AF767864

ADVANCED MEMBRANE TECHNOLOGY AND APPLICATIONS

ADVANCED MEMBRANE TECHNOLOGY AND APPLICATIONS

Edited By

Norman N. Li, Anthony G. Fane,
W. S. Winston Ho, and T. Matsuura

WILEY

A JOHN WILEY & SONS, INC., PUBLICATION

Copyright © 2008 by John Wiley & Sons, Inc. All rights reserved.

Published by John Wiley & Sons, Inc., Hoboken, New Jersey
Published simultaneously in Canada

No part of this publication may be reproduced, stored in a retrieval system, or transmitted in any form or by any means, electronic, mechanical, photocopying, recording, scanning, or otherwise, except as permitted under Sections 107 or 108 of the 1976 United States Copyright Act, without either the prior written permission of the Publisher, or authorization through payment of the appropriate per-copy fee to the Copyright Clearance Center, Inc., 222 Rosewood Drive, Danvers, MA 01923, (978) 750-8400, fax (978) 750-4470, or on the web at www.copyright.com. Requests to the Publisher for permission should be addressed to the Permissions Department, John Wiley & Sons, Inc., 111 River Street, Hoboken, NJ 07030, (201) 748-6011, fax (201) 748-6008, or online at http://www.wiley.com/go/permission.

Limit of Liability/Disclaimer of Warranty: While the publisher and author have used their best efforts in preparing this book, they make no representations or warranties with respect to the accuracy or completeness of the contents of this book and specifically disclaim any implied warranties of merchantability or fitness for a particular purpose. No warranty may be created or extended by sales representatives or written sales materials. The advice and strategies contained herein may not be suitable for your situation. You should consult with a professional where appropriate. Neither the publisher nor author shall be liable for any loss of profit or any other commercial damages, including but not limited to special, incidental, consequential, or other damages.

For general information on our other products and services or for technical support, please contact our Customer Care Department within the United States at (800) 762-2974, outside the United States at (317) 572-3993 or fax (317) 572-4002.

Wiley also publishes its books in a variety of electronic formats. Some content that appears in print may not be available in electronic format. For more information about Wiley products, visit our web site at www.wiley.com.

Library of Congress Cataloging-in-Publication Data:

Advanced membrane technology and applications/edited by Norman N. Li . . . [et al.].
p. cm.
Includes index.
ISBN 978-0-471-73167-2 (cloth)
1. Membranes (Technology) 2. Six sigma (Quality control standard) 3. Membrane industry.
I. Li, Norman N.
TP159.M4A38 2008
660′.28424—dc22 2007041577

MIX
Paper from responsible sources
FSC
www.fsc.org
FSC® C013604

CONTENTS

PREFACE

Since the last membrane book I published with the New York Academy of Sciences, I have attended several quite large membrane conferences including the one that I organized in the beautiful city of Irsee, Germany. I was struck by the fact that there had been very good progress made in the broad field of membranes science and technology. Also, membranes seem to be coming to the center of the water treatment and desalination technologies. Many parts of the world now are in critical need of clear water. Membrane technology is gaining increasing importance in treating and reusing wastewater and in producing potable water from seawater. It appears there is a timely need for a book that comprehensively reviews the up-to-date membrane technology and its many applications.

To undertake the task of publishing this book, I invited three of my colleagues, Tony Fan, Winston Ho, and Takeshi Matsuura to help, thus a team of four editors. Together we invited 35 chapters to cover membrane applications from gas to water separations. These chapters are now divided into six categories—membranes and applications in water and wastewater, membranes and applications in biotechnology and biomedical engineering, gas separations, membrane contactors and reactors, environmental and energy applications, and membrane materials and characterization. These six categories indeed cover a very broad field of applications.

I believe three somewhat unique features can be said about these chapters. One is that the percentage of contributors from industry is high. This is, of course, a relative comparison, in general, with the other published membrane books. As we know, most of the authors of the chapters in a membrane book are from academia, whereas many of the contributors from this book are from some of the major international membrane manufacturing companies. The other feature is that the chapters, in general, are more into applications than theories. The third feature is that a very strong coverage of water treatment and purification is presented for the reason mentioned above.

We are truly gratified to the strong response to contributing chapters. As a matter of fact, we still have quite many chapters that have been promised but have not been finished. This prompted me to consider publishing a second book in the near future. Meanwhile, we are indeed very pleased to have this book published and wish to thank all the reviewers and chapter contributors.

Norman N. Li

NL Chemical Technology, Inc.
Mount Prospect, Illinois

ABOUT THE EDITORS

Dr. Norman N. Li has about 40 years of working experience in the chemical and petroleum industries. He was a senior scientist with Exxon Research and Engineering Co, Director of Separation Science and Technology at UOP Co. and Director of Research and Technology at AlliedSignal Co. (now part of Honeywell). Since 1995, he is the president of NL Chemical Technology, Inc., which focuses on the development of membrane technologies. Dr. Li has more than 100 technical publications, 44 U.S. patents, and 13 books edited, all in the field of separation science and technology. He received the prestigious Award of Separation Science and Technology from the American Chemical Society, the Founders Award, Alpha Chi Sigma Award for Chemical Engineering Research, and the Award in Chemical Engineering Practice from the American Institute of Chemical Engineers and the Perkin Medal from the Society of Chemical Industry. The American Institute of Chemical Engineers held special symposia on membranes in his honor at its national meetings in 1995 and 2000. Dr. Li served as the president of the North American Membrane Society and the chair of the International Congress on Membranes and Membrane Processes (ICOM) in 1990. He is a member of the National Academy of Engineering, United States.

Dr. Tony Fane is a chemical engineer with a Ph.D. from Imperial College, London. He has been working on membranes since 1973 when he joined the University of New South Wales, in Sydney, Australia. His current interests are in membranes applied to environmental applications and the water cycle, with a focus on the sustainability aspects of membrane technology. He is a former director of the UNESCO Centre for Membrane Science and Technology at UNSW and recently Temasek Professor at Nanyang Technological University, Singapore. He is currently director of the Singapore Membrane Technology Centre at NTU. He is on the editorial board of the *Journal of Membrane Science and Desalination.* He is a fellow of the Australian Academy of Technological Sciences and Engineering, a recipient of the Centenary Medal in 2002 for services to Chemical Engineering and the Environment, and an honorary life member of the European Membrane Society.

Dr. W. S. Winston Ho is University Scholar Professor of Chemical and Materials Science and Engineering at the Ohio State University since 2002. Previously, he was a professor of chemical engineering at the University of Kentucky, after having more than 28 years of industrial R&D experience with Allied Chemical, Xerox, and Exxon, and serving as senior vice-president of technology at Commodore Separation Technologies. He was elected a member of the National Academy of Engineering, United States, in 2002. A New Jersey Inventor of the Year (1991), Dr. Ho holds more than 50 U.S. patents in separation processes. He is co-editor of *Membrane Handbook* and the recipient of the Professional and Scholarly Publishing Award for the most outstannding engineering work in 1993. He received the 2006 Institute Award for Excellence in Industrial Gases

Technology and the 2007 Clarence G. Gerhold Award from AIChE. He obtained his B.S. degree from National Taiwan University and his M.S. and Ph.D. degrees from the University of Illinois at Urbana–Champaign, all in chemical engineering.

Dr. Takeshi Matsuura received his B.Sc. and M.Sc. degrees from the Department of Applied Chemistry, University of Tokyo, and his doctoral degree from the Institute of Chemical Technology of the Technical University of Berlin in 1965. After working at the Department of Synthetic Chemisty of the University of Tokyo as a staff assistant and at the Department of Chemical Engineering of the University of California as a postdoc, he joined the National Research Council of Canada in 1969. He became a chair professor at the University of Ottawa in 1992. He also served as the director of the Industrial Membrane Research Institute until he retired in 2002. He is now a visiting professor at the National University of Singapore and the University Technology Malaysia, Skudai. Dr. Matsuura received the Research Award of International Desalination and Environmental Association in 1983. A symposium of membrane gas separation was held at the Eighth Annual Meeting of the North American Membrane Society, May 18–22, 1996, Ottawa, to honor him and Dr. S. Sourirajan. He received the George S. Links Award for Excellence in Research from University of Ottawa in 1998. He has published more than 300 articles in refereed journals, authored and co-authored 3 books, and edited 4 books.

CONTRIBUTORS

Fakhir U. Baig, Petro Sep Membrane Technologies Inc., Oakville, Ontario, Canada

Richard W. Baker, Membrane Technology and Research, Inc., Menlo Park, California 94025

Dibakar Bhattacharyya, Department of Chemical and Materials Engineering, University of Kentucky, Lexington, Kentucky 40506-0046

Bart Van Der Bruggen and Jeroen Geens, Department of Chemical Engineering, Laboratory for Applied Physical Chemistry and Environmental Technology, University of Leuven, Leuven, Belgium

G. Catapano, Department of Chemical Engineering and Materials, University of Calabria, Rende (CS), Italy

Tai-Shung Neal Chung, Department of Chemical and Biomolecular Engineering, National University of Singapore, Singapore 119260

P. Cornel, Technische Universität Darmstadt, Department of Civil Engineering, Institute WAR, Darmstadt, Germany

Pierre Côté and Mingang Liu, GE Water and Process Technologies, ZENON Membrane Solutions, Ontario, L6M 4B2, Canada

Zhanfeng Cui, Department of Engineering Science, Oxford University, Oxford, United Kingdom

Avijit Dey, Director – Application and Research, Omexell Inc., Stafford, Texas 77477

Enrico Drioli and Enrica Fontananova, Institute on Membrane Technology of the National Council Research (ITM-CNR), and Department of Chemical Engineering and Materials, University of Calabria, Rende (CS), Italy

Anthony G. Fane, UNESCO Centre for Membrane Science & Technology, University of New South Wales, Australia 2052 and Singapore Membrane Technology Centre, Nanyang Technological University, Singapore

Raja Ghosh, Department of Chemical Engineering, McMaster University, Hamilton, Ontario L8S 4L7, Canada

Alan R. Greenberg, Department of Mechanical Engineering, University of Colorado, Boulder, Colorado 80309-0427

Alexis M. W. Hillock, Shabbir Husain, and William J. Koros, School of Chemical & Biomolecular Engineering, Georgia Institute of Technology, Atlanta, Georgia, 30332

Sumod Kalakkunnath, Department of Chemical and Materials Engineering and Center for Manufacturing, University of Kentucky, Lexington, Kentucky 40506-0046

Douglass S. Kalika, Department of Chemical and Materials Engineering and Center for Manufacturing, University of Kentucky, Lexington, Kentucky 40506-0046

M. Kallioinen and M. Nyström, Laboratory of Membrane Technology and Technical Polymer Chemistry, Department of Chemical Technology, Lappeenranta University of Technology (LUT), Lappeenranta, Finland

Sujatha Karoor, Renal Division, Baxter Healthcare Corp., McGaw Park, Illinois, Massachusetts

Yoji Kase, UBE Industries Ltd., Ichihara, Chiba 290-0045, Japan

M. D. Kennedy, J. Kamanyi, S. G. Salinas Rodríguez, N. H. Lee, J. C. Schippers, and G. Amy, UNESCO–IHE Institute for Water Education, 2601 DA Delft, The Netherlands

Mohamed Khayet, Department of Applied Physics I, Faculty of Physics, University Complutense of Madrid, Madrid, Spain

William B. Krantz, Department of Chemical and Biomolecular Engineering, National University of Singapore, The Republic of Singapore, 117576

S. Krause, Microdyn-Nadir GmbH, Wiesbaden, Germany

N. Kubota, T. Hashimoto, and Y. Mori, Microza Research & Development Department, Specialty Products & Systems R&D Center, Asahi Kasei Chemicals Corporation, Fuji City, Shizuoka, 416-8501 Japan

A. Kumano and N. Fujiwara, Desalination Membrane Operating Department, Toyobo Co., Ltd., Osaka, Japan

Victor A. Kusuma, Benny D. Freeman, and Miguel Jose-YacamaN, Department of Chemical Engineering, University of Texas at Austin, Austin, Texas 78712

Haiqing Lin, Membrane Technology and Research, Inc., Menlo Park, California 94025

Chunqing Liu and Santi Kulprathipanja, UOP LLC, 25 East Algonquin Road, Des Plaines, Illinois, 60017

Yi Hua Ma, Center for Inorganic Membrane Studies, Department of Chemical Engineering, Worcester Polytechnic Institute, Worcester, Massachusetts 01609

M. L. Mottern, J. Y. Shi, K. Shgau, D. Yu, and Henk Verweiji, Department of Materials Science & Engineering, The Ohio State University, Columbus, Ohio 43210-1178

Norma J. Ofsthun, Clinical Science Department, Fresenius Medical Care, Lexington, Massachusetts 02420

Ho Bum Park and Young Moo Lee, School of Chemical Engineering, Hanyang University, Seoul, South Korea

Peter N. Pintauro and Ryszard Wycisk, Department of Chemical Engineering, Case Western Reserve University, Cleveland, Ohio 44106-7217

Raphael Semiat, Technion, Israel Institute of Technology, The Wolfson Chemical Engineering Department, Technion City, Haifa, Israel

P. Silva, L. G. Peeva, and A. G. Livingston, Department of Chemical Engineering, Imperial College, London SW7 2BY, United Kingdom

Kamalesh K. Sirkar, Otto H. York Department of Chemical Engineering, Center for Membrane Technologies, New Jersey Institute of Technology, Newark, New Jersey 07102

Steven Siverns, EnviroTower, Toronto, Ontario, M5V 1R7, Canada

Mitsuru Suzuki, Medical Membrane Department, Toyobo Corp., Osaka, Japan

Yit-Hong Tee, Department of Chemical and Materials Engineering, University of Kentucky, Lexington, Kentucky 40506-0046

R. L. Truby, Toray Membranes, Escondido, California 92026

Tadahiro Uemura and Masahiro Henmi, Global Environment Research Laboratories, Toray Industries Inc., Otsu Shiga, Japan

J. Vienken, Fresenius Medical Care, Bad Homburg, Germany

Nikolay Voutchkov, Poseidon Resources Corporation, Stamford, Connecticut

S. Ranil Wickramasinghe, Department of Chemical and Biological Engineering, Colorado State University, Fort Collins, Colorado 80523-1370

P. Jason Williams and William J. Koros, School of Chemical and Biomolecular Engineering, Georgia Institute of Technology, Atlanta, Georgia 30332

Jian Zou, Jin Huang, and W. S. Winston Ho, Department of Chemical and Biomolecular Engineering, Department of Materials Science and Engineering, The Ohio State University, Columbus, Ohio 43210-1180

PART I

MEMBRANES AND APPLICATIONS IN WATER AND WASTEWATER

CHAPTER 1

Thin-Film Composite Membranes for Reverse Osmosis

TADAHIRO UEMURA and MASAHIRO HENMI

Global Environment Research Laboratories, Toray Industries Inc., Otsu Shiga, Japan

1.1 INTRODUCTION

Because of vastly expanding populations, increasing water demand, and the deterioration of water resource quality and quantity, water is going to be the most precious resource in the world. Thus, the 21st century is called the "water century." In the 20th century, membrane technologies made great progress, and commercial markets have been spreading very rapidly and throughout the world. The key technologies fueling this progress are as follows:

1. *Materials*: Chemical design of high-performance materials suitable for each separation mode
2. *Morphology*: Morphological design of high-performance membranes
3. *Element/Module*: Element and module design for high-performance membranes
4. *Membrane Process*: Plant design and operation technology

In 21st century, to solve these water problems, membranes technology is going to be further expanded and new technology—further improvements of membrane performance, development of membrane systems, membranes stability such as antifouling membranes for wastewater treatment, and other highly qualified membranes—will be needed.

Among desalination technologies available today, reverse osmosis (RO) is regarded as the most economical desalination process. Therefore, RO membranes have played crucial roles in obtaining fresh water from nonconventional water resources such as seawater and wastewater.

1.2 APPLICATION OF RO MEMBRANES

Reverse osmosis membranes have been used widely for water treatment such as ultrapure water makeup, pure boiler water makeup in industrial fields, seawater and brackish water

Advanced Membrane Technology and Applications. Edited by Norman N. Li, Anthony G. Fane, W. S. Winston Ho, and T. Matsuura
Copyright © 2008 John Wiley & Sons, Inc.

TABLE 1.1 Application of Reverse Osmosis Membrane Process

Industrial Use	Drinking Water	Wastewater Treatment and Reuse
Ultrapure water, boiler water, process pure water, daily industries	Seawater desalination, brackish water desalination	Industrial water, agricultural water, indirect drinking water

desalination in drinking water production, and wastewater treatment and reuse in industrial, agricultural, and indirect drinking water production as shown in Table 1.1.

The expansion of RO membrane applications promoted the redesign of suitable membrane material to take into consideration chemical structure, membranes configuration, chemical stability, and ease of fabrication. And along with the improvements of the membranes, the applications are further developed.

1.3 MAJOR PROGRESS IN RO MEMBRANES

1.3.1 Cellulose Acetate Membrane

Reverse osmosis systems were originally presented by Reid in 1953. The first membrane, which could be used at the industrial level in actual water production plants, was a cellulose-acetate-based RO membrane invented by Loeb and Sourirajan in 1960. This membrane has a so-called asymmetric or anisotropic membrane structure having a very thin solute-rejecting active layer on a coarse supporting layer, as shown in Figure 1.1. The membrane is made from only one polymeric material, such as cellulose acetate, and made by the nonsolvent-induced phase separation method. After the invention by Loeb and Sourirajan, spiral-wound membranes elements using the cellulose acetate asymmetric flat-sheet membranes were developed and manufactured by several U.S. and Japanese companies. RO technologies have been on the market since around 1964 (Kurihara et al., 1987). They were widely used from the 1960s through the 1980s mainly for pure water makeup for industrial processes and ultrapure water production in semiconductor industries; and some are still used in some of these applications.

Figure 1.1 SEM photograph of CA asymmetric membrane.

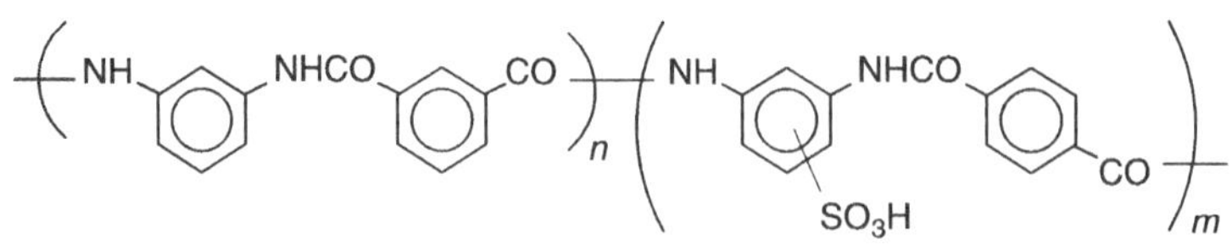

Figure 1.2 Representative chemical structure of linear polyamide membrane (B-10).

TABLE 1.2 Summary of Membrane Materials for RO

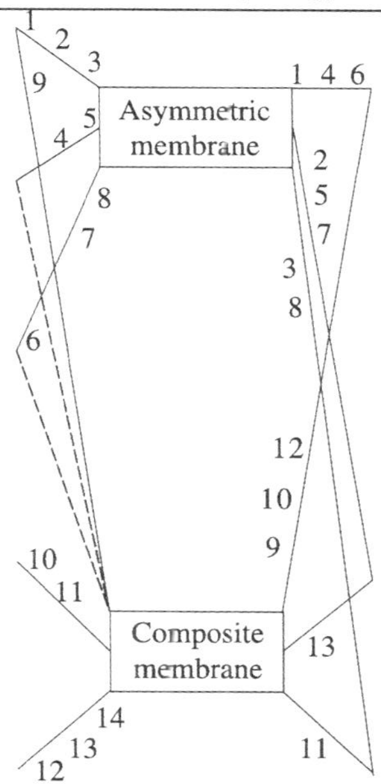

Membrane Material	Membrane Morphology	Module Configuration	Examples of Membrane and Module, Membrane Suppliers
Cellulose acetate Polyamide Heterocyclic polymer		Spiral	1. Toray, UOP, environgenics, Osmonics, Desalination, Ajax, Hydranautics, Daiseru 4. Toray-Polyamidic acid, Du Pont-DP-1, Monsanto 6. Cellanese-Polybenzimidazole 9. UOP-CTA 10. North Star-NS-100, UOP-PA-300, -100, LP-300, RC-100 12. North Star-NS-200, Osmonics-NS-200 Environgenics-SPFA (NS-200), Desalination-NS-200(?) 14. North Triangle Inst.-Plasma Polym. Toray-PEC-1000, Film Tec-FT-30 Asahi Glass-MVP, Nihon Syokubai
Cross-linked water-soluble polymer		Hollow fiber	2. Dow, Monsanto, Toyobo 5. Du Pont 7. Cellanese-Polybensimidazole 13. FRL-NS-200, Gulf South Research Inst.-NS-100
Polymerizable monomer (cross-linking)		Tubular	3. UOP, Environgenics, Universal Water Co. Raypak, Abcor, PCI, Nitto, Daiseru 8. Teijin-PBIL 11. North Star-NS-100, Others Sumitomo-PAN-Composite Memberane

1.3.2 Aromatic Polyamide Hollow Fiber Membrane

Since then, there has been intensive and continuous R&D efforts mainly around the United States and Japan to meet the demands from commercial markets, and there exist many inventions and breakthroughs in membrane materials and configurations to improve the performances of membranes.

To overcome the problems of cellulose acetate membranes, many synthetic polymeric materials for reverse osmosis were proposed, but except for one material, none of them proved successful. The only one material, which could remain on the market, was the linear aromatic polyamide with pendant sulfonic acid groups, as shown in Figure 1.2. This material was proposed by DuPont, which fabricated very fine hollow fiber membranes; the modules of this membrane were designated B-9 and B-10. They have a high rejection performance, which can be used for single-stage seawater desalination. They were widely used for mainly seawater or brackish water desalination and recovery of valuable materials such as electric deposition paints, until DuPont withdrew them from the market in 2001.

1.3.3 Composite Membrane

Another approach to obtain a high-performance RO membrane was investigated by some research institutes and companies in the 1970s. Many methods to prepare composite membranes have been proposed, as shown in Table 1.2. In the early stage, very thin films of a cellulose acetate (CA) polymer coating on a substrate, such as a porous cellulose nitrate substrate, was tried. However, in spite of their efforts, this approach did not succeed in industrial membranes manufacturing.

The next approach used the interfacial polycondensation reaction to form a very thin polymeric layer onto a substrate. Morgan first proposed this approach (Morgan, 1965), and then Scala et al. (1973) and Van Heuven (1976) actually applied this approach to obtain an RO membrane. But it was Cadotte who invented the high-performance membrane using the in situ interfacial condensation method (Cadotte, 1985). In his method, interfacial condensation reactions between polymeric polyamine and monomeric polyfunctional acid halides or isocyanates takes place on a substrate material to deposit a thin film barrier onto a substrate. Some of the composite membranes were succeeded in industrial fabrication by another method, which was designated as PA-300 or RC-100.

Another preparation method for composite membrane is an in situ monomer condensation method using the monomeric amine and monomeric acid halide, which was also invented by Cadotte. Then, many companies succeeded in developing composite membranes using this method, and the membrane performance has been drastically improved up to now. Now, composite membrane of cross-linked fully aromatic polyamide is regarded as the most popular and reliable material in the world. Permeate flow rate and its quality have been improved 10 times greater than that of the beginning (Kurihara et al., 1987, 1994b).

1.4 TRENDS IN RO MEMBRANE TECHNOLOGY

Figure 1.3 shows recent trends in RO membrane technology with two obvious tendencies. One is a tendency toward low-pressure membranes for operating energy reduction in the field of brackish water desalination. The other is a tendency toward high rejection with high-pressure resistance in the large seawater desalination market.

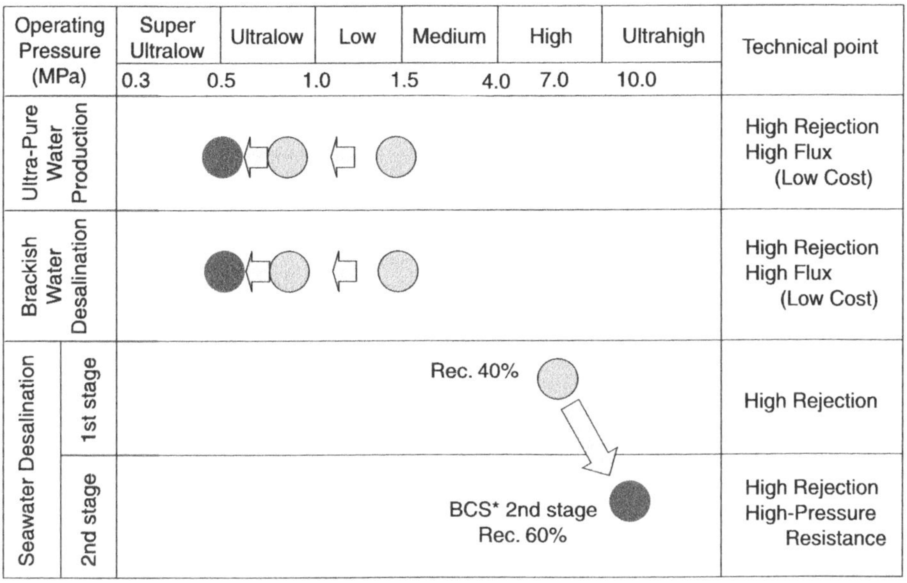

*Brine Conversion System

Figure 1.3 Technology trends in RO membrane.

1.4.1 Progress of Low-Pressure Membrane Performance in Brackish Water Desalination

Figure 1.4 shows the progress of low-pressure membrane performance trends in RO membrane on brackish water desalination from the 1970s to the 1990s, including industrial water treatment such as ultrapure water production. In the 1970s much effort was devoted to

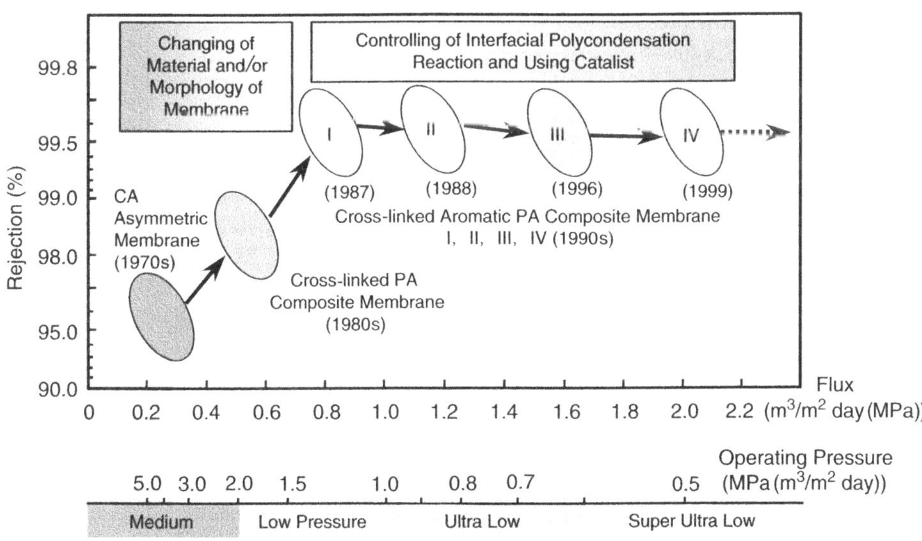

Figure 1.4 Performance trends in RO membrane for brackish water desalination.

TABLE 1.3 Typical Performance of Toray's Brackish Water RO Elements

	Low Pressure		Ultralow Pressure	Super-ultralow Pressure
Type of Membrane	I	II	III	IV
Name of membrane element (in market: year)	SU-720 (1987)	SU-720L (1988)	SUL-G20 (1996)	SUL-H20 (1999)
Performance				
Salt rejection (%)	99.4	99.0	99.4	99.4
Water permeability (m^3/day)	26.0	22.0	26.0	26.0
Test condition				
Operating pressure (MPa)	1.5	1.0	0.75	0.5
Temperature (°C)	25	25	25	25
Feed concentration (mg/L)	1500	1500	1500	1500
Brine flow rate (L/min)	80	80	80	80

developing high-performance membrane materials and improving the membrane performance. As a result performance was improved with a new developed material of cross-linked aromatic polyamide and by developing membrane morphology and fabrication technology. The cross-linked fully aromatic polyamide composite membrane developed in 1987 has four or five times larger water flux and five times higher product water quality than those of the CA membrane (Kurihara et al., 1987). Since 1987, membrane performance has been drastically developed. On the basis of the development of cross-linked fully aromatic polyamide composite membranes, RO membrane performance of brackish water desalination has improved very rapidly. Typical performances of the RO elements for brackish water desalination are shown in Table 1.3. The ultralow-pressure membrane, which can be used at ultralow pressures such as 0.75 MPa, has been developed, which saves on the operating cost (Ikeda et al., 1996). And now the super-ultralow-pressure membrane elements, which can be used at super-ultralow-pressure, such as 0.5–0.3 MPa, have been developed (Fusaoka, 1999). This membrane has three times the water permeability than the ordinary low-pressure RO membrane. This membrane can operate with one-third the pressure of a low-pressure membrane.

1.4.2 Progress of RO Membranes for Seawater Desalination

The progress of RO membranes for seawater desalination is shown in Figure 1.5 (Kurihara et al., 1994a). It is very important to increase the water recovery ratio on seawater desalination systems to achieve further cost reduction. Most seawater RO desalination systems in use today are confined to approximately 40% conversion of the feed water (salt concentration 3.5%), since most of commercially available RO membrane do not allow for high-pressure operation of more than around 7.0 MPa.

Recent progress on high-pressure–high-rejection spiral wound (SW) RO elements, combined with proven and innovative energy recovery and pumping devices, has opened new possibilities to reduce investment and operating cost. The progress of RO seawater desalination from a point of view of water recovery is shown in Table 1.4 (Moch, 2000).

Toray has developed a new low-cost seawater desalination system called the Brine Conversion Two-Stage (BCS) system, as shown in Figure 1.6, which provides 60%

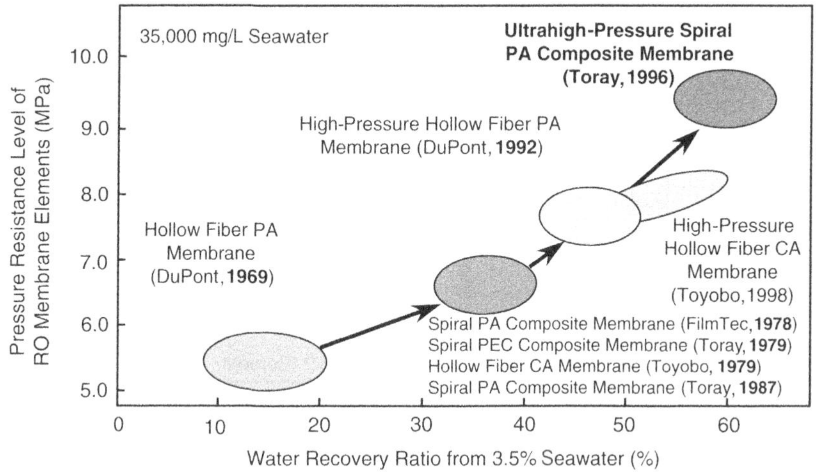

Figure 1.5 Performance trends in RO membranes for seawater desalination.

water recovery of freshwater (Yamamura et al., 1996). Ohya et al. (1996) and Nakao (1996) also suggest that higher recovery of RO seawater desalination by the BCS system is most effective in saving energy yet keeping a low operating cost.

As for achieving the 60% RO seawater desalination system, it is absolutely necessary to make the RO membrane element, which can be operated under very severe operating conditions, with high pressure and high feed water concentration such as 9.0 MPa and 5.8%. Toray has developed a high-performance membrane (BCM element) that can be operated at high pressure and high concentration conditions, as shown in Table 1.5.

1.4.3 High Boron Rejection SWRO Membrane

The removal of boron is a significant problem in SWRO desalination processes (Fukunaga et al., 1997). Boron exists as boric acid in the natural water, and boric acid mainly shows the

TABLE 1.4 Typical Performance of Toray's Seawater RO Membranes

	Ultrahigh Pressure Seawater RO Membrane SU-820BCM	Seawater RO Membrane SU-820
Membrane material	Cross-linked fully aromatic polyamide	
Membrane morphology	Composite membrane	
Membrane substrate	Flat unwoven fabric	Unwoven fabric or taffeta
Feed water spacer	Special spacer	Normal spacer
Permeate spacer	Ultrahigh-pressure resistant	High-pressure resistant
Performance		
Rejection (%)	99.70	99.75
Water permeability (m^3/day)	16.0	16.0
Test condition		
Feed concentration (mg/L)	58,000	35,000
Operating pressure (MPa)	9.0	5.5
Max. pressure (MPa)	10.0	7.0

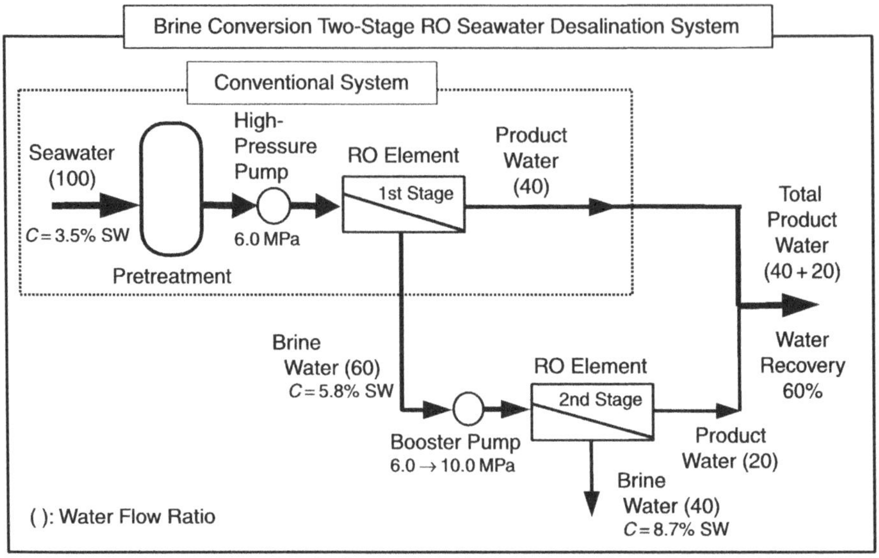

Toray's Patent: Japanese Patent Application No. 94-245184(1994), Application No. 96-108048(1996) priority.
Granted: US, CA, RC(1998), AU(1998); Pending - JP, EP, KR, CH

Figure 1.6 Typical flow diagram of BCS system.

male reproductive tract when administered orally to laboratory animals. The World Health Organization (WHO) proposes that boron concentration in drinking water be below 0.5 mg/L as a provisional guideline value (WHO, 2004). However, especially in SWRO desalination fields, this is not an easy goal to meet because boron concentration in seawater is comparatively high. Although conventional SWRO membrane elements have shown a little more than 90% of boron rejection, it is still inadequate. It is difficult for RO membranes to remove boric acid in water for the following reasons: First, the molecular size of boric acid is so small that it is difficult to remove by size exclusion. Second, since boric acid has a pK_a of 9.14–9.25, it is not ionized in natural seawater with a pH of 7.0–8.0 and dissociates at pH 9 or more (Rodriguz et al., 2001). The boron rejection by the electric repulsive force between boric acid and the membrane cannot be expected in a neutral condition. Therefore, some posttreatment processes are necessary to meet the WHO proposal.

The conventional SWRO membrane element TM820, which is typical with Toray, has exhibited 91–93% boron rejection, which was the highest level achieved by commercialized SWRO membrane elements (Toray, 2004; Redondo et al., 2003; Hiro and Hirose, 2000). This membrane element series has been installed in a large number of SWRO

TABLE 1.5 Progress in RO Seawater Desalination Plant

	Unit	1980s	1990s	2000s
Recovery	%	25	40–50	55–65
Operating pressure	MPa (psig)	6.9 (1000)	8.25 (1200)	9.7 (1400)
Product water TDS	mg/L	500	300	<200
Plant energy consumption	kWh/m^3 (kWh/kgal)	12 (45)	5.5 (21)	4.6 (17.4)

Source: Moch (2000).

TABLE 1.6 SWRO Products Lineup Released from Toray[a]

Type of Membrane	Standard Use	High Permeability	High Pressure	Ultrahigh Pressure
Name of membrane element	TM820-370	TM820L-370	TM820H-370	SU820BCM
Salt rejection, %	99.75	99.70	99.75	99.83
Product flow rate, m^3/day (gpd)	23 (6000)	34 (9000)	21 (5600)	23 (6000)
Boron rejection, %	91–93	88–90	91–93	91–93
Membrane area, m^2 (ft^2)	34 (370)	34 (370)	34 (370)	29 (320)
Max. operating pressure, MPa (psi)	6.9 (1000)	6.9 (1000)	8.3 (1200)	10.0 (1450)

[a]Testing conditions were: applied pressure, 800 psi (5.52 MPa) for others; recovery ratio, 8%; feed solution: 32,000 ppm, NaCl with 5.0 mg/L boron for others; pH = 8; temperature, 25°C.

plants. And Toray has commercialized many types of SWRO membrane elements, which are for different pressure ranges due to total dissolved solids (TDS) concentration and temperature of the seawater, as shown in Table 1.6. However, the highest boron rejection in those membrane elements is 91–93%, which is the same as TM820. This means that the improvement of boron rejection by membrane material had been sluggish for a while. Meanwhile, the new membrane element TM820A was developed based on the following two concepts: (1) reduction of affinity with boric acid by control of hydrophobic property and functional groups may reject boric acid selectively, and (2) molecular structure design was considered as blocking the boric-acid-permeable large pore (Taniguchi et al., 2004). TM820A exhibited 94–96% boron rejection with high TDS rejection and high water productivity. The specification and typical performance of TM820A is shown in Table 1.7.

Seen from various viewpoints, a single SWRO system is the most ideal. Therefore, to evaluate the performance of TM820A, the amount of boron that TM820A could remove by a single-stage operation was estimated. Table 1.8 shows the permeate boron concentration that corresponds to the boron rejection performance used by membrane elements when each region of seawater is treated by a single-stage SWRO operation considering the aging factor. According to Table 1.8, TM820A meets the Japanese guidelines of below 1 mg/L of boron concentration by a single-stage operation. But in severe conditions, for the WHO guideline grade and the Middle East seawater treatment, certain posttreatment processes are still needed. If 97% of boron rejection performance is gained, the WHO grade will be enabled until the Southeast Asia seawater treatment. Furthermore, at 99% boron

TABLE 1.7 Specifications and Typical Performances of TM820A[a]

Name of Membrane Element	TM820A-370
Salt rejection, %	99.80
Product flow rate, m^3/day (gpd)	21 (5500)
Boron rejection, %	94–96
Membrane area, m^2 (ft^2)	34 (370)

[a]Testing conditions were: applied pressure, 800 psi (5.52 MPa); recovery ratio, 8%; feed solution: 32,000 ppm, NaCl with 5.0 mg/L boron; pH = 8; temperature, 25°C.

TABLE 1.8 Permeate Boron Concentration by a Single-Stage SWRO Operation (Calculated)[a]

Seawater (Temp., TDS conc., Boron conc.)	Permeate Boron Concentration (mg/L) — Boron Rejection Performance of Used Membrane Element: 90%	95% (TM820A)	97%	99%
Japan (25°C, 3.5%, 5 mg/L)	1.5	0.9	0.4	0.2
Southeast Asia (32°C, 3.5%, 5 mg/L)	1.6	1.0	0.5	0.2
Middle East (38°C, 4.5%, 7 mg/L)	3.0	2.0	1.1	0.4

[a]Assumed conditions: Plant: 7 elements/vessel, 14 lmh; operation: 25°C, pH 8.0, 800 psi, 3.5 L/min, recovery ratio 40%, after 3 years. Japanese law grade: <1.0 mg/L, WHO guideline grade: <0.5 mg/L.

rejection performance, the WHO guideline grade will be enabled even in the Middle East seawater treatment.

Recently, Toray has been investigating SWRO membranes that focus on the removal of boron by the improvement of membrane performance. The history and future prospects of the boron removal at Toray and other companies are shown in Figure 1.7. Until 2000, although the boron rejection was also improved as various membranes were developed in each company, it was 90% at best. In the next period, from 2000 to 2003, the membranes in which a little more than 90% of boron rejection was shown were released, and these serve as main items for each company now.

From 2003 to 2005, Toray developed and released TM820A, whose performance was appreciably improved, and offered the membrane that showed around 95% boron rejection prior to other companies. However, the supportive systems are still required to meet the WHO proposal even by using TM820A as above. Thus, the next target is 97 or 99% boron rejection performance of renovative membrane. The further development of a new

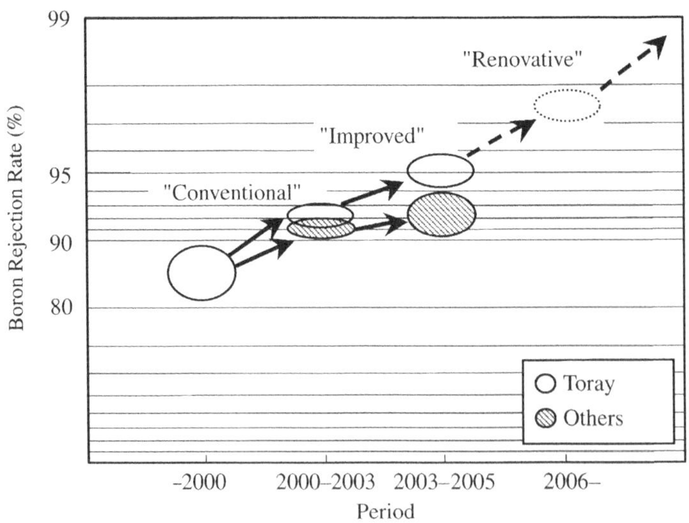

Figure 1.7 History and prospect of boron rejection performance of SWRO membrane element in Toray and comparable companies.

TABLE 1.9 Large Seawater Desalination Plants Utilizing RO Process (as of July, 2005)

No.	Country	Location	Capacity (m^3/d)	In Operation (year)	Membrane Supplier
1	Israel	Ashkelon	272,520	2005	Dow
2	UAE	Taweelah	227,300	2006	Undecided
3	Algeria	Hamma	200,000	2006	Undecided
4	UAE	Fujairah	170,000	2003	Hydranautics
5	Trinidad & Tobago	Point Lisa	136,000	2002	Toray
5	Singapore	Tuas	136,000	2005	Toray
7	Saudi Arabia	Tanbu	128,000	1995	Toyobo
8	Spain	Carponeras	120,000	2001	Hydranautics
9	Israel	Palmachim	92,250	2006	Toray
10	Saudi Arabia	Al Jubail III	90,900	2000	DuPont/Toray

renovative membrane that can meet the WHO proposal for every seawater continues (Tomioka et al., 2005). Table 1.9 summarizes large seawater RO desalination plants around the world. TM820A is installed in a large seawater RO desalination plant in Singapore.

1.5 REVERSE OSMOSIS/BIOFOULING PROTECTION

Biofouling has been regarded as the most serious problem in the operation of SWRO plants. The usual method to prevent biofouling is continuous chlorine dosing to intake seawater with sodium bisulfate (SBS) dosing at the RO portion. However, membranes performance deterioration occurred by oxidation in case of both polyamide and cellulose acetate membranes, and biofouling has not been solved yet. Toray has developed a new method that is effective to prevent biofouling on SWRO membranes and verified its effectiveness at actual plants.

First of all, by measuring the viable counts of bacteria at a plant, in case of the continuous chlorine/SBS dosing method, it was found that a number of bacteria drastically increased immediately after SBS dosing, as shown in Figure 1.8, and most of these bacteria were quite different from those found in raw seawater. Currently, the addition of SBS to feed water at relatively high concentration has been used for sterilization of RO membranes. However, when SBS was added to seawater, the pH was just dropped to 6, and most of the bacteria harbored in water were still alive. This result indicates that the sterilization ability of SBS is due to lower pH, and oxygen consumption with SBS only plays a role to repress the cell growth. Finally, Toray has developed a new agent, MT-901, which is effective in preventing biofouling on RO membranes. Adding MT-901 to seawater instead of SBS effectively killed bacteria in a few samples of seawater within a short time.

Finally, the effect of this method was verified at an SWRO plant. In this plant, when feed water was chlorinated and dechlorinated with SBS continuously and RO membrane modules were treated with SBS intermittently, differentiation pressure of the module increased gradually. MT-901 was used for membrane module treatment in place of SBS and the differential pressure decreased within 10 days. Moreover, using an intermittent chlorination method was effective to maintain the initial differential pressure with less concentration of MT-901 (Kallenberg et al., 1999).

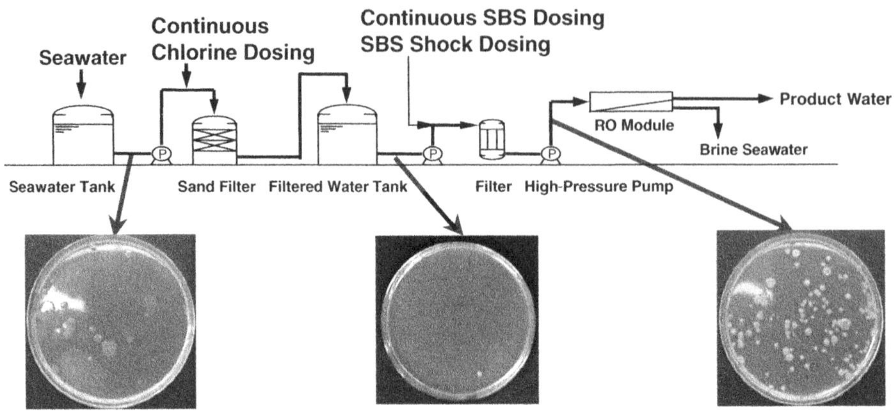

Sample	Seawater	Filtered Seawater	Feed Water of RO Module
Viable Cells (×1000/mL)	0.3	<0.1	1.5

Figure 1.8 Viable cell count assessment in RO plant.

1.6 LOW-FOULING RO MEMBRANE FOR WASTEWATER RECLAMATION

Wastewater reclamation and reuse plants have been constructed and operated around the world. Table 1.10 shows large wastewater reuse plants. RO membrane is necessary for wastewater reclamation to make the water quality reusable. The largest RO plant is operated in Kuwait since 2005.

For RO membrane modules, stable operation is very important. Many organizations, universities, and companies have reported many kinds of operation troubles. Fouling, membrane deterioration, and hardware problems have mainly caused these troubles, and the major troubles, which occupy 80%, are fouling problems.

As described above, it is important to consider the (1) proper RO membrane elements with low-fouling property, (2) proper pretreatment technology before the RO membrane, and (3) suitable sterilization methods and cleaning technology (Kurihara et al., 2003).

TABLE 1.10 Large Wastewater Reuse Plants (as of July, 2005)

No.	Country	Location	Capacity (m^3/d)	In Operation (year)	RO Supplier
1	Kuwait	Sulaibiya	310,000	2005	Toray
2	USA	Fountain Valley, CA	220,000	2007	Hydranautics
3	Singapore	Ulu Pandan	140,000	2006	Undecided
4	USA	West Basin, CA	75,000	1997–2001	Unidentified
5	Singapore	Kranji	40,000	2003	Hydranautics
6	Singapore	Bedok	32,000	2003	Hydranautics
7	Singapore	Seletar	24,000	2003	Toray
8	USA	Scottsdale, AZ	22,710	1998	Koch

The reasons for fouling of RO membrane are reported as consisting of chemical fouling, biological fouling, and scale precipitation.

It is estimated that chemical fouling is caused by the adsorption of organic materials such as humic substances and surfactants in the feed water or on the membrane surface. Humic substances have various chemical structures depending on the water origin, such as land water or seawater, and regions in the world. However, it has both hydrophobic groups of aromatic and linear structure and ionic groups of amino acid and carboxylic acid. The material of RO membrane is polyamide with hydrophobic and ionic properties. As mentioned above, chemical fouling depends on hydrophobic interaction and electrostatic interaction between organic materials in the feed water and membrane surface.

On the other hand, in case of biological fouling, the following estimations are reported: (1) microbe adsorption by hydrophobic or electrostatic interaction, (2) propagation of microbe with nutrition in the feed water, and (3) deposition of exhaust material of biological metabolism. Case 1 is a reversible phenomenon; however, cases 2 and 3 are irreversible phenomena, which are difficult to remove by simple chemical cleaning.

As a result of R&D activities, Toray has developed low-fouling RO membrane for wastewater reclamation. The low-fouling RO membrane has the same level of pure water permeability as conventional RO membranes, SU-700 and SUL-G, and also has low-fouling property with keeping water permeability against chemical and biological fouling during the operation (Yamamura et al., 2002).

The low-fouling property of membranes is evaluated with a nonionic surfactant aqueous solution. Test result shows that, in operation, low-fouling RO membrane has a relatively small permeability declaration ratio of 27%, compared with initial pure water permeability and shows stable operation. On the other hand, conventional fully aromatic polyamide membranes, SU-700 and SUL-G, show 36–47% declaration ratio, even if they show stable operation. And concerning the chemical cleaning properties, low-fouling RO membrane shows better recovery of permeability after chemical cleaning.

To evaluate the fouling property against microbes, adsorption property of a certain hydrophobic microbe and other hydrophilic microbes were measured. The hydrophobic microbe was severely adsorbed to conventional RO membranes and caused biological fouling of RO membranes. In case of low-fouling RO membranes, the adsorption property of the hydrophobic microbe is quite low, which is less than one-tenth of conventional RO membranes. Initial performance of low-fouling RO membrane element TML-20 is described in Table 1.11.

A test of wastewater reclamation using low-fouling RO membranes and conventional low-pressure RO membrane SUL-G10 has been carried out in a wastewater treatment facility in Japan, as shown in Figure 1.9. In this test, secondary effluent was directly filtered by ultrafiltration (UF) membrane and permeate was fed to both of the RO membranes. In case

TABLE 1.11 Initial Performance of Low-Fouling RO Membrane Element TML-20[a]

Items	Performance Data		
Salt rejection (%)	99.5	99.5	99.5
Permeability (m^3/day)	36.0	41.0	48.0

[a]Conditions: Pressure = 1.5 MPa, temperature = 25°C, feed water conc. = 1500 (NaCl mg/L), pH 6.5.

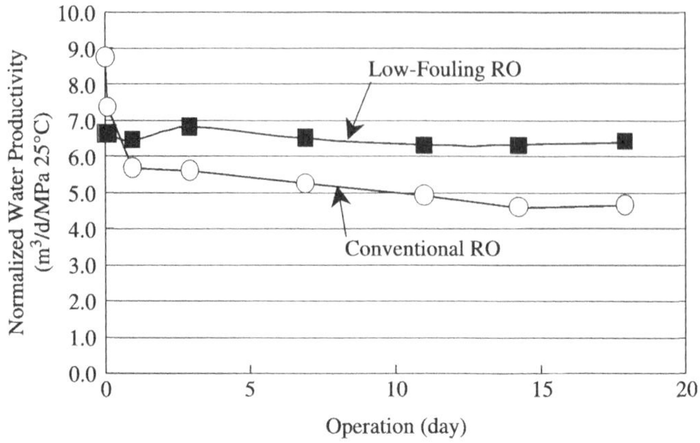

Figure 1.9 Water productivity of low-fouling RO compared with conventional RO.

Figure 1.10 Wastewater reclamation and reuse plant in Singapore (24,000 m^3/day).

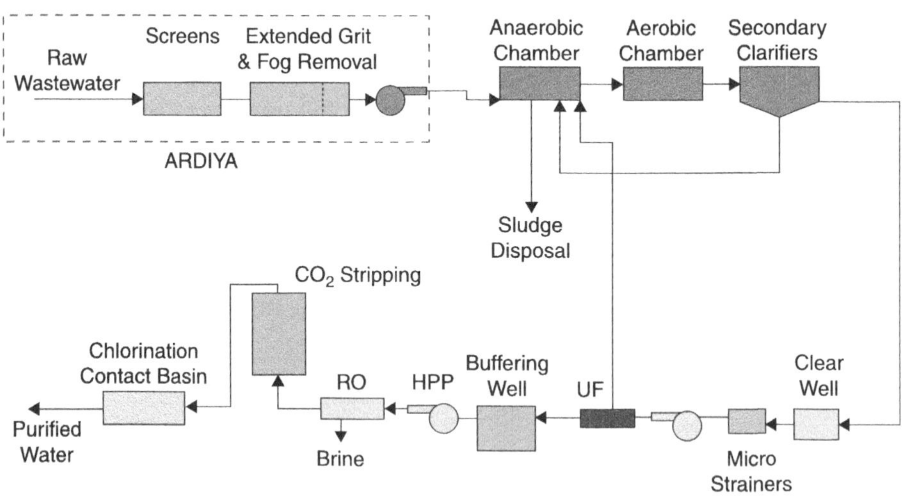

Figure 1.11 Process flow of wastewater reclamation and reuse plant in Kuwait (375,000 m^3/day).

Figure 1.12 RO portion of wastewater reclamation and reuse plant in Kuwait.

of the SUL-G membrane, water permeability dropped to 60% of initial permeability in a day due to biological fouling; however, the permeability drop of low-fouling RO TML20 was smaller than that of SUL-G and the stable operation has been performed.

The low-fouling RO membrane is strongly required for the stable operation of the wastewater reclamation plant. Two large water reclamation and reuse plants have been operated in Singapore and Kuwait, as shown in Figures 1.10, 1.11, and 1.12.

1.7 CHLORINE TOLERANCE OF CROSS-LINKED AROMATIC POLYAMIDE MEMBRANE

Chlorine tolerance is a very important characteristic to design an RO membrane process because chlorine dosing to water process is commonly used as a disinfection for microorganisms. Many authors have studied chlorine tolerance of RO membranes as listed in Table 1.12.

In our literature, we studied three kinds of RO membranes: a cellulose acetate asymmetric membrane (SC-3000), a cross-linked N substituted polyamide composite membrane

TABLE 1.12 Studies on Chlorine Tolerance of RO Membrane

Active Chlorine		Membrane Material		
HClO	ClO^-	Cellulose Acetate	Cross-linked N-substituted Polyamide	Cross-linked Aromatic Polyamide
—	yes	yes	—	—
yes	yes	yes	—	yes
yes	yes	—	yes	yes
yes	yes	—	yes	yes
yes	yes	—	yes	yes
yes	yes	yes	yes	yes

TABLE 1.13 Chlorine Degradation Rate Obtained from Each Membrane[a]

Membrane	$(D_{AM}/k\delta)/(D_{AM}/k\delta)_0 =$	$A = A_0 +$
UTC-70	$\exp(1.6 \times 10^{-3} \times C^{0.5} \times t)$	$(7 \times 10^{-13}) \times C^{0.5} \times t$
UTC-60	$\exp(5 \times 10^{-6} \times C^{1.2} \times t)$	$(4 \times 10^{-15}) \times C^{1.2} \times t$
SC-3000	$\exp(5 \times 10^{-6} \times C^{1.5} \times t)$	$(9 \times 10^{-16}) \times C^{1.5} \times t$

[a] C = chlorine conc. (ppm), t = chlorine exposure time (h).

(UTC-60), and a cross-linked aromatic polyamide composite membrane (UTC-70) using immersion and operating test conditions (Uemura and Kurihara, 2003). The degradation was observed as the increase in both solute and water permeation coefficients, which can be expressed as functions of a quantity of chlorine concentration to the Xth power times the chlorine exposure time. It was found that the values of the X are in the range of 0.5–0.7 for cross-linked aromatic composite, 1.2 for cross-linked N-substituted composite, and 1.5 for cellulose acetate asymmetric. The value of X seems closely related to the degradation mechanism. In the case of rapid degradation, the degradation might be mainly caused by chlorination reaction, and the value of X should be close to 0.5. On the other hand, in the case of slow degradation, the degradation might be caused by oxidation reaction, and value of X should be close to 2. The morphological and structural changes due to the chlorination degradation were observed using electron microscopy and electron spectroscopy for chemical analysis (ESCA). It was clarified that, as the degradation reaction progresses, the membrane thickness is reduced and its looseness and fixed charge density are increased. The results are summarized in Table 1.13.

Using the equations in Table 1.13, the membrane performance of both solute permeation coefficient ($D_{AM}/k\delta$) and water permeate coefficient (A) after certain chlorine concentration (C) and exposure time (t) can be roughly predicted for each membrane. But some catalytic acceleration reactions, which may be caused by the iron ion and other heavy-metal ions in water, must be taken into account in actual cases.

REFERENCES

Cadotte, J. E. (1985). Evolution of composite reverse osmosis membrane. In *Materials Science of Synthetic Membranes*. ACS Symposium Series 269. American Chemical Society, Washington, DC, p. 273.

Fukunaga, K., Matsukata, M., Ueyama, K., and Kimura, S. (1997). Reduction of boron concentration in water produced by a reverse osmosis sea water desalination unit. *Membrane* 22(4), 211.

Fusaoka, Y. (1999). Super ultra low pressure composite reverse osmosis membrane elements for brackish water desalination and ultrapure water production. *Membrane* 24(6), 319–323.

Hiro, A., and Hirose, M. (2000). Development of the high boron removal reverse osmosis membrane element for seawater desalination. *Nitto Giho* 40, 36.

Ikeda, T., Fusaoka, Y., Uemura, T., Tonouchi, T., and Fujino, H. (1996). Advanced ultra low pressure reverse osmosis membrane elements having a high water flux and a high solute rejection. In *Preprints of International Congress on Membranes and Membrane Processes*, Yokohama, Japan, Aug. 18–23, p. 182.

Kallenberg, K., Jose, P. V., Yamamura, H., and Kurihara, M. (1999). Brine conversion (BCS) enhances SWRO desalination case histories, operating data, novel design features. In *Preprints of IDA World Congress*, San Diego, USA, Vol. II, Aug. 29–Sept. 3, pp. 101–107.

Kurihara, M., Fusaoka, Y., Sasaki, T., Bairinji, R., and Uemura, T. (1994a). Development of cross-linked fully aromatic polyamide ultra-thin composite membranes for seawater desalination. *Desalination* 96, 133.

Kurihara, M., Himeshima, Y., and Uemura, T. (1987). In *Preprints of ICOM*, p. 428.

Kurihara, M., Matsuka, N., Fusaoka, Y., and Henmi, M. (2003). Newly developed wastewater treatment systems using separation membranes. In *Proceedings Water Reuse & Desalination Conference*, Suntec Singapore, Singapore, Feb. 25–27.

Kurihara, M., Uemura, T., Himeshima, Y., Ueno, K., and Bairinji, Y. (1994b). Development of crosslinked aromatic polyamide composite reverse osmosis membrane. *Nippon Kagaku Kaishi* 1994(2), 97–107.

Moch, I. (2000). The case for and feasibility of very high recovery sea water reverse osmosis plants. In *Preprints of ADA Conference*, Lake Tahoe, USA.

Morgan, P. W. (1965). Condensation polymers: By interfacial and solution methods. In *Polymer Reviews*, Vol. 10. Wiley, New York.

Nakao, S. (1996). Sea water desalination process for high recovery of fresh water by reverse osmosis. *Bull. Soc. Sea Water Sci. Japn.* 50(6), 406–412.

Ohya, H., Suzuki, T., and Nakao, S. (1996). Proposal and technological breakthrough of an integrated system for the complete usage of sea water. *Bull. Soc. Sea Water Sci. Japn.* 50(6), 389–395.

Redondo, J., Busch, M., and Witte, J. D. (2003). Boron removal from seawater using FILMTEC™ high rejection SWRO membranes. *Desalination* 156, 229.

Rodriguez, M., Ruiz, A. F., Chilon, M. F., and Rico, D. P. (2001). Influence of pH in the elimination of boron by means of reverse osmosis. *Desalination* 140, 145.

Scala, R. C., Ciliberti, D. F., and Berg, D. (1973). Interface condensation desalination membrane. U.S. Patent 3,744,642.

Taniguchi, M., Fusaoka, Y., Nishikawa, T., and Kurihara, M. (2004). Boron removal in RO seawater desalination. *Desalination* 167, 419.

Tomioka, H., Taniguchi, M., Okazaki, M., Goto, S., Uemura, T., and Kurihara, M. (2005). Milestone of high boron rejection seawater RO membrane. In *Proceedings of IDA World Congress on Desalination and Water Reuse*, Singapore, Sept. 11–14.

Toray (2004). Brochure of TM800. Tokyo, Japan.

Uemura, T., and Kurihara, M. (2003). Chlorine resistance of reverse osmosis membranes and changes in membrane structure and performance caused by chlorination degradation. *Bull. Soc. Sea Water Sci. Jpn.* 57, 498.

World Health Organization (WHO) (2004). *Guidelines for Drinking Water Quality*, 3rd ed. WHO, Geneva.

Van Heuven, J. W. (1976). Dynamic membrane. U.S. Patent 3,996,318.

Yamamura, H., Henmi, M., and Inoue, T. (2002). In Development innovative MF and RO membrane for wastewater treatment and reclamation. In *Proceedings of the 2nd International Conference on Application of Membrane Technology*, Beijing, China, Sept. 27–29.

Yamamura, H., Kurihara, M., and Maeda, K. (1994, 1996). Japanese Patent Applications H06-245184 and H08-108048.

CHAPTER 2

Cellulose Triacetate Membranes for Reverse Osmosis

A. KUMANO and N. FUJIWARA

Desalination Membrane Operating Department, Toyobo Co., Ltd., Osaka, Japan

2.1 INTRODUCTION

The reverse osmosis (RO) seawater desalination process has many advantages in the areas of saving energy, lower capital cost, short startup and shutdown time, short construction period, less installation space, and less total water cost. RO technology is becoming the key technology for obtaining freshwater from the sea, especially in the Middle East. Membrane manufacturers are working to develop membranes offering higher product water recovery, lower energy consumption, and lower installation costs in order to enable the RO process to be adopted as the most popular method for supplying freshwater around the world.

The commercialized RO modules consist of cellulose triacetate hollow-fiber type and polyamide spiral-wound type. The cellulose triacetate hollow-fiber RO modules are used around the world for seawater desalination mainly because of excellent features such as a chlorine tolerance and fouling resistance.

This summary describes the history of cellulose triacetate RO membrane, description of cellulose triacetate hollow-fiber RO modules of Toyobo for seawater desalination, actual operation results, and recent RO modules of cellulose triacetate including most recently developed advanced modules.

2.2 HISTORY OF CELLULOSE ACETATE MEMBRANE

2.2.1 Development of Loeb-Sourirajan Membrane

The Saline Water Act was enacted in the United States on July 3, 1952. Then, the Office of Saline Water (OSW) was installed in the Department of the Interior in order to study the method of obtaining freshwater from seawater and brackish water economically. This effort accelerated development of RO membranes. The RO process was proposed as one

Advanced Membrane Technology and Applications. Edited by Norman N. Li, Anthony G. Fane, W. S. Winston Ho, and T. Matsuura
Copyright © 2008 John Wiley & Sons, Inc.

of the methods of desalination by Reid of Florida University in the beginning of 1953. Workers at Florida University studied various commercial polymer films in 1957 and announced a cellulose diacetate film as the outstanding semipermeable membrane in which salt rejection was 96% or more. However, the permeability of water was very small. After that, Loeb and Sourirajan of UCLA succeeded in developing a method of manufacturing a new asymmetric membrane in 1960. The obtained membrane had high permeability and consisted of a heat-treated cellulose diacetate asymmetric membrane. These improvements in advanced performance of a membrane led to a practical application of the RO membrane module that was promoted (Breton, 1957; Reid and Breton, 1959; Loeb and Sourirajan, 1964).

2.2.2 Development of Commercial Cellulose Acetate Membrane Modules

Gulf General Atomic was funded by OSW and developed a spiral–wound-type module. Moreover, the company applied for the patent of the basic structure in 1964. In the patent, the example that used the cellulose diacetate membrane of Loeb and Sourirajan was indicated (Westmoreland, 1968). The improvement was performed after that in each company, and the spiral-wound-type module using a cellulose acetate membrane was put on the market by many companies, such as UOP, Hydranautics, Envirogenics, Toray Industries, and Daicel.

As for tubular-type modules, from the 1960s, various models of tubular-type modules were developed and put on the market by many companies. The RO plant using the module that Loeb and others developed worked in 1965 (McCutchan and Johnson, 1970).

A hollow-fiber-type module was developed and a fundamental patent was filed by Dow Chemical in 1960. An RO module using the cellulose triacetate hollow fiber is indicated there (Mahon, 1966). A significant portion of research and development at Dow Chemical was carried out based on its research contract with OSW, and the development results of the RO module for brackish water was published in 1970 and for seawater in 1974. The RO module using cellulose triacetate hollow fiber for brackish water was marketed in 1974 (Dance et al., 1971; Ammons and Mahon, 1974).

Research and development of the hollow-fiber-type module using a cellulose acetate membrane was conducted by Monsanto, Toyobo, and others, in addition to Dow Chemical. Toyobo announced an RO module for one-pass desalination of seawater that used the cellulose triacetate hollow-fiber membrane module in 1979 (Orofino, 1970; Ukai et al., 1980).

2.3 TOYOBO RO MODULE FOR SEAWATER DESALINATION

2.3.1 Hollow-Fiber RO Membrane for Seawater Desalination

Seawater desalination by reverse osmosis is the most effective method for the production of freshwater among various desalination technologies. Hollow-fiber RO membranes and flat-sheet membranes have been developed for brackish water and seawater desalination by a two-pass process since 1976 (Ohya, 1976). In spite of a satisfactory result of two-pass seawater desalination processes, the one-pass process has the advantages of simple and compact plant, simple operation, easy maintenance, and the lowest energy consumption. Although several one-pass seawater desalination systems by reverse osmosis have been developed so far, the membrane performance, especially salt rejections, have not been satisfactory and not stable in long-term operation (Macgowan et al., 1976; Ohya, 1978). In 1979, Toyobo succeeded in developing the high-performance, especially high-salt rejection, hollow-fiber RO membrane of cellulose triacetate for seawater desalination by one-pass process for the first time at a practical level (Ukai et al., 1980). This hollow-fiber membrane had high-pressure resistance in addition to high-salt rejection. In general, the salt rejection of membranes varies inversely as permeate water flux. In flat-sheet membrane modules, such as spiral-wound modules, high membrane permeate water flux is required to obtain practical product water flux due to narrow membrane area in the modules. The spiral-wound module configuration may also result in insufficient salt rejection. To solve this conflict, thin-film composite membranes were developed. In contrast, the hollow-fiber membrane module usually has a large membrane surface area, which makes it possible to design the module performance without any special method such as thin-film composite membranes.

The cellulose triacetate (CTA) hollow-fiber RO membranes are prepared by spinning a doped solution of CTA polymer followed by soaking and annealing. An outline of hollow-fiber preparation is shown in Figure 2.1 (Aptel and Buckley, 1996).

Optimization of preparation conditions such as hollow-fiber spinning technology using high concentrations of polymers, micropore control technology in manufacture, and post-treatment by high-temperature treatment made it possible to increase the permeate water flux of the hollow-fiber membrane without additional steps. CTA–hollow-fiber RO membrane itself has outstanding pressure resistance retention, and the hollow-fiber design and selection of suitable dimensions provide high-pressure resistance. Outer diameter is about 165 μm and inner diameter is about 70 μm. Resistance of this hollow-fiber membrane against high pressure is a critical characteristic to achieve a practical performance level. A microscopic view of hollow-fiber membrane for seawater desalination is shown in Figure 2.2.

The membrane made from CTA, which give an improved membrane performance, continue to be used widely today because of high performance and long-term reliability. CTA–hollow-fiber features superior chlorine tolerance compared with polyamide membrane as shown in Figure 2.3.

In biologically active seawaters sterilization by chlorine is considered a very effective solution to prevent biological fouling from occurring in the RO process. Material properties

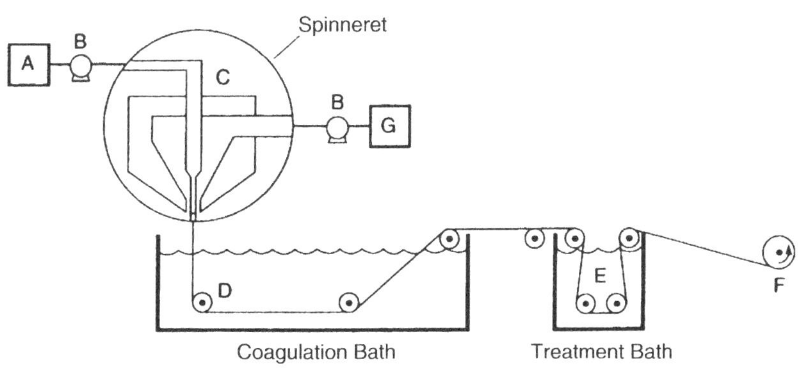

Figure 2.1 Outline of hollow-fiber preparation.

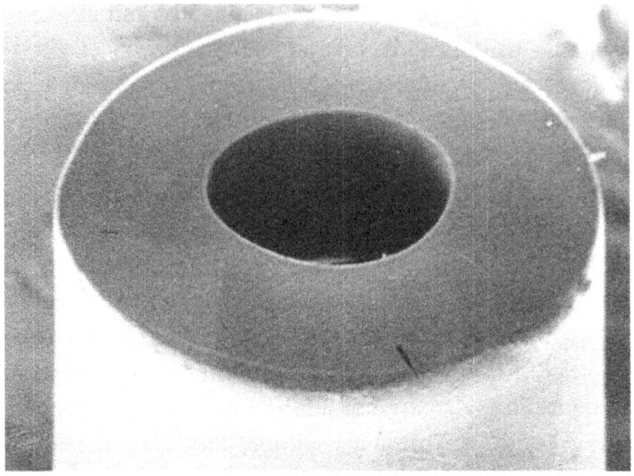

Figure 2.2 Microscopic view of hollow-fiber membrane.

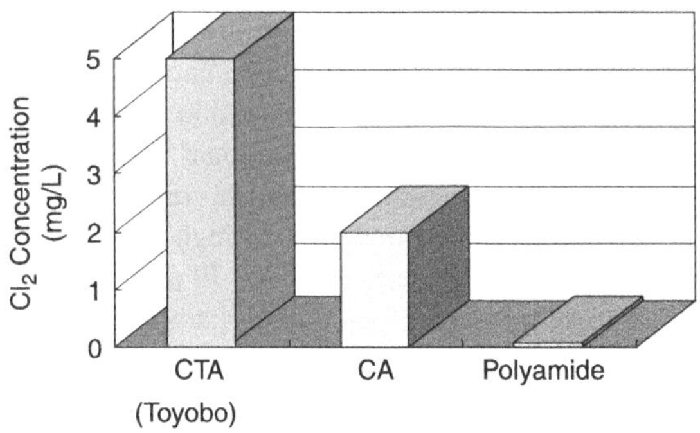

Figure 2.3 Comparison of chlorine resistance.

of CTA chemistry make it possible that the CTA–hollow-fiber membrane RO process can be simply controlled by only chlorine injection, which prevents biological fouling. The sterilization operation mode can be optimized by continuous or intermittent chlorine injection.

2.3.2 Seawater Desalination RO Module

The bundle of several tens of thousands of hollow fibers is fabricated into an element and installed in a module. Hollow fibers in an element are arranged in a mutually crossed configuration without any kinds of supporting materials between hollow-fiber layers.

The hollow-fiber membrane is wound into a bundle with a layered, cross arrangement, as shown in Figure 2.4, and moderately sized. Regular intervals are left between the hollow fibers to minimize pressure differentials and allow uniform flow and achieve small pressure drop.

These features minimize concentration polarization and extend the allowance of fouling index of feed water up to SDI = 4. This structure is also less prone to being clogged by fouling matter. From the microscopic photograph shown in Figure 2.5, the space of Toyobo's hollow fiber is remarkably large, and an accordingly rapid increase of differential

Figure 2.4 Cross arrangement of hollow fiber.

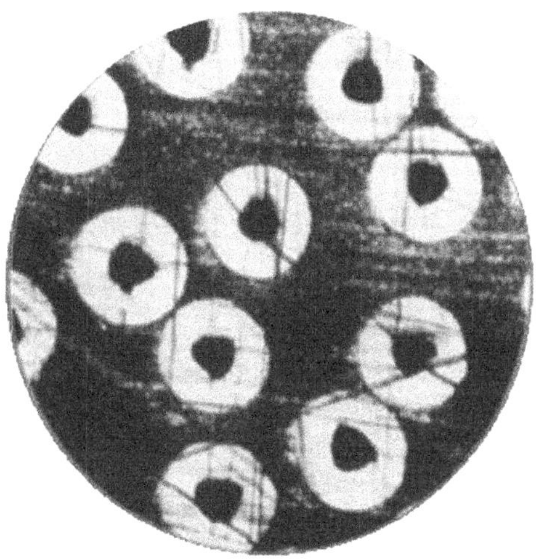

Figure 2.5 Cross section of hollow fibers.

pressure is not observed in the RO process. This characteristic allows an RO plant to be operated with easy maintenance.

Conventional hollow-fiber RO module consisted of a single element inserted in one pressure vessel. However, economical and technical considerations demanded modules with a number of elements inserted in one pressure vessel. Toyobo successfully produced double-element modules on a commercial basis for the first time in 1979 and was followed by various types of double-element modules based on their flow path design.

Toyobo double-element modules, with many advantages, were developed in 1981. This module structure and flow pattern is shown in Figure 2.6.

Feed water enters the feed center pipe of the feed-side element, feed water flows radially from the center pipe, past the hollow fibers in the bundle, and away from the element as concentrated brine of the feed-side element. Feed flow is uniformly distributed in the RO module throughout the cross-wound hollow-fiber bundle. The brine water of feed-side element flows on the periphery of the element, passes through the narrow space between the pressure vessel and element, and flows to the periphery of the brine-side element and

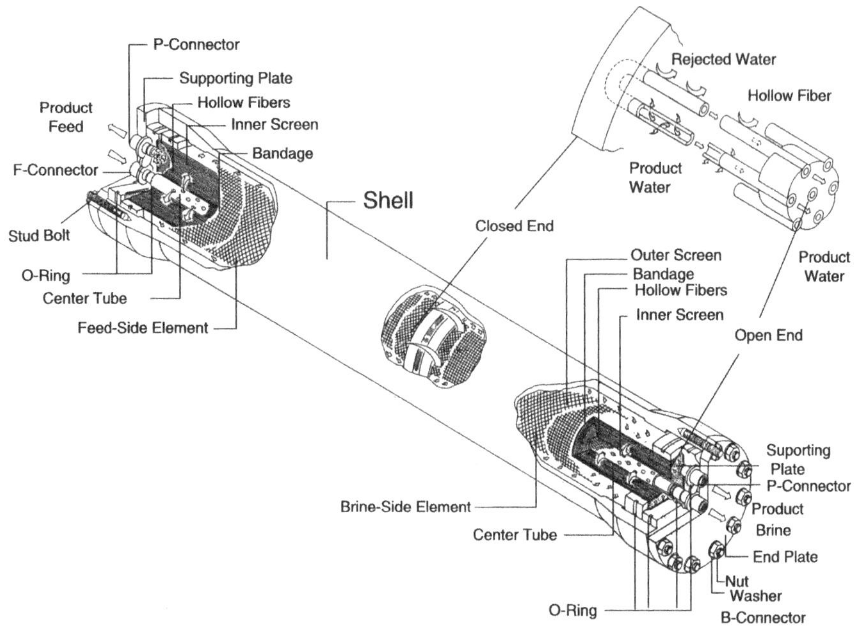

Figure 2.6 Structure of the hollow-fiber RO module (HM type).

enters and passes hollow fibers in the bundle radially from outside to inside. Brine water flows to and passes through the center tube of the brine-side element to brine port and away from the module. Permeate water is collected with supporting plates and passes through the permeate pipe of each side element.

2.3.3 Features of Toyobo RO Module for Seawater Desalination

2.3.3.1 Hollow-Fiber Configuration Hollow-fiber modules offer greater surface membrane area in the module than spiral-wound modules. This allows a high water production of the module and a smaller footprint of module banks in a desalination plant owing to high compactness. Figure 2.7 illustrates Toyobo's hollow-fiber array fixed in epoxy resin at both ends.

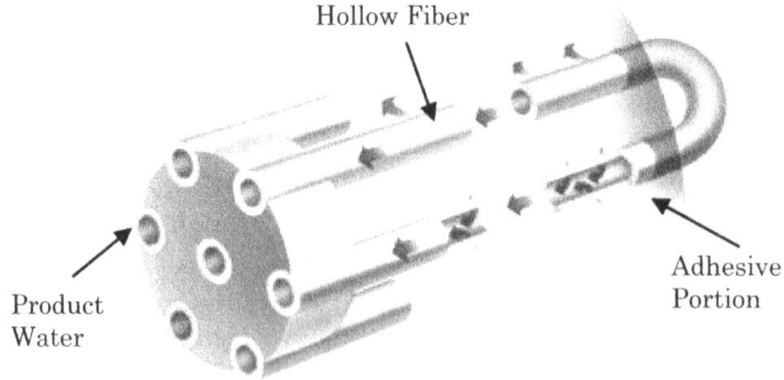

Figure 2.7 Hollow fibers with epoxy resin.

This provides mechanical stability to the fiber array in the RO element. The fibers at one end of the element are precisely cut so that product water can be discharged from the bore of the fibers. Toyobo RO modules for seawater desalination have configurations in the double-element type as shown above. Because the permeate water of each element in a pressure vessel can be obtained directly, the quality of each permeate water can also be measured directly. This feature allows easier maintenance of an RO plant (Kannari, 1995).

2.3.3.2 Stable RO Performance

Membrane Physical Structure The physical structure of the hollow-fiber membrane in Toyobo RO modules is optimized to provide mechanical strength resistance to collapse in a hydrostatic pressure environment. This is achieved by spinning the hollow-fiber geometry at dimensions consistent with the mechanics of a tiny hollow tube or pipe and the material properties of a strong CTA polymer.

2.3.3.3 Superior Fouling Resistance

Chlorine Resistance Chlorine resistance is significantly better with the CTA membrane chemistry than with membrane chemistry based on polyamide. This allows the biological control and stable operation of the RO module by the intermittent chlorine injection (ICI) disinfectant method.

Optimum Permeability Continuous flux (permeability or flow per membrane surface area) across membranes can cause performance difficulties in a desalination process due to deposition of fouling materials on the RO membrane surface. Generally, the higher the flux is, the sooner the membrane surface will become coated with fouling material. Spiral-wound membrane elements offer relatively low membrane surface area, and hence high flow must be more restricted to prevent fouling. Hollow-fiber membrane elements offer about 10 times greater membrane surface area than spiral-wound membrane elements. Figure 2.8 shows a comparison of fouling tendency.

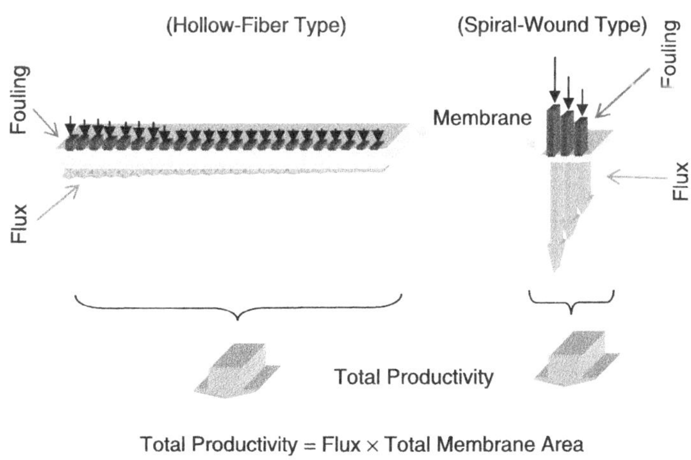

Figure 2.8 Comparison of fouling tendency in membrane type.

The larger membrane surface area advantage of the hollow-fiber element allows the same quantity of permeate water to be produced at a permeability about one-tenth that for spiral-wound elements. This results in much fewer restrictions in operation and less frequent cleaning.

Cross-Winding Style A hollow-fiber membrane manufactured via a multifilament spinning process allows element construction with parallel fiber arrays or filament winding, that is, cross-layered arrays. This unique fabrication allows regular open spacing between the hollow fibers, which minimizes element pressure drop and assures uniform flow throughout the element.

2.4 ACTUAL PERFORMANCE OF TOYOBO RO MODULE FOR SEAWATER DESALINATION

2.4.1 Actual Performance Around the World

The main features of a CTA membrane are summarized in following two points:

1. Chlorine is an effective disinfectant of CTA RO module and chlorine allows the RO membrane module to be directly sterilized.
2. A CTA membrane has low adhesion of the fouling matters due to its hydrophilic property to an organic matter.

These characteristics are most important for a seawater desalination plant because seawater has high potential for microbial growth in the module since there are many sources of nutrients. If the seawater RO module cannot be sterilized by chlorine, high microbe multiplication occurs and propagates on the membrane surface. Microbe propagation within an RO module will cause serious decrease of product water quantity and deterioration of water quality. Moreover, frequent membrane cleaning is required, plant downtime increases, and the amount of chemicals increases due to cleaning. However, such performance deterioration and trouble does not occur in the case of an RO module made from a CTA membrane, which can be sterilize directly by the chlorine as a disinfectant. Therefore, many RO modules made from a CTA membrane are adopted in the Arabian Gulf countries where the possibility of microbe multiplication is high due to surface intake unlike the beach-well intake and high temperature. The main desalination plants that have adopted RO modules made from CTA of Toyobo are listed in Table 2.1. Table 2.1 shows large-scale seawater desalination RO plants using Toyobo RO membrane modules.

2.4.2 Jeddah 1 RO Plant in Saudi Arabia

The Jeddah 1 RO seawater plant is introduced as an example of a Red Sea coast plant on the Arabian Peninsula. The Jeddah 1 RO Phase I plant, which has a capacity of 15 MGD (56,800 m^3/day), went into operation in 1989. The same sized Phase II plant came on stream in 1994, giving the plant a total capacity of 30 MGD (113,600 m^3/day) (Al-Badawi et al., 1995). The specifications of the Jeddah Phase I and Phase II RO plants are shown in Table 2.2.

The two plants are of almost the same construction, but the membrane at the Phase II plant was guaranteed for 5 years. Product water is blended with that from the MSF plant

TABLE 2.1 Main Supply Record of Toyobo Hollow-Fiber RO Module

Plant	Country	Capacity (m^3/day)	Startup
Rabigh	Saudi Arabia	205,000	2008
Medina-Yanbu	Saudi Arabia	128,000	1998
Jeddah 1 Phase I	Saudi Arabia	56,800	1989
Jeddah 1 Phase II	Saudi Arabia	56,800	1994
MARAFIQ-Yanbu	Saudi Arabia	50,400	2004b
Fukuoka	Japan	50,000	2005
AdDur	Bahrain	22,750 (45,500)	2005
Tanjun-Jati B	Indonesia	10,800	2005
Florida	United States	11,400	2005
Tanajib	Saudi Arabia	6,000	2001
Duba	Saudi Arabia	4,400	1989
Haql	Saudi Arabia	4,400	1989

TABLE 2.2 Plant Specifications of Jeddah 1 RO Plants

	Phase I	Phase II
Number of trains	10	10
Capacity	1.5 MGD × 10	1.5 MGD × 10
Permeate quality Cl^-	Less than 625 mg/L	Less than 625 mg/L

and distributed to Jeddah. The tight water supply in Jeddah just meets demand, and so product water from the plant is of crucial importance. The maintenance period must be kept to a minimum and the plant continuously operated at full capacity.

Toyobo double-element-type hollow-fiber RO modules (HM10255) are used for both Jeddah Phase I and Phase II plants. Figure 2.9 illustrates the schematic flow of the Jeddah 1 Phase II plant.

Raw seawater is taken from the Red Sea and then disinfected by sodium hypochlorite produced by a chlorine generation plant using filtered seawater as feed. Then ferric chloride as coagulant is added to the seawater feed ahead of dual-media filter (DMF) to help in reducing SDI values to suit the recommended SDI required by the membrane manufacturer. The filtered feed water is then collected in a clear well, and particles greater than 10 μm will not pass a cartridge filter in order to avoid membrane plugging. Sulfuric acid is added to the feed water to adjust pH to about 6.5. Sodium bisulfite (SBS) is injected for 7 h every

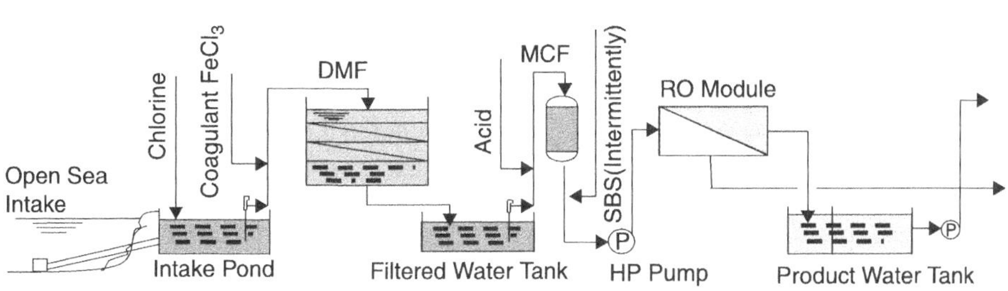

Figure 2.9 Schematic flow of Jeddah Phase II plant.

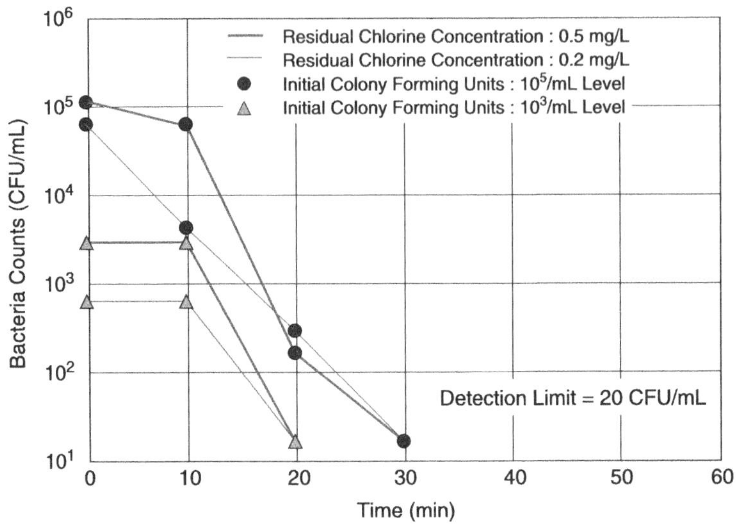

Figure 2.10 Disinfecting power of chlorine.

shift (8 h) ahead of the high-pressure pump to nullify residual chlorine and protect the membrane from oxidation by residual chlorine in the presence of heavy metals. To avoid biological fouling, 0.2 mg/L residual chlorine is allowed to pass through the membrane for 1 h every shift (8 h) intermittently by the ICI method instead of a conventional continuous chlorine injection (CCI) method.

The disinfecting power of chlorine was tested in Japan by cultivating bacteria taken from the Jeddah seawater. The results were as shown in Figure 2.10 (Fujiwara, 1999a).

It was confirmed that even when the bacteria count was high, disinfection with a chlorine concentration of 0.2 mg/L reduced it almost to zero in 30 min.

To determine whether the disinfection was sufficient, the concept of "lead time" was considered (Nada et al., 1994). Lead time is the time from when injection of SBS is halted and chlorine is fed into the module to when a certain chlorine concentration is detected in filtered water in the module. Except certain times when the SDI of feed seawater exceeded 4.5 and seawater was polluted, lead time remained stable at 10 min. This signifies that a 10-min injection of chlorine at a concentration of 0.2 mg/L was sufficient to disinfect the module. In fact, it was confirmed that differential pressure in the module was stable and no biofouling occurred during the test period.

2.4.2.1 Operation Results of Phase II Plant Due to the successful site testing by the ICI method at Jeddah, the operating conditions were kept the same for the Phase II plant as shown in Table 2.3.

This plant is of crucial importance to Jeddah, which is prone to water shortages, and barring times when the feed seawater is polluted and the SDI exceeds 4.5, it operates continuously at 99% capacity. It is important to note that the membrane has not been replaced at all in the 5 years since the plant went into operation.

Product Flow Rate and Quality Changes in the permeate flow rate are shown in Figure 2.11. The permeate flow rate remained stable at the rated 56,800 m^3/day throughout

TABLE 2.3 Plant Specification of Jeddah 1 RO Plants

Feed TDS	43,300 mg/L (Open-Sea Intake)
Feed temperature	24–34°C
Feed pressure	57–64 bars
Recovery	35%

the entire period. As is evident from the diagram, all 10 systems at the plant were in virtually continuous operation. Permeate total dissolved solids (TDS) are shown in Figure 2.11. Permeate TDS stayed below 500 mg/L without the membrane being replaced at all in 5 years, which is highly satisfactory (Fujiwara et al., 1999a).

Differential Pressure One advantage of ICI is the importance of preventing biofouling. While biofouling should not occur when the module is disinfected by CCI, intermittent disinfecting of the module necessitates caution. Disinfection tests were therefore conducted by cultivating bacteria taken from Jeddah seawater as described before, and extended testing was carried out using site test units to confirm that biofouling did not occur. As a result, differential pressure was stable at the Phase II plant as shown in Figure 2.12, and it was found that periodic cleaning with citric acid once a year reduced differential pressure to almost what it was at the start (Fujiwara et al., 1999a).

At the Jeddah Phase II plant, as a result of ICI adoption, product flow and product quality were excellent even with fact that the membrane was not replaced in 5 years.

Next we discuss the significant improvement of the performance of an existing plant by exchanging the RO membrane of the plant for a CTA membrane. Here, the operation test results at the AdDur plant in Bahrain is introduced as an example on the Arabian Gulf coast.

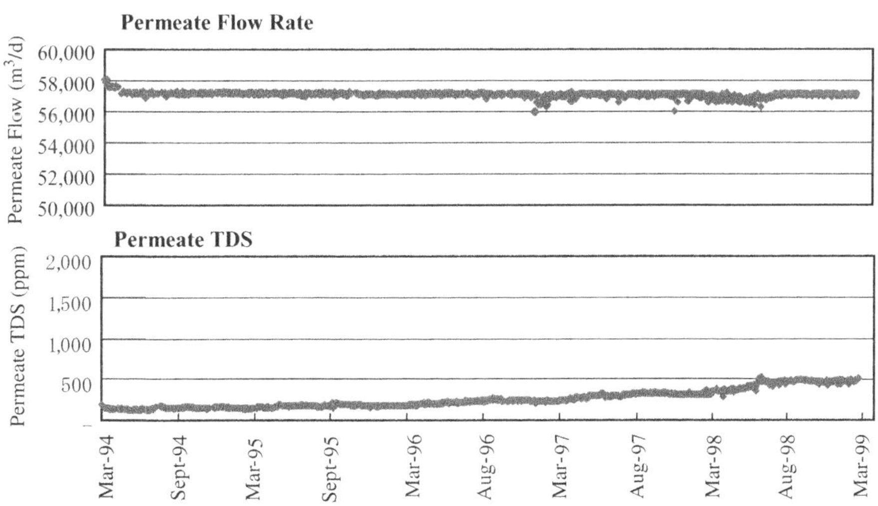

Figure 2.11 Permeate flow rate and permeate TDS of Jeddah Phase II plant.

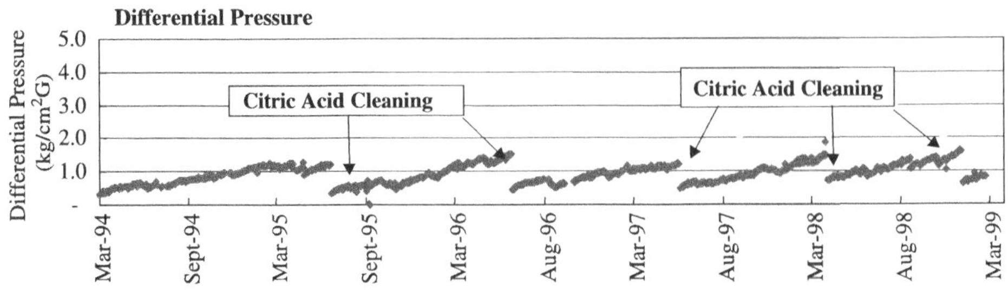

Figure 2.12 Differential pressure of Jeddah Phase II plant.

2.4.3 The AdDur Plant in Bahrain

The AdDur RO desalination plant was designed to produce 10 MGD (45,500 m^3/day) with polyamide RO hollow-fiber membrane modules from seawater containing up to 45,000 mg/L as TDS. This plant was commissioned in 1990. The AdDur plant process consisted of the dual-media filter (DMF) with coagulants, dechlorination by sodium bisulfate, high-pressure pumps, RO modules, and posttreatment by lime injection. The specifications of the AdDur plant are shown in Table 2.4 (Al-Badawi et al., 1995).

In 2000, rehabilitation work was conducted to obtain improved performance of the pretreatment by means of ultrafiltration (UF) membranes that proved most effective among the tests conducted. The work included modification of existing dual-media filter to single-media filter and additional installation of spiral-wound-type UF having a capacity of 130,000 m^3/day after the media filter in order to reduce SDI values. Chemical dose was stopped together with modification of media filter. The filtered seawater from the UF process was fed to the existing first pass 8 RO trains of the polyamide membrane module. However, even after installation of the UF system, the RO membranes required frequent cleaning. It was concluded that additional action was required to achieve more stable operation. It was suspected that the UF membrane had poor removal of natural organic materials in the raw seawater, so that such organic matter would act as a nutrient for biological activities in the RO membranes.

2.4.3.1 Schematic Flow Diagram of CTA Membrane Test Unit

Reverse osmosis feed water to the test plant is taken from acidified UF filtrate and is exactly the same as the actual RO feed water to the plant. SBS is dosed into the feed water to remove residual chlorine contained in pretreated water. The SBS dose is stopped three

TABLE 2.4 Plant Specification of AdDur RO Plant

Operation Conditions of Design	
Feed temperature	16–36°C
Feed pressure	Maximum 68 bars
Feed water TDS	45,000 mg/L
Recovery	35%
Total number of trains	8 trains, each divided into 4 banks
Specifications of Design	
Plant capacity	10 MIGD (45,500 m^3/day)
Product TDS	<500 mg/L

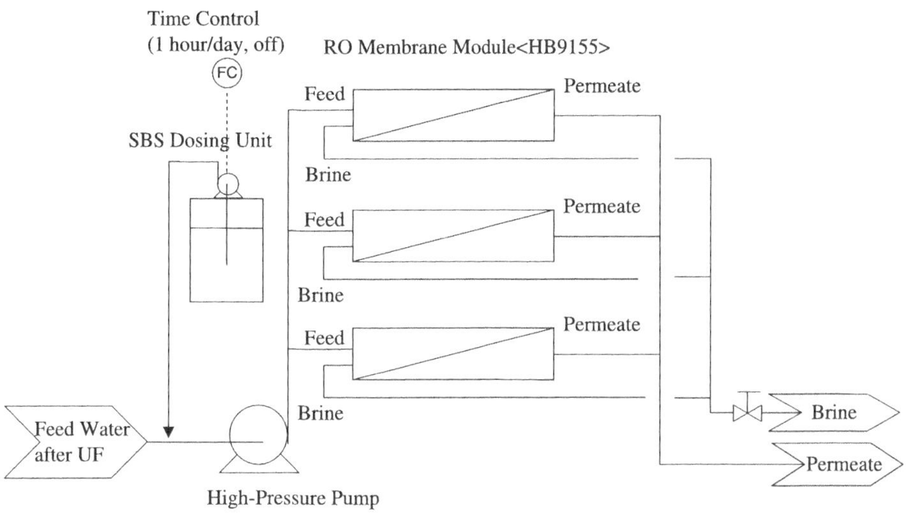

Figure 2.13 Schematic flow diagram of test unit.

times a day for 1 h to introduce residual chlorine into the RO membrane for sterilization purposes. Three Toyobo RO modules, HB9155 Model, were used in this test. A schematic flow diagram of the test unit is shown in Figure 2.13.

2.4.3.2 Test Conditions

Test conditions and test requirements are shown in Table 2.5. The test was aimed at satisfying the following plant specification. For prevention of biological fouling, it was shown that the ICI method provided effective chlorine injection.

2.4.3.3 RO Performance at Site Test

Permeate Flow Rate and Permeate Quality Actual permeate flow rate is shown in Figure 2.14. The permeate flow rate was gradually increased to about 7.8 L/min per module due to adjustment of operation conditions during about 10 days from startup. Then the permeate flow was set at about 7.5 L/min per module (10.77 m^3/day per module at design conditions). During the winter season the permeate flow rate was maintained by increasing feed pressure. The flow rate was stable and satisfied the plant requirements for 12 months. Actual permeate quality is shown in Figure 2.15.

TABLE 2.5 Test Conditions and Test Requirements

Test Conditions	Test Requirements
Feed pressure: 6.9 MPa max.	1. Permeate quality: Less than 500 mg/L as TDS
Feed temperature: 16–37°C	2. Permeate flow rate per module: 10.77 m^3/day
Feed TDS: 45,000 mg/L	3. Differential pressure: Normal increase or no change
Recovery: 35%	
Chlorine injection: ICI	

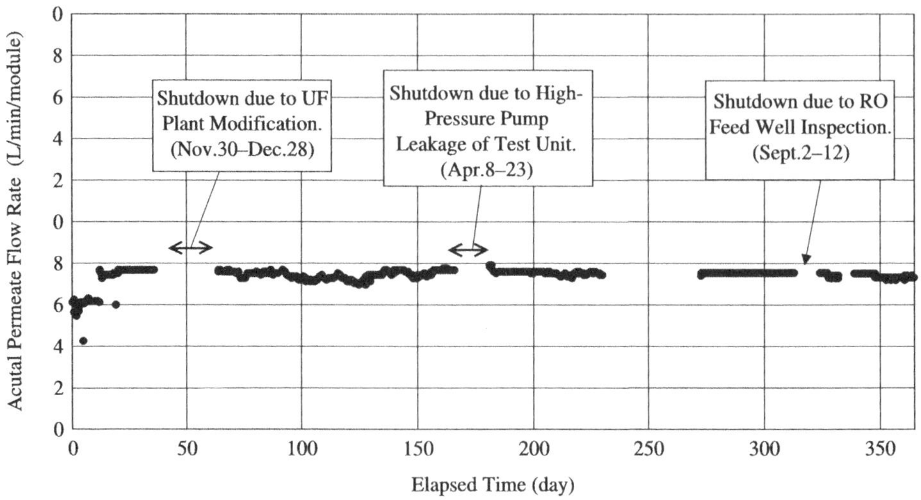

Figure 2.14 Permeate flow rate of test unit in AdDur plant.

The permeate quality was 764 μS/cm at the startup of operation and then gradually decreased to about 200 μS/cm due to natural membrane compaction. The permeate quality was stable at a level of 200 μS/cm (95 mg/L as TDS at design conditions) through the test period. This performance was much better than the plant requirement of 500 mg/L as TDS. RO performance (permeate flow rate and permeate quality) was very stable and satisfactory during the entire test period (Alawadhi et al., 2005).

Differential Pressure A differential pressure can be used as an indication of biological fouling growth in the RO module as shown in Figure 2.16. The differential pressure was stable at a low level of 20 kPa without the chemical cleaning. Therefore, it is considered that biological fouling had not occurred in RO modules for the 12-month test period. Also, it was concluded that the ICI method worked effectively to prevent biological fouling.

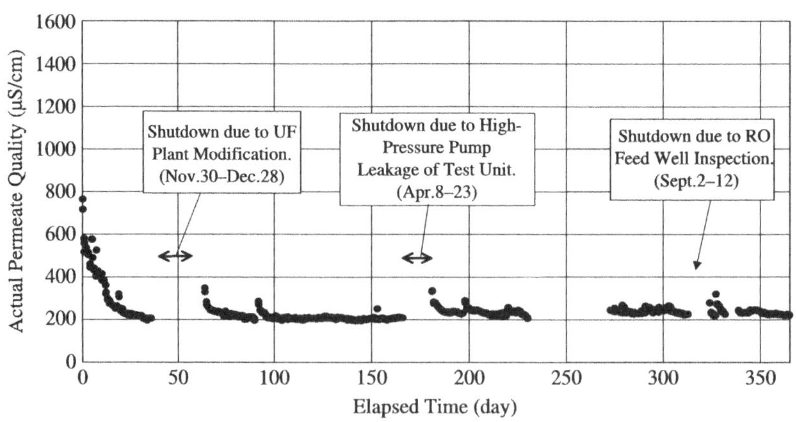

Figure 2.15 Permeate quality of test unit in AdDur plant.

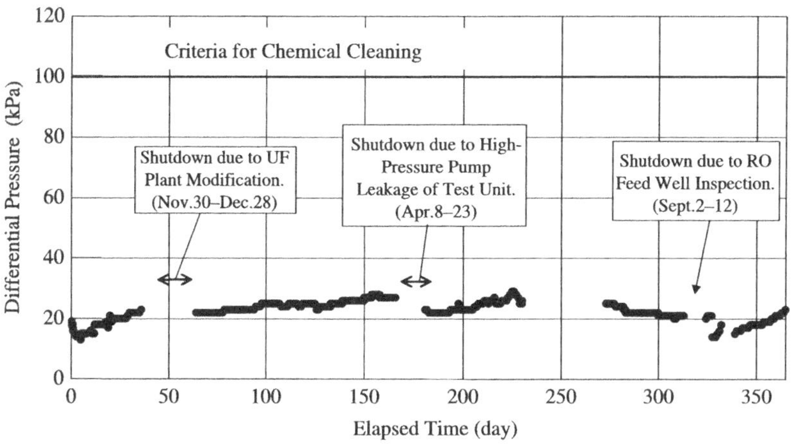

Figure 2.16 Differential pressure of test unit in AdDur plant.

For a 12-month test period, RO performance of HB9155 based on CTA material was very stable and satisfied the test requirements. Therefore, it was demonstrated that RO plant performance could be recovered by replacement of existing polyamide membranes with CTA membranes. Differential pressure was stable at a low level of about 20 kPa without the need for chemical cleaning during the test period. Therefore, it is concluded that the ICI method worked effectively to prevent biological fouling.

In general, the Arabian Gulf region is a difficult area for an RO plant operation due to its biological activity. Control of biological activity in the RO membrane is key to a stable performance in the region. The result obtained through this test is helpful to the RO plant operation in this region.

2.5 MOST RECENT RO MODULE OF CELLULOSE TRIACETATE

2.5.1 Development of RO Module for Higher Recovery

The reverse osmosis seawater desalination process has many advantages from the viewpoints of saving energy, lower capital cost, short startup and shutdown time, short construction period, less installation space, and less total water cost. In a seawater RO process, the seawater is subjected to disinfection, coagulation-filtration, and acidification processes in the pretreatment section and forwarded as feed water to RO section. When the recovery fraction (the ratio of product flow rate to feed flow rate) is high, the amount of feed water required for a desired production is lower and hence, the pretreatment system, chemical cost, equipment sizing, and energy costs are significantly reduced (Ohya et al., 1996; Nakao, 1996).

Membrane manufacturers are working to develop membranes offering higher water recovery, lower energy, and lower installation cost in order to enable the RO process to be recognized as the most popular method for supplying freshwater around the world.

In areas such as the Middle East where seawater has high salinity, commercial seawater RO desalination plants were normally designed to operate at approximately 35% recovery. This relatively low recovery was due to the very high osmotic pressure of the seawater, and most commercially available RO membranes did not allow operating pressures above

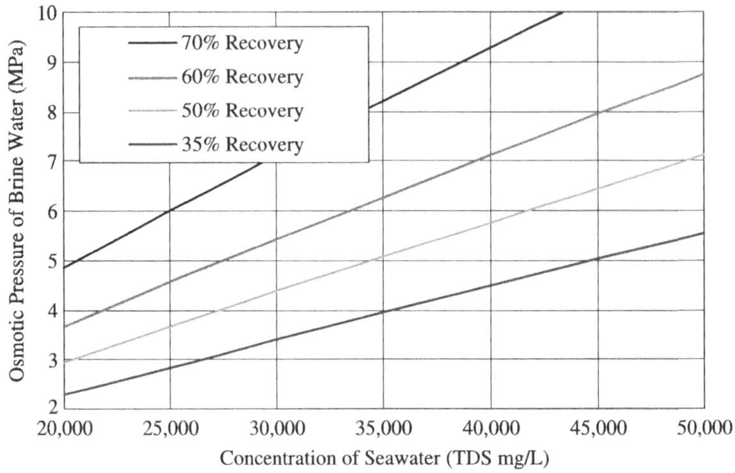

Figure 2.17 Osmotic pressure of brine water.

7.0 MPa. Osmotic pressure of seawater from the Red Sea is approximately 3.2 MPa. In the case of operation with recovery at 35%, the osmotic pressure of brine is about 4.8 MPa in the RO module, as shown in Figure 2.17.

If the recovery is increase to about 50%, the osmotic pressure of the brine increases to about 6.1 MPa. Therefore, a higher feed pressure (i.e., driving force) is required for a high-recovery operation in comparison with a conventional low-recovery operation. In order to enable the high-recovery operation, the RO module must be designed and manufactured to withstand the higher pressure.

Toyobo recently developed a new type of RO module to achieve higher product water recovery in order to further reduce the cost of RO desalination. Toyobo's hollow-fiber RO modules are widely used around the world in RO desalination plants. Based on the long operating experience and recent research efforts, Toyobo developed the high-pressure high-flux HB series modules. The HB series is an improved version of the conventional HM series type of module using the same materials (Ohnishi et al., 1997; Sekino, 1998; Fujiwara et al., 1999b). The hollow-fiber membrane in the HB series module is wound in a cross arrangement, designed to minimize pressure loss and allow uniform water flow in the module. The hollow fiber incorporated in the HB series has higher pressure resistance based on a change of the hollow-fiber outer diameter/inner diameter dimensions and optimization of manufacturing conditions. The specification of HB series modules are shown in Table 2.6.

The product flow rate of the new type improved by about 1.4 times compared with the conventional type. A high-pressure, single-pass desalination process of new HB series modules with high recovery was successfully conducted for the first time at an RO test plant on the Red Sea at the conditions of more than 52% recovery (Kumano et al., 2003).

2.5.2 Development of Both Open-Ended RO Module

Toyobo's newest innovative RO module builds on the proven reliability of the HM-type module. The new technology is based on both open-ended (BOE) hollow-fiber membrane structures versus single open-ended hollow-fiber membrane structures. The

TABLE 2.6 Specification of HB Series Modules

Model	Conventional Model HM10255	New Model HB Series HB10255	New Model HB Series HB9155
Element diameter (mm)	280	280	216
Product flow rate (m^3/day)	45	62	15
Product salt rejection[a] (%)	99.6	99.6	99.6
Number of elements (—)	2	2	1
Test conditions			
NaCl concentration (mg/L)	35,000	35,000	
Pressure (MPa)	5.4	5.4	
Temperature (°C)	25	25	
Recovery (%)	30	30	
Operation conditions (max.)			
Pressure (MPa)	6.9	8.2	
SDI (—)	4	4	
Temperature (°C)	40	40	
Residual chlorine[b] (mg/L)	1.0	1.0	

[a]Salt rejecton = (1 salt concentration in product water/salt concentration in feed water) × 100.
[b]Residual chlorine is limited by feed water quality.

BOE hollow-fiber membrane structure allows the reduction of the pressure drop along the hollow-fiber bore that leads to both greater salt rejection due to greater dilution effects and greater permeate water flow.

A comparison for the case where the effective fiber length, L_e, is reduced by opening both ends of the fiber to allow flow from each end compared to just one end is illustrated in Figure 2.18. Also shown in Figure 2.18 is a comparison of the average bore pressures for the BOE type versus the single open-ended type.

The maximum bore pressure in both cases is the pressure in the fiber at the furthest distance from the open end of the fiber. In the case of BOE type the maximum bore pressure is moved from the end of a longer fiber of length L to the midpoint of fiber length $L/2$, and the average pressure is reduced. The net effect of this change increases the amount of water flow since less bore pressure drop occurs for the same applied pressure. The BOE reverse osmosis element structure and flow pattern is illustrated in Figure 2.19.

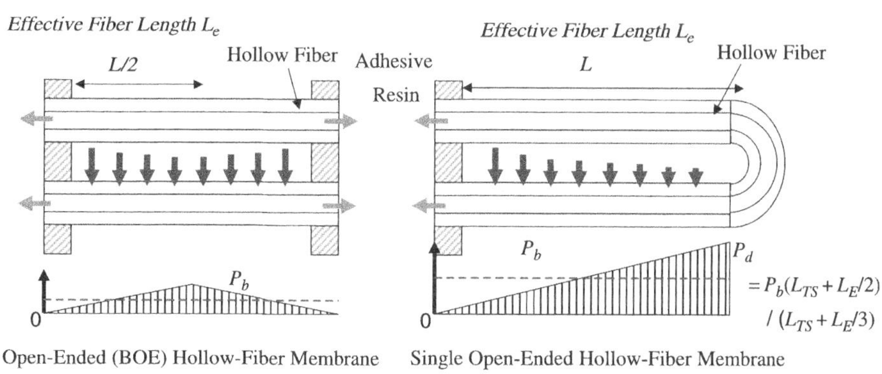

Figure 2.18 Comparison of both open-ended (BOE) vs. single open-ended hollow-fiber membrane.

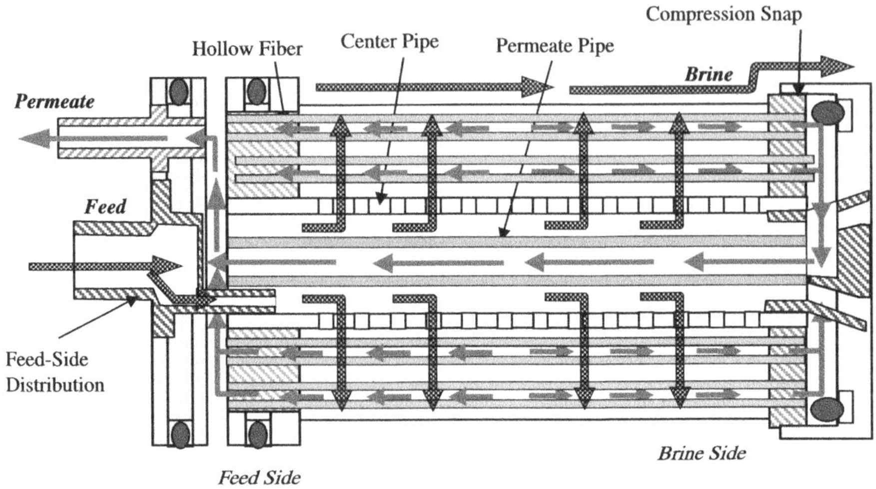

Figure 2.19 Cross section and flow pattern of both open-ended (BOE) hollow-fiber element.

The module structure also consists of two elements and is shown in Figure 2.20. The center pipe of the RO module is composed of a concentric pipe. Permeate flows in the inner tube of the concentric pipe, and feed water flows between the outer tube and the inner tube of the concentric pipe. Feed water enters the feed center pipe of the feed-side element through the feed-side distribution connector. The distribution connector divides intersecting flow of the feed and permeate. Feed water flows radially outward from the center pipe, past the hollow fibers in the bundle, and away from the element as concentrated brine flow. Feed flow is uniformly distributed in the RO module throughout the cross-wound hollow-fiber bundle.

Permeate water is recovered from both ends of the BOE reverse osmosis module and is collected with each supporting plate, respectively. At the opposite side of the feed water inlet (brine side), permeate water is collected through the supporting plate and passes through the permeate pipe in the inside of the concentric pipe. This permeate combines with the feed-side permeate water and flows to the permeate port of the pressure

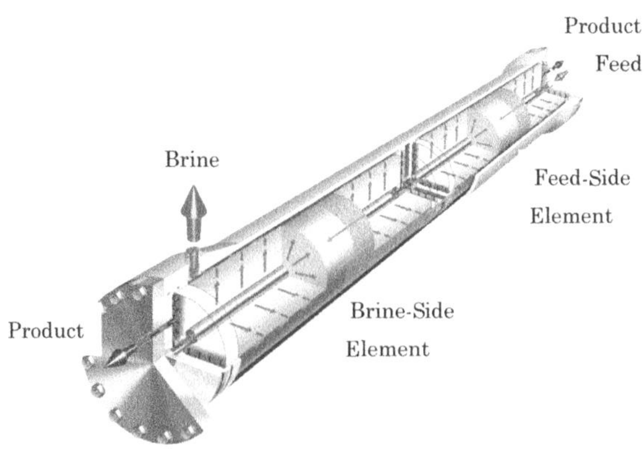

Figure 2.20 Both open-ended (BOE) hollow-fiber module.

vessel. The supporting plate is held in position using compression snaps and an O-ring seal on the face of open end, so that concentrate water cannot penetrate into permeate water area. The compression snap has another function, which is to center the element in the pressure vessel and form a narrow space between the pressure vessel and element for the brine water flow.

Brine water flows on the periphery of the RO element passes through the narrow space between pressure vessel and supporting plate and flows to the brine port of the pressure vessel. The specification of the HB series modules are shown in Table 2.7. The product flow rate of the BOE module improved by about 1.5 times compared with thc conventional type (Kumano, 2004).

2.5.3 Fukuoka Seawater RO Desalination Plant

The Fukuoka seawater desalination plant has a capacity of 50,000 m^3/day, the largest plant in Japan. The 60% recovery of the RO desalination system is the highest seawater recovery level in the world. To achieve this high recovery rate, Toyobo's advanced RO modules were adopted.

The specifications of the plant is shown in Table 2.8. The production capacity is 50,000 m^3/day and the permeate TDS is 200 mg/L. The process flow diagram and membrane arrangement is shown in Figure 2.21.

This plant has adopted an infiltration intake system, an ultrafiltration system, and a high-pressure RO system that provides 60% recovery. A low-pressure RO membrane partial second pass is used in the posttreatment system and helps to improve the quality of product water as needed.

TABLE 2.7 Specification of BOE Module

Model	Conventional Model HM10255 (HM10155) Single Open-Ended (SOE)	New Model HB10255FI (HD10155) Both Open-Ended (BOE)
Element diameter (mm)	280	280
Product flow rate (m^3/day)	45	67
Product Salt Rejection[a] (%)	99.6	99.6
Number of elements (—)	2	2
Test conditions		
NaCl concentration (mg/L)	35,000	35,000
Pressure (MPa)	5.4	5.4
Temperature (°C)	25	25
Recovery (%)	30	30
Operation conditions (max.)		
Pressure (MPa)	6.9	8.2
SDI (—)	4	4
Temperature (°C)	40	40
Residual chlorine[b] (mg/L)	1.0	1.0

[a]Salt rejection = (1 salt concentration in product water/salt concentration in feed water) × 100.
[b]Residual chlorine is limited by feed water quality.

TABLE 2.8 Specification of Fukuoka Plant

Production capacity:	50,000 m^3/day
Product TDS:	Less than 200 mg/L
System	
Intake:	Infiltration intake
Pretreatment:	Ultrafiltration
Desalination:	High-pressure RO
Posttreatment:	Low-pressure RO (partial 2 pass)
Operation condition	
Feed seawater TDS:	35,000 mg/L
Seawater temperature:	10–30°C
Recovery:	60%
Feed pressure:	High-pressure RO; max. 8.2 MPa
	Low-pessure RO; Max. 1.5 MPa

The specification of the high-pressure RO system is shown in Table 2.9, and a view of the high-pressure RO system is shown in Figures 2.22 and 2.23.

The high-pressure RO system consists of five units. Each RO unit has a capacity of 11,988 m^3/day. The recovery is controlled within 57.5–62.5% in response to water temperature changes (Kotera et al., 2005; Matsumoto et al., 2002).

The plant operation began on June 1, 2005. Depending upon water demand in Fukuoka City, the plant has produced up to the designed maximum capacity of 50,000 m^3/day, as needed. Since January, 2006, operation of the plant at full capacity of 50,000 m^3/day has continued up to date (April 2006).

Figure 2.24 shows mean performance of each high-pressure RO train. The pressure of feed was kept constant for several months and later showed a tendency to increase because of higher temperatures. After the middle of August, low temperatures affected the increase in feed pressure. The recovery is set according to the feed temperature. It was confirmed that the set value agreed with the real value. The performance of the high-pressure RO desalination was stable under the severe conditions of 60% average recovery in a wide range of 12–30°C temperatures in the feed.

2.5.4 Advanced Large-Sized RO Module

Toyobo developed an advanced, large-sized new-style module taking advantage of the technology of a both open-ended-type module. Since the bore pressure loss in a hollow fiber is smaller for a BOE type even if the fiber length is increased in a module, greater productivity performance is obtained. This new type advanced larger sized RO module has the same basic structure of the above-mentioned BOE-type module and has a length about 1.5 times. The specification of this module is shown in Table 2.10.

The amount of product water of one module has the capability of more than 2 times (i.e., double) or a 100-m^3/day capacity. The membrane material is the same CTA with the same characteristics of a cross-winding arrangement of the hollow fiber, the same excellent chlorine resistance, and stable operations are maintained (Marui et al., 2003; Hamano et al., 2006).

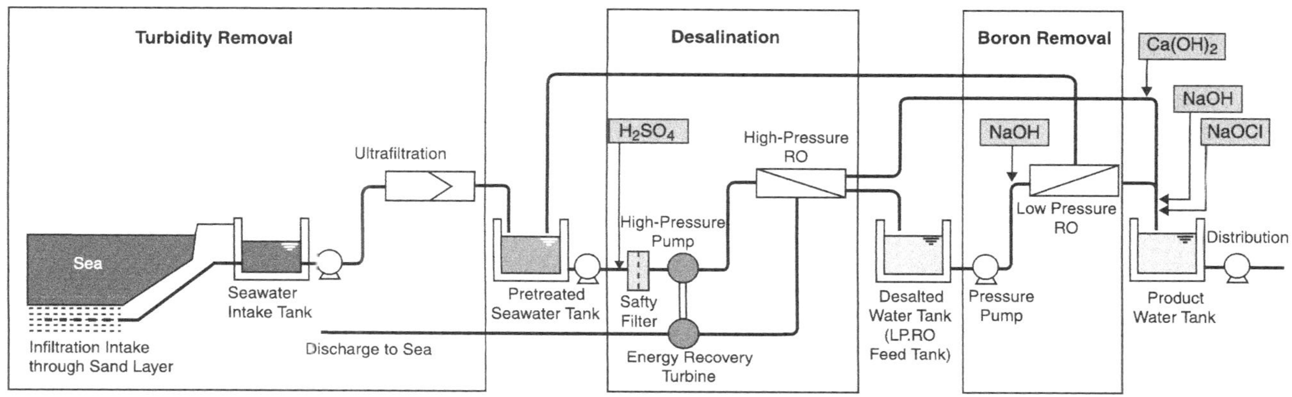

Figure 2.21 Process flow diagram and membrane arrangement.

TABLE 2.9 Specification of High-Pressure RO System (First Pass)

High-pressure RO system	
Number of units:	5
Production capacity:	11,988 m^3/day × 5 units
Product TDS:	less than 350 mg/L
RO membrane	
Model:	Toyobo HOLLOSEP HB10255FI
Type of membrane:	Hollow fiber
Material of membrane:	Cellulose triacetate
No. of modules:	200 pcs. × 5 units

Figure 2.22 High-pressure RO module unit (1).

Figure 2.23 High-pressure RO module unit (2).

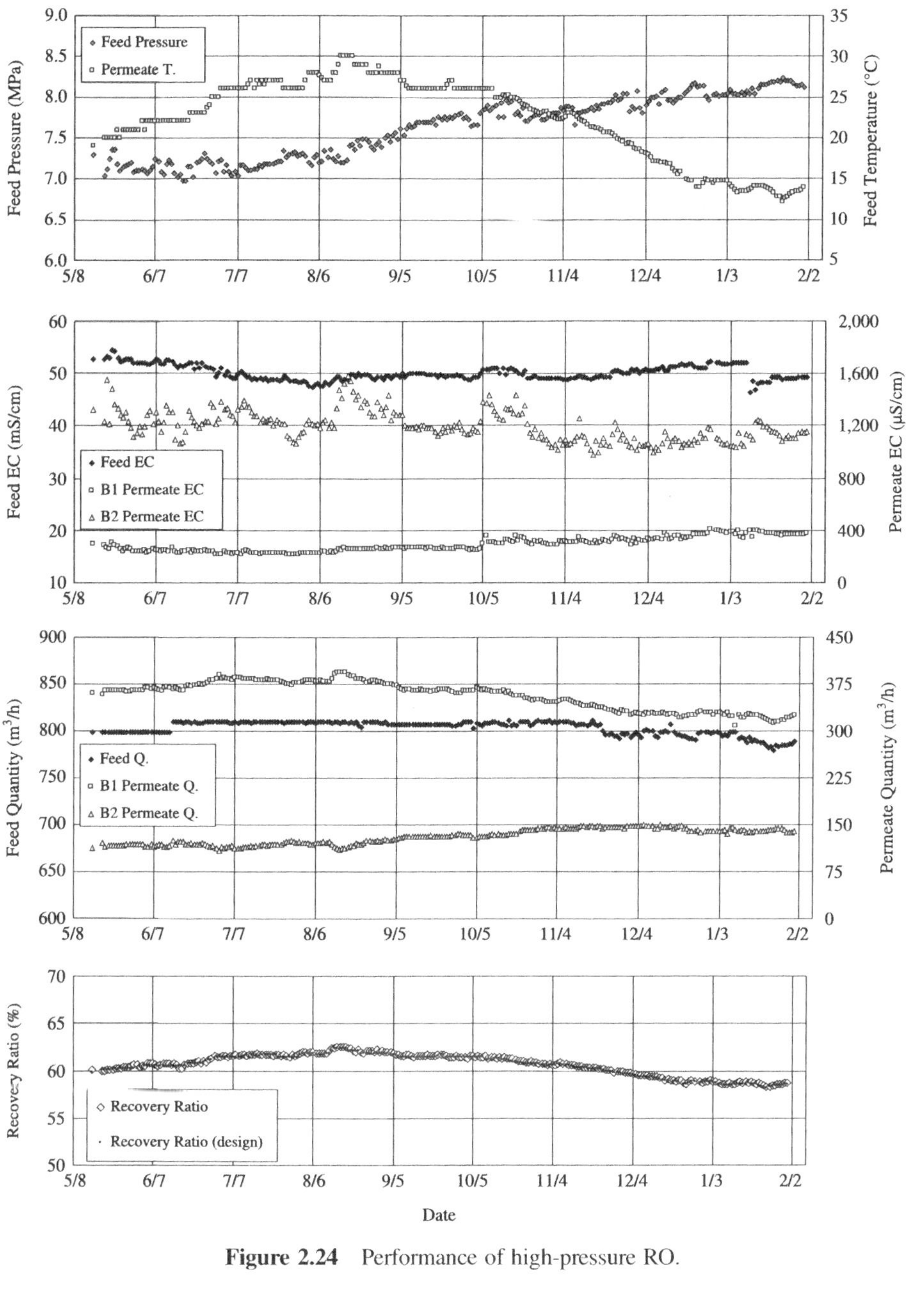

Figure 2.24 Performance of high-pressure RO.

2.6 CONCLUSION

Reverse osmosis membranes were first developed from cellulose acetate (CA) and a cellulose triacetate (CTA). These CA and CTA membranes have been in commercial use for many years. The CTA membrane, especially, is continuing to be used all over the world today. This wide adoption demonstrates that the CTA membrane is excellent as a RO

TABLE 2.10 Specification of Advanced Large-Sized BOE Module

Model	Conventional Model HM10255 Single Open-Ended (SOE)	Advanced Large-Sized HL10255 Both Open-Ended (BOE)
Element diameter (mm)	280	280
Element length (m)	1.35	2.0
Product flow rate (m^3/day)	45	100
Product salt rejection[a] (%)	99.6	99.6
Number of elements (—)	2	2
Test conditions		
NaCl concentration (mg/L)	35,000	35,000
Pressure (MPa)	5.4	5.4
Temperature (°C)	25	25
Recovery (%)	30	30
Operation conditions (max.)		
Pressure (MPa)	6.9	6.9
SDI (—)	4	4
Temperature (°C)	40	40
Residual chlorine[b] (mg/L)	1.0	1.0

[a]Salt Rejecton = (1 salt concentration in product water/salt concentration in feed water) × 100.
[b]Residual chlorine is limited by feed water quality.

membrane. As for the performance, the coexistence of high permeability and the high selectivity are enabled by outstanding balance of the hydrophilic and hydrophobic properties. In addition to the excellent basic characteristics of CTA membranes that other membranes also offer, CTA offers a very practical characteristic that makes it useful as a RO membrane. The CTA membrane is the only RO membrane among many RO membranes currently marketed that offers this high degree of chlorine resistance. The ICI method is effective with the CTA membrane to control biofouling growth and allows the stable operation of seawater desalination plants, especially in the Middle East where there is very high biofouling potential due to high seawater temperatures. In the case of RO membranes that do not have chlorine resistance biofouling becomes a problem, and these RO plants do not operate in stable mode because of frequent cleaning requirements. In those cases where CTA membranes have replaced the nonchlorine-tolerant RO membranes due to biofouling problems, stable operation has been attained.

Moreover, the material of CTA in hollow-fiber form is excellent in a general fouling-proof nature and has the characteristic of being difficult to become dirty with fouling matters. Furthermore, CTA in the form of a hollow-fiber membrane element in a module has a membrane area as large as about 10 times per unit volume compared with a spiral-wound membrane element and significantly reduces the flux per unit membrane area. Therefore, with CTA it becomes possible to decrease membrane load, and that makes it more difficult for fouling to occur. These key attributes lead to minimum chemical cleaning and long membrane life.

Furthermore, as described so far, development of the CTA hollow-fiber module continues, and by having realized further advanced features, RO membranes from CTA chemistry are adopted all over the world, especially in the seawater desalination field. CTA-based RO membranes contribute in a significant measure to meet the increase in water demand around the world. The commencement of commercial operation of new, large-scale seawater desalination plants in the Middle East now underway, future scheduled needs and installations in the Middle East, and an increase in future worldwide demand for fresh drinking water will be met by the hollow-fiber-type RO membranes and technology of Toyobo Co., Ltd. The ability to secure fresh drinking water by reverse osmosis surely leads to the possibility to remark: "The 21st century is the Age of Water."

REFERENCES

Alawadhi, A., Kannari, T., Katsube, M., Umemori, F., and Fujiwara, N. (2005). Elimination of biological fouling in seawater desalination plant in Bahrain. In *Proceedings of IDA World Congress on Desalination and Water Reuse*. IDA, Swissotel The Stamford, Singapore.

Al-Badawi, A. R., Al-Harthi, S. S., Imai, H., Iwahashi, H., Katsube, M., and Fujiwara, N. (1995). Operation and analysis of Jeddah 1—Phase 2 plant. In *Proceedings of IDA World Congress on Desalination and Water Reuse*, Vol. 3. IDA, Abu Dhabi, U.A.E., pp. 41–53 (November 18–24, 1995).

Ammons, R. D., and Mahon, H. I. (1974). Development of a one-pass hollow fiber seawater desalination module having a capacity of 2500–3000 gpd. Report No. 924. U.S. Office of Saline Water Res. Develop. Progr. U.S. Government Printing Office, Washington, D.C. 20402.

Aptel, P., and Buckley, C. A. (1996). Categories of Membrane Operations. In J. Mallevialle, P. E. Odendaal, and M. R. Wiesne (Eds.) *Water Treatment Membrane Processes*. McGraw-Hill, New York, p. 2.16.

Breton, E. J. (1957). Report No. 16. U.S. Office of Saline Water Res. Develop. Progr., New York.

Dance, E. L., Davis, T. E., Mahon, H. I., McLain, E. A., Skiens, W. E., and Spano, J. O. (1971). Development of cellulose triacetate hollow fiber reverse osmosis modules for brackish water desalination. Report No. 763. U.S. Office of Saline Water Res. Develop. Progr., New York.

Fujiwara, N., Tanaka, T., Katsube, M., Farhan, M. A., Al-Harthi, A. S. S., Imai, H., and Iwahashi, H. (1999a). Five-year operational performance of membrane at 15 MGPD Jeddah RO phase 2 plant. In *Proceedings of IDA World Congress on Desalination and Water Reuse*. IDA, San Diego, CA, pp. 89–98.

Fujiwara, N., Tanaka, T., Kumano, A., and Sekino, M. (1999b). New economical RO desalination system by a recovery ratio of 60%. In *Proceedings of IDA World Congress on Desalination and Water Reuse*. IDA, San Diego, pp. 101–109.

Hamano, T., Tsuge, H., and Goto, T. (2006). Innovations perform well in first year of operation. *Desalin. Water Reuse* 16(1), 31.

Kannari, T. (1995). RO module performance monitoring system. In *Proceedings of IDA World Congress on Desalination and Water Reuse*, Vol. 3. IDA, Abu Dhabi, U.A.E., pp. 3–11.

Kotera, H., Kumano, A., Marui, K., Fujiwara, N., Tanaka, T., and Sekino, M. (2005). Advanced RO module in the largest seawater desalination plant in Japan. In *Proceedings of IDA World Congress on Desalination and Water Reuse*. IDA, Swissotel The Stamford, Singapore.

Kumano, A. (2004). Focus on the high pressure and high recovery operation system and single stage RO module in Fukuoka plant. *Bull. Soc. Sea Water Sci. Jpn.* 58(3), 248.

Kumano, A., Fujiwara, N., Nishida, M., and Sekino, M. (2003). A single-stage RO seawater desalination process at higher recovery with improved hollow fiber membrane modules. In *Proceedings of IDA World Congress on Desalination and Water Reuse*. IDA, Atrantis Resort, Paradise Island, Bahamas.

Loeb, S., and Sourirajan, S. (1964). High flow porous membranes for separating water from saline solutions. U.S. Patent 3,133,132.

Macgowan, C. F., Ammons, D., Mahon, H., Wagener, E., and Davis, T. (1976). In A. Delyannis and E. Delyannis (Eds.), *Proceedings of the 5th International Symposium on Fresh Water from the Sea*, Vol. 4, Alghero, Italy, pp. 385–396.

Mahon, H. I. (1966), Permeability separatory apparatus, permeability separatory membrane element, method of making the same and process utilizing the same. U.S. Patent 3,228,876.

Marui, K., Kumano, A., Kotera, H., Fujiwara, N., Tanaka, T., and Sekino, M. (2003). Higher recovery with improved hollow fiber membrane modules. In *Proceedings of IDA World Congress on Desalination and Water Reuse*. IDA, Atrantis Resort, Paradise Island, Bahamas.

Matsumoto, Y., Kajiwara, T., Funayama, K., Sekino, M., Tanaka, T., and Iwahori, H. (2002). 50,000 m^3/day Fukuoka sea water RO desalination plant by a recovery ratio of 60%. In *Proceedings of IDA World Congress on Desalination and Water Reuse—Water for a Better Future*. IDA, Manama, Bahrain.

McCutchan, J. W., and Johnson, J. S. (1970). Reverse osmosis at Coalinga, California. *J. AWWA* 62, 346.

Nada, N., Iwahashi, H., and Umemori, F. (1994). Test result of the intermittent chlorine injection method in Jeddah 1 plant. *Desalination* 96, 283.

Nakao, S. (1996). Sea water desalination process for high recovery of fresh water by reverse osmosis. *Bull. Soc. Sea Water Sci. Jpn.* 50(6), 406.

Ohnishi, J., Nita, K., and Sekino, M. (1997). Development of high permeate flow rate and high pressure resistance RO module for seawater desalination. In *Proceedings of IDA World Congress on Desalination and Water Reuse*, IDA, Madrid, Spain, pp. 479–489.

Ohya, H. (1976). In *First Desalination Congress of the American Continent*, Vol. I, Session IV-5. Mexico City, Mexico.

Ohya, H. (1978). In *Proceedings of the 6th International Symposium on Fresh Water from the Sea*, Vol. 3, pp. 341–350.

Ohya, H., Suzuki, T., Nakao, S., Kato, S., Tsuji, M., and Sugi, J. (1996). Proposal and technological breakthrough of an integrated system for the complete usage of sea water. *Bull. Soc. Sea Water Sci. Jpn.* 50(6), 389.

Orofino, T. A. (1970). Development of hollow filament technology for reverse osmosis desalination systems. Report No. 549. U.S. Office Saline Water Res. Develop. Progr.

Reid, C. E., and Breton, E. J. (1959). Water and ion flow across cellulosic membranes. *J. Appl. Polym. Sci.* 1, 133.

Sekino, M. (1998). Important conditions and present state for seawater desalination RO module. *Bull. Soc. Sea Water Sci. Jpn.* 52(2), 107.

Ukai, T., Nimura, Y., Hamada, K., and Matsui, H. (1980). Development of one pass sea water reverse osmosis module, "HOLLOSEP". *Desalination* 32, 169.

Westmoreland, J. C. (1968). Spirally wrapped reverse osmosis membrane cell. U.S. Patent 3,367,504.

CHAPTER 3

Seawater Desalination

NIKOLAY VOUTCHKOV

Poseidon Resources Corporation, Stamford, Connecticut

RAPHAEL SEMIAT

Technion, Israel Institute of Technology, The Wolfson Chemical Engineering Department, Technion City, Haifa, Israel

3.1 INTRODUCTION

Seawater desalination is the production of fresh, low-salinity potable or industrial-quality water from a saline water source (sea, bay, or ocean water) via membrane separation or evaporation. Over the past 30 years, desalination technology has made great strides in many arid regions of the world such as the Middle East and the Mediterranean. Today, desalination plants operate in more than 120 countries worldwide, and some desert states, such as Saudi Arabia and the United Arab Emirates, rely on desalinated water for over 70% of their water supply. According to the 2004 desalination plant inventory report prepared by the International Desalination Association (Wagnick Consulting, 2004), by the end of 2003 worldwide there were over 17,000 desalination units with total installed treatment capacity of 37.8 million m^3/day. Seawater desalination plants contribute approximately 35% (13.2 million m^3/day) of this capacity.

Seawater is typically desalinated using two general types of water treatment technologies: thermal evaporation (distillation) and reverse osmosis (RO) membrane separation. Currently, approximately 56.5% (7.5 million m^3/day) of the world's desalination systems use RO membrane technologies. This percentage has been increasing steadily over the past 10 years due to the increasing popularity of membrane desalination, which is driven by remarkable advances in the membrane separation and energy recovery technologies and the associated reduction of the overall water production costs. Table 3.1 presents a list of the largest seawater reverse osmosis (SWRO) desalination plants built in the last 10 years. The total capacity of these facilities is approximately 1.5 million m^3/day.

Today, seawater desalination is mostly used to produce fresh potable water for human consumption and crop irrigation. Industrial applications of desalinated seawater are

Advanced Membrane Technology and Applications. Edited by Norman N. Li, Anthony G. Fane, W. S. Winston Ho, and T. Matsuura
Copyright © 2008 John Wiley & Sons, Inc.

TABLE 3.1 Large SWRO Plants Constructed from 1996 to 2005[a]

Plant Name/Location	Capacity (m^3/day)	In Operation Since
Ashkelon/Israel	325,000	2005
Tuas/Singapore	136,000	2005
Cartagena–Mauricia/Spain	65,000	2004
Fujairah/UAE	170,000	2003
Tampa Bay/United States	95,000	2003
Alikante/Spain	50,000	2003
Carboneras–Almeria/Spain	120,000	2003
Point Lisas/Trinidad	110,000	2002
Larnaca/Cyprus	54,000	2001
Al Jubail III/Saudi Arabia	91,000	2000
Muricia/Spain	65,000	1999
Bay of Palma/Palma de Mallorca	63,000	1998
Dhekelia/Cyprus	40,000	1997
Marbella–Mallaga/Spain	55,000	1997
Okinawa/Japan	40,000	1996

[a]This table includes only seawater RO desalination plants with a capacity of 40,000 m^3/day or higher.

typically limited to its use as a low-salinity power plant boiler water, process water for oil refineries, chemical manufacturing plants, and commercial fishing installations, canneries and other food industries. The limited industrial use of seawater desalination is related mainly to the high costs associated with production of high-purity or ultrapure water from seawater. Most industrial water supply facilities use low-cost groundwater or brackish water to produce high industrial-grade water for their specific applications.

3.1.1 Source Water Quality

Approximately 97.5% of the water on our planet is located in the oceans. Therefore, it is classified as seawater. Of the 2.5% of the planet's freshwater, approximately 70% is in the form of polar ice and snow; and 30% is groundwater, river and lake water, and air moisture. So even though the volume of Earth's water is vast, less than 10 million of the 1400 million m^3 of water on the planet are of low salinity and are suitable for use after applying conventional water treatment only. Seawater desalination provides means for tapping the world's main water resource—the ocean.

The mineral/salt content of the water is usually measured by the water quality parameter total dissolved solids (TDS), in milligrams per liter (mg/L) or parts per thousand (ppt). Natural water sources such as sea, bay, and ocean waters usually have TDS concentration higher than 15,000 mg/L. Seawater TDS and temperature are the two key source water quality parameters that have the most significant influence on the cost of seawater desalination. Table 3.2 presents typical TDS concentration and temperature for various seawater sources.

Source water quality has a key influence on the suitability of using seawater desalination for industrial water supply. The water quality parameters that have a significant impact on the desalination system design, operations, and cost of water production are the concentration of TDS, chlorides, turbidity, silt density index (SDI), organic content, nutrients, algae, bacteria, temperature, boron, silica, barium, calcium, and magnesium.

TABLE 3.2 Salinity and Temperature of Various Seawater Sources[a]

Seawater Source	Total Dissolved Solids Concentration (mg/L)	Temperature (°C)
Pacific/Atlantic Ocean	33,500	9–26 (avg. 18)
Caribbean	36,000	16–35 (avg. 26)
Mediterranean	38,000	16–35 (avg. 26)
Gulf of Oman, Indian Ocean	40,000	22–35 (avg. 30)
Red Sea	41,000	24–32 (avg. 28)
Arabian Gulf	45,000	16–35 (avg. 26)

[a]Seawater TDS and temperature may be outside the table ranges for a site-specific location.

3.1.2 Product Water Quality

Desalinated water quality is driven by its use. Typically, potable use of desalinated seawater is closely related to the levels of TDS, chlorides, boron, and bromides in this water. Drinking water regulations worldwide usually establish levels of TDS and chlorides in the product water below 500 and 250 mg/L, respectively. However, when using desalinated seawater, the importance of these parameters is often overshadowed by the health and irrigation-related water quality requirements in terms of boron and disinfection-related water quality targets in terms of bromides. The main reason boron and bromides are of specific importance for the overall quality of the desalinated water is the fact that their concentration in seawater is usually an order of magnitude higher than that of typical freshwater sources (rivers, lakes, groundwater, etc.). For example, typical river water has boron concentration of 0.05–0.2 mg/L, while the seawater boron levels are usually between 4.0 and 6.0 mg/L. Similarly, the bromide levels in freshwater sources are usually between 0.05 and 0.3 mg/L, while seawater has bromide concentration of 55–85 mg/L. While RO membranes typically remove over 70% of the boron and over 99% of the bromides in the source seawater, the remaining levels of these compounds are still several times higher than that in fresh surface water sources.

Usually, the boron level in the desalinated water is required to be less than 0.5 mg/L in order to alleviate problems associated with the use of this water for irrigation of sensitive crops (e.g., citrus trees, avocados, strawberries) or ornamental plants. To achieve this level of boron in the desalinated water, often the water TDS and chloride levels have to be reduced below 100 and 50 mg/L, respectively.

The bromide concentration of the desalinated seawater may also have a significant effect on the required level of removal of salts from the seawater, especially if this water will be disinfected using chloramines rather than chlorine, or it will be ozonated. While using chlorine only creates a stable chlorine residual that shows minimum decay over time, applying a combination of chlorine and ammonia to create chloramines (a practice widely used in the United States for example) to desalinated water with bromide levels above 0.4 mg/L, usually yields unstable chlorine residual that decays rapidly (within several hours) to unacceptably low levels. Although the effect of high levels of bromide in the desalinated water can be mitigated by superchlorination (i.e., applying initial chlorine at dosages of 4.0 mg/L or higher), this effect has to be accounted for especially when blending this water with other water sources that have low levels of bromides.

If desalinated seawater that contains bromide of levels above 0.2 mg/L is disinfected by ozonation, the ozonated water contains unacceptably high levels of bromate and is typically above the threshold of 10 mg/L, considered suitable for human consumption. Another important drawback of ozontaing desalinated water is the significant increase in the levels of brominated disinfection by-products (DBPs). Although currently individual brominated DBPs are not regulated, that is likely to occur in the near future. As a result, the target overall water quality of the desalinated seawater in some industrial applications, such as the production of bottled water where ozontaion is widely practiced, may be driven by the level of bromides in the water.

In addition to the potable uses discussed above, the desalinated water quality may be driven to even higher levels by the need of some industrial applications, especially those where ultrapure water quality is necessary.

3.2 SEAWATER DESALINATION PLANT CONFIGURATION

Seawater RO desalination plants typically consist of the following key components: intake; pretreatment system; filter effluent transfer pumps; high-pressure pumps, piping, and RO membrane system; energy recovery system; and permeate conditioning (posttreatment) facilities.

3.2.1 Seawater Intake Facilities

The seawater intake facilities are among the key components of every SWRO plant. Adequate and consistent flow and quality of source water over the entire useful life of the plant must be assured. The source water collection system for SWRO desalination plants could be an open-ocean intake or subsurface (beach well) intakes.

3.2.1.1 Open-Ocean Intakes Open-ocean intakes are commonly constructed for large seawater desalination plants. These surface water delivery systems include the following key components: off-shore intake structure; intake pipeline; intake chamber; trash racks; fine screens; source water intake pump station; electrical, instrumentation, and control equipment; and chemical feed equipment. The proper design of open-ocean intakes requires the collection of detailed source water quality data from the proposed site of the intake, characterization of aquatic life in the vicinity of the intake, and completion of detailed sanitary survey assessing the potential sources of SWRO plant source water quality contamination in the vicinity of the intake location (such as waste discharges of industrial and municipal wastewater plants, stormwater discharges or large port or marina activities, which may result in oil and gasoline spills, and other ocean water contamination).

The off-shore intake structure is usually a vertical concrete or steel well (vault) or pipe located at the ocean floor and submerged below the water surface, which is designed to reliably collect adequate amount of seawater that has a minimum content of debris and aquatic organisms. The exact location and depth of the off-shore intake structure must be determined based on a hydrological study to ensure that the intake will be adequately submerged at low tide; protected from the damaging orbital storm wave motion; and far enough off-shore to avoid the near-shore sediment transport zone where storms can cause suspension of large quantities of silt and sediment, and can ultimately damage the intake structure and the interconnecting piping. Diurnal and seasonal source water quality fluctuations should also be considered when determining the location of the intake structure. At minimum low-tide conditions, inlet mouth should be submerged at least 3 m below the water surface. In addition, the distance between the inlet mouth and the ocean floor should be no less than 3 m to prevent excessive sand carryover into the downstream intake facilities. The intake water supply can be protected against large aquatic organisms and large floating debris by installation of wire net across the intake mouth. A picture of the Larnaca, Cyprus, SWRO desalination plant intake structure is shown on Figure 3.1.

Typically, the open-ocean intake structure is located several hundred to several thousand meters off-shore. The best location of the intake structure in terms of source water quality is at ocean floor depths of 30 m or higher (deep-water intake). Debris load in the source water and algal content during red tides at such depths are typically 20 times lower than that in the surface water or the shallow waters of the tidally influenced near-shore area (Gille, 2003).

Depending on the plant location and ocean floor formation, installing the intake structure at a 30-m depth may require intake pipeline that is between 10 and 2000 m long. Because the construction cost for intake pipeline located on the ocean floor is usually very high (between 4 and 10 times higher than the cost of the same size pipe installed inland in the ground), the intake water quality benefits of locating the off-shore intake structure in deep waters have to be compared against the costs for construction of the intake structure and pipeline. The best location for an open-ocean intake from a life-cycle cost point of view is typically where the ocean floor depth of 30 m can be reached within 500 m from the shoreline. If such an ocean floor location is not available within a reasonably close vicinity of the SWRO desalination plant, usually it is more cost effective to collect source water of inferior water quality and build a more elaborate pretreatment system than to install a costly off-shore intake structure and a long intake pipeline.

Because of the high costs of deep intake structures and long pipelines, most of the existing SWRO desalination plants with open-ocean intakes are located in shallow near-shore areas where the ocean floor depth is typically between 3 and 10 m. As a result, plants with open-ocean intakes typically have source water with a high content of debris, solids, and aquatic organisms, which requires elaborate pretreatment prior to SWRO membrane separation. The construction material of the intake pipe should be chosen carefully, as possible release of minor constituents may harm the membranes, as the case of phthalates release in the Eilat project (Hasson et al., 1996).

3.2.1.2 Subsurface Intakes Subsurface (beach well) intakes are widely used for small and medium-size seawater desalination plants. Beach wells are typically located on the seashore, in close vicinity to the ocean. Beach well intakes include vertical or horizontal wells and associated intake pumping and electrical components. The subsurface intake

Figure 3.1 Open intake structure of Larnaka, Cyprus SWRO plant.

facilities are relatively simple to build, and the source seawater they collect is pretreated via slow filtration through the subsurface sand/seabed formations in the area of source water extraction. Therefore, source water collected using subsurface intakes is usually of better quality in terms of solids, slit, oil and grease, natural organic contamination, and aquatic microorganisms, as compared to seawater from open-ocean intakes.

Vertical intake wells consist of a nonmetallic casting [typically, fiberglass-reinforced pipe (FRP) or polyvinyl chloride (PVC)], proper-grade stainless steel well screen (if needed), and a stainless steel submersible or vertical turbine pump (see Fig. 3.2). These wells are usually less costly than the horizontal wells, but their yield is relatively small (typically 400–4000 m^3/day).

Horizontal collector intake wells are not as commonly used for SWRO source water supply as vertical wells because of their relatively higher installation costs. However, they may be the best choice for some applications. They consist of a caisson that extends below the ground surface with water well collector screens (laterals) projected out horizontally from inside the caisson into the surrounding aquifer (Fig. 3.3). Since the well screens in the collector wells are placed horizontally, a higher rate of source water collection is possible than with most vertical wells. This allows the same intake water quantity to be collected with fewer wells.

Individual horizontal intake wells are typically designed to collect between 2000 and 20,000 m^3/day of source seawater. The caisson is constructed of reinforced concrete that may be 2.5–6.0 m inside diameter with a wall thickness from approximately 0.5 to 1.0 m depending on the depth (Hunt, 1996). The caisson depth varies according to site-specific geologic conditions, ranging from approximately 10 to 45 m. The number, length, and location of the horizontal laterals are determined based on a detailed hydrogeological investigation. Typically, the diameter of the laterals ranges from 0.2 to 0.3 m, and their length

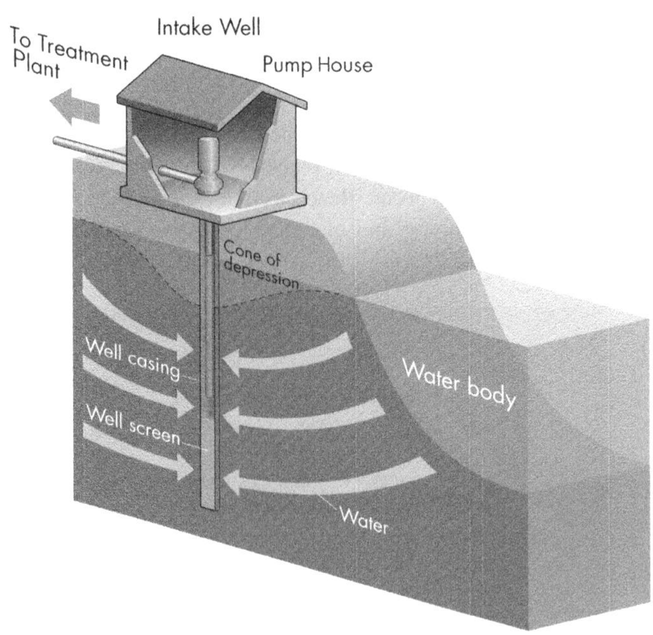

Figure 3.2 Vertical intake well.

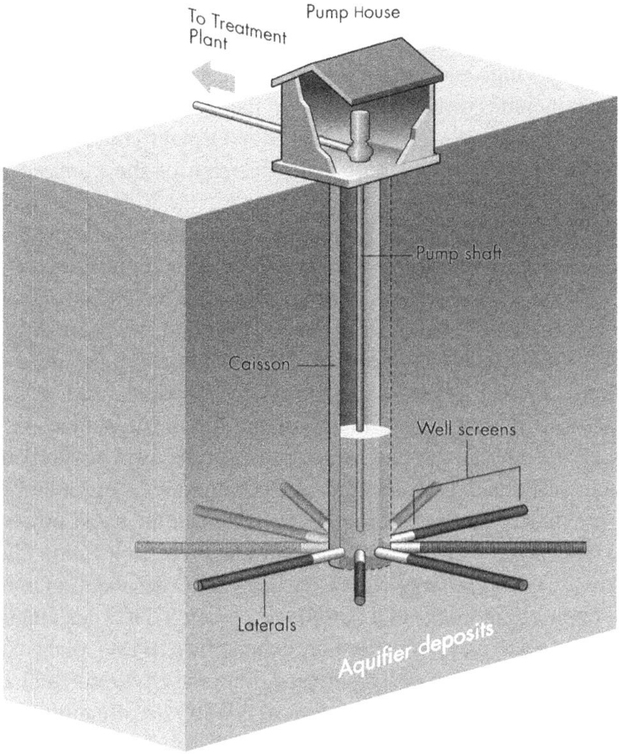

Figure 3.3 Horizontal Ranney collector intake well.

extends up to 60 m. The size of the lateral screens is selected to accommodate the grain size of the underground soil formation.

Although beach wells have proven to be quite cost competitive for plants with a capacity smaller than 4000 m^3/day, the open surface ocean intakes have found significantly wider application for large SWRO desalination plants. At present, worldwide there are only 4 operational SWRO facilities with capacities larger than 20,000 m^3/day using beach well intakes. The largest SWRO facility with beach wells is the 54,000-m^3/day Pembroke plant in Malta. This plant has been in operation since 1991. The 42,000-m^3/day Bay of Palma plant in Mallorca, Spain, has 16 vertical wells with capacities of 5600 m^3/day each. The third largest plant is the 24,000-m^3/day Ghar Lapsi SWRO in Malta. Source water for this facility is supplied by 15 vertical beach wells with unit capacities of 3800 m^3/day.

The largest SWRO plant in North America, which obtains source water from beach wells, is the 15,000-m^3/day water supply facility for the Pemex Salina Cruz refinery in Mexico. This plant also has the largest existing seawater intake wells—three Ranney-type horizontal collectors with capacities of 15,000 m^3/day each. Key considerations for the selection of the type of intake most suitable for the site-specific conditions of a given SWRO plant and guidelines for the development of subsurface intakes for seawater desalination plants are discussed elsewhere (AWWA, 2007; Wright and Missimer, 1997; Voutchkov, 2004a).

3.2.1.3 Colocation of Desalination Plant Intake with Power Plant Discharge Colocation of desalination plants with large power generation stations may yield significant cost savings and further reduce the cost of desalinated water. This alternative intake approach includes direct connection of the SWRO plant intake piping to the discharge outfall of the nearby power plant. The SRWO plant concentrate discharge line is connected to the same power plant outfall downstream of the point of the plant intake (see Fig. 3.4).

Typically, once-thru coastal power generation plants use large volumes of seawater for cooling purposes and screen the source seawater through a combination of coarse (100 mm or less) and fine (20 mm or less) screens in series. Since the SWRO desalination plant intake is connected to the discharge outfall of a power plant and the plant discharge is already screened, the use of the colocation approach allows eliminating the need for construction of a new separate SWRO plant intake structure, pipeline, and screening facilities. Microscreens, as those shown on the intake of the desalination plant in Figure 3.4, are only needed when membrane pretreatment facilities are used to prevent the intake of sharp objects such as shells, which could damage the pretreatment membranes.

As indicated previously, under the colocation approach the same power plant outfall is also used for the discharge of the RO concentrate. Therefore, colocation allows eliminating the need to construct a new discharge outfall for the SWRO desalination plant as well. Since the cost of a new open-ocean intake and outfall for a given SWRO desalination plant is typically 10–30% of the total plant construction expenditure, power plant colocation allows achieving significant construction cost savings. Additional environmental and cost benefits and feasibility considerations of the colocation of SWRO desalination plants with coastal power generation stations are described in detail elsewhere (Voutchkov, 2004b). A summary of key advantages and disadvantages of the colocation approach is presented in Table 3.3.

3.2.2 Pretreatment System

Depending on the type of SWRO plant intake and the source water quality, SWRO desalination plant pretreatment systems may include one or more of the following processes: screening, chemical conditioning, sedimentation, and filtration.

3.2.2.1 Intake Screens A typical open-ocean intake system for medium and large SWRO membrane plants includes a set of manually cleaned bar racks followed by automated traveling fine-bar screens and/or fine-mesh screens. The bar racks usually have 75–100 mm distance between the bars, and their purpose is to retain large-size debris and aquatic life in the source water. Fine self-cleaning bar screens typically have 3–10 mm openings between the bars. Because the main function of these screens is to protect the intake pumps from damage, the actual distance between the bars has to be selected to be smaller than the distance between the intake pump impellers. Fine-bar screens are usually used if the downstream pretreatment system consists of conventional granular media filters.

If the pretreatment system selected for the SWRO plant is of membrane type, conventional bar screens do not provide adequate removal of source water particles to protect the integrity of the microfiltration/ultrafiltration (MF/UF) membranes. One of the key issues of using a MF/UF pretreatment is that the MF/UF membrane fibers can be punctured by sharp objects in the source water, such as broken shells. Pilot testing experience at

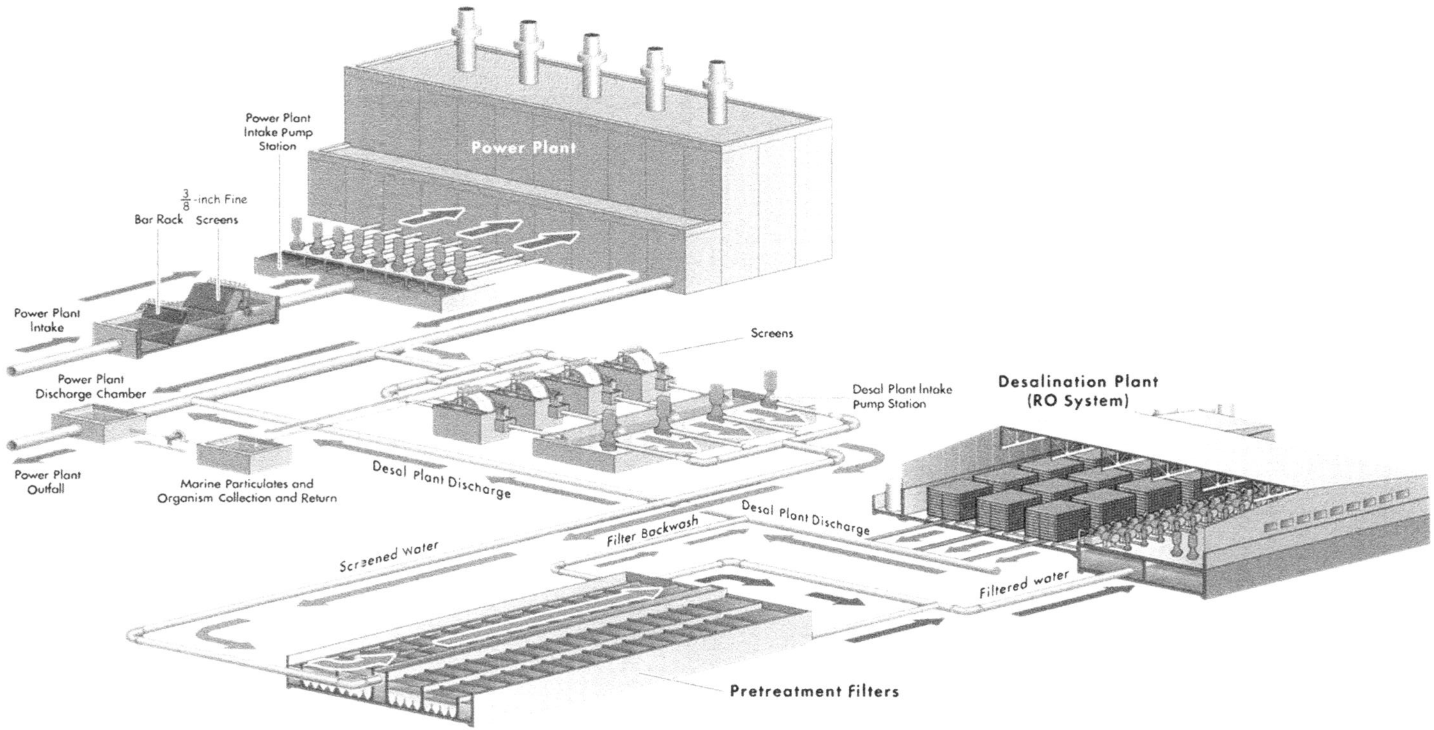

Figure 3.4 Colocation of desalination and power plant intake and discharge.

TABLE 3.3 Advantages and Disadvantages of Desalination Plant Colocation

Advantages	Disadvantages
• Capital cost savings by avoiding the construction of separate intake pipeline and structure and new discharge outfall.	• Use of warmer seawater may accelerate membrane biofouling, especially if the source water is rich in organics.
• Decrease of the required RO system feed pressure and power cost savings as a result of using warmer water.	• RO membranes may be exposed to iron, copper, or nickel fouling if the power plant condensers and piping are built of low-quality materials.
• Reduction of unit power cost by connecting directly to power plant generation facilities and avoiding power transmission charges.	• Source seawater has to be cooled if its temperature increases above 40°C in order to protect RO membrane integrity.
• Accelerated permitting process as a result of avoidance of construction of new intake and discharge outfalls in the ocean.	• Permeate water quality diminishes slightly with the increase of source water temperature.
• Reduction of marine organism impingement and entrainment because the desalination plant does not take additional seawater from the ocean.	• Use of warmer water may result in lower boron rejection and require feed water pH adjustment to meet stringent boron water quality targets.
• Reduction of the impact on marine environment as a result of faster dissipation of thermal plume and concentrate.	• RO plant source water screening may be required if the power plant disposes of its screenings through their outfall and the point of disposal is upstream of the desalination plant intake.
• Reduction of the power plant thermal discharge to the ocean because a portion of this discharge is converted to potable water.	• Desalination plant operations may need to be discontinued during periods of heat treatment of the power plant facilities.
• Use of already disturbed land at the power plant minimizes environmental impact.	

several locations indicates that if MF/UF pretreatment is used, the surface intake system should include fine-mesh screens of 120 μm or smaller ahead of the pretreatment filter to protect the membrane elements from damage and premature loss of integrity. Typically, microscreens or disk filters can be used for this application. The main disadvantage of the use of microscreens is that they add to the plant construction and operation and maintenance costs. Usually they are 20–50% more costly than conventional fine-bar screens. For comparison, the use of conventional granular media pretreatment filters does not require the installation of microscreens ahead of the filters because the granular media is not susceptible to damage by sharp objects in the source water and effectively retains these objects.

Smaller size SWRO desalination plants with open intakes use strainers instead of bar screens to protect the downstream intake pumps and pretreatment systems. For plants that have conventional granular media pretreatment systems, 500 to 900-μm strainers are usually adequate prescreening device. Plants equipped with membrane pretreatment filters would require the use of 80 to 120-μm strainers.

If the source seawater contains a significant amount of sand, which should be removed before processing through the downstream treatment facilities, strainers and cyclones are used upstream of the pretreatment facilities. Sand strainers operate by physical straining of the sand on a metallic or plastic mesh screen. Typically, strainers can achieve up to 95% removal of sand using a 150-mesh screen size. For larger membrane treatment plants, self-cleaning strainers are used. These units use a portion of the source water flow to wash the screen surface and remove the accumulated sand.

Disk filters operate using specially designed thin plastic disks that are diagonally grooved on both sides to a precise micrometer size. Disks are typically between 20 and 400 μm in size and are stacked and compressed on a specially designed spine by a spring. During the filtration mode, the disks are tightly compressed together by the spring and the differential pressure. Filtration occurs while water percolates from the outer diameter to the inner diameter of the element. During backwash, the disks are released by reducing the inlet hydraulic pressure. Multijet nozzles provide tangential spray on the loosened disks, causing them to spin and release the retained solids, which are than washed out to the drain. The disk filter technology has a proven track record for seawater applications.

Cyclones operate using the greater inertia of sand particles in the water stream in cyclonic motion to separate the sand from the source water stream. While cyclones can achieve up to 98.5% sand removal, the resulting pressure loss of 1–1.3 bars yields a high energy cost.

3.2.2.2 Pretreatment System Configurations As discussed previously, SWRO systems treating seawater from beach wells often require minimal filtration pretreatment—commonly just cartridge filtration (as a safety filtration device to remove suspended solids that can plug, foul, or damage membranes) and chemical addition (commonly acid and/or scale inhibitor).

For plants with open-ocean intakes, pretreatment facilities are usually more elaborate. In addition to the source water screening equipment, the SWRO plants with open intakes have to be equipped with pretreatment facilities to handle:

- Colloidal and particulate foulants (suspended solids and silt)
- Inorganic compounds that may precipitate and scale or foul the membranes (such as iron and manganese, calcium carbonate, calcium sulfate, barium sulfate, or silica)
- Organic foulants (soluble organic compounds that can serve as food to the microorganisms in the source water)

The most suitable pretreatment facility configuration for a given source seawater typically depends on how high the source water turbidity and silt content are above the acceptable feed water turbidity and SDI levels. Most spiral wound SWRO membrane manufacturers require membrane feed water turbidity not to exceed 1.0 Nephelometric Turbidity Units (NTU) and SDI not to exceed 4 or 5, depending on the membrane product. It is usually recommended that the feed water turbidity and SDI be less than 0.5 NTU and SDI less than 3.0, if economically practical.

Table 3.4 provides guidelines for a combination of treatment processes recommended for the effective pretreatment of the source seawater as a function of seawater turbidity and SDI levels. The pretreatment configurations shown in this table have to be used as a guideline only. Thorough water quality analysis and pilot testing are recommended to define an optimum pretreatment system for the site-specific source water quality of a given project.

TABLE 3.4 Guidelines for Selection of Pretreatment Based on Source Water Quality

	SDI < 4	SDI ≥ 4				
	Turbidity (NTU)					
Process Recommended to Be Included in the Pretreatment Configuration	<0.5	≥0.5 and <2	≥2 and <20	≥20 and <40	≥40 and <100	≥100
Coagulation and flocculation		×	×	×	×	×
Conventional sedimentation or dissolved air flotation				×	×	
Enhanced sedimentation or dissolved air flotation						×
Single-stage granular media filtration	×[a]	×[a]				
Two-stage granular media filtration			×[a]	×[a]	×[a]	×[a]
Single-stage MF or UF membrane filtration	×[a]	×[a]	×[a]	×[a]	×[a]	×[a]

[a]Select either granular media or membrane filtration. For source water turbidity of 20 NTU or higher consider a combination of single-stage coarse granular media filtration followed by MF or UF membrane filtration.

A basic description and key design criteria for the pretreatment facilities included in Table 3.4 are presented below. A more detailed reference for design of these pretreatment facilities can be found elsewhere (AWWA, 2007).

3.2.2.3 Coagulation and Flocculation The coagulants most frequently used for membrane plant source water conditioning are ferric salts (ferric sulfate and ferric chloride). Aluminum salts (such as alum or polyaluminum chloride) are rarely used because it is difficult to maintain aluminum concentrations at low levels in the filtered water and in the dissolved form because aluminum solubility is very pH dependent.

Large residual amounts of aluminum in the RO feed water may cause irreversible fouling of the membranes. Optimum coagulant type and dosage for a given source water are typically determined based on jar and/or pilot testing. Overdosing of coagulant and/or flocculant may have negative effects on membrane performance. Coagulation may continue after passage through the pretreatment system and cause large particles to increase filter effluent turbidity and SDI and to ultimately result in accelerated fouling of the RO membrane elements.

In general, it is best not to use polymers in SWRO pretreatment. However, if used, only nonionic or anionic polymers should be considered because most SWRO membrane elements carry a negative surface charge. Use of cationic polymer is likely to form a polymer film on the membrane surface that will foul the membrane elements. The type and dosage of polymer (nonionic or anionic) that is most suitable for a given application has to be determined by jar and/or pilot testing. Typically, polymer is added at a very low dosage (less than 1 mg/L). Adding elevated polymer dosages should be avoided because it usually results in a high content of unused polymer in the filtered seawater, which in turn plugs the cartridge filters and deposits on the SWRO membrane elements—thereby shortening the cartridge filter useful life and expediting the need for membrane cleaning.

3.2.2.4 Conventional and Enhanced Sedimentation Sedimentation is typically used upstream of granular media filters when the membrane plant source water has daily average turbidity higher than about 40 NTU (see Table 3.4—Conventional and Enhanced Sedimentation section). The sedimentation basins should be designed to produce settled source water of less than approximately 2.0 NTU and measurable SDI (15-min SDI below 6). To achieve this level of turbidity removal, the sedimentation basin is typically equipped with coagulant (most frequently iron salt) and flocculant (polymer) feed systems. The design coagulant and flocculent dosages should be established based on jar and/or pilot testing.

Typically, conventional sedimentation basins are designed for a surface loading rate of 30–50 m^3/m^2 day, weir loading rate of 250 m^3/linear m day, and hydraulic detention time of 2–4 h. Surface loading rate is used to calculate the surface area requirements of the sedimentation basin. Detention time is used in conjunction with the surface loading rate to calculate the volume and the side depth of the sedimentation basin.

If the source water turbidity exceeds 100 NTU, than conventional sedimentation basins are often inadequate to produce turbidity of the desired target level of less than 2.0 NTU. Under these conditions, the sedimentation basin has to be designed for enhanced solids removal by installing inclined "lamella-type" plates or using sedimentation technologies that combine lamella and fine granular media (microsand or settled residuals) for enhanced solids removal.

Typically, the use of enhanced sedimentation technologies is needed for treating seawater collected using open-ocean intake, which is under a strong influence of high-turbidity/high organics river water. This condition would occur when a seawater desalination plant intake is located in a river delta area or is strongly influenced by a seasonal runoff. For example, during the rainy season, the intake of the Point Lisas seawater desalination plant in Trinidad is under the influence of the Orinoco River, which carries a large amount of alluvial solids, and under the river water influence, the desalination plant intake turbidity could exceed 200 NTU (Irwin and Thompson, 2003). To handle this high solids load, the plant source seawater is settled in a lamella sedimentation tank prior to conventional single-stage dual-media filtration.

3.2.2.5 Granular Media Pretreatment Filtration Granular media filtration is the most commonly used source water pretreatment process for open-ocean intake SWRO plants today (other than cartridge filtration). This process includes gravity or pressure filtration of the source water through one or more layers of granular media (e.g., anthracite coal, silica sand, garnet).

Since the purpose of the pretreatment filters for SWRO plants is not only to remove over 99 percent of all suspended solids in the source water but also to reduce the content of the much finer silt particles with several orders of magnitude, the design of these pretreatment facilities is usually governed by the filter effluent SDI target levels and particle count than by the effluent turbidity.

The granular media filters are backwashed using filtered water or, if acceptable, concentrate from the SWRO membrane system. The backwash rate is typically designed to provide 30–50% media bed expansion. The number of individual filter units is usually selected such that the filtration system can operate at full production capacity with one filter out of service in backwash and one out of service for maintenance.

The two types of granular media filters most widely used for pretreatment of SWRO plants are pressure and gravity filters. The major differences between pressure and gravity filters are the head required to convey the water through the media bed and the type of vessel used to contain the filter unit. Gravity filters usually require 2–3 m of head and are housed in open concrete or steel tanks. Pressure filters have greater driving pressure and are contained in enclosed vessels. Because of the cost of constructing large steel pressure vessels with the proper wetted surfaces for corrosion resistance, pressure filters are typically used for small and medium-size capacity SWRO plants. Gravity pretreatment filters are typically concrete structures and are used both on small and large membrane plants.

3.2.2.6 Membrane Pretreatment Filtration The most common membrane pretreatment process uses hollow-fiber MF or UF membranes to separate suspended solids and particles (including microorganisms) from the source seawater. MF and UF membrane systems have been shown to be very effective for turbidity and silt removal. Turbidity can be lowered consistently below 0.1 NTU, and filter effluent typically has SDI levels below 3 over 90% of the time.

It is recommended that pilot testing be used in developing design criteria for the site-specific conditions and MF/UF product selection. Both pressure-type MF/UF systems, where the membranes are encased in pressure vessels, and vacuum-type systems, where the membrane are immersed in tanks open to atmosphere and use filtrate/permeate pumps to create the driving force, may be used for seawater pretreatment.

3.2.2.7 Comparison of Granular Media and Membrane Pretreatment Filtration Systems Membrane filtration technologies have a number of advantages as compared to conventional granular media filtration systems. Granular media filtration, however, is a well understood and widely used SWRO pretreatment technology with a proven track record, which has a number of features that may render it cost competitive under specific circumstances. Therefore, the selection of filtration technology for membrane pretreatment has to be based on a thorough life-cycle cost–benefit analysis. Side-by-side pilot testing of the two types of systems is also highly recommended to develop background system performance information for an objective technology evaluation. The following issues have to be taken into consideration when selecting between granular media and membrane pretreatment filtration for a specific application.

Effect of Source Seawater Quality MF and UF systems have a wider spectrum of particle removal capabilities than granular media filtration. Single-or dual-media filters usually have lesser removal efficiency in terms of source water organics, disinfection by-product precursors, fine particles, silt, and pathogens. Membrane filtration technologies are also less prone to upsets caused by seasonal changes in source seawater temperature, pH, turbidity, color, microbial contamination, and size and type of water particles because their primary treatment mechanism is mechanical particle removal through fine-pore membranes. Therefore, the upstream chemical coagulation and flocculation of the influent water particles is of a lesser importance for their consistent and efficient performance. In contrast, the pretreatment efficiency of the granular media filtration technologies is very dependent on how efficient chemical coagulation and flocculation of the source seawater is ahead of the filtration process. Typically, coagulation and flocculation water chemistry is more sensitive to changes in seasonal water quality than the mechanically driven membrane particle separation processes.

Therefore, for applications where intake water quality experiences significant seasonal variations and presents a challenge in terms of high microbial, fine particle, and organic contamination, membrane filtration technologies are likely to be the technology of choice, assuming it is cost effective for the specific application. However, for seawater sources of high quality and limited seasonal variations, granular media filtration may offer an efficient and cost-effective pretreatment alternative to membrane filtration.

Another condition under which the granular media filtration may have certain benefits is for a source of seawater that is very likely to be exposed to sudden and unpredictable changes of water quality such as very high or low pH chemical spills, large oil and grease spills, frequent exposures to very high water temperature, or other contaminants that may damage the MF or UF pretreatment membranes irreversibly, if they are used for this application. If the membrane elements are permanently damaged, the cost of their replacement could be significant, especially for large SWRO treatment plants. Typically, granular filter media can handle a wider range of extreme source seawater quality conditions before irreversible damage. In addition, the cost of media replacement is significantly lower than that for replacing all membrane elements for the same size plant.

This issue is very significant for pretreatment systems for seawater desalination plants with open-ocean intakes. Often the source seawater contains small sharp objects (such as shell particles), which can easily puncture the pretreatment membranes and result in a very quick loss of their integrity, unless the damaging particles are removed upstream of the membrane pretreatment system. As discussed previously, to remove sharp seawater particles that can damage the membranes from the source water, the SWRO plant intake system has to incorporate a microscreening system that can remove particles larger than 120 μm.

The installation and operation of a microscreening system is significantly more costly than the use of conventional traveling fine screens, and therefore its cost has to be taken into consideration when comparing granular and membrane filtration pretreatment. The performance and reliability of the granular pretreatment systems are not sensitive to the content of sharp objects in the seawater and do not require more elaborate and costly screening ahead of the filters. Typically, fine traveling screens of 3- to 10-mm openings provide adequate protection of granular media pretreatment systems.

Footprint Membrane technologies are very space efficient as compared to granular media filtration. The smaller footprint benefits of membrane filtration are usually of greater importance when upgrading existing water treatment plants of limited site area availability or where the cost of new land acquisition is significant.

Depending on the type and size of the membrane modules and the intake water quality characteristics, the membrane filtration system may have 20–60% smaller footprint than a granular filtration system. The space benefits of membrane filtration are more significant for high-turbidity waters where two-stage granular media filtration may be required to achieve comparable performance to a single-stage membrane system. For a more difficult-to-treat intake water, which requires the granular media filtration system to be designed for surface loading rates of less than $10\,m^3/m^2$ h or where two-stage granular media filtration is needed to produce comparable filter effluent, the membrane filtration systems may have up to 60% smaller footprint. As a rule of thumb, under typical surface water quality conditions, the footprint of granular media filters, designed at a surface loading rate of $10-12\,m^3/m^2$ h, is approximately 50% larger than that of MF or UF systems producing similar filtered water quality. For better-than-average influent water quality where granular media filters can perform adequately at $15-20\,m^3/m^2$ h of hydraulic surface loading rate, the footprint difference is usually 20–40% in the benefit of membrane filtration.

Waste Stream Quantity and Quality Granular and membrane pretreatment systems differ significantly by the type, quality, and amount of the generated waste streams. Typically, granular media filtration systems generate only one waste stream—waste filter backwash. The volume of this stream in a well-designed plant varies between 4 and 6% of the total plant intake source water volume. In addition to the solids that were originally in the source water, this waste stream also contains coagulant (typically iron salt) and polymer (if used). Currently, most often the pretreatment filter waste backwash water is treated, blended with the SWRO plant concentrate, and discharged to the ocean.

The membrane pretreatment systems typically generate two large-volume waste streams: waste membrane wash water and membrane cleaning solution. The volume of the membrane wash water stream is typically 5–10% of the plant intake source volume—that is, approximately 2 times larger than the waste filter backwash generated by granular pretreatment systems. The waste stream difference is even larger, taking into account that the microscreens required to be installed to protect the pretreatment membrane filters will generate additional waste discharge for their cleaning. While conventional traveling fine-bar screens use less than 0.5% of the intake source water for cleaning, microscreens would require a wash volume that equals 1.5–3% of the intake flow. The relatively larger waste stream volume of the membrane pretreatment system would need the collection of proportionally larger intake source volume, which in turn would result in increased size and construction costs for the SWRO membrane plant intake facilities and higher operation and maintenance costs for source water pumping to the pretreatment facilities. In addition to daily membrane washing and monthly membrane cleaning, cost-competitive design and operation of membrane pretreatment systems requires short daily chemically enhanced membrane backwash (CEB) using high dosage of chlorine and base and acid over a short period of time. This performance-enhancing CEB wastewater has to be discharged to the wastewater collection system and also adds to the volume of the waste streams generated at the SWRO membrane plant and to the overall cost of source water pretreatment.

One advantage of the main membrane waste wash stream is that it typically does not contain source water conditioning chemicals (coagulant and polymer), and, therefore, it is more environmentally friendly—that is, contains only solids that already have been in the source water. However, the other two waste streams: the spent chemical waste generated during the CEB and the monthly pretreatment membrane cleaning waste are not suitable for surface water discharge and have to be pretreated on-site in a neutralization tank, prior to their discharge to the sanitary sewer. The additional treatment and disposal costs of the waste membrane cleaning chemicals have to be taken into consideration when comparing the use of membrane pretreatment and granular media pretreatment systems.

Chemical Use Typically, granular media pretreatment systems use source water conditioning chemicals for effective solids separation. This adds to the plant chemical costs. However, they do not use any chemicals for media cleaning (outside of the occasional addition of chlorine). The membrane pretreatment systems use significant amounts of membrane cleaning chemicals. The cost of these cleaning chemicals has to be considered in the cost–benefit analysis of the plant pretreatment system. Another factor that has to be accounted for in the overall plant chemical use and cost analysis is that the SWRO system cleaning frequency, and therefore the SWRO membrane cleaning costs, may be reduced by using membrane pretreatment due to the typically better solids and silt removal capabilities of this type of pretreatment.

Power Use Granular pretreatment systems use a limited amount of power to separate particulates in the source water. Depending on the type of membrane system (pressure or vacuum driven), the membrane pretreatment systems typically consume several times larger amounts of power. More power is not only used to create a flow-driving pressure through the membranes, but also for membrane backwash, CEB, chemical membrane cleaning and feed water pumping, and screening. The total power use has to be taken into consideration when completing a life-cycle cost comparison of granular versus membrane pretreatment system for a given application.

3.2.2.8 *Frequency of Filtration Media Replacement* Typically, well-operating granular media filters loose less than 5% of filter media per year, which periodically needs to be replaced to maintain consistent performance. The costs of granular media replacement are usually predictable and relatively low. At present, a well-designed and operated MF/UF facility can have membrane life greater than 5 years. Assuming 5 years of useful life, on average approximately 20% of the membrane elements would need to be replaced per year to maintain system production capacity and performance.

Diversity of Membrane Elements and Configurations Currently, all membrane manufacturers offer their own design, size, and configuration of membrane elements and systems. The membrane systems differ by the type of filtration driving force (pressure versus vacuum), the size of the individual membrane elements, the size of the membrane vessels, the configuration of the membrane modules, the type of membrane element backwash, and the type of membrane integrity testing method and other factors.

The current diversity of membrane element sizes and configurations, and the lack of standardization and commoditization, may have a number of disadvantages for the membrane plant owner. For example, if an existing membrane manufacturer discontinues the production of membrane elements or a given type of membrane system or goes out of business, the plant owner would incur additional costs if suitable membrane replacements were not available and a new pretreatment system would need to be acquired.

3.2.2.9 *Cartridge Filters* The purpose of the cartridge filters is to protect the downstream RO membrane system from fouling and from mechanical plugging and damage caused by particulate matter, especially sand and other pretreatment filter media. Since membrane pretreatment systems do not use granular media, use of cartridge filters for seawater pretreated by UF or MF membranes is optional and only serves a strictly protective role to accommodate circumstances where large seawater particles pass through the pretreatment membranes in the case of loss of membrane integrity or defects.

Cartridge filters are typically designed to remove at least 90% of particles larger than their nominal rated size. For all but the smallest capacity SWRO systems, many filter cartridges, up to 100 cm (40 inches) long, are installed in stainless-steel or fiberglass-reinforced plastic (FRP) pressure housing vessels (see Fig. 3.5).

The vessels housing the cartridge filters may be oriented vertically or horizontally. Cartridges are rated for removal of particle sizes of 1, 2, 5, 10, and 25 μm, with the most frequently used size being 5 μm. Polypropylene wound cartridges are commonly used; although other types such has melt-blown or pleated cartridges have also found widespread use.

Cartridge filters are usually designed for a maximum SWRO feed water flow of 0.3 lps (liters per second) per equivalent 250 mm of cartridge length. Additional filtration capacity

Figure 3.5 Cartridge filters located horizontally in vessel.

is normally provided to allow cartridges to be replaced without the need to interrupt production. Pressure vessels are typically constructed of duplex stainless steel for seawater RO installations. Individual cartridge filter housing vessels that have capacity to process feed flow of up to 20,000 m^3/day per housing are commercially available at present.

The clean cartridge filter pressure drop is usually specified as less than 0.2 bar. Commonly, cartridges are replaced when the filter differential pressure reaches 0.7 or 1 bar. The operational time before replacement depends on source water quality and the degree of pretreatment. Typically, a cartridge filter replacement is needed once every 6–8 weeks. However, for high-quality source water cartridge filter useful life may exceed one year.

For SWRO systems whose source water may contain sand, rigid melt-blown cartridges or cartridge filters with single open ends and dual O-rings on the insertion nipple (rather than granular dual open-end cartridges) are commonly used. The single open-end insertion filters have positive seating and an insertion plate, which do not allow deformation of the filter cartridge under pressure caused by sand packing. Double open-end cartridge filters are held in place by a spring-loaded pressure plate. Under pressure from packed sand, wound-type cartridge filter can bend, causing the ends of the filter to unseat and allow direct entry of sand into the SWRO system feed line.

3.2.3 SWRO Desalination System

3.2.3.1 Key System Elements Figure 3.6 shows a typical configuration of a large SWRO system. The filtered water produced by the plant's pretreatment system is conveyed by transfer pumps from a filtrate water storage tank through cartridge filters and into the suction pipe of the high-pressure RO feed pumps. The main purpose of the cartridge filters is to protect the RO membranes from damage. The high-pressure feed pumps are designed to deliver the source water to the RO membranes at pressure required for membrane separation of the freshwater from the salts, which typically is 55–85 bars. The actual required feed pressure is site specific and is mainly determined by the source water salinity and the configuration of the RO system.

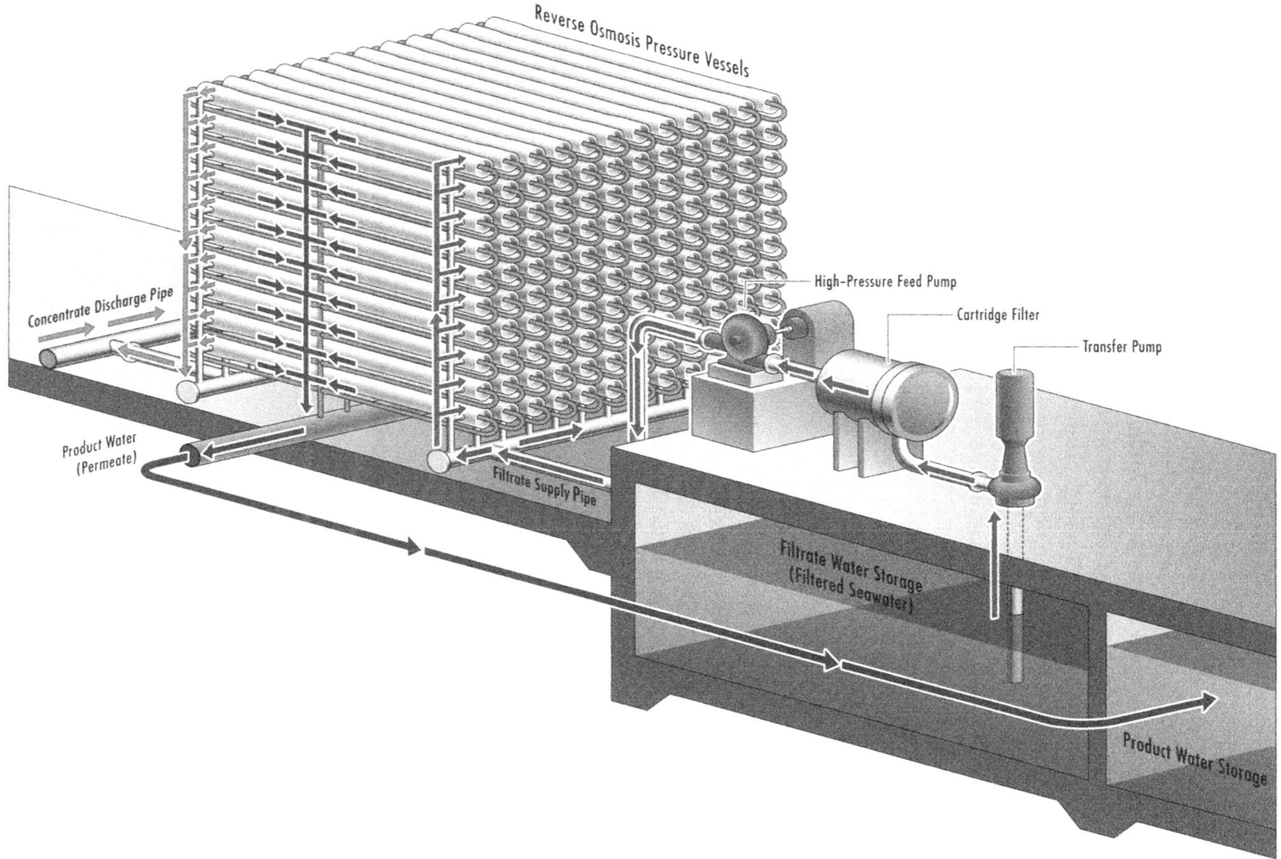

Figure 3.6 RO membrane system configuration.

3.2.3.2 SWRO Membrane Elements—Performance Parameters and Types The "engine" of every SWRO plant is the RO membrane element. The two primary materials used to manufacture SWRO membrane elements today are various polyamides referred to as thin-film composites, and cellulose acetate and its derivatives. Key advantages and disadvantages of the thin-film composite and cellulose acetate SWRO membranes are presented in Table 3.5.

Mainly because of their higher membrane rejection and lower operating pressures, the thin-film composite membranes are the choice for most SWRO membrane installations today. Exceptions are applications in the Middle East and Persian Gulf where the source water is rich in organics, and the cellulose acetate membranes offer significant benefits in terms of limited membrane biofouling, cleaning, and pretreatment needs, which there, because of the significantly lower unit power costs, are an acceptable trade-off for higher operating pressures/power demand.

The most widely used thin-film composite SWRO membrane elements consist of two membrane sheets glued together in a "leaf" that is spirally wound around a perforated central tube through which the desalinated water exits the membrane element (Fig. 3.7). The first membrane sheet, which actually retains the source water minerals on one side of the membrane surface, is typically made of thin-film composite polyamide material and has a microscopic porous structure that can retain compounds of size smaller than 200 Da. This sheet, however, is usually less than 0.2 μm thin, and in order to withstand the high pressure required for salt separation, it is supported by a second thicker membrane sheet, which is typically made of higher porosity polysulfone material that has several orders of magnitude larger membrane openings.

The membrane leaves are separated by approximately 0.7-mm-thick feed spacer that forms feed channels and facilitates the mixing and conveyance of the feed concentrate stream along the length of the membrane element. The most widely used and commercially available SWRO elements have a diameter of 20 cm (8 inches), a length of 100 cm (40 inches), and produce 11–14 m^3/day of permeate. As shown on Figure 3.7, the SWRO membrane elements are connected in series inside the pressure vessel. Typically, one pressure vessel houses from 6 to 8 SWRO membrane elements. A recent design trend is

TABLE 3.5 Comparison of Thin-Film Composite and Cellulose Acetate Membranes

Parameter	Thin-Film Composite Membranes	Cellulose Acetate Membranes
Salt rejection,%	Higher (>99.6%)	Lower (Up to 99.5%)
Net driving pressure, bars	Lower (10–15 bars)	Higher (15–30 bars)
Surface charge	Anionic (limits use of cationic pretreatment coagulants)	Neutral (no limitations on pretreatment coagulants)
Chlorine tolerance	Poor (Up to 1000 ppm-h) feed dechlorination needed	Good; continuous feed of 1–2 ppm of chlorine is acceptable
Cleaning frequency	High (weeks to months)	Lower (months to years)
Pretreatment requirements	High (SDI < 4)	Lower (SDI < 5)
Organics removal	High	Relatively lower
Biogrowth on membrane surface	May cause performance problems	Limited—not a cause of performance problems
pH tolerance	High (1–13)	Limited (4–6)

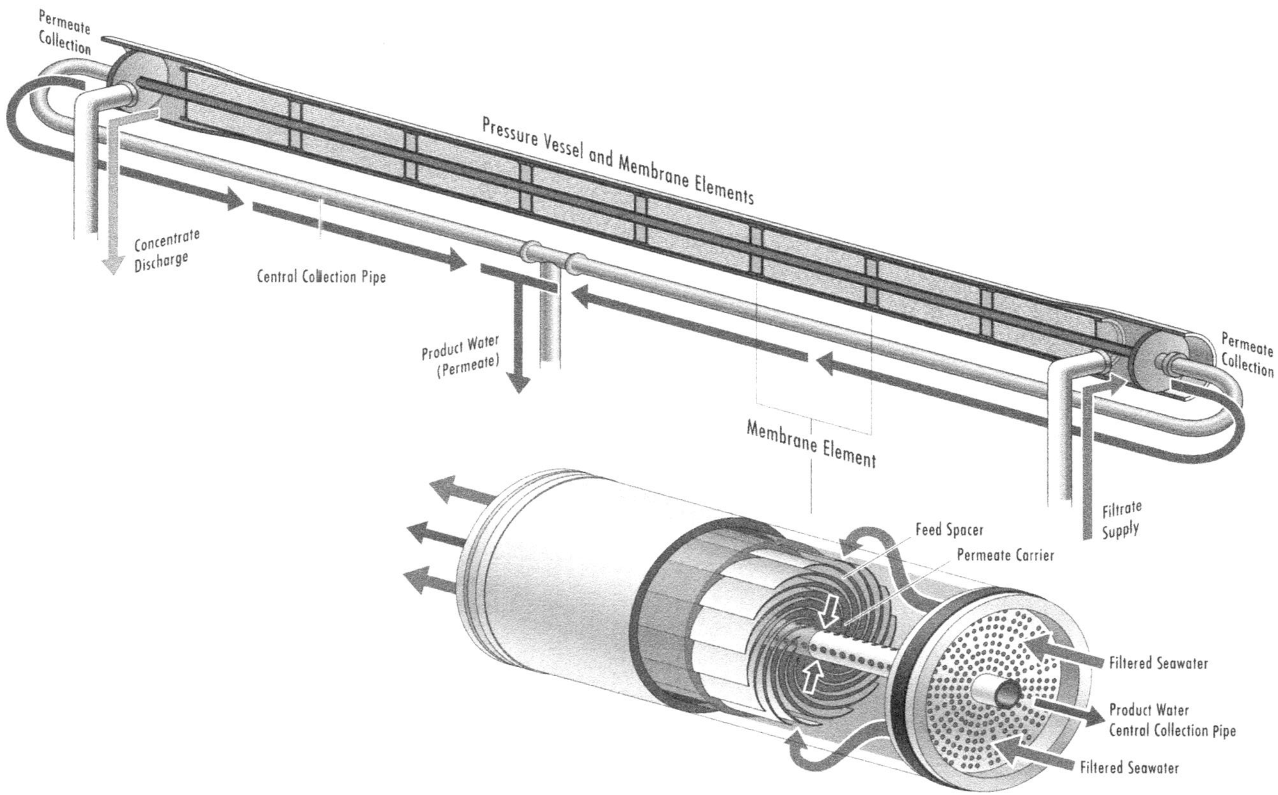

Figure 3.7 Spiral-wound thin-film-composite SWRO membrane element.

to use 8 elements per vessel. Typically, the higher the source seawater TDS and the lower the design membrane flux, the less elements are used per vessel. SWRO systems designed around lower salinity feed water and higher membrane flux are conducive to the use of 8 membrane elements per vessel.

The commercially available membrane RO elements today are of standardized diameters and length and salt rejection efficiency. Standard membrane elements have limitations with respect to a number of performance parameters such as feed water temperature (45°C), pH (minimum of 2 and maximum of 10), silt density index (less than 4), chlorine content (not tolerant to chlorine in measurable amounts), and feed water pressure (maximum of 80–100 bars).

Some membrane manufacturers currently are developing spiral-wound RO elements of 16–18 inches (40.64–45.72 cm) in diameter. Those membranes are suitable at the moment for brackish water desalination due to pressure limitations associated with the large diameter.

During the RO process the water molecules permeate through the RO membranes at a rate of permeation per unit area commonly referred to as membrane flux. Membrane flux is expressed in cubic meters per second per square meter ($m^3/s\ m^2$) or gallons per day per square foot (gfd) of active membrane surface area. For example, a typical seawater membrane RO element is operated at 8–10 gfd.

The ratio between the volume of the product water produced by the membrane desalination system and the volume of the source water used for its production is commonly defined as recovery and is presented in percent of the plant RO system feed water volume. The maximum recovery that can be achieved by a given pressure-driven membrane desalination system mainly depends on the source water salinity and is limited by the magnitude of the osmotic pressure to be overcome by the RO system high-pressure feed pumps and by the scaling potential of the source water. Scaling occurs when the minerals left behind on the rejection side of the RO membrane are concentrated to a level at which they begin to form precipitates (crystalline compounds), which in turn plug the membrane pores and interfere with freshwater transport through the membrane. Typically, seawater desalination plants can only turn 40–60% of the source water into low-salinity permeate.

Membrane performance tends to naturally deteriorate over time due to a combination of material wear-and-tear and irreversible fouling of the membrane elements. Typically, membrane elements have to be replaced every 5–7 years to maintain their performance in terms of water quality and power demand for salt separation. Improvements of membrane element polymer chemistry and production process have made the membranes more durable and have extended their useful life. Use of elaborate granular media pretreatment technologies and ultra and microfiltration (UF and MF) membrane pretreatment systems prior to RO desalination is expected to allow extending the membrane useful life to 7 years and beyond, thereby reducing the costs for their replacement and the overall cost of water. Detailed guidelines for designing SWRO plants are provided elsewhere (AWWA, 2007).

While water permeates through the membrane, the rejected matter accumulates behind the membrane and forms a layer of high concentration of salts that increases the osmotic pressure and reduces the permeability. Also over time, organic and suspended matter adsorb on the membrane and reduce membrane permeability. These materials may also serve as food for bacteria that attach on the membrane surface. The excessive bacterial growth may cause membrane biofouling. Sparingly, soluble salts may precipitate on the membrane surface and impact membrane performance. The phenomenon of the formation of a concentrated layer close to the membrane is called concentration polarization. The ratio of salt concentration close to the wall over the concentration in the bulk flow is referred as the β parameter. Membrane producers usually restrict the flux through the membrane such that the β value is below 1.15. This was chosen to limit membrane fouling.

While permeate is produced and collected along the membrane length, membrane flow and velocity are reduced and the concentration polarization/fouling effects typically increase. On the other hand, increasing flow along the membrane causes increased pressure drop and loss of energy. Proper design is needed to maintain proper pressure–flow configuration.

3.2.3.3 Classification of SWRO Membranes Based on Performance As indicated previously, thin-film composite SWRO membranes are the most widely used types of membranes today. The three most important SWRO membrane performance parameters are salt rejection, flux/productivity, and operating pressure. Currently, there are a number of commercially available SWRO membrane elements designed with special features allowing to optimize their performance around one or more of these three key performance parameters. Commercially available RO and nanofiltration (NF) elements at present can be classified in the following key groups:

1. Standard rejection membranes
2. High-rejection membranes
3. High-productivity (or low-energy) membranes
4. High-pressure membranes

Standard Rejection SWRO Membrane Elements Standard rejection membrane elements are designed to remove up to 99.6% of the salts in the source seawater. These membrane elements are most widely used today and have found applications in a variety of RO system configurations.

High-Rejection SRWO Membrane Elements High-rejection membrane elements are designed with tighter membrane structure allowing to increase the mass of rejected ions and to reject smaller size ions, such as boron, for example. The higher rejection membrane capabilities of 99.75–99.85% come at a price—10–20% higher operating pressure.

High-Productivity (Low Energy) Membrane Elements High-productivity membrane elements are designed with features to yield more product water per membrane element. These features are: (1) higher surface area and (2) denser membrane packing. Increasing active membrane leaf surface area allows to gain significant productivity using the same size (diameter) membrane element. Active surface area of the membrane leaf is typically increased by improving and automating the membrane production process.

The total active surface area in a membrane element can also be increased by increasing membrane size/diameter. Although 100-cm (8-inch) SWRO membrane elements are still a "standard" size for those most widely used in large full-scale applications, larger size membrane elements have been used in the past and are currently under development.

Another alternative for improving membrane productivity is increasing the number of membrane leaves packed into the same size (diameter) membrane. This is accomplished either by the use of thinner feed channel spacers or by improving element construction. Using thinner feed spacers typically increases the membrane pressure drop. As a result, higher productivity membrane elements using this approach also have higher operational pressure requirements for the same salt rejection level and flux.

Denser membrane leaf packing makes membranes also more susceptible to fouling, and their use requires high-quality source water and more elaborate pretreatment. To address this issue, the newest high-productivity membrane elements actually use wider spacers to compensate for the increased fouling potential and pressure.

The dynamics of the high-productivity (or low-energy) membrane element development is illustrated by an example of the development of seawater membranes. In the second half of 1990s the typical 100-cm (8-inch) SWRO membrane element had a standard productivity of 5000–6000 gpd at salt rejection of 99.6%. In 2003, several membrane manufacturers introduced high-productivity seawater membrane elements that are capable of producing 7500 gpd at salt rejection of 99.75%. Just one year later, even higher productivity (9000 gpd at 99.7% rejection) seawater membrane elements were released on the market. Newest membrane elements provide flexibility and choice and allow to trade productivity and pressure/power costs. The same water product quality goals can be achieved either by (1) reducing the system footprint/construction costs by designing the system at higher productivity or by (2) reducing the system overall power demand by using more membrane elements, designing the system at lower flux and recovery, and taking advantage of newest energy recovery technologies that further minimize energy use if the system is operated at lower (35–45%) recoveries.

High-Pressure SWRO Elements The main purpose of this type of SWRO element is to produce freshwater from concentrated seawater with salinity of 50,000–60,000 mg/L and are used to maximize water recovery from a given source water volume. While a standard membrane element can only allow to recover up to 50% of the source seawater, the high-pressure SWRO are suitable to obtain recoveries of 60% and higher. The high-pressure membrane elements are specifically designed to operate at 20–40% higher pressure than that of the other types of membrane elements listed above and to treat high-salinity concentrate produced by the first stage of a two-stage SWRO system. A more detailed discussion of the key features and benefits of the two-stage SWRO system is provided in the next section of this chapter.

3.2.3.4 Alternative SWRO Membrane System Configurations Reverse Osmosis membrane elements are installed in pressure vessels that usually house 6 to 8 elements per vessel (see Fig. 3.7). Multiple pressure vessels are arranged on support structures (racks) that form RO trains. Each RO train is typically designed to produce between 10 and 20% of the total amount of the membrane desalination product water flow. Figure 3.6 depicts one RO train.

Single-Stage SWRO Systems Single-stage SWRO systems are designed to produce desalinated seawater (permeate) in one step using only a single set of RO trains operating in parallel (see Fig. 3.6). In general, between 800 and 900 SWRO membrane elements installed in 100–150 vessels are needed to produce 10,000 m^3/day of permeate suitable for potable use in a single-stage SWRO system.

Under a typical single-stage SWRO system configuration, each RO train has a dedicated system of transfer pumps for pretreated seawater followed by a high-pressure RO feed pump. The high-pressure feed pump motor/operation is coupled with that of energy recovery equipment. The alternative energy recovery systems commonly used today are discussed further in this chapter.

Single-stage SWRO systems are widely used for production of drinking water. However, these systems have found limited industrial application mainly because of the water quality limitations of the produced permeate. Even if using the highest rejection RO membrane elements commercially available today (nominal minimum rejection of 99.75%), the single-stage SWRO desalination systems typically cannot consistently yield permeate with TDS concentration lower than 200 mg/L, chloride level of less than 100 mg/L, and boron concentration lower than 0.5 mg/L.

Two-Pass SWRO Systems Two-pass SWRO systems are typically used when either the source seawater salinity is relatively high (i.e., higher than 35,000 mg/L) and/or the product water quality requirements are very stringent. For example, if high-salinity (4.2% salts)/high-temperature source water (such as Red Sea and Persian Gulf seawater) is used in combination with standard rejection (99.6%) SWRO membranes, then single-stage SWRO systems may not be able to produce permeate suitable for drinking water use. In this case, two-pass systems have proven to be a very efficient and cost-effective configuration for potable water production. RO systems with two or more passes are also widely used for production of high-purity industrial water.

The two-pass SWRO systems typically consist of a combination of a single-stage SWRO system and a single-stage brackish water RO (BWRO) system connected in series (Fig. 3.8). Permeate from the SWRO system (i.e., pass one) is directed for further treatment to the BWRO system (i.e., pass two) to produce a high-quality TDS permeate. The concentrate form the pass-two BWRO system (indicated as "concentrate–pass two" on Fig. 3.8) is returned to the feed of the pass-one SWRO system to maximize the overall desalination system production capacity and efficiency.

A variation of the two-pass SWRO system is the single-stage/partial second-pass system where only a portion of the first-stage SWRO permeate is processed through a second-pass

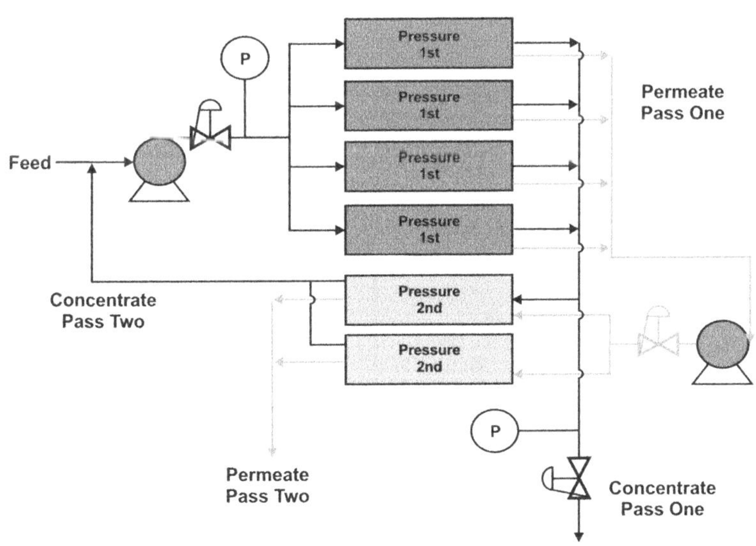

Figure 3.8 Two-pass SWRO system.

SWRO system. For example, a partial second-pass configuration is used at the 95,000-m^3/day Tampa Bay seawater desalination plant. The second pass at this facility is designed to treat up to 30% of the permeate produced by the first-pass SWRO system as needed in order to maintain the concentration of chlorides in the plant product water always below 100 mg/L. The partial second pass at the Tampa Bay seawater desalination plant was installed to provide operational flexibility and to accommodate the wide fluctuations of source water salinity (16,000–32,000 mg/L) and temperature (18–40°C). Typically, the product water quality target chloride concentration of 100 mg/L at this plant is achieved by only operating the first pass of the system. However, when source water TDS concentration exceeds 28,000 mg/L and/or the source water temperature exceeds 35°C, the second pass is activated to maintain adequate product water quality. The percent of first-pass permeate directed for additional treatment through the second pass is a function of the actual combination of source water TDS and temperature and is adjusted based on the plant product water chloride level.

Two-Stage SWRO Systems Two-stage SWRO membrane systems are mainly used to maximize the overall desalination plant recovery and reduce the volume of concentrate discharged by the desalination plant. A general schematic of a two-stage RO system is shown on Figure 3.9. In these SWRO systems, typically the entire volume of the concentrate generated by the first-stage SWRO system is directed to a second-stage SWRO system for further treatment and enhanced recovery. Permeate from both systems is blended prior to final use.

The main advantage of such SWRO system configurations is that it allows achieving very high levels of use (recovery) of the available source seawater and the energy used by the first-stage RO system. For example, while a single-stage SWRO system configuration typically allows turning 35–50% of the source seawater to potable water, the two-stage SWRO system recovery may reach 60–65%. Designing the SWRO plant around higher recovery allows minimizing the size of the plant intake and pretreatment facilities and the capital expenditures for their construction and operation.

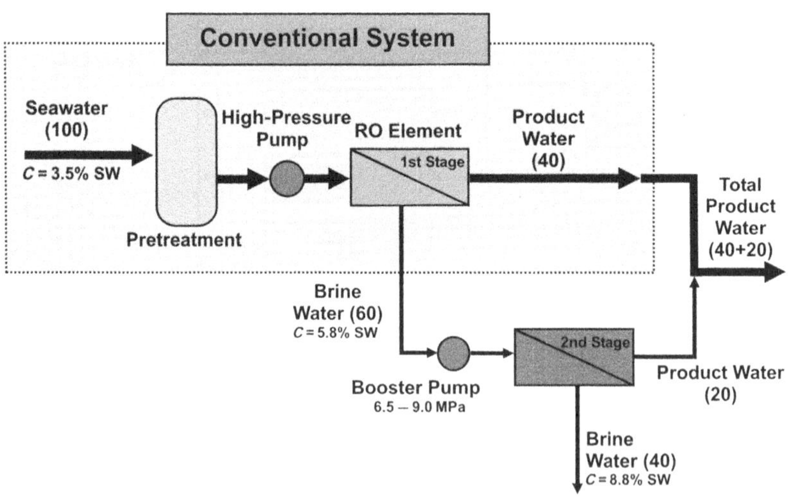

Figure 3.9 Two-stage SWRO system. (From Kurihara et al., 1999.)

However, because of the high salinity of the first-stage concentrate (typically above 55,000 mg/L), the practical implementation of two-stage SWRO systems requires the use of high-pressure SWRO membranes, membrane vessels, piping, and auxiliary equipment that can withstand and perform well at very high pressures (up to 98 bars/1400 psi). The cost of this equipment is usually higher than the cost of the same size and type of equipment built to operate at more "standard" pressures (i.e., below 70 bars/1000 psi). Therefore, the viability of using a two-stage SWRO system has to be carefully assessed based on a comprehensive life-cycle cost analysis for the site-specific conditions of a given project. To date, two-stage systems have been mainly used to upgrade the capacity and improve energy use of existing conventional single-stage SWRO plants (Kurihara et al., 1999).

Hybrid RO System Configurations The two-pass and two-stage RO system configurations may be combined to achieve an optimum plant design and tailor desalination plant operation to the site-specific water source water quality conditions and product water quality goals.

An example of a full-scale two-pass, two-stage RO system application is the 170,500-m^3/day Fujairah seawater desalination plant (Rovel, 2003). A general treatment process schematic of this plant is depicted on Figure 3.10. The plant uses Gulf of Oman, Indian Ocean, seawater (see Table 3.2 for water quality characteristics).

The first pass of the Fujairah plant consists of 17 duty and 1 standby RO trains using standard rejection SWRO membranes producing permeate of TDS concentration of 400–500 mg/L. The overall recovery of the first-pass SWRO system is 43%. The second pass has 8 BWRO trains with a total recovery rate of 90%. The second-pass BWRO trains have two stages and treat approximately 80% of the first-pass permeate to TDS concentration of 10–20 mg/L. The rest (i.e., 20%) of the first-pass permeate is blended with the second-pass permeate to produce finished water of TDS concentration of less than 120 mg/L. The concentrate produced by the second-pass RO system has salinity lower than that of the source seawater and is recycled to the feed of the first-pass RO system

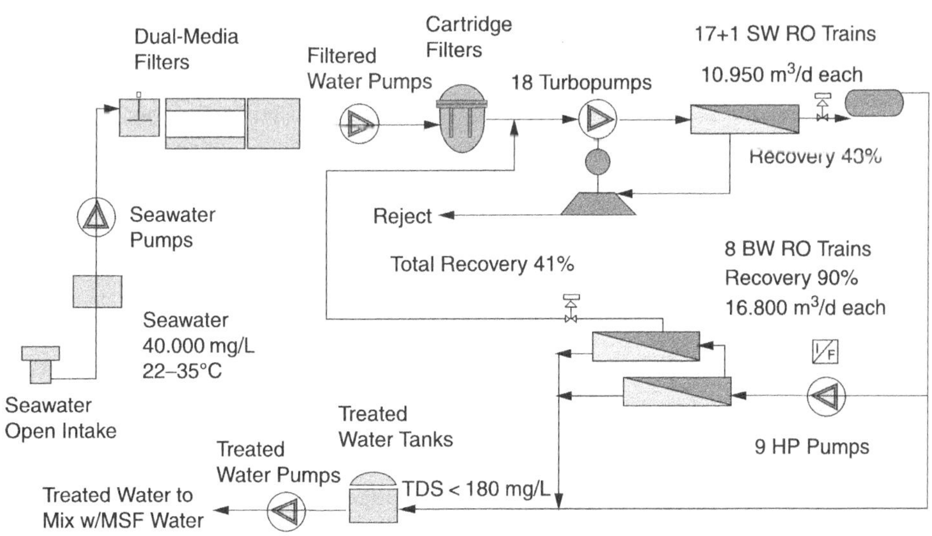

Figure 3.10 Fujairah seawater desalination plant schematic. (From Rovel, 2003.)

(see Fig. 3.10). Despite its complexity, the two-pass, two-stage RO system in Fujairah performs very reliably and has demonstrated plant availability of over 97% since its startup in 2003.

An example of a two-pass/two-stage SWRO plant is the 110,000-m^3/day Point Lisas facility in Trinidad. This plant produces high-quality desalinated seawater of TDS concentration of 85 mg/L or less, which is predominantly used for industrial applications. The first pass of the Point Lisas SWRO system consists of six two-stage RO units. Each of the first-stage RO trains uses SWRO membranes and is coupled with energy recovery devices. The second-stage trains of the first pass are equipped with brackish water RO elements.

The entire volume of permeate from the first pass of the Point Lisas SWRO system is further treated in a second-pass RO system to meet the final product water quality specifications. The second-pass system also consists of two stages—each equipped with BWRO membranes. The Point Lisas seawater desalination plant has the same number of first-pass and second-pass RO membrane trains.

Three-Center RO System Configuration As indicated previously, a typical conventional SWRO system is configured in individual equally sized RO trains each of which is serviced by a separate transfer pump, cartridge filter vessel, high-pressure feed pump, and energy recovery equipment dedicated to this RO train (see Fig. 3.6). The size of the individual RO trains depends on the overall production capacity of the SWRO plant and typically varies between 2000 and 20,000 m^3/day. The main advantage of this RO train-based configuration is that it is modular and allows for relatively easy flow distribution and service of the individual trains. Since typically the size of the individual RO trains does not exceed 10–20% of the total plant production capacity, train shutdown for maintenance (membrane cleaning and replacement and equipment service) is handled either by using a standby RO train or by temporary increase of the production capacity of the RO trains remaining in service.

The RO-train-based configuration is very suitable when the SWRO plant is designed and intended to operate at a constant production output. At present, most of the existing large municipal SWRO plants worldwide are designed to supplement existing conventional water supply sources rather than to be the primary or the only source of water supply for a given area. Therefore, the operation of these SWRO plants does not need to have the flexibility to follow the actual diurnal and monthly fluctuations in product water demand and most of the existing plants are designed to operate at constant production capacity.

In the future, the SWRO is likely to become a prime rather than a supplemental source of water supply for many coastal communities pressured by population growth, especially for large urbanized or industrial areas with limited traditional local sources of freshwater supply (i.e., groundwater or river or lake water). The SWRO systems servicing such areas have to be designed to have the operational flexibility of matching desalination plant production with the product water demand fluctuations.

Shift of the SWRO plant operational paradigm from constant to variable production flow requires a change of the typical SWRO configuration from one that is most suitable for constant production output to one that is most cost effective for delivery of varying permeate production. A response to such a shift of the desalination plant operational paradigm is the three-center RO system configuration used for the first time at the Ashkelon seawater desalination plant in Israel (see Fig. 3.11). Under this configuration, the RO membrane vessels, high-pressure pumps, and energy recovery equipment are no longer separated in individual RO trains but are rather combined in three functional centers—a high-pressure RO feed

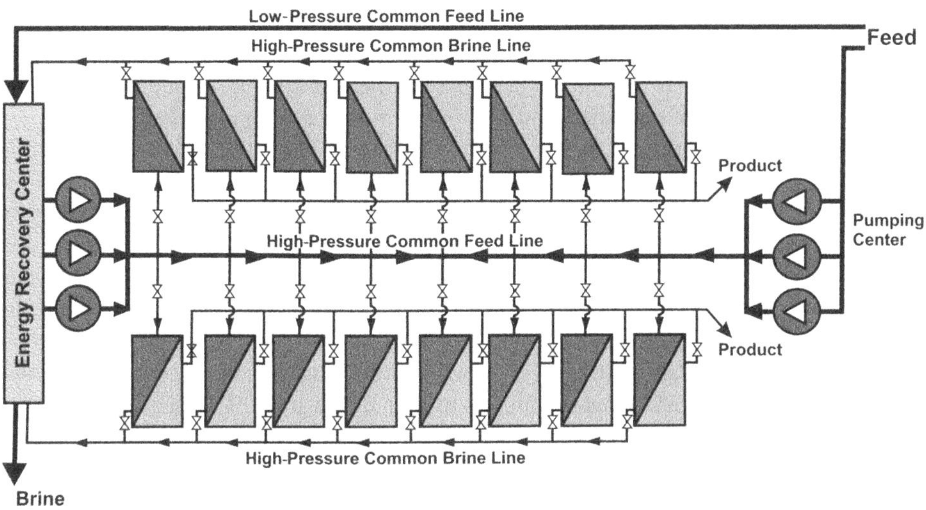

Figure 3.11 Three-center SWRO system. (From Liberman, 2002.)

pumping center, a membrane center, and an energy recovery center (Liberman, 2002). The three functional centers are interconnected via service piping.

The RO feed pumping center includes only a few large-capacity high-pressure pumps that deliver seawater to the RO membrane center. The main benefit of using a few large-capacity high-pressure pumps rather than a large number of small-capacity units is the gain in overall pumping efficiency. Typically, the smaller the ratio between the pressure and the flow delivered by a given pump, the better the pump efficiency and the "flatter" the pump curve (i.e., the pump efficiency is less dependent on the variation of the delivered flow). Therefore, pump efficiency can be improved by either reducing the pressure delivered by the pump or by increasing pump flow. Since the pump operating pressure decrease is limited by the RO system target salt separation performance, the main approach to improve pump efficiency is to increase unit pump flow. While a conventional size high-pressure RO feed pump of small capacity would typically have maximum total energy use efficiency of 80–85%, the use of 10 times larger size pump may allow to increase the pump efficiency to 88–92%, especially for large SWRO plants. This beneficial feature of the three-center design is very valuable in the case of systems delivering varying flow.

While in a conventional RO train design the membrane vessels are typically grouped in 100–200 units per train and in 2–20 RO trains, the membrane center configuration contains 2–4 times larger number of RO vessel groups (banks) and a smaller number of membrane vessels per bank. Under this configuration the individual vessel banks are directly connected to the high-pressure pump feed lines and can be taken off service one at a time for membrane replacement and cleaning. Although the feed water distribution piping for such membrane center configuration is more elaborate and costly than that use for individual RO trains that contain 2–3 times more vessels per train, what is lost in capital expenditure is gained in overall system performance reliability and availability. A reliability analysis completed for a 95,000-m^3/day (25-MGD) SWRO plant (Liberman, 2002) indicates that the optimum number of vessels per bank for this scenario is 54 and the number of RO banks per plant is 20. A typical RO-train-based configuration would include 2–4 times more (108–216) vessels per RO train and 2–4 times less (5–10) RO trains. According to this analysis, the use of the three-center configuration instead of the conventional RO-train-based approach allows to improve RO system availability from 92–96% (avg. 95%) to 98%, which is a significant benefit in terms of additional amount of water delivered to the customers and improvement in water supply reliability.

The centralized energy recovery system included in the three-center configuration (Fig. 3.11) uses high-efficiency pressure-exchanger-based energy recovery technology. The proposed configuration allows to improve the overall energy recovery efficiency of the RO system and to reduce system power, equipment, and construction costs. While typically the energy recovery of the conventional Pelton wheel systems drop significantly when the reduction of the overall SWRO plant recovery, the energy recovery efficiency of the pressure exchanger systems improve with lowering the plant recovery rate. This allows operating the SWRO plant cost effectively while delivering variable product water flow. For example, if the SWRO plant output has to be reduced by 40% to accommodate low diurnal demand, an SWRO system with RO-train-based configuration has to shut down 40% of its trains, and, if this low demand persists, it has to flush these trains to prepare them for the next startup. An RO system with three-center configuration would only need to lower its overall recovery to achieve the same reduction of the diurnal demand. Although temporary operation at lower recovery would result in elevated costs for pumping and pretreatment of larger volumes of source water, these extra operation expenses are typically compensated for by the improved energy recovery efficiency that results from operating the SWRO the system at lower water recovery ratio.

3.2.4 Energy Recovery Systems

The concentrate leaving the last membrane in the pressure vessel has a great portion of the feed water energy applied through the high-pressure RO pumps that can be recovered and reused to minimize the overall energy cost for seawater desalination. Dramatic improvements of the membrane element materials and energy recovery equipment over the last 20 years coupled with enhancements in the efficiency of RO feed pumps and reduction of the pressure losses through the membrane elements have allowed to reduce the use of power to desalinate seawater to less than 3.5 kWh/m^3 (13.5 kWh/1000 gal) of produced freshwater today. The main reduction in energy use is in the RO process itself. Today, the energy use of the RO system is as low as 2.2–2.7 kWh/m^3 of produced freshwater on a basis of 50% recovery. The balance of the desalination plant energy use is associated with the operation of the desalination plant intake and discharge pumps and facilities, the energy requirements of the pretreatment stage, ventilation, controls, lights, and the like. Taking into consideration that the cost of power is typically 25–30% of the total cost of desalinated water, these technological innovations contributed greatly to the reduction of the overall cost of seawater desalination.

Currently, several different devices are widely used to reduce the overall energy use by the SRWRO system. All of these devices are designed to recover energy from the concentrate generated by the SWRO system. When using turbines for energy recovery (Fig. 3.12 and 3.13), the pressure of the concentrate turns into water velocity that rotates the blades of the turbines. The use of this energy recovery equipment results in somewhat lower energy recovery (87–92%), in comparison with pressure exchangers (Figs. 3.14–3.17) and requires the location of the energy recovery turbine on the same shaft as the high-pressure SRWO system feed pump— that is, each pump is equipped with a separate energy recovery turbine. Compensation is needed for the pressure loss in the pressure vessel, which may be

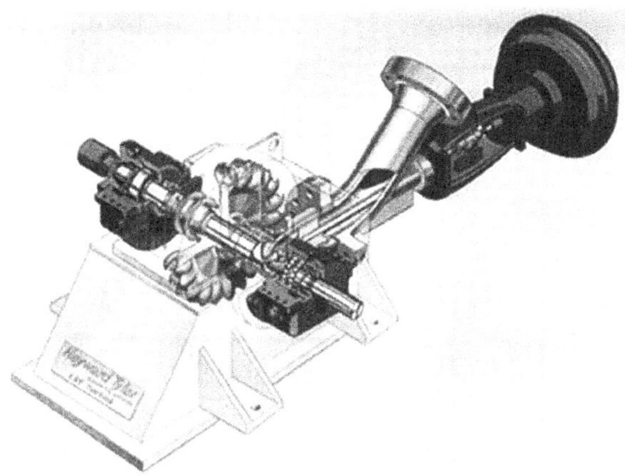

Figure 3.12 Schematic of Pelton turbine.

covered by a low-pressure pump operated at high water pressure to raise the feed pressure with approximately 2 bars. Separate turbines can be applied to use the SWRO concentrate energy and drive a second stage at elevated pressure, as shown in Figure 3.18.

Novel energy recovery systems working on the pressure exchange principle (pressure exchangers) are currently available on the market and allow to achieve higher energy recovery efficiency than turbines. The pressure exchangers transfer the high pressure of the concentrated seawater directly into the RO feed water, by using a piston moving in a pressure vessel and a set of automatic valves to change flow directions (Figs. 3.14 and 3.15). Two such vessels are needed to work in parallel, as shown in Figure 3.15, where one is filled with pretreated seawater, while the other pushes the pressurized seawater into the system, using the high pressure of the SWRO system concentrate. The efficiency of these devices may exceed 94%. Optimization is needed between pressure vessel diameters and the number of such devices needed per number of pressure vessels. A similar pressure-exchanger-type system that uses a rotating shaft has also found wide application for SWRO system energy recovery.

Figure 3.13 Pelton turbine connected to RO train motor and pump.

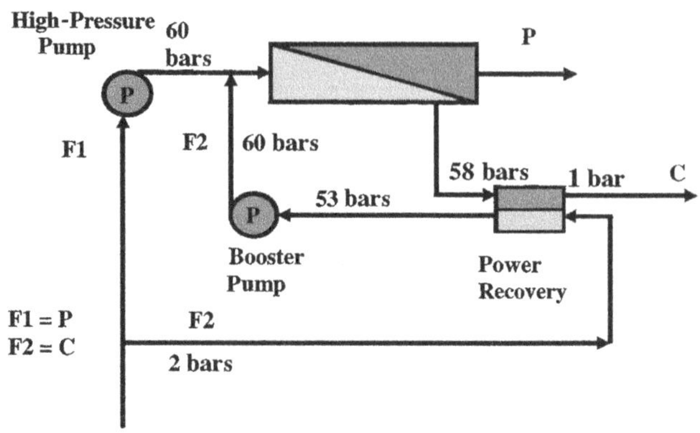

Figure 3.14 Pressure exchanger system with booster pump.

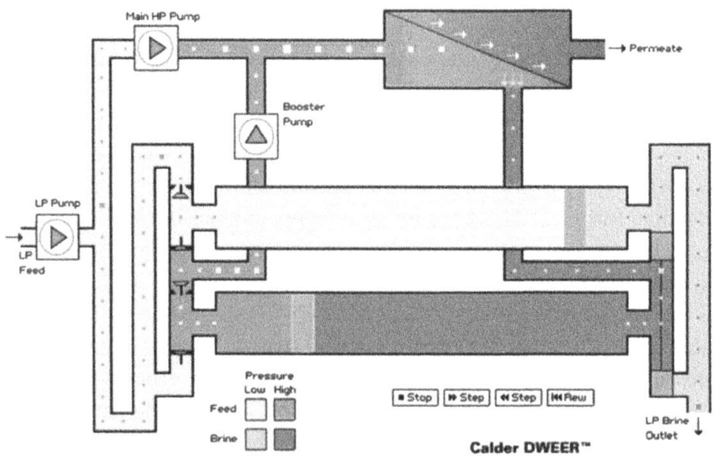

Figure 3.15 Schematic of pressure exchanger system.

Figure 3.16 DWEER pressure exchanger.

Figure 3.17 ERI pressure exchanger.

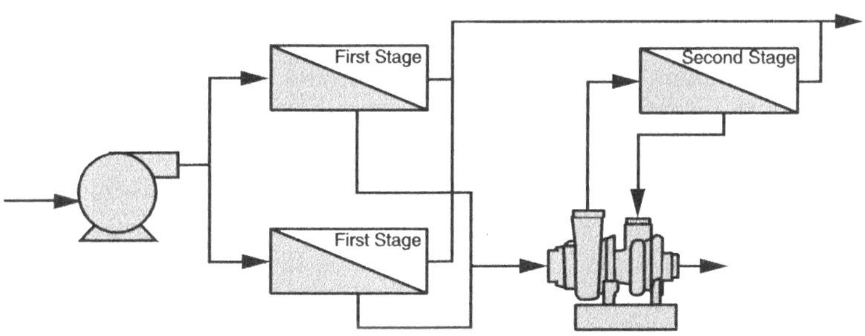

Figure 3.18 Two stage RO system with turbine running on first-stage concentrate.

Use of pressure exchangers instead of turbines for energy recovery has its trade-offs: better energy recovery is traded for a higher complexity system that has more moving parts and therefore higher maintenance costs. The use of energy recovery devices turns energy cost into equipment cost. Lower energy RO membrane elements (high flux) are expected to operate at even lower pressures and to continue to yield further reduction in the cost of desalinated water.

3.2.5 Permeate Conditioning

The quality of the permeate leaving the plant depends on the plant source water quality and configuration. Single-stage SWRO systems would typically produce permeate that has boron concentration close to 1 mg/L. This boron level is acceptable according to the regulations in most countries, such as the European Community (boron level limit of 1 mg/L), Canada (drinking standard for boron of 5 mg/L), and the United Startes (no drinking water limit for boron, except the State of California, where the boron action level is 1 mg/L). For comparison, the United Nations World Health Organization (WHO) guidelines have a recommendation for boron level in drinking water of 0.5 mg/L. Achieving this or even lower levels of boron content if needed for irrigation of boron-sensitive crops would require the use of multistage/multipass SWRO systems.

Boron reduction is currently possible by applying one or more of the following treatment approaches:

- Increase pH in feed seawater to the SWRO system, which increases its rejection by the membranes.
- Exchange borate ion in an ion exchanger.
- Use two- or three-stage SWRO systems with or without ion exchange or pH increase to reduce boron content.
- Use of electrodialysis membranes.

In addition, as indicated previously, membrane manufacturers have been focusing recently on the development of SWRO membranes with improved boron rejection.

Salt content of the desalinated water depends on the SWRO plant recovery ratio and the quality of the membranes. The product may contain NaCl at a level below 100 mg/L and up to 1000 mg/L. That can be reduced by a second-stage membrane treatment of permeate as explained in the previous sections of this chapter. Using two- or three-stage SWRO systems to remove boron to lower levels will further reduce the sodium chloride salt content as well. As the water quality is improved, however, the produced water would have very low mineral content (especially calcium hardness) and therefore would be more aggressive and may dissolve matter from the piping system and increase pipe material corrosion rate.

Therefore, it is advisable to treat the SWRO permeate in order to increase its calcium carbonate content. Since the saturation of this salt in water is relatively low, water may maintain supersaturation of this salt. Based on WHO guidelines, it is suggested to increase the calcium concentration slightly above the saturation level. This can be done by dissolving CO_2 or acid in the produced water and passing the water through a bed of limestone (Hasson and Bendrihem, 2003). Adjustment of pH may increase slightly the level of supersaturation referred to as Langelier index (LGI). Values close to $LGI = 0.5$ are recommended, as water at this LGI tends to precipitate $CaCO_3$ on the mains and pipes, and thereby to prevent water aggressiveness. Stabilizing SWRO permeate using calcium-based salts reduces corrosion and prevents a phenomena called "red water" denoted by release of corrosion products into the water system.

While desalinated seawater may be considered very suitable for industrial and potable water supply, this water may be deficient in some mineral constituents needed for agricultural irrigation. For example, deficiency of magnesium and other constituents may affect certain crops.

3.2.6 Concentrate Management

Concentrate leaving the desalination plant may reach up to twice the original seawater concentration based on 50% recovery. The problem with this high concentration is its high osmotic pressure, in the range of approximately 50 bars, which is approximately twice the osmotic pressure of the source seawater. Since most species living in seawater are adjusted to the osmotic pressure related to seawater concentration, significant change resulting from the concentrate discharge may cause damage to those species.

The most environmentally suitable way to mitigate concentrate impact on the marine environment is to dilute concentrate with seawater prior to its discharge to the ocean. This, for example, can be completed when a desalination plant is colocated with a large power station and the cooling seawater from the power plant is mixed with the desalination plant concentrate. For example, in the case of the Ashkelon seawater desalination project in Israel, the ratio of the cooling water to the concentrate flow is 8 : 1, which reduces the TDS concentration of the concentrate to approximately 11% above the ambient seawater concentration. This TDS concentration increment is within the salinity naturally occurring in the ocean and therefore is tolerable by the marine environment. In some cases it is possible to reduce SWRO discharge salinity by mixing plant concentrate with brackish water or with wastewater treatment plant effluent (Masaaki et al., 2003). When such co-disposal alternatives are not available, it is advisable not to release the concentrate along the bottom of the sea, where the flow is very slow, but instead to accelerate concentrate mixing with the ambient seawater using diffuser system similar to that shown in Figure 3.19. In this case the concentrate flow is divided into a large number of smaller volume streams, which are discharged upward to improve mixing and salinity discharge plume dissipation. As seen in Figure 3.19, such a solution can be adopted at a single spot or along the concentrate pipe line at different distances to minimize the possible environmental damage.

Another issue that should be taken into consideration is the effect of the treatment chemicals on the aquatic life. There are two types of chemicals used in the membrane SWRO desalination: chemicals applied continuously, such as source water conditioning coagulants, antiscalants, and antioxidants, and chemicals used intermittently—mostly during SWRO or pretreatment membrane cleaning. These chemicals have to be of high purity and low content of phosphorus in order to minimize their impact on marine life. All spent membrane cleaning chemicals have to be neutralized prior to discharge to the ocean or discharged to the nearby wastewater collection system if possible. Development of improved pretreatment backwash and membrane cleaning techniques, such as osmotic backwash, for example, are expected to reduce the quantity of membrane cleaning chemicals in the future and thereby to minimize the impact of these chemicals on aquatic life. New techniques that allow washing the residuals generated during seawater pretreatment may allow using these solids for land application (Gasson and Allison, 2004).

In addition to the concentrate discharge methods discussed above, other alternative concentrate disposal methods are deep-well injection, evaporation ponds, mechanical

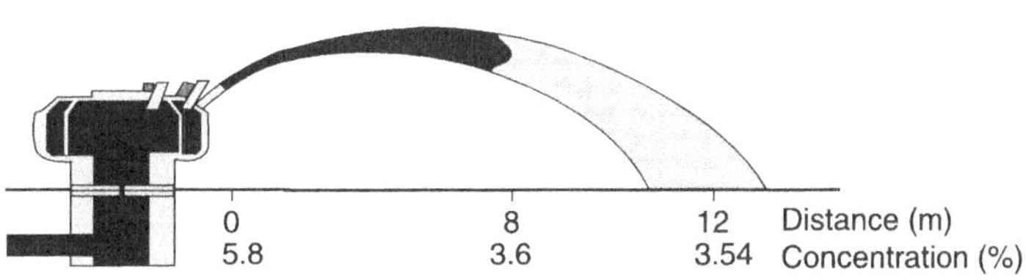

Figure 3.19 Concentrate disposal device. (From Gasson and Allison, 2004.)

evaporation, and sanitary sewer discharge. Most of these methods, however, have found application for small seawater desalination plants.

3.3 WATER PRODUCTION COSTS

The cost of SWRO desalination has decreased dramatically over the last 20 years, as seen from the example projects referenced in Figure 3.20. For this period the cost of seawater has been reduced over two times (i.e., with over US$0.5/m^3). One of the key reasons for this cost decrease is the reduction of the unit costs of the membrane elements combined with the increase in membrane element productivity (flux) and rejection. Another significant cost reduction factor is the decrease of energy costs due to the development of new energy recovery devices. In addition, improved pretreatment technologies and comprehensive operational experience allowed increasing membrane useful life and thereby reducing costs for membrane replacement and cleaning.

A recent conference on desalination costing held in Cyprus, in December 2004, organized by the Middle East Desalination Research Center (MEDRC) provides additional information of seawater desalination costs and trends (Wilf, 2004; Glueckestern, 2004; Velter, 2004). Table 3.6 presents a summary of water produced cost components for a large seawater desalination plant. It is important to note that the over cost of production of desalinated water depends not only on state-of-the-art technological solutions but on the well-structured project financing and implementation as well. Electric power costs are typically based on purchasing power from the electrical grid, yet power self-generation in an electrical power generation station dedicated to the desalination project may reduce significantly the unit cost of power and the overall desalination water production costs if

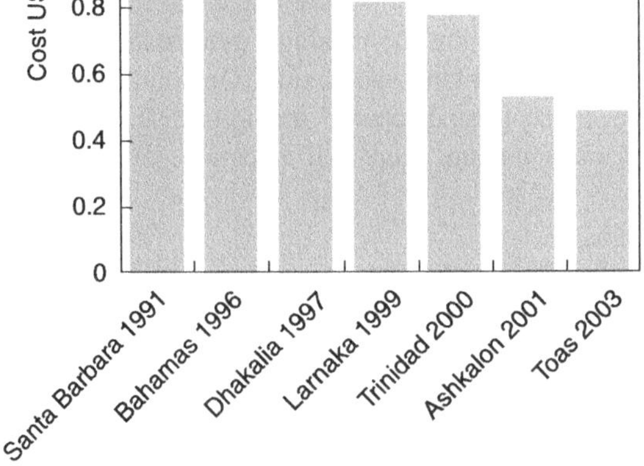

Figure 3.20 Seawater desalination cost reduction during last 20 years. (Taken from papers in Semiat et al., 2004.)

TABLE 3.6 Product Water Cost Components of 200,000 m^3/day SWRO Plant

Product Water Cost Component	US$/$m^3$
Capital cost, including land (25 years @ 6.0% interest)	0.203–0.338
Electric power ($0.060/kWh)	0.180–0.240
RO membrane replacement (5 years membrane life)	0.025–0.035
MF membrane replacement (7 years membrane life)	0.019–0.030
Chemicals	0.020–0.025
Maintenance and spare parts	0.023–0.038
Labor	0.030–0.044
Total product water cost	0.50–0.75

a low-cost source of power generation fuel is available. The energy consumption of a typical large seawater desalination plant is shown in Figure 3.21. As can be expected, the high-pressure pumping consumes most of the energy in such a plant.

As previously mentioned, the key SWRO project construction expenditures are associated with building the plant intake, the pretreatment system, procurement and installation of the plant pumps and piping, the SWRO membranes and pressure vessels, the energy recovery system, the water posttreatment facilities, and the concentrate disposal system. It is difficult to compare the investment and construction costs of existing desalination projects because projects may differ significantly in one or more of the cost-related parameters listed above. Based on previous experience, however, it can be estimated, for example, that the seawater pretreatment costs are in the range of 6–8 US cents/m^3; and the costs of water conditioning and boron and chloride removal are between 4 and 8 US cents/m^3.

Key Operating and maintenance (O&M) cost elements for the Ashkelon SWRO project are shown in Table 3.7 (Velter, 2004). The difference between the first 5 years is attributed to the use of new equipment at the first years of the plant operations and used equipment after that time. An O&M cost breakdown based on experience with seawater desalination in Spain is shown in Figure 3.22 (Medina, 2004).

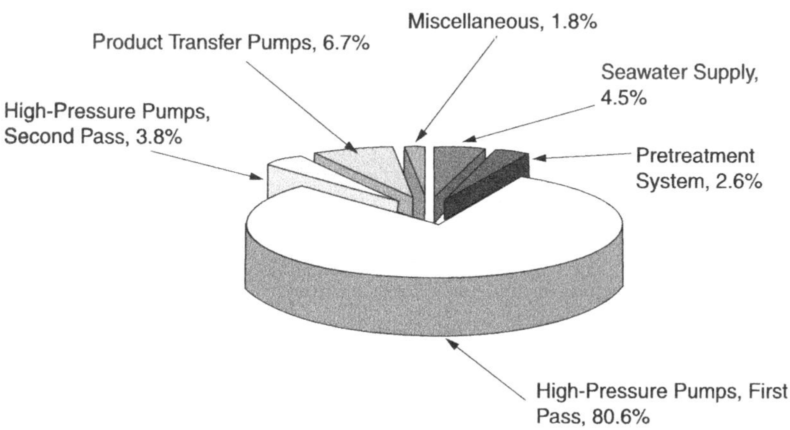

Figure 3.21 Power use in the large SWRO plant with a partial second stage.

TABLE 3.7 O&M Cost Breakdown for Ashkelon Seawater Desalination Project

O&M Cost Component	First 5 Years (%)	Subsequent Years (%)
Chemicals	25	22
Membranes	19	24
Maintenance	23	33
Labor	25	16
Other costs	8	5
Total	0.110	0.174
	In US$ 2002 per m^3	In US$ 2002 per m^3

Source: (Velter, 2004).

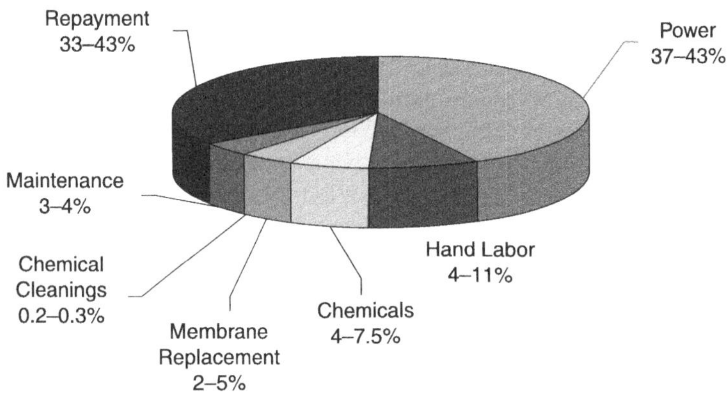

Figure 3.22 O&M cost breakdown based on Spanish SWRO projects (Medina, 2004).

3.4 FUTURE TRENDS

The accelerated development of new SWRO membrane elements of higher productivity and lower operating pressures is projected to continue in the future. Even that the membrane cost is a low fraction of the total investment in the system, membrane flux improvement is a key factor for the most significant cost reduction of desalinated water. Such improvement will allow achieving current production rates at lower pressure and at lower recovery ratio. This will allow to reduce the energy costs as well as the costs of pressure vessels, pumps, piping and flow devices, and significant overall reduction of production costs. The forecasted improvements of the SWRO membrane technologies are to encompass:

- Development of membranes of higher flux with high salt rejection that will eventually allow lower transmembrane pressure. This may influence significantly the cost of water production, due to the possibility to operate at lower pressures, meaning lower energy consumption and lower cost equipment.
- Development of membranes of increased pathogen rejection, as well as higher anti-fouling potential. This would require a better understanding of the fouling mechanisms and hence the ways to reduce all types of fouling.
- Improvement of membrane resistance to oxidants, elevated temperature, and compaction. Achieving this goal may allow to extend the useful life of the membranes.

- Integration of membrane pretreatment, advanced energy recovery, and SWRO systems.
- Integration of brackish and seawater desalination systems. In many places brackish water can be found close to the sea. Integrating the two types of desalination may reduce the cost of the final product.
- Development of new generation of high-efficiency pumps for SWRO applications.
- Reduction of membrane costs by complete automation of the entire production and testing process.
- Development of methods for low-cost continuous membrane cleaning, such as osmotic-pressure-induced backwash, which would allow to reduce downtime and chemical cleaning costs.
- Development of methods for low-cost membrane concentrate treatment, in-plant and off-site reuse, and disposal.
- Development of techniques that will allow monitoring the operation or failure of each membrane in each pressure vessel, to allow fast replacement and to maintain consistent water quality.

These technology advances are expected to ascertain the position of SWRO treatment as viable and cost-competitive processes for potable water production. Above all, the key for proper operation is to maintain proper pretreatment and to pay attention to all changes in the plant. This requires skilled operations staff and, therefore, building qualified manpower capacity is essential.

3.5 CONCLUSION

The advance of the reverse osmosis desalination technology is closest in dynamics to that of the computer technology. While conventional technologies, such as sedimentation and filtration have seen modest advancement since their initial use for potable water treatment several centuries ago, new more efficient seawater desalination membranes and membrane technologies, and equipment improvements are released every several years. Similar to computers, the SWRO membranes of today are many times smaller, more productive, and cheaper than the first working prototypes. Over the last 20 years, the cost of desalinated water dropped more than twofold. Although no major technology breakthroughs are expected to bring the cost of seawater desalination further down dramatically in the next several years, the steady reduction of desalinated water production costs coupled with increasing costs of water treatment driven by more stringent regulatory requirements, are expected to accelerate the current trend of increased reliance on the ocean as an environmentally friendly and competitive water source. This trend is forecasted to continue in the future and to further establish ocean water desalination as a reliable drought-proof alternative for many communities worldwide.

REFERENCES

American Water Works Association (AWWA) (2007). *Manual of Water Supply Practices—M46, Reverse Osmosis and Nanofiltration*, 2nd ed. AWWA, Denver, Colorado, p. 65.

Gasson, C., and Allison P. (2004). Desalination markets 2005–2015. In *A Global Assessment and Forecast, a Global Water Intelligence Publication*. Media Analytics, Oxford.

Gille, D. (2003). Seawater intakes for desalination plants. *Desalination* 156, 249.

Glueckstern, P. (2004). History of desalination cost estimations. In Semiat et al. (Ed.) *Proceedings of International Conference on Desalination Costing,* Limassol, Cyprus, The Middle East Desalination Research Center, Muscat, Oman.

Hasson, D., and Bendrihem, O. (2003). Modeling remineralization of desalinated water by limestone dissolution. In *Proceedings of IDA World Congress on Desalination & Water Reuse,* Paradise Island, Bahamas, International Desalination Association, Topsfield, Massachusetts, USA.

Hasson, D., Limoni-Relis, B., Semiat, R., and Tob, Ph. (1996). Fouling of RO membranes by phthalate ester contamination. *Desalination* 105, 13.

Hunt, H. C. (1996). Filtered seawater supplies—Naturally. *Desalin. Water Reuse Q.* 6(2), 32.

Irwin, K. J., and Thompson, J. D. (2003). Trinidad SWRO—Orinoco fluctuations fail to make filters falter. *Desalin. Water Reuse Q.* 13(3), 12.

Kurihara, M., Yamamura, H., and Nakanishi, T. (1999). High recovery/high pressure membranes for brine conversion SWRO process development and its performance data. *Desalination* 125, 9.

Liberman, B. (2002). The importance of energy recovery devices in reverse osmosis desalination. In *The Future of Desalination in Texas*, Vol. 2, Technical Papers, Case Studies Desalination Technology Resources (Report No. 363). Texas Water Development Board, Austin, Texas.

Masaaki, A., Mitsuharu, F., Hiroshi, I., Ryouichirou, N., Shinji, T. and Toru, Y. (2003). Seven year operation and environmental aspects of 40,000 m^3/day seawater RO plant in Okinawa, Japan. In *Proceedings of IDA World Congress on Desalination and Water Reuse—Desalination: The Source of Sustainable Water Supplies*, Paradise Island, Bahamas, International Desalination Association, Topsfield, Massachusetts, USA.

Medina, J. A. (2004). 20 years evolution of desalination costs in Spain. In Semiat et al. (Ed.) *Proceedings of International Conference on Desalination Costing*, Limassol, Cyprus, The Middle East Desalination Research Center, Muscat, Oman.

Rovel, J. (2003). Description of the largest SWRO ever built. In *Proceedings of IDA World Congress on Desalination and Water Reuse—Desalination: The Source of Sustainable Water Supplies*, Paradise Island, Bahamas, International Desalination Association, Topsfield, Massachusetts, USA.

Semiat, R., Chapman, M., Price, P., and Hasson, D. (Eds) (2004). Desalination project costs. In *Proceedings of International Conference on Desalination Costing*, Limassol, Cyprus, The Middle East Desalination Research Center, Muscat, Oman.

Velter, G. (2004). Case Studies: Ashkelon 100 MC/Year BOT Project. In Semiat et al. (Ed.) *Proceedings of International Conference on Desalination Costing*, Limassol, Cyprus, The Middle East Desalination Research Center, Muscat, Oman.

Voutchkov, N. (2004a). Thorough study is key to large beach well intakes. *Desalin. Water Reuse Q.* 14(1), 16.

Voutchkov, N. (2004b). Seawater desalination costs cut through power plant co-location. In *Filtration + Separation*, 41(7), 24.

Wagnick Consulting (2004). IDA worldwide desalination plant inventory. Report No. 18. International Desalination Association, Topsfield, Massachusetts, USA.

Wilf, M. (2004). Fundamentals and cost of RO-NF technology. In Semiat et al. (Ed.) *Proceedings of International Conference on Desalination Costing*, Limassol, Cyprus, The Middle East Desalination Research Center, Muscat, Oman.

Wright, R., and Missimer, T. (1997). Alternative intake systems for seawater membrane water treatment plants. In *Proceedings of International Desalination Association*, Congress on Desalination and Water Reuse, Madrid, Spain, International Desalination Association, Topsfield, Massachusetts, USA.

CHAPTER 4

Seawater Desalination by Ultralow-Energy Reverse Osmosis

R. L. TRUBY

Toray Membranes, Escondido, California 92026

4.1 INTRODUCTION

4.1.1 Early Research

In the 1950s and 1960s research by Loeb and Sourirajan at UCLA and Reid at the University of Florida demonstrated that a semipermeable reverse osmosis (RO) membrane could be produced that would allow water to pass but selectively reject impurities including dissolved ions such as sodium and chloride. These membranes were thick-film, asymmetric, cellulose acetate and required up to 105 bars pressure to desalt seawater (Buros, 1980). Furthermore, the selectivity of the early membranes and osmotic pressure of the seawater limited the volume of drinking quality permeate that could be extracted from a typical spiral-wound RO (SWRO) system to 25%. In this case the remaining 75% of the seawater that was pressurized to 105 bars would be discharged back to the ocean as a brine or concentrate stream. The energy required to produce drinking-quality permeate from seawater at this pressure and recovery was approximately $10\,kWh/m^3$.

The success of these early membranes stimulated a great deal of additional research to optimize and improve the RO process. The stated goal of much of this research was to develop a cost-effective technology to desalt large volumes of seawater for human consumption.

4.1.2 Thin-Film Composite Membranes

In the 1970s researchers such as Robert Riley at Fluid Systems Division of UOP Inc. and John Cadotte at Midwest Research Institute developed thin-film composite membranes using polymeric materials such as polyetherurea and polyamide. These thin films offered reduced resistance to the flow of water across the membrane layer and allowed production of drinking-quality permeate at a pressure of less than 70 bars and 35% recovery. At these operating parameters the energy requirement was lowered to approximately $6.6\,kWh/m^3$.

Advanced Membrane Technology and Applications. Edited by Norman N. Li, Anthony G. Fane, W. S. Winston Ho, and T. Matsuura
Copyright © 2008 John Wiley & Sons, Inc.

4.1.3 First Large Commercial SWRO: Jeddah, Saudi Arabia

In 1978 the first large SWRO desalination system for the commercial production of drinking water from seawater for a municipality was installed in Jeddah, Kingdom of Saudi Arabia, for the Saline Water Conversion Corporation (SWCC). This 12,000-m^3/day system was designed to operate at 56 bars and 30% recovery. The Jeddah RO produced 1000 mg/L total dissolved solids (TDS) permeate (Hassan et al., 1989). This permeate was indirectly blended with less than 50 mg/L TDS condensate from evaporators also operated by SWCC in the Jeddah distribution system. Energy consumption of this first SWRO was approximately 6.1 kWh/m^3. The Jeddah RO was not designed using any type of energy recovery technology.

4.2 SWRO ENERGY REDUCTION USING ENERGY RECOVERY TECHNOLOGY

4.2.1 Need for Energy Recovery

Before the Jeddah RO, most SWRO systems had a capacity of less than 1000 m^3/day, and the brine/concentrate stream was too small for economic use of energy recovery technology. As SWRO system capacities increased, it became obvious to engineers and designers that the brine/concentrate stream represented an opportunity to reduce energy consumption by using this discharge to help pressure the incoming raw seawater.

4.2.2 Pelton Wheels

The first energy recovery technology designed into SWRO systems was based upon a well-proven 100-year-old engineering concept—the Pelton wheel. The use of Pelton wheels dates back to the late 1800s when they were widely employed in the mining industry and elsewhere. Pelton wheels were easily and quickly adapted to SWRO high-pressure pump/motor assemblies beginning as early as 1982.

The Pelton wheel reduces energy by directing the brine/concentrate stream at full pressure into buckets on a wheel that turns a shaft. This shaft is directly connected to the high-pressure SWRO pump/motor assembly. The shaft reduces the load on the main pump/motor and thus lowers the energy required to drive the process.

While the Pelton wheel reduces energy consumption it is still necessary to size the high-pressure pump/motor assembly to pump the full feedwater flow required for the SWRO system. Furthermore, the Pelton device must convert the hydraulic energy into rotational shaft power that again must be converted back to hydraulic energy through the SWRO main feed pump. This results in a double efficiency penalty of the Pelton and SWRO pump device and limits their best overall net transfer efficiencies to approximately 88% (Villa Sallangos, 2004). Figure 4.1 shows a flow scheme incorporating a typical Pelton wheel energy recovery device. Figure 4.2 is a picture of a Pelton wheel.

Pelton Wheel Characteristics

- Low capital cost
- Directly connected to SWRO pump/motor
- Requires full-sized SWRO pump/motor

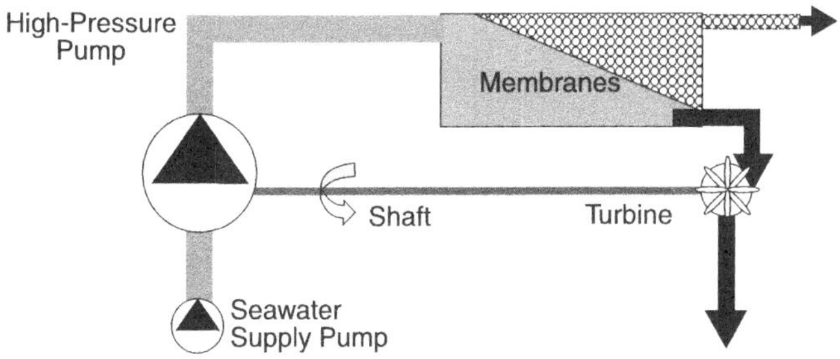

Figure 4.1 Pelton wheel RO flow scheme. (From Stover, 2006, p. 13.)

Figure 4.2 Picture of a Pelton wheel. (From Kuendig, 2006a, p. 19.)

- Operates in a centrifugal mode and follows flow and pressure curves
- Overall net transfer efficiencies follow a curve and can reach 88%
- Metallic construction

Pelton wheels have been modified and improved specifically to optimize them for use with SWRO systems. This innovation has resulted in efficiency and other improvements, and Pelton wheels continue to be used with SWRO systems today. Current Pelton wheel devices have overall net transfer efficiencies of 60–88% (Villa Sallangos, 2004). Figure 4.3 is a typical Pelton wheel efficiency curve and shows the device efficiency at 80%. Optimally, a Pelton wheel can reach 88% efficiency, but this efficiency must be multiplied by the SWRO pumping device it is connected to in order to produce the overall net transfer efficiency. For example, an 85% efficient Pelton wheel operating with an 82% efficient SWRO pump would have an overall net transfer efficiency of the reject hydraulic flow to the feed water hydraulic flow of approximately 70%.

4.2.3 Turbochargers

The success of Pelton wheels and reverse running pumps encouraged pump companies to seek innovations that would improve the economy and applicability of energy recovery technology. The "turbocharger" was developed as a stand-alone device, not connected to

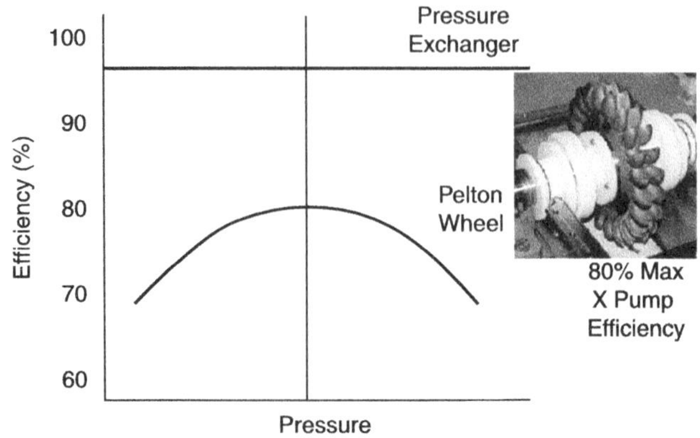

Figure 4.3 Typical Pelton wheel efficiency curve. (From Stover, 2006, p. 23.)

the high-pressure pump/motor, and can be employed in a variety of ways. Figure 4.4 shows a typical flow scheme using a turbocharger in an SWRO system to recover energy. Figure 4.5 shows an alternative flow scheme wherein the turbocharger serves as a booster pump between stages. Figure 4.6 is a photograph of a typical turbocharger.

Turbocharger Characteristics

- Moderate capital cost compared to Pelton wheel
- Not connected to SWRO pump/motor assembly
- Requires full flow but at a reduced outlet pressure for the SWRO pump/motor
- Operates in a centrifugal mode and follows flow and pressure curves
- Overall net transfer efficiency follows a curve and can reach 83%
- Metallic construction
- Can be used in a variety of flow schemes

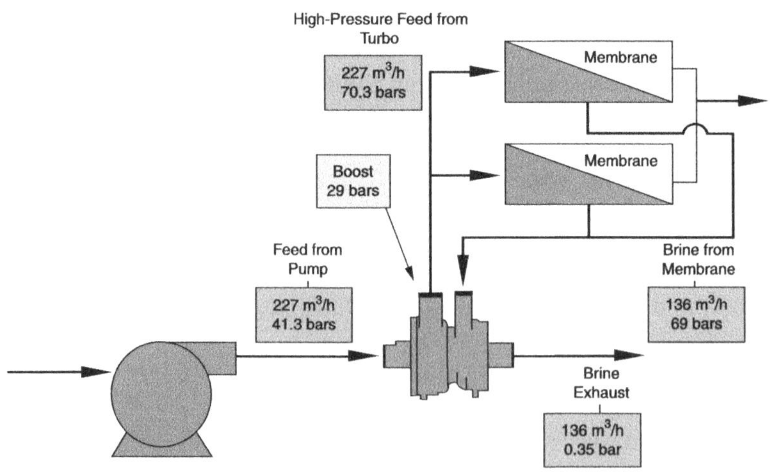

Figure 4.4 Typical turbocharger RO flow scheme. (From Oklejas, 2006, p. 32.)

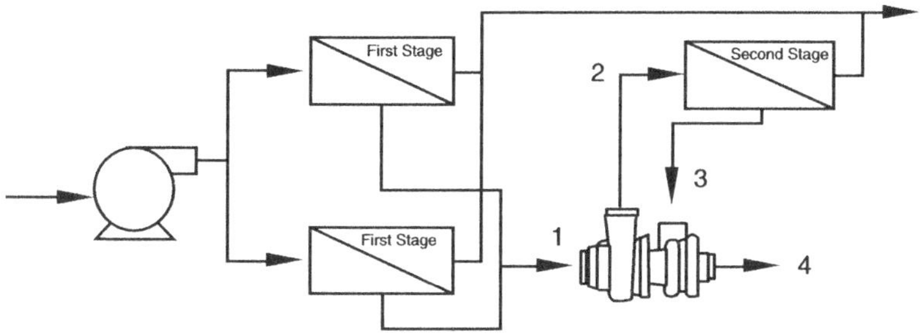

Figure 4.5 Alternative turbocharger RO flow scheme where the turbocharger is used as a booster pump. (From Oklejas, 2006, p. 22.)

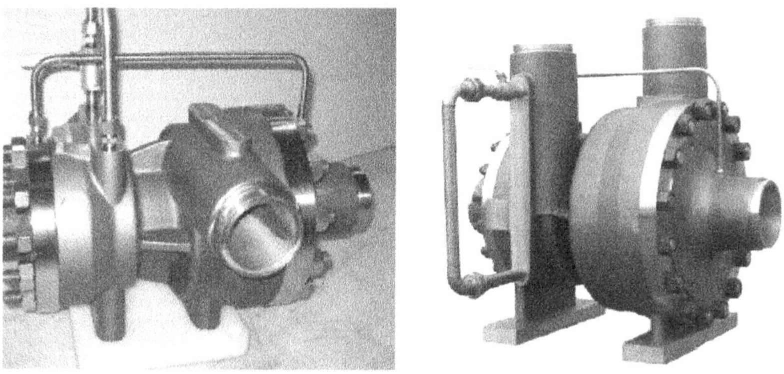

Figure 4.6 Photograph of a typical turbocharger. (From Oklejas, 2006, p. 8.)

Turbochargers are also centrifugal devices and have an efficiency curve much like the Pelton wheel shown in Figure 4.3. Similar to a Pelton wheel, turbochargers are connected to an integrated pump, but the turbocharger's pump operates in series with the SWRO pump. The turbocharger's pump receives feed water at some intermediate pressure from the SWRO pump and boosts the RO feed water to the full system pressure. The SWRO pump runs at an outlet pressure that is reduced by the turbocharger's boost pressure and thereby saves energy. Figure 4.7 shows a typical turbocharger overall net transfer efficiency curve. The highest overall net transfer efficiency achieved by a turbocharger is 83% (Moch et al., 2005).

4.2.4 Isobaric or Direct Energy Recovery Devices

In the late 1990s researchers introduced a new energy recovery concept to the marketplace that was unlike the Pelton wheel and turbocharger technology in that it operated in the mode of a positive displacement pump. This isobaric energy recover technology performs at higher overall net transfer efficiencies, typically around 95%, and the operating curve is essentially flat (Villa Sallangos, 2004). Figure 4.8 shows a typical efficiency curve for an isobaric energy recovery device.

The flow scheme for a typical isobaric energy recovery system is quite different from Pelton wheels or turbochargers and is shown in Figure 4.9. The isobaric energy recovery device is not connected to the high-pressure pump/motor assembly. Additionally, the

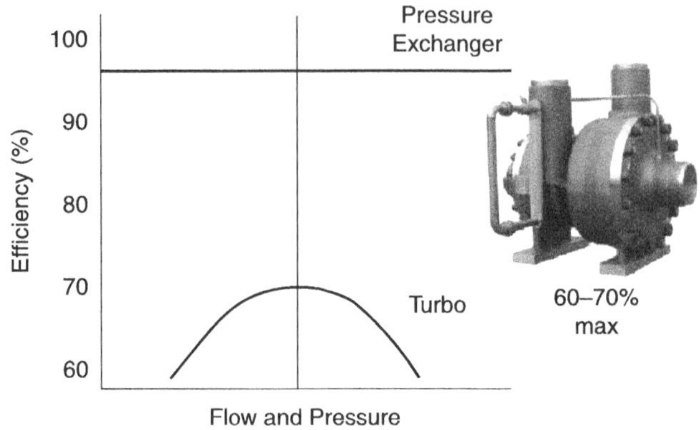

Figure 4.7 Typical turbocharger overall net transfer efficiency curve. (From Stover, 2006, p. 24.)

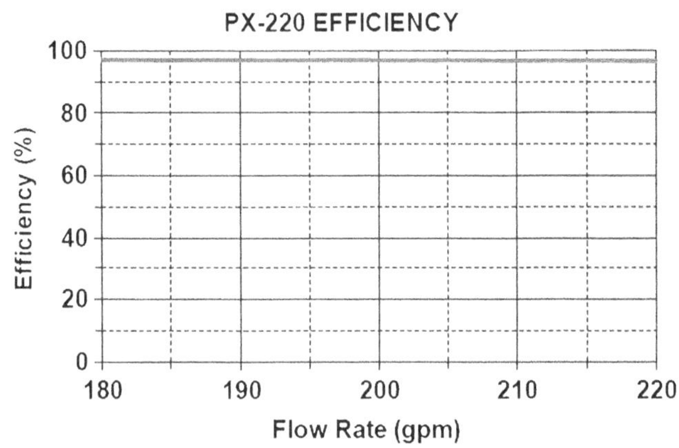

Figure 4.8 Typical efficiency curve for an isobaric energy recovery device. (From Stover, 2006, p. 22.)

high-pressure pump/motor does not need to be sized to pump the full flow required by the SWRO system. In fact when isobaric energy recovery devices are used, the high-pressure pump/motor assembly is approximately sized to match the volume of the permeate produced by the SWRO system.

Figure 4.10 is a depiction of a DWEER isobaric energy recovery system. DWEER is a work exchanger that uses the concentrate stream to pressurize the feed stream through parallel stainless steel pipes that alternatively fill and empty. Figure 4.11 shows a different isobaric device known as a pressure exchanger, or PX. The PX consists of a ceramic rotor as the only moving part, spinning on a thin film of water inside a ceramic sleeve. The ceramic end plates divide the high-pressure side from the incoming seawater. The entire device is housed in a typical fiberglass pressure vessel. The DWEER and PX devices are physically different in many ways, however, the end result and performance characteristics are quite similar (Stover, 2006; Kuendig, 2006b).

Isobaric energy recovery systems allow a small portion of the concentrate stream to mix with the raw seawater as part of the process. This volume of bypass water is necessary to lubricate the moving parts. This leakage results in a slight increase in the TDS of the seawater exposed to the membranes and must be included in the SWRO system design

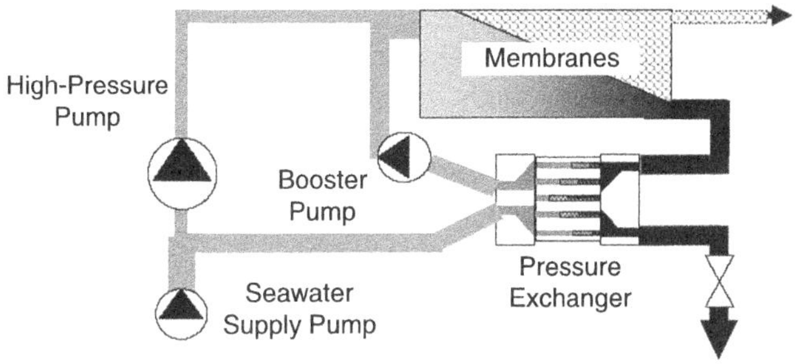

Figure 4.9 Flow scheme for a typical isobaric energy recovery system. (From Stover, 2006, p. 14.)

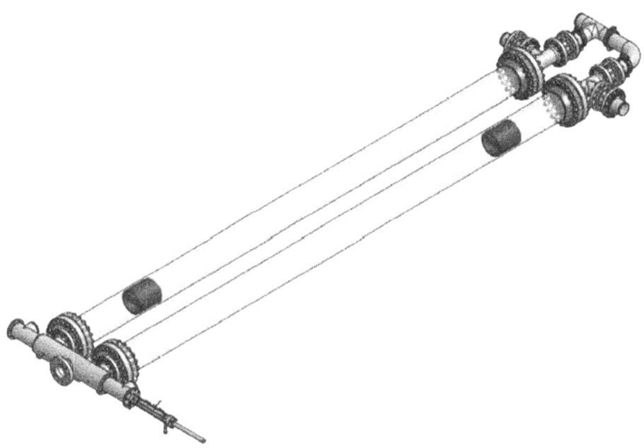

Figure 4.10 A DWEER isobaric energy recovery system. (From Kuendig, 2006b, p. 4.)

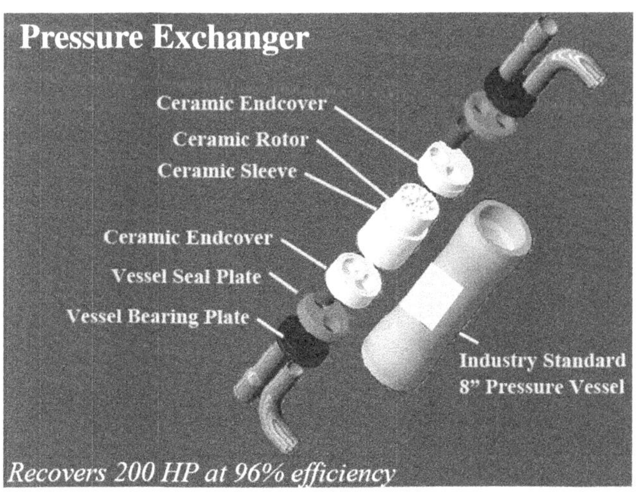

Figure 4.11 A pressure exchanger (PX) isobaric energy recovery system. (From Stover, 2006, p. 21.)

calculations. A typical isobaric system design might result in approximately a 1% TDS increase in the seawater concentration (Seacord et al., 2006).

Isobaric systems also develop a slight differential pressure loss from the brine/concentrate stream across the device to the pressurized seawater stream. To make up for this slight loss of pressure plus piping and friction losses, a small booster pump is necessary. The booster pump can be seen in Figure 4.9.

Isobaric Energy Recovery Characteristics

- Higher capital cost (approximately 250% higher than the net capital cost for a Pelton wheel with smaller SWRO pump/motor savings included in calculation)
- Not connected to the SWRO pump/motor assembly
- SWRO pump/motor smaller; sized to pump permeate volume only
- Operates in a positive displacement mode
- Efficiency curve is flat and operates at up to 95%
- Available with nonmetallic construction for corrosion resistance
- Requires a small booster pump/motor
- Increases in seawater TDS by approximately 1%

4.2.5 Energy Recovery Utilization Since 1980

The isobaric energy recovery devices are the most recent in a series of technological innovations that have made SWRO more cost effective. All of the energy recovery technologies described above are in wide use today. No single technology is optimum for all applications. Figure 4.12 shows an estimate of the number of energy recovery systems installed

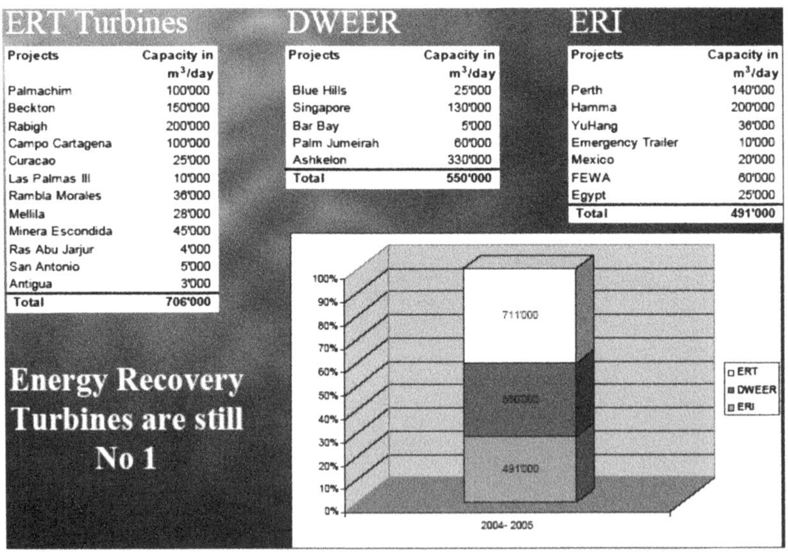

Figure 4.12 Estimate of number of energy recovery systems installed or under contract in 2004/05. (From Kuendig, 2006a, p. 4.)

Figure 4.13 Picture of the DWEER system installed at Ashkelon. (From Kuendig, 2006b, p. 35.)

HAMMA, ALGERIA 2007 – 53 MGD

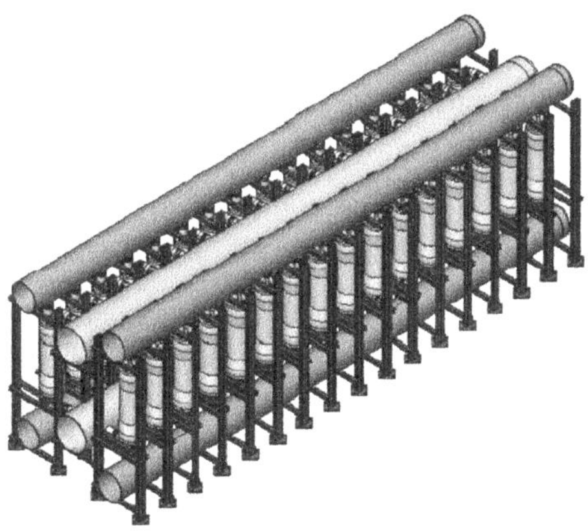

Figure 4.14 Picture of proposed PX system that will be installed at Hamma, Algeria, in 2007. (From Stover, 2006, p. 8.)

or under contract from 2004 to 2005. It is clear that the isobaric devices, which were only introduced in the late 1990s, have been well received (Kuendig, 2006a).

A picture and graphic representation of a typical DWEER energy recovery system and typical PX installation are shown in Figures 4.13 and 4.14, respectively.

4.3 SWRO ENERGY OPTIMIZATION

Advances in three distinct areas of SWRO system design have resulted in a significant reduction in energy consumption from 10 kWh/m^3 in the late 1970s to approximately 3.5 kWh/cm^3 by 2000.

The three areas of innovation are:

1. Membrane/element design
2. Large-capacity high-pressure pump/motor efficiency
3. Energy recovery technology improvement

4.3.1 Improved Membrane/Element Design

In 1978 a typical SWRO membrane/element had 15 m^2 of membrane and produced 7.5 m^3 of permeate per day with 98.6% salt rejection. Today a typical SWRO element contains 40 m^2 of membrane and produces 28–34 m^3 of permeate per day with 99.75% salt rejection. The improvement in membrane performance allows system engineers to select an optimum design point and also to plan for a wider range of operating conditions without capital cost penalties. This flexibility makes it easier for municipalities to meet the needs of their customers and gives them the option to run at lower pressures and/or recoveries, which in turn can reduce energy consumption.

4.3.2 Improved High-Pressure Pump/Motor Efficiency

Pump manufacturers have introduced larger capacity pumps that operate at higher efficiencies. This has given SWRO designers more options with respect to train sizes. Additionally, the energy consumption of these pump/motors designed specifically for SWRO use is lower than was available in the past.

4.3.3 Isobaric Energy Recovery Technology

The higher efficiency (95%) and flat efficiency curve of the isobaric device allows system designers and operators to cope with a wider range of variables. This makes it easier to design for common fluctuations such as temperature change, loss of flow due to normal fouling/cleaning cycles, or other events that might necessitate operation at other than the design conditions. For example, if permeate quality deteriorated to the point where it might be inappropriate for distribution, the system recovery could be lowered by as much as 10% with no increase in energy consumption. (This assumes there would be enough pretreatment capacity to accommodate such a change.) Lower recovery would have an impact reducing the TDS of the permeate, and the flat curve of the isobaric device would maintain energy consumption at the design point.

4.4 AFFORDABLE DESALINATION COLLABORATION (ADC)

4.4.1 Organization of ADC

The Affordable Desalination Collaboration was conceived in California by a group of public agencies and private companies. It was established as a nonprofit organization. ADC existed through the completion of the project work and then plans called for it to either be disbanded or refunded to conduct additional studies. Members of ADC contributed funds, services, and/or equipment used in the demonstration. The organization had a five-member board of directors and a CEO/managing director who was the only paid employee.

The founders of ADC were stakeholders in the production of safe drinking water. They recognized that the growth in global population had strained available water supplies to the point where seawater desalination must be included in future planning. One of the concerns expressed by the public and other interested parties about the planned use of seawater desalination was the high consumption of energy required to drive the process (Seacord et al., 2006).

4.4.2 Theoretical Lower Limit of Energy Consumption for Seawater Reverse Osmosis

The ADC conducted its study on Pacific Ocean seawater in central California. The typical salt content of seawater at the test location is 34,000 mg/L TDS and the temperature ranges from 12–20°C.

It is possible to calculate the lower limit to energy consumption using software provided by the pump and energy recovery manufacturers. Applying this software and assuming (1) 99% pump/motor/energy recovery device efficiencies and (2) a seawater element capable of producing 12,000 gal per day (gpd) or 45.5 m^3 per day (m^3/day) with 99.8% salt rejection, allows a calculation of the absolute lowest possible energy consumption. At 13 and 20°C, respectively, the SWRO high-pressure pump, booster pump, and isobaric energy recovery device would consume 1.38 kWh/m^3 of permeate and 1.33 kWh/m^3 of permeate.

If the seawater intake pump is added to this calculation, the energy consumption would increase to 1.94 and 1.88 kWh/m^3 of permeate at 13 and 20°C, respectively.

The theoretical lower limit of energy consumption that SWRO could achieve using the conditions at the site of the ADC study was estimated at 1.4 kWh/m^3 of permeate.

4.4.3 Goals

The ADC members were aware that the energy requirements to desalt seawater using SWRO had been declining in recent years. At the same time the energy cost of transporting water from northern California to the south in the State Water Project (SWP), and from the Colorado River to southern California in the Colorado River Aqueduct (CRA) represents the largest use of energy in the state.

Figure 4.15 shows the decrease in SWRO energy consumption since the late 1970s and the target set by ADC of approximately 1.6 kWh/m^3. Figure 4.16 shows the energy cost in the SWP and the CRA compared to the goals of ADC (Seacord et al., 2006).

The ADC ran the demonstration equipment over a range of parameters seeking the optimum energy point. This information was then used as part of a parametric study of a 50-mgd SWRO desalination system. Capital and operating costs were estimated at the optimum design point to calculate the lowest water cost (Seacord et al., 2006).

4.4.4 Equipment and Demonstration Protocol

The SWRO system constructed under the guidance of ADC used all commercially available components. No prototype materials or components were employed. The components were selected to easily scale-up to a much larger size SWRO. For example, the

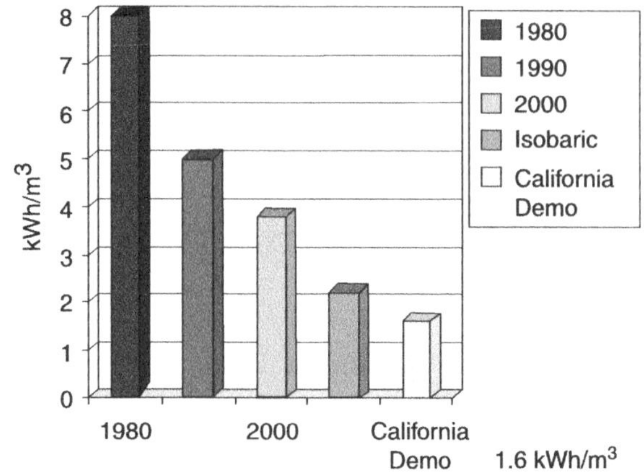

Figure 4.15 Historical cost of desalination as compared to the goal of the ADC demo (1.6 kWh/m^3). (From Stover, 2006, p. 75.)

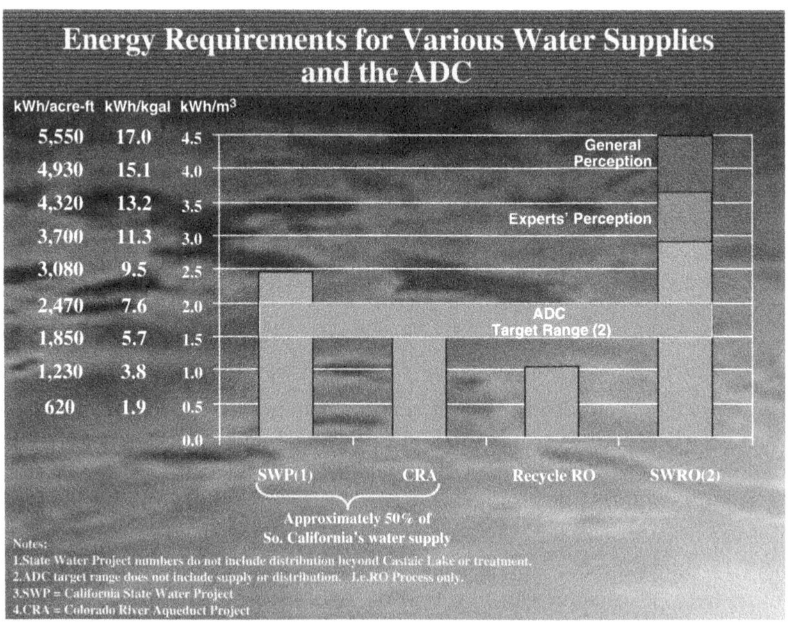

Figure 4.16 Energy cost of the State Water Project (SWP) and the Colorado River Aqueduct (CRA) as compared to the goals of the ADC test (1.6 kWh/m^3). (From MacHarg, 2005, p. 4.)

membrane/elements selected were typical 8×40-inch spiral elements containing approximately 40 m^2 of membrane. The system was designed around 21 of these elements installed in three typical fiberglass-reinforced plastic (FRP) pressure vessels all in parallel. The pump/motor assembly had an efficiency that is also available in much larger capacities. The isobaric energy recovery device was modular and is used in multimillion gallon per day systems. Figure 4.17 shows a simple flow scheme of the equipment.

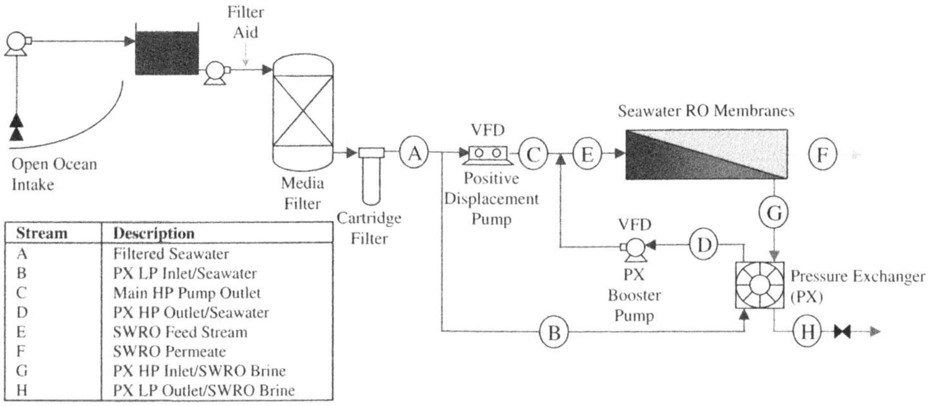

Stream	Description
A	Filtered Seawater
B	PX LP Inlet/Seawater
C	Main HP Pump Outlet
D	PX HP Outlet/Seawater
E	SWRO Feed Stream
F	SWRO Permeate
G	PX HP Inlet/SWRO Brine
H	PX LP Outlet/SWRO Brine

Figure 4.17 Simple flow scheme of the equipment used in the ADC test. (From Seacord et al., 2006, p. 4.)

The SWRO system was installed in a container at the U.S. Navy Test Station at Pt. Hueneme, California, about 80 miles north of Los Angeles. It operated over a range of capacities of 182–286 m^3/day. Pretreatment consisted of conventional media filtration.

Testing was carried out over a period of time from October 2005 through April 2006. The system was operated with three sets of membrane/elements representing differing membrane areas and flow capacities. Each membrane type was tested for 2 weeks to reach steady-state performance and then the system was operated over various flux and recovery rates. At the end of each trial the membrane/elements were operated at the same conditions as during the run-in period [7.5 gfd (12.75 lmh) and 42.5% recovery]. Data was collected including water samples for later analysis. Tests were run at fluxes of 6, 7.5, and 9 gfd (10.2, 12.75, and 15.3 lmh, respectively), while varying the recovery from 35 to 42.5–50%.

4.4.5 Results

The ADC has concluded its work and the final report was published in August 2006 at the American Membrane Technology Association (AMTA) conference in Anaheim, California (to read the full manuscript go to: www.membranes-amta.org).

The data has shown that the lowest process energy consumption was achieved at 6.0 gfd (10.2 lmh) and 42.5% recovery. Under these conditions the SWRO required 1.58 kWh/m^3 to drive the process. Permeate quality was also monitored. Over the entire range of operating conditions the permeate TDS varied from a low of 110 mg/L TDS to a high of 350 mg/L TDS. At the optimum energy point the permeate TDS was under 200 mg/L (Seacord et al., 2006).

The ADC has demonstrated that energy consumption can be lowered by as much as 40% by selecting energy-efficient components and designing around optimized operating conditions (Seacord et al., 2006).

4.5 CONCLUSION

The SWRO system has benefited from innovation beginning with the first application of the technology in the late 1970s to the present. This innovation has resulted in many improvements. One of the most significant changes is the reduction in the energy required to drive the process. Energy recovery technology and the introduction of isobaric devices that operate at 95% efficiency have had a major impact. Organizations like ADC continue to try to reduce the cost of producing reliable, affordable drinking water from desalted seawater.

The cost of energy has increased over time. This trend is expected to continue. The energy reduction innovations described above make SWRO more cost effective, affordable, and more valuable to consumers with every increase in energy cost.

ACKNOWLEDGMENTS

The author would like to acknowledge John MacHarg for his assistance and advice in reviewing sections of the manuscript regarding energy recovery devices and Peter F. Metcalfe for his assistance calculating the theoretical lower limit of energy consumption at ADC.

REFERENCES

Buros, O. K. U.S. Agency for International Development (1980). *The U.S.A.I.D. Desalination Manual.* CH2M Hill International, Washington, DC, Chapter 5, p. 3.

Hassan, A. M., Al-Jarrah, S., Al-Lohibi, T., Al-Hamdan, A., Bakheet, L. M., and Al-Amri, M. I. (1989, Aug.). Performance evaluation of SWCC SWRO plants. In *IDA 1989 World Conference*, S.A.S. Hotel, Kuwait, p. 10.

Kuendig, E. (2006a). Pelton wheel energy recovery turbine designed for reverse osmosis applications. In *Proceedings AMTA/SCDA Joint Technology Transfer Workshop*, Corpus Christi, TX, Feb. 8–9.

Kuendig, E. (2006b). Work exchanger DWEER product description and design. In *Proceedings AMTA/SCDA Joint Technology Transfer Workshop*, Corpus Christi, TX, Feb. 8–9.

MacHarg, J. (2005). The affordable desalination collaboration. In *2005 IDA World Congress*, Singapore, Sept. 13.

Moch, I., Jr., Oklejas, M., Terrasi, K., and Oklejas, R. A. (2005). Advanced high efficiency energy recovery. In *2005 IDA World Congress*, Singapore, p. 12, Sept. 13.

Seacord, T. F., Coker, S. D., and MacHarg, J. (2006). Affordable desalination collaboration. In *Proceedings AMTA Biennial Conference and Exposition*, Anaheim, CA, p. 20, July 31–Aug. 2.

Stover, R. (2006). Pressure exchanger. In *Proceedings AMTA/SCDA Joint Technology Transfer Workshop*, Corpus Christi, TX, Feb. 8–9.

Villa Sallangos, O. L. (2004). Operating experience of the Dhekelia seawater desalination plant using an innovative energy recovery system. In *Proceedings Euromed 2004*, Morocco, pp. 17–19.

Oklejas, M. (2006). Turbo charger. In *Proceedings AMTA/SCDA Joint Technology Transfer Workshop*, Carpus Christi, TX, p. 32, Feb. 8–9.

CHAPTER 5

Microfiltration and Ultrafiltration

N. KUBOTA, T. HASHIMOTO, and Y. MORI

Microza Research & Development Department, Specialty Products & Systems R&D Center, Asahi Kasei Chemicals Corporation, Fuji City, Shizuoka, 416-8501 Japan

5.1 INTRODUCTION

5.1.1 Overview and Major Trends in Microfilter/Ultrafilter Technology

Microfilter (MF) and ultrafilter (UF) membranes range from several nanometers to several micrometers in screening pore size (Fig. 5.1). Filtration membranes having a screening pore size ranging from several angstroms to several micrometers can be obtained by forming a polymer material into a porous body using the proper technique. Since filtration separation by membrane uses pressure difference as the driving force for separation and is not accompanied by phase change, it has the following advantages:

1. Separation measure is a low-energy consumption type.
2. Target to be separated is scarcely denatured or decomposed due to separation under mild conditions.

Therefore, since Loeb and Sourirajan innovated an industrially usable filtration membrane having a low permeation resistance in 1960, the filtration separation using a membrane has been used in various industrial fields around the world, such as the automobile industry (closed system for recovery of electrodeposition paint), the electronics industry (production of ultrapure water for semiconductor production), and the pharmaceutical industry (concentration and purification of enzymes and antibiotics, production of pyrogen-free water, etc).

As mentioned, the industrial use of MF/UF has progressed in various industrial fields, and its use has been further widely expanded in this decade. A major factor in this rapid expansion is that the MF/UF technology has been used as a clarification procedure in water supply and sewerage fields, where sand filtration had been conventionally used. Because of this, the leading factor of the progress of MF/UF technology is the use of membranes for water supply and sewerage. Therefore, recent innovations in MF/UF technology

Advanced Membrane Technology and Applications. Edited by Norman N. Li, Anthony G. Fane, W. S. Winston Ho, and T. Matsuura
Copyright © 2008 John Wiley & Sons, Inc.

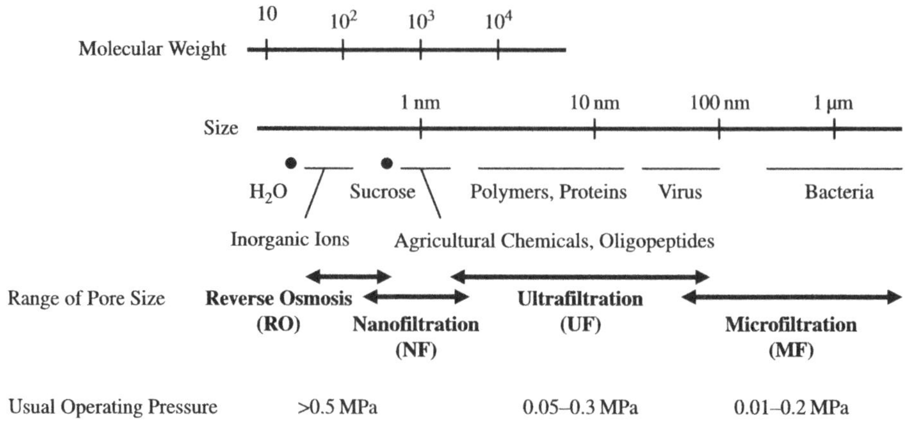

Figure 5.1 Pore size range of various membranes.

will be described centering around membranes for water supply and sewerage, including topics on application technology.

5.1.2 Use and Performance of MF/UF

Microfiltration and UF are used for solid–liquid separation and polymer/low-molecular-weight compound separation. Use of MF/UF is mostly for aqueous fluid treatments and is roughly classified as follows (Table 5.1):

1. Rejection of impurities (purification of water, etc.)
2. Retention and concentration of valuables (concentration and purification of enzymes, etc.)
3. Permeation and purification of valuables (bacteria elimination and purification of fermentation fluid, etc.)

In an industrial use, the following two characteristics are important. First, pore size must be appropriate for the purpose of separation. Smaller pore size is not necessarily better because it may be necessary to concentrate a valuable (useful component such as an enzyme) and also necessary to purify (separation of impurities that are larger than a valuable) by permeating a valuable. Furthermore, a narrow pore size distribution is also important. Thus, the membrane preparation method used in the production of MF/UF must capable of the selective production of a specified range of pore sizes from a wide range of pore sizes. Second, economic efficiency must be adequate to allow for the variations in pore sizes needed. The economic efficiency can be represented by the formula:

$$\text{Economical efficiency} = \frac{\text{Water permeability} \times \text{Life}}{\text{Price}}$$

The success of these methods relates to:

1. Reluctance against clogging
2. Mechanical strength (pressure resistance, etc.; reasonable required performance for forced filtration over a long period)

TABLE 5.1 Typical Application Examples of MF/UF Technology

					Separation Targets	
Types	Examples	Industry	UF	MF	Rejection	Permeation
Removal of impurities	Production of drinking water, wastewater treatment	Water supply	○	○	Bacteria, colloids	Water
	Portable water purifier	Water supply	○	○	Bacteria	Water
	Production of ultrapure water	Semiconductor	○	△	Microparticles	Water
	Purification of chemical fluid	Semiconductor		○	Microparticles	Chemical fluid
	Treatment of condensed water of power plant	Electric power		○	Microparticles	Water
	Preparation of sterile water	Medical, pharmaceuticals	○		Pyrogen	Water
Concentration of valuables	Recovery of electrodeposition paints	Automobile	○		Electrodeposition paints emulsion	Water
	Purification of proteins	Pharmaceuticals, chemicals	○		Proteins	Low-molecular-weight (ions, sugars, etc.)
	Purification of inorganic colloids	Chemicals	○		Colloids	Water
Permeation purification of valuables	Purification of antibiotics	Pharmaceuticals	○		Bacteria, proteins	Antibiotics
	Purification of fermentation fluid	Pharmaceuticals, chemicals		○	Bacteria, cells	Proteins, etc.
	Clarification of beverages	Beverages	○	○	Bacteria, dregs	Beverage

△ – Some actual examples.
○ – Many actual examples.

3. Chemical strength (chemical resistance: usually chemical cleaning with oxidant, acid, alkali, etc. is carried out when clogging becomes serious)

Since selective formation of a specified pore size has been achieved at a considerable level in the present membrane preparation technology, one important key to further extend spread of the membrane filtration in the future is economic efficiency. The key to increase the economic efficiency will be reduction of clogging and increasing mechanical and chemical strength. In this connection, a membrane superior in thermal resistance is advantageous because it can expand the range of application. In the field of pharmaceutical application, for example, steam sterilization resistance may be required.

Regarding the shape of the membrane, there are three types:

1. A sheetlike "flat membrane"
2. A "hollow fiber" with a diameter smaller than several millimeters
3. A hollow tube with a diameter of around several centimeters; "tubular"

The membrane is used as a "module" in which a membrane with a large membrane area is packed compactly in a housing. It is a hollow fiber that can provide the largest packed membrane area per unit volume.

5.2 RECENT TRENDS AND PROGRESS IN MF/UF TECHNOLOGY

5.2.1 Trends and Progress of Membrane Preparation Technology

Typical membrane preparation methods and features and examples of materials of membranes obtained by each of the methods are shown in Table 5.2. A list of certified MF/UF membranes for water supply based on the certified membrane modules for water supply

TABLE 5.2 Typical Preparation Process of MF/UF Membranes

Membrane Preparation Process		Examples of Membrane Materials	UF	MF	Properties
Phase separation	Nonsolvent-induced phase separation	Polysulfone, polyethersulfone, polyacrylonitrile, cellulose derivatives, polyvinylidene fluoride	○	○	Asymmetric structure High permeability High porosity
	Thermally induced phase separation	Polyethylene, polypropylene, polyvinylidene fluoride		○	Generally high strength High porosity
Stretched semicrystalline		Polyethylene, polypropylene, polytetrafluoroethylene		○	Slitlike pore
Nucleation track		Polycarbonate, polyester		○	Uniformly and straight pore
Sintered particles		Ceramics	○	○	

○ – Many actual examples.

TABLE 5.3 Examples of MF/UF Membranes for Water Purification Application

Membrane	Nominal Cut of Molecular Weight	Nominal Pore Size (μm)	Membrane Materials	Membrane Shape	Membrane Preparation Process
UF	13,000	—	Polysulfone	Hollow fiber	Phase separation
UF	80,000	—	Polyacrylonitrile	Hollow fiber	Phase separation
UF	100,000	0.01	Polyacrylonitrile	Hollow fiber	Phase separation
UF	100,000	0.01	Cellulose derivative	Hollow fiber	Phase separation
UF	150,000	0.01	Cellulose acetate	Hollow fiber	Phase separation
UF	150,000	—	Polyvinylidene fluoride	Flat sheet	Phase separation
MF	—	0.1	Polyvinylidene fluoride	Hollow fiber	Phase separation
MF	—	0.1	Hydrophilic polyethylene	Hollow fiber	Stretched semicrystalline
MF	—	0.1	Hydrophilic polysulfone	Hollow fiber	Phase separation
MF	—	0.1	Ceramics	Tubelar	Sintered particles
MF	—	0.1	Ceramics	Monolith	Sintered particles
MF	—	0.2	Polypropylene	Hollow fiber	Phase separatuion

catalog (version 2001) (Association of Membrane Separation Technology of Japan, 2001) is shown in Table 5.3.

Based on Table 5.3, technical trends of MF/UF membranes will be reviewed from the following two viewpoints: (1) membrane materials and (2) membrane preparation technology.

5.2.1.1 *Membrane Materials*

Although a variety of materials are used, as is apparent from Table 5.3, they are roughly classified into the following two groups:

1. Materials characterized by hydrophilicity; cellulose based, polyacrylonitrile, hydrophilized polyethylene, hydrophilized polysulfone, and so forth
2. Materials characterized by high strength and high durability; polyvinylidene fluoride (PVDF), and so forth

Group 1 materials are aimed at a stable and high level filtration rate by inhibiting membrane fouling due to organic substances in raw water. On the other hand, the group 2 materials are aimed at a stable and high level filtration rate over a long period by preventing mechanical breakdown and intensifying chemical cleaning (in kind of chemical and frequency) by using materials enhanced in mechanical strength and chemical resistance (chemical resistance in cleaning). At present, there is no material that satisfys both requirements at the same time. Therefore a material is usually selected that satisfys either group 1 or 2. Actually, in the last 2 years, products made of PVDF have become more numerous than those shown in Table 5.3. These membrane products have a life span of 3–5 years for the conventional applications, but many are still functional for 5–10 years in the water supply field. This is considered one reason why membranes made of PVDF have increased.

In this connection, it should be noted that the physical properties of membranes can be significantly varied depending on the membrane preparation method used even in membranes of the same material, and correlations between membrane materials and physical properties of membranes should not be considered unchanging (Kubota et al., 2003; Kubota, 2004). This may be easily understood if it is imagined that physical properties—in particular, mechanical strength of a polymer membrane—are influenced by a first-order structure of the material polymer but more intensely by its high-order structure, and that the membrane preparation process for a polymer membrane is usually "a process of deletion of the high-order structure of a raw polymer bulk (usually powder) followed by restructuring of the high-order structure" (The Society of Chemical Engineers Japan (2005), Advance of Chemical Engineering 39, Maki–syoten, Tokyo, p. 146), where a raw polymer is once dissolved by some measure and the high-order structure is built up again by coagulation, for example.

5.2.1.2 Membrane Preparation Technology As is apparent from Table 5.3, the most important method in membrane preparation is phase separation. The phase separation method has been widely utilized due to the following advantages:

1. A wide range of pore size variation is possible.
2. A high level of filtration rate can be easily obtained because a high percentage of void is possible.
3. Many types of polymers can be applicable.

In the phase separation method, the following two patterns are used:

1. Nonsolvent induced-phase separation process
2. Thermally induced-phase separation process

The nonsolvent-induced phase separation process is a process for obtaining a porous membrane by contacting a polymer solution with a nonsolvent to decrease dissolving power and induce a phase separation by diffusion and penetration of the nonsolvent into the polymer solution (Kesting, 1985). Usually, water is used as a nonsolvent. A technique generally employed is that a polymer solution is discharged into a water bath through a die with the proper shape to cause phase separation and coagulation to prepare a membrane. A hollow-fiber-like porous membrane can be obtained by using a double-ring nozzle as a die, and flowing a nonsolvent (e.g., water) from inside of the double rings (Fig. 5.2) (Watanabe and Ohya, 1993).

In the nonsolvent induced-phase separation process, a nonsolvent penetrates into a polymer solution by diffusion. Since phase separation occurs quickly in the surface where the penetration starts, and slowly in the inner part where the penetration proceeds comparatively slowly, the membrane has an "asymmetrical structure," which has a smaller pore size in the surface and a larger pore size in the inner part. An example of UF having the asymmetrical structure is shown in Figure 5.3.

The asymmetrical structure tends to provide a membrane that has a small permeation resistance despite the small pore size, and a high filtration rate despite the fine membrane. It was this asymmetrical structure membrane that was developed by Loeb et al. in 1964 as described earlier.

The thermally induced-phase separation process produces a porous membrane by mixing a polymer and a "potential solvent," which is a nonsolvent at room temperature

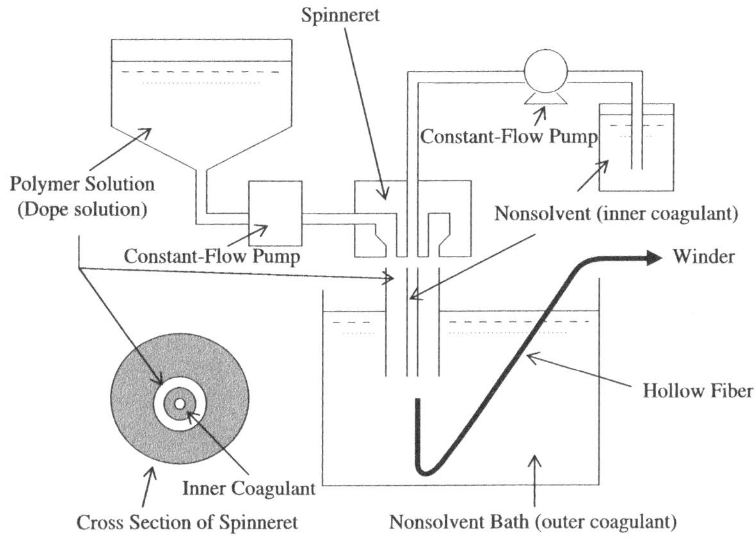

Figure 5.2 Illustration of a preparation system of hollow-fiber membranes by nonsolvent induced-phase separation process.

but becomes a solvent at a high temperature, to dissolve in one phase. It then cools to decrease dissolving power of the solvent to induce phase separation, and further cooling to a temperature at which the polymer solidifies to fix the structure, followed by extracting and removing the solvent (Lloyd et al., 1990, 1991). The cooling, which

Figure 5.3 SEM image of cross section of a polyacrylonitrile hollow-fiber UF membrane from nonsolvent phase separation process.

is a driving force of the phase separation, diffuses and proceeds from the surface toward the inner part, similarly to the nonsolvent in the nonsolvent induced-phase separation process. In any process of the nonsolvent induced process and the thermally induced process, the finest layer tends to be formed at the surface part, where the phase separation starts to occur. In the case of the nonsolvent induced-phase separation process, once the fine layer is formed on the surface, the layer disturbs further diffusion and permeation of the nonsolvent toward the inner part after that, and the asymmetrical structure is more easily formed. On the other hand, in the case of the thermally induced-phase separation process, even if the fine layer is formed on the surface in the beginning, the structure does not become so asymmetric as in the nonsolvent induced-phase separation process and tends to become comparatively uniform along the membrane thickness direction because the fine layer does not disturb thermal diffusion. Since the dissolving temperature of the polymer is usually higher than its melting point, when a crystalline polymer is used as a polymer component, crystallization tends to occur during the cooling process, giving a high-strength membrane due to a high degree of crystallinity. In this connection, as a modified method of this process, there is a process for obtaining a porous membrane by mixing a polymer, a potential solvent, and an inorganic filler at a high temperature and removing the solvent and the inorganic filler after cooling and solidifying (Doi and Matsumura, 1991). Pore structure of MF obtained by the thermally induced-phase separation process is shown in Figure 5.4.

Due to the prolongation of the useful life required for membrane products, the thermally induced-phase separation process has been attracting attention in recent years because the process has an advantage that a high-strength membrane with a high degree of crystallinity

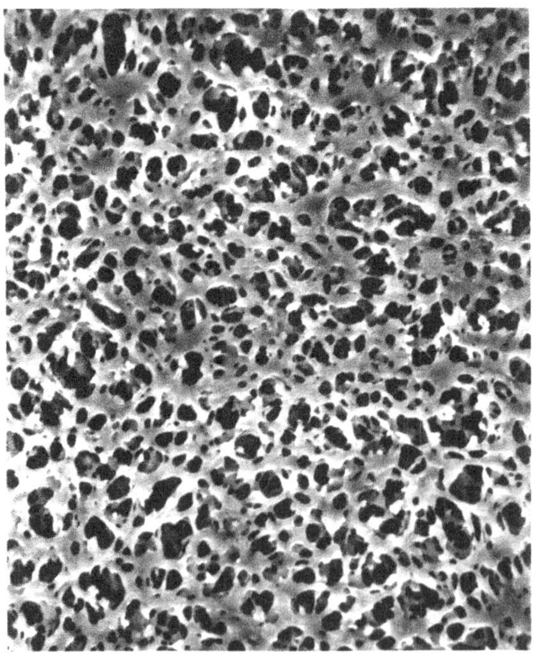

Figure 5.4 SEM image of surface of a polyvinylidene fluoride hollow-fiber MF membrane from thermally induced-phase separation process.

can be easily obtained because the membrane is prepared by heating at a temperature higher than the melting temperature of the polymer used. In particular, a membrane that has a fine seamless network structure, as shown in Figure 5.4, is likely to have high strength and high durability, differing from the structure in which particles are assembled and connected to each other, frequently seen in the conventional UF, for example. Improvements of product performances based on progress in technology, including development and putting to practical use of such a high-strength structure membrane, are moving forces in the expansion of membrane utilization to water supply and sewerage fields.

5.2.2 Topics in Application Technology: Ozone-Resistant Membrane and Ozone Microfiltration Systems

5.2.2.1 Introduction to Ozone Microfiltration System Ozone has high oxidation performance and has high effectiveness for decolorlation, odor removal, and disinfection ability. Ozone is now more widely used in advanced water treatment or wastewater treatment using its high oxidation performance to deal with deteriorating feed water quality or desire to improve the quality of water produced.

On the other hand, membrane filtration technology has been highly effective in the removal of turbidity and bacteria. Membrane filtration technology is a simple technology that provides high reliability, and it is approved for tap water treatment. Furthermore, membrane filtration technology is being considered for wastewater treatment in municipal wastewater or plant wastewater.

To use the current membrane technology for a large drinking water plant or secondary treated municipal wastewater treatment, there is always concern about the fouling of the membrane and the stability of flux. To prevent the fouling of the membrane, an excessive pretreatment or a very conservative design flux needs to be used. Consequently, the initial capital investment is high and results in less competitiveness against the existing treatment technology, thus limiting the application of the membrane technology to small-scale plants or to treatment of relatively clean feed water.

By combining the oxidation characteristics of ozone and membrane technology, a better quality of water is expected to be produced more easily and at lower cost. Combination of ozone with membrane filtration is effective for the prevention of membrane fouling. Thus, the process is expected to be simplified and requires less space than other methods. In addition, an excessive pretreatment is not required even with high turbid feed water.

Moulin et al. (1991a, 1991b) have shown a high permeate flux using ceramic membranes in conjunction with ozone. Ozone decomposes or denatures a part of organic materials in the raw water and reduces fouling of the membrane surface. But ceramic membranes face many problems such as small surface area per volume, high membrane cost, high energy loss because it needs high linear velocity, and cannot be easy cleaning by such methods as air scrubbing.

On the other hand, organic hollow-fiber membranes have superior characteristics, for example, large surface area per volume and relatively inexpensive cost, but they could not be used in conjunction with ozone because, generally, organic materials have no resistance to ozone so far.

However, we have succeeded in the development of ozone-resistant filtration modules by developing PVDF hollow-fiber membranes and potting materials having a high resistance to ozone, and assembling technology (Mori et al., 1998).

In addition, we have designed a new filtration system using our new ozone-resistant MF module in conjunction with ozone (Hashino et al., 2000). In this system, filtration was carried out at low pressure and low linear velocity and with air scrubbing to remove cake layer on the membrane. This system can provide a high permeate flux consistently for various types of water, especially as in the case of highly turbid raw water and secondary treated municipal wastewater.

5.2.2.2 Structure of Ozone-Resistant Microfiltration Modules

Module Structure Figure 5.5 shows an ozone-resistant module. Filtration takes place from outside to inside of the hollow-fiber membrane. The module is designed to feed raw water under high linear velocity, and air can be fed for the air scrubbing. This module is constructed by hollow-fiber membranes, potting material, and housing, and ozone resistance is required for all of these materials.

Hollow-Fiber Membrane Material We selected PVDF as hollow-fiber membrane material. Ozone resistance of PVDF hollow-fiber membrane depends on PVDF crystallinity. Higher crystallinity results in higher ozone resistance. Therefore, we have developed manufacturing technology to obtain high-crystalline PVDF. This hollow-fiber membrane has ozone resistance of over 5 years under exposure to 1.0 mg/L of dissolved ozone. Figure 5.6 shows the tensile elongation retention of PVDF hollow-fiber membranes when immersed in pure water in which ozone was dissolved at 35 mg/L concentration. The tensile elongation of high-crystalline PVDF hollow fiber did not change even in the concentration time (CT) value of 1800 mg/L·day, but the tensile elongation of low-crystalline PVDF hollow fiber decreased gradually. The CT value of 1800 mg/L·day equals to 5 years under the exposure to 1.0 mg/L of dissolved ozone. Figure 5.7 shows the scanning electron microscopy (SEM) image of high-crystalline PVDF hollow fiber. The lamella structure was clearly observed.

Potting Material We have studied the ozone resistance of various materials for potting, for example, fluoride materials, inorganic material, or others. We have selected silicone potting material because of its ozone resistance and ease of handling. The new silicone potting material has at least 5 years ozone resistance under the exposure to 1.0 mg/L of dissolved ozone.

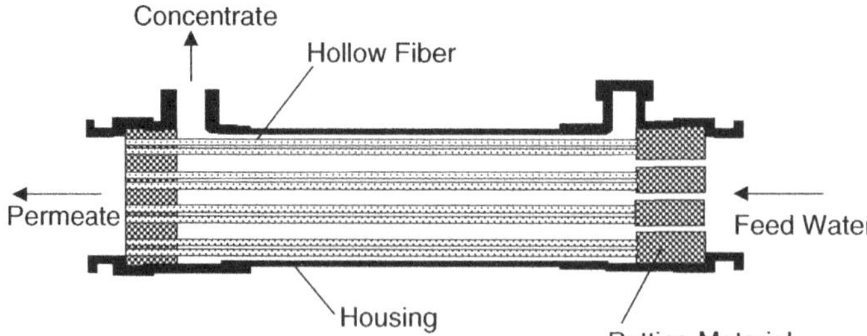

Figure 5.5 Module structure.

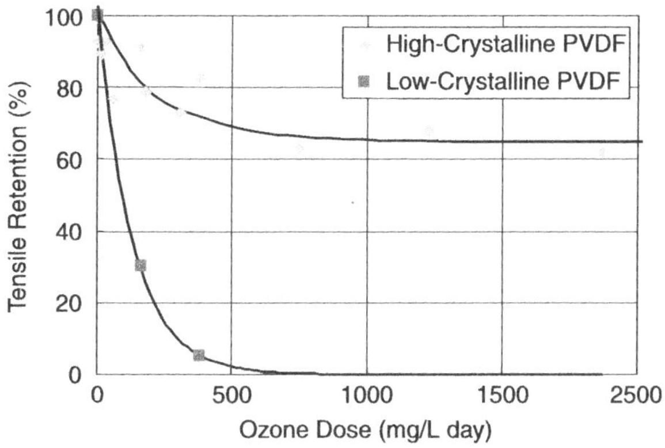

Figure 5.6 Ozone resistance of PVDF hollow fiber.

Housing Material Stainless steel housing is desirable for long-term ozone resistance.

5.2.2.3 Experimental Results and Discussion of Evaluation on Performance of Membrane Filtration

Experimental Outline Experimental site: Kitachiba Water Purification Plant, Kitachiba Water Supply Authority (raw water–surface water of the Edo River in Japan). Experimental duration: February 1999 to November 2001.

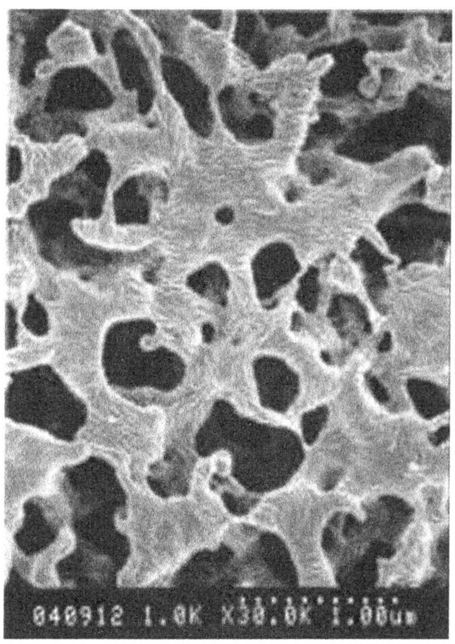

Figure 5.7 SEM image of high-crystalline PVDF hollow fiber.

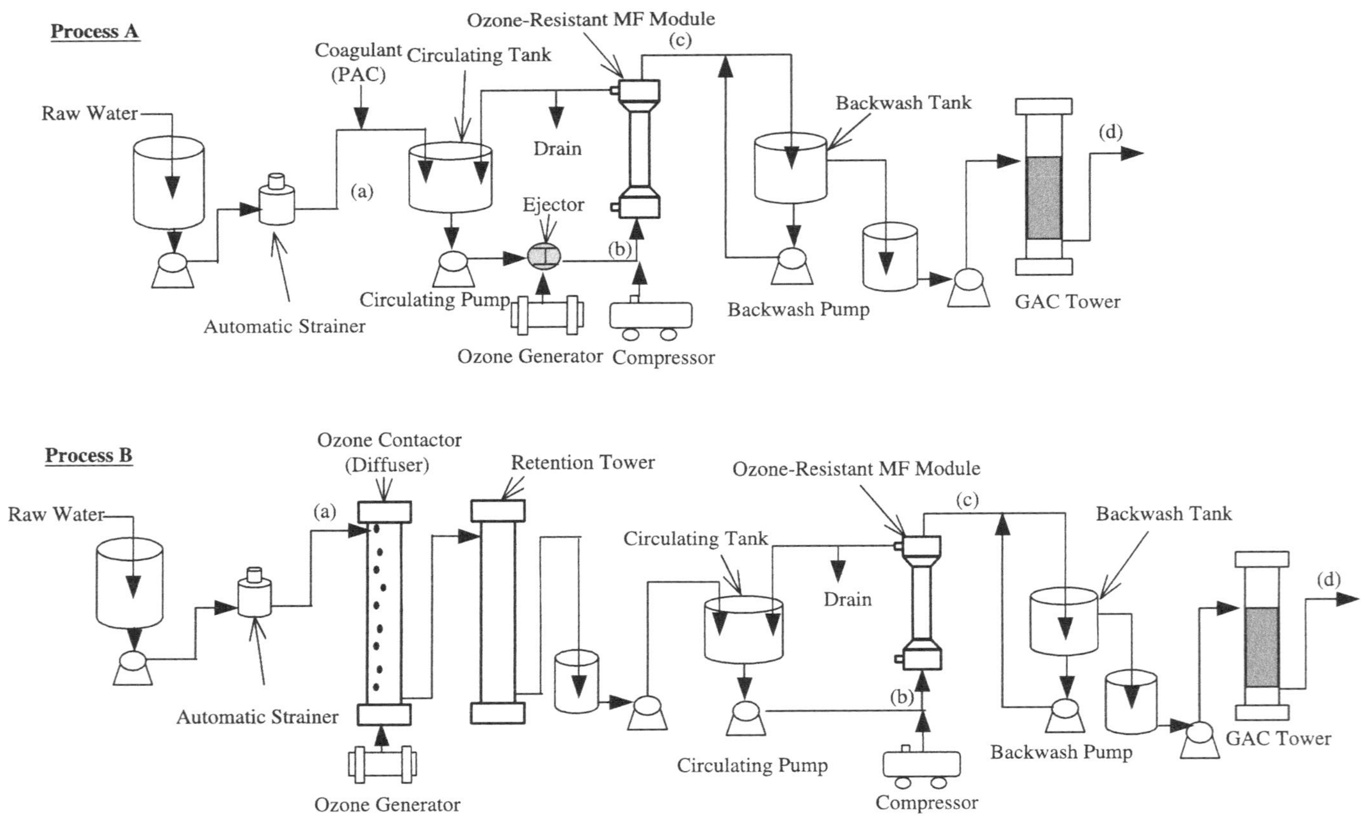

Figure 5.8 Experimental flow.

TABLE 5.4 Equipment Specifications of Treatment Units

Ozone Generator	Membrane Filtration Unit	GAC Unit
Capacity: 20 g O_3/h	Membrane material: PVDF hollow fiber Nominal pore size: 0.1 μm Membrane area: 6.9 m^2 Flow path: outside-in	Coal class: granular activated carbon Thickness: 2 m Flow method: gravity downward flow

EXPERIMENTAL FLOW The pilot plant, which consisted of an ozonation facility, an ozone-resistant MF module, and a granular activated carbon (GAC) tower, had a capacity of treating about 35 m^3/day of water. The experimental flow, as shown in Figure 5.8, had two trains in which ozone was injected before the MF module using an ejector (process A) and a diffuser (process B). The raw water taken from the Edo River through a raw water receiving tank was treated by an automatic strainer with a mesh size of 100 μm. Equipment specifications of treatment units are shown in Table 5.4.

The study has two experimental objectives as follows:

1. Examine the effects of ozone dosage, ozone dosage method, and membrane filtration method on membrane filtration flux and treated water quality through a short-term experiment, and determine conditions of a long-term experiment.
2. Conduct a long-term experiment to verify the treatment performance as well as the operating stability against raw water and seasonal variations, and establish a process management method such as determining a method to control ozone dosage and identifying the frequency of chemical cleaning.

The experimental conditions are shown in Table 5.5. The operating conditions for a GAC unit in the long-term experiment were 250 m/day of filtration rate [linear

TABLE 5.5 Experimental Conditions

	Short-Term Experiment			Long-Term Experiment		
Run No.	1–7	8–10	11–12	13	14	15
Operating method	Constant pressure operation (70 kPa)			Constant flow operation (5 m^3/m^2/day)		
Filtration method	Cross flow	Cross flow	Cross flow or dead end	Cross flow	Dead end	Dead end
Circulation/filtration flow rate	1/1	0.1/1	0/1–1/1	1/1	0/1	0/1
Operating pattern	(20-min filtration–20-s backwash) × 6–2 min air scrubbing			20-min filtration–1-min air scrubbing & backwash–Flushing		
Ozone dosage method	Ejector	Ejector or diffuser	Ejector	Ejector	Ejector	Ejector
Ozone dosage	0–4 mg/L	2 mg/L	2.5–3 mg/L	Ozone dosage was controlled to keep constant dissolved ozone conc. 1 mg/L in MF permeate		

velocity (LV)], 5.2 h^{-1} of space velocity (SV) and 48 h of backwash interval. Furthermore, water quality analysis was conducted in compliance with the Standard Methods for the Examination of Water (by Japan Water Works Association), sampling (a) raw water, (b) membrane feed water, (c) membrane filtrated water, and (d) water treated with activated carbon, as shown in Figure 5.8.

Effects of Ozone Dosage on Membrane Filtration Flux Figure 5.9 shows a relationship between ozone dosage and membrane filtration flux determined from the experiment of Run 1–7. When the ozone dosage was changed from 0 to 4 mg/L, the membrane filtration flux increased accordingly. With 4 mg/L, the flux reached 4.3 m^3/m^2/day (at 70 kPa 25°C), being 2.4 times as high as that without ozone (1.8 m^3/m^2/day). Based on this result, we have confirmed that high membrane filtration flux is possible with preozonation.

Figure 5.10 showed a relationship between dissolved ozone concentration of MF permeate and membrane filtration flux. Membrane filtration flux substantially increased with up to 0.3 mg/L dissolved ozone concentration, while rising gradually with 0.3–1 mg/L of the concentration. Based on this result, it became obvious that it is necessary to maintain dissolved ozone concentration more than 0.3 mg/L.

Effects of Ozone Dosage Method on Membrane Filtration Flux Figure 5.11 showed the changes in ozone dosage, dissolved ozone concentration of MF permeate, and membrane filtration flux when ozone was injected using an ejector (Run 8), a diffuser (Run 9), and a combination of a diffuser and a retention tower (Run 10). The contact time of ozone and raw water before reaching the ozone-resistant MF membrane was a few seconds for the ejector, 10 min for the diffuser, and 20 min for the diffuser plus retention tower. There was no major difference between the solubility of ozone into raw water between these three methods, reaching more than 80%. When the ozone dosage was 2 mg/L, dissolved ozone concentration of MF permeate was 0.6 mg/L with the ejector and 0 mg/L with the diffuser, and the membrane filtration flux for the ejector with high dissolved ozone concentration was higher than that of the latter. Based on this result, we found that the high interaction of dissolved ozone with foulant on the membrane surface is important to obtain high membrane filtration flux.

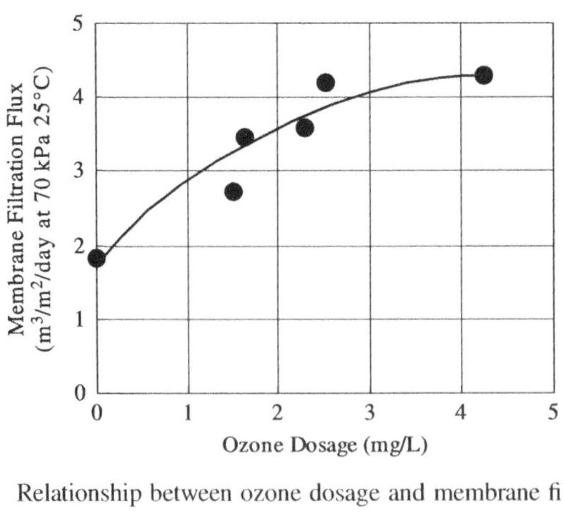

Figure 5.9 Relationship between ozone dosage and membrane filtration flux.

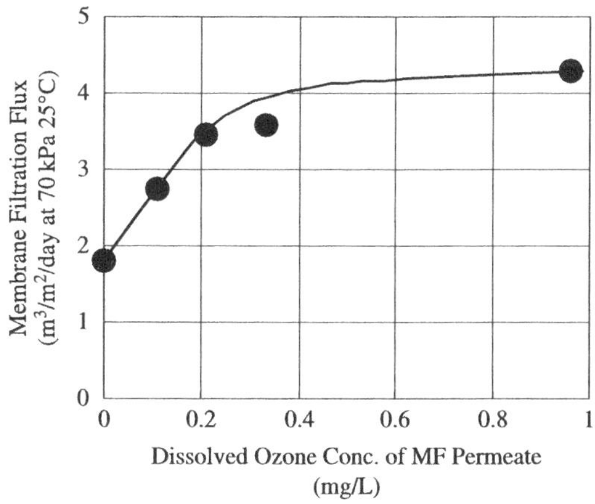

Figure 5.10 Relationship between dissolved ozone concentration of MF permeate and filtration flux.

Effects of Membrane Filtration Method on Membrane Filtration Flux Figure 5.12 shows the changes in ozone dosage, dissolved ozone concentration of MF permeate, and membrane filtration fluxes when ozone was injected with an ejector and water was filtrated with cross-flow (Run 11) and dead-end (Run 12) methods. Membrane filtration flux for the dead-end method was higher than that of the cross-flow method, and high membrane filtration flux operation (about 6 $m^3/m^2/day$) was achieved with 3 mg/L ozone dosage.

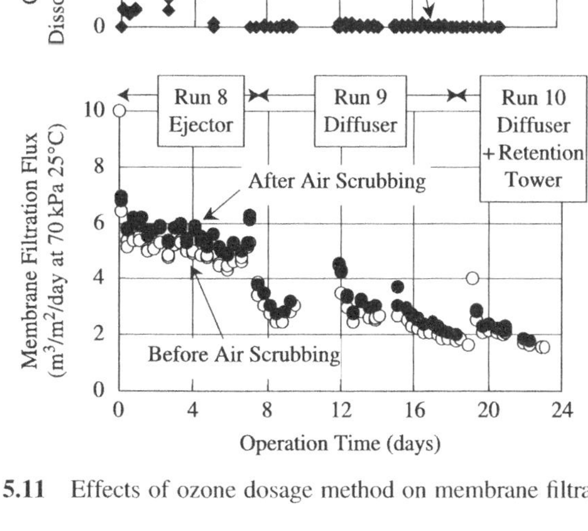

Figure 5.11 Effects of ozone dosage method on membrane filtration flux.

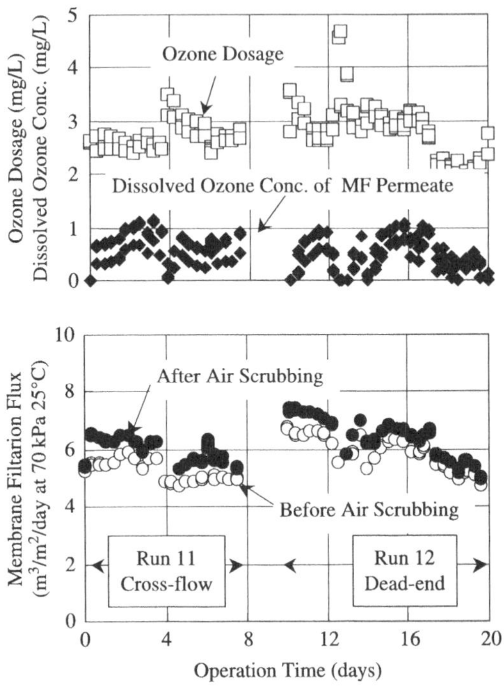

Figure 5.12 Effects of membrane filtration method on membrane filtration flux.

Based on these results, we found that higher membrane filtration flux could be obtained by injecting ozone just below the MF module, with the ejector and filtrating water with ozone left in MF permeate.

Figure 5.13 shows the changes in transmembrane pressure and raw water temperature when constant filtration operation was conducted on 5 m^3/m^2/day of membrane filtration

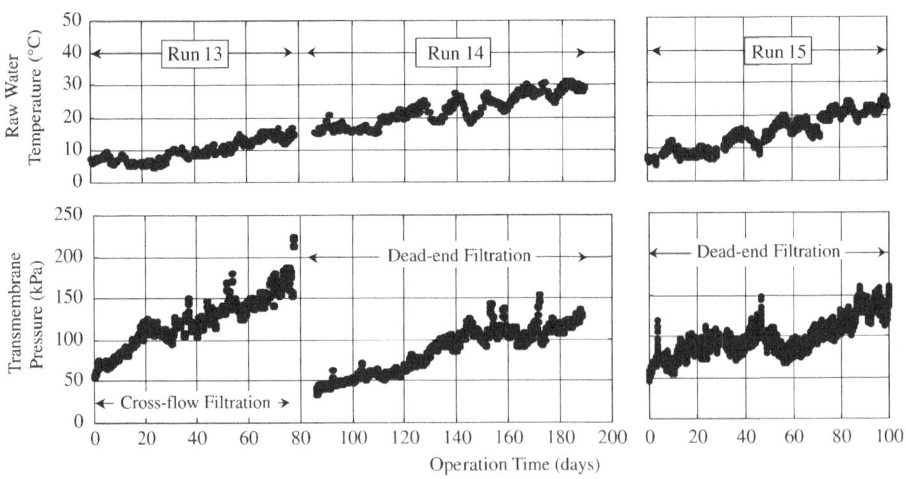

Figure 5.13 Changes in transmembrane pressure and raw water temperature.

flux (actual flux) with in-line injection of 20 mg/L of polyaluminum chloride (PAC) using Run 13 (cross-flow), Run 14 (dead-end), and Run 15 (dead-end). PAC was added to ensure stable membrane filtration during the period of low water temperature. During the operating period, raw water temperature was in the range of 5–30°C. By filtrating the water with approximately 1 mg/L ozone dissolved in MF permeate, continuous operation at high membrane filtration flux could be maintained for about 3 months. During this period, ozone dosage changed to between 2 and 6 mg/L.

Figure 5.14 shows the changes in raw water turbidity, dissolved ozone concentration of MF permeate, and transmembrane pressure. In this pilot plant, we continuously monitored the dissolved ozone concentration of MF permeate and controlled the ozone dosage so that the dissolved ozone concentration was kept at the given value (ca.1 mg/L). As a result, raw water turbidity rose up to about 150°, but stable filtration could be maintained during the period.

Based on these results, by controlling ozone dosage to have dissolved ozone concentration of MF permeate at around 1 mg/L, continuous high membrane filtration flux operation could be maintained for about 3 months even during the low water temperature period and the high turbidity period.

Treated Water Quality Table 5.6 shows the results of treated water quality analysis on raw water, MF feed, MF permeate, and water treated with GAC. Water quality after being treated with GAC met the requirements of the water quality standards of the Water

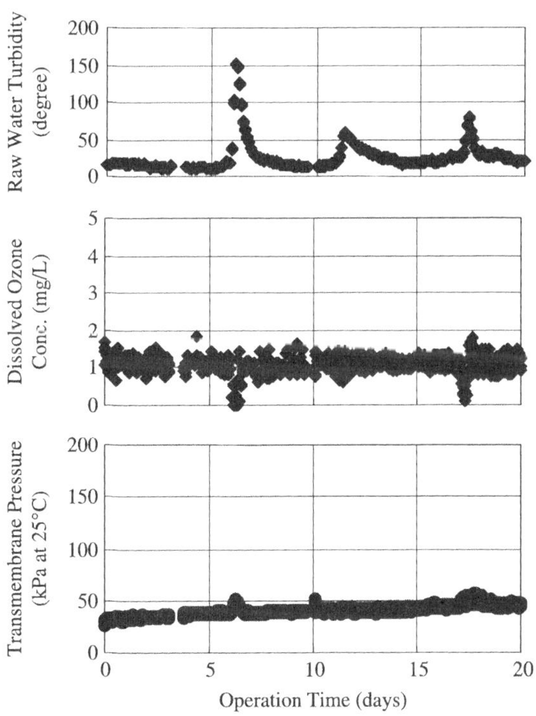

Figure 5.14 Changes in raw water turbidity, dissolved ozone concentration of MF permeate and transmembrane pressure, data corresponding to Run 14 of Table 5.5.

TABLE 5.6 Average Water Quality for Each Treatment Process

		Run 13 (Cross-flow Filtration)				Run 14 (Dead-end Filtration)			
Items	Unit	Raw Water	MF Feed	MF Permeate	GAC Permeate	Raw Water	MF Feed	MF Permeate	GAC Permeate
Turbidity	degree	7.6	7.8	<0.1	<0.1	19.6	19.8	<0.1	<0.1
Color	degree	5	1	<1	<1	6	1	<1	<1
pH	—	6.7	6.9	6.9	6.8	7.1	7.2	7.2	6.9
$KmnO_4$	mg/L	5.9	4.7	2.7	1.5	7.2	6.2	2.7	1.3
E260	—	0.161	0.057	0.038	0.028	0.155	0.057	0.043	0.026
NH_4-N	mg/L	0.27	0.28	0.28	<0.04	0.07	0.07	0.07	<0.04
Total Mn	mg/L	0.047	0.045	0.03	<0.005	0.069	0.064	0.019	<0.005
Dissolved Mn	mg/L	0.040	0.034	0.029	<0.005	0.019	0.020	0.019	<0.005
Fe	mg/L	0.61	0.58	<0.01	<0.01	1.1	1.18	<0.01	<0.01
Al	mg/L	0.50	1.25	0.02	<0.01	1.29	2.40	0.02	<0.01
THMFP	mg/L	0.034	0.030	0.018	0.012	0.039	0.029	0.016	0.010
DOC	mg/L	1.58	1.50	1.38	0.76	1.05	1.20	1.03	0.44
General bacteria	cfu/mL	6440	5	0	0	5550	28	0	1
Total coliform	cfu/m	69	0	0	0	74	0	0	0
BrO_3^-	mg/L	<0.01	<0.01	<0.01	<0.01	<0.01	<0.01	<0.01	<0.01

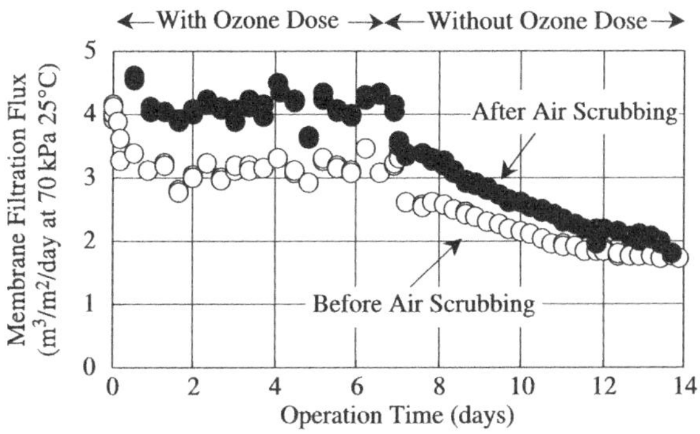

Figure 5.15 Changes in membrane filtration flux before and after air scrubbing.

Supply Law in all operating methods. Total manganese could be reduced to a detection limit or less by GAC treatment, although the efficiency of removing the manganese through membrane filtration was low. We presume that this was because the ozone oxidizes the manganese to be septivalent. Furthermore, ammoniacal nitrogen was hardly removed through membrane filtration but could be reduced to a detection limit or less, benefiting from the effects of biological activated carbon.

Discussion about Mechanism of High Membrane Filtration Flux with Ozone Dosage Figure 5.15 shows the changes over time in membrane filtration flux after and before air scrubbing with and without ozone dosage. Membrane filtration flux was substantially recovered with ozone dosage, while a degree of recovery of the flux was gradually reduced without ozone dosage. This result shows ozone dosage causes high membrane filtration flux as membrane surfaces can be kept clean by effective air scrubbing due to the interaction of foulant on the membrane with ozone.

Figure 5.16 shows the changes over time in dissolved organic carbon/extinction exponent at 260 nm (DOC/E260) of raw water and membrane feed water in the long-term consecutive experiment for Run 13 and Run 14. During the experiment, DOC/E260 of membrane feed water that was coagulated and ozonated represented a higher value than that of raw water. We presume this is because higher molecular weight humic substances were removed by PAC and became lower molecular weight substances and/or hydrophilic by ozone.

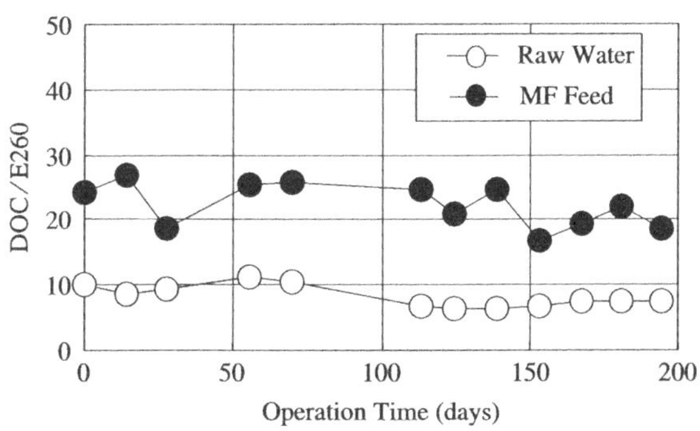

Figure 5.16 Changes in DOC/E260.

Figure 5.17a shows the distribution of molecular weights of raw water and MF permeate, while Figure 5.17b shows the results after measuring the distribution of molecular weights of raw water, ozonated water, and MF permeate. We found from Figure 5.17a that the substance removed with MF membrane had a molecular weight of about 25,000, and from Figure 5.17b that ozone interacted with that substance. The result showed that the interaction of ozone with a substance with a molecular weight of about 25,000 to be important for ozone-based high membrane filtration flux.

5.2.2.4 Conclusions From the above experiments, we were able to understand why such a high flux can be achieved by filtration using ozone-resistant module under the presence of residual ozone. This system has following superior features:

1. An economical membrane filtration system can be designed based on higher filtration flux with stable flux and lower filtration pressure.
2. Space saving (higher filtration flux needs less of a footprint).
3. Simple and highly reliable process based on the use of membrane.
4. Better quality of produced water by the use of ozone and membranes.
5. Cryptosporidium in concentrate as well as in filtrate may be disinfected by ozone.
6. Thorough pretreatment is not necessary.
7. Membrane contamination is cleaned immediately by ozone.

The application for sewage reuse was investigated and was applied to a sewage reuse process in Tokyo in 2003 (Sugimoto and Ogasawara, 2005). The applications of water treatment for various raw water can be expected.

5.2.3 Further Topics in Application Technology: Membrane Bioreactor System

5.2.3.1 Features of Membrane Bioreactor Membrane bioreactor (MBR) is a wastewater treatment system that has realized miniaturization of a plant by performing a solid–liquid separation process to separate activated sludge and treated water with a membrane module dipped in an activated sludge tank, and gives clear and high-quality treated water irrespective of the properties of the activated sludge.

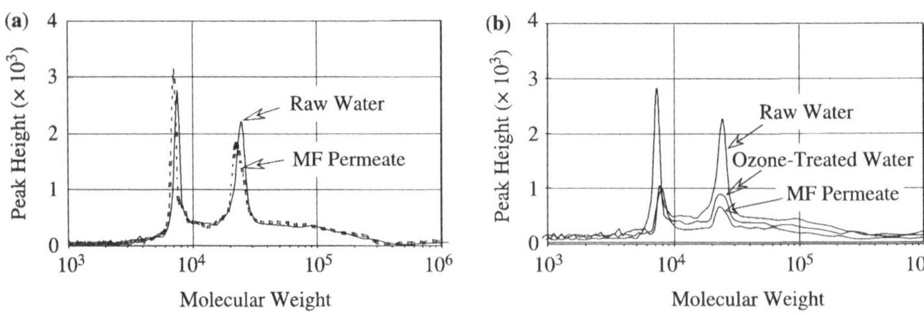

Figure 5.17 Changes in molecular weight distribution for each treatment process.

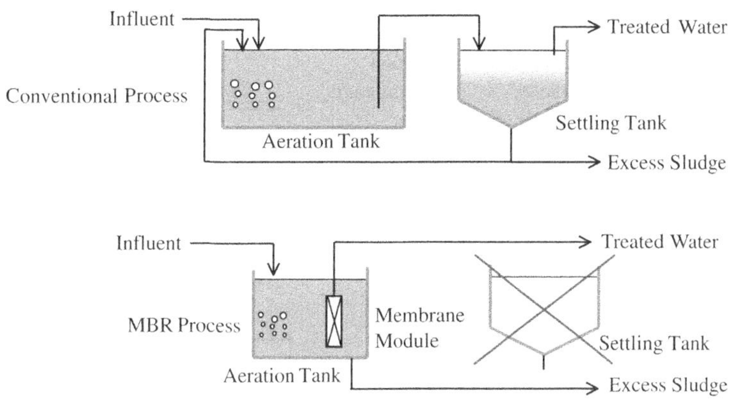

Figure 5.18 Comparison of the conventional activated sludge process and the MBR process.

Features of the wastewater treatment system by MBR are summarized as follows:

1. Final settling tank is not required, enabling space saving.
2. Biochemical oxygen demand (BOD) volumetric loading can be increased due to a high concentration of activated sludge [mixed liquor suspended solids (MLSS)] being possible. This enables miniaturization of activated sludge tank, reduction of treatment time, and the like.
3. Excess sludge can be directly removed from the activated sludge tank and subjected to dehydration treatment.
4. Maintenance and management are easy because solid–liquid separation can be done irrespective of the properties of the sludge. Particularly, release from the bulking problem may greatly contribute to reduction of workload of an operator at site.
5. High quality of treated water free of suspended solids (SS) and *E. coli* (*Escherichia coli*) can be obtained. Further, general service water can be reused for boiler water, for example, by combining a simple sterilization treatment and reverse osmosis (RO).

5.2.3.2 Necessary Conditions of MBR Module Requirements for the MBR module include cost reduction (membrane module cost and operation cost), prolongation of life, high reliability (safety operation, leak problem), easy handling, and the like.

In particular, with regard to differences from the module for water purification, first, the MBR module must be physically and chemically strong because aeration is always carried out during operation to avoid membrane blockage, and chemical cleaning is regularly required due to fouling by soluble organic substance, for example.

Second, a module structure, by which membrane blockage by activated sludge hardly occurs, is required. In the case of the MBR module, it is important that blockage on the membrane surface is prevented by shear force and oscillation generated by air bubbles of the aeration rising in the activated sludge. For this, it is required that the aeration works uniformly and effectively, and a flow of the activated sludge generated by the aeration goes out from the system through the upper part of module without stagnation.

5.2.3.3 Overview of Commercialized MBR Modules The modules for MBR have a variety of types and shapes of membranes and modules depending on manufacturers,

TABLE 5.7 Representative Commercialized Modules for MBR

Supplier	Asahi Kasei Chemicals Co., Ltd.	Zenon Environmental Inc.	Mitsubishi Rayon Co., Ltd.		Kubota Co., Ltd.
Membrane shape	Hollow fiber	Hollow fiber	Hollow fiber		Flat sheet
Nominal pore size (μm)	0.1	0.1	0.4		0.4
Membrane material	PVDF	PVDF on braid	Hydrophilic PE	PVDF on braid	Chlorinated PE
Element shape	Cylindrical	Skein	Skein	Skein	Flat sheet
Element length (m)	2	2	1.4	2	1
Element, stacked	1	1	2	1	2
Membrane surface area (m^2)	25	60	15	25	0.8
Standard average flux (m/day)	0.72	0.85	0.43	0.8	0.7
Air influent ratio, total	13.3	10.4	23	13–15	19.3
Air influent ratio for membrane	6.7	8.9	23	13–15	15.4

but any of them is well designed to satisfy the necessary conditions. A worldwide trend in membrane material is a shift from polyethylene to polyvinylidene fluoride (PVDF), which has high durability. As for shape of membrane, there are two types: hollow-fiber membranes and flat membranes.

Membrane manufacturers adopting the flat membrane include Kubota, Yuasa, and Toray. The flat membrane is hardly ever tangled with the fibrous substance contained in activated sludge, but it tends to require more installation space and a larger volume of aeration used for the membrane module. On the other hand, membrane manufacturers adopting the hollow-fiber membrane include ZENON, Mitsubishi Rayon, and Asahi-Kasei Chemicals. The hollow-fiber membrane has features such as a high degree of integration, a compact size, and less volume of aeration required. However, the activated sludge tends to accumulate unless its structure is optimized because space among membranes is narrow. Representative modules for MBR are shown in Table 5.7 prepared based on open information (Li et al., 2003; Kishino et al., 2003; Hashimoto et al., 2004; Fujii, 2004).

5.2.3.4 Design Technology of Membrane and Module for MBR The application of the MF/UF technology to MBR and design technology of membrane and module for MBR will be exemplified using the MBR modules made by Japan-based Asahi-Kasei Chemicals Co., Ltd (hollow-fiber type). The concept for design of the MBR module includes the following points:

1. Physically and chemically tough membrane: Since filtration membrane in MBR is required to be physically and chemically tough, Asahi-Kasei Chemicals employs the original PVDF membrane, which is superior in mechanical (physical) strength as well as

resistance to oxidants (sodium hypochlorite, etc.) and alkalis commonly used as a membrane cleaner.

2. High water permeability: Enhancement of water permeability of the filtration membrane is effective to reduce cost of equipment by reducing the number of modules used. The original PVDF hollow-fiber membrane has high water permeability, and a filtration operation at 0.8 m/day is possible in the sewage treatment field. Further, the membrane has a structure capable of back-washing as a feature of the products of Asahi-Kasei Chemicals. Stable operation is possible even under a high MLSS condition such as MLSS not less than 2% and a methane fermentation sludge that is extremely difficult to be filtered because the back-washing works effectively.

3. Operation is possible with a small volume of aeration for membrane cleaning: Since the cost of aeration occupies a large part of the electricity cost, reduction of aeration volume for membrane cleaning had been a target in the development. Asahi-Kasei Chemicals has prevented clogging by oscillating the membrane effectively with about half a volume of aeration compared with the conventional membrane module, by employing a cylindrical structure to increase a degree of integration of hollow-fiber membrane.

4. Design for compact module: An installation space per unit filtration membrane area is extremely small, about one third of the conventional module. This enabled miniaturization and weight reduction of an unit and simplification of its fitting up works. Further, the tank in which the membrane is dipped can vary in its dimensions and shape.

We will now describe the MBR module in more detail.

Hollow-Fiber Membrane A photograph of the cross section of the PVDF hollow-fiber membrane and an SEM image of the surface of the PVDF hollow-fiber membrane are shown in Figures 5.19a and 5.19b, respectively. On the external surface of the PVDF hollow-fiber membrane, a number of uniform and fine pores open, through which raw water is filtered.

Figure 5.20 shows the rejection rate in filtration of uniform latex particles. Particles having a particle size less than 0.1 μm pass through, but particles having a particle size not less than 0.1 μm can be rejeced perfectly. Therefore, treatment can be performed by the filtration without contamination of bacteria such as *E. coli* in activated sludge into treated water. Further, from the rapid rise of the rejection rate, pore size is found to be uniform.

Resistances to chemicals are shown in Figure 5.21. Figure 5.21a shows retention of elongation at break of the membrane after dipping in an oxidant (0.5% sodium hypochlorite aqueous solution) for 2 months, and Figure 5.21b shows that after dipping in an alkali (4% sodium hydroxide aqueous solution) for 2 months. Even after contact with chemicals for a long period, no mechanical deterioration is observed.

Membrane Module Membrane modules have a feature of cylindrical structure and are constructed of a header and a skirt to bundle the hollow-fiber membranes both having cylindrical shapes (Fig. 5.22). It has the following dimensions: diameter ca. 0.2 m, length ca. 2 m, and membrane outer surface area 25 m^2.

Further, since the membrane module has a high degree of integration of hollow-fiber membrane, it can be expected that clogging can be prevented by effectively oscillating the membrane with a small volume of aeration (Fig. 5.23). The aeration of membrane

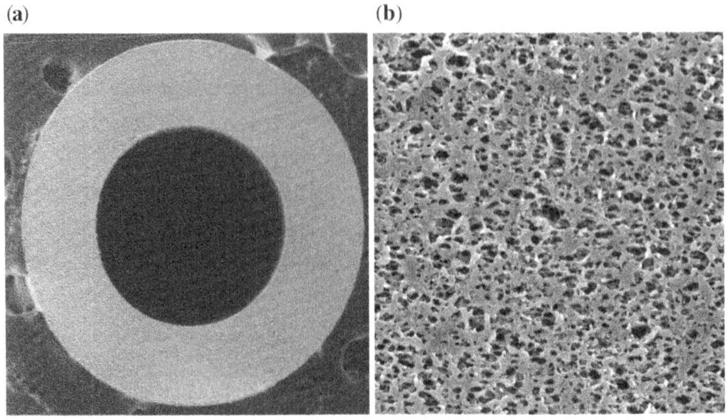

Figure 5.19 SEM images of the PVDF hollow-fiber membrane: (a) cross section and (b) outer surface.

module is performed by feeding air from the lower part of the skirt, and the air bubbles pass through holes equipped in the sealing part of the skirt, rise among the hollow-fiber membranes to oscillate them.

5.2.3.5 Examples of Applications Application examples of the above-mentioned MBR modules (hollow fiber and cylindrical type made by Asahi-Kasei Chemicals Co., Ltd.) will be exemplified.

Sewage Application of the MBR module to sewage has been studied at the Technology Development Experiment Center, Japan Sewage Works Agency in Mooka City, Tochigi Prefecture, aiming at cost reduction of the membrane separation activated sludge process, jointly with Japan Sewage Works Agency since the 2002 fiscal year (Fig. 5.24).

The main purpose of this study was reduction of maintenance cost by reducing air influent ratio, and finally a long-run operation was successfully demonstrated with a total air

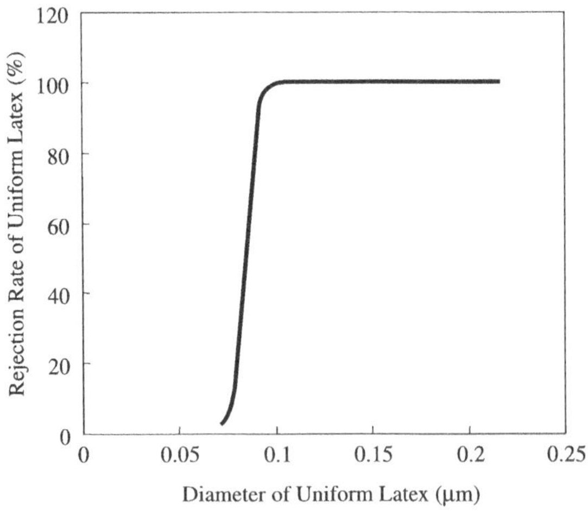

Figure 5.20 Rejection rate of uniform latex of the PVDF hollow-fiber membrane.

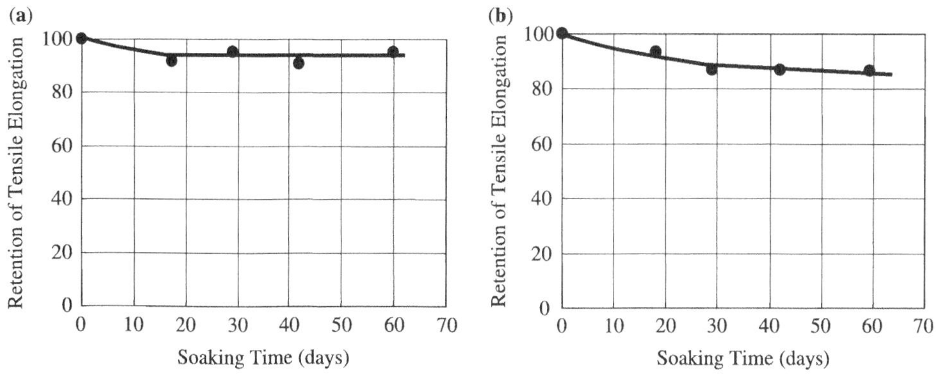

Figure 5.21 Chemical resistance of the PVDF hollow-fiber membrane: (a) soaked in NaOCl aqueous solution containing 5000 ppm of free Cl at room temperature and (b) soaked in 4 wt% NaOH aqueous solution at room temperature.

influent ratio of 13.3 times (aeration for membrane module cleaning: 6.7 times plus aeration for auxiliary diffusing air: 6.7 times), which corresponded to about a 40% reduction from the general level of 2 years before. In addition, stable operation was secured at a flux of 0.8 m/day during operation. Further, with regard to quality of treated water, good results were shown in all items such as BOD 0.7 mg/L, total nitrogen (TN) 5.37 mg/L, total phosphorus (TP) 0.2 mg/L, and SS $<$ 0.4 mg/L in average values during the experimental period. Little deterioration in strength of the hollow-fiber membrane was observed after 11 months operation, indicating that little deterioration in performances occurs during long-run use.

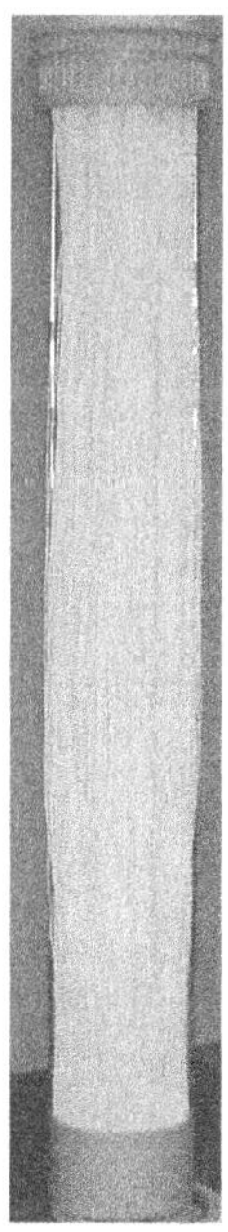

Figure 5.22 Appearance of the cylindrical MBR module.

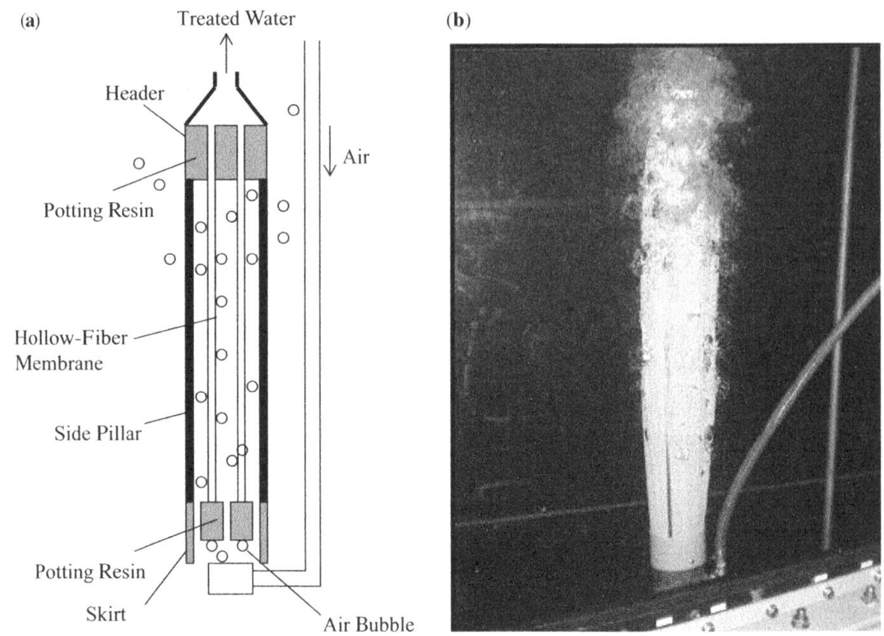

Figure 5.23 The MBR module in aeration.

Effluent from a Food Factory (1) A study on effluent from a sugar factory was carried out using an MBR pilot plant. In the existing plant, the effluent (BOD 1000–2000 mg/L) was subjected to a pretreatment to reduce BOD to 600–800 mg/L prior to a treatment in an activated sludge tank, whereas in this study, the raw water was directly treated in the activated sludge tank. In the MBR process, it was confirmed that the plant was able to be operated under 2 times of BOD volumetric loading compared with that of the conventional process, providing good quality of treated water. Also, the plant was able to be stably operated at a flux of 0.6 m/day.

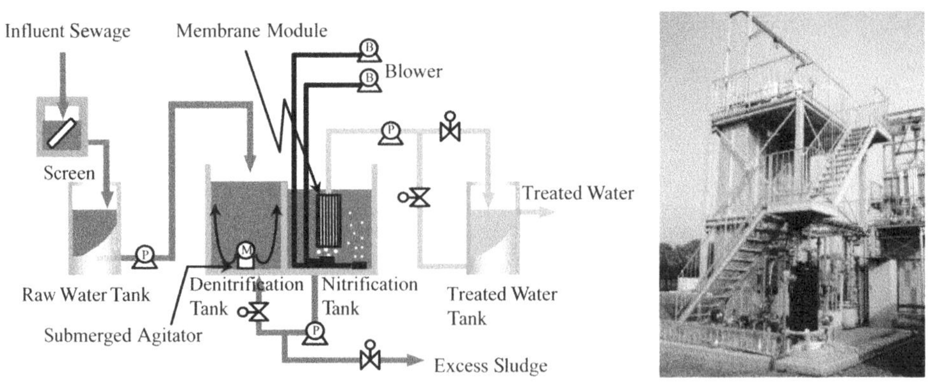

Figure 5.24 Schematic of a pilot plant in the joint study with Japan Sewage Works Agency and photograph.

Effluent from a Food Factory (2) A food factory has been operated using a membrane module system of our company installed in their effluent process. Despite of the raw water having very high values such as BOD 1000–2000 mg/L, MLSS 8000–10,000 mg/L, sludge viscosity about 300 Pa·s, the system has been operated stably showing a satisfactory level of quality such as BOD of around 1 mg/L and SS of not higher than the detection limit.

Effluent from a Liquid-Crystal Factory A study was conducted on effluent containing an organic solvent from a liquid-crystal factory. Two kinds of effluent containing different components were studied, and good results were obtained both in operation and quality of treated water. In addition, a stable operation was possible even with raw water having MLSS of 17,000 mg/L, proving that high loading operation was possible.

5.3 FUTURE PROSPECTS

Use of membrane filtration will continue to expand around water supply and sewerage fields, including the electronics industry and the pharmaceutical industry. One of the big factors controlling the extension is economic efficiency, and hence an advance on the technical side to improve the economic efficiency is important. In particular, a technology to suppress the clogging (fouling resistance technology), which is a fatal problem of filtration, is an old problem. To realize the further spread of membrane use, the construction of a total "membrane engineered for fouling resistance" is necessary that includes (1) membrane preparation technology (material design, structure design, etc.) and (2) filtration operation technology (module design, filtration conditions design, cleaning technology, etc.) is important.

The following three topics were exemplified in this chapter:

1. A PVDF porous membrane having a seamless network structure and high crystallinity for high strength and high durability, as shown in Figure 5.4
2. A unique membrane filtration system using ozone for high permeation flux
3. A unique cylindrical module design for high performance on MBR system

These are examples of the construction of a total membrane engineered for fouling resistance.

Abbreviations

BOD	Biochemical oxygen demand
CT value	Concentration time value
DOC	Dissolved organic carbon
E260	Extinction exponent at 260 nm
GAC	Granular activated carbon
LV	Linear velocity
MBR	Membrane bioreactor
MF	Microfilter
MLSS	Mixed-liquor suspended solids
NH_4-N	Ammoniac nitrogen

NF	Nanofilter
PAC	Polyaluminum chloride
PVDF	Polyvinylidene fluoride
RO	Reverse osmosis
SEM	Scanning electron microscopy
SS	Suspended solids
SV	Space velocity
THMFP	Trihalomethane formation potential
T-N	Total nitrogen
T-P	Total phosphorus
UF	Ultrafilter

REFERENCES

Association of Membrane Separation Technology of Japan (AMSTJ) (2001). Certified membrane modules for water supply catalogue book 2001. AMSTJ, Tokyo, Japan.

Doi, Y., and Matsumura, H. (1991). Polyvinylidene fluoride porous membrane and a method for producing the same. U.S. Patent 5,022,990.

Fujii, W. (2004). Waste water treatment using PVDF hollow fiber membrane. In *Abstracts of the 21th New Membrane Technology Symposium*. The Membrane Society of Japan and Japan Management Association, Tokyo, Japan, pp. 5/1/2–5/1/8.

Hashimoto, T., Okamura, D., Murakami, T., and Ohta, S. (2004). Running cost reduction of membrane bioreactor system using a newly cylindrical membrane module. In *Abstracts of the 41th Sewage Technology Conference*. Japan Sewage Works Association, Yokohama, Japan, pp. 762–764.

Hashino, M., Mori, Y., Fujii, Y., Motoyama, N., Kadokawa, N., Hoshikawa, H., Nishijima, W., and Okada, M. (2000). Pilot plant evaluation of an ozone-microfiltration system for drinking water treatment. *Water Sci. Technol.* 41, 10.

Kesting, R. E. (1985). *Synthetic Polymeric Membranes: A Structural Perspective*, 2nd ed. Wiley, Hoboken, NJ, pp. 237–261.

Kishino, H., Ohi, Y., Murakami, T., and Ohta, S. (2003). Running cost reduction of membrane bioreactor system. In *Abstracts of the 40th Sewage Technology Conference*. Japan Sewage Works Association, Tokyo, Japan, pp. 703–705.

Kubota, N. (2004). Fundamental of MF/UF technology. In *Abstracts of the 10th Membrane Separation Technology Seminar*. Association of Membrane Separation Technology of Japan, Tokyo, Japan, pp. 1–8.

Kubota, N., Takegawa, H., and Takamura, M. (2003). An unique membrane formation technology based on thermally induced phase separation technique and applications of the resulting membrane. In *Abstracts of the 36th Autumn Meeting of Society of Chemical Engineers*. Tohoku University, Sendai, Japan, p. M2P04.

Li, T., Sakamoto, Y., Murakami, T., and Ohta, S. (2003). Running cost reduction of membrane bioreactor system using intermittent aeration technique. In *Abstracts of the 40th Sewage Technology Conference*. Japan Sewage Works Association, Tokyo, Japan, pp. 697–699.

Lloyd, D. R., Kim, S. S., and Kinzer, K. E. (1991). Microporous membrane formation via thermally induced phase separation. 2. Liquid-liquid phase separation. *J. Membr. Sci.* 64, 1.

Lloyd, D. R., Kinzer, K. E., and Tseng, H. S. (1990). Microporous membrane formation via thermally induced phase separation. 1. Solid-liquid phase separation. *J. Membr. Sci.* 52, 239.

Loeb, S., and Sourirajan, S. (1964). US Patent 3,133,132.

Mori, Y., Oota, T., Hashino, M., Takamura, M., and Fujii, Y. (1998). Ozone-microfiltration system. *Desalination* 117, 211.

Moulin, C., Bourbigot, M. M., Tazi-Pain, A., and Faivre, M. (1991a). Potabilization of surface waters by crossflow ultra- and microfiltration on mineral membranes: Interest of ozone. In *Proceedings AWWA Membrane Processes Conference*, Orland, FL, pp. 729–738.

Moulin, C., Bourbigot, M. M., and Faivre, M. (1991b). Interest of the ozone/coagulant combination for the potabilization of surface waters by crossflow microfiltration on mineral membranes. *Key Eng. Mater.* 61/62, 229.

Sugimoto, K., and Ogasawara, H. (2005). Development of a sewage reuse process using ozone-resistant membranes. In *Proceedings International Forum on Water Industry Qingado 2005*, Qingado, China, pp. 271–279.

Watanabe, K., and Ohya, H. (1993). *Membrane Experiments Series 3; Artificial Membranes.* Membrane Society of Japan, Kyoritsu Shuppan, Tokyo, Japan, pp. 5–9.

CHAPTER 6

Water Treatment by Microfiltration and Ultrafiltration

M. D. KENNEDY, J. KAMANYI, S. G. SALINAS RODRÍGUEZ, N. H. LEE,
J. C. SCHIPPERS, and G. AMY

UNESCO–IHE Institute for Water Education, 2601 DA Delft, The Netherlands

6.1 INTRODUCTION

Advanced treatment processes are required to remove microorganisms, disinfection by-product (DBP) precursors, synthetic organic chemicals (SOCs), suspended and colloidal particles, natural organic matter, and salts from drinking water supplies. Microfiltration (MF) and ultrafiltration (UF) are low-pressure membrane processes that can be applied to remove microorganisms and suspended and colloidal particles. Classification of membrane technologies based on their pore size and the size of particles and molecules retained is illustrated in Figure 6.1.

Since the mid-1990s, the use of low-pressure hollow-fiber MF and UF processes has exploded. The most recent global study (Furukawa, 2002) of MF and UF technology showed accelerated market growth from 1996 onward, with an increase in the contracted/installed capacity from ca. 1000 m^3/day in 1997 to an estimated 10,000,000 m^3/day, by the end of 2003 (Fig. 6.2). One of the main drivers for the increase in MF and UF growth was the increasingly stringent environmental legislation implemented in the last decade (e.g., *Giardia* and *Cryptosporidium* removal guidelines of the Surface Water Treatment Rule, USA 1989, and the directive for the quality of water for human consumption, EC 1998), requiring the use of advanced technology, and in particular MF and UF, to comply with new disinfection standards for drinking water.

Furthermore, water scarcity has led to the widespread use of desalination to supplement freshwater resources. While reverse osmosis (RO) membranes are used to desalinate seawater and brackish water, MF/UF can be applied as a pretreatment to RO, in an *integrated membrane system* (IMS). The level of pretreatment, in terms of turbidity and silt density index (SDI), provided by MF/UF was found to be significantly better and more consistent than conventional pretreatment to RO (e.g., coagulation/sedimentation/filtration) for RO feed water. In recent years, the demand for MF/UF as a pretreatment to RO has grown as the need to augment our

Advanced Membrane Technology and Applications. Edited by Norman N. Li, Anthony G. Fane, W. S. Winston Ho, and T. Matsuura
Copyright © 2008 John Wiley & Sons, Inc.

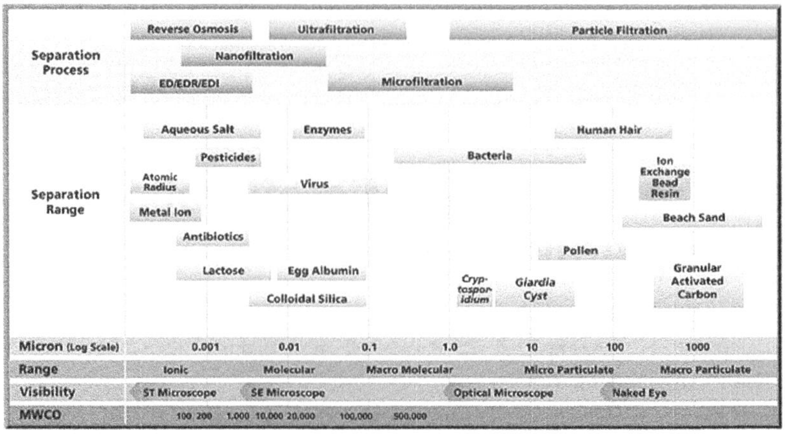

Figure 6.1 Classification of pressure-driven membrane filtration processes based on the size of particles and molecules removed. (From Applied Water Solution, 2006.)

freshwater supplies with nonconventional water resources such as seawater and industrial and municipal wastewater increased.

Other important triggers behind the recent surge in MF/UF growth were the development of hollow-fiber MF/UF membranes and the switch from cross-flow to dead-end filtration (direct filtration). Hollow-fiber membranes can be backwashed automatically with permeate, thus reducing the need for extensive pretreatment. Adopting dead-end filtration (instead of cross-flow filtration where feed water is recirculated across the membrane to control fouling) for large-scale MF and UF drinking water plants reduced the energy consumption considerably. In dead-end filtration, membrane fouling is controlled by a very short automatic backwash performed at regular time intervals with UF/MF permeate, or, if necessary, *an enhanced backwash* can be applied where a low dose of disinfectant/oxidant is added to the backwash water to remove foulants and restore membrane permeability. As a result of these innovations, MF and UF technology has developed into a cost-competitive and viable alternative to conventional methods of drinking and industrial water treatment and water reuse.

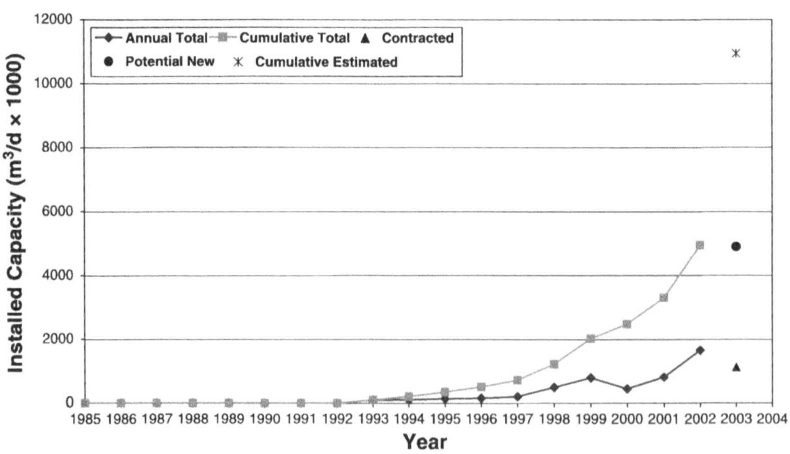

Figure 6.2 MF/UF global capacity. (From Furukawa, 2002.)

6.2 MATERIALS, MODULE CONFIGURATIONS, AND MANUFACTURERS

6.2.1 Materials

Microfiltration and ultrafiltration membranes can be made from organic polymers or inorganic materials such as ceramic, glass, or metal or organic polymers. Materials used in MF and UF membrane fabrication are shown in Table 6.1. A number of different techniques are employed to prepare synthetic MF/UF membranes; the most important are phase inversion, coating, sintering, and track etching.

6.2.1.1 Polymeric Membranes Synthetic polymeric membranes can divided into two classes, that is, hydrophobic and hydrophilic. Hydrophilic polymers such as cellulose and its derivatives have been used widely for the manufacture of MF and UF membranes. However,

TABLE 6.1 Materials for MF and UF Membrane Fabrication

Membrane	MF	UF
Organic		
Acrylonitrile polymer	✓	✓
Cellulose acetate (cellulose-2-acetate, cellulose-2,5-diacetate, cellulose-3-acetate)	✓	✓
Cellulose nitrate (CN)	✓	
Mixed cellulose esters	✓	
Regenerated cellulose	✓	✓
Nylon	✓	
Polyamide (aromatic polyamide, copolyamide, polyamide hydrazide)	✓	✓
Polyacrylonitrile (PAN)		✓
Polysulfone (PS)	✓	✓
Hydrophilic polysulfone		✓
Polyelectrolyte complexes		✓
Polyester		✓
Polyether sulfone (PES)	✓	✓
Polycarbonate (track etched)	✓	✓
Polyethylene terephthalate (PET) (track etched)	✓	✓
Polyimide		✓
Polyethylene (PE)	✓	
Polypropylene (PP)	✓	
Polytetrafluoroethylene (PTFE)	✓	
Polyvinylidene fluoride (PVDF)	✓	✓
Polytetrafluoroethylene (Teflon)	✓	
Polyvinylchloride (PVC)	✓	✓
Inorganic		
Alumina		✓
Aluminum oxide (Al_2O_3)	✓	
Zirconia (ZrO_2)–carbon		✓
Zirconia (ZrO_2)–polyacrylic acid		✓
Titania	✓	✓
Ceria (CeO_2)	✓	
Glass (SiO_2)	✓	✓
Stainless steel	✓	
Palladium (PD) and its alloy	✓	

cellulose acetate is sensitive to acid or alkaline hydrolysis, temperature, and biological degradation. Polysulfone (PSf) and polyethersulfone (PES) are also hydrophilic and used for UF membranes and a support for composite RO membranes. Hydrophobic membranes such as polytetrafluoroethylene (PTFE), polyvinylidene fluoride (PVDF), polyethylene (PE), or isotactic polypropylene (PP) are commonly used for MF membranes. These membranes are usually modified to reduce membrane fouling by blending with hydrophilic polymers.

6.2.1.2 Ceramic Membranes Ceramics [e.g., alumina (γ-Al_2O_3, α-Al_2O_3), zirconia (ZrO_2), titania (TiO_2), ceria (CeO_2), glass (SiO_2), and metal (e.g., sintered steel fibers or powders or thin- or thick-film deposits on various support media)] are widely used for the manufacture of inorganic membranes. Inorganic membranes have outstanding stability at high temperatures (over 100°C) and at extreme pH but are mechanically weak (i.e., brittle).

6.2.2 Membrane Properties

6.2.2.1 Pore Size or Molecular Weight Cutoff Microfiltration membranes have pore sizes (micrometer) ranging from 0.05 to 5 μm and UF membranes range from 1000 to 100,000 Da. Pore size is determined by bubble point analysis, porosimery (gas–liquid for MF or liquid–liquid for UF), or microscopic analysis for pore size and pore size distribution. However, the pore size measured is not absolute as the membrane pores are interconnected (networked) and not cylindrical capillaries.

Ultrafiltration membrane manufacturers frequently characterize their membranes using the "cutoff" concept rather than pore size. The nominal molecular weight cutoff (MWCO) is a performance-related parameter, defined as the lower limit of a solute molecular weight (e.g., dextran) for which the rejection is 95–98% (Boerlage, 2001). As the MWCO decreases, the mean pore diameter for most UF membranes has been found to decrease (Kim et al., 1990). However, the MWCO may be sharp or diffuse, that is, there is a range of MWCO, and in reality MWCO is only a rough indication of the membrane's ability to remove a given compound as molecular shape, polarity, and interaction with the membrane affect rejection (Mulder, 1996). Moreover, membrane surface characteristics (e.g., surface porosity and pore size distribution) may influence the apparent size of particles retained.

6.2.2.2 Porosity (Pore Density) The porosity is the pore volume divided by the volume of the material. Porosity can be measured by analyzing processed images obtained from microscopic analyses such as scanning electron microscopy (SEM), transmission electron microscopy (TEM), or atomic force microscopy (AFM) as well as pore size and pore size distribution. However, this simple definition obscures two difficulties, which can be significant in pores of atomic dimension.

6.2.2.3 Pure Water Permeability (PWP) In 1856, Darcy observed that the rate of flow of water through a bed of thickness Δx could be related to the driving pressure ΔP by the simple expression:

$$\frac{dV}{A\,dt} = J = \frac{\Delta P}{\eta R_m} \tag{6.1}$$

where J is the linear fluid velocity (of volume V flowing in time t through cross-section area A, $L/m^2 \cdot h$) for the pressure gradient (ΔP, N/m^2) and the viscosity of the fluid (η, Ns/m^2);

R_m refers to the permeability of the clean filter media. The resistance model is based on Darcy's law, which states that water flux through a membrane is proportional to the pressure gradient across the medium and the permeability of the medium.

In the situation of no fouling (clean membrane) and feed water completely free of any solutes, and assuming laminar flow through capillary tubes of radius r, the Hagen–Poiseuille law was obtained:

$$J = \frac{\varepsilon\, r^2}{8\,\eta\,\tau}\frac{\Delta P}{\Delta x} \tag{6.2}$$

where ε = porosity ($n\pi r^2$/surface area)
n = number of pores
r = pore radius (m)
η = viscosity (Pa·s)
τ = tortuosity factor
ΔP = transmembrane pressure (TMP, N/m^2)
Δx = membrane thickness (m)

Flux is proportional to porosity, pore size, and TMP.

6.2.2.4 *Hydrophobicity (Contact Angle)* Based on the chemical characteristics of the membrane material, membrane surfaces show hydrophilic or hydrophobic properties. Hydrophobicity is estimated by the sessile drop method using a goniometer. A goniometer measures the angle of a droplet of 1 μL of pure water on a membrane surface (Fig. 6.3). Hydrophobic membrane surfaces are often modified by blending with hydrophilic materials. The fouling potential of hydrophobic membrane is high due to the high binding affinity of proteins and humic substances.

6.2.2.5 *Surface/Pore Charge (Zeta Potential or Isoelectric Point)* Zeta potential (ζ) indicates the surface charge of a membrane and can be observed by measuring the streaming potential across a fluid shear plane at the surface. The isoelectric point is the pH at which the surface charge is zero. A streaming potential is generated when an ionic solution is forced to flow between two parallel membranes, and electrodes detect the difference in streaming potential. Zeta potential can be derived by the Helmholtz–Smoluchowski equation:

$$\frac{\Delta\phi}{\Delta P} = \frac{\varepsilon\zeta}{\mu k} \tag{6.3}$$

where $\Delta\phi$ = streaming potential (mV)
ΔP = forced pressure (Pa)

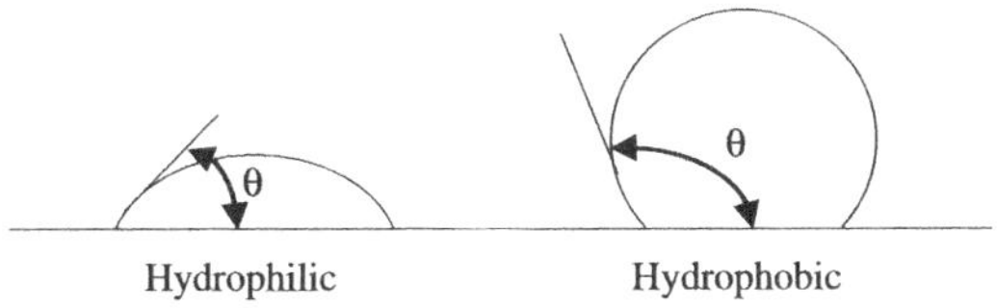

Figure 6.3 Contact angle measurement.

ε = permittivity of the solution (s/m)
μ = viscosity (Pa·s)
k = electrical conductivity of the solution (mS/m).

The surface charge implies different fouling tendencies. Generally, membrane materials carry a negative charge or are modified to have a negative charge because natural organic matter (NOM) in water is negatively charged at neutral pH, due to phenolic and carboxylic functional groups. A negatively charged membrane, therefore, prevents rapid deposition of NOM foulants on the membrane surface by charge repellence.

6.2.2.6 Roughness With AFM, the three-dimensional structure of a membrane surface can be observed directly. The image is generated by probing (tapping) on the membrane surface with a very fine tip. Roughness is calculated from Eq. (6.4). A change in roughness is associated with fouling. Generally, MF membranes have higher roughness than UF membranes. Roughness is reduced when foulants block pores and increased when foulant are deposited sporadically on a smooth membrane surface (Mulder, 1996).

$$R_a = \frac{\sum_{i=1}^{N} |Z_i - Z_{cp}|}{N} \tag{6.4}$$

where R_a (nm) is the arithmetic average of the deviations from the center plane, Z_i (nm) is current Z (nm) value, Z_{cp} is the Z value of the center plane, and N is the number of points within a given area.

6.2.3 Module Configurations

6.2.3.1 Hollow Fiber The hollow-fiber configuration is the most common configuration for MF and UF membranes. The hollow fibers are 0.5–1.5 mm (less than 5 mm) in diameter, and several thousands of hollow fibers are packed into a module (Fig. 6.4). The most important merit of hollow-fiber membranes in water treatment applications is that extensive pretreatment is not needed as the membrane can be backwashed (backflushed) automatically. Backwashing is carried out by changing the direction of flow of the permeate stream so that it flushes out the deposited particles that formed a cake layer on the membrane surface during the filtration cycle.

Inside-out, Outside-in Filtration in Hollow-Fiber Membranes Hollow-fiber membrane modules can be operated in two different flow modes—"inside-out" and "outside-in"—based on the direction of filtration flow. In the inside-out configuration, pressurized feed water flows through the bore of a hollow fiber, and permeate is collected on the outside of the membrane fibres. In the outside-in configuration, the pressurized feed stream flows from the outside of a hollow fiber, and permeate is collected inside the bore of the hollow fiber.

Immersed/submerged membranes operate in the outside-in mode, and a partial vacuum is created within the hollow fibers by the operation of a centrifugal pump. The feed water to be treated passes through the membrane (outside-in configuration), entering the hollow

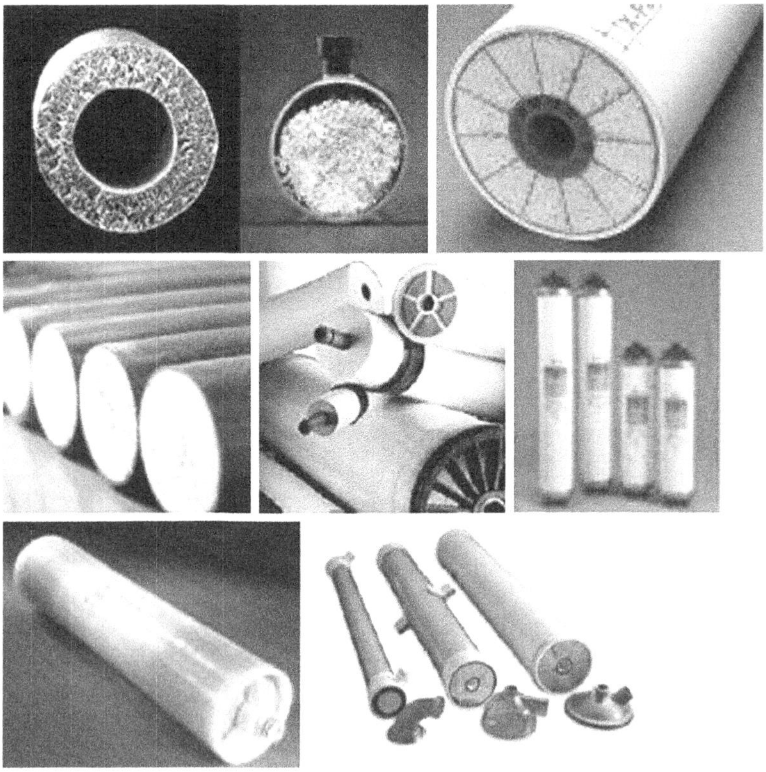

Figure 6.4 Profile of a hollow-fiber-type membrane and modules (Pall, X-flow, Inge AG, GEwater, Hydracap, Lenntech, Koch).

fibers, and is pumped to storage. Air is introduced either intermittently or continuosly at the bottom of the membrane module to minimize the thickness of the dynamic boundary layer on the outside of the membrane fibers, allowing them to operate at high flux. The clean filtered water is extracted by means of a permeate pump (Zenon, 2006). The operation of the system is highly automated and fibers can be easily cleaned with a clean-in-place backwashing process that forces permeate water back through the membranes, and, when necessary, membrane cassettes can be removed from service and placed in a separate cleaning tank to dissolve solids from the fibers. An immersed membrane system is illustrated in Figure 6.5. More discussion of immersed/submerged membrane systems can be found in Chapter 10, Submerged Membranes.

6.2.3.2 Tubular The tubular configuration is an extension of the hollow fiber, but with larger diameter (up to 25 mm) tubes (Fig. 6.6). The flow is inside-out similar to hollow-fiber membranes. The advantage of tubular membrane is that feed water with a high level of suspended solids can be processed and the system is easy to clean mechanically. However, tubular membranes are more costly than hollow-fiber membrane modules. The tubular configuration is widely used in the food and beverage industries and for industrial wastewater treatment.

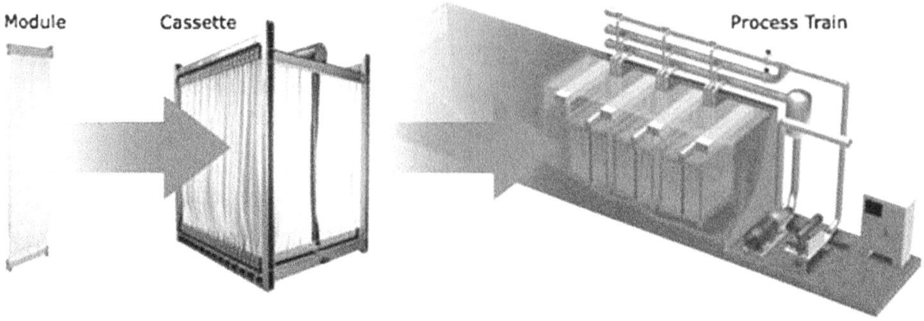

Figure 6.5 Immersed/submerged membrane system (Zenon 500 process).

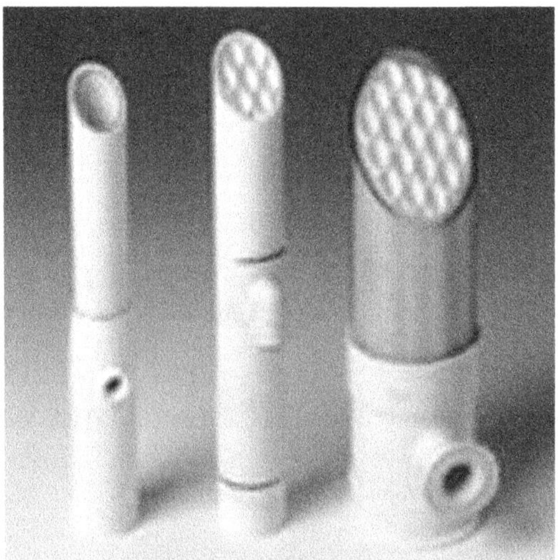

Figure 6.6 Tubular-type membranes.

6.2.3.3 Spiral Wound The spiral-wound membrane geometry in MF and UF is not widely used in surface water treatment because spiral-wound membranes cannot be backwashed, and extensive pretreatment is therefore required to keep the membrane from clogging. However, there are some applications in groundwater treatment where surface water UF membranes are applied to remove color (Nystrom et al., 1999). In the spiral-wound configuration, two flat sheets of membrane are separated with a permeate collector channel material to form a leaf. This assembly is sealed on three sides, with the fourth side left open for permeate to exit. A feed/brine spacer material sheet is also included in the leaf assembly. A number of these assemblies or leaves are wound around a central plastic permeate tube. This tube is perforated to collect the permeate from the multiple leaf assemblies (Hydranautics, 2001). Most commercial lengths are 1 or 1.5 m and 20 cm in diameter (Fig. 6.7).

6.2.3.4 Monolith The monolith configuration is a type of tubular configuration used in inorganic membrane systems (Fig. 6.8) and comprises a ceramic support (or hexagonal log)

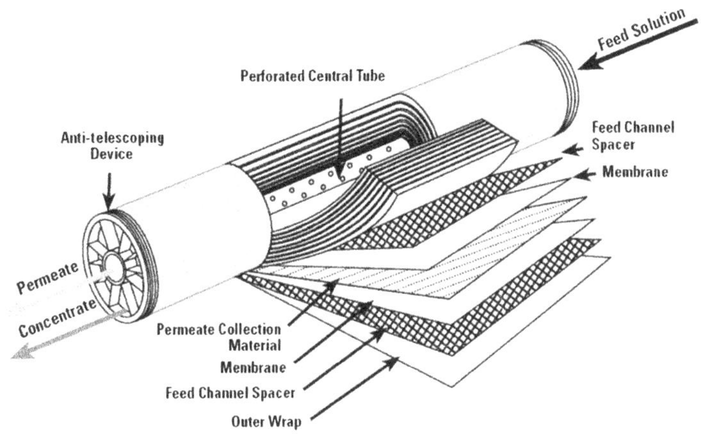

Figure 6.7 Schematic diagram of spiral wound membrane (Lenntech, 2006).

with several parallel-flow channels providing an extensive effective surface area for filtration.

6.2.4 Process Configuration

The application of MF and UF technology in water treatment is still relatively new, as many full-scale plants are less than 10 years old. Consequently, the design of MF and UF systems varies from plant to plant as no single design has proved itself to be the best (see Table 6.2). For example, MF/UF systems may be positioned horizontally or vertically, operate in dead-end or cross-filtration mode, the membranes may be immersed/submerged in a feed tank where permeate is sucked (via vacuum) into the inside of the hollow fiber (outside-in filtration) or the membranes may be housed in modules where pressurized feed water is forced through the fiber and permeate is collect on the outside (inside-out filtration).

6.2.4.1 Constant Flux, Constant Pressure Membrane systems operate in one of two possible ways: constant transmembrane water flux (flow rate per unit membrane area) with variable pressure or constant transmembrane pressure with variable water flux (Fig. 6.9). An increase in transmembrane pressure is required as the membrane fouls to maintain a particular water flux (constant-flux operation). In constant-pressure systems,

TABLE 6.2 Characteristics of Different Pressure-Driven Membrane Processes

Membrane Process	Applied Pressure [psi (kPa)]	Minimum Particle Size Removed	Application
Microfiltration	4–70 (30–500)	0.1–3 μm	Particle/turbidity removal Removal of bacteria and protozoa
Ultrafiltration	4–70 (30–500)	0.01–0.1 μm > 1000 D	Particle/turbidity removal Disinfection

Source: AWWA, 2004.

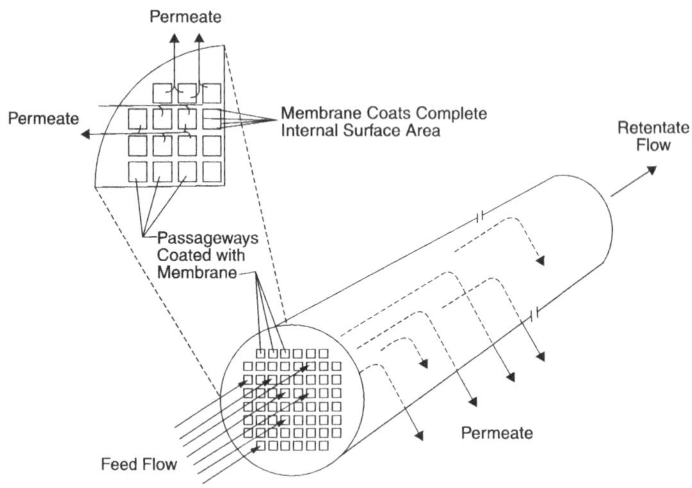

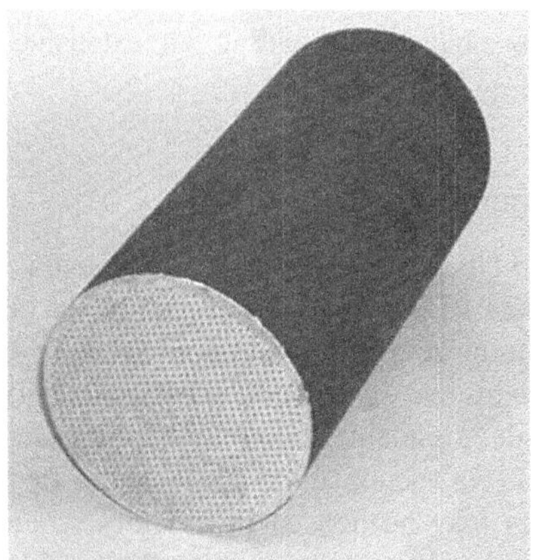

Figure 6.8 Schematic diagram of monolith-type membrane (CeraMem Corporation).

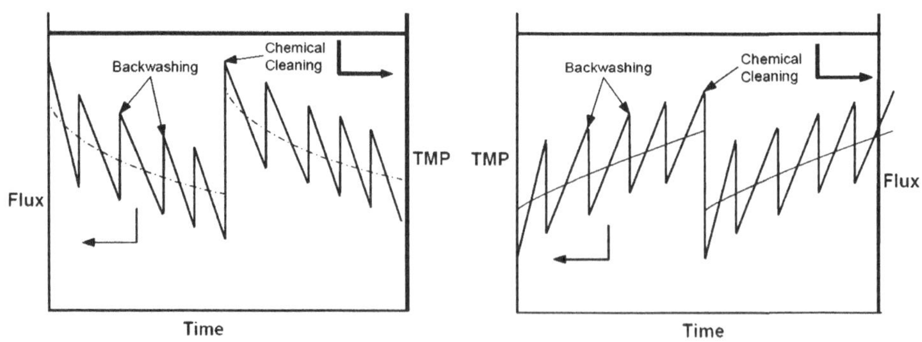

Figure 6.9 (*Left*) Constant-pressure and (*right*) constant-flux operation.

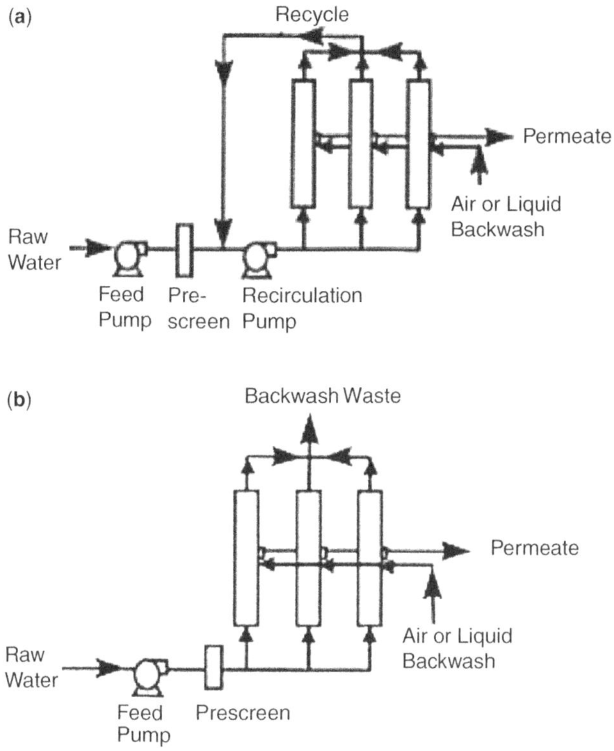

Figure 6.10 (a) Cross-flow configuration MF or UF and (b) dead-end filtration for MF or UF. (Adapted from Jacangelo et al., 1998.)

the pressure is constant and therefore the membrane fouls; the productivity (flux) decreases. Constant-flux operation is more common, particularly in full-scale plants.

6.2.4.2 *Cross-Flow, Dead-End Configurations*

Microfiltration and UF systems are operated in two possible filtration modes. Figure 6.10 shows the cross-flow configuration in which the feed water is pumped tangential to the membrane. Clean water passes the membrane while the water that does not permeate is recirculated as concentrate and combined with additional feed water. To control the concentration of the solids in the recirculation loop, a portion of the concentrate is discharged at a specific rate. In dead-end or direct filtration, all the feed water passes through the membrane. Therefore, the recovery is 100%, and a small fraction is used periodically for backwash in the system (5–15%).

Nowadays, most MF/UF plants operate in dead-end filtration mode, as the energy required is lower compared to cross-flow systems, as the high cross-flow velocity required to prevent fouling results in increased head loss and energy consumption.

6.2.5 Manufacturers and Commercially Available Membranes

The manufacturers of many commercially available membranes are given in Table 6.3.

TABLE 6.3 List of Manufacturers and Their Membranes

Manufacturer	Membrane Type	Commercial Name
Amicon (now Millipore)	UF	YM
Infilco Degremont (Suez)	UF	Aquasource
Nitto Denko (Hydranautics)	UF	Hydracap
Koch	MF, UF	Abcor, Fluid systems, ROMICON
Inge AG	UF	Multibore
Novasep	MF, UF	Persep. Carbosep, Kerasep, Pleiade
Norit Osmonics (now GE Water)	MF, UF	X-flow
Pall Siemens [U.S. Filter (Veolia)]	MF, UF	Microza, Memcor
Lenntech (TriSep)	UF	SpiraSep
Zenon	MF	Zeweed
NGK Insulators, Ltd.	MF, UF	NGK ceramic membrane filter
Lenntech	MF, UF	Ceramic membrane
CeraMEM	MF, UF	Ceramic membrane
Jiangsu JiuWu Hi-tech Co.	MF, UF	Ceramic membrane

6.3 MICROFILTRATION/ULTRAFILTRATION PRETREATMENT

Membranes are susceptible to fouling; therefore, pretreatment of the feed water is required to control colloidal, organic, and biological fouling, as well as scaling. The pretreatment scheme must be capable of controlling UF and MF fouling to such an extent that a practical cleaning frequency can be achieved. For low-pressure membranes, a number of pretreatment methods are currently used.

6.3.1 Direct Filtration (No Pretreatment)

Ultrafiltration can adequately produce disinfected water directly from surface water for different applications. MF can also be used for disinfection, although not all viruses are removed. However, direct membrane filtration is limited by fouling, which, during constant-flux filtration, leads to a continuous increase in transmembrane pressure. In addition, UF and MF membrane treatment alone cannot effectively and consistently remove organic material, measured as total organic carbon (TOC), and THM (tri-halo-methane) precursors, measured as chloroform formation potential (Berube et al., 2002).

6.3.2 Precoagulation

Precoagulation involves the addition of chemicals (e.g., coagulants such as $FeCl_3$ or $FeSO_4$, alum, polyaluminum chloride, etc.) to increase the size of suspended and colloidal particles in the feed water prior to UF or MF. This method of pretreatment can be problematic where coagulant residuals and an increased solids loading can reduce membrane fluxes and increase cleaning requirements. Moreover, the relatively "clean" membrane plant will now produce a chemical sludge that requires disposal.

Less extensive work has been done on coagulation and UF/MF process. To date research has been conducted in coagulation with other pretreatment processes such as powdered activated carbon (PAC) addition. Lahoussine-Turcaud et al. (1990) evaluated the effect of coagulation pretreatment on UF of surface water using a hydrophobic polysulfone membrane. They found that short-term reversible fouling was slowed, but the extent or the rate of irreversible fouling was uncharged. This was attributed to the fact that the coagulation process removed the larger size foulants (that usually cause reversible fouling) but not the smaller sized ones (usually responsible for irreversible fouling). Factors affecting membrane fouling included coagulant dosage, pH, nature of dissolved organic matter, as well as the Ca^{2+} content of the feed water.

Carroll et al. (2000) found that following coagulation pretreatment, most membrane fouling was due to the smaller sized hydrophilic NOM. This finding is consistent with the fact that most metal-based coagulants are known to preferentially remove hydrophobic rather than hydrophilic substances. Farahbakhsh and Smith (2002) concluded that coagulation reduced the rate of membrane fouling by minimizing pore plugging and increasing the efficiency of membrane backwashing.

6.3.3 In-line Coagulation

In-line coagulation (IC) refers to the dosing of coagulant into feed water, rapid mixing, allowing the flocs to form (but not settle), and finally feeding the resulting water (with the flocs) into the next process (e.g., MF or UF). In-line coagulation therefore, involves the use of coagulants without removal of coagulated solids (no sedimentation or prefiltration step) prior to UF or MF.

Despite the larger fouling load in terms of suspended matter, IC may improve membrane performance due to a change in the filtration mechanism (cake filtration). Once a cake is built, performance stabilizes, and operation is stable when the cake can be removed by backwashing.

Pearce and Hanks (1993) showed that MF offered good possibilities to remove *Cryptosporidium* oocysts from water after sedimentation, after in-line coagulation and from backwash water. The maximum removal measured was 99.5%. Bos et al. (1998) reported an improved UF membrane performance when IC or precoat was used. In other researches, however, iron sulfate dosage of 20–30 mg Fe/L were not capable of adequately controlling UF fouling, and in some cases even increased it.

Guigui et al. (2002) studied the IC (without settling)/UF process to improve membrane performance and water quality for surface water treatment. Employing coagulation before UF increased permeate quality, the extent of dissolved organic matter removal was controlled by the coagulation step. When optimized coagulation conditions for a coagulation/settling process were applied for the IC/UF process, membrane fouling was reduced. The possible reason proposed was that the (floc) cake resistance was lower than the resistance resulting from the unsettled flocs and the uncoagulated organics.

Doyen (2003) observed that due to in-line coagulation, the influences of membrane polymer nature and structure disappeared, cleaning could be postponed and cleaning aggressiveness could be lowered. In addition, water recovery and flux rate could be increased, and the influence of seasonal water quality variations were easier to control.

It has also been reported that aggregates produced under sweep floc conditions were more compressible than for charge neutralization conditions, resulting in compaction when the membrane filtration system was pressurized (Antelmi et al., 2001; Cabane et al., 2002). Lee et al. (2000) also reported that the specific resistance was lower with charge neutralization than with sweep floc, attributed to the formation of a less compressible and more porous cake. Judd and Hillis (2001) suggested that flocs need to reach a certain critical floc size prior to MF, otherwise membranes would be irreversibly fouled by the coagulant solids.

6.3.4 Coagulation–Sedimentation Process

In this pretreatment method, a coagulant is applied and the formed flocs are settled out by sedimentation. The supernatant is then fed to the membranes. At East St. Loius, when UF was applied after coagulation–sedimentation (CS) for 400 h, no fouling was observed at all (Kruithof et al., 2004). Minegishi et al. (2001) investigated the fouling mechanism of hollow-fiber UF membranes with pretreatment by the CS process. The experiments were carried out on three different feed water conditions: river/surface water, coagulated water containing microflocs, and clarified water from the CS process. The coagulation or CS pretreatment process was very effective in increasing UF membrane life because this process removed the primary foulants such as high molecular weight (MW) humics.

6.3.5 Coagulation–Adsorption Process

Coagulation–adsorption refers to the use of adsorbent such as PAC between the coagulation step and MF or UF. Enhancement of the removal of DBP (or DBP precursors) can be achieved by addition of PAC (Heneghan and Clark, 1991). PAC/UF has been used for the removal of SOCs—mainly pesticides.

Abdessemed and Nezzal (2002) treated wastewater with a COD of 165 mg O_2/L and turbidity of 90 NTU with 120 mg $FeCl_3$/L at a pH of 5.5. They obtained product water of a COD of 23 mg O_2/L and a turbidity of 12 NTU. When a further treatment step of adsorption with PAC was used, the COD dropped further to 7 mg O_2/L.

Berube et al. (2002) investigated the use of adsorption (PAC) and coagulation (alum and polyaluminum chloride) as pretreatment steps prior to membrane filtration to remove organic material. Coagulation, prior to membrane treatment, significantly improved the removal of organic material and THM precursors. MF treatment, with precoagulation (0.3 mg Al/L), consistently removed about 75% of the organic material and THM precursors. No flocculation step was required. However, preadsorption (PAC) alone moderately improved the removal of organic material or THM precursors, but increased the rate of membrane fouling. Farahbakhsh and Smith (2002) and Carroll et al. (2000) had obtained similar results. PAC has been shown to remove the smaller sized dissolved organic matter (DOM).

6.3.6 Preflocculation/Flocculated Water Filtration

Flocculation is another pretreatment method that can improve the permeate flux and remove particles and colloids. It is used to achieve three objectives: eliminating the penetration of colloidal particles into the membrane pores, increasing the critical flux, and modifying the characteristics of the deposits.

Guo et al. (2005) evaluated the effect of pretreatment of two types of wastewater (synthetic—with persisting organic compounds—and biologically treated wastewater from a wastewater treatment plant) on the critical flux of cross-flow MF. They found that (1) pretreatment by flocculation ($FeCl_3$) alone could effectively remove the large molecular weight organic matter (30–60 kDa); (2) pretreatment by flocculation, together with adsorption (PAC) could remove both large and small molecular weight organics; (3) flocculation as a pretreatment was significantly better than adsorption in improving the critical flux; and (4) critical flux increased fivefold when flocculation–adsorption was applied.

Thiruvenkatachari et al. (2002) used flocculation (hematite + $CaCl_2$) as a pretreatment to a cross-flow MF (CFMF) system. The flocculant was rapidly mixed with feed water and later slow mixed before finally feeding the resulting solution to the CFMF. For the same influent turbidity, the effluent turbidity was lower when rapid mixing was introduced prior to slow mixing, compared to CFMF with slow mixing alone, and there was a greater flux drop when the retentate was not recycled. When retentate was recycled, the flux remained fairly constant (especially when rapid mixing was also done) for the entire filter run.

The use of flocculation prior to membrane filtration reduced clogging of the membrane by aggregating smaller particles, thereby retaining them on the surface of the membrane. The larger flocs retained by the membrane surface are washed off by the retentate due to the tangential force (cross-flow) of the incoming solution, thus preventing membrane clogging. In the case of CFMF, employing slow mixing only (no rapid mixing), internal clogging of the membrane pores by smaller organic particles occur (due to the inefficient floc formation). This resulted in higher fouling rates and lower fluxes.

6.3.7 Magnetic Ion Exchange

Both MF and UF provide very little NOM removal. Pretreatment with coagulant or PAC can provide NOM removal, but drawbacks include reduction in membrane flux rates and increased backwash requirements in the short-term. The requirement to optimize coagulant dosage adds to the complexity of the UF or MF system.

A possible alternative is magnetic ion exchange (MIEX). MIEX was developed in Australia specifically for the removal of DOC from drinking water sources. This resin is a microsize, macroporous strong base, magnetic ion exchange resin, developed for the removal of negatively charged organic ions. In Australia, trials showed that MIEX resin, used in a continuous ion exchange process, was highly effective at removing low and medium MW NOM and could achieve greater removals of NOM than enhanced coagulation. MIEX preferentially removes the low-MW fraction that is not removed by inorganic coagulants even at very high concentrations (Slunjski et al., 2000). When used as a pretreatment step, up to 80% of raw water NOM can be removed prior to MF or UF, resulting in removals of DBP precursors. A full-scale plant utilizing this combined process was constructed in South Australia (Bourke et al., 2000). Bourke et al. (2000) further demonstrated that the fouling rate was decreased by MIEX pretreatment and there was no evidence of irreversible fouling.

Drikas et al. (2003) compared a number of DOC removal methods: alum coagulation (without pH control), alum coagulation (with pH controlled at 6), ion exchange using MIEX resin, and combined treatment of alum coagulation and MIEX. They found that each pretreatment removed different fractions of NOM. The relative effectiveness of each pretreatment for DOC removal was in the order alum/MIEX > MIEX > alum pH 6 > alum (no pH control). For both waters, the DOC concentration remaining after alum coagulation at pH 6 was approximately double that remaining after combined alum/MIEX and MIEX alone, which suggested that a significant DOC removal was achieved by including MIEX in the treatment process.

The emergence of immersed membrane technology has made it possible to utilize plant infrastructure, keeping capital costs down. Water treatment plants can be retrofitted with an MIEX process and immersed membranes into the existing infrastructure (Fig. 6.11).

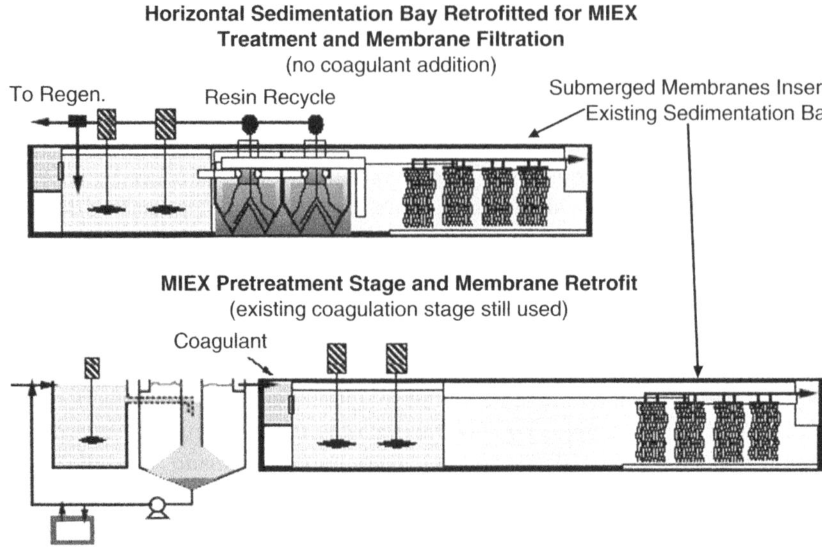

Figure 6.11 Potential MIEX/membrane configurations. (Adapted from Kruithof et al., 2004.)

6.3.8 Ozonation

Ozone can be used as a pretreatment to UF or MF. Ozone breaks up large organic molecules into smaller ones. Park (2002) studied the effect of ozonation on permeate flux using a polysulfone UF membrane. It was found that the ozonation increased the flux by 10%, and oxidation by ozone and hydrogen peroxide was even more effective. They indicated that ozone prolonged the period required to reach appreciable fouling rather than eliminated it.

6.4 MEMBRANE APPLICATIONS

6.4.1 Product Water Quality

6.4.1.1 Disinfection and Removal of Pathogenic Microorganisms Microfiltration is a partial removal barrier for pathogenic microorganisms, whereas UF is an absolute barrier to microorganisms, including viruses. MF will remove particles with sizes down to 0.1 μm, depending on the pore size. Theoretically, this process can remove bacteria (sizes 1–10 μm), but viruses, which are much smaller (sizes 0.01–0.1 μm), can pass. These results were confirmed by particle counting. Particles above 2 μm were ca. 98% removed, and up to 99.8% removal was achieved for 5-μm particles. Vickers et al. (1993) found that MF reduced turbidity to 0.5 NTU while Coffey et al. (1993) reported 0.24 NTU. In the Coffey et al. study, heterotrophic bacteria were 64–99% removed. They attributed the low removal efficiency to regrowth. *Escherichia coli* and *Giardia* cysts were removed to a large extent (>6 log and >4.4 log, respectively). In the Vickers et al. study, coliforms and *Giardia* cysts were removed to below the detection limit. MS2 bacteriophages could pass through the UF membrane to a certain extent with removal ranging between 97.8 and 99.9%.

Ultrafiltration membranes have pore sizes in the range 0.01–0.1 μm, much smaller than those of MF. UF membranes are capable of removing cysts [*Cryptosporidium* oocysts (4–6 μm) and *Giardia* cysts (8–14 μm)], bacteria, and viruses. Jacangelo et al. (1989) continuously obtained turbidities below 0.04 NTU, and no coliforms could be detected in the permeate of the UF membranes used. Like in other studies heterotrophic bacteria were detected, probably caused by regrowth. Jacangelo et al. (1991) spiked UF feed water with *Giardia muris* and MS2 phages and obtained log reductions of more than 4 and 6, respectively. Again heterotrophic bacteria were detected in the permeate. However, there is no guarantee of membrane integrity, and at least one extra disinfection barrier is required. Laine et al. (1993) indicated very low removal efficiencies for DBP precursors due to the high MWCO of the UF membranes. Other dissolved substances (e.g., inorganic salts, small organic molecules, etc.) can pass the membranes as well.

6.4.2 Integrity Testing

Disinfection capacity depends also on the integrity of the membrane element and the pressure vessels. Because of this, assessment of the membrane integrity is very important at the beginning and throughout the operation of MF and UF systems (Kruithof et al., 2004). The fibers must be intact and seals correctly seated to provide an impermeable barrier between influent water and permeate. Testing of the integrity of the system may then become the method of evaluating membrane performance. Membrane integrity testing can be classified as direct or indirect.

6.4.2.1 Direct Integrity Tests Direct integrity tests directly measure a breach in a membrane or membrane system. They monitor gas passing through a breach, filtrate flows, pressure changes, or sound that can be detected.

Pressure Decay Test (PDT)—Filled This test involves applying pressurized air to the feed side at a predetermined level below the bubble point and then isolating. When the pores of a membrane are filled with liquid and air pressure is applied to one side of the membrane, surface tension prevents the liquid in the pores from being blown out by the air pressure below a certain minimum pressure known as the bubble point. The predetermined pressure directly relates to the size of defect under investigation [Drinking Water Inspectorate (DWI), 2001].

The air pressure is maintained for a period of time (2–10 min) to observe the rate of decay. A small decrease of 0.1–0.2 psi/min is considered acceptable and is due to diffusion of the air across the microporous membrane structure (DWI, 2001). A faster decrease in pressure indicates a faulty membrane. As the membrane system is open to atmospheric pressure on the filtrate side, the airflow can be observed to confirm the location of any breach (defects, holes, or tears in the membrane). The system can be automated, but faults will only be detected in a membrane unit. If the unit is comprised of several membrane housings (larger systems may contain in excess of 100 membrane modules), additional diagnoses will be necessary to identify the failed membrane fibers.

Pressure Decay Test—Drained This is very similar to the PDT–filled, but in this case the module is drained on both sides of the element so that the membrane is only wetted.

Airflow Technique An alternative method, the airflow technique, measures the flow of air passing through the membrane surface. Several variations of this test method are available from the various membrane suppliers, and the test is usually automated. This test may involve applying a constant feed-side gas pressure below the bubble point of the selected size hole. The operator then measures the diffused gas filtrate flow or the displaced water from the membrane.

Advantages of this technique are that it can prove to be more sensitive than PDT, and the risk of damage to the membranes is reduced because lower air pressures are typically used. Depending on how the test system is engineered, it is possible to isolate cartridges and test each membrane or series of membranes, thus speeding the locating of failed fibers in a membrane or compromised seals. Further testing is necessary to identify broken fibers (Fig. 6.12). The test is conducted with the membrane unit off-line.

Acoustic Sensor This test is based on hydrophonic sensor technology; the acoustic monitoring technique consists of measuring the noise due to a compromised membrane. Its advantage is that integrity of the membranes is monitored continuously during filtration. The acoustic sensor is able to detect a noncut compromised fiber and, therefore, guarantees more than 6 log removal of viruses (DWI, 2001).

Audible Test—Stethoscope In this test, an operator employs a stethoscope to manually listen to working modules to identify any sound produced by a breach in membrane integrity.

6.4.2.2 Indirect Integrity Tests These tests measure the results of a breached membrane system. Instruments monitor water quality by measuring inherent particles, introduced particles, or turbidity.

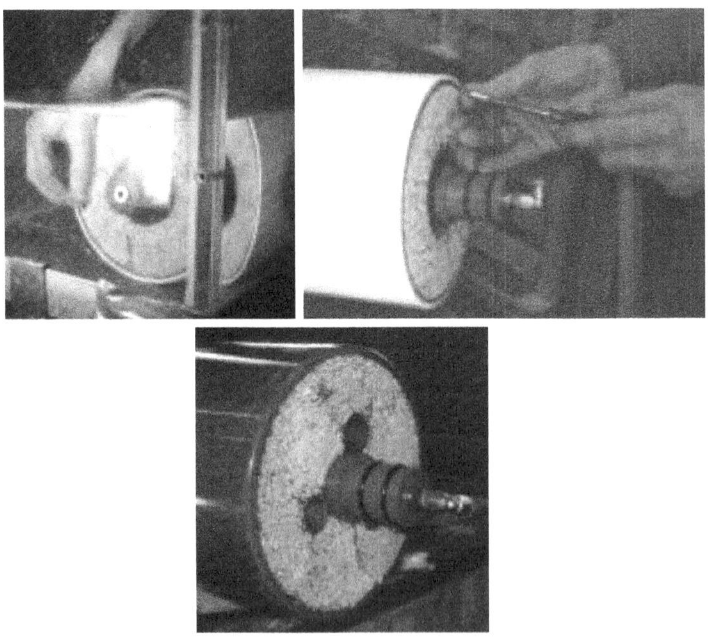

Figure 6.12 Integrity test: identifying and plugging of broken fibers.

Turbidity Monitoring The turbidity of feed water and filtrate is monitored. An intact membrane would be expected to show a 90% reduction of turbidity from feed to filtrate (DWI, 2001).

Particle Challenge Testing Particle challenge testing can be conducted in three ways as follows:

1. Measuring particles (in the permeate) of a given size that are naturally occurring in the source water.
2. Introducing inert particles into the membrane feed water, such as PAC or latex spheres.
3. Introducing indicator organisms into the membrane feed water such as MS2 phage.

Particle counting can seem attractive as monitoring can be conducted on-line. However, there are drawbacks associated with this method. First, a sufficient number of particles must be present in the feed water to demonstrate a high rejection. This may not be problematic for surface-derived waters, but some groundwater supplies contain far less particles, and a log rejection greater than 2 may not be achievable. Second, the instruments used for particle counting are susceptible to air bubbles and false high readings that may create unnecessary nuisance alarms. Such instruments are sensitive and require regular maintenance such as de-scaling and recalibration. Cost can also be a barrier if sub-1-μm particles need to be monitored, as such instruments may cost $100,000 or more. Third, only rejection of the particles measured can be proved. If the instrument measures only to 2 μm, the removal of viruses and coliforms cannot be demonstrated as these organisms are much smaller than 2 μm.

Spiked Integrity Monitoring System This test doses high concentrations of PAC into the feed and measures the particulate levels in the permeate. A log removal can then be calculated. Introducing particles into the water enables demonstration of a higher level of rejection. PAC is available in a tight particle size distribution around 1 μm, but for rejection of particles only the same size or greater than the PAC can be demonstrated. Viruses, for example, will generally be smaller than PAC, so this technique will not be suitable for demonstration of virus removal. Tests with PAC can, in theory, be conducted on-line because of the small risk of passage of PAC into the finished water.

MS2 Bacteriophage and Oocyst Challenge Test This test doses high concentrations of microorganisms into the feed and measures the particulate levels in the product. This is basically a research tool and permits the calculation of log removal values for the specific organism (DWI, 2001).

6.5 MEMBRANE FOULING AND CLEANING

Over long periods of operation, membrane fouling is generally not totally reversible by the hydraulic backwash procedure. As the number of filtration cycles increases, the irreversible fraction of membrane fouling also increases (see Fig. 6.13). In order to obtain the desired production flow rates (or flux), an increase in TMP is required. When this pressure reaches a maximum allowable mechanical resistance of the membrane, chemical cleaning is required for the membrane to regain most of its permeability. Regardless of the membrane system

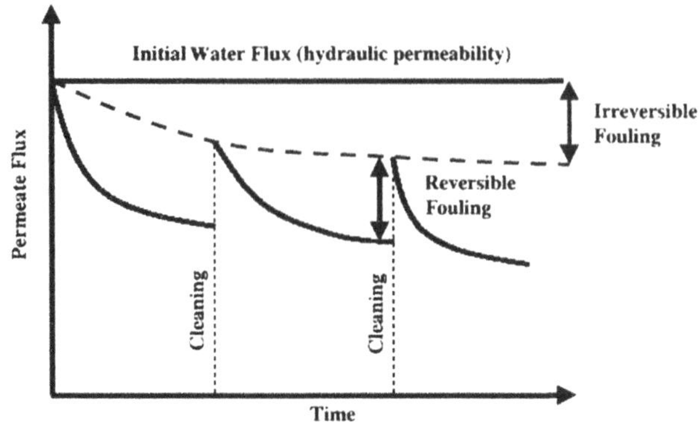

Figure 6.13 Illustration of reversible and irreversible permeate flux decline in membrane UF. (Adapted from Faibish and Cohen, 2001.)

used, chemical cleaning is typically cumbersome and requires shutdown of the unit for several hours. This results in a reduction of the overall plant capacity, and produces a waste that may be difficult to dispose of. There are also concerns that repeated chemical cleaning might affect the membrane life. Chemical cleaning should thus be limited to a minimum or avoided. This can be done through optimization of the operating conditions (pressure, flux, and recovery) and backwash conditions (pressure, flux, frequency, and duration), improving the efficiency of chemical cleaning so as to reduce the number of cleanings, pretreatment, and so forth.

6.5.1 Characterization of Feed Water

According to Laine et al. (2003), several water quality parameters can be used to assess and predict the fouling potential of a natural water: (1) turbidity of the particle content, (2) total organic carbon (TOC) for the total NOM concentration, and (3) the ultraviolet (UV) absorbance at 254 nm to characterize the aromaticity of the NOM as well as the polyhydroxyaromatic (PHA) content. In addition to these parameters, other researchers have found other approaches to indicate the NOM fouling potential of a water such as specific UVA_{254} (SUVA), permanganate ($KMnO_4$) value, assimilable organic carbon (AOC), biofilm formation rate (BFR), color, liquid chromatography–organic carbon detection (LC–OCD), high–performance size exclusion chromatography (HPSEC) values, colloid size, humic content, and properties such as SEC-UV absorbance ratio index (SECURI), and SEC fluorescence (Anselme et al., 1994; Matilainen et al., 2002; Kabsch-Korbutowicz, 2005; Edzwald and van Benschoten, 1990; Vrouwenvelder and van der Kooij, 2003; Cho and Amy, 1999; Lahoussine et al., 1990; Fan et al., 2001; Schäfer et al., 2000). However, none of the above water quality parameters or analytical techniques were found sufficient to fully characterize or predict fouling during membrane operation.

6.5.2 Fouling

Fouling refers to the blockage of membrane pores during filtration caused by the combination of sieving and adsorption of particulates and compounds onto the membrane surface or within the membrane pores. This blockage of the pores causes a flux decline over time when all operating parameters such as pressure, flow rate, temperature, and feed concentration are kept constant. Fouling is the most important issue affecting the development of membrane filtration—as it worsens membrane performance and shortens membrane life (Boerlage, 2001).

6.5.2.1 *Types of Fouling* Fouling can be broadly classified into (non-) backwashable and ir(reversible). Backwashable fouling can be removed by backwashing (reversing the direction of permeate flow through the pores of the membrane) at the end of each filtration cycle. Nonbackwashable fouling is that fouling that cannot be removed by normal hydraulic backwashing in between filtration cycles. In nonbackwashable fouling, the membrane can be returned to its original flux by other means (e.g., chemical cleaning). Irreversible fouling is that kind of fouling that cannot be removed with flushing, backwashing, chemical cleaning, or any other means, and the membrane cannot be restored to its original flux. Figure 6.14 summarizes the above explanation.

Inorganic Fouling/Scaling Fouling can also be classified according to the type of fouling material. Four categories of membrane fouling are generally recognized. They are: (a) inorganic fouling/scaling, (b) particle/colloidal fouling, (c) microbial/biological fouling, and (d) organic fouling.

Inorganic fouling or scaling is caused by the accumulation of inorganic precipitates, such as metal hydroxides, and "scales" on membrane surface or within pore structure. Precipitates are formed when the concentration of these chemical species exceeds their saturation concentrations. Limiting salts can be identified from solubility products of potential limiting salts in the raw feed water. Since ionic strength increases on the feed side of the membrane, the effect of ionic strength upon the solubility products should be considered. Some limiting salts can be controlled by the addition of acid or scale inhibitor or both to the feed water. Typical sparingly soluble salts that may limit recovery in pressure-driven membranes include, but are not limited to, $CaCO_3$, $CaSO_4$, $Ca_3(PO_4)_2$, $BaSO_4$, $SrSO_4$, CaF_2, and SiO_2 [U.S. Environmental Protection Agency (EPA), 2002].

Scaling is a major concern for reverse osmosis (RO) and nanofiltration (NF) since these membranes reject inorganic species. Those species form a concentrated layer in the vicinity of membrane–liquid interface—a phenomenon referred to as *concentration polarization*.

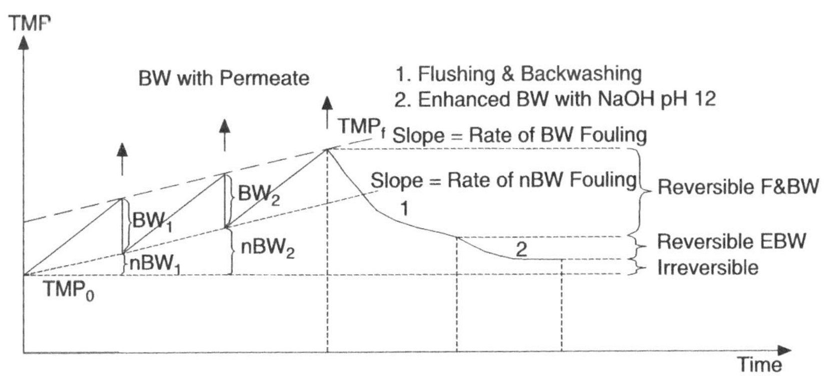

Figure 6.14 Backwashable, nonbackwashable, reversible and irreversible fouling. Bw_i = backwashable fouling after the *i*th filtration cycle and nBw_i = nonbackwashable fouling after the *i*th filtration cycle. (Adapted from Peavy et al., 1984.)

For microfiltration (MF) and ultrafiltration (UF), inorganic fouling due to concentration polarization is much less profound but can exist most likely due to interactions between ions and other fouling materials (i.e., organic polymers) via chemical bonding. Some pretreatment processes for membrane filtration such as coagulation and oxidation, if not designed or operated properly, may introduce metal hydroxides on membrane surface or within pore structure. Also cleaning methods such as enhanced backwash (EBW) could cause scaling in UF, if performed incorrectly. For example, enhanced backwash at a high pH with water initially containing Ca^{2+} and high HCO_3^- can result in $CaCO_3$ precipitation on the permeate side of the membrane during the cleaning cycle.

Particulate/Colloidal Fouling Algae, bacteria, and some natural organic matter fall into the size range of particles and colloids. However, they are different from inert particles and colloids such as silts and clays. To distinguish the different fouling phenomena, particles and colloids here are referred to as biologically inert particles and colloids that are inorganic in nature and originate from weathering of rocks. In most cases, particles and colloids do not really foul the membrane because the flux decline caused by their accumulation on the membrane surface is largely reversible by hydraulic cleaning measures such as backwash and air scrubbing. These particles form a cake, which eventually may be compressed and reduces the flow through the membrane. Initially, cake formation does not significantly reduce productivity. However, after cake compression, the productivity decreases and the compressed cake must be removed. MF and UF membranes can be backwashed to remove this cake. Also cross-flow filtration can be used to control colloid fouling (EPA, 2002).

Microbial/Biological Fouling/Biofouling Microbial fouling is a result of formation of biofilms on membrane surfaces. Such films (bacterial, algal, or fungal) grow and release biopolymers (polysaccharides, proteins, and amino sugars) as a result of microbial activity. For example, once bacteria attach to the membrane, they start to multiply and produce extracellular polymetric substances (EPS) to form a viscous, slimy, hydrated gel. EPS typically consists of heteropolysaccharides and have high negative charge density. This gel structure protects bacterial cells from hydraulic shearing and from chemical attacks of biocides such as chlorine. Severity of microbial fouling is greatly related to the characteristics of the feed water (Boerlage, 2001).

Organic Fouling Several studies have shown that NOM is a major UF membrane foulant, and that different components of NOM cause different forms of fouling (Lee et al., 2003; Aoustin et al., 2001; Makdissy et al., 2003; Jucker and Clark, 1994). Other studies have shown that increased concentration of Ca^{2+} and Mg^{2+} caused more fouling (Hong and Elimelech, 1997; Yangali-Quintanilla, 2005; Lee et al., 2005). Based on the analysis of the extracted solution during chemical cleaning, Mo and Huang, (2003) found that the majority of soluble organic foulants were of low molecular weight, and calcium was the major inorganic element in the foulants. These studies did not differentiate between reversible and irreversible fouling. Kimura et al. (2004), however, differentiated between the fouling and concluded that neither aluminum nor calcium contributed to irreversible fouling.

Although several researches have tried to identify the components of NOM that cause fouling of UF membranes, the results are confusing and conflicting. Weisner et al. (1992) identified four NOM categories that are strong foulants: proteins, amino sugars, polysaccharides, and polyoxyaromatics. Makdissy et al. (2003) partially agreed and also showed that the organic colloidal fraction (bacterial peptidoglycan—bacterial cell wall residue rich in sugars and amino sugars) caused the most significant fouling. However, Cho and Amy (1999) and Kimura et al. (2004) identified polysaccharides as dominant foulants. Speth et al. (1996) revealed that the NOM components of conventionally treated feed water fouled membranes in the order of polyhydroxyaromatics, proteins, polysaccharides, and amino sugars.

Whereas Jucker and Clark (1994) concluded that humic acids (the largest and most hydrophobic fraction of the humic material) caused irreversible fouling while fulvic acids caused less fouling than humic acids, later on Aoustin et al. (2003) found that fulvic acids were responsible for reversible fouling of UFs via concentration polarization. Schafer (2001) found that organic substances have a tendency to clog the membrane, and organic carbon generally concentrates on the internal surface of the membrane.

In other studies, Nilson and DiGiano (1996) reported most fouling was caused by hydrophobic NOM components. However, Carroll et al. (2000) found neutral hydrophilic NOM components were the major foulants, while Cho et al. (2000) and Fan et al. (2001) reported NOM components as the major foulants on the order neutral hydrophilics > hydrophobic acids > transphilic acids > charged hydrophilics. Due to conflicting opinions, further research on NOM in UF/MF is required.

6.5.3 Fouling Indices

The fouling potential can be estimated using fouling indices. The most commonly used fouling indices include silt density index (SDI), modified fouling index (MFI), miniplugging factor index (MPFI), and MFI-ultrafiltration (MFI-UF). Fouling indices are simple measurements that provide an estimate of the fouling potential of a membrane feed water. They are used to qualitatively estimate the pretreatment requirements and possibly predict membrane fouling (EPA, 2002).

6.5.3.1 Silt Density Index The SDI is a commonly used test to predict a feed water's potential to foul a membrane by colloidal particles smaller than 0.45 μm. It is only a guide for pretreatment and is not an indication of adequate pretreatment. The SDI is a static measurement of resistance, which is determined by samples taken at the beginning and at the end of the test (EPA, 2002).

The SDI test is performed by timing the flow through a 47 mm diameter, 0.45 μm membrane filter at a constant pressure of 30 psi. The time required for 500 mL of the feed water to pass through the filter is measured when the test is first initiated and is also measured at time intervals of 5, 10, and 15 min after the start of the test.

According to ASTM D-4189-82, the value of SDI can be calculated from

$$\mathrm{SDI} = \frac{1 - \frac{t_i}{t_f}}{T} \tag{6.5}$$

where t_i = time to collect initial 500-mL sample

t_f = time to collect 500-mL sample at time $t = T$

T = total running time of the test; 5, 10, or 15 min

If the SDI is below 3, the impact of colloidal fouling is minimized (EPA, 2002). It should be noted that even when the guidelines are not compromised serious fouling may occur. The reasons could be that a different type of fouling (e.g., biofouling) occurred, or more fundamentally that SDI is not based on any filtration mechanism and has no linear correlation with colloidal matter concentration (Schippers, 2005).

6.5.3.2 Modified Fouling Index ($MFI_{0.45}$) and MFI-UF The $MFI_{0.45\mu m}$ is determined using the same equipment and procedure used for the SDI, except that the volume is recorded every 30 s over a 15-min filtration period (Schippers and Verdouw, 1980). Unlike SDI, MFI is based on cake filtration mechanisms, has a linear relation with colloidal matter concentration, and can utilize membranes of different sizes (Schippers, 2005).

However, MFI depends strongly on pore size of the membrane, particle diameter, and particle concentration. MFI increases with decreasing particle and pore diameters and increasing particle concentration (Schippers, 2005).

The development of the MFI is consistent with Darcy's law in that the thickness of the cake layer formed on the membrane surface is assumed to be directly proportional to the filtrate volume. The total resistance is the sum of the filter and cake resistance.

The MFI is defined graphically (for constant pressure feed) as the slope of an inverse flow (t/v) versus cumulative volume (v) as shown in Eq. (6.6) (EPA, 2002):

$$\frac{t}{v} = \frac{\mu V R_f}{\Delta P A} + \frac{\mu V^2 I}{2\Delta P A^2} \qquad \frac{1}{Q} = (a + \text{MFI})V \tag{6.6}$$

where R_f = resistance of a clean filter
a = constant
Q = average flow (L/s)
I = measure of the fouling potential
A = area of filter
ΔP = transmembrane pressure
V = cumulative volume passing through the membrane

For constant flux feed, the fouling index I can be determined from the slope of the linear region in a plot of transmembrane pressure (ΔP) versus time using Eq. (6.7). The MFI can then be calculated from Eq. (6.8):

$$\Delta P = J\eta R_m + J^2 \eta I t \tag{6.7}$$

$$\text{MFI} = \frac{\eta_{20^\circ\text{C}} I}{2\Delta P_o A_o^2} \tag{6.8}$$

where R_m = resistance of a clean filter
η = viscosity
$\eta_{20^\circ\text{C}}$ = viscosity at 20°C
J = permeate water flux (m^3/m^2 s)
ΔP_o = reference applied TMP (2 bars)

A_o = reference surface area of a 0.45-μm membrane filter (13.8×10^{-4} m^2)
t = time (s)
I = fouling index

To incorporate the smaller particles in the MFI, Boerlage (2001) examined the use of ultrafiltration membranes for measuring the MFI using ultrafiltration membranes (MFI-UF). MFI-UF can be used to measure and monitor the fouling potential of a feed water. It is according to the cake filtration theory calculated from the ratio of filtration time and filtration volume as a function of the filtration volume. MFI-UF can be calculated from both experiments with constant transmembrane pressure (TMP) and experiments with constant flux. Differences in fouling potential can be measured for various feed waters using the same membrane type and for various membrane types using the same feed water (Roorda and van der Graaf, 2001).

6.5.4 Microfiltration/Ultrafiltration Membrane Cleaning

Membrane chemical cleaning is an integral part of the operation for MF and UF systems in the water industry and has significant impact on process operations. However, membrane fouling is poorly understood and related to site-specific water quality issues. Fouling is very much needed to advance on knowledge of membrane cleaning.

Certain fouling materials can be removed by hydraulic means such as backwashing; most can be removed by chemical means such as enhanced backwash (EBW), cleaning-in-place (CIP), or off-line chemical cleaning (or soaking). Backwashing involves reversing the flow across the membrane, using product water (permeate) to remove the foulants accumulated at the membrane surface and clogging the membrane pores. EBW involves adding a cleaning chemical to the backwash water and recirculation for a short period of time (10–15 min). In CIP the membrane module is cleaned without removing it from the installation, while in off-line cleaning the module is removed from the system and soaked in a chemical. Chemical cleaning is an integral part of membrane process operation that has a profound impact on the performance and economics of membrane processes. Currently, practices of chemical cleaning are mostly based on recommendations from membrane manufacturers. Some of them supply proprietary cleaners while others use commercial chemicals.

The effects of various operating strategies against different types of fouling are summarized in Table 6.4. As indicated in Table 6.4, chemical cleaning is an effective control strategy for a majority of membrane fouling.

TABLE 6.4 Effects of Operating Strategies on Membrane Fouling[a]

	Effects of Operating Strategy			
Type of Fouling	Hydraulic Cleaning/ Backwashing	Feed Chlorination	Feed Acidification	Chemical Cleaning
Inorganic	–	–	++	++
Particulate	++	–	–	++
Microbial	+	++	+*	++
Organic	–	+	–	++

[a]'–', No effects/negative effects; '+', some positive effects; '++', Positive effects; '*', together with feed chlorination.
Source: Adapted from Pall (2006).

6.5.4.1 Backwashing One of the methods of removing foulants in cross-flow and dead-end UF/MF systems is to disturb the onset of the mass transfer boundary layer near the membrane surface. Several techniques can be used. Increasing the cross-flow velocity is the simplest but results in an increase in energy consumption in cross-flow system. Other methods such as pulsative and reversed flow/backwash aim at creating flow instabilities (Finnigan and Howell, 1990).

The disturbance of the mass transfer boundary layer in dead-end operated UF and MF systems occurs by a periodic backwash. Hydraulic cleaning by backwash is performed by reversing the flow of permeate from the permeate storage tank into the membrane. It is generally effective in removing particle cakes from the membrane surface, and it can also remove foulants from the membrane interior, particularly when performed with a chemical cleaning solution (enhanced backwash). However, due to backwashing, the overall system recovery is reduced as permeate is used. Therefore, backwashing is often combined with a cross flush (or forward flush). A backwash lifts the accumulated material from the membrane surface, while a cross flush transports material out of the membrane module. The efficiency of backwashing is dependent on the frequency, duration, flux, applied pressure, and type of fouling.

The ideal situation regarding backwash flux and frequency is to use as high a flux, and as frequently as possible. However, such a practice would result in a very low net flux. It is therefore desirable to optimize the backwash flux and frequency. The frequency and duration of the backwashes are adjusted as a function of the raw water quality, and typically both are increased when the raw water solids load increases (Crozes et al., 1997).

Pressure, Flux, Frequency, and Duration of Backwash Backwashing conditions such as pressure, flux, frequency, and duration, in practice, are usually obtained by trial and error. Kennedy et al. (1998) studied the backwash conditions in order to maximize the net flux per filtration cycle. They found that increasing the backwash to filtration pressure ratio (P_b/P_f) above 2.5 did not result in any significant increase in flux restoration within the range of backwash pressures tested (0.2–1.6 bars) (Fig. 6.15). Their conclusion that applying high backwash pressures (8 times the filtration pressure) cannot restore irreversible flux decline was in agreement with membrane suppliers' recommendations.

Increasing the backwash duration from 0.5 to 1 min had almost negligible effect on flux restoration. However, when the backwash duration was increased to 2 min, a significant effect on flux restoration was observed for all backwash pressures tested. Increasing the backwash pressure increases the water consumption, which reduces the system recovery. It is, therefore, important to optimize the backwash pressure and duration to achieve as high a degree of flux restoration and system recovery as possible.

Net flux serves as a tool to determine the optimal backwashing conditions since it reflects the operation efficiency by taking into account water production and consumption, including the time spent for backwashing. Kennedy et al. (1998) observed that, for all backwash pressures and durations tested, the net flux increased up to a pressure ratio of 2.5, then decreased, even though flux restoration increased (Fig. 6.15, right). The reason is that above a pressure ratio of 2.5, the flux decline that was recovered by backwashing was small compared to the quantity of water consumed to realize the flux restoration.

Crozes et al. (1997) demonstrated a decrease in backwash effectiveness as the TMP applied in the previous filtration cycle increased. Backwash efficiency in terms of

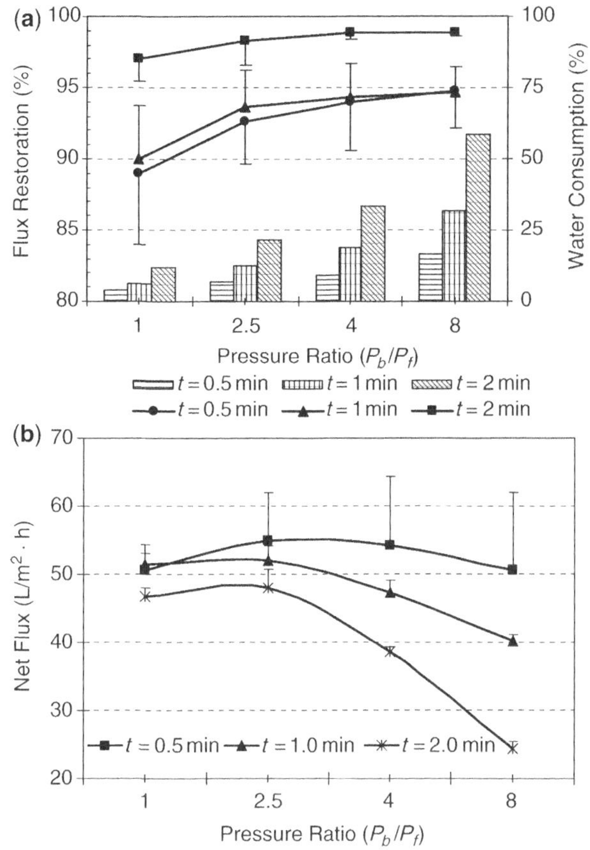

Figure 6.15 (a) Effect of backwash pressure on flux restoration and water consumption, and (b) effect of backwash pressure on net flux (*t* = *backwash duration*, P_b/P_f = *backwash : feed pressure ratio*). (Adapted from Kennedy et al., 1998.)

membrane flux recovery after hydraulic backwash was reduced by 50% when the TMP was increased from 0.4 to 1.4 bars.

Nakatsuka and Ase (1995) found backwash most effective if the backwash pressure is more than double the operating pressure. An increase in cross-flow velocity also led to higher flux. Hagmeyer et al. (1996) optimized the backwash interval to 30 s every 30 min. Efficiency could be further increased with the duration and frequency of backwash but at the expense of recovery.

Sanjeev and Davis (2004) designed a high-frequency reverse-filtration strategy to maximize the flux for washed yeast suspensions through 0.2-μm cellulose acetate flat-sheet membranes. Several experiments were conducted with reverse-filtration times ranging 1–40 s. They found that for every back-filtration time, there exists an optimum forward-filtration time that gives the maximum global average flux. The optimum average flux increases with decreasing back filtration and feed concentrations. However, the optimum average flux shows little dependence on cross-flow velocity and reverse-filtration transmembrane pressure. The optimum flux with back flushing was 20–30 times higher than the long-term flux in the absence of back flushing.

Role of Air in MF/UF Cleaning In MF, gas (air, nitrogen, etc.) can be injected in the permeate side of the membrane. By periodically pulsing the lumen with gas and backwashing with a gas–permeate mixture, the flux decline can be controlled. The gas pulse expands the fibers and opens the pores, allowing fouling material to be flushed out. However, this technique can only be used when the membrane has a low bubble point and an easy rewettability (Fane and Fell, 1987).

Verberk (2005) showed that combined water and airflow resulted in a much more intensive cleaning than a forward flush performed at a water velocity equal to the sum of the individual air and water velocities. Optimal values for the water and air velocities obtained in the study were 0.2 and 0.3 m/s, respectively. Equal distribution of air and water over the cross-sectional area of a membrane module is important to have the same cleaning conditions in every membrane fiber in a module. Verberk (2005) found that water was well distributed over the cross section of the module when a forward flush was performed. When a water–air flush was performed, there was an even distribution of water and air in the capillary membrane module membranes. However, this was not the case for the tubular modules, as stagnant water levels were observed. Intermittent application of water–air flush was recommended as a probable solution to this maldistribution problem. Chemical consumption is likely to be low, and stable operation was achieved by making use of the AirFlush cleaning method (Verberk, 2005).

In cross-flow UF or MF, air can be added continuously to the feed stream, resulting in a stable cake layer of a certain thickness and porosity (Cabassud et al., 1990; Laborie et al., 1998). In dead-end UF (or MF) the air is dosed intermittently into the forward flush (only when a cleaning is performed). Guigui et al. (2003) also showed that using an air–fluid mixture resulted into an increase in the quantity of particles removed from hollow fibers. They concluded that using air–fluid mixing in the backwash sequence prevented long-term fouling and reduced the rate of fouling during the filtration period.

6.5.4.2 *Enhanced Backwash* In an enhanced backwash, a low dose of oxidant/disinfectant is automatically injected into the permeate during a backwash (with permeate) in order to enhance the efficiency of cleaning. While a normal backwash (with permeate) is very short (ca. 15–30 s) and is performed automatically (every 15–45 min), an enhanced backwash usually takes long (i.e., approx 10–15 min), and the frequency is usually every 4–6 h.

The enhanced backwash comprises three steps: first, a backwash with permeate (approx 30 s) to remove accumulated particles from the hollow fiber. Second, a short soak (10–15 min) with a low dose of oxidant/disinfectant to remove adsorbed foulants from the membrane, and finally another short backwash (with permeate) to remove the chemicals from the systems.

6.5.4.3 *Chemical Cleaning* Chemical cleaning is performed when flushing and/or backwashing cannot restore the flux. In a chemical cleaning, the chemical dose is usual higher than during the enhanced backwash, and the frequency of chemical cleaning is usual lower (approximately 1 per week). Moreover, the enhanced backwash can be fully automated, but the chemical cleaning involves labor.

Proper selection of chemical cleaning agents, conditions for their application, and understanding their performance are important. The chemicals used depend on the foulant in question and the resistance of the membrane to the cleaning agent. Once the cause of membrane fouling is identified, various cleaning chemicals can be used to remove the fouling materials and restore the membrane flux. Chemicals commonly used for

TABLE 6.5 Major Categories of Membrane Cleaning Chemicals

Category	Major Functions	Typical Chemicals
Caustic	Hydrolysis, solubilization	NaOH
Oxidants/ disinfectants	Oxidation, disinfection	NaOCl, H_2O_2, peroxyacetic acid
Acids	Solubilization	Citric, nitric, hydrochloric acid
Chelating agents	Chelation	Citric acid, EDTA
Surfactants	Emulsifying, dispersion, surface conditioning	Surfactants, detergents

Source: Adapted from Pall (2006).

cleaning MF and UF membranes in the water industry fall into five categories, as summarized in Table 6.5.

Caustic Caustic is often used to clean membranes fouled by organic and microbial foulants. The function of caustic is twofold: (1) hydrolysis and (2) solubilization. Organic materials, including polysaccharides and proteins, can be hydrolyzed by caustic. Fats and oils also react with caustic through saponification, generating water-soluble soap micelles (Pall, 2006).

A very important function of caustic is to increase the negative charge of humic substances. Therefore, they become easier to remove from membranes. Phenolic groups are typically associated with the most hydrophilic portion of NOM and likely to have strong adhesion to membranes. Hydrophilizing this portion of organic matter undoubtedly weakens the bond between membrane and those fouling materials. In addition, the molecules of NOM are likely to have a stretched, linear configuration due to the repulsion between negatively charged functional groups (Hong and Elimelech, 1997). This change in molecule configuration creates a loose fouling layer and allows easier access of chemicals to penetrate the inner portion of the fouling layer, therefore, facilitating mass transfer, and enhancing the efficiency of cleaning. Maartens et al. 1998 determined that a combination of caustic solutions with detergents were effective in removing organic foulants in UF.

Oxidants The most common oxidants used for membrane cleaning include chlorine (sodium hypochlorite) and hydrogen peroxide. The oxidation of organic polymers generates more oxygen-containing functional groups such as ketones, aldehydes, and carboxylic acids. The existence of these functional groups generally increases hydrophilicity of their parent compounds. The increase in abundance of carboxylic groups on aromatic rings increases negative charges of natural organic matter at alkaline pH conditions, due to the dissociation of these acids (Pall, 2006). Therefore, oxidation reduces the adhesion of fouling materials to membranes. Oxidants are often mixed with caustic to form a cleaning "cocktail" to, first, enhance cleaning efficiency. The mixture provides a synergy for NOM-dominated fouling because the fouling layer tends to have more open structure in caustic conditions. This synergy allows chlorine to reach the inner layer of foulant materials, facilitating mass transfer and reactions between chlorine and foulants, and enhancing the cleaning efficiency.

Acids and Chelating Agents Acids are used primarily to remove scales and metal dioxides from fouling layers. When a membrane is fouled by iron oxides, citric acid is very effective because it not only dissolves iron oxides precipitates, but also complexes with iron (Pall, 2006). In addition, some of the organic compounds such as polysaccharides and proteins also hydrolyze at low pH. If there is coexistence of divalent cations (e.g., Ca^{2+}) and natural organic matter, *salt bridges* can be formed between the divalent cation and NOM, causing a denser and more adhesive fouling layer. The removal of divalent cations by either acids or chelating reagent such as EDTA can also improve the cleaning of membranes fouled by organic foulants (Hong and Elimelech, 1997). Speth et al. (1996) found acid cleaning very effective in removing inorganic foulants.

Surfactants Surfactants are compounds that have both hydrophilic and hydrophobic structures. They can form micelles with fat, oil, and proteins in water and help to clean the membranes fouled by these materials. Some surfactants may also interfere with hydrophobic interactions between bacteria and membranes. In addition, surfactants can disrupt functions of bacteria cell walls. They therefore affect fouling dominated by the formation of biofilms (Pall, 2006).

Kimura et al. (2004) cleaned fouled membranes with different individual chemicals and also consecutively used two different chemicals. For the individual chemicals, they found NaClO (500 ppm) as the best cleaner followed by NaOH (pH 12). Oxalic acid (pH 2)/NaClO (500 ppm) was found to be a better cleaning combination than HCl (pH 2)/NaClO (500 ppm) and NaOH (pH 12)/NaClO (500 ppm). The combinations were better cleaners than the individual chemicals—see Figure 6.16.

Factors Affecting Cleaning Efficiency During Chemical Cleaning Because membrane cleaning is essentially conducted through chemical reactions between cleaning chemicals and fouling materials, factors that affect the mass transfer and chemical reactions such as concentration, temperature, length of the cleaning period, and hydrodynamic conditions all affect cleaning efficiency (Pall, 2006). The *concentration* of cleaning chemicals can affect both the equilibrium and the rate of the reaction. The concentration of cleaning chemicals not only needs to maintain a reasonable reaction rate (kinetics need), but also needs to overcome mass transfer barriers imposed by the fouling layer. In practice, the concentrations of cleaning chemicals are usually high enough to satisfy the kinetic need. It is mass transfer that sets the lower boundary for the concentration of cleaning chemicals (Pall, 2006).

Temperature can affect membrane cleaning by (1) changing the equilibrium of a chemical reaction, (2) changing the reaction kinetics, and (3) changing the solubility of fouling materials and/or reaction products during the cleaning. Generally, membrane cleaning is more efficient at elevated temperatures, but (Pall, 2006) compatibility of the membrane and other filter components regarding temperature should also be checked.

Membrane cleaning involves mass transfer of chemicals to the fouling layer and the reaction products back to the bulk liquid phase. Therefore, *hydrodynamic conditions* that promote contact between cleaning chemicals and fouling materials during cleaning are required. From a mass transfer point of view, dynamic cleaning involving circulating cleaning solutions through the system can be more effective than simply static cleaning such as soaking.

6.6 INTEGRATED MEMBRANE SYSTEMS (MF OR UF + RO OR NF)

With increasingly stringent regulations to guarantee that drinking water presents minimal health risks, nanofiltration (NF) and low-pressure reverse osmosis (RO) processes are nowadays considered for surface water treatment. However, NF and RO membranes are sensitive

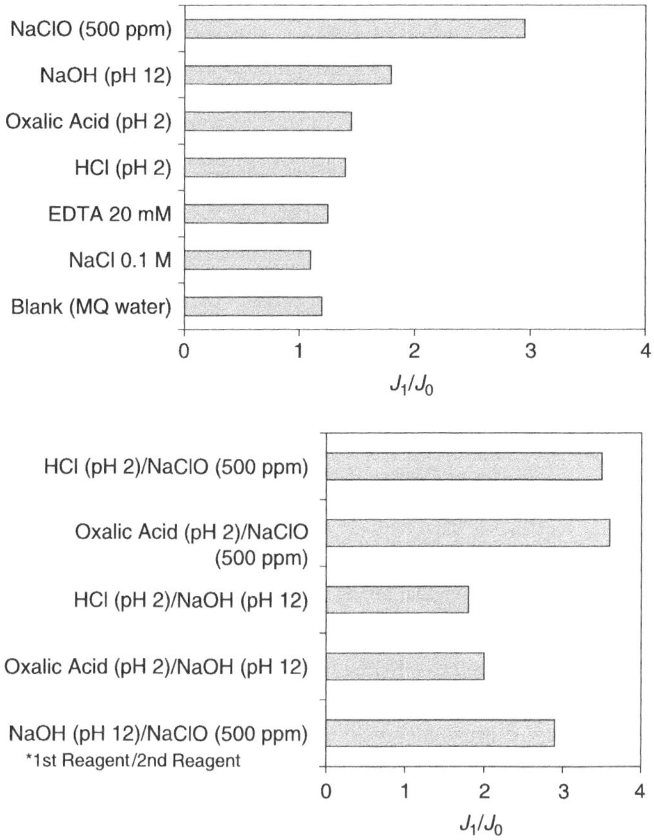

Figure 6.16 Effect of chemical membrane cleaning membrane cleaning with (a) single solution and (b) two different reagents (J_0, pure water flux before chemical cleaning; J_1, pure water flux after chemical cleaning). (Adapted from Kimura et al., 2004.)

to fouling and advanced pre-treatment such as MF and UF are often used to control productivity. Therefore, in addition to the removal of suspended and colloidal particles, including microorganisms, low-pressure membrane systems such as ultrafiltration (UF) and microfiltration (MF) can be used as a pretreatment for high-pressure membrane systems like RO and NF. In such cases, the UF or MF pretreatment processes combined with NF or RO form an integrated membrane system (IMS). The primary purpose of pretreatment systems in IMSs is to reduce and control membrane fouling. With the membrane process in these IMSs, multiple treatment objectives can be achieved. If the IMS is used for treatment of groundwater, the objectives include DBP precursor removal, color removal, SOC removal, and inorganics removal. For treatment of surface water, the membrane plays an additional role of disinfection. See Figures 6.17 and 6.18.

Before an IMS can be implemented successfully, strategies to control membrane fouling must be evaluated. If membranes are scaled or fouled, the productivity of the membrane system is reduced, and the membranes have to be chemically cleaned to restore productivity. The cleaning frequencies of surface water IMSs range from once per week to once per 3 months, while those for RO/NF systems treating groundwater range from 3 months to 2 years and average about 6 months (Taylor et al., 1989).

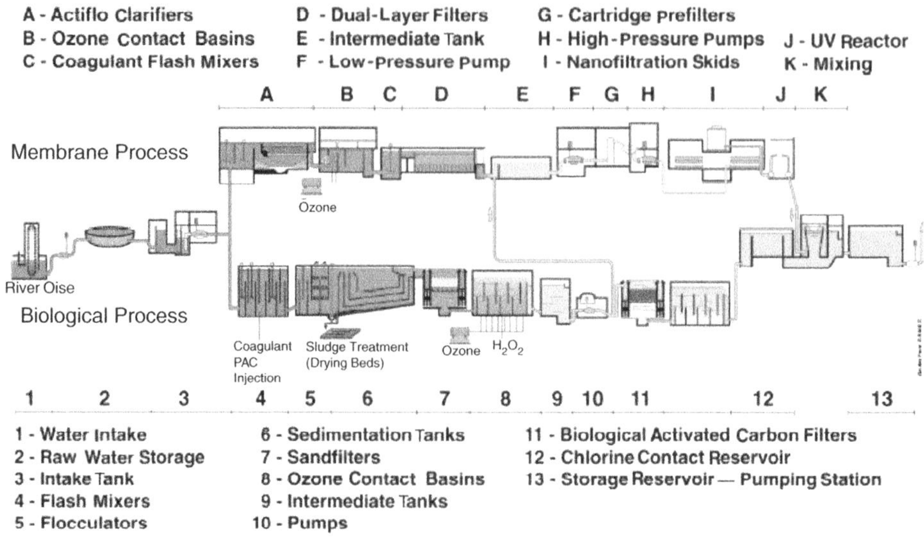

Figure 6.17 Process scheme of the new membrane facility of Mery-sur-Oise Water Treatment Plant: membrane process (train 2) and biological process (train 1). (Adapted from Ventresque et al., 2000.)

Glucina et al. (2000) evaluated two IMSs: UF/RO and conventional train/RO. Both IMSs produced water of a quality that exceeded the most stringent water quality requirements that were proposed in the new European directives. However, in case of blending with the pretreated water for remineralization, clarified water has to be disinfected. In terms of hydraulic performance, both treatments were found to be effective in controlling RO fouling.

Both MF and UF have been presented as extremely effective for the removal of particles, turbidity, and bacteria and cysts without the use of disinfectants. However, when high-organic-content feed water is used, neither MF nor UF can guarantee high water quality (Jacangelo et al., 1991). Therefore, in order to economically achieve the water quality objectives, MF or UF have to be combined with other processes such as adsorption (PAC), coagulation, or even tight membranes (RO, NF). While RO and NF are able to remove a wide range of contaminants, their applications are hydraulically limited by the feed water quality. NF and RO systems involved in surface water treatment generally require extensive pretreatment in order to control fouling of the spiral-wound membranes (Table 6.6).

The resulting IMSs combine the advantages of MF or UF for particle removal with the selectivity of RO or NF, thereby minimizing NF/RO fouling while controlling DBPs and SOCs. The IMS design approach to water treatment systems has some significant advantages over RO/NF systems designed with conventional pretreatment:

1. MF/UF filtrate quality is better, and, therefore, the colloidal fouling load to the RO is reduced with significantly lower SDI and turbidity.
2. MF/UF filtrate quality remains much more constant since it is an absolute membrane barrier.
3. RO/NF cleaning frequencies due to colloidal fouling are reduced.
4. MF/UF concentrate streams are easier to dispose of relative to chemically enhanced conventional pretreatment processes.

(a)

(b)

Figure 6.18 (a) UF/RO plant in Heemskerk, The Netherlands. (b) MF/NF plant in Paris (Mery-Sur-Oise). (Adapted from Veoliawater, 2006.)

TABLE 6.6 Effect of Pretreatment Processes Preceding NF/RO to Control Membrane Fouling[a]

Pretreatment	Particulate	Organic	Biological	Scaling
Groundwater	++	+/−	++	+/−
Riverbank filtration	++	+	++	−
Slow sand filtration	++	+	++	−
Biological activated carbon	+	+/−	+	−
In-line coagulation	+/−	+	−	−
Coagulation–sedimentation	+	+	−	−
Coagulation–sedimentation–filtration	++	+	+/−	−
Microfiltration	++	−	−	−
Ultrafiltration	++	+/−	+/−	−
Scale inhibitor and/or acid addition	−	−	−	++
Chloramine dosage	−	−	+	−

[a] '−', No effects/negative effects; '+', some positive effects; '++', positive effects.

Source: Adapted from Kruithof et al. (2004).

5. Capital and operating costs are competitive and in some schemes less.
6. Future expandability is easier to design in.

6.7 BACKWASH WATER REUSE, TREATMENT, AND DISPOSAL

6.7.1 Introduction

Microfiltration and UF processes may generate backwash water with chemicals (acids such as citric, HNO_3, HCl, or bases such as NaOH, chelating agents such as EDTA, oxidants/disinfectants such as NaOCl, H_2O_2, detergents, and surfactants), especially after chemical cleaning or enhanced backwash. Such residues may have to be neutralized prior to subsequent treatment such as sedimentation and dewatering. Backwash water may also contain high concentrations of viruses, bacteria, and protozoans that may have to be inactivated. Also depending on the type of pretreatment used, inorganics such as coagulants, acids, bases (used in pH correction), and flocculants may be present. In cases where enhanced backwash has been used, a high level of basic, acidic materials or even surfactants may be present. In addition to the composition of the waste described above, the volume generated is also important in which case reuse or recovery becomes an issue. A number of backwash water reuse/recovery options, disposal, and treatment methods are available. The choice of a disposal option is constrained by the legislation that is intended to clean up the environment.

The waste problem could also be handled from the generation side, where the main aim is to reduce the amount of waste generated in the first place, by adopting plant designs that generate little or reusable/recyclable wastes. At all times, the goal of water treatment plant waste management is to dispose of or treat the waste in an economical and environmentally acceptable manner.

6.7.2 Backwash Water Reuse—Recovery of Chemicals

In many countries the regulations for disposal are very exigent, and in this way treatment plants must treat the backwash water that is used during the normal operation. In the United Kingdom, there is a drinking water plant that used two units of ultrafiltration in its process. The first one is a pretreatment for RO units, and the second one is to treat the water used in the cleaning of the membranes.

In addition to minimizing residue quantity by choice of process, the quantity may also be reduced by recovering some of the constituents of the waste. For example, some coagulants (e.g., aluminum, iron) and lime-softening residues (e.g., CaO, $CaCO_3$) can be recovered (Doe, 1990). Chemical recovery is achieved by performing chemical reactions with the sludge, thereby altering sludge characteristics. The cost recovery is an important consideration and must be carefully evaluated when considering chemical recovery. A comparison must be made between the cost of chemical recovery plus the cost of makeup chemical and the cost of new chemical (without recovery) plus the cost of disposing of a portion of the sludge.

6.7.3 Recirculation and Reuse of Backwash Water

About 90% of source water suspended solids will be removed by sedimentation under normal circumstances, leaving the remaining 10% of lighter floc to be passed to the MF or UF system. During filter backwashing, the dirty water (mainly consisting of fines carried over from the sedimentation process) is moved to settling basins. A growing tendency is for backwash water to be returned to the head of the plant where it can mix and settle with the heavier floc in the sedimentation basins. Particles in the backwash water may serve as nuclei for flocculation of the source water. Chemical composition of the backwash water and the possibility of recirculating contaminants must be investigated before deciding to recirculate.

6.7.3.1 Treatment Suspended solids from the backwash process may be removed by sedimentation. Sedimentation may be accomplished by using sedimentation basins devoted to this purpose or by using impounding reservoirs. Solids deposited in impounding reservoirs must be dredged out from time to time. Suspended solids that do not settle in the sedimentation basins (where provided) are removed by coagulation, flocculation, and sedimentation. Sludge characteristics may be affected by choices made in the operation of the coagulation process. Studies by Knocke et al. (1987) found that increases in the rate and extent of dewatering occurred when the coagulation pH was reduced, the coagulant to influent turbidity ratio was reduced, or the coagulation mechanism was adjusted from enmeshment (sweep) to adsorption–charge neutralization.

6.7.4 Ultimate Disposal of Backwash Water

A number of major disposal areas for backwash water exist: The source water from which it came, the nearest wastewater pollution control plant, and a sanitary landfill or other means of land disposal. The choice of ultimate disposal option will dictate the type of treatment required and the degree of concentration needed to prepare it for that method of disposal.

REFERENCES

Abdessemed, D., and Nezzal, G. (2002). Treatment of primary effluent by coagulation-adsorption-ultrafiltration for reuse. *Desalination* 152, 367–373.

Antelmi, D., Cabane, B., Meireles, M., and Aimar, P. (2001). Cake collapse in pressure filtration. *Langmuir* 17, 7137–7144.

Aoustin, E., Schäfer, A. I., Fane, A. G. and Waite, T. D. (2001). Ultrafiltration of natural organic matter. *Separation and Purification Technology* 22–23, 63–78.

Applied water solution (2006). Separation technologies. Available: http://www.appliedwatersolutions.com/technicalgallery.html, accessed Sept. 14, 2006.

Berube, P. R., Mavinic, D. S., Hall, E. R., Kenway, S. E., and Roett, K. (2002). Evaluation of adsorption and coagulation as membrane pre-treatment steps for the removal of organic material and disinfection-by-product precursors. *J. Environ. Eng. Sci.* 1, 465–476.

Boerlage, S. F. E. (2001). Scaling and particulate fouling in membrane filtration systems. Ph.D. Thesis, Wageningen University and UNESCO-IHE.

Bos, R. H. M., Broens, L., Drost, S., van Hoof, S., Van der Maas, P., and Oosterom, H. A. (1998). Membrane filtration for greenhouse horticulture. *Desalination* 118, 93–100.

Bourke, M., and Slunjski, M. (1999). MIEX® DOC Process Launched in Western Australia – A New Australian Development, *Water*, Nov/Dec, 17–20.

Cabane, B., Meireles, M., and Aimar, P. (2002). Cake collapse in frontal filtration of colloidal aggregates: Mechanisms and consequences. *Desalination* 146, 155–161.

Cabassud, C., Anselme, C., Bersillon, J. L. and Aptel, P. (1991). Ultrafiltration as a nonpolluting alternative to traditional clarification in water treatment. *Filtration & Separation* 28, 194–198.

Cabassud, C., Laborie, S. and Laine, J. M. (1997). How slug flow can improve ultrafiltration flux in organic hollow fibres. *J. Membr. Sci.* 128, 93–101.

California Water Desalination Task Force (CWDTF) (2003). Issue paper on concentrate management associated with desalination facilities. Office of Water Use Efficiency, CWDTF. Available: http://www.owue.water.ca.gov/recycle/desal/docs/concentratemgmtissues.doc, accessed Mar. 25, 2006.

Carroll, T., King, S., Gray, S. R., Bolto, B. A., and Booker, N. A. (2000). The fouling of microfiltration membranes by NOM after coagulation treatment. *Water Res.* 34, 2861–2868.

Cho, J., and Amy, G. (1999). Interactions between natural organic matter (NOM) and membranes: Rejection and fouling. *Water Sci. Technol.* 40(9), 131–139.

Cho, J., Amy, G., and Pallegrino, J. (1999). Membrane filtration of natural organic matter: Initial comparison of rejection and flux decline characteristics with ultrafiltration and nanofiltration membranes. *Water Res.* 35(18), 4455–4463.

Cho, J., Amy, G., and Pallegrino, J. (2000). Membrane filtration of natural organic matter: Factors and mechanisms affecting rejection and flux decline with charged ultrafiltration (UF) membrane. *J. Membr. Sci.* 164, 89–110.

Coffey, B. M., Stewart, M. H., Wattier, K. L., and Wale, R. T. (1993). Evaluation of microfiltration for metropolitan's small domestic water systems. In *Proceedings of the Membrane Technology Conference.* American Water Works Association, Baltimore, MD.

Crozes, G. F., Jacangelo, J. G., Anselme, C., and Laine, J. M. (1997). Impact of ultrafiltration operating conditions on membrane irreversible fouling. *J. Membr. Sci.* 124, 63–76.

Doe, P. W. (1990). Water treatment plant waste management. In: F. W. Pontius, (Ed.), *Water Quality and Treatment: A Handbook of Community Water Supplies.* McGraw-Hill, New York, pp. 955–966.

Doyen, W. (2003). Latest developments in ultrafiltration for large-scale drinking water applications. *Desalination* 113, 165–177.

Drikas, M., Christopher, W., Chow, K., and Cook, D. (2003). The impact of recalcitrant organic character on disinfection stability, trihalomethane formation and bacterial regrowth: An evaluation of Magnetic Ion Exchange Resin (MIEX) and alum coagulation. *J. Water Supply Res. Technol. Aqua* 52(7), 475–487.

Drinking Water Inspectorate (DWI) (2001). Membrane integrity testing—Membrane company level. In *Review of the Adequacy of Existing Proposals for Membrane Integrity Monitoring.* Available: http://www.dwi.gov.uk/regs/pdf/mim.pdf, accessed May 22, 2006.

Edzwald, J. K., and Van Benschoten, J. E. (1990). Aluminum coagulation of natural organic matter. In: R. Klute and H. H. Hahn (Eds.), *Chemical Water and Wastewater Treatment.* Springer-Verlag, New York, pp. 341–359.

Faibish, R. S., and Cohen, Y. (2001). Fouling and rejection behavior of ceramic and polymer-modified ceramic membranes for ultrafiltration of oil-in-water emulsions and micro-emulsions. *Colloids Surfaces A: Physicochem. Eng. Aspects* 191, 27–40.

Fan, L., Harris, J. L., Roddick, F. A. & Booker, N. A. (2001). Influence of the characteristics of natural organic matter on the fouling of microfiltration membranes. *Water Research* 35, 4455–4463.

Fane, A. G., and Fell, C. J. D. (1987). A review of fouling and fouling control in ultrafiltration. *Desalination* 62, 117–136.

Farahbakhsh, K., and Smith, D. W. (2002). Performance comparison and pre-treatment evaluation of three water treatment membrane pilot plants treating low turbidity water. *J. Environ. Eng. Sci.* 1, 113–122.

Finnigan, S. M., and Howell, J. A. (1990). The effects of pulsatile flow on UF fluxes in a baffled tubular membrane system. *Desalination* 79, 181–202.

Furukawa, D. (2002). Global status of microfiltration and ultrafiltration membrane technology. *Watermark* (17). MEDRC (Newsletter). Muscat, Oman.

Glucina, K., Alvarez, A., and Laine, J. M. (2000). Assessment of an integrated membrane system for surface water treatment. In *Proceedings of Membranes in Drinking Water and Industrial Water Production*, Vol. 2, pp. 113–122.

Guigui, C., Mougenot, M., and Cabassud, C. (2003). Air sparging backwash in ultrafiltration hollow fibres for drinking water production. *Water Sci. Technol: Water Supply* 3(5), 415–422.

Guigui, C., Rouch, J. C., Durand-Bourlier, L., Bonnelye, V., and Aptel, P. (2002). Impact of coagulation conditions on the in-line coagulation/UF process for drinking water production. *Desalination* 147, 95–100.

Guo, W. S., Vigneswaran, S., Ngo, H. H., and Ben Aim, R. (2005). Effect of flocculation and/or adsorption on the critical flux of microfiltration. *Desalination* 172, 53–62.

Hagmeyer, G., Panglisch, S., and Gimbel, R. (1996). Ultrafiltration for drinking water treatment of reservoir water. In *Proceedings of WaterTECH,* AWWA, Sydney, 621–627.

Heneghan, K. S., and Clark, M. M. (1991). Surface water treatment by combined ultrafiltration/PAC adsorption/coagulation for removal of natural organics, turbidity and bacteria. In *Proceedings of the Seminar on Membrane Technologies in the Water Industry.* American Water Works Association, Orlando, FL.

Her, N., Amy, G., Foss, D., and Cho, J. (2002). Variations of molecular weight estimation by HP—Size exclusion chromatography with UVA versus on-line DOC detection. *Environ. Sci. Technol.* 36(15), 3393–3399.

Her, N., Amy, G., McKnight, D., Sohn, J., and Yoon, Y. (2003). Characterisation of DOM as a function of MW by fluorescence EEM and HPLC-SEC using UVA, DOC and fluorescence detection. *Water Res.* 37, 4295–4303.

Her, N., Amy, G., Park, H., and von Gunten, V. (2004). UV absorbance ratio index with size exclusion chromatography (URI-SEC) as a NOM property indicator.

Hong, S., and Elimelech, M. (1997). Chemical and physical aspects of natural organic matter (NOM) fouling of NF membranes. *J. Membr. Sci.* 132, 159–181.

Innocenti, P. (1988). Techniques for handling water treatment sludge. *Opflow* 14(2), 1.

Jacangelo, J. G., Aieta, E. M., Carns, K. E., Cummings, E. W., and Mallevialle, J. (1989). Assessing hollow fibre ultrafiltration for particle removal. *J. AWWA* 83(11), 68–75.

Jacangelo, J. G., Laine, J. M., Carns, K. E., Cummings, E. W., and Mallevialle, J. (1991). Low-pressure membrane filtration for removing *Giardia* and microbial indicators. *J. AWWA* 83(9), 97 106.

Jucker, C., and Clark, M. M. (1994). Adsorption of aquatic humic substances on hydrophobic ultrafiltration membranes. *Journal of Membrane Science*, 97, 37–52.

Judd, S. J., and Hillis, P. (2001). Optimisation of combined coagulation and microfiltration for water treatment. *Water Res.* 35(12), 2895–2904.

Kabsch-Korbutowicz, M. (2005). Application of ultrafiltration integrated with coagulation for improved NOM removal. *Desalination* 174, 13–22.

Kim, K. J., Fane, A. G., Fell, C. J. D., Suzuki, T., and Dickson, M. R. (1990) Quantitative microscopic study of surface characteristics of ultrafiltration membranes. *J. Membr. Sci.* 54, 89–102.

Kimura, K., Hane, Y., Watanabe, Y., Amy, G., and Ohkuma, N. (2004). Irreversible membrane fouling during ultrafiltration of surface water. *Water Res.* 38, 3431–3441.

Kennedy, M., Kim, S-M., Mutenyo, I., Broens, L., and Schippers, J. (1998). Intermittent crossflushing of hollow fiber ultrafiltration systems. *Desalination* 118, 175–188.

Knocke, W. R., Hamon, J. R., and Dulin, B. E. (1987). Effects of coagulation on sludge thickening and dewatering. *J. AWWA* 79(6), 89–97.

Kruithof, J. C., Nederlof, M. M., Hoffman, J. A. M. H., and Taylor, J. S. (2004). *Integrated Membrane Systems*. AWWA Research Foundation and American Water Works Association. Elbert, Colorado.

Laborie, S., Cabassud, C., Durand-Bourlier, L., and Laine, J. M. (1998). Fouling control by air sparging inside hollow fibre membranes—Effects on energy consumption. *Desalination* 118, 189–196.

Lahoussine, V., Weisner, M. R., and Bottero, J. (1990). Fouling tangential-flow ultrafiltration: The effect of colloid size and coagulation pre-treatment. *J. Membr. Sci.* 52, 173–190.

Lahoussine-Turcaud, V., Wiesner, M. R., Bottero, J. Y., and Mallevialle, J. (1990). Coagulation pretreatment for ultrafiltration of a surface water. *J. AWWA* 81, 76–81.

Laine, J. M., Cummings, E. W., Carns, K. E., and Mallevialle, J. (1993). Influence of bromide on low-pressure membrane filtration for controlling DBPs in surface waters. *J. AWWA* 85(6), 87–99.

Lee, J. D., Lee, S.-H., Jo, M.-H., Park, P.-K., Lee, C.-H., and Kwak, J.-W. (2000). Effect of coagulation conditions on membrane filtration characteristics in coagulation-microfiltration process for water treatment. *Environ. Sci. Technol.* 34(17), 3780–3788.

Lee, S., Cho, J. and Elimelech, M. (2005). Combined influence of natural organic matter (NOM) and colloidal particles on nanofiltration membrane fouling. *J. Membr. Sci.* 262, 27–41.

Maartens, A., Swart, P., and Jacobs, E. P. (1998). Humic membrane foulants in natural brown water: Characterization and removal. *Desalination* 115, 215–227.

Mace, R. E., Nicot, J.-P., Chowdhury, A. H., Dutton, A. R., and Kalaswad, S. (2005). Please pass the salt: Using oil fields for the disposal of concentrate from desalination plants. Texas Water Development Board. Available: http://www.twdb.state.tx.us/desalination/thefutureofdesalinationintexas-volume 202/documents/c10.pdf, accessed Mar. 25, 2006.

Makdissy, G., Croue, J. P., Buisson, H., Amy, G., and Legube, B. (2003). Organic matter fouling of ultrafiltration membranes. *Water Sci. Technol. Water Supply* 3(5–6), 175–182.

Mallevialle, J., Anselme, C., and Marsigny, O. (1989). Effects of humic substances on membrane processes. In I. H. Suffet and P. MacCarthy (Eds.), *Aquatic Humic Substances' Influence on Fate and Treatment of Pollutants*. American Chemical Society, Washington, DC, pp. 749–767.

Matilainen, A., Linqvist, N., Korhonen, S., and Tuhkanen, T. (2002) Removal of NOM in the different stages of the water treatment process. *Environ. Int.* 28, 457–465.

Minegishi, S., Jang, N.-Y., Watanabe, Y., Hirata, S., and Ozawa, G. (2001). Fouling mechanism of hollow fibre ultrafiltration membrane with pre-treatment by coagulation/sedimentation process. *Water Sci. Technol. Water Supply* 1(4), 49–56.

Mo, L., and Huanga, X. (2003). Fouling characteristics and cleaning strategies in a coagulation-microfiltration combination process for water purification. *Desalination* 159, 1–9.

Mulder, M. (1996), *Basic Principles of Membrane Technology*. Kluwer Academic, The Netherlands.

Nakatsuka, S., and Ase, T. (1995). Ultrafiltration of river water for drinking water production. In *Proceedings of AWWA Membrane Technology Conf.*, Reno, NV, 621–639.

Nilson, J., and Digiano, F. A. (1996). Influence of NOM composition on nanofiltration. *Am. Water Works Assoc.* 88, 53–66.

Office of Technology Assessment (OTA) (2006). Environmental considerations: Waste concentrates. Office of Technology Assessment, Congress of the US. Available: http://www.wws.princeton.edu/ota/disk2/1988/8842/884207.pdf, accessed Mar. 25, 2006.

Pall Corporation (2006). Membrane chemical cleaning: From art to science. Available: http://www.pall.com/pdf/mtcpaper.pdf, accessed Sept. 5, 2006.

Park, Y. G. (2002). Effect of ozonation for reducing membrane fouling in the UF membrane. *Desalination* 147, 43–48.

Pearce, G. K., and Hanks, G. (1993). Unique backwashable microfilter offers efficient compliance with SWTR requirements. In *Proceedings of the Membrane Technology Conference*. American Water Works Association, Baltimore, MD.

Peavy, H. S., Rowe, D. R., and Tchobanoglous, G. (1984). *Environmental Engineering*. McGraw-Hill, New York.

Roorda, J. H., and Van Der Graaf, J. H. J. M. (2001). New parameter for monitoring fouling during ultrafiltration of WWTP effluent. *Water Sci. Technol.* 43(10), 241–248.

Safe Drinking Water Act (SDWA) (2005). What do I do with this stuff? Options for inland membrane concentrate disposal. Safe Drinking Water Act Newsletter. Available: http://hdrinc.com/assets/documents/publications/SDWA/august2005/membraneconcentratedisposal.pdf, accessed Mar. 25, 2006.

Sanjeev, G. R., and Davis, R. H. (2004). Cross-flow microfiltration with high frequency reverse filtration. *AIChE J.* 41(3), 501–508.

Schäfer, A. I. (2001). Natural organics removal using membranes. Ph.D. Thesis, UNESCO-IHE, UNSW, Australia.

Schäfer, A. I., Fane, A. G., and Waite, T. D. (2000). Fouling effects on rejection in the membrane filtration of natural waters. *Desalination* 131, 215–224.

Schippers, J. C., and Verdouw, J. (1980). The modified fouling index, a method of determining the fouling characteristics of water. *Desalination* 32, 137–148.

Schippers, J. C., and Buiteman, J. P. (2005). *Conventional Water Treatment Technology*. UNESCO-IHE, Delft, The Netherlands.

Slunjski, M., Bourke, M., and O'Leary, B. (2000). MIEX DOC process for removal of humics in water treatment. In *Proceedings of IHSS-Australian Branch Symposium: Humic Substances—Science and Commercial Applications*. Monash University, Melbourne, pp. 22–27.

Speth, T. F., Summers, R. S., and Gusses, A. M. (1996). Evaluation of membrane foulants from conventionally-treated drinking waters. In *Proceedings of Natural Organic Matter Workshop*, Poitiers, France, p. 44.

Taylor, J. S., Mulford, S. J., Duranceau, S. J., and Barrett, W. M. (1989). Cost and performance of membrane processes. *J. AWWA* 81(11), 52–60.

Thiruvenkatachari, R., Ngo, H. H., Hagare, P., Vigneswaran, S., and Ben Aim, R. (2002). Flocculation-cross-flow microfiltration hybrid system for natural organic matter (NOM) removal using hematite as a flocculant. *Desalination* 147, 83–88.

Thurman, E. M. (1985). *Organic Geochemstry of Natural Waters*. Martinus Nijhoff/Dr W Junk Publishers, Dordrecht, Netherlands.

U.S. Environmental Protection Agency (EPA) (2002). Chapter 2, Equipment verification testing plan for the removal of synthetic organic chemical contaminants by membrane filtration process. Available: http://www.epa.gov/etvprgrm/pdfs/testplan/01_tp_111299_soc.pdf, accessed May 22, 2006.

Ventresque, C., Gisclon, V., Bablon, G., and Chagneau, G. (2000). An outstanding feat of modern technology: The Mery-Sur-Oise Nanofiltration Treatment Plant (340,000 m^3/d). In *Proceedings of the Conference on Membranes in Drinking and Industrial Water Production*, Vol. 1, Desalination Publications, L'Aquila, Italy, pp. 1–16.

Veoliawater (2006). 1999—Méry-sur-Oise—Nanofiltration. Available: http://www.veoliawater.com/history/1999/, accessed May 22, 2006.

Verberk, J. Q. J. C. (2005). Application of air in membrane filtration. Ph.D. Thesis, Technical University of Delft.

Vickers, J. C., Johnson, P. E., and Willingham, G. A. (1993). Meeting the surface water treatment rule using continuous microfiltration. In *Proceedings of the Membrane Technology Conference*. American Water Works Association, Baltimore, MD.

Viessman, W., and Hammer, M. J. (1993). *Water Supply and Pollution Control*. HarperCollins College Publishers, New York.

Vrouwenvelder, J. S., Kappelhof, J. W. N. M., Heijman, S. G. J., Schippers, J. C., and Van Der Kooij, D. (2003). Tools for fouling diagnosis of NF and RO membranes and assessment of the fouling potential of feed water. *Desalination* 157, 361–365.

Weisner, M. R., Clarke, M. M., Jacangelo, J. G., Lykins, B. W., Marinas, B. J., O'Melia, C. R., Ritmann, B. E., and Semmens, M. J. (1992). Committee report: Membrane processes in portable water treatment. *J. AWWA* 84(1), 59–67.

Yangali-Quintanilla, V. A. Y. (2005). Colloidal and non-colloidal NOM fouling of ultrafiltration membranes: Analysis of membrane fouling and cleaning. M.Sc. Thesis, UNESCO-IHE, Delft, The Netherlands.

Yoo, R. S., Brown, D. R., Pardini, R. J., and Bentson, G. D. (1995). Microfiltration: A case study. *J. AWWA* 87(3), 38–49.

Zenon membrane bioreactors. (2006). GE Water & Process Technologies, www.GEwater.com.

CHAPTER 7

Water Reclamation and Desalination by Membranes

PIERRE CÔTÉ and MINGANG LIU

GE Water and Process Technologies, ZENON Membrane Solutions, Ontario, L6M 4B2, Canada

STEVEN SIVERNS

EnviroTower, Toronto, Ontario, M5V 1R7, Canada

7.1 INTRODUCTION

In populated, water-scarce regions, two sources of water are normally available abundantly: seawater and sewage. Although brackish groundwater is easier to treat than seawater, its availability is limited in populated areas, and it is not considered further in this chapter. Both seawater and sewage pose significant treatment challenges but are being increasingly used as a source of freshwater for a wide variety of uses.

Seawater can be desalinated using thermal processes and reverse osmosis (RO). While approximately half of the worldwide installed desalination capacity is based on thermal processes, there is a clear trend that favors RO, primarily because it consumes much less energy.

In the last decade, the reclamation of effluents has developed rapidly as an alternative to seawater desalination for industrial uses, irrigation, and indirect potable water reuse. For water reclamation, contaminants that require treatment over and above what is provided by conventional biological treatment include suspended solids, microbes, nutrients (nitrogen and phosphorus, trace organic compounds (e.g., pesticides, endocrine disruptor compounds), and in certain cases dissolved salts.

Microfiltration and ultrafiltration (MF/UF) membranes are flexible water treatment tools that can be used in a number of process configurations to meet advanced effluent treatment objectives. MF/UF membranes, when used by themselves, are limited to the removal of particulate and colloidal contaminants; however, they can be combined with biological or chemical treatment to remove dissolved contaminants. Furthermore, they represent the ideal pretreatment to reverse osmosis by addressing their main weakness, fouling by particulate materials.

Advanced Membrane Technology and Applications. Edited by Norman N. Li, Anthony G. Fane, W. S. Winston Ho, and T. Matsuura
Copyright © 2008 John Wiley & Sons, Inc.

Whether in water reclamation or seawater desalination, membrane processes offer the following benefits:

- Small footprint
- Simplification of treatment (reduced number of unit processes)
- Reduced use of treatment chemicals and generation of residuals
- Ease of operation through automation

The purpose of this chapter is to compare membrane treatment processes and their costs for seawater desalination and wastewater reclamation.

7.2 WATER RECLAMATION AND SEAWATER DESALINATION

A comparison of the feed water quality for each of the two sources (many parameters shown in Table 7.1) indicates that they are markedly different. Seawater is high in total dissolved solids (TDS), mostly sodium chloride but low in organic materials [total organic carbon (TOC)] and nutrients (nitrogen, phosphorus); it contains trace amounts of sparingly soluble salt that are close to their saturation concentration and represent a scaling potential. Biologically treated sewage has a low TDS but a relatively high TOC and nutrients. Total nitrogen compounds could be reduced much further through nitrification/denitrification in the activated sludge process as compared to the example given in Table 7.1. Biologically treated sewage contains trace amounts of organic compounds (pesticides, endocrine disrupting compounds) that are considered contaminants.

Figure 7.1 shows the two major treatment options to obtain RO-quality water from sewage and seawater. The key in water reclamation is to first treat the sewage biologically and use MF/UF membrane filtration to remove suspended solids. Two membrane filtration alternatives are available for water reclamation: tertiary filtration (TF) of the effluent from a conventional activated sludge (CAS) process and an integrated membrane bioreactor (MBR). For seawater desalination, pretreatment must be provided if the source is open seawater. The current practice involves multimedia filtration, but membrane filtration has also been considered.

The RO permeate quality for the two water sources described in Table 7.1 are presented in Table 7.2; they are very similar. For seawater, RO process conditions are selected

TABLE 7.1 Comparison of Feed Water Quality for Reclamation and Desalination

Parameter	Units	Seawater[a]	Biologically Treated Sewage[b]
Alkalinity	ppm ($CaCO_3$)	135	261
Ammonia	ppm	<0.02	19.2
Boron	ppm	5.0	0.45
Chloride	ppm	22,080	225
Nitrate + nitrite	ppm	<1.0	1.8
Silica	ppm	4.0	21
Sulfate	ppm	3,100	237
Total dissolved solids	ppm	41,000	986
Total organic carbon	ppm	2.6	10.6

[a]Mediterranean seawater at Ashdod, Israel, May 23, 2001.
[b]Water Factory 21, Demonstration testing, May 2001 to April 2002 (Daugherty et al., 2005).

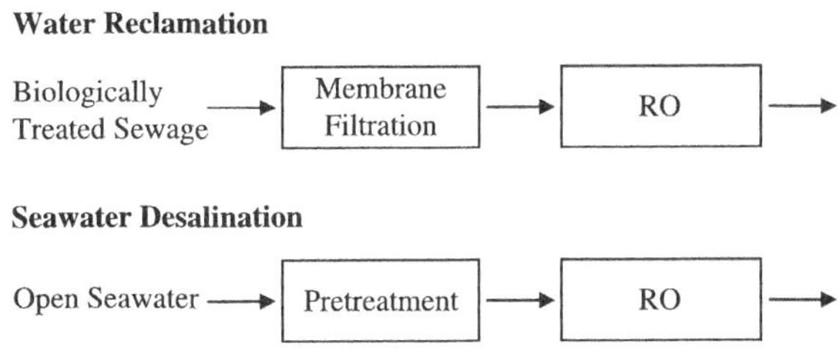

Figure 7.1 Water reclamation versus seawater desalination.

TABLE 7.2 Comparison of RO Permeate Quality for Reclamation and Desalination

Parameter	Units	Seawater[a]	Biologically Treated Sewage[b]
Alkalinity	ppm ($CaCO_3$)	<5	15.7
Ammonia	ppm	<0.02	1.4
Boron	ppm	1.0	0.3
Chloride	ppm	174	6.1
Nitrate + nitrite	ppm	0.07	0.11
Silica	ppm	0.05	<0.1
Sulfate	ppm	2.4	4.7
Total dissolved solids	ppm	300	420
Total organic carbon	ppm	<0.5	0.5

[a]Mediterranean seawater at Ashdod, Israel, May 23, 2001. Simulated permeate quality using ROSA (Dow Filmtec).
[b]Water Factory 21, demonstration testing, May 2001 to April 2002 (Daugherty et al., 2005).

to reduce TDS to below 500 ppm. Most other parameters are reduced to close or below their detection limit. The RO permeate from water reclamation meets all drinking water guidelines. As discussed by Daugherty et al. (2005), the reported TOC value of 0.5 mg/L is questionable as there is interference from inorganic carbon in the measurement; separate measurements indicated that the real value should be between 0.2 and 0.3 ppm. This is confirmed by Van Houtte and Verbauwhede (2005) who reported a TOC < 0.2 mg/L for a similar treatment train. A number of researchers have reported that trace organic contaminants potentially present in biologically treated sewage are essentially removed to below detection limit by RO (Daugherty et al., 2005; Van Houtte and Verbauwhede, 2005).

7.3 COST ESTIMATION

Costing was done using the following tools. CapdetWorks was used for the design and cost estimation of conventional activated sludge plants. CapdetWorks is a preliminary design and costing program available from Hydromantis, Inc. (www.Hydromantis.com). WTCost was used for the design and cost estimation of RO plants. WTCost is a preliminary design and costing program for membrane process plants developed with the support of the U.S. Bureau of Reclamation, available from imoch@aol.com. ZeeCost was used for the design and cost estimation of the immersed membrane systems used as tertiary filtration with the conventional activated sludge plants and for the membrane bioreactor plants.

ZeeCost is ZENON's proprietary design and costing program for ZeeWeed-immersed membrane systems. Total plant costs were considered, including land to reflect the benefit provided by smaller footprint, except where indicated. While the cost estimations presented in this chapter allow comparing options and drawing conclusions on a relative basis, the absolute precision is considered to be $\pm 25\%$.

7.4 PROCESS OPTIONS FOR WATER RECLAMATION

In this section, the options for sewage treatment are investigated, including MF/UF filtration, but without RO. The two primary options shown are conventional activated sludge (CAS) followed by tertiary filtration (TF) and integrated membrane bioreactor (MBR). These options are developed and costed for plants ranging in size between 3800 m^3/day (1 MGD) and 76,000 m^3/day (20 MGD).

Process flow diagrams for the CAS-TF option are presented in Figure 7.2. Two levels of screening are provided, a coarse screen similar to a CAS plant and a fine screen to protect the immersed membranes from accumulation of trash and hair. For small plants [$<$19,000 m^3/day (5 MGD)], there is no primary clarification and the thickened sludge is digested aerobically (Fig. 7.2a). For large plants [$>$19,000 m^3/day (5 MGD)], a primary clarification step was added, and the thickened sludge is digested anaerobically (Fig. 7.2b).

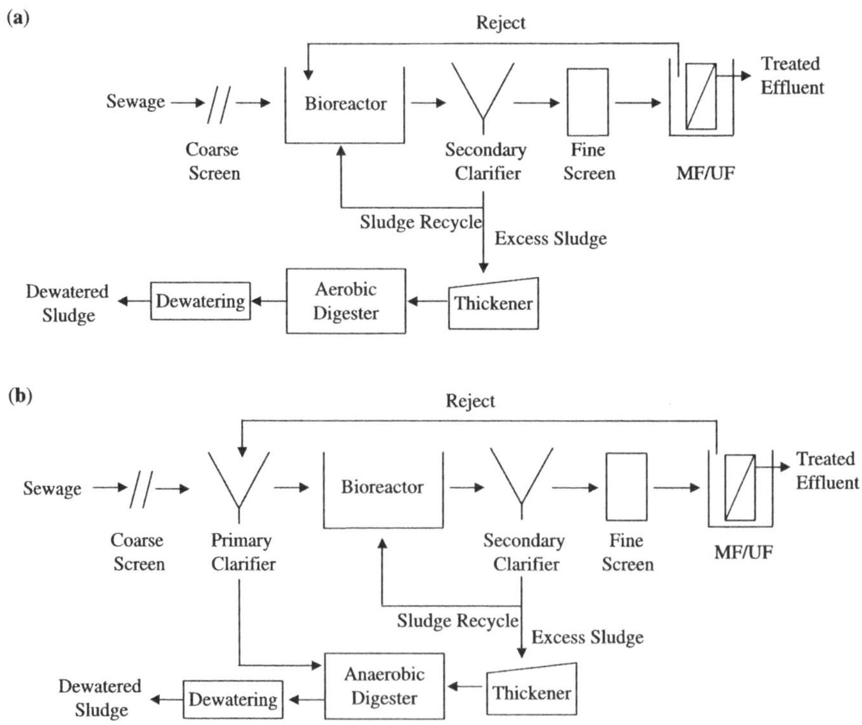

Figure 7.2 Process flow diagram for the conventional activated sludge–tertiary filtration (CAS-TF) option: (a) small plants and (b) large plants.

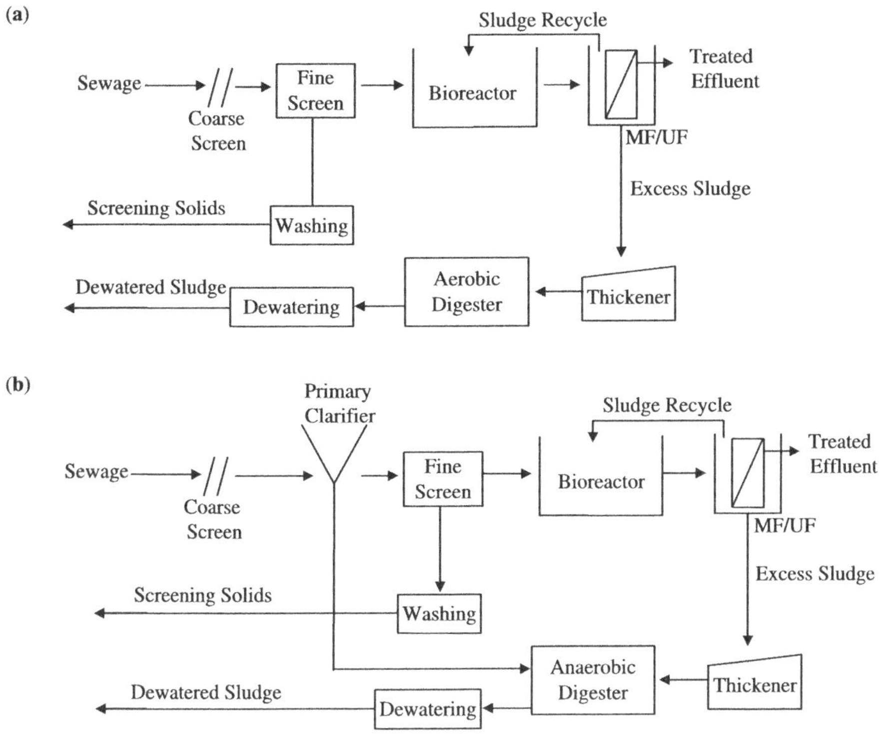

Figure 7.3 Process flow diagram for the membrane bioreactor option (MBR): (a) small plants and (b) large plants.

Process flow diagrams for the MBR option are presented in Figure 7.3. The inclusion of primary clarification and aerobic or anaerobic digestion as a function of flow rate is identical to the CAS-TF case. Also, for all but very small plants, the membranes are immersed into separate tanks from the main bioreactors. This provides more flexibility in isolating membranes for cleaning or maintenance without handling.

Typical sewage concentrations were used as input to the models in order to design the biological treatment steps and estimate sludge production [biochemical oxygen demand (BOD) and suspended solids (SS)= 240 mg/L; total Kjeldahl nitrogen (TKN) = 40 mg/L; total phosphorus (TP) = 8 mg/L]. Simulations were run for average flows of 3800 m^3/day (1 MGD), 19,000 m^3/day (5 MGD), 38,000 m^3/day (10 MGD), and 76,000 m^3/day (20 MGD). Typical peaking factors of 2 twice the average flow were used for all plants.

The process conditions used to design the plants are summarized in Table 7.3. For the conventional activated sludge system, typical values for North American design were selected by using most of the default values suggested by CapdetWorks. The same sludge age [solids retention time (SRT)] was used for CAS and MBR, but hydraulic retention times and mixed liquor concentrations were significantly different.

The membrane system used for the MBR is ZENON's ZeeWeed 500d (Crossley et al., 2003). Membrane cassettes containing 48 modules of 31.6 m^2 each (Fig. 7.4a) were arranged into separate membrane tanks, shown in Figure 7.4c (2, 6, 8, and 12 tanks for the four flow rates simulated). The continuous mixed liquor recirculation flow rate between the membrane tanks and the main bioreactor was set at $5 \times Q_{ave}$ to deconcentrate

TABLE 7.3 Design Process Conditions for Sewage Treatment Plants

			Small Plants $\leq$19,000 m^3/day		Large Plants >19,000 m^3/day	
Unit Process	Parameter	Units	CAS-TF	MBR	CAS-TF	MBR
Coarse screen	Size	mm	10	10	10	10
Primary clarifier	Loading rate	m/h	No	No	1.7	1.7
Fine screen		mm	2	2	2	2
Bioreactor	HRT	h	23	6.5	12	3.6
	MLSS	g/L	3	10	3	10
	SRT	day	15	15	15	15
Secondary clarifier	Loading rate	m/h	1.4	No	1.4	No
ZeeWeed filtration	Average flux	L/m^2/h	22	20	22	20
Gravity thickener	Discharge TSS	g/L	60	60	60	60
Sludge digestion			Aerobic	Aerobic	Anaerobic	Anaerobic
	HRT	day	15	15	35	35

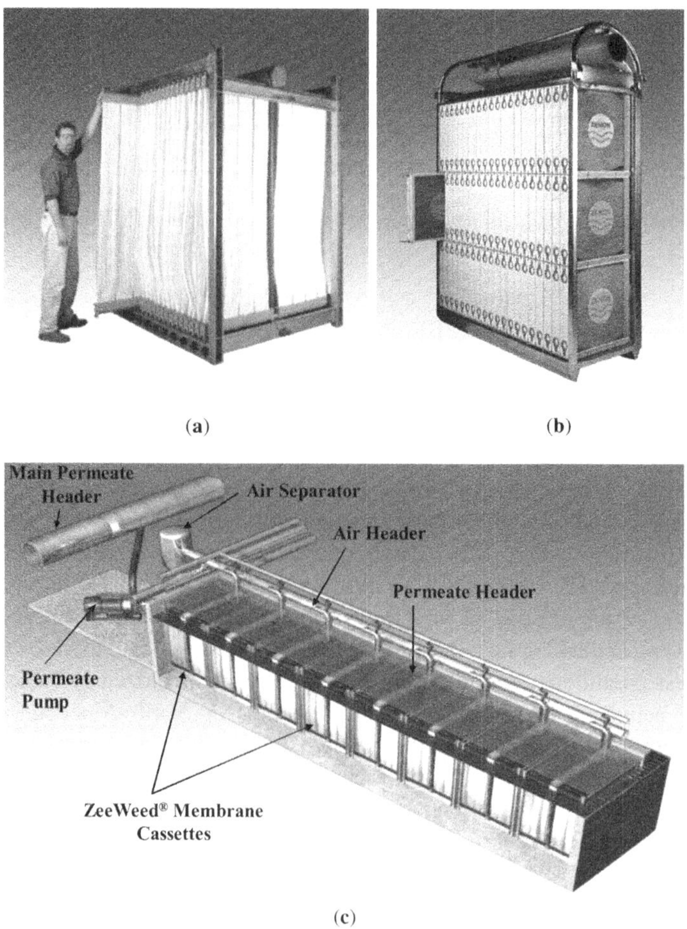

Figure 7.4 ZeeWeed filtration system for MBR: (a) 500d cassette, (b) 1000 cassette, and (c) membrane tank.

the membrane tank and provide for nitrification/denitrification. The design was based on an average flux of 20 $L/m^2/h$. A set of blowers for membrane scouring was provided independently from the biological process blowers and was sized to provide an average of 0.26 $m^3/h/m^2$ of membrane surface area.

The membrane system used for the CAS tertiary filtration is ZENON's ZeeWeed 1000, as shown in Figure 7.4b (Côté et al., 2001). Membrane cassettes containing 60 modules of 37 m^2 each were arranged into separate membrane tanks shown in Figure 7.4c (2, 6, 8, and 12 tanks for the four flow rates simulated). The ZeeWeed 1000 is operated as a dead-end filtration system with backpulses and deconcentration every 20 min. It was assumed that the backwash reject water was returned to the head of the plant. The design was based on an average flux of 22 $L/m^2/h$. The scouring aeration for the ZeeWeed 1000 is an average of 0.02 $m^3/h/m^2$ of membrane surface area, applied only during the backwashing sequence.

Energy consumption for each of the two membrane systems included suction pumps assuming an average transmembrane pressure of 35 kPa and scouring aeration as described above.

7.5 COST OF WATER RECLAMATION

7.5.1 Plant Footprint

Membrane bioreactor plants are much smaller than CAS-TF plants. This is shown by comparing the total hydraulic retention time (HRT) in Figure 7.5 and the total plant surface area in Figure 7.6.

The total HRT (the volume of *all* tanks divided by the average flow rate) for the CAS-TF plants is 28 h in small plants, decreasing to 20 h for the larger plants; HRT decreases with flow rate as the net impact of adding primary clarifiers and reducing the biological tanks HRT in larger plants is positive regardless of anaerobic digester. For MBR plants, the total HRT increases from 7.5 h in small plants to 9.5 h in large plants; for large plants, the addition of the primary clarifiers is not compensated by the reduction of the biological tank's HRT, and the total HRT increases with flow rate as a result of using an anaerobic

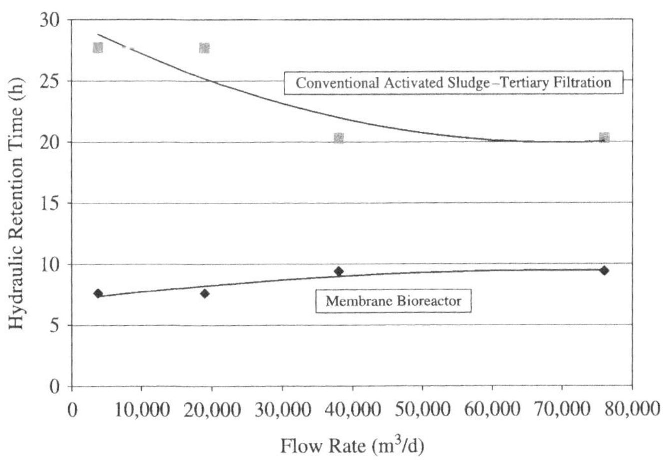

Figure 7.5 Hydraulic retention time.

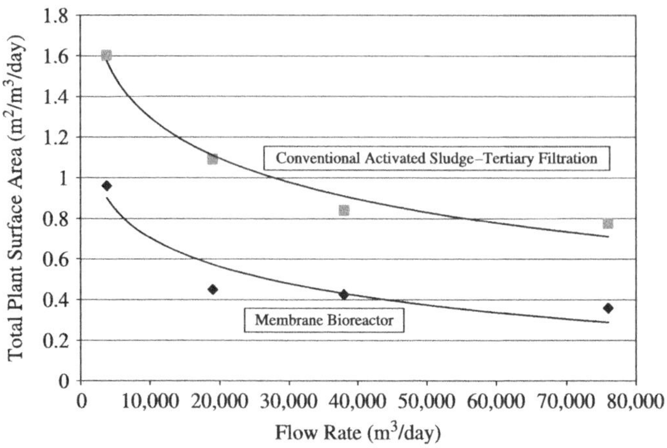

Figure 7.6 Total plant surface area.

digester. Overall, the MBR plant's HRT is 75% (small plants) to 50% (large plants) smaller than those of the CAS-TF plants.

Figure 7.6 shows the total surface area occupied by the plant as a function of flow rate. The total surface area was estimated based on the total process tanks footprint (proportional to flow rate) and the area required for roads, parking, laboratory, and the like (not directly proportional to flow rate). The land required for an MBR plant is about half of that required for a CAS-TF plant.

7.5.2 Capital Costs

The total capital costs are presented in Figure 7.7 for three plants, CAS, CAS-TF, and MBR, including the following components for all unit processes shown in Figures 7.1 and 7.2:

- Direct costs (equipment for all unit processes, mobilization, site preparation, site electrical, yard piping, instrumentation and control, laboratory, and administration buildings)
- Indirect costs (legal fees, engineering design, inspection, contingency, and miscellaneous)
- Land cost

The capital costs of CAS plants are shown as a reference in Figure 7.7a. For all plant sizes considered, MBR plants are less expensive than CAS plants. This is because the savings associated with eliminating secondary clarifiers, reducing the size of the aeration tanks, and reducing footprint are larger than the added costs for the membrane system and the fine screens. The capital costs of CAS-TF plants are higher that the costs of CAS or MBR plants. The cost of a tertiary filtration membrane system is basically added to that of a CAS plant, without eliminating anything. There are, however, situations where the cost of a CAS-TF plant can be lower than the cost of an MBR. A typical case is a water reclamation system that treats a fraction of the effluent produced by a CAS plant. In this case, the TF system does not have to be sized for peaking, resulting in significant savings.

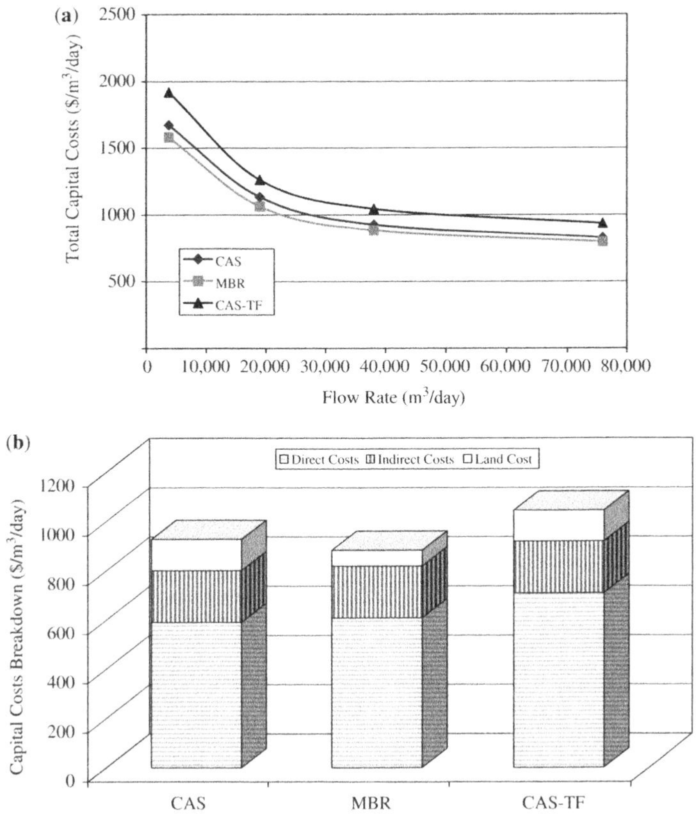

Figure 7.7 (a) Total capital costs and (b) capital costs breakdown (38,000 m^3/day).

The breakdown of capital costs between direct, indirect, and land costs is shown in Figure 7.7b for a 38,000-m^3/day (10-MGD) plant size. Direct costs represent about two thirds of total capital costs. The cost of land, which represents 13.6 and 12.1% for the CAS and CAS-TF plants, shrinks to 7.3% for the MBR plants; this fraction is indeed sensitive to land unit cost ($150/$m^2$ in this case).

7.5.3 Operation and Maintenance Costs

The total operation and maintenance (O&M) costs are presented in Figure 7.8 for the three plants CAS, CAS-TF, and MBR, including the following components for all unit processes shown in Figures 7.1 and 7.2:

- Labor
- Materials (renewal of equipment, membrane replacement)
- Energy
- Chemicals (membrane cleaning)

The O&M costs of the plants including membranes (MBR and CAS-TF) are higher than the CAS plant by 20–30% for all flows (Fig. 7.8a). The O&M cost of an MBR plant is

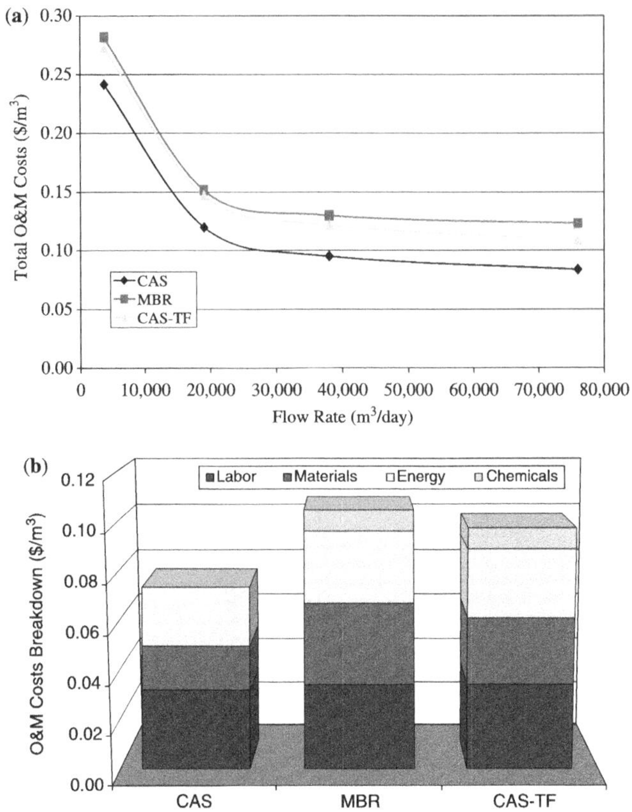

Figure 7.8 (a) Total operation and maintenance costs and (b) operation and maintenance costs breakdown (38,000 m^3/day).

slightly higher than the CAS-TF plant because the MBR membrane system requires a higher scouring aeration rate and the membrane replacement cost is higher.

Figure 7.8b shows that all four categories of O&M costs are higher for the MBR system. However, these were based on conservative assumptions. For labor, it was assumed that the use of membranes did not reduce labor requirement for the conventional part of the plant; however, there is evidence that a fully automated MBR plant requires less labor than a CAS plant. For materials, a membrane life of 8 years was assumed. For energy, a relatively high unit cost of \$0.10/kWh was used.

7.5.4 Total Life-Cycle Costs

The total life-cycle costs presented in Figure 7.9 were generated using a present value (PV) factor of 14.32 (20 years, 6% interest rate, and 2.5% inflation rate). They vary from \$0.45/m^3 for small plants to \$0.20/m^3 for large plants. Total life-cycle costs are smallest for the CAS plants, followed by MBR and CAS-TF plants. The premium for a membrane-filtered wastewater over CAS is 5–20% and increases with plant size.

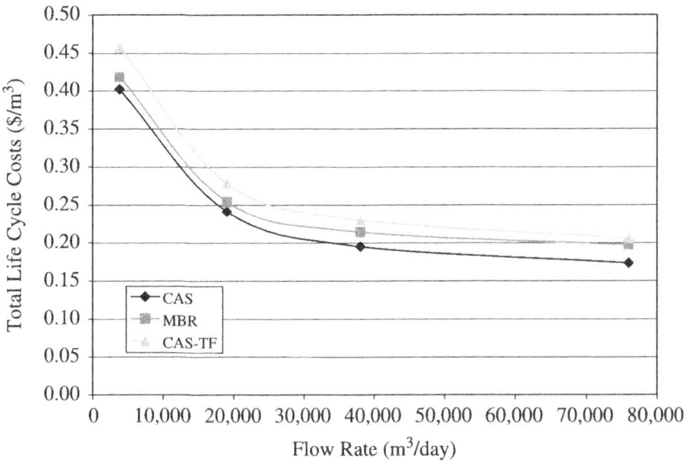

Figure 7.9 Total life-cycle costs.

7.6 PROCESS OPTIONS FOR DESALINATION

Options for pretreatment of seawater include beach wells, media filtration, and ultrafiltration.

Beach wells use the natural filtration ability of beach sand to remove particulate matter. While beach wells provide excellent feed water quality, their use is normally limited to small plants where sufficient beach area is available.

The most common pretreatment for open seawater is multimedia filtration. It is sometimes possible to get away with a single filtration stage, if the feed water is *constantly* of high quality. However, double-stage filtration is required if the source is degraded (i.e., harbors receiving river water or effluent) or there is a risk of quality degradation during storms or algae blooms.

Ultrafiltration membranes have been considered as an alternative to multimedia filtration for seawater RO pretreatment in recent years. There have been pilot studies at numerous sites. The benefits of UF over media filtration pretreatment, lower fouling rate, has been demonstrated in long-term tests (2 years) by Jamaluddin et al. (1998). UF has been recently selected as pretreatment for a large-scale desalination plants in Saudi Arabia, Japan, and China (Wolf et al., 2005).

A comparison of typical design conditions for the two alternatives, including reverse osmosis is presented in Table 7.4. It was assumed the pretreatment system would have to treat (at least occasionally) degraded water, and that two stages of media filtration with coagulant addition were required.

7.7 COST OF DESALINATION

Capital and total life-cycle costs for 75,000-m^3/day plants with each of the two pretreatment options are presented in Table 7.5. The comparison shows a premium for ultrafiltration of 6.5 and 3.6%, respectively, for capital costs and total life-cycle costs.

TABLE 7.4 Process Design Conditions for Seawater Desalination with Media Filtration and Ultrafiltration Pretreatment

			Pretreatment Type	
Unit Process	Parameter	Units	Media Filtration	Immersed UF
Two-stage media filtration	Velocity 1	m/h	7.3	—
	Recovery 1	%	96	—
	Velocity 2	m/h	10.2	—
	Recovery 2	%	97	—
ZeeWeed 1000 ultrafiltration	Flux	$L/h/m^2$	—	42.5
	Recovery	%		92
	Pressure	bar		0.7
Two-pass RO	Flux 1	$L/h/m^2$	13.9	13.9
	Recovery 1	%	50	50
	Pressure 1	bar	54.3	54.3
	Flux 2	$L/h/m^2$	26.0	26.0
	Recovery 2	%	90	90
	Pressure 2	bar	10	10

The breakdown of costs for the ultrafiltration option is presented in Figure 7.10. Figure 7.10a shows that UF pretreatment only represents about 10% of the total life-cycle costs and that energy accounts for one third of the costs. Figure 7.3b shows a breakdown of the total energy cost (3.46 kWh/m^3), with UF accounting for only 5% of the total.

In the comparison presented above, no credit was taken for the better pretreatment provided by ultrafiltration. However, the potential benefits provided by ultrafiltration, when compared to multimedia filtration include:

1. Higher RO flux
2. Increased RO recovery
3. Longer RO membrane life

TABLE 7.5 Costs for 75,000-m^3/day Seawater Desalination Plant with Media Filtration and Ultrafiltration Pretreatment

	Capital Cost ($\$/m^3/day$)			Total Life-Cycle Cost ($\$/m^3$)	
	Media Filtration	Immersed UF		Media Filtration	Immersed UF
Infrastructure (power supply, intake, brine pipes, building, fee/permeate/reject tanks)	196	196	Capital	0.251	0.254
Pretreatment (incl. cartridge filters)	124	138	Fixed O&M	0.122	0.132
Reverse osmosis	558	558	Variable O&M	0.253	0.252
Membranes (UF and RO)	66	113			
Total	944	1005	Total	0.615	0.637

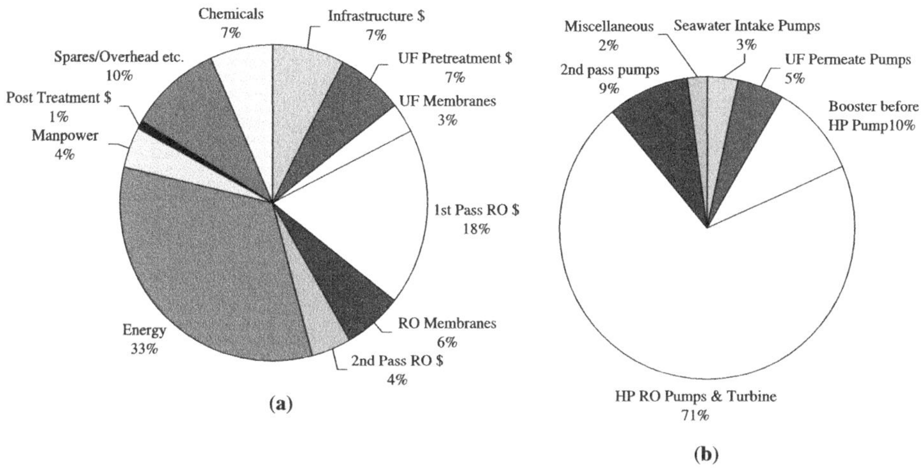

Figure 7.10 Cost breakdown for seawater desalination with ultrafiltration pretreatment: (a) Total life-cycle cost ($ indicates a capital cost) and (b) energy cost (total 3.46 kWh/m^3).

4. Increased plant availability
5. Smaller plant footprint

An attempt was made at quantifying these benefits by doing cost sensitivity simulations using the costs presented above as a base case.

Higher RO Flux The benefit of designing the plant at a higher RO flux is illustrated in Figure 7.11. While running at higher flux also means running at higher pressure, there is a benefit to increasing RO flux over the base case of 14 $L/m^2/h$ all the way up to 20 $L/m^2/h$. This means that the savings in capital cost (smaller RO) and membrane replacement cost outweigh the increased energy cost. This is true for energy cost of 5.5 ¢/kWh and 10.0 ¢/kWh. This benefit becomes quite significant if credit can be taken for improved permeate quality that results from running at higher flux, and permeate from the first pass is blended with that of the second pass (resulting in a smaller second pass). This

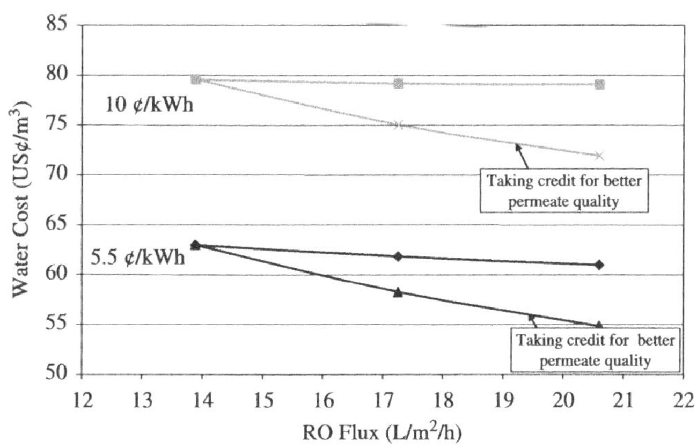

Figure 7.11 Impact of RO flux on total life-cycle costs of a seawater desalination system.

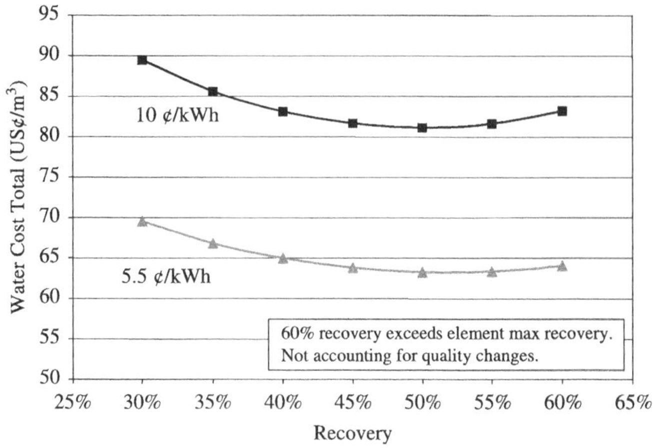

Figure 7.12 Impact of RO recovery on total life-cycle costs of a seawater desalination system.

situation would arise for high-salinity seawater or if a particular contaminant like boron controls the design.

Increased RO Recovery Figure 7.12 shows that total life-cycle costs decrease as recovery increases all the way to 50%. The benefit is derived from building a smaller intake and pretreatment system. At recovery higher than 50%, higher energy costs outweigh these benefits.

Longer RO Membrane Life Ultrafiltration pretreatment has the potential of increasing RO membrane life by protecting membranes from particulate fouling. Figure 7.13 shows the impact of membrane life on water cost.

Ultrafiltration pretreatment positively protects an RO desalination system, and thus has the potential to increase plant availability and reliability. In addition, the analysis presented above shows that UF pretreatment offers the possibility to optimize the design and operation of RO systems to reduce total life-cycle cost below current practice.

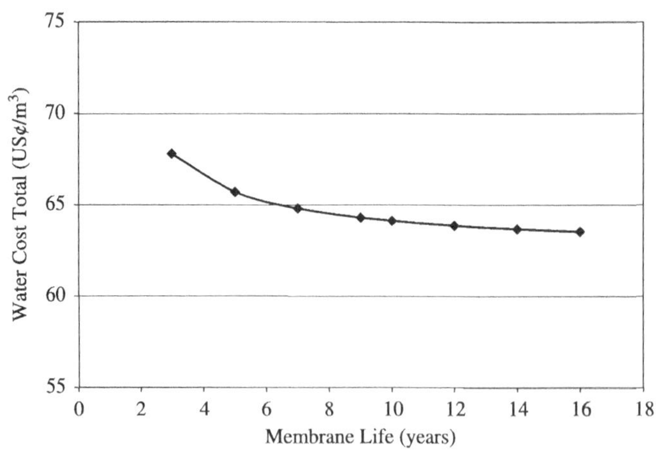

Figure 7.13 Impact of membrane life on total life-cycle costs of a seawater desalination system.

7.8 WATER REUSE VERSUS DESALINATION

Two treatment plants were designed and costed to produce RO-quality water from sewage and from seawater, using current best practice. The comparison is based on the process flow diagram in Figure 7.1, assuming that biologically treated sewage is available as a source for water reclamation (the biological treatment cost is not considered because it is assumed that sewage has to be treated for discharge). For seawater desalination, conventional multimedia filtration was assumed. The key design conditions for each system are shown in Table 7.6. Key differences, expressed as benefits for water reclamation, include:

- No need for coagulant addition
- Single pass versus two-pass RO
- Higher RO recovery
- Higher RO flux
- Lower RO feed pressure

The total costs estimated for producing RO water from secondary effluent and from seawater are compared in Table 7.7. It is assumed that the RO concentrate from both plants could be disposed of at no cost.

The capital costs for a plant producing water from seawater are about twice the costs of a plant reclaiming secondary sewage. Both the pretreatment costs and RO costs are higher.

TABLE 7.6 Process Design Conditions for Comparison of Water Reclamation and Seawater Desalination

Unit Process	Parameter	Units	Water Reuse	Desalination
Coagulation	$FeCl_3$ dose	mg/L	No	5
Pretreatment			Membrane filtration	Multimedia filtration
Reverse osmosis	Stages	Number	1	2
	Recovery	%	75	50
	Flux	$L/m^2/h$	20	14
	Feed pressure	bar	13.6	54

TABLE 7.7 Costs for Producing Water from Secondary Effluent and Seawater

Component	Units	A: From CAS Effluent	B: From Seawater	Ratio (B/A)
		Capital Costs		
Infrastructure & pretreatment	US$/$m^3$/day	161	320	1.9
RO	US$/$m^3$/day	321	624	1.9
Total	US$/$m^3$/day	482	944	2.0
		Total Life-Cycle Costs		
Capital	US$/$m^3$	0.07	0.25	3.4
O&M	US$/$m^3$	0.21	0.37	1.8
Total	US$/$m^3$	0.28	0.62	2.2

In the case of pretreatment, this is due to the difference in recovery (75% for secondary effluent; 50% for seawater), which results in a larger seawater pretreatment system. The capital cost for the seawater RO process is higher than for the secondary effluent RO as it is operating at a much higher pressure, lower permeate flux, lower recovery, and must be made of materials that resist corrosion in seawater.

Similarly, the O&M costs for producing RO water from seawater are about twice the costs of reusing secondary sewage. The higher pretreatment costs are due to continuous dosage of a coagulant. The higher RO costs are due primarily to energy (the operating pressure is 4 times higher and the feed flow is 1.5 times higher) but also to membrane replacement.

The total life-cycle costs for producing RO water from secondary effluent and seawater are 0.28US$/m^3 and 0.62US$/m^3, respectively, a ratio of 2.2.

7.9 CONCLUSIONS

The typical quality of a water reclamation plant after reverse osmosis was compared to that of desalinated water and found to be similar and acceptable for many water reuse applications.

Microfiltration and Ultrafiltration are the best available technology for water reuse. Two options are available: conventional activated sludge followed by tertiary filtration and an integrated membrane bioreactor. Both provide effluent of high quality suitable for treatment by reverse osmosis. The cost of tertiary filtration can be lower than a membrane bioreactor if the water reclamation plant is designed for constant flow and is located at a different site.

Ultrafiltration for seawater desalination is currently slightly more expensive than conventional media filtration. However, if the benefits of a better/safer pretreatment are taken into account (increased RO flux, operation of RO at higher recovery, and longer RO membrane life), UF has the potential to reduce the cost of seawater desalination below current levels.

The costs to treat sewage to indirect potable reuse standards are only a fraction of the costs to desalinate seawater. When total life-cycle costs are considered, the cost of treating secondary effluent by membrane filtration and RO is 0.28US$/m^3, as compared to 0.62US$/m^3 for seawater desalination.

REFERENCES

Côté, P., Cadera, J., Adams, N., and Mende, U. (2001). A new immersed membrane for water filtration. In *Proceedings of IWA 2nd World Water Congress: Efficient Water Management—Making it Happen*, Berlin, Germany, Oct. 15–18, 2001.

Crossley, I., Pedersen, S., and Janson, A. (2003). ZENON introduces new reinforced hollow fiber membrane system. In *Proceedings of AWWA Membrane Technology Conference and Exhibition*, Atlanta, GA, Mar. 2–5, 2003.

Daugherty, J., Alexander, K., Cutler, D., Patel, M., and Deshmukh, S. (2005). Applying advanced membrane technology for Orange County's water reuse treatment facilities. In *Proceedings of AWWA Membrane Technology Conference*, Phoenix, AZ, Mar. 6–8, 2005.

Jamaluddin, A. T. M., Hassan, A. M., and Farooque, A. M. (1998). Selection of membranes for hybrid systems, Saline Water Conversion Corporation (SWCC), Saudi Arabia, Report 8, pp. 1418–1419.

Van Houtte, E., and Verbauwhede, J. (2005). Artificial recharge of treated wastewater effluent enables sustainable groundwater management of a dune aquifer in Flanders, Belgium. In *Proceedings of the 5th International Symposium on Management of Aquifer Recharge*, Berlin, Germany, June 11–16, 2005.

Wolf, P., Siverns, S., and Monti, S. (2005). UF membranes for RO desalination pretreatment. Paper present at EDS Conference, Cyprus, May, 2005.

CHAPTER 8

Chitosan Membranes with Nanoparticles for Remediation of Chlorinated Organics

YIT-HONG TEE and DIBAKAR BHATTACHARYYA*

Department of Chemical and Materials Engineering, University of Kentucky, Lexington, Kentucky 40506-0046

8.1 INTRODUCTION

The unique chemical and physical properties of nanosized metals as compared to their bulk particles have increased the studies of nanoparticle synthesis for specific optical, magnetic, electronic, and catalytic purposes. Polymeric systems have been used in the preparation of nanoscaled particles due to the presence of specific functional groups on the backbone of the polymer chain. These groups are often ionic in nature or have lone-pair electrons that can served as a chelating agent as well as imposing a stabilizing effect on the synthesized nanoparticles. Polymers such as polyacrylic acid (PAA) (Wang et al., 2003; Shuetz and Caruso, 2004; Si et al., 2004; Kidambi and Brueing, 2005), poly(vinyl alcohol) (PVA) (Temgire and Joshi, 2004; Korchev et al., 2004; Porel et al., 2005), poly(vinyl pyridine) (P2VP or P4VP) (Bronstein et al., 1998; Lin et al., 1999; Malynych et al., 2002), polyaniline (PA) (Hable and Wrighton, 1991; Gan et al., 1995; Drelinkiewicz et al., 1998), poly(vinyl pyrrolidone) (PVP) (Suslick et al., 1996; Teranishi and Miyake, 1999; Isabel and Luis, 2002; Kapoor and Mukherjee, 2003; Machulek et al., 2003), and polyamines such as polyethyleneimine (PEI) (Sidorov et al., 1999; Bao et al., 2003; Kidambi and Brueing, 2005) that have either negatively charged carboxylic acid group or posses lone-pair electrons of amine functional groups are most commonly used in this area of studies.

Various methods have been used for the synthesis and studies of the chemical and physical properties of the polymeric membrane reactor with embedded nanoparticles. These methods include sonochemical, ultraviolet (UV), and γ irradiation, and chemical reduction by reducing agents such as borohydrides and hydrazine. Studies have revealed the successful synthesis of cellulose acetate thin film with embedded bimetallic nanoparticles such as Ni/Fe, Pd/Co, and monometallic Pd for the catalytic dechlorination and hydrogenation processes (Liu et al., 1997; Liu et al., 2000; Meyer et al., 2004). Catalytically active

*Corresponding author, email: db@engr.uky.edu

Advanced Membrane Technology and Applications. Edited by Norman N. Li, Anthony G. Fane, W. S. Winston Ho, and T. Matsuura
Copyright © 2008 John Wiley & Sons, Inc.

nanoparticles and bimetallic nanoclusters incorporated into poly(amide imide) film have been reported by Tröger et al. (1997), Fritsch and Peinemann (1995), and French et al. (2005). Formation of catalytically active nanoparticles embedded in the poly(dimethylsiloxane) matrix for the hydrodechlorination of chlorobenzene have also been reported by Fritsch et al. (2003).

Chitosan, poly(β-[1-4]-2-amino-2-deoxy-D-glucopyranose), a biodegradable, nontoxic polysaccharide, as depicted in Figure 8.1, consists of a linear chain of a six-membered ring structure with both the hydroxyl and amine functional groups on its backbone. The unique chemical and physical properties of chitosan have been the subject of numerous studies. These include the use of chitosan as clinically applicable biomaterials (Patel and Amigi, 1996; Schmuch and Krajicek, 1998; Denuziere et al., 1998), separation polymers such as pervaporation and protein separation membranes (Zeng and Ruckenstein, 1998; Jegal and Li, 1999; Nam and Lee, 1999; Kittur et al., 2005), as well as ion exchange beads (Kawamura et al., 1993; Guibal et al., 1998; Ngah and Liang, 1999; Beppu et al., 2004). The presence of hydroxyl and amine functional groups as possible chelating/ion exchange agent for metal ions has also attracted interest to study the use of chitosan for the synthesis of nanoparticles. Reports have shown that silver and other precious metals such as gold, palladium, and platinum have been successfully coated and stabilized by polymeric chitosan forming nanosized particles (Ishizuki et al., 1991; Adlim et al., 2004; Huang et al., 2004). These kinds of chitosan-supported nanoparticles have shown to be catalytically active.

The use of chitosan–Pd nanocomposites for heterogeneous catalytic reactions is the most extensively studied system and reported in the literature. For example, allylic substitution reactions of (*E*)-cinnamyl ethyl carbonate by morpholine was reported by Quignard et al. (2000) using a chitosan-supported palladium complex. Dechlorination reaction of chlorophenol to phenol, degradation of nitrophenol, and reduction of chromate were studied by Vincent et al. (2002, 2003, 2004). Hydrogenation of nitrobenzene to aniline by chitosan-impregnated palladium composites was also reported by Jin et al. (1994). Excellent review of chitosan-based materials used in the heterogeneous catalytic reactions

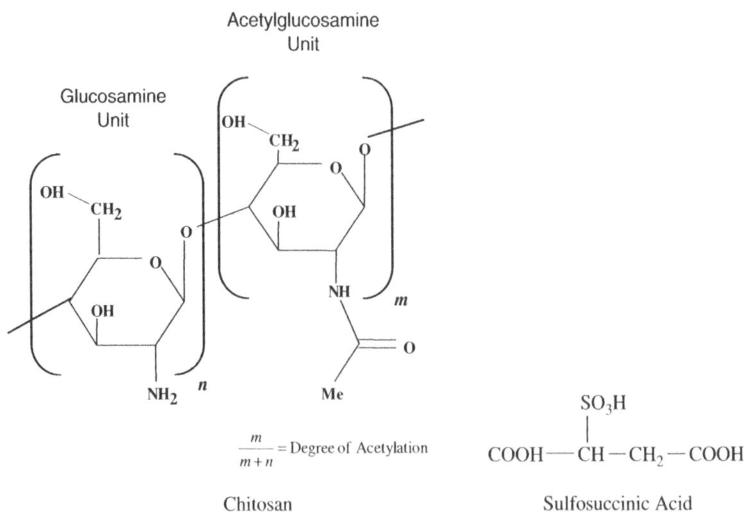

Figure 8.1 Chemical structure for chitosan and sulfosuccinic acid.

is detailed by Guibal (2005). Most of these catalytic systems were in the form of beads or flakes supported on silica or zeolites. Yet very few studies were conducted using chitosan in the form of membrane with active nanoparticles embedded in the polymer matrix for catalytic reactions. These were largely due to the relatively dense structure of the final membrane formed by chitosan. The low porosity of the thin film increases the diffusion resistance of the target compound from contacting the nanoparticles embedded in the membrane if the compound has a low partition coefficient with chitosan. These will render the nanoparticles in the polymer matrix "inactive" and limit the catalytic reaction on the external surface of the membrane. The diffusion limitation of chitosan film can be reduced by inducing porosity during or after the formation of the membrane such as the use of silica as the pore-inducing agent reported by Zeng and Ruckenstein (1996) for the formation of macroporous chitosan membranes.

In view of the film-forming properties of chitosan and its unique chelating characteristic, studies were conducted to explore the feasible use of chitosan as the polymeric precursor to prepare membranes with embedded nanoparticles. The main objective of this work is to synthesize and use the bimetallic Ni/Fe and Pd/Fe nanoparticles in the chitosan membrane for the degradation study of hazardous chlorinated compounds: trichloroethylene (TCE) and 2,2′-dichlorobiphenyl (DCB). The first part of the study involved the synthesis of chitosan-embedded bimetallic nanoparticle membranes using the chemical reduction technique by borohydride. The membrane is then characterized by conventional methods such as X-ray diffraction (XRD) spectroscopy, thermogravimetric analysis (TGA), Fourier transform infrared (FTIR) spectroscopy, and scanning and transmission electron microscopy (SEM and TEM). This is followed by the catalytic degradation study of TCE and DCB using the mixed-matrix membrane reactor conducted in the batch and pressure-induced experiments, respectively. The kinetic results of TCE degradation obtained are quantitatively compared with the solution phase studies reported in the literature (Schrick et al., 2004; Tee et al., 2005). A long-term study is conducted for DCB degradation under pressure-induced condition to investigate the applicability of the chitosan-embedded Pd/Fe nanoparticles under continuous operation. To the best of our knowledge, this is the first study reporting the use of chitosan to synthesize porous membranes with embedded nanoparticles for the degradation of TCE and DCB.

8.2 EXPERIMENTAL SECTION

8.2.1 Chemicals

High-molecular-weight chitosan flakes ($M_w \approx 1{,}000{,}000$ with >75% deacetylation), sulfosuccinic acid (70 wt% in aqueous solution), granular sodium borohydride ($NaBH_4 = 99.99\%$), nickel chloride ($NiCl_2 \cdot 6H_2O = 99.99\%$), palladium(II) acetate [$(CH_3COO)_2Pd = 99.98\%$], silica gel (Davisil Grade 710, pore size 60 Å, pore volume 0.75 cm^3/g, particle size 4–20 μm), and ethanol (denatured reagent grade) were purchased from Aldrich Chemical Company Inc. Trichloroethylene (TCE: ACS reagent grade), Glacial acetic acid (HPLC) [high pressure liquid chromatography grade], ferrous chloride (assay as $FeCl_2 \cdot 4H_2O = 102.0$), nitric acid (trace-metal grade), and deionized ultrafiltered water (DIUF) were from Fischer Scientific. 2,2′-Dichloroethylene (DCB: 99.9% purity), 2-chlorobiphenyl (99.9% purity), and biphenyl (99.9% purity) were from Ultra Scientific. All of the mentioned chemicals and materials were used without further treatment. DIUF was deoxygenated by heating at ~60°C and bubbling with nitrogen gas overnight before use in the study.

8.2.2 Synthesis of Chitosan Membranes

8.2.2.1 Chitosan Membrane with Silica Gel Chitosan membrane was synthesized by dissolving 1.0 g of chitosan flakes in a 100-mL total volume of 1 vol% acetic acid solution. The mixture was vigorously stirred for 2 h with a mechanical stirrer to dissolve the chitosan. This was followed by the addition of 1.0 mL of 70 wt% sulfosuccinic acid in the solution and stirred for another 2 h. Silica gel was then added slowly into the mixtures and allowed to mix for 30 min under intense stirring for homogeneity. The final pH of the solution was about 2.2. A specific weight of the solution was poured on a Petri dish (diameter = 7.5 cm) and dried in the hood. The dried chitosan film was soaked in 500-mL aqueous solution of 0.2 M $NaBH_4$ twice for a total immersion time of 3 h. Chitosan membrane was formed and detached easily from the Petri dish once being soaked in the $NaBH_4$ solution. The resulting membrane was thoroughly washed with deoxygenated water followed by rinsing with ethanol before further analysis.

8.2.2.2 Chitosan-Embedded Fe Nanoparticle Membrane with Silica Gel Solution with a mixture of chitosan, sulfosuccinic acid, and silica gel was prepared by the same procedure outlined above (Section 8.2.2.1). A 10-mL 0.18 M Fe^{2+} aqueous solution was added to the chitosan mixture and mechanically stirred for 2 h. The final pH of the solution mixture was 3.00. Specific amount of the mixture was poured on a Petri dish (diameter = 7.5 cm) and allowed to dry overnight in the hood. The synthesis of chitosan-embedded Fe nanoparticle membrane was achieved by immersing the dried film in 500 mL of 0.4 M $NaBH_4$ solution (pH $\approx$ 11.0). Intense reaction with bubbling accompanied by the gradual formation of a dark and rigid membrane was observed at the beginning. Reduction reaction of Fe^{2+} to zero-valent Fe nanoparticles and the dissolution of silica gel in alkaline medium continued as the membrane immersed in the BH_4^- solution. This process is repeated a second time for a total of 3 h immersion of the membrane in the borohydride solution. After the immersion process, the membrane was placed in a pressure cell with a filter paper as support. The cell was filled with 50 mL of 0.4 M $NaBH_4$ (50 vol% in ethanol) and pressurized to 5.8 bars using nitrogen gas. This second step is to ensure the complete reduction of Fe^{2+} in the membrane matrix to zero-valent Fe^0 nanoparticles as the BH_4^- convectively passed through the membrane in the pressure cell. This process is also repeated twice with a total volume of 100 mL of 0.4 M $NaBH_4$. Four batches of 50 mL ethanol were added and convectively passed through the membrane to wash the membrane thoroughly in the cell. The synthesis flowchart is shown in Figure 8.2.

8.2.2.3 Chitosan-Embedded Ni/Fe and Pd/Fe Nanoparticle Membrane with Silica Gel The doping of chitosan-embedded Fe nanoparticle membrane with Ni was conducted using 60 mL of 0.50×10^{-3} M Ni^{2+} in ethanol solution (Ni^{2+} from stock of 12,000 mg/L aqueous solution prepared from $NiCl_2 \cdot 6H_2O$). The Ni^{2+} was allowed to convectively pass through the membrane in the cell under pressurized condition (5.8 bars). After the doping process, 100 mL of 0.2 M $NaBH_4$ solution (50 vol% ethanol) was added in the cell and convectively passed through the membrane at 5.8 bars. The chitosan-embedded Ni/Fe nanoparticle mixed-matrix membrane was washed with copious amounts of ethanol in the cell followed by rinsing with deoxygenated water

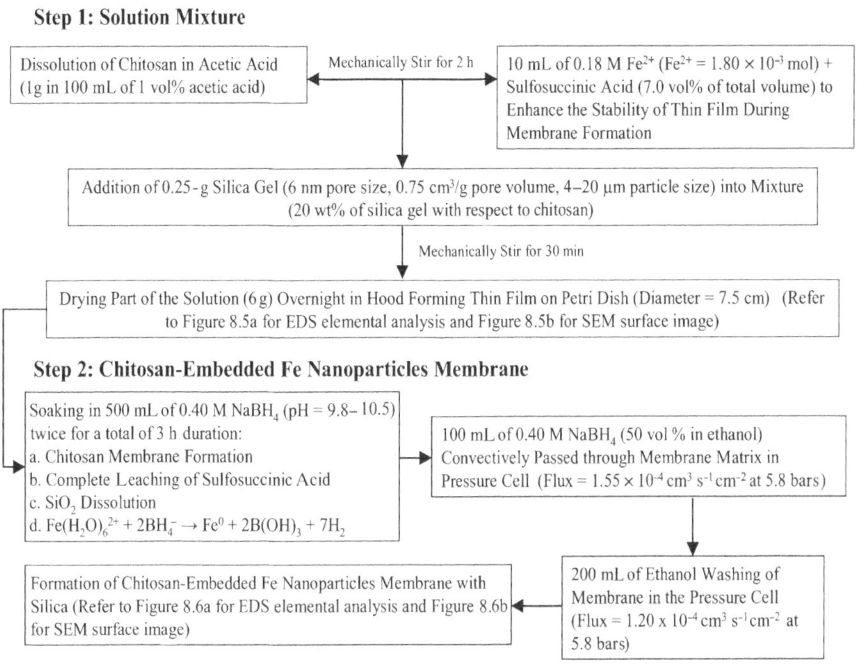

Figure 8.2 Flowchart for the synthesis of chitosan-embedded Fe nanoparticles membrane with silica.

before being used in the TCE degradation study. Postcoating of the chitosan-embedded Fe nanoparticles with palladium is similar as described above. In brief, the membrane in the pressure cell was filled with 6 mL of 0.05×10^{-3} M Pd^{2+} [550 mg/L Pd^{2+} from palladium(II) acetate] in acetone. The cell was pressurized and postcoating of Fe nanoparticles with Pd was achieved as the Pd^{2+} convectively passed through the membrane. No $NaBH_4$ was used after the postcoating of Fe nanoparticles with Pd. This is due to the ability of zero-valent Fe to reduce Pd^{2+} based on the difference in the reduction potential:

$$Pd^{2+} + 2e^- \rightarrow Pd^0 \qquad E^0 = +0.98 \tag{8.1}$$

$$Fe^{2+} + 2e^- \rightarrow Fe^0 \qquad E^0 = -0.44 \tag{8.2}$$

$$Fe^0 + Pd^{2+} \rightarrow Fe^{2+} + Pd^0 \tag{8.3}$$

TABLE 8.1 Characteristics and Metal Loading of Chitosan-Embedded Bimetallic Ni/Fe Nanoparticles Mixed-Matrix Membranes with Silica for TCE Degradation Study

Membrane weight	65.00 ± 4.50 mg
Membrane diameter	6.50 cm
Membrane surface area	33.20 cm^2
Membrane thickness	60.50 ± 5.50 μm
Fe weight	3.60 ± 0.25 mg
Ni weight	1.20 ± 0.15 mg
Silica weight	3.50 ± 0.15 mg

TABLE 8.2 Characteristics and Metal Loading of Chitosan-Embedded Bimetallic Pd/Fe Nanoparticles Mixed-Matrix Membranes with Silica for DCB Degradation Study

Membrane weight	39.00 ± 2.00 mg
Membrane diameter	4.15 cm
Membrane surface area	13.50 cm^2
Membrane thickness	58.50 ± 6.50 μm
Fe weight	2.80 ± 0.15 mg
Pd weight	0.015 ± 0.005 mg
Silica weight	2.20 ± 0.05 mg

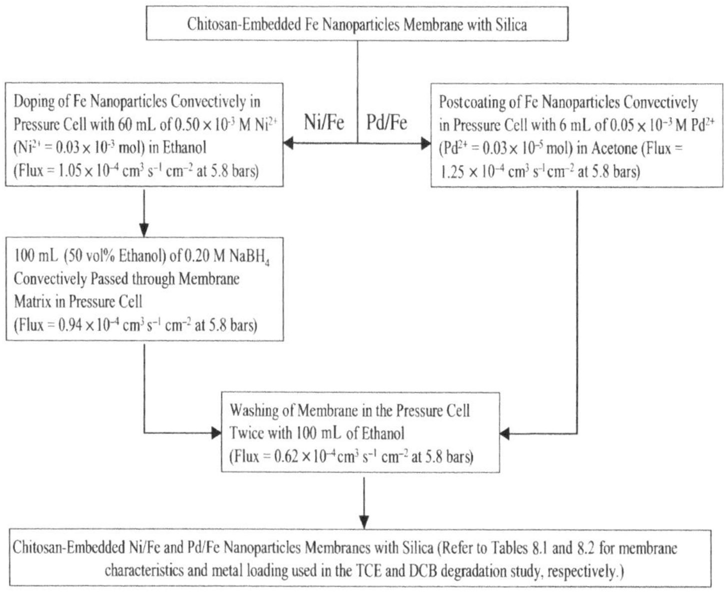

Figure 8.3 Flowchart for the synthesis of chitosan-embedded bimetallic (Ni/Fe and Pd/Fe) nanoparticles mixed-matrix membrane with silica.

The chitosan-embedded Pd/Fe nanoparticle membrane was thoroughly washed with ethanol in the cell. This was followed by rinsing with deoxygenated water once the membranes were taken out from the cell and used immediately in the DCB degradation study. Tables 8.1 and 8.2 show the characteristics of the membranes used in the TCE and DCB degradation studies, respectively. Figure 8.3 shows the doping and postcoating of Ni and Pd, respectively.

8.2.3 Methods of Degradation Study

8.2.3.1 Batch Degradation Study of TCE Batch experiments were conducted in 42-mL total volume serum-seal glass vials using two pieces of chitosan-embedded Ni/Fe nanoparticle membranes with an initial TCE concentration of 10 mg/L. The metal loading and Ni/Fe ratio for each of the membrane was determined by atomic adsorption (AA) analysis (metal loading = 5.20 mg, Ni/Fe ratio = $\frac{1}{3}$). Initial and final pH of the solution was 6.5 and ~7.2, respectively. A headspace of 2 mL was left in the vial for gas-phase analysis. The reaction vials were clamped to a shaker and samples were collected at different time intervals. Control experiments were performed using chitosan membrane without the Ni/Fe nanoparticles. The total TCE balance in the liquid phase was 96% of the initial TCE concentration. This showed that TCE has negligible partition coefficient in the chitosan, and the volatility of TCE in the headspace can be neglected in the mass balance calculation. Reproducibility of the TCE degradation study was checked by using different membranes synthesized from three different batches and doping with Ni^{2+} under the same experimental conditions.

8.2.3.2 Degradation Study of 2,2′-dichlorobiphenyls (DCB) under Pressure-Induced Operation A high-pressure cell with an inner diameter of 4.15 cm (active surface area = 13.50 cm^2) was used for the degradation study of DCB under pressure-induced (convective flow) operation. The metal loading and Pd wt% for the chitosan-embedded Pd/Fe nanoparticle membranes were determined by ion-coupled plasma (ICP) analysis (metal loading = 2.30 mg, Pd = 0.50 wt% with respect to the total metal loading). The pressure cell was filled with DCB with an initial concentration and total volume of 15 mg/L and 100 mL, respectively. Transmembrane pressures (using nitrogen gas pressure) ranged from 0.35 to 11.0 bars. For each of the pressures studied, the first 10 mL of the sample was collected and the collection time was recorded to determine the volumetric flux of the membrane. This was followed by the collection of different sample volumes at different time intervals. The volumetric fluxes were determined and compared with the initial value to study the flux fluctuation. DCB [gas chromatography–mass spectrometer (GC–M)] was analyzed as a function of time to establish steady-state behavior of the reaction. It was found that the volumetric fluxes determined at different time intervals for each of the pressures studied were within ±15% deviation. The DCB's GC–MS analysis also showed that the reaction had reached its steady-state value as the results showed that the first 10-mL sample was ±10% to the sample collected thereafter. Control experiment was conducted using chitosan membranes with silica gel, and the results showed that the adsorption of DCB in the membrane phase was 20.00 ± 4.5% relative to the initial concentration of DCB. The adsorption of DCB in the membrane was taken into account when reporting the degradation results.

8.2.4 Methods of Analysis

8.2.4.1 Trichloroethylene Analysis A similar analytical method, reported previously in the degradation of TCE by bimetallic Ni/Fe nanoparticles, was used in this study (Tee et al., 2005). In brief, headspace volume of 0.4 mL was drawn from the reaction vial and injected manually into the sampling port of a cryogenic-capable HP 5890 Series II gas chromatography (column = Supelco – 1.4 μm SPB-624) equipped with a Series 6150 mass spectrometer (Hewlett-Packard) for gas-phase analysis. The gas chromatography was cooled to −5°C using liquid CO_2 and increased to 200°C at 8°C/min. Quantitative ethane analysis was conducted by correlation using a five-point calibration curve (correlation coefficient, $R^2 = 0.986$) constructed by standards obtained from Supelco. The reproducibility of the calibration curve was checked periodically by known standards that fall within the concentration range of the curve.

For liquid-phase analysis, a sample volume of 0.4 mL was taken directly from the reaction vial and diluted with 40 mL DIUF water in another 42-mL serum-seal glass vial. An analytical purge-and-trap instrument (OI, model 4560) interfaced with the Hewlett-Packard 5890 (column = Supelco–1.4 μm SPB-624) gas chromatography machine was used for the TCE analysis. Initial GC temperature of 35°C was increased to 200°C with a ramping rate of 6°C/min. An internal standard of methylene chloride (0.1 mg/L) was mixed with the diluted liquid sample during the purging stage of the purge-and-trap analysis. Control experiments showed that the error associated with the GC–MS analysis was 8% relative to the initial TCE concentration. Reproducibility of the liquid-phase analysis was checked by TCE standard from Ultra Scientific (HC-270). The calculated TCE standard, relative to the actual concentration, was 96%.

8.2.4.2 Analysis of DCB A Varian CP-3800 gas chromatography (column = Varian EZ-Guard capillary column CP-9012) coupled with Varian Saturn-2200 GC–MS was used to measure the concentration of DCB and its degradation products of 2-chlorobiphenyl and biphenyl. The GC was initially held at 100°C and increased to 210°C with a ramping rate of 12°C/min. This is followed by a final 30°C/min increase to 300°C. For each of the samples collected at different transmembrane pressures and time intervals, 2 mL was added into an 8-mL vial containing 2 mL of hexane for extraction. The vials were clamped to a shaker for 2 h of extraction process. At the end of the extraction, 1-mL aliquot from the hexane phase was transferred to a 1.5 mL GC autosampler vial. This was followed by the addition of 10 μL of naphthalene d-8 (5000 mg/L in methylene chloride from Ultra Scientific) into the autosampler vial as internal standard for GC–MS analysis. At the end of the degradation study, the used membrane was removed from the convective cell and placed into a 20-mL vial with 10 mL of hexane for membrane phase extraction. The same extraction and sampling procedures described previously were conducted for the membrane's GC–MS analysis. The concentration of DCB, 2-chlorobiphenyl, and biphenyl were correlated using a calibration curve ranging from 0.5 to 20 mg/L constructed from external standards obtained from Ultra Scientific. A periodic check was conducted to ensure the reproducibility of the calibration curve using 5- and 10-mg/L standards from dilution of the premade 100 mg/L DCBs, 2-chlorobiphenyl, and biphenyl (Ultra Scientific).

8.2.5 Characterization of Membranes

8.2.5.1 Fourier Transform Infrared Spectroscopy Analysis A ThermoNicolet Nexus 470 spectrometer was used for FTIR analysis to detect the presence of functional groups in the membranes. Thin membrane samples were attached to a polyethylene substrate FTIR card from Thermo Electron Corp. and analyzed at ambient temperature. All spectra were obtained from 100 scans at 4.0 cm^{-1} resolution with spectra ranging from 400 to 4000 cm^{-1}.

8.2.5.2 X-ray Diffraction Analysis Thin-films from X-ray diffraction for all chitosan membranes were characterized using a Siemen diffractometer interfaced with DACO-Kristalloflex using Cu $K_{\alpha 1}$ ($\lambda = 0.1541$ nm) as radiation source generated at 40 kV and 30 mA. The XRD data were collected using a scanning rate of 1.0° min^{-1} and a scanning mode with a step size of 0.08° in the 2θ range from 5 to 85° for chitosan–silica gel membranes and 10–90° for all other samples.

8.2.5.3 Thermogravimetric Analysis The thermal properties of the membranes were evaluated using a Perkin-Elmer thermogravimetric analyzer (TGA-7) interfaced with Pyris Thermal Analysis software Version 7. The membrane samples were freeze-dried and crushed into pieces using liquid nitrogen. This was followed by weighing approximately 5.0 mg of sample and heating it from 40 to 800°C at 10°C/min under a nitrogen atmosphere.

8.2.5.4 Atomic Absorption Analysis and Inductively Coupled Plasma Analysis The total metal loading of Fe and Ni in the membrane was analyzed using Varian Fast Sequential AA spectrometer (SpectraAA 220/FS) interfaced with SpectraAA V-3.0 for quantitative analysis. The Pd content in the membrane was determined using Varian Vista-Pro Simultaneous ICP-OES machine. The use of ICP for Pd analysis is due to the expected low Pd content in the membrane that required an analytical method with high sensitivity at low detection limit. Five-point linear calibration curves for Fe, Ni, and Pd were constructed and the results were correlated accordingly. The calibration curves were periodically checked with solutions appropriately diluted using standards obtained from Ultra Scientific (Pd^{2+} = 1000 mg/L) and Fisher Scientific (Fe^{2+} and Ni^{2+} = 1000 mg/L).

8.2.5.5 Scanning Electron Microscopy and Energy Dispersive X-ray Spectroscopy (SEM–EDS) Analysis High-resolution SEM and SEM–EDS analyses were performed on a Hitachi S-900 operated at 3 kV and a Hitachi S-3200 at 20 kV, respectively. Both machines were pumped to near vacuum at ~50 mTorr. Samples for high-resolution SEM imaging were sputter-coated with a thin layer of gold coating with a thickness of ~2–5 nm (~20 s sputter time) by an EmScope SC 400 sputter coater to reduce electrostatic charging. SEM–EDS elemental analysis was conducted using a Noran Voyager EDS detector coupled with Noran Voyager Easy Spectra V-4.2.2 for qualitative and quantitative analyses.

8.2.5.6 High-Resolution Transmission Electron Microscopy and Energy Dispersive X-ray Spectroscopy (HRTEM–EDS) Analysis A JEOL 2010F FasTEM field emission electron microscopy equipped with EDS detector operated at 200 kV was used to carry out the HRTEM–EDS analysis. Electron beam with point-to-point resolution of 0.5 nm was used for the EDS elemental analysis of Fe, Ni, and Si at different locations of the membrane. Lacey carbon copper grid (Lacey Carbon Type-A, Ted Palla, Inc.) was coated with membrane solution and dried under vacuum. This is followed by the reduction of Fe^{2+} and the dissolution of silica gel using borohydride solution. After the reduction process, the Fe nanoparticles was doped with Ni and redried under vacuum. The vacuum-dried sample was dipped into another borohydride solution. The copper grid with chitosan-embedded Ni/Fe nanoparticles is washed with ethanol and dried in the hood before the HRTEM–EDS analysis.

8.3 RESULTS AND DISCUSSIONS

8.3.1 Synthesis and Characterization of Chitosan Membranes

8.3.1.1 Membrane Synthesis The synthesis of chitosan membranes with embedded Fe nanoparticles is achieved by the simultaneous pore formation utilizing silica particles as porogen and the reduction of Fe^{2+} by sodium borohydride, as described in Figure 8.2.

The polymeric chitosan formed a hydrophilic yet mechanically stable thin film with nanoparticles randomly distributed on the surface and in the membrane. The use of silica gels to induce porosity in the chitosan membrane has been previously reported by Zeng and Ruckenstein (1996). They reported that stable macroporous membranes can be synthesized when the silica particles were extracted out of the dried chitosan membrane by immersion in an alkaline medium. The synthesis of porous chitosan membrane has also been studied and used as an ion exchange agent in the adsorption of Cu(II) (Beppu et al., 2004) and the separation of protein (Zeng and Rukenstein, 1998). In this study, only aqueous sodium borohydride as the alkaline medium was used to form porous chitosan membrane and simultaneously reducing Fe^{2+} to zero-valent iron. This is different from the approach used by Zeng and Ruckenstein (1996) where dissolution of silica was attained using a warm bath of aqueous NaOH. Partial dissolution of silica gel was observed at the end with approximately 60% loss of silica weight. SEM analysis (image not shown) of the membrane without the Fe nanoparticles showed that the silica particle size was 60–70 nm versus the initial value of 4–20 μm. The silica weight loss was 65% relative to the initial value. This is close to the silica weight loss in the chitosan-embedded Fe nanoparticle membrane, and the assumption was made that the silica had a roughly similar particle size in both of the membranes with and without the Fe nanoparticles. Table 8.1 shows the structural characteristics and metal loading of the chitosan-embedded Fe nanoparticle membrane with silica before the doping and postcoating with Ni and Pd, respectively.

8.3.1.2 Fourier Transform Infrared Analysis The FTIR analysis in Figure 8.4 shows the spectrums for the dried polymer solution and the chitosan-embedded Fe nanoparticle membrane with silica. The results in Figure 8.4 for both the dried polymer solution and the membrane exhibit quite similar stretching bands and are outlined with its wavenumber (cm^{-1}) as follows:

1. Broad peaks range from 3450 to 3390 cm^{-1} (region in line a): OH and NH group stretching
2. 2960 cm^{-1} (region in line b): CH stretching of CH_3
3. 1590 cm^{-1} (line d): NH_2 in amino group
4. 1020 to 1150 cm^{-1} (region in line f): C–O–C ether linkage and Si–O–Si stretching vibrations
5. 800 cm^{-1} (line g): CH_3OH group

In Figure 8.4a of the dried polymer solution, the characteristic bands correspond to the C=O and SO_3^- groups from sulfosuccinic acid and were observed at 1750 cm^{-1} (line c in Fig. 8.4b) and 1250 cm^{-1} (line e in Fig. 8.4a), respectively. A small shoulder at 1620 cm^{-1} appeared next to the NH_2 stretching band at 1590 cm^{-1} was attributed to the symmetric deformation of NH_3^+. This indicated that the SO_3^- group of the sulfosuccinic acid can form a complex with NH_3^+ in chitosan via electrostatic interaction. This will increase the stability of thin film during membrane formation where an intense reduction reaction of Fe^{2+} and dissolution of silica occurred in the borohydride solution. Due to the small molecule size of the sulfosuccinic acid, gradual leaching occurred during the synthesis process. This is confirmed by the FTIR spectra in Figure 8.4b and the SEM image coupled with EDS analysis in Figures 8.5 and 8.6 showing the presence and complete leaching of sulfosuccinic acid before and after the synthesis, respectively.

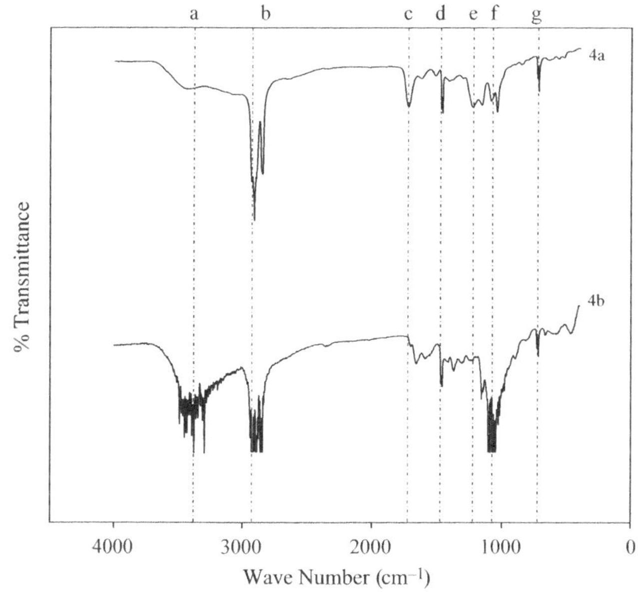

Figure 8.4 FTIR analysis of (a) dried polymer solution of chitosan, silica, Fe^{2+}, and sulfosuccinic acid; (b) chitosan-embedded Fe nanoparticles with silica membrane.

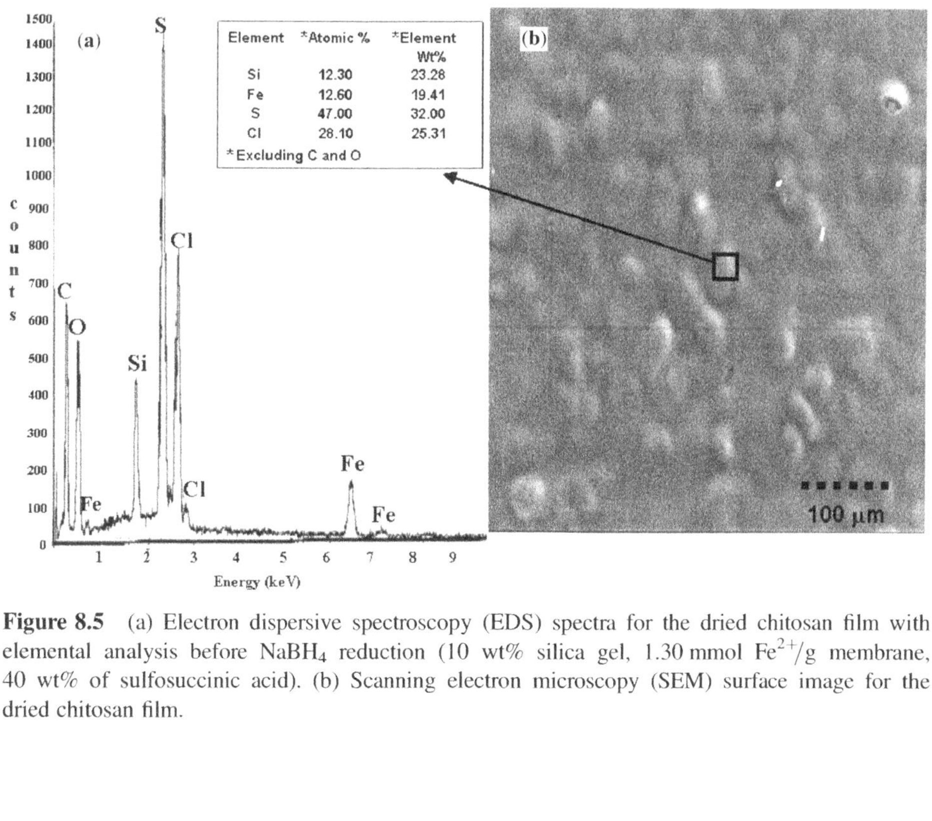

Figure 8.5 (a) Electron dispersive spectroscopy (EDS) spectra for the dried chitosan film with elemental analysis before $NaBH_4$ reduction (10 wt% silica gel, 1.30 mmol Fe^{2+}/g membrane, 40 wt% of sulfosuccinic acid). (b) Scanning electron microscopy (SEM) surface image for the dried chitosan film.

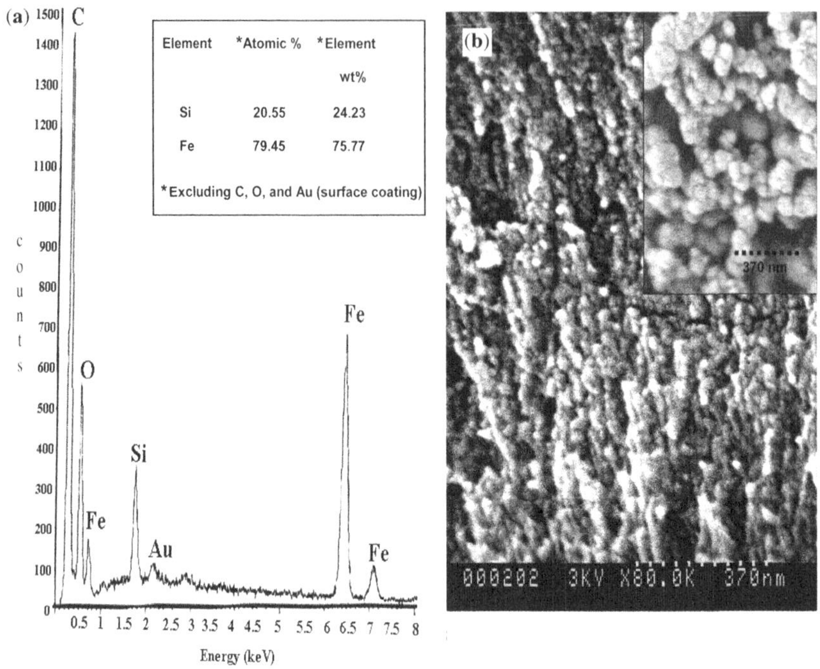

Figure 8.6 (a) Electron dispersive spectroscopy (EDS) spectra with elemental analysis for the cross section of the chitosan-embedded Fe nanoparticles with silica (6.0 wt% silica gel, 73 mg Fe/g membrane). (b) Scanning electron microscopy (SEM) of cross-section image and surface image (shown as insert) for the membrane.

8.3.1.3 X-ray Diffraction Analysis For chitosan membranes with and without the embedded Fe nanoparticles XRD analysis was performed and shown in Figure 8.7a–c. It was observed that the chitosan membrane in Figure 8.7a shows two relatively broadened diffraction peaks at approximately 10° (2θ) and 20° (2θ), which were attributed to the hydrated and anhydrous crystals, respectively. This is consistent with the observations made by Ogawa and Yui (1993) and Nunthanid et al. (2001). They reported the loss of crystallinity of the chitosan membranes by the observed peak broadening at the diffraction angles of $2\theta = 10°$ and $20°$. The loss of crystalline structure together with the formation of the amorphous phase is due to the reprecipitation and restructuring of the chitosan after the dissolution in acidic medium. The same observations are made for the chitosan membranes with embedded nanoparticles with similar diffraction patterns, as shown in Figure 8.7b and 8.7c. The increased intensity of the hydrated phase accompanied by weaker and broader diffraction of the anhydrous region can be attributed to the morphological changes of the chitosan membrane. These changes can be attributed to the dissolution of silica and Fe nanoparticle formation in the membrane when it is treated with $NaBH_4$. This kind of phenomenon has been reported by Ogawa and Yui (1993). Based on the XRD analysis, it is concluded that the chitosan-embedded Fe nanoparticle membrane has more amorphous and less oriented structure than the chitosan thin film.

The XRD analysis for Fisher electrolytic Fe and Fe nanoparticles synthesized by the solution method of $NaBH_4$ reduction were also obtained for comparison purpose. Diffraction patterns for the Fisher electrolytic Fe in Figure 8.7e were confirmed by the

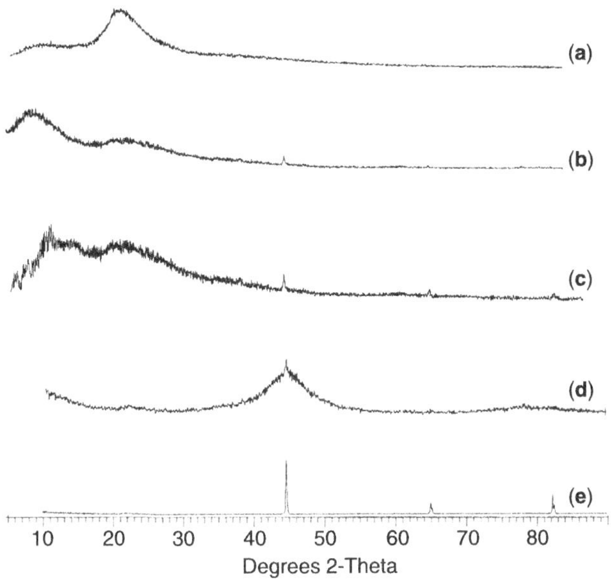

Figure 8.7 XRD analysis for (a) chitosan membrane, (b) 1.30 mmol Fe^{2+}/g chitosan membrane, (c) 3.25 mmol Fe^{2+}/g chitosan membrane, (d) Fe nanoparticles by $NaBH_4$, and (e) Fisher electrolytic Fe.

data files of the International Centre for Diffraction Data (ICDD) [formerly Joint Committee on Powder Diffraction Standards, (JCPDS)] (ICDD card number: 6-696). The Fe nanoparticles synthesized by the solution method show a strong and broadened peak at a diffraction angle of approximately 45° in Figure 8.7d, which is similar to the most intense diffraction peak ($2\theta = 44.788°$, $d_{spacing} = 2.022$, RI = 100%) of the electrolytic Fe. The peak broadening is due to the amorphousity and the different solid phases detected on the particles. Neglecting peak broadening due to microstrain in the structure and stripping undesirable peaks caused by Cu $K_{\alpha 2}$ radiation using TracersV6 software from Diffraction Technology, the solid phases with respective diffraction patterns were examined. The results showed the presence of iron oxides [$Fe_{0.98}O$—ICDD: 39-1088, FeO (wüstite)—ICDD: 6-615, Fe_2O_3 (maghemite)—ICDD: 25-1402, and Fe_3O_4 (magnetite)—ICDD: 19-629] in addition to Fe^0. The presence of oxides in the sample is assumed to be attributed to the oxidation of surface Fe atoms during the handling and sample preparation. The appearance of similar diffraction angle with less peak broadening is observed for the chitosan-embedded Fe nanoparticle membrane. With the presence of polymeric chitosan, fewer solid oxide phases were found on the surface of membrane-embedded nanoparticles. The only oxide phase detected was $Fe_{0.98}O$ and the nanoparticles on the surface were predominantly Fe^0. When the metal loading on the membrane increases, the protective and stabilizing effect of the polymeric chitosan gradually decreases. This is confirmed by the detection of more oxide phases in membranes with a higher ratio of Fe per unit mass of membrane (mmol Fe/g membrane) (data not shown).

8.3.1.4 Thermogravimetric Analysis Thermal stability and the amount of nonvolatile components in the chitosan membranes could be inferred from the TGA thermograms in Figure 8.8. The initial weight loss from 50 to 140°C for all of the membranes was attributed to the dehydration of water in the chitosan. This agrees with the observation made by

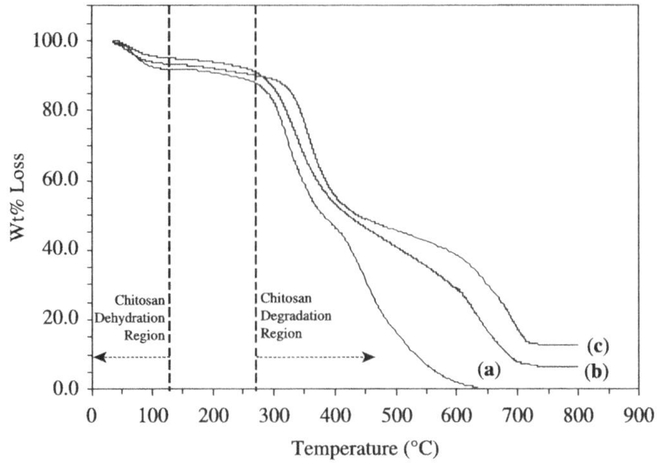

Figure 8.8 TGA analysis for (a) chitosan membrane, (b) chitosan membrane with silica (6 wt% silica relative to membrane weight), and (c) chitosan-embedded Fe nanoparticles with silica (6 wt% silica and 7.5 wt% Fe relative to membrane weight).

Nunthanid et al. (2001). A rapid weight loss at a higher temperature of 260°C was due to the chitosan chain degradation. Degradation of chitosan continued with the increase in temperature, and complete decomposition was observed at 640 and 700°C for the chitosan (Fig. 8.8a) and chitosan mixed-matrix membranes (Fig. 8.8b and 8.8c), respectively. It is observed in Figure 8.8 that with the addition of nanoparticles in the membrane matrix, the physical and mechanical properties of chitosan was slightly improved by decomposing at a higher temperature. The final weight percentage of the nonvolatile silica in the membrane shown in Figure 8.8b was around 6.0 wt% relative to the initial membrane mass. The combined weight of the silica and Fe nanoparticles retained in the mixed-matrix membrane in Figure 8.8c was approximately 14.0 wt%. Both of the results are consistent with the AA and ICP analyses (data not shown) by digestion technique conducted using the same mixed-matrix membranes (chitosan–silica and chitosan-embedded Fe with silica membranes).

8.3.1.5 Analysis by SEM, HRTEM, and EDS The formation of Fe nanoparticles in the membrane domain with silica was observed by the SEM as shown in Figure 8.6. The distinction between the Fe and silica in the image was not possible since both particles are in close contact with each other. Based on the previously reported results for the membrane systems with embedded nanoparticles (Meyer et al., 2004; Xu et al., 2005) and the solution-phase synthesis studies using borohydride as the reducing agent (Schrick et al., 2004; Tee et al., 2005), bimetallic nanoparticles obtained in this study are in the range of 30–70 nm in diameter.

$$Fe^{2+} + 2BH_4^- + 6H_2O \rightarrow Fe(0) + 3B(OH)_3 + 7H_2 \quad (8.4)$$

The presence of polymers prevent the agglomeration of the metal nanoparticles in the final membrane. In order to investigate the effect of membrane on nanoparticle formation, HRTEM was used to study the structural characteristic of the synthesized Fe nanoparticles. The HRTEM image in Figure 8.9 shows that nanoparticles shown in the SEM image

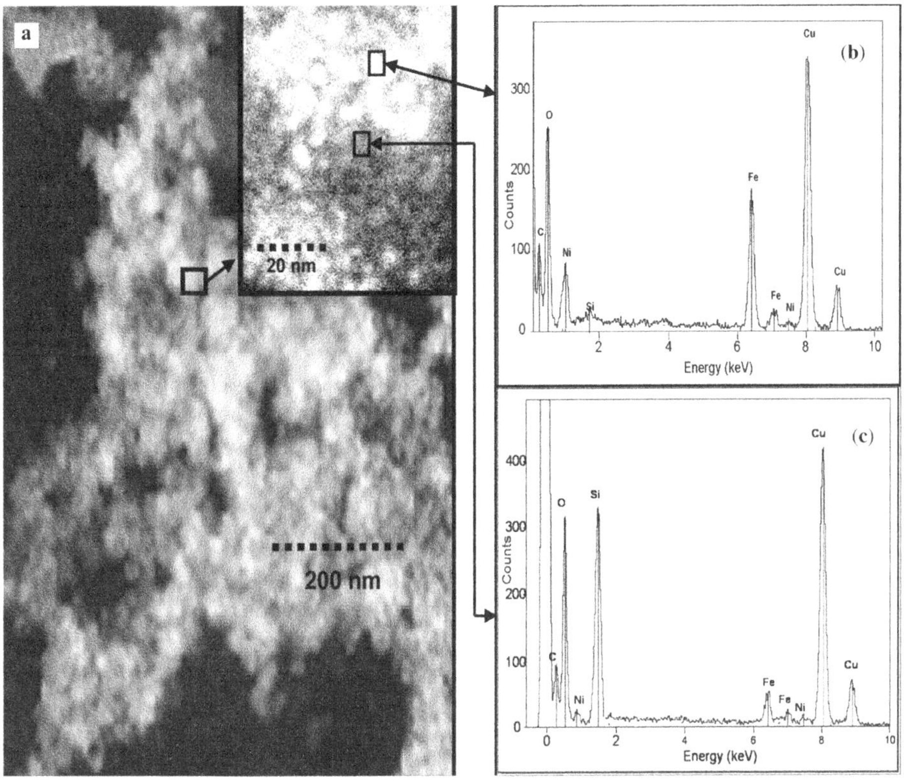

Figure 8.9 (a) High-resolution transmission electron microscopy (HRTEM) image of chitosan-embedded bimetallic Ni/Fe nanoparticles (Ni/Fe weight ratio = 1/4) mixed-matrix membrane (insert shows enlarged image for the squared area of the same membrane). (b) and (c) Electron dispersive spectroscopy (EDS) spectra with elemental analysis (shown in Table 8.3) of the membrane with silica at different positions.

actually consisted of nanoclusters with smaller silica and Ni/Fe particles with diameter of 5–10 nm in the membrane domain.

The bimetallic Ni/Fe nanoparticles in Figure 8.9 synthesized by doping the Fe nanoparticles with stoichiometric amount of Ni followed by $NaBH_4$ treatment (Ni/Fe weight ratio = 1/4) seem to be deposited on the larger silica particle surface. Elemental analysis of the nanoclusters in Table 8.3 shows that the bimetallic Ni/Fe compositions of the 5 to 10-nm size particles was indeed close to 1/4 weight ratio with high concentration of Si in the background.

TABLE 8.3 Elemental Analysis of Figure 8.9b and 8.9c for Chitosan-Embedded Bimetallic Ni/Fe Nanoparticles (Ni/Fe Weight Ratio = 1/4) Mixed-Matrix Membranes with Silica

	Si		Fe		Ni	
Element	Atomic %	Element wt%	Atomic %	Element wt%	Atomic %	Element wt%
Figure 8.9b	2.65	3.22	77.10	74.70	20.25	22.08
Figure 8.9c	85.26	81.67	12.08	15.52	2.66	2.81

8.3.2 Degradation Study of TCE

8.3.2.1 Batch Degradation of TCE The results of the batch degradation experiments by chitosan-embedded Ni/Fe nanoparticles in Figure 8.10 show the simultaneous dechlorination of TCE and formation of ethane as the main product, consistent with the results obtained in the reported literature using colloidal bimetallic Ni/Fe nanoparticles (Schrick et al., 2004; Tee et al., 2005) and different systems of membrane-embedded Ni/Fe nanoparticles (Meyer et al., 2004; Xu et al., 2005).

This shows that the bimetallic Ni/Fe nanoparticles of the chitosan membranes are catalytically active for the degradation of hazardous chlorinated organics. In this study where the catalytically active nanoparticles are embedded in the membrane matrix, considerations have to be taken to account for the mass transfer phenomena such as the sorption and diffusion of TCE and other degraded products in the membrane phase. In order to check that the decrease of TCE concentration in the aqueous-phase GC–MS analysis is actually due to the degradation reaction by the Ni/Fe nanoparticles and not the sorption of TCE in the chitosan membrane, control experiments were conducted using chitosan membrane without the nanoparticles. The aqueous-phase GC–MS analysis for the control study showed that the TCE concentration was $92.00 \pm 0.53\%$ relative to the initial total TCE used. Hexane extraction of the chitosan recovered approximately $4.0 \pm 0.65\%$ of TCE in the membrane phase. The combined GC–MS analysis and extraction study showed that the TCE balance obtained was approximately 96.00% of the TCE used in the control experiments. The volatility and loss of TCE during the sampling procedure might account for the difference in the TCE balance. Based on the results obtained in the control study, the sorption of TCE in the membrane and the loss due to the volatility of TCE is neglected in the kinetic analysis. The decrease of TCE in the aqueous phase is expected to be the

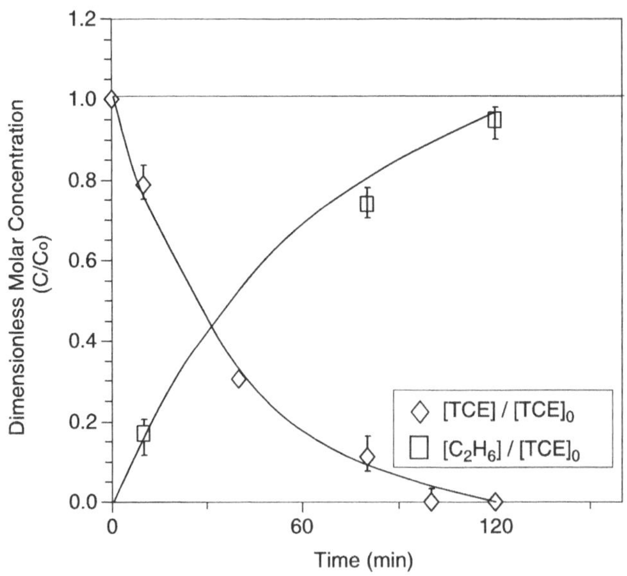

Figure 8.10 Batch TCE degradation study of chitosan-embedded Ni/Fe nanoparticles mixed-matrix membrane. Initial TCE concentration = 10 mg/L, reaction volume = 35 mL + 7 mL headspace, initial pH = 6.5.

degradation reaction by the active Ni/Fe nanoparticles in the chitosan membrane. Quantitative gas-phase analysis showed that the mass balance based on C_2 is $90.0 \pm 2.5\%$ relative to the initial value of GC–MS analysis.

8.3.2.2 *Kinetic Analysis* Kinetic studies were conducted by measuring the rate of TCE degraded over time and is modeled based on the pseudo-first-order reaction rate law:

$$-\frac{dC_{\mathrm{TCE}}}{dt} = k_{\mathrm{obs}} C_{\mathrm{cat}} C_{\mathrm{TCE}} \tag{8.5}$$

where C_{TCE} and C_{cat} represent the TCE (mol L^{-1}) and catalyst concentration (g_{cat} L^{-1}), respectively, and k_{obs} is the observed pseudo-first-order reaction rate constant (L $\mathrm{g_{cat}}^{-1}$ h^{-1}). It is noted that most of the degradation study by bulk or nanoparticles involved the use of surface-area-normalized rate constant, k_{SA} (L m^{-2} h^{-1}), to characterize the reaction kinetic:

$$-\frac{dC_{\mathrm{TCE}}}{dt} = k_{\mathrm{SA}} a_s \rho_m C_{\mathrm{TCE}} \tag{8.6}$$

where a_s (m^2 g^{-1}) and ρ_m (g L^{-1}) is the specific surface area and mass concentration of metal, respectively. To compare our results with others, k_{SA} values reported in the literature were converted to k_{obs} as followed:

$$k_{\mathrm{obs}} = k_{\mathrm{SA}} a_s (C_{\mathrm{cat}} = \rho_m) \tag{8.7}$$

Table 8.4 shows the observed reaction rate of the batch TCE degradation study using Ni/Fe nanoparticles (Tee et al., 2005) and membrane-embedded nanoparticles of different polymeric systems reported in the literature (Meyer et al., 2004; Xu et al., 2005).

It is observed that the peudo-first-order reaction rate in our study has the same order of magnitude with the reported result in Xu et al. (2005) and three times higher than those in Tee et al. (2005) and Meyer et al. (2004). The difference in the rate constants of the TCE degradation study can be caused by different factors. These include the synthesis method and the use of membrane systems with different polymeric properties that might affect the transport of TCE to the nanoparticles. In addition, the presence of membrane may play a role in the reactivity of the immobilized nanoparticles. The nonagglomerated and more uniform-sized nanoparticles may have better electronic and enhanced catalytic properties and structurally favored in the adsorption and degradation reactions.

TABLE 8.4 Pseudo-First-Order Reaction Rate Constants of TCE Degradation by Ni/Fe Nanoparticles

Reference	Ni/Fe Ratio	k (L/g · h)
Tee et al. (2005)	1/4 (nanoparticles)	1.40
Meyer et al. (2004)	1/4 (cellulose acetate–immobilized nanoparticles)	1.20
Xu et al. (2005)	1/4 (polyacrylic acid–immobilized nanoparticles supported on PVDF)	3.50
This study	1/4 (chitosan-immobilized nanoparticles with silica)	3.75

8.3.2.3 Compositional Effect of Bimetallic Ni/Fe Nanoparticles on TCE Degradation The Ni/Fe weight ratio of 1/4 used in this study is based on the previously established optimum compositions showing the highest TCE degradation rate with the formation of ethane as the main product (Tee et al., 2005). It is assumed that the Ni/Fe nanoparticles synthesized by the solution reduction method and the bimetallic Ni/Fe nanoparticles in the chitosan membrane synthesized in situ by doping the Fe nanoparticles with Ni possess similar degradation characteristics toward chlorinated organics. To verify that Ni loading higher than 25 wt% in the bimetallic Ni/Fe will have a lower TCE degradation kinetic as reported in the nanoparticles study [Figure 8.7 of Tee et al. (2005)], chitosan-embedded Ni/Fe nanoparticles with 40 wt% Ni was synthesized and used to degrade the TCE under the same experimental conditions. As anticipated, bimetallic Ni/Fe nanoparticles with 40 wt% Ni showed lower degradation rate for the complete conversion of TCE to ethane (data not shown). The kinetic results characterized by the pseudo-first-order reaction constant for the 25 and 40 wt% Ni was 9.31 L $g_{cat}{}^{-1}$ h^{-1} ($R^2 = 0.9969$) and 5.52 L $g_{cat}{}^{-1}$ h^{-1} ($R^2 = 0.9900$), respectively. The lower reaction kinetic with higher Ni content can be explained by the high Ni surface composition that will hinder the electrochemical process of the zero-valent Fe that generates electrons and assists in the production of hydrogen needed for the reduction of TCE:

$$Fe^0 \rightarrow Fe^{2+} + 2e^- \tag{8.8}$$

$$Fe^0 + 2H^+ \rightarrow Fe^{2+} + H_2 \tag{8.9}$$

$$Fe^0 + 2H_2O \rightarrow Fe^{2+} + H_2 + 2OH^- \tag{8.10}$$

$$Cl_2C{=}CHCl + \frac{3}{2}H_2 + 3e^- \xrightarrow{Ni} CH_4 + 3Cl^- \tag{8.11}$$

The presence of Ni as dopant is hypothesized to be able to strengthen the metal–carbon bonds on the metal surfaces. This will enhance the chemisorptions and subsequently increase the dissociation of C=C intermediates on the surface (Burke and Madix, 1991; Agrawal and Tratnyek, 1996; Park and Baker, 1998). The observed optimum degradation was reported previously by Tee et al. (2005) and Xu et al. (2005) at approximately 25 wt% Ni loading. Beside the observed compositional effect of the bimetallic Ni/Fe nanoparticles on TCE degradation shown in this study, Grenier et al. (2004) also reported the reaction kinetics in the vapor-phase reduction of organohalides (*cis*-1,2-dichloroethylene) by Ni/Fe as a function of the Ni loading of the bimetallic system. Different studies showing the dependence of reaction rates on the compositions of the bimetallic systems were also reported in the literature. These include the hydrogenation of nitrobenzene derivatives by Ni/Pd bimetallic nanoclusters in Lu and Toshima (2000), reduction of chlorophenols by Pd/Fe in Liu et al. (2005), and TCE hydrodechlorination by Pd/Au nanoparticles in Nutt et al. (2005). All of this studies, including our results, have indicated the importance of the second dopant metal and the synergistic compositions of the bimetallic systems in achieving the optimum physical and chemical properties for the dechlorination reactions.

8.3.3 Degradation Study of DCB

8.3.3.1 Solvent Permeation through the Membrane The membrane solvent permeation property for the chitosan-embedded Pd/Fe nanoparticles was determined using water and ethanol solutions (data not shown). In this study, water has the highest permeability (1.42×10^{-4} cm^3 s^{-1} cm^{-2} bar^{-1}) followed by ethanol (1.00×10^{-4} cm^3 s^{-1} cm^{-2} bar^{-1}) and mixture of 1/1 volume ratio of ethanol and water solution (0.50×10^{-4} cm^3 s^{-1} cm^{-2} bar^{-1}). It is reported that the transport of solvent through polymeric porous membrane is influenced both by the membrane characteristics and the properties of solvents (Bhanushali et al., 2001). This includes the hydrophilicity and porosity of the membrane, and the viscosity and polarity of the solvent. High water permeability is expected due to the high degree of hydrophilicity of the chitosan membrane. The ethanol flux through the hydrophilic chitosan would be lower than the water as anticipated due to the limited hydrogen bonding of the organic polar solvent and lower polymer–solvent interaction. The lowest solvent permeation for the 1/1 volume mixture of ethanol and water can be attributed to the highest viscosity of the solution (2.50×10^{-3} Pa · s at 25°C) with respect to water (0.85×10^{-3} Pa · s at 25°C) and ethanol (1.10×10^{-3} Pa · s at 25°C). The addition of ethanol in the water also decreases the mixture polarity and subsequently increases the resistance at the membrane–solvent interphase (Geens et al., 2005). Based on this observation, the solvent effect is more important on the permeation properties for the mixture of ethanol and water solution (1/1 volume ratio), which is the solvent used in the DCB degradation study.

The simplest models describing the volume flux through porous membrane are the viscous flow based on the Hagen–Poiseuille equation:

$$J_w = \left(\frac{\varepsilon r^2}{8l\tau}\right)\left(\frac{\Delta P}{\mu}\right) = \frac{1}{R_m}\left(\frac{\Delta P}{\mu}\right) \tag{8.12}$$

where J_w is the solvent flux, ΔP the applied transmembrane pressure, μ is the solvent viscosity, and R_m is the intrinsic membrane resistance that includes the membrane thickness (l), surface porosity (ε), pore radius (r), and tortuosity factor (τ). It assumes an ideal situation where the solvent flow through a uniformly distributed and evenly sized pores in the membrane, without concentration polarization, fouling and the like (Mulder, 1995). Based on these assumptions, the permeation of solvent should be dependent only on the membrane properties, the inverse of membrane resistance, $1/R_m$. Figure 8.11 shows the expected linear relation for the different solvent fluxes versus the viscosity-normalized transmembrane pressure with different observed membrane resistances.

Based on the permeation data, water showed the lowest membrane resistance of 57.47 cm^{-1} ($R^2 = 0.9789$), followed by ethanol with R_m of 87.00 cm^{-1} ($R^2 = 0.9712$). The highest resistance observed was the mixture of 1/1 volume ratio of ethanol and water with R_m of 131.60 ($R^2 = 0.9893$). The difference in the membrane resistances for different solvents shows that the flux through the chitosan-embedded nanoparticle membranes is affected not only by the polymeric materials used, but also by the solvent properties (viscosity, molar volume, polarity) and polymer–solvent interactions (surface tension, etc.). The influence of these parameters is being studied and reported in the literature (Machodo et al., 1999; Robinson et al., 2004).

8.3.3.2 Degradation of DCB under Pressure-Induced Operation Degradation of DCB is conducted under pressure-induced condition. As the solution convectively passed through the membrane by applying pressure at the cell reservoir, mass transfer of

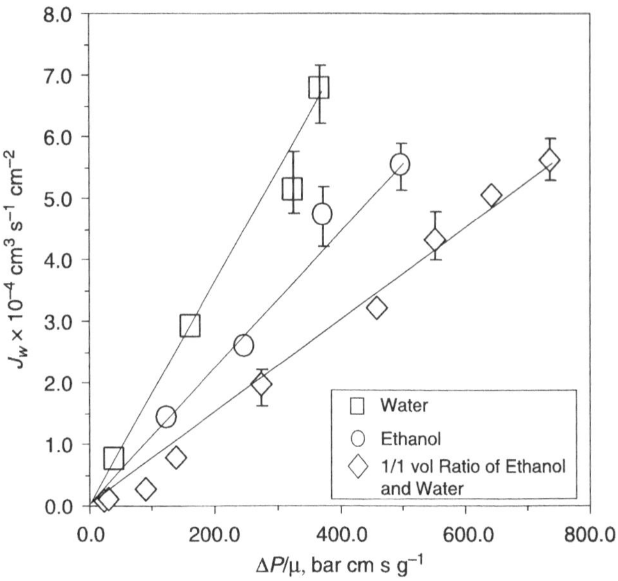

Figure 8.11 Volumetric flux vs. transmembrane pressure normalized by solution viscosity of different solvents for chitosan-embedded Pd/Fe nanoparticles under pressurized condition. Each data point represents flux after 96 h of continuous operation.

the chlorinated DCB to the active Pd/Fe nanoparticles in the membrane is enhanced. It is observed that the DCB degradation and biphenyl formation as end product increased with the reciprocal of volumetric flux, $1/J_w$ (data not shown). As expected, lower volumetric flux (higher $1/J_w$) corresponds to longer residence time of DCB in the membrane phase containing the bimetallic Pd/Fe nanoparticles. This will increase the contact time of the targeted DCB with the catalytically active Pd/Fe and subsequently lead to higher DCB degradation and formation of biphenyl as shown in Figure 8.12.

Based on the flux data with the 60-μm-thick membrane with a diameter of 4.15 cm (membrane volume = 0.081 cm^3), the recorded residence time for 80% DCB degradation was 12.5 min. This corresponds to a flux of 0.12×10^{-4} cm^3 s^{-1} cm^{-2} at transmembrane pressure of 0.5 bar. DCB degradation of 50% was observed at 1.5 min residence time that corresponds to applied pressure of 2 bars with solvent flux of 0.78×10^{-4} cm^3 s^{-1} cm^{-2}.

8.3.3.3 Degradation Mechanism of DCB The hypothesized reductive hydrodechlorination reaction process of DCB with chitosan-embedded bimetallic Pd/Fe nanoparticles at room temperature is shown in Figure 8.13. A previous dechlorination study of DCB congeners suggested a stepwise reaction pathway (Korte et al., 2002; Kim et al., 2004) and is expected to be the same in this investigation. The palladized Fe nanoparticles in the chitosan are in close electronic contact with each other, and dechlorination is assumed to occur at the Pd/Fe interface, not on the iron or palladium surfaces alone. This is illustrated in Figure 8.13 showing the corrosion (oxidation) of elemental iron generating H_2 [reactions in (8.9) and (8.10)] and H^+, which are adsorbed by palladium forming the powerful reducing agent of surface-bound hydride, Pd–H*:

$$Fe^{2+} + H_2O \rightarrow Fe(OH)^+ + H^+ \tag{8.13}$$

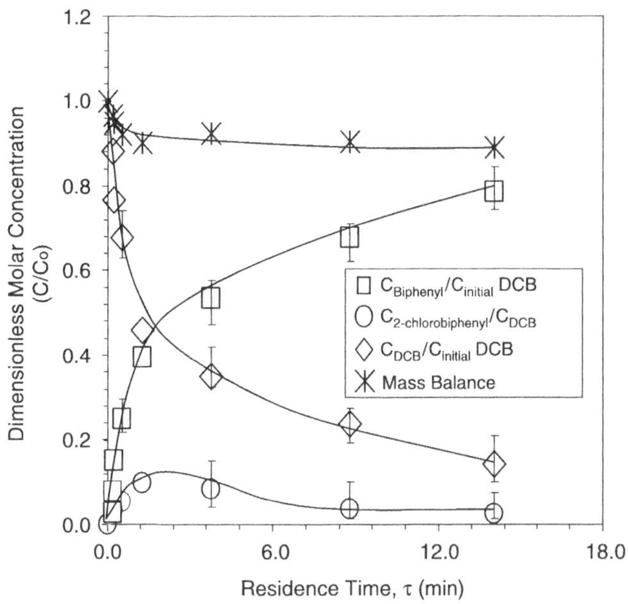

Figure 8.12 Degradation study of 2,2′-dichlorobiphenyl (DCB) vs. residence time under pressure-induced operation. Top right shows the applied transmembrane pressure vs. residence time. Initial DCB concentration = 15 mg/L, solvent = 1/1 (v/v) of ethanol and water, solvent permeability = 0.50×10^{-4} cm^3 s^{-1} cm^{-2} bar^{-1}.

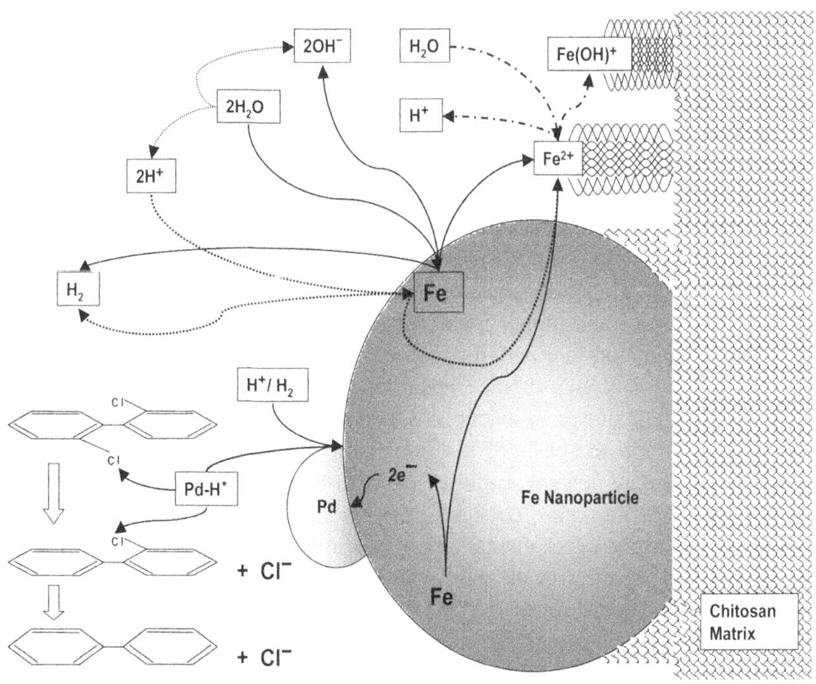

Figure 8.13 Illustration of the hypothesized reductive hydrodechlorination process of DCB on the bimetallic Pd/Fe surface and recapturing of Fe ions by chitosan.

$$H_2O \rightarrow H^+ + OH^- \quad (8.14)$$

$$H^+ + e^- \xrightarrow{Pd} Pd\text{-}H^* \quad (8.15)$$

$$H_2 \xrightarrow{2Pd} 2Pd\text{-}H^* \quad (8.16)$$

$$R\text{-}Cl_2 + 2Pd\text{-}H^* + 2e^- \rightarrow R\text{-}H_2 + 2Cl^- + 2Pd \quad (8.17)$$

In addition, the possible formation of strong palladium–chlorine bonds as reported in the literature (Coq et al., 1986; Bodnariuk et al., 1989; Muftikian et al., 1995) may also enhance the removal of atomic Cl from DCB.

Based on the hypothesized reductive dechlorination mechanism of DCB on the Pd/Fe surface, palladium as the catalytic agent is not consumed in the reaction, and iron as the electron donor and reductant for water is transformed to soluble ionic species [Fe^{2+}, $Fe(OH)^+$]. In contrast to the gradual dissolution of ionic species into the aqueous medium, as is observed by the solution-phase dechlorination study of colloidal nanoparticles, leaching is minimized by embedding the bimetallic Pd/Fe nanoparticles in the membrane matrix. This can be explained in the recapturing properties of the polymers with chelating functional groups. Chitosan with hydroxyl and amine functional groups on its backbone is expected to be able to recapture the water-soluble Fe^{2+} ions that are being "consumed" in the dechlorination process. The elemental analysis (AA for Fe ions and ICP for Pd ion) of the permeates and the remaining solution in the cell reservoir indeed showed negligible leaching of metals with respect to the initial loading of the bimetallic Pd/Fe nanoparticles in the membrane, with less than 1% for Fe^{2+} and below the ICP detection limit for Pd^{2+}. The total metal mass balance of the used membranes was $98.00 \pm 0.05\%$ relative to the fresh membranes by elemental analysis of digestion study.

8.3.3.4 Long-Term Study Further study is conducted to investigate the long-term usage of the chitosan-embedded Pd/Fe nanoparticles in the degradation of DCB. Figure 8.14 shows the degradation of DCB versus reaction time at a fixed applied transmembrane pressure of 0.5 bar.

At the initial stage of the study, the degradation reaction achieved steady state after continuous operation for 2.5 h. At the steady-state condition, the conversion of DCB to biphenyl was around 80% with 2-chlorobiphenyl accounting for the rest of the total carbon balance. With increasing elapsed time, DCB degradation gradually decreased with declining biphenyl conversion and higher 2-chlorobiphenyl formation at 50 h of operation and thereafter. This attenuation in reactivity was assumed to be attributed by several factors. The major deactivation is due to the consumption of Fe as the electron donor and H_2 as the generating agent shown in reaction (8.8) and (8.9, 8.10), respectively, in the DCB degradation process. On the basis of reaction in (m), we hypothesized that 2 mol of Fe are needed to completely degrade 1 mol of DCB to biphenyl, assuming that Pd remained catalytically active for the entire course of the reaction. Based on the initial Fe loading ($\sim$0.04 mmol Fe) and the experimental conditions used in this study, the calculated membrane reactor lifetime usage will be about 160 h of continuous operation at 0.5-bar transmembrane pressure [permeability $= 0.50 \times 10^{-4}$ $cm^3/(s\ bar\ cm^2)$]. In view of the relatively shorter lifetime usage of the membrane, as shown in Figure 8.14 with declining DCB conversion after 50 h of operation, we hypothesized that the formation of iron oxides/hydroxides on the particle surface that act as a passivating layer to be the cause of the activity loss. This inhibitory effect has been studied and reported in the literature (Farrell et al., 2000;

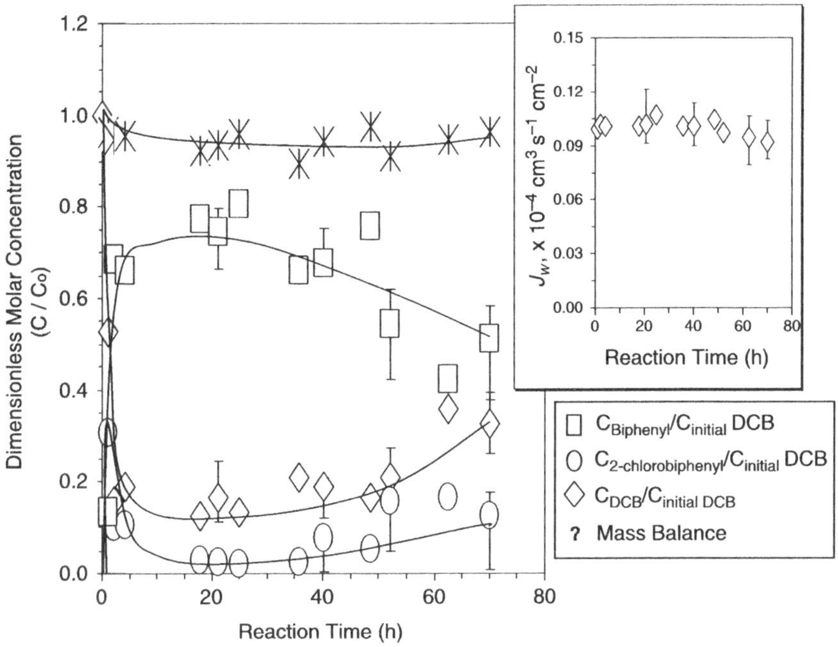

Figure 8.14 Long-term study for degradation of 2,2′-dichlorobiphenyl (DCB) under pressure-induced operation. Top right shows the volumetric flux vs. reaction time. Each data point represents successive collection of permeate volumes with continuous degradation at transmembrane pressure of 0.5 bar. Initial DCB concentration = 15 mg/L, solvent = 1/1 (v/v) of ethanol and water.

Furukawa et al., 2005; Kohn et al., 2005). Even though the reported results were obtained using commercial Fe^0 filling as the permeable reactive barrier for underground water treatment, it is expected that a similar inhibitory effect of the passivating iron layer will be observed in systems where Fe is used as the reductant and electron donor in the degradation reaction. The activity loss due to the dissolution of Pd or Fe into the reaction medium as reported in the degradation study by colloidal nanoparticles (Jovanovic et al., 2005) is assumed to be negligible in this case. This is supported by the previous results (Section 3.3.3) showing negligible loss of metals due to the recapturing/chelating properties of the chitosan membrane.

A slight decrease in solvent flux was also observed after 50 h of continuous operation at transmembrane pressure of 0.5 bar. This observation can be explained by the internal pore clogging and external surface fouling due to insoluble iron oxide/hydroxide precipitation that reduced the pore size of the membrane. Such phenomena have been extensively studied and reported in the literature (Tansel et al., 2000). Taking into account the adverse effects on the permeation property that occurred for long duration of membrane operation, the Hagen–Poiseuille equation is written as follow:

$$J_w = \frac{1}{R_m + R_d}\left(\frac{\Delta P}{\mu}\right) \tag{8.18}$$

where R_d is the external resistances caused by fouling, clogging, and the like that lowered the solvent flux. Based on these results, further study is needed to optimize the membrane in order to prolong the activity of the embedded bimetallic Pd/Fe nanoparticles.

8.4 CONCLUSIONS

Chitosan membrane-embedded bimetallic nanoparticles of Ni/Fe and Pd/Fe were effective in the degradation of toxic chlorinated organics. In the case of TCE degradation by Ni/Fe nanoparticles, the observed pseudo-first-order reaction rate was two times higher than the reported values in the studies using colloidal and other membrane systems with immobilized nanoparticles. Complete TCE degradation was achieved using milligram quantities of the Ni/Fe nanoparticles with the formation of ethane as the final product. DCB degradation with the formation of biphenyl and less chlorinated product by chitosan-embedded Pd/Fe nanoparticles under pressure-induced operation was related to the applied transmembrane pressure (residence time). As expected, lower applied pressure that corresponds to longer residence time will lead to higher DCB degradation and biphenyl formation, and vice versa. The long-term study of DCB degradation showed the occurrence of deactivation of the bimetallic Pd/Fe and declining solvent flux after 50 h of continuous operation. This was due to the oxidation of zero-valent Fe that subsequently led to the formation of inactive iron oxides/hydroxide on the surface of other active Pd/Fe nanoparticles and caused the fouling/pore clogging of the membrane. Optimization of the membrane is thus needed to increase the lifetime usage of the chitosan-embedded bimetallic nanoparticles.

ACKNOWLEDGMENT

The funding provided by the U.S. EPA-STAR program and NIEHS-SBRP program is greatly appreciated. We also thank the analytical support by the Environmental Research and Training Laboratory (ERTL) facility and the Electron Microscopy Center (EMC) at the University of Kentucky.

REFERENCES

Adlim, M., Bakar, M. A., Liew, K. Y., and Ismail, J. (2004). Synthesis of chitosan-stabilized platinum and palladium nanoparticles and their hydrogenation activity. *J. Mol. Catal. A Chem.* 212, 141–149.

Agrawal, A., and Tratnyek, P. G. (1996). Reduction of nitro aromatic compounds by zero valent iron metal. *Environ. Sci. Technol.* 30, 153–160.

Bao, C. Y., Jin, M., Lu, R., Xue, P. C., Zhang, T. R., Tan, C. H., and Zhao, Y. Y. (2003). Synthesis of hyperbranched poly(amine-ester)-protected noble metal nanoparticles in aqueous solution. *J. Mater. Res.* 18, 1392–1398.

Beppu, M. M., Arruda, E. J., Vieira, R. S., and Santos, N. N. (2004). Adsorption of Cu(II) on porous chitosan membranes functionalized with histidine. *J. Membr. Sci.* 240, 227–235.

Bhanushali, D., Kloos, S., Kurth, C., and Bhattacharyya, D. (2001). Performance of solvent-resistant membranes for non-aqueous systems: Solvent permeation results and modeling. *J. Membr. Sci.* 189, 1–21.

Bodnariuk, P., Coq, B., Ferrat, G., and Figueras, F. (1989). Carbon-chlorine hydrogenolysis over PdRh and PdSn bimetallic catalysts. *J. Catal.* 116, 459–466.

Bronstein, L. M., Sidorov, S., Valetsky, P. M., Hartmann, J., Cölfen, H., and Antonietti, M. (1998). Investigation of polymer-protected noble metal nanoparticles by transmission electron microscopy: Control of particle morphology and shape. *Colloid Polym. Sci.* 276, 769–779.

Burke, M. L., and Madix, R. J. (1991). Hydrogenation and hydrogen-deuterium exchange reactions of C3-C6 substituted alkenes on iron(100). *J. Am. Chem. Soc.* 113, 4151–4157.

Coq, B., Ferrat, G., and Figueras, F. (1986). Conversion of chlorobenzene over palladium and rhodium catalysts of widely varying dispersion. *J. Catal.* 101, 434–445.

Davis, R. H., and Grant, D. H. (1992). Microfiltration—Theory for deadend microfiltration. In W. S. W. Ho and K. K. Sirkar (Eds.), *Membrane Handbook*, Chapman & Hall, New York, p. 461.

Denuziere, A., Ferrier, D., Damour, O., and Domarel, A. (1998). Chitosan–chondroitin sulfate and chitosan–hyaluronate polyelectrolyte complexes: Biological properties. *Biomaterials* 19, 1275–1285.

Drelinkiewicz, A., Hasik, M., and Choczyński, M. (1998). Preparation and properties of polyaniline containing palladium. *Mater. Res. Bull.* 33, 739–762.

Farrell, J., Kason, M., Melitas, N., and Li, T. (2000). Investigation of the long-term performance of zero-valent iron for reductive dechlorination of trichloroethylene. *Environ. Sci. Technol.* 34, 514–521.

French, B. L., Davis, L. M., Munzinger, E. S., Slavin, J. W. J., Christy, P. C., Thompson, D. W., and Southward, R. E. (2005). Palladium-polyimide nanocomposite membranes: Synthesis and characterization of reflective and electrically conductive surface-metalized films. *Chem. Mater.* 17, 2091–2100.

Fritsch, D., Kuhr, K., Mackenzie, K., and Kopinke, F. D. (2003). Hydrodechlorination of chloroorganic compounds in ground water by palladium catalysts. Part 1. Development of polymer-based catalysts and membrane reactor tests. *Catal. Today* 82, 105–118.

Fritsch, D., and Peinemann, K. V. (1995). Novel highly permselective 6F-poly(amide-imide)s as membrane host for nano-sized catalysts. *J. Membr. Sci.* 99, 29–38.

Furukawa, Y., Kim, J.-W., Watkins, J., and Wilkin, R. T. (2005). Formation of ferrihydrite and associated iron corrosion products in permeable reactive barriers of zero-valent iron. *Environ. Sci. Technol.* 36, 5469–5475.

Gan, L. M., Zhang, L. H., Chan, H. S. O., and Chew, C. H. (1995). Preparation of conducting polyaniline-coated barium sulfate nanoparticles in inverse microemulsions. *Mater. Chem. Phys.* 40, 94–98.

Geens, J., Peeters, K., Van der Bruggen, B., and Vandecasteele, C. (2005). Polymeric nanofiltration of binary water-alcohol mixtures: Influence of feed composition and membrane properties on permeability and rejection. *J. Membr. Sci.* 255, 255–264.

Grenier, A. C., McGuire, M. M., Fairbrother, D. H., and Roberts, A. L. (2004). Treatment of vapor-phase organohalides with zero-valent iron and Ni/Fe reductants. *Environ. Eng. Sci.* 21, 421–435.

Guibal, E. (2005). Heterogeneous catalysis on chitosan-based materials: A review. *Prog. Polym. Sci.* 30, 71–109.

Guibal, E., Milot, C., and Tobin, J. M. (1998). Metal-anion sorption by chitosan beads: Equilibrium and kinetic studies. *Ind. Eng. Chem. Res.* 37, 1454–1463.

Hable, C. T., and Wrighton, M. S. (1991). Electrocatalytic oxidation of methanol by assemblies of platinum/tin catalyst particles in a conducting polyaniline matrix. *Langmuir* 7, 1305–1309.

Huang, H. Z., Yuan, Q., and Yang, X. R. (2004). Preparation and characterization of metal-chitosan nanocomposites. *Colloids Surf. B Biointerf.* 39, 31–37.

Isabel, P. S., and Luis, M. L. M. (2002). Formation of PVP-protected metal nanoparticles in DMF. *Langmuir* 18, 2888–2894.

Ishizuki, N., Torigoe, K., Esumi, K., and Meguro, K. (1991). Characterization of precious metal particles prepared using chitosan as a protective agent. *Colloids Surf.* 55, 15–21.

Jegal, J. G., and Lee, K. H. (1999). Chitosan membranes crosslinked with sulfosuccinic acid for the pervaporation separation of water/alcohol mixtures. *J. Appl. Polym. Sci.* 71, 671–675.

Jin, J.-J., Chen, G.-C., Huang, M.-Y., and Jiang, Y.-Y. (1994). Catalytic hydrogenation behaviours of palladium complexes of chitosan-polyacrylic acid and chitosan polymethacrylic acid. *React. Funct. Polym.* 23, 95–100.

Jovanovic, G. N., Plazl, P. Z., Sakrittichai, P., and Al-Khaldi, K. (2005). Dechlorination of *p*-chlorophenol in a microreactor with bimetallic Pd/Fe catalyst. *Ind. Eng. Chem. Res.* 43, 1–7.

Kapoor, S., and Mukherjee, T. (2003). Photochemical formation of copper nanoparticles in poly(*N*-vinylpyrrolidone). *Chem. Phys. Lett.* 370, 83–87.

Kawamura, Y., Mitsuhashi, M., Tanibe, H., and Yoshida, H. (1993). Adsorption of metal ions on polyaminated highly porous chitosan chelating resin. *Ind. Eng. Chem. Res.* 32, 386–391.

Kidambi, S., and Brueing, M. L. (2005). Multilayered polyelectrolyte films containing palladium nanoparticles: Synthesis, characterization, and application in selective hydrogenation. *Chem. Mater.* 17, 301–307.

Kim, Y.-H., Shin, W. S., and Ko, S.-O. (2004). Reductive dechlorination of chlorinated biphenyls by palladized zero-valent metals. *J. Environ. Sci. Heal. A.* A39, 1177–1188.

Kittur, A. A., Kulkarni, S. S., Aralaguppi, M. I., and Kariduraganavar, M. Y. (2005). Preparation and characterization of novel pervaporation membranes for the separation of water–isopropanol mixtures using chitosan and NaY zeolite. *J. Membr. Sci.* 247, 75–86.

Kohn, T., Livi, K. J. T., Roberts, A. L., and Vikesland, P. J. (2005). Longevity of granular iron in groundwater treatment processes: Corrosion product development. *Environ. Sci. Technol.* 39, 2867–2879.

Korchev, A. S., Bozack, M. J., Slaten, B. L., and Mills, G. (2004). Polymer-initiated photogeneration of silver nanoparticles in SPEEK/PVA films: Direct metal photopatterning. *J. Am. Chem. Soc.* 126, 10–11.

Korte, N. E., West, O. R., Liang, L., Gu, B., Zutman, J. L., and Fernando, Q. (2002). The effect of solvent concentration on the use of palladized-iron for the step-wise dechlorination of polychlorinated biphenyls in soil extracts. *Waste Manag.* 22, 343–349.

Lin, B. H., Morkved, T. L., Meron, M., Huang, Z. Q., Viccaro, J., Jaeger, H. M., Williams, S. M., and Schlossman, M. L. (1999). X-ray studies of polymer/gold nanocomposites. *J. Appl. Phys.* 85, 3180–3184.

Liu, C., Xu, Y., Liao, S., and Yu, D. (2000). Selective hydrogenation of cyclopentadiene in mono- and bimetallic catalytic hollow-fiber reactors. *J. Mol. Catal. A. Chem.* 157, 253–259.

Liu, C., Xu, Y., Liao, S., Yu, D., Zhao, Y., and Fan, Y. (1997). Selective hydrogenation of propadiene and propyne in propene with catalytic polymeric hollow-fiber reactor. *J. Membr. Sci.* 137, 139–144.

Liu, Y. L., Su, Y. H., Lee, K. R., and Lai, J. Y. (2005). Crosslinked organic–inorganic hybrid chitosan membranes for pervaporation dehydration of isopropanol–water mixtures with a long-term stability. *J. Membr. Sci.* 251, 233–238.

Lu, P., and Toshima, N. (2000). Catalysis of polymer-protected Ni/Pd bimetallic nano-clusters for hydrogenation of nitrobenzene derivatives. *Bull. Chem. Soc. Jpn.* 73, 751–758.

Machodo, D. R., Hasson, D., and Semiat, R. (1999). Effect of solvent properties on permeate flow through nanofiltration membranes: Part I: Investigation of parameters affecting solvent flux. *J. Membr. Sci.* 163, 93–102.

Machulek, A., Jr., Oliveira, H. P. M., and Gehlen, M. H. (2003). Preparation of silver nanoprisms using poly(*N*-vinyl-2-pyrrolidone) as a colloid-stabilizing agent and the effect of silver nanoparticles on the photophysical properties of cationic dyes. *Photochem. Photobiol. Sci.* 2, 921–925.

Malynych, S., Luzinov, I., and Chumanov, G. (2002). Poly(vinyl pyridine) as a universal surface modifier for immobilization of nanoparticles. *J. Phys. Chem. B* 106, 1280–1285.

Meyer, D. E., Wood, K., Bachas, L. G., and Bhattacharyya, D. (2004). Degradation of chlorinated organics by membrane-immobilized nanosized metals. *Environ. Prog.* 23, 232–242.

Muftikian, R., Fernando, Q., and Korte, N. (1995). A method for the rapid dechlorination of low molecular weight chlorinated hydrocarbons in water. *Water Res.* 29, 2434–2439.

Mulder, M. V. H. (1995). Polarization phenomena and membrane fouling. In R. D. Noble and S. A. Stern (Eds.), *Membrane Separations and Technology Principles and Applications*. Elsevier, Amsterdam, The Netherlands, p. 45.

Nam, S. Y., and Lee, Y. M. (1999). Pervaporation separation of methanol/methyl *t*-butyl ether through chitosan composite membrane modified with surfactants. *J. Membr. Sci.* 157, 63–71.

Ngah, W. S. W., and Liang, K. H. (1999). Adsorption of gold(III) ions onto chitosan and *N*-carboxymethyl chitosan: Equilibrium studies. *Ind. Eng. Chem. Res.* 38, 1411–1414.

Nunthanid, J., Puttipipatkhachorn, S., Yamamoto, K., and Peck, G. E. (2001). Physical properties and molecular behavior of chitosan films. *Drug Dev. Ind. Pharm.* 27, 143–157.

Nutt, M. O., Hughes, J. B., and Wong, M. S. (2005). Designing Pd-on-Au bimetallic nanoparticles catalysts for trichloroethene hydrodechlorination. *Environ. Sci. Technol.* 39, 1346–1353.

Ogawa, K., and Yui, T. (1993). Crystallinity of partially *N*-acetylated chitosan. *Biosci. Biotechnol. Biochem.* 57, 1466–1469.

Park, C., and Baker, R. T. K. (1998). Carbon deposition on iron–nickel during interaction with ethylene–hydrogen mixtures. *J. Catal.* 179, 361–374.

Patel, V. R., and Amigi, M. M. (1996). Preparation and characterization of freeze-dried chitosan-poly(ethylene oxide) hydrogels for site-specific antibiotic delivery in the stomach. *Pharm. Res.* 13, 588–593.

Porel, S., Singh, S., Harsha, S. S., Rao, D. N., and Radhakrishnan, T. P. (2005). Nanoparticle-embedded polymer: In situ synthesis, free-standing films with highly monodisperse silver nanoparticles and optical limiting. *Chem. Mater.* 17, 9–12.

Quignard, F., Choplin, A., and Domard, A. (2000). Chitosan: A natural polymeric support of catalysts for the synthesis of fine chemicals. *Langmuir* 16, 9106–9108.

Robinson, J. P., Tarleton, E. E., Millington, C. R., and Nijmeijer, A. (2004). Solvent flux through dense polymeric nanofiltration membranes. *J. Membr. Sci.* 230, 29–37.

Schmuch, A. B., and Krajicek, M. E. (1998). Mucoadhesive polymers as platforms for peroral peptide delivery and absorption: Synthesis and evaluation of different chitosan–EDTA conjugates. *J. Control. Release* 50, 215–223.

Schrick, B., Blough, J. L., Jones, A. D., and Mallouk, T. E. (2004). Hydrodechlorination of trichloroethylene to hydrocarbons using bimetallic nickel-iron nanoparticles. *Chem. Mater.* 14, 5140–5147.

Shuetz, P., and Caruso, F. (2004). Semiconductor and metal nanoparticle formation on polymer spheres coated with weak polyelectrolyte multilayers. *Chem. Mater.* 16, 3066–3073.

Si, S., Kotal, A., Mandal, T. K., Giri, S., Nakamura, H., and Kohara, T. (2004). Size-controlled synthesis of magnetite nanoparticles in the presence of polyelectrolytes. *Chem. Mater.* 16, 3489–3496.

Sidorov, S. N., Bronstein, L. M., Valetsky, P. M., Hartmann, J., Cölfen, H., Schnablegger, H., and Antonietti, M. (1999). Stabilization of metal nanoparticles in aqueous medium by polyethyleneoxide–polyethyleneimine block copolymers. *J. Colloid Interface Sci.* 212, 197–211.

Suslick, K. S., Fang, M., and Hyeon, T. (1996). Sonochemical synthesis of iron colloids. *J. Am. Chem. Soc.* 118, 11960–11961.

Tansel, B., Bao, W. Y., and Tansel, I. N. (2000). Characterization of fouling kinetics in ultrafiltration systems by resistances in series model. *Desalination* 129, 7–14.

Tee, Y.-H., Grulke, E., and Bhattacharyya, D. (2005). Role of Ni/Fe nanoparticle composition on the degradation of trichloroethylene from water. *Ind. Eng. Chem. Res.* 44, 7062–7072.

Temgire, M. K., and Joshi, S. S. (2004). Optical and structural studies of silver nanoparticles. *Radiat. Phys. Chem.* 71, 1039–1044.

Teranishi, T., and Miyake, M. (1999). Novel synthesis of monodispersed Pd/Ni nanoparticles. *Chem. Mater.* 11, 3414–3416.

Tröger, L., Hünnefeld, H., Nunes, S., Oehring, M., and Fritsch, D. (1997). Structural characterization of catalytically active metal nanoclusters in poly(amide imide) films with high metal loading. *J. Phys. Chem. B* 101, 1279–1291.

Vincent, T., and Guibal, E. (2002). Chitosan-supported palladium catalyst. 1. Synthesis procedure. *Ind. Eng. Chem. Res.* 41, 5158–5164.

Vincent, T., and Guibal, E. (2004). Chitosan-supported palladium catalyst. 5. Nitrophenol degradation using palladium supported on hollow chitosan fibers. *Environ. Sci. Technol.* 38, 4233–4240.

Vincent, T., Spinelli, S., and Guibal, E. (2003). Chitosan-supported palladium catalyst. II. Chlorophenol dehalogenation. *Ind. Eng. Chem. Res.* 42, 5968–5976.

Wang, T. C., Rubner, M. F., and Cohen, R. E. (2003). Manipulating nanoparticle size within polyelectrolyte multilayers via electroless nickel deposition. *Chem. Mater.* 15, 299–304.

Xu, J., Dozier, A., and Bhattacharyya, D. (2005). Synthesis of nanoscale bimetallic particles in polyelectrolyte membrane matrix for reductive transformation of halogenated organic compounds. *J. Nanopart. Res.* 7, 449–467.

Zeng, X., and Ruckenstein, E. (1996). Control of pore sizes in macroporous chitosan and chitin membranes. *Ind. Eng. Chem. Res.* 35, 4169–4175.

Zeng, X., and Ruckenstein, E. (1998). Cross-linked macroporous chitosan anion-exchange membranes for protein separations. *J. Membr. Sci.* 148, 195–205.

CHAPTER 9

Membrane Bioreactors for Wastewater Treatment

P. CORNEL

Technische Universität Darmstadt, Department of Civil Engineering, Institute WAR, Darmstadt, Germany

S. KRAUSE

Microdyn-Nadir GmbH, Wiesbaden, Germany

9.1 INTRODUCTION

The membrane bioreactor (MBR) technology for wastewater treatment has attracted enormous attention worldwide over the past 10–15 years. The technology is based on the combination of the activated sludge process with biomass separation by membrane filtration. The first step involves the biological treatment of the wastewater using microorganisms. The elimination of organics, nitrification, and denitrification are due to microbial activities and require an adequate process design to stimulate the microorganisms to do the required job. Separation of the biomass (activated sludge) from the treated wastewater is achieved with the help of membranes as against gravity separation in the conventional activated sludge (CAS) process. Micro- and ultrafiltration membranes are used in different modules and configurations. Compared to the conventional activated sludge process, the MBR produces a significantly better effluent quality as it removes all suspended, colloidal solids and bacteria including attached viruses or adsorbed compounds. Also, the MBR process can be operated at a much higher mixed-liquor suspended solids (MLSS) concentrations so that higher volumetric loads are feasible, resulting in a small footprint, all the more so as there is no need for a secondary sedimentation tank.

9.2 PRINCIPLE OF THE MEMBRANE BIOREACTOR PROCESS

In the MBR process, the membrane separates the MLSS into a particle-free "permeate" phase (filtrate, effluent) and a "concentrate" phase (biomass, return sludge; MLSS) which remains in the bioreactor (Fig. 9.1).

Advanced Membrane Technology and Applications. Edited by Norman N. Li, Anthony G. Fane, W. S. Winston Ho, and T. Matsuura
Copyright © 2008 John Wiley & Sons, Inc.

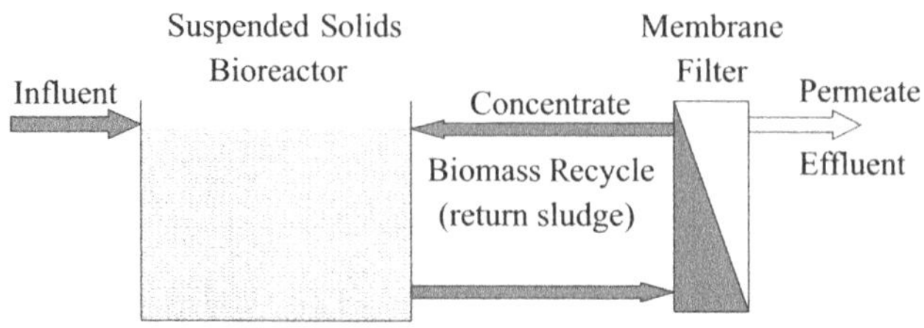

Figure 9.1 Operating principle of membrane bioreactor (MBR).

For the use of membrane technology in biological wastewater treatment applications, several aspects have to be considered:

- Membrane materials and modules
- Configuration of MBR
- Membrane fouling phenomena
- Adaptation of the design of the activated sludge process as well as the pretreatment

9.2.1 Membrane Materials for MBRs

The selection of membrane materials for MBR applications is governed by the need for a material of appropriate mechanical strength and chemical resistance combined with high throughput, narrow range of pore size, and minor fouling tendency. All MBRs use porous membranes, that is, solids separation occurs by sieving and/or surface filtration. While ceramic membranes are unsurpassed in terms of chemical, thermal, and fouling resistance, they are currently far too expensive to be competitive in wastewater treatment applications. This is the reason polymeric membranes, which are generally less resistant, are widespread in wastewater treatment (WWT). Not only are they much cheaper [price per meter squared is approximately 1/20th of that of ceramic membranes in 2005 (Judd, 2005)], but they are also versatile with respect to membrane configuration and module types. An overview of polymer-based membrane materials published by Stephenson et al. in 2000 is given in Table 9.1.

9.2.2 Membrane Configuration

The term *membrane configuration* is not well defined. It describes the geometry of the membranes, for example, planar, tubular, and hollow fiber, but also how the membranes

TABLE 9.1 Materials of Polymer-Based Membranes

Polymer	Characteristic
Polypropylene (PP)	Chemically resistant
Polyethylene	Chemically resistant
Polytetrafluoroethylene (PTFE)	Highly hydrophobic, excellent organic and chemical resistance
Polyethersulphone (PES)	Excellent chemical resistance, hydrophilic
Polyvinylidenefluoride (PVDF)	Excellent chemical resistance, good chlorine resistance

Source: Adapted from Stephenson et al. (2000).

are fashioned into elements and modules and the way in which the membrane modules are integrated into the process.

9.2.2.1 *Membrane Modules for MBRs* For the implementation of the membranes into the process, they are fixed in elements and housed in technical units, so-called modules. Membrane modules for biomass separation should exhibit the following properties (Stephenson et al., 2000; ATV-DVWK, 2002; Vossenkaul, 2005; Judd 2005):

- No dead zones where sludge can accumulate
- High degree of turbulence at the feed side to enhance mass transfer and reduce fouling
- Mechanical, chemical, and thermal stability
- Low pressure drop [low transmembrane pressure (TMP) in the range of 50 to 250 mbar]
- High packing density (meter squared membrane area to meter cubic bulk volume of the module)
- Ease of installation and membrane replacement
- Modularization, retrofit
- Ease of cleaning
- Low energy requirements per unit volume of treated water

As no configuration satisfies all these requirements, three types of modules have become established in wastewater treatment applications: hollow, fiber modules, planar, for example, plate-and-frame modules as well as tubular modules.

In hollow-fiber modules, the flow is usually from outside to inside. The diameter of the hollow fibers can range from a few hundred micrometers up to a few millimeters. The fixed support may be provided either on both sides or on one side only. The fibers are bundled in so-called racks and may be aligned vertically or horizontally. The modules are installed either horizontally or vertically. These modules feature a high packing density and are submerged in the biomass. Fouling control is usually achieved by aerators installed underneath the membrane module. Often coarse bubble aeration acts as the source of scour at the membrane surface, but fine bubble aerators are also employed. A schematic view of a hollow-fiber module is depicted in Figure 9.2. Membrane operation may include periodic relaxation (pressure release) and/or back-flushing for removing the fouling layer from the membrane surface.

In plate-and-frame membrane modules, the planar membranes are arranged in parallel and supported by a plate. Different implementations of flat membranes in the modules include:

- Submerged vertically arranged rectangular plates
- Submerged rotating plate modules (comparable to a rotating disk reactor)
- Flat membranes outside the aeration tank (flow alongside the membrane surface)

Fouling control in submerged modules is achieved by coarse bubble aeration. (External) horizontal modules are operated in the side-stream mode. Most systems are relaxed periodically; some can be back pulsed (at very low pressure). A schematic view is depicted in Figure 9.3.

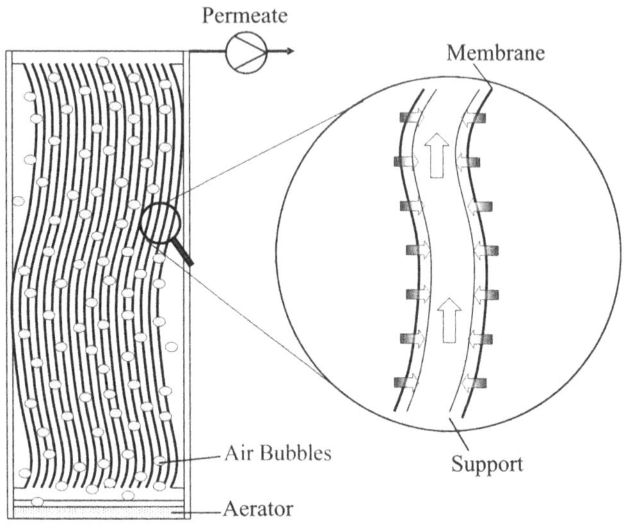

Figure 9.2 Schematic view of a hollow-fiber membrane module.

Commercially available tubular modules are mainly operated in the "inside-out" mode. Over the years, the diameter has been progressively reduced from more than 20 mm down to 5 mm in order to increase the packing density. Tubular modules can have installation lengths of up to 6 m. These side-stream modules are operated at flow velocities of 1–4 m/s. Arrangement may be vertical or horizontal. Some systems are operated with additional aeration for fouling control. Tubular modules provide more direct hydrodynamic control at the membrane surface (e.g., Baker, 2004; Chang et al., 2002). Compared to submerged modules, the flux per surface area is higher, however, at the cost of a higher specific energy demand (kWh/m^3). The modules can be back pulsed or flushed with permeate to clean the membrane surface. A schematic view of a tubular module is illustrated in

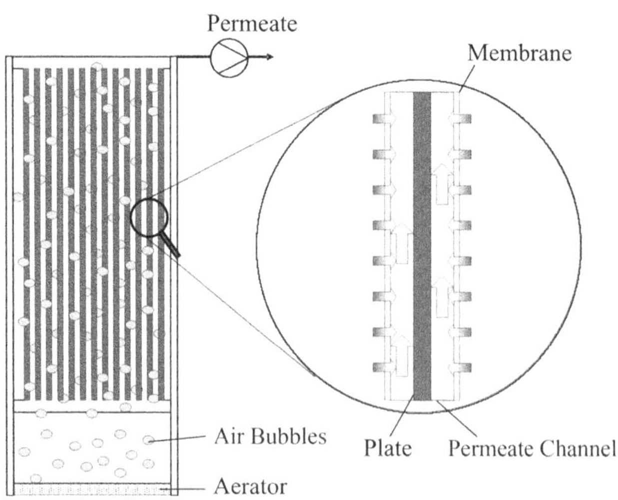

Figure 9.3 Schematic view of a vertically arranged submerged plate membrane module.

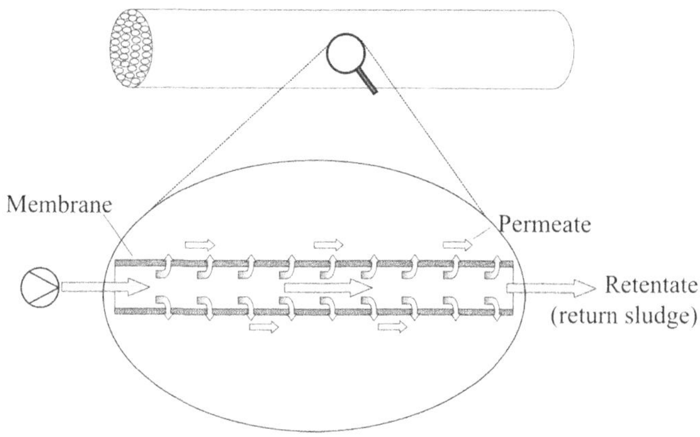

Figure 9.4 Schematic view of a tubular membrane module (side-stream module).

Figure 9.4. Some characteristics of membrane configurations for MBR applications are summarized in Table 9.2.

9.2.2.2 Integration of Membrane Modules into the Process Generally, there are two options of integrating the membranes into the activated sludge process: the submerged configuration and the side-stream configuration. In the first case, the membranes are submerged in the mixed liquor, and permeate is sucked mechanically or by gravity flow. Due to the fouling potential of the mixed liquor, the membranes need some fouling control. In submerged systems, this is usually accomplished by an air scour at the membrane surface using a coarse bubble aeration system that generates a "cross flow" at the membrane surface. Figures 9.5 and 9.6 depict submerged membrane configurations inside the aeration tank (Fig. 9.5) and in a separate filtration tank (Fig. 9.6), respectively. In the latter design, the activated sludge has to be recycled in order to maintain a constant sludge concentration. According to Daigger (2003) and Judd (2006), submerged MBRs have proven to be more cost- and energy-effective than tubular side-stream modules.

TABLE 9.2 Membrane Configurations

Configuration	Packing Density[a] (m^2/m^3)	Invest Cost[c] (€/m^2)	Turbulence Promotion[b]	Advantages[b]	Disadvantages[b]
Hollow fiber	600–1200	Low	Poor	Backflush capability	Fouls readily (braiding)
Planar	100–250	Low–high	Fair	Can be dismantled for cleaning	Large installation, no backflush
Tubular	20–90	High	Very good	Good cleaning performance, tolerates high MLSS	High cost & energy demand

[a]Judd (2005).
[b]ATV-DVWK (2002).
[c]This study.

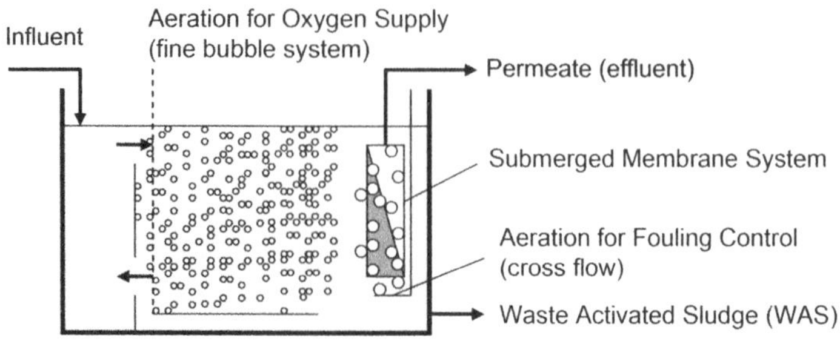

Figure 9.5 Submerged membrane system (inside aeration tank).

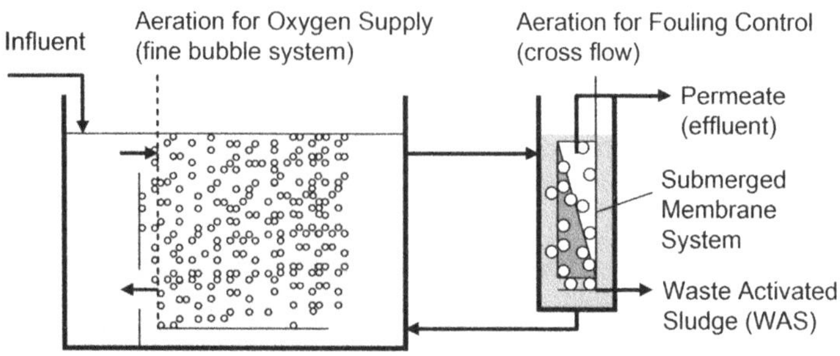

Figure 9.6 Submerged membrane system (in an external filtration tank).

In the side-stream configuration, the MLSS is pumped through the membrane module. Side-stream systems typically use tubular membranes. Fouling is controlled by a well-defined flow velocity in the range of 1–4 m/s, generating a turbulent crossflow. Figure 9.7 shows a schematic view of a side-stream system.

Due to fouling control by aeration or by pumping the MLSS through the tubes, the use of membranes results in a higher energy demand than the conventional sedimentation process for sludge–water separation. Reasonable estimates are about 0.4 kWh/m^3 for submerged systems and about 3 kWh/m^3 for tubular modules (Krause, 2005; Cornel and Krause, 2006).

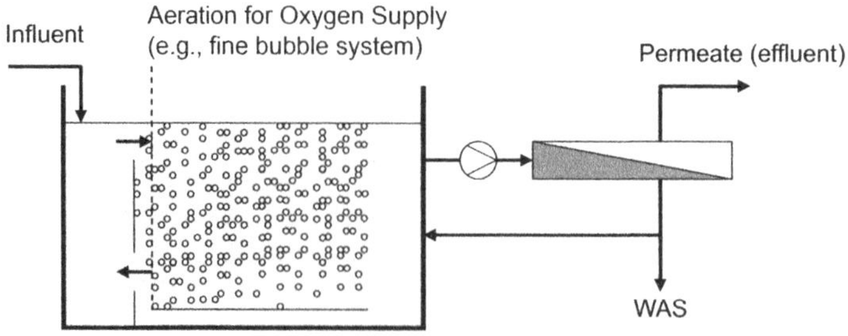

Figure 9.7 Side-stream system.

9.2.3 Membrane Fouling and Cleaning

9.2.3.1 Membrane Fouling Fouling is the reduction in membrane permeability [= flux divided by pressure expressed in $L/(m^2 \cdot h \cdot bar)$]. Typically, the transmembrane pressure (TMP) has to be increased to keep the flux at a constant level (Fig. 9.8). Thus, membrane fouling reduces productivity by increasing TMP, which in turn increases maintenance and operational costs.

Membrane fouling may be physical or biological (Judd, 2004). It is caused by the deposition of biosolids, colloidal species, scalants, or macromolecular species on the membrane surface and leads to a flux and permeability decline. An excellent overview is given by Le-Clech et al. (2006). According to Chang et al. (2002), *reversible* fouling is defined as fouling on the membrane surface that can be removed by physical washing, whereas *irreversible* fouling designates internal fouling into the membrane pores, which can be removed by chemical cleaning only. A description of membrane fouling is difficult due to the heterogeneity of the activated sludge. According to Chang et al. (2002) and Judd (2004), there are three factors influencing the overall fouling behavior in MBRs: (1) biomass characteristics, (2) membrane characteristics, and (3) operating conditions.

1. Biomass characteristics affecting the fouling behavior:
   - According to Rosenberger (2001), the MLSS can affect the fouling rate.
   - The particle size distribution of the biosolids and/or the so-called extracellular polymeric substances (EPS)—in particular the dissolved portion of the EPS—influences the loss in permeability (Judd, 2004; Rosenberger, 2001; Kim et al., 2001).
   - Bulk characteristics such as viscosity and hydrophobicity (Judd, 2005).
2. Membrane and module characteristics affecting the fouling behavior:
   - Pore size and shape
   - Surface characteristics such as polarity, hydrophobicity, surface topography, and porosity of the membrane as well as
   - Module geometry and dimensions

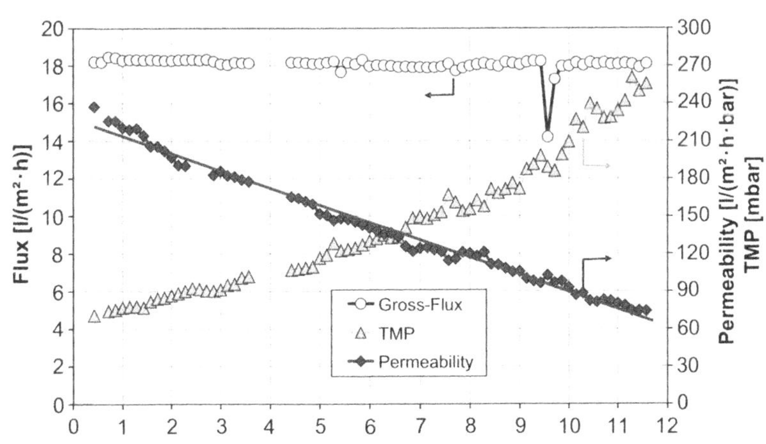

Figure 9.8 Flux, transmembrane pressure (TMP), and permeability vs. time (Municipal MBR, pilot plant) (Krause, 2005).

3. Operating conditions affecting the fouling behavior:
   - Hydraulics, the most significant factor being the flux
   - Hydrodynamics: that is, the cross-flow velocity or aeration rate (submerged membrane modules), flow rate, pulse rate, relaxation time, and soforth
   - Cleaning: for example, back-flush, physical cleaning, chemical cleaning, cleaning intervals

The interaction between all these parameters is largely unknown. As these factors can be controlled by operational measures, operational fouling control will be discussed in greater detail next.

9.2.3.2 Operational Fouling Control Fouling control measures include: (1) reducing the flux, (2) increasing the shear, (3) back-flush the membrane, (4) modify the mixed liquor, and (5) apply chemical or physical cleaning. Another fouling control strategy, the applying of an appropriate pretreatment of the feed water (Judd, 2006) in order to prevent membrane clogging, is described in Sections 9.2.5.2 and 9.3.2.

1. *Reducing the Flux* Reducing the flux requires a proportionally larger membrane surface area, resulting in a higher energy demand for fouling control and hence higher cost (see also Chapter 10). Differentiation between gross flux and net flux as well as peak and average flux must be made (Judd, 2006). In particular, in municipal applications the peak flux, usually during storm water flow, impacts the MBR design.
2. *Increasing the Shear*
   a. Enhancing the Air Scouring of Submerged Modules The shear velocity can be increased by applying a higher airflow. Typical values range from about 0.2 up to 0.8 m^3N per m^2 membrane surface area per hour. As the specific airflow rate is more or less directly linked to the energy demand, a higher aeration rate translates into higher energy cost.
   b. Increasing the Flow Velocity (in tubular modules) In side-stream membrane systems, fouling control is accomplished by a high sludge velocity at the membrane surface in order to remove accumulated particles. The shear velocity of about 1 up to 4 m/s is generated by pumps. Some side-stream systems use an additional air flush or frequent back-pulsing for particle removal. Higher flow velocity results in an increased pressure drop, translating into a higher energy demand and, hence, higher operational cost.
3. *Back-flush of the Membrane* Cyclical back-flushing with permeate and/or relaxation helps slow down membrane fouling. Back-flush or relaxation cycles are automatically initiated. Depending on the membrane/module type, back-flush or relaxation cycles can range from a few minutes to several hours. According to Judd (2006) less frequent, longer back-flushing (e.g., 600 s filtration, 45 s back-flush) are more efficient than more frequent but shorter back-flushes (e.g., 200 s filtration, 15 s back-flush). Temporarily enhanced air bubbling, intermittent aeration, back-flushing, and relaxation of the membranes are very effective in removing the cake layer from the membrane (van der Roest et al., 2002). At higher back-flush

rates, the net flux decreases, requiring a larger membrane surface area with a resultant impact on cost (changing of gross/net flux ratio).

4. *Modifying the Mixed Liquor* The mixed liquor can be changed by adjustment of sludge retention time (SRT) or through the addition of chemicals. Coagulants/flocculants as ferric chloride or aluminum sulfate show both an improvement of membrane hydraulic performance according to Judd (2006). Addition of adsorbents [powdered activated carbon (PAC)] reduces EPS level and the effects the filtration performance (Kim and Lee, 2003). Polymer-based filtration enhancer for MBR application are recently developed (e.g., MPE50 by Nalco Company) and lead to performance enhancement (Yoon et al., 2005).
5. *Employing Physical or Chemical Cleaning* (see next section).

9.2.3.3 Membrane Cleaning Regardless of operational fouling control, the membranes require regular cleaning to remove membrane fouling and keep the permeability loss within a given range. As the wastewater composition varies greatly from plant to plant, different wastewater streams require different cleaning strategies. Overall, two main cleaning methods can be distinguished: mechanical cleaning and chemical cleaning. Mechanical cleaning involves the physical removal of solids from the membrane material. For chemical cleaning, the following two cleaning procedures exist (Fig. 9.9):

- *Chemically Enhanced Back-flush or Maintenance Cleaning (in situ)* A back-flush with chemicals such as acids or oxidizing agents is typically activated in an automatic control mode, for example, daily or weekly. This cleaning procedure is applied in situ [cleaning in place (CIP)], that is, cleaning of the (submerged) membranes inside the tank or pumping chemicals through the pipes (side-stream modules). Key parameters are duration, frequency, and back-flush flux. With submerged membranes, it is also possible to remove the sludge first (pumping the sludge out of the aeration tank) and back pulse the membranes in such a way that the chemicals remain in the pores for a while [cleaning in air (CIA)]. This is a special cleaning procedure for

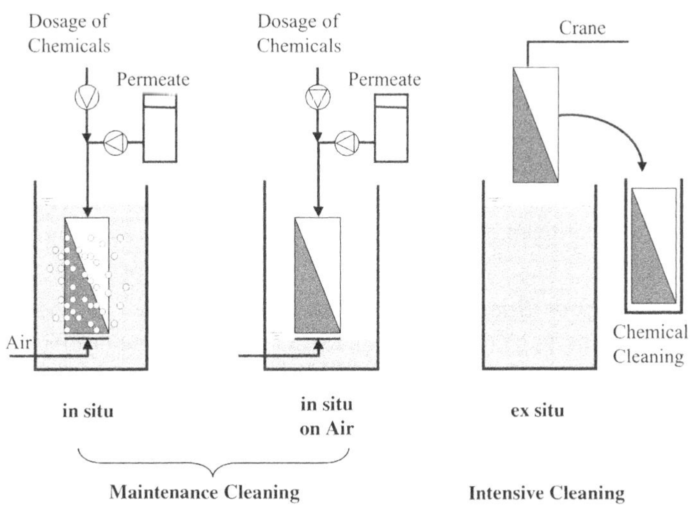

Figure 9.9 Membrane cleaning strategies for submerged modules.

submerged membranes that are back-flushed without direct contact with the activated sludge (air contact). Chemical maintenance cleaning can be accomplished using a 500-ppm NaOCl (Cl) solution, for instance, followed by cleaning with water, citric acid (0.5%) at a pH of 2.5–3, and a final cleaning step with water.

- *Intensive Cleaning Outside the MBR (ex situ)* For membrane recovery, the (submerged) membranes have to be removed from the aeration tank and cleaned outside the tank (need for membrane craning). This procedure uses higher chemical concentrations, for example, with citric acid and NaOCl 1000 ppm at 35°C (e.g., 1–2 times per year).

Typical chemicals used for membrane cleaning include strong acids such as nitric, sulfuric, or hydrochloric acid or weak acids such as citric acid to remove scaling or oxidizing agents such as sodium hypochlorite or peroxide, in some cases in combination with caustic soda for pH adjustment. The cleaning frequency as well as the types of chemicals and their concentration depend strongly on the wastewater composition and the membrane and module type. So far, no standard cleaning strategies exist. Currently, cleaning strategies are a focus of research with regard to avoiding the use of chlorinated products [adsorbable organic halogen compounds (AOX) formation!] and reducing the so-called aging of membranes caused by the use of oxidizing chemicals that corrode the membranes. Some commercial cleaning agents of unknown formulation are offered as well. Attention should be paid to minimizing chemical usage through maximum reuse and to the use of alternative, less environmentally harmful chemicals.

9.2.4 Comparison of MBR and CAS Processes

9.2.4.1 Sludge Characteristics The characteristics of the sludge formed in MBRs may be different from those of conventional activated sludge. While only biomass capable of forming flocs big enough to be settled in the sedimentation units can exist in the CAS process, MBR sludge may contain even single bacteria. Especially in side-stream modules where the sludge is subject to high shear forces, the existence of single bacteria is possible (e.g., Rosenberger, 2001) and the microbial floc size is reduced (Kim et al., 2001). Microscopically significant changes in the floc structure and free suspended cells were observed by Witzig et al. (2002). Bulking sludge, as may occur in the activated sludge process, cannot affect the quality of the MBR effluent and is no problem in MBR units.

In MBRs using side-stream modules, in particular, the biocenosis may be "stressed" by pressure variations and high flow velocities. This may lead to a release of EPS from the cells in the activated sludge (Kim et al., 2001). In combination with the higher aeration rate (m^3 air/m^3 tank) resulting from the smaller reactor volumes, this may lead to foaming problems, which has to be considered in the system design. Flat tanks with a large surface area or physical or chemical foam destruction may be advisable.

9.2.4.2 Mixed-Liquor Suspended Solids and Viscosity The MBR technology allows the aeration tank to be operated at higher MLSS concentrations. While in the conventional activated sludge (CAS) process the MLSS concentration is typically limited to 3–5 g MLSS/L by the sludge sedimentation properties, membranes overcome these limitations. Practical aspects such as reduced mass transfer, reduced oxygen transfer, increased viscosity (Fig. 9.10), membrane fouling, and sludge handling problems with increasing MLSS concentrations have led to typical MLSS ranges of between 10 and 15 g MLSS/L in submerged systems and up to 30 g/L in tubular side-stream applications.

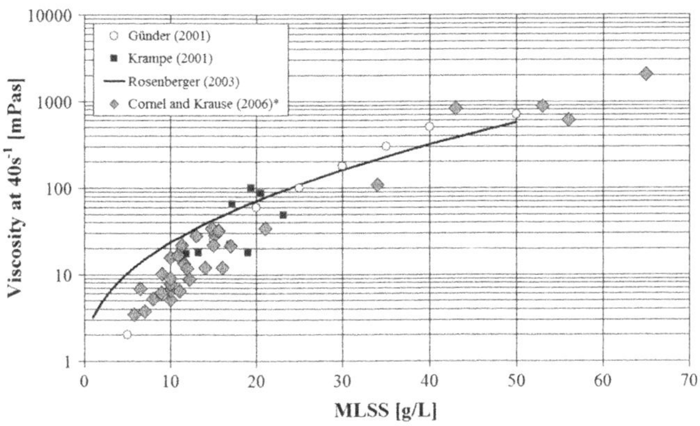

Figure 9.10 Viscosity vs. MLSS concentration. (*Own measurements performed in full-scale MBRs from 2001 to 2004.)

Figure 9.10 depicts the results of viscosity tests on different MBR sludges measured with a rotational viscometer at a shear stress of 40 s^{-1}. Although the results from laboratory-scale test, pilot plants and full-scale municipal plants gathered in the plot show some variation, the strong relation between viscosity and MLSS is obvious.

The impact of a higher viscosity was investigated by numerous research groups (e.g., Slatter, 1997; Günder, 2001; Rosenberger, 2001; Cornel et al., 2003). With increasing viscosity, the mass transfer decreases, resulting in a less efficient oxygen supply and a reduced oxygen capacity. Also, the energy demand for mixing and pumping the activated sludge rises with increasing viscosity. For side-stream membrane modules, in particular, the energy demand for pumping sludge at high viscosity is a factor that has to be considered.

9.2.4.3 Surplus Sludge Production Surplus sludge production results from the biomass growth, the inert portion of the suspended solids in the influent, and the inert material from the decay of biomass. It is well known that the yield decreases with decreasing food-to-microorganism (F/M) ratio [i.e., increasing sludge retention time (SRT), decreasing inflowing solid matter, and rising temperature].

The low or even zero sludge production frequently reported in the literature (e.g., Günder, 2001; Rosenberger, 2001) is mainly observed at high to very high sludge ages (calculated values of up to several hundred days) in laboratory- or pilot-scale plants.

Today, there is broad agreement that the yields of MBR and CAS are comparable as long as the sludge retention time is similar and in the normal technical range of <30 days. The amount of surplus sludge can be calculated by established models (e.g., Henze et al., 1987). In the design and operation of MBRs, a trade-off has to be made between the following two extremes:

- Small aeration tanks at the same F/M ratio and similar surplus sludge production as in CAS process
- Operation at extremely long sludge ages, that is, very low F/M ratio, minimizing the amount of surplus sludge but maximising biochemical oxidation and thus the specific oxygen consumption and the energy demand.

There is no operating mode to get both positive effects, that is, low-energy consumption and zero sludge production rates.

9.2.5 Experience with MBR Technology

9.2.5.1 Effluent Quality The quality of the MBR effluent is mainly determined by the proper design and operation of the activated sludge process. To a first approximation, the biological degradation in MBRs is equivalent to that of the CAS process as long as SRT is the same. However, due to the complete solids removal by the membranes, the MBR effluent quality is superior to that of the CAS process, not only in respect of suspended solids but also of other parameters such as chemical oxygen demand (COD), biochemical oxygen demand, total nitrogen (N), and phosphorus (P) as 1 mg/L MLSS correspond to about:

1.3 mg/L COD	(0.75 g MLSS/g COD)
0.18 mg/L N	(0.14 $gN/gCOD_{part}$) (Gujer et al., 1995, 1999)
0.03 mg/L P	(WAS 0.03 %P) (e.g., Schaum et al., 2005)

Accordingly, a solids-free effluent leads to measurably lower COD, N, and P levels compared to the effluent of secondary sedimentation with suspended solids in the range of 5–10 mg/L. No need to say that the biological removal rates including nitrogen removal at the low F/M ratio, for example, about 0.10 kg COD/(kg MLSS · day), and long SRTs of >25–50 days reported in the literature for most of the municipal plants are excellent (e.g., van der Roest et al., 2002; Cicek et al., 1999). Also, stable biological P removal was achieved (e.g., Adam et al., 2001; Gnirss et al., 2003, van der Roest et al., 2002). Of course, enhanced biological phosphorus removal (EBPR) requires the withdrawal of excess sludge.

Hygienic parameters were investigated and extensive disinfection reported (e.g., Churchhouse and Brindle, 2002). Van der Roest et al. (2002) reported the complete removal of coliform bacteria. The effluent quality meets the European quality standard for bathing water [Council of European Communities (CEC), 1975]. The bacteriophage concentration in MBR permeate is 100–1000 times lower than in CAS effluent (van der Roest, 2002).

Dutch STOWA (Foundation for Applied Water Research, NL) also investigated heavy metals, polycyclic aromatic hydrocarbons (PAH), and extractable organic halogens (EOX) in the MBR effluent. Not surprisingly it was found that the MBR technology offers superior removal efficiency for all heavy metals having a tendency to be sorbed by the sludge such as Cu, Hg, Pb, and Zn as well as for organics such as PAH, which is almost completely attached to sludge. However, high-peak EOX concentrations were measured after cleaning with NaOCl (van der Roest, 2002).

9.2.5.2 Specific Features of the Modules

Hollow-Fiber Modules In hollow-fiber modules, in particular, sludging and braiding of the membranes can be a major problem. Braiding is caused by hair and/or long fibers (e.g., additive cellulose fibers) that may loop around the membrane bundle, typically in the upper part of the module. Braiding is to be avoided by carefully screening the influent or the recirculated mixed liquor. In the lower part of the module, sludge may accumulate due to insufficient water and/or airflow (Vossenkaul, 2005) (Fig. 9.11). Especially when reducing the air scour to save energy, it should be kept in mind that a sufficient flow is crucial to

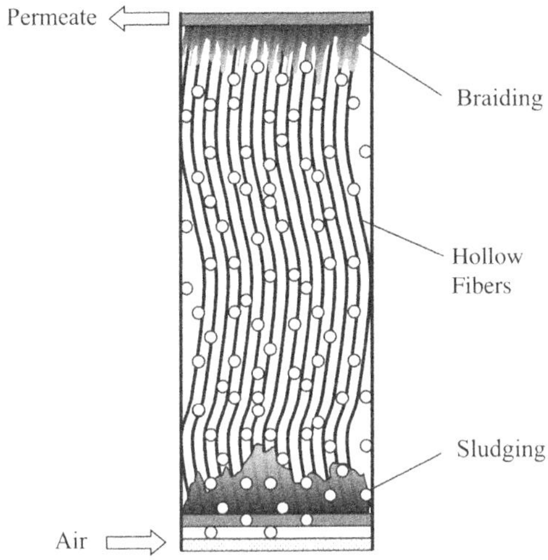

Figure 9.11 Potential operational problems with hollow-fiber modules (according to Vossenkaul, 2005).

preventing sludging of hollow-fiber bundles. Both effects are likely to reduce the active membrane surface area and mechanical cleaning is necessary.

Plate-and-Frame Membrane Modules Plate-and-frame membrane modules are less prone to braiding and sludging. On the one hand, long fibers and hair cannot loop around the membrane, and on the other hand, the high upward flow in the bottom part counteracts sludging. In plate-and-frame modules, fouling may occur in the peripheral area. Also, blocking of the gaps occurs. At present, plate-and-frame modules are operated at high airflow rates in order to minimize these phenomena. Most of the commercially available plate-and-frame modules cannot be back-flushed, which restricts the cleaning strategy (Fig. 9.12).

Tubular Modules In tubular modules, hair and/or long fibers may accumulate in or even block the inflow region where the mixed liquor enters the tubes. This results in a higher energy demand and/or a flux decline.

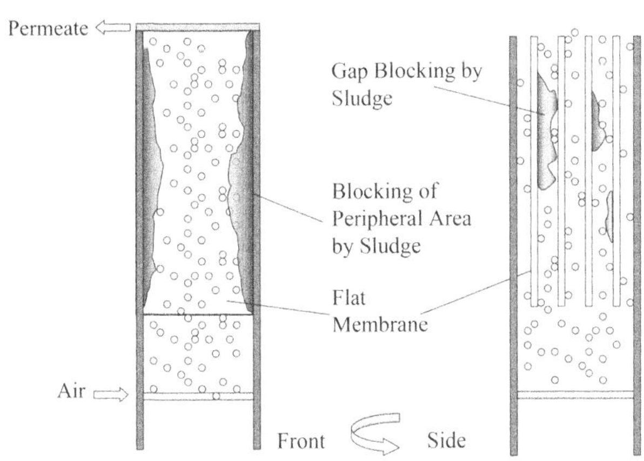

Figure 9.12 Potential operational problems with plate-and-frame modules (according to Vossenkaul, 2005).

9.3 MBR DESIGN CONSIDERATIONS

9.3.1 Introduction

More than 10 years' operating experience with full-scale MBRs has shown that the use of membranes in the activated sludge process requires an adaptation of the overall process—consisting of (1) pretreatment, (2) aeration tank, and (3) filtration—to the specific strengths and weaknesses of membranes for biomass separation. As the specific measures required may differ with the module design, the various options will be discussed in general in the following sections.

9.3.2 Pretreatment

9.3.2.1 Screening As shown before, fibers and hair may cause disturbance to membrane filtration. Moreover, abrasive or sharp-edged materials likely to damage the membrane surface should be kept clear of the membranes. Small-mesh sieves or, even better, microscreens with mesh sizes <0.5 mm have proved to be suitable for this purpose (Cornel and Krause, 2006). Van der Roest et al. (2002) report that a one-step screening process is inadequate and can result in sludging of the membrane modules and therefore suggest two-step screening. According to Frechen et al. (2006), a mesh screen is preferable as it removes around twice as much suspended solids, COD, N, and P as a bar screen. Van der Roest et al. (2002) reports that a discontinuously cleaned screen produces the largest amount of screenings. No hair was observed in a discontinuously cleaned drum filter screen.

9.3.2.2 Oil and Grease As nondegradable oil and grease may reduce the membrane flux, proper design of the oil and grease traps is crucial (Cornel and Krause, 2006). Screens remove only about 50% of the mineral oil present (van der Roest et al., 2002).

9.3.2.3 Equalization Tank As membranes are costly and the membrane surface area has to be designed for the maximum flow, it has to be checked whether hydraulic peaks can be leveled out. Equalization tanks and/or an aeration tank operation mode with alternating water levels are measures to be considered in the design phase. At municipal plants, the maximum stormwater flow at low temperatures has to be taken in account.

9.3.2.4 Inorganic Chemicals (Calcium and Precipitants) High calcium concentrations (>200 mg/L Ca^{2+}) may cause scaling problems (ATV-DVWK, 2002), depending on aeration and pH. With the high specific aeration at the membrane surface of submerged membranes, CO_2 is stripped out, which leads to an increase in the pH and, as a consequence, may cause $CaCO_3$ precipitation at or in the membrane; pH control helps avoid precipitation as has been demonstrated at an industrial site in Germany. In tubular membrane systems without scour aeration, the pressure drop across the module and/or the membrane can shift the solubility equilibrium and cause $CaCO_3$ precipitation. Typically, only industrial applications are affected by calcium precipitation. The membranes can be cleaned with (weak) acids.

Oxidized forms of iron or aluminum salts—which are used in the wastewater treatment process for phosphate precipitation—have no significant influence on the membrane filtration performance. Dissolved forms that can be oxidized under aerobic conditions may deposit on the membrane surface, causing a cohesive coating (ATV-DVWK, 2002).

9.3.3 Aeration Tank

9.3.3.1 Sizing of the Aeration Tank As the biological processes do not differ much from those in the conventional activated sludge process, common design rules apply for the aeration tank. As in the CAS process, the tank size is determined by the F/M ratio. Higher MLSS (see Section 9.2.4.2) concentrations at the same F/M ratio result in smaller aeration tank volumes. The volume of the filtration tank in the configuration with external filtration tanks (Fig. 9.6) can likewise be addressed as an aeration tank. Full-scale municipal MBRs in Europe are operated as extended aeration plants with SRTs of above 25 days. Predenitrification and intermittent denitrification can be applied. Recirculation of the oxygen-saturated return sludge from the filtration tank to the denitrification area should be avoided.

The hydraulic retention time (HRT) should be long enough to ensure hydrolysis and avoid short cuts. According to van der Roest et al. (2002), the use of a selector tank to improve the sludge characteristics may be beneficial.

9.3.3.2 Aeration Capacity The aeration system must be adequate to satisfy the oxygen needs of the biomass. In municipal MBRs, diffused-air aeration is the predominant method. Jet aerators and mechanical (surface) aerators have been tested and shown to be less efficient in MBR applications (Krampe and Krauth, 2003; Krause, 2005).

Mention should be made of two features specific to MBRs:

1. Due to the higher MLSS concentration, the specific oxygen transfer rate (SOTR) (kg/h) will be higher. Mixed liquors that tend to foam might foam even more in these intensively aerated MBR tanks.

2. The α correction factor is defined as the ratio of the oxygen transfer coefficient (k_La) in process water to k_La in clean water [American Society of Civil Engineers (ASCE), 2007]. The α value decreases with increasing MLSS, as has been shown by many researchers. Figure 9.13 depicts some published results from municipal plants. As the α value decreases with decreasing oxygen transfer, the amount of air supply has to be increased, which in turn increases the energy demand. This interrelationship is illustrated by the following example. While at an MLSS concentration of 5 g/L as typically used in the CAS process, the α value of a fine-bubble diffused-air aerator will be 0.7 ± 0.1, it will be as low as 0.5 ± 0.1 at 12 g MLSS/L in an MBR tank. Consequently, the aerators and blowers have to be designed for a correspondingly larger capacity, and the energy demand will be higher by a factor of $0.7/0.5 = 1.4$. [As the correction factor might be influenced by salt and electrolytes (ASCE, 2007), alcohol, organic acids, and surfactants (Mueller et al., 2002), all of which may vary more strongly in industrial wastewaters, measured α values of industrial MBRs are not included in Fig. 9.13.]

The main reason for the decreasing α value with increasing MLSS may be the viscosity increase with higher MLSS, which directly affects the mass transfer. Figure 9.14 depicts this relationship by plotting α versus viscosity.

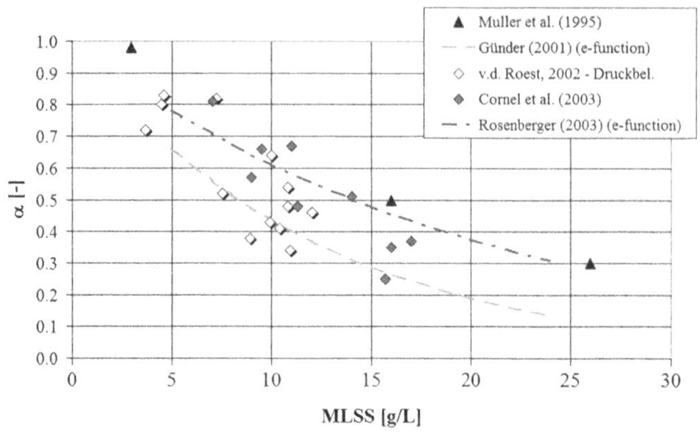

Figure 9.13 The α values vs. MLSS concentration in municipal MBRs.

9.3.3.3 Return Sludge If the membrane modules of submerged systems are placed in a separate filtration tank as shown in Figure 9.6, the thickened sludge has to be recirculated to the aeration tank. Typically, a recycle rate of 4–5 is applied, allowing the sludge to be thickened by 16–20% in the filtration chamber. As the membranes in the filtration tank are air-scoured, the return sludge will be virtually saturated with oxygen and should not be recycled to an anoxic denitrification zone or an anaerobic P-release zone in EBPR configurations.

9.3.3.4 Sludge Treatment Treatment of MBR sludge need not be significantly different from that of CAS sludge. Van der Roest et al. (2002) report that centrifuges yield better results compared to a filter press in laboratory tests. The digestion of the sludge does not differ from the CAS process (DWA, 2005). As part of the polymers added for dewatering may be recycled to the aeration tank with the filtrate, tests should be performed to ensure that the polymer does not interfere with or block the membrane (ATV-DVWK, 2002).

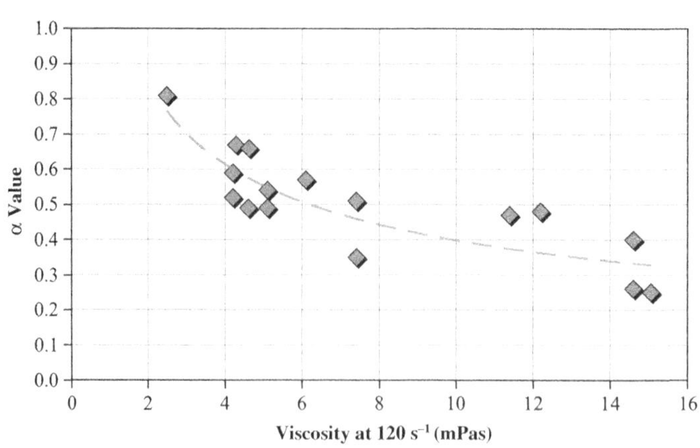

Figure 9.14 The α values vs. viscosity in municipal MBRs (Krause, 2005).

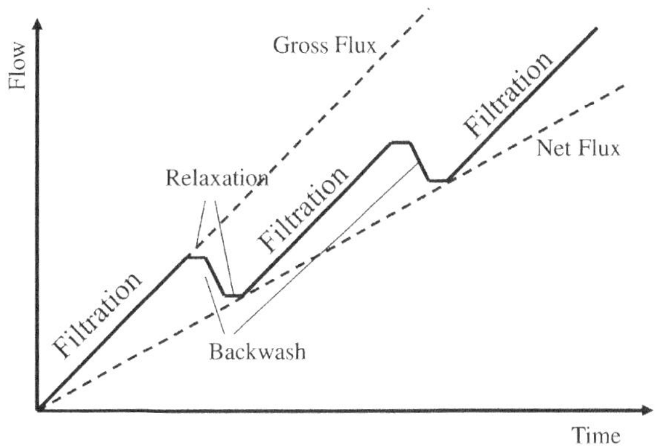

Figure 9.15 Net flux and gross flux.

9.3.4 Membrane Filtration

The required membrane surface area is a function of the hydraulic load and the membrane flux. For municipal applications with combined sewers, the stormwater runoff has to be taken into account. Concerning the flux, a distinction has to be made between gross flux and net flux (Fig. 9.15). The gross flux characterizes the actual filtration rate while withdrawing the effluent. A more important parameter for the design of the MBR is the net flux that characterizes the average flow rate including phases of relaxation and backflushing. Some membranes allow a temporarily increased net flux to handle peak flows.

Typical values for the maximum fluxes of submerged MBRs in municipal wastewater treatment are in the range of about 25–30 $L/(m^2 \cdot h)$ (e.g., DWA, 2005). The net flux will be about 20% lower, depending on membrane type and system. For industrial wastewater, pilot tests are recommended to estimate realistic fluxes. The test duration may depend on the variation of the wastewater composition over time. As the flux typically decreases with time due to fouling and/or scaling phenomena (see Section 9.2.3 and Fig. 9.8), it is important not only to record the initial flux, pilot tests offer the chance to test different cleaning strategies and to reconcile fouling control measures, cleaning procedures, and required flux. Frequently, the fluxes applied in submerged modules in industrial applications are as low as 5–15 $L/(m^2 \cdot h)$. For tubular membranes in industrial applications, fluxes of up to 100 $L/(m^2 \cdot h)$ have been reported (Cornel and Krause, 2006). Besides surface area, the design of the filtration tank may be of importance. The inflow and outflow pipes should be installed over the full tank length in order to attain constant hydraulic flows and avoid membrane fouling (van Bentem et al., 2005). A pH gradient should be avoided when precipitation is to be expected. A proper amount of recycle and a proper flow regime can minimize sludging of the membrane modules.

9.4 APPLICATIONS AND COST

According to van der Roest et al. (2002), more than 1000 MBRs were in operation worldwide in 2002, most of them in Japan (approximately 66% of the installations). Many of the Japanese installations consist of an in-building treatment (office and domestic). In Europe, most applications relate to the treatment of landfill leachate. About 98% of the systems accounts for a combination of aerobic wastewater treatment with membrane separation and 2% for a combination with an anaerobic wastewater treatment. According to van der Roest et al. (2002), 55% of the installed MBRs are of the submerged type, whereas 45% of the installations use external side-stream modules. Nearly all of the large municipal MBRs in Europe use submerged membranes. Side-stream modules are used for industrial wastewater treatment and predominate in the treatment of landfill leachate.

9.4.1 Applications of MBR in Municipal WWT

Membrane bioreactors are an option for municipal wastewater treatment when high effluent water quality is required, for example, bathing water quality, or when the receiving water body is very sensitive or when the water is to be treated for reuse. As mentioned before (see Section 9.2.5.1), the effluent quality is superior to that of secondary sedimentation. To attain a similar effluent quality by conventional treatment, effluent filtration and disinfection would be required in addition. This needs to be taken into account when comparing the cost of MBR and conventional activated sludge treatment.

Especially for an extension of wastewater treatment plants (WWTPs) where the capacity of the existing plant is insufficient and floor area is limited or where the buildings are in good condition, so-called hybrid MBRs are a viable solution (e.g., de Vente et al., 2005).

The small footprint resulting from the smaller aeration units and the elimination of secondary sedimentation allows MBRs to be housed in a building, which might be of interest in urban areas as well as in tourist locations. This not only facilitates emission control but also eliminates the visibility of the wastewater treatment plant, a technical advantage that is not to be underestimated in populated areas.

9.4.2 Applications of MBR in Industrial Water Management

In industrial wastewater treatment, MBRs offer an attractive alternative when the treated effluent is to be routed to reuse. MBRs are a perfect pretreatment solution when further treatment by nanofiltration or reverse osmosis (RO) is considered.

Other typical applications are retrofits to existing plants combined with a capacity increase or installation where the small footprint is advantageous. According to ATV-DVWK (2002), MBRs are operated in more than 14 industries in western Europe, representing more than 120 plants ranging up to 18,000 m^3/day in capacity. MBRs are operated in particular in the chemical and pharmaceutical industries and also in leachate treatment applications. Installations on cruise ships are an increasing market as both the small footprint and superior effluent quality are beneficial.

9.4.3 Cost Comparison

9.4.3.1 Capital Cost Cost data are difficult to obtain. The answer to the question of how the capital cost of MBRs is related to that of conventional activated sludge treatment depends on who is doing the analysis, what will be covered by the evaluation, and how the better effluent quality and the smaller land use will be figured in.

As a matter of fact, membrane prices have been steadily decreasing, for example, in the United Kingdom from 400 US\$/m^2 (320 €/m^2) in 1992 to 100 US\$/m^2 (80 €/m^2) in 2000 (Judd, 2005). However, cheap membranes do not mean cheap elements. The aeration devices, automated control, cleaning devices, pipes, and fittings as well as additional pretreatment have to be taken into account in balancing the additional costs. Beneficial are (a) the smaller required tank volume, which is about 150 L per population equivalent (PE). for municipal applications with full nutrient removal (compared to 450 L/PE for CAS), (b) the smaller footprint, which may be a cost factor when space is at a premium, and (c) the better effluent quality. For the CAS effluent to be comparable, it would require additional effluent filtration and disinfection.

9.4.3.2 Operating Costs When comparing operating costs, the higher energy demand of MBRs is a main factor. Besides the lower α value, which results in a higher aeration energy demand in the range of 0.1–0.15 kWh/m^3 for normal municipal wastewater in Europe (40% of approximately 0.25–0.4 kWh/m^3 aeration energy in the CAS process) (Krause, 2005), the greater part of the additional energy demand is accounted for by fouling control. Aeration in submerged membrane systems consumes about 0.4 kWh per m^{3}effluent (Krause, 2005). The greater part of the energy demand in side-stream systems relates to pumping. Depending on the tube diameter and flow velocity, cross-flow generation requires about 3 kWh/m^3.

Membrane cleaning consumes chemicals. In municipal applications, the cost of membrane cleaning ranges between 0.2 and 1 € per m^2 membrane surface area per year (0.25–1.25 US$) (Wedi et al., 2005).

Membrane replacement cost can be considered as operational cost as well. Up to now, there is little reliable information on the lifetime of membranes. Although some membrane systems have been in operation for 6 years without failure or with an annual replacement rate of less than 3% (Churchhouse and Brindle, 2002), others had to be replaced after 2–3 years because of serious fouling, braiding, sludging, or even mechanical destruction. It is evident that lifetime is correlated with the composition and pretreatment of the wastewater, the applied cleaning strategies, as well as with the type of membrane and module construction.

In case the MBR effluent can be reused, the savings on freshwater must appear on the balance sheet.

9.5 CONCLUSIONS AND SUMMARY

Membrane bioreactor technology has become "mature" and can be applied in both municipal and industrial wastewater treatment. The process may play a major role in future wastewater treatment and water reuse applications.

Membrane bioreactors combine the activated sludge process for wastewater treatment with biomass separation from the mixed liquor by ultra- or microfiltration membranes. Advantages are the superior effluent quality characterized by complete solids removal and "disinfection," the small footprint of the plant resulting from more compact aeration tanks, the absence of a secondary sedimentation tank, and the modular construction.

Attention should be paid to the pretreatment, which typically has to be more advanced, depending on the type of membrane and membrane module. In municipal applications, a two-step screening process (fine screening with a small mesh size) is favorable. Due to fouling and scaling, the permeability of the membrane will decrease with time. The fouling behavior depends on the wastewater composition, the biomass characteristics, as well as the operating conditions. Measures to minimize fouling are reducing the flux, increasing the shear or relaxation and back-flushing, and, ultimately, chemically enhanced membrane cleaning. All these measures have an impact on cost and energy demand. Further research and development is required to improve fouling control and cleaning strategies.

The overall energy demand of membrane bioreactors is 2–4 times higher compared to conventional activated sludge plants. This is mainly due to operational fouling control (either aeration or pumping energy). As against the CAS process, only a small portion of the additional energy is needed for oxygen transfer due to the lower α values (approximately 30–40% more energy). The lower α value is due to the higher viscosity of the mixed liquor resulting from the higher suspended solids (biomass) concentration.

Nomenclature

Parameter	Unit	Definition
α		Ratio of the oxygen transfer coefficient (k_La) in clean water and in wastewater
AOX	μg/L	Adsorbable organic halogen compounds
CA		Cellulose acetate
CAS		Conventional activated sludge
CIA		Cleaning in air
CIP		Cleaning in place
COD	mg/L	Chemical oxygen demand
EBPR		Enhanced biological phosphorus removal
EOX		Extractable organic halogens
EPS		Extracellular polymeric substances
F/M ratio	g substrate/g biomass · day	Food-to-microorganism ratio
HRT	h	Hydraulic retention time
k_La	L/h	Overall mass transfer coefficient (in water or as αk_La in activated sludge)
MBR		Membrane bioreactor
MF		Microfiltration
MLSS	g/L	Mixed-liquor suspended solids
NaOCl		Sodium hypochlorite
PAH		Polycyclic aromatic hydrocarbons
PE	PE	Population equivalent
PES		Polyethersulfone
PP		Polypropylene
PTFE		Polytetrafluorethylene
PVDF		Polyvinylidenfluoride
SOTR	kg/h	Standard oxygen transfer rate
SRT	day	Sludge retention time (sludge age)
TKN	mg/L	Total Kjeldahl nitrogen
TMP	bar	Transmembrane pressure
UF		Ultrafiltration
WAS	kg/day	Waste activated sludge
WWT		Wastewater treatment
WWTP		Wastewater treatment plant

REFERENCES

Adam, C., Kraume, M., and Gnirss, R. (2001). Vermehrte biologische Phosphorelimination in Membranbioreaktoren, Proceedings A11 at 4. Aachener Tagung Siedlungswasserwirtschaft und Verfahrenstechnik. In *Membrantechnik*, Vol. 4. ISBN 3-921955-25-4, A11-1–A11-16.

American Society of Civil Engineers (ASCE) (2007). Measurement of oxygen transfer in clean water. ASCE Standard No. ASCE/EWRI 2-06, Michael Stenstrom (Editor), ISBN 10 0-7844-0848-3.

ATV-DVWK (2002). Aufbereitung von Industrieabwasser und Prozesswasser mit Membranverfahren und Membranbelebungsverfahren—Teil 2: aerobe Membranbelebungsverfahren. *KA Wasserwirtschaft, Abwasser, Abfall* 49(11), 1563–1571.

Baker, R. (2004). *Membrane Technology and Applications*, 2nd ed., Wiley, Hoboken, NJ.

Chang, I.-S., Le Clech, P., Jefferson, B., and Judd, S. (2002). Membrane fouling in membrane bioreactors for wastewater treatment. *J. Environ. Eng.* 128(11), 1018–1029.

Churchouse, S., and Brindle, K. (2002). Operational experience of full scale membrane bioreactor sewage treatment plants, proceedings. Paper presented at the 3rd IWA World Water Congress, Melbourne.

Cicek, N., et al. (1999). Characterization and comparison of a membrane bioreactor and a conventional activated sludge system in the treatment of wastewater containing high molecular weight compounds. *Water Environ Res.* 71(1), 64–71.

Cornel, P. and Krause, S. (2006). Membrane bioreactors in industrial wastewater treatment—European experiences, examples and trends. *Water Sci. Technol.* 53(3), 37–44.

Cornel, P., Wagner, M., and Krause, S. (2003). Investigation of oxygen transfer rates in full scale membrane bioreactors. *Water Sci. Technol.* 47(11), 313–319.

Council of European Communities (CEC) (1975). Council directive of 8 December 1975 concerning the quality of bathing water, 76/160/EEC. CEC.

Daigger, G. T. (2003). State of the art of membrane bioreactors in North America. In *Proceedings of the International Conference Application and Perspectives of MBRs in Wastewater Treatment and Reuse*, Cremona, Apr. 28–29.

De Vente, D., Geraats, B., Boenders, H., and Futselaar, H. (2005, Apr.). The Ootmarsum hybrid MBR project, *H_2O*, MBR Special III.

DWA (2005). Membranbelebungsverfahren, 2. Arbeitsbericht des DWA-Fachausschusses KA-7, Fassung vom 19.01.2005. Available: www.dwa.de (in German).

Frechen, F.-B., Schier, W., and Wett, M. (2006). Pre-treatment of municipal MBR applications in Germany—Current status and treatment efficiency. In *Proceedings, IWA World Water Congress*, Peking.

Gnirss, R., Lesjean, B., and Buisson, H. (2003). Biologische Phosphorentfernung mit einer nachgeschalteten Denitrifikation im Membranbelebungsverfahren, Proceedings A17 at 5. Aachener Tagung Siedlungswasserwirtschaft und Verfahrenstechnik. In *Membrantechnik*, Vol. 5. ISBN 3-921955-28-9, A17-1–A17-13.

Gujer, W., Henze, M., Mino, T., Matsuo, T., Wentzel, M. C., and Marais, G. v. R. (1995). Activated sludge model no. 2. *Water Sci. Technol.* 31(2), 1–11.

Gujer, W., Henze, M., Mino, T., and van Loodsrecht, M. (1999). Activated sludge model no. 3. *Water Sci. Technol.* 39(1), 183–193.

Günder, B. (2001). *The Membrane-Coupled Activated Sludge Process in Municipal Wastewater Treatment.* Technomic Publishing Lancaster, PA.

Henze, M., Grady, C. P. L., Gujer, W., Marais, G. V. R., and Matsuo, T. (1987). Activated sludge model No. 1. Scientific and Technical Report No. 1. International Association on Water Pollution Research and Control (IAWPRC), London, U.K.

Judd, S. (2004). A review of fouling of membrane bioreactors in sewage treatment. *Water Sci. Technol.* 49(2), 229–235.

Judd, S. (2005). Membrane technology. In *Proceedings MBR5 Short Course*, Cranfield University, July 18–19, 2005.

Judd, S. (2006). *The MBR Book: Principles and Applications of Membrane Bioreactors in Water and Wastewater Treatment.* Elsevier, New York.

Kim, J. S., and Lee, C. H. (2003). Effect of powdered activated carbon on the performance of an aerobic membrane bioreactor: Comparison between cross-flow and submerged membrane systems. *Water Environ. Res.* 75, 300–307.

Kim, J.-S., Lee, C.-H., and Chang, I.-S. (2001). Effect of pump shear on the performance of a cross-flow membrane bioreactor. *Water Res.* 35(9), 2137–2144.

Krampe, J. (2001). Das SBR-Membranbelebungsverfahren, Stuttgarter Berichte zur Siedlungswasserwirtschaft, No. 163, Ph.D. Thesis, Universität Stuttgart, Germany.

Krampe, J., and Krauth, K. (2003). Oxygen transfer into activated sludge with high MLSS concentrations. *Water Sci. Technol.* 47(11), 297–303.

Krause, S. (2005). Untersuchungen zum Energiebedarf von Membranbelebungsanlagen, Ph.D. Thesis, Technische Universitaet Darmstadt.

Le-Clech, P., Chen, V., and Fane, A. G. (2006). Fouling in membrane bioreactors used in wastewater treatment. *J. Membr. Sci.* 284(1–2), 17–53.

Mueller, J. A., Boyle, W. C., and Pöpel, H. J. (2002). *Aeration: Principles and Practice.* CRC Press, Boca Raton, FL.

Muller, E. B., Stouthamer, A. H., van Verseveld, H. W., and Eikelboom, D. H. (1995). Aerobic domestic waste water treatment in a pilot plant with complete sludge retention by cross flow filtration. *Water Res.* 29(4), 1179–1189.

Rosenberger, S. (2003). Charakterisierung von belebtem Schlamm in Membranbelebungsreaktoren zur Abwasserreinigung. Fortschr. Ber. VDI Reihe 3 Nr. 769. Düsseldorf: VDI Verlag. ISBN 3-18-376903-4.

Schaum, C., Cornel, P., and Jardin, N. (2005). Possibilities for a phosphorus recovery from sewage sludge ash. In *Proceedings Conference on the Management of Residues Emanating from Water and Wastewater Treatment*, Johannesburg, South Africa, Aug. 9–12, 2005.

Slatter, P. T. (1997). The rheological characterisation of sludges. *Water Sci. Technol.* 36(11), 9–18.

Stephenson, T., Judd, S., Jefferson, B., and Brindle, K. (2000). *Membrane Bioreactors for Wastewater Treatment.* IWA Publishing, London.

Van Bentem, A., Petri, C., Mulder, J. W., Evenblij, H., de Vente, D., and Geraats, B. (2005, Apr.). Dutch MBR grows maturiti, *H_2O*, MBR Special III.

Van der Roest, H. F., Lawrence, D. P., and van Bentem, A. G. N. (2002). *Membrane Bioreactors for Municipal Wastewater Treatment.* IWA Publishing, London.

Vossenkaul, K. (2005). Grundlagen des Membranbelebungsverfahrens. In *Proceedings DWA-DVGW Membrantage*, Osnabrück, June 21–23, 2005.

Wedi, D., Wild, W., Resch, H., and Bleisteiner, S. (2005). Betriebsergebnisse der MBR Monheim—Abwasserreinigung und Erhalt der Permeabilitäten mittels chlorfreier chemischer Reinigung, Proceedings A9 at 6. Aachener Tagung Siedlungswasserwirtschaft und Verfahrenstechnik. In *Membrantechnik*, Vol. 6. ISBN 3-86130-775-8, A9-1–A9-12.

Witzig, R., Manz, W., Rosenberger, S., Krüger, U., Kraume, M., and Szewzyk, U. (2002). Microbiological aspects of a bioreactor with submerged membranes for aerobic treatment of municipal wastewater. *Water Res.* 36, 394–402.

Yoon, S. H., Collins, J. H., Musale, D., Sundarajan, S., Tsai, S. P., Hallsby, G. A., Kong, J. F., Koppes, J., and Cacuia, P. (2005). Effects of flux enhancing polymer on the characteristics of sludge in membrane bioreactor process. *Water Sci. Technol.* 51, 151–157.

CHAPTER 10

Submerged Membranes

ANTHONY G. FANE

UNESCO Centre for Membrane Science & Technology, University of New South Wales, Australia 2052 and Singapore Membrane Technology Centre, Nanyang Technological University, Singapore

10.1 INTRODUCTION

Submerged (or immersed) membranes are now commonplace in advanced water and wastewater applications. However, this is a relatively recent development; for example, the subject does not rate a mention in *Membrane Handbook* (Ho and Sirkar, 1992) and only has the briefest mention in *Water Treatment Membrane Processes* (Mallevialle et al., 1996). This chapter provides an overview of submerged membranes, from basic features to practical aspects and applications. We will start with the background to this development.

10.1.1 History and Development

The traditional approach to the pressure-driven liquid-phase membrane technologies has been to apply significant positive pressure to maximize flux and pumped cross flow to control concentration polarization. As a result, the first two or three decades of microfiltration (MF) and ultrafiltration (UF) were based on contained pressure vessels in pumped circulation loops. At the same time the typical applications were for medium to high value products, such as in the food and beverage industries. Water production by membranes involved reverse osmosis (RO) desalination at high cost in niche areas. Waste treatment by membranes usually included resource recovery to offset the "costly" membranes. However, around the late 1980s and early 1990s a number of significant trends were occurring:

- The nonsaline water industry became a target for membrane manufacturers, driving down the cost of low-pressure membranes.
- A major driver for application of low-pressure membranes in water treatment was the need for improved removal of pathogens, particularly chlorine-tolerant protozoa, such as *Cryptosporidium*.

Advanced Membrane Technology and Applications. Edited by Norman N. Li, Anthony G. Fane, W. S. Winston Ho, and T. Matsuura
Copyright © 2008 John Wiley & Sons, Inc.

- For low solids feeds, such as raw water and industrial clarification, it was realized that cross flow was not essential and that the "batch-continuous" filtration approach with intermittent backwash was feasible.
- It was accepted that practical fluxes could be obtained with MF and UF membranes using pressure driving forces of a fraction of an atmosphere.
- Bubbly (two-phase) flow was being studied and shown to be very effective at controlling concentration polarization in membrane processes, including membrane bioreactors.

These trends set the scene for a paradigm shift in membrane module design.

The early references to submerged membranes came from Japan. Ohkubo et al. (1988) obtained a patent describing hollow fibers in a vertical bundle in a vessel with air scour to vibrate the fibers to remove the cake. The first reported use of submerged hollow fibers in a wastewater membrane bioreactor (MBR) was by Yamamoto et al. (1989), who used fibers in a bundle and air bubbles for aeration, mixing, and induced liquid flow. Permeate was removed by suction. At that time the concept was more of a curiosity, but within a decade the submerged membrane has become the dominant approach for low-pressure membrane processing in the water and wastewater industry.

10.1.2 Concept and Features

Submerged membranes, as the name implies, are an assembly of membranes positioned in a "flooded" tank at atmospheric pressure typically open at the top. The liquid to be filtered is fed to the tank and permeate is removed under suction, either by a pump or by a gravity effect. Concentrate is removed continuously or intermittently from the tank. Figure 10.1

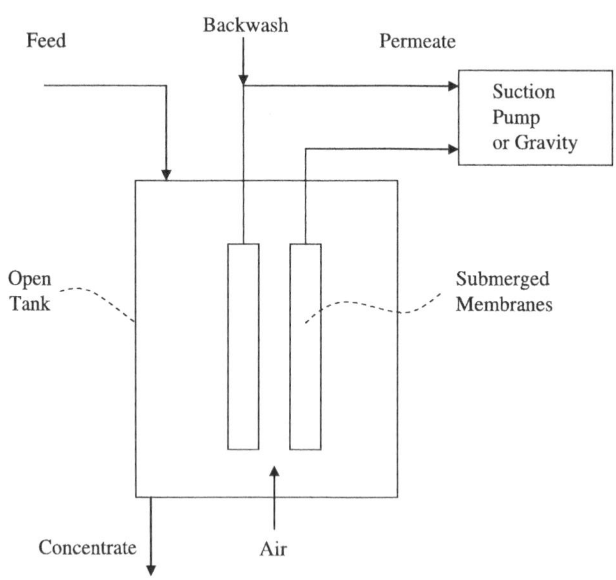

Figure 10.1 Schematic showing common features of submerged membrane system.

depicts the general features of a submerged membrane system. There are several possible modes of operation (Section 10.2) and various module geometries (Section 10.3).

The common features of submerged membrane systems are:

- Open tank (no pressure vessel)
- Membranes in bundles of fibers or arrangements of flat plates
- Permeate removed by suction
- Transmembrane pressures (TMP) <1 atm
- Polarization control by backwash and/or two-phase bubbly flow (depending on the module geometry and type of feed)

10.2 MODES OF OPERATION OF SUBMERGED MEMBRANES

Some of the phenomena and governing relationships for membrane operation are discussed below, with emphasis on submerged membranes, although not limited to them.

10.2.1 Polarization

Accumulation of retained species at the membrane surface is a consequence of the membrane separation. In the low-pressure membrane processes the retained species will be particulate, colloidal, and macromolecular, depending on the membrane pore size. The localized accumulation on the membrane is polarization, either particle polarization or concentration polarization (applies to dissolved species). The basic equation describing the relationship between flux (J) and driving force is

$$J = \frac{\Delta P - \Delta \Pi}{\eta(R_m + R_d)} \tag{10.1}$$

where R_d represents the "deposit" resistance due to polarization, some of which could be fouling. The net driving force is TMP minus any osmotic effects. In MF and UF, as applied in submerged membranes, it is convenient to assume $\Delta\Pi$ is zero because much of the deposit is particulate. In fact retained macrosolutes and colloids, particularly using UF membranes, could exert small osmotic effects, which may be significant when TMP is a fraction of an atmosphere. Thus, for submerged membrane filtration we assume:

$$J = \frac{\Delta P}{\eta(R_m + R_d)} \tag{10.2}$$

Equation (10.2) applies for both steady-state and unsteady-state conditions. The key feature is how the mode of operation controls the magnitude of deposit resistance, R_d. One other point of interest from Eqs. (10.1) and (10.2) is the role of permeate (filtrate) viscosity, η, which can have a strong influence on production. For example, the effect of temperature on water viscosity would mean at 10°C the flux would only be about 60% of that at 30°C for the same TMP. It is important to note that η refers to permeate and not feed viscosity, which can be much higher than water, as found in a typical membrane bioreactor.

Submerged membranes are typically operated at fixed (imposed) flux (J_i). This means that changes due to deposition/fouling are reflected in changes to TMP. So we can write

$$\Delta P = \eta J_i R_m + \eta J_i R_d \tag{10.3}$$

or as a rate of change,

$$\left(\frac{d\,\Delta P}{dt}\right) = \left(\frac{dTMP}{dt}\right) = \eta J_i \frac{dR_d}{dt} \tag{10.4}$$

Equation (10.4) shows for constant flux processing how the driving force (TMP) will increase as deposition progresses. For submerged membranes the feed side is atmospheric pressure; so this means the permeate-side pressure becomes increasingly subatmospheric.

The rate of resistance rise, dR_d/dt, can be related to the "net" flux of solids and their specific resistance, α_c, (assuming incompressible solids); so,

$$\frac{dR_d}{dt} = \frac{dm}{dt}\alpha_c = J_{\mathrm{Net}} C_b \alpha_c \tag{10.5}$$

where m is mass load per unit surface of membrane, C_b is feed concentration, and J_{Net} is the imposed flux minus backtransport flux (see Section 10.2.2). Equations (10.4) and (10.5) are easiest to visualize for incompressible particle cakes, no membrane fouling (pore closure, etc.), and in flat geometry. With some modification they can be applied to conditions where R_m changes with usage, for compressible or dynamic deposits and in cylindrical geometry (hollow fibers, etc.).

10.2.2 Cross-Flow Filtration

An obvious approach to limiting the effect of polarization and the magnitude of R_d or dR_d/dt is to apply shear stress to the accumulating layer. This is the principle of cross flow, or tangential flow, where fluid is forced across the membrane surface and provides a velocity gradient (shear rate) adjacent to the membrane surface. In submerged membrane processes cross flow is usually induced by bubbling and is applied when the feed to the membrane has a high fouling tendency, typically with high feed solids such as the MBR. We can anticipate the effect of shear stress (τ) and feed concentration (C_b) by the relationship

$$\frac{dTMP}{dt} = A_1 f(\tau) g(C_b) \tag{10.6}$$

where f and g imply functions of shear stress and concentration, respectively. In common with other membrane systems, there is evidence (see Section 10.4.3) for submerged membrane systems that $dTMP/dt$ is increased as C_b increases and decreases as τ increases. The use of bubbling to induce flow across submerged membranes provides an unsteady-state flow condition, and this also helps as $dTMP/dt$ appears to decrease with the standard deviation of shear stress (see Section 10.4.1).

A feature of cross flow, whether pumped, stirred, or bubbled, is that flux-convected deposition is mitigated by shear-induced "back transport." In principle, for a given feed and cross-flow hydrodynamics, a flux can be identified at which convection and back transport are balanced. Under these conditions it is possible that $dTMP/dt \rightarrow 0$, that is, fouling is negligibly slow. This is the *critical flux*, J_{crit}, (Section 10.4.3) for that set of conditions. For complex feeds such as mixed liquor in the MBR, the critical flux is difficult to define because the limiting foulant needs to be specified, and it is more practical to consider the *sustainable flux*, for which $dTMP/dt$ is acceptably low. However, the critical flux concept is useful to aid our understanding of fouling. For example, we can rewrite Eq. (10.5) as

$$\frac{dR_d}{dt} = (J_i - J_{\text{crit}})\,C_b \alpha_c \tag{10.7}$$

In other words, the net flux of foulant is reduced by the magnitude of the critical flux. More discussion of critical flux in submerged systems with bubbly flow is given in Section 10.4.3.

The characteristic TMP history for a submerged membrane with cross flow (bubbling) is depicted in Figure 10.2a. Due to deposition TMP rises slowly, and the rate of rise is less with more imposed shear or lower solids. In an idealized situation TMP would remain unchanged at subcritical flux conditions. However, for various reasons (see Section 10.5.1) some degree of TMP rise tends to occur at all fluxes; so the interest is in the acceptable rate of rise and the sustainable flux. Figure 10.2a also includes the potential for a sudden TMP "jump" that can be observed in prolonged operation at constant flux (discussed further in Section 10.5.1).

10.2.3 Dead-end Filtration and Batch-Continuous Operation

For relatively low solids content feeds, it is feasible to operate without continuous cross flow or surface shear, and this can reduce energy costs. Such applications include water treatment and pretreatment for RO desalination and reclamation. This mode of operation is called dead-end filtration or frontal filtration, and the key feature is that the deposition of retained species is allowed to grow. A typical cycle commences with a "clean" membrane (after backwash), and at constant flux the TMP rises according to Eq. (10.4). After a specified period (t_c), or at a predetermined ΔP_{max}, the flux is stopped and the deposit is removed by backwashing and (usually) vigorous aeration. The contents of the membrane

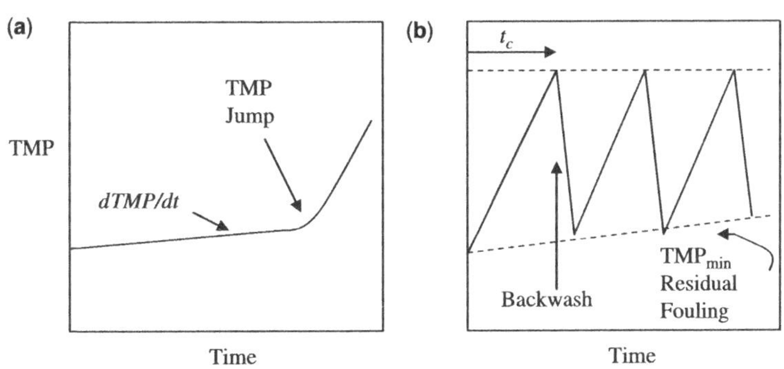

Figure 10.2 Transmembrane pressure (TMP) histories for constant-flux operation, typically applied to submerged membranes: (a) cross flow filtration and (b) dead-end filtration.

tank are then discharged along with the backwashed solids before the tank is recharged and the next filtration cycle commences. The cycle times would typically be about 30 min and the backwash and recharge <5 min. This mode of operation is batch continuous, and net flux would be slightly less than the imposed flux; that is, for t_c of 30 min and t_{BW} for backwash and recharge of 5 min the net flux is about 86% of imposed flux.

The idealized TMP cycle is depicted in Figure 10.2b for a cycle that is limited by $\Delta P_{\max}$, rather than a specified cycle time. In reality the backwash efficiency will be <100% so that $\Delta P_{\min}$ (the "cleaned" membrane TMP) may slowly rise over time, eventually requiring chemical cleaning (see Section 10.5.2). During a filtration cycle the deposit resistance grows so that at time t it is given by

$$R_d = m\alpha_c = (J_i t C_b)\,\alpha_c \tag{10.8}$$

The TMP variation is then, from Eq. (10.3), given by

$$\Delta P = \eta J_i R_m + \eta (J_i)^2 t C_b \alpha_c \tag{10.9}$$

So for a maximum TMP ($\Delta P_{\max}$), the cycle time, t_c, is

$$t_c = \frac{\Delta P_{\max} - (\eta J_i R_m)}{\eta J_i^2 C_b \alpha_c} \tag{10.10}$$

Equation (10.10) shows that t_c is very sensitive to the imposed flux J_i and inversely proportional to the feed concentration C_b. Thus, a 40% increase in flux could almost halve the cycle time.

10.2.4 Submerged versus Contained Modules

As noted in Section 10.1.1 submerged membrane modules have been developed and gained popularity for a number of reasons—the targeting of the water industry, the recognition that dead-end filtration worked for low solids feeds and that bubbled cross flow worked for more fouling feeds, and an acceptance of modest fluxes and low TMPs. In a water treatment application a tank containing submerged membranes can achieve a superficial velocity (volumetric flow rate/footprint area) about 10× that of a rapid sand filter with a product water quality far superior. When compared with contained (pressurized) membrane modules, the benefits are not so clear-cut, and both approaches have their pros and cons. Table 10.1 provides a qualitative comparison of submerged and contained membrane modules, with the following qualifications:

Packing Density A high packing density can have benefit in terms of plant footprint, but it can be a compromise in terms of fluid management and fouling control, and this can be influenced by feed properties. For example, submerged hollow-fiber bundles for MBRs tend to have lower packing densities than those for water treatment. Similarly, contained modules for MBRs tend to have membranes of moderate diameter (several millimeters), viewed as either small-bore tubes or large-bore hollow fibers, with in-to-out feed. For flat-plate modules the footprint can be decreased by stacking one set of cassettes above another, although this adds to required aeration pressure (see Section 10.4.2).

TABLE 10.1 Comparison of Contained and Submerged MF and UF Systems

Characteristic Feature	Contained Tubular	Contained Hollow Fiber	Submerged Flat Sheet	Submerged Hollow Fiber
Packing density	Low	High	Moderate	High
Mode of operation (10.2.2, 10.2.3)	Cross flow	Cross flow and deadend (in/out, out/in)	Cross flow	Cross flow and deadend (out/in)
Piping and valves	High	Modest	Low	Low
Capital cost	High	Modest to low	Modest to low	Low
Energy usage	High (turbulent)	Low (laminar or deadend)	Low (bubbly flow)	Low (bubbly flow or deadend)
Fluid management	Good	Moderate to good	Moderate to good	Moderate to poor
Standardization of module	No	No	No	No
Replacement and repair	Tubes or element	Element	Element	Element or bundle
Cleaning	Good—incl. physical	Backflush; smaller vol. for chemical	Poor (low backflush)	Backflush possible
Turn-up/down	Good	Good	Limited by TMP	Limited by TMP

Mode of Operation The option for dead-end as well as cross-flow situations determines the range of applications. Dead-end operation is now widely used for low solids feed, such as water treatment and pretreatment to RO. This application favors the use of hollow fibers, either contained or submerged. The submerged fibers operate from out-to-in, but the contained fibers can be operated from in-to-out and out-to-in. Out-to-in filtration usually involves air scouring during permeate backwash, and this tends to be easier in a submerged system.

Piping and Valves Submerged modules, particularly hollow fibers, come in "blocks" that can be connected with less piping and valving than contained modules.

Capital Costs The capital costs have dropped for all membrane systems. However, there is evidence that submerged hollow-fiber modules have marginally lower costs, particularly for larger plants, due to the absence of pressure vessels, simpler plumbing and simplicity of the hollow-fiber bundle.

Energy Usage For applications involving pumped cross flow, turbulent conditions (typical of large-diameter channels) tend to be energy expensive. Two phase-bubbly flow also creates unsteady flows (see Section 10.4.1) but appears to be more efficient. Thus, the use of bubbly flow in submerged membrane systems offers a potential energy saving. For low solids feeds the application of dead-end operation (Section 10.2.3) can reduce energy usage; this mode of operation is applicable to contained as well as submerged hollow fibers. It will be apparent that submerged flat sheets would not be applicable to dead-end operation with backwash, due to physical limitations of negative TMP on the flat-sheet structure.

Fluid Management This influences attainable flux via control of polarization (Section 10.2.1) and fouling. While large-diameter tubes in contained modules provide very good fluid management, this comes at a capital and operating cost penalty. Contained hollow fibers can provide good fluid management, providing the fluid flow distribution is effective; any incipient blockage can accelerate as flow is locally reduced. Submerged membranes can also give effective fluid management, but they also suffer from the self-accelerating blockage phenomenon, particularly in hollow-fiber bundles. This characteristic is important for system selection, and manufacturers should be able to show the effectiveness of their design.

Standardization Unlike the high-pressure processes that use spiral-wound modules as a standard, the low-pressure systems are not standardized. The benefit of standardization comes at membrane replacement time when alternative suppliers can be used. However, in the contained hollow-fiber market, there is some attempt to use standard pressure vessels, and in the submerged membrane market it would be feasible (but possibly difficult) to retrofit an alternative design.

Replacement and Repair Influences maintenance and labor costs. The most convenient modules to repair use hollow fibers rather than tubes or flat plates. Both contained and submerged hollow fibers can be monitored, and the individual damaged fiber eventually identified and plugged. The procedure tends to be labor intensive in either system.

Cleaning Maintenance cleaning by backwash or chemically enhanced backwash is an option for contained and submerged modules and easier for hollow fibers. Clean-in-place may be easier for contained systems and smaller volumes of chemicals may be required.

Turn-up/Turn-down Operation below design flux is not a problem, but, if short-term operation above design flux is required, it may be easier for the contained systems operating at positive TMP. However, high-flux operation risks rapid fouling and would be a limited strategy, for example, for unsteady flow to an MBR.

In summary, submerged systems compete favorably with contained systems and may have a marginal capital cost advantage. Hollow fibers, submerged or contained, can operate at both dead-end and cross flow with backwash and are suited to water treatment through to MBRs. Flat-sheet submerged modules are not suited to dead-end operation and are only suited to MBRs. Characteristics where contained modules have an advantage include fluid management, cleaning, and turn-up/turn-down.

10.3 SUBMERGED MEMBRANE MODULE GEOMETRIES

As mentioned above there is no standardization in submerged membranes, and there are many commercial suppliers. For example, Judd (2006) describes 12 different MBRs using submerged membranes. The majority of submerged membranes are either hollow fibers (HF) or flat sheets (FS), as depicted in Figure 10.3. This section gives a brief description of the different submerged membrane concepts.

10.3.1 Hollow Fibers

Submerged hollow fibers are used for both low solids content feed, such as water treatment, and high solids contents, such as MBRs. Hollow fibers allow the use of backwash, which

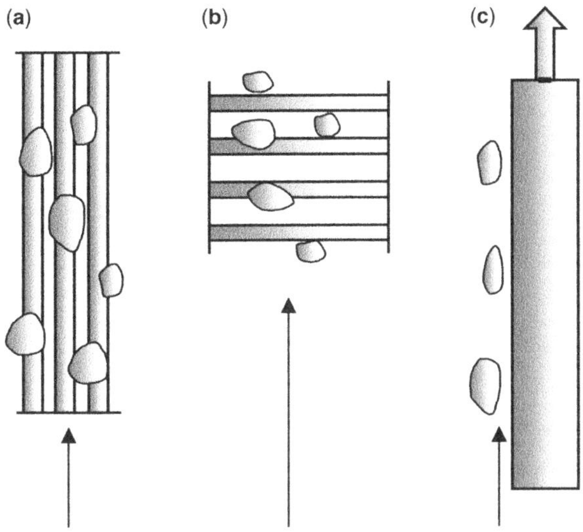

Figure 10.3 Submerged membrane concepts: (a) vertical HF, (b) horizontal HF, and (c) flat sheet.

is required for dead-end (nonbubbled) filtration in water treatment or pretreatment for RO. The major difference between submerged HF for water treatment and the MBR is the higher packing density allowable for low solids contents and possibly the arrangement for air sparging. Some submerged HF modules are vertically aligned (Fig. 10.3a) with fibers potted at both ends and suction applied either from one or both ends (this often helps to overcome the permeate flow issues—see Section 10.4.4). The fibers are not held tightly but with a small degree of looseness to facilitate fiber movement (discussed further in Section 10.4.1). One manufacturer (see below) has fibers only fixed at one end to allow more movement.

Other submerged HF are potted in a frame and aligned horizontally in a curtain arrangement (Fig. 10.3b) and permeate is withdrawn from both ends. Again the fibers have a degree of looseness that promotes a swaying movement to control fouling. Table 10.2

TABLE 10.2 Some Submerged Hollow-Fiber Modules

Supplier	MF or UF[a]	Fiber Size (OD) (mm)	Alignment[b]	Specific Area (m^2/m^3)[e]
Asahi-Kasai	MF/UF	1.3	V	710 (66)
GE-Zenon	UF	1.9	V	300
	UF	0.8	H[c]	750
KMS-Puron	UF	2.6	V[d]	314 (125)
Mitsubishi	MF	0.54	H	485 (131)
	MF	2.8	V	333 (71)
Siemens–Water Technologies	UF	1.3	V	334
	UF	0.8	V[c]	

[a]MF > 0.1 μm pore; UF < 0.1 μm pore.
[b]V = vertical; H = horizontal.
[c]Water treatment only.
[d]Potted at bottom only.
[e]Membrane area per volume of element (per volume of module).
Source: Adapted from Judd (2006).

Figure 10.4 Zenon's ZeeWeed 500d submerged HF cassette.

lists some of the submerged HF modules. It is evident that most use vertically aligned fibers. Special features of several submerged HF modules are briefly described below. More detailed descriptions can be found in Chapter 4 of Judd (2006) and the report of DeCarolis et al. (2006).

10.3.1.1 GE–Zenon The Zenon system uses PVDF-reinforced hollow-fiber membranes with a pore size of 0.04 μm, fiber diameter of 1.9 mm, and length about 1.7 m for its MBRs. Membranes are assembled into cassettes with a membrane area of about 30 m^2 and held in a frame that provides support and connection to aeration. Zenon has evolved the ZeeWeed concept from the ZW500a to the current ZW500d, which has higher packing density and lower air-sparging requirements (Benedek and Cote, 2003). Figure 10.4 shows a view of the ZeeWeed 500d cassette. For water treatment Zenon has developed the ZeeWeed 1000, which has a higher packing density (2 ×), smaller diameter fibers, and is horizontally aligned.

10.3.1.2 KMS–Puron The Koch Membrane Systems (KMS) Puron uses polyether sulfone membranes of 0.05-μm pore size and fiber diameter of 2.6 mm. A standard module contains about 30 m^2. The fibers are vertically aligned and potted only at the base. The top end of each fiber is individually sealed, and this unique arrangement allows a limited range of free movement that avoids blocking of the bundle (see Section 10.5.1.2). The module depicted in Figure 10.5.

10.3.1.3 Mitsubishi The most common Mitsubishi HF module uses a polyethylene MF membrane of 0.4-μm pore size and fiber diameter of 0.54 mm and length about 0.5 m. The membranes are horizontally aligned and potted at both ends in elements of up to 3 m^2 area combined into modules of 100–200 m^2. The modules are either installed as single or double deck, with the latter requiring less airflow per unit membrane surface (see Section 10.4.2 and discussion of specific air demand). In a recent development

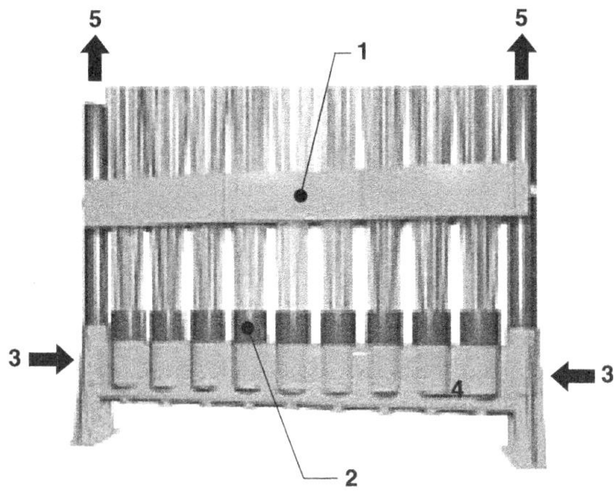

Figure 10.5 KMS Puron submerged HF module: (1) fiber support, (2) module element, (3) air supply, (4) module row, and (5) filtrate.

Mitsubishi has introduced a vertically aligned hollow fiber based on a PVDF MF membrane of 0.4-μm pore size, fiber diameter of 2.8 mm, and almost 2 m long. The arrangement is not unlike the Zenon ZeeWeed MBR module. A typical element has 25 m^2 and is combined into units of about 500 m^2. The reasons for the change are not known but could be due to improved bubble distribution with vertical HF and greater membrane integrity for larger fibers. Figure 10.6 depicts the Mitsubishi system with horizontal fibers.

Figure 10.6 Mitsubishi submerged HF module.

Figure 10.7 Siemens Water Technologies—MEMCOR CS submerged (water treatment): module; view of submerged modules; drained tank.

10.3.1.4 Siemens Water Technologies Siemens Water Technologies has developed submerged membranes for both low solids applications (water treatment, RO pretreatment) and MBRs. For low solids feed the MEMCOR CS uses PVDF hollow fibers, 0.04-μm nominal pore size, 0.8-mm diameter, and about 1.1 m long, vertically aligned. The modules are about 28 m^2 and assembled in isolatable units of four, giving 112 m^2, which are then connected to a permeate header. Figure 10.7 shows the MEMCOR CS module and views from above the submerged modules and the drained tank; the stainless steel permeate headers and air supply hoses are evident. These membranes are used in dead-end operation with backwash and air scour at set cycle times.

For MBR applications Siemens Water Technologies has developed the MEMCOR Memjet system, which uses PVDF hollow fibers of 0.04-μm pore size, 1.3-mm diameter, and vertically aligned. The modules come in sizes of 10 m^2 (the B10R) and 38 m^2 (the B30R), the latter combines four B10R modules in a square array. Up to 16 B30R modules can be combined in racks of 608 m^2. A unique feature of the design is the use of pumped two-phase flow into the fiber bundle to improve flow distribution and provide homogeneous surface shear for fouling control. Figure 10.8 depicts the MEMCOR Memjet submerged MBR system.

10.3.2 Flat Sheets

The submerged flat-sheet systems are principally used for high solids content MBR applications. Table 10.3 summarizes the products from three suppliers, which are briefly

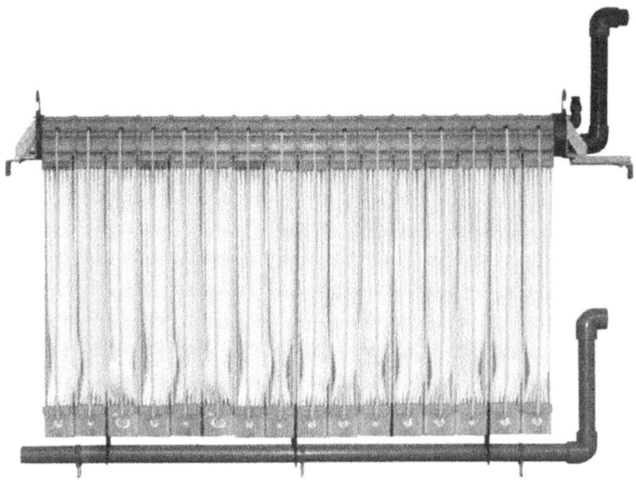

Figure 10.8 MEMCOR Memjet submerged (MBR).

discussed below. More detailed descriptions can be found in Chapter 4 of Judd (2006) and the report of DeCarolis et al. (2006).

10.3.2.1 Kubota Kubota has been involved in submerged membrane MBRs since 1990 and is now the major supplier of flat-sheet MBRs. Its membranes are hydrophilized polyethylene with a nominal pore size of 0.4 μm. A typical cartridge has a permeable support with membrane layers on both sides, and dimensions of 0.5 m (width) × 1.0 m (height) giving 0.8 m^2 active area; larger cartridges (1.5 × 0.5 m) are also available. The cartridges are aligned vertically and stacked with a gap of 7–8 mm between cartridges, which obviates blocking in most cases. Coarse aeration is provided by diffusers beneath the cartridge stack, and improved air demand is obtained by use of a double-decker (DD) arrangement. The DD system also reduces the footprint of the MBR. Although the flat-sheet design does not allow vigorous backwash, it is allowable to use low-flow chemically enhanced backwash for cleaning (see Section 10.5.2). Figure 10.9 depicts the Kubota flat-sheet concept. The Toray flat-sheet MBR (Table 10.3) is similar in concept to the Kubota MBR but is based on an oxidant-resistant hydrophilized PVDF membrane with a 0.08-μm nominal pore size.

10.3.2.2 Huber The Huber MBR is radically different from all other commercial submerged membrane systems, being based on a mechanically moved membrane. The Huber VRM (vacuum rotating membrane) uses flat-sheet membranes of polyethersulfone

TABLE 10.3 Some Submerged Flat-Sheet Modules

Supplier	MF or UF	Arrangement	Specific Area (m^2/m^3)
Kubota	MF	Vertical, single, or double decker; 7- to 8-mm spacing	115
Toray	UF	Vertical 7-mm spacing	135
Huber	UF	Rotating vertical plates, 6-mm spacing	160

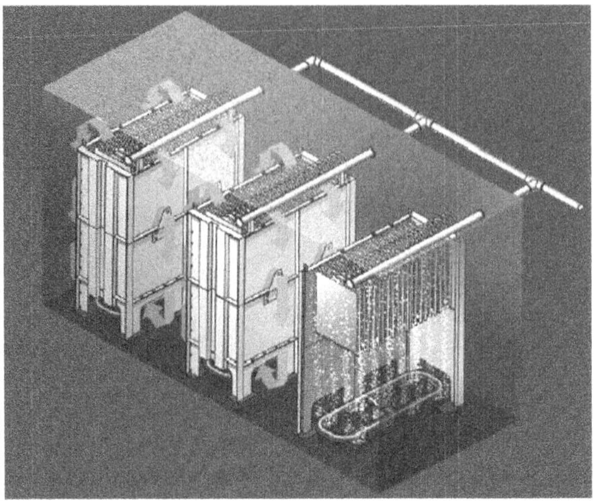

Figure 10.9 Kubota submerged flat-sheet module (MBR).

of 0.04-μm nominal pore size. They are assembled into vertical sector-shaped panels that are connected around a horizontal collection tube to form a donut disk. The tube rotates slowly (1–2 rpm) and moves the membranes through the air-sparged zone; this arrangement achieves relatively low specific aeration demand (see Figure 10.13). Four panels form one "circular" module, and up to 60 modules form one VRM unit with a maximum area of 2880 m^2. Although this is a rather more complex design than fixed submerged membranes, it is reported to have good fouling control at low-energy demand. Figure 10.10 is a schematic of the VRM MBR.

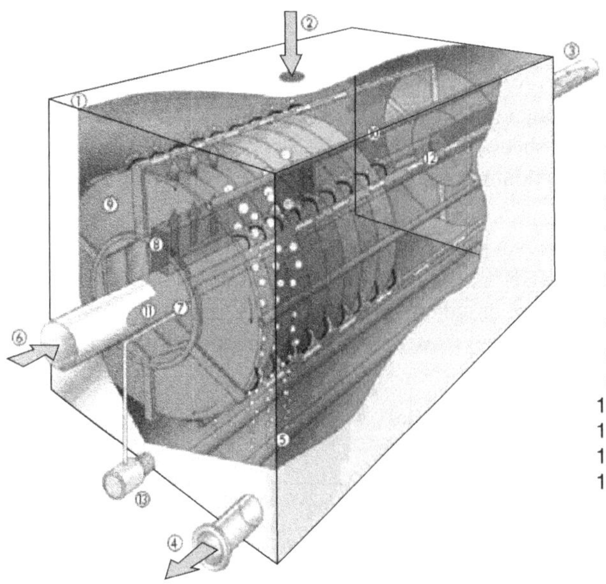

Figure 10.10 Huber VRM rotating submerged flat-sheet module (MBR).

10.4 BUBBLING AND HYDRODYNAMIC CONSIDERATIONS

In membrane processing the application of surface shear is required to control concentration polarization and fouling for high solids feeds or to assist cake removal for batch membrane filtration of low solids feeds. For submerged membranes the common practice is to use two-phase bubbly flow to induce surface shear. This section deals with the role of bubbles as well as other hydrodynamic aspects of submerged membranes.

10.4.1 Role of Bubbles

The use of bubbly flow has been reviewed by Cui et al. (2003). Figure 10.11 depicts the role of bubbles in the enhancement of filtration by submerged hollow fibers. For submerged flat sheets similar mechanisms apply except for the lack of induced membrane movement.

In the majority of submerged membrane designs the bubbles are generated by sparging into the liquid-filled tank. However, the liquid contents in the tank are not stagnant due to liquid recirculation induced by the bubble flow, rather like a draft-tube effect. If the system can be considered as having a *riser* region (bubbles and liquid moving upward) and a *downcomer* region (liquid moving down), then Liu et al. (2003) show that the analysis of Yusuf and Moo-Young (1993) can be used to estimate the upward liquid velocity U_L. The relevant equation takes the form

$$U_L = \left[\frac{2gh_L\varepsilon_r}{K_B(1-\varepsilon_r)}\right]^{0.5} \tag{10.11}$$

where h_L is the liquid height and K_B is the friction loss coefficient (typically 7–10); this assumes the area of the riser and downcomer are similar. Parameter ε_r is the fractional gas holdup in the dispersion (on the order of 0.05) and depends on the superficial air velocity U_G in the riser; so,

$$\varepsilon_r = \frac{U_G}{0.24 + 1.35(U_G + U_L)^{0.93}} \tag{10.12}$$

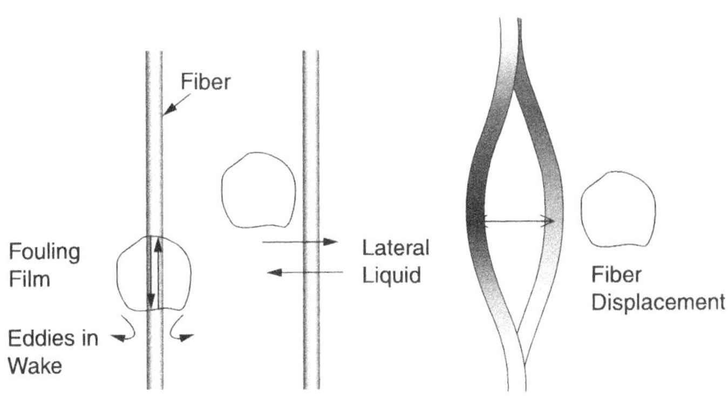

Figure 10.11 Fouling and cake control mechanisms by bubbles outside fibers. (Adapted from Cui et al., 2003).

There have been few experimental measurements of U_L in submerged systems, but the reported values are in the range 0.1–0.5 m/s (Liu et al., 2003; Madec, 2004). A decrease in fouling rate ($dTMP/dt$) as U_L increased was reported by Liu et al. (2003). However, this is probably only part of the story, as decreased fouling rate appears to be linked to bubble-induced instabilities as well as average shear stress effects.

One design of a submerged hollow-fiber system used for MBRs [Siemens-Memcor Memjet (Fig. 10.8)] provides additional liquid flow by pumping a two-phase (air–liquid) dispersion into the base of the fiber assemblies in the tank. This may give improved control of flow distribution into the bundle with greater potential for turn-up/turn-down. However, as we discuss below, a major benefit of bubbling flow is the unsteady or transient shear stress induced by bubbles that helps to mitigate fouling, and this "unsteady" flow may be dampened if the imposed background liquid flow is too high.

For submerged flat-sheet systems the bubbles induce surface shear, and there is evidence (Ndinisa et al., 2006) that this is favored by relatively large bubbles that are bigger than the dimension of the channel separating the flat membranes, typically 7 mm. Under these conditions the bubbles behave rather like "slugs" rising up a tube where the surface shear is generated by a falling film between the bubble and the surface and by eddies in the bubble wake. Results by Ducom et al. (2002) showed that the flux enhancement due to bubbling across a vertical flat-plate membrane could be related to shear stress transients (characterized by the maximum and minimum shear stress). They correlated the flux enhancement E due to bubbling by

$$E = 1 + 0.16\left[\frac{\text{average}\,(\tau_{\text{max}} - \tau_{\text{min}})}{\text{average}\,\tau_{\text{single phase}}}\right] \tag{10.13}$$

where E is [flux (bubbling)/flux (no bubbling)], and τ is the shear stress.

For hollow fibers the bubbles also generate liquid shear stress at the membrane surface. These effects have been observed in laboratory-scale systems using the laser-driven particle image velocimetry (PIV) technique (Yeo et al., 2007) that allows measurement of shear stress in real time. This gives temporal and spatial data. Figure 10.12 shows how fouling rate (characterized by $dTMP/dt$) for a single submerged hollow fiber decreased as the standard deviation of the surface shear stress induced by bubbling was increased. The unsteady character of shear stress again provides a reasonable approach to characterizing the effect of bubbling.

With submerged hollow fibers it is common to allow fiber movement, rather than having them tightly constrained. The benefit of allowing some fiber movement was recognized early in the development of the submerged hollow-fiber module (Cote et al., 1994; Futamura et al., 1994). This movement is achieved by a degree of fiber looseness [here, looseness (%) = 100 − tightness (%), where tightness = distance between fixed ends of fiber/fiber length]. A looseness of 1–5% is recommended, but excessive looseness can lead to fiber damage as the fibers experience rapidly changing stresses and strains. The degree of movement of loose fibers is substantial even for a 1% looseness. For example, a 100-cm fiber with 1% looseness could have a maximum midpoint displacement of about 14 cm, although in practice it would be less (50–80% of this) and attenuated by the viscosity of the surrounding medium (Wicaksana et al., 2006). Fiber movement improves fouling control due to increased surface shear stress and also fiber–fiber interaction (Berube and Lei, 2006). Wicaksana et al. (2006) compared the fouling rate

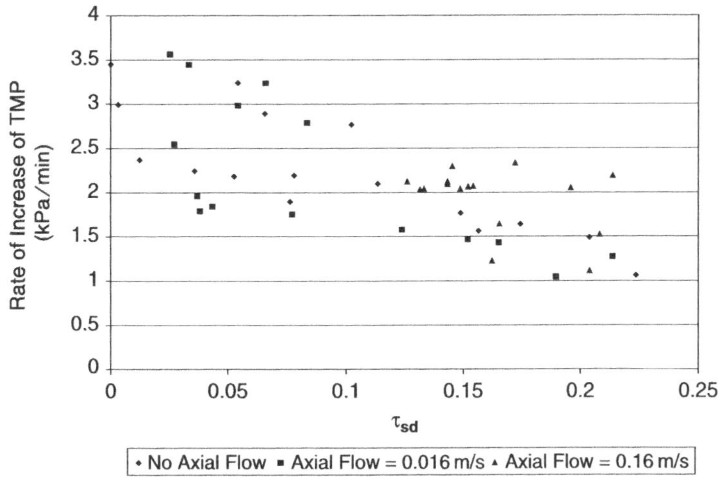

Figure 10.12 Effect of standard deviation of shear stress induced by bubbles on the rate of fouling, $dTMP/dt$, of submerged fiber (Yeo et al., 2007).

($dTMP/dt$) for a tight submerged fiber with one at 1% looseness, and the benefit from bubble-induced movement per se was of the order of 30% of the fouling control.

In MBRs bubbling also has a role in mixing the reactor contents, presumably through the induced liquid circulation [Eq. (10.11)]. There is some evidence from tracer studies that mixing induced by bubbles may be a little more effective with submerged hollow fibers than flat-sheet membranes because the latter act as baffles in the vessel (Wang et al., 2008).

10.4.2 Bubble Production

The bubble flow in submerged membrane systems is usually produced by air pumped through diffusers or spargers located below the membrane modules. The diffuser or sparger design tends to be proprietary information and is not discussed in the open literature. Judd (2006, p. 128 et seq.) provides some information, and Mayer et al. (2006) compare various aeration devices for nonsubmerged membrane applications. Aeration applied to MBRs serves two purposes: (i) to provide oxygen transfer to the biomass and (ii) to produce shear stress on the membrane surface to control fouling (Section 10.4.1). To achieve effective oxygen transfer, it is necessary to use fine-bubble aeration; and to provide shear, it is common practice to use coarse-bubble aeration. For these reasons both fine and coarse bubbling tend to be used in MBRs. (For information on the relative oxygen transfer efficiencies of fine and coarse bubbles, the reader should refer to Chapter 9, Membrane Bioreactors for Wastewater Treatment). However, recent studies at lab scale (Yeo et al., 2007) suggest that for submerged hollow fibers, well-distributed small bubbles can be as effective as, or better than, coarse bubbles for a given airflow rate. In effect many small transients (bubble events) may be more effective than few large transients. More work is required to identify the optimal bubble size for a given MBR design and operation. Another factor that has to be incorporated in diffuser/sparger design is the need to minimize biofouling, and this may limit the use of fine diffusers.

The amount of air sparging required to control fouling in submerged membrane systems is an important operating cost. This is because the power required, $\overline{P_A}$, is a function of the airflow rate, Q_A, and also the ratio of the blower delivery pressure to the blower inlet pressure (p_2/p_1). The procedure for calculation of $\overline{P_A}$ (kW) is given by Judd (2006) and a simplified expression is

$$\overline{P_A} = k_B \left[\left(\frac{p_2}{p_1} \right)^{0.29} - 1 \right] Q_A \tag{10.14}$$

where k_B incorporates temperature, ratio of specific heats of air (constant pressure to volume; about 1.4), and blower efficiency. The blower inlet pressure, p_1, is usually atmospheric, and the outlet pressure, p_2, must allow for the sparger (orifice) pressure drop, (h_0), the static liquid head (h_L) above the sparger, and piping losses (h_P); so $p_2 = p_1 + h_0 + h_L + h_P$. Equation (10.14) shows that power required for sparging is directly proportional to the amount of air used and increases for fine orifice diffusers (h_0 increased) and deeper tanks (h_L increased).

Specifications for airflow rate requirements for MBRs are based on experience rather than theoretical analyses, although there is a growing literature on the effect of bubble flow on fouling control [see reviews by Cui et al. (2003) and Le Clech et al. (2006)]. Judd (2006) has summarized a wide range of available data for submerged MBRs as the specific air demand (SAD) and defines two parameters:

$$\text{SAD}_m = \frac{Q_A}{A_m} = \text{airflow rate per unit area } (\text{Nm}^3/\text{h} \cdot \text{m}^2) \tag{10.15}$$

$$\text{SAD}_p = \frac{Q_A}{JA_m} = \text{airflow per unit permeate } (\text{Nm}^3/\text{m}^3) \tag{10.16}$$

Values of SAD_m range from 1.0 down to about 0.2 Nm/h, and data are plotted in Figure 10.13 versus membrane permeability (L/m^2 h bar). SAD_p ranges from about 60 down to as low as 10, with typical values from 20 to 15. The permeabilities in Figure 10.13 are based on long-term operation and can be considered as *sustainable flux*

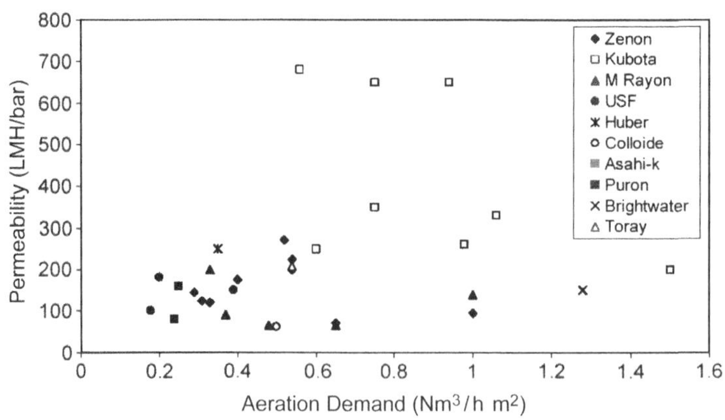

Figure 10.13 Relationship between specific air demand and "sustainable" permeability for submerged membrane MBRs (Judd, 2006).

data (see Section 10.4.3). Although the data are very spread out, the higher permeabilities tend to involve greater aeration demand. The data also indicate rather lower permeabilities for the submerged hollow fibers (typically <200 L/m^2 h bar), with lower SAD$_m$ values possibly due to intermittent aeration. In this context an important observation was made by Guibert et al. (2002) working with the Zenon ZeeWeed submerged modules. By using sequential aeration, or "air cycling," on either side of a fiber bundle, they were able to substantially reduce the total air demand. Comparing fouling rates ($dTMP/dt$) at different imposed fluxes for continuous aeration versus air cycling, they were able to halve total air demand for a cycle of 10 s on and 10 s off. This approach is now used in large-scale plants.

10.4.3 Critical and Sustainable Flux Phenomena

The concept of critical flux was introduced in Section 10.2.2, and in simple terms it refers to a flux beyond which deposition and/or fouling occurs and $dTMP/dt$ [Eq. (10.4)] becomes nonzero. Critical flux can be viewed as a condition where the convection and back transport of foulant are in balance; so the net flux of solids to the membrane [Eq. (10.5)] is zero. The situation can be observed in Figure 10.14, which shows the deposits on hollow fibers operated at different fluxes in a bubbled yeast suspension. The flux at 10 L/m^2 h is clearly subcritical and ≥ 15 L/m^2 h is supercritical.

In practical applications the feed tends to be a mixture in which each component (particle, colloid, solute) has a specific critical flux determined by back transport (dependent on size, charge, etc.). MBR mixed liquor is a classic example of this type of feed, and as a result $dTMP/dt$ tends to be nonzero even at very low flux. The approach to this is to define a sustainable flux, which is a more subjective concept and depends on an "acceptable" value of $dTMP/dt$ and the cleaning frequency.

Even with a complex feed, such as MBR mixed liquor, it is usually possible to identify the critical flux of the dominant species by measuring the flux where $dTMP/dt$ starts to rise

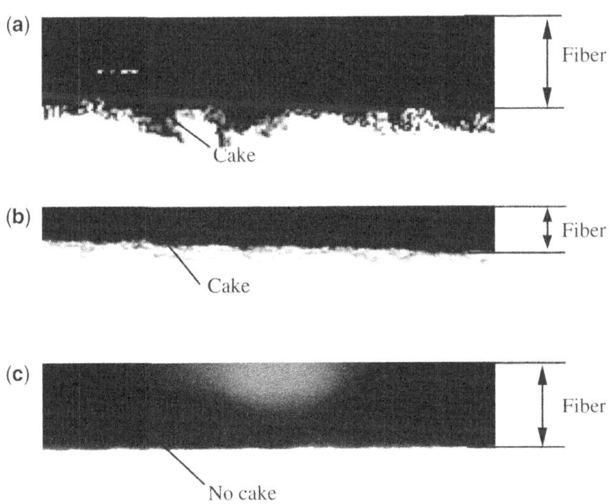

Figure 10.14 Images of fiber surface after long-term filtration of yeast suspension. Fiber 2.7 mm OD, $U_G = 0.34$ m/s, $U_L = 0.19$ m/s. Fluxes (a) = 30, (b) = 15, and (c) = 10 L/m^2 h (Chang and Fane, 2000).

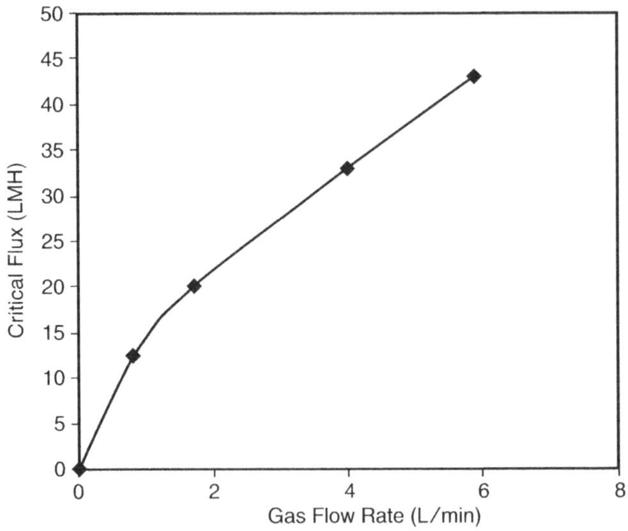

Figure 10.15 Effect of bubbling rate on critical flux of bentonite suspension with submerged hollow-fiber membranes.

rapidly (Le-Clech et al., 2003). The sustainable value will probably be 60–80% of this dominant species critical flux, depending on applications.

Increasing bubbling increases both average and unsteady-state shear stresses, and this increases critical flux in submerged membrane systems as illustrated in Figures 10.15 and 10.16. Figure 10.15 is for a lab-scale submerged hollow-fiber module with bentonite suspension. The effect is not linear but shows a gradually diminishing effect at higher gas rate. This "plateau" effect is more evident in Figure 10.16, which was obtained using a 6-g/L mixed-liquor suspension and a submerged flat-sheet system (Ndinisa et al., 2006). Also evident in Figure 10.16 are the higher critical fluxes when the distribution of air bubbles was improved by vertical baffles placed between the vertical membranes. For example, the critical flux at 16 L/min airflow without baffles was achieved with 50% of this (8 L/min) with baffles. This emphasizes the importance of delivering the bubble-induced scour to all the membrane surface. It can also be seen that the trends in critical flux versus airflow are reflected in the "sustainable" permeabilities versus specific air demand for submerged membrane MBRs shown in Figure 10.13.

As pointed out by Cui et al. (2003) there are two features of submerged membranes in a bubbling system that complicate the critical flux concept for such systems. These features relate to the spatial and temporal variation of filtration conditions, as indicated below.

10.4.3.1 Flux Distribution along a Fiber Submerged hollow fibers are operated under suction and usually at constant permeate flow, that is, constant average flux. The pressure losses due to permeate flow down the lumen reduce the net driving force, and this decreases with distance from the suction end. As a result, the flux varies axially along the fiber. This phenomenon is discussed below (Section 10.4.4). In terms of critical flux analysis the problem is that while the length average flux may be below critical the condition near the suction end could have an elevated flux that exceeds critical. This means that deposition will occur near the suction end and as cake resistance increases the local flux

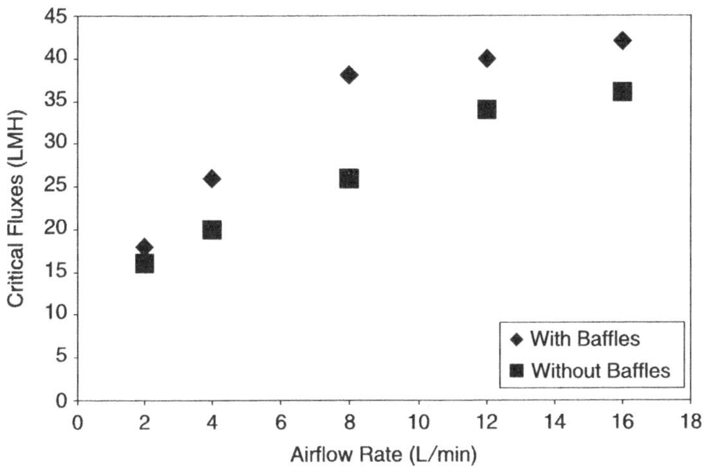

Figure 10.16 Effect of bubbling rate on critical flux of mixed-liquor suspension with submerged flat-sheet membranes. Also shown is improvement due to baffles (Ndinisa et al., 2006).

drops, shifting more flux down the length of the fiber. This would cause an increase in TMP. If the average flux is not too high, the flux distribution could reach a steady state where no point along the fiber is above critical flux, and at this condition the TMP could stabilize. However, if average flux is too high, local fluxes will remain above critical and TMP will continue to rise until a limit is reached (see Section 10.4.4). Figure 10.17 shows suction pressure (TMP) profiles versus time, illustrating subcritical conditions, partial subcritical (rises and stabilizes), and supercritical (rises and does not stabilize).

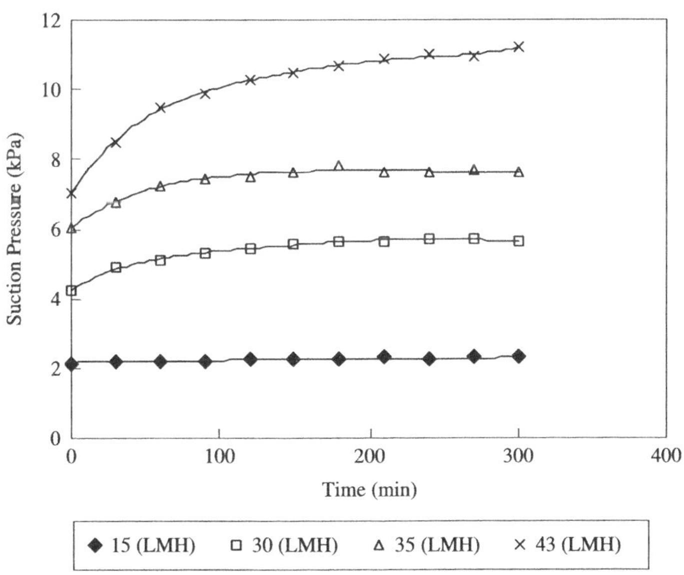

Figure 10.17 Effect of imposed flux on TMP profiles for submerged hollow fibers. Yeast 5 g/L with bubbling (Chang and Fane, 2002).

Kim and DiGiano (2006) have also analyzed and measured the effect of fiber length on measured critical flux, they state: "Fouling will occur in full-scale if the critical flux test is established in tests with fibers that are much shorter than in full-scale."

10.4.3.2 Transient Effects The concept of critical flux considers a steady-state balance between convection and back transport. However, for bubbled systems, such as submerged membranes in MBRs, the depolarization mechanisms are transient, and the fouling control relates to unsteady-state shear stresses [recall Eq. (10.13) and Fig. 10.12]. In effect the filtration cycles from conditions of low shear to high shear or supercritical to subcritical flux. If the deposit is reversible, the time-averaged TMP will remain constant, leading to the conclusion that the overall condition is subcritical, even though the instantaneous flux may be supercritical. If the deposit is only partially reversible, the time-average TMP will slowly rise and the condition appears supercritical.

In summary, the spatial and temporal features of submerged membranes in bubbly flow complicate the definition of critical flux in these systems. However, in practice it is found that sustainable flux can be achieved and that it is improved by bubbling.

10.4.4 Permeate-Side Flow

Submerged membranes are operated with the feed in a vessel at atmospheric pressure. To obtain a TMP, the permeate side has to be below atmospheric pressure, and this is achieved by suction provided by permeate pumping. The pressure on the permeate side will be determined by the TMP defined by Eq. (10.3) as well as a pressure drop due to permeate-side (lumen) flow, which may be significant for hollow fibers. This situation has been analyzed for vertical submerged hollow fibers closed at the bottom and with suction at the top. The situation is depicted in Figure 10.18a for a clean water feed, and the axial flux distribution can be estimated from (Chang et al., 2000)

$$J_i(x) = \lambda L J_{mi}\left(\frac{e^{\lambda x} + e^{-\lambda x}}{e^{\lambda L} - e^{-\lambda L}}\right) \tag{10.17}$$

where $J_i(x)$ is the local flux at position x, based on the inner radius, J_{mi} is the imposed average flux, L is fiber length, λ is a function of the fiber radius, r_i, and resistance R_m,

$$\lambda = 4\, r_i^{-3/2} R_m^{-1/2} \tag{10.18}$$

Equations (10.17) and (10.18) relate to the conditions where there is no deposit and describe the flux distribution along a clean fiber with water feed (or for the initial moments with contaminated feed).

Figure 10.19 shows the local flux distribution with different imposed average fluxes and highlights the potentially much greater flux near the suction end. The ratio of local flux to average flux, $J_i(x)/J_{mi}$, is also shown and is independent of J_{mi}. For the typical conditions chosen $J_{i,\max}/J_{mi}$ is rather high at about 4.5, so that for an average flux of $20\ \mathrm{L/m^2\,h}$ the flux at the suction end could be $90\ \mathrm{L/m^2\,h}$. This could easily lead to a local condition where $J_i(x) > J_{\mathrm{crit}}$, as depicted in Figure 10.18b. An implication of this analysis is that as fiber radius reduces and length increases the initial flux distribution becomes more inhomogeneous. Other implications of this model are discussed in Chang and Fane (2001),

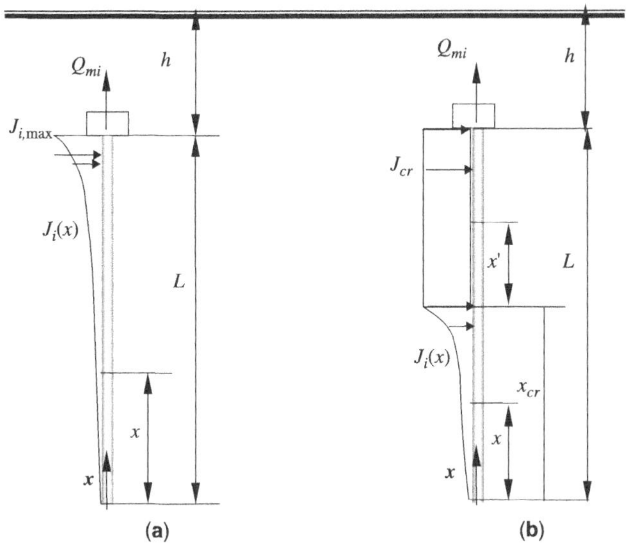

Figure 10.18 Flux distribution at steady state for different imposed flux conditions: (a) $J_{i,\max} < J_{\mathrm{Crit}}$ and (b) $J_{mi} < J_{\mathrm{Crit}}$, $J_{i,\max} > J_{\mathrm{Crit}}$ (Chang and Fane, 2002).

Chang et al. (2002), and Kim and DiGiano (2006). For example, simulations based on the model show that for fiber lengths of 0.5–3.0 m there is an optimal fiber inner radius of 0.2–0.35 mm based on water productivity per unit pressure loss.

The need to use suction to produce TMP means that the permeate side of the membrane is under subatmospheric pressure. This has the potential to cause bubble formation as dissolved air comes out of solution. If the bubbles so formed are free to flow with the permeate, the effect on pressure drops will be negligible. However, if bubbles become stagnant and/or attach to the lumen wall, they will present a restriction to flow causing an added resistance.

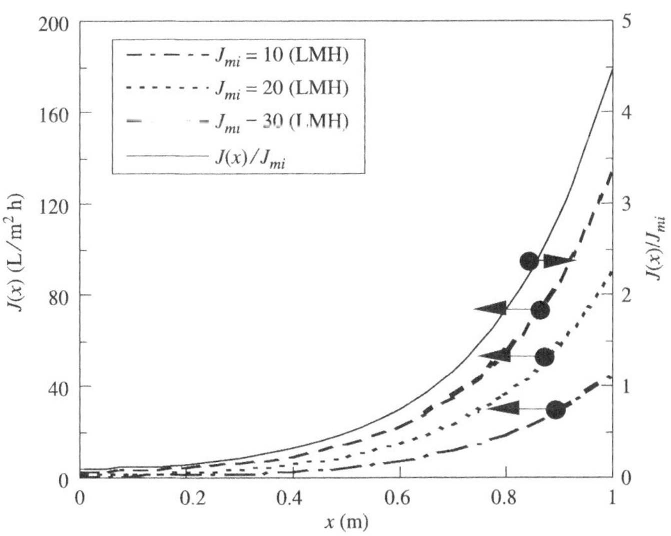

Figure 10.19 Simulated flux distribution along a fiber (Chang and Fane, 2002).

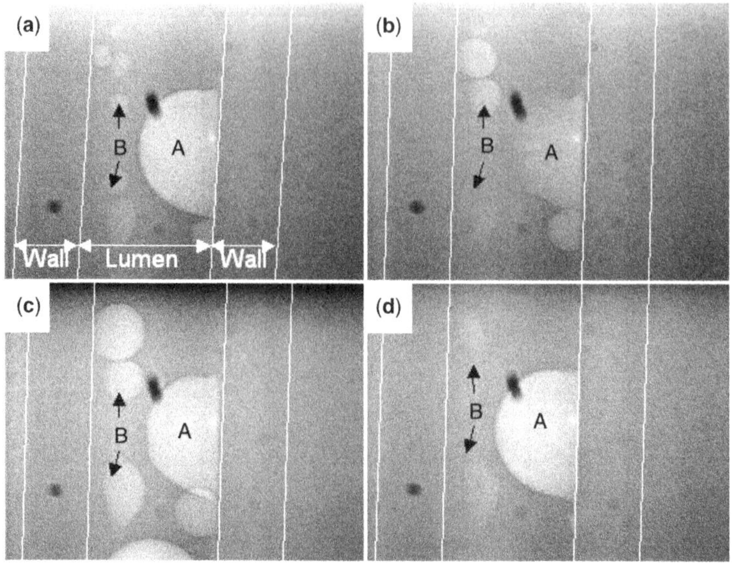

Figure 10.20 Stagnant and mobile bubbles in the lumen of a hollow-fiber membrane at a flux of $26\,L/m^2\,h$ at (a) 5 min, (b) 6 min, (c) 7 min, and (d) 8 min from the start of filtration (Chang et al., 2007).

The membrane region below this will then deliver less flux and the region above will be required to produce more (equalized by the net flow of all fibers). Figure 10.20 provides a view inside a hollow-fiber lumen under considerable suction. The observation was made using X-ray microimaging of a single fiber and clearly shows a stagnant bubble (A) and smaller mobile bubbles (B). This behavior is likely to be specific to the membrane wall morphology and the material used for fiber fabrication, and more likely at high values of TMP.

10.5 PRACTICAL ASPECTS

10.5.1 Fouling and Blocking

In common with all membrane processes submerged membranes are prone to fouling. The key factors are feed characteristics, membrane properties, and hydrodynamic environment. Submerged membranes are predominantly micro- and ultrafiltration, which means they are microporous. For such membranes fouling is either internal (pore plugging or restriction) or surface (cake formation). Various texts describe membrane fouling and control (e.g., Ho and Sirkar, 1992, Chapter 26; Mallevialle et al., 1996, Chapters 10 and 11). Submerged membranes tend to be applied to water treatment and MBRs, for which there are detailed fouling reviews in Fane et al. (2006) and Le-Clech et al. (2006), respectively. However, it is pertinent to discuss several aspects of fouling that are particularly germane to submerged membranes.

10.5.1.1 Low Solids Feed by Dead-end Filtration For applications such as water treatment the solids content is low and it is common to use submerged membranes in dead-end filtration mode (see Section 10.2.3). In this case permeate is withdrawn in cycles of filtration and backwash; during backwash it is common to also apply air scouring. Flux (J_i) is controlled and the cycle time is either fixed (say 30 min) or determined by the TMP, which is given by Eq. (10.9), rewritten as

$$\text{TMP} = (\text{TMP})_{\text{membrane}} + (\text{TMP})_{\text{deposit}} \tag{10.19}$$

If removal of cake, after backwash, is incomplete, the membrane resistance R_{m1} (start of cycle 1) rises to R_{m2} (start of cycle 2). This increase in R_m is due to irreversible (or not easily reversible) fouling. It is manifest as a steady rise in TMP_{min} [Fig. 10.2b]. The fouling can have two consequences:

1. *Raised TMP* The additional fouling resistance (ΔR_m) causes the TMP to commence at a higher value at the start of the next cycle. This either causes a higher $\Delta P_{\max}$ for a fixed cycle time or a shorter cycle time [t_c in Eq. (10.10)] for a fixed $\Delta P_{\max}$.
2. *Raise Effective Flux* If ΔR_m is due to loss of pores or active surface due to plugging or coverage, the effective flux will have to increase for a fixed permeate rate, because

$$J_{\text{effective}} = J_{\text{average imposed}} \left(\frac{\text{Total area}}{\text{Effective active area}} \right) \tag{10.20}$$

Thus, if incomplete cake removal due to fouling decreases the effective area to 90%, this will increase local flux, $J_{\text{effective}}$, by 10% and increase R_m by 10%. The importance of this can be seen by inspection of Eq. (10.10) where the numerator is increased and the denominator is decreased. The net effect on cycle time would be more than a 20% reduction. The problem is exacerbated because a rise in local flux increases irreversible fouling making backwash even less effective.

Strategies to mitigate fouling under these conditions include:

- *Pretreatment* Limits fine colloids, for example, by coagulation because larger particles are easier to filter and less likely to plug pores.
- *Sustainable Flux Operation* Balances the capital cost of lowering flux with the benefits of less fouling and cleaning.
- *Membrane Selection* There may be benefits in hydrophilic and relatively neutral membranes. UF membranes may be less susceptible to pore pluggung than MF, but this would be application specific.
- *Module Design* As discussed in Section 10.4.4 long submerged fibers with small diameters tend to have a significant flux distribution due to pressure losses, and this can lead to fouling at the suction end and difficulties in backwashing this region. The other issue relates to module blocking where solids fill spaces between fibers so that backwashing and bubble scour become less effective (see below).

10.5.1.2 High Solids Feed by (Bubbled) Cross-Flow Filtration For applications, such as MBRs, the particulate solids content is relatively high [suspended solids (SS) up to 10–15 g/L] with additional colloids and macrosolutes. Under these conditions bubbled cross flow is essential to limit concentration polarization and subsequent fouling.

Figure 10.2a depicts a typical TMP history for an MBR. Even below the critical flux of the biomass floc, the TMP rises slowly due to colloidal and biopolymer foulants. The fouling mechanisms in MBRs are complex and have been reviewed by Le-Clech et al. (2006). Strategies to mitigate fouling in submerged MBRs include:

- *Feed Manipulation* Can involve selection of "optimal" biological conditions (sludge retention time, etc.), additives (polymers, adsorbents), and so forth.
- *Sustainable Flux Operation* Typical fluxes are chosen in the modest range of 10–30 L/m^2 h to avoid cake formation of biomass.
- *Intermittent Operation* To avoid or delay the rapid TMP rise (the jump) illustrated in Figure 10.2a, it is usual to operate with short rest periods to allow relaxation and removal of deposits as air scouring continues without flux.
- *Membrane Selection and Module Design* Points made above (Section 10.5.1.1) also apply. For MBRs the design of the module and choice of operating mode to deliver shear stress to the total membrane surface is crucial. The various commercial designs described in Section 10.3 have evolved to meet this challenge.

Blocking (or clogging) of submerged membrane modules is when regions of the module, such as within a hollow-fiber bundle, become filled with solid material. This obstructs the flow of liquid and bubbles (if used) and causes maldistribution and stagnation of the feed liquor. The problem can be particularly serious in MBRs where fouling (a membrane effect) is controlled by cross flow. Blocking will reduce local cross flow and shear stress, and this accelerates fouling. (Note that module blocking is not fouling per se but can lead to fouling.)

To mitigate blocking it is necessary to consider:

- *Preatreatment* Particularly the use of adequate prescreening for MBRs to eliminate debris, hair, lint and the like (see, e.g., Le-Clech et al., 2001).
- *Module Design* Where the packing density needs to be sufficiently low for high solids feeds. Modules most prone to blocking could be vertical hollow fibers, rather than horizontal fibers or vertical flat-plate cassettes. One design that overcomes this uses vertical hollow fibers attached (to suction) only at the base and closed at the top end, which is free to move (see Section 10.3.1).
- *Flow Distribution* Avoids dead spots and bypassing. This requires attention to module geometry and delivery of two-phase flow in a homogeneous fashion.

10.5.2 Cleaning and Backwashing

Fouled membranes can usually be restored by cleaning. The cleaning can be physical or chemical, and the frequency varies depending on convenience, complexity, and cost. Details of cleaning for MBRs are given in the Chapter 9 (Section 9.2.3.2, point 3 and Section 9.2.3.3). Table 10.4 summarizes the various cleaning strategies applied to submerged membrane MBRs, indicating the frequency of a particular cleaning regime. Table 10.5 (based on Le-Clech et al., 2001) gives more details of the intensive cleaning protocols used to restore well-fouled membranes. It should be noted that submerged membranes can either be cleaned in situ, which requires draining the operating vessel, or

TABLE 10.4 Cleaning Methods Typically Used with Submerged Membranes

Method	Period
Air scour	Continuous (MBR); with backwash for low solids
Backwash with permeate; intermittent filtration	<1 h; such as 10 min (on)/1 min (off)
Chemically enhanced backwash	Daily (or few days)
Maintenance chemical	Weekly
Intensive chemical (details in Table 10.5)	Once/twice per year

ex situ, which requires moving the membranes to a separate cleaning vessel (see Fig. 9.9 in Chapter 9 by Cornel and Krause).

The importance of efficient backwash has been explained in Section 10.5.1.1 for dead-end operation, and it is also important for similar reasons for cross-flow (bubbled) systems. In particular, unrestored membrane area means higher local fluxes in the remaining "active" area with acceleration of fouling; this may be the cause of the characteristic TMP "jump" (Fig. 10.2a). Backwash frequency needs to be high enough to avoid consolidated deposits that are difficult to remove. The backwash will preferentially flow through clean or easily cleaned regions and not where consolidated deposits lie. Similarly, the backwash flux or imposed pressure also needs to be high enough to stress the more tightly bound foulants. Vigorous backwash is not possible with flat-sheet membranes due to structural issues. For dead-end filtration with submerged membranes, the backwash is frequently combined with a period of air scour. After the solids have been displaced from the membranes, the tank has a relatively high solids content; so it is common to drain the tank to waste recovery before the next cycle begins.

10.5.3 Integrity Testing

Submerged membranes are applied to both water and wastewater treatment, and in both applications it is vital that damage to the membranes can be monitored. The issue is probably more important for drinking water where passage of pathogens (such as *Cryptosporidium*) into the final product water is a public health risk. Regulations are

TABLE 10.5 Intensive Cleaning Protocols for Various MBR Systems

MBR Supplier	Type	Chemical Agent	Conc. (%)	Protocol[a]
Mitsubishi	CIL[b]	NaOCl; citric acid	0.3 0.2	Backflow through membrane (2 h) + soak (2 h)
Zenon (GE)	CIT[c]	NaOCl; citric acid	0.2 0.2–0.3	Backpulse and recirculate
Memcor (Siemens)	CIT	NaOCl; citric acid	0.01 0.2	Recirculate through lumens; and in-tank air manifolds
Kubota	CIL	NaOCl	0.5	Backflow and soaking (2 h)

[a]Exact protocol varies plant to plant.
[b]CIL: cleaning in-line where chemical solutions are allowed to backflow (under gravity) inside the module.
[c]CIT: clean in tank, either in situ where membrane tank is isolated and drained or ex situ where membrane is removed from MBR and cleaned in a separate vessel.
Source: Based on Le-Clech et al. (2001).

specific about these risks, as evident in the U.S. Environmental Protection Agency (EPA) Long Term 2 Enhanced Surface Water Treatment Rule (LT2ESWTR) (EPA, 2006). The LT2 rule specifies membrane filtration as one of the "microbial toolbox" treatment options for surface water systems. Similar concerns about treatment integrity apply for wastewater MBRs particularly where the product water is for reuse.

Reasons for membrane damage include:

- Chemical oxidation (incompatibility with cleaning agents)
- Faulty installation
- Presence of foreign bodies (emphasizes importance of prescreening)
- Faulty membrane or module assembly (such as inadequate potting of hollow fibers)
- Operating stresses (e.g., excessive movement due to air sparging)

Some insights into membrane failure can be found in the literature (Childress et al., 2005; Gijsbertsen et al., 2006). The frequency of failure will depend on applications, but Gijsbertsen et al. (2006) estimate annual rates in water treatment of 1–10 per million fibers. Higher rates would be anticipated in MBRs where membranes are subject to more fouling conditions and greater operational stresses. Even flat-sheet membranes used in MBRs are not immune to occasional failure (Cornel and Krause, 2003).

Integrity tests are designed to assess whether the installed membranes can treat the feed water with a high retention of particles >1 μm. The target objective is typically to maintain a log removal value [$\text{LRV} = \log_{10}(C_{\text{Feed}}/C_{\text{Permeate}})$] of 4, which is a 99.99% removal. The available integrity test methods are either *indirect* (measuring filtrate quality) or *direct* (measuring a defect in the membrane). Table 10.6 summarizes the various types of integrity test for low-pressure membranes, based on information in Mallevialle et al. (1996, Chapters 10 and 11). The latest guidance in these matters can be found in the American Society for Testing and Materials (ASTM) Standard D-6908 (ASTM, 2006) and a review of the adequacy of various methods is available (DWI, 2001). It should be noted that the air pressure tests are not continuous and require the modules to be off-line for a few minutes during the test. The methods are the pressure decay test and the diffusive airflow test. Continuous monitoring is possible using particle counters that employ laser diffraction, but the method is costly and "counts" microbubbles as particles. Turbidity monitoring is not sufficiently sensitive for water treatment plants but could be used for MBRs. It should be noted that flat-sheet submerged membranes cannot be monitored by air pressure tests and rely on particle counters or turbidity monitoring.

TABLE 10.6 Summary of Membrane Integrity Test Methods for Submerged Membranes

Method	On-line	Sensitivity	Comment
Particle counting	Yes	High	Indirect method; high cost; several needed for a large plant
Particle monitoring	Yes	Modest	Indirect method; lower cost; several needed for a large plant
Turbidity	Yes	Low	Indirect method; low cost
Air pressure test	No	High	Direct method; integrated into system; not for flat-sheet membranes

10.6 APPLICATIONS

Submerged membranes are now an established feature in the water industry for water treatment, water process pretreatment, and wastewater treatment. The following sections provide brief comments on these applications.

10.6.1 Water Treatment

The introduction of submerged membranes coincided with growth in membrane filtration applications to drinking water after the major cryptosporidiosis outbreak in Milwaukee, Wisconsin, in 1993. From a very low level of installed capacity at that time, the installed capacity of MF/UF in the water industry 10 years later was on the order of 12,000 ML/day (Furukawa, 2006). Of this the major part would be water treatment and more than 50% of that would be submerged membranes. The growth in this application exceeds 15% per annum. It should be noted that contained/pressurized systems are also popular and continue to compete strongly with submerged systems. A typical application is in Coliban, Australia, where a submerged MF plant processes about 140 ML/day of surface water. In this process following drum screening the feed is dosed with a few parts per million of coagulant (polyaluminum chloride) and filtered by submerged MF hollow fibers (Memcor CFM-S; Fig. 10.7) housed in multiple tanks operating in parallel. The plant runs in dead-end mode with backwash and air scour after about 30 min. The membranes remove essentially all particulates but only about 20% of the natural organic matter, which is removed by posttreatment (ozone and biological carbon). Significantly larger plants are now being built (such as the Zenon plant at Chestnut in Singapore at 273 ML/day) with variations in terms of pre- and post-treatment. Costs are now close to conventional water treatment with an improved quality water product.

10.6.2 Water Process Pretreatment—Dual Membranes

Low-pressure MF and UF are now state-of-the-art as pretreatment to RO in water reclamation processes and increasingly considered as an option for seawater desalination by RO (see Chapter 3, Seawater Desalination for details). In these dual-membrane processes the MF and UF play a crucial role in protecting the RO. Both submerged and contained membranes are being used in these applications. For example, in Singapore the NeWater reclamation plants that produce nearly 300 ML/day of high-quality product employ submerged membranes prior to RO. The Water Factory 21 Ground Water Recharge facility in Orange County now also uses this arrangement. More details of this type of system can be found in Chapter 7, Water Reclamation and Desalination by Membranes.

10.6.3 Membrane Bioreactors

Membrane bioreactors are growing rapidly in application, and the majority of commercial MBRs are based on submerged membranes, due to issues of energy usage and footprint. Initially MBRs have been relatively small scale with capacities of <10 ML/day, but increasingly the trends are to large plants with capacities >100 ML/day. However, the inherent characteristics of submerged-membrane MBRs means they are also of considerable interest for small decentralized systems (DiGiano et al., 2004). Another trend is that the dual-membrane reclamation process could be superseded by the MBR + RO combination, converting wastewater to indirect potable quality water. Further discussion of MBRs can be found in Chapter 9.

10.6.4 Future Prospects

Submerged membranes have come from being a "curiosity" to a mainstream membrane technology in less than a decade. The concept continues to evolve and can be expected to have an increasing major role in the water industry. Potential developments could include:

- Lower energy demand as operational issues (flux, fouling control) are optimized.
- Lower capital costs as submerged units are further simplified and modularized.
- More retrofitting in existing wastewater tanks to increase capacity and quality.
- More usage in hybrid processes, including sorbents and catalysts, which can be suspended in the tank with the membranes.
- Application to other membrane separations, such as membrane distillation and membrane contactors.

It is quite possible that a future review of submerged membranes will grow out of this research, noting that the subject of submerged membranes applied to (some as yet undefined) technology that does not rate a mention in this book.

10.7 CONCLUSIONS

Submerged membranes have evolved rapidly since the early 1990s. The advantages of the concept have lead to increasingly important roles in the water industry. For low solids content feeds submerged membranes are used in dead-end batch filtration mode with membrane regeneration by backwashing and intermittent air sparging. For more concentrated feeds, such as the MBR, submerged membranes are continuously scoured by air bubbles in order to control fouling. Performance loss due to fouling can be mitigated by identifying the sustainable flux where TMP rise is modest. Submerged membranes also bring new challenges, including permeate side flow issues and the need for effective integrity testing. Overall, the use of submerged membranes can be expected to grow in the foreseeable future.

REFERENCES

American Society for Testing and Materials (ASTM). (2006). Standard practice for integrity testing of water filtration membrane systems, revision 06. ASTM D 6908. ASTM, West Conshohocken, PA.

Benedek, A., and Cote, P. L. (2003). Long term experience with hollow fiber membrane bioreactors. In *Proceedings IDA World Congress*, Sept 28–Oct 3, 2003, Paradise Island, Bahamas, CDROM, ISBN 0-9745019-1-3.

Berube, P. R., and Lei, E. (2006). The effect of hydrodynamic conditions and system configurations on the permeate flux in a submerged hollow fiber system. *J. Membr. Sci.* 271(1–2), 29–37.

Chang, S., and Fane, A. G. (2000). Characteristics of microfiltration of suspensions with inter-fiber two-phase flow. *J. Chem. Technol. Biotechnol.* 75, 533–540.

Chang, S., and Fane, A. G. (2001). The effect of fibre diameter on filtration and flux distribution-relevance to submerged hollow fibre modules. *J. Memb. Sci.* 184(2), 221–231.

Chang, S., and Fane, A. G. (2002). Filtration of biomass with laboratory-scale submerged hollow fiber modules—Effect of operating conditions and module configuration. *J. Chem. Technol. Biotechnol.* 77, 1030–1038.

Chang, S., Fane, A. G., and Vigneswaran, S. (2002). Modeling and optimizing submerged hollow fiber membrane modules. *AIChE J.* 48(10), 2203–2212.

Chang, S., Yeo, A., Fane, A. G., Cholewa, M., Ping, Y., and Moser, M. (2007). Observation of flow characteristics in a hollow fiber lumen using non-invasive X-ray microimaging (XMI). *J. Memb. Sci.* 304(1–2), 181–189.

Childress, A., Le-Clech, P., Daugherty, J. L., Chen, C., and Leslie, G. (2005). Mechanical analysis of hollow fiber membrane integrity in water reuse applications. *Desalination* 180(1–3), 5–14.

Cornel, P., and Krause, S. (2003). State of the art of MBRs in Europe. Paper presented at Conference on Application and Perspective of MBRs in Wastewater Treatment and Reuse, Cremona, Italy, Apr. 28–29.

Cote, P. L., Smith, B. M., Deutschmann, A. A., Rodrigues, C. F., and Pedersen, S. K. (1994). Frameless array of hollow fiber membranes and method of maintaining clean fiber surfaces while filtering a substrate to withdraw a permeate, PCT WO 94/11094.

Cui, Z. F., Chang, S., and Fane, A. G. (2003). The use of gas bubbling to enhance membrane processes *J. Membr. Sci.* 221, 1–35.

DeCarolis, J. F., Hirani, Z., Adham, S. S., Tran, N., and Lagos, S. (2006). Commercially available membrane bioreactor systems. In S. S. Adham (Ed.), *Proceedings of Short Course on Membrane Bioreactors*. National Water Research Institute, Anaheim, pp. 11–17.

DiGiano, F. A., Andreottola, G., Adham, S. et al. (2004). Membrane bioreactor technology and sustainable water. *Water Environ. Res.* 76(3), 195–196.

Ducom, G., Puech, F. P., and Cabassud, C. (2002). Air sparging with flat-sheet nanofiltration: A link between wall shear stresses and flux enhancement. *Desalination* 145, 97–102.

DWI (2001). Review of the adequacy of existing proposals for membrane integrity monitoring. DWI Report 43/2/159. Available: www.dwi.gov.uk/reg/pdf/min.

Fane, A. G., Wei, X., and Wang, R. (2006). Membrane filtration processes and fouling. In G. Newcombe and D. Dixon (Eds.), *Interface Science in Drinking Water Treatment: Fundamentals and Applications*. Academic, New York, pp. 109–132.

Furukawa, D. (2006). Microfiltration and ultrafiltration: What's next? Paper presented at Microfiltration 4 Conference, National Water Research Institute, Anaheim, Mar. 20–21.

Futamura, D., Katoh, M., and Takeuchi, K. (1994). Organic waste water treatment by activated sludge process using integrated type membrane separation. *Desalination* 98, 17–25.

Gijsbertsen-Abrahamse, A. J., Cornelissen, E. R., and Hofmann, J. A. M. H. (2006). Fiber failure analysis and causes of hollow fiber integrity loss. *Desalination* 194, 251–258.

Guibert, D., Ben Aim, R., Rabie, H., and Cote, P. (2000). Aeration performance of immersed hollow-fiber membranes in a bentonite suspension. *Desalination* 148, 395–400.

Ho, W. S., and Sirkar, K. K. (1992). *Membrane Handbook*. van Nostrand Reinhold, New York.

Judd, S. (2006). The MBR Book: Principles and Applications of Membrane Bioreactors in Water and Wastewater Treatment. Elsevier, New York.

Kim, J., and DiGiano, F. A. (2006). Defining critical flux in submerged membranes: Influence of length-distributed flux. *J. Membr. Sci.* 280, 752–761.

Le-Clech, P., Chen, V., and Fane, A. G. (2006). Fouling in membrane bioreactors used in wastewater treatment—A review. *J. Membr. Sci.* 284(1–2), 17–53.

Le-Clech, P., Fane, A. G., Leslie, G., and Childress, A. (2005, June). MBR focus: The operators perspective. *Filtration and Separation*, pp. 20–23.

Le-Clech, P., Jefferson, B., Chang, I. S., and Judd, S. J. (2003). Critical flux determination by the flux-step method in a submerged membrane reactor. *J. Membr. Sci.* 227(1–2), 81–93.

Liu, R., Huang, X., Wang, C., Chen, L., and Qin, Y. (2000). Study on hydraulic characteristics in a submerged membrane bioreactor process. *Process Biochem.* 36, 249–254.

Madec, A. (2000). Influence d'un ecoulement diphasique sur les performances de filtration d'un procede a membranes immerges. PhD Thesis, L'Institut National Des Sciences Appliques De Toulouse.

Mallevialle, J., Odendaal, P. E., and Wiesner, M. R. (1996). *Water Treatment Membrane Processes.* McGraw-Hill, New York.

Mayer, M., Braun, R., and Fuchs, W. (2006). Comparison of various aeration devices for air sparging in crossflow membrane filtration. *J. Memb. Sci.* 277(1–2), 258–269.

Ndinisa, N. V., Fane, A. G., and Wiley, D. E. (2006). Fouling control in a submerged flat sheet membrane system, Part I—Bubbling and hydrodynamic effects. *Separation Sci. Technol.* 41(7), 1383–1410.

Ohkubo, K., Hayashi, T., and Nasgai, H. (1988). Hollow fiber filter device. U.S. Patent 4,876,006.

U.S. Environmental Protection Agency (EPA) (2006). Long term 2 enhanced water treatment rule. Available: www.epa.gov/safewater/disinfection.

Wang, Y., Brannock, M. W. D., Leslie, G. L., and Ong, K. W. (2008). Evaluation of membrane bioreactor performance via residence time distribution: effects of membrane configuration and mixing. *Water Sci. Technol.* 57(3), 353–359.

Wicaksana, F., Fane, A. G., and Chen, V. (2006). Fibre movement induced by bubbling using submerged hollow fibre membranes. *J. Membr. Sci.* 271(1–2), 186–195.

Yamamoto, K., Hiasa, M., Mahmood, T., and Matsuo, T. (1989). Direct solid-liquid separation using hollow fiber membrane in an activated sludge aeration tank. *Water Sci. Technol.* 21, 43–54.

Yeo, A. P. S., Law, A. W. K., and Fane, A. G. (2007). The relationship between performance of submerged hollow fibers and bubble-induced phenomena examined by particle image velocimetry. *J. Memb. Sci.* 304(1–2), 125–137.

Yusuf, C., and Moo-Young, M. (1993). Improve the performance of airlift reactors. *Chem. Eng. Prog.* 6, 38–45.

CHAPTER 11

Nanofiltration

BART VAN DER BRUGGEN and JEROEN GEENS

Department of Chemical Engineering, Laboratory for Applied Physical Chemistry and Environmental Technology, University of Leuven, Leuven, Belgium

11.1 INTRODUCTION

The history of nanofiltration (NF) dates back to the 1970s when efforts started to develop reverse osmosis (RO) membranes with a reasonable water flux at relatively low pressures. The high pressures used in reverse osmosis resulted in a considerable energy cost, but, on the other hand, the quality of the obtained permeate was very good, and often even too good. Thus, membranes with lower rejections of dissolved components, but with a higher water permeability, would be a great improvement for separation technology. Such low-pressure RO membranes became known as nanofiltration membranes. By the second half of the 1980s, nanofiltration slowly started to come of age, and the first applications were reported (Eriksson, 1988; Conlon and McClellan, 1989).

In comparison with ultrafiltration (UF) and reverse osmosis, nanofiltration has always been a difficult process to define and to describe. Tight NF membranes are in some ways similar to RO membranes, and loose NF membranes could probably be classified as UF membranes. The specific features of NF membranes are mainly the combination of very high rejections for multivalent ions (>99%) with low to moderate rejections for monovalent ions (0–70%), and the high rejection (>90%) for organic compounds with a molecular weight above the molecular weight of the membrane, which is usually in the range of 150–300. However, nanofiltration is still a gray zone in terms of physicochemical interactions and transport mechanisms; a transition zone with features of both UF and RO, but with its own particular characteristics as well. Understanding these phenomena is a challenge for researchers and product developers.

In the second half of the 1990s, research on nanofiltration increased (see Fig. 11.1). As a consequence, scientists and industrialists nowadays feel more confident about what can be expected from a nanofiltration membrane, and more and more applications proved to be successful.

By 2000, the installed capacity was about 6000 ML/day, which is 10 times higher than in 1990 (Schäfer et al., 2005). As research continues, membranes become better defined,

Advanced Membrane Technology and Applications. Edited by Norman N. Li, Anthony G. Fane, W. S. Winston Ho, and T. Matsuura
Copyright © 2008 John Wiley & Sons, Inc.

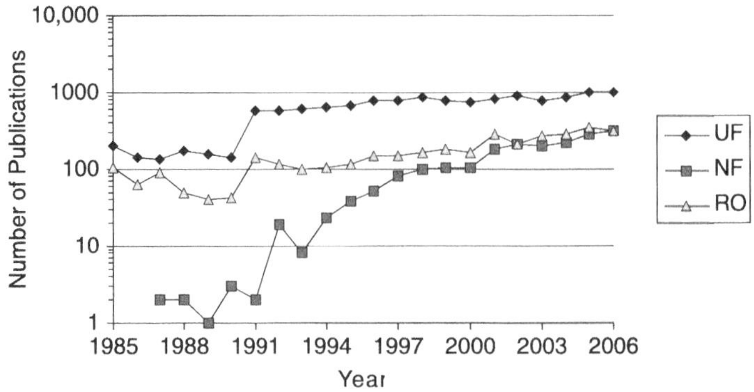

Figure 11.1 Publications on ultrafiltration, nanofiltration, and reverse osmosis (Web of Knowledge, 2005—values for 2005 are extrapolated, apps.isiknowledge.com).

less prone to fouling, and more resistant to harsh conditions. Examples are the development of ceramic NF membranes and polymeric solvent-resistant nanofiltration (SRNF) membranes. Taking the number of possible applications in, for example, the chemical and pharmaceutical industry into account, in addition to the applications that are still to be implemented but can be considered as state-of-the-art, it can be assumed that the increase of installed capacity will continue for many more years.

This chapter gives a short overview of process principles in nanofiltration, with a focus on the aspects that distinguish NF from RO and UF. This should allow us to determine which applications are theoretically possible, based on the composition of the liquid to be treated. Some of the most important applications of NF that are already realized will be discussed as well. The use of NF for drinking water production, which is historically the first and to date still the most important application, will be described in detail; other industrial applications with proven performance will be reviewed, and a glimpse of the growing market of separations in organic solvents will be given.

11.2 PROCESS PRINCIPLES

11.2.1 Membranes and Process Design

The traditional materials used for NF membranes are organic polymers. NF membranes are made by phase inversion or by interfacial polymerization (Vankelecom et al., 2005). Phase inversion membranes are homogeneous and asymmetric and often made of cellulose acetate or poly(ether)sulfone. Membranes made by interfacial polymerization are heterogeneous: They consist of a thin-film composite layer on top of a substrate UF layer. Typical polymers are (aromatic) polyamides, polysulfone/poly(ether sulfone)/sulfonated polysulfone, polyimide, and poly(piperazine amide); other polymers or blends can be used as well. Recent trends are the use of highly cross-linked polymers in order to obtain enhanced membrane stability at low or high pH, at high temperature or in organic solvents. NF membranes contain functional groups that can be charged, depending on the pH of the solution in contact with the membrane. Typically, NF membranes are negatively charged at neutral pH, with the isoelectric point around pH 3–4.

The production of ceramic NF membranes is also possible, but to date the pore size of most ceramic NF membranes is still relatively high. The molecular weight cutoff, the molecular weight of a component retained by 90%, is usually above 500 (Siewert et al., 2000; Tsuru et al., 2001a, b; Voigt et al., 2001). Molecular weight cutoff (MWC) values of 200 and below were recently reported for Al_2O_3/TiO_2 membranes (Van Gestel et al., 2002). These membranes were obtained by a careful preparation of each sublayer. The macroporous substrate consisted of α-Al_2O_3; the intermediate layers were prepared from TiO_2, α-Al_2O_3, γ-Al_2O_3, or mixtures or these components; the top layer is a fine textured polymeric TiO_2 top layer.

Most NF membranes are packed into spiral-wound elements; however, tubular, hollow-fiber, and flat-sheet or plate-and-frame modules are also available (Yacubowicz & Yacubowicz, 2005). Tubular membranes with diameter around 1 mm, denoted as capillary membranes, are interesting in view of fouling control (Verberk et al., 2002).

11.2.2 Performance

Three parameters are crucial for the operation of a (nano)filtration unit: solvent permeability or flux through the membrane, rejection of solutes, and yield or recovery.

The flux J or the permeability (flux per unit of applied pressure) of a membrane is, similarly to other pressure-driven membrane processes, a crucial parameter. Most NF membranes—except some used for solvent applications—are hydrophilic. If the Hagen–Poiseuille equation can be assumed (although this equation is, in fact, only valid for porous membranes), the other parameters influencing the permeability are obvious:

$$J = \frac{\varepsilon r^2}{8\eta\tau}\frac{\Delta P}{\Delta x}$$

A large membrane surface porosity (ε), large pore radii (r), and a low tortuosity (τ), together with a low membrane thickness Δx, are advantageous. The influence of the viscosity (η) is important when the temperature is varied: a lower viscosity is obtained at higher temperatures, which results in higher fluxes. An increase of the temperature by 1°C corresponds to a flux increase of 2–2.5%. For concentrated solutions or solutions with high salinity, the osmotic pressure $\Delta\pi$ should be substracted from the applied pressure ΔP. The flux equation then becomes

$$J = L_p(\Delta P - \sigma\Delta\pi),$$

where L_p is the solvent permeability and σ is the (maximal) rejection of the solute. The osmotic pressure can be calculated by using the Van't Hoff equation (Mulder, 1996) or, with more precision, by using the Pitzer model (Pitzer, 1973).

A similar transport equation can be written for the solute:

$$J_s = P_s\Delta x\frac{dc}{dx} + (1 - \sigma)Jc$$

where c is the concentration of the solute and P_s is the permeability of the solute. Transport by diffusion is represented by the first term in this equation; the second term stands for the contribution of convection to the transport of (uncharged) molecules.

The rejection of component i is defined as

$$R_i(\%) = \left(1 - \frac{c_{p,i}}{c_{f,i}}\right)100$$

where $c_{p,i}$ is the permeate concentration and $c_{f,i}$ is the feed concentration of component i; R is a dimensionless parameter and its value normally varies between 100% (complete rejection of the solute) and 0% (solute and solvent pass freely through the membrane). Negative rejections can sometimes be observed when the solute passes favorably through the membrane, for example, in salt mixtures.

The rejection of a given molecule can be calculated from the equations above as

$$R = \frac{\sigma(1 - F)}{1 - \sigma F}$$

$$F = \exp\left(-\frac{1 - \sigma}{P_s}J\right)$$

The rejection in NF is mainly determined by molecular size, hydrophobicity, and charge (Van der Bruggen et al., 1999; Kiso et al., 2001; Bowen and Welfoot, 2005; Braeken et al., 2005b), but effects of, for example, molecular shape and dipole moment, might play a role as well. The pore/void dimensions are statistically distributed and can be described by a log-normal distribution (Bowen and Welfoot, 2005). This explains the smooth transition from no rejection to complete rejection in a typical S-shaped curve when molecular size is varied. The MWC value is often used to indicate the lower limit of molecules that are (almost completely) retained, similar to UF membranes. For NF membranes, with MWC values between 150 and 1000 (but often in the range 150–300), this concept should be used with care: Hydrophobic molecules larger than the MWC, for example, often have a low rejection; the pH of the solution might change the membrane's surface charge as well as the charge of the solute, so that the rejection of this solute can be higher or lower than expected.

The third important parameter is the recovery or yield. This is a parameter for the design of an industrial application rather than a membrane characteristic. The recovery is the ratio of the permeate stream to the feed stream; its value ranges from 40 to 90%.

11.2.3 Membrane Fouling in NF

A problem often encountered in practical applications of NF is the decrease of the water flux for real feed solutions in comparison to the pure water flux. Flux decline can be caused by membrane fouling (Wiesner and Chellam, 1999). Fouling is caused by precipitation of inorganic components such as $CaCO_3$ or $CaSO_4$ (Lee and Lee, 2000; Her et al., 2000), deposition of organic compounds (Her et al., 2000), or possibly growth of bacteria on the membrane surface (biofouling) (Vrouwenvelder et al., 1998). Fouling can be defined as irreversible flux decline that can only be removed, for example, by chemical cleaning. When flux decline disappears by simply changing the feed solution to pure water, the phenomenon is reversible and should therefore not be considered as fouling.

Both reversible flux decline and fouling cause practical problems in the application of nanofiltration: For a given membrane surface, the yield of permeate decreases; the energy consumption increases because higher pressures are needed to obtain the same flow rate; cleaning procedures need additional chemical reagents; and the lifetime of the membrane decreases. Furthermore, the rejection of different components might change. It might be expected that rejections generally increase when flux decline occurs, for example, because of pore narrowing, but this is not always the case.

Flux decline due to the presence of organic compounds in the feed solution was, for example, encountered in nanofiltration of surface water containing high concentrations of natural organic matter (NOM) (Nilsson and DiGiano, 1996; Braghetta et al., 1998), where interactions between organic compounds and the membrane material (in a hollow-fiber module) even lead to the formation of a cake layer. This was also reported for groundwaters and for surface waters during nanofiltration with spiral-wound membranes (Baker et al., 1995).

It is usually accepted that flux decline in aqueous solutions containing organic molecules is mainly caused by adsorption, possibly enhanced by pore blocking (Van der Bruggen and Vandecasteele, 2001; Braeken et al., 2005a). Adsorption on NF membranes has been related to high-performance liquid chromatography (HPLC) characteristics (Kiso et al., 1999a, b). Similar problems have been reported for UF membranes (Lindau et al., 1998) and for RO membranes (Matsuura and Sourirajan, 1972, 1973a, b). For ultrafiltration, it was found that molecular size is the most important factor determining flux decline, whereas in reverse osmosis different factors reflecting hydrophobicity play a role.

11.2.4 Concentrates

One of the main problems still to be solved for nanofiltration, and for pressure-driven membrane filtration in general, is the further treatment of the concentrate fraction (Van der Bruggen et al., 2003). The relative volume of the concentrate may range from 40 to 90% of the feed volume; its composition is similar to the feed, but the concentration factor (CF) of rejected compounds is higher by a factor CF calculated as

$$\mathrm{CF} = \frac{C_{r,i}}{C_{f,i}} = \frac{Q_f}{Q_r}\left[1 - \mathrm{REC}\left(\frac{C_{p,i}}{C_{f,i}}\right)\right]$$

where Q is the volumetric flow (L/h) and C is the concentration (mg/L); the subscripts r, f, p, and i refer to the concentrate (or retentate), the feed, the permeate, and the component used, respectively, and REC is the relative fraction of the permeate compared to the feed. For components that are completely rejected, this equation simplifies to

$$\mathrm{CF} = \frac{1}{1 - \mathrm{REC}}$$

As a consequence of this large variation, the further environmental fate of the concentrate is unpredictable; a large variation in possibilities for reuse, further treatment, or discharge exists. Cost factors and legal aspects also play an important role. Generally, all methods for concentrate processing can be classified into one of the following categories (Van der Bruggen et al., 2003): (1) reuse, (2) further treatment by removal of contaminants, (3) incineration, (4) direct or indirect discharge in surface water, (5) direct or indirect discharge in groundwater, and (6) landfilling. Reuse is the most attractive option but only applicable in a few cases where the concentrated fraction is actually the desired product, such as in the food industry (e.g., dairy products, starch processing). The permeate is then a side product, which can be reused as a rinsing water or discharged. If reuse of the concentrate is not possible, further treatment may be necessary before discharge. Two options for further treatment can be distinguished: (a) water removal from the concentrate and (b) removal of specific components by a proper choice of a selective treatment method. The first option leads to a sludge or solid waste that is subsequently reused (if possible), landfilled (if necessary after solidification/stabilization or a similar pretreatment to avoid leaching of contaminants), or incinerated in a rotating kiln furnace (hazardous waste) or a grate furnace (nonhazardous waste). The second option leads to a (treated) wastewater, which has to be reused (if possible) or discharged in surface water (direct or indirect via sewage systems) or in groundwater.

Other factors than the volume and composition that have to be taken into account for selecting a proper treatment process are similar to reverse osmosis (Squire et al., 2000; Ahmed et al., 2001): legal requirements such as permits and conditions; cost of further treatment; local factors such as the proximity and size of a wastewater treatment plant, the presence of surface water or open land, soil characteristics, and geological structure; flexibility of the disposal method in case of an expansion of the existing plant; and public acceptance.

In the mid-1990s a survey on the use of membrane techniques in the drinking water industry in the United States was conducted (Truesdall et al., 1995), for installations with a capacity above 25,000 gal/day (95 m^3/day). Of the 137 installations 73% were RO installations for desalination of brackish water; 11% were NF installations; another 11% were electrodialysis installations; the remaining 5% were RO plants for seawater desalination. In 48% of the installations the concentrate was discharged in surface water; in 23% the concentrate was treated in a wastewater treatment plant; in 13% the concentrate was reused on the land for, for example, irrigation; in 10% the concentrate was discharged to groundwater by deep injection; and in the remaining 6% the concentrate was discharged to evaporation ponds.

11.3 APPLICATION OF NANOFILTRATION FOR PRODUCTION OF DRINKING WATER AND PROCESS WATER

11.3.1 Softening

Today still the most described application of NF is softening of surface water and groundwater. As a softening process, NF is in competition with traditional water-softening processes such as inorganic and organic ion exchange systems, as well as processes such as cold and hot lime softening and pellet softening (Schneider, 1994; Sombekke et al., 1997). The NF permeate contains fewer hardness ions, ranging from almost no hardness ions for low-pressure RO membranes or tight NF membranes, to a slight decrease in hardness for open porous NF membranes. Typical rejection values are 70–99% (Schaep et al., 1998; Bannoud, 2001).

Groundwater generally has a stable composition, but it is often too hard for use as process water or drinking water. Softening is necessary in many cases. Nanofiltration is here a competitive alternative for lime softening; a comparison between both for application in Florida by Bergman (1995) showed that for lime softening operation and maintenance

(O&M) costs were lower than for membrane softening, but the relative difference in costs decreased with larger facilities from a factor 2 for a production of 4000 m^3/d to 15% for a production capacity of about 50,000 m^3/d. If additional treatment processes are added to lime softening to match the better membrane softening permeate quality, or if some water can be bypassed around the membranes and blended to produce water comparable to the finished water in the lime softening plant, the cost of NF is much lower than for lime softening.

A comparison between nanofiltration and pellet softening [combined with granular activated carbon (GAC) adsorption for organics removal] was made by Sombekke et al. (1997), based on a life-cycle analysis (LCA). Both treatment schemes were found to have a comparable impact, except that NF was advantageous for quality and health aspects. Other positive aspects for NF were the investments and costs, and the impact on the landscape. The main environmental impact for both alternatives is caused by the use of energy. The Water Supply Company of Overijssel (WMO), The Netherlands, decided to extend current treatment capacity with nanofiltration on the basis of this LCA.

The most important advantage of NF is the product quality, which is superior to lime treatment because of the additional removal of color and turbidity. Other issues that are favorable for nanofiltration are the process flexibility, the smaller land requirements, the absence of sludge to dispose, and even site aesthetics. Moreover, newer developments have resulted in improved performance characteristics of NF membranes, including lower operating pressure requirements, which favorably affects membrane system and overall plant construction and O&M costs (Bergman, 1996). This should make NF even more cost effective in the future. Another trend is the use of membranes such as NF 200 (Filmtec), with high rejection of organics but low hardness rejection (De Witte, 1997). Energy consumption is still low, and the membrane performance remains good after repetitive cleaning. In contrast to other membranes, this new type of NF membranes is capable to maintain minimum hardness requirements without remineralization of the permeate. It was succesfully used at Debden Road water works, Saffron Walden, England (Wittmann et al., 1998), although hardness rejection was insufficient when the plant was run at a feed flow rate much below its design capacity.

Possible problems related to the use of NF are the further treatment of the concentrate fraction (see above), and the consequent loss of water. Because groundwater sources can be protected, or regulated by permits or taxes, this might be a legal and/or economical obstruction.

11.3.2 Organics Removal

Surface waters, in comparison with groundwaters, often have a frequently changing chemistry or composition due to seasonal changes or by dilution with rain. This is a disadvantage for softening by precipitation processes, which require an enhanced system control, but not for NF. The permeate quality in NF depends largely on the type of membrane used and is not very susceptible to changes in feed water quality. Therefore, NF is a reliable option for surface water treatment, although the focus is here rather on removal of organics than on softening. As a result of new regulations such as the U.S. Environmental Protection Agency Disinfectant/Disinfection By-Product Rule, several utilities have been evaluating NF for surface water applications (Jacangelo et al., 1997). The nature of the water source does not affect the filtration characteristics significantly, but a more extensive pretreatment is necessary for surface waters compared to groundwaters (Chellam, 2000). For groundwaters, pretreatment may be limited to depth filtration. For surface waters, coagulation/flocculation followed by sedimentation is usually necessary; ultrafiltration or microfiltration is a valuable alternative. Surface waters may even show better results, on condition that the right pretreatment is used (Allgeier and Summers, 1995).

A typical NF application is the treatment of water from a lake in Taiwan (Yeh et al., 2000), where softening had to be combined with a solution for taste and odor problems. Compared to conventional processes, NF produced water with the best quality measured by turbidity, dissolved organics, biostability, and organoleptic parameters.

Removal of organics usually focuses on disinfection byproduct (DBP) precursors and NOM. A sufficient but incomplete organics rejection is obtained with membranes with a relatively high molecular weight cut-off (MWCO) (Visvanathan et al., 1998; Levine et al., 1999). Because NOM consists of molecules from a large range of molecular weights, the smallest molecules are not rejected by NF membranes. Agbekodo et al. (1996) studied the organics in NF permeate in detail and concluded that about 60% of the remaining dissolved organic carbon (DOC) is caused by amino acids, with lower fractions of fatty aromatic acids and aldehydes. Around 80% of the total DOC of 0.15 mg/L that was found in the permeate is thus identified. It was concluded that the NF permeate appears to be a high-quality drinking water; should chlorination be required to maintain a disinfectant residual in the distribution network, the produced permeate would likely be low in DBPs and would present little or no risk of bacterial regrowth.

Cho et al. (1999) found that NOM rejection is influenced by size exclusion, electrostatic repulsion, and NOM aromaticity. Given the broad range of molecular sizes found in NOM, this confirms the need for a membrane with a low MWCO for complete organics removal. Visvanathan et al. (1998) mention the influence of the presence of ions on rejection properties and conclude that rejections are somewhat lower at higher ionic strength. This can be explained by charge effects and the interaction with charged NOM compounds.

Tuhkanen et al. (1994) studied nanofiltration as an alternative to ozonation. Ericsson and Trägårdh (1997) compared nanofiltration of a highly colored water from Lake Muskan, Sweden, with conventional flocculation–sedimentation–filtration and activated carbon treatment. The removal of color and organic matter was almost complete (undetectable levels) with NF. Because of the superior quality of the NF permeate, a pilot plant was built for design of a full-scale production unit of 6000 m^3/day (Ericsson et al., 1996). The membranes used were DS-5 DL (Osmonics) and NF45 (Dow-Filmtec). Several other successful studies on the removal of DBP precursors and NOM by NF have been reported (Alborzfar et al., 1998; Escobar et al., 2000; Everest and Malloy, 2000; Khalik and Praptowidodo, 2000); the best results were obtained with membranes with an MWCO around 200.

11.3.3 Micropollutants

A growing concern nowadays is the presence of small concentrations of pollutants thought to have a high (long-term) impact on human health, generally denoted as micropollutants but also known as POPs (persistant organic pollutants), new or emerging pollutants, or, more specifically, PCPs (personal care products), PhACs (pharmaceutically active compounds), EDCs (endocrine disruptors), and pesticides.

The use of nanofiltration for the removal of pesticides has been investigated by many authors (Montovay et al., 1996; Berg et al., 1997; Hofman et al., 1997; Van der Bruggen et al., 1998, 2001b; Boussahel et al., 2000; Kiso et al., 2000). The removal efficiencies largely depend on the membranes used and on the pesticides that have to be removed. Montovay et al. (1996) found an 80% removal of atrazine and a 40% removal of metazachlor, which is insufficient. Kiso et al. (2000) studied the removal of 12 pesticides (acaricides, fungicides, insecticides, herbicides, and rodenticides), among which were atrazine and simazine, together with pyridine and a chlorinated pyridine compound. Four different (unspecified) membranes made by Nitto-Denko were used. Rejections obtained with three of these membranes were too low; the rejections with the fourth membrane were very high (over 95%), but this membrane appears to be an RO membrane, given the high NaCl rejection. Van der Bruggen et al. (1998) obtained good results for the removal of atrazine, simazine, diuron, and isoproturon. Rejections were all over 90% with NF70 (Dow/Filmtec), but for two other membranes (NF45, a Dow/Filmtec membrane, and UTC-20, a Toray membrane) relatively low rejections were found for diuron and isoproturon. This can be explained by the high polarity of these compounds, which causes interactions with the charged membrane. Further studies (Van der Bruggen et al., 2001b) proved that there was no significant effect of concentration on the rejection of pesticides. Comparable results were reported by Berg et al. (1997) for bank filtrate of the Elbe River (Germany). Removal of atrazine, terbutylazine, simazine, and metazachlor was tested with seven different NF membranes. Rejection of diuron was always lower than rejection of the other pesticides. The best result was found for the PVD1 membrane (Hydranautics), where diuron rejection is over 80%. A slight influence of pH was found, but at neutral pH values, which are normal in drinking water treatment, a nearly complete rejection of pesticides was observed.

Ducom and Cabassud (1999) studied the removal of trichloroethylene, tetrachloroethylene, and chloroform by nanofiltration. The removal efficiency of the former two compounds was satisfactory, but chloroform rejection was significantly lower. This effect was attributed to the selective sorption of chloroform into the membrane structure, which causes higher concentrations in the permeate. However, good removal efficiencies of chloroform were obtained by Waniek et al. (2002).

Rejection of PhACs and other compounds with EDCs in nanofiltration is also gaining attention (Nghiem et al., 2004). Bellona and Drewes (2005) investigated the rejection of the organo-acidic pharmaceutical residue ibuprofen. Ibuprofen has hydrophobic properties below its pK_a value of 4.91 and thus acts the same way in nanofiltration as the hormones (high initial rejection due to absorption, which decreases in time). Bellona and Drewes (2005) described an initial rejection of 95%, which dropped to about 30% over time for a Dow/Filmtec NF-200 membrane. Above the pK_a value, ibuprofen is negatively charged and the rejection remains constant at a 70% value. Kimura et al. (2003) investigated a number of PhACs and EDCs as a function of their physicochemical properties. For a Hydranautics ESNA nanofiltration membrane, high rejections of over 90% were reached for a number of charged compounds (salicylic acid, diclofenac, respectively, 92 and 93%). Rejection values of the noncharged compounds bisphenol-A (an industrial EDC), phenacetine and primidone was lower: 45, 19, and 87%, respectively. Rejection was measured after a 24-h operation time, except for phenacetine (8 h), for adsorption equilibrium. Experiments were performed at 100 μg/L concentrations. When initial concentrations were decreased to a 100-ng/L level, rejection efficiency dropped to 14 and 72% for phenacetine and primidone, respectively. This shows that other factors (diffusion/partitioning) than steric hindrance are also important for rejection of PhACs, after all, if steric hindrance would be the only mechanism of removal, rejection would be independent of concentration. Rejection of primidone was also tested by Xu et al. (2005), together with other PhACs, on different Dow/Filmtec and Koch NF membranes. For the Koch membrane, rejection ranged from 10% for primidone in an water matrix with effluent organic matter (EfOM), up to 70% for diclofenac. Xu et al. (2005) also investigated the pharmaceutical carbamazepine, which was rejected with 90%. This information shows that NF is not always an effective technique to remove pharmaceutical residues. Kimura et al. (2004) concluded that membrane material is an important parameter as different membranes can exhibit different rejection trends.

Rejection of PCPs was examined by Lee et al. (2005). Four PCPs (DEET, Galaxolide, Tonalide, and Musk ketone, molecular weights ranging from 190 to 295) were filtered on two different NF membranes. Rejection was obtained by adsorption/partitioning/diffusion and steric hindrance. Rejection of DEET was around 87% and followed the same pattern as CF (as investigated by Ducom and Cabassud), i.e. rejection decreased for increasing operating pressure.

Another study on removal of organic micropollutants was reported by Hofman et al. (1997). In this study, rejection of four different mixtures of micropollutants was measured with ultra-low-pressure RO membranes, comparable to tight NF membranes. The authors mention maximal concentrations of these components found in IJssel Lake and the Rhine River in the Netherlands and calculate the maximal concentrations for which a membrane plant would still be able to fulfill the European standards for drinking water. Surprisingly, there is no significant difference between pesticides and other organic micropollutants, although the pesticides are larger and should show a higher rejection. Low-pressure reverse osmosis or nanofiltration produce an effluent that meets all regulations for pesticides and other organic micropollutants. The results of the Méry-sur-Oise surface water treatment plant (Ventresque and Bablon, 1997) support this conclusion: Pesticides, together with other compounds, are rejected efficiently in a large-scale NF plant producing 140,000 m^3/day of water in Paris, France. The NF200 membrane, mentioned above, was used for this industrial application.

11.4 WASTEWATER POLISHING AND WATER REUSE

Nanofiltration permeates have a very good quality because the process is capable of removing, in many cases, most inorganic as well as organic foulants. This makes NF a good candidate for wastewater treatment when a high quality is necessary, that is, for wastewater polishing and water reuse. The list of possible applications is very long, but we will limit ourselves to some of the most important.

11.4.1 Treatment of Effluents from the Textile Industry

Effluents from the textile industry are a challenge: They are visually polluted by the presence of dyestuff after exhaustion, they contain large salt concentrations (depending on the dyeing technique, but up to 100 g/L), have a high chemical oxygen demand (COD) (additives, dyes) and have a high temperature and pH. Conventional treatment is insufficient, not in the least because many new dye components are not biodegradable. NF can be a solution because most dyes currently used in the textile industry are in the range 700–1000 g/mol, well above the MWC of the average NF membrane (Tang and Chen, 2005). Salts can be retained (in case Na_2SO_4 is used) or not (in case NaCl is used). The remaining salt concentration is essential if the water is to be reused because it determines the fastness and evenness of the product.

If NF follows an existing secondary treatment, concentrations are usually lower due to dilution with other streams (e.g., diluted rinsing water) (Van der Bruggen et al., 2001a). This reduces membrane fouling, which can be a problem for treating textile effluents (Tang and Chen, 2005). On the other hand, breakdown products from the secondary treatment, which are fed to the NF unit, are smaller and thus more difficult to retain (Van der Bruggen et al., 2001b).

Another option is direct application of NF on textile effluents without a secondary treatment step. The advantage of this strategy is that the filtration can be carried out at enhanced temperature, which results in a significant energy saving. However, the complexity of the solution makes filtration difficult in many ways. Membrane fouling is a logical problem because of the reactivity of the dyestuff, although in some cases the problem is smaller than expected (Chen et al., 1997). Biofouling can occur when easily biodegradable components are present; this is to date still an unsolved problem. A possible solution is the combination of NF with photocatalytic oxidation, a process that was proposed by Tang and Chen (2005) for a better removal of recalcitrant compounds but could possibly deactivate biofouling in situ.

11.4.2 Treatment of Leachates from Landfills and Composting Plants

Leachates from landfills for municipal or industrial waste contain a complex mixture of organic and inorganic pollutants in high concentrations. Treatment is extremely difficult with classical technologies due to the complex and variable composition and the high concentrations to be removed. Even the combination of biological treatment (aerobic or anaerobic) with activated carbon adsorption or oxidation by ozonization was not sufficient (Peters, 1998), although the efficiency depends on composition and age of the landfill (Trebouet et al., 1999). Wastewater from composting plants can be treated in a similar way as leachates from landfills, although the composition is significantly different (larger organic fraction). It can be assumed that a similar approach can be used.

Reverse osmosis is used full scale for the treatment of landfill leachates (Linde et al., 1995). The RO retentate can be recycled to the landfill either directly or after evaporation. If this is not allowed, as is the case in Germany, more expensive solutions are to be found for the concentrate (Melin et al., 2005). A possible treatment method is activated carbon adsorption/oxidation, patented as Biomembrat-Plus, or even the combination of high-pressure RO, a crystallizer, and NF (Melin et al., 2005).

The use of RO can be improved by replacing the first RO unit by a nanofiltration unit (Peters, 1998), as shown in Figure 11.2. The rejection of dissolved components (inorganic and organic) by NF is lower than by RO; but when NF is combined with RO in the second membrane unit, a sufficient removal is obtained. The energy consumption in nanofiltration,

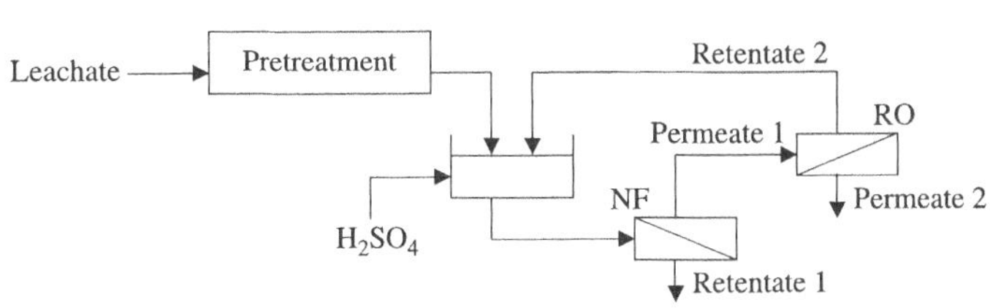

Figure 11.2 Treatment of landfill leachates by using a combination of NF and RO (Peters, 1998).

which operates typically at 10 bars, is much lower than in reverse osmosis, which operates at 40–80 bars. This decreases the operating cost of the plant.

Another combination that was suggested (Melin et al., 2005) is the use of biology and NF, extended with powdered activated carbon (PAC) adsorption.

11.4.3 Reclamation of Municipal Wastewater and Groundwater Remediation

If the additional cost of extending existing wastewater treatment plants toward tertiary treatment is taken into account, water reclamation is limited to low-quality purposes such as irrigation. In many cases a sand filtration is already sufficient for attaining reuse standards. Taking cost considerations into account, the majority of reclamation projects has modest ambitions: 60% of all reclamation projects is aimed at irrigation for agricultural purposes; 30% is aimed at industrial reuse, most of which is cooling.

However, the full potential of reclaimed water as an alternative water supply is realized when municipal wastewater is treated by membranes to potable standards (Fane et al., 2005). RO is frequently mentioned for this purpose; the Groundwater Replenishment System in Orange County, California, uses RO for injection into a potable aquifer to prevent the ingress of seawater into the groundwater basin (Leslie et al., 1999). NF is mentioned as an interesting alternative requiring less energy (Fane et al., 2005). However, the rejection of small toxic compounds may be too small for potable reuse; appropriate reuse applications should be defined. If the permeate is not reused, NF is possibly the best solution for, for example, groundwater remediation (Fane et al., 2005). The additional advantage of NF is the possibility to separate monovalent and bivalent ions, which is of interest when ammonium sulfate needs to be removed from groundwater (Macintosh et al., 2003). In many (abandoned) industrial locations in former Eastern Europe, this type of groundwater pollution exists.

11.4.4 Water Reuse in the Pulp and Paper Industry

Nanofiltration was evaluated for treatment of effluents from the paper and pulp industry as an alternative to UF, in view of a better process water quality and, eventually, closed water circuits (Nyström et al., 2005). Circulation and effluent waters contain large concentrations of fibers, colloids, and resins and are colored; therefore, pretreatment is essential to avoid fouling (Mänttäri et al., 1999). Spiral-wound modules can be used but require a pretreatment sequence of various methods; a better option are tubular modules or special configurations such as the cross-rotational module (Zaidi et al., 1992), where rotors between flat-sheet membranes increase the turbulence of the feed flow, or the vibratory shear enhanced processing (VSEP) module, in which a vibrating movement of the membrane stack decreases the effect of fouling (Culkin, 1990).

Similar to the application of NF for textile effluents, the operating temperatures in the pulp and paper industry are high—ideally, NF should operate at 70°C and withstand sudden increases up to 90°C (Nyström et al., 2005). In some cases, very alkaline conditions occur as well. Another similarity with textile effluents is the occurrence of (bio)fouling when easily degradable components (lignins, polysaccharides) are present (Nyström et al., 2005).

Given the huge volumes of water used in the pulp and paper industry, it is surprising how few existing mills incorporate NF. Nyström et al. (2005) give an overview of industrial plants and pilot-scale plants. From these experiences, it is clear that the permeate quality in NF is satisfactory, although a two-stage process can be required when a high recovery is needed. Technical issues being solved, it is expected that more stringent demands will be the trigger for application of NF on a large scale in the paper and pulp industry.

11.4.5 Nanofiltration in the Leather Industry

A fine example of a closed water cycle is the application of nanofiltration in the leather industry (Cassano et al., 2001). Wastewater contains large concentrations of tannins or chromium; after pretreatment by ultrafiltration, the chemicals used in the baths can be recovered in the concentrate fraction (Molinari et al., 2004), whereas the permeate can be reused for other purposes, for example, in the pickling step (Cassano et al., 1999).

11.5 OTHER APPLICATIONS

11.5.1 Application of NF in the Food Industry

Among the most mentioned industries where NF can be used are the dairy industry, the sugar industry. and the beverage industry. Concentration and demineralization of whey in the dairy industry is one of the oldest applications of NF (Eriksson, 1988). Whey is a by-product of cheese production and needs to be concentrated to produce whey powders, used in chocolate, bakery products, ice cream, and frozen dairy desserts (Bargeman et al., 2005). Monovalent ions contribute to a negative sensoric perception, but divalent ions are thought to have a health effect. In NF (diafiltration mode), 90–95% of the sodium chloride can be removed from salty whey in a three-step process (Kelly et al., 1992); usually, ultrafiltration is used prior to nanofiltration. Acid whey can be treated as well. Demineralization of sweet whey is a more recent application and requires a careful membrane selection in order to prevent lactose leakage (Bargeman et al., 2005).

In the sugar industry, NF is used for concentration of dextrose syrup originating from starch and for demineralization of colored brine from anion exchange resin elution solutions (Bargeman et al., 2005). The purification of alternative (low-calorie) sweeteners such as stevioside might also require a nanofiltration step (Zhang et al., 2000).

In the beverage industry, NF can be used for concentration of juices (Warczok et al., 2004) and for water reclamation in breweries (Braeken et al., 2004).

Finally, the reclamation of water used in cleaning-in-place (CIP) operations, used in many branches of the food industry, can be done by using NF, on condition that the removal of organics is sufficient, and that membranes are used that can operate in extreme conditions of pH.

11.5.2 Dual-Step Desalination

Due to the low rejection of monovalent salts, osmotic pressures in NF are lower than in RO. Thus, lower pressures need to be applied, and the energy consumption is proportionally lower. RO membranes operable at ultralow pressures have been developed (Semiat, 2000; Matsuura, 2001) that allow the desalination of brackish water at pressures comparable to those applied in NF. These membranes operate at the interface of NF and RO and might be helpful in optimizing the desalination process. Hassan et al. (1998) reported the use of NF in an integrated desalination system NF-SWRO (sea water reverse osmosis) and

NF-MSF (multistage flash). The NF plant received noncoagulated filtered seawater and reduced turbidity, microorganisms, and hardness. The concentration of monovalent salts was reduced by 40%, and the overall concentration of TDS (total dissolved salts) was reduced by 57.7%. The permeate thus obtained was far superior to seawater as a feed to SWRO or MSF. This made it possible to operate a SWRO and MSF pilot plant at a high recovery (resp. 70 and 80%, respectively). The MSF could be operated at a top brine temperature of 120°C without any scaling problem (Al-Sofi et al., 1998). The high water output in both integrated desalination systems, combined with a reduction of chemicals and energy (by about 25–30%), allows producing freshwater from seawater at a 30% lower cost compared to conventional SWRO (Al-Sofi, 2001). A demonstration plant was built at Umm Lujj, Saudi Arabia, consisting of six spiral-wound NF modules (8 in. × 40 in.) followed by three SWRO elements (Hassan et al., 2000).

11.5.3 Removal of Specific Contaminants from Water

Partial removal of nitrates is frequently reported, usually as a side effect in other applications (Ratanatamskul, 1998; Lee and Lueptow, 2001; Van der Bruggen et al., 2001b). Bohdziewicz et al. (1999) also remark that a combination NF/RO for nitrate removal would suffer less from scaling than a single RO because of the $CaSO_4$ and $CaCO_3$ removal in the NF step. Redondo and Lanari (1997) suggest RO for nitrate reduction in drinking water production; rejection of nitrates with NF membranes is too low. The reduction of hardness and essential minerals would, however, be too large, so that remineralization of the water is needed; this is an important disadvantage. Nanofiltration is also one of the methods that can be used to meet regulations for lowered arsenic concentrations in drinking water (Kartinen and Martin, 1995). Arsenic removal from synthetic freshwater and from surface water sources (including Colorado River water) by NF and RO was studied by Waypa et al. (1997). Both As(V) and As(III) were effectively removed from the water by RO and tight NF membranes (NF70, Dow/Filmtec) over a wide range of operational conditions. Rejections of 99% could easily be obtained with both NF and RO membranes. Removal of As(V) (negatively charged) and As(III) (uncharged) was comparable, with no preferential rejection of As(V) over As(III). This suggests that size exclusion governed their separation behavior and not the charge interaction. Seidel et al. (2001) used loose (porous) NF membranes to study the difference in rejection between As(V) and As(III). The rejection of As(III) was here indeed much lower than the rejection of As(V); the latter was pH dependent and varied between 60 and 90%, whereas As(III) rejection was always below 30%. Pilot studies (Brandhuber and Amy, 1998) confirmed the difference between As(V), which can be effectively removed, and As(III), which is far more difficult to remove, except when NF membranes with tight pores are used. Comparable results were found by Urase et al. (1998) with relatively porous membranes. The removal of arsenic by NF can even be applied in rural areas without electricity supply by using a manually operated bicycle pump (Oh et al., 2000).

Other applications that have been reported are the removal of fluoride and aluminum (Kettunen and Keskitalo, 2000; Pervov et al., 2000) and uranium removal from natural waters (Raff and Wilken, 1999).

11.6 SOLVENT-RESISTANT NANOFILTRATION

In contrast to the use in aqueous media, the first applications studied in nonaqueous media were not very successful. Membranes showed important performance loss, due to chemical instability of the polymeric materials in organic solvents. Different problems occurred: zero flux due to membrane collapse (Raman et al., 1996), "infinite" flux due to membrane swelling (Raman et al., 1996), membrane deterioration (Bridge et al., 2002), poor separation quality (Subramanian et al., 1998), and the like. In the beginning of the twenty-first century, new solvent-stable polymeric membranes came on the market, and ceramic NF membranes were developed. Researchers gained more and more insight in the interactions between solvents and membranes, which result in a performance that is totally different from the performance in water (Machado et al., 1999, 2000; Bhanushali et al., 2002; Robinson et al., 2004; Geens et al., 2005). In addition, the first successful applications of solvent-resistant nanofiltration (SRNF) were started.

The largest industrial SRNF plant is installed in the petrochemical industry (Bhore et al., 1999). Wax is a monoester of fatty acids that severely modifies the properties of lube oil and must therefore be removed (Hart et al., 1995). The traditional process of dewaxing involved the cooling of a hydrocarbon mixture in solvent or solvent mixtures (methyl ethyl keton, acetone) to temperatures typically ranging from -5 to -18°C. In this chilling section, waxy components coagulated and were precipitated or filtered; the solvent in the filtrate was removed by evaporation and reused in the process (Cuperus and Ebert, 2002).

In a new approach, a membrane unit replaces or supports the distillation step. White and Nitsch (2000) incorporated nanofiltration membranes in the conventional process, so that 25% of the solvent was removed by filtration prior to further purification. A great benefit of this process is that filtered solvent does not need to be heated. Hence, for an initial plant with 300 m^2 of membrane could handle 25% of the feed of 160 m^3, resulting in 20% reduction in size and capacity for the recovery section, 25% reduction in heat, and 10% in refrigeration requirements. Nowadays, an industrial plant for an ultimate feed rate of 11,500 m^3/day is running at the Exxon-Mobil refinery in Beaumont, Texas. The throughput in 2001 was 5800 m^3/day, consistent with existing demand. The average base oil production increased by over 25% and the improved dewaxed oil yields by 3–5%. The net uplift for additional lube and wax margin, attributed only to the use of membranes, was estimated at $6.1 million/year. The reduction in fuel oil consumption was calculated to 5800 m^3/year, equating to a reduction of greenhouse gas emissions of about 20,000 tons/year. Cooling water requirements are reduced by approximately 250 Mm^3/year. The diffuse loss of dewaxing solvents, which are VOCs, into the environment could be decreased by 50–200 tons/year.

A typical application of SRNF is found in the vegetable oil industry. It is estimated that more than 2 million tons of extraction solvent is used in the United States alone. Starting from seed from plants as soybean or sunflower, oil is obtained by solvent extraction, eventually in combination with mechanical extraction. Hexane is by far the most common extraction solvent. Currently, evaporation is used to recover these solvents and reuse them in the process, which requires a considerable amount of energy, approximately 530 kJ/kg oil. In addition, the elevated temperatures hold a risk on thermal damage, and explosive vapors may create safety problems (Raman et al., 1996). These limitations can be partially overcome by membrane technology.

Oil–micelle mixtures are formed during hexane extraction and consist of triglycerides (oils), phospholipids, and solvent. Due to its polarity, phospholipids form very loose conglomerates that can be filtered off by membranes. The process results in a filtrate with clear oil and hexane and a phospholipid fraction that can be worked up more easily. Although hexane has to be removed by stripping, potential savings are found in the reduction of chemicals and improved quality of oil. Raman et al. (1996) reported that a mixture containing 20% of oil could be concentrated to 45% with a commercial SRNF membrane, resulting in a 50% reduction of the energy consumption for the evaporation unit. Stafie et al. (2003) reported oil rejections of over 90% with a poly(dimethylsiloxane) membrane.

In organic synthesis of complex molecules, homogeneous catalysis has become increasingly important. Examples of such catalysts are the Heck and the Co-Jacobsen catalysts (Nair et al., 2001; Aerts et al., 2004). Transition metal catalysis (TMC) is used for the synthesis of enantiomerically pure components. Phase transfer catalysis (PTC) is applied to reactions involving a water-soluble electrophilic reagent. Two immiscible phases are contacted, and a phase transfer catalyst is used to transfer a reactant from one of these phases into the other, so that the reaction can occur (Luthra et al., 2002). Typically, postreaction, the catalyst remains in the organic phase. Applications are found in pharmaceuticals, agricultural chemicals, and fine chemicals. The advantages include increased productivity and quality, improved environmental performance, avoidance of need to use polar aprotic solvents (dimethyl sulfoxide, DMF, dimethyl acetamide), enhanced safety, and reduction of other manufacturing costs.

However, beside the relatively high cost, one of the major technical drawbacks for the use of TMC and PTC in industrial applications is the need to separate the product and the phase transfer catalyst. The most commonly used methods are extraction by water washing and distillation, but problems with toxic waste streams and degradation may then occur. A mild separation technique could facilitate the use of homogeneous catalysis in larger industrial organic synthesis processes. A straightforward approach to accomplish this, involves the use of membrane filtration (Luthra et al., 2001). Different research groups investigated the potential of nonaqueous membrane technology for the separations in TMC and PTC, and it has become clear that for industrially relevant catalysts, the membranes and the catalysts should be adopted to each other. Scarpello et al. (2002) reported catalyst rejections of more than 95%, coupled with good solvent fluxes, which makes this technology very promising. Nair et al. (2002) reported rejections up to 97% in methyl tert butyl ether and THF. It must be pointed that for this application, the membrane unit is not used as a "debottlenecking tool" but as a full part of a hybrid production process, as can be seen in the Figure 11.3.

A large number of fine chemicals or pharmaceutically active ingredients are synthesized with complex reaction paths. It often occurs that different steps require different solvents as reaction medium. In such cases, a solvent exchange must take place. A traditional procedure to transport relatively large, nonvolatile components to another solvent is a "put-and-take distillation," in which the original solvent is stepwise boiled out and replaced by similar volumes of the second solvent. The technique is only effective if the first solvent has a much lower boiling temperature than the second solvent, for example, the replacement of methanol by toluene. Azeotropic mixtures may cause additional problems.

Nanofiltration can become a strong alternative for traditional solvent exchange methods, without the drawbacks of elevated temperatures or azeotropic mixtures (Livingston et al., 2003). The principle requires a membrane that shows high permeation rates for the (first) solvent, whereas the component should be rejected. The feed stream is then filtered until nearly 10–30% of the volume is retained (and thus severely concentrated). Afterwards,

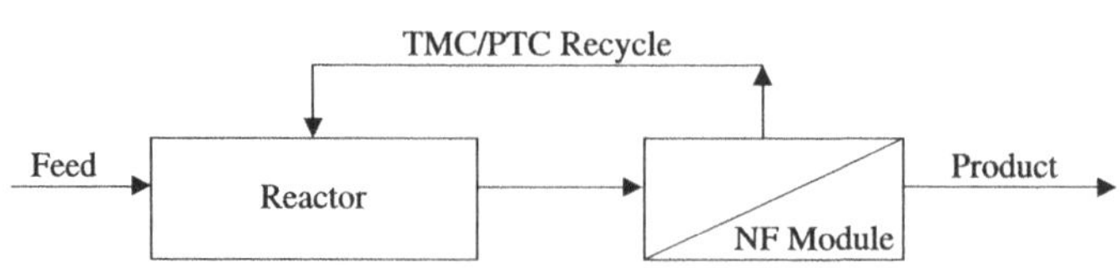

Figure 11.3 Scheme of a hybrid process in synthesis with homogeneous catalysis.

the second solvent is added to the retentate and the filtration process can restart. In a stepwise procedure, a concentration of less than 1% of the original solvent can be reached. No external heating is required and the permeate of the first membrane step can directly be reused in the production process.

This procedure was tested for the exchange of toluene by methanol, with tetraoctylammonium bromide (TOABr) as a reference solute. After five filtration steps, only 0.3% of toluene was found in the methanol fraction. The solute rejection in all steps was 100%.

The difference in energy consumption between put-and-take distillation and the membrane setup is difficult to estimate. Both technologies result in waste streams that require further treatment. Moreover, the required pressure for the membrane setup is hard to define on the base of pilot-scale experiments. It is, however, clear that the membrane setup shows a much lower mass intensity than the distillation. Mass intensity is a commonly used concept in the context of green energy and is defined as the ratio of the total mass of the reactants and the mass of the reaction products (Curzons et al., 2001). Solvent exchange by distillation typically shows a mass intensity of 5–10 kg/kg, whereas the use of membrane technology results in a value of 2–3 kg/kg. Both values are valid for a 95% exchange, which is often the required value in industrial applications.

11.7 CONCLUSIONS

Nanofiltration, a technology conceived in the 1980s, with a childhood in the first half of the 1990s and a rapid growth in the second half of the 1990s, is a mature process that can be strategically used in many applications. The combination of a pure permeate, in combination with the relatively low energy requirements, make NF attractive for applications where multivalent ions have to be separated from monovalent ions, and where relatively small organics have to be removed.

Although it is safe to say that NF nowadays can be implemented without problems in many known applications, it is expected that the evolution in NF will continue. Some of the expected progresses include modeling of rejection in aqueous solutions, understanding of the performance in organic solvents, prediction and control of fouling, membranes with controlled pore size, possibly in hollow-fiber configuration, development of hydrophobic ceramic NF membranes, and the optimization of hybrid processes with NF.

List of Symbols

c	concentration (kg/m^3)
CF	concentration factor
$c_{f,i}$	concentration of component i in the feed (mol/L)
$c_{p,i}$	concentration of component i in the permeate (mol/L)
J	flux (L/h m^2)
J_s	solute flux (mol/h m^2)
L_p	permeability (L/h m^2 bar)
P_s	solute permeability (m/s)
Q_f	feed flow (L/h)
Q_p	permeate flow (L/h)
r	pore radius (m)
R	rejection (%)

REC	permeate recovery
ε	surface porosity (m^2/m^2)
η	dynamic viscosity ($Pa \cdot s$)
σ	reflection coefficient
τ	tortuosity
ΔP	pressure difference (bar)
Δx	membrane thickness (m)
$\Delta\pi$	osmotic pressure (bar)

ACKNOWLEDGMENT

The authors wish to thank collaborators Leen Braeken, Katleen Boussu, Arne Verliefde, Ben Bettens, Bart Verrecht, and Kathleen Moons for valuable input.

REFERENCES

Aerts, S., Weyten, H., Buekenhoudt, A., Gevers, L. E. M., Vankelecom, I. F. J., and Jacobs, P. A. (2004). Recycling of the homogeneous Co-Jacobsen catalyst through solvent-resistant nanofiltration (SRNF). *Chem. Commun.* 6, 710–711.

Agbekodo, K. M., Legube, B., and Cote, P. (1996). Organics in NF permeate. *J. AWWA* 88(5), 67–74.

Ahmed, M., Shayya, W. H., Hoey, D., and Al-Handaly, J. (2001). Brine disposal from reverse osmosis desalination plants in Oman and the United Arab Emirates. *Desalination* 133, 135–147.

Alborzfar, M., Escande, K., and Allen, S. J. (1998). Removal of natural organic matter from two types of humic ground waters by nanofiltration. *Water Res.* 32(10), 2983–2970.

Allgeier, S. C., and Summers, R. S. (1995). Evaluating NF for DBP Control with the RBSMT. *J. AWWA* 87(3), 87–99.

Al-Sofi, M. A. K. (2001). Seawater desalination—SWCC experience and vision. *Desalination* 135(1–3), 121–139.

Al-Sofi, M. A. K., Hassan, A. M., Mustafa, G. M., Dalvi, A. G. I., and Kither, M. N. M. (1998). Nanofiltration as a means of achieving higher TBT of ≥120 degrees C in MSF. *Desalination* 118(1–3), 123–129.

Baker, J., Stephenson, T., Dard, S., and Cote, P. (1995). Characterisation of fouling of nanofiltration membranes used to treat surface waters. *Environ. Technol.* 16(10), 977–985.

Bannoud, A. H. (2001). Elimination of hardness and sulfate content in water by nanofiltration. *Desalination* 137(1–3), 133–139.

Bargeman, G., Timmer, M., and Van der Horst, C. (2005). Nanofiltration in the food industry. In A. I. Schäfer, A. G. Fane, and T. D. Waite (Eds.), *Nanofiltration—Principles and Applications*. Elsevier, Oxford.

Bellona, C., and Drewes, J. E. (2005). The role of membrane surface charge and solute physico-chemical properties in the rejection of organic acids by NF. *J. Membr. Sci.* 249, 227–234.

Berg, P., Hagmeyer, G., and Gimbel, R. (1997). Removal of pesticides and other micropollutants by nanofiltration. *Desalination* 113(2–3), 205–208.

Bergman, R. A. (1995). Membrane softening versus lime softening in Florida—A cost comparison update. *Desalination* 102(1–3), 11–24.

Bergman, R. A. (1996). Cost of membrane softening in Florida. *J. AWWA* 88(5), 32–43.

Bhanushali, D., Kloos, S., and Bhattacharyya, D. (2002). Solute transport in solvent-resistant nanofiltration membranes for non-aqueous systems: Experimental results and the role of solute-solvent coupling. *J. Membr. Sci.* 208, 343–359.

Bhore, N. A., Gould, R. M., Jacob, S. M., McNally, S. D., Smiley, P. H., and Wildemuth, C. R. (1999). *Oil Gas J.*, Nov. 15, pp. 67–74.

Bohdziewicz, J., Bodzek, M., and Wasik, E. (1999). The application of reverse osmosis and nanofiltration to the removal of nitrates from groundwater. *Desalination* 121(2), 139–147.

Boussahel, R., Bouland, S., Moussaoui, K. M., and Montiel, A. (2000). Removal of pesticide residues in water using the nanofiltration process. *Desalination* 132(1–3), 205–209.

Bowen, W. R., and Welfoot, J. S. (2005). Modelling the performance of nanofiltration membranes. In A. I. Schäfer, A. G. Fane, and T. D., Waite (Eds.), *Nanofiltration—Principles and Applications*. Elsevier, Oxford.

Braeken, L., Boussu, K., Van der Bruggen, B., and Vandecasteele, C. (2005a). Modeling of adsorption of organic compounds on polymeric nanofiltration membranes in solutions containing two compounds. *Chem. Phys. Chem.* 6(8), 1606–1612.

Braeken, L., Ramaekers, R., Zhang, Y., Maes, G., Van der Bruggen, B., and Vandecasteele, C. (2005b). Influence of hydrophobicity on retention in nanofiltration of aqueous solutions containing organic compounds. *J. Membr. Sci.* 252(1–2), 195–203.

Braeken, L., Van der Bruggen, B., and Vandecasteele, C. (2004). Regeneration of brewery waste water using nanofiltration. *Water Res.* 38(13), 3075–3082.

Braghetta, A., DiGiano, F. A., and Ball, W. P. (1998). NOM accumulation at NF membrane surface: Impact of chemistry and shear. *J. Environ. Eng. ASCE* 124(11), 1087–1098.

Brandhuber, P., and Amy, G. (1998). Alternative methods for membrane filtration of arsenic from drinking water. *Desalination* 117(1–3), 1–10.

Bridge, M. J., Broadhead, K. W., Hlady, V., and Tresco, P. A. (2002). Ethanol treatment alters the ultrastructure and permeability of PAN-PVC hollow fiber cell encapsulation membranes. *J. Membr. Sci.* 195, 51–64.

Cassano, A., Molinari, R., and Drioli, E. (1999). Saving of water and chemicals in tanning industry by membrane processes. *Water Sci. Technol.* 40(4–5), 443–450.

Cassano, A., Molinari, R., Romano, M., and Drioli, E. (2001). Treatment of aqueous effluents of the leather industry by membrane processes—A review. *J. Membr. Sci.* 181(1), 111–126.

Chellam, S. (2000). Effects of nanofiltration on trihalomethane and haloacetic acid precursor removal and speciation in waters containing low concentrations of bromide ion. *Environ. Sci. Technol.* 34(9), 1813–1820.

Chen, G., Xijun, C., Yue, P.-L., and Mi, Y. (1997). Treatment of textile desizing wastewater by pilot scale nanofiltration membrane separation. *J. Membr. Sci.* 127(1), 93–99.

Cho, J. W., Amy, G., and Pellegrino, J. (1999). Membrane filtration of natural organic matter: Initial comparison of rejection and flux decline characteristics with ultrafiltration and nanofiltration membranes. *Water Res.* 33(11), 2517–2526.

Conlon, W. J., and McClellan, S. A. (1989). Membrane softening: Treatment process comes of age. *J. AWWA* 81(11), 47–51.

Culkin, B. (1990). Device and method for filtering a colloidal suspension. US Patent No. 4952317.

Cuperus, P., and Ebert, K. (2002). Non-aqueous applications of NF. In A. I. Schäfer, A. G. Fane, and T. D. Waite (Eds.), *Nanofiltration—Principles and Applications*. Elsevier Advanced Technology, Oxford, pp. 521–536.

Curzons, A. D., Constable, J. C., Mortimer, D. N., and Cunningham, V. L. (2001). So you think your process is green, how do you know? Using principles of sustainability to determine what is green—A corporate perspective. *Green Chem.* 3, 1–6.

De Witte, J. P. (1997). Surface water potabilisation by means of a novel nanofiltration element. *Desalination* 108(1–3), 153–157.

Ducom, G., and Cabassud, C. (1999). Interests and limitations of nano-filtration for the removal of volatile organic compounds in drinking water production. *Desalination* 124(1–3), 115–123.

Ericsson, B., Hallberg, M., and Wachenfeldt, J. (1996). Nanofiltration of highly colored raw water for drinking water production. *Desalination* 108(1–3), 129–141.

Ericsson, B., and Trägårdh, G. (1997). Treatment of surface water rich in humus—Membrane filtration vs. conventional treatment. *Desalination* 108(1–3), 117–128.

Eriksson, P. (1988). Nanofiltration extends the range of membrane filtration. *Environ. Prog.* 7(1), 58–62.

Escobar, I. C., Hong, S., and Randall, A. (2000). Removal of assimilable and biodegradable dissolved organic carbon by reverse osmosis and nanofiltration membranes. *J. Membr. Sci.* 175(1), 1–17.

Everest, W. R., and Malloy, S. (2000). A design/build approach to deep aquifer membrane treatment in Southern California. *Desalination* 132(1–3), 41–45.

Fane, A. G., Macintosh, P., and Leslie, G. (2005). Water reclamation, remediation and cleaner production with nanofiltration. In A. I. Schäfer, A. G. Fane, and T. D. Waite (Eds.), *Nanofiltration—Principles and Applications*. Elsevier, Oxford.

Geens, J., Peeters, K., Van der Bruggen, B., and Vandecasteele, C. (2005). Polymeric nanofiltration of binary water-alcohol mixtures: Influence of feed composition and membrane properties on permeability and rejection. *J. Membr. Sci.* 255(1–2), 255–264.

Hart, H., Hart, D., and Craine, L. E. (1995). *Organic Chemistry: A Short Course*, 9th ed. Houghton Mifflin, Boston.

Hassan, A. M., Al-Sofi, M. A. K., Al-Amoudi, A. S., Jamaluddin, A. T. M., Farooque, A. M., Rowaili, A., Dalvi, A. G. I., Kither, N. M., Mustafa, G. M., and Al-Tisan, I. A. R. (1998). A new approach to thermal seawater desalination processes using nanofiltration membranes (Part 1). *Desalination* 118(1–3), 35–51.

Hassan, A. M., Farooque, A. M., Jamaluddin, A. T. M., Al-Amoudi, A. S., Al-Sofi, M. A., Al-Rubaian, A. F., Kither, N. M., Al-Tisan, I. A. R., and Rowaili, A. (2000). A demonstration plant based on the new NF-SWRO process. *Desalination* 131(1–3), 157–171.

Her, N., Amy, G., and Jarusutthirak, C. (2000). Seasonal variations of nanofiltration (NF) foulants: Identification and control. *Desalination* 132(1–3), 143–160.

Hofman, J. A. M. H., Beerendonk, E. F., Folmer, H. C., and Kruithof, J. C. (1997). Removal of pesticides and other micropollutants with cellulose-acetate, polyamide and ultra-low pressure reverse osmosis membranes. *Desalination* 113(2–3), 209–214.

Jacangelo, J. G., Trussell, R. R., and Watson, M. (1997). Role of membrane technology in drinking water treatment in the United States. *Desalination* 113(2–3), 119–127.

Kartinen, E. O., and Martin, C. J. (1995). An overview of arsenic removal processes. *Desalination* 103(1–2), 79–88.

Kelly, P. M., Horton, B. S., and Burling, H. (1992). Partial demineralisation of whey by nanofiltration. *IDF Special Issue* 1, 130–140.

Kettunen, R., and Keskitalo, P. (2000). Combination of membrane technology and limestone filtration to control drinking water quality. *Desalination* 131, 271–283.

Khalik, A., and Praptowidodo, V. S. (2000). Nanofiltration for drinking water production from deep well water. *Desalination* 132(1–3), 287–292.

Kimura, K., Amy, G., Drewes, J. E., Heberer, T., Kim, T.-U., and Watanabe, Y. (2003). Rejection of organic micropollutants (disinfection by-products, endocrine disrupting compounds and pharmaceutically active compounds) by NF/RO membranes. *J. Membr. Sci.* 227, 113–121.

Kimura, K., Toshima, S., Amy, G., and Watanabe, Y. (2004). Rejection of neutral endocrine disrupting compounds (EDCs) and pharmaceutical active compounds (PhACs) by RO membranes. *J. Membr. Sci.* 245, 71–78.

Kiso, Y., Kitao, T., and Nishimura, K. (1999a). Adsorption properties of cyclic compounds on cellulose acetate. *J. Appl. Polym. Sci.* 71, 1657–1663.

Kiso, Y., Kitao, T., and Nishimura, K. (1999b). Adsorption properties of aromatic compounds on polyethylene as a membrane material. *J. Appl. Polym. Sci.* 74, 1037–1043.

Kiso, Y., Nishimura, Y., Kitao, T., and Nishimura, K. (2000). Rejection properties of non-phenylic pesticides with nanofiltration membranes. *J. Membr. Sci.* 171, 229–237.

Kiso, Y., Sugiura, Y., and Nishimura, K. (2001). Effects of hydrophobicity and molecular size on rejection of aromatic pesticides with nanofiltration membranes. *J. Membr. Sci.* 192, 1–10.

Lee, S., and Lee, C. H. (2000). Effect of operating conditions on $CaSO_4$ scale formation mechanism in nanofiltration for water softening. *Water Res.* 34(15), 3854–3866.

Lee, S., and Lueptow, R. M. (2001). Membrane rejection of nitrogen compounds. *Environ. Sci. Technol.* 35(14), 3008–3018.

Lee, S. Y., Park, G., Park, N., Yoon, Y., and Cho, J. (2005). Evaluation of personal care products (PCP) removal efficiency: NF membranes, granular activated carbon (GAC) and ozone process. AWWA Membrane Technology Conference, Phoenix, AZ.

Leslie, G., Dawes, T. M., Snow, T. S., Mills, W. R., and Macintyre, D. (1999). Meeting the demand for potable water in Orange County in the 21st century: The role of membrane processes. AWWA Membrane Technology Conference, Long Beach, CA, Mar. 3.

Levine, B. B., Madireddi, K., Lazarova, V., Stentrom, M. K., and Suffet, M. (1999). Treatment of trace organic compounds by membrane processes: At the Lake Arrowhead water reuse pilot plant. *Water Sci. Technol.* 40(4–5), 293–301.

Lindau, J., Jönsson, A. S., and Bottino, A. (1998). Flux reduction of ultrafiltration membranes with different cut-off due to adsorption of a low-molecular-weight hydrophobic solute—Correlation between flux decline and pore size. *J. Membr. Sci.* 149(1998), 11–20.

Linde, K., Jonsson, A. S., and Wimmerstedt, R. (1995). Treatment of 3 types of landfill leachate with reverse osmosis. *Desalination* 101(1), 21–30.

Livingston, A., Peeva, L., Han, S., Nair, D., Luthra, S. S., and Freitas dos Santos, L. M. (2003). Membrane separation in green chemical processing—Solvent nanofiltration in liquid phase organic synthesis reactions. *Ann. N. Y. Acad. Sci.* 984, 123–141.

Luthra, S. S., Yang, X., Freitas dos Santos, L. M., White, L. S., and Livingston, A. G. (2001). Phase-transfer catalyst separation and reuse by solvent resistant nanofiltration membranes. *Chem. Commun.* 16, 1468–1469.

Luthtra, S. S., Yang, X., Freitas dos Santos, L. M., White, L. S., and Livingston, A. G. (2002). Homogeneous phase transfer catalyst recovery and re-use using solvent resistant membranes. *J. Membr. Sci.* 201, 65–75.

Machado, D. R., Hasson, D., and Semiat, R. (1999). Effects of solvent properties on permeate flow through nanofiltration membranes. Part I: Investigation of parameters affecting solvent flux. *J. Membr. Sci.* 163, 93–102.

Machado, D. R., Hasson, D., and Semiat, R. (2000). Effects of solvent properties on permeate flow through nanofiltration membranes. Part II: Transport model. *J. Membr. Sci.* 166, 63–69.

Macintosh, P. D., Fane, A. G., and Papazoglou, D. (2003). Recolamation of contaminated ground waters by a multiple membrane process. *Water Sci. Technol.: Water Supply* 3(5/6), 353–360.

Mänttäri, M., Martin, H., Nuortila-Jokinen, J., and Nyström, M. (1999). Using a spiral wound nanofiltration element for the filtration of paper mill effluents: Pretreatment and fouling. *Adv. Environ. Res.* 3(2), U9–214.

Matsuura, T. (2001). Progress in membrane science and technology for seawater desalination—A review. *Desalination* 134, 47–54.

Matsuura, T., and Sourirajan, S. (1972). Reverse osmosis separation of phenols in aqueous-solutions using porous cellulose-acetate membranes. *J. Appl. Polym. Sci.* 16(10) 2531.

Matsuura, T., and Sourirajan, S. (1973a). Reverse osmosis separation of hydrocarbons in aqueous-solutions using porous cellulose-acetate membranes. *J. Appl. Polym. Sci.* 17(12), 3683–3708.

Matsuura, T., and Sourirajan, S. (1973b). Reverse osmosis separation of organic acids in aqueous-solutions using porous cellulose-acetate membranes. *J. Appl. Polym. Sci.* 17(12), 3661–3682.

Melin, T., Meier, J., and Wintgens, T. (2005). Nanofiltration in landfill leachate treatment. In A. I. Schäfer, A. G. Fane, and T. D. Waite (Eds.), *Nanofiltration—Principles and Applications*. Elsevier, Oxford.

Molinari, R., Buonomenna, M. G., Cassano, A., and Drioli, E. (2004). Recovery and recycle of tannins in the leather industry by nanofiltration membranes. *J. Chem. Technol. Biotechnol.* 79(4), 361–368.

Montovay, T., Assenmacher, M., and Frimmel, F. H. (1996). Elimination of pesticides from aqueous solution by nanofiltration. *Magyar Kemiai Folyoirat* 102(5), 241–247.

Mulder, M. (1996). *Basic principles of membrane technology*, 2nd ed. Kluwer Academic, Dordrecht, The Netherlands.

Nair, C., Luthra, S. S., Scarpello, J. T., White, L. S., Freitas dos Santos, L. M., and Livingston, A. G. (2002). Homogeneous catalyst separation and re-use through nanofiltration of organic solvents. *Desalination* 147, 301–306.

Nair, D., Scarpello, J. T., White, L. S., Freitas dos Santos, L. M., Vankelecom, I. F. J., and Livingston, A. G. (2001). Semi-continuous NF-coupled Heck reactions as a new approach to improve productivity of homogeneous catalysts, *Tetrahedr. Lett.* 42, 8219–8222.

Nghiem, L. D., Schäfer, A. I., and Elimelech, M. (2004). Removal of natural hormones by nanofiltration membranes: Measurement, modelling and mechanisms. *Environ. Sci. Technol.* 38, 1888–1896.

Nilson, J. A., and DiGiano, F. A. (1996). Influence of NOM composition on nanofiltration. *J. AWWA* 88(5), 53–66.

Nyström, M., Nuortila-Jokinen, J. M. K., and Mänttäri, M. J. (2005). Nanofiltration in the pulp and paper industry. In A. I. Schäfer, A. G. Fane, and T. D. Waite (Eds.), *Nanofiltration—Principles and Applications*. Elsevier, Oxford.

Oh, J. I., Yamamoto, K., Kitawaki, H., Nakao, S., Sugawara, T., Rahman, M. M., and Rahman, M. H. (2000). Application of low-pressure nanofiltration coupled with a bicycle pump for the treatment of arsenic-contaminated groundwater. *Desalination* 132(1–3), 307–314.

Pervov, A. G., Dudkin, E. V., Sidorenko, O. A., Antipov, V. V., Khakhanov, S. A., and Makarov, R. I. (2000). RO and NF membrane systems for drinking water production and their maintenance techniques. *Desalination* 132(1–3), 315–321.

Peters, T. A. (1998). Purification of landfill leachate with reverse osmosis and nanofiltration. *Desalination* 119(1998), 289–293.

Pitzer, K. S. (1973). Thermodynamics of electrolytes I. Theoretical basis and general equations. *J. Phys. Chem.* 77, 268–277.

Raff, O., and Wilken, R. D. (1999). Removal of dissolved uranium by nanofiltration. *Desalination* 122(2–3), 147–150.

Raman, L. P., Cheryan, M., and Rajagopalan, N. (1996). Solvent recovery and partial deacidification of vegetable oils by membrane technology. *Fett-Lipid* 98(1), 10–14.

Ratanatamskul, C. (1998). Description of behavior in rejection of pollutants in ultra low pressure nanofiltration. *Water Sci. Technol.* 38(4–5), 453–462.

Redondo, J. A., and Lanari, F. (1997). Membrane selection and design considerations for meeting European potable water requirements based on different feedwater conditions. *Desalination* 113(2–3), 309–323.

Robinson, J. P., Tarleton, E. S., Millington, C. R., and Nijmeijer, A. (2004). Solvent flux through dense polymeric nanofiltration membranes. *J. Membr. Sci.* 230, 29–37.

Scarpello, J. T., Nair, D., dos Santos, L. M. F., White, L. S., and Livingston, A. G. (2002a). The separation of homogeneous organometallic catalysts using solvent resistant nanofiltration. *J. Membr. Sci.* 203(1–2), 71–85.

Scarpello, J. T., Nair, D., Freitas dos Santos, L. M., White, L. S., and Livingston, A. G. (2002b). The separation of homogeneous organometallic catalysts using solvent resistant nanofiltration. *J. Membr. Sci.* 203, 71–85.

Schaep, J., Van der Bruggen, B., Uytterhoeven, S., Croux, R., Vandecasteele, C., Wilms, D., Van Houtte, E., and Vanlerberghe, F. (1998). Removal of hardness from groundwater by nanofiltration. *Desalination* 119, 295–302.

Schäfer, A. I., Fane, A. G., and Waite, T. D. (Eds.) (2005). *Nanofiltration—Principles and Applications*. Elsevier, Oxford.

Schneider, B. M. (1994). Membranes—Part 1—Nanofiltration compared to other softening processes. *Ultrapure Water* 65, 65–74.

Seidel, A., Waypa, J. J., and Elimech, M. (2001). Role of charge (Donnan) exclusion in removal of arsenic from water by a negatively charged porous nanofiltration membrane. *Environ. Eng. Sci.* 18(2), 105–113.

Semiat, R. (2000). Desalination: Present and future. *Water Int.* 25(1), 54–65.

Siewert, C., Richter, H., Piorra, A., and Tomandl, G. (2000). Development of ceramic nanofiltration membranes. *Ind. Cer.* 20(1), 31–33.

Sombekke, H. D. M., Voorhoeve, D. K., and Hiemstra, P. (1997). Environmental impact assessment of groundwater treatment with nanofiltration. *Desalination* 113(2–3), 293–296.

Squire, D., Murrer, J., and Holden, P. (2000). Disposal of reverse osmosis membrane concentrate. *Desalination* 132, 47–54.

Stafie, N., Stamatialis, D. F., and Wessling, M. (2003). Insight into the transport of hexane-solute systems through tailor-made composite membranes. *J. Membr. Sci.* 228, 103–116.

Subramanian, R., Nakajima, M., and Kawakatsu, T. (1998). Processing of vegetable oils using polymeric composite membranes. *J. Food Eng.* 38(1), 41–56.

Tang, C., and Chen, V. (2005). Nanofiltration of textile dye effluent. In A. I. Schäfer, A. G. Fane, and T. D., Waite (Eds.), *Nanofiltration Principles and Applications*. Elsevier, Oxford.

Trebouet, D., Schlumpf, J. P., Jaouen, P., Maleriat, J. P., and Quemeneur, F. (1999). Effect of operating conditions on the nanofiltration of landfill leachates: Pilot-scale studies. *Environ. Technol.* 20, 587–596.

Truesdall, J., Mickley, M., and Hamilton, R. (1995). Survey of membrane drinking water plant disposal methods. *Desalination* 102, 93–105.

Tsuru, T., Hironaka, D., Yoshioka, T., and Asaeda, M. (2001a). Titania membranes for liquid phase separation: Effect of surface charge on flux. *Separ. Purif. Technol.* 25(1–3), 307–314.

Tsuru, T., Sudoh, T., Yoshioka, T., and Asaeda, M. (2001b). Nanofiltration in non-aqueous solutions by porous silica-zirconia membranes. *J. Membr. Sci.* 185(2), 253–261.

Tuhkanen, T. A., Kainulainen, T. K., Vartiainen, T. K., and Kalliokoski, P. J. (1994). The effect of preozonization, ozone hydrogen-peroxide treatment, and nanofiltration on the removal of organic-matter from drinking-water. *Ozone-Sci. Eng.* 16(5), 367–383.

Urase, T., Oh, J., and Yamamoto, K. (1998). Effect of pH on rejection of different species of arsenic by nanofiltration. *Desalination* 117(1–3), 11–18.

Van der Bruggen, B., Daems, B., Wilms, D., and Vandecasteele, C. (2001a). Mechanisms of retention and flux decline for the nanofiltration of dye baths from the textile industry. *Separ. Purif. Technol.* 22–23, 519–528.

Van der Bruggen, B., Everaert, K., Wilms, D., and Vandecasteele, C. (2001b). Application of nanofiltration for the removal of pesticides, nitrate and hardness from ground water: Retention properties and economic evaluation. *J. Membr. Sci.* 193(2), 239–248.

Van der Bruggen, B., Lejon, L., and Vandecasteele, C. (2003). Reuse, treatment and discharge of the concentrate of pressure driven membrane processes. *Environ. Sci. Technol.* 37(17), 3733–3738.

Van der Bruggen, B., Schaep, J., Maes, W., Wilms, D., and Vandecasteele, C. (1998). Nanofiltration as a treatment method for the removal of pesticides from ground waters. *Desalination* 117(1–3), 139–147.

Van der Bruggen, B., Schaep, J., Wilms, D., and Vandecasteele, C. (1999). Influence of molecular size, polarity and charge on the retention of organic molecules by nanofiltration. *J. Membr. Sci.* 156, 29–41.

Van der Bruggen, B., and Vandecasteele, C. (2001). Flux decline during nanofiltration of organic components in aqueous solution. *Environ. Sci. Technol.* 35(17), 3535–3540.

Van Gestel, T., Vandecasteele, C., Buekenhoudt, A., Dotremont, C., Luyten, J., Leysen, R., Van der Bruggen, B., and Maes, G. (2002). Alumina and titania multilayer membranes for nanofiltration: Preparation, characterization and chemical stability. *J. Membr. Sci.* 207, 73–89.

Vankelecom, I. F. J., De Smet, K., Gevers, L. E. M., and Jacobs, P. A. (2005). Nanofiltration membrane materials and preparation. In A. I. Schäfer, A. G. Fane, and T. D., Waite (Eds.), *Nanofiltration—Principles and Applications*. Elsevier, Oxford.

Ventresque, C., and Bablon, G. (1997). The integrated nanofiltration system of the Méry-sur-Oise surface treatment plant (37 mgd). *Desalination* 113(2–3), 263–266.

Verberk, J. Q. J. C., Post, J., van der Meer, W. G. J., and van Dijk, J. C. (2002). Direct capillary nanofiltration for ground water and surface water treatment. *Water Sci. Technol. Water Supply* 2(5–6), 277–283.

Visvanathan, C., Marsono, B. D., and Basu, B. (1998). Removal of THMP by nanofiltration: Effects of interference parameters. *Water Res.* 32(12), 3527–3538.

Voigt, I., Stahn, M., Wohner, S., Junghans, A., Rost, J., and Voigt, W. (2001). Integrated cleaning of coloured waste water by ceramic NF membranes. *Separ. Purif. Technol.* 25(1–3), 509–512.

Vrouwenvelder, H. S., van Paassen, J. A. M., Folmer, H. C., Hofman, J. A. M. H., Nederlof, M. M., and van der Kooij, D. (1998). Biofouling of membranes for drinking water production. *Desalination* 118(1–3), 157–166.

Waniek, A., Bodzek, M., and Konieczny, K. (2002). Trihalomethanes removal from water using membrane processes. *Polish J. Environ. Stud.* 11(2), 171–178.

Warczok, J., Ferrando, M., Lopez, F., and Guell, C. (2004). Concentration of apple and pear juices by nanofiltration at low pressures. *J. Food Eng.* 63(1), 63–70.

Waypa, J. J., Elimelech, M., and Hering, J. G. (1997). Arsenic removal by RO and NF membranes. *J. AWWA* 89(10), 102–114.

White, L. S., and Nitsch, A. R. (2000). Solvent recovery from lube oil filtrates with a polyimide membrane. *J. Membr. Sci.* 179, 267–274.

Wiesner, M. R., and Chellam, S. (1999). The promise of membrane technologies. *Environ. Sci. Technol.* 33(17), 360A–366A.

Wittmann, E., Cote, P., Medici, C., Leech, J., and Turner, A. G. (1998). Treatment of a hard borehole water containing low levels of pesticide by nanofiltration. *Desalination* 119(1–3), 347–352.

Xu, P., Drewes, J. E., Bellona, C., Amy, G., Kim, T., Adam, M., and Heberer, T. (2005). Rejection of emerging organic micropollutants in nanofiltration/reverse osmosis membrane applications. *Water Environ. Res.* 77(1), 40–48.

Yacubowicz, H., and Yacubowicz, J. (2005). Nanofiltration: Properties and use. *Filtrat. Separ.* 9, 16–21.

Yeh, H. H., Tseng, I. C., Kao, S. J., Lai, W. L., Chen, J. J., Wang, G. T., and Lin, S. H. (2000). Comparison of the finished water quality among an integrated membrane process, conventional and other advanced treatment processes. *Desalination* 131(1–3), 237–244.

Zaidi, A., Buisson, H., Sourirajan, S., and Wood, H. (1992). Ultra- and nanofiltration in advanced effluent treatment schemes for pollution control in the pulp and paper industry. *Water Sci. Technol.* 25(19), 263–276.

Zhang, S. Q., Kumar, A., and Kutowy, O. (2000). Membrane based separation scheme for processing sweeteners from stevia leaves. *Food Res. Int.* 33, 617–620.

CHAPTER 12

Membrane Distillation

MOHAMED KHAYET

Department of Applied Physics I, Faculty of Physics, University Complutense of Madrid, Madrid, Spain

12.1 INTRODUCTION TO MEMBRANE DISTILLATION

12.1.1 Brief Introduction and Membrane Distillation Configurations

Membrane processes have become competitive with the conventional separation methods in a wide variety of applications such as distillation. Different membrane separation processes have been developed during the past half century, and new membrane applications are constantly emerging from industries or from academic and government laboratories. Most of membrane transport processes are isothermal processes that depend on either hydrostatic pressure, concentration, or electrical potential as a driving force. The isothermal membrane processes include reverse osmosis, nanofiltration, ultrafiltration, microfiltration, electrodialysis, and membrane gas separation. Membrane distillation (MD) is an emerging nonisothermal membrane technology for separations that are traditionally accomplished by conventional separation processes such as distillation and pressure-driven membrane processes. MD is a process mainly suited for applications in which water is the major component present in the feed solution to be treated and refers to a thermally driven transport of vapor through nonwetted porous hydrophobic membranes, the driving force being the partial pressure difference between each side of the membrane pores.

The liquid feed to be treated by MD is maintained in direct contact with one side of the membrane without penetrating its dry pores unless a transmembrane pressure higher than the membrane liquid entry pressure [i.e., liquid entry pressure or breakthrough pressure (LEP)] is applied. In fact, the hydrophobic nature of the membrane prevents liquid solutions from entering its pores due to the surface tension forces. As a result, liquid–vapor interfaces are formed at the entrances of the membrane pores. Various MD modes differing in the technology applied to establish the driving force can be used. The differences between them are localized only in the permeate side, as illustrated in Figure 12.1.

Advanced Membrane Technology and Applications. Edited by Norman N. Li, Anthony G. Fane, W. S. Winston Ho, and T. Matsuura
Copyright © 2008 John Wiley & Sons, Inc.

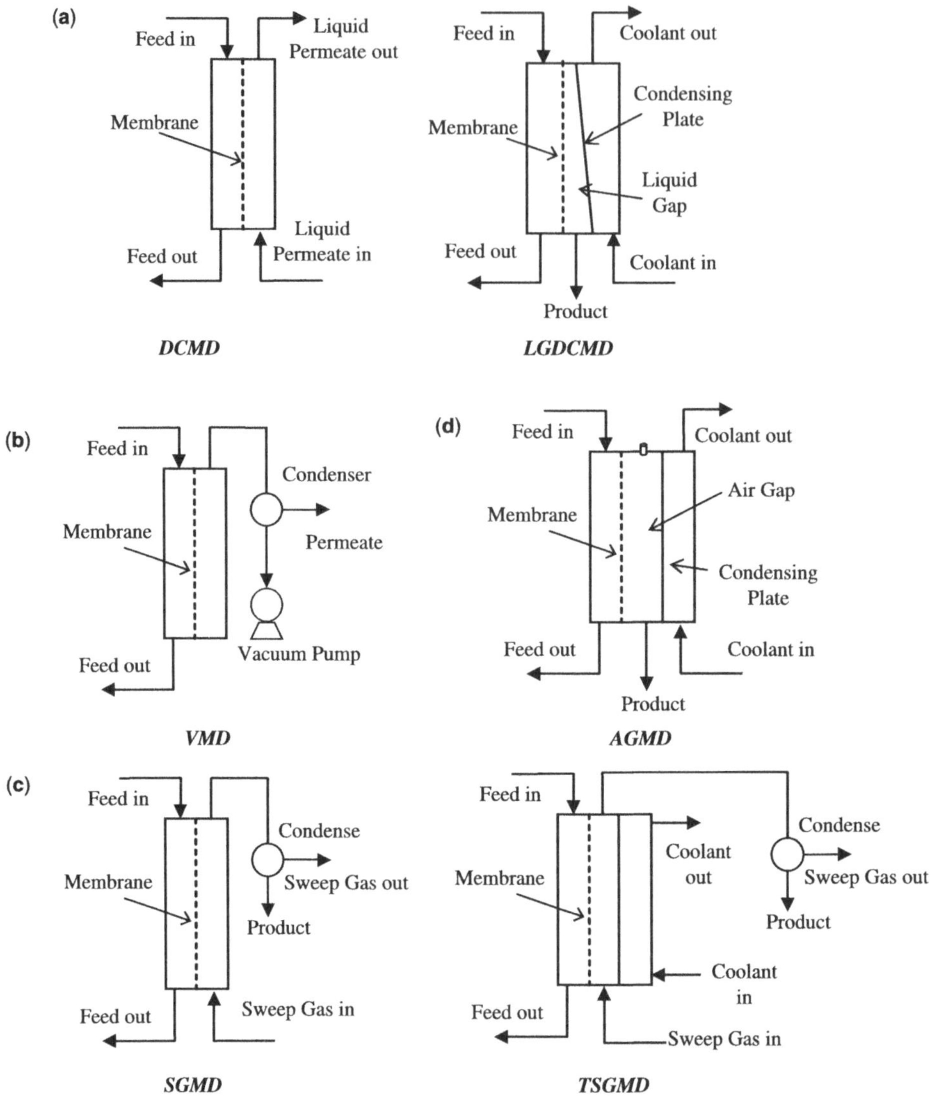

Figure 12.1 Types of MD process: (a) DCMD and DCMD with liquid gap, (b) VMD, (c) SGMD and thermostatic SGMD, and (d) AGMD.

The MD driving force may be maintained with any one of the four following possibilities applied in the permeate side:

- An aqueous solution colder than the feed solution is maintained in direct contact with the permeate side of the membrane, giving rise to the configuration known as direct-contact membrane distillation (DCMD). Both the feed and permeate aqueous solutions are circulated tangentially to the membrane surfaces by means of circulating pumps or are stirred inside the membrane cell by means of a magnetic stirrer. In this case the transmembrane temperature difference induces a vapor pressure difference. Consequently, volatile molecules evaporate at the hot liquid–vapor interface, cross the membrane pores in vapor phase, and condense in the cold liquid–vapor interface inside the membrane module. A DCMD with liquid gap is an other DCMD variant in which a stagnant cold liquid, frequently distilled water, is kept in direct contact with the permeate side of the membrane (Fig. 12.1).

- Vacuum is applied in the permeate side of the membrane module by means of a vacuum pump. The applied vacuum pressure is lower than the saturation pressure of volatile molecules to be separated from the feed solution. In this case, condensation takes place outside of the membrane module. This MD configuration is termed vacuum membrane distillation (VMD).
- A stagnant air gap is interposed between the membrane and a condensation surface. In this case, the evaporated volatile molecules cross both the membrane pores and the air gap to finally condense over a cold surface inside the membrane module. This MD configuration is called air gap membrane distillation (AGMD).
- A cold inert gas sweeps the permeate side of the membrane carrying the vapor molecules, and condensation takes place outside the membrane module. This type of configuration is termed sweeping gas membrane distillation (SGMD). In this configuration, due to the heat transferred from the feed side through the membrane, the sweeping gas temperature in the permeate side increases considerably along the membrane module length. A SGMD variant, termed thermostatic sweeping gas membrane distillation (TSGMD), has been proposed recently. In this mode of SGMD the increase of the gas temperature is minimized by using a cold wall in the permeate side.

Each of the above MD configurations has its advantages and disadvantages for a given application. The involved simultaneous heat and mass transfer phenomena through the membrane, the different MD configurations, and the various applications in desalination, environmental/waste cleanup, water reuse, food, medical, and the like make MD attractive within the academic community as a kind of didactic application. Furthermore, the lower operating temperatures than the conventional distillation (feed solutions having temperatures much lower than its boiling point can be used in MD), the lower operating hydrostatic pressures than the pressure-driven membrane processes (i.e., reverse osmosis, nanofiltration, ultrafiltration, microfiltration), the less demanding membrane mechanical properties, and the high rejection factor achieved makes MD more attractive than other popular separation processes. Additionally, the possibility of using waste heat and/or alternative energy sources, such as solar and geothermal energy, enable an MD technique to cooperate with other processes in integrated systems, making it a more promising separation technique.

It must be mentioned here that various authors involved in MD investigations often abbreviate DCMD to MD. However, as stated above, there are various MD configurations. To avoid misconceptions and complications, it is better to use the adequate term following the note given below.

12.1.2 Nomenclature in MD

The term MD comes from the similarity of the MD to conventional distillation as both processes are based on the vapor–liquid equilibrium for separation, and both processes require heat to be supplied to the feed solution in order to achieve the required latent heat of vaporization. Before the Workshop on Membrane Distillation held in Rome on

May 5, 1986, various terms had been used to identify what is nowadays known as MD, including transmembrane distillation, thermo-pervaporation, pervaporation, membrane evaporation, and capillary distillation. The terminology for MD was first discussed by the committee formed during the 1986 workshop by six members: V. Calabro (University of Calabria, Calabria, Italy), A.C.M. Franken (Twente University, Enschede, Netherlands), S. Kimura (University of Tokyo, Tokyo, Japan), S. Ripperger (Enka Membrana, Wuppertal, Germany), G. Sarti (University of Bologna, Bologna, Italy), and R. Schofield (University of New South Wales, Kensington, Australia). Terms, definitions, and symbols related to MD have been discussed, standardized, and lately reported in the literature (Franken and Ripperger, 1988; Smolders and Franken, 1989). Mainly, MD should be applied for nonisothermal membrane operations in which the driving force is the partial pressure gradient across the membrane that has the following characteristics: (1) porous, (2) not wetted by the process liquids, (3) does not alter the vapor–liquid equilibrium of the involved species, (4) does not permit condensation to occur inside its pores, and (5) is maintained in direct contact at least with the hot feed liquid solution to be treated. Following the 1996 International Union of Pure and Applied Chemistry (IUPAC) recommendations (Koros et al., 1996), the 21st term *membrane distillation* was defined as a "distillation process in which the liquid and gas phases are separated by a porous membrane, the pores of which are not wetted by the liquid phase."

12.1.3 A Historical Survey of MD

Membrane distillation has been used for more than 40 years, but for the present it is still necessary to be developed for its industrial implementation. The first patent was filed by Bodell on June 3, 1963 (Bodell, 1963), and the first MD article was published 4 years later by Findley in the journal *Industrial & Engineering Chemistry Process Design Development* (Findley, 1967). Within the 16 cited references throughout the article, Findley did not mention the first patent made by Bodell (1963). However, the well-known Loeb–Sourirajan's study (Loeb and Sourirajan, 1960) was cited. Findley used the DCMD configuration using various types of coated and uncoated membrane materials such as paper hot cup, gum wood, aluminum foil, cellophane, glass fibers, paper plate, diatomaceous earth mat, and nylon. Silicon and Teflon have been used as coating materials to achieve the required membrane hydrophobicity. Some of the proposed membranes discussed by Findley (1967) appeared to have internal condensation or adsorption of moisture between layers or inside surfaces. Based on experimental observations, Findley outlined, throughout his article, the most suitable membrane characteristics needed for an MD membrane and stated in the abstract the following: "Calculations indicate possible economical performance, especially at high temperatures, if high temperature, long life and low cost membrane are obtainable." Findley also suggested the possibility of using infinite-stage flash evaporation through porous membranes. During the same year, 1967, a U.S. patent filed by Weyl on May 14, 1964, appeared (Weyl, 1967). This patent claimed an improved method and apparatus for the recovery of demineralized water from saline waters using also DCMD. It was stated that the two bodies of water may be stationary or moving, that is, passing with respect to the membrane, and the process may be effected in a single stage or may be multistaged. The membrane used was poly-(tetrafluoroethylene) (PTFE) membrane having a thickness of 3175 μm, an average pore size of 9 μm, and a porosity of 42%. In this patent, the use of other suitable hydrophobic membranes made of polyethylene (PE), polypropylene (PP), and polyvinyl chloride (PVC) was suggested. Weyl also stated that membranes may be constructed of nonhydrophobic materials coated with a hydrophobic substance, for example, by liquid or vapor impregnation (Weyl, 1967). In this patent, an alternate geometrical form for multistage operation was presented by having the membrane coiled up into a cylinder giving rise to the actual known DCMD with liquid gap in a spiral-wound module. A year later, a second U.S. patent by Bodell (1968) was made partly as a continuation of his first patent (Bodell, 1963). Bodell described a system and a method to convert impotable aqueous solutions to a potable water using a parallel array of tubular silicone membranes having 0.30-mm inner diameter and 0.64-mm outer diameter. No membrane characteristics such as pore size and porosity were presented. Air was circulated through the lumen side of the tubular membranes, and condensation was carried out in an external condenser, giving rise to the actual known sweeping gas membrane distillation (SGMD) configuration. The important object of the invention was to provide novel apparatus and methods for desalting seawater in an economical manner. The improved apparatus was also provided for extracting potable water from brine, sewage, urine, wastewater, bacteria-containing water, and other non-potable water sources. Bodell recommended the water vapor pressure in the air side of the SGMD system be at least 4 kPa below that of the aqueous medium. Moreover, Bodell suggested, for the first time, an alternative means of providing low water vapor pressure in the tubes by applying vacuum, leading to the vacuum membrane distillation (VMD) configuration (Bodell, 1963, 1968). A second MD article was published by Findley and co-authors (Findley et al., 1969) without mentioning any of the previously cited U.S. patents (Bodell, 1963, 1968; Weyl, 1967). The study concerns heat and mass transfer of water vapor from a hot salt solution through a hydrophobic porous membrane to a cooled water condensate. Their experimental studies indicated that the major factor influencing the rates of transfer was the diffusion through the stagnant gas (i.e., air) in the membrane pores. The first theoretical calculations have been reported taking into account the membrane thermal conductivity and the film heat transfer coefficients (Findley, 1967). An empirical correction related to the possible internal condensation and diffusion along the surfaces has been considered to perform their calculations (Findley et al., 1969). In Europe, a seawater desalination SGMD process using dry air was also proposed by van Haute and Henderyckx at the second European Symposium on Fresh Water from the Sea held in Athens (Greece) in May of 1967 (van Haute and Henderyckx, 1967; Henderyckx, 1967). The authors stated that the proposed apparatus can utilize waste hot water and should be possible to use it for solar distillation; but this has not been further developed.

After this short period of time, interest in MD process faded quickly, loosing its brightness due partly to the observed lower MD production compared to that of the reverse osmosis process. Rodgers (1972, 1974), in his successive patents related to distillation, presented a system and method of desalination using a stack of flat-sheet membranes separated by nonpermeable corrugated heat transfer films and working under DCMD configuration. A temperature gradient is applied over the membrane stack, and the latent heat passing from feed to distillate is recovered by heat transfer to a lower temperature feed. Thus, the latent heat of vaporization may be used several times, as with multiple effect evaporation. The feed liquid supplied was subjected to treatment including heating and deaeration. The main object was to provide an improved economical desalination system and method of desalination. It was disclosed in the aforementioned patents that the distillation unit comprises a multiplicity of sheetlike elements all substantially rectangular, having the same dimensions and arranged with their edges in alignment so that the distillation unit takes the form of a relatively thin parallelepipedon. It was also indicated that the suitable materials for the membranes are those that permit the formation of microporous membranes having high porosity, that is 70–80%, uniform pore size distribution, and must be either poorly wettable or nonwettable (i.e., hydrophobic) by the used liquids or can be treated to render them nonwettable. The cited polymers were polycarbonates, polyesters, polyethylene, polypropylene, and the halogenated polyethylenes, particularly the flurocarbons. Particular mention was made to polyvinylidene fluoride (PVDF) as a preferred membrane material, and the so-called solvent–nonsolvent casting process was the preferred method of forming the membranes. The use of cellulose nitrate, cellulose acetate, and cellulose triacetate microscopic porous filter media coated with a silicone water repellant to provide a nonwetting porous membrane was also mentioned.

The MD process garnered much interest within the academic community in the early of the 1980s when novel membranes with better characteristics and modules became available (Esato and Eiseman, 1975; Cheng, 1981; Cheng and Wiersma, 1982, 1983a, 1983b; Gore, 1982; Carlsson, 1983; Enka, 1984; Schneider and van Gassel, 1984; Andersson et al., 1985; van Gassel and Schneider, 1986). Gore-Tex membrane (i.e., expanded PTFE membrane 50 μm thick with a 0.5-μm pore size) was used first by Esato and Eiseman (1975) as a biologically inert membrane oxygenator and later proposed by Gore & Associated Co. under the name Gore-Tex Membrane Distillation for application in MD in a spiral-type module using the AGMD or the liquid gap DCMD configuration (Fig. 12.1) (Gore, 1982). The proposed Gore-Tex membranes are made of PTFE having a thickness as thin as 25 μm, a porosity up to 80%, and a pore size of 0.2–0.45 μm. Other types of MD membranes, methods, and apparatus have been proposed by Cheng and Wiersma in a series of patents (Cheng, 1981; Cheng and Wiersma, 1982, 1983a, 1983b). The object of the first patent (Cheng, 1981), filed by Cheng on February 14, 1979, was to provide an improved thermal membrane distillation process with continuous distillate production over a prolonged period of time. A desalination system with three cell stages was presented. Multiple-layered (i.e., composite) membranes have been proposed comprising a thin hydrophobic microporous layer or membrane and a thin hydrophilic layer or membrane. The hydrophilic layer was maintained adjacent to the feed (i.e., saltwater), whereas the hydrophobic layer was kept adjacent to the distillate (i.e., freshwater). In this case, the evaporation and condensation phenomena take place within the micropores of the hydrophobic layer while the hydrophilic layer prevents intrusion of feed into the pores of the hydrophobic layer. It was reported that, in the case of saltwater distillation, generally higher distillate production rates have been observed in composite membranes with the smaller pore sizes in the hydrophilic layer than in the hydrophobic layer. The best results have been obtained with the hydrophobic layer having a mean pore size smaller than 0.5 μm. It was also reported that the hydrophilic layer can be nonporous. The proposed hydrophobic materials for the composite membrane include PTFE and PVDF, whereas the proposed hydrophilic materials included cellulose acetate, cellulose nitrate, mixed esters of cellulose, and polysulfone. In fact, the proposed composite porous membrane in Cheng (1981) was formed by clamping the hydrophobic/hydrophilic layers closely together to form a cell with a suitable support backing to maintain the integrity of the composite membrane.

The following patents (Cheng and Wiersma, 1982, 1983a, 1983b) filed by Cheng and Wiersma claimed the use of composite membranes having a thin microporous hydrophobic layer and one or two thin hydrophilic layers with the hydrophobic layer sandwiched between the two hydrophilic layers. It was stated that the two hydrophilic layers may be of different materials (cellulose acetate, mixed esters of cellulose, polysulfone, and polyallylamine), and the composite membrane could be formed by coating the hydrophilic layers on the hydrophobic layer. It was found that the freshwater production rate for a distilland bulk temperature of 62.8°C and distillate temperature of 56.7°C was $75.2\ kg/d \cdot m^2$, and for the same distilland and distillate temperature conditions a composite membrane of similar structure, except with nonhydrophilic layer on the distillate side of the hydrophobic membrane, yielded a freshwater production rate of $51.3\ kg/d \cdot m^2$. In other words, it was reported that the addition of the hydrophilic layer of the distillate side of the hydrophobic membrane increased the freshwater production rate by almost 50%. The fourth patent (Cheng and Wiersma, 1983b), filed on March 4, 1982, proposed an improved apparatus and method for MD using a composite membrane comprising a microporous hydrophobic layer having deposited thereon an essentially nonporous hydrophilic coating. The hydrophobic layer of the membrane had either asymmetrical or symmetrical shaped micropores. Fluoro-substituted vinyl polymers, which are suitably hydrophobic, were proposed as ideal materials for the microporous hydrophobic layer of the composite membrane. The coating material and method was selected, taking into account the adequate adhesion to the hydrophobic substrate, its resistance to both mechanical abrasion and chemical damage from the distilland, its ability to be coatable as a thin continuous layer on the surface of the porous substrate, and to allow certain liquids to pass through to the hydrophobic layer of the membrane. Examples were plasma-polymerized allylamine, dehydrated polyvinyl alcohol, and polyacrylic acid. The latter coating is deposited on the surface of the hydrophobic layer, which has been modified by electrical discharge to accept a graft of the acrylic acid monomer to form the polyacrylic acid layer. Generally, the coating was allowed to become about 0.6 μm thick. The proposed system can be operated at temperatures as low as 15.6°C to as high as 148.9°C if the system is pressurized. Preferably, it was operated between 48.9 and 100°C for a system not requiring pressurization.

At the same time, the Swedish National Development Co. (Svenska Utvecklings AB) developed plate and frame membrane modules applying the AGMD configuration (Carlsson, 1983; Andersson et al., 1985). The German company Enka AG presented polypropylene hollow-fiber membranes in tubular modules at the Europe–Japan Joint Congress on Membranes and Membrane Processes, held at Stresa (Italy) in 1984 (Enka, 1984). Their experiments using the DCMD process with heat recovery took a more academic orientation by publication of their results in journals (Schneider and van Gassel, 1984). In the same congress, other papers on MD have also been published (Sarti and Gostoli, 1986; van Gassel and Schneider, 1986). More MD papers were presented at the Second World Congress on Desalination and Water Reuse in 1985, held in Bermuda. This renewed interest is a result of the development of various porous hydrophobic membranes used in different MD configurations (Drioli and Wu, 1985; Sarti et al., 1985; Jonsson et al., 1985; Hanbury and Hodgkiess, 1985; Kimura and Nakao, 1987). Since then, various studies have been carried out having a more academic interest rather than industrial. Most of those studies are published in journals such as *Journal of Membrane Science* and *Desalination*, as will be shown throughout this chapter.

Being an attractive separation process, MD has been the subject of worldwide academic studies by many experimentalist and theoreticians. Unfortunately, from the commercial standpoint, MD has gained only little acceptance and as yet not implemented in industry. The major barriers include MD membrane and module design, membrane pore wetting, low permeate flow rate, and flux decay as well as uncertain energy and economic costs. Recently, interest in MD has increased significantly, as can be seen in Figure 12.2 presenting the number of published papers in journals per year in the MD field.

It must be stated that the number of MD papers referenced in a (1997) MD review by Lawson and Lloyd (1997) is below 53. However, the number of MD articles

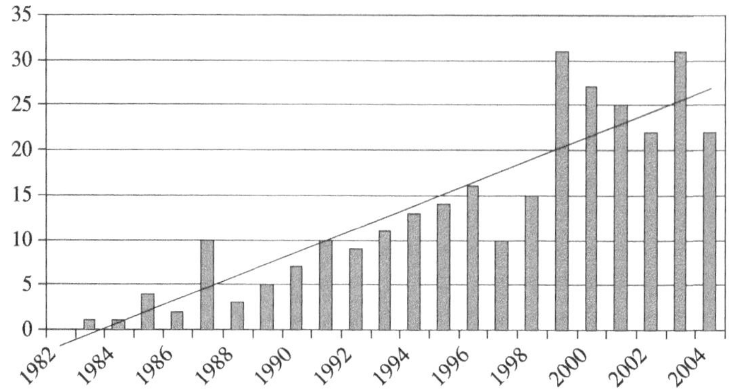

Figure 12.2 Growth of research activity of MD represented as a plot of number of articles published in refereed journals per year.

published in journals from 1997 to date is more than 200. In fact, DCMD is the most studied MD configuration, as can be observed in Figure 12.3, although the heat loss by conduction through the membrane matrix, which is considered heat loss as is not utilized to evaporate components in DCMD, is higher than in the other MD configurations. More than 60% of the MD studies are focused in DCMD. This is attributed mainly to the fact that the condensation step is carried out inside the membrane module, leading in this way to a simple operation mode. In contrast, SGMD and VMD configurations require external condensers to collect the permeate, which complicates the system design and increases its cost.

Most of the publications listed in Figures 12.2 and 12.3 are concerned with theoretical models of MD and experimental studies on the effects of the operating conditions. More than 56% of the MD publications dealt with theoretical models (i.e., 50.7% for DCMD, 62.1% for AGMD, 64.5% for VMD, and 80% for SGMD), whereas less than 8% focused on the preparation of membranes for MD. Few authors have considered the possibility of manufacturing new membranes and membrane module designs specifically for MD

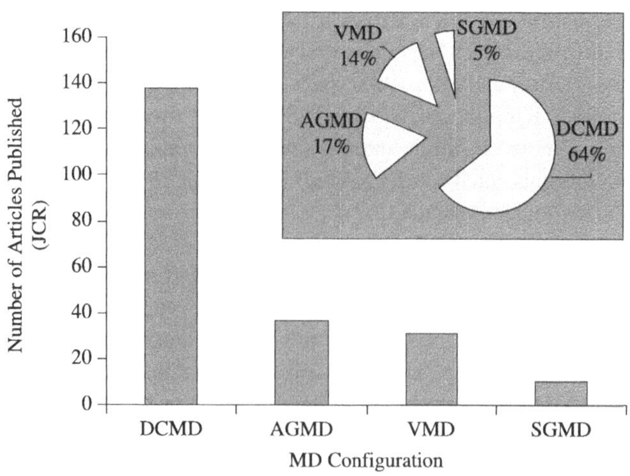

Figure 12.3 Number of articles published in refereed journals for each MD configuration.

applications (Schneider et al., 1988; Ohta et al., 1991; Kong et al., 1992; Wu et al., 1992; Fujii et al., 1992a, 1992b; Ortiz de Zárate et al., 1995; Tomaszewska, 1996; Khayet and Matsuura, 2001, 2003a; Wang et al., 2002; Belleville et al., 2002; Brodard et al., 2003; Li et al., 2003; Feng et al., 2004a, 2004b; Larbot et al., 2004; Xu et al., 2004; Li and Sirkar, 2004, 2005; Khayet et al., 2005a, 2005b; Peng et al., 2005). As a matter of fact, commercial microporous hydrophobic membranes available in capillary or flat-sheet forms have been used in MD experiments, although these membranes were prepared initially for microfiltration purposes. This chapter will also focus on MD membrane engineering and highlight the areas within the MD field that are either usually or rarely studied, helping new researchers in the field of MD to quickly be updated to avoid repetition of already existing studies.

12.2 MEMBRANE DISTILLATION MEMBRANES AND MODULES

12.2.1 Membranes Used in MD and MD Membrane Design

As stated previously, the membranes used in MD must be porous and hydrophobic. It can be a single hydrophobic layer (i.e., conventional and most used membrane), a composite porous bilayer hydrophobic/hydrophilic membrane or a composite trilayer hydrophilic/hydrophobic/hydrophilic or hydrophobic/hydrophilic/hydrophobic porous membranes. Both supported and unsupported membranes can be used in this process. The pore size of the membranes frequently used in MD lies between 10 nm and 1 μm.

It is agreed that in MD the membrane itself acts only as a barrier to hold the liquid–vapor interface at the entrance of the membrane pores and is not involved in the transport phenomenon (Lawson and Lloyd, 1997). Both water and the other volatile compounds present in the feed solution are transferred through the membrane pores according to the vapor–liquid equilibrium (VLE) principle, and the separation performance is predominantly determined by the VLE. In other words, it is believed that the membrane is not necessary to be selective as required in other membrane processes such as pervaporation (PV). The solution–diffusion flow through the nonporous portion of the membrane is not considered in spite of the possible existence of a high "affinity" (i.e., close solubility parameters) between the species to be separated and the membrane material. Franken et al. (1987a) stated that only selectivity lower than the vapor–liquid equilibrium can be expected because the membrane itself has no selectivity to any components in the solution. However, the effect of the membrane and the properties of polymers used for preparation of fine porous hollow-fiber membrane on the DCMD selectivity have been studied more than 15 years ago by Fujii et al. (1992a, 1992b). When using ethanol–water mixtures, it was found that the selectivity for ethanol varied according to the properties of the polymers used and the membranes as well as the DCMD operation conditions. Recently, Khayet and Matsuura (2004) considered the solution–diffusion contribution when using various PVDF flat-sheet membranes with different pore sizes and porosities in VMD modes. Both experiments and modeling were performed for PV and VMD processes. Rivier et al. (2002) and García-Payo et al. (2002) indicated that selectivity in MD depends on both the differences in volatility and diffusion rates of the components across the membrane and gas gap. They explained that the separation performance in MD is not only based on VLE of the involved components in the feed mixture but also on the thermodynamics (operating temperature) and kinetic effects (diffusion of the components through the membrane and the gas gap in SGMD and AGMD, for example). Generally, we must point out that this aspect is not

TABLE 12.1 Commercial Membranes Commonly Used in MD (Membrane Thickness, δ; Mean Pore Size, d_p; Porosity, ε; Liquid Entry Pressure of Water, LEP_w).

Membrane Type	Membrane Trade Name	Manufacturer	Material	δ (μm)	d_p (μm)	ε (%)	LEP_w (kPa)
Flat-sheet membranes	TF200	Gelman	PTFE/PP[a]	178	0.20	80	282
	TF450				0.45		138
	TF1000				1.00		48
	GVHP	Millipore	PVDF[b]	110	0.22	75	204
	HVHP			140	0.45		105[i]
	FGLP		PTFE/PE[a]	130	0.20	70	280
	FHLP			130	0.5	70	124
	Gore		PTFE	64	0.2	90	368[j]
			PTFE	77	0.45	89	288[j]
			PTFE/PP[a]	184	0.2	44	463[j]
	Enka		PP	100	0.1	75	—
				140	0.2		
	3MA	3M Corporation	0.29	91	0.29[k]	66	—
	3MB		0.40	81	0.40[k]	76	
	3MC		0.51	76	0.51[k]	79	
	3MD		0.58	86	0.58[k]	80	
	3ME		0.73	79	0.73[k]	85	
	G-4.0-6-7[c]	GoreTex Sep GmbH	PTFE	100	0.20	80	—

Capillary membranes	Accurel S6/2 MD020CP2N[d]	AkzoNobel Microdyn	PP	450	0.2	70	140
	Accurel BFMF 06-30-33[e]	Enka A.G. Euro-Sep		200	0.2		
	Sartocon-Mini SM 3031 750701W[f]	Sartorius	Polyolefine		0.22	—	—
	POREFLON[g]	Sumitomo Electric	PTFE	550	0.8	62	
	TA001[h]	Gore-tex	PTFE	400	2[l]	50	

[a]Flat-sheet poly(tetrafluoroethylene) (PTFE), membranes supported by polypropylene (PP) or polyethylene (PE).
[b]Flat-sheet polyvinylidene fluoride membranes.
[c]Spiral-wound module, SEP Gesellschaft für Technische Studien, Entwicklung, Planmung GmbH; filtration area: 4 m^2.
[d]Shell-and-tube capillary membrane module: filtration area: 0.1 m^2, inner capillary diameter: 1.8 mm; length of capillaries: 470 mm.
[e]Shell-and-tube capillary membrane module: filtration area: 0.3 m^2, inner capillary diameter: 0.33 mm, length of capillaries: 200 mm.
[f]Plate-and-frame module, dimensions: 138/117/7 mm, filtration area: 0.1 m^2.
[g]PTFE hollow fiber, inner/outer diameters: 0.9/2 mm (Lee and Hong, 2001).
[h] PTFE hollow fiber, inner diameters: 1 mm (Calibo et al., 1987).
[i]Measured value (Khayet et al., 2004a).
[j]Measured value (Izquierdo-Gil et al., 1999b).
[k]Maximum pore size (Lawson and Lloyd, 1996b).
[l]Maximum pore size.

fully investigated yet, and more experiments should be done using membranes with different pore sizes and different configurations and materials.

As stated earlier, the main requirements for the MD process are that the membrane must not be wetted by the aqueous solutions in contact with it, and only vapor and noncondensable gases are present within its pores. As water is the major component used, the membrane must be hydrophobic and therefore have to be made by polymers or inorganic materials with low surface energy. Capillary or flat-sheet commercial microporous hydrophobic membranes, made of PP, PVDF, and PTFE have been used in MD experiments despite the fact that these membranes were prepared primarily for microfiltration purposes. The choice of membrane for MD applications is a compromise between low thermal conductivity achieved by thicker membranes and high flux achieved by thin membrane, large pore size, and porosity. Table 12.1 summarizes the commercial membranes commonly used in MD studies together with their principal characteristics as specified by the manufacturers. Among the membranes studied are commercial module units, formed in shell-and-tube, plate-and-frame, and spiral-wound configurations that were tested in laboratory and pilot plant experiments.

Hydrophobic porous membranes can be prepared by different techniques depending on the properties of the materials to be used. The useful materials should be selected according to criteria that include compatibility with the liquids involved, cost, ease of fabrication and assembly, useful operating temperatures, and thermal conductivity. Microporous membranes can be made by sintering, stretching, phase inversion, or thermally induced phase separation (TIPS) (Mulder, 1992; Matsuura, 1994; Pinnau and Freeman, 2000; Lloyd et al., 1990). For example, PVDF membranes are made by the phase inversion method; PP membranes are made generally by stretching and thermal phase separation; PTFE membranes are made by a sintering or stretching process. Among these techniques the phase inversion is the most used one. As mentioned in the previous section, hydrophilic membrane surface modification and possible applications in MD have also been attempted. Various physical and chemical techniques were carried out for membrane surface modification including coating, grafting, plasma polymerization, and the like (Mulder, 1992; Matsuura, 1994; Pinnau and Freeman, 2000; Lloyd et al., 1990).

Compared to other membrane separation processes such as pervaporation, reverse osmosis, and gas separation, only a few authors have considered the possibility of designing and manufacturing new membranes for MD processes (Khayet et al., 2006a). Moreover, looking into the extended published articles in the MD field, only a few authors have considered the possibility of designing and manufacturing membranes for MD processes. The MD membranes have to meet several requirements simultaneously:

- Good thermal stability: long-term stability is required for MD membranes to temperatures as high as 100°C.
- Excellent chemical resistance to various feed solutions, acids, and bases especially if the membrane has to be cleaned.
- High liquid entry pressure (LEP): This is the minimum transmembrane hydrostatic pressure that must be applied before liquid solutions overcome the hydrophobic forces of the membrane and penetrate into the membrane pores. This pressure is characteristic of each membrane and permits to prevent wetting of the membrane pores during MD experiments. High LEP may be achieved using a membrane material with high hydrophobicity (i.e., large water contact angle) and a small maximum pore

size at its surface. However, a small maximum pore size parallels small mean pore size, leading to low membrane permeability.

- High permeability: The MD flux will *increase* with an increase in the membrane pore size and porosity and with a decrease of the membrane thickness and pore tortuosity. In other words, to obtain a high MD permeability, the surface layer that governs the membrane transport must be as thin as possible so that the vapor transport distance across the membrane is as short as possible and its surface porosity as well as pore size must be as large as possible.
- Narrow pore size distribution: To avoid wetting of the maximum pore sizes or a fraction of the number of pores, leading to a decrease of both the membrane surface area for MD and the membrane rejection factor.
- Low thermal conductivity: Heat transferred by conduction through both the pores and the membrane matrix from the feed to the permeate side is heat loss in MD. This conductive heat loss is greater for thinner membranes. Various possibilities may be applied to diminish the conductive heat loss by using: (1) membrane material with low thermal conductivity. However, most of the hydrophobic polymers have similar conductive heat transfer coefficients, at least with the same order of magnitude. The thermal conductivity of commercial membranes lies between 0.04 and 0.06 $W/m \cdot K$ (Schofield et al., 1987). (2) Membranes with high porosity, since the conductive heat transfer coefficients of the gases entrapped within the membrane pores are an order of magnitude smaller than that of the membrane matrix. This possibility is parallel to the need of high MD permeability as the available surface area of evaporation is enhanced. The thermal conductivity decreases with the porosity. (3) Thicker membranes are used, which contradicts the achievement of high permeability. (4) Use composite porous membranes with hydrophobic and hydrophilic layers. In this case, the hydrophobic layer responsible for the mass transfer must be very thin, whereas thick hydrophilic layer can be used to prevent heat loss through the whole membrane. This last option has been selected by Khayet and Matsuura (2003a) and Khayet et al. (2005a, 2005b) for the preparation of membranes for DCMD. It seems to be a relatively simple solution that fulfills all the above conditions for achieving high permeability and low thermal conductivity.

Furthermore, supported membranes can be used in MD. However, the membrane support must be chosen so that it does not significantly increase the heat and mass transfer resistances, but it must be strong enough to prevent deflection or rupture of the membrane.

Other than the membranes mentioned in the previous section either patented, reported in journals, or communicated in meetings and congresses (Bodell, 1963; Findley, 1967; Weyl, 1967; Rodgers, 1972, 1974; Cheng, 1981; Gore, 1982; Cheng and Wiersma, 1982, 1983a, 1983b; Enka, 1984), very few other studies have been performed on the preparation and modification of membranes for the MD process.

Ohta et al. (1991) looked at using a partially hydrophilic dense fluorocarbon composite membrane and tested it for seawater desalination. The authors used the term MD for both porous and dense membranes. DCMD configuration was used and the obtained fluxes ($<6\ kg/m^2 \cdot h$) were of similar magnitude to those obtained with porous hydrophobic membranes. The effects of the DCMD operation parameters were similar to those observed using a single porous hydrophobic layer. It was also found that the permeability and thermal efficiency when using fluorocarbon membrane were superior than those of silicon membrane. However, no details were given on the transport mechanism through this type of membrane.

Kong et al. (1992) employed a hydrophilic microporous cellulose nitrate membrane surface-modified via plasma polymerization using two monomer systems: octafluorocyclobutane (OFCB) and vinyltrimethylsilicon/carbon tetrafluoride (VTMS/CF_4). A trilayer membrane (a hydrophilic layer sandwiched between two hydrophobic layers) has been prepared and tested in a DCMD system of a 0.3–0.5 M NaCl feed solution. The effects of various polymerization conditions on the MD performance and structure of the membrane have been investigated. The trilayer membrane exhibited similar MD behavior as the typical MD behavior frequently observed when using single-layer porous hydrophobic membranes. Moreover, it was found that the MD performance of the plasma-modified cellulose nitrate membrane by VTMS/CF_4 was better than that prepared using OFCB, and the DCMD flux of both membranes decreased while the rejection factor increased when the discharge time was getting higher. This was attributed to the gradual decrease of the pore size with increasing discharge time until the formation of a dense layer on the membrane surfaces. The same group, Wu et al. (1992), extended their study presenting other composite membrane for MD prepared via radiation polystyrene grafting of cellulose acetate to achieve the required hydrophobicity. The morphological structure of the membranes was investigated by various techniques, including scanning electron microscopy (SEM), contact angle measurements, and X-ray photoelectron spectroscopy (XPS). Compared to the asymmetric PVDF membrane having a permeability of $0.68 \times 10^{-3}\ kg/m^2 \cdot h \cdot Pa$ and a rejection factor of 99.9%, it was found that the permeability of the polystyrene grafting modified the cellulose acetate membrane ranging from 0.14×10^{-3} to $0.96 \times 10^{-3}\ kg/m^2 \cdot h \cdot Pa$ with a rejection factor decreasing from 99.1 to 66.7%; the OFCB plasma-modified cellulose nitrate membrane exhibited higher permeability (0.82×10^{-3} to $1.32 \times 10^{-3}\ kg/m^2 \cdot h \cdot Pa$) and a lower rejection factor (99.5–92.1%), whereas the reported values of the permeability of the VTMS/CF_4 plasma-modified cellulose nitrate were higher than the permeability of the PVDF membrane (1.81×10^{-3} to $2.22 \times 10^{-3}\ kg/m^2 \cdot h \cdot Pa$) with a rejection factor varying from 99.9 to 96.4%.

Fujii et al. (1992a, 1992b) prepared fine porous hollow-fiber membranes for DCMD from different polymers, PVDF, polysulfone (PSF), poly(phenylene oxide) (PPO), polyacrylonitrile (PAN), and cellulose triacetate (CTA) using the dry-jet wet spinning technique. Dimethyl sulfoxide (DMSO) was used as solvent for PVDF and CTA, while *N*-methyl-2-pyrrolidone was used as solvent for PS and PPO. It must be pointed out that CTA is a relatively hydrophilic polymer, but the authors still called the process DCMD. No differences were given about the transport mechanism through the hydrophobic membrane such as PVDF and the hydrophilic one such as CTA, and no details were given on why the authors thought that the process used could be termed DCMD when using hydrophilic porous membranes. In any case, the pore sizes of the prepared membranes are smaller than those of the microfiltration process. For example, PVDF hollow fibers were prepared with 4.0–24.8 nm mean pore size, 56–73% porosity, 0.675–0.844 mm internal diameter, and 0.982–1.071 mm external diameter. When ethanol aqueous solutions were considered, the DCMD fluxes of ethanol and water were found to be directly proportional to their partial vapor pressure difference and the separation factors varied according to the membrane polymers and the DCMD operation conditions. Ethanol permeability in the DCMD experiments was fairly constant, but that of water varied widely with different operation conditions and membranes. PPO membranes exhibited an ethanol selectivity of 7.3 with a partial ethanol flux of $0.168\ kg/m^2 \cdot h$ and a partial water flux of $0.422\ kg/m^2 \cdot h$. The ethanol selectivity of PVDF membranes was found to be lower (i.e., 3.7–5.2) with an ethanol flux of 0.239–0.640 $kg/m^2 \cdot h$ and a water flux of 0.878–2.314 $kg/m^2 \cdot h$. PSF membranes exhibited very low ethanol selectivity (1.6–3.1) with an ethanol flux of 0.272–1.21 $kg/m^2 \cdot h$ and water flux of 1.637–10.34 $kg/m^2 \cdot h$. The ethanol selectivity of PAN membranes was 1.3–3.2 with an ethanol flux of 0.045–0.199 $kg/m^2 \cdot h$ and water flux of 0.307–2.737 $kg/m^2 \cdot h$. CTA exhibited the lowest ethanol selectivity, 1.1, with an ethanol flux of 0.177 $kg/m^2 \cdot h$ and water flux of 2.962 $kg/m^2 \cdot h$.

Moreover, both temperature and concentration polarization effects and heat transport characteristics have been studied. The effect of coating polymers and heat treatment for PVDF hollow-fiber membranes have been investigated by the same authors (Fujii et al., 1992b). Silicone rubber (Si-LTV, Toray-Dow-Corning Silicone Co. Ltd.), poly(1-trimethylsilyl-1-propyne) (PMSP), and polyketone (Honshu Kagaku Co. Ltd.) have been used as coating material for the bores of the PVDF hollow-fiber membranes. It was observed that ethanol permeability through both silicone-coated and uncoated PVDF hollow-fiber membranes were similar but that of water decreased remarkably after coating treatment. Thus, the ethanol selectivities for the coated membranes were higher and in many results exceed the vapor–liquid equilibrium. The PMSP-coated PVDF membranes yielded decreased ethanol and water permeability; however, the decrease in water permeability was much greater and, thus, the selectivity was increased. Similar results were obtained with the polyketone-coated PVDF membranes. Moreover, when using acetone, acetonitrile, and *n*-butanol aqueous solutions, the selectivity of the same coated PVDF membranes were found to be higher than their relative volatility. This was attributed to the decrease of the water flux. On the other hand, the heat treatment of the PVDF membrane had the effect of increasing the hydrophobicity of the membrane. The contact angles increased from 94.3° to 102° after heat treatment at 353 K for 89 h. Therefore, the water flux was decreased while the ethanol flux was maintained stable, and thereby the selectivity was increased to be higher than the relative volatility. These results may be explained by the fact that the ethanol molecule has a hydrophobic ethyl group and a hydrophilic OH group, and a change of hydrophobicity has little effect on the diffusion of this molecule as the effects on both groups may be compensated. In contrast, the water molecule has only hydrophilic OH groups, and a change of hydrophobicity has stronger effects on the sorption parameter while the structural changes have less effect because of its small molecular size.

Ortiz de Zárate et al. (1995) reported asymmetric PVDF flat-sheet membranes prepared by the phase inversion technique from binary solutions of PVDF/dimethylacetamide (DMAC) or PVDF/dimethylformamide (DMF) formed by different polymer concentrations (10–25 wt%). The membranes were tested in DCMD configuration using distilled water as feed. They observed that pore diameters and porosity increased as the PVDF content decreased, and there was no improvement in flux compared to commercial membranes (GVHP in Table 12.1). A maximum appeared when plotting the transmembrane mass transfer coefficient against the polymer concentration in the casting solution. This maximum was found to depend on the solvent used for preparation of casting solutions (i.e., 13.2 wt% PVDF for DMAC and 15 wt% PVDF for DMF). This was attributed to the fact that the membrane top layer became more dense and so the fluxes were reduced. Also it was mentioned that for the PVDF concentration lower than 15 wt% the formed membrane matrix was inconsistent with big holes, which became wetted and therefore do not contribute to the MD process.

Tomaszewska (1996) studied the effect of the salt LiCl added to the casting polymer solution on the structural properties and permeability of PVDF flat-sheet membranes prepared for MD.

The effects of the casting solution composition, solvent evaporation time prior to coagulation, and temperature of coagulation bath on the properties of the PVDF membranes were investigated. Two types of solvents were employed, DMF and DMAC. The PVDF concentration in the casting solutions varied from 8 to 15 wt%. It was observed that both the porosity (>79%) and the pore size (0.0698–0.349 μm) increased with the concentration of LiCl varied from 0 to 3 wt%, and the liquid entry pressure of water (LEP_w) decreased from 290 to 45 kPa. The addition of only 0.5 g LiCl per 8 g of PVDF in DMAC solution increased the permeate flux very strongly from 90 to about 180 $L/m^2 \cdot day$ and additional amounts of LiCl in a casting solution caused a gradual increase of permeate flux from 180 to 240 $L/m^2 \cdot day$, when 1–2% NaCl feed solution was used with feed and permeate temperatures of 333 and 293 K, respectively. A separation factor higher than 99% has been obtained. The inconvenience of using LiCl as additive is the drastic decrease of the mechanical resistance of the PVDF membranes. This was the result of the presence of bigger cavities in the membrane structure. It was therefore concluded that the characteristics and properties of the membrane were affected by the composition of the casting solution and by the temperature of the coagulation bath used in the phase inversion. The same type of membranes were prepared previously by Bottino et al. (1988) for ultrafiltration applications. An increase of the porosity and permeation flux of PVDF membranes was also observed with increasing the content of LiCl in the casting solution.

Khayet and Matsuura (2001) used pure water as a nonsolvent additive for the preparation of PVDF flat-sheet supported and unsupported PVDF membranes from 15 wt% polymer in the solvent DMAC. The support was a nonwoven polyester backing material (Osmonics, Inc.). Water was selected as a nonsolvent additive to improve the MD permeability and to reduce the cost of the membranes, which also must be taken into account as an important parameter during membrane design. The concentration of water in the casting solution was varied from 0 to 6.8 wt%. The prepared membranes were tested for the removal of organic compounds (i.e., chloroform) from water by VMD configuration. Various membrane characterization techniques have been applied to determine the porosity, the pore size, the pore size distribution and the LEP_w. It was observed that the porosity (26.8–79.6%) and pore size (0.02–0.7 μm) both increased with increasing the water content in the casting solution for both the supported and unsupported membranes, whereas the LEP_w decreased. The VMD flux increased exponentially with the water content in the PVDF casting solution for both the supported and the unsupported membranes (1–14 $kg/m^2 \cdot h$ for the supported membranes and 0.6–16 $kg/m^2 \cdot h$ for the unsupported membranes). Both the VMD flux and the mass transfer coefficient of the supported membranes were found to be higher than those of the unsupported membranes when the water amount in the casting solution was lower than 4.25 wt%. However, at the 4.3 and 5.1 wt% water concentrations, the overall mass transfer coefficients were lower for the supported membranes. This was attributed to the resistance of the backing material. Furthermore, it was found that the separation factor decreased with increasing concentration of water in the PVDF casting solution for both supported and unsupported membranes and is generally lower for the supported membranes. The decrease of the separation factor was explained by the increase of the permeation rate of water through the membrane pores.

Other MD membranes have been prepared from the copolymer poly(vinylidene fluoride-*co*-tetrafluoroethylene) by Feng et al. (2004a, 2004b) using the phase inversion technique. F2.4 was the name given to the copolymer membranes. The DCMD performance of the membranes were compared to that of PVDF membranes prepared under similar conditions. It was found that the F2.4 membranes prepared from the copolymer and LiCl exhibited better mechanical performances, higher contact angles, and lower DCMD flux than those of the PVDF membrane (i.e., the DCMD flux was 6.5 and 3.5 $kg/m^2 \cdot h$ for the PVDF and F2.4 membrane, respectively; when the feed and permeate temperatures were 328 and 293 K, respectively). The low permeate flux observed for the F2.4 membrane was attributed to its smaller pore size ($<2.4 \times 10^{-2}$ μm) and porosity ($<80\%$) compared to those of PVDF membranes (Feng et al., 2004a). However, the F2.4 membranes exhibited excellent mechanical properties (i.e., stretching strain and extension ratio at break approximately 6–8 times higher than those of PVDF membranes) and higher hydrophobicity. For all tested membranes, close to 100% separation factors were observed when aqueous salt solutions were used. Moreover, the same authors (Feng et al., 2004b) used $LiClO_4 \cdot 3H_2O$/trimethyl phosphate (TMP) as a pore-forming additive. The pore size, porosity, and DCMD fluxes of the prepared F2.4 membranes were higher than those prepared using LiCl.

It is worth quoting that a series of studies has been carried out in an effort to improve the properties of PVDF membranes by introducing nonsolvent additives in the PVDF polymer dope, although these membranes were not prepared for MD purposes. Deshmuck and Li (1998) and Wang et al. (1999) introduced poly(vinylpyrrolidone) (PVP) in the PVDF casting solution as an additive in order to obtain highly porous PVDF hollow-fiber membranes. However, trace quantities of PVP in the membrane affect the hydrophobicity of the PVDF membrane. Uragami et al. (1980) tested the effect of the addition of polystyrene sulfonic acid in the PVDF dope as well as the influence of poly(ethylene glycol) (PEG). The permeability of the prepared flat-sheet membranes was improved, as expected. In contrast, the strength of the membranes was reduced. Khayet et al. (2002a, 2002b) also used the additive PEG for the preparation of PVDF hollow-fiber membranes for ultrafiltration. Glycerol and phosphoric acid were employed by Benzinger and Robinson (1982) as pore-forming agents in PVDF membranes to increase the membrane permeability. Shih et al. (1990) used ethanol to increase the gas permeability through dried PVDF membranes. Small molecular additives such as water, ethanol, and i-propanol were used by Wang et al. (2000) for preparation of PVDF asymmetric hollow-fiber membranes. The wet prepared fibers exhibited high water permeability while the dry fibers had a high gas permeability, good mechanical strength, and excellent hydrophobicity.

Li et al. (2003) proposed the use of polyethylene (PE) and polypropylene (PP) hollow-fiber membranes for desalination by DCMD and VMD. The hollow-fiber membranes were prepared by the melt-extruded/cold-stretching method. Compared to PP hollow-fiber membranes, higher water fluxes have been obtained for the PE membranes in both DCMD and VMD. This was attributed to the larger pore size of the PE membranes. The highest permeate flux reported was 0.8 $L/m^2 \cdot h$ in DCMD and about 4 $L/m^2 \cdot h$ in VMD.

As stated in the previous section, the use of composite membranes in MD was described first by Cheng (1981) and Cheng and Wiersma (1982, 1983a, 1983b) in a series of patents. Recently, Khayet and Matsuura (2003a) and Khayet et al. (2005a, 2005b) proposed a new type of porous composite hydrophobic/hydrophilic flat-sheet membrane for DCMD application. As presented in Figure 12.4, the hydrophobic side of the membrane was brought into contact with the hot feed solution, while the hydrophilic layer of the membrane was maintained in contact with cold water, which penetrates into the pores of the hydrophilic layer. The composite porous hydrophobic/hydrophilic membranes are promising for desalination by DCMD as they combine the low resistance to mass flux achieved by the diminution of the water vapor transport path through the hydrophobic thin top layer and a low conductive heat loss through the membrane obtained by using a thicker hydrophilic sublayer. These types of membranes were prepared by the phase inversion technique in a single casting step from polyetherimide (PEI) polymer solutions containing fluorinated surface

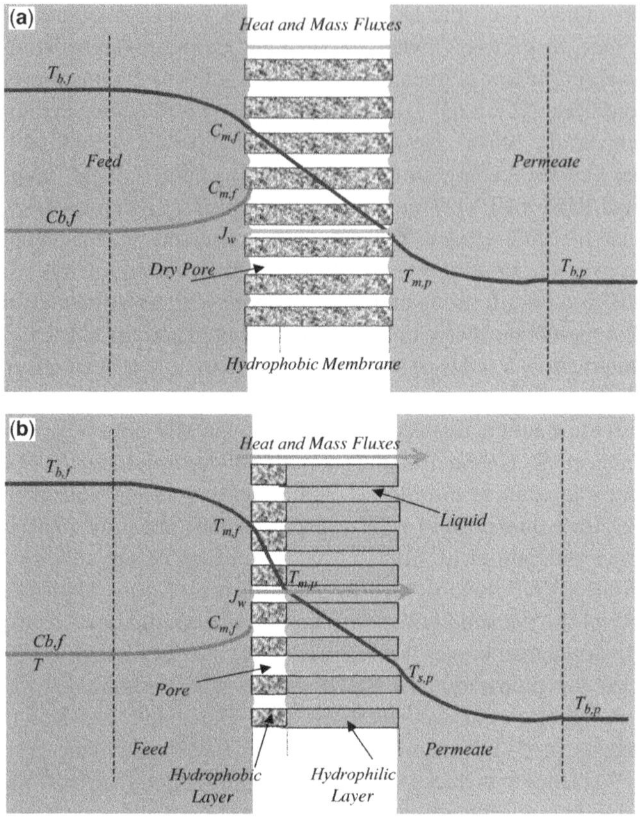

Figure 12.4 Schema of DCMD mechanism of transport through porous membranes: (a) homogenous hydrophobic and (b) composite hydrophobic/hydrophilic. The concentration profiles correspond to nonvolatile solutes. (Adapted from Khayet et al., 2005b.)

modifying macromolecules (SMMs), the solvent *N,N*-dimethylacetamide, and the non-solvent γ-butyrolactone (GBL). Only 2 wt% SMM was added to the casting solution. These SMMs are oligomeric fluoropolymers synthesized by polyurethane chemistry and tailored with fluorinated end groups. A complete detail on SMM preparation and characterization was reviewed by Khayet et al. (2003b). During membrane formation, these SMMs migrate toward the top air–polymer interface rendering the membrane hydrophobic. This fact was confirmed by contact angle measurements and X-ray photoelectron spectroscopy (XPS) analysis, which indicated fluorine concentration gradient along the membrane thickness, resulting from the migration of fluorinated blocks to the air side surface during membrane formation.

The effects of the PEI amount in the casting solution on the permeate flux and on the membrane characteristics were investigated. For each PEI concentration, it was found that the LEP_w of the SMM-modified membrane, was higher than that of the unmodified membrane, while the pore sizes were smaller and decreased as the concentration of PEI in the casting solution was increased. The pore sizes of the SMM-modified PEI membranes determined from the gas permeation test were 12–23 nm, lower than the pore sizes of the membranes shown in Table 12.1. However, their DCMD permeabilities were found to be of the same order of magnitude to those of the commercial PTFE membranes (TF200 and

TF450 in Table 12.1) having higher porosity and an order of magnitude higher pore sizes. Furthermore, the DCMD flux of the SMM/PEI membrane prepared with 12 wt% PEI was found to be higher than that of the commercial PTFE membranes, up to 30.5% compared to the membrane TF450 and up to 66.6% compared to the membrane TF200. Very high separation factor (>99.7%) was observed for the SMM-modified PEI membranes when NaCl aqueous solutions were considered. A theoretical model has been developed to estimate the thickness of the two layers of these composite membranes using the DCMD experiments together with other membrane parameters and the heat and mass transfer equations. Thicknesses lower than 8 μm have been found for the hydrophobic layer (Khayet et al., 2005b). Thus, the SMM blended membranes exhibit hydrophobic layer thickness, responsible for mass transport in DCMD, an order of magnitude lower than that of the PTFE layer of the commercial membranes.

Larbot et al. (2004) used grafted ceramic hollow-fiber membranes by fluoroalkylsilanes for water desalination by DCMD configuration. The ceramic membranes were alumina (Al_2O_3) and zirconia (ZrO_2) of pore size 200 and 50 nm, respectively. The measured water contact angles of the membranes were higher after grafting, indicating that the grafted membranes were more hydrophobic. It was stated that when using the grafted Al_2O_3 membrane having pore size of 200 nm, at low NaCl concentrations (10^{-3}–10^{-2} M), the rejection factor ranged from 90 to 96%, whereas for higher NaCl concentrations in feed higher rejection factors up to 99% were obtained with DCMD fluxes up to 163.2 $L/m^2 \cdot day$ when the feed temperatures were up to 95°C. No indications were given on the cause of the lower rejection factor observed when using lower NaCl concentrations. When using the grafted ZrO_2 membrane having lower pore size (i.e., 50 nm), lower DCMD fluxes ($\approx$95 $L/m^2 \cdot day$) and near 100% rejection factor were obtained; whereas higher fluxes ($\approx$202 $L/m^2 \cdot day$) with lower rejection factors (99.5%) were found for the grafted ZrO_2 having higher pore size (i.e., 200 nm). Finally, it was concluded that the use of ceramic hollow fibers would be of great interest in MD. A similar type of membranes has been used by Belleville et al. (2002) and Brodard et al. (2003) in the osmotic evaporation process. Tubular microporous alumina membranes, provided by Pall Corporation (Exekia Division) with mean pore sizes of 0.2 and 0.8 μm, were grafted by siloxane compounds. It was reported that the permeate flux was independent of the membrane pore size, concluding that a limiting water vapor diffusion occurred through the membrane pores. Xu et al. (2004) proposed coating hydrophobic microporous membranes used in MD and osmotic distillation by hydrophilic sodium alginate hydrogel for protection against pore wetting by surface-active agents such as oils, fats, and detergents. A microporous PTFE membrane (Poreflon 020-40, Sumitomo Electric Fine Polymer, Osaka, Japan) of pore size 0.2 μm was alginate coated and then cross-linked by a water-soluble carbodiimide. The observed reduction of the overall mass transfer coefficient due to coating effect was less than 5%, and the coated membranes were resistant to wetting during at least 300 min when exposed to orange–oil–water mixture solution.

Li and Sirkar (2004, 2005) reported on novel hollow-fiber membranes and modules for use in both DCMD and VMD configurations. The presented new types of membranes were commercial porous PP hollow fibers (Accurel MEMBRANA, Wuppertal, Germany) of different dimensions and thicknesses coated with a variety of ultrathin microporous silicone–fluoropolymer layer on their external surface by plasma polymerization at Applied Membrane Technology (AMT) Inc. (Minnetonka, MN). The fibers were arranged in a rectangular cross-flow module design for the hot feed to flow over the outside surface of the fibers and to reduce the temperature polarization effect. Both the DCMD and VMD experiments were carried out at feed temperatures ranging from 60 to 90°C with 1% NaCl aqueous solution. Higher fluxes (41–79 $kg/m^2 \cdot h$), complete absence of membrane pore wetting, higher temperature polarization coefficients (93–99% in VMD) than in other MD apparatuses, and heat transfer coefficients in the membrane module feed side have been reported.

Peng et al. (2005) looked at casting a dense hydrophilic polymer solution on porous PVDF (GVHP, Table 12.1) membrane. The polymer solution was a blend of polyvinyl alcohol (PVA) and polyethylene glycol (PEG) cross-linked by aldehydes and sodium acetate. The composite hydrophilic/hydrophobic membranes were tested for desalination by DCMD configuration using 3.5 wt% NaCl solution. The effects of the feed temperature and salt concentration were investigated and the obtained behavior for the coated membranes was similar to the uncoated PVDF membrane. More than 99% separation factor was obtained, and the DCMD flux of the coated membrane is only 9% lower than that of the uncoated membrane, which was 23.7 $kg/m^2 \cdot h$ at 70°C feed temperature and 22°C permeate temperature. It was supposed that the feed liquid is first dissolved in the hydrophilic layer of the membrane, whose molecular network has a suitable swelling, followed by diffusion through a continuous pathway along the wriggling polymer chain mechanism and finally evaporates at the interface between the hydrophobic and the hydrophilic layers. It was concluded that a hydrophilic layer can prevent wetting, which is a problem for the single hydrophobic layer membranes in MD. In fact, 25% ethanol was added to the salt feed solution, and the DCMD flux was found to be higher at the beginning of the experiment when ethanol permeated with water, and then it remained constant indicating that wetting did not occur. However, the authors (Peng et al., 2005) did not measure the LEP of water or their feed solutions for this type of membrane.

Despite all the above-cited works, detailed studies concerning the design of membranes for MD are still lacking. MD membranes with different pore size, porosity, thickness, and materials are needed to help understanding the physical nature of mass transport in different MD configurations.

12.2.2 Membrane Distillation Membrane Modules

Two membrane configurations have been tested so far for MD: (a) flat-sheet membranes and (b) capillary or hollow-fiber membranes. These membrane types have been packed in a large variety of membrane module configurations and tested in MD systems. A majority of the laboratory-scale modules are designed for use with flat-sheet membrane, either in Lewis test cells [Fig. 12.5(a)] or plate-and-frame modules working under tangential flows by means of circulating pumps [Fig. 12.5(b)]. In both systems, the flat-sheet membranes can be easily replaced, changed, examined, or cleaned. Generally, Lewis cells have been employed for DCMD experiments in which both feed and permeate liquids are stirred inside the cell by graduated magnetic stirrers without the need of circulation pumps (Sakai et al., 1988; Mengual and Peña, 1997; Sudoh et al., 1997; Khayet et al., 2005a). Khayet and Matsuura (2001) conducted VMD experiments in a device where the feed solution is also stirred inside the cell by a magnetic stirrer while vacuum was applied in the permeate side. Lewis cells are only used for laboratory-scale studies. A thin channel device shown in Figure 12.5(d) was used for DCMD by Schofield et al. (1987). In fact, the plate-and-frame modules are being used for all MD configurations (Gostoli and Sarti, 1989; Martínez-Díez and Vázquez-González, 1996; Lawson and Lloyd, 1996a; Khayet et al., 2000a). The only inconvenience of using flat-sheet membranes

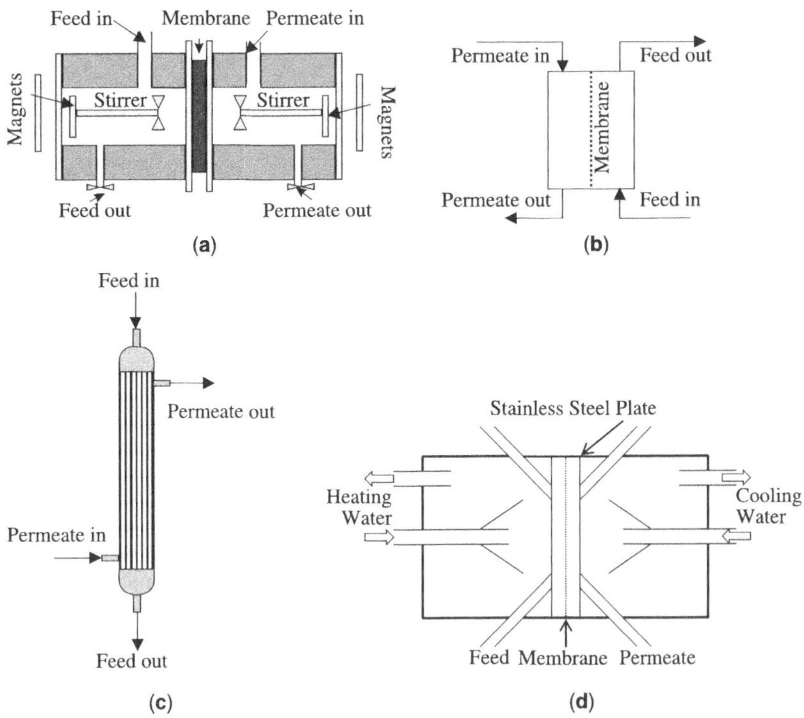

Figure 12.5 Different membrane modules used in MD: (a) a Lewis cell, (b) flat-sheet or tangential-flow module, (c) hollow-fiber module, and (d) a thin channel module.

in plate-and-frame modules is the requirement of using supports to hold the membrane, specially when the membrane surface area exposed to flow is large. On the other hand, capillary or hollow-fiber membranes do not require a support but are an integrated part of the module and cannot be replaced easily [Fig. 12.5(c)]. In this case, if the membrane pores are wetted by the liquid solutions, the whole module is useless for MD applications. In fact, from a commercial standpoint, tubular membrane modules are more attractive than the plate-and frame modules as a much higher membrane surface area to module volume ratio can be reached. Tubular modules have also been applied in all MD configurations (Drioli and Wu, 1985; Rivier et al., 2002; Khayet et al., 2003c; Guijt et al., 2005). Actually, the availability of the industrial MD modules is one of the limitations for MD process implementation. Most of the used tubular modules are prepared commercially for other separation processes (Table 12.1).

It must be mentioned that capillary membranes were also assembled in plate-and-frame membrane modules in cross-flow mode to reduce the temperature polarization effect by increasing the heat transfer coefficients (Li and Sirkar, 2004, 2005). Two face boxes and face plates were assembled with a rectangular membrane module channel to constitute the whole device, as can be seen in Figure 12.6. The modules have surface areas varying between 113 and 257 cm^2 (internal dimensions of the module frame: length 6.4 cm, width 2.5 cm, height 1.8 cm). The effective membrane area used was between 119 and 491 cm^2 with packing fractions between 12 and 30%. Considerably enhanced water vapor flux and high module productivity in both DCMD and VMD configurations were achieved. Scale-up of this device for similar flow conditions would be of great interest.

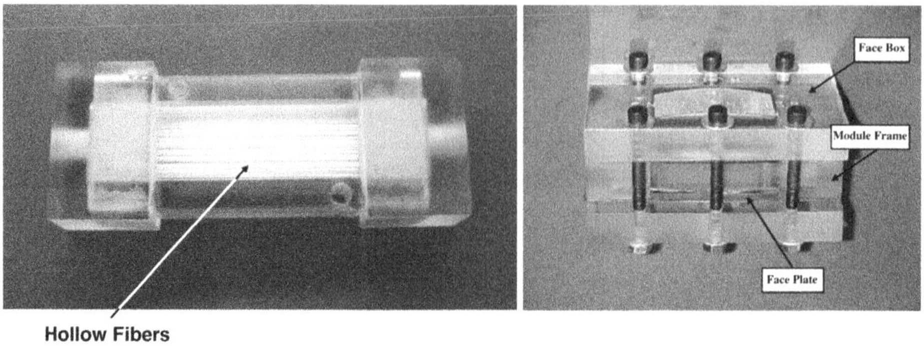

Figure 12.6 Rectangular cross-flow module used in Li and Sirkar (2004, 2005).

In SGMD, the gas temperature, the heat transfer rate, and the mass transport through the membrane change during the gas progression along the membrane module. In order to overcome this problem, a novel tubular device was proposed by Rivier et al. (2002) for SGMD in which the increase of the sweeping gas temperature along the membrane module was minimized by using a cold wall in the permeate chamber. A schematic representation of the module is shown in Figure 12.7. It must be mentioned here that part of the permeating vapors condense inside the module depending on the operating conditions, while the rest is collected in an external condenser. The liquid feed was pumped through the internal side of the membrane tubes and air through their outer side. The membrane used to construct tubes of 5 mm internal diameters and 0.2 m length was flat-sheet membrane TF200 (Table 12.1). These tubes were subsequently thermowelded to connections of the module, which were made of PP. A spiral turbulent promoter was placed around the tubular membranes in order to reduce the polarization effect in the gas side.

Flat-sheet membranes were also arranged in spiral-wound modules as shown in Figure 12.8. Commercial spiral-wound modules were used in DCMD experiments by Zakrewska-Trznadel et al. (1999) in a pilot plant illustrated in Figure 12.8(a). Koschikowski et al. (2003) proposed the use of a spiral-wound module for AGMD or

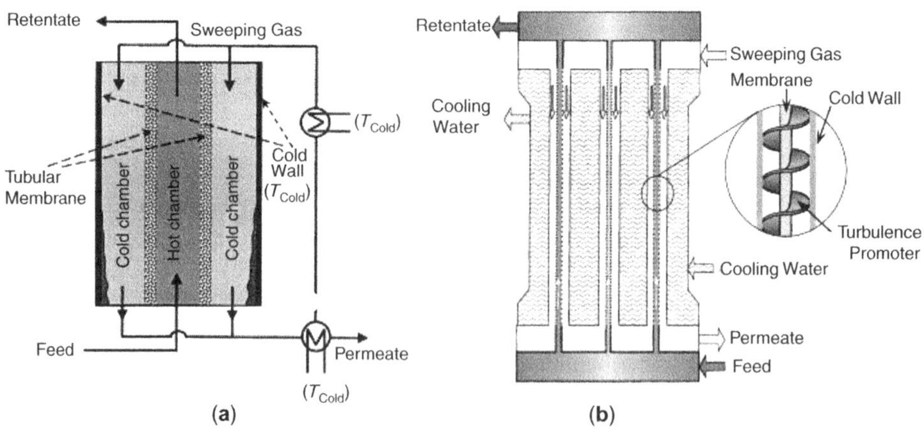

Figure 12.7 Schematic representation of TSGMD module with detail of the turbulence promoter. (Adapted from Rivier et al., 2002.)

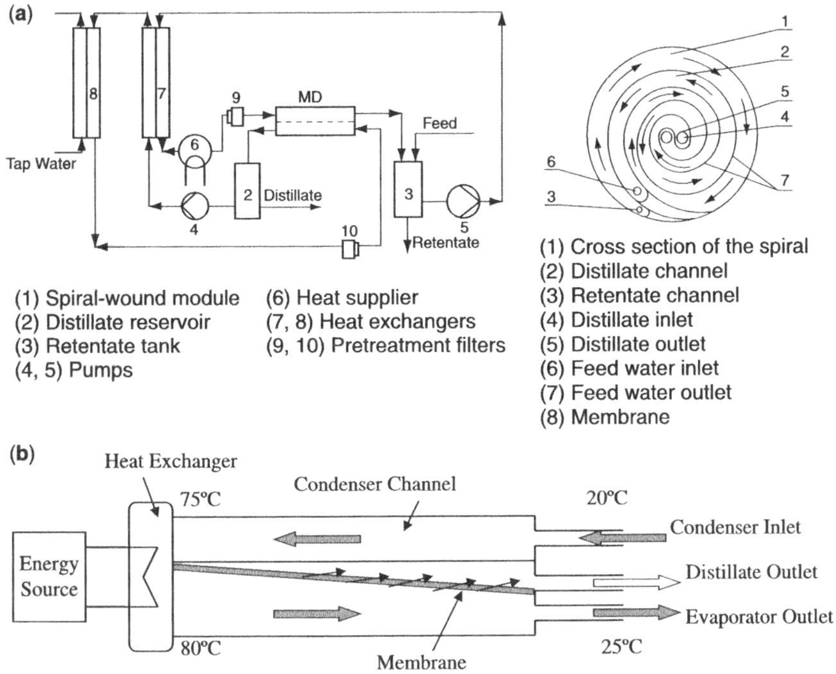

Figure 12.8 Application in spiral-wound module in (a) DCMD from Zakrewska-Trznadel et al. (1999) and in (b) AGMD or liquid gap DCMD from Koschikowski et al. (2003).

the liquid gap DCMD configuration with an integrated heat recovery for the design of solar-powered desalination systems as shown in Figure 12.8(b). However, the same design has been previously communicated (more than 20 years before) by Gore & Associated Co. (Gore, 1982) and Hanbury and Hodgkiess (1985).

A module to be used in MD must exhibit a high packing density (i.e., high membrane surface area) and must provide both high feed and permeate flow rates tangentially to the membrane or in cross-flow mode with high turbulence in order to reduce the temperature polarization effect. A well-designed membrane module should provide high rates of heat and mass transfer between the bulk solution and the solution at the membrane interface. A good MD module design must provide high heat and mass transfer coefficients, which in turn will lower both the temperature and concentration polarization effects. Moreover, the MD module should satisfy low pressure drop along the membrane module length to prevent excessively high transmembrane hydrostatic pressure that may cause flooding of membrane pores. The MD module should not only provide good flow conditions but also must guarantee low heat loss to the environment and if possible a good heat recovery system (i.e., internal heat exchanger). A uniform temperature of the liquid solutions must be maintained along the module length being accomplished with high heat transfer coefficients. The possibility of using plastic equipment also reduces or avoids erosion problems. Additionally, mass transfer rates at the feed boundary layer must be sufficient to prevent excessive concentration polarization that can lead to membrane wetting by scaling and building up of salt crystals or minerals on the membrane surface. Some authors considered the use of spacer-filled channels of the plate-and-frame membrane modules in both the feed or permeate side of the membrane. However, this increases the heat transfer coefficients but may lead to a decrease of the membrane surface area and therefore to the MD flux. A membrane support must be chosen that does not significantly increase the heat and mass transfer resistance, but it must be strong enough to prevent deflection or rupture of the membrane.

It must be pointed out that although one of the major advantages of hollow-fiber membrane modules over others is its high packing density and compactness, normally the fibers are randomly packed inside the module and care must be taken to avoid flow maldistribution caused by the polydispersity of fiber inner diameter at the lumen side and the nonuniformity of fiber packing at the shell side of the membrane modules. These facts have been analyzed by Ding et al. (2003). Zander et al. (1989) also observed that the air pressure drops at the shell side across hollow-fiber membrane modules used in the membrane air stripping (MAS) process were much higher than those for packed towers. The packing density of the hollow fibers in the modules was thought to be the cause for this high pressure drop. Proper design of the bundle to prevent fiber blockage of the exit port and avoiding major head loss across the air ports may help reduce the pressure drop.

Other authors (Martínez-Díez et al., 1998; Phattaranawik et al., 2001) have considered the use of spacer-filled channels of plate-and-frame membrane modules used in DCMD. It was observed that DCMD permeate fluxes were higher when using a screen separator than when an open separator was used. The spacers changed the flow characteristics and promoted regions of turbulence, leading to a decrease of the temperature polarization. These effects were found to be higher for the coarse spacer than for the fine spacer.

12.3 MEMBRANE DISTILLATION MEMBRANE CHARACTERIZATION TECHNIQUES

As occurred in other membrane processes, the successful application of MD may be aided by a good knowledge of the different membrane parameters. Before conducting MD experiments, various techniques must be used for characterization of the MD membranes in order to avoid pore wetting. As was discussed by the committee formed during the Workshop on Membrane Distillation held in Rome on May 5, 1986, and later reported (Franken and Ripperger, 1988; Smolders and Franken, 1989), the membranes used in MD should be characterized by the following five membrane "performance" parameters: "polymer" or other membrane material, thickness, porosity, nominal pore size, and liquid-entry-pressure of water (LEP_w). Here one can add the thermal conductivity of the membrane, its pore tortuosity, and pore size distribution, which includes the maximum pore size related directly to the membrane pore wetting. In fact, pore tortuosity factor and pore size distribution convey much more information, which can be used to predict the permeate MD fluxes as will be shown later on.

The characterization techniques of MD membranes are physical methods, which can be divided in two main groups: (i) the techniques related to membrane permeation such as liquid and gas flow tests and (ii) the techniques permitting to obtain directly the morphological properties of the membranes, including scanning electron microscopy (SEM), field emission scanning electron microscopy (FESEM), and atomic force microscopy (AFM).

12.3.1 Liquid Entry Pressure

To avoid membrane pore wetting, the membrane material must be hydrophobic with high water contact angle and small maximum pore size. However, pore wetting may occur and permeate quality may be affected if solutions with surface-active components are brought in direct contact with the membrane surface and/or if the transmembrane hydrostatic pressure exceeds the liquid entry pressure (LEP) of the aqueous solutions present in the membrane module. LEP is the pressure that must be applied onto the aqueous solution to be treated before it goes into the pores. This depends on each membrane. It decreases with the increase of the maximum pore size and/or the decrease of the contact angle of the feed solution to be treated. It must be pointed out that various authors, by measuring only the water contact angle and the LEP_w, exposed their membranes to different aqueous solutions containing lower surface tension compounds than water without paying attention to the possible pore flooding of their membranes. Thus, a difference exists between the LEP_w of water and that of the feed and permeate solutions when they are brought in direct contact with the membrane. Care must also be taken when using DCMD for removal of organic compounds from aqueous solutions. The organic compounds are more volatile than water, and after the DCMD process, the permeate becomes more concentrated in organics, increasing the risk of membrane pore wetting by the permeate aqueous solution. Figure 12.9 shows a schematic system to measure the LEP. Generally, a slight pressure ($\approx 0.3 \times 10^5$ Pa) is first applied to the feed solution maintained in direct contact with one side of the membrane for at least 10 min; then the pressure is increased stepwise until a first drop of the feed solution appears in the permeate side. This procedure was applied for both flat-sheet (Franken et al., 1987b; Gostoli and Sarti, 1989; Tomaszewska, 1996; Izquierdo-Gil et al., 1999a;

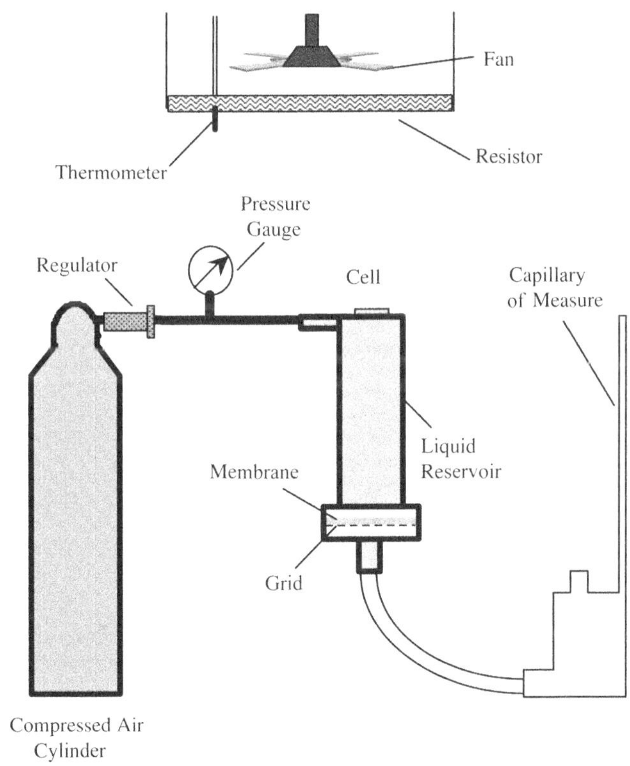

Figure 12.9 Schema of LEP system used in García-Payo et al. (2000a).

García-Payo et al., 2000a; Khayet and Matsuura, 2001, 2003a; Khayet et al., 2004a) and hollow-fiber membranes (Khayet et al., 2002a).

The LEP_w of commercial flat-sheet membranes prepared with different materials, PVDF (GVHP and HVHP) and PTFE (Gore membranes and TF200) was measured by Izquierdo-Gil et al. (1999a) and Khayet et al. (2004a). The results of the PVDF membranes are reported in Table 12.1, while the value of TF200 membrane was found to be slightly lower (i.e., 2.2%) than that given by the manufacturer reported in Table 12.1. Gostoli and Sarti (1989) measured the LEP of ethanol–water mixtures for the membrane TF200 (Table 12.1). As can be seen in Figure 12.10, the LEP decreases linearly with ethanol concentration and vanishes at nearly 75 wt%. The authors (Gostoli and Sarti, 1989) finally concluded that the membrane TF200 has adequate barrier properties for dilute ethanol–water mixtures.

Franken et al. (1987b) established wetting criteria for the application of MD. Various membrane types and liquids, including ethanol, acids (formic acid, acetic acid, propionic acid, and butyric acid), solvents (dimethylacetamide, dimethylsulfoxide, and dimethylformamide), acetone, and 1–4-dioxane were tested. For each membrane, different LEP values were obtained. It was concluded that the maximum allowable concentration of organic compound in water cannot be calculated but has to be measured.

García-Payo et al. (2000a) looked at the study of the effects of alcohol type, alcohol concentration, temperature, and type of membrane on the LEP value. It was observed that the LEP decreased with the increase of the concentration of alcohol (i.e., methanol, ethanol, isopropanol) in water, the length of the alcohol hydrocarbon chain, the temperature, which was varied in the range 25–50°C, and the membrane pore size. Moreover, for membranes with similar pore sizes, the LEP of PTFE membranes was found to be higher than that of the PVDF membranes. Figure 12.11 presents as an example the increase of the mass flux with the applied hydrostatic transmembrane pressure on distilled water at 25°C for PVDF and PTFE membranes, after the LEP was exceeded. However, with the decrease of the transmembrane hydrostatic pressure, the measured fluxes showed linear trends, which could not be observed for the PTFE membrane (Gore flat-sheet membranes having 0.2 μm mean pore size in Table 12.1) because of the system limitations.

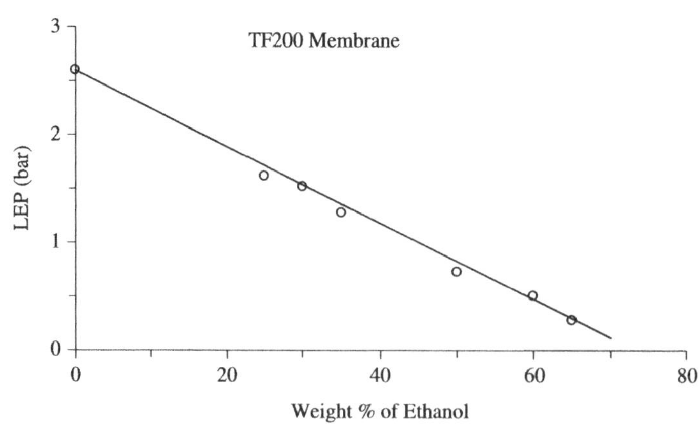

Figure 12.10 LEP of ethanol–water mixtures of the TF200 membrane. (Adapted from Gostoli and Sarti, 1989.)

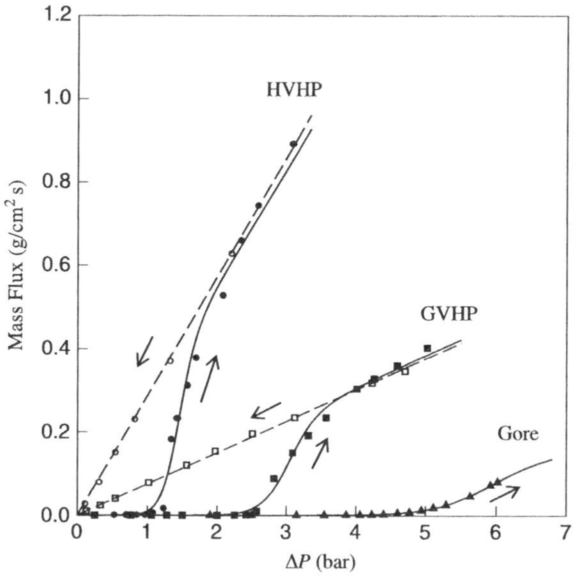

Figure 12.11 Water flux vs. applied transmembrane pressure, ΔP, for PVDF (GVHP and HVHP) and PTFE (Gore, 0.2-μm mean pore size) membranes at 25°C. (Adapted from García-Payo et al., 2000a.)

Tomaszewska (1996) studied the variation of the LEP_w of the laboratory-made PVDF membranes with the LiCl concentration in the polymer casting solution. As can be seen in Figure 12.12, the PVDF membranes prepared with higher amounts of LiCl exhibited lower values of LEP_w due to their larger pore size. Similar results were obtained by Khayet and Matsuura (2001) for the supported and unsupported PVDF membranes prepared with the same solvent (DMAC) and water as a nonsolvent additive.

Khayet and Matsuura (2003a) found that the LEP_w values of both the modified hydrophilic polyetherimide (PEI) membranes by surface modifying macromolecules (SMMs) and the unmodified ones increased with the increase of the PEI concentration in

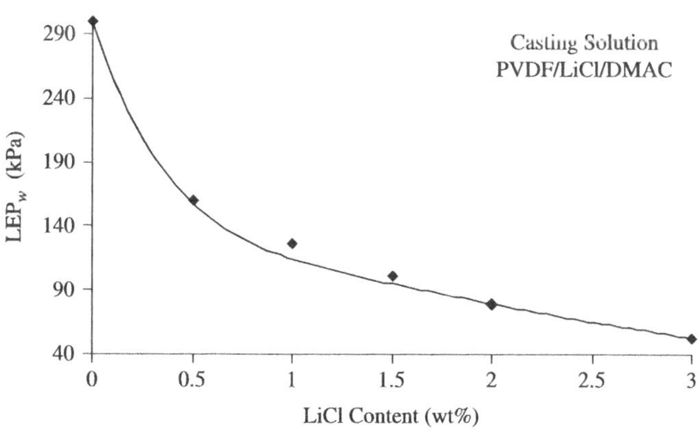

Figure 12.12 Effect of LiCl content in 8 wt% PVDF solution in dimethylacetamide (DMAC) on the membrane LEP_w. (Adapted from Tomaszewska, 1996.)

the casting solution used for the preparation of the membranes, and for the same PEI concentration the LEP_w of the SMM-modified membrane was higher than that of the unmodified membrane. These results were attributed to the higher hydrophobicity of the SMM-modified PEI membranes and to the larger pore size of the membranes prepared with lower PEI concentration in the casting solution.

12.3.2 Contact Angle Measurements

The relative wettability of the MD membranes can be studied by measuring their liquid contact angles. Both the advancing and receding water contact angles on both sides of the MD membranes must be determined. It must be highlighted here that the contact angle measurement should be taken at different locations of the membrane sample and the effect of the membrane pore size and roughness must be considered to determine the true contact angle of the MD membrane (Franken et al., 1987a; Mittal, 1993; Wu, 1982; Matijevic, 1989). In fact, measurements were determined at room temperature using different contact angle meters (Franken et al., 1987a; Hoffman et al., 1987; Wu et al., 1992; García-Payo et al., 2000a; Khayet and Matsuura, 2003a; Feng et al., 2004a, 2004b), and drops of different liquids of about 2 μL were deposited on the membrane surface employing tight syringes. It was seen that the measured advancing and receding contact angles of the PTFE membranes were higher than those of the PVDF membranes. As an example, the reported advancing contact angles were $110.2 \pm 3.3^\circ$ and $113.6 \pm 2.7^\circ$ for GVHP and TF200 membranes, respectively; while their receding contact angles were lower, $95.9 \pm 2.5^\circ$ and $100.7 \pm 3.6^\circ$, respectively.

Wu et al. (1992) measured the advancing water contact angles on the top surface of modified membranes by plasma surface polymerization of polystyrene grafting of cellulose acetate membranes to test their hydrophobicity. The water contact angles were found to increase with the radiation time and presented a minimum with discharge power.

Feng et al. (2004a, 2004b) also measured the advancing water contact angles of membranes prepared from PVDF and the copolymer poly(vinylidene fluoride-*co*-tetrafluoroethylene) under the same conditions using the phase inversion technique. Both asymmetric porous and symmetric dense membranes were considered for each polymer. It was found that the membranes prepared from the copolymer poly(vinylidene fluoride-*co*-tetrafluoroethylene) exhibited larger pore size than the PVDF membranes, and the water contact angles of the porous membranes were lower than those of the dense membranes. This may be attributed to the pore effect.

Khayet and Matsuura (2003a) measured both the advancing and the receding water contact angles of the SMM-modified and unmodified PEI membranes. Both contact angles increased when blending SMM with the polymer PEI solution, indicating that the SMM-modified PEI membranes were more hydrophobic than the unmodified ones prepared with the same PEI polymer concentration in the casting solution. This result was due to the hydrophobic character of the SMM fluorine tails associated to the SMM. Moreover, the increase of the receding contact angle of the SMM-modified membranes indicated that hydrophobic segments of the SMM remain at the membrane surface even after the surface was brought in contact with water.

Hoffmann et al. (1987) measured the contact angles between PTFE foils and binary mixtures of methanol–water, ethanol–water and *n*-propanol–water of different concentrations. A decrease of the contact angle with increasing alcohol concentration was observed, and the contact angles were lower when using larger length of the hydrocarbon chain of the alcohol.

García-Payo et al. (2000a) measured the advancing contact angle of various membranes of different pore sizes and materials (PVDF, PTFE) using distilled water and different alcohol–water mixtures. The researchers reported that, for the same membrane, the water contact angles were independent of membrane pore size, and the membrane support had no effect on the measured water contact angles. Mean values for PVDF ($111 \pm 3^\circ$) and PTFE ($123 \pm 2^\circ$) were reported. These values are higher than those stated previously. This may be attributed to the fact that in the previous measurements the receding contact angles were considered. Accurate measurements on contact angles and the corresponding corrections due to pore size and membrane roughness effects are needed for MD membranes.

12.3.3 Thermal Conductivity

The thermal conductivity of the MD membrane is contributed from both polymer (k_p) and gas (k_g) present in the membrane pores:

$$k_m = \varepsilon k_g + (1 - \varepsilon)k_p \tag{12.1}$$

Following the standard ASTM C177 method, the reported thermal conductivities of PVDF material were 0.17–0.19 W/m · K at 296 K and 0.21 W/m · K at 348 K. Lower values were reported for PP (0.11–0.16 W/m · K at 296 K and 0.2 W/m · K at 348 K), while those of PTFE were higher (0.25–0.27 W/m · K at 296 K and 0.29 W/m · K at 348 K) (Phattaranawik et al., 2003a). In fact, both the void volume and the structure of the pore affect the thermal conductivity prediction of MD membranes.

The thermal conductivity of various commercial porous and hydrophobic membranes was measured by Izquierdo-Gil et al. (1999a, 1999b) and García-Payo and Izquierdo-Gil (2004) using a modified Less method shown in Figure 12.13. Superpositions of three or more membranes were needed to conduct their experiments, leading to large error measurements due to the presence of the air layer between membranes. However, the authors (García-Payo and Izquierdo-Gil, 2004) claimed that their apparatus was simple and accurate for determination of the thermal conductivity.

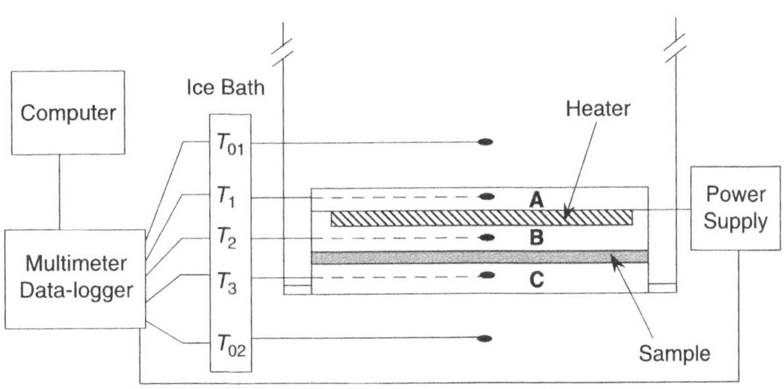

Figure 12.13 Schematic of the modified Less system used for the thermal conductivity measurement (García-Payo and Izquierdo, 2004).

The obtained thermal conductivity of the PVDF membranes, GVHP and HVHP, were 0.041 and 0.040 W/m · K, respectively, while that of PTFE membranes supplied by Gore shown in Table 12.1 were 0.043 W/m · K for the supported PTFE/PP membrane, 0.031 and 0.027 W/m · K for the unsupported PTFE membranes with 0.2 and 0.45 μm pore sizes, respectively. Phattaranawik et al. (2003a) found that these values were close to those calculated using the isostress model, Eq. (12.2), rather than the usually used isostrain model, Eq. (12.1), in most MD studies. Various models for calculation of thermal conductivity were reviewed in García-Payo and Izquierdo-Gil (2004):

$$k_m = \left(\frac{\varepsilon}{k_g} + \frac{1 - \varepsilon}{k_p} \right)^{-1} \tag{12.2}$$

12.3.4 Porosity

The ratio between the volume of the pores and the total volume of the membrane is the membrane porosity, also called membrane void volume. The symbol of the porosity is often ε as defined in Franken and Ripperger (1988) and Smolders and Franken (1989). For hydrophobic porous membranes, the porosity can be determined by measuring the density of the membrane using two types of liquids. A wetting liquid such as isopropyl alcohol (IPA), which penetrates inside the membrane pores, and a nonwetting liquid such as water, which does not enter the pores because of the hydrophobicity of the membrane material. In this method, a pyknometer and a balance are necessary (Franken and Ripperger, 1988; Smolders and Franken, 1989; Izquierdo-Gil et al., 1999a; Khayet and Matsuura, 2001; García-Payo and Izquierdo-Gil, 2004), and the following equation can be used to calculate the value of the porosity:

$$\varepsilon = 1 - \frac{\rho_m}{\rho_P} \tag{12.3}$$

where ρ_m and ρ_P are the densities obtained using water and IPA, respectively.

By using this method, Khayet and Matsuura (2001) obtained experimental values of 70.1 ± 2.7% for the membrane GVHP (Table 12.1), 71.3 ± 3.4% for the membrane HVHP (Table 12.1), and 68.7 ± 5.4% for the membrane TF200 (Table 12.1). These values are lower than the values given by the manufacturer (i.e., 75% for GVHP and HVHP membranes and 80% for TF200). This may be attributed to the different measurement techniques used. However, when using the same technique, the obtained values by Khayet and Matsuura (2001) were higher than the ones given by Izquierdo-Gil et al. (1999a, 1999b) (62 ± 2% for the GVHP membrane and 66 ± 2% for the HVHP membrane).

12.3.5 Mean Pore Size, Pore Size Distribution, Effective Porosity, Roughness, and Pore Tortuosity

12.3.5.1 Gas Permeation Test Gas permeation test has been used by various groups (Deshmuck and Li, 1998; Wang et al., 1999; Khayet and Matsuura, 2001) to determine both the mean pore size (d_p) and the effective porosity (ε/L_p), defined as the ratio of the porosity and the effective pore length (L_p) that takes into account the tortuosity (τ) of the membrane pores. Air and nitrogen can be employed as a standard gas. The gas permeation flux of each dry membrane is measured at various transmembrane pressures. The permeate pressure can be allowed at atmospheric pressure or under lower pressure. In general, the gas permeance contains both the diffusive term and the viscous term that depends on the mean pressure. By plotting a linear dependence between the permeance and the mean pressure, the intercept and the slope can be determined, and consequently the pore size and the effective porosity can be calculated as stated in the literature (Deshmuck and Li, 1998; Wang et al., 1999; Khayet and Matsuura, 2001). It must be mentioned that one of the limitations of this method is the determination of the pore size distribution. However, recently, Kong and Li (2001) reported a procedure to determine the pore size distribution using the gas permeation test. Both standard normal distribution and log-normal distribution functions were used to represent the pore size distribution.

It must be pointed out that if the effective membrane porosity (ε/L_p), the membrane thickness (δ), and porosity (ε), are known, the tortuosity of the membrane pores can be calculated (Khayet et al., 2004a, 2004b):

$$\tau = \frac{\varepsilon}{\delta(\varepsilon/L_p)} \tag{12.4}$$

In MD studies, a value of 2 is frequently assumed for the tortuosity factor to predict the MD fluxes (Schofield et al., 1987). The calculated tortuosity factor by Khayet et al. (2004a, 2004b) was 1.59, 2.14, and 2.12 for the membranes TF200, GVHP, and HVHP, respectively.

Khayet et al. (2004a, 2004b) determined the mean pore size and the effective porosity by the gas permeation test of three commercial membranes (TF200, GVHP, and HVHP). The mean pore size of the membrane TF200 (198.96 nm) was found to be almost the same as the value given by the manufacturer (Table 12.1), while those obtained for the membranes GVHP (283.15 nm) and HVHP (463.86 nm) were slightly higher.

12.3.5.2 Wet/Dry Flow Method The bubble point together with the gas permeation test, known also as the wet/dry flow method or liquid displacement method, can be employed for determining the maximum pore size, the mean pore size, and the pore size distribution of MD membranes. First, the gas permeation is measured through a dry membrane. Generally, a straight line is observed between the gas permeation and the transmembrane pressure difference. Subsequently, the membrane is wetted by a liquid with low surface tension such as isopropyl alcohol (IPA), which is assumed to fill all the pores, and again the gas permeation is measured at increasing transmembrane pressures. In this case, the dependence of the gas flux with the applied transmembrane pressure is not linear as the membrane pores are flooded with IPA. As the pressure increases, it will reach a point where it can overcome the surface tension of the liquid in the largest pores and will push the liquid out. At low transmembrane pressure, the pores remain filled with IPA and the gas flux is practically zero, whereas for higher pressure than the bubble point of the membrane, the largest pores will be empty and the gas flux starts to increase with the pressure because smaller pores are opened with the increase of the pressure until all pores become empty at the pressure corresponding to the minimum pore size. Both the wet and dry steps must be carried out under the same temperature.

Generally, the experiment is taken at room temperature, maintaining the downstream side of the system at atmospheric pressure. The method suggested by both Kesting (1985) and Khayet et al. (2004a) can be used to obtain the pore size distribution. The cumulative pore size and the probability density curves of different membranes (TF200, TF450, TF1000, GVHP, and HVHP indicated in Table 12.1) are illustrated in Figure 12.14. A detailed explanation of the method used was given by Khayet et al. (2004a).

It must be pointed out that a Coulter Porometer II, manufactured by Coulter Electronics Ltd., is based on the same procedure. In this system the membrane sample is wetted with a Coulter Porofilm of low surface tension, low vapor pressure, and low reactivity, which is assumed to fill all the membrane pores.

It was found that the mean pore sizes of the membranes determined from the wet/dry flow method were more than 2.7% lower than those determined by means of the gas permeation test (Khayet et al., 2004a). Martínez et al. (2002, 2003), by using a similar wet/dry flow method, obtained pore size distributions of various porous hydrophobic membranes (TF200, TF450, TF100, GVHP, and HVHP) with mean pore sizes larger than those

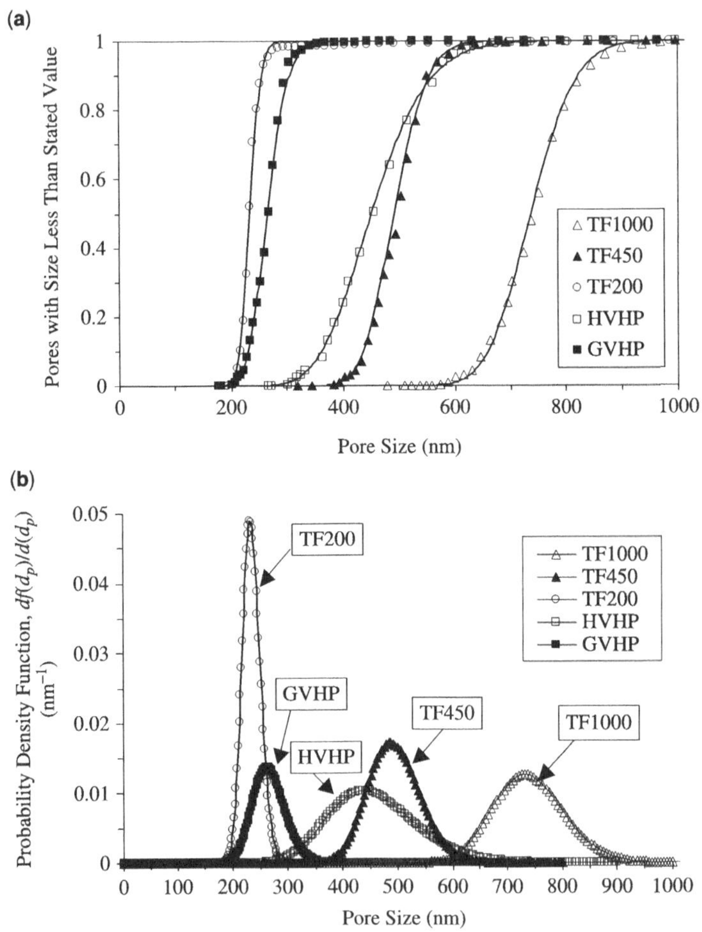

Figure 12.14 (a) Cumulative pore size and (b) probability density curves of different membranes indicated in Table 12.1.

reported by Khayet et al. (2004a). They were 1.32, 1.20, and 1.46 times higher for TF200, GVHP, and HVHP membranes, respectively.

Mercury porosimetry and measurements of gas–liquid displacement porometry have been carried out by Martínez-Díez and Vázquez-González (1996) to determine the pore size distribution of the membranes TF200, TF450, and HVHP (Table 12.1). The mercury intrusion was carried out using a mercury porosimeter Quanta-Chrome 33000. In this case a weighed amount of membrane was intruded into the porosimeter chamber and evacuated to a certain pressure during about 10 h. Subsequently, mercury was put into the chamber, and the pressure was then increased to fill the pores of the membrane with mercury. The pore radius was calculated using the known Washburn equation. By monitoring the volume change of mercury and the corresponding pressure, the pore size distribution of the membrane was calculated by a computer program. One of the disadvantages of this technique is the membrane compaction caused by the high pressures used in the mercury intrusion measurement as well as the formation of microfractures with a consequent alteration of pore shape and size. Moreover, mercury cannot differentiate between opened pores and dead-ended pores since it intrudes from all directions. All these shortcomings can be avoided by using the wet/dry flow method.

12.3.5.3 Scanning Electron Microscopy This technique permits the direct visualization of the morphological structure of porous membrane (top and bottom surfaces as well as membrane cross section). SEM was used in different MD studies (Sakai et al., 1988; Wu et al., 1992; Fujii et al., 1992b; Ortiz de Zárate et al., 1995; Tomaszewska, 1996; Feng et al., 2004a, 2004b), and the mean pore sizes were calculated. Before the SEM test, the membrane samples must be first frozen in liquid nitrogen and then broken to obtain small fragments. However, one of the limitations of this technique is the heavy-metal coating required for membrane sample preparation, which gives some artefacts and tends to damage the membrane surface.

Phattaranawick et al. (2003b) by using the field emission scanning electron microscopy (FESEM) reported mean pore sizes of 251 nm for GVHP membranes and 414 nm for HVHP membranes, which are slightly lower than those obtained from the gas permeation test and the wet/dry flow method. As stated previously, this difference may be attributed to pore contraction during metal coating of the membrane sample, which is required to take the FESEM images. The authors (Phattaranawik et al., 2003b) assumed the log-normal distribution to plot the pore size distribution of each membrane.

12.3.5.4 Atomic Force Microscopy Atomic force microscopy (AFM) is a newly developed high-resolution technique to study the surface morphology of the membranes. Three-dimensional images of the MD membrane surface can be obtained directly without special sample preparation. In other words, a more accurate surface structure of an MD membrane can be observed by AFM. In addition, the AFM may be used in a number of different modes: contact mode, noncontact mode, and tapping or intermittent mode. The mean pore size, the pore size distribution, the surface porosity as well as its roughness parameters (mean roughness, root mean square, average difference in height) can be determined.

Khayet et al. (2004c) were the first to characterize MD membranes by AFM. The mean pore size, pore size distribution, surface porosity, pore density, and roughness parameters of the prepared PVDF membranes for MD with the phase inversion method were determined. The membranes were prepared with different amounts of nonsolvent additive

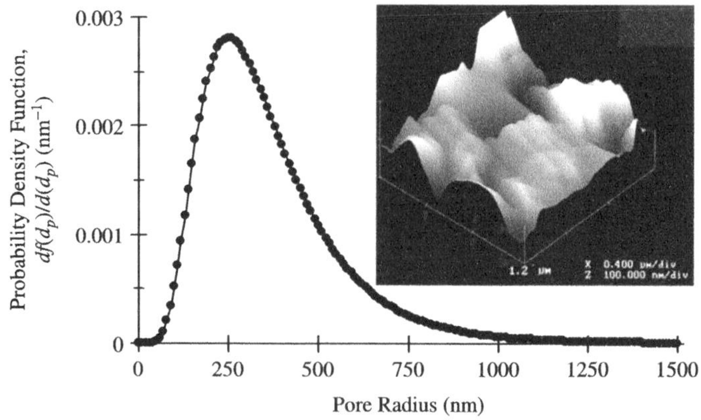

Figure 12.15 AFM image of HVHP (Table 12.1) membrane together with its pore size distribution.

(i.e., water) in the casting solution. Also, the two commercial PVDF membranes (GVHP and HVHP) were studied by AFM. Figure 12.15 shows as an example the AFM images of HVHP membrane together with its pore size distribution. A detailed analysis carried out to obtain the pore size distribution from AFM images was reported by Khayet et al. (2004c). Generally, the mean pore sizes determined by AFM were 1.2–2.1 times larger than those determined from the gas permeation test. This was attributed to the fact that the pores measured by AFM might have maximum openings at the surface entrance and a few small pores could easily be misinterpreted as one large pore when they are amalgamated, resulting in an overestimation of the pore sizes of the MD membranes.

Khayet and Matsuura (2003b) compared the surface and bulk pore size of various laboratory-made flat-sheet and hollow-fiber membranes obtained by different methods including the gas permeation test, the solute transport method, and the AFM analysis. The pore sizes of both the internal and external surfaces of the hollow-fiber membranes as well as the pore sizes of the top and bottom surfaces of flat-sheet membranes were evaluated using AFM analysis. All techniques showed the same tendency in the pore size with the changes of the membrane preparation conditions. The pore sizes at the membrane surfaces were found to be larger than those in the bulk of the membrane. The pores at the bottom surfaces of flat-sheet membranes were 3.7–9.8 times larger than those at the top membrane surface, while the pores at the top surface were 2.1 times larger than the bulk pore sizes determined from the gas permeation test. For the PVDF hollow-fiber membranes, the pore size at the outer surface was 1.7 times larger than the bulk pore size determined from the gas permeation test, and the pore size of the inner surface was 1.1–1.4 times larger than the pore size of the outer surface.

Other characterization techniques such as the X-ray photoelectron spectroscopy (XPS) have also been used to analyze the compositional gradient of different species near the membrane surface when porous composite membranes are considered for MD applications (Khayet and Matsuura, 2003a). The relative atomic percentages of fluorine, nitrogen, oxygen, and carbon can be determined as a function of the depths into the membrane using the variable photoelectron take-off angle method.

Additionally, although the MD membranes are not required to be mechanically resistant like the membranes used in the pressure-driven membrane separation processes, the mechanical properties are of interest for the sake of comparison. Feng et al. (2004a, 2004b) studied the mechanical properties of phase inversion membranes prepared from PVDF and the copolymer poly(vinylidene fluoride-*co*-tetrafluoroethylene) under the same conditions using an Instrom 1121 test. Parameters such as the maximal load during tensile, the stretching strength, stretching strain, elastic tensile modulus, the load at break, for example, were determined. The researchers found better mechanical properties of the membranes prepared from the copolymer poly(vinylidene fluoride-*co*-tetrafluoroethylene) than those of PVDF membrane.

Recently, Khayet et al. (2005b) presented a method to determine the thickness of the hydrophobic layer of composite porous hydrophobic/hydrophilic membranes used in MD. The results of different DCMD experiments together with a theoretical model involving the structural characteristics of the membranes and the heat and mass transfer mechanisms were considered. The proposed method can be used to measure the layer thickness of porous composite hydrophobic/hydrophilic membranes with random uncertainties less than 5% and maximum deviation of each individual thickness from the corresponding average value less than 9%. The DCMD parameters were found to exert no effect on the calculated thickness.

12.4 TRANSPORT MECHANISMS IN MD: TEMPERATURE POLARIZATION, CONCENTRATION POLARIZATION, AND THEORETICAL MODELS

12.4.1 Mass Transfer through the Membrane

Transport of gases and vapors through porous media has been extensively studied, and theoretical models have been developed based on the kinetic theory of gases to predict the MD performance of the membranes depending on the MD configuration used (Present, 1958; Mason and Malinauskas, 1983; Lawson and Lloyd, 1997; Khayet et al., 2000c, 2001; Martínez et al., 2002, 2003; Phattaranawik et al., 2003b). The different types of mechanisms proposed for the mass transport through MD membranes are the Knudsen flow model, the viscous flow model, the ordinary molecular diffusion model, and/or the combination thereof often summarized as the dusty gas model. In MD theoretical studies, generally a membrane of uniform and noninterconnected cylindrical pores is assumed. Recently, the pore size distribution of MD membranes rather than their uniform pore size have been considered in DCMD configuration by Martínez et al. (2002, 2003), Laganà et al. (2000), Phattaranawik et al. (2003b), and Khayet et al. (2004a) and in VMD configuration by Khayet and Matsuura (2004). A three-dimensional network of interconnected cylindrical pores with a pore size distribution was considered for Monte Carlo simulation of DCMD by Imdakem and Matsuura (2004, 2005). Knudsen and viscous flows were considered in DCMD ignoring both the ordinary diffusion type of flow, which must be applied whenever air is present in the membrane pores, and the transition (i.e., combined) flow when the pore size is close to the mean free path of the transported molecules through the membrane pores. More elaborated theoretical model and comparisons to the experimental results are needed before confirming the goodness of the proposed three-dimensional network model.

It must be pointed out that the governing quantity, which provides a guideline in determining which mechanism is operative in a given pore under given experimental conditions, is the Knudsen number (Kn) defined as the ratio of the mean free path, λ, of the transported molecules to the pore size of the membrane. The mean free path, λ_i, for specie i can be calculated using the following expression (Matsuura, 1994; Lawson and Lloyd, 1997):

$$\lambda_i = \frac{k_B T}{\sqrt{2}\pi \bar{p} \sigma_i^2} \tag{12.5}$$

where σ_i is the collision diameter (2.641 Å for water vapor), k_B is the Boltzmann constant, $\bar{p}$ the mean pressure within the membrane pores, and T the absolute temperature.

For the binary mixture (i and j) in air, the mean free path can be evaluated by the following equation:

$$\lambda_{i/j} = \frac{k_B T}{\pi \bar{p}[(\sigma_i + \sigma_j)/2]^2} \frac{1}{\sqrt{1 + M_j/M_i}} \tag{12.6}$$

where σ_i and σ_j are the collision diameters and M_i and M_j the molecular weight of the components i and j, respectively.

In DCMD, the mean free path for water vapor at 50°C under atmospheric pressure is approximately 0.14 μm, which is around the pore sizes of the membranes used in MD. However, in VMD, the mean free path value is higher due to the low pressure in the permeate side. This indicates that the physical nature of mass transport may be different when using the same membrane under different MD configurations. Furthermore, for the membrane having a pore size distribution, different mechanisms may occur simultaneously.

12.4.1.1 *Direct-Contact Membrane Distillation Configuration* If the mean free path of the transported molecules is large in comparison with the membrane pore size (i.e., $\mathrm{Kn} > 1$ or $d_p < \lambda_i$), the molecule–pore wall collisions are dominant over the molecule–molecule collisions, and the Knudsen type of flow is responsible for mass transfer through the membrane pore. In this case the permeability through each pore in the Knudsen region was expressed as follows (Matsuura, 1994; Khayet et al., 2004a):

$$B_i^K = \frac{2\pi}{3} \frac{1}{RT} \left(\frac{8RT}{\pi M_i} \right)^{1/2} \frac{r_k^3}{\tau \delta} \tag{12.7}$$

where r_k is the pore radius in the Knudsen region, M_i is the molecular weight of specie i, R is the gas constant, and δ is the membrane thickness. It must be mentioned here that Matsuura (1994) stated that the Knudsen type of flow is predominant when the ratio of the pore radius to the mean free path (i.e., r_p/λ_i) is lower than 0.05.

If $\mathrm{Kn} < 0.01$ (i.e., $d_p > 100\ \lambda_i$), molecular diffusion was used to describe the mass transport in the continuum region caused by the virtually stagnant air trapped within each membrane pore due to the low solubility of air in water, which is the major component in feed and permeate solutions used in DCMD. Schofield et al. (1987, 1990a, 1990b, 1990c) stated that the air flux across the membrane is extremely small relative to the flux of water, and viscous flux can be neglected unless the liquid solutions are degassed. The following relationship was used to determine the MD permeability through a pore having an area of πr_D^2 in the ordinary diffusion region (Kimura and Nakao, 1987; Khayet et al., 2004a):

$$B_i^D = \frac{\pi}{RT}\frac{PD_i}{p_a}\frac{r_D^2}{\tau\delta} \tag{12.8}$$

where D_i is the diffusion coefficient, P is the total pressure inside the pore, and p_a is the air pressure in the membrane pore. For water–air, PD_i (Pa m^2/s) can be calculated from the following equation (Khayet et al., 2004a; Phattaranawik et al., 2003b):

$$PD_i = 1.895 \times 10^{-5} T^{2.072} \tag{12.9}$$

In transition region, $0.01 < \text{Kn} < 1$ (i.e., $\lambda_i < d_p < 100\,\lambda_i$), the mass transport takes place via a combined Knudsen/ordinary diffusion mechanism. In this case, the following model was used to determine the permeability through the membrane pores (Khayet et al., 2004a):

$$B_i^C = \frac{\pi}{RT}\frac{1}{\tau\delta}\left[\left(\frac{2}{3}\left(\frac{8RT}{\pi M_i}\right)^{1/2} r_t^3\right)^{-1} + \left(\frac{PD_i}{p_a} r_t^2\right)^{-1}\right]^{-1} \tag{12.10}$$

where r_t is the pore radius in the transition region.

The dusty gas model (Mason and Malinauskas, 1983; Lawson and Lloyd, 1997), considered to be a more complete model for all MD configurations, has been applied by Lawson and Lloyd (1996b) to predict the DCMD flux. This model combines all transport mechanisms through the membranes: Knudsen diffusion, molecular diffusion, and viscous flow. For DCMD mode, the more complete equations were reduced to the Knudsen/ordinary molecular diffusion transition form (Lawson and Lloyd, 1996b). Due to the fact that in DCMD configuration both feed and permeate aqueous solutions are brought into contact with the membrane under atmospheric pressure, the total pressure is maintained constant at $\approx$1 atm, resulting in negligible viscous flow (Lawson and Lloyd, 1996b).

Khayet et al. (2000c) studied the physical nature of mass transport in DCMD. A uniform pore size (i.e., mean pore size) was assumed for the membranes TF200, TF100, FGLP, and FHLP indicated in Table 12.1. The DCMD experiments were performed under different operation conditions. They concluded that the mass transport mechanism takes place via the combined Knudsen/molecular diffusion mechanism for the studied membranes having mean pore sizes between 0.2 and 1 μm. It must be pointed out that the pore tortuosity was taken by the authors as an adjustment parameter.

It must be mentioned here that for given experimental conditions, the calculated DCMD flux considering Knudsen mechanism is higher than that considering the combined Knudsen/molecular diffusion mechanism. This fact indicates that when the pore size is near the critical pore size (i.e., mean free path of the transported molecules through the membrane pore), the DCMD flux will not necessarily increase with the increase of the pore size. Therefore, under some operating conditions, it would be better to use membranes with lower pore size than the corresponding mean free path so that Knudsen type of flow will take place, leading to higher DCMD flux than the membranes having larger pore sizes where the combined Knudsen/molecular diffusion flux is that responsible for mass transfer. Therefore, care must be taken to choose the appropriate membrane pore size, taking into account the value of the mean free path of the molecules transported from the feed solution and trying to drive the membrane to work under the Knudsen type of flow. This fact must be considered during the design of DCMD membranes.

As stated previously, for MD membranes having pore size distributions, more than one mechanism of mass transport can occur simultaneously through the entire membrane. The following equation was proposed by Khayet et al. (2004a) to determine the total DCMD permeability of the membrane:

$$B_i^m = \frac{N}{\delta}\left[\sum_{j=1}^{m(r=0.5\lambda)} G_i^K f_j r_j^3 + \sum_{j=m(r=0.5\lambda)}^{p(r=50\lambda)} \left(\frac{1}{G_i^K r_j} + \frac{1}{G_i^D}\right)^{-1} f_j r_j^2 + \sum_{j=p(r=50\lambda)}^{n(r=r_{\max})} G_i^D f_j r_j^2\right] \tag{12.11}$$

where

$$G_i^K = \left(\frac{32\pi}{9M_iRT}\right)^{1/2} \tag{12.12}$$

$$G_i^D = \frac{\pi}{RT}\frac{PD_i}{p_a} \tag{12.13}$$

and f_j is the fraction of pores with pore radius r_j, N is the total number of pores per unit area, m is the last class of pores in the Knudsen region, and p is the last class of pores in the transition region. It is to be noted that in Eq. (12.11) the upper limit of each summation is altered depending on the maximum pore radius ($r_{\max}$). For example, if $r_{\max} < 0.5\ \lambda_i$, only the Knudsen mechanism can be applied. If $0.5\ \lambda_i < r_{\max} < 50\ \lambda_i$, both Knudsen and transition mechanisms are applicable; and finally if $r_{\max} > 50\ \lambda_i$, all mechanisms are operative simultaneously.

Furthermore, when a uniform pore radius (r) is assumed for MD membrane, Eq. (12.11) can be simplified to:

$$B_i^m = \frac{2}{3RT}\frac{\varepsilon(r)}{\tau\delta}\left(\frac{8RT}{\pi M_i}\right)^{1/2} \tag{12.14}$$

when the Knudsen model is applied ($(r) < 0.5\lambda$), and to the following expression:

$$B_i^m = \frac{1}{RT\delta}\left[\frac{3\tau}{2\varepsilon(r)}\left(\frac{\pi M_i}{8RT}\right)^{1/2} + \frac{p_a\tau}{\varepsilon PD_i}\right]^{-1} \tag{12.15}$$

when the transition flow dominates ($0.5\lambda_i < (r) < 50\lambda_i$).

Laganà et al. (2000) reported that the flux of commercial porous membranes, calculated assuming all pores having the same size and the one calculated with a Gaussian (symmetric) function are similar and the predicted fluxes were lower than the experimental ones. Phattaranawik et al. (2003b) by using commercial membranes (GVHP, HVHP in Table 12.1) also concluded that the influence of pore size distribution on the predicted DCMD flux was insignificant. However, in their calculations an adjustment factor (i.e., pore tortuosity) was assumed, and mean pore sizes in each mass transfer region (i.e., Knudsen region, continuum region, and transition region) were considered. Khayet et al. (2004a) found slightly higher membrane permeability when calculated including pore size distribution than that predicted from mean pore sizes. The authors explained that this slight effect was attributed to the low values of the geometric standard deviations of the commercial membrane pore size distribution. They stated that larger discrepancy may be detected if laboratory-made membranes with broader pore size distributions are used. By using their theoretical model, Khayet et al. (2004a) found an increase of the predicted water vapor permeability with an increase of the geometric standard deviation of the pore size distribution.

It must be mentioned that the presence of air within the membrane pores between the feed and permeate liquid–vapor interfaces hinders the mass transfer resulting in a reduction of the DCMD flux. Deaerated DCMD systems were proposed by Schneider and van Gassel (1984), Schofield et al. (1990a, 1990b), and Fane et al. (1987). For membrane having small pore sizes, Knudsen flow is predominant, and the removal of air results only in small increase of the DCMD flux; however, for membranes having larger pores a substantial increase in DCMD flux can be achieved by deareation. This can be carried out by lowering the pressure of the liquid streams, hence controlling the maximum pressure of gas within the membrane pores.

12.4.1.2 Vacuum Membrane Distillation Configuration In VMD, the driving force is maintained by applying vacuum at the permeate side below the equilibrium vapor pressure. In this configuration, the ordinary molecular diffusion resistance is neglected because it depends on the partial pressure of air in the membrane pores, and in VMD only traces of air are present within the membrane pores. Thus, a Knudsen type of diffusion was established theoretically by various authors (Lawson and Lloyd, 1997; Bandini et al., 1997; Couffin et al., 1998; Urtiaga et al., 2000; Khayet and Matsuura, 2001; Bandini and Sarti, 2002; Mengual et al., 2004; Banat et al., 2005). However, in VMD, because of the applied transmembrane hydrostatic pressure and for membranes with large pore size compared to the mean free path of the transported molecules through the membrane pores (i.e., $r_p > 50\,\lambda_i$) the molecule–molecule collisions will dominate and viscous flow occurs (i.e., Poisseuille flow) (Lawson and Lloyd, 1997; Khayet and Matsuura, 2004; Khayet et al., 2004c). In this case, permeability of specie i through a single pore was expressed as

$$B_i^v = \frac{\pi r_v^4}{8\eta_i}\frac{\bar{p}}{RT}\frac{1}{\tau\delta} \tag{12.16}$$

where r_v is the pore radius in Poisseuille region, η_i the viscosity of specie i, and $\bar{p}$ the average pressure in the pore. When the pore radius, r_p, is between $0.05\,\lambda_i$ and $50\,\lambda_i$, both molecule–molecule and molecule–pore wall interactions have to be considered, and the pores contribute to the total mass transport by a mechanism operative in the Knudsen–viscous transition region. In this case, the VMD permeability of specie i

through a single pore was described by the following expression (Khayet and Matsuura, 2004; Khayet et al., 2004c):

$$B_i^t = \frac{\pi}{RT\tau\delta}\left[\frac{2}{3}\left(\frac{8RT}{\pi M_i}\right)^{1/2} r_t^3 + \frac{r_t^4}{8\eta_i}\bar{p}\right] \tag{12.17}$$

where r_t is the pore radius in the transition region. Lawson and Lloyd (1996a) stated that for VMD configuration, the more complete equations of dusty gas model reduced to the Knudsen–viscous transition flow [Eq. (12.17)].

For a membrane with a distribution of pore size applied in the VMD system, all mechanisms can occur simultaneously, depending on the operating condition. The total VMD permeability of the membrane was calculated using the following equation (Khayet and Matsuura, 2004; Khayet et al., 2004c):

$$B_i^m = \frac{N}{\tau\delta}\left(\sum_{j=1}^{m(r=0.05\lambda)} G_i^k f_j r_j^3 + \sum_{j=m(r=0.05\lambda)}^{p(r=50\lambda)} (G_i^k f_j r_j^3 + G_i^v f_j r_j^4 \bar{p}) + \sum_{j=p(r=50\lambda)}^{n(r=r_{\max})} G_i^v f_j r_j^4 \bar{p}\right) \tag{12.18}$$

where

$$G_i^k = \left(\frac{32\pi}{9M_i RT}\right)^{1/2} \tag{12.19}$$

$$G_i^v = \frac{\pi}{8\eta_i}\frac{1}{RT} \tag{12.20}$$

and f_j is the fraction of pores with pore radius r_j, N is the total number of pores per unit area as stated earlier, m is the last class of pores in the Knudsen region, and p is the last class of pores in the transition region. It is to be noted also here that in Eq. (12.18) the upper limit of each summation may be altered by the relative values of the maximum pore radius ($r_{\max}$).

In MD processes, the transport of molecules through the membrane matrix (i.e., surface diffusion) is neglected due to the fact that the diffusion area of the membrane matrix is small compared to the pore area (Lawson and Lloyd, 1997). For hydrophobic MD membranes, the "affinity" between water and the membrane material is very low, and it may be allowed to neglect the contribution of transport through the membrane matrix, especially for porous membranes with large pore sizes and high porosities. Nevertheless, when other compounds are present in the feed solution, especially for compounds having strong affinity with the membrane material, the transport mechanism through the matrix of the membrane may have a significant effect. Systematic studies are needed to clarify this point in MD, although Fujii et al. (1992a, 1992b) reported that for membranes with small pore sizes (<0.02 μm), surface diffusion may affect MD performance. Recently, Khayet and Matsuura (2004) in their theoretical model considered mass transport through the membrane matrix by the solution–diffusion mechanism and realized an extensive comparative study between pervaporation (PV) and VMD.

12.4.1.3 SGMD Configuration The physical nature of the mass transport through microporous and hydrophobic membranes was analyzed for the SGMD process by Khayet et al. (2000a; 2000c) using two PTFE membranes (TF200 and TF450). A theoretical model that permits determining the local SGMD flux from which the overall flux for the complete membrane module was developed. A comparison between the theoretical predictions of the SGMD fluxes and the experimental results permitted researchers (Khayet et al., 2000c) to conclude that the combined Knudsen–molecular diffusive flux is responsible for the water transport through the membrane as occurred for DCMD. The same equations presented previously for DCMD configuration can be used in SGMD. Studies on the effect of pore size distribution and the nature of fluxes through laboratory-made membranes with broader pore sizes are not conducted yet and may help understand the nature of mass transport through membrane pores in SGMD configuration.

12.4.1.4 Air Gap MD and Thermostated SGMD Configurations In AGMD configuration, Udriot et al. (1994) assumed the transport of vapors across the membrane to be described by the theory of molecular diffusion, admitting the air inside the pores of the membrane and in the gap width as a stagnant film. Banat and Simandl (1998, 1999) used Stefan diffusion and binary-type relations (i.e., Fick's equation of molecular diffusion) to describe the multicomponent mass transfer in AGMD system. Gostoli and Sarti (1989) looked at using Stefan–Maxwell equations. In all these models, to account for membrane parameters, the diffusion coefficient was multiplied by the porosity and divided by the effective length of the membrane pores (i.e., the product of pore tortuosity and membrane thickness). However, the pore size was not considered in their equations, although experimental studies proved the dependence of the AGMD flux on this parameter (Izquierdo-Gil et al., 1999b). In fact, systematic studies on the effect of pore size and pore size distribution on the AGMD process performance has not been performed yet.

Moreover, Fickian approach assumes that the rate of diffusion of species depends only on its concentration gradient, and coupling interactions between the diffusing species, which may occur, were neglected. Banat et al. (1999a, 1999b, 1999c) used the Stefan–Maxwell formalism, taking into consideration all diffusional interactions between the diffusing species as well as the temperature and concentration polarization effects. The predictions based of the Stefan–Maxwell model were closer to the experimental data than those based on the Fickian mass transfer mathematical model. Rivier et al. (2002) also used the Stefan–Maxwell equations to predict the performance of the thermostated SGMD configuration. The model predictions were compared with the experimental results using one membrane, and a good agreement between fluxes were found (García-Payo et al., 2002). The Stefan–Maxwell model must be checked for other membranes of different materials and parameters. Recently, attempts were made by Guijt et al. (2005) to predict the AGMD performance process using the dusty gas model that takes into account all membrane parameters to describe the simultaneous Knudsen diffusion, molecular diffusion, and viscous flow. The results seem promising.

12.4.2 Heat Transfer through the MD Membrane

In MD, the heat transfer within the membrane is due to the latent heat accompanying vapor flux and the heat transferred by conduction across both the membrane material and the gas-filled membrane pores. The following equation was applied for DCMD and SGMD configurations (Schofield et al., 1987, 1990a, 1990b, 1990c; Martínez-Díez and

Vázquez-González, 1996; Gryta et al. 1997; Mengual and Peña, 1997; Gryta and Tomaszewska, 1998; Khayet et al., 2000c, 2001, 2003c, 2004a; Phattaranawik et al., 2003a).

$$Q_m = \frac{k_m}{\delta}(T_{m,f} - T_{m,p}) + \sum_{i=1}^{s} J_i^t \, \Delta H_{v,i} \tag{12.21}$$

where k_m is the thermal conductivity of the membrane, which can be calculated by Eqs. (12.1) and (12.2), $\Delta H_{v,i}$ is the evaporation enthalpy of specie i of the transmembrane flux J_i^t, s is the number of permeated components, and $T_{m,f}$ and $T_{m,p}$ are the feed and permeate temperatures, respectively.

It is worth quoting that the total heat flux transferred through the membrane, typically 50–80% is consumed as the latent heat for permeate production, while the remainder is lost by thermal conduction. In addition to the different methods of minimizing heat loss through the membrane commented previously in Section , the heat loss becomes less significant by letting the MD system to work under higher operating temperatures.

In VMD, the boundary layer resistance in the permeate side and the contribution of the heat transported by conduction through the membrane are negligible (Lawson and Lloyd, 1996a; Bandini et al., 1997; Lawson and Lloyd, 1997). This makes VMD of pure water useful to determine the temperature of the feed solution at the membrane surface ($T_{m,f}$), and therefore the boundary layer heat transfer coefficients in the membrane module can be evaluated (Mengual et al., 2004). This helps in selecting the adequate empirical heat transfer correlation of a given MD system, which is a complex task when developing theoretical models to determine the temperature polarization coefficients.

12.4.3 Heat and Mass Transfer Boundary Layers: Temperature and Concentration Polarization Effects

In various MD studies, the permeate flux of a specie i, J_i, is linearly related to its transmembrane partial pressure, Δp_i (Lawson and Lloyd, 1997):

$$J_i^t = B_i^m \, \Delta p_i \tag{12.22}$$

On the other hand, the total heat flux, Q, transferred from the feed to the permeate side was expressed as (Lawson and Lloyd, 1996b; Khayet et al., 2004d)

$$Q = \left(\frac{1}{h_f} + \frac{1}{(k_m/\delta) + \sum_{i=1}^{s} J_i^t \Delta H_{v,i}/(T_{m,f} - T_{m,p})} + \frac{1}{h_p} \right)^{-1} (T_f - T_p) \tag{12.23}$$

where h_f and h_p are the heat transfer coefficients on the feed and permeate sides of the membranes, respectively; and T_f and T_p are the feed and permeate temperatures at the bulk fluids, respectively.

Equation (12.22) seems simple and trivial. However, Δp_i depends on both temperatures and concentrations at the membrane surfaces, which are not directly measurable and are different from those at bulk phases due to the simultaneous heat and mass transfers through the membrane. These phenomena are called temperature polarization and concentration polarization and are a major problem for MD. Therefore, the overall MD process rate is controlled by the heat and mass transfer through both the membrane and fluid phases. Figure 12.4a depicts as an example the temperature and concentration profiles for single-layer hydrophobic porous membrane when nonvolatile solutes are present in the feed side for DCMD and SGMD configurations. For the other MD configurations, a slight change may be applied in the permeate side. However, when volatile solutes are present in the feed side, the concentration of the components having the highest vapor pressure will be lower at the feed membrane surface and higher at the permeate membrane surface than in the corresponding bulk phases. When other types of membranes, such as composite porous membranes, are considered, changes must be applied on the temperature polarization profile across the membrane (Fig. 12.4b).

The temperature polarization coefficient (TPC) in MD is frequently defined as the ratio of the feed and permeate temperature difference at the membrane surfaces to that in the bulk phases (Schofield et al., 1987). Similarly to what occurs to the temperature, the concentration cannot be directly measured at the membrane surfaces but may only be obtained using some approximations. The concentration polarization coefficient (CPC) was defined as the ratio of the feed concentration at the membrane surface to that at the bulk phase. The Nernst film model that neglects the eddy and thermal diffusions in relation to the ordinary diffusion was frequently used in MD to relate the solute concentration at the membrane surface to that at the bulk solution (Schofield et al., 1990a; Tomaszewska et al., 1995; Lawson and Lloyd, 1996a; Mengual and Peña, 1997; Banat and Simandl, 1998; Martínez-Díez and Vázquez-González, 1999). Furthermore, as the driving force in MD is the vapor pressure difference, which depends on both temperature and concentration, Martínez-Díez and Vázquez-González (1999) and Khayet et al. (2004d) defined a general coefficient termed vapor pressure polarization coefficient (VPPC) as the fraction of the externally applied driving force that contributes to the mass transfer. When water was used as feed, the temperature and vapor pressure polarization coefficients were found to differ in less than 0.6%, but the difference was higher when salt aqueous solutions were considered (Martínez-Díez and Vázquez-González, 1999). Additionally, when nonvolatile solutes were considered, concentration polarization was found to be insignificant (i.e., CPC $\approx$ 1.1) compared to temperature polarization (Schofield et al., 1990a; Lawson and Lloyd, 1996b; Sudoh et al., 1997; Laganá et al., 2000; Khayet et al., 2004d).

Different theoretical approaches and a large number of studies have been undertaken to analyze the influence of temperature and concentration polarization on the performance of the different MD configurations (Schofield et al., 1987, 1990a; Velázquez and Mengual, 1995; Martínez-Díez and Vázquez-González, 1996; Lawson and Lloyd, 1996b; 1997; Mengual and Peña, 1997; Martínez-Díez and Vázquez-González, 1999; Khayet et al., 2002c, 2004d). The TPC of an ideal MD system should be equal to unity; however, due to the unavoidable presence of boundary layers adjoining the membrane surfaces, most of the MD systems exhibit lower TPC values and the MD process is heat transfer limited. In contrast, for MD systems having TPC values as close to unity as possible, the MD process is mass transfer controlled only by the permeability of the membrane (Fane et al., 1987; Schofield et al., 1990c). It is worth mentioning that the TPC was considered a tool to design MD systems (Schofield et al., 1987; Lawson and Lloyd, 1996a; Xu et al., 2004). The closer the TPC was to unity, the better the heat transfer in the module, and the more suitable for MD applications. By estimating the TPC for a variety of MD geometries Schofield et al. (1987) concluded that tubular membrane systems with turbulent liquid flow, and hollow-fiber membranes under laminar flow regime, are the most thermally efficient MD systems exhibiting low TPC values at the internal side of the membrane,

0.9 and 0.85, respectively. Further, they stated that the temperature polarization at the external side of the membrane tubes may be excessive leading to lower TPC (0.62–0.65). Lawson and Lloyd (1996b) reported a TPC value as high as 0.85 for a DCMD laboratory plate and frame membrane module, although the commonly observed values were within 0.4 and 0.7 (Fane et al., 1987).

In SGMD configuration, lower TPC coefficients (<0.44) were found by Khayet et al. (2000a, 2002c) when using plate-and-frame PTFE membrane modules. The researchers attributed the lower TPC values to the predominant effect of the gas boundary layer. In VMD a value as high as 1 was recently reported by Banat et al. (2005). Phattaranawik et al. (2001) observed a substantial increase of the TPC approaching unity in DCMD when spacers were used in the channels. In general, it is agreed upon that the temperature and concentration polarization coefficients depend strongly on fluid dynamics, membrane characteristics, and temperature. These coefficients can be reduced by increasing the flow rates of the fluids (i.e., feed and/or permeate velocities), with the inclusion of turbulence promoters in the membrane module channels, by decreasing the section of fluid channels, by using spacer-filled channels, and so forth.

In DCMD and SGMD, various MD researchers found that the TPC is dependent on temperature and flow rate. It becomes more significant at higher temperature (i.e., higher MD fluxes) and at lower flow rates (Velázquez and Mengual; 1995; Lawson and Lloyd, 1996b; Khayet et al., 2000a, 2002c; Phattaranawik et al., 2001). However, in VMD, Lawson and Lloyd (1996a) reported that the TPC values were maintained constant or increased with the feed temperature, depending on the membrane used. It must be mentioned that the same authors (Lawson and Lloyd, 1996b) by using the same membrane module but under DCMD mode, found a decrease of the TPC with the feed temperature. Khayet and Matsuura (2004) examined the effect of membrane pore size on the VPPC using VMD configuration. Smaller VPPC values were determined for membranes having larger pore size when water and chloroform-diluted aqueous solutions were used. Furthermore, it must be pointed out that in most of the reported MD studies, particularly in DCMD mode (Schofield et al., 1987, 1990a; Velázquez and Mengual, 1995; Mengual and Peña, 1997; Martínez-Díez and Vázquez-González, 1999), the temperature polarization effects were supposed to be similar on both sides of the membrane in spite of the fact that the adjoining fluids have different physicochemical properties. Khayet et al. (2002c, 2004d) took into account for the first time the different properties of the fluids adjoining the membrane surfaces when DCMD and SGMD were considered. They proved the existence of an asymmetric temperature polarization and vapor pressure polarization in both MD configurations by determining the TPCs and the VPPCs of the feed and permeate sides individually.

The effects of the different operating parameters have been studied and reported (Khayet et al., 2002c, 2004d). In SGMD, for example, the TPC increased slightly with the feed flow rate, increased more clearly with the air circulation velocity, decreased slightly with the gas (i.e., humid air) inlet temperature, decreased more clearly with the water inlet temperature, and a slight difference was observed between the TPCs of two PTFE membranes of different pore sizes (TF200 and TF450 indicated in Table 12.1), concluding that the mass transport is predominantly controlled by heat transfer through the boundary layers adjacent to the membrane surfaces, especially the gas boundary layer, as the heat transfer coefficient through the liquid phase is very large in comparison with the heat transfer coefficient in the gas phase. In DCMD (Khayet et al., 2002c), it was found that the temperature polarization in the feed side was higher than that in the permeate side; however, the VPPC in the feed side was lower than the corresponding one in the permeate side. This result was attributed to the exponential increase of the vapor pressure with temperature.

In a large number of MD studies carried out under the different known MD configurations, the TPC was related to the heat transfer coefficients in the feed and permeate boundary layers, and the heat transfer coefficients were further estimated by means of the empirical heat transfer correlations developed for different systems under different flow regimes (Tomaszewska et al., 1995; Lawson and Lloyd, 1996b; Bandini et al., 1997; Sudoh et al., 1997; Gryta and Tomaszewska, 1998; Banat and Simandl, 1998, 1999; Izquierdo-Gil et al., 1999a, 1999b; Martínez-Díez and Vázquez-González, 1999; Martínez and Florido-Díaz, 2001; Phattaranawik et al., 2001, 2003b; Khayet et al., 2003a, 2003c, 2004d; Khayet and Matsuura, 2004). The mass transfer analogy of the empirical heat transfer correlations was also applied by various MD investigators to determine the mass transfer coefficient through the boundary layers (Bandini et al., 1992, 1997; Tomaszewska et al., 1995; Sudoh et al., 1997; Banat and Simandl, 1998, 1999; Martínez-Díez and Vázquez-González, 1999; Bandini and Sarti, 1999; Urtiaga et al., 2000; Martínez and Florido-Díaz, 2001; Khayet et al., 2003c; Khayet and Matsuura, 2004). In fact, for some MD systems, the selection of the adequate empirical heat transfer correlation and the calculation of the dimensionless involved parameters (i.e., Reynolds numbers) are difficult especially in the presence of supports and channel spacers.

A critical review of the most frequently used empirical heat transfer correlations in MD studies was presented by Mengual et al. (2004), concluding that special care must be taken into account when the empirical correlations, developed originally for nonporous heat exchangers, are used in MD for prediction of the MD fluxes and calculation of the TPC. The characteristic constants in the empirical heat and mass transfer correlations must be reevaluated for application in MD systems as made in previous studies reported by Camera-Roda et al. (1994), Sudoh et al. (1997), and Lawson and Lloyd (1996a). A summary of the correlations commonly used for the estimation of the heat transfer coefficients under laminar flow was reported by Gryta and co-workers (Gryta et al., 1997; Gryta and Tomaszewska, 1998). They concluded that the correlations developed for heat exchangers can be used successfully for heat transfer calculations in MD modules; however, care must be exercised in selecting the most suitable correlation.

12.5 MEMBRANE DISTILLATION APPLICATIONS

12.5.1 Direct-Contact Membrane Distillation

The DCMD process is currently applied mostly at laboratory scale. Recently, some pilot plant applications have been proposed but still under test at laboratory stage (Zakrewska-Trznadel et al., 1999; Koschikowski et al., 2003). As stated previously, DCMD is the most studied MD configuration (Fig. 12.2), although the air entrapped within the membrane pores, which leads to a mass transfer inefficiency, and the heat loss by conduction through the membrane, which is considered to be heat loss and therefore a heat transfer inefficiency, are higher compared to the other MD configurations. Additionally, the performance of a specific DCMD system is highly dependent on the membrane used, the design of the membrane module, and the specific separation being performed. One of the inconveniencies of DCMD is the high risk of membrane pore wetting from the permeate side of the membrane during the separation of volatile organic compounds having lower surface tensions than water.

Generally, it is agreed upon the exponential increase of the DCMD flux with the feed temperature and/or mean temperature. The DCMD flux is higher at higher feed and permeate flow rates and decreases with the increase of the permeate temperature and nonvolatile solute concentration (Lawson and Lloyd, 1997). The major DCMD application has been in desalination for production of high-purity water (Weyl, 1967; Drioli and Wu, 1985; Hanbury and Hodgkiess, 1985; Drioli et al., 1986, 1987; Schneider et al., 1988; Schofield et al., 1990a; Ohta et al., 1990; Ortiz de Zárate et al., 1992; Lawson and Lloyd, 1996b; Sudoh et al., 1997; Izquierdo-Gil et al., 1999a; Martínez and Florido-Díaz, 2001; Karakulski et al., 2002; Hsu et al., 2002; Li and Sirkar, 2004). Near 100% rejection of nonvolatile elyctrolytes (i.e., NaCl, KCl, LiBr, etc.) and nonelectrolytes (i.e., glucose, sucrose, fructose, etc.) solutes in aqueous solutions was achieved. As the permeate product is very pure, it is suitable for use in the medical and pharmaceutical sectors. In fact, in the case of a solution with nonvolatile components, only water molecules flow through the membrane pores. Karakulski et al. (2002) reported a quality water as low as 0.8 μS/cm with 0.6 ppm TDS (total dissolved solids) when using Accurel S6/2 membrane module (Table 12.1).

Weyl (1967) was the first to conduct desalination by DCMD. However, the obtained permeate fluxes were up to 1 $kg/m^2 \cdot h$, which were lower than the reverse osmosis (RO) fluxes (20–75 $kg/m^2 \cdot h$). Drioli and Wu (1985) and Drioli et al. (1986) applied DCMD for the treatment of NaCl and sugar aqueous solutions at different concentrations and temperature gradients using both flat sheet (PTFE) and capillary PP and PVDF membranes. The rejection factor was shown to increase with the decrease of membrane pore size from 96.5 to 100%, and membranes having nominal pore sizes near 0.2 μm exhibited rejection factors always higher than 99% with permeate fluxes lower than 2.5 $kg/m^2 \cdot h$. Ohta et al. (1990), by performing experimental studies on seawater desalination using a plate-and-frame module containing 3.12 m^2 of nonporous silicone/polysulfone composite membrane, obtained DCMD fluxes lower than 2.5 $kg/m^2 \cdot h$. Martínez and Florido-Díaz (2001) were able to predict fluxes about 65 $kg/m^2 \cdot h$ for the DCMD desalination of seawater by HVHP and GVHP (Table 12.1) membranes at high feed temperature (i.e., 85°C). DCMD permeate fluxes as high as those of RO systems were obtained by Schneider et al. (1988), Schofield et al. (1990a), and recently by Li and Sirkar (2004). Additionally, Schofield et al. (1990a) stated that DCMD of aqueous salt solutions (up to 25%) and aqueous sucrose solutions (up to 30%) was able to give fluxes of 60–70% of that of pure water used as feed. Their analysis revealed that the flux reduction for aqueous NaCl salt solutions was largely due to vapor pressure reduction with small effect of increased viscosity, whereas for the aqueous sucrose solutions the viscosity was the major factor in reducing the DCMD flux.

Lawson and Lloyd (1996b) by using a microporous PP flat-sheet membrane (3ME in Table 12.1) reported DCMD fluxes two to three times higher (i.e., up to 130 $kg/m^2 \cdot h$ at a feed temperature of 78°C and a permeate temperature of 20°C) than those reported in the DCMD literature or RO with nearly total rejection of NaCl solute. Figure 12.16, which is adapted from the experimental along with the predicted results in Lawson and Lloyd (1996b), illustrates the effect of the feed temperature and salt concentration ranging from 0 to 1.3 mol % for the two membranes 3ME and 3MA. At this range of concentrations, it was found that the effect of feed concentration was almost negligible due to the high mass and heat transfer coefficients across the feed and permeate boundary layers

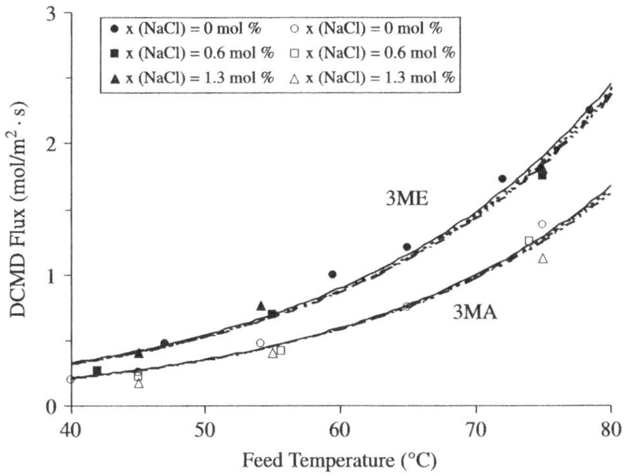

Figure 12.16 DCMD flux of NaCl aqueous solutions as a function of feed temperature for the membranes 3ME and 3MA indicated in Table 12.1. The permeate temperature was maintained at 20°C. (Adapted from Schofield et al., 1990a.)

(i.e., minimum effect of concentration and temperature polarization). It must be pointed out that other than the permeate flux, the advantage of DCMD desalination is the observed high rejection factor, which may not be accomplished by RO at high fluxes.

On the other hand, due to the fact that DCMD can be conducted at relatively low feed temperatures, it was successfully tested in many areas where high-temperature applications lead to degradation of the process fluids. It has been shown that DCMD is effective in the concentration of many types of juices, including orange juice (Kimura and Nakao, 1987; Drioli et al., 1992; Calabró et al., 1994), apple juice (Laganá et al., 2000; Lukanin et al., 2003), sugarcane juice (Nene et al., 2002), and the like. The utilization of DCMD in the food industry for concentration or separation is promising, especially at high feed concentrations.

Direct-Contact MD was also applied for the concentration of biological solutions. Sakai et al. (1986, 1988) have applied the process to the concentration of bovine plasma and bovine blood using PTFE membranes of different pore sizes and thicknesses. It was found that the permeate flux was directly proportional to the vapor pressure difference and exhibited only a slight decrease during blood treatment, concluding that PTFE membranes, which possess outstanding properties of biocompatibility, were suitable for stable removal of solute-free water from blood with a hematocrit of 45% by DCMD. Ortiz de Zárate et al. (1998) looked at the concentration of protein (0.4 and 1% bovine serum albumin at pH 7.4) aqueous solutions by DCMD at low temperatures and found that fouling effects were practically absent, while the limiting factor of the process was the temperature polarization. Capuano et al. (2000) and Criscuoli et al. (2002) suggested DCMD as an innovative tool to ameliorate treatment of uremia by allowing purification of the blood ultrafiltrate and the reinjection of the purified water to the patients. Tests have been made on artificial solutions and plasma ultrafiltrate obtained from a hemofiltration unit. The researchers proposed the clinical application of DCMD for patient treatment by operating online with hemofiltration or hemodialysis. Optimization of the whole process in terms of energetic requirements, membrane area, and operating conditions has been made.

Direct-contact MD has also been applied successfully to wastewater treatment, either to produce a permeate less hazardous to the environment or to recover valuable compounds. Among the different studies carried out, we include pharmaceutical waste- water containing taurine (Wu et al., 1991), textile wastewater contaminated with dyes (Calabró et al., 1991), heavy metals (Zolotarev et al., 1994), coolant liquid (i.e., glycols) (Rincón et al., 1999), oil–water emulsions (Gryta and Karakulski, 1999), humic acid (Khayet et al., 2004b; Khayet and Mengual, 2004; Srisurichan et al., 2005), acid solutions rich in specific compounds (Tomaszewska, 1993, 2000a, 2000b; Tomaszewska et al., 1995, 1998, 2001a; Ugrosov and Elkina, 1998), and radioactive wastewater solutions (Chmielewki et al., 1995; Zakrewska-Trznadel et al., 1996, 1999, 2001; Khayet et al., 2006b). Recently, DCMD was proposed for wastewater reclamation in space in a combined direct osmosis system (Cath et al., 2005). The perspectives for the future of this type of dual system appear to be promising.

It is worth noting that few studies appeared on the DCMD treatment of aqueous solutions containing volatile organic components (VOCs). Fujii et al. (1992a, 1992b) studied the removal of low-concentration organics (ethanol, acetone, acetonitrile, *n*-butanol) from water using various coated and uncoated hollow-fiber fine porous membranes. It was found that the selectivity varied according to the properties of the type of polymer used, the membrane characteristics, and the DCMD operation conditions. Detailed results were given previously in Section 12.2.1. It must be pointed out here that in DCMD, the permeate must be dilute to prevent membrane pore wetting, while in the other MD configurations concentration of the condensed permeate is not a concern as it does not come into contact with the membrane. Franken et al. (1987a) studied the possible use of tubular PP membrane modules for separating dilute ethanol (5–6 wt%) from ethanol–water mixture. They investigated the influence of time and hydrodynamic conditions such as feed and permeate flow rates upon the permeate flux and selectivity. After 30 days of operation, the researchers found that the flux was reduced by 30% and the selectivity by 9%. This was attributed to the wetting of some membrane pores. DCMD has also potential applications in biotechnology for the removal of toxic products from culture broths. Udriot et al. (1989) described the application of a DCMD unit connected to a laboratory bioreactor for the selective recovery of ethanol from the culture medium. The experiments were run at a constant temperature of 38°C on an anaerobic cultures of *Fragilis* using a PTFE membrane. The researchers found that continuous extraction of ethanol using DCMD resulted in an increase in ethanol productivity by 87%. Gryta et al. (2000) also investigated the batch fermentation combined with the removal of ethanol from the broth by means of DCMD and an increase in productivity, and rate of conversion of sugar to ethanol was also observed.

The application of DCMD for breaking azeotropic mixtures was first proposed by Udriot et al. (1994), who looked at separating hydrochloric acid–water and propionic acid–water azeotrope mixtures. In fact, the azeotropic mixtures are impossible to separate by simple distillation. Retention selectivities of the solute between 0.6 and 0.8 were achieved instead of unity as implied by vapor–liquid equilibrium (VLE). In this case DCMD may be used to shift the selectivity above or below the one obtained by the VLE.

The pilot plant depicted in Figure 12.8a employing DCMD configuration was proposed by Zakrewska-Trznadel et al. (1999) to clean low-level radioactive wastes from a nuclear center at a throughput of about 0.05 m^3/h. A spiral-wound module equipped with a PTFE membrane of an effective surface area of 4 m^2 was used. The pilot plant laboratory experiments were able to produce very pure water with about 30 mg/L total solute concentration and activity on the level of natural background, and permeate fluxes between 30 and 50 L/h (i.e., 180–300 $L/m^2 \cdot day$) using feed inlet temperatures of 45–80°C, permeate inlet temperatures of 5–20°C, and feed and distillate flow rates of up to 1500 L/h. Moreover, it was proved that DCMD is feasible to process low- and medium-level radioactive wastes giving a high decontamination factor in only one stage (Zakrewska-Trznadel et al., 1996, 1999, 2001; Chmielewki et al., 1995).

It must be mentioned that up to now, MD is not applied in the nuclear industry. The process applications are still under evaluation due to its high-energy consumption and difficulties with long-time operation together with the membrane wettability. In fact, nuclear power industry should be a very convenient place for MD implementation because a lot of waste heat can be recovered in many places around the nuclear cycle.

12.5.2 Sweeping Gas Membrane Distillation

As stated earlier, SGMD involves the evaporation of water and volatile molecules at the hot feed side, transport of water vapor and volatile molecules through dry pores of hydrophobic membranes due to transmembrane vapor pressure, which is the driving force, collection of the permeating molecules by an inert cold sweeping gas, and finally condensation out of the membrane module. Compared to DCMD, SGMD has received only little attention (Fig. 12.3). In fact, in different series of studies (Zander et al., 1989; Semmens et al., 1989; Mahmud et al., 1998, 2000), membrane air stripping (MAS) was maintained instead of SGMD. The same devices and the same membranes (i.e., microporous hydrophobic) are used by both processes. Moreover, SGMD is similar to the well-known sweeping gas pervaporation (PV), the only difference is the use of porous hydrophobic membrane instead of dense selective membranes in PV.

The SGMD configuration has a great perspective for the future because it combines a relatively low conductive heat loss through the membrane with a reduced mass transfer resistance. As in the AGMD configuration, there is a gas barrier, which results in reduction in heat loss by conduction through the membrane. However, the gas in SGMD is not stationary and sweeps the membrane, resulting in higher mass transfer coefficients leading to higher permeate fluxes than in AGMD. Khayet et al. (2003c), by comparing SGMD and DCMD configurations using the same membrane module and the same feed operating conditions, found higher fluxes (1.4 times) and lower internal heat loss by conduction in SGMD. As occurred in DCMD, the SGMD mass fluxes are strongly dependent on the feed temperature because of the exponential increase of the vapor pressure with temperature, since the vapor pressure difference is the driving force in MD (Fig. 12.17). However, in SGMD (Khayet et al., 2000a, 2002c), it was found that the effect of the feed flow rate on the SGMD flux was practically negligible, while in the other MD configurations the mass flux increased and in most cases tended to approach asymptotic values.

Khayet et al. (2000a, 2002c, 2003a, 2003c) presented a series of theoretical and experimental SGMD studies using flat-sheet PTFE (TF200 and TF450 in Table 12.1) and PP capillary (Accurel S6/2 MD020CP2N, Table 12.1) membranes. As observed in DCMD, the permeate flux was found to be higher for membranes having larger pore sizes. The temperature of the liquid feed and the sweeping gas flow rate were the important parameters controlling the SGMD flux. Moreover, it was concluded that the major disadvantage of the SGMD configuration is the very high increase of the gas temperature along the membrane module compared to the decrease of the feed liquid temperature as shown in Figure 12.18.

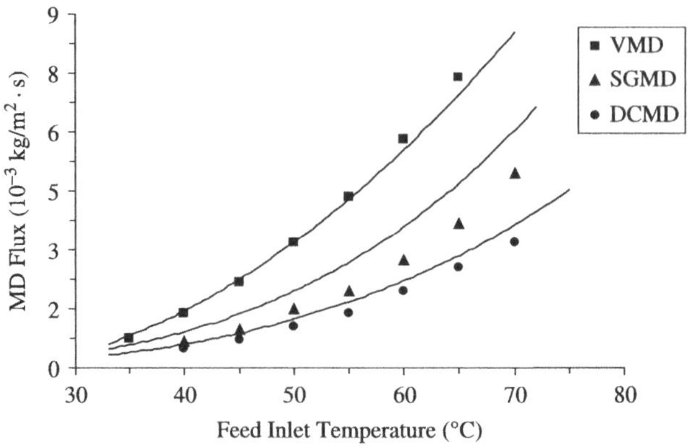

Figure 12.17 Effect of feed inlet temperature on MD flux obtained using PP capillary (Accurel S6/2 MD020CP2N, Table 12.1) with feed flow velocity 0.8 m/s. In DCMD (0.8 m/s permeate flow velocity, 20°C permeate inlet temperature); in SGMD (11.3 m/s humid air velocity, 20°C air inlet temperature); in VMD (3500 Pa downstream pressure). The solid lines are theoretical prediction curves. (Adapted from Khayet et al., 2003c.)

This temperature gradient along the membrane module length affects the local driving force and consequently the SGMD flux.

Additionally, Khayet et al. (2002c) found the TPC in the feed side to be very close to unity, whereas the main temperature polarization in SGMD was located in the air phase, concluding that the mass flux in the SGMD process is mostly controlled by the heat transfer through the air boundary layer.

Sweeping gas MD was applied successfully for desalination of aqueous solutions, first by Basini et al. (1987) and later by Khayet et al. (2003a, 2003c). Rejection of practically 100% was achieved. SGMD was used also for concentration of aqueous sucrose solutions by Korngold and Korin (1993) and separation of alcohol water mixtures by Calibo et al. (1987) and Lee and Hong (2001).

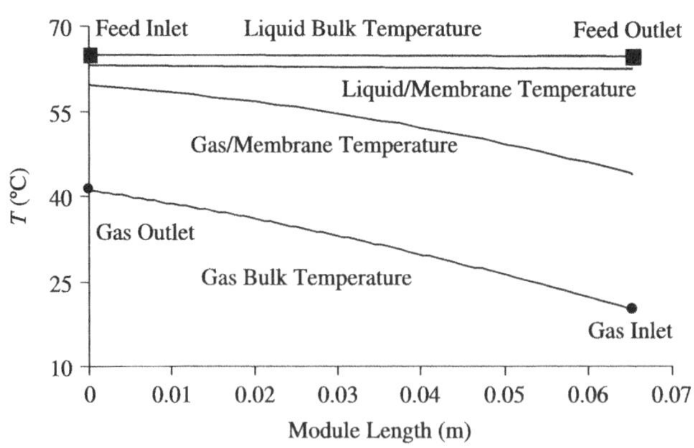

Figure 12.18 Temperature profile along a plate-and-frame TF200 membrane module working under SGMD configuration. (Feed liquid velocity 0.15 m/s; air velocity 1.5 m/s). (Adapted from Khayet et al., 2000a.)

Basini et al. (1987) used dry air as sweeping gas, and the experiments were conducted on both flat-sheet PTFE and tubular PP membranes. A significant increase of the SGMD flux at low air velocity was plotted, reaching a maximum value after which it decreased. For the same operation conditions, the PTFE membrane exhibited higher SGMD performance than the PP capillary membrane due to its larger pore size and porosity. Basini et al. (1987) did not indicate the value of the membrane area to compare their results to other SGMD studies. They concluded that the mass transfer resistance in the gas phase was the rate-determining step when low gas flow rates were considered, while at higher gas flow rates the resistance exerted by the membrane itself became dominant. An iterative theoretical model was presented assuming a tortuosity factor of 3 (adjustment factor) for PP membrane. A good agreement was obtained between the predicted SGMD fluxes and the experimental ones for the shell-and-tube PP capillary membrane module. However, some deviations appeared for the plate-and-frame PTFE membrane module, and the authors attributed them to the small module for which no standard heat and mass transfer correlation were available.

As stated previously, theoretical and experimental desalination cases by SGMD were also studied by Khayet et al. (2003a) using a shell-and-tube PP membrane module (Accurel S6/2 MD020CP2N, Table 12.1). Humid air was employed as the vapor carrier gas, and the feed solution was circulated in countercurrent through the lumen side of the membrane module. The effects of various process parameters including salt concentration on the distillate flux have been investigated. Only a slight decrease of the SGMD flux with the NaCl concentration in the aqueous feed solution was detected, and high-purity water (<9 μS/cm) was obtained, indicating a solute rejection of more than 99.9%. It was also found that the predicted SGMD fluxes were slightly higher than the experimental ones due partly to the limitation of the condenser efficiency.

Calibo et al. (1987) tested PP and PTFE hollow-fiber membranes for ethanol removal from an 8 wt% ethanol–water solution. Nitrogen was used as carrier gas circulating in countercurrent mode through the shell side of the membrane module. They found that the overall mass transfer coefficient of the membrane was affected more by the feed flow rate than by the sweeping gas flow rate, and no temperature effect was observed in the range of 21–32°C. The overall mass transfer coefficient of the PTFE membrane was found to be higher than that of PP hollow fiber having smaller pore size.

Lee and Hong (2001) studied the effects of different operating parameters on the SGMD flux and selectivity when using dilute aqueous isopropanol (IPA) solutions and PTFE hollow-fiber membrane module (Table 12.1). Nitrogen was employed as sweeping gas through the lumen side of the membrane module. The upper IPA concentration in feed was 10 wt% and the feed temperature was varied from 20 to 50°C. It was found that the IPA selectivity was not affect by the feed flow rate and decreased with the increase of the sweeping gas flow rate to a saturation point at high sweep gas velocity. Moreover, as can be seen in Figure 12.19, the total SGMD flux increased and the IPA selectivity decreased with the IPA concentration. IPA selectivity was found to increase continuously with the feed temperature at lower IPA concentrations, whereas for higher IPA concentrations an optimum value was observed at a feed temperature near 40°C. This fact was explained by the negative effect of the viscosity. In contrast, at lower IPA concentration, the diffusion coefficient for IPA in water is higher, leading to an increase of the mass transfer coefficient and therefore better selectivity was achieved. The addition of the salt, $MgCl_2$, to the feed solution was found to increase the IPA selectivity significantly with a slight

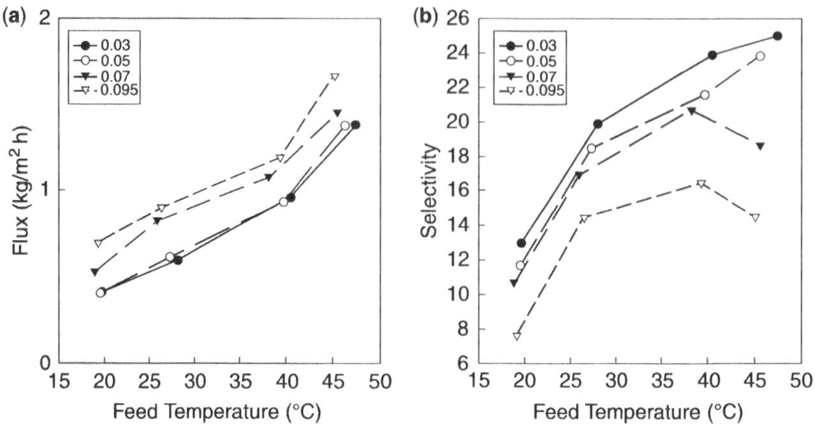

Figure 12.19 (a) SGMD flux and (b) IPA selectivity as a function of feed temperature at various feed concentrations using the PTFE membrane Poreflon indicated in Table 12.1. 12.9 kg/h feed flow rate, 8.7 m/s gas velocity; 25°C gas temperature (Lee and Hong, 2001).

decrease in total permeate flux. This was attributed to the reduction in water vapor pressure, leading to a decrease in the water mass transfer through the membrane.

As shown in Figure 12.18, one of the inconveniencies of the traditional SGMD is the steep increase of the gas temperature along the membrane module. Rivier et al. (2002) and García-Payo et al. (2002) proposed a thermostatic sweeping gas membrane distillation (TSGMD) in which the increase of the sweeping gas temperature was minimized by using a cold wall in the cold chamber (Fig. 12.7) for the treatment of aqueous formic acid solutions. A Stefan–Maxwell-based model, including vapor–liquid equilibria and heat and mass transfer relations, was developed, and the effects of the mean temperature and temperature difference on both the TPC and energy efficiency (EE) for both TSGMD and DCMD configurations were studied. It was observed that the TSGMD attained much higher TPC and EE values than those obtained in DCMD due to a large heat transfer resistance of the boundary layer in the permeate side. This means that the TPC for DCMD is significantly higher than that for TSGMD. Based on simulation results, higher membrane performance was found when using the TSGMD configuration.

García-Payo et al. (2002) applied the TSGMD tubular module for separation of the azeotropic mixture of formic acid–water. The effect of concentration on the entire range of formic acid mass fraction was studied, and no PTFE membrane wetting was observed. It was seen that the permeate flux increased with the temperature difference, with the mean temperature and with the sweeping air velocity, and formic acid fluxes were higher than water fluxes. It must be pointed out here that a value of selectivity higher than 1 indicates formic acid enrichment in the permeate, while a value lower than 1 means water enrichment in the permeate. Gacía-Payo et al. (2002) found selectivity to be always lower than unity when the formic acid mass fraction in the feed was 0.7; in contrast, the selectivity for 0.8 mass fraction was greater than 1, depending on the applied temperature.

Possible SGMD applications can be the removal/concentration of organics from dilute organic/water mixtures such as esters, ethers, chlorinated hydrocarbons, and aromatic compounds. These applications are very appropriate for environmental, chemical, petrochemical, and biotechnology industries, as they need removal or recovery of organics from dilute solutions. In fact, MAS has shown the potential for the removal of VOCs from aqueous streams over a wide range of concentration levels and demonstrated to be usefully applied for water pollution reduction, groundwater cleanup, and for organic recovery and reuse from industrial and petroleum wastewater streams. Several MAS studies have been published in different journals, indicating that MAS is a promising technique for the removal of organic compounds from aqueous mixtures such as chloroform, trichlroethane, trichloroethylene, tetrachloroethylene, carbon tetrachloride, bromoform, and bromodichloromethane (Zander et al., 1989; Semmens et al., 1989; Mahmud et al., 1998, 2000).

12.5.3 Vacuum Membrane Distillation

As shown earlier in Figures 12.3 and 12.17, the vacuum membrane distillation (VMD) configuration exhibiting higher permeate flux and practically negligible heat transfer by conduction through the membrane is the least used configuration next to SGMD. The effects of the feed temperature, flow rate, and nonvolatile solute concentration in VMD are similar to their effects in DCMD and SGMD. Generally, in VMD, membranes of smaller pore size (i.e., less than 0.45 μm) than in the other MD configurations are used because in VMD vacuum is applied and the risk of pore wetting is high.

Vacuum MD has been proposed for the extraction of VOCs from dilute aqueous solutions (Bandini et al., 1992; Sarti et al., 1993; Banat and Simandl, 1996; Couffin et al., 1998; Bandini and Sarti, 1999; Urtiaga et al., 2000; Khayet and Matsuura, 2001, 2004). Banat and Simandl (1996) investigated the removal of benzene traces from contaminated water using tubular PP membrane module. Among the three resistances involved in the mass transport (liquid layer, membrane, and gas layer), the liquid layer resistance was the predominant. Sarti et al. (1993) showed that the overall flux was high relative to the benzene flux, at low downstream pressures, resulting in a relatively poor selectivity, whereas at permeate pressures higher than the vapor pressure of water, the overall flux was found to be similar to that of benzene, resulting in a high concentrated benzene permeate. Couffin et al. (1998) applied VMD to remove halogenated VOCs, namely chloroform, trichloroethylene, and tetrachloroethylene, from drinking water at very low concentration (400 μg/L) using a VMD pilot plant designed to test both flat-sheet and hollow-fiber modules. The partial VMD flux was found to be higher for chloroform followed by that of trichloroethylene and then tetrachloroethylene. Trichloroethylene was removed from water at very high selectivity varying from 9 to 860 depending on feed temperature and downstream pressure. The total flux decreased by increasing the downstream pressure and decreasing the temperature (30–50°C), while the selectivity was found to increase.

The same tendencies were observed by Bandini et al. (1997) when using other dilute binary aqueous solutions containing acetone, ethanol, isopropanol, ethylacetate, methylacetate or methyl tertiary butyl ether in the concentration range 2–10 wt%. Urtiaga et al. (2000) looked at removing chloroform from dilute aqueous solutions (500–2012 mg/L) using porous PP hollow-fiber membranes at different VMD operating parameters. It was concluded by means of a kinetic analysis that only under turbulent flow regime the resistance to mass transfer in the membrane affected the overall mass transfer coefficient, whereas in the laminar regime, the diffusion coefficient of chloroform in the feed liquid phase was the only parameter describing the separation performance. Similar experiments were conducted by Khayet and Matsuura (2001, 2004) using PVDF flat-sheet membranes of different pore sizes and different dilute aqueous solutions containing chloroform.

The chloroform selectivity was found to decrease from 72.5 to 7.9 with the increase of membrane pore size (0.01–0.2 μm), while both the chloroform and water fluxes increased.

Vacuum MD has been applied for the treatment of dilute alcohol solutions. Hoffman et al. (1987), by performing VMD experiments of the binary mixtures of methanol–water, ethanol–water, and *n*-propanol–water of different concentrations, stated that selectivities, when using membranes with 0.45 μm nominal pore radius, were smaller than those calculated from the corresponding equilibrium data. On the contrary, the maximum values of the selectivity, higher by more than a factor of 3 compared to the calculated ones on the basis of the vapor–liquid equilibrium data, were achieved by the PTFE membranes having lower nominal pore radius, which was 0.2 μm. Bandini et al. (1992) investigated VMD for the removal of ethanol using PTFE and PP membranes. They found that the separation was limited by concentration polarization, and the increase in permeate flux was usually accompanied by the decrease in selectivity. They reported that the liquid mass transfer resistance greatly affected the separation factor of a diluted (5 wt%) ethanol–water mixture, which was found to be lower than 10 and increased with the mass transfer coefficient in the liquid boundary layer.

Izquierdo-Gil and Jonsson (2003) studied the factors affecting VMD flux and ethanol separation. They reported that the separation factor increased with the feed circulation velocity from 5 to 7 for the PTFE and PVDF membranes, whereas it was maintained practically constant around 7–8 for the PP membranes, although the membranes have similar pore sizes (i.e., 0.1–0.2 μm). No explanations on this aspect were given by the authors. Bandini et al. (1997) tested various dilute binary mixtures by VMD and found that the VOC permeate composition was lower than the corresponding equilibrium value with the feed, and at low downstream pressures and feed temperatures high permeate fluxes and VOC permeate concentrations up to the equilibrium values were achieved. In Figure 12.20, the overall flux as well as the fluxes of water and organic components are plotted against the downstream pressure for the case of methyl acetate–water mixture. At low permeate pressure, the water flux was higher than that of the organic flux approaching the total flux, so a low separation factor was obtained. In contrast, at permeate pressure higher than the equilibrium water vapor pressure, the water flux became lower, and the permeate had high organic concentration in the permeate.

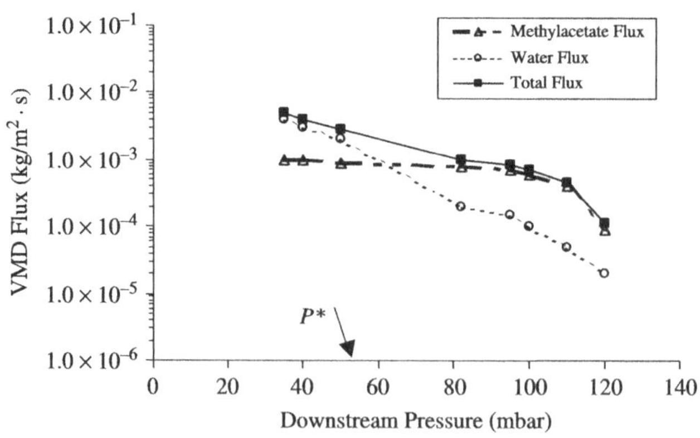

Figure 12.20 Total water and organic VMD fluxes vs. downstream pressure for TF200 membrane: 2.8 L/min feed velocity, 35°C feed temperature, 3wt% methylacetate–water mixture. (Adapted from Bandini et al., 1997.)

As in the other MD configurations, VMD has been successfully applied for desalination (Wirth and Cabassud, 2002; Li et al., 2003; Khayet et al., 2003c; Cabassud and Wirth, 2003; Li and Sirkar, 2005). Fluxes as high as 71 $kg/m^2 \cdot h$ from 85°C were achieved (Li and Sirkar, 2005), and higher fluxes than those of the other MD configurations were obtained (Fig. 12.17). Moreover, Cabassud and Wirth (2003), based on a developed computational analysis, stated that for a membrane area of 100 times larger than that of the experimental membranes at low temperatures (25°C), VMD can compete with RO on energy consumption (<2 kWh/m^3) with approximately the same water production.

Vacuum MD was also investigated as a tool for textile wastewater colored with dyes such as methylene blue (Banat et al., 2005) and in the food industry for concentration of must (i.e., the juice obtained from grape pressing containing sugars and a wide variety of aroma compounds) (Bandini and Sarti, 2002) and for the recovery of volatile aroma compounds from black currant juice (Bagger-Jorgensen et al., 2004). In all these application areas, VMD seems to be a promising technique; however, further studies and economical analysis together with detailed comparative works within the different MD configurations and the pressure-driven separation processes are needed.

12.5.4 Air Gap Membrane Distillation

To solve the problem of heat loss by conduction through the membrane, which leads to relatively low efficiency of the MD process, an air gap was placed inside the membrane module between the permeate side of the membrane and the condensing surface. This reduces considerably both the heat loss by conduction and temperature polarization, thereby improving the separation effect. However, the permeate flux has to overcome the air barrier, and, therefore, it is drastically reduced, depending on the effective air gap width. On the other hand, because permeate is condensed on a cold surface rather than directly on membrane surface, air gap MD (AGMD) can be applied in fields where the DCMD is limited such as the removal of organic compounds from aqueous solutions.

In the 1980s, the Swedish National Development Co. (Svenska Utvecklings AB) developed plate-and-frame membrane modules applying the AGMD configuration (Carlsson, 1983; Andersson et al., 1985), and Gore & Associated Co. proposed the use of a spiral-type module using the AGMD (or the liquid gap DCMD) configuration (Gore, 1982). Since then, research studies related to the use of AGMD in many applications have increased (Kimura and Nakao, 1987; Kubota et al., 1988; Gostoli and Sarti, 1989; Kurokawa et al., 1990; Banat and Simandl, 1994, 1998, 1999, 2000; Udriot et al., 1994; Izquierdo-Gil et al., 1999b; Banat et al., 1999d; García-Payo et al., 2000b; Bouguesha and Dhahbi, 2002; Koschikowski et al., 2003; Kim et al., 2004).

The effects of feed temperature, feed flow rate, and solute concentration on the AGMD process performance are similar to their effects in the other MD configurations. In addition, it was observed that the AGMD fluxes decreased with the increase of the air gap thickness and is practically independent on the cold side flow rate (Jonsson et al., 1985; Kimura and Nakao, 1987; Banat and Simandl, 1994, 1998). Moreover, as far as the separation rate is concerned, the optimum thickness of the air gap was found to be dependent on the salt concentration in the feed membrane side. In other words, the higher the salt concentration is the larger the gap needs to be (Gostoli and Sarti, 1989).

Air Gap MD has been applied for water desalination (Kimura and Nakao, 1987; Kubota et al., 1988; Kurokawa et al., 1990; Banat and Simandl, 1994,1998; Izquierdo-Gil et al., 1999b; Bouguesha and Dhahbi, 2002; Koschikowki et al., 2003). Kimura and Nakao (1987) carried out AGMD tests of 3.8% NaCl aqueous solution using PTFE membranes of different pore sizes and observed that the permeate concentration was independent on the pore size, while the AGMD flux did depend to some extent on the pore size and increased linearly with the porosity. Banat and Simandl (1994) showed that the AGMD flux of PVDF and PTFE membranes of 0.45 and 0.5 μm pore sizes, respectively, was maintained steady over more than a 6-week run, and it was affected only slightly by an increase in salt concentration. Kurokawa et al. (1990) found a linear increase of the AGMD flux with the vapor pressure difference between the feed and the condensation surface and observed a decrease with increasing concentration of LiBr and H_2SO_4 because of the temperature and concentration polarization effects, with rejection factors higher than 99.9% (i.e., electrical conductivities below 10 μS/cm). The maximum concentration of LiBr in aqueous solution was 56% while that of H_2SO_4 was 83%.

Experiments on seawater desalination by AGMD were conducted by Kubota et al. (1988) using various plate-and-frame membrane modules containing porous PTFE membrane and silicone/polysulfone composite dense membranes with an effective area of 1.92 and 2.93 m^2, respectively. For both membrane modules, the product water quality was very good with an electric conductivity of about 10 μS/cm, while the fluxes of the PTFE membranes (i.e., maximum flux 10 kg/h) were higher than those of the silicone–polysulfone composite dense membrane (i.e., maximum flux 8 kg/h). Banat and Simandl (1998) obtained very pure water with less than 5 ppm total solid content when treating a simulated seawater used as a feed solution by AGMD employing PVDF (GVHP, Table 12.1) flat-sheet membranes. Fluxes as high as 10 $kg/m^2 \cdot h$ were obtained with a 0.19-cm air gap thickness and lower than 1 $kg/m^2 \cdot h$ with a 0.99-cm air gap thickness. The combined use of AGMD and solar energy was investigated by Koschikowski et al. (2003) in a pilot plant (Fig. 12.8b) using spiral-wound PTFE membrane modules working at 60–80°C feed temperature. The maximum distillate reported was 15 kg/h, and the gained output ratio (GOR) was 5.5 at 350 L/h flow rate and 75°C evaporator inlet temperature. Their simulation results during summer in a southern country (Eilat, Israel) indicated that the plant was able to distillate 120–160 L/day with a solar collector area less than 6 m^2 and without heat storage. Bouguesha and Dhahbi (2002) looked at coupling fluidized-bed crystallizers and AGMD as possible solution to geothermal waste desalination. Preliminary experiments were conducted, and more investigations are needed to prove the viability and efficiency of the combined system.

Air gap MD was applied for the concentration of nonvolatile solutes and food processing (Kimura and Nakao, 1987; Izquierdo-Gil et al., 1999b). Kimura and Nakao (1987) tested the use of AGMD for the concentration of mandarin juice. It was seen that at high feed concentration ratios (>1.5), AGMD fluxes at a feed temperature of 48°C and a condensation temperature of 10°C were higher than those of the RO system. AGMD was demonstrated to be used for the concentration of milk (Kimura and Nakao, 1987). However, the membranes tend to be fouled by the adhesion of fat, resulting in very low fluxes. Concentration of sugar and gelatin was also investigated (Kimura and Nakao, 1987; Izquierdo-Gil et al., 1999b). It was seen that the permeate flux was maintained at the same level after 30 days of AGMD operation of 150 g/L sucrose solutions. Kimura and Nakao (1987) also investigated the concentration of aqueous solutions of H_2SO_4 and NaOH at different pH values and noticed comparable AGMD flux and electrical conductivity to those of the obtained permeate when using aqueous NaCl solutions. Moreover, they looked at the separation of volatile solutes such as HNO_3 and HCl and noticed similar trends for both components, different from that of the nonvolatile solutes. The permeate concentration increased with the feed concentration until both coincided at a concentration higher than 1 mol/L. When testing acetic acid (0.02–0.85 mol/L) and formic acid (0.14–1.1 mol/L) aqueous solutions at 50°C feed temperature and 18°C condensation temperature, the measured AGMD fluxes were 9.5–10 $kg/m^2 \cdot h$ with rejection factors of 40–44% for acetic acid–water solutions and higher AGMD performance for formic acid–water solutions (12.8–13.0 $kg/m^2 \cdot h$ permeate fluxes and 57–59% rejection factors) (Kimura and Nakao, 1987).

The potential advantage for ethanol recovery from fermentation broth was discussed by Gostoli and Sarti (1989). They used TF200 (Table 12.1) membrane for ethanol–water separation by AGMD instead of direct contact in order to prevent the ossible membrane wetting by ethanol, which may occur on the permeate side. The researchers (Gostoli and Sarti, 1989) found that the separation factor increased with the increase of the temperature difference across the membrane. When using 3–7 wt% feed ethanol concentrations, the measured total AGMD flux was practically the same as that of pure water used as feed, and the distillate was more concentrated in ethanol, being the separation factor up to 2, far below the observed values for a single distillation. Additionally, when 23.8 wt% ethanol–water solution was considered, the separation factor was extremely low, close to unity. In other words, the process was ethanol selective for low ethanol content in the feed, while it became water selective for high ethanol content. An exact Stefan–Maxwell formulation of flux equations for an ethanol–water–air system was used without incorporating the concentration polarization effect. Based on their mathematical model, Gostoli and Sarti (1989) found the separation factor to be highly sensitive to the feed composition at constant average temperature and temperature difference.

Banat and Simandl (1999) investigated the removal of ethanol from aqueous solutions by AGMD using the HVHP membrane (Table 12.1). The upper ethanol feed concentration tested was varied from 0.83 to 10 wt% within the feed temperature range of 40–70°C. The effects of various AGMD operating parameter on permeate flux and ethanol selectivity were investigated. Generally, the ethanol selectivity of 2–3.5 was achieved with fluxes lower than 9 $kg/m^2 \cdot h$. Figure 12.21 shows the effect of air gap width on the ethanol flux and

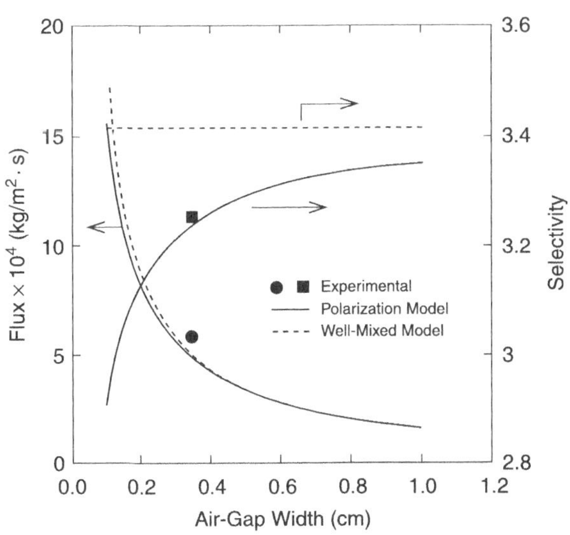

Figure 12.21 Effect of air gap width on AGMD flux and ethanol selectivity of HVHP (Table 12.1). 50°C feed temperature, 20°C cold temperature, 0.0155 weight fraction ethanol in feed (Banat and Simandl, 1999).

selectivity. As can be seen, the permeate flux is inversely proportional to the air gap thickness, and increasing the air gap width increased the ethanol selectivity. Furthermore, it was found that the Fickian mass-transfer-based mathematical model version, which neglected the concentration and temperature polarization effects, could not adequately predict the experimental results. Banat and Simandl (1999) also examined the effect of salt on the AGMD process performance during the concentration of ethanol aqueous solutions and found better ethanol selectivity with only a slight decrease in total permeate flux. The exact Stefan–Maxwell and the approximate Stefan–Maxwell mass-transfer-based models were analyzed by Banat et al. (1999b), and they noticed that the results of these models agreed well, whereas some differences existed between those models and the Fickian-based model. Moreover, the authors (Banat and Simandl, 1999) concluded that the temperature and concentration variation along the membrane module length must be considered at low air gap widths and high feed temperatures.

García-Payo et al. (2000b) also looked at the removal of alcohol (methanol, ethanol, isopropanol) from water binary mixtures using various membranes (PTFE, PVDF) of different parameters. The effects of the relevant AGMD parameters were investigated. For membranes prepared with the same material, the AGMD fluxes were found to be higher when the membrane pore size was larger and the porosity was higher. It was also noticed that generally the AGMD process performance was better when using PTFE membranes. Moreover, for the PTFE membranes having the same pore size, the AGMD flux was lower for the supported membrane. It must be pointed out that García-Payo and co-authors (2000b) provided neither detailed alcohol selectivity values of each membrane nor the permeate partial fluxes.

Air gap MD is of potential interest in breaking azeotropic mixtures. This was proposed first by Udriot et al. (1994) who conducted experiments in a plate-and-frame PTFE membrane module with azeotropic mixtures of hydrochloric acid–water and propionic acid–water. Selectivities between 0.6 and 0.8 were achieved instead of unity implied by the vapor–liquid equilibrium. Additionally, it was seen that the hydrochloric acid–water azeotrope was shifted to higher acid concentration, whereas the azeotropic point of the propionic acid–water system disappeared. This phenomenon was attributed to the differences in the acid–air and water–air diffusion rates across the membrane and the air gap of the different components of the azeotropic mixtures. Udriot et al. (1994) reported that the selectivity was reduced by about 12% for the propionic acid–water system when going from a DCMD operation to an AGMD of 7-mm air gap width. Antagonistic effects of the gas gap length on the permeate flux and on the selectivity were pointed out. Banat et al. (1999d) examined the effect of the inert gases—helium, air and sulfur hexafluoride—in breaking the formic acid–water azeotropic mixtures. The selectivity was found to be larger when using helium (around 0.96), followed by that in air (about 0.9) and then in sulfur hexafluoride (0.85–0.86). They found that both the AGMD flux and selectivity were governed by the vapor–liquid equilibrium relations and the different diffusivities in the inert gas, showing that the heavy inert gases such as sulfur hexafluoride helped more in eliminating totally the azeotropic point than the lighter ones such as air and helium. Banat and Simandl (2000) also used AGMD for the selective removal of propanone from aqueous solutions using PVDF membranes, within feed temperature range of 40–70°C and propanone concentration up to 6 wt%. Selectivities of 2–6 were achieved.

As stated previously in DCMD section, MD can be applied for radioactive wastewater treatment and for nuclear desalination. AGMD was also used for the ^{18}O isotopic water separation by Kim et al. (2004) using PTFE membranes of different pore sizes. They noticed that the AGMD flux and the degree of ^{18}O separation were higher for large temperature gradient across the membrane.

12.6 LONG-TERM MD PERFORMANCE AND MEMBRANE FOULING IN MD

In the MD literature, few studies were reported on long-term MD performance (Drioli and Wu, 1985; van Gassel and Schneider, 1986; Schneider et al., 1988; Banat and Simandl, 1994, 1998; Gryta, 2001, 2003a; Karakulski et al., 2002; Khayet and Mengual, 2004; Li and Sirkar, 2004; Khayet et al., 2004b). However, the MD researchers seem to disagree in this point. A permeate flux decline up to 66% during over 4 days was reported by Drioli and Wu (1985) when using 0.1 M NaCl aqueous solution at different temperatures. Schneider et al. (1988) observed around a 20% drop of the permeate flux after 18 weeks using tap water with 350 μS/cm electrical conductivity. It was further observed that the original flux was restored by a short time treatment with hydrochloric acid. No flux decay was detected during 10 days AGMD desalination experiments by Banat and Simandl (1998). Many more experimental works are needed in this MD area using the different MD configurations and the different applications outlined in Section 12.5.

Recently, Gryta (2001) described the results of much longer term DCMD experiments, of over 3 years, carried out for production of demineralized water using Accurel PP S6/2 capillary PP membrane (0.5-μm maximum pore size, 0.22-μm nominal pore size, 72% porosity, 0.0889-m^2 effective membrane surface area: Table 12.1). The feed was circulated through the lumen side of the membrane module, the inlet feed and permeate temperatures were 80 and 20°C, respectively, and the inlet pressures were around 52 kPa. It was found that the membrane was thermally stable, maintaining its morphology and its good separation characteristics throughout the 3 years of DCMD operation (i.e., electrical conductivity was around 1 μS/cm). When using the permeate of the RO system as DCMD feed solution, membrane pore wetting was not observed; and, irrespective of the duration of the experiment, the DCMD flux was found to be similar to the initial permeate flux. However, when using tap water as DCMD feed solution, deposition of calcium carbonate ($CaCO_3$) on the membrane surface occurred, and a partial wetting of the membrane was observed, resulting in an increase of the electrical conductivity of the permeate from 0.9 to 2.5 μS/cm and a decrease of the permeate flux from 700 to 550 $L/m^2 \cdot day$. However, the formed deposit was removed every 40–80 h by rinsing the membrane module first with a 2–5 wt% HCl aqueous solution and then with pure water, permitting the recovery of the initial permeate flux. It must be pointed out here that the multiple repetition of the cleaning operation caused a gradual DCMD flux decline.

Membrane fouling (i.e., deposition of particles, colloids, emulsions, suspensions, macromolecules, etc.) and microorganism growth on membrane surface (i.e., membrane biofouling) may be also a problem in MD systems (Sakai et al., 1986, 1988; Kimura and Nakao, 1987; Calabró et al., 1994; Ortiz de Zárate et al., 1998; Gryta, 2000, 2002b, 2005; Gryta et al., 2001a; Khayet et al., 2004b; Khayet and Mengual, 2004; Srisurichan et al., 2005), although some MD researchers detected very low DCMD fouling during the treatment of humic acid solutions, extraction of solute-free water from blood (i.e., bovine plasma and bovine blood), and concentration of protein (i.e., bovine serum albumin, BSA) aqueous solutions (Sakai et al., 1986; Ortiz de Zárate et al., 1998; Khayet and Mengual, 2004; Khayet et al., 2004b; Srisurichan et al., 2005). They concluded that problems resulting from fouling were significantly lower than those encountered in other pressure-driven membrane separation processes (Gryta, 2002b). Fouling of the membrane surfaces and depositions of salts at the membrane pore surface can lead to wetting of the pores and a decrease of the effective membrane area resulting in an MD flux decline and low separation factors. In fact, fouling is an important issue in all membrane applications including MD and needs to be investigated further in MD process.

12.7 HYBRID MD SYSTEMS

The combination of MD with other membrane systems such as reverse osmosis (RO), ultrafiltration (UF), and nanofiltration (NF) as well as conventional distillation systems (i.e., multieffect distillers) and renewable energy sources proved to be a very advantageous option improving final product quality, reducing both energy consumption of the whole installation and the discharged brine to a lower environmental impact (Hogan et al., 1990; Calabró et al., 1990, 1994; Morrison et al., 1991; de Andrés et al., 1998; Drioli et al., 1999; Criscuoli and Drioli, 1999; Khayet et al., 2000b; Gryta et al., 2001b; Zakrewska-Trznadel et al., 2001; Criscuoli et al., 2002; Banat et al., 2002; Koschikowski et al., 2003; Cath et al., 2005; Ding et al., 2005). In fact, integrated membrane systems or hybrid processes are beneficial when they combine the processes that supplement one another or when such a combination reduces the disadvantages, which are associated with a single process.

Calabró et al. (1990) looked at integrating RO with MD using simulated textile aqueous solutions. The RO permeate water was used as a cold water stream in the DCMD unit while the reject brine from the RO system was employed as a feed solution since the MD process is weakly influenced by nonvolatile feed solutes at high concentration. Integrating MD with RO was also carried out by Drioli et al. (1999) for desalination operation to increase the water recovery factor, which was found to be up to 87.6%. The RO–MD combined system can produce more than twice as much water as the RO plant at the same water cost, whereas the MD plant alone can produce as much water as the RO–MD plant, but at a water cost about 5% higher. Criscuoli and Drioli (1999) presented a detailed energetic and exergetic analysis of both RO–MD and NF–RO–MD integrated systems. They observed an improvement in the performance of the integrated system by introducing NF as a water pretreatment for the RO–MD system with almost the same energy. Zakrewska-Trznadel et al. (2001) proposed the combination of one RO unit and two MD units for liquid radioactive wastewater treatment to improve the parameters of effluent discharged and to produce high-purity water. One MD unit was used for the concentration of the RO brine and production of pure water, and the other MD unit for final purification of RO permeate.

The combination of UF and DCMD units was examined by Calabró et al. (1994) for the concentration of orange juice, by Gryta et al. (2001b) for the purification of oily wastewater, and by Criscuoli et al. (2002) for purification of human plasma. In fact, UF is a useful pretreatment technique for heavy fouling feeds to remove the larger particles that could increase the viscosity of the stream through the MD process. For example, Gryta et al. (2001b) noticed that UF generally reduced the content of oil to less than 5 ppm, whereas further purification of the UF permeate by DCMD, as a supplementary second stage, resulted in a complete removal of oil pollutants from wastewater with a very high reduction of the total organic carbon (99.5%) and the total dissolved solids (99.9%).

In addition, the combination of evaporation (i.e., single and multieffect distillers) and the MD process was proposed for desalination by de Andrés et al. (1998) using the DCMD configuration and Khayet et al. (2000b) using the SGMD configuration, and was also proposed by Zakrewska-Trznadel et al. (2001) for processing liquid radioactive wastes to produce high-purity water in nuclear power stations.

The possibility of using waste heat and renewable energy sources was investigated by various authors (Koschikowski et al., 2003; Morrison et al., 1991; Banat et al., 2002; Ding et al., 2005). Hogan et al. (1990) and Morrison et al. (1991) looked at solar-powered MD desalination pilot plants. A simulation model of MD combined with the TRNSYS solar simulation program was developed by Morrison et al. (1991) who observed that the solar-heated DCMD pilot plant would be economically feasible if a 60–80% heat was recovered. An optimum design of 50 L/day pilot plant was shown with a 3-m^2 solar collector area, 1.8-m^2 membrane area, and a 0.7-m^2 total heat exchange area with a capital cost of $3500 (Australian scaled to 1991). More recently, Banat et al. (2002) also evaluated the technical feasibility of producing potable water from simulated water by an integrated solar still and MD module. The hot brine of the solar still was employed as the feed to the MD module. A sensitivity of the permeate flux to the brine temperature, flow rate, salt concentration, and solar irradiation was evaluated. The contribution of the solar still was found to be no more than 20% of the total flux in the outdoor experiments and less than 10% in the indoor experiments. The researchers obtained maximum pure water fluxes 2–3 h after the solar irradiation peak. Koschikowski et al. (2003) developed a solar thermally driven stand-alone AGMD desalination system with a capacity ranging from 0.2 to 20 m^3/day (Fig. 12.8b). Details of this pilot plant, which is still under laboratory test, were given earlier in Section 12.5.4. An extensive theoretical analysis of solar-heated MD systems using a computational program solved numerically was given by Ding et al. (2005). They found that plant productivity can be improved by increasing the heat exchanger capacity to an optimum value, decreasing the feed and permeate flow rates or by increasing the membrane effective area to an optimum value depending on the heat exchanger capacity. They concluded that both the membrane area and the heat exchanger capacity should be optimized in the design of a solar-heated MD pilot plant.

It is worth quoting that few cost estimations for some MD-specific cases were made by MD researchers (Fane et al., 1987; Sarti et al., 1993; Drioli et al., 1999). This is due to the fact that the MD process has not been implemented commercially yet. The cost of various principal components of an MD plant such as large-scale MD modules and membranes, pretreatments, optimum flow conditions, long-term MD performance, which may be associated with flux decline and influence the costs, fouling, and membrane life are not sufficiently known yet.

12.8 CONCLUDING REMARKS AND FUTURE DIRECTIONS IN MD

Membrane distillation technology is still in the laboratory development stage, although the concept has been known for more than 40 years and used successfully in numerous application areas such as desalination and radioactive waste treatments. Different opinions exist concerning MD industrialization. The MD researchers discuss the high energy consumption in MD units, the difficulties with long-term operation with the simultaneous risk of membrane wetting and fouling, the lack of research on MD membranes and modules, and the uncertain energy and economic costs. Different propositions to improve product quantity and quality and reduce energy consumption have been reported, including integrated MD units to other conventional processes such as distillation, pressure-driven membrane processes, to alternative energy sources such as solar and geothermal energy, and also around nuclear installations where waste heat can be recovered.

On the other hand, as observed throughout the chapter, MD technology has taken a more academic role, and most of the publications in the MD field are concerned with experimental studies on the effects of the process operating conditions and theoretical models including heat and mass transfer mechanisms. In fact, although the effect of some MD operating parameters on commercial membranes performance are generally agreed upon and well known, still some studies are repeated and appear with almost the same results, and some areas within the MD field still ignored or rarely studied. Among the areas that are roughly studied are preparation of novel membranes and membrane module designs specifically for MD applications, adequate methods for MD membrane characterization, studies on the effect of pore size near the critical region (i.e., mean free path of the transported molecules), effect of pore size distribution and nature of fluxes through laboratory-made membranes with broader pore sizes than the commercial membranes, a systematic study on the effect of membrane pore size and pore size distribution in the AGMD mode, effect of membrane material on MD separation when using volatile organic compounds with high affinity to membrane material, effect of membrane pore structure on MD performance especially in certain applications such as in liquid food processing and separation of VOCs from water by VMD, MD fouling, MD membrane aging, long-term MD performance of the different MD configurations, economical analysis together with detailed comparative works within the different MD configurations and the pressure-driven separation processes, and the like. An intensive and focused research effort in these fields is needed, both in experimental and modeling studies, where a central issue is the construction of pilot plants for scale-up studies.

Nomenclature

AFM	Atomic force microscopy
AGMD	Air gap membrane distillation
Al_2O_3	Alumina
AMT	Applied Membrane Technology
BSA	Bovine serum albumin
$CaCO_3$	Calcium carbonate
CP	Concentration polarization
CPC	Concentration polarization coefficient
CTA	Cellulose triacetate
DCMD	Direct-contact membrane distillation
DMAC	Dimethylacetamide
DMF	Dimethylformamide
DMSO	Dimethyl sulfoxide
EE	Energy efficiency
FESEM	Field emission scanning electron microscopy
FGLP	Membrane supplied by Millipore
FHLP	Membrane supplied by Millipore
F2.4	Membrane made by the copolymer poly(vinylidene fluoride-*co*-tetrafluoroethylene)

GBL	γ-Butyrolactone
GOR	Gained output ratio
GVHP	Membrane supplied by Millipore
HCl	Hydrochloric acid
HNO_3	Nitrate acid
HVHP	Membrane supplied by Millipore
IPA	Isopropyl alcohol
IUPAC	International Union of Pure and Applied Chemistry
LEP	Liquid entry pressure or breakthrough pressure
LEP_w	Liquid entry pressure of water
LiBr	Lithium bromide
LiCl	Lithium chloride
MAS	Membrane air stripping
MD	Membrane distillation
NaCl	Sodium chloride
NaOH	Sodium hydroxide
NF	Nanofiltration
OFCB	Octafluorocyclobutane
PAN	Polyacrylonitrile
PEG	Poly(ethylene glycol)
PE	Polyethylene
PEI	Polyetherimide
PMSP	Poly(1-trimethylsilyl-1-propyne)
PP	Polypropylene
PPO	Poly(phenylene oxide)
PSF	Polysulfone
PTFE	Poly(tetrafluoroetylene)
PV	Pervaporation
PVC	Polyvinyl chloride
PVDF	Polyvinylidene fluoride
PVP	Poly(vinylpyrrolidone)
RO	Reverse osmosis
SEM	Scanning electron microscopy
SGMD	Sweeping gas membrane distillation
SMMs	Surface modifying macromolecules
TDS	Total dissolved solids
TF1000	Membrane supplied by Gelman Company
TF200	Membrane supplied by Gelman Company
TF450	Membrane supplied by Gelman Company
TIPS	Thermally induced phase separation
TMP	Trimethyl phosphate
TP	Temperature polarization
TPC	Temperature polarization coefficient
TSGMD	Thermostatic sweeping gas membrane distillation
UF	Ultrafiltration
VLE	Vapor–liquid equilibrium
VMD	Vacuum membrane distillation

VOCs	Volatile organic compounds
VPPC	Vapor pressure polarization coefficient
$VTMS/CF_4$	Vinyltrimethylsilicon/carbon tetrafluoride
XPS	X-ray photoelectron spectroscopy
ZrO_2	Zirconia

ACKNOWLEDGMENTS

M. Khayet would like to thank Juan I. Mengual from the Department of Applied Physics I, Faculty of Physics, University Complutense of Madrid (Spain) for introducing him to the fascinating field of MD and Takeshi Matsuura for his guidance in membrane design.

REFERENCES

Andersson, S. I., Kjellander, N., and Rodesjo, B. (1985). Design and field tests of a new membrane distillation desalination process. *Desalination* 56, 345.

Bagger-Jorgensen, R., Meyer, A. S., Varming, C., and Jonsson, G. (2004). Recovery of volatile aroma compounds from black currant juice by vacuum membrane distillation. *J. Food Eng.* 64, 23.

Banat, F. A., Abu Al-Rub, F., Jumah, R., and Al-Shannag, M. (1999a). Application of Stefan-Maxwell approach to azeotropic separation by membrane distillation. *Chem. Eng. J.* 73, 71.

Banat, F. A., Abu Al-Rub, F., and Shannag, M. (1999b). Modeling of dilute ethanol-water mixture separation by membrane distillation. *Sep. Purif. Technol.* 16, 119.

Banat, F. A., Abu Al-Rub, F., Jumah, R., and Shannag, M. (1999c). Theoretical investigation of membrane distillation role in breaking the formic acid–water azeotropic point: Comparison between Fickian and Stefan-Maxwell-based models. *Int. Commun. Heat Mass Transfer* 26, 879.

Banat, F. A., Abu Al-Rub, F., Jumah, R., and Shannag, M. (1999d). On the effect of inert gases in breaking the formic acid–water azeotrope by gas-gap membrane distillation. *Chem. Eng. J.* 73, 37.

Banat, F., Al-Asheh, S., and Qtaishat, M. (2005). Treatment of waters colored with methylene blue dye by vacuum membrane distillation. *Desalination* 174, 87.

Banat, F., Jumah, R., and Garaibeh, M. (2002). Exploitation of solar energy collected by solar stills for desalination by membrane distillation. *Renewable Energy* 25, 293.

Banat, F. A., and Simandl, J. (1994). Theoretical and experimental study in membrane distillation. *Desalination* 95, 39.

Banat, F. A., and Simandl, J. (1996). Removal of benzene traces from contaminated water by vacuum membrane distillation. *Chem. Eng. Sci.* 51, 1257.

Banat, F. A., and Simandl, J. (1998). Desalination by membrane distillation: A parametric study. *Sep. Sci. Technol.* 33, 201.

Banat, F. A., and Simandl, J. (1999). Membrane distillation for dilute methanol: Separation from aqueous streams. *J. Membr. Sci.* 163, 333.

Banat, F. A., and Simandl, J. (2000). Membrane distillation for propanone removal from aqueous streams. *J. Chem. Technol. & Biotechnol.* 75, 168.

Bandini, S., Gostoli, C., and Sarti, G. C. (1992). Separation efficiency in vacuum membrane distillation. *J. Membr. Sci.* 73, 217.

Bandini, S., Saavedra, A., and Sarti, G. C. (1997). Vacuum membrane distillation: Experiments and modeling. *AIChE J.* 43(2), 398.

Bandini, S., and Sarti, G. C. (1999). Heat and mass transport resistances in vacuum membrane distillation per drop. *AIChE J.* 45, 1422.

Bandini, S., and Sarti, G. C. (2002). Concentration of must through vacuum membrane distillation. *Desalination* 149, 253.

Basini, L., D'Angelo, G., Gobbi, M., Sarti, G. C., and Gostoli, C. (1987). A desalination process through sweeping gas membrane distillation. *Desalination* 64, 245.

Belleville, M. P., Dornier, M., Sánchez, J., Vaillant, F., Rynes, M., and Rios, G. (2002). Procedé d'extraction para diffusion membranaire. French Patent 0205959.

Benzinger, W. D., and Robinson, D. N. (1982). Porous polyvinylidene fluoride membrane and process for its preparation. U.S. Patent 4,384,047.

Bodell, B. R. (1963). Silicone rubber vapor diffusion in saline water distillation. U.S. Patent 285,032.

Bodell, B. R. (1968). Distillation of saline water using silicone rubber membrane. U.S. Patent 3,361,645.

Bottino, A., Capannelli, G., Munari, S., and Turturro, A. (1988). High performance ultrafiltration membranes cast from LiCl doped solutions. *Desalination* 68, 167.

Bouguesha, S., and Dhahbi, M. (2002). Fluized bed crystalliser and air gap membrane distillation as a solution to geothermal water desalination. *Desalination* 152, 237.

Brodard, F., Romero, J., Belleville, M. P., Sánchez, J., Combe-James, C., Dornier, M., and Rios, G. M. (2003). New hydrophobic membranes for osmotic evaporation process. *Sep. Purif. Technol.* 32, 3.

Cabassud, C., and Wirth, D. (2003). Membrane distillation for water desalination: How to choose an appropriate membrane? *Desalination* 157, 307.

Calabró, V., Drioli, E., and Matera, F. (1991). Membrane distillation in the textile wastewater treatment. *Desalination* 83, 209.

Calabró, V., Jiao, B. L., and Drioli, E. (1994). Theoretical and experimental study on membrane distillation in the concentration of orange juice. *Ind. Eng. Chem. Res.* 33, 1803.

Calabró, V., Pantano, G., Kang, M., Molinari, R., and Drioli, E. (1990). Experimental study on integrated membrane processes in the treatment of solutions simulating textile effluents. Energy and exergy analysis. *Desalination* 78, 257.

Calibo, R. L., Matsumura, M., Takahashi, J., and Kataoka, H. (1987). Ethanol stripping by pervaporation using porous PTFE membrane. *J. Ferment. Technol.* 65, 665.

Camera-Roda, G., Boi, C., Saavedra, A., and Sarti, G. C. (1994). Heat and mass transfer boundary layers in radial creeping flow. *Int. J. Heat Mass Transfer* 37, 2145.

Capuano, A., Memoli, B., Andreucci, V. E., Criscuoli, A., and Drioli, E. (2000). Membrane distillation of human plasma ultrafiltrate and its theoretical applications to haemodialysis techniques. *Int. J. Artif. Organs* 23, 415.

Carlsson, L. (1983). The new generation in sea water desalination: SU membrane distillation system. *Desalination* 45, 221.

Cath, T. Y., Adams, D., and Childress, A. E. (2005). Membrane contactor processes for wastewater reclamation in space, II. Combined direct osmosis, osmotic distillation, and membrane distillation for treatment of metabolic wastewater. *J. Membr. Sci.* 257, 111.

Cheng, D. Y. (1981). Method and apparatus for distillation. U.S. Patent 4,265,713.

Cheng, D. Y., and Wiersma, S. J. (1982). Composite membrane for a membrane distillation system. U.S. Patent 4,316,772.

Cheng, D. Y., and Wiersma, S. J. (1983a). Composite membrane for a membrane distillation system. U.S. Patent 4,419,242.

Cheng, D. Y., and Wiersma, S. J. (1983b). Apparatus and method for thermal membrane distillation. U.S. Patent 4,419,187.

Chmielewski, A. G., Zakrewska-Trznadel, G., Miljevic, N. R., and Van Hook, W. A. (1995). Membrane distillation employed for separation of water isotopic compounds. *Sep. Sci. Technol.* 30, 1653.

Couffin, N., Cabassud, C., and Lahoussine-Turcaud, V. (1998). A new process to remove halogenated VOCs for drinking water production: Vacuum membrane distillation. *Desalination* 117, 233.

Criscuoli, A., and Drioli, E. (1999). Energetic and exegetic analysis of an integrated membrane desalination system. *Desalination* 124, 243.

Criscuoli, A., Drioli, E., Capuano, A., Memoli, B., and Andreucci, V. E. (2002). Human plasma ultrafiltrate purification by membrane distillation: Process optimisation and evaluation of its possible application on-line. *Desalination* 147, 147.

de Andrés, M. C., Doria, J., Khayet, M., Peña, L., and Mengual, J. I. (1998). Coupling of a membrane distillation module to a multieffect distiller for pure water production. *Desalination* 115, 71.

Deshmuck, S. P., and Li, K. (1998). Effect of ethanol composition in water coagulation bath on morphology of PVDF hollow fiber membranes. *J. Membr. Sci.* 150, 75.

Ding, Z., Liu, L., El-Bourawi, M. S., and Ma, R. (2005). Analysis of a solar-powered membrane distillation system. *Desalination* 172, 27.

Ding, Z., Liu, L., and Ma, R. (2003). Study on the effect of flow maldistribution on the performance of the hollow fiber modules used in membrane distillation. *J. Membr. Sci.* 215, 11.

Drioli, E., Calabró, V., and Wu, Y. (1986). Microporous membranes in membrane distillation. *Pure Appl. Chem.* 58, 1657.

Drioli, E., Jiao, B. L., and Calabró, V. (1992). The preliminary study on the concentration of orange juice by membrane distillation. *Proc. Int. Soc. Citriculture* 3, 1140.

Drioli, E., Laganá, F., Criscuoli, A., and Barbieri, G. (1999). Integrated membrane operations in desalination processes. *Desalination* 122, 141.

Drioli, E., and Wu, Y. (1985). Membrane distillation: An experimental study. *Desalination* 53, 339.

Drioli, E., Wu, Y., and Calabró, V. (1987). Membrane distillation in the treatment of aqueous solutions. *J. Membr. Sci.* 33, 277.

Enka, A. G. (1984, June), Catalogue presented at Europe-Japan Joint Congress on Membranes and Membrane Processes, Stresa, Italy.

Esato, K., and Eiseman, B. (1975). Experimental evaluation of Gore-Tex membrane oxygenator. *J. Thorac. Cardiovasc. Surg.* 69, 690.

Fane, A. G., Schofield, R. W., and Fell, C. J. D. (1987). The efficient use of energy in membrane distillation. *Desalination* 64, 231.

Feng, C., Shi, B., Li, G., and Wu, Y. (2004a). Preliminary research on microporous membrane from F2.4 for membrane distillation. *Sep. Purif. Technol.* 39, 221.

Feng, C., Shi, B., Li, G., and Wu, Y. (2004b). Preparation and properties of microporous membrane from poly(vinylidene fluoride-*co*-tetrafluoroethylene) (F2.4) for membrane distillation. *J. Membr. Sci.* 237, 15.

Findley, M. E. (1967). Vaporization through porous membranes. *Ind. Eng. Chem. Process Des. Dev.* 6, 226.

Findley, M. E., Tanna, V. V., Rao, Y. B., and Yeh, C. L. (1969). Mass and heat transfer relations in evaporation through porous membranes. *AIChE J.* 15, 483.

Franken, A. C. M., Nolten, J. A. M., Mulder, M. H. V., and Smolders, C. A. (1987a). Ethanol-water separation by membrane distillation: Effect of temperature polarization. In: B. Sedlacek and J. Kahovec (Eds.) *Synthetic Polymeric Membranes*, Walter de Gruyter, Berlin, p. 531.

Franken, A. C. M., Nolten, J. A. M., Mulder, M. H. V., Bargeman, D., and Smolders, C. A. (1987b). Wetting criteria for the applicability of membrane distillation. *J. Membr. Sci.* 33, 315.

Franken, A. C. M., and Ripperger, S. (1988, Jan.), *Terminology for Membrane Distillation*. European Society of Membrane Science and Technology. (Today European Membrane Society, Toulouse, France).

Fujii, Y., Kigoshi, S., Iwatani, H., and Aoyama, M. (1992a). Selectivity and characteristics of direct contact membrane distillation type experiment: I. Permeability and selectivity through dried hydrophobic fine porous membranes. *J. Membr. Sci.* 72, 53.

Fujii, Y., Kigoshi, S., Iwatani, H., Aoyama, M., and Fusaoka, Y. (1992b). Selectivity and characteristics of direct contact membrane distillation type experiment: II. Membrane treatment and selectivity increase. *J. Membr. Sci.* 72, 73.

García-Payo, M. C., and Izquierdo-Gil, M. A. (2004). Thermal resistance technique for measuring the thermal conductivity of thin microprous membranes. *J. Phys. D Appl. Phys.* 37, 3008.

García-Payo, M. C., Izquierdo-Gil, M. A., and Fernández-Pineda, C. (2000a). Wetting study of hydrophobic membranes via liquid entry pressure measurements with aqueous alcohol solutions. *J. Colloid Interface Sci.* 230, 420.

García-Payo, M. C., Izquierdo-Gil, M. A., and Fernández-Pineda, C. (2000b). Air gap membrane distillation of aqueous alcohol solutions. *J. Membr. Sci.* 169, 61.

García-Payo, M. C., Rivier, C. A., Marison, I. W., and von Stockar, U. (2002). Separation of binary mixtures by thermostatic sweeping gas membrane distillation: II. Experimental results with aqueous formic acid solutions. *J. Membr. Sci.* 198, 197.

Gore, D. W. (1982). Gore-Tex membrane distillation. In Proceedings of the Tenth Annual Convention of the Water Supply Improvement Association, Honolulu, July 25–29, 1982.

Gostoli, C., and Sarti, G. C. (1989). Separation of liquid mixtures by membrane distillation. *J. Membr. Sci.* 41, 211.

Gryta, M. (2000). Concentration of saline wastewater from the production of heparin. *Desalination* 129, 35.

Gryta, M. (2001). The fermentation process integrated with membrane distillation. *Sep. Purif. Technol.* 24, 283.

Gryta, M. (2002a). Concentration of NaCl solution by membrane distillation integrated with crystallization. *Sep. Sci. Technol.* 37(15), 3535.

Gryta, M. (2002b). The assessment of microorganism growth in the membrane distillation system. *Desalination* 142, 79.

Gryta, M. (2005). Long-term performance of membrane distillation process. *J. Membr. Sci.* 265, 153.

Gryta, M., and Karakulski, K. (1999). The application of membrane distillation for the concentration of oil-water emulsions. *Desalination* 121, 23.

Gryta, M., and Tomaszewska, M. (1998). Heat transport in the membrane distillation process. *J. Membr. Sci.* 144, 211.

Gryta, M., Tomaszewska, M., and Morawski, A. W. (1997). Membrane distillation with laminar flow. *Sep. Purif. Technol.* 11, 93.

Gryta, M., Morawski, A. W., and Tomaszewska, M. (2000). Ethanol production in membrane distillation bioreactor. *Catalysis Today* 56, 159.

Gryta, M., Tomaszewska, M., Grzechulska, J., and Morawski, A. W. (2001a). Membrane distillation of NaCl solution containing natural organic matter. *J. Membr. Sci.* 181, 279.

Gryta, M., Karakulski, K., and Morawski, A. W. (2001b). Purification of oily wastewater by hybrid UF/MD. *Water Res.* 35(15), 3665.

Guijt, C. M., Meindersma, G. W., Reith, R., and Haan, A. B. (2005). Air gap membrane distillation: 1. Modelling and mass transport properties for hollow fiber membranes. *Sep. Purif. Technol.* 43, 233.

Hanbury, W. T., and Hodgkiess, T. (1985). Membrane distillation: An assessment. *Desalination* 56, 287.

Henderyckx, Y. (1967). Diffusion doublet research. *Desalination* 3, 237.

Hoffman, E., Pfenning, D. M., Phillippsen, E., Schwahn, P., Sieber, M., When, R., and Woermann, D. (1987). Evaporation of alcohol/water mixtures through hydrophobic porous membranes. *J. Membr. Sci.* 34, 199.

Hogan, P. A., Sudjito, Fane, A. G., and Morrison, G. L. (1990). Desalination by solar heated membrane distillation. *Desalination* 81(1–3), 81.

Hsu, S. T., Cheng, K. T., and Chiou, J. S. (2002). Seawater desalination by direct contact membrane distillation. *Desalination* 143, 279.

Imdakem, A. O., and Matsuura, T. (2004). A Monte Carlo simulation model for membrane distillation processes: Direct contact (MD). *J. Membr. Sci.* 237, 51.

Imdakem, A. O., and Matsuura, T. (2005). Simulation of heat and mass transfer in direct contact membrane distillation (MD): The effect of membrane physical properties. *J. Membr. Sci.* 262, 117.

Izquierdo-Gil, M. A., García-Payo, M. C., and Fernández-Pineda, C. (1999a). Direct contact membrane distillation of sugar aqueous solutions. *Sep. Sci. Technol.* 34, 773.

Izquierdo-Gil, M. A., García-Payo, M. C., and Fernández-Pineda, C. (1999b). Air gap membrane distillation of sucrose aqueous solutions. *J. Membr. Sci.* 155, 291.

Izquierdo-Gil, M. A., and Jonsson, G. (2003). Factors affecting flux and ethanol separation performance in vacuum membrane distillation (VMD). *J. Membr. Sci.* 214, 113.

Jonsson, A. S., Wimmerstedt, R., and Harrysson, A. C. (1985). Membrane distillation: A theoretical study of evaporation through microporous membranes. *Desalination* 56, 237.

Karakulski, K., Gryta, M., and Morawski, A. (2002). Membrane processes used for potable water quality improvement. *Desalination* 145, 315.

Kesting, R. E. (1985). *Synthetic Polymeric Membranes*, 2nd ed., Wiley, New York.

Khayet, M., and Matsuura, T. (2001). Preparation and characterization of polyvinylidene fluoride membranes for membrane distillation. *Ind. Eng. Chem. Res.* 40, 5710.

Khayet, M., and Matsuura, T. (2003a). Application of surface modifying macromolecules for the preparation of membranes for membrane distillation. *Desalination* 158, 51.

Khayet, M., and Matsuura, T. (2003b). Determination of surface and bulk pore sizes of flat-sheet and hollow-fiber membranes by atomic force microscopy, gas permeation and solute transport methods. *Desalination* 158, 57.

Khayet, M., and Matsuura, T. (2004). Pervaporation and vacuum membrane distillation processes: Modeling and experiments. *AIChE J.* 50, 1697.

Khayet, M., and Mengual, J. I. (2004). Effect of salt concentration during the treatment of humic acid solutions by membrane distillation. *Desalination* 168, 373.

Khayet, M., Godino, P., and Mengual, J. I. (2000a). Theory and experiments on sweeping gas membrane distillation. *J. Membr. Sci.* 165, 261.

Khayet, M., Peña, L., and Mengual, J. I. (2000b). Utilización de un módulo de destilación en membranas con gas de barrido para mejorar el rendimiento de un destilador multiefecto de una sola etapa. *Anal. Fís.* 95, 169.

Khayet, M., Godino, P., and Mengual, J. I. (2000c). Nature of flow on sweeping gas membrane distillation. *J. Membr. Sci.* 170, 243.

Khayet, M., Godino, M. P., and Mengual, J. I. (2001). Modelling transport mechanism through a porous partition. *J. Non-Equilib. Thermodyn.* 26, 1.

Khayet, M., Feng, C. Y., Khulbe, K. C., and Matsuura, T. (2002a). Preparation and characterization of polyvinylidene fluoride hollow fiber membranes for ultrafiltration. *Polymer* 43, 2879.

Khayet, M., Feng, C. Y., Khulbe, K. C., and Matsuura, T. (2002b). Study on the effect of a non-solvent additive on the morphology and performance of ultrafiltration hollow-fiber membranes. *Desalination* 148, 321.

Khayet, M., Godino, M. P., and Mengual, J. I. (2002c). Thermal boundary layers in sweeping gas membrane distillation. *AIChE J.* 48, 1488.

Khayet, M., Godino, P., and Mengual, J. I. (2003a). Theoretical and experimental studies on desalination using sweeping gas membrane method. *Desalination* 157, 297.

Khayet, M., Suk, D. E., Narbaitz, R. M., Santerre, J. P., and Matsuura, T. (2003b). Study on surface modification by surface-modifying macromolecules and its applications in membrane-separation processes. *J. Appl. Polym. Sci.* 89, 2902.

Khayet, M., Godino, M. P., and Mengual, J. I. (2003c). Possibility of nuclear desalination through various membrane distillation configurations: A comparative study. *Int. J. Nucl. Desalin.* 1, 30.

Khayet, M., Velázquez, A., and Mengual, J. I. (2004a). Modelling mass transport through a porous partition: Effect of pore size distribution. *J. Non-Equilib. Thermodyn.* 29, 279.

Khayet, M., Velázquez, A., and Mengual, J. I. (2004b). Direct contact membrane distillation of humic acid solutions. *J. Membr. Sci.* 240, 123.

Khayet, M., Khulbe, K. C., and Matsuura, T. (2004c). Characterization of membranes for membrane distillation by atomic force microscopy and estimation of their water vapor transfer coefficients in vacuum membrane distillation process. *J. Membr. Sci.* 238, 199.

Khayet, M., Godino, M. P., and Mengual, J. I. (2004d). Study of asymmetric polarization in direct contact membrane distillation. *Sep. Sci. Technol.* 39, 125.

Khayet, M., Mengual, J. I., and Matsuura, T. (2005a). Porous hydrophobic/hydrophilic composite membranes: Application in desalination using direct contact membrane distillation. *J. Membr. Sci.* 252, 101.

Khayet, M., Matsuura, T., and Mengual, J. I. (2005b). Porous hydrophobic/hydrophilic composite membranes: Estimation of the hydrophobic-layer thickness. *J. Membr. Sci.* 266, 68.

Khayet, M., Matsuura, T., Mengual, J. I., and Qtaishat, M. (2006a). Design of novel direct contact membrane distillation membranes. *Desalination* 192, 105.

Khayet, M., Mengual, J. I., and Zakrzewska-Trznadel, G. (2006b). Direct contact membrane distillation for nuclear desalination. Part II. Experiments with radioactive solutions. *Int. J. Nucl. Desalin.* 2, 56.

Kim, J., Park, S. E., Kim, T. S., Jeong, D. E., and Ko, K. H. (2004). Isotopic water separation using AGMD and VEMD. *Nukleonika* 49, 137.

Kimura, S., and Nakao, S. (1987). Transport phenomena in membrane distillation. *J. Membr. Sci.* 33, 285.

Kong, J., and Li, K. (2001). An improved gas permeation method for characterizing and predicting the performance of microporous asymmetric hollow fiber membranes used in gas adsorption. *J. Membr. Sci.* 182, 271.

Kong, Y., Lin, X., Wu, Y., Cheng, J., and Xu, J. (1992). Plasma polymerization of octafluorocyclobutane and hydrophobic microporous composite membranes for membrane distillation. *J. Appl. Polym. Sci.* 46, 191.

Korngold, E., and Korin, E. (1993). Air sweep water pervaporation with hollow fiber membranes. *Desalination* 91, 187.

Koros, W. J., Ma, Y. H., and Shimadzu, T. (1996). Terminology for membranes and membrane processes (IUPAC recommendations 1996). *J. Membr. Sci.* 120, 149.

Koschikowski, J., Wieghaus, M., and Rommel, M. (2003). Solar thermal-driven desalination plants based on membrane distillation. *Desalination* 156, 295.

Kubota, S., Ohta, K., Hayano, I., Hirai, M., Kikuchi, K., and Murayama, Y. (1988). Experiments on seawater desalination by membrane distillation. *Desalination* 69, 19.

Kurokawa, H., Ebara, K., Kuroda, O., and Takahashi, S. (1990). Vapor permeate characteristics of membrane distillation. *Sep. Sci. Technol.* 25, 1349.

Laganà, F., Barbieri, G., and Drioli, E. (2000). Direct contact membrane distillation: Modelling and concentration experiments. *J. Membr. Sci.* 166, 1.

Larbot, A., Gazagnes, L., Krajewski, S., Bukowska, M., and Kujawski, W. (2004). Water desalination using ceramic membrane distillation. *Desalination* 168, 367.

Lawson, K. W., and Lloyd, D. R. (1996a). Membrane distillation. I. Module design and performance evaluation using vacuum membrane distillation. *J. Membr. Sci.* 120, 111.

Lawson, K. W., and Lloyd, D. R. (1996b). Membrane distillation: II. Direct contact MD. *J. Membr. Sci.* 120, 123.

Lawson, K. W., and Lloyd, D. R. (1997). Membrane distillation: Review. *J. Membr. Sci.* 124, 1.

Lee, C. H., and Hong, W. (2001). Effect of operating variables on the flux and selectivity in sweep gas membrane distillation for dilute aqueous isopropanol. *J. Membr. Sci.* 188, 79.

Li, B., and Sirkar, K. K. (2004). Novel membrane and device for direct contact membrane distillation-based desalination process. *Ind. Eng. Chem. Res.* 43, 5300.

Li, B., and Sirkar, K. K. (2005). Novel membrane and device for vacuum membrane distillation-based desalination process. *J. Membr. Sci.* 257, 60.

Li, J., Xu, Z., Liu, Z., Yuan, W., Xiang, H., Wang, S., and Xu, Y. (2003). Microporous polypropylene and polyethylene hollow fiber membranes: Part 3. Experimental studies on membrane distillation for desalination. *Desalination* 155, 153.

Lloyd, D. R., Kinzer, K. E., and Tseng, H. S. (1990). Microporous membrane formation via thermally-induced phase separation: I. Solid-liquid phase separation. *J. Membr. Sci.* 52, 239.

Loeb, S., and Sourirajan, S. (1960). Sea water demineralization by means of a semi-permeable membrane, UCLA-ENG-60-60, UCLA-SEAS Report. University of California, Los Angeles, CA.

Lukanin, O. S., Gunko, S. M., Bryk, M. T., and Nigmatullin, R. R. (2003). The effect of content of apple juice biopolymers on the concentration by membrane distillation. *J. Food Eng.* 60, 275.

Mahmud, H., Kumar, A., Narbaitz, R. M., and Matsuura, T. (1998). Membrane air stripping: A process for removal of organics from aqueous solutions. *Sep. Sci. Technol.* 33, 2241.

Mahmud, H., Kumar, A., Narbaitz, R. M., and Matsuura, T. (2000). A study of mass transfer in the membrane air-stripping process using microporous polypropylene hollow fibers. *J. Membr. Sci.* 179, 29.

Martínez, L., and Florido-Díaz, F. J. (2001). Theoretical and experimental studies on desalination using membrane distillation. *Desalination* 139, 373.

Martínez, L., Florido-Díaz, F. J., Hernández, A., and Prádanos, P. (2002). Characterization of three hydrophobic porous membranes used in membrane distillation: Modelling and evaluation of their water vapor permeabilities. *J. Membr. Sci.* 203, 15.

Martínez, L., Florido-Díaz, F. J., Hernández, A., and Prádanos, P. (2003). Estimation of vapor transfer coefficient of hydrophobic porous membranes for applications in membrane distillation. *Sep. Purif. Technol.* 33, 45.

Martínez-Díez, L., and Vázquez-González, M. I. (1996). Temperature polarization in mass transport through hydrophobic porous membranes. *AIChE J.* 42, 1844.

Martínez-Díez, L., and Vázquez-González, M. I. (1999). Temperature and concentration polarization in membrane distillation of aqueous salt solutions. *J. Membr. Sci.* 156, 265.

Martínez-Díez, L., Vázquez-González, M. I., and Florido-Díaz, F. J. (1998). Study of membrane distillation using channel spacers. *J. Membr. Sci.* 144, 45.

Mason, E. A., and Malinauskas, A. P. (1983). *Gas Transport in Porous Media: The Dusty-Gas Model*. Elsevier, Amsterdam.

Matijevic, E. (1989). *Surface and Colloid Science*. Wiley, New York.

Matsuura, T. (1994). *Synthetic Membranes and Membrane Separation Processes*. CRC Press, Boca Raton, FL.

Mengual, J. I., Khayet, M., and Godino, M. P. (2004). Heat and mass transfer in vacuum membrane distillation. *Int. J. Heat Mass Transfer* 47, 865.

Mengual, J. I., and Peña, L. (1997). Membrane distillation. *Colloid Interface Sci.* 1, 17.

Mittal, K. L. (1993). *Contact Angle, Wettability and Adhesion*. VSP. Utrecht, The Netherlands.

Morrison, G. L., Sudjito, Fane, A. G., and Hogan, P. (1991). Solar heated membrane distillation. In *Proceedings of the Twenty-Sixth International Congress of the International Solar Energy Society (ISES)*, Solar World Congress, Denver, Colorado, August 19–23, 1991.

Mulder, M. (1992). *Basic Principles of Membrane Technology*. Kluwer Academic, Dordrecht, The Netherlands.

Nene, S., Kaur, S., Sumod, K. K., Joshi, B., and Raghavarao, K. S. M. S. (2002). Membrane distillation for the concentration of raw-cane sugar syrup and membrane clarified sugarcane juice. *Desalination* 147, 157.

Ohta, K., Hayano, I., Okabe, T., Goto, T., Kimura, S., and Ohya, H. (1991). Membrane distillation with fluoro-carbon membranes. *Desalination* 81, 107.

Ohta, K., Kikuchi, K., Hayano, I., Okabe, T., Goto, T., Kimura, S., and Ohya, H. (1990). Experiments on sea water desalination by membrane distillation. *Desalination* 78, 177.

Ortiz de Zárate, J. M., Peña, L. and Mengual, J. I. (1995). Characterization of membrane distillation membranes prepared by phase inversion. *Desalination* 100, 139.

Ortiz de Zárate, J. M., Rincón, C., and Mengual, J. I. (1998). Concentration of bovine serum albumin aqueous solutions by membrane distillation. *Sep. Sci. Technol.* 33, 283.

Ortiz de Zárate, J. M., Velázquez, A., Peña, L., García-López, F., and Mengual, J. I. (1992). Non-isothermal solute transport through PTFE membranes. *J. Membr. Sci.* 69, 169.

Peng, P., Fane, A. G., and Li, X. (2005). Desalination by membrane distillation adopting a hydrophilic membrane. *Desalination* 173, 45.

Phattaranawik, J., Jiraratananon, R., and Fane, A. G. (2003a). Heat transport and membrane distillation coefficients in direct contact membrane distillation. *J. Membr. Sci.* 212, 177.

Phattaranawik, J., Jiraratananon, R., and Fane, A. G. (2003b). Effect of pore size distribution and air flux on mass transport in direct contact membrane distillation. *J. Membr. Sci.* 215, 75.

Phattaranawik, J., Jiraratananon, R., Fane, A. G., and Halim, C. (2001). Mass flux enhancement using spacer filled channels in direct contact membrane distillation. *J. Membr. Sci.* 187, 193.

Pinnau, I., and Freeman, B. D. (2000). *Membrane Formation and Modification*. ACS Symposium Series 744. American Chemical Society, Washington, DC.

Present, R. D. (1958). *Kinetic Theory of Gases*. McGraw-Hill, New York.

Rincón, C., Ortiz de Zárate, J. M., and Mengual, J. I. (1999). Separation of water and glycols by direct contact membrane distillation. *J. Membr. Sci.* 158, 155.

Rivier, C. A., García-Payo, M. C., Marison, I. W., and von Stockar, U. (2002). Separation of binary mixtures by thermostatic sweeping gas membrane distillation: I. Theory and simulations. *J. Membr. Sci.* 201, 1.

Rodgers, F. A. (1972). Stacked microporous vapor permeable membrane distillation system. U.S. Patent 3,650,905.

Rodgers, F. A. (1974). Compact multiple effect still having stacked impervious and previous membranes. U.S. Patent 3,497,423 (Re. 27,982).

Sakai, K., Koyano, T., Muroi, T., and Tamura, M. (1988). Effects of temperature and concentration polarization on water vapor permeability for blood in membrane distillation. *Chem. Eng. J.* 38, B33.

Sakai, K., Muroi, T., Ozawa, K., Takesawa, S., Tamura, M., and Nakane, T. (1986). Extraction of solute-free water from blood by MD. *Am. Soc. Artif. Intern. Organs* 32, 397.

Sarti, G. C., and Gostoli, C. (1986). Use of hydrophobic membranes in thermal separation of liquid mixture: Theory and experiments. In *Proceedings of the Europe-Japan Congress on Membranes and Membrane Processes*, Stresa, Italy, June 18–22, 1984. In E. Drioli and M. Nagaki (Eds.), *Membranes and Membrane Processes*. Plenum, New York, p. 349.

Sarti, G. C., Gostoli, C., and Bandini, S. (1993). Extraction of organic components from aqueous streams by vacuum membrane distillation. *J. Membr. Sci.* 80, 21.

Sarti, G. C., Gostoli, C., and Matulli, S. (1985). Low energy cost desalination processes using hydrophobic membranes. *Desalination* 56, 277.

Schneider, K., Holz, W., and Wollbeck, R. (1988). Membranes and modules for transmembrane distillation. *J. Membr. Sci.* 39, 25.

Schneider, K., and van Gassel, T. J. (1984). Membrane distillation. *Chem. Ing. Technol.* 56, 514.

Schofield, R. W., Fane, A. G., and Fell, C. J. D. (1987). Heat and mass transfer in membrane distillation. *J. Membr. Sci.* 33, 299.

Schofield, R. W., Fane, A. G., Fell, C. J. D., and Macoun, R. (1990a). Factors affecting flux in membrane distillation. *Desalination* 77, 279.

Schofield, R. W., Fane, A. G., and Fell, C. J. D., and Macoun, R. (1990b). Gas and vapor transport through microporous membranes: I. Knudsen-Poisseuille transition. *J. Membr. Sci.* 53, 159.

Schofield, R. W., Fane, A. G., Fell, C. J. D., and Macoun, R. (1990c). Gas and vapor transport through microporous membranes: II. Membrane distillation. *J. Membr. Sci.* 53, 173.

Semmens, M. J., Qin, R., and Zander, A. (1989). Using a microporous hollow-fiber membrane to separate VOCs from water. *J. Am. Water Works Assoc.* 81, 162.

Shih, H. C., Yeh, Y. S., and Yasuda, H. (1990). Morphology of microporous poly(vinylidene fluoride) membranes studied by gas permeation and scanning electron microscopy. *J. Membr. Sci.* 50, 299.

Smolders, C. A., and Franken, A. C. M. (1989). Terminology for membrane distillation. *Desalination* 72, 249.

Srisurichan, S., Jiraratananon, R., and Fane, A. G. (2005). Humic acid fouling in the membrane distillation process. *Desalination* 174, 63.

Sudoh, M., Takuwa, K., Iizuka, H., and Nagamatsuya, K. (1997). Effects of thermal and concentration boundary layers on vapor permeation in membrane distillation of aqueous lithium bromide solution. *J. Membr. Sci.* 131, 1.

Tomaszewska, M. (1993). Concentration of the extraction fluid from sulphuric acid treatment of phosphogypsum by membrane distillation. *J. Membr. Sci.* 78, 277.

Tomaszewska, M. (1996). Preparation and properties of flat-sheet membranes from polyvinylidene fluoride for membrane distillation. *Desalination* 104, 1.

Tomaszewska, M. (2000a). Concentration and purification of fluosilicic acid by membrane distillation. *Ind. Eng. Chem. Res.* 39, 3038.

Tomaszewska, M. (2000b). Mass tansfer of HCl and H_2O across the hydrophobic membrane during membrane distillation. *J. Membr. Sci.* 166, 149.

Tomaszewska, M., Gryta, M., and Morawski, A. W. (1995). Study on the concentration of acids by membrane distillation. *J. Membr. Sci.* 102, 113.

Tomaszewska, M., Gryta, M., and Morawski, A. W. (1998). The influence of salt in solutions on hydrochloric acid recovery by membrane distillation. *Sep. Purif. Technol.* 14, 183.

Tomaszewska, M., Gryta, M., and Morawski, A. W. (2001). Recovery of hydrochloric acid from metal pickling solutions by membrane distillation. *Sep. Purif. Technol.* 22–23, 591.

Udriot, H., Ampuero, S., Marison, I. W., and von Stokar, U. (1989). Extractive fermentation of ethanol using membrane distillation. *Biotechnol. Lett.* 11, 509.

Udriot, H., Araque, A., and von Stockar, U. (1994). Azeotropic mixtures may be broken by membrane distillation. *Cheng. Eng. J.* 54, 87.

Ugrosov, V. V., and Elkina, I. B. (1998). Concentration of binary aqueous solutions by the method of membrane distillation. *Theor. Fund. Chem. Eng.* 32, 97.

Uragami, T., Fujimoto, M., and Sugihara, M. (1980). Studies on syntheses and permeabilities of special polymer membranes. 28. Permeation characteristics and structure of interpolymer membranes from poly(vinylidene fluoride) and poly(styrene sulfonic acid). *Desalination* 34, 311.

Urtiaga, A. M., Ruiz, G., and Ortiz, I. (2000). Kinetic analysis of the vacuum membrane distillation of chloroform from aqueous solutions. *J. Membr. Sci.* 165, 99.

van Gassel, T. J., and Schneider, K. (1986). An energy-efficient membrane distillation process. In *Proceedings of the European-Japan Congress on Membranes and Membrane Processes*, Stresa, Italy. In E. Drioli and M. Nagaki (Eds.), *Membranes and Membrane Processes*. Plenum, New York, p. 343.

van Haute, A., and Henderyckx, Y. (1967). The permeability of membranes to water vapor. *Desalination* 3, 169.

Velázquez, A., and Mengual, J. I. (1995). Temperature polarization coefficients in membrane distillation. *Ind. Eng. Chem. Res.* 34, 585.

Wang, D., Li, K., and Teo, W. K. (1999). Preparation and characterization of polyvinylidene fluoride (PVDF) hollow fiber membranes. *J. Membr. Sci.* 163, 211.

Wang, D., Li, K., and Teo, W. K. (2000). Porous PVDF asymmetric hollow fiber membranes prepared with the use of small molecular additives. *J. Membr. Sci.* 178, 13.

Wang, D. M., Chang, C. Y., Wang, C. Y., Wu, T. T., and Lai, J. Y. (2002). Preparation and application of microporous TPX membranes. *Macromol. Symp.* 188, 49.

Weyl, P. K. (1967). Recovery of demineralized water from saline waters. U.S. Patent 3,340,186.

Wirth, D., and Cabassud, C. (2002). Water desalination using membrane distillation: Comparison between inside/out and outside/in permeation. *Desalination* 147, 139.

Wu, S. (1982). *Polymer Interface and Adhesion*. Marcel Dekker, New York.

Wu, Y., Kong, Y., Lin, X., Liu, W., and Xu, J. (1992). Surface-modified hydrophilic membranes in membrane distillation. *J. Membr. Sci.* 72, 189.

Wu, Y., Kong, Y., Liu, J., Zhang, J., Xu, J. (1991). An experimental study on membrane distillation-crystallization for treating waste water in taurine production. *Desalination* 80, 235.

Xu, J. B., Lange, S., Bartley, J. P., and Johnson, R. A. (2004). Alginate-coated microporous PTFE membranes for use in the osmotic distillation of oily feeds. *J. Membr. Sci.* 240, 81.

Zakrewska-Trznadel, G., Chmielewski, A. G., and Miljevic, N. R. (1996). Separation of protium/deuterium and oxygen-16/oxygen-18 by membrane distillation. *J. Membr. Sci.* 113, 337.

Zakrewska-Trznadel, G., Harasimowicz, M., and Chmielewski, A. G. (1999). Concentration of radioactive components in liquid low-level radioactive waste by membrane distillation. *J. Membr. Sci.* 163, 257.

Zakrewska-Trznadel, G., Harasimowicz, M., and Chmielewski, A. G. (2001). Membrane processes in nuclear technology–application for liquid radioactive waste treatment. *Sep. Purif. Technol.* 22–23, 617.

Zander, A. K., Semmens, M. J., and Narbaitz, R. M. (1989). Removing VOCs by membrane stripping. *J. Am. Water Works Assoc.* 81, 76.

Zolotarev, P. P., Urgozov, V. V., Volkina, I. B., and Nikulin, V. N. (1994). Treatment of waste water for removing heavy metals by membrane distillation. *J. Hazardous Mater.* 37, 77.

CHAPTER 13

Ultrapure Water by Membranes

AVIJIT DEY

Director – Application and Research, Omexell Inc., Stafford, Texas 77477

13.1 INTRODUCTION

Membrane-based technologies have become the industry standard for the ultrapure water systems in the semiconductor, pharmaceutical, and power industries. The seminal discovery that changed membrane separation from a laboratory to an industrial process was the development, in the early 1960s, of the Loeb–Sourirajan process for making defect-free, high-flux, anisotropic reverse osmosis membranes (Loeb and Sourirajan, 1963). The most important development in the 1980s was the innovation of industrial membrane gas separation process.

There has been a large demand for ultrapure water (UPW) in various industrial uses. The purification of aqueous streams using membrane-based technologies has become of immense interest in many industries. Membrane-based technologies are the answer to applications that require reliable and economical removal of particulate, organic, ionic, and gaseous contaminants from aqueous streams without the application of hazardous chemicals. Membrane-based technologies have been extensively used in the semiconductor, pharmaceutical, and power industries for UPW production. Production of UPW in a cost-effective manner is becoming more and more important in these industries. Effective application of membrane-based systems can in many cases provide a critical competitive edge.

13.1.1 Semiconductor Industry

Large quantities of UPW are required for wafer production; the demand of UPW even increases with the wafer size. For instance, the production of one 200-mm wafer with a 16-MB dynamic random access memory (DRAM) needs 4–5 m^3 of high-purity water (Nagel and Will, 1999). It is recommended to give adequate consideration to the Semiconductor Equipment and Materials International (SEMI) F63 2001 guidelines, chip manufacturer's specification, and to the International Technology Roadmap for Semiconductors (ITRS) 2004 guidelines to decide the specification for UPW for a given system. These guidelines should be used as a starting point for setting water quality

Advanced Membrane Technology and Applications. Edited by Norman N. Li, Anthony G. Fane, W. S. Winston Ho, and T. Matsuura
Copyright © 2008 John Wiley & Sons, Inc.

specifications. Although few studies have been conducted that actually relate specific water quality to various types of device performance, specifications could be set by compiling the information gained from microcontamination studies where yield loss was dramatic and caused by water (Air Liquide, 2004). As geometries shrink to 0.09 μm and beyond, water quality discussions by the active organizations continue to help identify process needs. However, each site and process will need to develop a set of specifications based on their own specific internal requirements. Maximum allowable impurity levels in semiconductor rinse water was specified in ASTM D-5127-99: Standard Guide for Ultra Pure Water used in the electronics and semiconductor industry. Six types of electronic-grade water are described in this standard. Type E-1.2 water is of ultimate practical purity produced in large volumes and is intended for most critical uses. This water is classified as microelectronic water to be used in the production of devices having line widths between 0.18 and 0.25 μm. Type E-1.1 water is classified as microelectronic water to be used in the production of devices having line widths between 0.25 and 0.5 μm. Table 13.1 indicates the requirements for water according to Type E-1 and above-grade specification.

A typical ultrapure water system for the semiconductor industry is illustrated in Figure 13.1. The reverse osmosis (RO)/electrodeionization (EDI) system is gaining more and more in importance in a typical UPW system due to its contamination-free design.

TABLE 13.1 Requirement for Water in E-1 and above Grades

Parameter	Type E-1	Type E-1.1	Type E-1.2
Resistivity, 25°C	18.2	18.2	18.2
TOC, ppb	5	2	1
DO, ppb	1	1	1
On-line particle count/liter (μm range)	500	500	100
0.05–0.1	300	300	50
0.1–0.2	50	50	20
0.2–0.3	20	20	10
0.3–0.5	4	4	1
>0.5			
Dissolved silica, ppt	1000	100	50
Ions and metals, ppt	100	100	50
Ammonium	100	50	20
Chloride	100	50	30
Fluoride	100	50	20
Nitrate	100	50	20
Sulfate	50	20	5
Aluminum	50	20	5
Boron	50	20	2
Calcium	50	20	2
Chromium	50	20	2
Copper	50	20	2
Iron	50	20	2
Magnesium			
Sodium	50	20	5
Potassium	50	20	5
Zinc	50	20	2

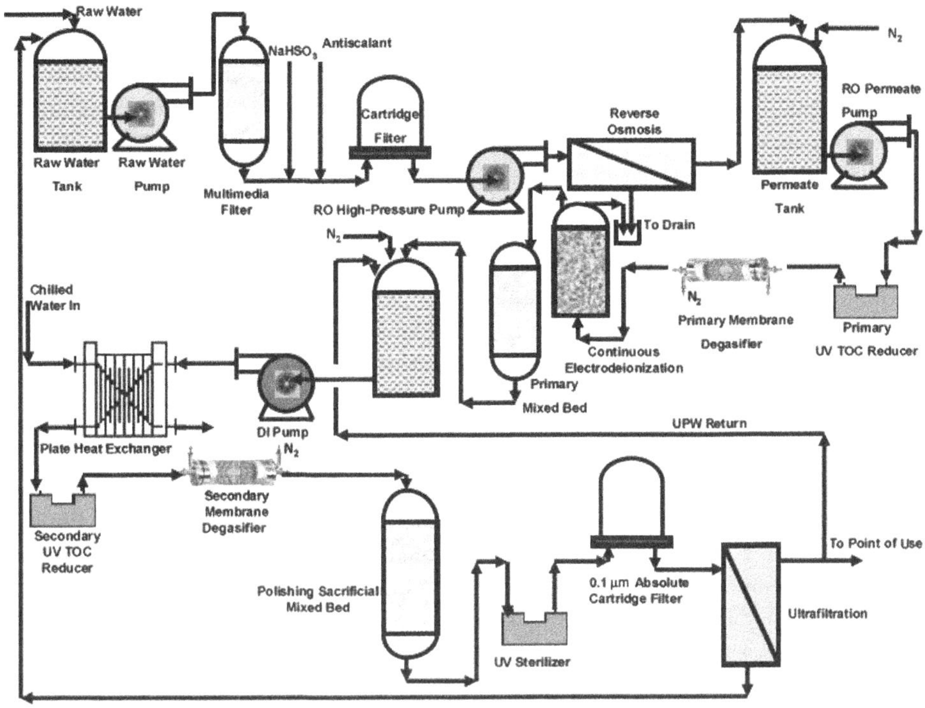

Figure 13.1 Typical ultrapure water system in the semiconductor industry.

The popularity of EDI systems in UPW systems can also be ascribed to physical space constraints needed to be taken into consideration in the UPW solution. However, in most cases a sacrificial polishing mixed-bed deionizer has to be considered downstream of the EDI to achieve ASTM D-5127-99 E-1.2: Standard Guide for Ultra Pure Water used in the Electronics and Semiconductor Industry. While the RO systems remove most of the organic and ionic impurities, downstream EDI devices can easily achieve a product water resistivity greater than 16 MΩ-cm at 25°C. Recent developments in EDI units offer more than 99% rejection of weakly ionized silica and boron and thus drastically reduce the regeneration frequency of the downstream mixed-bed deionizers. Dissolved oxygen in the UPW used in the wet cleaning process forms a native oxide film on a silicon wafer surface. This film has an impact on the quality of semiconductor devices, particularly as the level of integration becomes higher (Sato et al., 1991). Formation of native oxide layers can be effectively suppressed if the amount of dissolved oxygen is reduced to very low levels. Users in the UPW market are continuously searching to find innovative, cost-effective solutions to achieve low levels of oxygen in the product water.

Several methods for removal of dissolved oxygen from UPW are currently available. The most conventional ones are the thermal and vacuum degassing systems, which have inherent drawbacks in terms of both operating costs and bulky construction. Also, with these physical methods, it is difficult to reduce the dissolved oxygen concentration from the parts-per-million (ppm) level down to a few parts-per-billion (ppb) levels (Sato et al., 1991; Kasama et al., 1990; Imaoka et al., 1991). Most of the recent installations use microporous hydrophobic membrane modules for the removal of dissolved oxygen from UPW. Such membrane modules also simultaneously achieve an ultralow level of carbon dioxide in the treated water, which improves the performance of downstream EDI modules significantly. Another major benefit of membrane modules is their ability to remove low-molecular-weight organic impurities like trihalomethane (THM) compounds, which are poorly rejected by conventional polyamide RO membranes. The ultrafiltration (UF) unit placed in the final stage of the polishing loop has an important function as a final filter to remove particles up to 50 nm. The use of UF modules in the RO pretreatment is also becoming popular. As RO has become a widely accepted technology for the removal of ionic impurities from water bodies, the water sources that are being used to feed RO systems have become more and more challenging. Poor pretreatment can lead to a failure to control scale formation, organic fouling, and biogrowth. Bacteria, filamentous fungi, and yeasts frequently contribute to membrane biofouling and are a major component of the biomass that builds up at the membrane surface. Humic acid, which is commonly present in the surface water, may provide a food source for bacteria and may cause severe biofouling. RO manufacturers recommend that a silt density index (SDI) of lower than 3 is necessary for trouble-free RO operation. Conventional media filters typically undergo a ripening period when they are first put into service after a backwash, during which time water quality is inferior. In a similar manner, termination of the filtration cycle also sees a leakage of turbidity. Long-term performance can deteriorate due to mudball formation. A well-designed media filter typically achieves a SDI value in the vicinity of 3. In contrast, most UF modules remove all particulate matter, bacteria ($>$6 log reduction), most viruses, and colloids. In addition, the large molecular weight fraction of the humic acid can be removed by such UF systems. Further discussion of the UF system shall be limited only to the point-of-use application in the polishing loop because abundant literature is available on the RO pretreatment application.

13.1.2 Pharmaceutical Industry

Pharmaceutical water treatment plants are designed to meet the United States Pharmacopeia (USP) standards required for compendial waters. The term *compendial waters* represents any water intended to be used for final drug dosage forms, including sterile purified water (PW), sterile water for injection (WFI), sterile bacteriostatic water for injection, and sterile water for inhalation (Paul, 2002). High-purity water for application in the pharmaceutical industry essentially has to meet the chemical requirements laid down in the pharmacopeias. Consequently, the application of chemical-free RO/EDI systems has become increasingly popular in this industry (Fig. 13.2). The USP publishes standards for the pharmaceutical industry, including those for water quality. USP guidelines are listed in Table 13.2 for WFI and PW.

13.1.3 Power Industry

The thermoelectric power generation industry is the largest consumer of water in the United States. Deregulation of government-controlled power markets is compelling utilities and independent power producers to take a closer look at their production costs. To be competitive in the deregulated power market, authorities are looking for ways to lessen operating costs while maintaining an unfailing supply (Smith and Hyde, 2000). The heart of most electric power plants that use coal, gas, oil, or nuclear fuel is to create steam that will turn a turbine to generate electricity. Industrial boiler operating pressures continue to increase, and more independent combined cycle and co-generation facilities are being

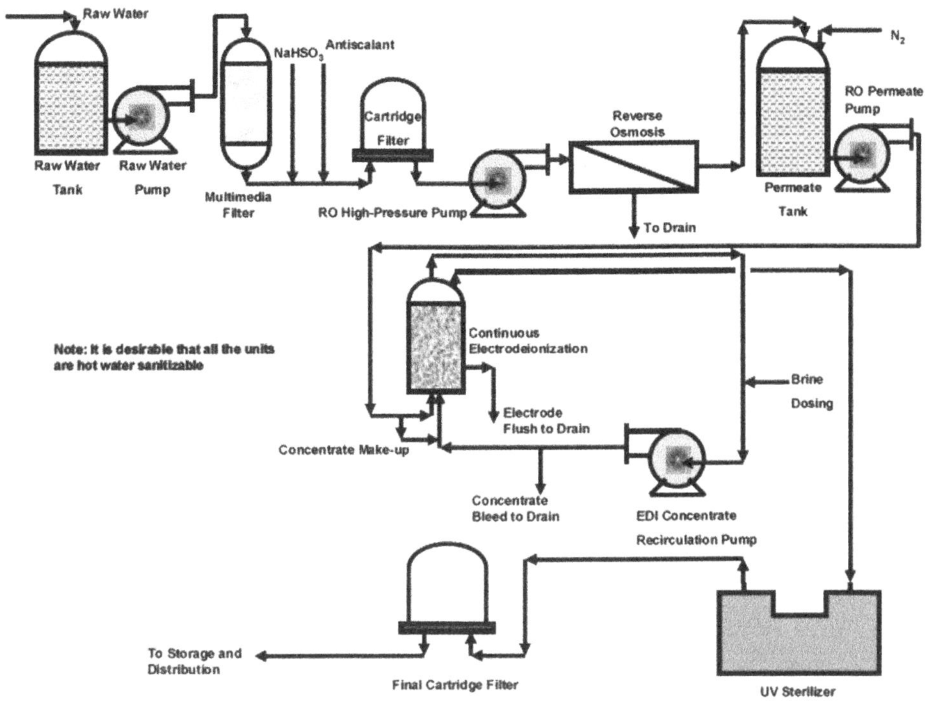

Figure 13.2 Typical ultrapure water system in the pharmaceutical industry.

built every day. High-purity feed water is required in any steam generation system. To minimize the overall water consumption, the makeup is combined with the purified condensed steam to become the boiler feed water. Freshwater added to replace water lost by blowdown, evaporation, wind drift, leaks, or withdrawal in these systems is referred to as makeup. The makeup water quality requirements vary greatly with the type of boiler system. There are many documents published by various authorities, but one extensively used is that published by the American Society of Mechanical Engineers (ASME, 1994). Typical makeup water specification for high-pressure boilers is given in Table 13.3. In boiler and turbine systems, the deposition of silica is often associated with temperature, pressure, and phase state changes that occur. Silica exhibits enough volatility that high-pressure boilers will have silica carryover to the vapor partition. When the steam pressure is reduced in the low-pressure turbines, the silica will precipitate on the turbine blade. Organic impurities will decompose in the steam/water circuit to produce CO_2 and organic acids such as acetic, formic, and glycolic. These impurities have potential corrosion

TABLE 13.2 USP Guidelines for PW and WFI

Parameters	USP PW	USP WFI
pH	5.0–7.0	5.0–7.0
Conductivity at 25°C, μS/cm	<1.3	<1.3
TOC, ppb	500	500
Bacteria, cfu/mL	100 cfu/mL	10 cfu/100 mL
Endotoxins, EU/mL	—	0.25

TABLE 13.3 Typical Makeup Water Specification for High-Pressure Boilers

Parameter	Value
Conductivity, μS/cm at 25°C	≤0.1
Silica, ppb	5–10
Sodium, ppb	2–5
Chloride, ppb	2–5
Sulfate, ppb	2–5
TOC, ppb	≤100

implications. Low-molecular-weight chlorinated organic impurities are especially troublesome and produce chloride.

The conventional three-bed ion exchange system in the thermoelectric power generation industry employs a cation, a forced-draft degassifier, an anion, and a mixed-bed polisher. The noteworthy market penetration of ultra-low-pressure RO technology in the front end of boiler feed water systems demonstrates a definite trend to reduce dependence on ion exchange processes in order to minimize the use of regenerant chemicals and improve the yield of product water in a cost-effective manner. The breakeven point in total dissolved solids (TDS) above which it is more economical to use RO technology over the ion exchange depends on a number of aspects that are discussed in the literature (Coker et al., 1994). It is imperative that readers should realize that since this study was conducted, the cost of RO membranes has declined significantly. In general, one can conclude from this study that it is more economical to use RO as the feed water TDS increases. Moreover, the power industry seldom has a dedicated water chemist because water is one of its raw materials to produce energy. A combination of RO and EDI does not necessitate continuous monitoring, whereas the ion exchange processes are operated in batch mode and entail careful monitoring by operators (Fig. 13.3). The conventional chemical methods for the

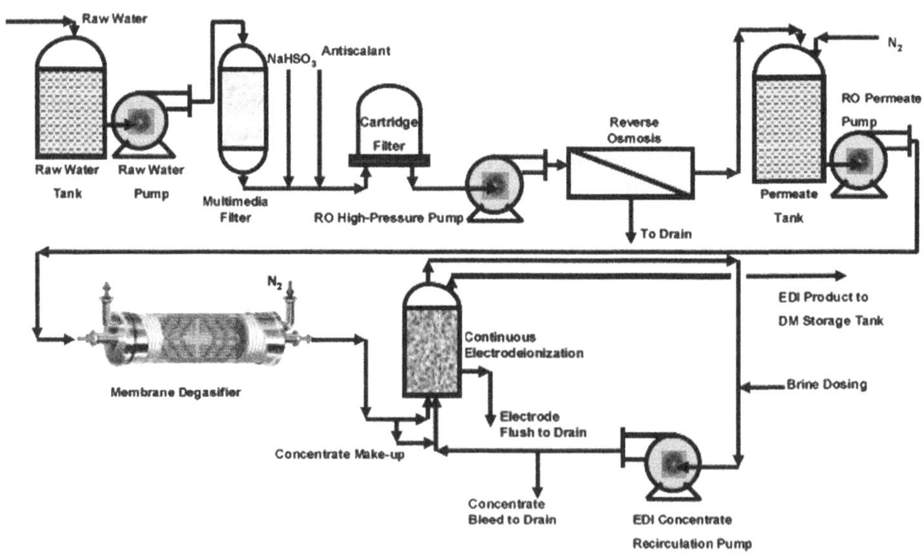

Figure 13.3 Typical ultrapure water system in the power industry.

removal of dissolved oxygen using hydrazine or sodium bisulfite are undesirable because of the toxicity of hydrazine and because the addition of sodium bisulfite will result in an increase of the solid content of the water. Many end users have recently installed microporous hydrophobic membrane modules for the removal of dissolved oxygen from UPW.

13.2 INTEGRATED MEMBRANE TECHNOLOGY IN UPW SYSTEMS

An integrated membrane approach in UPW systems consists of four major membrane-based water treatment components: ultrafiltration (UF), reverse osmosis (RO), electrodeionization (EDI), and membrane degasification. Each process is unique and contributes particular advantages to the system design. As the need increases and the costs become more acceptable, these technologies will become lynchpins of UPW systems.

13.2.1 Reverse Osmosis

Osmosis is a natural process involving fluid flow across a semipermeable membrane barrier. The phenomenon of osmosis occurs when pure water flows from a dilute saline solution through a membrane into a higher concentration saline solution. Its chemical potential, which is a function of pressure, temperature, and concentration of dissolved solids, determines the direction of solvent flow. This process can be reversed by applying pressure to the concentrate solution, causing water to flow from the concentrated solution into the dilute solution (Fig. 13.4). These membranes permit the passage of water molecules but are a barrier to most of the dissolved solids in water.

Membrane separation utilizes cross-flow filtration in which feed water flows over the membrane surface, separating the feed water into two streams: product water and concentrated water. The driving force for this filtration process is the pressure differential. Reverse osmosis membranes reject ionic species and operate at pressures of 700–4200 kPa (100–600 psig) for brackish water applications. Reverse osmosis is the process of forcing water through a semipermeable membrane against the natural osmotic gradient. When water is

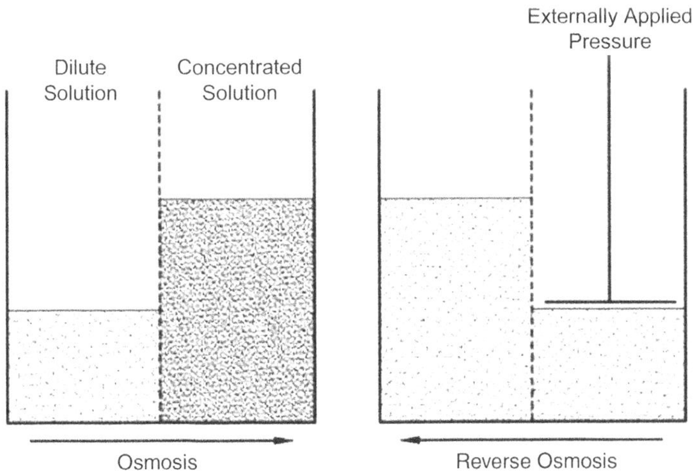

Figure 13.4 Principle of reverse osmosis.

forced through the membrane, a large percentage of the dissolved salts, and other material in the water, are removed from the water with the permeate being relatively pure water. The nominal pore size of the RO membrane is less than 1 nm. Water is a small molecule and passes through the membrane, reducing the solute concentration in the product stream. Water molecules, whose radius is about one tenth of 1 nm, can pass through the membrane freely.

There are two major groups of polymeric materials that can be used to produce satisfactory RO membranes: Cellulose acetate (CA) and polyamide (PA). Most of the industrial systems utilize PA membranes due to lower feed pressure and superior salt rejection. PA membranes are manufactured in two distinct steps. First, a polysulfone support layer is casted on a nonwoven polyester fabric. The polysulfone layer is very porous and not semipermeable. In a separate manufacturing step, a semipermeable membrane skin is formed on the polysulfone substrate by interfacial polymerization of monomers containing amine and carboxylic acid chloride functional groups. This manufacturing process enables independent optimization of the distinct properties of the membrane support and salt-rejecting skin. The resulting composite membrane is characterized by higher specific water flux and lower salt passage than CA membranes. Primarily, the surface layer governs the permeation rate since the pore size of the porous layer is very large. Membrane rejection of solutes and flux will deteriorate over time due to compaction, fouling, scaling, and biodegradation of the membrane surface. The normal life of membranes is 3–5 years.

The surface charge of the membrane is the result of ionization of particular functional groups existing on the membrane surface. It is important to note that ionization of a functional group depends on solution pH, and as a result surface charge of a particular membrane is also pH dependent. Most membranes have a neutral to negative net surface charge in the pH range of typical natural waters. Neutral surface charge for a wide feed water pH range alleviates the fouling potential. Some popular membrane suppliers have already developed low fouling membranes with a neutral surface charge over a wide feed water pH for water sources with higher fouling potential.

Typically the spiral-wound configuration offers significantly lower replacement cost, simpler plumbing systems, easier maintenance, and greater design freedom than other module configurations (Dow, 1995). The basic components of the spiral modules are the membrane, feed spacer, permeate spacer (carrier), leaf adhesive, permeate tube, antitelescoping device, and brine seal (Winter, 1995). Fiberglass housings with side-entry ports are commonly used in brackish water applications. The fiberglass has the advantage of being relatively inert to corrosion.

Based on long-term field experience, a relative fouling potential can be associated with the source of feed water. Accordingly, the recommended flux range has been defined in Table 13.4 for RO systems processing different types of feed water.

TABLE 13.4 Permeate Flux for Different Feed Sources

Feed Water Source	Flux Rate (GFD)
Seawater	8–10
Surface water (river, lake, or ocean)	8–12
City water	12–14
Well water	14–18
RO permeate	18–25

In a manner similar to ultrafiltration systems, concentration polarization effects need to be accommodated in designing RO systems. The maximum allowable value of the concentration polarization factor is 1.13–1.2 as recommended by different RO membrane suppliers. The concentration polarization factor can be defined as the ratio of salt concentration at the membrane surface to the bulk concentration.

Computer models easily perform many of the calculations necessary to design an RO system. Such programs are currently available from the RO membrane manufacturers for predicting RO permeate quality and monitoring and graphing normalized RO performance. The particular array (staging) is entered, along with the characteristics of the system (e.g., product water flow rate and feed solute concentration). The program then predicts the required feed pressure to obtain the desired product water flow rate and permeate quality. The limiting values of various parameters have been incorporated into these computer programs. Design of the system in excess of the guidelines results in a warning message on the printout. However, computer programs do not specify RO system designs. The user proposes an RO system configuration, and the computer program helps the user to evaluate the optimized system.

13.2.1.1 RO Membrane Pore Size At a given set of design and operating conditions, the feed pressure is determined by the net driving pressure (NDP) required to produce the design value of average permeate flux. The design feed pressure (P_f) is equal to the sum of NDP required to produce the average permeate flux, osmotic pressure corresponding to an average feed salinity (P_o), average pressure drop (P_d) along the RO system and permeate pressure (P_p) as follows.

$$P_f = \text{NDP} + P_o + P_d + P_p \tag{13.1}$$

(Mathematical notation used throughout chapter are defined in the Notation section at the end of the chapter.) Consequently, the only way to reduce the RO feed water pressure is to reduce the NDP at a given permeate flux rate. Commercially available ultra-low-pressure RO membranes are manufactured with a larger pore diameter to achieve lower NDP at a given permeate flux rate. This can be further explained by the Hagen–Poiseuille equation as follows:

$$J = \frac{\varepsilon \times d_p^2 \times \Delta P}{32 \times \delta \times \mu} \tag{13.2}$$

One can easily infer from Eq. (13.2) that to maintain the same flux rate (J) at a given temperature, one needs to increase d_p to reduce ΔP (NDP), which is related to the feed pressure by Eq. (13.1). Larger pore diameters will allow a higher passage of ionic impurities into the permeate water, and as a result one can see a deterioration in the permeate quality. Designers of UPW systems should pay adequate attention to this issue when selecting the RO membrane modules. Especially, one needs to ensure that the RO permeate quality should be acceptable for the downstream EDI devices in any UPW system.

13.2.1.2 Chemistry of Ionic Rejection Multivalent ions can be rejected better than univalent ions. Wiesner and Aptel (1996) have reported that the rejection of ionic solutes by RO membranes have been observed to an approximation to follow the lyotrophic series (increasing rejection with increasing hydrated radius). This is also consistent with being

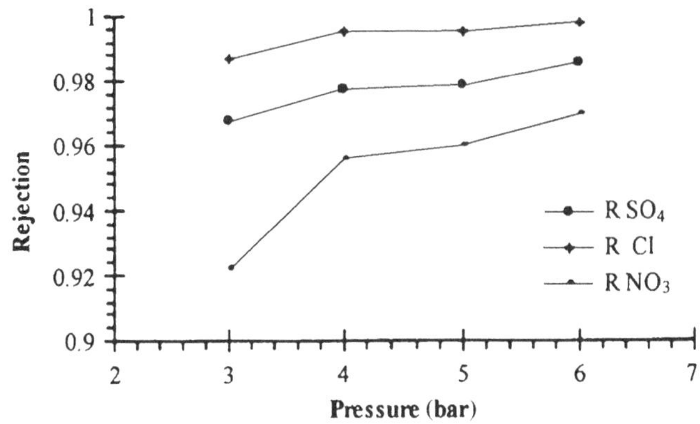

Figure 13.5 Rejection of anions (Koyuncu et al., 2001).

less for ions of lower valences. The lyotrophic series predicts that the rejection of anions by RO membrane should obey the following order: $SO_4^{2-} > Cl^- > Br^- > NO_3^- > I^-$. This theory is further supported in the following experimental study (Fig. 13.5). Figure 13.5 also shows the impact of the feed water pressure on the rejection characteristics of various anions. From the slope of the different curves, one can infer that the rejection of nitrate ions is highly dependent on the feed water pressure. Consequently, the use of ultra-low-pressure membranes may result in a raised nitrate level in the RO permeate stream. Such rejections are also impacted by ion pairing, complexation, and other solute–solute interactions.

13.2.1.3 Interstage Caustic Dosing in Double-Pass RO It was observed that the retention of weak acids by RO systems, such as boric acid at neutral or acidic pH values, is inadequate. If the degree of dissociation of boric acid is increased by increasing the interstage pH, the rejection in the second pass improves tremendously. The effect on retention for total organic carbon (TOC) of increasing the interstage pH value is also significant.

In a recent experimental study by Dey et al. (2001a,b), a series of experiments were conducted at various interstage pH conditions. The permeate water from the first pass was adjusted to various pH conditions with 45% caustic soda. Raw water boron concentration varied from 45 to 55 ppb. Boron rejection in the first-pass RO ranges between 40 and 50%. Figure 13.6 shows the boron concentrations in the second-pass feed and permeate as a function of interstage pH values. Percentage of boron rejection as a function of interstage pH values is also shown in Figure 13.6. The data in Figure 13.6 indicate that the interstage pH value strongly affects the boron rejection. The improvement in the boron rejection at higher interstage pH values can be attributed to the higher degree of ionization of boron at higher pH.

In the same experimental work, raw water TOC concentration varied from 2.9 to 3.1 ppm. TOC concentration was in the range of 1.4–1.9 ppm after the carbon filter. TOC concentrations in the first-pass RO permeate ranges between 150 and 180 ppb. Figure 13.7 shows the TOC concentrations in the second-pass feed and permeate as a function of interstage pH values. Percentage of TOC rejection as a function of interstage pH values is also shown in Figure 13.7. The data in Figure 13.7 indicate that the interstage pH value strongly affects the TOC rejection. The improvement in the TOC rejection at higher interstage pH values can be attributed to the higher degree of ionization of weakly ionized organic acids at higher pH. This is because the dissociation of the solute molecule tends to favor the electrostatic repulsion of the solute at the membrane–solution interface.

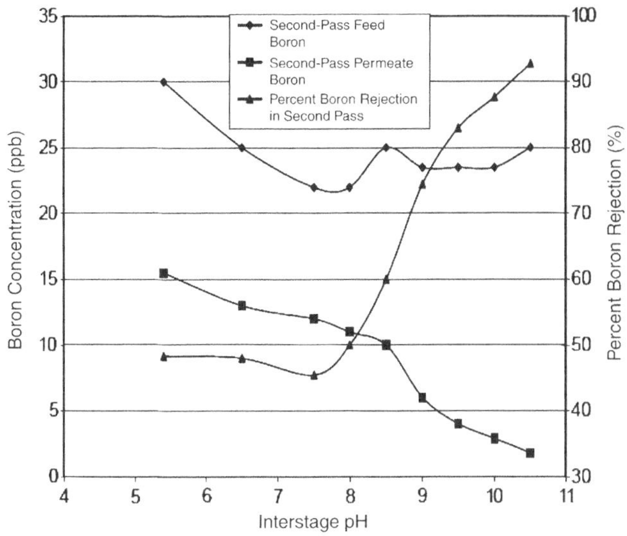

Figure 13.6 Effect of interstage pH on rejection of boron by double-pass RO (Dey et al., 2001a,b).

13.2.2 Electrodeionization

Electrodeionization is a separation process combining electrodialysis and conventional ion exchange; the resulting hybrid process does not entail regenerant chemicals. In the case of competing regenerable mixed beds, it is essential that distinct separation of cation and anion resins occurs upon backwashing to assure complete regeneration of the resins and avoid contamination by the other regenerant. However, the very nature of a regenerable mixed bed requires compromise: The resins must be able to separate for regeneration and to remain intimately mixed during exhaustion (Dey and Thomas, 2003).

The terminal deionization step in the membrane-based system includes the use of EDI technology. Industrial-scale EDI devices are available in two major

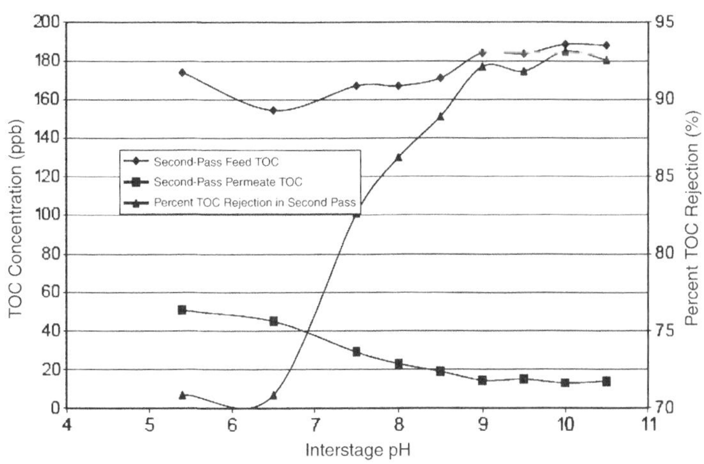

Figure 13.7 Effect of interstage pH on TOC rejection by double-pass RO (Dey et al., 2001a,b).

configurations: plate-and-frame and spiral-wound configurations. While the majority of these systems employ first-generation plate-and-frame designs, the spiral-wound version has turned out to be the fastest growing technology in the EDI field.

The plate-and-frame type EDI devices are similar in construction to a plate heat exchanger, with multiple fluid chambers sandwiched between a set of end plates (and electrodes) that are held in compression by bolts or threaded rods. Plate-and-frame devices are large in size and characteristically suffer from leaks because of the difficulty of sealing large vessels. Edges of the ion exchange membranes in the plate-and-frame modules are normally exposed to the ambient air. Any leakage from concentrate compartments results in the buildup of salt deposits due to subsequent water evaporation. They can form a bridge between the cell pairs and the metallic frame of the modules. Such salt bridges can lead to arcing, which leads to the module damage (Wood and Gifford, 2004).

The spiral-wound EDI membranes with central cathode are similar to that of RO membranes in that the membranes and spacers are rolled to form a cylindrical element (Fig. 13.8). The EDI element is manufactured by placing a stainless steel concentrate pipe on a rolling machine and winding the membrane and spacers around the pipe. The element is then placed into a fiberglass pressure vessel, and dilute chamber spacers are filled with resin. The unit is sealed inside the pressure vessel. As a result, these devices are not susceptible to salt bridging. The central stainless steel pipe acts as the concentrate distributor/collector and the cathode. A titanium anode lines the inside of the fiberglass pressure vessel and becomes the anode.

13.2.2.1 Electromigration Through Ion Exchange Membranes Ion exchange membranes are made of ion exchange resins manufactured in sheet form. The connection between the ion exchange properties of a membrane material and its permselectivity, that is, its ability to transfer only certain kinds of ions, in particular cations or anions, had been known even before synthetic ion exchange resins became available and was discussed quantitatively by many researchers (Teorell, 1951). When a membrane made of ion exchange material is placed between two solutions and a potential is applied, exchangeable ions are transferred from one solution to the other. If a cation exchange membrane is placed between dilute solutions and an electrical potential applied across it, the current is carried

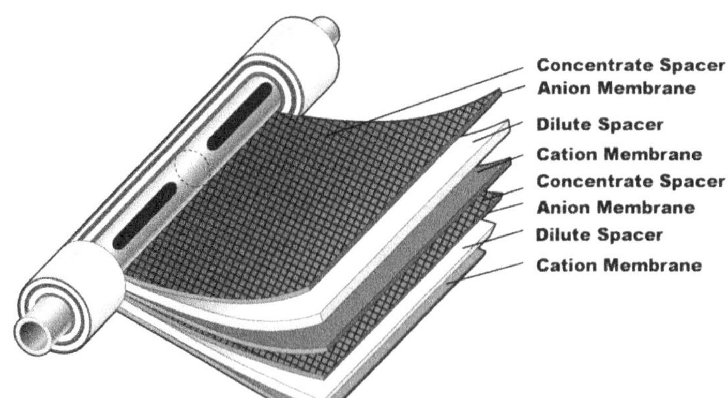

Figure 13.8 Spiral-wound EDI with central cathode. Photo of Omexell spiral-wound EDI (SWEDI) module.

mainly by the cations because the anions from the solution are almost completely excluded from the resin, and the fixed anion groups of the resins are immobile. Thus, in dilute solutions, cation exchange membranes are almost impermeable to anions, acting as "ionic sieves" that screen out anions. Similarly, anion exchange membranes are impermeable to cations (Spiegler, 1953). However, permselectivity declines with increasing electrolyte concentrations and improves with increasing capacity of the ion exchange resin. The average transference number is the quantitative measure for the permselectivity of the membrane. In general, it was found that the transference number of the exchangeable ion equals unity at low concentrations of the solutions and decreases with increasing solution concentration. This fact is due to the diffusion of electrolyte into the resin, which causes a certain proportion of the current to be carried by the nonexchangeable ions (Bauman et al., 1952). As explained earlier, ion exchange resins of high capacity are particularly suitable as permselective materials. The diffusion of electrolytes into the membranes and electroosmotic water transport fall with increasing cross-linking of the membrane material. In contrast, the higher electrical resistance of highly cross-linked ion exchange resins requires a compromise in the extent of cross-linking (Bauman et al., 1952).

The deionization range will be limited by the maximum current density that the membrane can withstand. The current density will vary directly with the solution conductivity, and thus a series of cell pairs will be necessary to achieve deionization over a large range (Walters et al., 1955). The transfer of ions proceeds in two stages. First, they are transferred from the solution to the resin particles by the ion exchange process. This exchange process is diffusion controlled. Film diffusion is influenced by solution flow rate and hence the Reynolds number (Gittens and Glueckauf, 1965). Second, the ions are transported inside the resin particles by the electrical current flowing through the resin bed until they reach the membrane walls of the diluting chamber. Because the concentration of ions in the resin phase is so high, the process that dictates the overall removal of ions is their rate of diffusion from the solution to the surface of the ion exchange resins. This rate of diffusion is a function of three factors: directly proportional to resin surface area, inversely proportional to the thickness of liquid film, and proportional to the difference in concentration of ions in the bulk and their concentration adjacent to the ion exchange resins. The use of small particle size resins commensurate with avoidance of excessive pressure drop satisfies the first two conditions to improve the rate of diffusion. Theoretically, the pressure drop is inversely proportional to the square of particle diameter of the ion exchange resin. Moreover, resins with broad size distributions tend to pack more densely in the diluting chamber and smaller resins have more narrow interstitial flow passages yielding larger pressure drops. Pressure drop may be easily correlated by the Ergun equation. To satisfy the third condition, one needs to maximize the electroregeneration of the ion exchange resins (Leon, 2001). In general, the better the current utilization, the slower the diffusion transport and the larger equipment size. Optimum conditions thus have to be established for the most economical deionization, which forms a compromise between current consumption on the one hand and size of the equipment on the other (Glueckauf, 1959).

13.2.2.2 Mechanism of Ion Removal in EDI The EDI system is composed of an arrangement of flow-directing spacers separated by anion and cation semipermeable membranes. Spacers are provided between alternating cation and anion exchange membranes to maintain separation of associated membranes. Spacers support even distribution of liquid through the chamber. Spacers are usually manufactured from thermoplastic materials. Permselective ion exchange membranes are arranged to form parallel-flow chambers.

In EDI systems, ions are removed from water by a combination of ion exchange resin, ion exchange membranes, and a direct current (dc) electric current. The electric field creates a driving force for the motion of ionic impurities present in the EDI feed water. It drives cations toward the cathode and anions toward the anode. An electric current passes at right angles to the liquid flow in such a way that both cations and anions are removed from the mixed bed through the membranes. Ions in the diluting chambers flow through the resin-filled electrolytic solution toward the electrodes, and the circuit is completed by electrons flowing through the external conductor. The feed water entering the EDI apparatus is divided into at least three parts. A small percentage flows over the electrodes, a majority of the feed passes through the diluting chambers, and the remainder passes along the concentrating chambers. Diluting chambers are filled with ion exchange resins. A conductive path is developed through the resin beads that are much lower in electrical resistance compared to the path through the bulk solution. The chambers bounded by the anion membrane facing the anode and the cation membrane facing the cathode become exhausted of ions and are thus called diluting chambers. The chambers bounded by the cathode-facing anion membrane and anode-facing cation membrane are called concentrating chambers. The concentrating chambers will then "trap" ions that have electrically migrated in from the diluting chambers. Consequently, the ionic concentration of the water will decrease in the diluting chamber with a corresponding increase in the ionic concentration of the water in the concentrating chamber. These flow chambers are hydraulically in parallel but electrically in series.

The current also splits water molecules into hydrogen and hydroxyl ions. The EDI systems operate in two different modes: electrodeionization and electroregeneration (Ganzi, 1988). The system is working in the electrodeionization mode when the feed water salinity is high (Ganzi and Parise, 1990). Polyvalent cations and anions are discharged to the concentrating compartments with relative ease at the upstream flow path (Fig. 13.9). Monovalent cations and anions are next to receive treatment. On the contrary,

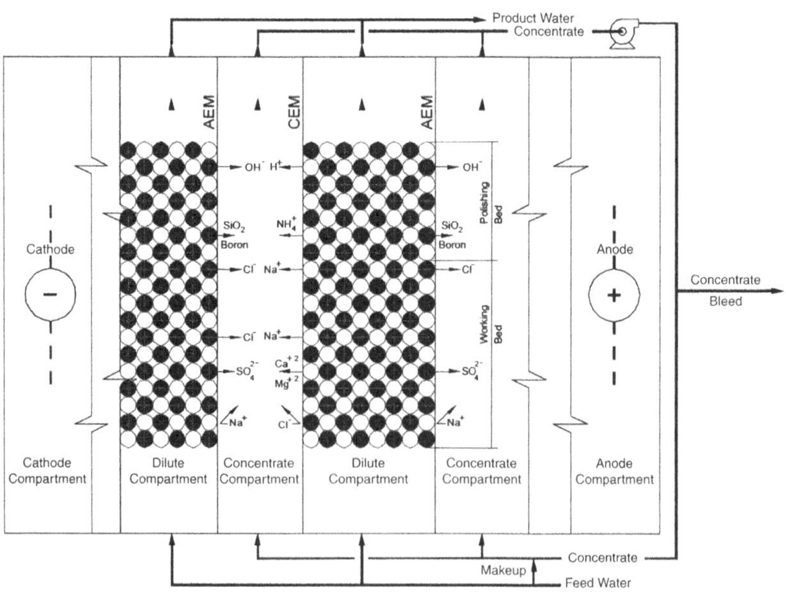

Figure 13.9 Ion transport in an electrodeionization device.

the system is working in the electroregeneration mode when the feed water salinity is very low due to the passage of strongly ionized species to the concentrating chamber in the upstream flow path. This allows a portion of the resins in the EDI to always be in the fully regenerated state. This will form a highly conductive path through the resin beads. Moreover, a higher degree of regeneration of resin beads will reduce the leakage of all the ionic species including boron and silica.

13.2.2.3 *Electrodeionization System Material Balance*

The percentage of the feed water that becomes product is referred to as the recovery of the system. The recent EDI units are typically operated at very high system recovery. With the addition of a concentrate recycle system, the new generation EDI modules are routinely operated at 80–95% feed water recovery in post-RO application. Recovery rate is decided by EDI feed water hardness. Recovery for the EDI units can be expressed as follows:

$$\text{Recovery} = \frac{P}{P + \text{CB} + \text{EL} \times N_M} \times 100\% \tag{13.3}$$

The following guidelines for the water recovery based on the feed water total hardness are established for the spiral-wound EDI modules with the central cathode (Table 13.5). Unfortunately, vendors of plate-and-frame EDI devices suggest that the maximum total hardness in the feed to the EDI be limited to 0.5–1 ppm as $CaCO_3$.

With this design, the concentrate is recycled to the concentrate feed of the module with a small bleed to maintain the desired recovery. Concentrate bleed off can be estimated from the following relationship:

$$\text{CB} = P \times \left(\frac{1}{\alpha} - 1\right) - \text{EL} \times N_M \tag{13.4}$$

If high overall water treatment system recovery is desired, it is feasible to recycle some or all of the concentrate bleed flow. In many situations, concentrate bleed is purer than the raw water. However, when the EDI feed stream contains high levels of CO_2, its buildup may prevent recycling without de-carbonation of the EDI feed stream. De-carbonation of the EDI feed water using membrane degasification units shows a lot of promise. Alternatively, pH of the upstream RO feed water is increased by the addition of a small amount of sodium hydroxide, which converts carbon dioxide to sodium bicarbonate, which can be rejected by conventional polyamide RO membranes as follows:

$$CO_2 + NaOH = NaHCO_3 \tag{13.5}$$

TABLE 13.5 Water Recovery versus EDI Feed Water Hardness

EDI Feed Water Hardness (ppm as $CaCO_3$)	Water Recovery (%)
0.0–0.5	95
0.5–1.0	90
1.0–1.5	85
1.5–2.0	80

Electrodeionization vendors recommend injecting brine in the concentrate loop of their systems to maintain the conductivity of the concentrate stream between 250 and 600 μS/cm. The use of Canners evaporated salt or equivalent is recommended for this purpose. At lower concentrate water conductivity, a rise in the electrical resistance will augment the power consumption.

13.2.2.4 Electrodeionization Feed Water Requirements Appropriate pretreatment of the water is a basic prerequisite for optimum performance of the EDI process to remove hardness, particulate matter, and chlorine. A major limitation of this process is the tendency of the membranes to become fouled by hard-water constituents. Because of their requirements for softened pretreated water, EDI devices are always coupled with upstream RO units (Meltzer, 1997). RO membrane is used for removal of organic and ionic impurities. Chlorine will attack ion exchange resins and cause de-cross-linking, which results in reduced capacity. Particulate matter, organics, and colloids can cause the plugging and fouling of membrane and resin beds. Since carbon dioxide is typically present at significant concentrations in the RO permeate water and is removed in a combination of anionic bicarbonate and carbonate forms, the total exchangeable anion (TEA) load is usually greater than the cation load (total exchangeable cation, TEC) and determines the required EDI operating conditions (Barber et al., 2000). If free carbon dioxide is present in the EDI feed, as it generally is, it will also be present in the concentrate. Since it is not ionic, it will diffuse without restraint through the cation membrane back to the diluting chamber. It cannot diffuse through the anion membrane because most of the anion membrane is alkaline, and the carbon dioxide would be converted into bicarbonate in the membrane and forced back into the concentrating chamber by the voltage gradient. Bicarbonate or carbonate ions are forced by the voltage gradient within the concentrating chamber toward the cation membrane. The boundary layer next to this membrane is acidic, as is the membrane itself. This converts both bicarbonate and carbonate into carbon dioxide, which can diffuse without restraint into the diluting chamber through the cation membrane (Hidaka, 2001). Consequently, occurrence of carbon dioxide and alkalinity at more than the recommended values are mostly accountable for the degradation of EDI product water quality. The feed water requirements for commercial EDI modules are as follows:

TEA $\leq$ 25 ppm as $CaCO_3$
pH 5.0–9.0
Hardness $\leq$ plate and frame: 1 ppm as $CaCO_3$; spiral wound: 2 ppm as $CaCO_3$
Dissolved silica $\leq$ 1000 ppb as SiO_2
TOC $\leq$ 500 ppb
Free chlorine $\leq$ 0.05 ppm
Fe, Mn $\leq$ 0.01 ppm
Carbon dioxide $\leq$ 10 ppm
Alkalinity $\leq$ 20 ppm as $CaCO_3$
Temperature 10–38°C

At this point it is important to realize that feed water conductivity does not show a complete picture of the total ionic load in a water system. Conductivity measurement devices do not detect the full amount of weakly ionized species such as CO_2, silica, and boron (Liqui-Cel, 2007). Consequently, total ionic load is more accurately described by TEA, TEC, and FCE (feed water conductivity equivalent). TEA represents the concentration of all the anions present in the feed water, including contributions from silica, carbon dioxide, and the hydroxide ion. In a similar manner, TEC represents the concentration of all the cations present in the feed water, including contributions from ammonia (NH_3) and the hydrogen ion.

$$\begin{aligned}\text{TEA (ppm as CaCO}_3) = {} & CO_3^{2-} \text{ as ppm CaCO}_3 \\ & + 1.7 \times HCO_3^- \text{ as ppm CaCO}_3 + 50{,}000 \times 10^{(\text{pH}-14)} + SO_4^{2-} \\ & + F^- + Cl^- + NO_3^-) \text{ as ppm CaCO}_3 + 0.83 \times SiO_2 \text{ as ppm ion} \\ & + \frac{1.7 \times 50}{44} \times CO_2 \text{ as ppm ion}\end{aligned} \tag{13.6}$$

$$\begin{aligned}\text{TEC (ppm as CaCO}_3) = {} & (Ca^{2+} + Mg^{2+} + K^+ + NH_4^+ \\ & + Ba^{2+} + Sr^{2+}) \text{ as ppm CaCO}_3 \\ & + \frac{NH_4^+ \text{ as ppm CaCO}_3}{10^{[(0.0902 - \text{pH} + 2730/273.2 + t]}} \\ & + 50{,}000 \times 10^{-\text{pH}}\end{aligned} \tag{13.7}$$

$$\begin{aligned}\text{FCE } (\mu\text{S/cm}) = {} & \text{feed water conductivity } (\mu\text{S/cm}) + \text{ppm } CO_2 \times 2.66 \\ & + \text{ppm } SiO_2 \times 1.94\end{aligned} \tag{13.8}$$

All the EDI units have an upper operating temperature limit of approximately 38°C. This limit is established by the spacer material, the membrane material, and the anion exchange resin present in the diluting chamber.

13.2.2.5 Estimation of Current in EDI The applied dc current, not the voltage, is the main parameter for the design of EDI units. The amount of current used by an EDI module is a function of the flow rate and the amount of salt being removed. Faraday's law states that 96,485 A of electric current is required for one second to move one mole of ionic charge between electrodes. This is same as 26.8 A for one hour. One can also write that one Faraday of charge is required to transfer one-gram equivalent of salt.

$$1 \text{ Faraday} = 96{,}485 \text{ A-s} = 26.8 \text{ A-h}$$

$$I_M = \frac{96{,}485 \times \text{MAX (TEA, TEC)} \times Q_d \times 3.785/60}{\eta_c \times N_{cp} \times 50{,}000} \tag{13.9}$$

Equation (13.9) indicates that higher impurity concentrations and higher flow rate will entail a higher applied dc current in order to achieve the desired product water quality. Current efficiency (η_c) is defined as the ratio of the current that transfers salt to the total amount of current applied. Water splitting occurs when the unit is operated at low current efficiency. Water splitting continuously regenerates the resin in the diluting chamber, enabling the EDI to remove weakly ionized contaminants such as CO_2, boron, and silica.

More precisely, current density is the driving force of the process. Running at a high current density reduces the required ion exchange membrane surface area of the EDI modules making the process more attractive. Current density is the flux of charge, that is, the rate of flow of charge per unit area perpendicular to the direction of flow. The behavior of electrochemical systems is determined more by the current density than by the total current, which is the product of the current density and the cross-sectional area.

Cell designs may be classified into two main groups: rectangular tank cells in plate-and-frame modules and cylindrical electrode cell in spiral-wound modules. Spiral-wound EDI modules employ two concentric electrodes. Current flows only in the radial direction and is uniform in the angular and axial direction. If a total current I_M is applied between two concentric electrodes, then the current density $i(r)$ in solution will vary with radial position by the following equation:

$$i(r) = \frac{I_M}{2 \times \pi \times r \times H_D} \tag{13.10}$$

Current distribution on each electrode is uniform. However, it is different on the two electrodes in a spiral-wound EDI module, being larger on the inner electrode (cathode in this case).

$$i(r_a) = \frac{I_M}{2 \times \pi \times r_a \times H_D} \tag{13.11}$$

$$i(r_c) = \frac{I_M}{2 \times \pi \times r_c \times H_D} \tag{13.12}$$

Average current density (i_{av} in A/ft^2) in a spiral-wound module can be determined using a formal integration technique as follows:

$$i_{av} = \frac{\int_{r_c}^{r_a} \frac{I_M}{2 \times \pi \times r \times H_D} dr}{r_a - r_c} = \frac{I_M}{2 \times \pi \times H_D \times (r_a - r_c)} \times \ln\left(\frac{r_a}{r_c}\right) \tag{13.13}$$

13.2.2.6 E Factor In EDI devices, H^+ and OH^- ions are formed by dissociation of the water molecules to continuously regenerate the ion exchange resin beads in the dilute compartment. Water splitting, cross-leakage, and back diffusion can cause current loss in the operation of EDI devices. Such losses can be estimated using current efficiency (η_c) as discussed before. Some authors prefer to use the term E-factor (Tessier, 1998). The combined effects of impurity concentration, feed flow rate, and applied current can be represented by a dimensionless parameter termed E factor. This is defined as the ratio of the applied current to the theoretical current and is, therefore, the reciprocal of the current efficiency. The value of E factor for plate-and-frame EDI devices can be conveniently determined using the following simple equation:

$$E = \frac{I_M \times N_{CP} \times 50{,}000}{96{,}485 \times \text{MAX (TEA, TEC)} \times Q_d \times (3.785/60)} \tag{13.14}$$

Let us assume that a plate-and-frame EDI module is operating at a current level of 4.5 A to produce 10 gpm deionized water from a feed with TEA of 25 ppm as $CaCO_3$ that has 38 cell pairs. The E factor can then be calculated simply as shown below:

$$E = \frac{4.5 \times 38 \times 50{,}000}{96{,}485 \times 25 \times 10 \times (3.785/60)} = 5.62 \tag{13.15}$$

It has been determined that an E factor of at least 1 is required for acceptable operation of an EDI device. Post-RO EDI modules are normally operated at 3–20 E factor to achieve essentially complete deionization, including the desired level of boron and silica removal.

One needs to account for the variation in the current density in the radial direction of the spiral-wound EDI modules in determining its current efficiency or E factor. To achieve desired deionization performance at the low current density region near the outer wall of the EDI module, spiral-wound E factor (SWE) calculation is performed based on the current density at the inner electrode (cathode in this case):

$$\text{SWE} = \frac{I_M \times N_{\text{CP}} \times L_D \times 50{,}000}{96{,}485 \times 2 \times \pi \times r_c \times \text{MAX (TEA, TEC)} \times Q_d \times (3.785/60)} \tag{13.16}$$

Let us assume that a spiral-wound EDI module is operating at a current level of 8 A to produce 8.8 gpm deionized water from a feed stream with TEA of 25 ppm as $CaCO_3$ that has the following characteristics:

$$N_{\text{CP}} = 6$$
$$L_D = 2.886\ \text{ft}$$
$$r_c = 0.136\ \text{ft}$$

The spiral-wound E factor can then be estimated as follows:

$$\text{SWE} = \frac{8 \times 6 \times 2.886 \times 50{,}000}{96{,}485 \times 2 \times \pi \times 0.136 \times 25 \times 8.8 \times (3.785/60)} = 6.05 \tag{13.17}$$

13.2.2.7 Electrodeionization Module Resistance The current is supplied by a power source capable of automatically increasing or decreasing voltage in response to a change in the electrical impedance of the EDI module to maintain constant current. Over time, the electrical impedance of all the modules increase, resulting in a gradual decrease in current at a given voltage level. Electrical resistance of the module controls how much current can be passed through the units. Since the direct current determines how much feed water of a given conductivity can be processed for a given product quality, it is important to optimize the electrical resistance of the module. Early in the nineteenth century George Ohm discovered that, as long as the temperature is constant, the magnitude of the current is proportional to the applied voltage. Ohm's law indicates that the current flowing through an EDI module is directly proportional to the voltage applied, and inversely proportional to the overall resistance of the module under ideal operating conditions:

$$I_M = \frac{V}{R} \tag{13.18}$$

Equation (13.18) indicates that lower overall EDI module resistance will entail a lower voltage in order to pass a fixed amount of current through the module. The overall resistance is equal to the sum of the individual resistances, just as is the case in the flow of electric current through a series of resistances. The overall resistance of an EDI module is equal to the sum of that offered by individual membranes, resins, concentrate stream, anolyte, and catholyte at a given feed water temperature and ionic composition.

The structural design of the spiral-wound EDI modules reduces the distance between anode and cathode, resulting in less overall module resistance when compared to conventional plate-and-frame modules. It is interesting to note that the distance between anode and cathode in the spiral-wound EDI modules is smaller than its radius, while anode and cathode are separated by the overall length of the module in the conventional plate-and-frame configuration. Lower overall module resistance in spiral-wound EDI configuration leads to reduction of necessary voltage when compared to conventional plate-and-frame counterpart to obtain a given current through the module.

13.2.2.8 Estimation of EDI Voltage Factors such as desired current through the module, water temperature, and C-loop conductivity will influence the minimum necessary voltage required for the operation of an EDI system. The multiple regression equation for the calculation of voltage for central cathode spiral-wound EDI modules offered by Omexell is given in Eq. (13.19) (Dey, 2005a).

$$V = \frac{404.88}{t} + 18.18 \times \ln\left(\frac{2 \times I}{N_M}\right) + 1334 \times \frac{(2 \times I)/N_M}{T_C \times \mu_C} - 10.74 \qquad (13.19)$$

$$T_C = 1 + 0.02 \times (t - 25) \qquad (13.20)$$

13.2.2.9 Electrodeionization Power Consumption The current is supplied by a power source capable of automatically increasing or decreasing voltage in response to a change in the electrical impedance of the EDI module to maintain constant current. Over time, the electrical impedance of all the modules increase, resulting in a gradual decrease in current at a given voltage level. All EDI manufacturers limit the maximum dc rectifier voltage to 350–600 Vdc, in order to avoid the need for the more expensive wiring construction that is necessary for higher voltages. Actual power consumption can be computed by the following simple equation:

$$\text{Power consumption} = \frac{I \times V}{1000 \times \eta_R \times Q_P} \frac{\text{KW-h}}{\text{m}^3 \text{ water}} \qquad (13.21)$$

13.2.2.10 Electrodeionization Desalination Performance A full-size EDI system is normally operated on varying feed water quality to evaluate its desalination efficiency. Module desalination performance is evaluated by constructing a desalination performance curve as shown in Figure 13.10 (Dey and Li, 2004). Desalination performance of the spiral-wound EDI modules with a central cathode was evaluated at various applied direct currents and feed water conductivities. The dilute flow rate was kept constant per EDI module. During such experiments water temperature was maintained at 25 ± 1°C. The curves all have the same characteristic shape, in the beginning showing a high resistivity plateau as the feed conductivity increased, which eventually gives way to deterioration in product resistivity at higher feed conductivity values. The rudimentary requirement is to provide sufficient current to attain the desired product water quality for a given feed water conductivity.

13.2.2.11 Rejection of Weakly Ionized Species by EDI The presence of weakly ionized species at high concentration levels in the feed water interferes with the

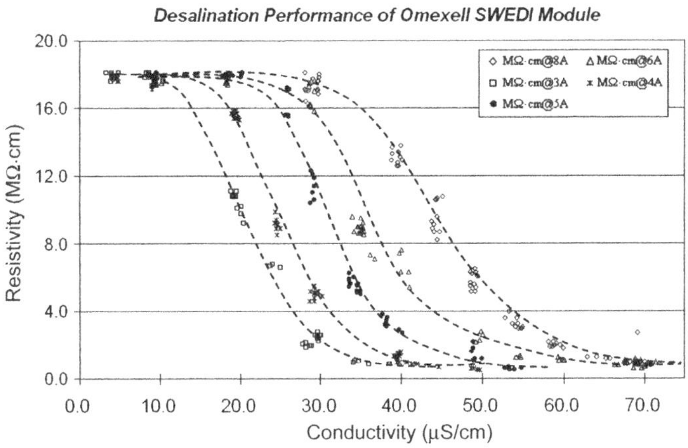

Figure 13.10 Product resistivity versus feed water conductivity at varying currents (Dey and Li, 2004).

performance of ultrapure water systems. For removal of weakly ionized species such as carbon dioxide, silica, ammonia, and boron, it is necessary to ionize these species. Dissociation of weak acids to ionized form at different pH is related by the following simple equation:

$$pK_a = pH + \log\left(\frac{1-\alpha}{\alpha}\right) \tag{13.22}$$

Dissociation of the above-mentioned weakly ionized species occurs as shown below:

$$\begin{aligned} CO_2 + OH^- &= HCO_3^- & pK_a &= 6.38 \\ HCO_3^- + OH^- &= CO_3^{2-} & pK_a &= 10.37 \\ SiO_2 + OH^- &= HSiO_3^- & pK_a &= 9.50 \\ H_3BO_3 + OH^- &= B(OH)_4^- & pK_a &= 9.27 \\ NH_3 + H^+ &= NH_4^+ & pK_a &= 9.24 \end{aligned}$$

Conventional EDI devices can offer high removal rates for weakly ionized species having a low dissociation constant (pK_a) such as CO_2 by increasing the applied current to dissociate enough water. However, the removal rate for weakly ionized species having high dissociation constant such as silica and boron is limited across conventional EDI devices. Careful selection of ion exchange resins in the dilute chambers is necessary to achieve high rejection rate of silica and boron across EDI modules.

Silica rejection across any EDI module strongly depends upon the applied current, dilute flow rate, water temperature, and inlet silica concentration. Pilot data can be easily fitted into the following empirical correlation (Dey, 2005b):

$$\%SiO_2 = [A + B \times \ln(SWE)] \times [C \times \ln(t) + D] \tag{13.23}$$

where, $\%SiO_2$ indicates percentage rejection rate of silica (%), and A, B, C, and D are empirical constants unique to a given EDI module.

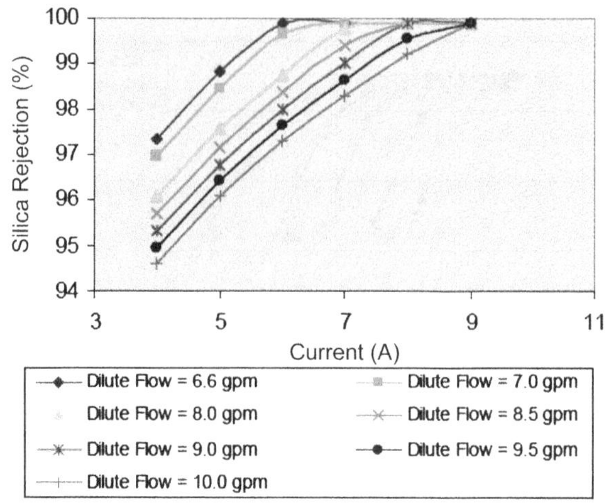

Figure 13.11 Silica rejection versus applied current at constant temperature of 23°C (Dey, 2005b).

A test unit of cross-flow spiral-wound EDI module with a working range of 6.6–10 gpm product flows was operated on varying applied currents at various dilute flow rates and water temperature to evaluate its silica rejection efficiency (Dey, 2005b). The inlet silica concentrations in all such experimental studies were maintained at 250 ± 10 ppb. Module silica removal performance was evaluated by constructing silica rejection performance curves as shown in Figures 13.11 through 13.13. Silica rejection performance of the cross-flow spiral-wound EDI test modules was evaluated at various dilute flow rates and applied direct currents at a constant water temperature of 23°C in Figure 13.11. Figure 13.11 clearly shows that decrease in dilute flow rate and/or increase in applied current improves the silica rejection performance of the EDI modules.

Silica rejection performance of the cross-flow spiral-wound EDI test modules was also evaluated at various water temperatures and applied direct currents at a constant dilute flow rate in Figure 13.12. The dilute flow rate was kept constant at 8.5 gpm per EDI test module.

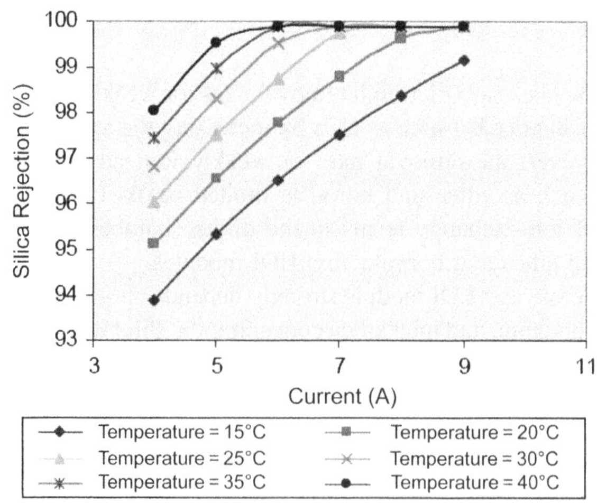

Figure 13.12 Silica rejection versus applied current at constant dilute flow of 8.5 gpm (Dey, 2005b).

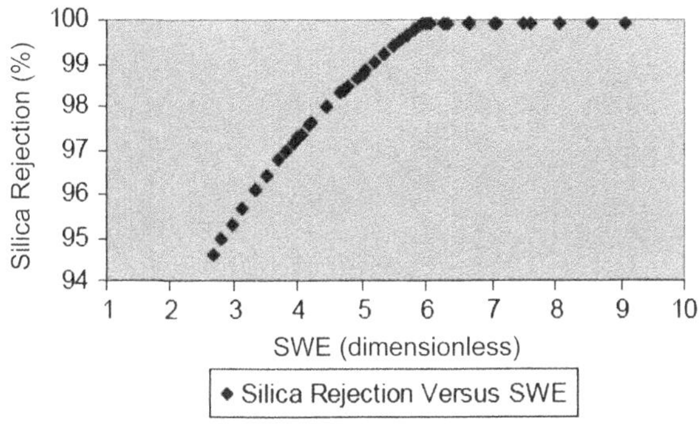

Figure 13.13 Silica rejection versus SWE at 23°C (Dey, 2005b).

Figure 13.12 clearly shows that increase in water temperatures and/or increase in applied current improves the silica rejection performance of the EDI modules.

The curves between silica rejection efficiency and SWE in the beginning show a high silica rejection plateau as the SWE decreased, which eventually gives way to deterioration in product silica at lower SWE values (Fig. 13.13). The rudimentary requirement is to operate the EDI modules at high SWE (low current efficiency) to attain the desired product water quality in terms of resistivity and silica rejection performance for a given feed water conductivity and temperature.

13.2.3 Membrane Degasification

Membrane degasification units are devices that can be used to permit mass transfer between a gaseous phase and a liquid phase without dispersing one phase into another. The gas layer is stabilized within the pores of a hydrophobic microporous filter in membrane degasification units. Solutes that are volatile can pass across these membranes, but nonvolatile solutes and aqueous liquids such as electrolytes are completely retained. High flux occurs when the solute is volatile and when it is relatively insoluble in water.

The membrane degasification technology has three major advantages and one potential disadvantage over conventional equipment based on packed towers. The advantages are (Reed et al., 1995):

- High surface area per volume
- Complete loading
- No flooding

Additionally, in membrane degasification technology liquid and gas phases can be maintained at very different pressures, the advantage being that there may be no need for a booster pump. However, the disadvantage is that the mass transfer is slower. In packed towers, two resistances to mass transfer are present, one in the aqueous feed phase and one in the gas phase. In membrane degasification units, the membrane provides a third resistance. In most cases the resistance of the porous membrane, which is gas filled, is negligible. Consequently, the use of membrane degasification units in place of conventional vacuum degasification units is becoming popular. A typical membrane degasification unit installation is shown in Figure 13.14.

Figure 13.14 Typical membrane degasification unit installation.

Membrane degasification units generally use hydrophobic hollow-fiber polypropylene or polytetrafluoroethylene (PTFE) (Teflon) microporous membrane (Wiesler, 1996). Hollow-fiber modules are made by potting the desired number of fibers into an external shell. The potting compound may be polyurethane, epoxy, polyolefin, or fluorinated polymers. Since the membranes are hydrophobic and have small pores (Fig. 13.15), water will not easily pass through the pores.

The pressure required to force water to enter the pore is called the breakthrough pressure. Bubble point as measured and reported in the literature is the air pressure needed to push out liquid imbibed in the pore of the membrane. The procedure for a bubble point test is described in ASTM Method F-316. The relationship between pore size and bubble point pressure is based on the application of the Young–Laplace equation. The smaller the

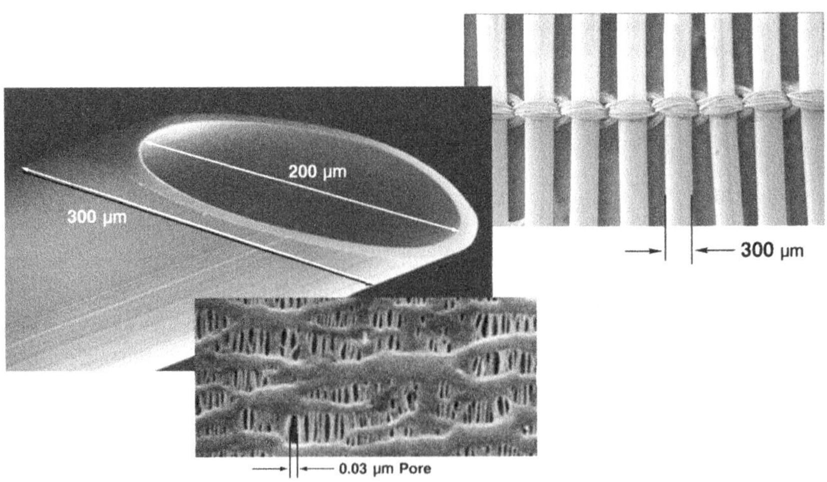

Figure 13.15 Membrane morphology in membrane degasification unit.

pore diameter, the greater the bubble point pressures. A commonly used form of this equation is (Goel et al., 1992)

$$P_{\mathrm{BP}} = \frac{4 \times \gamma \times \cos \theta}{d} \tag{13.24}$$

As an example, the above equation predicts that if water at 20°C (surface tension of 73 dyn/cm, contact angle of zero) were used, then the equation can be simplified to (Brock, 1983):

$$P_{\mathrm{bp}} = \frac{41.6}{d} \tag{13.25}$$

The high breakthrough pressure typically offered by commercially available membrane degasification units prevents water from passing through the membrane at pressures below 150 psig (Wiesler, 1996).

Extensive research both in industry and academia has resulted in the innovation of porous membranes, which are gas filled, with much smaller mass transfer resistance. Parallel research on microporous membranes has adjusted the pore size and membrane hydrophobicity, again yielding a much smaller mass transfer resistance. However, modules with different geometries perform differently. Flow outside of, but perpendicular to, the fiber bundle offers reasonably fast mass transfer. Not surprisingly, this geometry is chosen in most of the commercial membrane degasification units.

Henry's law describes the stripping of various dissolved gases from water. If we reduce the partial pressure of the volatile solute, some quantity of dissolved solute will come out of the aqueous phase until the system reaches equilibrium. Partial pressure of a solute is equal to the product of the total gas-phase pressure and the concentration (mole fraction) of the solute in the gas phase. The partial pressure of the solute in the nitrogen gas phase is reduced in the membrane degasification units by applying a strong vacuum and using a 99.995% pure nitrogen gas.

13.2.3.1 Mass Transfer in Membrane Degasification

Commercial membrane degasification units involve mass transfer in hollow-fiber modules where the vacuum and sweep gases are applied inside the fibers, and the water flows outside the fibers in cross flow perpendicular to the fiber axis. The mass transfer involves three sequential steps. First, dissolved gas diffuses out of the water to the membrane surface. Second, it diffuses into vapor pores in the walls of the hydrophobic hollow fibers. Third, when the dissolved gas reaches the other wall of the fibers, it diffuses into the surrounding nitrogen sweep gas in a high vacuum condition.

This overall process corresponds to mass transfer across three resistances in series. Diffusional resistances in series are additive just as electrical resistances in series are additive. The overall mass transfer coefficient is merely the reciprocal of the overall resistance. The overall mass transfer coefficient is given by

$$\frac{1}{k} = \frac{1}{k_L} + \frac{P_e}{\delta} + \frac{1}{k_G H} \tag{13.26}$$

The resistance of the vacuum-side boundary layer (k_G) is negligible compared to the others. The gas-filled membrane wall offers much smaller mass transfer resistance across the membrane. The resistance of the membrane has dominated overall mass transfer resistance in the past. This was because the permeability of the membrane was low and because the membrane was thick. The small resistance observed across the new-generation membranes is also the consequence of the choice of a hydrophobic fiber. The water does not wet this fiber, so its pores remain filled with nitrogen gas. Diffusion through the nitrogen gas is fast, making the membrane resistance unimportant. This conclusion would be different had we used a hydrophilic fiber. This implies that the key to the mass transfer is diffusion in the liquid. The overall mass transfer coefficients of oxygen were observed to be dominated by the individual mass transfer coefficient in the liquid film (Tai et al., 1994).

Membrane degasification unit design contains thousands of microporous polypropylene hollow fibers knitted into an array that is wound around a distribution tube with a central plug. Often a baffle design is also employed on the shell side to direct liquid flow on the shell side (Fig. 13.16). Sweep gas flows through the lumen of each fiber. Liquid enters into the porous central tube, is forced radially outward across the fibers because of the presence of the plug, and then flows back inward into the other half of the central tube before exiting the module out of the central tube. This geometry, involving flow outside of, but perpendicular to, a fiber bundle, offers faster mass transfer.

Owing to the complexity of mass transfer, calculations of actual mass transfer equipment are often based on empirical methods and equations derived by dimensional analysis and semitheoretical analogies. The behavior of mass transfer in membrane degasification units is most commonly simulated in terms of dimensionless correlations. A number of important dimensionless groups have been found by dimensional analysis or by other means. Many are important enough to justify names and symbols. The numerical value of a dimensionless group for a given case is independent of the units chosen for the primary quantities, provided consistent units are used within that group. The units chosen for one group need not be consistent with those used for another.

The mass transfer coefficient is normally reported as a Sherwood number (Sh). Sherwood number is the mass transfer analog to the Nusselt number of heat transfer. Sherwood number is defined as follows:

$$\mathrm{Sh} = \frac{k \times d_e}{D_f} \tag{13.27}$$

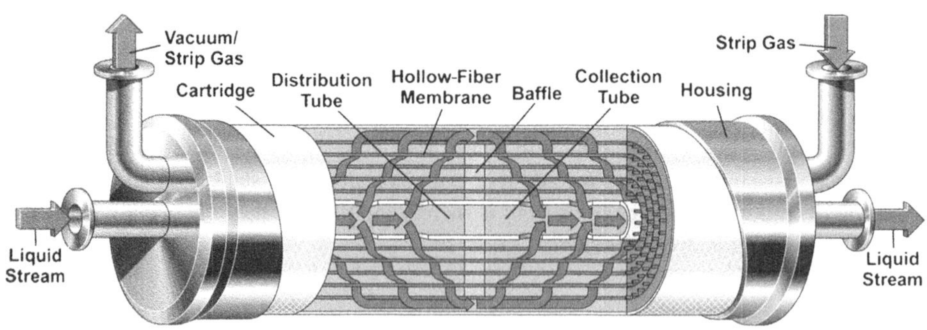

Figure 13.16 Membrane degasification module. (Courtesy of Membrana.)

Fluid flow can be classified by the ratio of the inertial forces to the viscous forces. The Reynolds number (Re) represents this ratio. The variation in water velocity is reported as a Reynolds number (Re). Reynolds number is defined as follows:

$$\mathrm{Re} = \frac{d_e \times V_L}{\nu} \tag{13.28}$$

In fluid motion where the frictional forces interact with the inertia forces, it is important to consider the ratio of the viscosity (μ) to the density (ρ). This ratio is known as the kinematic viscosity (ν). The kinematic viscosity has the same dimension as diffusivity, length^2/time.

Data reported for flow perpendicular to the fiber by numerous investigators are well reproduced by the following dimensionless equation:

$$\mathrm{Sh} = C \times \mathrm{Re}^m \tag{13.29}$$

The above form of the mass transfer equation is adopted in the design of membrane degasification units.

13.2.3.2 Dissolved Oxygen Removal Performance of Membrane Degasifier Membrane degasifiers offer high removal rates of dissolved oxygen from pure water when operated in the combo mode (Fig. 13.17). Combo mode refers to the application of both vacuum and nitrogen sweep gas for its operation. In the semiconductor industry, two stages of membrane degasification are typically used, one in the makeup and one in the polishing loop. The membrane degasifier in the makeup loop is commonly designed to achieve less than 50 ppb dissolved oxygen from feed water that is saturated with atmospheric oxygen. It is imperative to note that the secondary ultraviolet (UV) TOC reducer in the polishing loop requires a minimum dissolved oxygen level to work efficiently because dissolved oxygen in the water stream primarily functions as the electron scavenger in UV TOC reducers. The membrane degasifier units in the polishing loop can easily achieve less than 1 ppb dissolved oxygen with a standard 30–40% recirculation flow in

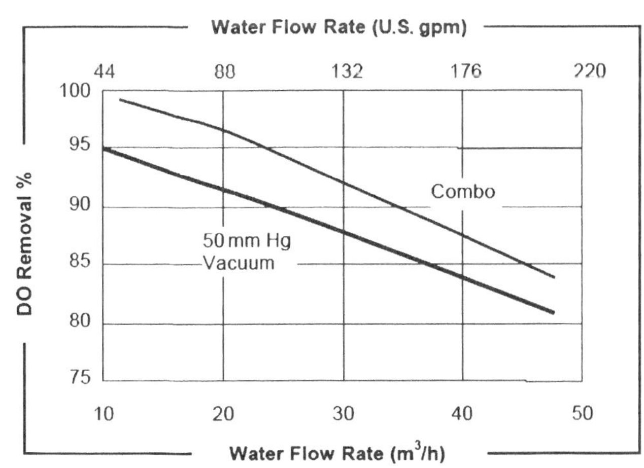

Figure 13.17 DO removal performance of 10 × 28-inch degasifier unit with 130 m^2 surface area (Membrana, 2005). (Courtesy of Membrana.)

the final polishing loop of the ultrapure water system. Performance curves presented in the Figure 13.17 are based on 50 mm Hg vacuum and a nitrogen gas sweep rate of 0.25 scfm per unit at a water temperature of 20°C. Performance may change under different operating conditions. For example, dissolved oxygen removal efficiency will improve at a higher water temperature, vacuum, and nitrogen gas sweep rate. Empirical mass transfer equations are developed by the equipment manufacturers to address this issue.

13.2.3.3 TOC and THMs Removal Performance of Membrane Degasifier On a weight basis, THMs are the largest class of disinfection by-products (DBP) found to be present. THMs are produced when chlorine reacts with naturally occurring humic acids, fulvic acids, and other natural organic material (NOM). The four common THMs are trichloromethane ($CHCl_3$: chloroform), dibromochloromethane ($CHBr_2Cl$), dichlorobromomethane ($CHCl_2Br$), and bromoform ($CHBr_3$). Total trihalomethanes (TTHM) is measured as the sum concentration of these four components.

Trihalomethanes are volatile organic compounds (VOCs). VOC separation, especially removal at trace levels, poses a challenge to the pharmaceutical and electronic industries where such impurities cannot be tolerated. Their low molecular weight allows free passage through RO membranes. It is generally accepted that RO membranes can effectively reject TOC of larger (over 200 Da) molecular weight. Chloroform cannot be modified by UV 185 treatment to yield an ionic entity capable of ion exchange removal. The carbon–chlorine bond is more stable than the chlorine–bromine bond. The THMs containing bromine are decomposed by the action of UV radiation. The resulting bromide ions lower the resistivity of the water (Meltzer, 1997). Their boiling points are close enough to that of water to render them difficult to separate from water by distillation, particularly at trace levels. Vacuum degasification is promising for the lighter THMs, those with fewer bromine atoms, but less encouraging for their heavier analogs. THMs and other halogenated organics can be reduced by adsorption with high iodine value granular activated carbon (GAC) filters to a fair extent. Use of ozonation has been reported to be an effective means of controlling THM levels. Unfortunately, these technologies are less preferable in the ultrapure water systems.

Dey and Thomas 2003 studied the performance of membrane degasification units in terms of organic removal, including THMs. The effect of variation in flow rate and applied vacuum level on the removal rate of TOC and chloroform are, respectively, given in Tables 13.6 and 13.7.

The improvement in the removal of TOC and chloroform at higher vacuum levels and lower flow rates can be attributed to the issues related to the efficiency of mass transfer operation. It was observed in the above study that the effect of the level of applied vacuum influences the TOC including chloroform removal efficiency more than the flow rate through the unit. Consequently, it is recommended to apply a high vacuum level (−700 mm Hg gauge or higher) to strip the TOC including chloroform from the ultrapure water stream with reasonably high efficiency.

13.2.4 Point-of-Use Ultrafiltration in Semiconductor UPW Systems

Particles present in the water if not removed may be deposited on the wafer surface during rinsing. These can interfere with the photomasking process or in later steps, are trapped within the circuitry and cause line interrupts. As a rule of thumb, circuit sensitivity to particle contaminant size is roughly one-tenth of the line width size. The economic justification of UF is usually found in improved integrated circuit (IC) yields due to better water quality.

TABLE 13.6 Effect of Flow Rate and Vacuum Conditions on the Removal Rate of TOC by 4 × 28-inch Membrane Degasifier with a 20-m^2 Effective Membrane Area

		TOC Concentration (ppb)		
Feed Flow Rate (gpm)	Vacuum Level (mm Hg gauge)	Membrane Contactor Inlet	Membrane Contactor Outlet	% TOC Removal
14	−700	12.45	8.470	31.97
	−650	13.42	10.59	21.09
	−550	14.45	12.49	13.56
	No vacuum	14.23	14.03	1.410
12	−700	12.79	7.910	38.15
	−650	13.98	10.11	27.68
	−550	14.12	12.90	8.640
	No vacuum	14.56	14.20	2.470
10	−700	13.14	7.210	45.13
	−650	14.83	10.06	32.16
	−550	15.36	12.59	18.03
	No vacuum	14.64	13.19	9.900
8.5	−700	14.20	7.020	50.56
	−650	15.79	10.01	36.61
	−550	15.62	12.36	20.87
	No vacuum	17.11	15.74	8.000
7.0	−700	12.65	5.540	56.21
	−650	15.32	10.09	34.14
	−550	16.09	12.89	19.89
	No vacuum	16.68	14.36	13.91

Source: Dey and Thomas (2003).

One significant difference between U.S. and Japanese semiconductor plants is that most large Japanese plants are using UF. Currently, only some U.S. plants have UF.

For many years the standard final filtration step in high-purity water was dead-end cartridge filtration. Various attempts with membrane filtration had mixed success. Recent installation of double-skin, hollow-fiber UF membranes appears to be successful; this approach is quickly becoming the industry standard (Payne et al., 1999). External pressurized UF membranes with 6000–10,000 nominal molecular weight cutoff are recommended for the removal of particulates at the point-of-use. Ultrafiltration membranes perform better in polishing units than the traditional cartridge microfiltration membranes in regards to removal of submicron particles and stability of performance. Ultrafiltration removes virtually all particles, colloidal materials, bacteria, and other microorganisms. The capillary configuration of the ultrafiltration membrane has been shown to give better performance than the spiral-wound configuration. Capillary configuration strikes an excellent balance between high surface area to volume and resistance to plugging and is typically the configuration of choice for point-of-use application. The outside/in capillary modules perform better than inside/out design. In both designs the membrane fibers have asymmetric double-skin layers with a tight UF layer on the inside of the fiber.

Number of particles per unit volume of feed water determines the UF permeate flux. To obtain better particulate removal efficiency and higher flux rate, it is recommended to use a 0.1- to 0.2-μm absolute cartridge filter at the upstream of the UF unit. Due to their higher

TABLE 13.7 Effect of Flow Rate and Vacuum Conditions on the Removal Rate of THMs by 4 × 28-inch Membrane Degasifier with a 20-m^2 Effective Membrane Area

		Chloroform Concentration (ppb)		
Feed Flow Rate (gpm)	Vacuum Level (mm Hg gauge)	Membrane Contactor Inlet	Membrane Contactor Outlet	Chloroform Removal (%)
14	−700	7.15	4.6	35.66
	−650	7.57	5.9	22.06
	−550	12.4	10.3	16.94
	No vacuum	12.1	11.1	8.26
12	−700	12.3	7.3	40.65
	−650	11.0	8.27	24.82
	−550	8.52	7.0	17.84
	No vacuum	8.78	8.0	8.88
10	−700	13.8	7.5	45.65
	−650	13.9	9.84	29.21
	−550	13.9	11	20.86
	No vacuum	13.8	12.5	9.42
8.5	−700	13.77	6.8	50.61
	−650	14.1	9.8	30.50
	−550	15.5	11.8	23.87
	No vacuum	16.4	14.6	10.98
7.0	−700	11.1	4.9	55.86
	−650	11.1	7.2	35.14
	−550	16.8	12.7	24.40
	No vacuum	16.9	14.7	13.02

Source: Dey and Thomas (2003).

particle retention efficiency, absolute-rated filters exhibit superior particle control performance in circulating flow systems compared with normal-rated filters.

Unlike RO, or EDI, ultrafiltration does not remove ions but removes colloids, macromolecules, or fine suspended matter. Ultrafilters reject solutes ranging in size from 0.003 to 10 μm. UF resembles RO in that the energy requirement is supplied as pumping power. UF plants are usually well automated and require little labor.

The UF unit placed in the final stage of the polishing system has an important function as a final filter to remove particles up to 50-nm size. The UF unit also completely eliminates bacteria from the system. The spiral-type UF unit has a complicated structure that allows particle leakage. On the other hand, the hollow-fiber structure is simple and capable of removing particles completely (Yabe, 1993). Consequently, hollow-fiber-type UF units are widely used in the polishing section of the ultrapure water systems. This ultrafiltration unit should possess the following necessary characteristics (Ishikawa, 1993):

- Less elution of ionic and organic substances into the water stream
- Structure without stagnation zones

Theoretically, particles larger than 50 nm should be completely removed by this UF unit. In actual operation, particles are still found in the permeate water. These are the particles that are dissociated from the surface of the permeate waterside of the module. It is therefore critical to minimize the dissociation of such particles (Ishikawa, 1993).

Both the membrane and the housing in such UF units are made of polysulfone material. Epoxy-based adhesives are commonly used in such units. Such materials have enough heat resistance to be processed by hot water sterilization. External pressurized UF is used to avoid particle contamination in the filtered water. However, higher cross flow inside the hollow fibers cannot be created in the external pressurized-type UF units. Higher cross flow inside the fibers minimizes the number of particles adhering to the inside surface of the fibers. Fortunately, clogging of fibers due to the presence of particles in external pressurized ultrafiltration units is not a serious problem with the level of particle contamination found in conventional makeup systems used in the industry.

Dead space is minimized in a capillary UF unit due to its design configuration. The original design of capillary UF modules was internal pressurizing. An internal pressurizing system allows the water to flow from inside the fiber to the outside. This design yields low particle counts in the product water under steady-state conditions. However, any system disturbance results in an unacceptably high particle count. In external pressurized UF systems, the feed water flows around the membrane fibers and the permeate is collected on the inside of the fibers. The larger membrane surface area on the feed side of the fibers for the external pressurized modules also results in about twice the flow rate per module as compared to the internal pressurized design. External pressurized modules drastically reduce such spikes in the particle count values (Maki et al., 1992).

Particle contamination arising from the manufacturing process normally occurs on the outside of the fibers. The external pressurized configuration provides lower particle count values because the product water only contacts the inside surface of the fibers. Moreover, the product water never contacts the housing material or potting resin, and consequently contamination arising from such sources can be completely eliminated.

The permeate flow rate per unit membrane area is termed flux. Higher particle counts in the feed water can result in severe flux decline. Cake filtration theory provides a theoretical basis for the estimation of such flux decline due to particulate fouling. Consequently, design flux should be selected based on the feed water particle count levels (Table 13.8). Consider a 225-m^3/h UPW generation system that utilizes Nitto Denko NTU–3306–K6R UF modules with 30-m^2 effective membrane area where the quality of the point-of-use UF influent water in terms of resistivity and particle count are, respectively, 18.0 MΩ-cm and less than 100 particles/mL. Based on a design flux of 500 (L/h)/m^2 area (Table 13.8), one can easily determine that at least 15 $[225/(0.5 \times 30) = 15]$ of these modules are required for this system. These UF modules can easily achieve <100/L (50- to 100-nm particles), <50/L (100- to 200-nm particles), <20/L (200- to 300-nm particles), <10/L (300- to 500-nm particles), and <1/L (particles larger than 500 nm) where the influent stream contains a TOC of less than 20 ppb and particle count <100/mL (larger than 100 nm) and <20/mL (larger than 200 nm).

TABLE 13.8 Design Flux for NTU–3306–K6R UF Module

Number of Particles per mL	Design Flux [(L/h)/m^2]
~100	500
100–500	470
500–1000	420
1000–2000	400

Source: Boon (2002).

With continued filtration, the bacteria count on the membranes increases exponentially. Regular sterilization of the polishing loop is recommended to avoid biofouling. Commercially available external pressurized UF modules can withstand both hydrogen peroxide and hot water up to 98°C for the sterilization. Hot water sterilisation is preferred as it does not introduce any chemical into the polishing loop, and it provides a quicker rinse up time compared to hydrogen peroxide. It is important to control the ion exchange resin leachables in the UF feed water. The presence of such leachables can result in severe flux decline in polysulfone UF modules (Maki et al., 1992).

Notation

C	empirical constant, dimensionless
CB	concentrate bleed off, gpm (L/s)
d	pore diameter in microporous membranes, μm
d_e	external fiber diameter, ft (μm)
D_f	diffusion coefficient, $ft^2/h(m^2/s)$
d_p	pore diameter, length unit
E	E factor for plate-and-frame EDI modules, dimensionless
EL	Electrode flush per module, gpm/module
F	Faraday's constant, A-s/g-equivalent
H	Henry's constant for solute gas at operating temperature, $atm/(mol/m^3)$
H_D	height of the ion exchange membrane in dilute chamber, ft (m)
I	rectifier current, A
$i(r)$	current density at radial distance r, A/ft^2
I_M	direct current per module, A
J	permeate flux, GFD (L/m^2 h)
k	overall mass transfer coefficient, ft/h (m/s)
k_G	mass transfer coefficient in the gas, ft/h (m/s)
k_L	mass transfer coefficient in the liquid, ft/h (m/s)
L_D	length of the membrane leafs, ft (m)
m	empirical constant, dimensionless
N_{cp}	number of cell pairs, dimensionless
NDP	Net driving pressure, psig (kPa)
N_M	minimum number of EDI modules in the complete system, dimensionless
P	EDI product flow, gpm (L/s)
P_{BP}	Bubble point pressure (dyn/cm × μm)
P_{bp}	Bubble point pressure, psig (kPa)
P_d	average pressure drop along RO system, psig (kPa)
P_e	permeability of the membrane, h
P_f	RO feed pressure, psig (kPa)
pK_a	dissociation constant of weak acids, dimensionless
P_o	osmotic pressure corresponding to an average feed salinity, psig (kPa)
P_p	RO permeate backpressure, psig (kPa)
Q_d	flow rate through dilute chambers, gpm (L/s)
R	overall resistance of the EDI module, Ω

r	radial distance, ft (m)
r_a	radius of outer anode, ft (m)
r_c	radius of inner cathode, ft (m)
Re	Reynolds number, dimensionless
Sh	Sherwood number, dimensionless
SWE	E factor for spiral-wound EDI modules, dimensionless
t	temperature, °C
T_C	temperature coefficient
TEA	total exchangeable anion, ppm as $CaCO_3$
TEC	total exchangeable cation, ppm as $CaCO_3$
V	Voltage per module, V
v_L	actual velocity of water in the membrane degasification module, ft/h (m/s)
ΔP	net driving pressure in RO modules
α	EDI water recovery, fraction; dissociation of a weakly ionized species at any given pH, fraction
θ	contact angle between water and membrane surface
γ	surface tension of the water, dyn/cm
μ	Viscosity, lb/(h × ft)
μ_C	C-loop conductivity, μS/cm at 25°C.
ρ	density, lb/ft^3 (kg/m^3)
δ	membrane thickness, ft (μm)
ν	kinematic viscosity, ft^2/h (m^2/s)
ϵ	porosity of the membrane
η_c	current efficiency, dimensionless
η_R	rectifier efficiency (fraction)

REFERENCES

Air Liquide—Balazs Analytical Services (2004). Ultrapure water monitoring. Guidelines for facility and fabrication engineers, pp. 11–13.

American Society of Mechanical Engineers (ASME). Research and Technology Committee on Water and Steam in Thermal Power Systems (1994). Consensus on operating practices for the control of feedwater and boiler water chemistry in modern industrial boilers. ASME I 00367. ASME, Washington, DC.

American Society for Testing and Materials (ASTM) (1999). Standard Guide for Ultrapure Water Used in the Electronics Industry. ASTM D5127.

American Society for Testing and Materials (ASTM) (2003). Standard test methods for pore size characteristics of membrane filters by bubble point and mean flow pore test. ASTM F316-03.

Barber, J. H., Towe, I. G., and Tessier, D. F. (2000). EDI operation for removal of weakly ionized impurities. Paper presented at the International Water Conference, IWC-00-46, Pittsburgh, PA.

Bauman, W. C., Anderson, R. E., and Wheaton, R. M. (1952). *Ann. Rev. Phys. Chem.* 3, 109.

Boon, I. (2002). Nitto Denko. Personal communication.

Brock, T. D. (1983). Membrane Filtration: A User's Guide and Reference Manual. Science Tech Publishers, Madison, WI.

Coker, S. D., Beardsley, S. S., and Whipple, S. S. (1994). An economical comparison of demineralization with reverse osmosis and ion exchange technology. Power-Gen Americas. Las Vegas, Nevada.

Dey, A. (2005a). Spiral wound electrodeionization: An emerging energy-efficient deionization process in high purity water systems. Paper presented at the Semiconductor Pure Water Chemical Conference (SPWCC), Santa Clara, CA, Feb. 14–16.

Dey, A. (2005b). Removal of weakly ionized species by cross-flow spiral wound EDI modules. Paper presented at the 66th International Water Conference, IWC-05-76, Orlando, FL, Oct. 9–13.

Dey, A., and Li, G. (2004). Introduction of spiral wound EDI—Exclusive design and its application. Paper presented at the 65th International Water Conference, IWC-04-51, Pittsburgh, PA, Oct. 17–21.

Dey, A., and Thomas, G. (2003). *Electronics Grade Water Preparation*. Tall Oaks Publishing, Littleton, CO.

Dey, A., Thomas, G., Kekre, K. A., and Tao, G. (2001a). Part 1: Effect of inter-stage caustic dosing on silica, boron, and organic removal using double pass RO. *Ultrapure Water*, July/Aug., pp. 52–58.

Dey, A., Thomas, G., Kekre, K. A., and Tao, G. (2001b). Part 2: Impact of caustic dosing on contaminant removal using double pass RO. *Ultrapure Water*, Sept., pp. 43–49.

Dow Chemical Company (DOW) (1995). *FILMTEC Membranes Technical Manual*. DOW, Middlesex, England.

Ganzi, G. C. (1988). Electrodeionization for high-purity water production. In K. K. Sirkar and D. R. Lloyd (Eds.), New membrane materials and processes for separation. AIChE Symposium series, No. 261, Vol. 84, AIChE, New York, pp. 73–83.

Ganzi, G. C., and Parise, P. L. (1990). The production of pharmaceutical grades of water using continuous deionization post-reverse osmosis. *J. Parenter. Sci. Technol. Parenter. Drug Assoc.*, July/Aug, Volume 44, Issue 4, pp. 231–241.

Gittens, G. J., and Glueckauf, E. (1965). The application of electrodialysis to demineralization. AIChE–I. Chem. E Symposium Series, No. 9. AIChE, London.

Glueckauf, E. (1959). Electro-deionization through a packed bed. *Br. Chem. Eng.* 646–651, December.

Goel, V., Mauro, A. A., DiLeo, A. J., Meiser, A. P., and Pluskal, M. (1992). Dead-end microfiltration: Applications, design, and cost. In W. S. W. Ho and K. K. Sirkar (Eds.), *Membrane Handbook*. Van Vostrand Reinhold, New York.

Hidaka, M. (2001). Electrodeionization apparatus comprising sub-desalination chamber. U.S. Patent 6,274,018.

Imaoka, T., Yagi, Y., Kasama, Y., Sugiyama, I., Isagawa, T., and Ohmi, T. (1991). Advanced ultrapure water systems for ULSI processing. In M. K. Balazs (Ed.), *Proceedings of the Tenth Annual Semiconductor Pure Water Conference*, Balazs, Santa Clara, CA, Feb. 26–28, pp. 128–146.

International Technology Roadmap for Semiconductors (ITRS) (2007). Executive summary (http://www.itrs.net/Links/2007ITRS/ExecSum2007.pdf).

Ishikawa, H. (1993). Membranes. In T. Ohmi (Ed.), Ultraclean Technology Handbook, Vol. 1, *Ultrapure Water*. Marcel Dekker, New York.

Kasama, Y., Yagi, Y., Imaoka, T, Kawakami, M., and Ohmi, T. (1990). Advanced DI water system with low dissolved oxygen for ULSI processing. *Proc. Instit. Environ. Sci.* 344–349.

Koyuncu, I., Yazgan, M., Topacik, D., and Sarikaya, H. Z. (2001). Evaluation of the low pressure RO and NF membranes for an alternative treatment of Buyukcekmece Lake. *Water Sci. Technol.* Water Supply 1(1), 107–115.

Leon, M. (2001). Electrodeionization apparatus with fixed ion exchange materials. U.S. Patent 6,241,866.

Liqui-Cel® (2003). Liqui-Cel® membrane contactors improve water quality and EDI performance. Technical Brief, Membrana, Charlotte, NC, Vol. 46, Aug.

Loeb, S., and Sourirajan, S. (1963). Sea water demineralization by means of an osmotic membrane. In *Saline Water Conversion—II*, Advances in Chemistry Series Number 28. American Chemical Society, Washington DC, pp. 117–132.

Maki, M. L., Katsura, T., Tasaka, K., Inoue, K., and Kamiyama, Y. (1992). New external pressurising capillary ultrafiltration module for ultrapure water systems. Paper presented at the Sixth Annual Ultrapure Water Expo, Philadelphia, PA, May.

Meltzer, T. H. (1997). *High-Purity Water Preparation: For The Semiconductor, Pharmaceutical, and Power Industries*, 2nd ed. Tall Oaks Publishing, Littleton, CO.

Membrana (2005). 10 × 28 Extra-flow product data Sheet—Liqui-Cel® Membrane Contactor, Membrana, Charlotte, NC.

Nagel, R., and Will, T. (1999). Membrane processes for water treatment in the semiconductor industry. *Ultrapure Water*, Oct., pp. 35–39.

Paul, D. H. (2002). Electrodeionization in pharmaceutical water treatment. *Pharma. Technol.*, July, pp. 36–42.

Payne, J., Sommer, K., and Williams, R. (1999). Microelectronics water system evolution through 300 mm. *Ultrapure Water*, May/June, pp. 11–13.

Reed, B. W., Semmens, M. J., and Cussler, E. L. (1995). Membrane contactors. In R. D. Noble and S. A. Stern (Eds.), *Membrane Separation Technology: Principles and Applications*. Elsevier Science, Amsterdam.

Sato, H., Hashimoto, N., Shinoda, T., and Takino, K. (1991). Dissolved oxygen removal in ultrapure water for semiconductor manufacturing. Paper presented at the 10th Annual Semiconductor Pure Water Conference, Santa Clara, CA, Feb. 26–28.

Semiconductor Equipment and Materials International (SEMI) (2001). Guide for ultrapure water used in semiconductor processing. SEMI F63-0701.

Smith, B., and Hyde, B. (2000). Short-bed demineralization: An alternative to electrodeionization. Paper presented at the 6th International Conference on Cycle Chemistry in Fossil Plants (EPRI), Columbus, OH, June.

Spiegler, K. S. (1953). On the electrochemistry of ion exchange resins: A review of recent works. *J. Electrochem. Soc.* 100, 303C–316C.

Tai, M. S. L., Chua, I., Li, K., Ng, W. J., and Teo, W. K. (1994). Removal of dissolved oxygen in ultrapure water production using microporous membrane modules. *J. Membr. Sci.* 87, 99–105.

Teorell, T. (1951). *Z. Elektrochem.* 55, 460.

Tessier, D. F. (1998). Apparatus for the purification of liquids and a method of manufacturing and operating the same. U.S. Patent 5,762,774.

Walters, W. R., Weiser, D. W., and Marek, L. J. (1955). Concentration of radioactive aqueous wastes—Electro-migration through ion exchange membranes. *Ind. Eng. Chem.* 47(1), 61–67.

Wiesler, F. (1996). Membrane contactors: An introduction to the technology. *Ultrapure Water*, May/June, pp. 27–31.

Wiesner, M. R., and Aptel, P. (1996). Mass transport and permeate flux and fouling in pressure-driven processes. In J. Malleviella (Ed.), *Water Treatment Membrane Handbook*. McGraw-Hill, New York, Chapter 9.

Winter, D. D. (1995). Membrane construction and evaluation. Paper presented at the Membrane Technology/Separations Planning Conference, Business Communications Company, Newton, MA, Oct. 23–25.

Wood, J., and Gifford, J. (2004). Process and system design for reliable operation of RO/CEDI systems. Paper presented at the International Water Conference, IWC-04-47, Pittsburgh, PA.

Yabe, K. (1993). Ultrapure water system. In T. Ohmi (Ed.), *Ultraclean Technology Handbook*, Vol. 1, *Ultrapure Water*. Marcel Dekker, New York.

PART II

MEMBRANES FOR BIOTECHNOLOGY AND CHEMICAL/BIOMEDICAL APPLICATIONS

CHAPTER 14

Tissue Engineering with Membranes

ZHANFENG CUI

Department of Engineering Science, Oxford University, Oxford, United Kingdom

14.1 INTRODUCTION

14.1.1 Regenerative Medicine

Regenerative medicine uses its unique approach to regenerate cells, tissues, and even organs to prevent and cure diseases and to repair, restore, and enhance functions of damaged and diseased tissues and organs. Common approaches in regenerative medicine include stem cell transplantation, cell therapy, gene therapy and tissue engineering, and their combinations, with tissue engineering and stem cell transplantation being the newest and most exciting approaches.

Tissue engineering, as an identified field, was established in the late 1980s, although it has to be pointed out that such an approach may have been practiced clinically for centuries. Its basic concept is to produce live and functional tissue by culturing appropriate cells on a three-dimensional scaffold for a certain period of time and then to implant the engineered tissue into the body to replace or repair the damaged or lost tissue. The concept has been proved and engineered skin products are available commercially.

Stem cells offer unprecedented opportunities and potentials for regenerative medicine. Bone marrow transplantation is an established clinical procedure, and the transplant of blood stem cells separated from umbilical cord blood, has also been established. It is, however, the potential of using embryonic and adult stem cells to regenerate all the tissues and organs in the body that has attracted attention and imagination all over the world. If stem cell potentials materialize, or even a fraction of their potentials are materialized, the health care industry will be revolutionized.

The ethical and social issues surrounding stem cell research and applications are well known—for example, where the stem cells are from and who will pay for the treatment. It must also be emphasized that stem cell research and application is still in "embryo" stage. There are many huge obstacles to be overcome before clinical application becomes a reality.

The research and development in tissue engineering and stem cell therapy need an integrated effort by multidisciplinary teams. Clinical relevance and technology development play equally important roles, as shown in Figure 14.1.

Advanced Membrane Technology and Applications. Edited by Norman N. Li, Anthony G. Fane, W. S. Winston Ho, and T. Matsuura
Copyright © 2008 John Wiley & Sons, Inc.

Due to its enormous impact, both economically and socially, huge R&D efforts have been poured into this emerging and exciting field. In 2001, the total R&D spending was about $580 million, and the estimated market was $15 billion just for connective structural tissues and $10 billion for soft tissue repair worldwide (RMD report) (Bassett, 2001).

14.1.2 Technical Challenges to Tissue Engineering and Stem Cell Transplantation

However, apart from skin that is avascular and of thin-layered structure, there is no major breakthrough, both clinically and commercially, regarding any other tissues, despite continuous and huge effort into research and development of engineered bone, cartilage, small-diameter blood vessels, pancreas, liver, and so forth. The reason for this includes:

1. Lack of understanding of biological systems at all levels from the subcellular, cellular, tissue, organ to the whole-body level. Important fundamental questions remain unanswered such as:

- What cells to use (e.g., differentiated cells vs. stem cells, banked cells vs. patient own cells, etc.)
- How to culture them (medium, physical, chemical, mechanical conditions, growth factors)
- How to prevent host rejection
- How to integrate the engineered tissue with the host

2. Technical challenges and lack of an engineering approach. There have not been established engineering descriptions or even rules of thumb for the design and operation of the tissue engineering process. Many laboratory tests followed the trial-and-error approach, which, not surprisingly, had a low success rate. The technical and engineering challenges cannot be underestimated, particularly considering clinical applications and commercialization. These include:

Scaffold Materials Most materials tested in research and development were medically approved biomaterials, designed and approved for other applications such as sutures,

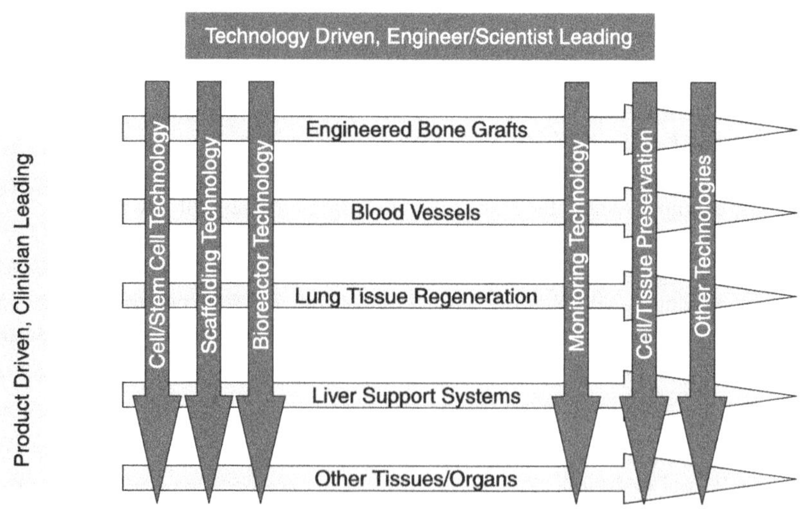

Figure 14.1 Multidisciplinary approach to tissue engineering.

wound dressing, or implants. It is attempting to use these biomaterials for scaffold materials for their readiness for regulatory approval, but they are not designed for tissue engineering scaffolds, and hence the chance to succeed is greatly reduced. Purpose-designed biomaterials, specific to engineered tissue types, will need to be developed to fulfill this requirement.

Bioreactor Technology The engineered tissue needs to be cultured in a bioreactor at a well-defined and well-controlled environment including the microenvironment around cells at different locations. The bioreactor is central to the tissue culture process and provides the predefined chemical, biochemical, physical, and mechanical environments for the seeded scaffolds, in which cells proliferate and differentiate to form neotissues. Various types of bioreactors have been investigated for tissue culture, including the spinner flask bioreactor, the rotating vessel bioreactor (slow-turning lateral vessel and high-aspect-ratio vessel), the rotating wall perfused vessel bioreactor, and the perfused column bioreactor, reviewed by Freed and Vunjak-Novakovic (Seidel et al., 2004). Also, perfused chambers have been used to encourage culture medium to reach the center of the bioconstruct through microchannels. This may function well in the initial stages when cells begin to grow and proliferate on the scaffold, but with time as the cell densities increase and an extracellular matrix (ECM) accumulates, most of the microchannels may be occupied by cellular materials and finally may be occluded. In addition, especially in culture of bone tissues, any mineralization of the ECM significantly restricts nutrient diffusion.

Designing bioreactors is not an easy task. Important parameters include nutrient and oxygen supply, metabolic waste removal, hydrodynamic and mechanical stimuli, and the like. Also, tissue culture takes a certain period of time, typically weeks, and is a costly process. Monitoring the process and adequate control is essential. Unfortunately, monitoring cell status within a three-dimensional tissue remains an unsolved problem, and hence sensible control strategy cannot be implemented.

Bioprocess Design How to design a process to manufacture 1000 pieces of live tissue per week? Nobody has come up with a well-thought-out design methodology yet. The process must be scalable, economical, and more importantly, as well as good manufacture practice (GMP) compliant.

Preservation and Storage of the Engineered Tissue Products Off-the-shelf availability is essential for commercial viability, although patient-specific, hospital-based tissue engineering is the low-hanging fruit. Cryopreservation is the process of choice, but it brings with it a great technical challenge of minimizing freezing damage to the product and could increase the distribution cost significantly.

Similar to stem-cell-based therapies, many technical challenges need to be tackled. Biological challenges include, for example, how to identify stem cells, how to culture stem cells, how to control their differentiation, and how to follow their fate after implantation. Engineering challenges must also be addressed, which are essential to the clinical and commercial success. These include:

How to Expand Stem Cells Stem cells are rare and of very limited supply. For any therapy, a certain number of stem cells are required. How to efficiently expand stem cells in a controlled manner is a key issue. This becomes more important for clinical applications of stem cells. Using passages in cell culture dishes cannot meet this need. Bioreactor technology is essential.

How to Differentiate Stem Cells Controlled differentiation of stem cells to the desired cell type is the final step of tissue regeneration. First, one needs to know exactly the quantitative relationship between cell local microenvironment and cell proliferation and differentiation. New research tools are required to assist determining such links. BioMEMS and the lab-on-chip engineering approach could provide the required new tools. Second, how to achieve and deliver the required microenvironmental conditions to the local stem cells. For in vitro tissue culture using stem cells, bioreactor technology again is the key, coupled with scaffold design. For in vivo cell transplantation on tissue regeneration, design and selection of the materials injected together with the cells may hold the key, which may have to be based on a detailed mathematical modeling of the local environment.

How to Preserve the Stem Cells Stem cells are precious. Stem cell storage and banking are essential steps for their clinical and commercial applications. Cryopreservation, that is, storing them in liquid nitrogen, is the obvious choice. However, cryopreservation, even with the best available protecting chemicals, could cause the death of a significant portion of cells. This is particularly true for human embryonic stem cells. Even for those cells that survive the procedure, cryopreservation acts like a selection step—the fittest survive. Those that survived may not be the ones required. What is more, it is not clear that the surviving stem cells still have all the functions and potency. Hence, there is an urgent need for new and efficient protocols for stem cell preservation to ensure maximum viability and functionality. In parallel, noninvasive nondestructive techniques for testing and monitoring the quality of stem cells in storage need to be developed. If the stem cells are already dead, there is no point to storing them for another 20 or more years.

14.1.3 Membrane Technology as an Enabling Technology

The term *membrane* can have a wide range of meanings, and hence it is necessary and important to clarify at the beginning that the discussion in this chapter is only limited to microporous membranes, and further the pores having the function of transmembrane transport. Application of membrane-like porous structures for engineering of thin layered tissues including skin (Dai et al., 2004, 2005), bladder (Danielsson et al., 2006), and cornea are excluded from the discussion, as are the dense films, for example, for wound dressing (Mi et al., 2003; Wright et al., 1998) and for providing physical barrier to preventing tissue adhesion (Lee et al., 2005). Hence the porous membranes discussed in this chapter have the conventional function of selective separation.

Membrane technology, even in this narrowed definition, can play an important role in tissue engineering and stem cell technology. The following sections will discuss three applications in depth, and Section 14.4 will briefly discuss additional applications, to demonstrate that membrane technology can indeed be an enabling technology.

14.2 HOLLOW-FIBER MEMBRANE BIOREACTORS FOR THREE-DIMENSIONAL TISSUE CULTURE

14.2.1 Mass Transfer Limitation

As briefly discussed in the previous section, new developments in the tissue engineering field are extensive with almost all tissues and organ systems under consideration. However, currently only avascular (cartilage) or thin sheets of tissues (skin) are capable of being successfully engineered (Marler et al., 1998). The mass of three-dimensional (3D) tissue that may be produced in vitro, especially dense tissue, is very limited. To develop a 3D tissue, the problems of oxygen and nutrient supply to the cells and waste removal from these cells, particularly in the center of the construct, must be addressed. Cell densities decrease with distance from the construct surface and no cells within a tissue mass can survive being more than approximately 100 μm away from a source of oxygen or nutrients in vivo (Tharakan et al., 1988). It has been suggested that the possible maximum thicknesses of engineered cartilaginous, bonelike and cardiaclike tissues grown in vitro are approximately 5.0, 0.5, and 0.18 mm, respectively (Freed and Vunjak-Novakovic, 1998).

Biological angiogenesis is the primary requirement for generation of an appreciable mass of most tissues, but initiation and control of angiogenesis remain a major technical challenge to tissue engineering. An engineering solution to supply oxygen and essential nutrients to the growing tissue in vitro is to use hollow-fiber membrane bioreactors so that the hollow-fiber membrane network mimics the blood capillary system.

The hollow-fiber membrane bioreactors (HFMB) with embedded hollow-fiber membranes within the scaffolding constructs can overcome the diffusional limitation (Ye et al., 2007). The wall of the hollow-fiber membranes are porous and selectively semipermeable to medium solutes. Depending on the pore sizes, passage of small molecules (nutrients, oxygen, and metabolic wastes) and macromolecules (e.g., growth factors) occurs, and cells are retained completely. The nutrients and oxygen flowing within the fibers will diffuse out through the membrane and be distributed more evenly within the tissue. Also, the spent media and metabolic waste products will permeate back into the fibers and be removed from the tissue. Therefore, a nutrient circulation system similar to the capillary system in the native tissue is created.

The concept of HFMB for use in cell culture has been investigated by numerous investigators. HFMBs have found application in monoclonal antibody production (Lowrey et al., 1994; Nagel et al., 1999; Tharakan et al., 1986), enzymatic reactions (Kawakita et al., 2002; Rony, 1972), and mammalian cell culture (Karlik et al., 1999; Nagel et al., 1999; Wang et al., 2005). Knazek first developed the HFMB system for cell culture in 1974 (Knazek, 1974; Knazek et al., 1972). The advantages of using hollow-fiber bioreactors for cell culture include: (a) the cultured cells are maintained in a more physiologically appropriate environment regarding nutrient supply, metabolic waste removal, and a stable pericellular microenvironment (Wang et al., 2005); (b) the cells can be grown to very high densities ($\geq 10^7 - 10^8$ cells/mL); and (c) the cultured cells are protected from major mechanical shear stresses. However, using the conventional HFMB, tissue structures could not be produced because of the lack of the necessary 3D scaffolds for cell attachment and proliferation. In vivo, cells are supported and maintained by ECM, and a similar environment is required for cells to form 3D tissue structures in vitro. A major difference between the design of HFMBs to be used for amplifying cell numbers and that for producing engineered tissues lies in the inclusion of biomaterial scaffolds in the latter to allow the growth of the required bulk of tissue (Ye et al., 2004).

14.2.2 Application into Bone Tissue Engineering

To prove the concept of HFMB for tissue engineering, Ye et al. (2004, 2007) conducted experiments using HFMB for growing engineered bone grafts. Ye et al. (2007) used collagen, a commonly used natural polymer, as the scaffolding material and seeded rat and human bone marrow fibroblastic cells attached to microcarriers and cultured the constructs for up to 4 weeks in vitro with bone-inducing culture medium, which is followed by a further 2 weeks in vivo study.

The hollow-fiber membrane bioreactor took the simple cylindrical geometry housing [dimension 13 mm inner diameter (ID) × 22 mm outer diameter (OD) × 40 mm L; see Fig. 14.2]. Cellulose acetate hollow-fiber membranes [200 μm ID, wall thickness of 14 μm and molecular weight cutoff (MWCO) of 10 kDa] derived from hemodialysers used to construct the HFMBs. The hollow-fiber membranes were fixed in the bioreactor by using molded silicon rubber. The effective length of the fiber in the reactor was 30 mm with approximately 200 fibers in each bioreactor. The distance between adjacent fibers was approximately 400 μm, of the order of the distance between natural blood capillaries in human bone. The volume external to the hollow fibers in each HFMB was approximately 3.5 mL, and this volume was available for the collagen gel together with the microcarriers with adherent cells.

After filling the spaces between fibers with the cells and scaffolding materials [for details see Ye et al. (2007)], the bioreactors were placed in a 37°C incubator together with the cell culture medium reservoir. Continuously gassed culture medium (5% CO_2 in air) was withdrawn from the medium reservoir (with 150 mL of medium) by the peristaltic pump, entered into the intracapillary space (ICS), and then recirculated back to the reservoir. Nutrients and oxygen diffuse through the porous wall of the membrane and are distributed within the bioreactor. Metabolic wastes within the construct diffuse into the membrane through the porous wall and flow out of the bioreactor. The medium in the reservoir was changed regularly. In comparison, HFMBs were operated without perfusion as controls for comparison with the perfused system, with medium changed manually on a daily basis.

Figure 14.3 shows the comparison of cell proliferation (Alamar Blue assay) between the perfused and control groups. There is no difference between the nonperfused and the perfused bioreactors on day 1. While the cell number in the perfused group increases up to 4 days and then stabilizes, the nonperfused group shows a decrease with time. The cell number in the perfused group is about 5 times that of the nonperfused group on both days 4 and 7 and the differences are statistically significant ($p < 0.05$). Experimental results also showed that daily lactate production (related to cell numbers and metabolic activities in the perfused group) increased with time while it decreased in the nonperfused group (Ye et al., 2007). The light and fluorescence microscopy of LIVE/DEAD™ cell

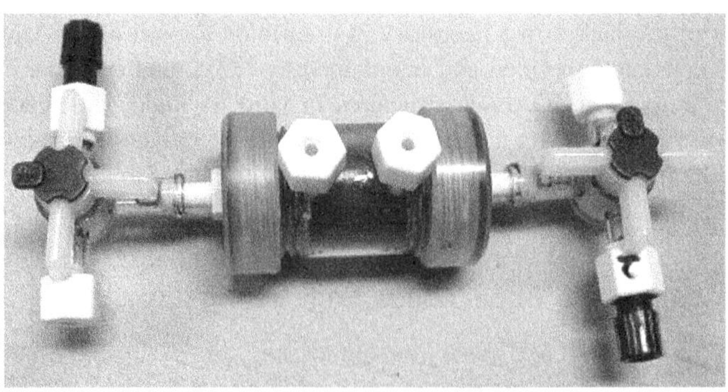

Figure 14.2 Hollow-fiber membrane bioreactors for bone tissue engineering testing (Ye et al., 2004, 2006c).

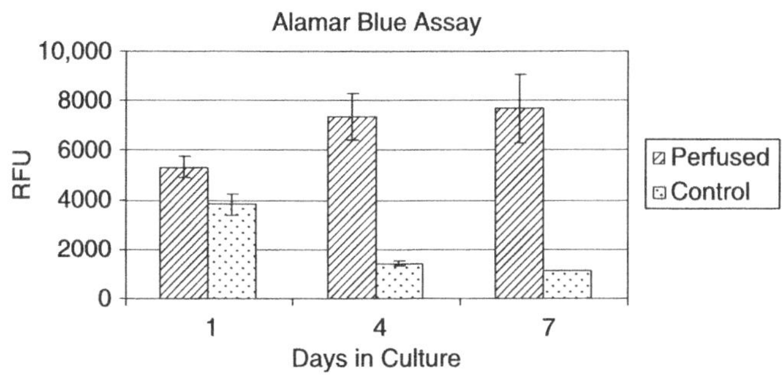

Figure 14.3 Comparison of cell proliferation in hollow-fiber membrane bioreactors. [Nutrient supply in the control group all come from external surface; hence results showing cell number decline due to mass transfer limitation (Ye et al., 2007).]

staining of the specimen removed from the HFMBs revealed that in the nonperfused group the overall cell number was lower compared with the perfusion group and there were many dead cells too. In contrast, in the perfusion group, the cells were viable, and there were no detectable nonviable cells in this group. There was clear evidence of bonelike tissue formation in the hollow-fiber membrane bioreactors as well (Ye et al., 2007).

Further development of hollow-fiber membranes with controlled biodegradable characteristics will be valuable for future work. The rates of fiber biodegradation could be adjusted according to the rate of tissue formation or to the vascular invasion after implantation in vivo. With time, this would result in increased pore sizes of the hollow-fiber membranes, thus providing more nutrient and oxygen supplies to meet the needs of increasing cell densities and tissue formation. It is very important to match the degradation rate of the hollow-fiber membrane with the growth of the cells and tissues. Cellulose acetate hollow-fiber membranes were used in the present studies as proof of concept. However, more advanced biomaterials with controllable degradation rates are being reported (Freed et al., 1994; Middleton and Tipton, 2000), and thus it is feasible that suitable biodegradable hollow-fiber membranes may become available in the near future (Ellis and Chaudhuri, 2007). In addition, the composition of the composite scaffold needs further investigation and development with possible substitution of biodegradable materials for adherent cell growth for the microcarrier particles and alternatives to the collagen scaffolds used here.

14.2.3 Mathematical Modeling of Tissue Engineering HFMB

In theory, a well-designed HFMB can reduce the diffusion path for nutrient and waste products to and from the cells and maintain a stable microenvironment similar to that existing in vivo. The perfused HFMB has shown significant advantages over conventional static culture in terms of cell proliferation and differentiation. Bonelike tissue has been detected after the cell–biomaterial constructs cultured in the HFMBs were implanted in vivo. These initial findings have proved the concept of using HFMB for bone tissue engineering and indicated that an implantable amount of bone tissue could be grown by using HFMBs in the future. However, it is necessary to develop a theoretical framework that elucidates the *uncertain* quantitative relationships between the cell environment and tissue behavior in order to guide the design of effective bone tissue engineering protocols.

These may include, for example, a framework to analyze the flow and transport behavior in relation to tissue and nutrient (solute) properties and the reaction rates in a changing tissue environment. This can then help to interpret the obtained results and direct further experimental work in an appropriate manner. It should also be noted that due to the nature of the HFMB system used for growing tissue (i.e., sterile operating conditions and the relatively small dimension compared to industrial processes), it does not allow precise online monitoring of most variables of interest. Instead, a mathematical model can be used to determine these variables.

As mentioned before, HFMB has been used extensively for various purposes, and a considerable amount of mathematical modeling work has been done to analyze the performance of HFMBs, including the ones for enzyme production and cell growth (e.g., Waterland et al., 1974). Based on these previous studies Ye et al. (2006) and Abdullah et al. (2006) have developed a mathematical description of the simplified HFMB for tissue engineering. The model has been developed to describe mass transfer and nutrient distribution profiles in HFMB for growing 3D bone tissues. Also, it has been used to simulate the effects of the important operation parameters on the nutrient and oxygen distribution in the system used for growing bone tissue.

In the model development, the Krogh cylinder approach was adopted (Krogh, 1919) for the hollow-fiber membrane bioreactor where straight fibers are arranged in a patterned manner (Labecki et al., 1996). This implies that the whole HFMB is composed of many identical fibers surrounded by a regular annulus of homogeneous cellular matrix (Krogh, 1918; Millard and Gorman, 1997). The interstitial space between the Krogh cylinders is ignored. A representative hollow-fiber model is schematically shown in Figure 14.4. As shown, region R_1 (from $r = 0$ to $r = R_1$) is the fiber lumen, region R_2 (from $r = R_1$ to $r = R_2$) is the fiber wall, and region R_3 (from $r = R_2$ to $r = R_3$) is the cellular matrix where scaffolding materials with cells are accommodated.

The continuity equations for each region and the boundary conditions used to solve these equations are given next (Ye et al., 2006). The lumen region:

$$2u\left(1 - \frac{r^2}{R_1^2}\right)\frac{\partial C}{\partial Z} = \frac{D_1}{r}\frac{\partial}{\partial r}\left(r\frac{\partial C}{\partial r}\right) \tag{14.1}$$

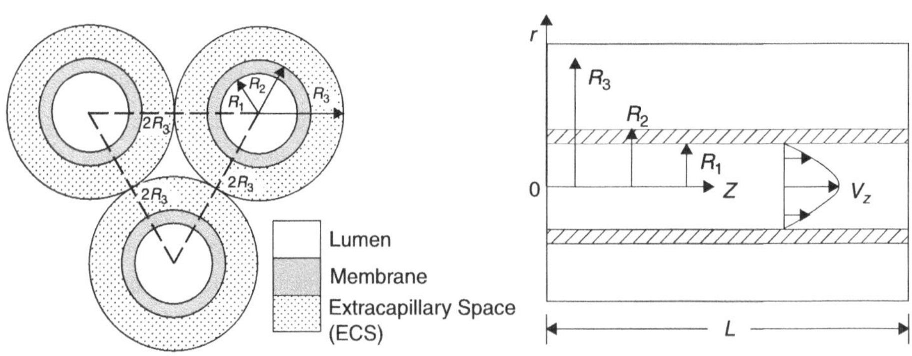

Figure 14.4 Schematic diagram of Krogh cylinder approximation: R_1, fiber lumen; R_2, fiber wall; R_3, cellular matrix.

with the boundary conditions (BCs):

$$C = C_0 \quad \text{at} \quad Z = 0, \qquad \frac{\partial C}{\partial r} = 0 \quad \text{at} \quad r = 0, \qquad D_1 \frac{\partial C}{\partial r} = D_2 \frac{\partial C}{\partial r} \quad \text{at} \quad r = R_1 \quad (14.2)$$

In the fiber lumen, axial convection is by laminar flow with an average steady fluid velocity of u. In this region, three BCs are imposed. The first condition is a Dirichlet-type BC and accounts for the substrate concentration at the entrance of the lumen region. The second is a Neumann-type BC and represents symmetry of radial substrate concentration gradient at the center of the fiber lumen. The third BC is also a Neumann-type BC and reflects continuity of flux at the fiber wall. In the above equations, parameter D with subscripts 1, 2, and 3 refers to the diffusivity values of a given species in the lumen (R_1), fiber membrane wall (R_2), and cellular matrix (R_3), respectively. The fiber wall:

$$\frac{D_2}{r} \frac{\partial}{\partial r}\left(r \frac{\partial C}{\partial r}\right) = 0 \quad (14.3)$$

with the BCs:

$$D_1 \frac{\partial C}{\partial r} = D_2 \frac{\partial C}{\partial r} \quad \text{at} \quad r = R_1 \quad \text{and} \quad D_2 \frac{\partial C}{\partial r} = D_3 \frac{\partial C}{\partial r} \quad \text{at} \quad r = R_2 \quad (14.4)$$

Both boundary conditions are of the Neumann-type BCs and account for the continuity of flux at the fiber walls. The cellular matrix:

$$\frac{D_3}{r} \frac{\partial}{\partial r}\left(r \frac{\partial C}{\partial r}\right) = V \quad (14.5)$$

In the cellular region, radial convection is ignored because the scaffolding materials in this region significantly reduce convective flow across the fiber wall. The term V is the overall cell consumption rate of the substrate, which can be expressed as $v \times d$, where v is the consumption/production rate of the species for a single cell and d is the cell density. In the analyses, v can take a different form of kinetics, for example, constant (zero-order), linear function of concentration (first-order), or depending on C^2 (second-order), or other types of kinetic expressions (e.g., Monod kinetics).

Equation (14.5) is solved subject to the following boundary conditions:

$$D_2 \frac{\partial C}{\partial r} = D_3 \frac{\partial C}{\partial r} \quad \text{at} \quad r = R_2 \quad \text{and} \quad \frac{\partial C}{\partial r} = 0 \quad \text{at} \quad r = R_3 \qquad Z = 0, Z = L \quad (14.6)$$

The final boundary condition reflects symmetry of the concentration gradient between adjacent fibers and both ends of the Krough cylinder (Ye et al., 2006).

The numerical solutions of the governing model equations have been obtained (Ye et al., 2006; Abdullah et al., 2006) using the software package FEMLAB (Comsol, Sweden). The base values of model parameters were taken from the experimental work of Ye et al. (2007) and are listed in Table 14.1.

Figure 14.5 shows the typical results on the concentration distribution within the HFMB. As can be seen, cell density is an important parameter. Furthermore, cell kinetics is also important and needs to be determined accurately, particularly when the cell density is high, as shown in Figure 14.6.

TABLE 14.1 Model Parameter Values Used to Determine Concentration Profiles of Glucose and Oxygen[a]

Model Parameter		Symbol	Unit	Basic Value
Diffusivity	Glucose	D_{1g}	$m^2\ s^{-1}$	5.4×10^{-10}
	Oxygen	D_{1o}	$m^2\ s^{-1}$	3×10^{-9}
Hindering factor	Membrane	$\alpha = D_1/D_2$		10
	Cell matrix	$\beta = D_1/D_3$		5
Concentration at entrance	Glucose	C_g	$mol\ m^{-3}$	5.55
	Oxygen	C_o	$mol\ m^{-3}$	0.22
Medium flow rate		u	$m\ s^{-1}$	7.45×10^{-3}
Fiber inner radius		R_1	m	1×10^{-4}
Fiber outer radius		R_2	m	1.2×10^{-4}
Krogh cylinder radius		R_3	m	3.2×10^{-4}
Effective fiber length		L	m	0.03
Cell seeding density		d	$cells\ m^{-3}$	2×1012
Cell metabolic rate	Glucose	v_g	$mol\ cell^{-1}\ s^{-1}$	3.83×10^{-16}
	Oxygen	v_o	$mol\ cell^{-1}\ s^{-1}$	3.75×10^{-17}

Source: Ye et al. (2007).

[a]List of Symbols

C	substrate concentration ($mol\ m^{-3}$)
C_0	substrate concentration at the inlet ($mol\ m^{-3}$)
d	cell density ($cells\ m^{-3}$)
D_1	diffusivity in water ($m^2\ s^{-1}$)
D_2	effective diffusivity in the membrane wall ($m^2\ s^{-1}$)
D_3	effective diffusivity in the cellular matrix ($m^2\ s^{-1}$)
n	number of hollow fibers
r	radial distance (m)
R	subscript 1—inner radius of hollow fiber; 2—outer radius of hollow fiber; 3—Krogh cylinder radius (m)
u	average velocity of substrate in lumen ($m\ s^{-1}$)
v	cell metabolic rate ($mol\ cell^{-1}\ s^{-1}$)
V	substrate consumption rate ($mol\ m^{-3}\ s^{-1}$)
Z	axial distance (m)
α	hindering factor of membrane ($=D_1/D_2$)
β	hindering factor of cellular matrix ($=D_1/D_3$)

Subscript:

g	glucose
o	oxygen

Of course, the diffusional distance is important, as shown in Figure 14.7. With the developed mathematical model, the influence of all individual parameters can be systematically studied; so can the sensitivity of each parameter on the concentrations of key nutrients. Such information can be used to determine design and operation criteria to prevent cell starvation and to guide experimental design to focus on the study of important and sensitive parameters in order to minimize the expensive laboratory measurements and testing.

Further study should focus on the determination of important parameters including the effective diffusivities of interested molecules in the scaffold, metabolic activities of different cells under different conditions, and matrix turnover measurements. For the model itself, the simulation of macromolecules transport needs to be address as most growth factors are proteins, including protein transport through membranes and diffusion within the scaffold. The electrostatic interaction is expected to be important as demonstrated in

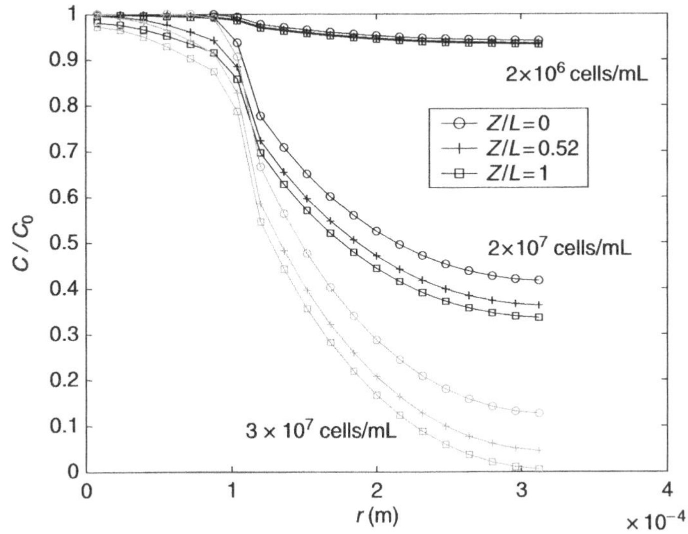

Figure 14.5 Radial concentration profiles of glucose at different cell densities and fiber length (zero kinetics, fiber length = 0.03 m, $Z/L = 0$: inlet; $Z/L = 0.52$: near middle; $Z/L = 1$: outlet) (Ye et al., 2006).

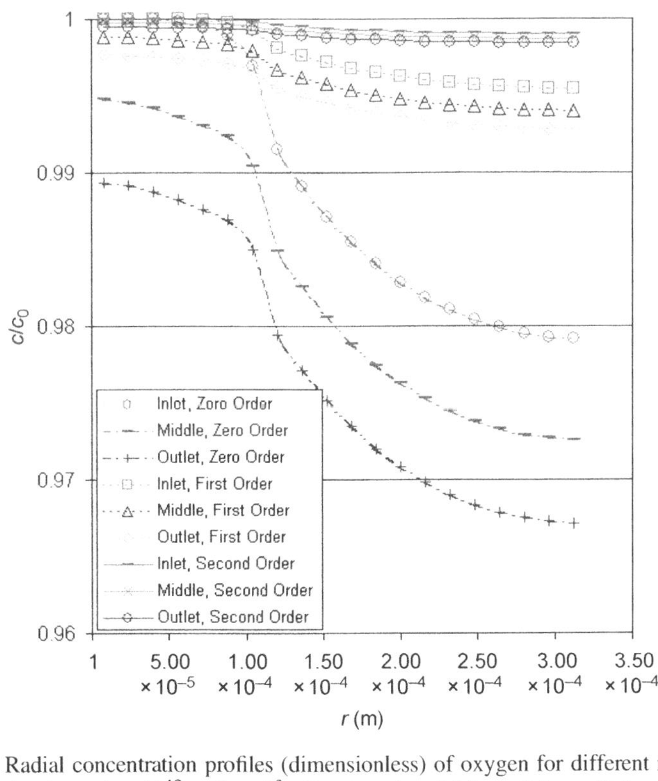

Figure 14.6 Radial concentration profiles (dimensionless) of oxygen for different reaction kinetic schemes (cell density = 2×10^{12} cells m^{-3}, fiber length = 0.03 m). Inlet, middle, and outlet positions are represented at Z/L ratios of 2.5×10^{-2}, 5.25×10^{-1}, and 9.75×10^{-1}, respectively (Abdullah et al., 2006).

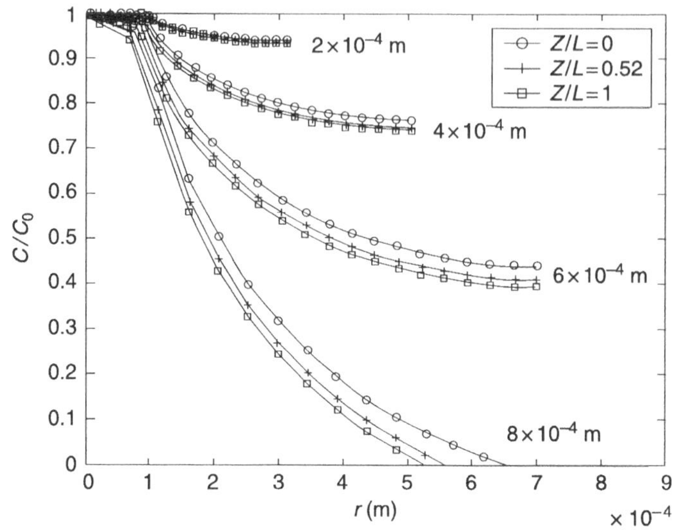

Figure 14.7 Glucose radial concentration profiles with different cellular matrix thickness. The numbers shown beside the curves are the thickness of cellular matrix. $Z/L = 0$: inlet; $Z/L = 0.52$: near middle; $Z/L = 1$: outlet (Ye et al., 2006).

protein fractionation work (Wan et al., 2004). The model should also be able to cope with changing properties of the hollow-fiber membrane when eventually biodegradable hollow fibers are used.

14.3 MICROMEMBRANE PROBES FOR TISSUE ENGINEERING MONITORING

14.3.1 Need for New Monitoring Techniques

Engineered tissue culture takes a relatively long period of time (e.g., 3 weeks for skin, even longer for other tissues) and is expensive due to the need for various chemicals and biochemicals. Hence the monitoring of the tissue culture process and early intervention if necessary is important and has high financial impact, apart from technical necessity. It is possible to measure the surrounding environment including feed concentrations and to monitor the effluents on-line or off-line in term of pH, glucose, lactate, and macromolecular markers [e.g., glycosaminoglycan (GAG)]. But it is difficult to monitor nutrient gradients and cell functions within a 3D-engineered tissue construct. And such monitoring is important to identify mass diffusion limitations and nonuniformity of tissue growth. For example, Constantinidis et al. (1999) showed the severe nonuniformity in cell distribution within β-cell encapsulating bead following 4 weeks of culture, with most of the cells located close to the shell (close to nutrient source) and no cells in the central core region. Such nonuniformity cannot be detected by monitoring the effluent, which can only give collective information.

Currently, the assessment of a developing construct is almost always carried out destructively using biochemical or histological methods to determine cell number, viability, and tissue growth throughout the construct (Martin et al., 2004; Obradovic et al., 2000).

Assessment methods using only changes in volume or shape have limited value as in many cases tissue and cell growth is uneven and occurs primarily at the construct boundary [reviewed by Martin et al. (2004)]. Since many of these experiments are long, taking weeks or even months to complete (Freyria et al., 2004; Seidel et al., 2004), simple and readily applicable nondestructive methods of monitoring changes in cell metabolism, viability, and tissue deposition within the construct would be invaluable; such methods could point out adverse responses during the early stages of culture. At present, nondestructive monitoring throughout the construct can be achieved using technologies such as magnetic resonance imaging (MRI) (Washburn et al., 2004; Williams et al., 2005) or optical methods (Xu et al., 2006). These techniques can provide useful information on spatial variations in tissue deposition but cannot monitor cellular activity or any biochemical parameters. Moreover their use is limited by cost and availability. Recently, a technique of using a micromembrane probe, based on the principle of microdialysis and ultrafiltration, has been successfully applied to tissue engineering monitoring.

14.3.2 Microdialysis and Ultrafiltration

Microdialysis is used to collect solutes present in the extracellular fluid via a microdialysis probe with a semipermeable membrane at its tip (see Fig. 14.8). The probe is perfused by a buffer (perfusate), and solutes from the environment surrounding the probe diffuse though the membrane into the perfused solution. Perfusate with the solutes (dialysate) is then collected for ex situ analysis. Microdialysis has been used in vivo to monitor the local concentrations of solutes in the extracellular fluids in a number of different tissues. The first microdialysis experiments were conducted on the brain and blood plasma (Bito et al., 1966). It has since been used to study metabolism in numerous tissues such as brain, muscles, tendons, subcutaneous adipose tissue, lungs, kidneys, and liver (Plock and Kloft, 2005; de la Pena et al., 2000; Jackson, 2005; Siddiqui and Shuaib, 2001). Microdialysis is widely used for pharmacokinetic research (Davies, 1999; de Lange et al., 2000; Verbeeck, 2000) and has also been used to monitor cell metabolites in cell culture medium (Wu et al., 2001).

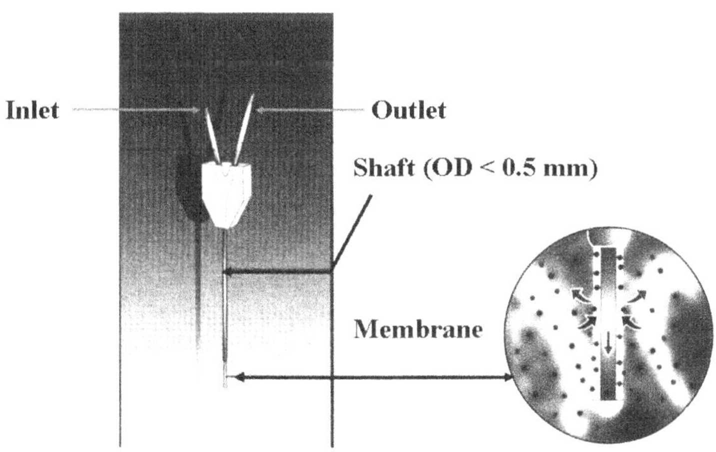

Figure 14.8 Illustration of microdialysis probe. (Adapted from http://www.microdialysis.se/probes.htm.)

To pick up macromolecules based on their diffusion into the probe would be too slow. This can be overcome by applying a negative transmembrane pressure, that is, via ultrafiltration. Microdialysis is commmonly conducted with one pushing pump where the pressure within the probe is slightly higher than its surrounding. The transmembrane pressure is usually insignificant due to the minute flow rate of the perfusate applied. Nevertheless, possible fluid loss due to this transmembrane prossure can be prevented by increasing the osmotic pressure of the perfusate to balance the transmembrane pressure (adding inert macromolecular solutes). In ultrafiltration operation, one "pull" pump can be used that generates a negative transmembrane pressure. In practice, it is more convenient to use two pumps, one pushing and one pulling, to conduct such measurements to increase the adjustability and controllability of the operation.

14.3.3 Fouling Detection and in situ Calibration

There are two main challenges related to the application of micromembrane probes: (1) how to quantify the true value of the measured parameters based on the assay of their concentrations in the collected samples and (2) whether this determined quantitative link would change with time due to membrane fouling.

To link the concentrations of the solute of interest in the dialysate (collected samples) and in the probe surrounding (to be measured), a parameter called relative recovery is often used. The concentration of solute in dialysate can be equal to the solute concentration in the probe surrounding if flow rate of perfusate equals 0 (in equilibrium). Otherwise the concentration of the solute in the dialysate will be lower than that in the surrounding probe. Relative recovery (RR) is expressed as a percentage of the concentration of the solute in the probe (the dialysate) relative to that in the surrounding external solution, that is,

$$\mathrm{RR} = C_e/C_d \times 100\% \tag{14.7}$$

where C_d is the concentration of solute in the dialysate, and C_e is the solute concentration in the environment (extracellular matrix). If the relative recovery is determined, C_e can be easily evaluated by the measurement of C_d.

The recovery of solutes from the interstitial fluid by the probe can be evaluated by introducing internal standards into the perfusate and measuring the relative loss (RL). Relative loss of a standard was estimated as a percentage of the standard solute that diffused from the probe perfusate into the surrounding interstitial fluid:

$$\mathrm{RL} = \left(\frac{C_p - C_d}{C_p}\right)100\% \tag{14.8}$$

where C_p is the concentration of the standard in the perfusate, and C_d is the concentration of the standard in dialysate collected.

Relative loss is assumed to be equal to a relative recovery (Ingvast Larsson, 1991; Peters et al., 2000; Scheller and Kolb, 1991) if the solute taken as internal standard has the same or similar molecular size and diffusional properties as a solute of interest. Diffusion of the marker from the perfusate and across the dialysis membrane is not dependent on the concentration of unlabeled molecules (Plock and Kloft, 2005).

Boubriak et al. (2006b) developed an in situ calibration method by measuring the RL. To determine relative loss of lactate and glucose, 0.4 μC/mL [^{14}C] lactate or [^{14}C] methyl-D-glucose was added to the perfusate as an internal marker for lactate and glucose, respectively. The concentration of radioactively labeled solutes in the dialysate was determined by radioactive counting. In situ probe relative recovery was also assessed on the basis of the percentage of phenol red (PhR) in the dialysate after equilibration of the construct with Dulbecco's Modified Eagle's Medium (DMEM) containing PhR. PhR is routinely added to cell culture media as a pH indicator and is hence a convenient indicator. For macromolecules, 40-kDa fluorescent dextran is used as the internal standard for measurement of RL and hence for the evaluation of the RR of proteins of similar size (Li et al., 2006).

Evaluation of possible fouling of the microdialysis probe membrane is important because it is necessary to distinguish a decrease of monitored molecules caused by a fall of metabolic cell activity from a decrease caused by fouling; this is particularly important for long-term monitoring. Probe membrane fouling can be assessed on the basis of tracer losses from the perfusate and phenol red gain in the dialysate over the time period of monitoring the construct.

Figure 14.9 showed the change in probe lactate relative recovery with time as a consequence of probe fouling (15-kDa probe, chondrocyte in alginate, see next sections for details). The recovery fell to 66.2% of its original value within the first 15 h and then was stable for 75 h. A loss of permeability may have occurred due to membrane fouling by macromolecules present in the probe's environment. For this system these would initially be mainly non-cross-linked alginate polymers because proteins secreted by the cells would be at a very low concentration during the early stages of culture. This change in relative recovery did not, however, influence the calculated level of lactate because individual recovery values were taken for each lactate value recorded in the dialysate.

14.3.4 Monitoring Nutrient Gradients and Metabolic Activities

Microdialysis probes can be used to monitor nutrient gradients and metabolites. Boubriak et al. (2006a) inserted eight probes into a cylindrical engineered cartilage construct to monitor the transient distribution of glucose and lactate. The bioreactor for culturing the

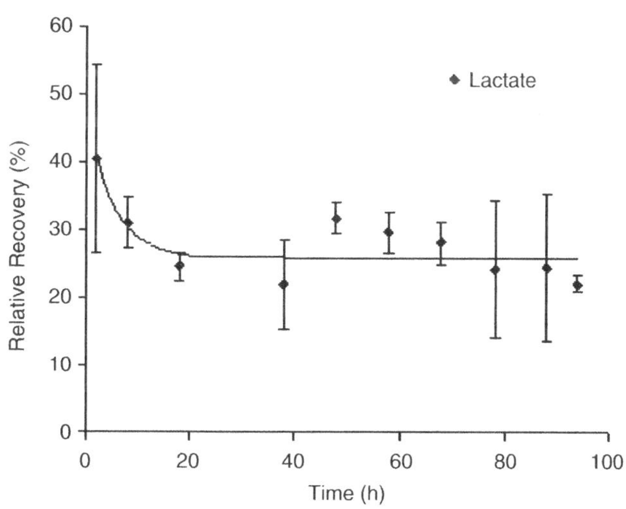

Figure 14.9 Change of probe relative recovery of lactate with time, assuming to be equal to relative loss of [^{14}C] lactate.

alginate gel containing the bovine cartilage chondrocytes consists of two identical cap-ground glass disks (ID 20 mm) for supplying culture medium to the cultured cells; two Isopore polycarbonate membranes retained the gel-cell construct and separated the alginate gel from the circulating solution. The spacer (ID 20 mm) of different heights determined the thickness of the construct.

Figure 14.10 shows a schematic of the experimental setup. Both the culture medium and the buffer (dialysate) for microdialysis probes were supplied by an 8-channel peristaltic pump that produced a flow of perfusate into the microdialysis probe at 0.56 ± 0.12 μL/min [polyvinyl chloride (PVC) tubing of ID 0.25 mm] and supplied the cultured cells in the bioreactor with DMEM culture medium at 30.0 ± 5.1 μL/min (silicone tubing ID 2.06 mm), the DMEM leaving the bioreactor not recirculated. The bioreactor itself and reservoirs with buffer for the probe and DMEM for the bioreactor were all kept in a bench incubator under atmospheric concentrations of oxygen and carbon dioxide at 37°C. The microdialysis probe was introduced into the spacer, and the bioreactor was assembled and autoclaved before injection of the cell–alginate mixture. For measurement of concentrations of low-molecular-weight solutes (<1 kDa) in alginate gels or alginate–chondrocyte constructs, autoclavable microdialysis probes formed from polyethersulfone (PES) dialysis membrane with 15-kDa cutoff and with an effective length of 4 mm, OD 0.6 mm and with a 35-mm flexible polyurethane shaft were used (Boubriak et al., 2006a).

Figure 14.11 shows the measured transient distribution of lactate concentration following its introduction at the boundary of the cell-free alginate scaffold at time = zero, measured by eight probes. These measurements agree well with theoretical predictions by solving the diffusion equation for diffusion into a plane sheet from a constant source (Crank, 1975; Boubriak et al., 2006a, 2006b).

Figure 14.12 shows a measurement of lactate at a defined location and, more important, the cellular metabolic response to the introduction of growth factors [fetal bovine serum (FBS)]. The results presented here demonstrate that microdialysis provides a nondestructive means of monitoring metabolic changes within cell-seeded three-dimensional constructs cultured in a bioreactor. Microdialysis was able to detect local changes in concentrations of metabolites within the construct and hence revealed differential responses to environmental perturbations (Boubriak et al., 2006a). Moreover microdialysis provided a

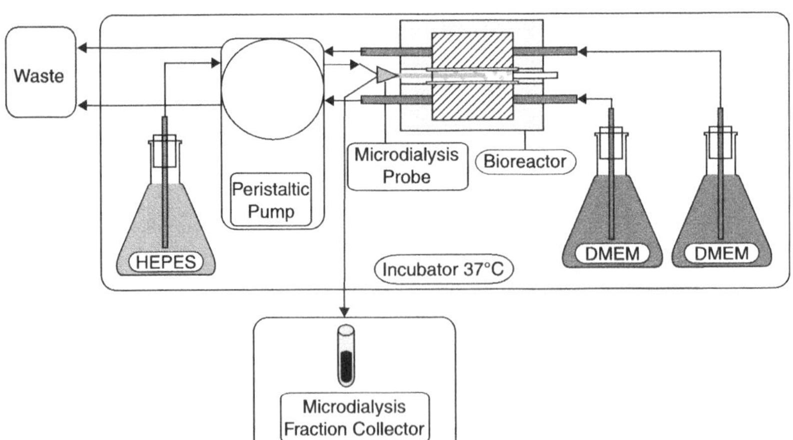

Figure 14.10 Microdialysis setup for monitoring engineered cartilage culture.

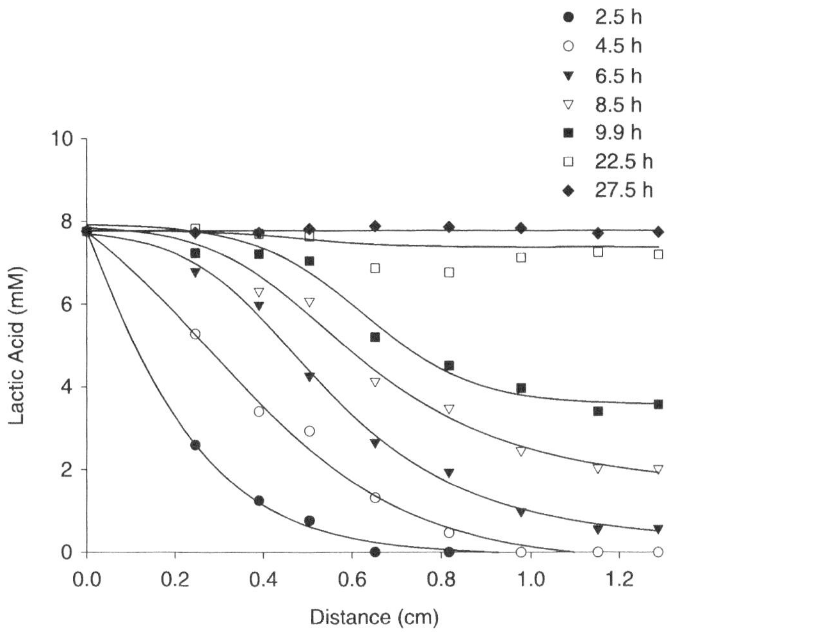

Figure 14.11 Measurement of solute diffusion through alginate gel using eight microdialysis probes at different locations.

nondestructive means of detecting changes in metabolite concentrations in the construct center, which were indicative of nutrient deprivation and cell death.

14.3.5 Monitoring Cell Functions

Many complex cell functions, apart from viability and metabolic activities, are assayed using various macromolecular markers. Hence it is important to be able to sample

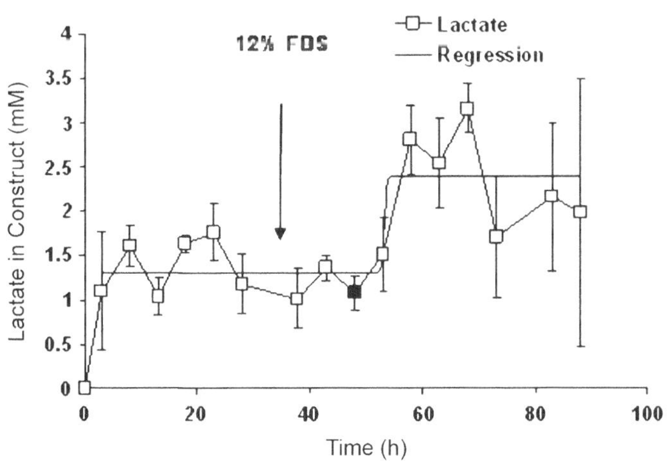

Figure 14.12 Monitoring of lactate concentration and the change of cell metabolic activity after stimulation by fetal bovine serum (FBS) (Boubriak et al., 2006a).

macromolecules to monitor cell functions and tissue formation. Microdialysis probes with the membranes of pore size ranged from 15 to 3000 kDa molecular weight (MW) cutoff are commercially available. It is hence feasible to monitor high-MW molecules such as products of matrix protein turnover and growth factors to use this technique. Micromembrane probes, in particular operated under ultrafiltration, have been shown to be capable of picking up protein markers (Ao and Stenken, 2006; Rosenbloom et al., 2005), and others have shown that a similar system can be used for measurement of soluble proteins such as collagen pro-peptides or growth factors in tissues in vivo.

Figure 14.13 shows the sample protein assay results of using a commericial 100-kDa microdialysis probe (PES dialysis membrane, an effective length 10 mm, 0.4 mm OD) to monitor bovine intervetebral disk (IVD) culture (Li et al., 2006). The IVD is cultured up to 7 days, and the probe is operated by either push, pull, or push-and-pull modes. Apart from glucose and lactate, the probe picked up at least three proteins with a molecular weight around 40 kDa. Three peaks of proteins (P1, P2, and P3) were evident on fast protein liquid chromatography (FPLC) outcome during the 7-day culture period. Three bands were also evident using 12% SDS PAGE electrophoresis (sodium dodecyl sulfate polyacrylamide gel electrophoresis) with silver stain (see Fig. 14.13). In these experiments, the relative recovery is easily determined using the developed in situ calibration using 40-kDa fluoresecent dextran as the internal standard. In the 7-day experiments, the relative loss of fluorescent dextran (40 kDa) stablized at 8–10%, which is equivalent to the protein transmission across the membrane (Li et al., 2006).

The change of these three identified proteins during the culture was sucessfully monitored, as shown in Figure 14.14. Now at least one of the proteins is identified as chitinase-3-like protein 1 (CHI3L1) by mass spectrometry and Western blotting (antibody was kindly provided by A. Recklies, Shriners Inst, Montreal), which appears as a sensitive marker for chondrocyte and cartilage metabolism. The detailed biological link of these protein markers to cell/tissue functions, and the reason why there is a sudden drop in protein 1 expression on day 4, merit further detailed studies.

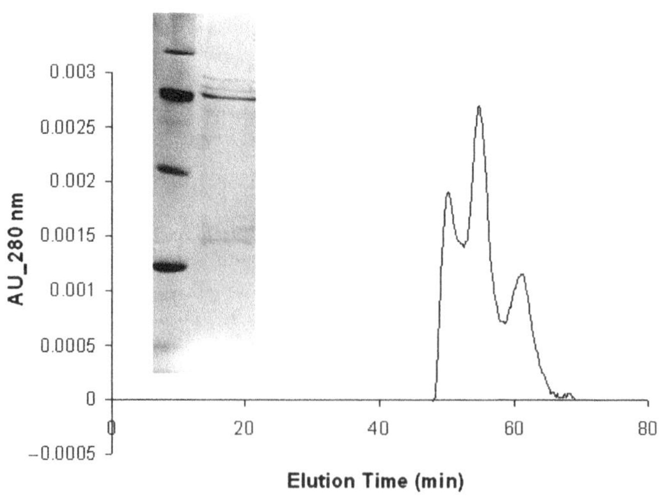

Figure 14.13 Protein assays using FPLC and and SDS-PAGE with silver stain (left inset). Lane 1, molecular mass standards (Bio-Rad); lane 2, crude dialysate.

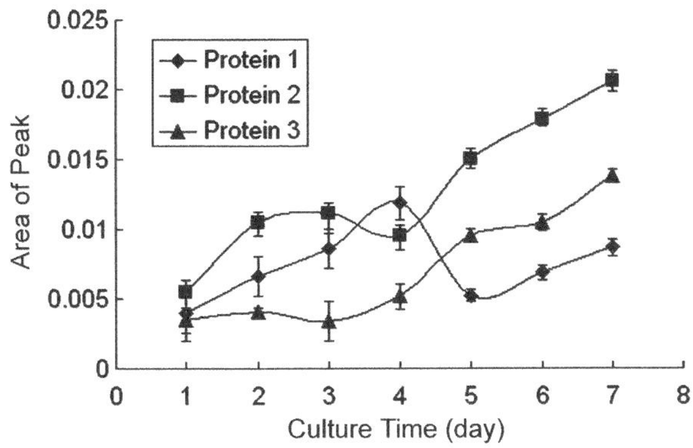

Figure 14.14 Changes of the three protein concentration in dialysate daily during 1-week IVD culture assayed by FPLC. Results were given as mean $\pm$ std ($n = 3$).

14.4 FUTURE OPPORTUNITIES

14.4.1 Membrane Bioreactors for Stem Cell Expansion

Efficient stem cell expansion is a key bottleneck for clinical application and commercialization of stem cell therapy. Membrane bioreactors may make a significant contribution due to its important features such as possibility for uniform chemical and biochemical conditions within the bioreactor, low or even zero hydrodynamic shears, large surface-to-volume ratios, and physical separation between two cell types but allowing biochemical signaling between them. For example, it may be possible to culture the feeder cells on one side of the membrane, while culturing human embryonic stem cells on the other. In this way human embryonic stem cells are not mixed with the feeder cells, which eliminates the need for later difficult separation, but get the biochemical signals from the feeder cells that are necessary to proliferate embryonic stem cells (e.g., Choo et al., 2006; Klimanskaya et al., 2005).

14.4.2 Bioartificial Organs

Membranes have been successfully applied to hemodialysis and artificial lung (e.g., Nagasc et al., 2005; Wickramasinghe et al., 2002). Bioaritificial organs are the next generation of devices involving the use of living cells, and membranes could be an essential part of these developments (Strathmann, 2001). Membrane devices incorporating kidney and lung cells are being studied (e.g., Saito et al., 1998), but the most studied device is probably the liver support systems (e.g., Flendrig et al., 1997; Krasteva et al., 2002; Borberg, 2006). In such devices, the advantages of offering large surface area to support cell growth of compact membrane device and providing a physical barrier to retain and confine cells are exploited.

14.4.3 Immunoisolation

To protect implanted tissues or functional cells, an immunoisolation membrane can be used (Dionne et al., 1996; Lanza and Chick, 1997). An example of such systems is the artificial pancreas (Desai et al., 2004; Figliuzzi et al., 2005; Ohgawara et al., 2000). An immunoisolation membrane, which can also be formed in place, is used to protect the implanted islets (or β-cell-based construct). The membrane allows nutrients to diffuse through the membrane and to feed the implanted tissue and cells and insulin to diffuse out but rejects host T-cells and antibodies.

14.4.4 Mass Transfer Control in Cryopreservation Procedures

Cryopreservation is likely to be the method to preserve engineered tissues to ensure the off-the-shelf availability to clinicians and a critical step for commercialization of the engineered tissues. In cryopreservation, cryoprotective agents (CPAs), which protect cells in the tissue against the damage caused either by intracellular ice formation or dehydration during freezing, are first added to the tissue to be preserved, and usually in the form of mixture of several CPA components. The subsequent freezing is achieved following a controlled freezing rate protocol to the desired temperature. Due to possible toxicity of CPAs to the end users, CPAs must be removed after thawing. Although CPAs can protect cells from freezing damage, they can also cause damage to cells by osmotic stress and toxicity, especially when used in high concentration, for example, in vitrification. To ensure the cell viability, the step of CPA addition and removal should be conducted in a controlled multistep and optimized manner.

Figure 14.15 illustrates the stepwise addition and removal of high concentrations of CPA mixtures. Such an operation particularly for the CPA removal, is laborious and tedious and subject to human errors within clinical settings.

One possible solution is to use porous membrane to control the CPA concentration around the tissue after direct exposure to the highest CPA concentration during addition or to water or cell culture medium for removal. The membrane provides a physical barrier to direct mixing but allows CPA diffusion through the pores. Xu and Cui (2001) and Xu and Cui (2003) have studied the feasibility of this approach and used Stafen–Maxwell modeling to simulate the co-transport of high concentration CPA mixtures (dimethyl sulfoxide, glycerol, etc.) across the membrane. Furthermore, the model was validated with experiments. The conclusion is that the desired CPA removal profile (Song et al., 2000) can only be achieved with membranes with changing properties. The porosity of the membrane needs to be increased significantly to match the required transport rate. Common polymeric membranes with fixed porosity cannot meet this need. High porous membrane with the pore filled with a fast degrading hydrogel may fulfill this need.

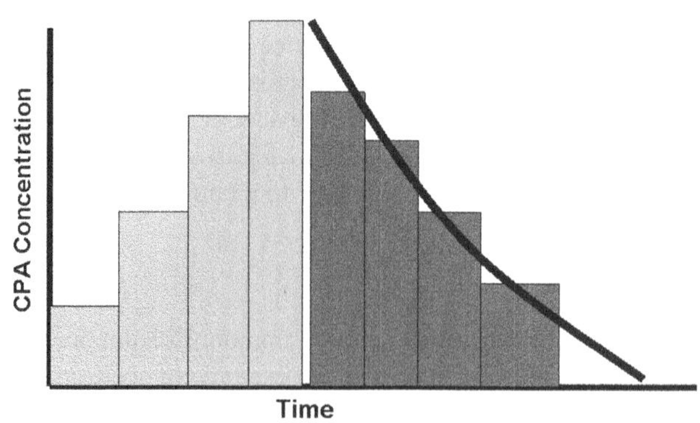

Figure 14.15 Stepwise CPA addition and removal (for vitrification the highest concentration is usually about 6 mol/L consisting of several chemicals).

14.4.5 Development of New Types of Membranes

It has become obvious that new types of membranes need to be developed for tissue engineering applications. For example, some applications will require the controllable and tunable biodegradation rate, so the degradation of the membrane and scaffold can match the neotissue formation and angiogenesis (Ye et al., 2007). Ellis and Chaudhuri (2007) developed poly(lactide) (PLA) and poly(lactide-co-glycolide) (PLGA) copolymer hollow-fiber membranes and cultured osteoblasts on these hollow fibers. Some applications may need controlled mass transfer rate (Dai and Barbari, 2000). Some applications may require a narrow pore size distribution, as it has been shown that pore size does affect cell adhesion and proliferation (Lee et al., 2004). One thing is common for all these applications: the surface properties are likely to be the most important parameter, and also these membranes are likely to be operated under ultra-low-permeate flux. For most tissue engineering applications, suitable membranes are to be developed, which provides a great opportunity for membrane research and development. The new membranes are likely to be sold by functions, instead of quantity, and with very high added values.

14.5 SUMMARY

Membrane technologies have been applied into tissue engineering or more generally regenerative medicine. As an enabling technology, application of membranes can help to overcome some key bottlenecks in tissue engineering and stem cell therapy. Examples given in this chapter include how to grow three-dimensional bulky tissues and how to monitor the cell functions within a three-dimensional tissue. Other important applications, such as membrane bioreactors for stem cell expansion, immunoisolation, mass transport control, and bioartificial organs, are also briefly discussed. However, the membranes currently commercially available cannot fulfill the requirements for tissue engineering applications and hence development of new membranes with novel functions such as biodegradability, biocompatability, and controlled release of active agents provides a good opportunity for membrane research. A new set of rules and parameters need to be defined to assess these new membranes as well.

ACKNOWLEDGMENTS

The author wishes to thank his research group for their contributions to the reported work, in particular, Hua Ye, Olga Boubriak, Xia Xu, Zhidao Xia, Zhaohui Li, N S Abdullah and his collaborators, Jill Urban, Diganta Das, and Jim Triffitt. He is grateful to funding agents, the UK Biotechnology and Biological Science Research Council, Engineering and Physical Research Council, for financial support for his research.

REFERENCES

Abdullah, N. S., Das, D. B., Ye, H., and Cui, Z. F. (2006). 3d bone tissue growth in hollow fibre membrane bioreactor: Implications of various process parameters on tissue nutrition. *Int. J. Artif. Organs* 29(9), 841–851.

Ao, X., and Stenken, J. A. (2006). Microdialysis sampling of cytokines. *Methods Anal. Cytokines* 38(4), 331–341.

Bassett, P. (2001). *Tissue Engineering: Technologies, Markets and Opportunities*, 3rd ed. D&MD Report.

Bito, L., Davson, H., Levin, E., Murray, M., and Snider, N. (1966). The concentrations of free amino acids and other electrolytes in cerebrospinal fluid, in vivo dialysate of brain, and blood plasma of the dog. *J. Neurochem.* 13(11), 1057–1067.

Borberg, H. (2006). Quo vadis haemapheresis: Current developments in haemapheresis. *Transfusion Apheresis Sci.* 34(1), 51–73.

Boubriak, O. A., Urban, J. P. G., and Cui, Z. F. (2006a). Monitoring of lactate and glucose levels in engineered cartilage construct by microdialysis. *J. Membr. Sci.* 273, 77–83.

Boubriak, O. A., Urban, J. P. G., and Cui, Z. F. (2006b). On-line monitoring of metabolite gradients in tissue-engineered constructs. *Proc. R. Soci. Interface* 3(10), 637–648.

Choo, A., Padmanabhan, J., Chin, A., Fong, W. J., and Oh, S. K. W. (2006). Immortalized feeders for the scale-up of human embryonic stem cells in feeder and feeder-free conditions. *J. Biotechnol.* 122(1), 130–141.

Constantinidis, I., Rask, I., Long Jr, R. C., and Sambanis, A. (1999). Effects of alginate composition on the metabolic, secretory, and growth characteristics of entrapped [beta]TC3 mouse insulinoma cells. *Biomaterials* 20(21), 2019–2027.

Crank, J. (1975). *Mathematics of Diffusion*. Oxford University Press.

Dai, W. S., and Barbari, T. A. (2000). Gel-impregnated pore membranes with mesh-size asymmetry for biohybrid artificial organs. *Biomaterials* 21(13), 1363–1371.

Dai, N.-T., Williamson, M. R., Adams, E. F., Coombes, A. G. A., and Khammo, N. (2004). Composite cell support membranes based on collagen and polycaprolactone for tissue engineering of skin. *Biomaterials* 25(18), 4263–4271.

Dai, N.-T., Yeh, M.-K., Chiang, C.-H., Yen, C.-Y., Shih, C.-M., Sytwu, H.-K., Chen, T.-M., Wang, H.-J., Adams, E. F., Liu, D. D., Williamson, M. R., and Coombes, A. G. A. (2005). A co-cultured skin model based on cell support membranes. *Biochem. Biophys. Res. Commun.* 329(3), 905–908.

Danielsson, C., Ruault, S., Basset-Dardare, A., and Frey, P. (2006). Modified collagen fleece, a scaffold for transplantation of human bladder smooth muscle cells. *Biomaterials* 27(7), 1054–1060.

Davies, M. I. (1999). A review of microdialysis sampling for pharmacokinetic applications. *Analy. Chim. Acta* 379(3), 227–249.

de Lange, E. C. M., de Boer, A. G., and Breimer, D. D. (2000). Methodological issues in microdialysis sampling for pharmacokinetic studies. *Adv. Drug Deliv. Rev.* 45(2–3), 125–148.

de la Pena, A., Liu, P., and Derendorf, H. (2000). Microdialysis in peripheral tissues. *Adv. Drug Deliv. Rev.* 45(2–3), 189–216.

Desai, T. A., West, T., Cohen, M., Boiarski, T., and Rampersaud, A. (2004). Nanoporous microsystems for islet cell replacement. *Adv. Drug Deliv. Rev.* 56(11), 1661–1673.

Dionne, K. E., Cain, B. M., Li, R. H., Bell, W. J., Doherty, E. J., Rein, D. H., Lysaght, M. J., and Gentile, F. T. (1996). Transport characterization of membranes for immunoisolation. *Biomaterials* 17(3), 257–266.

Ellis, M. J., and Chaudhuri, J. B. (2007). Development of PLGA hollow fibre membranes for use as a tissue engineering scaffold. *Biotechnol. Bioeng* 96(1), 177–187.

Figliuzzi, M., Cornolti, R., Plati, T., Rajan, N., Adobati, F., Remuzzi, G., and Remuzzi, A. (2005). Subcutaneous xenotransplantation of bovine pancreatic islets. *Biomaterials* 26(28), 5640–5647.

Flendrig, L. M., la Soe, J. W., Jorning, G. G. A., Steenbeek, A., Karlsen, O. T., Bovee, W. M. M. J., Ladiges, N. C. J. J., te Velde, A. A., and Chamuleau, R. A. F. M. (1997). In vitro evaluation of a novel bioreactor based on an integral oxygenator and a spirally wound nonwoven polyester matrix for hepatocyte culture as small aggregates. *J. Hepatol.* 26(6), 1379–1392.

Freed, L. E., and Vunjak-Novakovic, G. (1998). Culture of organized cell communities. *Adv. Drug Deliv. Rev.* 33(1–2), 15–30.

Freed, L. E., Vunjak-Novakovic, G., Biron, R. J., Eagles, D. B., Lesnoy, D. C., Barlow, S. K., and Langer, R. (1994). Biodegradable polymer scaffolds for tissue engineering. *Biotechnolog* (N. Y.) 12(7), 689–93.

Freyria, A.-M., Cortial, D., Ronzi, M.-C., Herbage, D., and Guerret, S. (2004). Influence of medium composition, static and stirred conditions on the proliferation of and matrix protein expression of bovine articular chondrocytes cultured in a 3-D collagen scaffold. *Biomaterials* 25(4), 687–697.

Ingvast Larsson, C. (1991). The use of an "internal standard" for control of the recovery in microdialysis. *Life Sci.* 49(13), PL73–PL78.

Jackson, M. J. (2005). Use of microdialysis to study interstitial nitric oxide and other reactive oxygen and nitrogen species in skeletal muscle methods in enzymology. In E. C.Lester Packer (Ed.) *Nitric Oxide, Part E.* Academic, pp. 514–525.

Karlik, W., Jozwiak, A., Wiechetek, M., and Werynski, A. (1999). A simple method for hepatocyte attachment in hollow fibre bioreactors. *Int. J. Artif. Organs* 22(8), 566–572.

Kawakita, H., Sugita, K., Saito, K., Tamada, M., Sugo, T., and Kawamoto, H. (2002). Production of cycloisomaltooligosaccharides from dextran using enzyme immobilized in multilayers onto porous membranes. *Biotechnol. Prog.* 18(3), 465–469.

Klimanskaya, I., Chung, Y., Meisner, L., Johnson, J., West, M. D., and Lanza, R. (2005). Human embryonic stem cells derived without feeder cells. *Lancet* 365(9471), 1636–1641.

Knazek, R. A. (1974). Solid tissue masses formed in vitro from cells cultured on artificial capillaries. *Fed. Proc.* 33(8), 1978–1981.

Knazek, R. A., Gullino, P. M., Kohler, P. O., and Dedrick, R. C. (1972). Cell culture on artificial capillaries. *Science 178*, 65–67.

Krasteva, N., Harms, U., Albrecht, W., Seifert, B., Hopp, M., Altankov, G., and Groth, T. (2002). Membranes for biohybrid liver support systems—Investigations on hepatocyte attachment, morphology and growth. *Biomaterials* 23(12), 2467–2478.

Krogh, A. (1918). The rate of diffusion of gases through animal tissue with some remarks on the coefficient of invasion. *J. Physiol. (Lond.)*, 52, 391–409.

Krogh, A. (1919). The number and distribution of capillaries in muscle with calculations of the oxygen pressure head necessary for supplying the tissue. *J. Physiol. (Lond.)* 52, 409–415.

Labecki, M., Bowen, B. D., and Piret, J. M. (1996). Two-dimensional analysis of protein transport in the extracapillary space of hollow-fibre bioreactors. *Chem. Eng. Sci.* 51(17), 4197–4213.

Lanza, R. P., and Chick, W. L. (1997). Immunoisolation: At a turning point. *Immunol. Today* 18(3), 135–139.

Lee, M.-W., Hung, C.-L., Cheng, J.-C., and Wang, Y.-J. (2005). A new anti-adhesion film synthesized from polygalacturonic acid with 1-ethyl-3-(3-dimethylaminopropyl)carbodiimide crosslinker. *Biomaterials* 26(18), 3793–3799.

Lee, S. J., Choi, J. S., Lee, H. B., Park, K. S., Khang, G., and Lee, Y. M. (2004). Response of MG63 osteoblast-like cells onto polycarbonate membrane surfaces with different micropore sizes. *Biomaterials* 25(19), 4699–4707.

Li, Z. H., Boubriak, O. A., Urban, J. P. G., and Cui, Z. F. (2006). Microdialysis for monitoring the process of functional tissue culture. *Int. J. Artif. Organs* 29(9), 858–865.

Lowrey, D., Murphy, S., and Goffe, R. A. (1994). A comparison of monoclonal antibody productivity in different hollow fiber bioreactors. *J. Biotechnol.* 36(1), 35–38.

Marler, J. J., Upton, J., Langer, R., and Vacanti, J. (1998). Transplantation of cells in matrices for tissue regeneration. *Adv. Drug Deliv. Rev.* 33, 165–182.

Martin, I., Wendt, D., and Heberer, M. (2004). The role of bioreactors in tissue engineering. *Trends Biotechnol.* 22(2), 80–86.

Mi, F.-L., Wu, Y.-B., Shyu, S.-S., Chao, A.-C., Lai, J.-Y., and Su, C.-C. (2003). Asymmetric chitosan membranes prepared by dry/wet phase separation: A new type of wound dressing for controlled antibacterial release. *J. Membr. Sci.* 212(1–2), 237–254.

Middleton, J. C., and Tipton, A. J. (2000). Synthetic biodegradable polymers as orthopaedic devices. *Biomaterials* 21(23), 2335–2346.

Millard, C. A., and Gorman, A. D. (1997). A model for substrate concentrations in tissue surrounding single capillaries. *Math. Computer Modelling* 25(11), 1–7.

Nagase, K., Kohori, F., and Sakai, K. (2005). Oxygen transfer performance of a membrane oxygenator composed of crossed and parallel hollow fibers. *Biochem. Eng. J.* 24(2), 105–113.

Nagel, A., Koch, S., Valley, U., Emmrich, F., and Marx, U. (1999). Membrane-based cell culture systems—An alternative to in vivo production of monoclonal antibodies. *Dev. Biol. Stand.* 101, 57–64.

Obradovic, B., Meldon, J. H., Freed, L. E., and Vunjak-Novakovic, G. (2000). Glycosaminoglycan deposition in engineered cartilage: Experiments and mathematical model. *AIChE J.* 46(9), 1860–1871.

Ohgawara, H., Hirotani, S., Edamura, K., Sasaki, S., and Miyazaki, J. I. (2000). Membrane immunoisolation of a diffusion chamber for a bioartificial pancreas. *Diabetes Res. Clin. Practice* 50(Suppl. 1) 180.

Peters, J. L., Yang, H., and Michael, A. C. (2000). Quantitative aspects of brain microdialysis. *Anal. Chim. Acta* 412(1–2), 1–12.

Plock, N., and Kloft, C. (2005). Microdialysis—Theoretical background and recent implementation in applied life-sciences. *Eur. J. Pharm. Sci.* 25(1), 1–24.

Rony, P. R. (1972). Hollow fiber enzyme reactors. *J. Am. Chem. Soc.* 94(23), 8247–8248.

Rosenbloom, A. J., Sipe, D. M., and Weedn, V. W. (2005). Microdialysis of proteins: Performance of the CMA/20 probe. *J. Neurosci. Methods* 148(2), 147–153.

Saito, A., Suzuki, H., Bomsztyk, K., and Ahmad, S. (1998). Regeneration of peritoneal effluent by Madin-Darby canine kidney cells-lined hollow fibers. *Mater. Sci. Eng. C* 6(4), 221–226.

Scheller, D., and Kolb, J. (1991). The internal reference technique in microdialysis: A practical approach to monitoring dialysis efficiency and to calculating tissue concentration from dialysate samples. *J. Neurosci. Methods* 40(1), 31–38.

Seidel, J. O., Pei, M., Gray, M. L., Langer, R., Freed, L. E., and Vunjak-Novakovic, G. (2004). Long-term culture of tissue engineered cartilage in a perfused chamber with mechanical stimulation. *Biorheology* 41(3–4), 445–458.

Siddiqui, M. M., and Shuaib, A. (2001). Intracerebral microdialysis and its clinical application: A review. *Methods* 23(1), 83–94.

Song, Y. C., Khirabadi, B. S., Lightfoot, F., Brockbank, K. G., and Taylor, M. J. (2000). Vitreous cryopreservation maintains the function of vascular grafts. *Nat. Biotechnol.* 18(3), 296–299.

Strathmann, H. (2001). Membrane separation processes: Current relevance and future opportunities. *AIChE J.* 47(5), 1077–1087.

Tharakan, J. P., Gallagher, S. L., and Chau, P. C. (1988). Hollow-fiber bioreactors for mammalian cell culture. *Adv. Biotechnol. Process.* 7, 153–184.

Tharakan, J. P., Lucas, A., and Chau, P. C. (1986). Hybridoma growth and antibody secretion in serum-supplemented and low protein serum-free media. *J. Immunol. Methods* 94(1–2), 225–235.

Verbeeck, R. K. (2000). Blood microdialysis in pharmacokinetic and drug metabolism studies. *Adv. Drug Deliv. Rev.* 45(2–3), 217–228.

Wan, Y., Ghosh, R., and Cui, Z. (2004). Separation of human serum albumin and human immunoglobulins using carrier phase ultrafiltration. *Biotechnol. Prog.* 20(4), 1103–1112.

Wang, D., Liu, W., Han, B., and Xu, R. (2005). The bioreactor: A powerful tool for large-scale culture of animal cells. *Curr. Pharm. Biotechnol.* 6(5), 397–403.

Washburn, N. R., Weir, M., Anderson, P., and Potter, K. (2004). Bone formation in polymeric scaffolds evaluated by proton magnetic resonance microscopy and X-ray microtomography. *J. Biomed. Mater. Res. A* 69(4), 738–747.

Waterland, L. R., Michaels, A. S., and Robertson, C. R. (1974). Theoretical model for enzymatic catalysis using asymmetric hollow fibre membranes. *AIChE J.* 20(1), 50–59.

Wickramasinghe, S. R., Garcia, J. D., and Han, B. (2002). Mass and momentum transfer in hollow fibre blood oxygenators. *J. Membr. Sci.* 208(1–2), 247–256.

Williams, J. M., Adewunmi, A., Schek, R. M., Flanagan, C. L., Krebsbach, P. H., Feinberg, S. E., Hollister, S. J., and Das, S. (2005). Bone tissue engineering using polycaprolactone scaffolds fabricated via selective laser sintering. *Biomaterials* 26(23), 4817–4827.

Wright, K. A., Nadire, K. B., Busto, P., Tubo, R., McPherson, J. M., and Wentworth, B. M. (1998). Alternative delivery of keratinocytes using a polyurethane membrane and the implications for its use in the treatment of full-thickness burn injury. *Burns* 24(1), 7–17.

Wu, Y.-S., Tsai, T.-H., Wu, T.-F., and Cheng, F.-C. (2001). Determination of pyruvate and lactate in primary liver cell culture medium during hypoxia by on-line microdialysis-liquid chromatography. *J. Chromatogr. A* 913(1–2), 341–347.

Xu, X., and Cui, Z. F. (2003). Modeling of the co-transport of cryoprotective agents in a porous medium as a model tissue. *Biotechnol. Prog.* 19(3), 972–981.

Xu, X., and Cui, Z. F. (2001). Maxwell-Stefan modelling of membrane transport in medical applications. Paper presented to *Engineering with Membranes*. June 3–6, 2001, Granada, Spain.

Xu, X., Smith, S., Urban, J., and Cui, Z. (2006). An in line non-invasive optical system to monitor pH in cell and tissue culture. *Med. Eng. Phys.* 28(6), 468–474.

Ye, H., Das, D. B., Cui, Z., and Triffitt, J. T. (2006). Modelling nutrient transport in hollow fibre membrane bioreactors for growing three-dimensional bone tissue. *J. Membr. Sci.* 272(1–2), 169–178.

Ye, H., Xia, Z., Ferguson, D. J. P., Triffitt, J. T., and Cui, Z. (2004). Hollow fibre membrane bioreactors for 3D tissue generation. *Cytotherapy* 6, 276.

Ye, H., Xia, Z., Ferguson, D. J. P., Triffitt, J. T., and Cui, Z. (2007). Studies on the use of hollow fibre membrane bioreactors for tissue generation by using rat bone marrow fibroblastic cells and a composite scaffold. *J. Mater. Sci. Mater. Med.* 18(40), 641–648.

CHAPTER 15

Biopharmaceutical Separations by Ultrafiltration

RAJA GHOSH

Department of Chemical Engineering, McMaster University, Hamilton, Ontario L8S 4L7, Canada

15.1 INTRODUCTION

The biopharmaceutical industry is a major user of membrane-based techniques, the main areas of application being biopharmaceutical purification, sterile filtration of pharmaceutical products and intermediates, endotoxin removal, water purification, and membrane-based biological analysis. The membrane-based technologies that are widely used are microfiltration, ultrafiltration, reverse osmosis, and membrane chromatography. The first ultrafiltration membranes ever used were developed for processing proteins. However, for a long time the use of membrane-based technologies was largely restricted to food processing, water purification, and environmental applications. Sterile filtration using microfiltration membranes was perhaps the only major biopharmaceutical application for quite a while. The widespread use of membranes in the biopharmaceutical industry did not really happen until recently. Membranes are now increasingly being used for a range of different applications. Membranes specifically designed for biopharmaceutical applications are now widely available, and we are likely to see a huge expansion in the use of membrane-based technologies in this sector.

The main focus in the biopharmaceutical industry in recent years has been in the area of protein-based therapeutics. Biopharmaceutical proteins such as monoclonal antibodies, plasma proteins, interleukins, interferon, growth factors, vaccines, and hyperimmune antibodies are increasingly being purified using microfiltration, ultrafiltration, and membrane chromatography (Przybycien et al., 2004). These techniques are generally less expensive, more reproducible, and more easily scalable than conventional purification techniques (van Reis and Zydney, 2001). These are also more desirable from a process validation point of view on account of the availability of relatively inexpensive disposable membrane products (Ghosh, 2002). Sterility is an important requirement in biopharmaceuticals, and by using membrane-based techniques, which are mostly based on barrier permeation, this attribute can be built into a purification process at multiple stages.

Advanced Membrane Technology and Applications. Edited by Norman N. Li, Anthony G. Fane, W. S. Winston Ho, and T. Matsuura
Copyright © 2008 John Wiley & Sons, Inc.

Ultrafiltration is traditionally used for selectively removing low-molecular-weight substances such as salts, peptide fragments, and impurities from protein solutions as well as for concentrating these solutions, that is, removing water from them. Research done in the past decade has clearly demonstrated that protein–protein fractionation using ultrafiltration is feasible (Saksena and Zydney, 1994; van Eijndhoven et al., 1995; van Reis et al., 1997; Nystrom et al., 1998; Ghosh and Cui, 1998). With the advent of new and improved membranes as well as new ways of carrying out ultrafiltration, this technique can easily compete with other high-resolution protein purification techniques (Ghosh, 2003a). For a long time sterile filtration mainly relied on microfiltration since this technique worked very well for efficient removal of bacterial and fungal contaminants. With virus removal from pharmaceutical products becoming mandatory, ultrafiltration is increasingly being used for sterile filtration since these membranes are better at retaining virus particles than microfiltration membranes (DiLeo et al., 1992). Ultrafiltration can also potentially capture infective prion proteins (van Holten et al., 2002).

Endotoxins, which are contaminants from bacteria sources, are traditionally removed from pharmaceutical products by packed-bed adsorption. Membrane adsorption, which involves the use of membrane stacks as adsorbent, is now increasingly being used for this application on account of several vital advantages over packed bed adsorption (Ghosh, 2002). When used to purify proteins, membrane adsorption is frequently referred to as membrane chromatography on account of similarities with column chromatography. Membrane chromatography is increasingly being used for protein purification, particularly where the target protein concentration is very low. This has been made possible due to the availability of a range of adsorptive membranes with different separation chemistries. Pharmaceutical-grade water is traditionally produced by distillation. With the availability of efficient reverse osmosis and microfiltration membranes, water used in the pharmaceutical industry is increasingly being produced using membranes. Another major use of membranes in the biopharmaceutical industry is for biological analysis. A range of membrane-based analytical techniques such as immunoblotting, Southern blotting, Western transfer blotting, and DNA (deoxyribonucleic acid) hybridization, all of which are based on the use of appropriate adsorptive membranes, are used for analyzing proteins and nucleic acids. Membranes are also extensively used in biological research laboratories for sample preparation and small-volume clarification.

This chapter will focus on the application of ultrafiltration in biopharmaceutical manufacturing.

15.2 ULTRAFILTRATION: AN OVERVIEW

Ultrafiltration is a pressure-driven, permeability-based membrane separation technique, mainly used for processing macromolecules such as proteins, nucleic acids, and polysaccharides. Ultrafiltration membranes have pore size in the 1- to 100-nm range. The main separation mechanism is size-based sieving, but other factors such as electrostatic solute–membrane and solute–solute interactions can significantly affect separation (Saksena and Zydney, 1994; van Eijndhoven et al., 1995; van Reis et al., 1997; Nystrom et al., 1998; Ghosh and Cui, 1998). Hence, the still widely held view of ultrafiltration as a purely size exclusion process is incorrect. In order to use ultrafiltration as an efficient bioseparation technique, all factors influencing solute transmission through the membrane need to be taken into account and properly utilized. The significant role of non-size-based factors in ultrafiltration tremendously increases its scope for application. However, the fact that separation depends on a large number of factors means that process optimization is essential (Ghosh, 2003a).

15.3 BASIC WORKING PRINCIPLES OF ULTRAFILTRATION

There are two main issues in an ultrafiltration process: productivity and selectivity. Productivity is quantified in terms of the permeate flux, this being defined as the permeation rate per unit membrane surface area. Factors that affect permeate flux are solute type, solute concentration, membrane type, solution pH, solution ionic strength, applied pressure (also called the transmembrane pressure), and the hydrodynamic conditions on the feed side. The volumetric permeate flux, which is the volume of permeate collected per unit time per unit membrane area is given by the following generalized equation, based on a resistance model:

$$J_v = \frac{\Delta P - \Delta\pi}{\mu(R_m + R_{\text{cp}} + R_g)} \tag{15.1}$$

Another equation, which is also widely used to obtain permeate flux in ultrafiltration, is the concentration polarization equation (Blatt, 1976):

$$J_v = k \ln\left(\frac{C_w - C_p}{C_b - C_p}\right) \tag{15.2}$$

High permeate flux can be achieved by optimizing all the parameters affecting permeate flux. Generally speaking, reducing solute accumulation near the membrane surface (also referred to a concentration polarization) and reducing adsorption and deposition of solutes and other feed components on the membrane (also referred to as membrane fouling) helps to maintain a high permeate flux. The following techniques have also been known to help achieve high permeate flux in ultrafiltration:

1. Periodic backflushing of the membrane using the permeate (Rodgers and Spark, 1992)
2. Creation of turbulence on the feed side by using flow inserts and baffles (Bellhouse et al., 2001)
3. Introduction of gas bubbles into the membrane module for membrane cleaning and creation of turbulence (Cui and Wright, 1994)

The selectivity of an ultrafiltration membrane depends on its ability to transmit different solutes to different extents. Factors that affect solute transmission are solute type, membrane type, solution pH, solution ionic strength, the permeate flux, and the hydrodynamic conditions on the feed side. The transport of a solute through an ultrafiltration membrane is best quantified in terms of the apparent sieving coefficient:

$$S_a = \frac{C_p}{C_b} \tag{15.3}$$

The apparent sieving coefficient depends on the permeate flux, the solute mass transfer coefficient, and a fundamental solute–membrane parameter called the intrinsic sieving coefficient:

$$\ln\left(\frac{S_a}{1-S_a}\right) = \ln\left(\frac{S_i}{1-S_i}\right) + \frac{J_v}{k} \tag{15.4}$$

For a solute passing through an ultrafiltration membrane without any physicochemical interaction with it other than hydrodynamic drag, the sieving coefficient can be estimated using the following equation (Opong and Zydney, 1991):

$$S_i = \frac{S_\infty \exp\left(\frac{S_\infty J_v \delta_m}{D}\right)}{S_\infty + \exp\left(\frac{S_\infty J_v \delta_m}{D}\right) - 1} \tag{15.5}$$

Where there is physicochemical interaction between the solute and the membrane, the intrinsic sieving coefficient can be obtained by (Zydney and van Reis, 2001)

$$S_i^* = S_i \exp\left(-\frac{E}{KT}\right) \tag{15.6}$$

The selectivity of a separation process is usually expressed in terms of the selectivity parameter:

$$\psi = \frac{S_{aT}}{S_{aR}}$$

The selectivity of an ultrafiltration process can be increased by optimizing the factors affecting solute transmission, mainly solution pH, ionic strength, permeate flux, and hydrodynamic conditions on the feed side.

15.4 ULTRAFILTRATION MEMBRANES AND DEVICES

Membrane manufacturers use a term called the molecular weight cutoff (or MWCO) for specifying the solute retention property of ultrafiltration membranes. The MWCO of a membrane refers to the molecular weight of the solute, which will have an apparent sieving coefficient of 0.1 when ultrafiltered using this membrane. This type of rating is quite arbitrary since there is no industrywide unanimously agreed upon testing protocol for determining MWCO. Generally, synthetic polymers such as polyethylene glycol (PEG) and natural polymers such as dextran are used as solutes for MWCO determination. However, two different solutes having the same molecular weight will not necessarily have the same sieving coefficient with respect to a given membrane. Equation (15.6) clearly shows that an interacting solute will have different sieving behavior relative to a noninteracting solute of the same molecular weight. Moreover, as shown in Eq. (15.4) and (15.5), the sieving coefficient of a particular solute will be affected by both the permeate flux and the mass transfer coefficient. The sieving coefficients of proteins are also significantly affected by solution pH and ionic strength (Saksena and Zydney, 1994;

van Eijndhoven et al., 1995; van Reis et al., 1997; Nystrom et al., 1998; Ghosh and Cui, 1998). It is therefore perhaps better to rate ultrafiltration membranes in terms of their pore sizes. Some membrane manufacturers are beginning to provide pore size data along with the MWCO rating.

Membranes used in the biopharmaceutical industry are usually made using polymers. Ultrafiltration membranes were traditionally made using cellulose and its derivatives. The main advantage of using cellulose-based membranes is that they are hydrophilic and hence tend to be less susceptible to membrane fouling. However, cellulose-based membranes can only be used over a limited pH range and can degrade at conditions used for sterilization. Hence, ultrafiltration membranes made of synthetic polymers are more commonly used these days. The two most commonly used polymers for making ultrafiltration membranes are polysulfone (PS) and polyethersulfone (PES). Other substances such as polyvinylidine fluoride (PVDF), polyamide (PA), and polyacrylonitrile (PAN) are also used to make membranes. Membranes made from synthetic polymers tend to be hydrophobic and need to be surface treated to make them hydrophilic, and thus less susceptible to fouling. Membranes used in biopharmaceutical manufacturing need to be tested for leachables, which should be below specified limits. These membranes should also be sterilizable. Temperature and steam-based sterilization of equipment is widespread in the industry, and hence membranes should ideally be able to withstand high temperatures for extended durations. Where this is not possible, membranes are sterilized using chemical agents such as ethylene oxide. Organic solvents such as ethanol are frequently used for sterilizing membranes. Therefore, stability in the presence of solvents is also a highly desirable property. Fouled membranes are usually cleaned using alkali (mainly sodium hydroxide), hypochlorite, detergents, and proprietary cleaning formulations. The membranes should therefore be able to withstand these cleaning agents.

The term *membrane element* refers to the basic form in which a membrane is produced. There are three types of membrane elements: flat sheets, tubular membranes, and hollow-fiber membranes. Flat-sheet and hollow-fiber membranes are more commonly used in biopharmaceutical manufacturing. A flat-sheet membrane, as the name suggests, looks like a sheet of paper. A hollow-fiber membrane looks like a fine tube, typically a millimeter or less in diameter. The wall of the hollow fiber acts as the membrane. The device within which a membrane element is housed is called the *membrane module*. The stirred cell ultrafiltration modules is the workhorse of laboratory-scale ultrafiltration. Figure 15.1 shows custom-designed stirred cell modules. These devices are fitted with flat sheet membrane disks. Table 15.1 lists some commercial flat-sheet ultrafiltration membranes available in the form of disks along with their properties. Many ultrafiltration processes are developed at the laboratory scale using stirred cell modules. The results obtained at this scale are usually translated to large-scale ultrafiltration through scale-independent parameters such as solution pH, ionic strength, transmembrane pressure, permeate flux, shear rate, and mass transfer coefficient.

Tangential-flow or cross-flow membrane modules are mainly used for biopharmaceutical manufacturing. This involves the flow of the feed solution parallel to the membrane surface (see Fig. 15.2). Using such flow arrangement, both concentration polarization and membrane fouling can be minimized. The basic flat-sheet membrane-based tangential-flow module consists of a shallow channel with rectangular flat-sheet membrane(s) on one or both sides of the channel. The feed is pumped into the channel from one end and the retentate is removed from the other. The transmembrane pressure is usually created by using a valve on the retentate line. In some cases, the permeate is drawn through the membrane

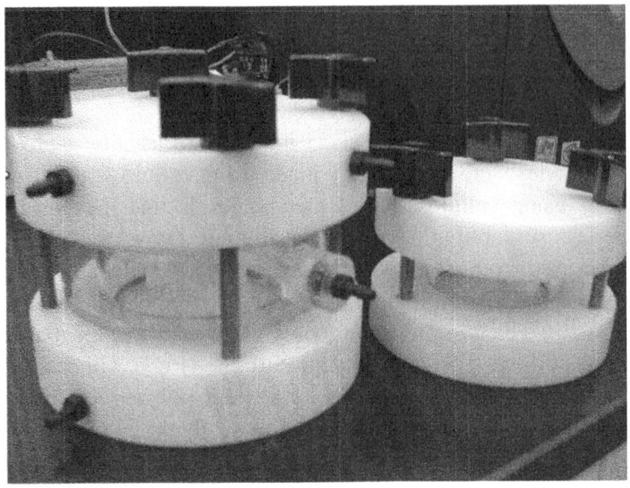

Figure 15.1 Stirred cell ultrafiltration modules.

TABLE 15.1 Ultrafiltration Membrane Disks[a]

Name	Membrane Material	Membrane Diameter (mm)	MWCO Availability (kDa)	Manufacturer
Omega	Modified polyethersulfone	25, 43, 47, 50, 62, 76, 90,150	1, 3, 5, 10, 30, 50, 100, 300	Pall
Biomax	Modified polyethersulfone	25, 44.5, 47, 63.5, 76, 90, 150	5, 8, 10, 30, 50, 100, 300, 500, 1000	Millipore
Ultracel-PL	Regenerated cellulose	25, 44.5, 47, 63.5, 76, 90, 150	1, 3, 5, 10, 30, 100	Millipore
Ultracel-YM	Regenerated cellulose	25, 44.5, 47, 63.5, 76, 90, 150	10, 30, 100	Millipore

[a]Data provided by manufacturers: Pall Corporation, www.pall.com; Millipore Corporation, www.millipore.com.

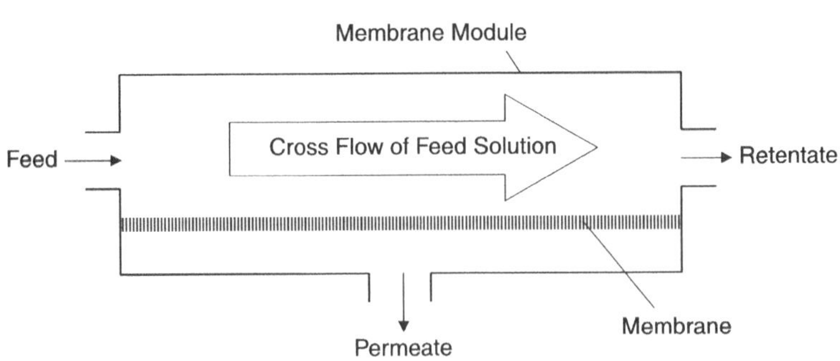

Figure 15.2 Tangential-flow ultrafiltration.

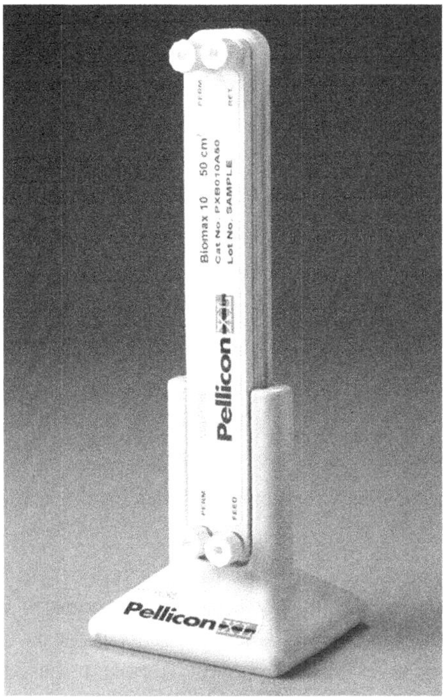

Figure 15.3 Small-scale tangential-flow ultrafiltration module. (Courtesy Millipore Corporation.)

by using a permeate pump, that is, by applying suction on the permeate side. The membranes are supported on grooved plates that facilitate collection of the permeate. Figure 15.3 shows a small-scale flat-sheet tangential-flow membrane module. Such devices are used to process up to 1500 mL of feed solution. For intermediate-scale processes, tangential-flow membrane modules, still based on the same basic design but having larger membrane areas are used. Figure 15.4 shows an intermediate-scale

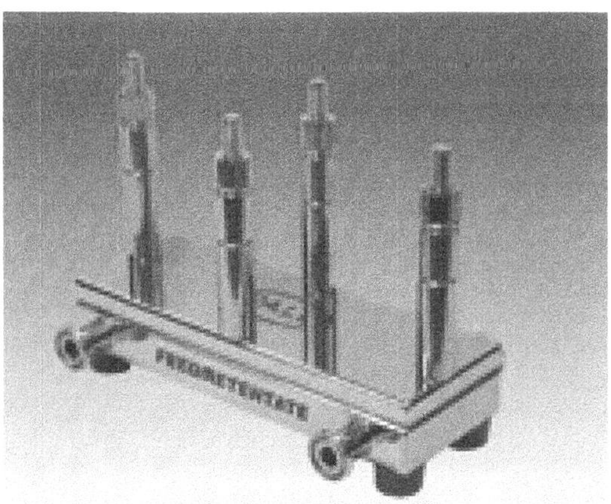

Figure 15.4 Medium-scale tangential-flow ultrafiltration module. (Courtesy Pall Corporation.)

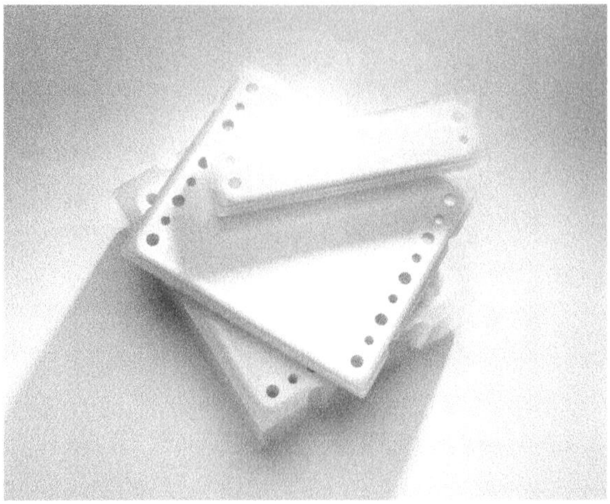

Figure 15.5 Tangential-flow ultrafiltration cassette. (Courtesy Millipore Corporation.)

flat-sheet tangential-flow membrane module that can process up to 15 L of feed solution. Most large-scale flat-sheet tangential-flow membrane modules are based on the use of membrane cassettes. A cassette (see Fig. 15.5) usually consists of several basic flat-sheet membrane units connected in series or in parallel. A cassette-based membrane module typically consists of several cassettes connected either in series or in parallel. Figure 15.6 shows one such assembled membrane module that can be used to process up to 150 L of feed solution. Table 15.2 lists some commercial flat-sheet tangential-flow membrane modules along with their specifications. The main advantages of the flat-sheet tangential-flow membrane modules are the ease of cleaning, the possibility of replacing defective membrane elements, and the ability to handle viscous feeds and suspensions. The main drawbacks are the low membrane area to module volume ratio and the high holdup volume of the feed side.

Hollow-fiber membranes are usually operated in the cross-flow mode with the feed entering one end of the fiber, the retentate leaving the other, and the permeate being forced through the fiber wall. Such a hollow-fiber membrane is called the inside-out type.

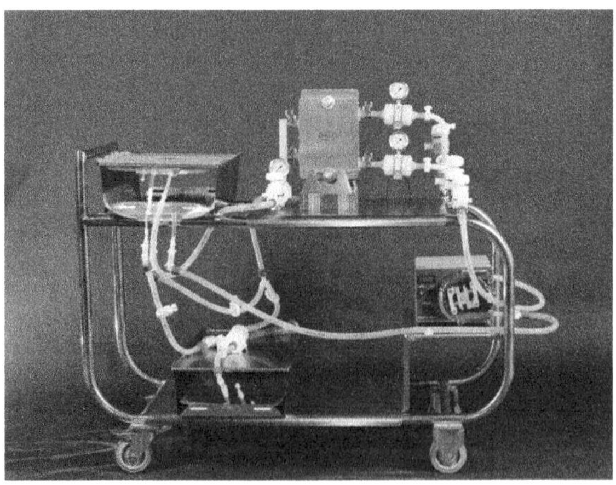

Figure 15.6 Tangential-flow ultrafiltration system. (Courtesy Pall Corporation.)

TABLE 15.2 Tangential-Flow Ultrafiltration Modules/Cassettes Suitable for Biopharmaceutical Processing[a]

Name	Membrane Material	Housing Material	Membrane Area (m^2)	MWCO Availability (kDa)	Manufacturer
Minimate TFF	Omega (modified, polyethersulfone)	Polypropylene	0.005	0.65, 1, 3, 5, 10, 30, 50, 70, 100, 300, 500, 1000	Pall
Centramate	Omega (modified polyethersulfone)	Polyester, polyolefin, polyurethane	0.09/cassette	1, 3, 5, 8, 10, 30, 50, 70, 100, 300, 500, 1000	Pall
Centrasette	Omega (modified polyethersulfone)	Polyester, polyolefin, polyurethane	0.46/cassette	1, 3, 5, 8, 10, 30, 50, 70, 100, 300, 500, 1000	Pall
Maximate	Omega (modified polyethersulfone)	Polyester, polyolefin, polyurethane	0.19/cassette	1, 3, 5, 8, 10, 30, 50, 70, 100, 300, 500, 1000	Pall
Maxisette	Omega (modified polyethersulfone)	Polyester, polyolefin, polyurethane	2.79/cassette	1, 3, 5, 8, 10, 30, 50, 70, 100, 300, 500, 1000	Pall
Pellicon XL 50	Biomax (polyethersulfone), Ultracel (regenerated cellulose)	Polypropylene	0.005	Biomax: 5, 8, 10, 30, 50, 100, 300, 500, 1000 Ultracel: 5, 10, 30, 300, 1000	Millipore
Pellicon 2	Biomax (polyethersulfone), Ultracel (regenerated cellulose)	Cassettes: polyurethane Cassette holder: stainless steel	0.11/cassette, 0.53/cassette 2.0/cassette, 2.65/cassette	Biomax: 5, 8, 10, 30, 50, 100, 300, 500, 1000 Ultracel: 5, 10, 30, 300, 1000	Millipore
Prostak UF	Polyethersulfone, regenerated cellulose	Plate: PVDF Holder: stainless steel	0.39, 0.93, 1.9	Polyethersulfone: 10, 30, 100, 300 Regenerated cellulose: 1, 3, 5, 10, 30, 100	Millipore

(*Continued*)

TABLE 15.2 *Continued*

Name	Membrane Material	Housing Material	Membrane Area (m^2)	MWCO Availability (kDa)	Manufacturer
Pellicon 3	Regenerated cellulose	Polypropylene, polyethylene, thermoplastic elastomer	0.0088/cassette, 0.11/cassette, 0.57/cassette, 1.14/cassette	10, 30	Millipore
ULTRAN Lab	Polyethersulfone	Silicon/stainless steel	0.07/cassette	5, 10, 30, 50, 100, 300, 500	Whatman
ULTRAN Pilot	Polyethersulfone	Silicon/stainless steel	0.45/cassette	5, 10, 30, 50, 100, 300, 500	Whatman
ULTRAN ProSlice	Polyethersulfone	Silicon/stainless steel	0.2/OC cassette, 0.23/SC cassette	5, 10, 30, 50	Whatman

[a]Data provided by manufacturers: Pall Corporation, www.pall.com; Millipore Corporation, www.millipore.com; Whatman, www.whatman.com.

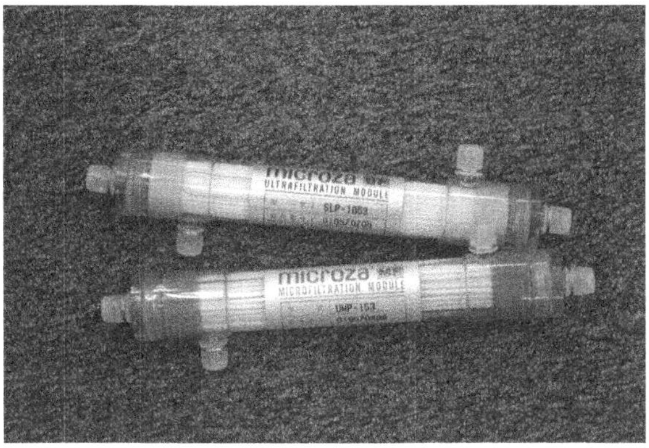

Figure 15.7 Hollow-fiber ultrafiltration module. (Courtesy Asahi Kasei Corporation.)

Outside-in-type hollow fibers are also available, but these are rarely used in the biopharmaceutical industry. The hollow-fiber membrane module is similar in design to a shell-and-tube heat exchanger. Hundreds of hollow-fiber membranes are bundled together, potted at both ends, and housed within a tubelike module. Figure 15.7 shows a typical hollow-fiber membrane module. These are available in different sizes. The membrane area can be increased by increasing the fiber length or simply by increasing the number of fibers bundled in parallel. Table 15.3 lists some commercial hollow-fiber membrane modules along with their specifications. The main advantages of hollow-fiber membrane modules include low pumping cost, high membrane area to module volume ratio, and low dead volume on the feed side. The main disadvantages include the possibility of fiber breakage and their tendency to get blocked by particles.

TABLE 15.3 Hollow-Fiber Ultrafiltration Modules Suitable for Biopharmaceutical Manufacturing[a]

Name	Membrane Material	Housing Material	Membrane Area (m^2)	MWCO Availability (kDa)	Manufacturer
Microza AV	Polyacrylonitrile	PVC, epoxy (potting)	3.1, 4.7, 7.8, 12.3	6, 13, 50, 80	Asahi Kasei
Microza SV	Polysulfone	PVC, epoxy (potting)	5.6, 12	3, 4, 6, 10	Asahi Kasei
Microza AP	Polyacrylonitrile	Polysulfone, epoxy (potting)	0.017, 0.19, 1, 4.7	6, 13, 50, 80	Asahi Kasei
Microza SP	Polysulfone	Polysulfone, epoxy (potting)	0.017, 0.2, 1, 4.7	3, 4, 6, 10	Asahi Kasei
Microza VP	Polysulfone	Polysulfone, epoxy (potting)	4.7	6	Asahi Kasei

[a]Data provided by manufacturer: Asahi Kasei Corporation, www.asahi-kasei.co.jp/.

15.5 ULTRAFILTRATION PROCESSES

Ultrafiltration is mainly used for the following types of separation:

1. Protein concentration: removal of water from protein solutions
2. Desalting: removal of salts and low-molecular-weight impurities from protein solutions
3. Fractionation: separation of one protein from another
4. Clarification: removal of particles from protein solutions

In concentration and desalting processes, the membrane should totally retain the protein, that is, its S_a should be equal to zero, while at the same time the membrane should allow high water flux as well as unhindered transport of the low-molecular-weight species. The type of membrane used for a particular application depends on the protein being concentrated or desalted. By the rule of thumb, the membrane MWCO should be one fifth to one tenth of the molecular weight of the protein being processed in order to ensure its total retention. In a fractionation process, the relative differences in sieving coefficients of different proteins are utilized for their separation. Fractionation is significantly more challenging than both concentration and desalting and needs very precise fine-tuning. The main objective in a clarification process is the retention of particulate matter. However, it is also desirable to have the highest possible protein transmission through the membrane. These objectives are usually at conflict with each other and if a choice has to be made, it is in favor of particle retention.

Specialized types of membranes are used in clarification processes. Figure 15.8 shows commercial hollow-fiber membrane modules suitable for filtering out virus particles. The efficiency of virus removal is measured in terms of the log removal value (LRV), which is the logarithm to the base 10 of the ratio of the particle concentration in the feed divided by that in the permeate. Virus filters are usually rated in terms of their pore sizes (these being in the nanometer range), thus clearly indicating their main objective, that is, particle retention. However, a good clarification membrane should allow greater than 95% protein transmission (i.e., S_a should be greater than 0.95). Table 15.4 lists some commercial virus filters with their specifications.

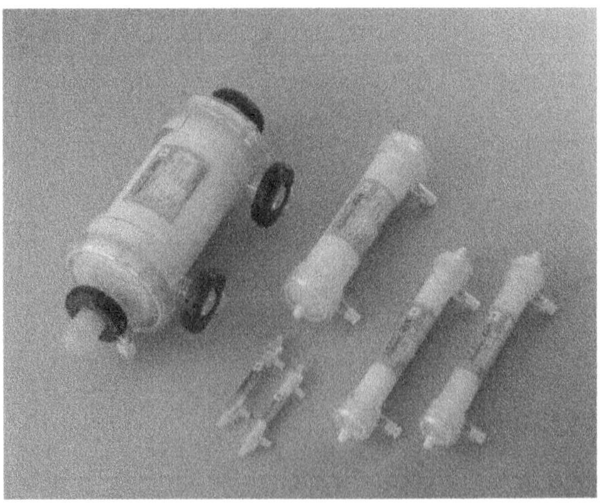

Figure 15.8 Hollow-fiber membrane modules for virus removal. (Courtesy Asahi Kasei Corporation.)

TABLE 15.4 Virus Removal Membrane Modules[a]

Name	Membrane Material	Membrane Module Type	Membrane Area (m^2)	Pore Size (nm)	LRV Rating with Test Virus	Manufacturer
Planova 15N	Hydrophilic cuprammonium regenerated cellulose	Hollow fiber	0.001, 0.01, 0.12, 0.3, 1.0, 4.0	15 ± 2	Porcine parvovirus >4.4 Poliovirus >7.8	Asahi Kasei Medical Co., Ltd.
Planova 20N	Hydrophilic cuprammonium regenerated cellulose	Hollow fiber	0.001, 0.01, 0.12, 0.3, 1.0, 4.0	19 ± 2	Porcine parvovirus >4.2 Encephalomyocarditis >5.4	Asahi Kasei Medical Co., Ltd.
Planova 35N	Hydrophilic cuprammonium regenerated cellulose	Hollow fiber	0.001, 0.01, 0.12, 0.3, 1.0, 4.0	35 ± 2	Bovine viral diarrhea virus >5.9 Human immunodeficiency virus >7.3	Asahi Kasei Medical Co., Ltd.
Planova 75N	Hydrophilic cuprammonium regenerated cellulose	Hollow fiber	0.001, 0.01, 0.3, 1.0	72 ± 4	Used as prefilter	Asahi Kasei Medical Co., Ltd.
Viresolve 70	Hydrophilic PVDF	Flat-sheet TF	1-stack 0.015 6-stack 0.1 10-stack 0.7 20-stack 1.4		Polio virus >3.5 Simian virus (SV-40) >5.6 Sindbis >7.4 Reovirus 3 >7.2 Murine leukemia virus >6.6 Human immunodeficiency virus >8.5	Millipore
Viresolve 180	Hydrophilic PVDF	Flat-sheet TF	1-stack 0.015 6-stack 0.1 10-stack 0.7 20-stack 1.4		Polio virus >2 Hepatitis A >4 Sindbis >4.5 Reovirus 3 >5.5 Murine leukemia virus >6	Millipore

[a]Data provided by manufacturers: Planova Division, Asahi Kasei Medical Co., Ltd., www.PlanovaFilters.com; Millipore Corporation, www.millipore.com.

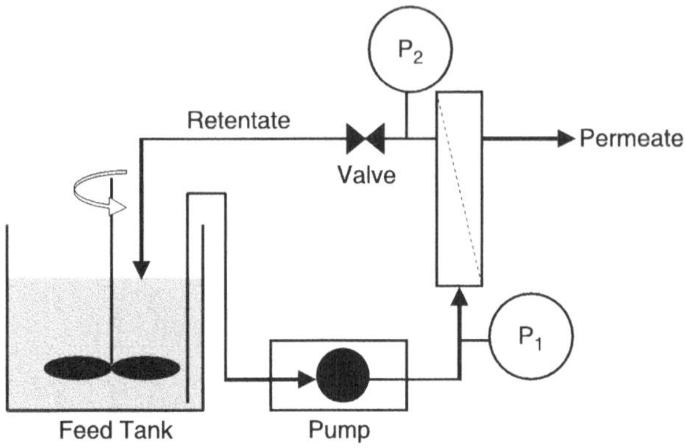

Figure 15.9 Batch clarification/concentration.

Batchwise operation is still widely used in the biopharmaceutical industry. The main reason for this is the small process volume, which is typically hundreds of liters at the most. Another reason for batchwise operation is that most biological processes such as fermentation and cell culture are carried out in batch mode. Batch concentration or clarification is usually carried out using the configuration shown in Figure 15.9. In batch concentration, the retentate is the product while in batch clarification, the permeate is the product. Batch desalting or fractionation is usually carried out using the configuration shown in Figure 15.10. As mentioned earlier, fractionation is significantly more demanding than other types of separation. The configuration shown in Figure 15.10 is not particularly well suited for a high-resolution fractionation process. Several researchers have suggested alternative ways of carrying out ultrafiltration that are better suited for fractionation (van Reis et al., 1997; Ghosh, 2003b). Also, a protein fractionation process needs to be optimized, taking into account the several factors that affect separation. Some recently developed techniques based on pulsed sample injection in the ultrafiltration device allow rapid optimization of membrane-based fractionation (Ghosh and Cui, 2000; Ghosh et al., 2003).

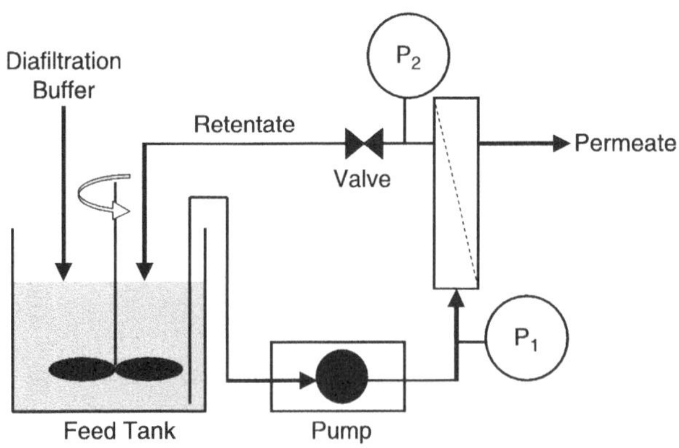

Figure 15.10 Batch desalting/fractionation.

15.6 CONCLUSION

Membranes are increasingly being used in the biopharmaceutical industry for a range of different applications. The main focus in the biopharmaceutical industry in recent years has been on the development and manufacture of protein-based therapeutic products. Therapeutic proteins such as monoclonal antibodies, plasma proteins, interleukins, interferon, growth factors, vaccines, and hyperimmune antibodies are processed using membrane-based techniques such as microfiltration, ultrafiltration, and membrane chromatography Ultrafiltration is primarily used for protein concentration, desalting, and clarification. In recent years there has been a significant amount of research work done on the use of ultrafiltration for protein fractionation, the overwhelming conclusion being that ultrafiltration can indeed be used for protein–protein separation. However, this would be possible only under highly optimized conditions. A huge range of commercial ultrafiltration membranes specifically developed for biopharmaceutical applications is now available. Configurations currently used in the industry are suitable for carrying out concentration, desalting, and clarification, but these are not suitable for fractionation. Several new configurations suitable for protein fractionation have been proposed in recent years. Also, techniques for rapidly optimizing fractionation processes are now available.

Symbols List

C_b	solute concentrate in feed (kg m^{-3})
C_p	solute concentration in permeate (kg m^{-3})
C_w	solute concentration on membrane surface (kg m^{-3})
D	diffusivity of the solute within the membrane (m^2 s^{-1})
E	activation energy (J)
J_v	volumetric permeate flux (m^3 m^{-2} s^{-1})
k	solute mass transfer coefficient (m s^{-1})
K	Boltzmann constant (J K^{-1})
ΔP	transmembrane pressure (Pa)
R_{cp}	resistance due to concentration polarization layer (m^{-1})
R_g	gel-layer resistance (m^{-1})
R_m	membrane resistance (m^{-1})
S_a	apparent sieving coefficient
S_{aR}	apparent sieving coefficient of retained solute
S_{aT}	apparent sieving coefficient of transmitted solute
S_i	intrinsic sieving coefficient of a noninteracting solute of same size
S_i^*	intrinsic sieving coefficient of interacting solute
S_∞	asymptotic sieving coefficient
T	absolute temperature (K)

Greek Symbols

μ	viscosity of permeate (kg m^{-1} s^{-1})
$\Delta\pi$	osmotic back pressure (Pa)
δ_m	thickness of the membrane (m)
Ψ	selectivity parameter

REFERENCES

Bellhouse, B. J., Costigan, G., Abhinava, K., and Merry, A. (2001). The performance of helical screw-thread inserts in tubular membranes. *Sep. Purif. Technol.* 22, 89.

Blatt, W. F. (1976). Principles and practice of ultrafiltration. In P. Meares (Ed.), *Membrane Separation Processes.* Elsevier Science, Amsterdam.

Cui, Z. F., and Wright, K. I. T. (1994). Gas-liquid 2-phase cross-flow ultrafiltration of BSA and dextran solutions. *J. Membr. Sci.* 90, 183.

DiLeo, A. J., Allegrezza, Jr., A. E., and Builder, S. E. (1992). High-resolution removal of virus from protein solutions using a membrane of unique structure. *Bio/Technol.* 10, 61.

Ghosh, R. (2002). Protein separation using membrane chromatography: Opportunities and challenges. *J. Chromatogr. A* 952, 13.

Ghosh, R. (2003a). *Protein Bioseparation Using Ultrafiltration: Theory, Applications and New Developments.* Imperial College Press, London.

Ghosh, R. (2003b). Novel cascade ultrafiltration configuration for continuous, high-resolution protein-protein fractionation: A simulation study. *J. Membr. Sci.* 226, 85.

Ghosh, R., and Cui, Z. F. (1998). Fractionation of BSA and lysozyme using ultrafiltration: Effect of pH and membrane pretreatment. *J. Membr. Sci.* 139, 17.

Ghosh, R., and Cui, Z. F. (2000). Analysis of protein transport and polarization through membranes using pulsed sample injection technique. *J. Membr. Sci.* 175, 75.

Ghosh, R., Wan, Y., Cui, Z. F., and Hale, G. (2003). Parameter scanning ultrafiltration: Rapid optimization of protein separation. *Biotechnol. Bioeng.* 81, 673.

Nystrom, M., Aimar, P., Luque, S., Kulovaara, M., and Metsamuuronen, S. (1998). Fractionation of model proteins using their physiochemical properties. *Colloids Surf. A.* 138, 185.

Opong, W. S., and Zydney, A. L. (1991). Diffusive and convective protein-transport through asymmetric membranes. *AIChE J.* 37, 1497.

Przybycien, T. M., Pujar, N. S., and Steel, L. M. (2004). Alternative bioseparation operations: Life beyond packed-bed chromatography. *Curr. Opin. Biotechnol.* 15, 469.

Rodgers, V. G. J., and Spark, R. E. (1992). Effect of transmembrane pressure pulsing on concentration polarization. *J. Membr. Sci.* 68, 149.

Saksena, S., and Zydney, A. L. (1994). Effect of solution pH and ionic strength on the separation of albumin and immunoglobulin (IgG) by selective filtration, *Biotechnol. Bioeng.* 43, 960.

van Eijndhoven, R. H. C. M., Saksena, S., and Zydney, A. L. (1995). Protein fractionation using electrostatic interactions in membrane filtration. *Biotechnol. Bioeng.* 48, 406.

van Holten, R. W., Autenrieth, S., Boose, J. A., Hsieh, W.T., and Dolan, S. (2002). Removal of prion challenge from an immune globulin preparation by use of a size exclusion filter. *Transfusion* 42, 999.

van Reis, R., and Zydney, A. (2001). Membrane separations in biotechnology. *Curr. Opin. Biotechnol.* 12, 208.

van Reis, R., Gadam, S., Frautschy, L. N., Orlando, S., Goodrich, E. M., Saksena, S., Kuriyel, R., Simpson, C. M., Pearl, S., and Zydney, A. L. (1997). High performance tangential flow filtration. *Biotechnol. Bioeng.* 56, 71.

Zydney, A. L., and van Reis, R. (2001). High-performance tangential-flow filtration. In W. K. Wang (Ed.), *Membrane Separation in Biotechnology.* Marcel Dekker, New York.

CHAPTER 16

Nanofiltration in Organic Solvents

P. SILVA, L. G. PEEVA, and A. G. LIVINGSTON

Department of Chemical Engineering, Imperial College, London SW7 2BY, United Kingdom

16.1 ORGANIC SOLVENT NANOFILTRATION MEMBRANES

The industrial development of membranes dates back to the 1960s with the implementation of the first water desalination plants based on reverse osmosis technology. Nowadays membranes play a key role in modern life and medicine. However, their liquid-phase applications are still mainly limited to separations in aqueous streams. The development of nanofiltration membranes for use in organic solvents has attracted much attention as they have potential applications in diverse areas, such as organometallic catalyst separation, solvent exchange, lube oil dewaxing, and the like. This creates a new opportunity for membrane technology, and a large expansion of membrane processes is expected in unexploited areas such as petroleum, electrochemical, pharmaceutical, and fine chemical synthesis. Unfortunately, currently available organic solvent nanofiltration (OSN) membranes have limited solvent compatibility and lifetime, and the central challenge for membrane technology is to produce membranes that are functionally stable in a broad range of organic solvents and at a variety of temperatures. Thus, the search for efficient and cost-effective solvent and thermally resistant materials continues. Both polymeric and inorganic materials have been used for the preparation of OSN membranes. In what follows we will present a brief summary of the currently commonly available OSN membranes.

16.1.1 Polymeric Membranes

Most polymeric OSN membranes have an asymmetric structure and are porous with a dense top layer. This asymmetry can be divided into two major types: the integral type, where the whole membrane is composed of the same material, and the thin-film composite (TFC), where the membrane separating layer is made of a different material.

Polymeric membranes generally fail to maintain their physical integrity in organic solvents because of their tendency to swell or dissolve. This is a major drawback since nonaqueous processes generally require polymers that are rigid and crystalline, thermally

Advanced Membrane Technology and Applications. Edited by Norman N. Li, Anthony G. Fane, W. S. Winston Ho, and T. Matsuura
Copyright © 2008 John Wiley & Sons, Inc.

stable, resistant to compaction, inert and nonswollen by solvents, and stable over long usage. Nevertheless, there are several polymeric materials that exhibit satisfactorily solvent resistance. McCarthy (1990, p. 147) states that "the chemical resistance of a polymeric material is its ability to withstand chemical attack with minimal change in appearance, dimensions, mechanical properties, and weight over a period of time" and is often interconnected with the thermal stability, that is, factors that favor thermal stability often favor chemical stability as well. As a general rule, the higher the glass transition temperature, the more rigid the polymer, and the higher its stability in solvents. Usually, the factors that promote chemical stability are (Gebben, 1988):

- Aromatic or heterocyclic backbone structures, that is, the presence of resonance structures
- Absence of "reactive" groups such as unsaturated bonds, —OH groups, free —NH groups, aliphatic groups
- Presence of high bond energies that cause strong chemical bonds, for example, C—F, C—Si, and C—P
- Polybonding: atoms are linked to the polymer chain with two or more bonds, which implies that chains cannot be broken by the rupture of one single bond, for example, in ladder polymers

Some examples of classes of highly resistant polymers are presented in Table 16.1 (Critchley et al., 1983).

16.1.1.1 Integral Asymmetric Polymeric Membranes Integral asymmetric polymeric membranes are prepared by the phase inversion immersion precipitation process. For this purpose, a solution of the polymer is cast as a thin film (usually on a nonwoven fabric), dried for a few seconds to create a dense top layer, and immersed in a coagulation bath that contains a nonsolvent for the polymer. The solvent starts to diffuse out of the homogeneous liquid polymer film, whereas the nonsolvent diffuses into the film. Due to the presence of nonsolvent, phase separation takes place in the polymer film, and the polymer precipitates as a solid phase to form a porous asymmetric membrane structure. The thermodynamic properties of the casting system and the kinetics involved in the exchange of solvent and nonsolvent affect the morphology of the membrane and consequently its permeability and solvent rejection (Park et al., 2000). More detailed information about membrane preparation techniques can be found elsewhere (Mulder, 1997).

Few of the highly resistant polymers mentioned in Section 16.1.1 are suitable for producing integral asymmetric OSN membranes, mainly due to the fact that to perform

TABLE 16.1 Classes of Highly Resistant Polymers

Polymer Class	Stability Promoted by	Examples
Thermosetting	High cross-linking density	Phenol-formaldehyde resin
Fluorinated	Strong chemical bonds	Polytetrafluorethylene
Inorganic	Strong chemical bonds	Polyphosphazene, polysiloxane
Aromatic	Resonance stability	Polyphenylene, aromatic polyamide
Heterocyclic	Resonance stability	Polyimide, polybenzimidazole
Ladder polymers	Polybonding, resonance	Polypyrrone

the phase inversion the polymer needs to be soluble in at least one solvent. Some soluble commercially available polymers are polysulfone (PSF), polyethersulfone (PES), polyacrylonitrile (PAN), polyvinylidenefluoride (PVDF), polyetherimide (PEI), and BTDA-based polyimide (PI) (BTDA = benzophenone tetracarboxylic dianhydride). Several researchers have compared the chemical resistance of these polymers, and the main conclusions are summarized in Table 16.2 (Beerlage, 1994).

The information presented in Table 16.2 suggests PAN and PI are solvent-resistant polymers. Both polymers can be processed by the phase inversion technique. Polyimides are commonly used for integral asymmetric nanofiltration (NF) membrane preparation and will be addressed in this section. PAN is mainly associated with the preparation of integral asymmetric ultrafiltration (UF) membranes and consequently serves as a support for composite NF membranes to be discussed in the next chapter. PIs are made by the reaction of diamines (DA) with dianhydrides to form the soluble polymer precursor known as poly(amic acid). This can then be processed into a useful shape by converting it into the final PI by cyclodehydration of the amic acid (imidization). Figure 16.1 shows the mechanism for the pyromellitic dianhydride (PMDA) case.

Some of the most common commercially available PIs used for forming integral asymmetric polymeric OSN membranes are shown in Table 16.3. There are several published examples of OSN membranes based on PIs: Strathmann (1978) developed solvent-stable membranes based on PDMA-ODA (4,4′-oxydianiline) (by reacting benzidine with pyromellitic anhydride). For increased solvent/thermal resistance these membranes were cyclized with *N,N*-dicyclohexylcarbodiimide. After all processing steps the membranes were able to withstand 50 days exposure to dichloromethane and cyclohexane without loss of mechanical stability. Alegranti (1978), developed OSN membranes by a similar procedure differing only in the cyclization step. At 5.5 MPa these membranes could perform hexane/ethanol (50%/50%) mixture separation (75% ethanol in the permeate).

Matrimid-5218-based OSN membranes were developed for lube oil separation (White et al., 1993) and were able to achieve 96% lube oil rejection and 12.9 L/m^2/h solvent flux for a blend of methyl ethyl ketone (MEK) and toluene at 4.1 MPa and 263 K. These are the first OSN membranes employed in a commercial plant installation at Mobil's Beaumont, Texas, refinery.

TABLE 16.2 Chemical Resistance of Soluble Commercially Available Polymers

Solvent Class	PSF	PES	PAN[a]	PVDF	PEI	PI[b]
Alcohols	+	+	+/−	+	+	+
Aliphatic hydrocarbons	+	+	+	+	+	+
Aromatic hydrocarbons	−	+	+	+	−	+
Ethers	−	+/−[c]	+	+	+/−[c]	+
Esters	−	−	+	−	+	+
Ketones	−	−	+	−	+	+
Aliphatic acids	+	+	+	+	−	+/−[d]
Amides	−	−	−	−	−	−
Halogenated hydrocarbons	−	−	+	+	−	+

[a]Pure polyacrylonitrile.
[b]Soluble BTDA-copolyimide: P84/PI 2080.
[c]Not stable in tetrahydrofuran.
[d]Stable in aqueous solutions; not stable in concentrated acid.
+ = stable, no visible change.

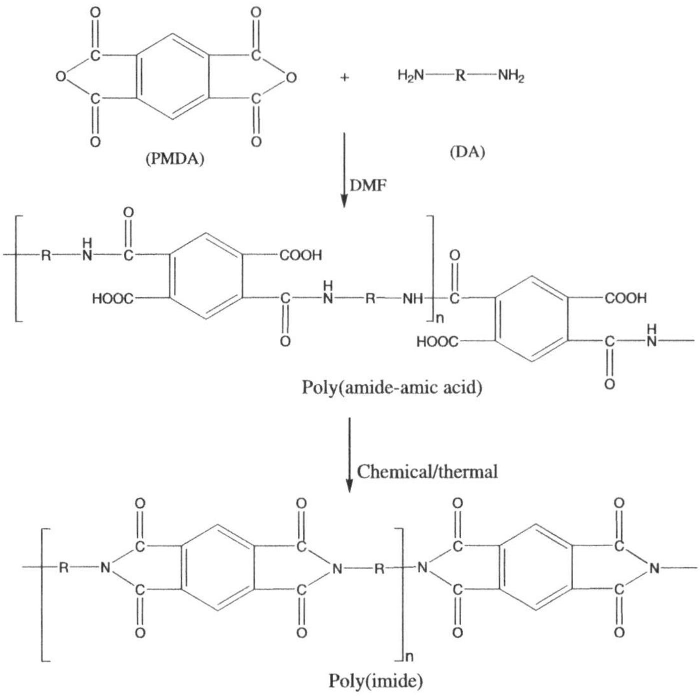

Figure 16.1 Polyimide synthesis schematic representation.

16.1.1.2 Thin-Film Composite Membranes Composite membranes consist of at least two different materials. Usually, a selective membrane material is deposited as a thin layer upon a porous sublayer, which serves as support. The advantage of this kind of membrane over the integrally skinned ones is that each layer can be optimized independently in order to achieve the desired membrane performance. There are several well-established techniques to apply a thin top layer upon a support: dip-coating, spray coating, spin coating, interfacial polymerization, in situ polymerization, plasma polymerization, and grafting. Details of these techniques can be found elsewhere (Mulder, 1997). Due to the large variety of preparation techniques, almost all polymeric materials can be used to produce these kind of membranes. The top layer and the support both contribute to the overall performance. High solvent resistance makes PAN a good support material for TFC membranes as can be seen from the following examples.

Preparation of cross-linked PAN supports has been reported in the literature by Peinemann et al. (2001), when an epoxidized PAN copolymer was subjected to ammonolysis providing highly solvent stable [including stability in dimethylformamide (DMF)] supports. A highly solvent-resistant UF membrane based on poly(acrylonitrile-*co*-glycidyl methacrylate) (PANGMA) was reported (Hicke et al., 2002). This membrane was prepared by phase inversion followed by ammonolysis and was also stable in DMF, thus being a very attractive support for preparation of OSN-TFC membranes.

Composite membranes made of a support layer of PAN (10–80 μm thick), and a top, thermally cross-linked, elastomeric barrier layer of an adduct of maleic acid anhydride and a poly(aliphaticterpene) have also been reported for lube oil dewaxing (Pasternak, 1993). These membranes were able to achieve 87.2% dewaxed oil rejection

TABLE 16.3 Commercially Available Soluble Polyimides[a]

Polymer	Chemical Structure
P 84 (Lenzing) BTDA-TDI/MDI soluble in amides	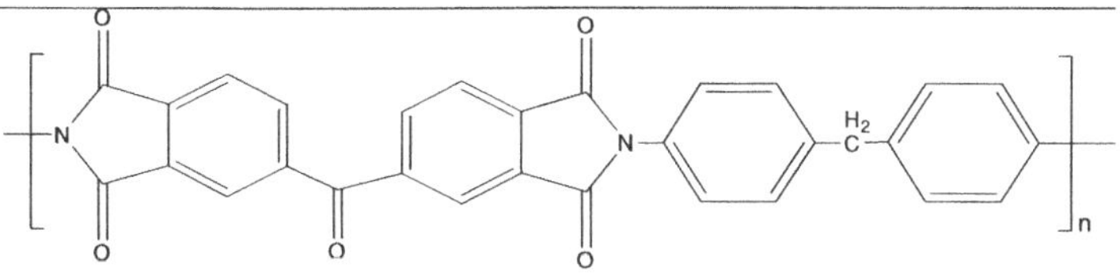 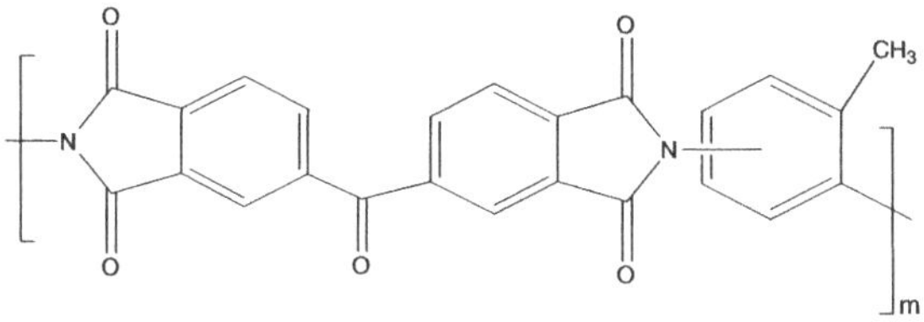
Matrimid 5218 (Ciba Geigy) BTDA-AAPTMI soluble in amides, chloroform and THF	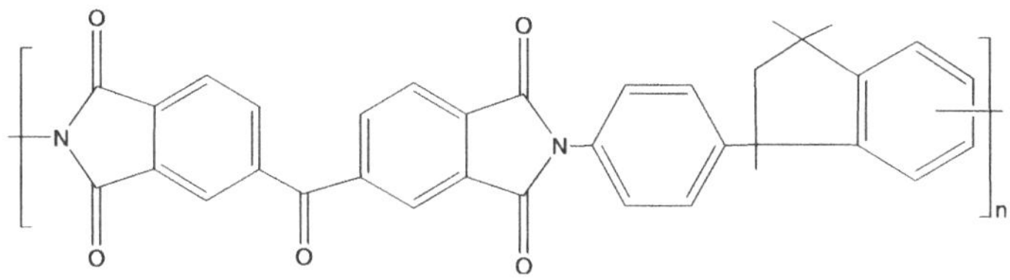
Sixef-44 (Hoechst Celanese) 6FDA-6FipDA soluble in amides, THF, chloroform and acetone	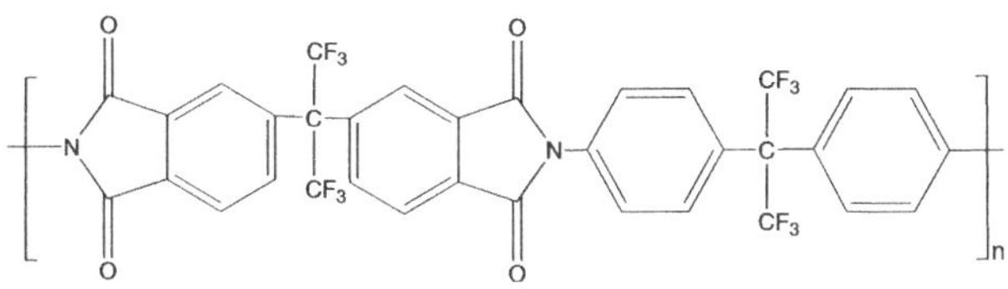
Kapton HN (DuPont) PMDA-ODA soluble in amides and THF	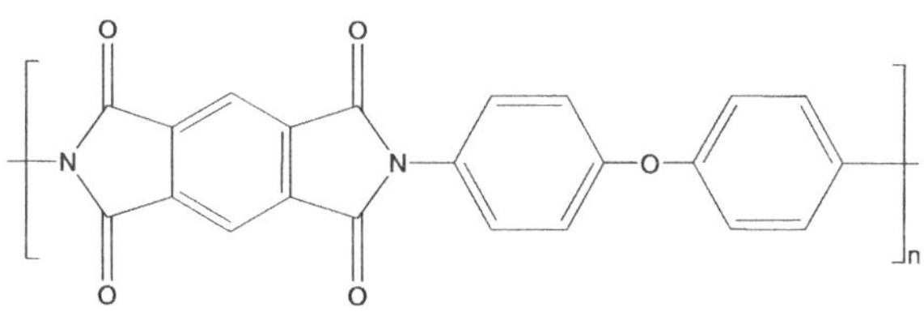

[a]BTDA, benzophenone-3,3′,4,4′-tetracarboxylic dianhydride; TDI, toluene diisocyanate; MDI, 4,4′-methylene bis(phenyl isocyanate); AAPTMI, 5(6)-amino-1-(4′-aminophenyl)-1,3-trimethylindane; 6FDA, 5,5-[2,2,2-trifluoro-1-(trifluoromethyl)-ethylidene]-bis-1,3-isobenzofuranedione; 6FipDA, hexafluoro-2,2-bis(4-aminophenyl)-propane; PMDA, pyromellitic dianhydride; ODA, 4,4′-oxydianiline.

and 58.3 $kg/m^2/h$ fluxes for a blend of MEK and toluene at 5.5 MPa and 298 K. PAN homopolymers or copolymers cross-linked with acids or bases, functionalized with amino, hydroxyl, or carboxylic groups, and coated with an additional hydrophilic or polyelectrolyte polymer layer have been synthesized (Linder et al., 1991). The resultant composite membrane was stable in solvents such as DMF, methylethyl ketone, and dichloromethane and could achieve fluxes of 147.5 $L/m^2/h$ DMF and 99% congo-red dye rejections at 2.9 MPa and room temperature. Polydimethylsiloxane is extensively used in OSN applications as the TFC membrane top layer, and dewaxing solvents were separated from dewaxed oils by nonporous silicone rubber membranes cross-linked with polyisocyanates, polyacidchlorides, or silanes (Pasternak, 1992). At 5.5 MPa and 298 K these membranes were able to achieve 89.7% dewaxed oil rejection and 37.1 $kg/m^2/h$ fluxes for a blend of MEK and toluene.

16.1.2 Inorganic Membranes

Due to the upper temperature constrain for polymeric-based membranes (Mulder, 1997), there is a growing market for membranes based on solvent-resistant materials able to withstand high temperatures. Ceramic materials (silicium carbide, zirconium oxide, titanium oxide) endure harsh temperature conditions and show stable performance in solvent medium and so are excellent materials for membrane preparation. On that basis a new generation of OSN membranes have been developed, the inorganic composite membranes.

Inorganic membranes with the stable, defect-free nanopore structure required for OSN are prepared by the sol–gel process. Two different routes are widely used (Fig. 16.2), the colloidal gel route and the polymeric gel route. Both of these routes start with an alkoxide precursor that is hydrolyzed and polymerized. The drying of the gel structures is regarded as the critical step and the calcination temperature determines the membrane pore size.

Porous OSN membranes (1 to 4 nm pore size) were prepared from silica-zirconia colloidal sols, by the sol–gel processes (Tsuru et al., 2001). Pore size control for these membranes was possible by appropriate choice of the colloidal particle size. These membranes were

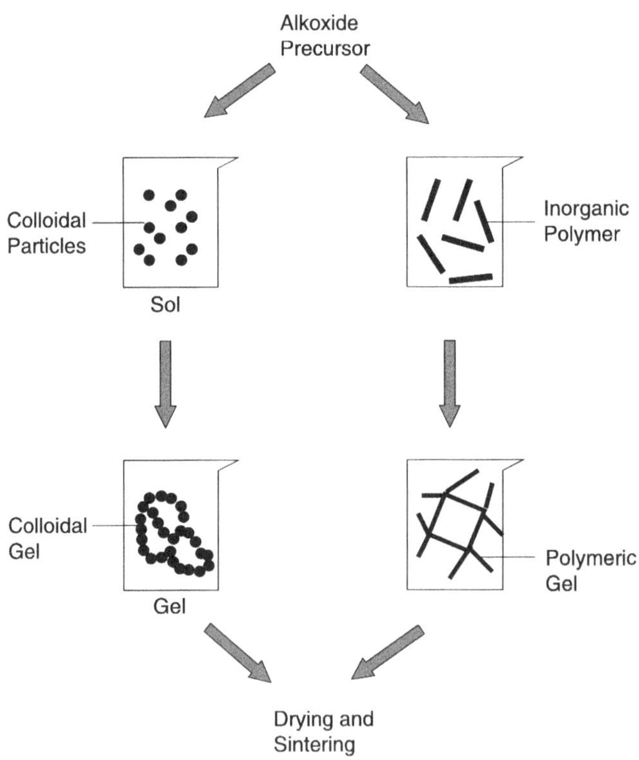

Figure 16.2 Inorganic membrane preparation by the sol–gel process.

TABLE 16.4 Literature OSN Performance Data

Membrane/ Material	Solute/Solvent System	Permeability (L m^{-2} h^{-1} bar^{-1})	Separation Factor*/ Rejection+	Reference
MPF-44 (Koch)	Safranin (0.01%)/ methanol	0.34	67.6%+	Whu et al. (2000)
	Solvent blue (35 mg/L)/ methanol	0.19	85.0%+	Yang et al. (2001)
MPF-60 (Koch)	Safranin (0.01%)/ methanol	0.50	86.9%+	Whu et al. (2000)
	Solvent blue (35 mg/L)/ methanol	0.13	81.0%+	Yang et al. (2001)
MPF-50 (Koch)	Vitamin B_{12} (0.01%)/ methanol	0.98	89.0%+	Whu et al. (2000)
	Methanol	1.24	—	Whu et al. (2000)
	Octane	11.63	—	Machado et al. (1999)
	Methanol	5.83	—	Machado et al. (1999)
	Pentanol	1.03	—	Machado et al. (1999)
	Acetone (40% molar)/ propanol	6.67	1*	Machado et al. (1999)
	Pentane (40% molar)/ acetone	16.67	1*	Machado et al. (1999)
	Ethanol	6.35	—	Van der Bruggen et al. (2002)
	n-Hexane	46.3	—	Van der Bruggen et al. (2002)
Matrimid-5218 (Ciba-Geigy)	Lube-oil (20%)/methyl ethyl ketone-toluene	0.11	99.0%+	White et al. (2000)
Lenzing P84 (HP polymers)	*n*-Decane (2%),	0.86	44.0%+	White et al. (2000)
	1-Methylnaphthalene (2%),		1.0%+	White et al. (2000)
	n-Hexadecane (2%),		79.0%+	White et al. (2000)
	1-Phenyl undecane (2%),		66.0%+	White et al. (2000)
	pristine (2%),		95.0%+	White et al. (2000)
	n-Docosane (2%)/ toluene (88%)		92.0%+	White et al. (2000)
STARMEM 122 (W.R. Grace)	Jacobsen catalyst (1.2 mM)/THF	2.73	96.0%+	Scarpello et al. (2002)
	Jacbsen catalyst (1.2 mM)/EA	4.24	99.0%+	Scarpello et al. (2002)
PDMS-PAN	*n*-Hexane	8.41	—	Robinson et al. (2004)
	n-Heptane	7.00	—	Robinson et al. (2004)

(Continued)

TABLE 16.4 *Continued*

Membrane/ Material	Solute/Solvent System	Permeability ($L\ m^{-2}\ h^{-1}\ bar^{-1}$)	Separation Factor*/ Rejection$^+$	Reference
	i-Hexane	7.80	—	Robinson et al. (2004)
	i-Heptane	6.25	—	Robinson et al. (2004)
	i-Octane	4.66	—	Robinson et al. (2004)
	Cyclohexane	3.66	—	Robinson et al. (2004)
	Xylene	4.90	—	Robinson et al. (2004)
	Sunflower oil (0%)/ hexane	3.70	—	Stafie et al. (2004)
	Sunflower oil (8%)/ hexane	2.40	88.0%$^+$	Stafie et al. (2004)
	Sunflower oil (19%)/ hexane	1.80	84.0%$^+$	Stafie et al. (2004)
	Sunflower oil (30%)/ hexane	1.10	82.0%$^+$	Stafie et al. (2004)
Silica-zirconia	Methanol	1.92	—	Tsuru et al. (2001)
	Polyethylene glycol (MW = 200)		40.0%$^+$	Tsuru et al. (2001)

tested in nonaqueous solutions of ethanol and methanol. At 323 K and 1.5 MPa the pure methanol flux was 28.8 $L/m^2/h$ and a 200 Da polyethylene glycol rejection of 40% was obtained.

One major drawback of the inorganic membranes has been their relatively high molecular weight cutoff (MWCO) > 1000 Da, which makes them unsuitable for nanofiltration-type separations. However, recently, inorganic membranes with a MWCO < 500 Da have been reported in the literature, although this rejection has been determined in aqueous solution for a mixture of PEGs (Puhlfürß et al., 2000; Weber et al., 2003). Table 16.4 summarizes some of the reported performance data for the OSN membrane types mentioned above.

16.2 OSN TRANSPORT MECHANISMS—THEORETICAL BACKGROUND

Although transport processes in aqueous NF systems have been studied for several years and much knowledge has been gained, OSN systems are not yet well understood. While some studies support the use of pore-flow models, others suggest using a solution-diffusion approach. The basic equations of these models are outlined below along with the major simplifying assumptions.

16.2.1 Pore-Flow Model

Stable pores are assumed to be present inside the membrane, and the driving force for transport is the pressure gradient across the membrane. Assuming a system at constant temperature where there are no external forces except pressure, one can derive, from the Stefan–Maxwell equations (Silva et al., 2005), the following equations describing the total volumetric flux through a membrane:

Hagen–Poiseuille equation—if the membrane is assumed composed of more or less cylindrical pores:

$$N_v = -\frac{d_{\text{pore}}^2}{32\eta}\frac{\varepsilon}{\tau}\nabla p \tag{16.1}$$

Carman–Kozeny equation—if the membrane consists of a packed bed of particles:

$$N_v = -\frac{d_{\text{particle}}^2}{180(1-\varepsilon)^2\eta}\frac{\varepsilon^3}{\tau}\nabla p \tag{16.2}$$

The use of pore flow to describe transport in OSN is common in the literature. Robinson et al. (2004) reported that their experimental data for the permeation of *n*-alkanes, *i*-alkanes, and cyclic compounds in a polydimethylsiloxane (PDMS) composite OSN membrane were consistent with the Hagen–Poiseuille pore-flow model of Eq. (16.1). Whu et al. (2000) also suggested this pore-flow model to explain the compaction and rejection behavior of Koch MPF series commercial OSN membranes. Van der Bruggen et al. (2002) described OSN rejection curves with log-normal pore size distributions in the commercially available membranes tested. Bhanushali et al. (2002) suggested that a pore-flow model including an interaction parameter between the membrane and the permeating species could qualitatively describe OSN rejection data. Finally, Machado et al. (2000) developed a resistances-in-series model and proposed that solvent transport through the MPF 50 membrane consists of three main steps: (1) transfer of the solvent into the top active layer, which is characterized by surface resistance, (2) viscous flow through NF pores, and (3) viscous flow through support layer pores, all expressed by viscous resistances, that is,

$$N_v = \frac{\Delta p}{R_s^0 + R_\mu^1 + R_\mu^2} \tag{16.3}$$

where R_s^0, R_μ^1, and R_μ^2 are the surface resistance and viscous resistances through NF active layer and support layers, respectively. The surface resistance is proportional to surface tension difference between the solvent and the OSN top layer, and viscous resistances are proportional to solvent viscosity.

16.2.2 Solution-Diffusion Model

In the solution-diffusion model, it is assumed that each permeating molecule dissolves in and diffuses through the membrane phase in response to a concentration gradient. There is no pressure gradient inside the membrane, and, starting again from the Stefan–Maxwell equation, the following equation can be derived (Silva et al., 2005) for the molar flux of each specie through the membrane:

$$N_i = P_{i,m}^{\text{molar}}\left[x_{i,F} - \frac{\gamma_{i,P}}{\gamma_{i,F}}x_{i,P}\ \exp\left(-\frac{\overline{V}_i\Delta p}{RT}\right)\right] \tag{16.4}$$

This equation assumes that the swelling of the membrane separating layer is negligible and is similar to the well-known equation presented by Wijmans and Baker (1995), differing only by the ratio of $\gamma_{i,P}/\gamma_{i,F}$, which has been shown to be important when there are significantly different concentrations on each side of the membrane. For cases where the membrane layer is a rubbery layer, the assumption of low swelling is unlikely to be true, and in fact the membrane will often be highly swollen—in that case the analysis developed by Paul through a series of studies in the 1970s and recently reformulated (Paul, 2004) is more appropriate.

The solution activity coefficients of the components in a mixture can be easily determined from the vapor–liquid equilibrium data. Vapor–liquid equilibrium relation assumes that:

$$y_i p = \gamma_i x_i p_{vpi} \tag{16.5}$$

So, from Eq. (16.5), knowing the pure liquid–vapor pressures (p_{vpi}) and the experimental p-x,y diagram at the required temperature, we can calculate the activity coefficients at several points and determine the molar excess Gibbs energy (g^E), for each of the points. For a binary mixture we have

$$g^E = RT[x_1 \ln(\gamma_1) + x_2 \ln(\gamma_2)] \tag{16.6}$$

Finally, in order to proceed we need to adopt some mathematical expression for the molar excess Gibbs energy. There are several models available in the literature (Reid et al., 1987) describing g^E, and all of them contain adjustable parameters whose best fitting values are obtained from Eq. (16.6).

The solution-diffusion mechanism is also often used in the literature to describe the transport in OSN. Bhanushali et al. (2001) suggested that solvent viscosity and surface tension are the dominant factors controlling solvent transport through NF membranes, and a solution-diffusion approach was proposed to predict pure solvent permeation. Stafie et al. (2004) employed the solution-diffusion model to describe sunflower oil/hexane and polyisobutylene/hexane permeation through a composite PDMS membrane with PAN support. White (2002) investigated the transport of normal and branched alkanes and aromatic compounds through a series of asymmetric PI OSN membranes. His experimental results were consistent with the solution-diffusion model presented by Wijmans and Baker (1995). Since PIs are reported to swell by less than 15%, and usually considerably less, in common solvents the simple solution-diffusion model can be used. However, we note that the solution-diffusion model assumes a discontinuity in pressure profile at the downstream side of the separating layer. When the separating layer is not a rubbery polymer coated onto a support material, but is a dense top layer formed by phase inversion, as in the PI membranes reported (White, 2002), it is not clear where this discontinuity is located or whether it will actually exist.

16.2.3 Concentration Polarization

Almost all reported OSN data has been mainly obtained at lab scale and with dilute solutions (<1 wt% solute in solvent), whereas in actual applications, solutes will be more concentrated (>5 wt%). Under these conditions, concentration polarization and osmotic pressure may contribute to the solvent flux, as they do in well-studied aqueous systems.

The use of film theory to describe solution mass transfer phenomena in pressure-driven membrane processes has a proven track record for aqueous systems. Under the flow conditions encountered in nanofiltration, the simplified film theory description of mass transfer has an accuracy close to solutions obtained by computational fluid dynamics (CFD) modeling (Zydney, 1997). The film theory, for component i, gives, for the total volumetric flux [see Peeva et al. (2004) for details]:

$$\frac{N_v}{k_i} = \ln\left(\frac{c_{i,FM} - c_{i,P}}{c_{i,F} - c_{i,P}}\right) \tag{16.7}$$

Concentration polarization can be coupled with a model for membrane transport [e.g., the solution diffusion model of Eq. (16.4)] (Nakao et al., 1986; Van der Berg and Smolders, 1989), to describe membrane transport in a mass transfer limited system.

Combined solution-diffusion film theory models have been presented already in several publications on aqueous systems; however, either 100% rejection of the solute is assumed or detailed experimental flux and rejection results are required in order to find parameters by nonlinear parameter estimation (Murthy and Gupta, 1997). Consequently, it is difficult to apply these models for predictive purposes. Peeva et al. (2004) presented the first consideration of concentration polarization in OSN. They coupled the solution-diffusion membrane transport model, Eq. (16.4), with film theory to describe flux and rejection of toluene/docosane and tolune/TOABr binary mixtures. This approach was able to integrate concentration polarization and nonideal solution behavior into OSN design models and predict fluxes over a wide range of solvent mixtures from a limited data set of the pure solvent fluxes. The only parameters to be estimated, other than physical properties, are the mass transfer coefficients, which may be measured, and the permeabilities, $P_{i,m}^{\text{molar}}$, which may be calculated from flux data.

16.3 APPLICATIONS OF ORGANIC SOLVENT NANOFILTRATION

Applications have been proposed for a variety of industries including fine chemical and pharmaceutical synthesis, food and beverage, and refining.

16.3.1 Fine Chemical and Pharmaceutical Synthesis

The separation of reaction products from catalysts is a recurrent problem in homogeneous catalysis. The major drawback of the common separation techniques applied in homogeneous catalysis is the extensive (and usually destructive) postreaction workup required. OSN membranes, being selective between high MW catalysts (>450 Da) and reaction products, are able to perform this separation. Nair et al. (2002) presented a membrane-based (STARMEM 122) process for the separation of a phase transfer catalyst (PTC) and a Heck reaction transition metal catalyst from the reaction media. For the PTC catalyst the process was so efficient that rejections superior to 99% were observed for both pre- and postreaction mixtures and no reaction rate decline was observed for two consecutive catalyst recycles.

Multistep organic-solvent-based pharmaceutical synthesis of larger organic solutes having MW in the range of 300–1000 Da, usually requires separation processes able to deal with thermally labile compounds. A series of works suggested that membrane technology could be a valid option. Sheth et al. (2003) demonstrated that OSN could be successfully applied for solvent exchange in multistep organic synthesis of bulk drugs where each reaction may be carried out in a different solvent. Ethyl acetate was reduced to the level of a low concentration impurity in methanol using OSN membranes MPF-50 and MPF-60 and solute, erythromycin, representing an active intermediate. Livingston et al. (2003) have also shown the solvent exchange potential of STARMEM 122 OSN membrane, for the methanol toluene system.

16.3.2 Food and Beverage

Membranes can be used in many areas of the vegetable oil industry; conceptually they can be applied to almost all stages of oil production and purification

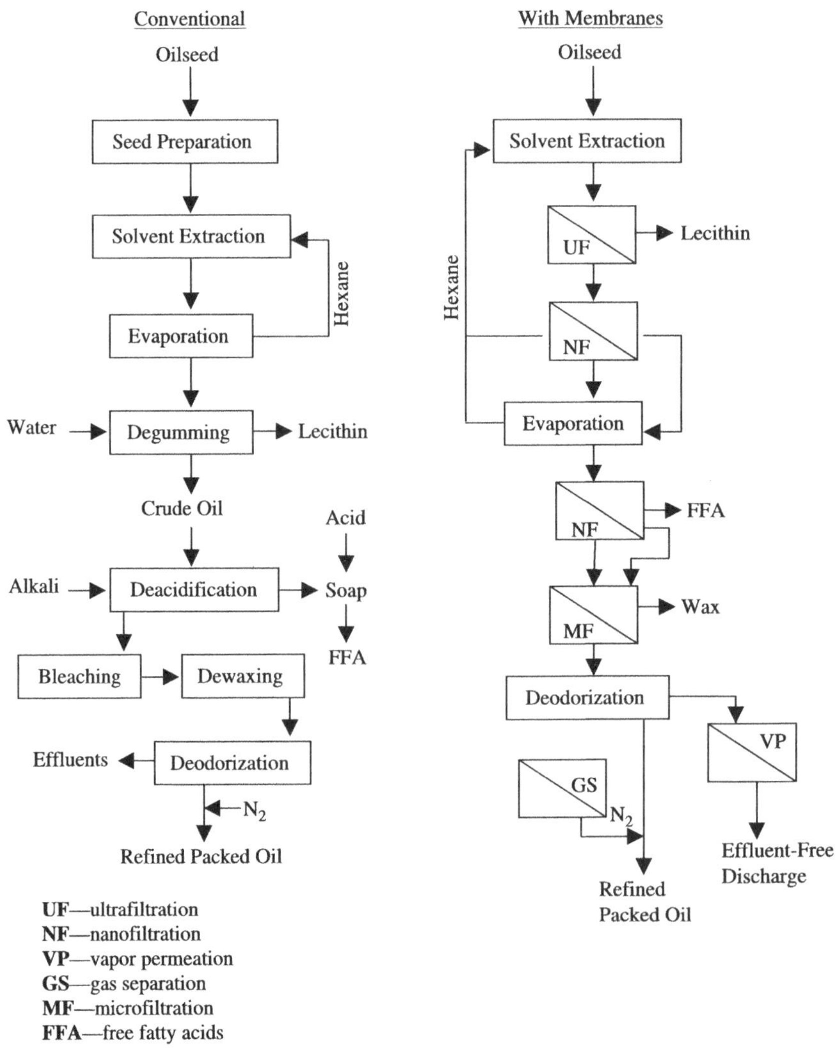

Figure 16.3 Vegetable oil processing, using a conventional approach and membrane technology.

(Fig. 16.3). In the conventional vegetable oil processing there are four major drawbacks (Cheryan, 2005):

- *High-energy usage:* After the oil extraction with solvent, the oil solvent mixture is then separated by evaporation. This requires a considerable amount of energy, and in addition safety problems might arise from the released explosive vapors.
- *Oil losses:* The saponification in the refining step traps a considerable amount of oil.
- *Resources:* Large amount of water and chemicals are used.
- *Effluents:* Heavily contaminated effluents are produced.

The use of solvent-resistant membranes in the vegetable oil industry would allow (Fig. 16.3): minimization of thermal damage, solvent recycle, lower emissions, lower energy consumption, reduced oil losses, and a decrease in bleaching earth requirements. Some experimental evidences: Pioch et al. (1998) observed that cross-flow filtration gives promising results for vegetable oil refining, while Raman et al. (1996) and Zwijnenberg et al. (1999) suggested that OSN could be successfully applied for deacidification of vegetable oil.

16.3.3 Refining

Organic solvent nanofiltration membranes can be used in the separation of petroleum fractions, recovery of components from petroleum streams, or fuel upgrading. Ohya et al. (1997) prepared a series of asymmetric PI membranes, and using ones with an MWCO of 170 Da succeeded in separating gasoline–kerosine mixtures, with a

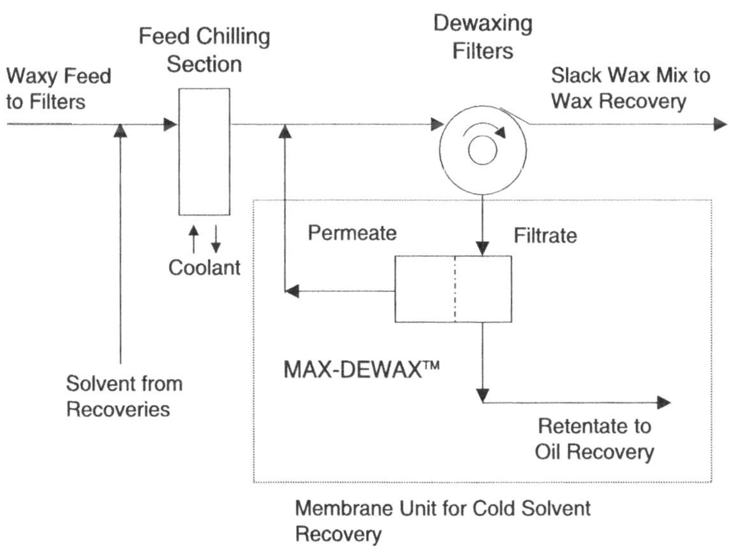

Figure 16.4 Schematic of the MAX-DEWAX™ process.

separating factor (for gasoline–kerosene) of 19.5 and a 40 kg/m^2/day flux at 423 K and 10 MPa. Separations of light gas–oil from crude oil and kerosene from lower MW mixtures were also possible with membranes presenting 380 and 270 Da MWCO, respectively. White and Nitsch (2000) showed that the commercial polymer known as Matrimid 5218 can be used to form asymmetric nanofiltration membranes with excellent chemical stability, able to separate light hydrocarbon solvents from lube oil filtrates. More specifically, in a 2-month high-pressure continuous test, this PI membrane was able to achieve a 99% purity chilled solvent (methyl ethyl ketone–toluene mixture at 263 K) recovery from lube oil filtrates with a steady permeate rate. The energy requirements for this membrane process is around 45% less than that for the usual distillation-based lube processing. This work led to the installation of a commercial OSN plant, located at Mobil's Beaumont, Texas, refinery. The process (trademarked as MAX-DEWAX) (Fig. 16.4) is able to deal with 11,500 m^3 of solvent per day. The MAX-DEWAX process, combined with selected ancillary equipment upgrades, increased base oil production by over 25 vol% and improved dewaxed oil yields by 3–5 vol%. The net annual uplift from the membrane unit is over \$6 million, making it a very attractive technology. The capital expenditure was paid back in less than 1 year by the increase in the net profitability of the lube dewaxing plant. This highly successful application at large scale clearly shows the potential for OSN to impact the energy and chemicals sectors.

Nomenclature

c	molar concentration (mol/m^3)
d_{pore}	pore diameter (m)
d_{particle}	particle diameter (m)
g^E	excess Gibbs energy (Pa/m^3/mol)
k	mass transfer coefficient (m/s)
N	molar flux (mol/m^2/s)
N_v	total volumetric flux (m/s)
p	pressure (Pa)
p_{vpi}	pure liquid vapor pressure (Pa)
$P_{i,m}^{\text{molar}}$	molar permeability (mol/m^2/s)
R	ideal gas constant (Pa·m^3/mol/K)
T	temperature (K)
$\overline{V}$	partial molar volume (m^3/mol)
x	molar fraction in the liquid phase (−)
y	molar fraction in the gas phase (−)

Greek Letters

ε	porosity (−)
γ	molar activity coefficient (−)
η	viscosity (Pa/s)
τ	tortuosity factor (−)
∇	gradient (m^{-1})

Subscripts	
$i,1,2$	species
$m,(m)$	membrane
F	feed side
FM	feed side, at the membrane–liquid interface
P	permeate side

REFERENCES

Alegranti, C. W. (1978). Assymetric polyimide membranes. U.S. Patent 4,113,628.

Beerlage, M. A. M. (1994). Polyimide ultrafiltration membranes for non-aqueous systems. Ph.D. Dissertation, University of Twente, The Netherlands.

Bhanushali, D., Kloos, S., and Battarcharyya, D. (2002). Solute transport in solvent-resistant nanofiltration membranes for non-aqueous systems: Experimental results and the role of solute-solvent coupling. *J. Membr. Sci.* 208, 343.

Bhanushali, D., Kloos, S., Kurth, C., and Battacharyya, D. (2001). Performance of solvent-resistant membranes for non-aqueous systems: Solvent permeation results and modelling. *J. Membr. Sci.* 189, 1.

Cheryan, M. (2005). Membrane technology in the vegetable oil industry. *Membr. Technol.* 2, 5.

Critchley, J. P., Knight, G. J., and Wright, W. W. (1983). *Heat Resistant Polymers.* Plenum, New York.

Gebben, B. (1988). Thermally stable and chemically resistant polymer membranes; aromatic polyoxadiazoles and polytriazoles. Ph.D. Dissertation, University of Twente, The Netherlands.

Hicke, H. G., Lehmann, I., Malsch, G., Ulbricht, M., and Becker, M. (2002). Preparation and characterization of a novel solvent-resistant and autoclavable polymer membrane. *J. Membr. Sci.* 198, 187.

Linder, C., Perry, M., Nemas, M., and Katraro, R. (1991). Solvent stable membranes. U.S. Patent 5,039,421.

Livingston, A., Peeva, L., Han, S., Nair, D., Luthra, S. S., White, L. S., and Dos Santos, L. M. F. (2003). Membrane separation in green chemical processing: Solvent nanofiltration in liquid phase organic synthesis reactions. *Ann. N. Y. Acad. Sci.* 984, 123.

Machado, D. R., Hasson, D., and Semiat, R. (1999). Effect of solvent properties on permeate flow through nanofiltration membranes. Part I: Investigation of parameters affecting solvent flux. *Journal of Membrane Science* 163(1), 93.

Machado, D. R., Hasson, D., and Semiat, R. (2000). Effect of solvent properties on permeate flow through nanofiltration membranes, Part II. Transport model. *J. Membr. Sci.* 166, 63.

McCarthy, R. A. (1990). *Encyclopaedia of Polymer Science and Engineering.* Wiley, New York.

Mulder, M. (1997). *Basic Principles of Membrane Technology*, 2nd ed. Kluwer Academic, Dordrecht, The Netherlands.

Murthy, Z. V. P., and Gupta, S. K. (1997). Estimation of mass transfer coefficient using a combined nonlinear membrane transport and film theory model. *Desalination* 109, 39.

Nair, D., Luthra, S. S., Scarpello, J. T., White, L. S., Freitas, L. M., and Livingston, A. G. (2002). Homogeneous catalyst separation and re-use through nanofiltration of organic solvents. *Desalination* 147, 301.

Nakao, S., Wijmans, J. G., and Smolders C. A. (1986). Resistance to the permeate flux in unstirred ultra-filtration of dissolved macromolecular solutions. *J. Membr. Sci.* 26, 165.

Ohya, H., Okazaki, I., Aihara, M., Tanisho, S., and Negishi, Y. (1997). Study on molecular weight cut-off performance of asymmetric aromatic polyimide membrane. *J. Membr. Sci.* 123, 143.

Park, J. S., Kim, S. K., and Lee, K. H. (2000). Effect of $ZnCl_2$ on formation of asymmetric PEI membrane by phase inversion process. *J. Ind. Eng. Chem.* 6, 93.

Pasternak, M. (1992). Membrane process for treating a mixture containing dewaxed oil and dewaxing solvent. U.S. Patent 5,093,002.

Pasternak, M. (1993). Membrane process for treating a charge containing dewaxing solvent and dewaxed oil. U.S. Patent 5,234,579.

Paul, D. R. (2004). Reformulation of the solution-diffusion theory of reverse osmosis. *J. Membr. Sci.* 241, 371.

Peeva, G. L., Gibbins, E., Luthra, S. S., White, L. S., Stateva, R. P., and Livingston, A. G. (2004). Effect of concentration polarization and osmotic pressure on flux in organic solvent nanofiltration. *J. Membr. Sci.* 236, 121.

Peinemman, K. V., Ebet, K., Hicke, H. G., and Schanagl, N. (2001). Polymeric composite ultrafiltration membranes for non-aqueous applications. *Environ. Prog.* 20, 17.

Pioch, D., Larguéze, C., Graille, J., Ajana, H., and Rouviere, J. (1998). Towards an efficient membrane based vegetable oils refining. *Ind Crops Products* 7, 83.

Puhlfürß, P., Voigt, A., Weber, R., and Morbé, M. (2000). Microporous TiO_2 membranes with a cut off <500 Da. *J. Membr. Sci.* 174, 123.

Raman, L. P., Cheryan, M., and Rajagopalan, N. (1996). Deacidification of soybean oil by membrane technology. *J. Am. Oil Chem. Soc.* 73, 219.

Reid, C. R., Prausnitz, J. M., and Poling, B. E. (1987). *The Properties of Gases and Liquids*, 4th ed. McGraw-Hill, New York.

Robinson, J. P., Tarleton, E. S., Millington, C. R., and Nijmeijer, A. (2004). Solvent flux through dense polymeric nanofiltration membranes. *J. Membr. Sci.* 230, 29.

Scarpello, J. T., Nair, D., Freitas dos Santos, L. M., White, L. S., and Livingston, A. G. (2002). The separation of homogeneous organometallic catalysts using solvent resistant nanofiltration. *J. Membr. Sci.* 203(1–2), 71.

Sheth, J., Qin, Y., Sirkar, K., and Baltzis, B. (2003). Nanofiltration-based diafiltration process for solvent exchange in pharmaceutical manufacturing. *J. Membr. Sci.* 211, 251.

Silva, P., Han, S., and Livingston, A. G. (2005). Solvent transport in organic solvent nanofiltration membranes. *J. Membr. Sci.* 262, 49.

Stafie, N., Stamatialis, D. F., and Wessling, M. (2004). Insight into the transport of hexane-solute systems through tailor made composite membranes. *J. Membr. Sci.* 228, 103.

Strathmann, H. (1978). Asymmetric polyimide membranes for filtration of non-aqueous solutions. *Desalination* 26, 85.

Tsuru, T., Sudoh, T., Yoshioka, T., and Asaeda, M. (2001). Nanofiltration in non-aqueous solutions by porous silica-zirconia membranes. *J. Membr. Sci.* 185, 253.

Van der Berg, G. B., and Smolders, C. A. (1989). The boundary-layer resistance model for unstirred ultrafiltration. A new approach. *J. Membr. Sci.* 40, 149.

Van der Bruggen, B., Geens, J., and Vandecasteele, C. (2002). Fluxes and rejections for nanofiltration with solvent stable polymeric membranes in water, ethanol and *n*-hexane. *Chem. Eng. Sci.* 57, 2511.

Weber, R., Chmiel, H., and Mavrov, V. (2003). Characteristics and application of new ceramic nanofiltration membranes. *Desalination* 157, 113.

White, L. S. (2002). Transport properties of a polyimide solvent resistant nanofiltration membrane. *J. Membr. Sci.* 205, 191.

White, L. S., and Nitsch, A. R. (2000). Solvent recovery from lube oil filtrates with a polyimide membrane. *J. Membr. Sci.* 179, 267.

White, L. S., Wang, I., and Minhas, B. S. (1993). Polyimide membrane for separation of solvents from lube oil. U.S. Patent 5,264,166.

Whu, J. A., Baltzis, B. C. and Sirkar, K. K. (2000). Nanofiltration studies of larger organic microsolutes in methanol solutions *J. Membr. Sci.* 170, 159.

Wijmans, J. G., and Baker, R. W. (1995). The solution diffusion model: A review. *J. Membr. Sci.* 107, 1.

Yang, X. J., Livingston, A. G., and Freitas dos Santos, L. M. (2001). Experimental observations of nanofiltration with organic solvents. *J. Membr. Sci.* 190, 45.

Zwijnenberg, H. J., Krosse, A. M., Ebert, K., Peinemann, K.-V., and Cuperus, F. P. (1999). Acetone-stable nanofiltration membranes in deacidifying vegetable oil. *J. Am. Oil Chem. Soc.* 76(1), 83.

Zydney, A. L. (1997). Stagnant film model for concentration polarization in membrane systems. *J. Membr. Sci.* 130, 275.

CHAPTER 17

Pervaporation

FAKHIR U. BAIG

Petro Sep Membrane Technologies Inc., Oakville, Ontario, Canada

17.1 INTRODUCTION

Pervaporation is a process in which organic solvent water mixture or organic solvent mixture can be separated by partial vaporization through a nonporous permeate selective membrane. In this process liquid feed mixture circulates in contact with the active nonporous side of the membrane while a vacuum is applied on the other side of the membrane. A phase change of membrane-selective permeate takes place in the membrane. The membrane-selective permeate diffuses through the membrane and desorbs on the posterior side of the membrane. Later, it evaporates with the help of a vacuum from the posterior side of the active nonporous membrane. The transport of the permeate through a nonporous permeate-selective membrane is quite complex. This could be explained in three steps, which are as follows:

1. Separation of the solvent mixture with a permeate-selective membrane on the surface of a nonporous permeate selective membrane
2. Diffusion of the permeate through the permeate-selective membrane
3. Permeate desorption on the posterior side of the membrane

Pervaporation has been known to the scientific world since the early 1900s. Kahlenberg (1906) reported some qualitative observations concerning the selective transport of hydrocarbon–alcohol mixtures through a thin rubber sheet. The term *pervaporation* was first introduced by Kober (1917) in a study published in 1917 in the *Journal of the American Chemical Society*. Schwob (1949) demonstrated dehydration of alcohols by using 20-μm-thin membranes during his scientific work. Binning et al. (1961) from The American Oil Company, Texas City, Texas, carried out several experiments to separate various hydrocarbons by using pervaporation experiments. He made a pilot plant consisting of 10 m^2 of membrane area to separate hydrocarbons. However, after several years of work, the technology was not commercialized.

Advanced Membrane Technology and Applications. Edited by Norman N. Li, Anthony G. Fane, W. S. Winston Ho, and T. Matsuura
Copyright © 2008 John Wiley & Sons, Inc.

Pervaporation research has continued in various parts of the world without any significant breakthrough in commercialization. In 1982 G.F.T., a German company, commercialized a pervaporation plant for alcohol dehydration. This plant could produce 1300 L of ethanol per day of 99.2% purity from predistilled ethanol. In 1987 G.F.T. was taken over by a French company, Carbone Lorraine. In 1994 Carbone Lorraine sold its pervaporation technology to Sulzer Chemtech. In Japan during the same time Mitsui, Sasakura Engineering, and Asahi Chemicals put a lot of effort into research and development to commercialize the pervaporation technology.

In 1999, based on 10 years of research by membrane scientists of Petro Sep Membrane Technologies Inc. with the Industrial Membrane Research Institute (IMRI) of the University of Ottawa, Petro Sep Membrane Technologies Inc. of Oakville, Canada, introduced a new type of pervaporation membranes. These membranes are very robust and chemically resistant and are available in hollow-fiber as well as flat-sheet configuration. The design is very user friendly and very economical. Petro Sep has built many pilot-scale plants and several full-scale plants for dehydration of solvents as well as a complete solvent recycling plant by using AZEO SEP and VOC SEP pervaporation technologies. Petro Sep continues to serve the industry by using its innovative membrane solutions for complex separation problems.

There are three kinds of pervaporation membranes: (a) hydrophilic membranes, (b) hydrophobic membranes, and (c) organophilic membranes.

17.1.1 Hydrophilic Membranes

Hydrophilic membranes can be used to dehydrate organic solvents or organic mixtures. The membranes are made of hydrophilic polymers. These membranes are cross-linked by using various types of cross-linking agents to provide mechanical strength, a better separation factor, as well as a higher flux as compared to non-cross-linked hydrophilic membranes. The membrane chemistry is designed to attract water molecules to the surface of the membrane. Pervaporation takes place in the membrane. Water molecules diffuse through the membrane and evaporate on the permeate side with the help of a vacuum. Later, vapors condense in a condenser. Examples of applications for these membranes are numerous. There are many organic solvents that form azeotropes with water. Using these hydrophilic membranes pervaporation can break the azeotropes in the solvents.

17.1.2 Hydrophobic Membranes

Hydrophobic membranes can be used to extract organic solvents or volatile organic compounds (VOCs) from water. The membranes are made of hydrophobic cross-linked polymers. The membrane chemistry is designed to attract VOC molecules to the surface of the membrane. Pervaporation takes place in the membrane. VOCs diffuse through the membrane and evaporate on the permeate side with the help of a vacuum. These VOCs later condense in a condenser. Examples of these membrane applications include extraction of numerous types of VOCs from water, extraction of aromatics from water, ketones from water, esters from water, and many more.

17.1.3 Organophilic Membranes

Organophilic membranes can be used to extract organic solvents from organic solvents. The membranes are designed to attract certain organic molecules to the membrane surface. The membranes can reject other types of molecules. The molecules travel toward the surface, diffuse through the membrane, and evaporate on the permeate side with the help of a vacuum. These organics later condense in a condenser. An example is the extraction of VOCs from glycols. Another example is the separation of aromatics from the aliphatics, and many more.

17.2 APPLICATIONS OF AZEO SEP AND VOC SEP

One of the major challenges that industries are facing today is the complex separation problems arising due to the azoetrope formation in the various processes in the chemical industry. Until 1999 membrane companies did not have a complete solution for such problems. However, after the breakthrough of Petro Sep Membrane Research Inc., a subsidiary of Petro Sep Membrane Technologies Inc., it became possible to provide a complete solution for the industries facing complex separation problems.

The two major complex separations were the acidic as well as the basic nature of organic mixtures forming azeotrope with water. It has been proven in the industry that mixtures of organics having a pH up to 1 can be dehydrated as well as separated by using innovative techniques of the AZEO SEP pervaporation membrane hybrid system. Similarly, organics having a pH up to 9 can be dehydrated as well as separated.

In the past it was impossible to dehydrate amines by using membrane technology. Various companies have tried to dehydrate amines by using membrane technology; however, none of the companies have succeeded. One of the major reasons was the failure of the membranes. Similarly, in the resin industry, it was impossible to recover reactants from wastewater. Wastewater was always the biggest issue in the chemical process industry, and it is still a major problem for the industry. Especially with very heavy loading of organics, the companies have no solution other than incineration. The AZEO SEP pervaporation system has proven that these reactants can be recovered and recycled. This process not only resolves the wastewater problem but can also provide significant savings for the chemical industry by recovering high concentrations of organic solvents from the wastewater.

It is quite common in the chemical process industry to use various kinds of salts to dehydrate complex mixtures of solvents to avoid azeotropic distillation. These salts can cause many serious problems in the whole process. They can also cause a rusting problem in the whole piping system. The disposal of such toxic wastewater is a colossal issue for the chemical industry. The AZEO SEP pervaporation hybrid system can eliminate the use of salt, and it can breakdown the azeotropes for further process or use.

In the semiconductor industry, companies use a lot of valuable organic solvents in various washing processes. If the concentration of organic solvents is high in the wastewater, it cannot be treated by biological methods, advanced oxidation, membrane bioreactor, or any other method. The only method available is incineration. Worldwide incineration regulations are becoming more stringent and difficult for the industries to implement. Therefore, in the future any wastewater with high organic loading will become a nightmare for the industry. The AZEO SEP membrane hybrid system can completely recycle the valuable solvents and can resolve the wastewater problems. At the same time valuable solvents can be purified to the highest purity level up to 99.90% and can be reused. The recovery of the valuable solvents can save significant amounts annually for these companies.

Small- to medium-size footprint ethanol plants are fading away from the industry due to the high operational cost and the high capital cost. AZEO SEP pervaporation hybrid systems offer alternatives for conventional four-column distillation. This includes only one column hybrid with AZEO SEP. This can reduce the operational cost by as much as 30%. (See Figs. 17.1 and 17.2.)

Esterification is quite a complex process; especially, separation steps to separate the ester, ethanol, water, and the catalyst (Fig. 17.3). In addition to a reactor, it involves six to eight unit operations to complete the separation steps in a conventional process. Therefore, the cost of ester is always much higher than the reactants. However, by using AZEO SEP pervaporation hybrid systems the process becomes very simple and economical. This process can reduce the operating and the capital cost significantly.

During esterification, the following reaction takes place:

$$\text{Acid} + \text{alcohol} \Longleftrightarrow \text{ester} + \text{water}$$

With the formation of ester and water the rate of reaction slows down with time. It takes 24–36 h to process 25,000–30,000 L/day. Then it goes through a complex process to separate ester from the water and catalyst. The AZEO SEP hybrid process can breakdown the azeotrope of ethyl ester, ethanol, and water. After breaking down the azeotrope, it sends ethyl ester and unreacted ethanol back to the reactor to complete the reaction. This increases the yield of reaction significantly, and it reduces the reaction time two to threefold.

In the resin manufacturing industry, resin manufacturers lose 30% of the reactants in the wastewater. The process is very similar to esterification. The reaction produces water with the formation of resins. At the end, 30% of the reactant remains unreacted due to the presence of water. Similar to esterification in the later stage, the reaction slows down. This step takes a lot of time to complete the batch. It has been demonstrated in a resin manufacturing application that if an AZEO SEP pervaporation hybrid system is incorporated with the conventional system, it increases the yield. It also reduces the reaction time and the wastewater problem. The savings in recovering the unreacted reactants could be phenomenal. It has

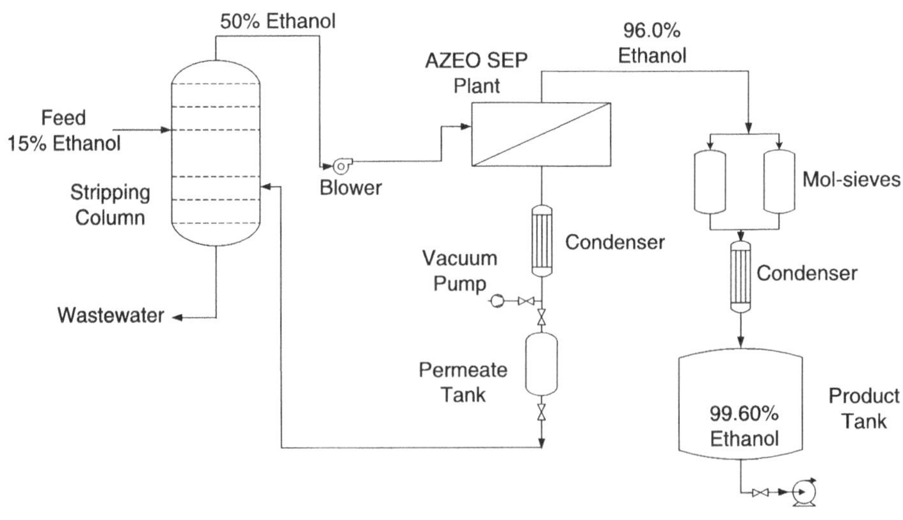

Figure 17.1 AZEO SEP hybrid ethanol dehydration plant (option 1).

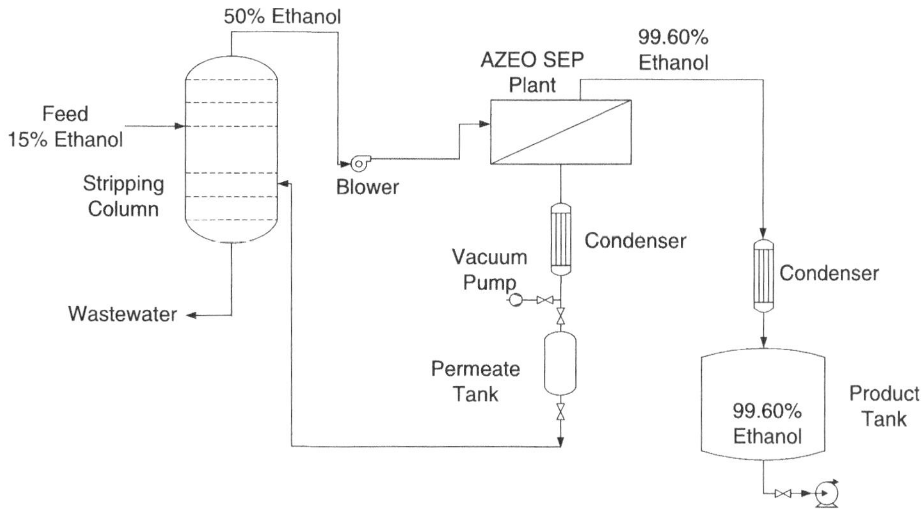

Figure 17.2 AZEO SEP hybrid ethanol dehydration plant (option 2).

been demonstrated in the field that AZEO SEP pervaporation membranes have been proven to be long lasting and robust for such a harsh environment.

In the solvent recycling industry, most of the solvents form azeotropes with water. Solvents come from the pharmaceutical industry, specialty chemicals, printing and paint industry, cosmetic industry, and petrochemical industry. These mixed solvents are very difficult to dry. Conventional methods cannot dehydrate complex solvent mixtures. However, AZEO SEP pervaporation membranes have demonstrated that even complex solvent mixtures can be dehydrated to a very high level of purity.

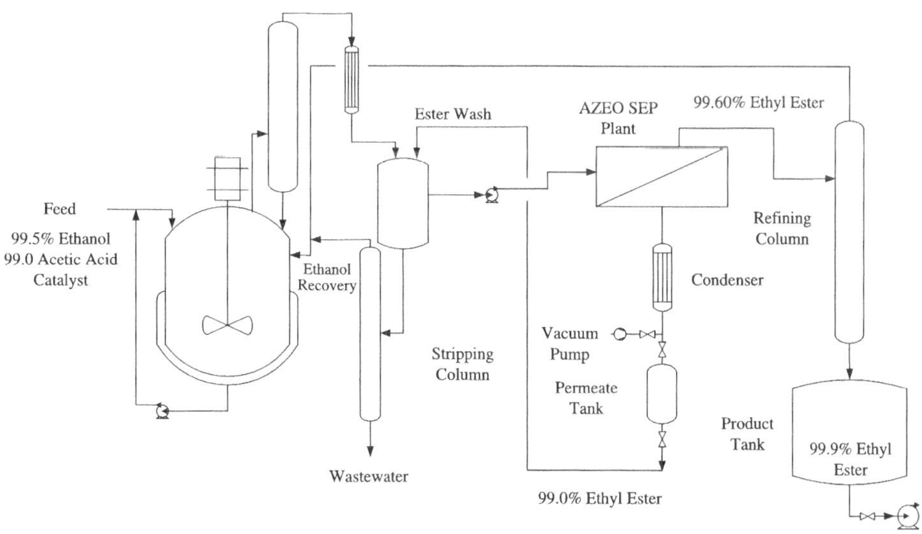

Figure 17.3 AZEO SEP hybrid esterificaton plant for ethyl ester.

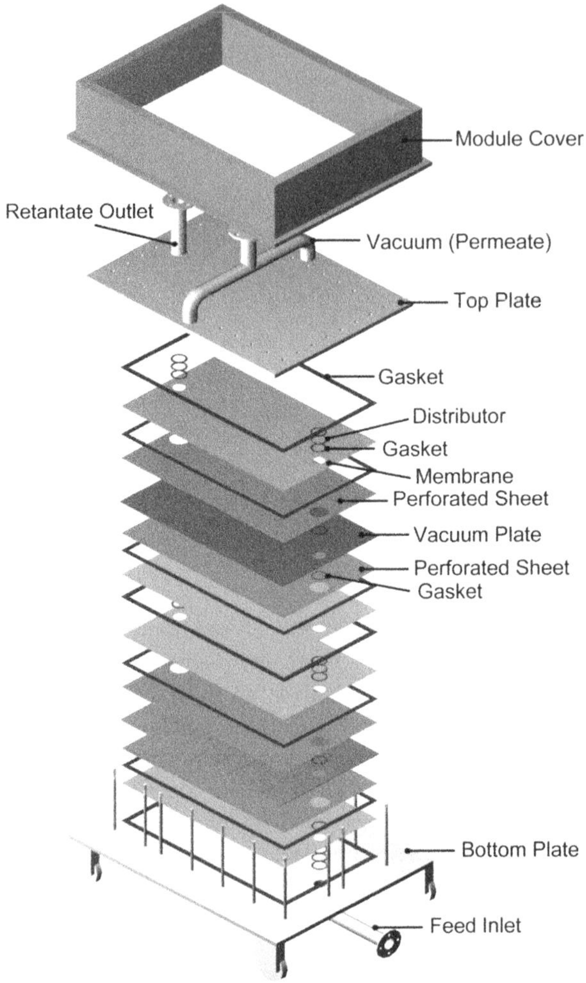

Figure 17.4 Flat-sheet membrane module.

Pervaporation membranes can be made in flat sheet (Fig. 17.4) or hollow fiber (Figs. 17.5 and 17.6). So far, flat-sheet membranes in plate-and-frame type of modules were the only type of membranes available for very few applications. There has been significant failure in spiral-wound modules and hollow-fiber modules due to membrane failure as well as lack of compatibility of epoxy and glue. Even though flat-sheet pervaporation membranes showed good performance in ethanol applications, there have been numerous failures in other applications.

Recent developments have changed the future of pervaporation. For the first time in the history of pervaporation, AZEO SEP pervaporation solvent-resistant membranes are now available in hollow-fiber configuration as well as flat sheets in plate-and-frame modules. Significant work has been done on the solvent-resistant epoxy and module design to use hollow-fiber pervaporation membrane configuration.

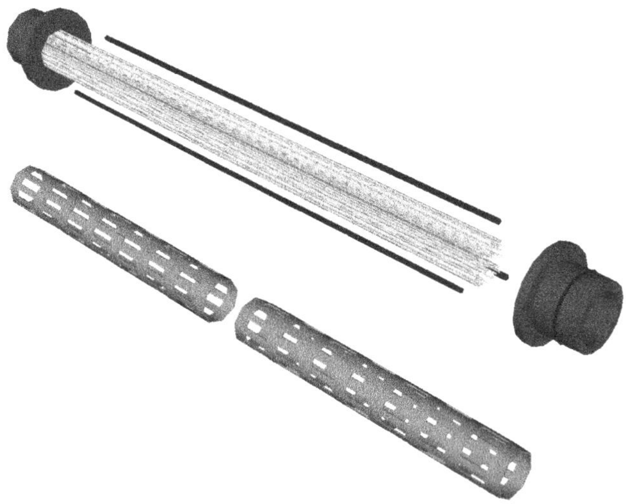

Figure 17.5 Hollow-fiber membrane module.

Figure 17.6 Hollow-fiber membrane module.

17.3 COMPUTER SIMULATION OF MODULE PERFORMANCE

An advanced simulation program based on a phenomenological/semiemprical approach in the pervaporation dehydration of solvents has been developed. This has been done to verify the actual and theoretical performance of the flat-sheet modules. An ethanol–water mixture was used as a model mixture. Some of the permeation parameters, which are used in the simulation model, were quantified from the real dehydration pervaporation of ethanol through a hydrophilic membrane. By adopting the coefficients determined from the real system, the simulation model has a more practical value. Its application to two basic processes, that is, continuous and batch processes, could highlight the importance of process parameters: for example, feed concentration, feed temperature, target feed quality in the process design, and optimization of pervaporation. In the following section, the theoretical development of the simulation using a finite elements-in-succession method is presented.

With respect to its design, a plate-and-frame module is characterized by: (a) dimensions of the plates, length l_{unit} and width w_{unit}, (b) height of the feed channel, and (c) number of membrane sheets in a unit and number of units making up a module. In this study, two arrangements of the feed channels are considered in a module unit: (1) series connection and (2) parallel connection (Fig. 17.7).

When a vacuum is applied to the permeate side of the membrane, a driving force and activity gradient can be created across the membrane thickness. Selective permeation takes place, and then feed flow rate as well as feed composition change through the differential volume. Since the heat of evaporation is supplied from the feed side, the feed temperature falls constantly, and so the flux through the membrane decreases. Therefore, three different balances over the differential volume are taken into account as follows:

Mass Balance:

$$\frac{d}{dz}F = -J(2w_{\text{unit}}) \tag{17.1}$$

Concentration Balance:

$$\frac{d}{dz}Fx = -J\,y(2w_{\text{unit}}) \tag{17.2}$$

Heat Balance:

$$\frac{d}{dz}Fh_F = -J\,\Delta h_v\,(2w_{\text{unit}}) \tag{17.3}$$

where F denotes feed flow rate, J is total flux, and w_{unit} the width of unit, x is the concentration of a selectively permeating component in the feed, y is the concentration of a selectively permeating component in permeate, h_F is the enthalpy of feed flow, and Δh_v is the heat of the evaporation of permeate. Equations (17.1)–(17.3) can be rewritten as follows, respectively:

$$dF = -2J\,w_{\text{unit}}\,dz \tag{17.4}$$

$$x\,dF + F\,dx = -2Jy\,w_{\text{unit}}\,dz \tag{17.5}$$

$$FC_p\,dT = -2J\,\Delta h_v\,w_{\text{unit}}\,dz \tag{17.6}$$

where C_p is the heat capacity of the feed liquid. From Eqs. (17.4) and (17.5), the following equation can be obtained:

$$dx = \frac{2Jw_{\text{unit}}}{F}\,(x - y)\,dz \tag{17.7}$$

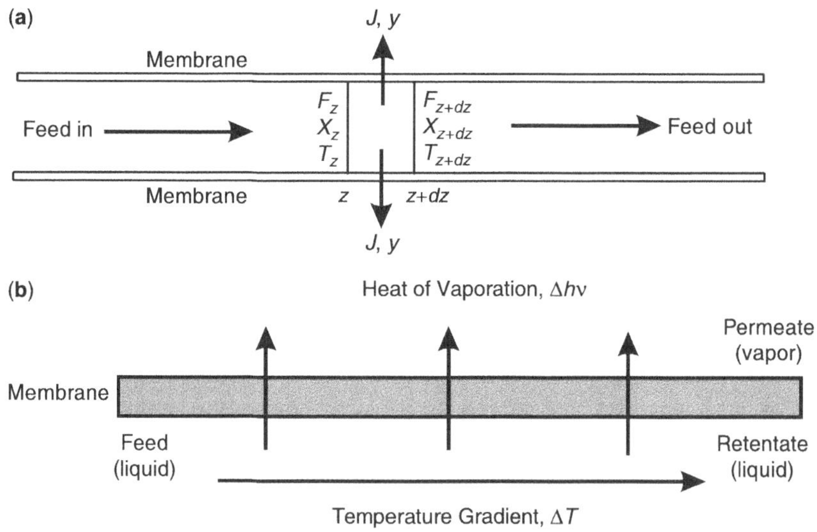

Figure 17.7 (a) Feed channel and flows through a differential element of volume (dz) and (b) temperature gradient along feed flow.

The changes in feed flow rate, feed composition, and feed temperature along the z direction can be calculated if the flux and permeate composition are expressed as a function of both feed composition and feed temperature:

$$J = f(x,\ T) \tag{17.8}$$

$$y = g(x,\ T) \tag{17.9}$$

Hence, flux, feed composition, and permeate composition should be expressed as a function of location along with z direction.

The simulation is carried out by an algorithm based on a numerical method, a so-called succession of elements method. This method involves dividing the unit length into finite elements. Within these elements it is assumed that the mass transfer driving force, which incorporates the activity coefficients on the permeate side and feed side, as well as the component properties, remain constant. The simulation of one finite element can then be completed. This approach has the advantage of different parameters influencing the overall process and can be implemented in modular form into the simulation program. By using the simulation model described above, the numerical method will be applied to simulate permeation performance in two processes, that is, a continuous process and a batch process. In a unit of plate-and-frame module, a number of membrane sheets are stacked. The feed channels are connected to each other in a series or arranged in parallel. Generally, the series connection of the channels is used in the continuous pervaporation process, while the parallel connection is used in the batch process.

17.3.1 Continuous Pervaporation Process

In the continuous process, feed mixture flows through a sequence of channels. Thus, the final retentate can be a product. If each membrane sheet is not large enough for permeation through it to significantly change the feed composition or feed temperature, each feed channel can be taken as an element volume in the numerical method. Then, the differential form of the parameters involved in Eqs. (17.4), (17.6), and (17.7) can be transformed into a different form, respectively, as follows;

$$\Delta F = F(k+1) - F(k) = -2J(k)A_m \tag{17.10}$$

$$\Delta T = T(k+1) - T(k) = -\frac{2J(k)\,\Delta h_v}{F(k)\,C_p} A_m \tag{17.11}$$

$$\Delta x = x(k+1) - x(k) = \frac{2J(k)}{F(k)} [x(k) - y(k)]\,A_m \tag{17.12}$$

where $F(k)$, $T(k)$, $J(k)$, $x(k)$, and $y(k)$ are the respective parameters at the kth feed channel and A_m is a membrane area of each membrane sheet. The initial condition for each parameter can be given:

$F(1) =$ initial feed flow rate
$T(1) =$ initial feed temperature
$x(1) =$ water concentration in initial feed
$J(1) = f[x(1), T(1)]$
$y(1) = g[x(1), T(1)]$

Now retentate temperature, retentate composition, retentate flow rate, flux, and permeate composition at a location in the module unit can be determined by the numerical method.

17.3.2 Batch Pervaporation Process

In the batch process, the feed channels in a module unit are arranged in a parallel connection so that feed mixture can feed each channel in a parallel mode. Feed mixture circulates from a feed tank through a membrane module for a given period of time. During the circulation of feed mixture through the membrane module, feed volume will decrease with time as much as the permeate amount permeating through the membrane sheets in the module. Therefore, total mass balance for feed volume in the system for a differential time interval, dt, is given as follows:

$$dM = -Jn\,A_m\,dt \tag{17.13}$$

where M denotes a feed mass and n is a number of sheets stacked in a membrane module unit. For a component permeating preferentially through a membrane, the mass balance equation is expressed by

$$d(Mx) = -y\,JnA_m\,dt \tag{17.14}$$

$$x\,dM + M\,dx = -yJnA_m\,dt \tag{17.15}$$

$$dx + x\frac{dM}{M} = -\frac{yJnA_m}{M} dt \tag{17.16}$$

Combining Eqs. (17.13) and (17.16) yields

$$\frac{dM}{M} = -\frac{dx}{x+y} \tag{17.17}$$

Thus, rewriting Eq. (17.16) gives

$$dx = -\left(\frac{x+y}{y}\right)\left(\frac{yJnA_m}{M}\right) dt \tag{17.18}$$

Feed mass at time t_k, M is expressed by initial feed mass M_0 taken by permeating amount M_k through membrane accumulated for a period of time, t_k:

$$M = M_0 - M_k = M_0 - nA_m \int_0^{t_k} J\, dt \tag{17.19}$$

To calculate feed composition and feed amount with permeating time, the same numerical method as used in the continuous process can be adopted. Total time period is divided equally into finite element time interval, Δt. Feed amount $M(k+1)$ and concentration $x(k+1)$ at the $(k+1)$th time interval are

$$M(k+1) = M_0 - nA_m \sum_{k=1}^{k} J(k)\,\Delta t \tag{17.20}$$

$$x(k+1) = x(k) - \left(\frac{x(k)+y(k)}{y(k)}\right)\left(\frac{y(k)\,J(k)\,n\,A_m}{M(k)}\right)\Delta t \tag{17.21}$$

In the batch process, since the heat of evaporation is supplied from a heater equipped in the feed tank, the feed temperature is assumed to be constant with permeating time. Therefore, flux and permeate concentration can be a function of feed composition in the batch process. The initial condition for each parameter can be given as

$M(0) = M_0$
$T =$ initial feed temperature
$x(1) =$ water concentration in initial feed
$J(1) = f[x(1), T]$
$y(1) = g[x(1), T]$

Now, retentate composition and retentate amount with permeating time can be determined by the numerical method.

AzeoSep-2002 commercial membrane (Petro Sep Membrane Technology Inc., Ontario, Canada) is a composite-type membrane. It is very effective for the dehydration of a wide range of organic solvents, such as, alcohols, ethers, acetates, and so on.

To produce permeation data for a simulation model, pervaporation experiments were carried out with a lab-scale permeation apparatus. Ethanol–water mixtures were used as a model mixture. The feed tank holds a volume of 4 L of feed mixture and has a function of temperature control. A permeation cell made of stainless steel holds a flat membrane of 154 cm^2. Feed composition ranged from 0.4 to 7 wt% water contents. Feed temperatures used were 60–90°C. Permeate pressure was kept below 5 Torr. In all measurements, feed flow rate was above 5 L/min to minimize a concentration polarization effect at feed mixture. The feed and permeate compositions were determined by Karl-Fisher titration. The permeate fluxes were determined by weighing the permeate samples in cold traps.

17.3.3 Determination of Permeation Parameters and Verification of Simulation Model

The permeation data of ethanol–water mixtures through the AzeoSep-2002 membrane, that is, flux and permeate concentration were obtained at different feed compositions. The curve fitting of the permeation data was performed by using software of Sigma Plot-2000 and produced the following dependencies of flux (J) and permeate concentration (y) on both feed composition and feed temperature temperatures, as shown in Eqs. (17.8) and (17.9):

$$J = A_0 \exp\left(-\frac{E_p}{T}\right)$$

$$A_0 = 295.5 \exp(2.8441x) \tag{17.22}$$

$$E_p = 3300.04 + 839.58x$$

$$y = \left[(ax)^{-2} + 98.3^{-2}\right]^{-0.5}$$

$$a = 440.9 - \frac{112{,}700}{T} \tag{17.23}$$

where y is water concentration in permeate (wt%), x a water concentration in feed (wt%), J a flux $[\mathrm{kg}/(\mathrm{m}^2 \cdot \mathrm{h})]$, A_0 a preexponential factor $[\mathrm{kg}/(\mathrm{m}^2 \cdot \mathrm{h})]$, and E_p a permeation activation energy (cal/mol). The preexponential factor and permeation activation energy were obtained as functions of feed composition. The determined permeation functions were in agreement with experimental data within $\pm 4.5\%$ in the given range of operating condition, as shown in Figure 17.8 (Yeom et al., 2002).

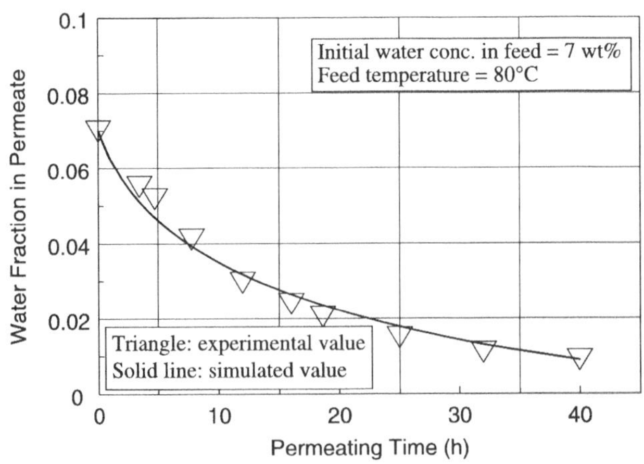

Figure 17.8 Comparison of simulated feed composition to experimental value with permeating time in dehydration of ethanol through AzeoSep-2002.

17.4 PERMEATION AND SEPARATION MODEL IN HOLLOW-FIBER MEMBRANE MODULE

C. K. Yeom, of Petro Sep Membrane Research Inc., demonstrated by simulation the permeation and the separation in the hollow-fiber membrane module. Feed mixture circulates from the feed tank through the hollow-fiber membrane module for a given period of time as described in Figure 17.9. During the circulation of the feed through the membrane module, selective permeation takes place across a hollow-fiber membrane thickness in the module so that the feed amount in the tank, M_F, decreases as much as the amount permeating through the membranes. Thus, the total mass change in the feed tank is balanced with the permeation amount through the membrane module for a differential time interval, dt, as follows:

$$dM_F = -J_M A_T \, dt \tag{17.24}$$

where J_M is an average flux through total membrane area, A_T, in the membrane module. The amount of water in the feed tank, which is a preferentially permeating component, can be affected by water-selective permeation through the membranes:

$$d(M_F x_F) = -J_M y_M A_T \, dt \tag{17.25}$$

where y_M is a water content in permeate. Equation (17.25) can be rewritten

$$x_F \, dM_F + M_F \, dx_F = -J_M y_M A_T \, dt \tag{17.26}$$

Combining Eqs. (17.24) and (17.26) gives

$$dx_F = \frac{(x_F - y_M) J_M A_T}{M_F} dt \tag{17.27}$$

Figure 17.9 A 20,000-L/day capacity AZEO SEP pervaporation plant for dehydration of solvents.

and M_F, a feed amount in the tank at time t, can be given by

$$M_F = (M_F)_0 - \int_0^t (A_T J_M)\, dt \tag{17.28}$$

where $(M_F)_0$ is an initial feed amount, and the integration term expresses an accumulated permeate amount for a permeating time t.

To calculate feed amount at a time t, the membrane performance over the hollow-fiber membrane module should be characterized with permeating time and location along with hollow-fiber length. The permeation through a single hollow fiber can be considered as a continuous process described in Figure 17.10. It schematically represents the feed flow through a differential volume (dz) inside the hollow fiber. When a driving force activity gradient is created across the membrane thickness, selective permeation takes place and then feed flow rate as well as feed composition changes through the differential volume. The feed temperature falls constantly because the heat of the evaporation of the permeate is supplied from the feed side, resulting in decreasing flux. Thus, three different balances

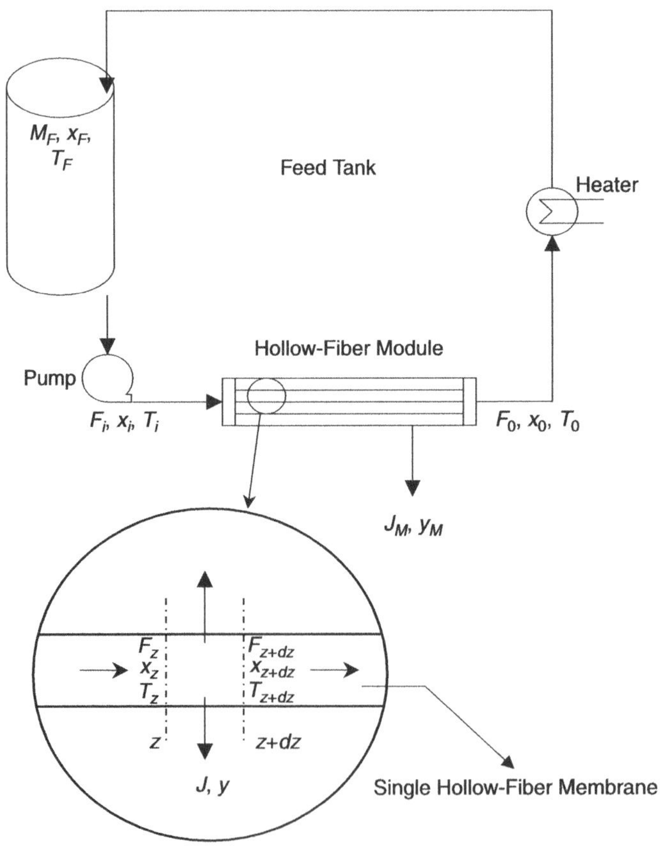

Figure 17.10 Model of single hollow-fiber membrane.

over the differential volume are taken into account as follows:

$$dF = -JD_i m\pi\, dz \tag{17.29}$$

$$d(Fx) = -JymD_i\pi\, dz \tag{17.30}$$

$$d(Fh_F) = -J\,\Delta h_v\, mD_i\pi\, dz \tag{17.31}$$

where F denotes the feed flow rate at z, D_i the inside diameter of the hollow fiber, J the total flux, m the number of hollow fibers in the module, x the concentration of a selectively permeating component in the feed, y the concentration of the component in the permeate, h_F the enthalpy of feed flow, and Δh_v the heat of the evaporation of permeate. Equation (17.30) and (17.31) can be rewritten as follows, respectively:

$$x\,dF + F\,dx = -JymD_i\pi\, dz \tag{17.32}$$

$$FC_p\,dT = -J\,\Delta h_v\, mD_i\pi\, dz \tag{17.33}$$

where C_p is the heat capacity of the feed liquid. From Eqs. (17.29) and (17.32), the following equation can be obtained:

$$dx = [(x - y)J]\frac{mD_i\pi}{F}dz \tag{17.34}$$

At a position z in hollow-fiber length,

$$F = F_i - mD_i\pi\int_0^z J\,dz \tag{17.35}$$

where F_i is an initial feed flow rate at the inlet of the module. The changes of feed flow rate, feed composition, and feed temperature along with z direction can be determined. This can be done when flux and permeate are expressed as functions of both feed composition and feed temperature as follows, respectively:

$$J = f(x, T) \tag{17.36}$$

$$y = g(x, T) \tag{17.37}$$

Hence, these three parameters can also be expressed as a function of location along with z direction. Finite difference schemes are employed to get a numerical solution. This method involves dividing permeating time and hollow-fiber length into finite elements, respectively. In the continuous process, feed mixture flows through a sequence of finite element volumes. This occurs to such an extent that the feed in an element volume closer to the outlet of the module has a longer residence time in the module at a given permeating time. A two-dimensional finite difference grid is applied to each to obtain a numerical solution, as shown in Figure 17.11. The horizontal directional parameter j denotes the jth element volume from the entrance of the module while vertical directional parameter i refers to feed introduced into the module at ith time interval. Thus, a coordinate (i, j) in the grid indicates a feed that is located at the jth element volume from the inlet of the

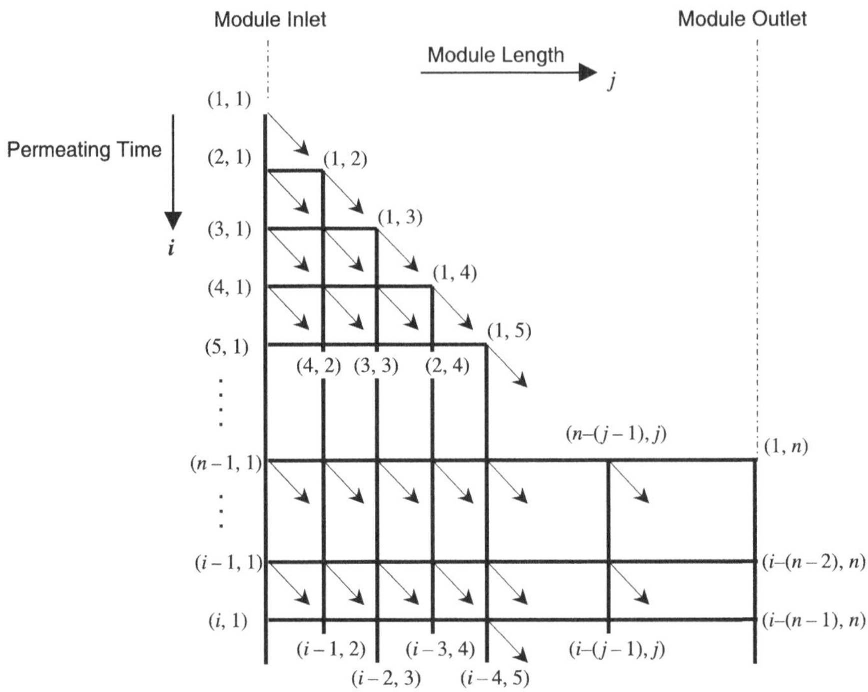

Figure 17.11 Finite difference grid.

module and introduced into the module at the ith time interval. The differential forms of the parameters involved in Eqs. (17.33) and (17.34) can be transformed into a different form, respectively, as follows:

$$T(i, j+1) = T(i, j) - \frac{J(i, j)\,\Delta h_v m D_i \pi}{F(i, j) C_p} \Delta z \tag{17.38}$$

$$x(i, j+1) = x(i, j) + [(x(i, j) - y(i, j))J(i, j)] \frac{m D_i \pi}{F(i, j)} \Delta z \tag{17.39}$$

where $T(i, j)$, $F(i, j)$, $J(i, j)$, $x(i, j)$, and $y(i, j)$ are the respective parameters at the jth element volume, which is introduced into the module at the ith time interval. The initial and boundary condition for each parameter can be given as

$x(1, 1) = x_0$: initial water concentration in feed
$x(i, 1) = xi$: water concentration to enter the membrane module
$T(1, 1) = T(i, 1) = T_F = Ti$: feed temperature in the tank
$F(1, 1) = F(i, 1)$: initial feed pumping speed

The feed temperature, feed composition, feed flow rate, flux, and permeate composition at a location in the module and a permeating time can be determined by the numerical method. When feed enters into the module at the ith time interval, the sequence of time interval in a respective element volume can be given as $i-(j-1)$ with a position, j, showing a residence time distribution along with membrane length as described in Figure 17.11. The permeation parameters at a coordinate (i, j) can be expressed as $J(i, j)$ and $y(i, j)$, which are flux and water content in the permeate, respectively. In Eq. (17. 28), $J_M A_T$ is a total permeation amount through the membrane module per unit time. Now the parameters in Eq. (17.28), that is, the total permeation amount and feed amount M_F can be calculated by using the finite difference scheme:

1. For $i < n$ that is a number of element volumes divided:

$$J_M(i)A_T = mD_i\pi \sum_{j=1}^{j} J(i-(j-1), j)\,\Delta z \tag{17.40}$$

$$M_F(i) = (M_F)_0 - mD_i\pi \sum_{i=1}^{i}\sum_{j=1}^{j} J(i-(j-1), j)\,\Delta z\,\Delta t \tag{17.41}$$

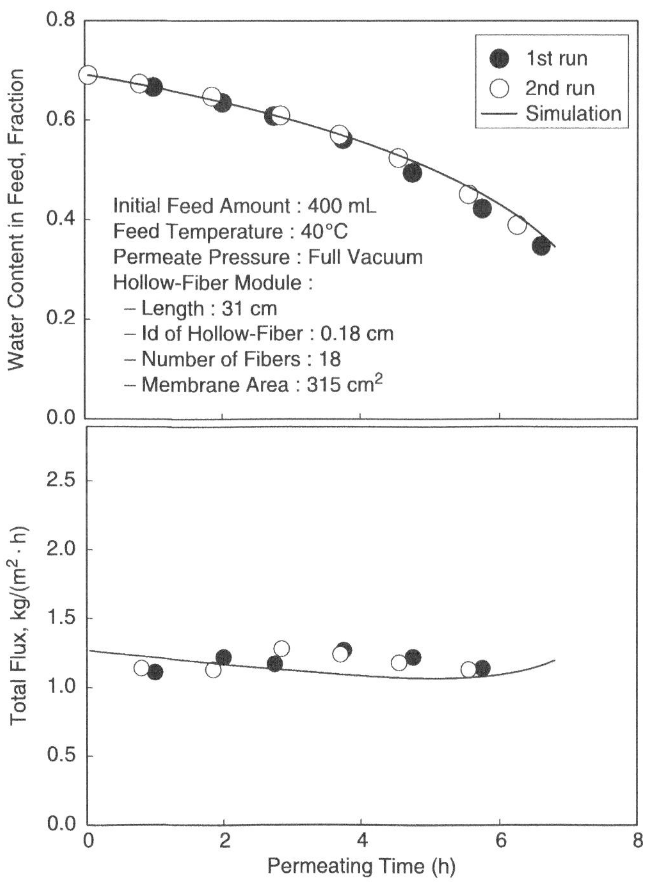

Figure 17.12 Comparison of simulated values with real values in permeation of aqueous organic solution.

2. For $i \geq n$:

$$J_M(i)A_T = mD_i\pi \sum_{j=1}^{n} J(i-(j-1), j)\,\Delta z \tag{17.42}$$

$$M_F(i) = (M_F)_0 - mD_i\pi \sum_{i=1}^{i}\sum_{j=1}^{n} J(i-(j-1), j)\Delta z\,\Delta t \tag{17.43}$$

Equation (17.27) can be transformed into a different form as follows:

$$x_F(k+1) = x_F(k) + \frac{(x_F(k) - y_M(k))J_M(k)A_T}{M_F(k)}\Delta t \tag{17.44}$$

where the parameter k refers to feed in the feed tank at the kth time interval in the finite difference grid applied to each to obtain a numerical solution. The initial condition for each parameter can be given as

$x(1) = x_0$: initial water concentration in feed
$M_F(1) = (M_F)_0$
$J(1) = f[x(1), T_F]$
$y(1) = g[x(1), T_F]$

Figure 17.13 A 10,000-L/day capacity pervaporation AZEO SEP plant for dehydration of solvents based on hollow-fiber membranes.

Figure 17.14 VOC SEP hollow-fiber pervaporation plant to treat 10,000 L of water per day.

Figure 17.12 presents a comparison of membrane performance simulated by using the model equations described above with measured real values. The comparison shows an excellent agreement between them.

17.5 CONCLUSION

Since the early twentieth century, there has been remarkable efforts worldwide to develop pervaporation membrane technology. Many researchers, scientists, and engineers tried to resolve the problems of the chemical industry in processing azeotropic solvent mixtures. Companies have invested millions of dollars to commercialize pervaporation membrane technology. However, there was no great success in processing azeotropic solvents other than in the alcohol industry for alcohol dehydration. The cost of these plants was quite prohibitive. Only a few major projects were completed worldwide. These plants were based on plate-and-frame modules for alcohol dehydration. Later, pervaporation technology was challenged by molecular sieve technology in the ethanol industry. This competition basically eliminated the use of pervaporation technology in the field of ethanol industry. On the other hand, manufacturers of pervaporation technology were not very successful in processing complex mixtures. Therefore, the use of pevaporation technology and the research started to decline by the early 1990s. Over the last 10 years, membrane researchers and membrane manufacturers once again put enormous efforts in the field of pervaporation membrane technology. They demonstrated that pervaportion membrane technology is the solution for processing azeotropes for various combinations of solvents and solvent mixtures with water. Especially in hollow-fiber applications, there have been extremely intricate challenges to develop solvent-resistant membranes as well as a solvent-resistant epoxy. For the first time in the history of pervaportion, it has been demonstrated commercially that hollow-fiber membrane modules can successfully process azeotropes and can dehydrate many solvent mixtures. This is possible even with an operating temperature of up to 120°C. The use of hollow fibers in the pervaporation technology makes this technology a very economical option for the process industry in the field of complex separation problems.

REFERENCES

Binning, R. C., Lee, R. J., Jenning, J. F., and Martin, E. C. (1961). Separation of liquid mixture by permeation. *Ind. Eng. Chem.* 53, 45–50.

Kahlenberg, L. (1906). On the nature of the process of osmosis and osmotic pressure with observations concerning dialysis. *J. Phys. Chem.* 10, 141.

Kober, P. A. (1917). Pervaporation, perstillation and percrystallization. *J. Am. Chem. Soc.* 39, 944–950.

Schwob, Y. (1949). Sur l'semipereabilite a l'ean, des membrane de cellulose regeneree. *Toulouse (France)*, May 23.

Yeom, C. K., Kazi, M., and Baig, F. (2002). Simulation and process design of pervaporation plate-and-frame modules for dehydration of organics solvents. *Membr. J.* 12, 226–239.

CHAPTER 18

Biomedical Applications of Membranes

G. CATAPANO

Department of Chemical Engineering and Materials, University of Calabria, Rende (CS), Italy

J. VIENKEN

Fresenius Medical Care, Bad Homburg, Germany

18.1 INTRODUCTION

Artificial membranes are used in medicine to treat blood for a broad variety of therapeutic purposes. In most treatments, blood is continuously withdrawn from the patient's blood circulation and brought in extracorporeal circulation into direct contact with the artificial membranes assembled in the device where solutes are permselectively removed from or supplied to it. The treated blood is then given back to the patient. Treatment sessions generally last a few hours and may have to be repeated a few times a week, sometimes for years. In Japan, more than 5000 uremic patients have been reported to be on hemodialysis for longer than 25 years (Patient Registration Committee, 2004).

In these treatments, membranes are used as permselective barriers to permit transport of selected solutes to/from the blood, while hindering the loss of essential blood constituents, and to regulate the rate at which solutes are transferred across the membrane so as to maintain the patient's homeostasis. For instance, waste metabolites are removed from renal failure patients by hemodialysis; excess water is removed from fluid overload patients by hemofiltration; plasma is separated from the blood cells of patients with immune-mediated diseases and removed by therapeutic apheresis; gaseous O_2 and CO_2 are supplied to and removed from the blood, respectively, in gas exchangers (often termed artificial lungs). A number of factors contribute to making the development of adequate membranes and devices for such processes rather complex, among which are the broad spectrum of molecular weights and physical-chemical properties of the species to be transported or rejected, the interactions between membranes and body fluids that continuously modify both membrane and blood properties, and the need for materials and processes that cause neither short-term nor long-term harm to the patient. It should also be recalled that membrane capacity to meet the set therapeutic objective, in terms of the solutes transported and the actual rate at which they cross the membrane, is determined by the intrinsic

Advanced Membrane Technology and Applications. Edited by Norman N. Li, Anthony G. Fane, W. S. Winston Ho, and T. Matsuura
Copyright © 2008 John Wiley & Sons, Inc.

mass transport properties of the membrane, but also by the conditions under which solutes and cells are transported from the bulk of the membrane-contacting fluids to the membrane surface. These effects and the process engineering aspects of the treatments are dealt with in other publications, some of which are referenced in Bronzino (1995), and are not discussed in detail here.

In this chapter, the state-of-the-art artificial membranes used in clinically established blood treatment processes is reviewed. Information on medical membrane preparation and surface modification is often not disclosed. For this reason, the typical features of commercial medical membranes will be reported as provided by the manufacturers. The criteria that have driven membrane development will also be discussed to inform membranologists about what is currently available commercially and the opportunities for further development in selected medical treatments. Proper consideration is given to the influence of membrane–blood interactions on membrane performance and the therapeutic success of a membrane-based treatment. Membranes used in accessories for the above treatments or for related pharmaceutical purposes are not considered.

18.2 MEMBRANE THERAPEUTIC TREATMENTS

Membranes have been used in clinics for therapeutic purposes since 1943, when Kolff successfully treated a uremic patient with his rotating drum membrane dialyzer (Kolff and Berk, 1943). Nowadays, hemodialysis, hemofiltration, hemodiafiltration, therapeutic apheresis, and gas exchange during cardiopulmonary bypass surgery have become clinically established treatments. In the following, features and therapeutic objectives of these treatments are briefly presented.

18.2.1 Hemodialysis

Hemodialysis (HD) is the most common therapeutic treatment using artificial membranes, with nearly 1.4 million uremic patients treated worldwide in the year 2004 (Grassmann et al., 2005). HD is a life-sustaining procedure for replacing, or augmenting, the normal renal excretory functions to clear the blood of low-molecular-weight (LMW) waste metabolites or toxins in the treatment of acute or chronic renal failure and in drug detoxification, respectively. HD is meant to simulate the excretory function of the one million nephrons in each kidney, which are the functional units in charge of the elimination of waste metabolites and of maintaining the body fluid balance. Schematically, the nephron consists of a negative charged filter, the *glomerulus*, where blood is depleted of waste metabolites and electrolytes, followed by the tubules, a series of reabsorbing units where essential species and electrolytes are reabsorbed. The concentrated waste leaving the tubules is removed in the urine. In commercial hemodialyzers, the blood flows in the lumen of hollow-fiber semipermeable membranes (ca. 0.4–2.6 m^2 surface area) that keep it separate from an essentially physiologic electrolyte solution (i.e., the dialysate) flowing in the dialyzer shell countercurrently, as shown in Figure 18.1a. In HD, unlike the nephron, LMW solutes (i.e., MW up to ca. 1500) are removed from the blood into the dialysate by diffusion driven by the existing transmembrane concentration gradient. Membranes act as barriers that retain blood proteins and cells and regulate the solute flux based only on its diffusivity in the membrane, hence its molecular weight (see Section 18.3). The size-selective nature of transport requires the addition of electrolytes to the dialysate to prevent their depletion

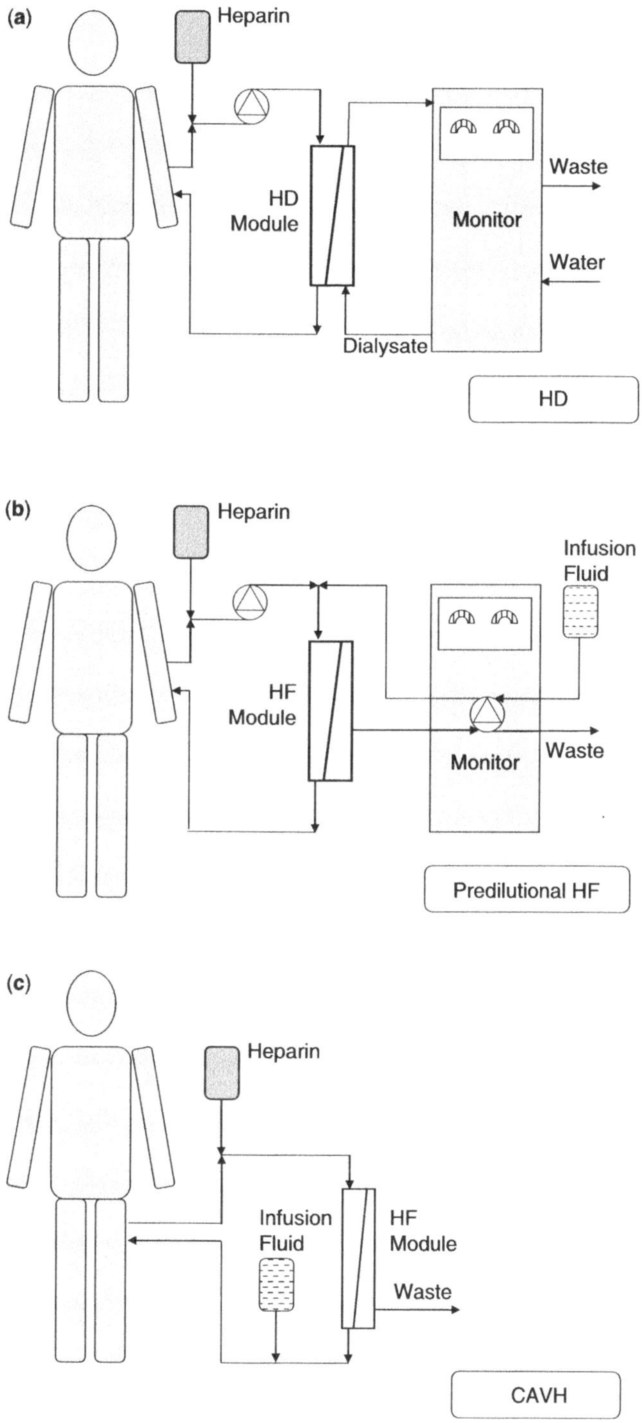

Figure 18.1 Schemes of some clinically established membrane blood treatments. For nomenclature see text.

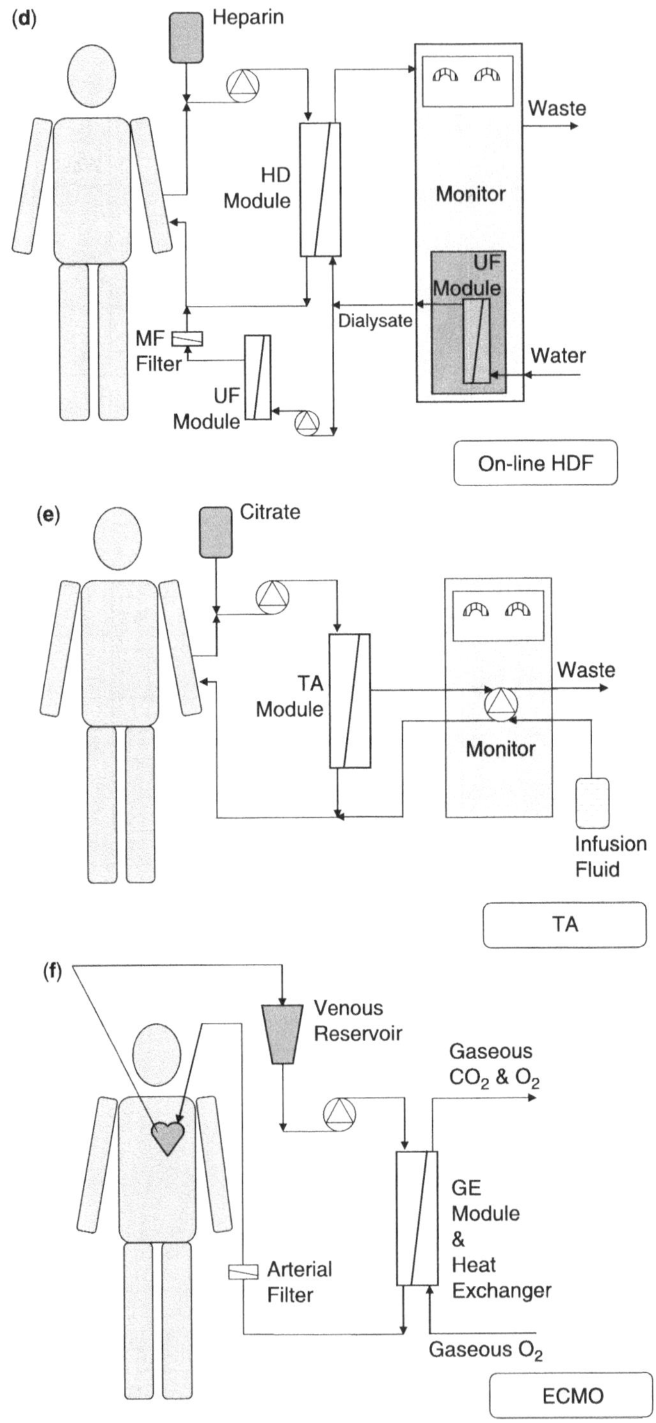

Figure 18.1 *Continued.*

and to equalize the transmembrane osmolarity. This causes essential LMW solutes to be lost in the dialysate together with LMW waste metabolites. While effective in the removal of LMW solutes, the typical HD membrane sieving properties (see Section 18.3) do not permit larger toxins to be removed as would occur in the *glomerulus*. In HD, the excess water accumulated during the interdialytic period is filtered out by transiently applying a transmembrane pressure (TMP) difference to maintain the patient's fluid balance. The low hydraulic permeability, L_p, typical of HD membranes (see Section 18.3) makes control of the removed fluid volume quite accurate and easy. The typical HD treatment generally lasts 3–5 h and is repeated 3 times per week.

18.2.2 Hemofiltration and Hemodiafiltration

Filtration processes, such as hemofiltration (HF), continuous arterovenous ultrafiltration (CAVU) or hemofiltration (CAVH), and hemodiafiltration (HDF), were introduced at the end of the 1970s to remove middle molecular weight (MMW) waste metabolites and toxins that are normally cleared by the kidneys but accumulate in chronic HD patients and have suspect pathological relevance (Babb et al., 1972). More like the nephron, in HF, CAVU, and CAVH plasma water and low-to-middle MW solutes (i.e., MW up to ca. 15,000) are removed from the blood flowing in cross-flow mode and transported across ultrafiltration-type membranes (ca. 0.5–2.4 m^2 surface area) by a convective mechanism (i.e., by solvent drag), as an effect of the applied transmembrane pressure gradient (see Fig. 18.1b–c). The last two treatments basically differ from HF for the access site and the use of smaller filters (ca. 0.07–1.8 m^2 surface area), less resistant to blood flow. HF and HDF membranes are about an order of magnitude more permeable to plasma water than HD membranes. Under typical operating conditions, such high permeability permits plasma water filtration across the membranes at up to ca. 80–100 mL/min with quick removal of the fluid overload but requires strict water flux control to maintain the patient's fluid balance. The membrane acts as a sieve that retains high-molecular weight (HMW) blood proteins and cells, and regulates the solute flux based only on its size, shape, and possible interactions with the membrane material. In particular, LMW solutes that do not interact with the membrane material and cross the membrane unhindered are all eliminated at the same rate, their plasma concentration being equal to that in the filtrate. HF membranes are typically more permeable to small proteins than HD membranes but have to retain larger essential proteins as albumin. Their separation properties more closely resemble those of the *glomerulus* (Streicher and Schneider, 1983). These features make HF very effective in removing MMW solutes but do not permit LMW solutes to be removed as efficiently as in HD (see Section 18.2.4). Additionally, in HF essential solutes (e.g., amino acids), and even some albumin, are discarded in the filtrate together with the waste metabolites. Fluid and electrolyte balance is maintained by infusing a sterile, nonpyrogenic electrolyte solution in the blood upstream (i.e., predilution mode) or downstream (i.e., postdilution mode) from the hemofilter at flow rates matching those of the filtrate.

In hemodiafiltration, diffusive solute removal is superimposed to purely convective transport across HF membranes by circulating a dialysate in the shell of a hemofilter counter currently with respect to the blood. Superimposition of diffusive transport, driven by a concentration gradient, to convection augments the poor LMW solute clearance typical of HF, while still retaining the good MMW solute clearance typical of HF (Legallais et al., 2000). Different techniques to deliver the replacement fluid have been proposed to contain the cost of the treatment (Sternby et al., 1992). Two solutions appear to be the most cost effective. In the first solution, the electrolyte-containing water lost in the first hemodiafiltration module is partially backfiltered from the dialysate into the blood in a second hemodiafiltration module in series with the first one. In the second solution, the replacement fluid is continuously produced by on-line filtration of the dialysate, as shown in Figure 18.1b. Patients on HF or HDF are typically treated for the same time and frequency as those on HD.

18.2.3 Therapeutic Apheresis (or Plasmapheresis)

Therapeutic apheresis (TA) (or plasmapheresis) is the treatment that aims at the nonspecific removal from the blood of pathogens (such as excess proteins, autoantibodies, macroproteins, immune complexes, HMW toxins, etc.) present at toxic (or pathogenic) concentrations to restore the function of the reticulo-endothelial system (Gurland et al., 1986). Apheresis is the process in which a "specific component of the blood is separated and removed with the remainder of the blood returned to the patient," and has been used since 1944 to safely collect plasma, plasma components, or cells from donor blood (Zydney, 1995). When pharmacological treatment is ineffective, apheresis is used therapeutically (hence TA) for the treatment of autoimmune, metabolic, and hematologic disorders in which soluble species are proven (or thought) to mediate progression of the disease, whose removal might be beneficial to the patient. Well over 50 diseases have been treated by plasmapheresis [American Society for Apheresis (ASFA), 2005]. However, the ASFA and the American Association of Blood Banks (AABB) endorse the use of TA as a standard and acceptable primary therapy or as valuable first-line adjunct therapy only for the treatment of Goodpasture syndrome, thrombotic thrombocytopenic purpura, chronic inflammatory demyelinating polyneuropathy, myasthenia gravis, and the Guillan Barré syndrome (ASFA, 2005). Japanese health insurance policy is reported to qualify TA as an acceptable therapy for at least another 15 diseases (AsahiKasei, 2008). Pathogen removal with the plasma, and return of the concentrated blood cell suspension to the patient, may be achieved either by continuous blood centrifugation or by cross-flow filtration of blood with artificial microporous membranes. In this chapter, reference is made only to the latter process.

In TA, the blood is filtered in cross-flow mode across microporous membranes (ca. 0.2–0.5 m^2 surface area) with 0.2–0.6 μm maximal pore size about one order of magnitude more permeable to plasma water than HF membranes (see Fig. 18.1e). A large volume of plasma (containing water, dissolved essential species, and pathogens) is convectively removed at a flow rate of 10–60 mL/min from the blood and across the membrane as an effect of the applied transmembrane pressure gradient and is discarded. The membrane acts as a sieve and retains only the cells in the concentrated bloodstream. Prior to returning to the patients, the blood is reconstituted downstream from the TA membrane module by the continuous infusion of an equivalent volume of fresh frozen plasma from donor blood or of a sterile nonpyrogenic buffered physiologic solution containing electrolytes and donor human albumin. Plasma proteins or albumin in the replacement fluid are needed to maintain the oncotic pressure and prevent fluid shifts from other body compartments with subsequent complications. In typical TA treatment, 2–3 L of plasma is removed up to 3–4 times a week. Treatment may be continued for several weeks, or even years but with a lower frequency, depending on the etiology of the disease and the patient's response to the treatment.

18.2.4 Gas Exchange

The earliest devices for gas exchange (GE) were introduced in clinics in the 1950s. They were designed to replace entirely the gas exchange functions of the natural lungs (i.e., O_2 supply to and CO_2 removal from the blood) to oxygenate the blood when the heart has to be stopped for surgery intervention (such as coronary artery bypass grafting) or to assist the malfunctioning lungs temporarily, in the hope that the treatment would heal the pulmonary tissue or even lead to a permanent cure. In the early devices, attention was focused on maximizing oxygen delivery to the blood in bubble oxygenators. In fact, physiological rates of oxygen transfer to the blood were obtained by directly contacting the blood with bubbles of an oxygen-rich gas small enough to give a large contact surface area per unit blood volume (i.e., several millimeters in diameter) (Galletti and Colton, 1995). Since then, gas exchangers equipped with gas-permeable membranes, yet impermeable to liquids, have mostly replaced the earlier oxygenators in clinics, and CO_2 removal has been acknowledged to be of therapeutic relevance in maintaining the physiological blood pH. In these devices, the membrane (ca. 0.3–5 m^2 surface area) physically separates the blood, flowing at rates equal to the cardiac output, from an O_2-rich, CO_2-poor gas while providing a large contact area, as shown in Figure 18.1f. In typical cardiopulmonary bypass surgery, gas is exchanged with the blood for 2–4 h. Interposition of membranes between the blood and the gas phase permits the reduction of plasma protein denaturation, thus preventing the subsequent blood foaming and trauma associated with earlier oxygenators, and prevents blood volume fluctuations in the expansion chambers, thus minimizing the risk of blood volume shifts during total body perfusion. However, the membrane itself and the associated fluid–membrane interfaces present additional resistances to gas transfer across the membrane. Nowadays, microporous hydrophobic membranes with maximal pore size lower than ca. 0.1 μm are generally used that have a high enough surface tension to prevent plasma water from crossing the membrane, and through which gas can freely diffuse. In currently available gas exchangers, gaseous oxygen flows into the membrane lumen and blood flows outside the membranes arranged in stacks and through the interstices among them. In this arrangement, gas is transported across the external surface area of hollow-fiber membranes, larger than the inner luminal area. Moreover, the membranes can be arranged in the blood flow channel at angles that induce an effective blood mixing, with angles increasing up to 90° generally causing increased mixing. Mixing thins down the average thickness of the stagnant blood layer at the blood–membrane interface and significantly reduces resistance to oxygen transport (Mockros and Leonard, 1985; Catapano et al., 1992, 2001). O_2 and CO_2 transmembrane transport is largely influenced by mass transport resistance at the blood–membrane interface that, in these gas exchangers, is minimized by their convenient fluid dynamics. However, water vapor transport and condensation in the membrane lumen, and/or liquid water breakthrough across the membrane caused by its hydrophilization, limits the use of gas exchangers equipped with microporous membranes for long-term assistance to patients with acute respiratory deficiency. In this case, gas exchangers equipped with dense silicon membranes are generally preferred. The therapeutic efficacy of long-term extracorporeal GE in the treatment of acute respiratory deficiency is still controversial. For this reason, membranes used in this application are not discussed here.

18.3 MEDICAL MEMBRANE PROPERTIES

In medical treatments, membranes are used to transfer toxic (or essential solutes) perm-selectively from the blood to a stripping fluid (or vice versa) while retaining all essential blood constituents and to regulate the rate at which solute transfer occurs. The molecular weight of the species of interest is rather broad, ranging from 18 for water to that of large lipoproteins, immune complexes, and cells, the latter being well in excess of 10^6. So are their physical-chemical properties, ranging from very hydrophilic to very hydrophobic. In all cases, the actual rate at which solutes are transferred across the membrane and the separation efficiency depend on both membrane transport and separation properties, and on the fluid dynamics of the blood and the stripping fluid compartment, hence the device geometry and operation. In the following, the parameters more commonly used to characterize the intrinsic membrane performance (i.e., in the absence of external resistance to transport) are presented and their typical values reported. The interactions between membranes and the body fluids, and their effects on membrane properties, are also briefly examined.

18.3.1 Transport Properties

In this Section, it is implicitly assumed that the mass transport resistance at the fluid–membrane interface on either side of the membrane is negligible. Also the following is information that is made available publicly by the membrane manufacturers, when not otherwise noted. As in technical processes, mass transport across semipermeable medical membranes is conveniently related to the concentration and pressure driving forces according to irreversible thermodynamics. Hence, for a two-component mixture the solute and solvent capacity to permeate a semipermeable membrane under an applied pressure and concentration gradient across the membrane can be expressed in terms of the following three parameters: L_p, hydraulic permeability; P_M, diffusive permeability; and σ, Staverman reflection coefficient (Kedem and Katchalski, 1958). All of them are more accurately measured experimentally because a limited knowledge of membrane structure means that theoretical models provide rather inaccurate predictions.

Parameter L_p is the reciprocal membrane hydraulic resistance to solvent (i.e., plasma water or plasma) flow, and is estimated as the initial slope of the solvent filtrate flux dependence on the applied TMP difference. When pressure drop along the module length on either membrane side cannot be neglected, TMP is often estimated by averaging out the inlet and outlet pressures as follows:

$$\text{TMP} \approx \frac{P_{Bi} + P_{Bo}}{2} - \frac{P_{Di} + P_{Do}}{2} \tag{18.1}$$

A finite TMP is needed to filter blood across HF membranes to counteract the oncotic pressure of the rejected species that accumulate at the membrane surface.

Transport models relate L_p to HD membrane structure, as follows (Klein et al., 1978):

$$L_p = \frac{A_p\, r_p^2}{8\, \mu\delta} \tag{18.2}$$

Equation (18.2) suggests that L_p increases for membranes featuring higher porosity and pore size but decreases when membrane thickness increases and fluids more viscous than water are filtered. Typical L_p values toward blood range from about 3–10 mL/(h/mm Hg m^2) for HD membranes, to about 30–50 mL/(h/mm Hg m^2) for HF and HDF membranes, to about 500 mL/(h/mm Hg m^2) for TA membranes.

Parameter P_M is the reciprocal resistance offered by the membrane to purely diffusive solute transport. Transport models relate P_M of HD membranes to their structure and solute mobility in the membrane as follows (Klein et al., 1978; Colton and Lowrie, 1981):

$$P_M = K \frac{A_p \, D(1 - q_m)^2}{K_{1,m} \, \delta} = K \frac{\varepsilon K_r D}{\tau \, \delta} = K \frac{D_M}{\delta} = \frac{D_{\text{eff}}}{\delta} \tag{18.3}$$

In Eq. (18.3), it is also assumed that the permeating solute partitions equally in the fluids on either membrane side. The unconstrained Brownian diffusivity in the liquid solvent (i.e., without accounting for pore density and tortuosity in the membrane) of a solute with radius r_s is obtained experimentally and decreases with increasing solute MW. In the absence of experimental estimates, D is related to the solute equivalent spherical radius, r_s, according to the Einstein–Stokes equation, as follows:

$$D = \frac{R\,T}{6\,\pi\,\mu\,r_s\,N_A} \tag{18.4}$$

If r_s is estimated as follows:

$$r_s = \left(\frac{3\ \text{MW}}{4\pi\,N_A\,\rho_h} \right)^{1/3} \tag{18.5}$$

where D is inversely proportional to the 0.33 power of the solute MW, at constant T. Increasing q_m values make the effective diffusivity in the membrane D_M decay with increasing solute MW much faster than the corresponding unconstrained Brownian diffusivity, as Eq. (18.3) suggests (Klein et al., 1977). The actual D_M value for a solute in a given membrane is better obtained from the experimental D_M/D_w curves reported in the literature once the solute MW is known (Colton et al., 1971; Klein et al., 1978). Parameter P_M could be directly estimated from the solute permeate flux under a given transmembrane concentration difference when transport across the membrane occurs by diffusion only (i.e., no solvent permeation). Techniques are available for estimating the intrinsic diffusive permeability toward different solutes of HD membranes with low L_p, which minimize the transport resistance at either fluid–membrane interface (Smith et al., 1968; Klein et al., 1977). The high L_p of most commercial membranes currently used for high-flux HD or HDF makes it difficult to prevent solvent permeation across the membrane even in the presence of small TMPs. Techniques have been proposed to estimate membrane P_M toward a given solute indirectly from the actual solute clearance of the whole device based on transport models in HD and HDF devices accounting for both convective and diffusive solute and solvent transport across the membrane (Legallais et al., 1998). Parameter P_M is the most relevant membrane transport property for gas exchange with the blood. In this case, only volatile species diffuse across the membrane wall (e.g., gaseous O_2 and CO_2 and water vapor), and P_M toward a given species is determined by both its Knudsen and Brownian diffusivity in the membrane. In practice, the transport of therapeutically relevant gases (e.g., O_2 and CO_2) across microporous hydrophobic membranes is mainly limited by the mass transport resistance of the stagnant layer at the membrane interface with the blood or the test liquid replacing it. Techniques have been proposed that provide membrane P_M estimates toward a given gas where the liquid mass transport resistance is minimized either by adding chemicals that bind oxygen to the test liquid (Keller and Shultis, 1979) or by flowing the liquid along the membrane under turbulent conditions (Catapano et al., 2004).

The Staverman reflection coefficient, σ, measures the extent to which the membrane rejects a given solute purely transported by convection. Solutes fully rejected by the membrane feature $\sigma = 1$. Solutes freely permeating the membrane feature $\sigma = 0$. Membrane rejection toward a given solute is experimentally assessed in the course of pure filtration experiments in terms of its rejection coefficient R, or its sieving coefficient S, with $S = 1 - R$ being the permeate-to-retentate solute concentration ratio. In fact, R is related to σ as follows (Spiegler and Kedem, 1996):

$$R = \frac{\sigma\,[e^{\beta} - 1]}{e^{\beta} - \sigma} \quad \text{with} \quad \beta = \frac{J_v\,(1 - \sigma)}{P_M} \tag{18.6}$$

and at very high filtrate fluxes $R\,(J_v \to \infty) = \sigma$. The membrane sieving coefficient is more commonly used for medical membranes and is estimated in pure filtration experiments as follows:

$$S_i = \frac{C_p}{C_B} \approx \frac{2\,C_p}{C_{Bi} + C_{Bo}} \tag{18.7}$$

where C_p and C_B are the solute concentrations in the permeate and in the retentate (i.e., on the blood side, in the plasma), respectively. The right-hand side of Eq. (18.7) provides an approximate expression for S when C_B significantly varies from the blood inlet to the oulet. Solutes freely permeating the membrane feature $S = 1$, whereas those fully rejected feature $S = 0$. Medical membrane sieving properties are often expressed in terms of the solute MW that is 90% rejected by the membrane, that is, its nominal molecular weight cut-off (NMWCO) (Mulder, 1991). Typical NMWCO of HD membranes is ca. 5000, whereas that of HF and HDF membranes is about an order of magnitude higher but generally lower than ca. 60,000 to retain albumin in the blood and is closer to that of the *glomerulus* as shown in Figure 18.2.

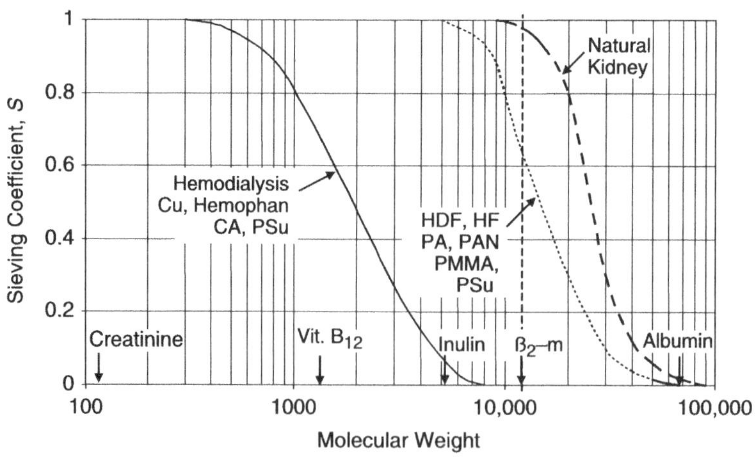

Figure 18.2 Typical sieving coefficient (S) spectrum of HD, HF, and HDF membranes compared to that of the natural kidney.

To estimate the membrane NMWCO, markers of LMW-to-HMW waste metabolites are generally used. The use of surrogate tracers with MW close to that of the presumed toxins, such as mixtures of maltodextrins, has been reported to minimize unwanted solute interactions with the membrane to affect the experimental outcome, thus providing fast information less dependent on the operating conditions than that obtained with marker proteins or metabolites (Feldhoff et al., 1984). However, medical membranes are not ideal separators, and essential blood constituents may be lost during the treatment, although their MW exceeds the membrane NMWCO. The membrane sieving coefficient or diffusive permeability curve as a function of the solute MW (i.e., the sieving or diffusive permeability spectrum) provides more effective information on the membrane pore size distribution and separation properties than its NMWCO only, as shown in Figure 18.2.

18.3.2 Adsorptive Properties

Membranes used in medicine are but inert permselective barriers. The presence of charged moieties, hydrophobic domains, hydrogen bonds, or specific ligands make them interact with dissolved species in the membrane-contacting fluids and may cause their adsorption on the membrane surface.

Adsorption of solutes whose MW far exceeds the membrane NMWCO may occur only at the blood-contacting membrane surface. It is well known that blood proteins adsorb on artificial membranes within a fraction of a second from contact with the blood (Basmadjian et al., 1997). The type of proteins adsorbed, their amount, and the binding strength vary depending on the hydrophobicity, the net charge, and the hydrogen bonds at the blood-contacting membrane surface; increasing hydrophobicity generally leads to the adsorption of larger amounts of proteins (Feldhoff et al., 1984). Drugs administered to the patient prior to, or during, the treatment may also adsorb on the membrane surface. Heparin systematically given as an anticoagulant in the treatment of renal failure has been reported to adsorb on positively charged membranes, such as the surface-modified AN69ST® polyacrylonitrile (PAN) membranes (Lavaud et al., 2003) or the *N,N*-diethylaminoethyl (DEAE)-presenting Hemophan® membranes (Gretz and Vienken, 1995). This may require adjustment of the injected heparin dose to effectively prevent blood coagulation in the extracorporeal circulation loop.

Species capable of permeating the membrane may also be adsorbed on the internal pore surface. In this case, the actual surface area available for adsorption is determined by the pore density, morphology, and size distribution in the membrane wall, as well as by the wall thickness. In fact, in the devices used for HDF and HD with highly permeable membranes, the pressure profiles that establish in the blood and dialysate compartment promote plasma water filtration from the blood to the dialysate at the blood inlet half of the device and dialysis fluid reabsorption into the blood at the blood outlet half of the device (Leypoldt et al., 1991). In filtration processes, such as HF or TA, membrane-permeating proteins may adsorb on the pore surface restricting its cross-sectional area or even clogging it. Together with the adsorption of fully rejected protein on the blood-contacting membrane surface, this may significantly worsen the membrane intrinsic transport and separation properties. Adsorptive phenomena may also be beneficial to the therapy. Adsorption of toxins capable of permeating the membrane boosts the device toxin clearance, as has been reported for β_2-microglobulin adsorption on polymethylmethacrylate (PMMA) or PAN membranes (Clark et al., 1999). Although the use of ultrapure dialysis water has been reported to improve the treatment biocompatibility and reduce the long-term mordibity of patients on dialysis, not all centers are equipped to provide it yet. In this case, adsorption on the membrane pore wall of bacterial degradation products, such as endotoxins and their fragments, present in contaminated dialysis fluid enhances membrane rejection toward these species and increases the safety of the treatment. Endotoxins effectively adsorb on polyamide (PA), PAN, and polysulfone (PSu) HF membranes. Adsorption on the former was reported to cause lower cytokine generation during dialysis with contaminated dialysate than with HD membranes made of regenerated cellulose in spite of their higher NMWCO (Pertosa et al., 1995). In fact, very hydrophobic membranes, in particular those with benzene rings in their molecular structure, may be expected to adsorb lipopolysaccharide endotoxins via their lipid part more effectively than very hydrophilic membranes.

18.3.3 Membrane Properties *ex vivo*

Blood is a cell suspension with high protein concentrations, whose MW ranges from about 10^4 to 10^6. As noted above, these proteins adsorb on (or foul) the membrane surface as soon as the blood contacts it and may disguise the membrane surface. In fact, this protein layer may change the hydrophilicity of the actual surface that contacts the blood, and its interactions with proteins and cells of the immune system of the host, to an extent that depends on the type and amount of the adsorbed proteins (Vroman et al., 1980; Norde, 1996).

In filtration processes for the extracorporeal treatment of renal failure, partially rejected proteins accumulate at the membrane separation layer [i.e., concentration polarization (CP) phenomena occur] to an extent that depends on protein concentration in the bulk blood, and the fluid dynamics of the blood compartment. Higher protein concentrations at the membrane surface cause the membrane sieving coefficient to be higher than that expected, based on the intrinsic membrane separation properties. However, once the latter are known, the actual sieving coefficient can be estimated from the operating conditions and module geometry as follows (Klein et al., 1978):

$$S_{\text{obs}} = \frac{Se^{\beta}}{(1 - S) + Se^{\beta}} \quad \text{with} \quad \beta = \frac{J_v\,(1 - \sigma)}{k} \tag{18.8}$$

The accumulation of fully rejected proteins and protein adsorption at the blood-contacting membrane surface may significantly reduce membrane L_p and sieving properties, thus reducing the rate at which toxins are removed in HF. In fact, protein adsorption on hydrophobic PA and PSu membranes was reported to shift their sieving coefficient spectrum and NMWCO toward lower MW values (Feldhoff et al., 1984). This effect is more evident at high filtrate and low blood flow rates and in postdilution mode. In the worst case, a protein layer may deposit on the membrane surface from the concentrated bloodstream that acts as an additional transport barrier, limiting water and toxin transport across the membrane (Dorson et al., 1975; Roeckel et al., 1986). Compaction of this layer might even reduce the toxin clearance potential of a hemofilter to less than that of a hemodialyzer (Dorson et al., 1978). Membranes used for TA are generally symmetric with pores far larger than those of HD or HF membranes and do not generally exhibit a dense separation layer. During the treatment, proteins and immune complexes enter the pores and may adsorb on their internal surface, or be trapped in them, thus fouling the membrane. The rejected blood cells accumulate at the blood-contacting membrane surface. Under a given TMP, both phenomena may strongly decrease membrane L_p causing plasma to be removed at a much lower rate than when filtering water under the same conditions. In long-term treatments (as is the case of acute liver failure patients), filtrate flow rate keeps decreasing during treatment and may require substitution of the device. When this happens, maintaining a predetermined plasma filtration rate (e.g., high enough to remove the preset volume of plasma within reasonable treatment times) requires high TMPs that may cause the unwanted lysis of the red blood cells. The extent of hemolysis generally increases with increasing TMPs and decreases with decreasing wall shear rate and membrane pore size (Zydney and Colton, 1982; Ding et al., 1986). Theoretical models are available that describe the formation and estimate the effect of the dynamic cell layer on the actual hydraulic resistance to the filtrate flow and that relate the operating conditions to the occurrence of cell lysis. They may be used to optimize membrane module geometry and operating conditions so as to minimize cell concentration polarization and its detrimental effects without inducing hemolysis (Zydney and Colton, 1982; Ding et al., 1986; Zydney, 1995).

18.4 MEDICAL MEMBRANE MATERIALS

Membranes are mostly used in medicine in the hollow-fiber configuration, arranged in the "shell-and-tube" configuration or in stacks of cross-wound HF membrane mats. Only in a few cases (e.g., in gas exchange with the blood for long-term pulmonary support) are flat-sheet membranes used, generally in spiral-wound modules. Since these treatments are not clinically established, in this chapter reference is made only to materials used for medical hollow-fiber membrane production. Hollow-fiber membranes are generally prepared by phase inversion by spinning the dissolved or molten polymer in a spinneret with a hollow core. Polymer precipitation is induced by immersion in a coagulation liquid, directly (i.e., wet spinning) or after a short exposure to a gas (i.e., dry-wet spinning), or by sudden cooling of the polymeric dope (i.e., melt spinning or thermal precipitation) (Mulder, 1991). Membrane precipitation is generally followed by a number of physical and chemical processing steps aimed at inducing or stabilizing a particular wall structure and washing out leachable species (e.g., plasticizers, solvents, stabilizers, additives) that could cause undesired reactions if extracted into the blood during the therapeutic treatment. All materials used in membrane production are medical grade or potential contaminants have to be removed prior to the membrane use, to comply with the safety standards set by national regulatory agencies. A large number of membranes and membrane modules are currently available for medical applications. The material a membrane is made of determines its physical and chemical properties, for example, its physical strength and workability; its resistance to temperature and chemicals, which determines the sterilization methods that can be used without altering its structure and separation properties; and its biocompatibility. However, membrane wall structure, transport, and separation properties are ultimately determined by the actual technique and conditions under which the membrane is produced (Mulder, 1991).

The earliest membranes used for medical purposes were prepared with cellulose by exploiting the spinning expertise of the textile industry. Only in the late 1960s, did techniques become available to prepare artificial membranes with the available technical polymers and to exploit the convenient properties of these polymers. Medical membranes have been traditionally classified as either natural or synthetic, depending on whether their backbone was made of cellulose or of a technical polymer, respectively (Klinkmann and Vienken, 1994). The reason for this was that, at that time, the material used was correlated with the membrane separation properties, cellulosic membranes generally being less permeable and featuring narrower pores than those prepared with technical polymers. Nowadays, this correlation does not hold any longer. However, in the following discussion, the polymers used for medical membrane preparation are presented according to the traditional classification. The information reported is made available publicly by the membrane manufacturers, when not otherwise noted.

18.4.1 Natural Polymers (i.e., Cellulose-Based Polymers)

The first membranes used in medicine were made of cellophane, a cellulose-based material that was used at that time as sausage casing (Kolff and Berk, 1943). To take advantage of the physical strength of cellulose membranes, but to improve on their diffusive permeability and biocompatibility, techniques have been developed over the years to produce cellulose-based membranes with walls as thin as 5 μm and featuring microdistributed hydrophobic side branches. Below, the main features of the commercial semipermeable medical membranes with a cellulosic backbone are briefly discussed.

18.4.1.1 Regenerated Cellulose Cellulose is a natural semicrystalline polymer consisting of repeating cellobiose (a sugar dimer) molecules. Semipermeable membranes are generally prepared starting from cellulose derivatives, molten or dissolved in suitable solvents, rather than the insoluble natural cellulose. The most widely used derivatives are obtained by acetylation to cellulose-2.5-acetate, or by the formation of amine complexes with copper (i.e., the Cuprammonium® process) (Ward et al., 1985).

In the former case, membranes are prepared by solvent precipitation or melt spinning followed by cellulose regeneration with mild alkali or saponification. In the Cuprammonium® process, membranes are spun from solutions of copper amine complexes with cellulose while using isopropylmyristate as the core fluid. Strong acids are used to precipitate the cellulose derivatives and regenerate the cellulose structure. Regenerated cellulose (RC) membranes are homogeneous hydrophilic hydrogels, with an average 40–50% water content at equilibrium. RC membranes are also produced that feature a wall structure with fins sticking out from the outer surface, with the intent to enhance mixing in the dialysate compartment during treatment. All RC membranes feature high diffusive permeability toward LMW solutes and low hydraulic permeability and nominal molecular weight cut-off, which make them feasible for HD. After drying, the membrane structure shrinks and does not recover entirely after rehydration. Shrinkage occurs at each drying/hydration cycle to an extent that depends on temperature, drying rate, and concentration of plasticizers in the starting polymeric dope, and reduces membrane permeability. In fact, the latter is related to the actual membrane hydration state. To reduce shrinkage and maintain membrane permeability unaltered, RC membranes are generally delivered with 5–40% glycerol of polymer weight (Ward et al., 1985). The hydrophilicity of RC membranes makes them very hemocompatible but has been related to a strong activation of the complement system that causes the transient leucopenia observed in HD treatments (see Section 18.5). RC membranes can be sterilized according to all available techniques (e.g., gaseous EtO, γ-ray irradiation, steam).

18.4.1.2 Cellulose Derivatives Cellulose derivatives are prepared by esterification or etherification to substitute some (or all) of the nucleophilic hydroxyl moieties (–OH) in the strongly hydrophilic cellulose backbone with more hydrophobic groups. This causes the formation at the membrane surface of hydrophilic-hydrophobic microdomains that reduce complement activation typical of RC membranes (Vienken et al., 1995).

Cellulose Esters Cellulose acetate (CA) is prepared by esterification of cellulose with acetic anhydride in the presence of small amounts of perchloric or sulforic acid, used as the catalyst. Cellulose acetate derivatives used to prepare medical membranes have a substitution degree from 2 to 3, most often of about 2.4. Commercial CA hollow-fiber membranes are prepared by both solvent-induced and thermal phase inversion, and feature symmetric (e.g., the melt–spun cellulose diacetate named Althane®, which is not commercially available any more), as well as asymmetric wall morphology (e.g., the cellulose triacetate membrane from Toyobo with an inner dense layer, made for high-flux HD and HDF). CA membranes generally feature high L_p and good selectivity. By changing the spinning conditions, CA membranes are produced that are suitable for processes from HF to AT. CA membranes are more hydrophobic than RC membranes. This causes reduced complement activation but makes them adsorb proteins easily, which affects their *ex vivo* separation properties. CA membranes have a low resistance to temperature and extreme pH values. In fact, CA degrades in concentrated alkalis and acids by deacetylation or by hydrolytic or oxidative scission of the glycosidic linkage in the cellulosic backbone, respectively. Release of hydrolysis products and the plasticizers mobilized by cellulose degradation has been reported to cause adverse reactions in patients. CA membranes cannot be steam sterilized.

Cellulose Ethers The introduction of hydrophobic moieties in the hydrophilic cellulose backbone is also achieved by etherification to substitute some of the hydroxyl groups with *N,N*-diethylaminoethyl (DEAE) or benzyl moieties. In the Hemophan® membranes (Baurmeister et al., 1987), about 1.5% of the hydroxyl groups are substituted with positively charged DEAE moieties. In the synthetically modified cellulose (SMC®) membranes, less than 1% of the hydroxyl groups are substituted with hydrophobic benzyl moieties. Both membranes feature the same good properties as RC membranes but elicit complement activation to a lower extent, SMC® less than Hemophan® membranes. Steric hindrance due to the large size of the hydrophobic substitute groups is suspected to reduce complement activation to values similar to CA membranes, in spite of the smaller number of substitute side groups.

18.4.2 Synthetic Polymers

Thermoplastic synthetic polymers are used for preparing semipermeable medical membranes owing to their good mechanical resistance and low density, which make it possible to easily prepare thin self-supporting hollow-fiber membranes, and their low energy requirement and processing costs. In the following, the main features of the commercial semipermeable medical membranes made of synthetic polymers are briefly discussed.

18.4.2.1 Acrylic Polymers Polyacrylonitrile was the first synthetic polymer used to prepare HD membranes. PAN used for preparing HD and HF membranes is a copolymer consisting of a mixture (about 85 : 15 by mole) of an acrylic and an acrylonitrile monomer and a copolymer that is either methallylsulfonate or methylmethacrylate. PAN membranes may exhibit a symmetric homogeneous (e.g., the Hospal AN69® membrane) or an asymmetric (e.g., the Asahi PAN membrane) wall structure. All feature high L_p and are used in high-flux HD, HDF, and HF. Today, PAN membranes are commercially available as hollow fibers and flat sheets. PAN membranes cannot be steam sterilized and are generally dried and stored with glycerol to prevent collapse of the membrane wall. The chemical backbone of PAN is hydrophobic and electronegative. This causes blood proteins and circulating drugs to adsorb easily, in particular those positively charged. Adsorption changes the membrane separation properties *ex vivo* and requires adjustment of the drug dose during treatment. In recent years, important anaphylactic reactions have been reported during treatment in association with the administration of angiotensin-converting-enzyme inhibitors that were related to contact-phase activation of the kallikrein-kinin system caused by the negative charges at the PAN membrane surface.

Rather hydrophobic hollow-fiber membranes are produced by melt spinning from blends of isotactic and atactic PMMA starting from the pure MMA monomers, or copolymerized with small amounts of *p*-styrene sodium sulfonate, dissolved in dimethylsulfoxide. Symmetric, almost homogeneous, PMMA membranes for HD, HF and TA are commercially available that are all sterilized by γ-ray irradiation. Their adsorptive properties have been reported to cause platelet adsorption with fibrin formation, possibly leading to high residual blood volumes in the filter (Locatelli et al., 2000) and one reported case of anaphylactic reaction (Hoenich, 1998).

Ethilenevinylalcohol (EVAL) is used to prepare hydrophilic symmetric low-flux membranes, usually sterilized by means of gaseous ethylene oxide (EtO) or γ-ray irradiation. Details on membrane preparation are not available. Their surface oxygen content has been reported to reduce protein adsorption and activation of the humoral system as compared to RC membranes.

18.4.2.2 Polyamides Polyamides are amorphous, rather hydrophobic, polymers where the repeating units are held together by amide (–CONH–) links. Asymmetric PA hollow-fiber membranes are produced by solvent phase inversion with an aliphatic-aromatic PA copolymer hydrophilized by blending with polyvinylpirrolidone (PVP). Membranes exhibit a wall with an inner thin dense layer supported by a spongy layer and an outermost layer with finger voids, and pores whose size rapidly increases outward. PA membranes feature high L_p, good diffusive permeability toward LMW solutes, and a sharp sieving coefficient spectrum with an NMWCO close to 40,000 (Goehl et al., 1992). These properties make them suitable for HD, HDF, and HF. Polyamide membranes are reported to be rather biocompatible, presumably for the presence of hydrophilic-hydrophobic microdomains a few 100 nm large, and to adsorb pyrogens effectively. They are sterilized with EtO.

18.4.2.3 Polycarbonate Polycarbonate (PC) is an unusually tough polymer, due to its chemical bonding, whose chemical resistance is not outstanding particularly to alkalis. Flat-sheet, low-flux membranes had been prepared with copolymers of PC and polyethylene glycol that exhibited alternating hydrophobic-hydrophilic microdomains at their surface (i.e., the Gambrane® membranes). PC–polyether membranes had good biocompatibility and were reported to clear middle molecules (e.g., β_2-microglobulin) in spite of their low L_p. These membranes are no longer commercially available owing to their scarce resistance to hydrolysis.

18.4.2.4 Polyolephines Semicrystalline low-density polypropylene (PP) and polyethylene (PE) are used to produce microporous hollow-fiber membranes by thermal phase inversion in the presence of plasticizers and solvents. Commercial PP and PE membranes feature a symmetric wall with pores whose maximal size is a fraction of a micron large; they are very inert but also very hydrophobic. Once solvents, plasticizers, and additives are washed out, membranes with pores smaller than about 0.1 μm are used for GE. In fact, their surface tension is high enough to prevent plasma water filtration at the blood pressures typical of therapeutic gas exchange devices. Microporous PP and PE membranes are also hydrophilized by special treatments with surfactants to permit their use in TA. Their thermal resistance is not outstanding. Thus, they cannot be steam sterilized.

Asymmetric hollow-fiber membranes with an outer thin dense layer supported by a thicker porous spongy layer are prepared with semicrystalline polymethylpentene (PMP) by melt spinning and quenching with a cold gas, followed by stretching and heat setting. PMP hollow-fiber asymmetric membranes are commercially available for gas exchange and are sterilized with gaseous EtO. The dense layer decreases membrane P_M toward O_2 and CO_2 with respect to that of microporous symmetric membranes but acts as a barrier to minimize the occurrence of plasma water breakthrough across the membrane occurring in microporous symmetric membranes possibly as an effect of polymer hydrophilization during treatment.

18.4.2.5 Polysulfones Polysulfone and its derivatives [e.g., polyethersulfones (PES)] are amorphous polymers widely used to prepare medical membranes. The repeating sequence of phenyl rings in the molecule makes the polymer very rigid, resistant to tensile, compression and thermal stresses, as well as dimensionally very stable. These polymers are glassy at room temperature and very hydrophobic. Thus, the polymers used to prepare membranes for blood treatment processes have to be hydrophilized either by chemically binding hydrophilic moieties to the polymeric backbone or by physically blending with a hydrophilic polymer (e.g., PVP). As compared to PSu membranes, PES membranes lack the isopropylidene groups in their chemical backbone and are slightly less hydrophobic. Blends of PES and polyarylate or of PES, PA, and PVP are also used to enhance membrane hydrophilicity. However, only the former do not require further addition of hydrophilization additives such as PVP. A large number of PSu and PES membranes are commercially available and the commercial terminology is often misleading. They differ for the hydrophilization and the preparation technique and exhibit different wall structure morphology. The wall morphology is generally asymmetric but differs for the number and thickness of the different layers and the pore size, shape, and distribution across the wall. This causes the membranes to exhibit different separation and adsorption properties (Streicher and Schneider, 1985). PSu-based membranes are available with low-to-high L_p and NMWCO from 1000 to 500,000 (Cheryan, 1986) and can be used in HD, HDF, HF, and TA. The hydrophobicity of their backbone makes PSu-based membranes susceptible to protein adsorption and fouling. This causes their separation properties to be strongly dependent on the presence of proteins but makes the membranes rather biocompatible. PSu membranes are resistant to temperature and pH. This makes it possible to sterilize them according to all available techniques and to clean and reuse them without significantly altering their separation and transport properties.

18.4.3 Surface-Modified Membranes

Materials used to make medical membranes were primarily developed for technical applications and were adapted to medical use only later on. In spite of the changes to polymer formulation and of the proposed polymer chemical modifications, the blood-contacting surface of commercial medical membranes bears but a pale resemblance to the endothelial cell lining of blood vessels in the body. During treatment, such nonphysiological surfaces are recognized by the blood as "foreign." This causes massive activation of the humoral and cellular immune defense systems that may cause important short-term syndromes and long-lasting morbidity (see Section 18.6). To reduce the surface-induced activation mechanisms, biomimetic techniques have been developed to make the blood-contacting surface of commercial membranes more closely resemble the natural endothelium, or simply to improve its interactions with the immunocompetent proteins and cells in the blood. The former strategy has been pursued in the development of biomimetic membranes for gas exchange. The latter for membranes used in the treatment of renal failure. The information reported in the following is made available publicly by the manufacturers, when not otherwise noted.

18.4.3.1 Surface Modifications in Gas Exchange In GE, biomimetic coatings resembling the natural endothelium were developed mostly to reduce activation of the coagulation cascade caused by the membrane and the whole extracorporeal loop (i.e., their thrombogenicity). On account of the different materials used for membranes, tubing, pump heads, and so forth coating attachment to the material surface generally relies on relatively nonspecific and weak bonds, such as hydrophobic or ionic interactions, and occurs in three stages. First, the biomaterial (e.g., the membrane) surface is functionalized. Second, hydrophilic spacer arms are grafted or coupled to the functions introduced at the surface. Lastly, anticoagulant or antiplatelet biochemical agents are attached to the spacer arms. The spacer arms form a hydrophilic layer at the surface that should reduce protein adsorption and cell adhesion. They should also be long enough to ensure proper interactions of the attached biomolecules with the cell receptors.

Most available commercial techniques aim at coating the membranes with heparin, a negatively charged anticoagulant that is known to simulate the antithrombogenic functions of heparan sulfate in the natural endothelium (Marcum and Rosenberg, 1989). In the Carmeda® Bioactive Surface technique, a layer of "priming" water-soluble polymeric cationic tenside containing primary amine groups is ionically adsorbed at the ionized membrane surface, and heparin molecules are covalently bound "endpoint" to it. In particular, diazotized heparin with a reactive aldehyde group at one end of the linear chain is reacted with the primary amino groups in the "priming" polymers, and the resulting unstable Schiff's base is eventually reduced with agents such as cyanoborohydride (Larm et al., 1983). In this treatment, the "active sites" on the heparin chains sticking out of the surface are not involved in the coupling and may freely interact with blood components and antithrombin III (ATIII) similarly to how heparin sulfate does on the endothelium. In the Bioline® coating, HMW heparin is ionically and covalently bound to a layer of cationic polypeptides adsorbed at the membrane surface. In the Trillium® Biosurface coating, a hydrophilic priming layer is adsorbed at the membrane surface on which a second functional layer is adsorbed containing negatively charged sulfate and sulfonate groups that are presumed to repel the negatively charged platelets. Leashes of polyethyleneoxide (PEO) are then attached to the functional layer to which heparin is covalently bound. PEO hydrophilicity is presumed to conceal the surface and prevent protein adsorption and cell adhesion (Han et al., 1995). These effects are presumed to act synergistically with the anticoagulant properties of heparin. Reports on the effects of the various heparin coatings are not very consistent. However, they all suggest that the heparin coating reduces the coated surface thrombogenicity and the general inflammatory response elicited by the extracorporeal treatment and does not influence significantly the gas exchange capacity of the coated membranes. The administration of protamine toward the end of treatment necessary to remove the high systemic heparin concentration given to prevent blood coagulation has been reported to worsen the hemocompatibility of heparin-coated surfaces (Muehrcke et al., 1995).

Alternatively membranes have been coated with nonthrombogenic materials, that is, materials that do not enhance protein binding nor activate platelets or white cells. In the X-coating® technique, poly(2-methoxyethyl acrylate) (PMEA) is adsorbed on the blood-contacting membrane surface. PMEA has a hydrophobic polyethylene backbone with pendant hydrophilic groups. Its chemical inertness and mild hydrophilicity is presumed to reduce blood protein and platelet adhesion on the coated surface, as well as complement activation (Suhara et al., 2001). Materials mimicking the outer surface of blood cells have also been used to stealth the membrane surface. Phosphorylcholine (PhC)-containing lipids are found in the outer membrane of blood cells that contribute to the nonthrombogenic properties of these cells (Chapman and Lee, 1987). In the Ph.Isio® technique, membranes are coated with a layer of strongly hydrated synthetic methacryloil-phosphorylcholine/lauryl-metacrylate copolymers (Yianni, 1992). PhC-containing coatings have been shown to possess good thrombogenic resistance and to hinder plasma protein deposition and platelet adhesion on the material surface (Campbell et al., 1994; Von Segesser et al., 1994).

18.4.3.2 Surface Modifications in the Treatment of Renal Failure In the AM-BIO® membranes (Asahi Medical, Japan), a hydrogel layer made of hydrophilic leashes protruding into the blood is formed at the surface of RC membranes by esterification of polyethyleneglycol acid through its carboxyl moiety with the hydroxyl groups in the cellulosic backbone. These membranes are available with different surface roughness in low- and high-flux devices and exhibit low platelet adhesion and complement activation (Fukuda et al., 1999).

The Excebrane® membranes (Terumo Ltd., Japan) are highly porous cellulosic membranes coated with the bioactive radical scavenger vitamin E (i.e., *d*-α-tocopherol) with the intent to improving membrane biocompatibility and reducing the oxidative stress by neutralizing the oxygen radicals where they form. The blood-contacting luminal membrane surface is modified by covalently binding a hydrophilic acrylic polymer via its epoxy group to the hydroxyl groups in the cellulosic backbone, to which a fluororesin polymer, and oleyl-alcohol chains are bound. These two hydrophobic moieties are reported to inhibit complement activation and platelet aggregation, respectively. Acrylic polymer, fluororesin polymer, and oleyl alcohol are all added to the fluid flowing in the inner spinneret cavity to maintain the membrane hollow, and both reactions occur during membrane spinning. Eventually, vitamin E is physically adsorbed to the oleyl-alcohol chains via hydrophobic bonding (Sasaki et al., 2000). The long-term membrane therapeutic properties and stability of the vitamin E bond are still under investigation.

In the AN69ST® membranes (Hospal, France), ca. 9 mg/m^2 of the polycationic polymer poly(ethyleneimine) (PEI) is adsorbed by ionic interactions at the negatively charged sulfonate groups of AN69® PAN membranes to minimize the anaphylactic reactions associated with the negative charges at PAN membrane surface (see Section 18.5.3). No anaphylactic reactions have been reported with these modified membranes (Maheut and Lacour, 2001). Adverse reactions and heparin adsorption were only occasionally reported to occur (Lavaud et al., 2003).

18.5 BIOCOMPATIBILITY OF MEMBRANE-BASED THERAPEUTIC TREATMENTS

Patients with kidney failure generally undergo hemodialysis, or related treatments, three times a week often for years. During open heart surgery, the whole patient's blood volume contacts the artificial surfaces in the extracorporeal loop every minute. Repeated blood contact with and massive exposure to artificial surfaces that the blood senses as foreign have raised serious concerns about the material bio(in)compatibility and its consequence on the patient's status. As a result, the bio(in)compatibility pathways that become activated when blood interacts with the materials of the extracorporeal loop were extensively investigated in the last two decades of the twentieth century. The importance of the problem may be easily envisioned by considering that, in his/her lifetime, the blood of an average kidney patient comes into contact with foreign surfaces in the extracorporeal loop whose areas have been estimated to add up to the size of a football field. All materials in the loop, that is, membranes in the dialyzer, filter, or gas exchanger, tubing, potting, and housing material, are thought to play a role. However, owing to the large surface area at which they are used, membranes are presumed to play the most important role. Indeed, larger membrane surface areas were shown to activate the immune system to a larger extent (Mahiout et al., 1987). On this account, membrane biocompatibility has also been used as a powerful marketing tool. In the last years, the alleged lack of biocompatibility of the classical cellulosic membranes has even oriented the dialysis market toward the use of synthetic membranes.

Inadequate membrane biocompatibility elicits a variety of responses, including activation of the coagulation and complement system, activation of platelets and leukocytes, production of cytokines and free oxygen radicals, and accumulation of bradykinin. Although they initiate locally, these responses cause systemic changes to the blood and the body behavior. They may also change the surface and transport properties of the membranes used. However, it should be recalled that the treatment of bio(in)compatibility is not solely related to the biomaterials used but rather depends also on the blood fluid dynamics in the membrane device and the whole extracorporeal blood circulation loop, as well as on how blood is handled in extracorporeal circulation.

Although often studied separately, a large body of evidence suggests that the pathways to bio(in)compatibility significantly interact in a complex and partly still unknown fashion. Biocompatibility parameters that provide information on membrane interactions with the organism, and that have been proven to have clinical *sequaelae*, are activation of the coagulation cascade (i.e., thrombogenicity); complement activation and hypersensitivity reactions; drug-mediated amplification of phase contact activation; and exposure to exogenous agents in the dialysate. Hereinafter, only these aspects will be very shortly discussed. For more detailed information, the interested reader may refer to specialized reviews, such as Klinkmann and Davison (1994) and Smith et al. (2002).

18.5.1 Activation of the Coagulation Cascade (Thrombogenicity)

Blood clotting at artificial surfaces has been, and still is, a major problem in extracorporeal blood treatments. Within milliseconds from blood contact with an artificial surface, plasma proteins (e.g., fibrinogen) adsorb on it (Basmadjian et al. 1997), the intrinsic coagulation pathway is activated, platelets adhere to the surface or aggregate and release factors that further activate the intrinsic coagulation pathway, thus promoting thrombogenesis.

Contact-phase activation (i.e., surface activation of the intrinsic pathway of coagulation) is thought to be the major stimulus for thrombin generation. In this complex process, factor XII adsorbs specially on negatively charged surfaces together with other plasma proteins (such as fibrinogen, albumin, HMW kininogen, prekallikrein, γ-globulin and the von Willebrand factor–factor VIII complex), undergoes a conformational rearrangement and subsequent proteolytic cleavage that, following a number of Ca^{2+}-dependent reaction steps, leads to the formation of thrombin, a proteolytic enzyme, that cleaves the HMW fibrinogen into fibrin monomers. These molecules polymerize to a soluble form of fibrin that is cross-linked by factor XIIIa and forms a mesh of fibrin fibrils that eventually traps platelets or red cells, thus forming "white" or "red" thrombi, respectively. In gas exchange in cardiopulmonary bypass, *in vitro* evidence suggests that coagulation may also occur by initiation of the extrinsic pathway caused by up-regulation of tissue factor expression in monocytes adherent to the biomaterial surfaces in the extracorporeal loop (Kappelmayer et al., 1993).

The mechanisms leading to the biomaterial-induced activation of coagulation are not yet fully understood. All biomaterials induce contact-phase activation and blood clotting, that is generally prevented by systemic administration of anticoagulants (Hofbauer et al., 1999). In particular, the presence at the membrane surface of crystallites, highly charged ionic moieties, and hydrophobic zones is associated with high thrombogenicity. Hydrogel and elastomeric materials above their glass transition temperature are associated with low thrombogenicity (Ward et al., 1985). Only recently, membranes have been made commercially available whose surface is modified to minimize activation of blood coagulation (see Section 18.4.3).

18.5.2 Complement Activation

Craddock and colleagues (1977) were the first to suggest that complement activation along the alternative pathway is implicated in the transient leukopenia and hypoxemia occurring in the initial phase of HD with cellulosic membranes. The complement system is a proteolytic cascade consisting of more than 20 blood proteins that protect the host against microorganisms in nonspecific fashion. Complement can be activated along three routes: the classical pathway, initiated by antigen–antibody complexes; the lectin pathway, initiated by terminal mannose containing glycoproteins or carbohydrates; and the alternative pathway, which defends the body against bacteria present in the circulating blood. All routes converge to the formation of C3 convertase, and through a common pathway lead to the formation of the membrane attack complex, that kills the undesired cells, the inflammatory anaphylatoxins C3a, C4a, and C5a, and the C3b fragment that attaches to the microorganism surface and attracts the phagocytes. A complex system of complement inhibitors, soluble and at the cell surface, controls the C3 convertase activity and prevents a complement system from attacking the host human cells. Artificial surfaces do not possess these inhibitory agents that prevent complement activation. This causes artificial membranes with surface properties similar to the bacterial cell wall to activate the alternative pathway and amplify the activation of the enzymatic activities. This leads to the accumulation in the blood of the anaphylatoxins C3a, C3b, C5a, and C5b and the formation of the membrane attack complex. Evidence of complement activation has been reported in all extracorporeal blood treatment processes, that is, hemodialysis, plasmapheresis, and blood oxygenation. The relationship between the physical-chemical properties of the membrane surface and complement activation is complex and still unclear. The presence of hydroxyl and amino groups at the membrane surface has been correlated with enhanced complement activation, as is the case of HD membranes made of RC (Cheung et al., 1989). Negatively charged groups, such as sulfate ions, appear to be anticomplementary (Chenoweth, 1988). However, the effective complement activation elicited by an artificial membrane is also related to its capacity to adsorb activated complement components and remove them from the bloodstream after they have been generated. For instance, HD membranes made of polyacrylonitrile and sodium methylsulfonate (i.e., AN69®) indeed activate complement but produce low levels of soluble anaphylatoxins because they are effectively adsorbed at the membrane surface (Momoi et al., 1995).

Complement activation has been associated with leukostasis in the pulmonary vasculature, granulocytopenia, anaphylactoid reactions, hypotension, pyrogenic reactions, amyloid formation in the joints of long-term renal patients, and cardiopulmonary complications (Chenoweth, 1988). This has led to the development of surface-modified RC membranes less complement activating (see Section 18.4.1.2). However, the controversial results obtained in many reported clinical trials comparing treatments of renal failure with membranes known to activate or not to activate complement have not yet shown a significant impact on patients mortality or morbidity.

18.5.2.1 Hypersensitivity Reactions Although rare today, hypersensitivity reactions are a dreadful complication in blood treatment processes, considering their potential for a deadly outcome following hypotension, bronchospasm, and upper airway angioedema (Grammer, 1994). These reactions are caused by an immunological response against a specific antigen that leads to the formation of antibodies or T cells that specifically recognize the antigen. Hypersensitivity reactions in HD were first reported in 1983 when dialyzers sterilized with ethylene oxide were used, an effect termed *first-use syndrome* (Ing et al., 1983). EtO is commonly used to sterilize medical membrane devices and equipment that cannot withstand heat or high-dose irradiation. Apparently, the gaseous EtO absorbed in the polyurethane potting of capillary dialyzers is slowly released into the blood when this enters the dialyzer. It has been postulated that EtO forms conjugates with human serum albumin (HSA) that act as antigens, based on the fact that two-thirds to three-quarters of the renal patients with hypersensitivity reactions have antibodies against the EtO–HSA conjugate (Bommer and Ritz, 1987). To prevent this response, rinsing techniques were developed to remove EtO prior to the dialyzer use (Ansorge et al., 1987). When possible, gaseous EtO has been replaced with heat treatment or high-dose irradiation to sterilize the dialyzers.

18.5.3 Drug-Mediated Amplification of Contact-Phase Activation

At the beginning of the 1990s, severe anaphylactoid reactions were reported in patients dialyzed with PAN AN69® membranes and taking drugs inhibiting the angiotensin-converting enzyme (ACE) (Tielemans et al., 1990). After extensive investigation, it was found that the positively charged factor XII adsorbs on negatively charged membranes (as is the case of the PAN membranes) and is activated to factor XIIa. Factor XIIa is a protease that cleaves prekallikrein to kallikrein, which in turn forms bradykinin from HMW kininogen (see Section 18.5.1). Bradykinin is a vasoactive peptide that causes vasodilation, increases the cardiac output, and decreases blood pressure. Under normal conditions, bradykinin is degraded by ACEs. However, in patients taking ACE inhibitors bradykinin is not degraded, and its concentration in the blood may rise and cause mild to severe anaphylactoid reactions in a membrane-dependent fashion (Krieter et al., 1998). Similar reactions have been also reported in patients treated by selective apheresis with columns loaded with beads of negatively charged dextran-sulfate (Kojima et al., 1991). Today, AN membranes are commercially available whose surface is modified to reduce their electronegativity and prevent drug-mediated anaphylactoid reactions (see Section 18.4.3.1).

18.5.4 Exposure to Exogenous Agents in the Dialysate

At the end of the 1980s it was reported that many dialysis centers were using dialysis water and dialysate contaminated with bacteria or fragments of their cell wall, such as lipopolysaccharides, peptidoglycans, and other proinflammatory products, that elicit leukocyte activation and cytokine generation (Klein et al., 1990). Owing to their broad MW spectrum, these pyrogenic contaminants virtually permeate all membranes and enter the blood-stream of renal patients at rates that decrease with decreasing membrane NMWCO and permeability and increase when high L_p membranes are operated in a way to induce backfiltration, that is, dialysate-to-blood convective transport. When this occurs, it causes leukocyte activation and cytokine release that add to the chronic inflammatory state of dialysis patients caused by complement activation (Pereira et al., 1995). In the presence of contaminated dialysate, membrane capacity to adsorb pyrogens most effectively prevents their transfer across the membrane and into the blood. In particular, very hydrophobic membranes (such those made of PA and PSu) with a thick finely porous membrane wall were reported to effectively adsorb the lipophilic pyrogens (Weber et al., 2004). However, in HDF with very permeable membranes the occurrence of dialysate backfiltration suggests the use of pure dialysate.

18.6 CONCLUSIONS

The broad variety of membranes commercially available for clinically established extracorporeal blood treatment processes testifies to the ingenuity of the membranologists who developed membranes and processes to treat otherwise deadly diseases. The successful treatment of about 1.2 million dialysis patients worldwide up-to-date would be impossible without adequate membranes, and represents one of the success stories of biomaterial application. However, in spite of these pioneers' efforts, commercial medical membranes still select the chemical species transported out of, or into, the bloodstream only based on their physical properties (i.e., size, charge, hydrophilicity, etc.) with an efficiency by far lower than that of any biological membrane. Membrane blood-contacting surfaces still bear but a pale resemblance to the endothelial cell lining of blood vessels and cause an often massive activation of the coagulation and immune defense of the host during treatment. It is now widely accepted that current membrane therapies often save lives but seldom cure the treated patients. This holds specially true for acute uremic patients on hemodialysis who become chronic patients with a different pathology than end-stage renal failure. This suggests that there is still room for the improvement of medical membranes. The better understanding of phenomena at the membrane interface with body fluids and tissues, and of the pathophysiology of the treated diseases, shall set new requirements for the new membranes that are also expected to foster the healing process. The criteria of medical membrane development in the last two decades reviewed in this chapter suggest that the next generation of medical membranes will not be crude passive separation media. They will rather incorporate functions that make it possible to actively remove only selected relevant pathogens from the bloodstream. Membranes actively interfering with the physiological cascades of the body in a controlled manner might even elicit a controlled immune response that can contribute to reestablishing the body homeostasis and promote healing of the treated patients. As discussed in this chapter, a number of complex and interrelated phenomena concur to determining the treatment outcome. They call for a multidisciplinary approach to new medical membrane development in which a new figure of membranologist will emerge who is capable of acting in close and synergic cooperation with experts in biology, medicine, physics, material science, and process engineering.

Nomenclature

A_p	ratio of area available for diffusive transport to cross-sectional membrane surface area
C	solute concentration, $\mathrm{mol/m^3}$
D	unhindered solute diffusion coefficient, $\mathrm{m^2/s}$
D_M	solute diffusion coefficient in the membrane, $\mathrm{m^2/s}$
D_{eff}	effective solute diffusion coefficient in the membrane, $\mathrm{m^2/s}$
k	blood-side solute mass transport coefficient, $\mathrm{m^3/(m^2\ s)}$
K	partition coefficient of a solute between the membrane and the neighboring fluid
K_r	fractional reduction in diffusion coefficient when the solute and pore size are comparable
$K_{1,m}$	$= (1 - 0.75857q_m^5)/(1 - 2.105q_m + 2.0685q_m^3 - 1.7068q_m^5 + 0.72603q_m^6)$
MW	molecular weight
$N_A = 6.023 \times 10^{23}$	Avogadro's number, molecules/mole
P	pressure, kPa
P_M	solute diffusive permeability in the membrane, m/s
$q_m = r_s/r_p$	solute to pore radius ratio
r_s	equivalent spherical solute radius, m
r_p	cylindrical pore radius, m
$R = 8.314 \times 10^7$	universal gas constant, dyn cm/(mol K)
R	membrane rejection coefficient toward the solute
S	membrane sieving coefficient toward the solute
T	temperature, K

Greek Symbols

δ	membrane wall thickness, m
ε	membrane void fraction
μ	bulk solution viscosity, kg/(m s)
Π	osmotic pressure, kPa
ρ_h	hydrated solute density, $\mathrm{kg/m^3}$
σ	Staverman reflection coefficient
σ	Staverman reflection coefficient toward the solute
τ	parameter related to tortuosity accounting for the fact that the diffusive path length is greater than membrane thickness

Superscripts and Subscripts	
B	refers to the device blood side
D	refers to the device side where the stripping fluid flows
i	refers to device inlet
o	refers to device outlet
obs	observed
p	refers to permeate or filtrate
w	refers to water

REFERENCES

American Society for Apheresis (ASFA), Nashville, TN (2005). *Therapeutic Apheresis. A Guide to Billing and Securing Appropriate Reimbursement.* ASFA.

Ansorge, W., Pelger, M., Dietrich, W., and Baurmeister, U. (1987). Ethylene oxide in the dialyser rinsing fluid: Effect of rinsing technique, dialyser storage time and potting compound. *Artif. Organs* 11, 118.

AsahiKasei (2008). Therapeutic apheresis and CRRT. Available: http://www.asahikasei.co.jp/medical/en/apheresis/department/index.html, accessed Feb. 2008.

Babb, A. L., Farrell, P. C., Uvelli, D. A., and Scribner, B. H. (1972). Haemodialyzer evaluation by examination of solute molecular spectra. *Trans. Am. Soc. Artif. Organs* 18, 98.

Basmadjian, D., Sefton, M., and Baldwin, S. (1997). Coagulation on biomaterials in flowing blood: Some theoretical considerations. *Biomaterials* 18, 1511.

Baurmeister, U., Brodowski, W., Diamantoglou, M., Dunweg, G., Henne, W., Pelger, M., and Schulze, H. (1987). Dialysis membrane of modified cellulose with improved biocompatibility. U.S. Patent 4,668,396.

Bommer, J., and Ritz, E. (1987). Ethylene oxide as a major cause of anaphylactoid reactions in dialysis. *Artif. Organs* 11, 111.

Bronzino, J. D. (1995). *The Biomedical Engineering Handbook.* CRC Press, Boca Raton, FL, pp. 1879–1951.

Campbell, E. J., O'Byrne, V., Stratford, P. W., Quirk, I., Vick, T. A., Wiles, M. C., and Yianni, Y. P. (1994). Biocompatible surfaces using methacryloylphosphorylcholine laurylmethacrylate copolymer. *ASAIO J.* 40, M619.

Catapano, G., Bergins, M., Wodetzki, A., Papenfuss, H. D., Rintelen, T., and Baurmeister, U. (2001). Mass and momentum transport in extra-luminal flow (ELF) membrane devices for blood oxygenation. *J. Membr. Sci.* 184, 123.

Catapano, G., Hornscheidt, R., Wodetzi, A., and Baurmeister, U. (2004). Turbulent flow technique for the estimation of oxygen diffusive permeability of membranes for the oxygenation of blood and other cell suspensions. *J. Membr. Sci.* 230(1–2), 131.

Catapano, G., Wodetzki, A., and Baurmeister, U. (1992). Blood flow outside regularly spaced hollow fibers: The future concept of membrane devices? *Int. J. Artif. Organs* 15(4), 327.

Chapman, D., and Lee, D. C. (1987). Dynamics and structure of biomembranes. *Biochem. Soc. Trans.* 15, 475.

Chenoweth, D. E. (1988). Complement activation produced by biomaterials. *Artif. Organs* 12, 508.

Cheryan, M. (1986). *Ultrafiltration Handbook.* Technomic Publishing, Lancaster, PA, p. 42.

Cheung, A. K., Parker, C. J., Wilcox, L., and Janatova, J. (1989). Activation of the alternative pathway of complement by cellulosic hemodialysis membranes. *Kidney Int.* 36, 257.

Clark, W. R., Hamburger, R. H., and Lysaght, M. J. (1999). Effect of membrane composition and structure on solute removal and biocompatibility in hemodialysis. *Kidney Int.* 56, 2005.

Colton, C. K., and Lowrie, E. G. (1981). Hemodialysis: Physical principles and technical considerations. In B. Rector (Ed.), *The Kidney*, 2nd ed. Saunders, New York, p. 2425.

Colton, C. K., Smith, K. A., Merrill, E. W., and Farrell, P. C. (1971). Permeability studies with cellulosic membranes. *J. Biomed. Mater. Res.* 5, 459.

Craddock, P., Fehr, J., Brigham, K., Kronenberg, R., and Jacob, H. (1977). Complement and leukocyte-mediated pulmonary dysfunction in hemodialysis. *New Engl. J. Med.* 296, 769.

Ding, L. H., Jaffrin, M. Y., and Gupta, B. B. (1986). A model of hemolysis in membrane plasmapheresis. *Trans. ASAIO* 32, 330.

Dorson, W. J., Jr., Cotter, D. J., and Pizziconi, V. B. (1975). Ultrafiltration of molecules through deposited protein layers. *Trans. ASAIO* 21, 132.

Dorson, W. J., Jr., Pizziconi, V. B., Ferdman, M. H., and Sizto, C. N. (1978). Quantitation of membrane-protein-solute interactions during ultrafiltration. *Trans. ASAIO* 24, 155.

Feldhoff, P., Turnham, T., and Klein, E. (1984). Effect of plasma proteins on the sieving spectra of hemofilters. *Artif. Organs* 8(2), 186.

Fukuda, M., Miyazaki, M., Hiyoshi, T., Iwata, M., and Hongou, T. (1999). Newly developed biocompatible membrane and effects of its smoother surface on antithrombogenicity. *J. Appl. Polym. Sci.* 72, 1249.

Galletti, P. M., and Colton, C. K. (1995). Artificial lungs and blood-gas exchange devices. In J. D. Bronzino (Ed.), *The Biomedical Engineering Handbook.* CRC Press, Boca Raton, FL, p. 1879.

Goehl, H., Buck, R., and Strathmann, H. (1992). Basic features of the polyamide membranes. In S. Shaldon and K. M. Koch (Eds.), *Contribution in Nephrology*, Vol. 96, *The Evolution of a Synthetic Membrane for Renal Therapy*. Karger, Basel, p. 1.

Grammer, L. C. (1994). Hypersensitivity. *Nephrol. Dial. Transplant.* 9(Suppl. 2), 29.

Grassmann, A., Gioberge, S., Moeller, S., and Brown, G. (2005). ESRD patients in 2004: Global overview of patient numbers, treatment modalities and associated trends. *Nephrol. Dial. Transplant.* 20, 2587.

Gretz, N., and Vienken, J. (1995). Heparin binding to dialysis membranes. *Semin. Dial.* 8, 128.

Gurland, H. L., Lysaght, M. J., and Samtleben, W. (1986). Immunomodulation: Critical aspects. *Artif. Organs* 10(2), 122.

Han, D. K., Lee, N. Y., Park, K. D., Kim, Y. H., Cho, H. I., and Min, B. G. (1995). Heparin-like anticoagulant activity of sulphonated poly(ethylene oxide) and sulphonated poly(ethylene oxide)-grafted polyurethane. *Biomaterials* 16(6), 467.

Hoenich, N. A. (1998). Platelet and leukocyte behaviour during hemodialysis. In C. Ronco (Ed.), *Contribution in Nephrology*, Vol. 125, *Polymethylmethacrylate. A Flexible Membrane for a Tailored Dialysis.* Karger, Basel, p. 76.

Hofbauer, R., Moser, D., Frass, M., Oberbauer, R., Kaye, A., Wagner, O., Kapiotis, S., and Druml, W. (1999). Effect of anticoagulation on blood membrane interaction during hemodialysis. *Kidney Int.* 56, 1578.

Ing, T. S., Daugirdas, J. T., Popli, S., and Gandhi, V. C. (1983). First-use syndrome with cuprammonium cellulose dialyzers. *Int. J. Artif. Organs* 6, 235.

Kappelmayer, J., Bernabei, A., Edmunds, L. H., Jr., Edgington, T. S., and Colman, R. W. (1993). Tissue factor is expressed on monocytes during simulated extracorporeal circulation. *Circ. Res.* 72(5), 1075.

Kedem, O., and Katchalsky, A. (1958). Thermodynamic analysis of the permeability of biological membranes to non-electrolytes. *Biochem. Biophys. Acta* 27, 229.

Keller, K. H., and Shultis, K. L. (1979). Oxygen permeability in ultrathin and microporous membranes during gas-liquid transfer. *Trans. ASAIO* 25, 469.

Klein, E., Holland, F. F., Donnaud, A., Leboeuf, A., and Eberle, K. (1977). Diffusive and hydraulic permeabilities of commercially available cellulosic hemodialysis films and hollow fibers. *J. Membr. Sci.* 2, 349.

Klein, E., Holland, F. F., and Eberle, K. (1978). Rejection of solutes by hemofiltration membranes. *ASAIO J.* 1, 15.

Klein, E., Pass, T., Harding, G. B., Wright, R., and Million, C. (1990). Microbial and endotoxin contamination in water and dialysate in the central United States. *Artif. Organs* 14, 85.

Klinkmann, H., and Davison, A. (1994). Consensus conference on biocompatibility. *Nephrol. Dial. Transplant.* 9(Suppl. 2).

Klinkmann, H., and Vienken, J. (1994). Membranes for dialysis. *Casopis Lekaru Ceskych* 11, 323.

Kojima, S., Harada-Shiba, M., Nomura, S., Kimura, G., Tsushima, M., Kuramochi, M., Yamamoto, M., and Omae, T. (1991). Effect of nafamostat mesilate on bradykinin generation during LDL-apheresis using a dextran sulfate cellulose column. *Trans. ASAIO* 37, 644.

Kolff, W. J., and Berk, H. T. J. (1943). De kunstmatige nier: een dialysator met groot oppervlak. *Ned. Tijdschr. Geneeskd.* 87, 1684.

Krieter, D. H., Grude, M., Lemke, H. D., Fink, E., Bönner, G., Schölkens, B., Schulz, E. and Müller, G. (1998). Anaphylactoid reactions during hemodialsysis in sheep are ACE inhibitor dose-dependent and mediated by bradykinin. *Kidney Int.* 53, 1026.

Larm, O., Larsson, R., and Olsson, P. (1983). A new non-thrombogenic surface prepared by selective covalent binding of heparin via a modified reducing terminal residue. *Biomater. Med. Devices Artif. Organs* 11, 161.

Lavaud, S., Canivet, E., Wuillai, A., Maheut, H., Randoux, C., Bonnet, J.-M., Renaux, J.-L., and Chanard, J. (2003). Optimal anticoagulation strategy in haemodialysis with heparin-coated polyacrylonitrile membrane. *Nephrol. Dial. Trasplant.* 18, 2097.

Legallais, C., Catapano, G., von Harten, B., and Baurmeister, U. (1998). Technique for the estimation of the diffusive permeability of high flux dialysis membranes. *Int. J. Artif. Organs* 21(10), 595.

Legallais, C., Catapano, G., von Harten, B., and Baurmeister, U. (2000). A theoretical model to predict the in vitro performance of hemodiafilters. *J. Membr. Sci.* 168, 3.

Leypoldt, J. K., Schmidt, B., and Gurland, H. J. (1991). Measurement of backfiltration rates during hemodialysis with highly permeable membranes. *Blood Purif.* 9, 74.

Locatelli, F., Andrulli, S., Pecchini, F., Pedrini, L., Agliata, S., Lucchi, L., Farina, M., La Milia, V., Grassi, C., Borghi, M., Redaelli, B., Conte, F., Ratto, G., Cabiddu, G., Grossi, C., and Modenese, R. (2000). Effect of high-flux dialysis on the anemia of hemodialysis patients. *Nephrol. Dial. Transplant.* 15, 1399.

Maheut, H., and Lacour, F. (2001). Using AN69ST membrane: A dialysis center experience. *Nephrol. Dial. Transplant.* 16(7), 1519.

Mahiout, A., Meinhold, H., Kessel, M., Schulze, H., and Baurmeister, U. (1987). Dialyzer membranes: Effect of surface area and chemical modification of cellulose on complement and platelet activation. *Artif. Organs* 11(2), 149.

Marcum, J. A., and Rosenberg, R. D. (1989). The biochemistry, cell biology and pathophysiology of anticoagulantly heparin-like molecules of the vessel wall. In D.A. Lane and U. Lindahl (Eds.), *Herapin: Clinical and Biological Properties, Clinical Applications*. CRC Press, Boca Raton, FL, p. 275.

Mockros, L. F., and Leonard, R. (1985). Compact cross-flow tubular oxygenators. *Trans. ASAIO* 31, 628.

Momoi, T., Ono, M., Takagi, T., Sugiura, S., Ogata, H., and Saito, A. (1995). The effects of hemodialysis (HD) membranes on interleukin 1-beta (IL-1 beta) production from peripheral blood mononuclear cells (PBMN). *Clin. Nephrol.* 44(Suppl. 1), S24.

Muehrcke, D. D., McCarthy, P. M., Stewart, R. W., Seshagiri, S., Ogella, D. A., Foster, R. C., and Cosgrove, D. M. (1995). Complications of extracorporeal life support systems using heparin-bound surfaces. The risk of intracardiac clot formation. *J. Thorac. Cardiovasc. Surg.* 110, 843.

Mulder, M. (1991). *Basic Principles of Membrane Technology*. Kluwer Acadaemic, Dordrecht, The Netherlands.

Norde, W. (1996). Driving forces for protein adsorption at solid surfaces. *Macromol. Symp.* 103, 5.

Patient Registration Committee (2004). An overview of regular dialysis treatment in Japan. *Ther. Apher. Dial.* 8, 358.

Pereira, B., Snodgrass, B., Hogan, P., and King, A. (1995). Diffusive and convective transfer of cytokine-inducing bacterial products across hemodialysis membranes. *Kidney Int.* 47, 603.

Pertosa, G., Gesualdo, L., Bottalico, D., and Schena, F. P. (1995). Endotoxins modulate chronically tumor necrosis factor a and interleukin 6 release by uraemic monocytes. *Nephrol. Dial. Transplant.* 10, 328.

Roeckel, A., Hertel, J., Fiegel, P., Abdelhamid, S., Panitz, N., and Walb, D. (1986). Permeability and secondary membrane formation of a high flux polysulfone membrane. *Kidney Int.* 30, 429.

Sasaki, M., Hosoya, N., and Saruhashi, M. (2000). Vitamin E modified cellulose membrane. *Artif. Organs* 24(10), 779.

Smith, B. R., Rinder, H. M., and Rinder, C. S. (2002). Interaction of blood and artificial surfaces. In J. Loscalzo and A. I. Schafer (Eds.), *Thrombosis and Hemorrhage*, 3rd ed. Lippicott, Williams & Wilkins, Baltimore, MD, p. 865.

Smith, K. A., Colton, C. K., Merrill, E. W., and Evans, L. B. (1968). Convective transport in a batch dialyzer: Determination of true membrane permeability from a single measurement. *Chem. Eng. Progr. Symp. Ser.* 64, 45.

Spiegler, K. S., and Kedem, O. (1996). Thermodynamics of hyperfiltration (reverse osmosis): criteria for efficient membranes. *Desalination* 1, 311.

Sternby, J., Jonsson, S., and Ledebo, I. (1992). Hemodiafiltration: Technical aspects. In S. Shaldon and K. M. Koch (Eds.), Vol. 96, *The Evolution of a Synthetic Membrane for Renal Therapy. The Contribution in Nephrology*, Karger, Basel, p. 86.

Streicher, E., and Schneider, H. (1983). Poysulphone membrane mimicking human glomerular basement membrane. *Lancet* II, 1136.

Streicher, E., and Schneider, H. (1985). The development of a polysulfone membrane. *Contr. Nephrol.* 46, 1.

Suhara, H., Sawa, Y., Nishimura, M., Oshiyama, H., Yokohama, K., Saito, N., and Matsuda, H. (2001). Efficacy of a new coating material, PMEA, for cardiopulmonary bypass circuits in a porcine model. *Ann. Thorac. Surg.* 71(5), 1603.

Tielemans, C., Madhoun, P., Lenaers, M., Schandene, L., Goldman, M., and Vanherweghem, J. (1990). Anaphylactoid reactions during hemodialysis on AN69 membranes in patients receiving ACE inhibitors. *Kidney Int.* 38, 982.

Vienken, J., Diamantoglou, M., Hahn, C., Kamusewitz, H., and Paul, D. (1995). Considerations on developmental aspects of biocompatible dialysis membranes. *Artif. Organs* 19, 398.

Von Segesser, L. K., Tonz, M., Leskosek, B., and Turina, M. (1994). Evaluation of phospholipidic surface coatings ex-vivo. *Int. J. Artif. Organs* 17, 294.

Vroman, L., Adams, A. L., Fischer, G. C., and Munoz, P. C. (1980). Interaction of high molecular weight kininogen, factor XII, and fibrinogen in plasma at interfaces. *Blood* 55(1), 156.

Ward, R. A., Feldhoff, P. W., and Klein, E. (1985). Membrane materials for therapeutic applications. In D. R. Lloyd (Ed.) *Materials Science of Synthetics Membranes*. American Chemical Society, Washington, DC, p. 99.

Weber, V., Linsberger, I., Rossmanith, E., Weber, C., and Falkenhagen, D. (2004). Pyrogen transfer across high and low flux hemodialysis membranes. *Artif. Organs* 28, 210.

Yianni, Y. P. (1992). Biocompatible surfaces based upon biomembrane mimicry. In P. J. Quinn and R. J. Cherry (Eds.) *Structural and Dynamic Properties of Lipids and Membranes*. Portland, London, p. 187.

Zydney, A. L. (1995). Therapeutic apheresis and blood fractionation. In J. D. Bronzino (Ed.), *The Biomedical Engineering Handbook*. CRC Press, Boca Raton, FL, p. 1936.

Zydney, A. L., and Colton, C. K. (1982). Continuous flow membrane plasmapheresis: Theoretical models for flux and hemolysis predictions. *Trans. ASAIO* 28, 408.

CHAPTER 19

Hemodialysis Membranes

NORMA J. OFSTHUN

Clinical Science Department, Fresenius Medical Care, Lexington, Massachusetts 02420

SUJATHA KAROOR

Renal Division, Baxter Healthcare Corp., McGaw Park, Illinois, Massachusetts

MITSURU SUZUKI

Medical Membrane Department, Toyobo Corp., Osaka, Japan

19.1 INTRODUCTION

19.1.1 Normal Kidney Function

The kidneys are two fist-sized organs whose primary function is to generate urine for excretion of water and metabolic waste products. The kidneys not only remove accumulated nitrogen products (urea, creatinine, uric acid, and others) but also maintain homeostasis of water and electrolytes (sodium, potassium, chloride, calcium, phosphate, magnesium) and regulate acid–base balance. In addition, human kidneys perform a few endocrine and metabolic functions, such as production of the hormone erythropoietin (a hormone that stimulates blood cell production) and conversion of vitamin D to its active form. Because of the tremendous overcapacity of normal kidney function, a person can live with only a fraction of normal kidney capacity, and the 0.1% of the population who are born with a single kidney often are not even aware of the missing kidney.

19.1.2 Kidney Failure

Acute renal failure (ARF) occurs when the kidneys fail due to an event such as trauma, poisoning, or surgery. Patients who recover from ARF typically do so within 10–14 days. Chronic kidney disease (CKD) is a degenerative process most often caused by diabetes or high blood pressure and less frequently as a result of genetic diseases. CKD patients whose kidneys function at less than 10% of normal capacity require regular dialysis treatment and are classified as CKD Stage V or end stage renal disease (ESRD) patients. In the United States, as of December 31, 2003, there were over 310,000 patients whose lives

Advanced Membrane Technology and Applications. Edited by Norman N. Li, Anthony G. Fane, W. S. Winston Ho, and T. Matsuura
Copyright © 2008 John Wiley & Sons, Inc.

are sustained by dialysis [U.S. Renal Data System (USRDS), 2005]. While most patients are treated by hemodialysis, which is the focus of this chapter, about 10% receive peritoneal dialysis, in which mass transport of water and toxins occurs across the patient's peritoneal membrane (into a periodically refreshed infused solution) rather than via an extracorporeal blood circuit. Each year, fewer than 15,000 Americans (USRDS, 2005) receive a transplant, though over 50,000 are listed on transplant waiting lists (USRDS, 2005).

It should be mentioned that while historically dialyzers have been called *artificial kidneys*, dialysis does not replace the kidneys' endocrine or metabolic functions. As a result, dialysis patients are given erythropoietin and intravenous vitamin D analogs to address their anemia and bone disease. This chapter will focus only on the excretory functions carried out by hemodialysis.

19.1.3 Hemodialysis Process

The process of hemodialysis is illustrated in Figure 19.1. Blood is pumped out of the body to the dialyzer through one of three types of vascular accesses—a fistula (natural vessel used to create a short-circuit between an artery and vein, usually in the arm or leg), a graft [polytetrafluoroethylene (PTFE) or other artificial vessel surgically implanted between an artery and vein], or a vascular catheter such as a central venous catheter. The blood flow rate to the dialyzer is often limited by quality of the vascular access rather than the pump setting. Prepump arterial pressure measurement is typically employed for monitoring the capacity of the vascular access, with a high negative pressure indicating that the vascular access is unable to supply the blood flow rate demanded by the pump, and that the actual blood flow rate is lower than the pump setting.

The dialyzer provides a membrane barrier that permits the passage of metabolic waste products such as urea, creatinine, uric acid, and inorganic phosphate to move from the bloodstream of the patient to the dialysate, while at the same time preventing the elimination of important blood proteins such as albumin and immunoglobulin.

Hemodialysis patients generally receive treatment 3 times per week. Over the past 30 years, with the development of more permeable membranes, average treatment times

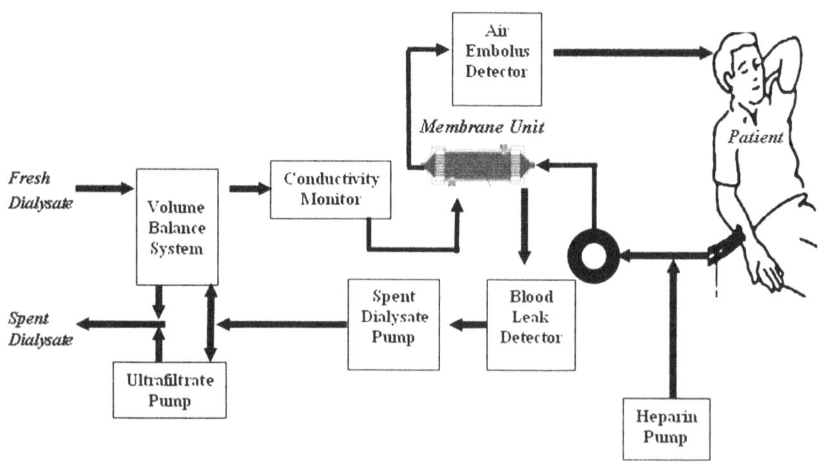

Figure 19.1 Schematic of hemodialysis process.

have dropped from 4 or 5 h to around 3.5 h. According to the Annual Report of the Clinical Performance Measures (CPM) Project, typical blood flow rates in the United States have risen from 300 mL/min to over 400 mL/min [Clinical Performance Measures (CPM), 2004a]. In the 1970s and 1980s, the standard dialysate flow rate was 500 mL/min. In the 1990s, dialysate flow rates of 800 mL/min became commonplace. To reduce unnecessary usage of dialysate solution at low blood flow rates, some dialysis machine manufacturers have added the option of setting the dialysate flow rate at a fixed ratio of 1.5 or 2.0 times the blood flow rate. Beyond such ratios, increasing the dialysate flow rate produces little to no improvement in solute removal rates.

19.2 TRANSPORT REQUIREMENTS

The terminology for characterizing dialysis membranes is somewhat unique to the dialysis field. Instead of being characterized in terms of hydraulic permeability, diffusive membrane permeabilities, and solute rejection coefficients, dialyzers are generally characterized in terms of an ultrafiltration coefficient (Kuf), solute clearances, and the product of the mass transfer coefficient times the surface area (KoA).

19.2.1 Hydraulic Permeability/Ultrafiltration Coefficient

As with all membranes, the hydraulic permeability of a dialysis membrane varies with thickness, pore size distribution, and pore density. Because exposure to blood affects the hydraulic permeability of a dialyzer, the intrinsic water permeability of a dialysis membrane is rarely reported. Instead, the *ultrafiltration coefficient* (Kuf) of a dialyzer is reported as the volumetric filtration rate (mL/h) per mm Hg transmembrane pressure (TMP) across the membrane when filtering blood. The Kuf of a dialyzer is usually derived from in vitro experiments using bovine blood in which the filtration rate is measured as a function of varying TMP. The filtration rate is linear with TMP at low TMP and reaches a plateau at high TMP (Ofsthun et al., 1991). The slope of the linear portion of the curve is defined as the Kuf of the dialyzer. Note that because the ultrafiltration coefficient is not normalized with respect to surface area, it is the property of a dialyzer, not a property of a membrane. Thus, a dialyzer containing membranes with relatively small pores can have a high Kuf if the surface area is large.

The Food and Drug Administration (FDA) classifies dialyzers as high flux if Kuf >8 mL/h mm Hg [Code of Federal Regulations (CFR), 2005]. This classification dates back to a period when dialyzers were operated in free filtrate mode, with the filtration rate controlled by varying the TMP. At that time, there was concern that a small error in setting the TMP could result in excessive fluid loss from a patient. High-flux dialyzers can only be used with hemodialysis (HD) machines that have volumetric control to prevent excessive fluid loss. With the possible exception of developing markets, all dialysis machines sold today employ ultrafiltration control. In the U.S. market, over 92% of patients are treated with high-flux dialyzers (CPM, 2004b). The Association for the Advancement of Medical Instrumentation (AAMI) Standards limit allowable lot-to-lot variability in ultrafiltration by requiring that dialyzer ultrafiltration coefficients be within 20% of the value reported on the package instructions.

19.2.2 Solute Clearance

The clearance rate of a solute, or solute clearance, is defined as the mass removal rate divided by the concentration of the solute in the blood and is expressed in units of milliliters/minute. Thus, the clearance represents the equivalent volume of blood fully cleared of the solute each minute and cannot exceed the blood flow rate to the dialyzer. The term *clearance* was originally used as a measure of the performance of the natural kidneys, which operate continuously and rely primarily on convection for solute removal.

Since dialyzers are rarely used in pure convective mode, and diffusion is the predominant mechanism of mass transport (Ofsthun and Zydney, 1994), a more appropriate measure of permeability would be a dialysance, which is defined as the mass removal rate divided by the concentration gradient across the membrane. Nevertheless, since physicians and nurses are the users of dialyzers, medical terminology had prevailed over engineering terminology in the characterization of dialyzers.

The AAMI Standards limit allowable lot-to-lot variability in clearance by requiring that reported clearances be within 10% of the value reported on the package instructions.

Another term used to characterize the transport properties of dialysis membranes is the so-called mass transfer area coefficient (MTAC), which is the product of the mass transfer coefficient (Ko) times the membrane surface area (A), or KoA. Usually, the terms MTAC and KoA reported are those for urea. While Ko should equal the maximum clearance obtained at high blood and dialysate flow rates, reports in the dialysis literature (Leypoldt et al., 1997) discuss the variation of KoA with dialysate flow rate. Such reports reflect the manufacturers' or others' inappropriate extrapolation of KoA from data obtained at typically clinically relevant flow rates, which are not high enough to minimize boundary layer resistance. While the measurement of in vivo rather than in vitro characteristics of dialyzers is meant to provide more accurate or realistic information, it can be misleading in this context.

Since mass transport rates depend on solute size and other characteristics, evaluation of dialyzer performance requires identification of solutes to be removed. While urea [molecular weight (MW) = 60 Da] has served as a marker solute for about 40 years, the full spectrum of solutes has yet to be identified.

The European Uremic Toxin Work Group (EUTox) is analyzing the uremic toxins, including each toxin's normal concentration, highest mean uremic concentration, highest single ever-reported uremic concentration, molecular weight, and the chemical class of each uremic retention compound (Vanholder et al., 2003). The uremic syndrome is characterized by accumulation of uremic toxins due to inadequate kidney function and new solutes are added to the list of uremic toxins every year (Vanholder et al., 2003). Uremic retention products differ in water solubility, protein binding, and molecular weight. Under normal kidney function, the glomerular membrane in the kidney allows passage of solutes with molecular weights up to approximately 35,000 Da (Pitts, 1968). Tubular secretion, readsorption, and metabolic breakdown are all altered when the renal function is reduced. While all substances that accumulate in renal failure can be considered uremic toxins, we do not yet understand which of these toxins and how much of them should be removed, nor the relationship between retention of a particular solute and a specific toxicity. Furthermore, the removal of these solutes is dependent on a variety of different factors such as compartmental distribution, intracellular concentration, rates of transport across cell membranes, protein binding, electrostatic charge, steric configuration, and molecular weight. For the purposes of discussion here, uremic toxins will be classified into three major categories based on their physiochemical properties that influence their dialytic removal: (1) small water-soluble compounds (MW $<$ 500 Da), (2) so-called middle molecules (500 Da $<$ MW $< \sim$15,000 Da), and (3) protein-bound molecules (both small and middle molecules).

19.2.2.1 Small Water-Soluble Molecules The water-soluble toxins generally include compounds with molecular weights less than 500 Da, such as urea and creatinine. The clearances of these molecules are primarily driven by diffusion (Ofsthun and Zydney, 1994), but other factors such as intercompartmental partition coefficients, intercompartmental mass transport rates, and protein binding can also play a role. For example, while urea is in local equilibrium between plasma and red cell water within the hemodialyzer, other solutes such as creatinine and uric acid remain partially trapped within red cells during passage through the hemodialyzer. Inorganic phosphate is a small-molecular-weight toxin that is removed relatively rapidly from the blood during the initial phase of the dialysis session, after which the transfer from the intracellular compartment(s) becomes rate limiting. Spalding et al. (2002) have considered four pools of phosphate, namely the extracellular space, intracellular space, bone, and finally as glycophosphates found in the intracellular space. For solutes with such compartmentalization effects, more frequent dialysis or longer treatment times offer greater promise than more permeable dialysis membranes. Studies sponsored by the National Institutes of Health (NIH) of short daily dialysis and nocturnal dialysis are currently underway.

19.2.2.2 Middle Molecular Solutes Several retrospective studies have provided suggestive evidence that middle molecule removal influences the outcomes in hemodialysis patients (Ward et al., 2001; Port et al., 1998; Leypoldt et al., 1997; Koda et al., 1997). Some of the examples of potential uremic toxins are β_2-microglobulin, advanced glycation end products, leptin, complement proteins, proinflammatory cytokines, factor D, and granulocyte-inhibiting protein (Vanholder et al., 2003). Middle molecules are distributed in multiple compartments, and their clearance by diffusion decreases as the molecular weight increases. Beta-2-microglobulin was the first so-called middle molecule to be linked to a specific clinical syndrome (dialysis-associated amyloidosis) occurring exclusively in hemodialysis patients. With a molecular weight of approximately 11,800 Da, β_2-microglobulin is largely removed by convection (Ofsthun and Zydney, 1994), but its sieving coefficient is >0 only with high-flux membranes. Because net ultrafiltration volumes removed during dialysis are typically small (1–3 L/treatment), the quantity of solutes removed by convection is limited. Recent studies have demonstrated that increasing the *internal filtration*, or Starling's flow (i.e., positive filtrate flux near the blood inlet and negative filtrate flux near the blood outlet), by increasing the length (Sato et al., 2003) or reducing the fiber diameter (Ronco et al., 2000), or increasing the dialysate-side pressure drop (Fujimura et al., 2004) resulted in improved clearances of middle molecules.

19.2.2.3 Protein-Bound Uremic Toxins Among the 20 protein-bound uremic toxins identified in the EUTox review (Vanholder et al., 2003), leptin and retinol binding protein were middle molecules, and 18 others were small solutes (MW $<$ 500 Da). Some of these protein-bound molecules exert a toxic effect at concentrations observed in dialysis patients. For example, the 3-carboxy-4-methyl propylfuranpropionic acid, which is 98% bound to albumin, inhibits erythropoiesis (Niwa et al., 1990). Similarly, P-cresol and indoxyl sulfate, both of which are nearly 100% protein bound, inhibit endothelial proliferation and wound repair (Dou et al., 2004). The flux of large solutes depends primarily on convective transport, and increasing the flux increases the removal of protein-bound toxins (Falkenhagen et al., 1999; Galli et al., 2003). Mathematical modeling by Meyer et al. (2004) showed that substantially higher clearances of protein-bound solutes could be achieved by combined increases of the dialyzer mass transfer area coefficient and dialysate flow rate (Meyer et al., 2004).

19.2.2.4 Retention of Plasma Proteins An important constraint on the maximum pore size of dialysis membranes is the need to retain plasma proteins such as albumin (MW = 66,000). Since patients are often malnourished, loss of significant plasma proteins is not clinically acceptable. Furthermore, the presence of even relatively small quantities of protein in dialysate waste streams creates a practical problem of foam being generated in drain lines and rising through floor drains.

19.2.2.5 Requirements for Adequate Dialysis The quantitative assessment of efficacy of dialysis therapy and renal function is based on the small solutes, even though the molecular weights range over 3 orders of magnitude. Urea and creatinine are considered to be representative or surrogates for the small molecules and easily measured. Three different methods of accessing urea removal rates are currently used:

1. *Urea Reduction Ratio (URR)*, which is defined in terms of pre- and post-blood-urea-nitrogen (BUN) values as (Owen et al., 1993):

$$\text{URR}(\%) = 100 \times (\text{pre-BUN} - \text{post-BUN})/\text{pre-BUN}$$

   For three times a week dialysis, the government CPM report considers a URR of 65% to be indicative of adequate dialysis (CPM, 2004c), but other reports suggest a target of 70% (Hakim et al., 1984).

2. Kt/V is defined as K = urea clearance, t = treatment time, and V = urea distribution volume. Using a simple mass balance, a single pool Kt/V can be calculated as

$$Kt/V = -\ln(\text{post-BUN}/\text{pre-BUN})$$

   To account for rebound due to intercompartmental gradients, the following so-called Daugirdas II calculation is often employed (Daugirdas, 1993):

$$\frac{Kt}{V} = \ln(R - 0.008 \times t) + (4 - 3.5 \times R) \times 0.55\,\frac{U_F}{W}$$

   where R = pre-BUN/post-BUN
   W = postdialysis body mass of patient (kg)
   V = body water (L)
   U_F = ultrafiltration volume per dialysis (L)
   t = dialysis treatment time (h)

   The K/DOQI (Kidney Disease Outcome Quality Initiative) national guidelines recommend a delivered Kt/V of at least 1.2 for a thrice-weekly hemodialysis (NKF-K/DOQI, 2001).

3. *Formal urea kinetic modeling (UKM)*, which involves iterative solution of the differential equations, can be used to calculate an equilibrated Kt/V, which takes into account postdialysis solute rebound as urea moves from into the blood compartment from other spaces. Equilibrated Kt/V's are typically about 0.2 units smaller than single pool Kt/V's calculated from the same BUN results.

A recently developed on-line clearance monitor (Gotch et al., 2004) has been demonstrated to be effective at predicting future mortality (Lowrie et al., 2005), indicating that it is a valid tool for assessing dialysis adequacy.

19.3 OTHER REQUIREMENTS

19.3.1 Membrane Biocompatibility

While transport properties play an important role in the selection of a dialyzer membrane, an equally important consideration in the evolution of the dialyzer technology has been biocompatibility, or the compatibility of the dialyzer with blood. Three aspects of biocompatibility that are important in dialysis are clotting, activation of the complement cascade, and cytokine generation.

19.3.1.1 Clotting Clotting is generally managed through the use of systemic heparinization, although a small percentage ($<5\%$) of dialysis treatments are performed heparin free due to allergic reactions to heparin. A recent report suggests that as many as 50% of dialysis procedures performed in the intensive care unit are performed heparin free to avoid bleeding complications (McGill et al., 2005). While heparin-coated membrane oxygenators and hemofilters were commercialized several years ago, heparin-coated dialyzers have only recently been introduced (Lavaud et al., 2003).

19.3.1.2 Complement Activation Complement proteins are so-named because they complement antibody activity to eliminate pathogens. The *alternate pathway* of the complement cascade is normally activated by bacterial surface molecules. Complement activation during dialysis was first identified by the rapid drop in white blood cell counts (neutropenia) during the first 30 min of dialysis. Regenerated cellulose membranes activate complement through the alternate pathway (Chenoweth et al., 1983). Modified cellulose membranes approach the biocompatibility profile of synthetic materials in terms of neutropenia and complement activation.

19.3.1.3 Adsorption of Endotoxins and Cytokines Whereas membrane materials used in downstream processing of biological products may be selected for their resistance to protein fouling, early dialysis membranes were found to be more blood compatible after adsorption of blood proteins. Endotoxins are bacterial products released from gram-negative bacteria upon death. Because endotoxins cause fever, they are also called pyrogens. Because endotoxin fragments fall in the middle molecule range, they may be inadvertently transported from dialysate to blood during high-flux dialysis, leading to cytokine generation.

Adsorption (membrane binding) is one mechanism by which hydrophobic compounds such as endotoxins, cytokines, peptides, growth factors, and proteins may be removed during HD. Although adsorption during HD is a relatively poorly understood phenomenon, certain membrane characteristics play an important role. The binding characteristics, distribution of hydrophobic and hydrophilic domains, and charge distribution on the surface and in the pores are important factors that govern the membrane biocompatibility. Renaux et al. (1999) proposed classification based on the ζ potential. Adsorption primarily occurs within the pore structure of the membrane rather than only at the luminal surface that contacts the blood. Therefore, the open-pore structure of high-flux membranes affords more adsorptive potential than do low-flux counterparts. Second, synthetic membranes, many of which are fundamentally hydrophobic, generally are much more adsorptive than hydrophilic cellulosic membranes (Clark et al., 1995).

While adsorption of endotoxins and cytokines is clearly desirable, one could argue that adsorptive properties of a dialyzer are not important for removal of solutes such as β_2-microglobulin because it does not matter to the patient whether a toxin goes down the drain or is adsorbed within the membrane wall.

19.3.1.4 Medication Interactions Potential adverse interactions between membranes and medications are difficult to predict. In 1990, reports of life-threatening anaphylactoid reactions with polyacrylonitrile membrane dialyzers and angiotensin converting enzyme (ACE) inhibitor blood pressure medications surfaced (Tielemans et al., 1990). These reactions were subsequently shown to be a result of bradykinin accumulation due to the combination of increased synthesis stimulated by interaction of blood with the negatively charged membrane, and reduced catabolism of bradykinin with ACE inhibitors (Schaefer et al., 1993; Deppisch et al., 1998). Other concerns with medications have arisen related to correct dosing of potentially dialyzable drugs such as vancomycin (Ariano et al., 2005).

19.3.2 Sterilizability

Since dialyzed blood is returned to patients, dialyzers and associated tubing must be sterilized. Historically, the most common form of sterilization used ethylene oxide. With the recognition that some patients develop allergic reactions to ethylene-oxide-altered human serum albumin, ETO-HSA (Grammer and Patterson, 1987), other sterilization methods such as γ irradiation, steam sterilization, and e-beam sterilization have been developed. Care must be taken to characterize dialyzers after sterilization and to ensure that any variability in sterilization regimen does not significantly alter dialyzer properties.

19.3.3 Dialyzer Reuse

In the 1980s and 1990s, the practice of dialyzer reuse became commonplace in the United States. At first, this was driven by both the medical benefit of improved biocompatibility and the financial benefit of reducing the per-treatment cost of the use of more expensive dialyzer membranes. With the use of more biocompatible polysulfone membranes, a recent retrospective study found a 5–10% mortality benefit associated with single-use dialyzers (Lowrie et al., 2004).

Vertical integration of dialyzer manufacturers with dialysis provider chains has enabled one major company (Fresenius) to offer single-use dialyzers to all patients. In 2003, this company reached a 50 million dialyzers/year production milestone.

19.4 MEMBRANE MATERIALS, SPINNING TECHNOLOGY, AND STRUCTURE

Dialyzer membrane performance depends on the biomaterial used, its thickness, and the hydraulic permeability, pore size and density, biocompatibility, and the hydrophilic/hydrophobic properties. Some of these properties are discussed below.

19.4.1 Membrane Materials

Current dialyzer membranes can be classified based on their chemical compositions as (1) cellulosic, (2) modified cellulosic, and (3) synthetic. Each of these membrane types will be discussed in detail.

19.4.1.1 Cellulosic Membranes Cellulosic membranes were exclusively used in the 1940s through 1960s. Regenerated cellulose membranes were produced using the cuprammonium process and were commonly known by the trade name Cuprophan or the term cuprammonium rayon. These are polysaccharide membranes derived from cotton linters, the short fibers left on cottonseed after long fibers have been removed. This natural cellulose is of high quality with minimal variation in its molecular chain length. Regenerated cellulose membranes can be manufactured to have very low wall thickness (6–15 μm), high pore density, and low cost. Cellulosic membranes are very effective in removing low-molecular-weight toxins, but their very low mean pore size results in poor middle molecule removal. The membrane has a high density of hydroxyl groups on their glucosan rings, which activates the complement cascade via the alternate pathway. Activation of the complement cascade makes these membranes bioincompatible (Hakim et al., 1984). Despite being considered the least biocompatible dialyzer material, these membranes are still used in some parts of the world primarily due to their lower cost.

19.4.1.2 Modified Cellulosic Membranes Modified cellulosic membranes are made more biocompatible by the substitution of the hydroxyl groups with other moieties or by coating the membrane with a biocompatible coating. Cellulose diacetate and cellulose triacetate differ in the degree of substitution of the hydroxyl groups with acetate groups. Other groups such as diethylaminoethyl (DEAE) and benzyl groups were added to make the membranes more biocompatible. These substituted membranes are more hydrophobic and also have a larger mean pore size, which results in higher water permeability and middle molecule clearances compared to unmodified cellulosics. This chemical modification influences membrane properties such as protein absorption, wettability, biocompatibility, and clearance of both small and middle molecules. Cellulose acetate membranes have been produced by melt spinning as well as solution-diffusion processes. Table 19.1 lists cellulose and modified cellulose dialyzer membranes used in hemodialyzers today.

19.4.1.3 Noncellulosic Synthetic Membranes In the dialysis field, the term *synthetic membrane* is used to denote all polymeric membranes that are not cellulose based. Table 19.2 lists the various synthetic membranes that are commercially available. Polymers such as polyacrylonitrile (PAN), polymethylmethacrylate (PMMA), and ethylene vinyl alcohol (EVAL) copolymer were adapted from the textile industry, while polymers such as polysulfone, polycarbonate, and polyurethane were developed as engineering plastics. Synthetic membranes with high water permeability were developed in the 1960s

TABLE 19.1 Cellulose and Modified Cellulose Membranes Used in Hemodialyzers Today

Membrane	Substitution of OH Groups	High/Low Flux	Common Trade Names
Regenerated cellulose	Not applicable	Primarily LF	Cuprophan Cuprammonium Rayon
Cellulose acetate	Acetate, nominal degree of substitution = 2.0	Primarily HF	CA
Cellulose 2. 5 acetate	Acetate, nominal degree of substitution = 2.5	Primarily HF	
Cellulose triacetate	Acetate, nominal degree of substitution = 3.0	High flux	CT
DEAE—modified cellulose	Diethylaminoethyl group	Primarily LF	Hemophan
Benzyl—modified cellulose	Benzyl group	Primarily LF	Synthetically modified cellulose (SMC), polysynthane (PSN)
PEG-coated cellulose	Not applicable	Primarily LF	
Vitamin-E-coated cellulose	Not applicable	Primarily LF	Excebrane

primarily for hemofiltration. These membranes are now manufactured with a range of permeabilities. These membranes have thicker walls (>20 μm) and are either symmetric or asymmetric. Figure 19.2 shows a typical asymmetric membrane.

The asymmetric membrane consists of a skin layer, which is 1 μm thick, and a support layer, which comprises the rest of the wall thickness (Sugaya and Sakai, 1999). The skin layer is in contact with the blood and controls the solute removal. The pore structure of the support layer is much more open and varies among the various synthetic membranes, and this layer dictates the thermal and mechanical properties of the membranes. The average pore size in the skin layer for low-flux membranes is around 10 Å and for the high-flux membranes is around 30–50 Å. The pore size in the support layer is greater

TABLE 19.2 Synthetic Membranes Used in Hemodialyzers Today

Membrane	Common Trade Names
Polysulfone (PSF)	Polysulfone (Althane, ASAHI, Fresenius) Helixone Toraysulfone Clirans PS
Polyethersulfone/polyarylethersulfone (PES)	DIAPES Synphan
Polyethersulfone + polyarylate (PEPA)	PEPA
Polyamide/polyarylethersulfone (PA/PES) (Bowry, 2002)	Polyamide-S
Polyacrylonitrile (PAN)	AN-69
Ethylene vinyl alcohol (EVAL)	
Polymethylmethacrylate (PMMA)	

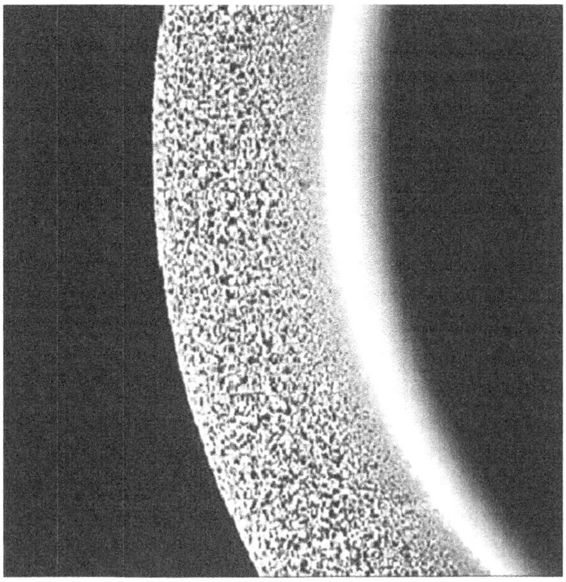

Figure 19.2 Asymmetric synthetic hemodialysis membrane.

than 100 Å. The synthetic membranes that were produced early on were hydrophobic and resulted in excessive protein losses (Gohl et al., 1992).

19.4.2 Hollow-Fiber Spinning Technology

Selected polymers are dissolved with solvents and spun through tube-in-orifice nozzles to form hollow-fiber structures in either dry-wet or dry spinning mode. Because the inner surface of the hollow fiber plays an important role for separation, the hollow fiber is usually spun with an inner liquid to control the pore structure of the lumen surface. Then, the fibers are introduced into a coagulation bath, where pores are formed by microphase separation, induced by thermal and/or solvent concentration differences. Finally, the porous hollow fiber is formed with the desired inner and outer diameters, for example, ID = 200, OD = 230 μm. As will be discussed in more detail below, membrane porosities up to around 75% may be employed, while the pore diameters are controlled in the nanometer range. The hollow fibers are wound on a spool or a reel, with the number of fibers in the bundle set by the number of revolutions of the spool. Spinning hollow-fiber membranes is very similar to regular fiber spinning but requires substantially greater quality control to ensure product safety. For example, dimensional uniformity, microphase separation, and material purities must be strictly controlled to ensure consistent transport performance.

19.4.3 Pore Size, Distribution, and Density

As described recently by Ronco and Bowry (2001), the nature of the pore size distribution may significantly influence a membrane's sieving properties. Desirable features for a high-flux membrane include a large number of relatively large pores (radius as large as 45 Å) having a narrow distribution of sizes. This type of distribution leads ideally to a solute-sieving coefficient versus molecular weight profile with a sharp cut-off at a molecular weight just below that of albumin, similar to that of the native kidney. In actual practice, all highly permeable membranes have measurable albumin sieving coefficient values, such that the design of this type of membrane involves striking a balance between optimized large-molecular-weight toxin removal and minimal albumin losses.

19.4.4 Hollow-Fiber Geometry

19.4.4.1 Inner Diameter The inner diameters of the current hollow fibers vary from about 180 to 220 μm. Recently, dialyzers have been produced with a decreased inner fiber diameter (Ronco et al., 2000). This preliminary study suggested that small changes of the inner diameter of the fiber could result in dramatic changes in removal of both urea and middle molecules. High shear rates were also obtained by lowering the inner diameter of the fiber at a given blood flow rate. This leads to a reduction of the protein boundary layer and improves the membrane permeability (Zydney and Colton, 1988). Decreasing hollow-fiber inner diameter improves diffusive mass transfer by shortening path length and attenuating boundary layer effects through higher shear rates (Clark and Gao, 2002). However, one factor constraining possible decreases in hollow-fiber inner diameter becomes evident with calculation of the axial pressure drop from the Hagen–Poiseuille equation (Bird et al., 1960), which can be rearranged to

$$\Delta P = 8\mu L Q_B / N D^4$$

where ΔP is the axial pressure drop, μ is the viscosity, L is the length, Q_B is blood flow rate, N is the number of fibers, and D is the hollow-fiber inner diameter. Because of the exponent on D, a small decrease in hollow-fiber inner diameter causes a large increase in axial pressure drop (at constant blood flow rate). Therefore, hollow-fiber lumen dimensions represent a compromise that reflects both mass transfer and hydrodynamic considerations.

19.4.4.2 Wall Thickness As mentioned previously, wall thicknesses of cellulosic fibers range from 6 to 15 μm. Noncellulosic membrane thicknesses are greater than 20 μm.

19.4.4.3 Axial Undulations Since wavy hollow fibers provide better dialysate flow distribution, hollow fibers are often axially undulated in a postspinning process. Other techniques for improving performance include radial variations in wall thickness, spacer yarns, and knitting of hollow fibers. All of these methods are designed to achieve uniform dialysate flow around all hollow fibers.

19.5 DIALYZER DESIGN AND PERFORMANCE

Dialyzer performance depends not only on the membrane properties but also on device properties.

19.5.1 Typical Dialyzer Dimensions

Figure 19.3 shows a typical hemodialyzer. Device properties such as the fiber length, membrane surface area, number of fibers, hollow fiber packing density, and header design all affect solute clearances.

Figure 19.3 Photograph of a typical hemodialyzer.

19.5.1.1 Length Dialyzer lengths are typically 20–24 cm and represent a trade-off between being long enough to allow virtual equilibration of dialysate and blood urea concentrations and short enough to have acceptable axial pressure drops. Recent reports have suggested increasing dialyzer lengths to improve middle molecule clearances via internal filtration (Sato et al., 2003).

19.5.1.2 Surface Area and Number of Fibers Dialyzers used for adult patients typically have 1–2 m^2 of membrane area (measured at the lumen surface) distributed among 8000–16,000 hollow fibers.

19.5.1.3 Hollow-Fiber Packing Density Fiber packing densities are optimized to provide uniform dialysate distribution. Typical dialyzers employ packing densities of roughly 50–75%.

19.5.2 Solute Clearances

The effectiveness of dialysis as a replacement for kidney function depends on the mass transfer characteristics of the membranes as well as device parameters. Table 19.3 provides a sampling of dialyzer characteristics. This table is meant to be illustrative of the many available dialyzers, not an exhaustive compilation.

19.5.3 Sieving Coefficients

Sieving coefficients of marker solutes such as vitamin B_{12} (MW = 1355 Da), inulin (MW = 5200 Da), and myoglobin (MW = 17,200 Da) are reported more often than sieving coefficients of known uremic toxins. Nevertheless, they serve as a useful tool in comparing potential middle molecule removal.

TABLE 19.3 Performance Comparison of Selected Hemodialyzers

						Performance Comparison											
						Clearances (mL/min) at Given Blood Flow Rate (mL/min)											
						Urea			Creatinine			Phosphate			Vitamin B_{12}		
Model	Size (m^2)	Wall (μm)	ID (μm)	Kuf (mL/h/mm Hg)	KoA (mL/min)	200	300	400	200	300	400	200	300	400	200	300	400
Exeltra 190 (cellulose triacetate)	1.9	15	200	36	1214	197	273	323	190	251	289	186	242	276	143	168	183
Polyflux 170H (polyamide)	1.7	50	200	48	1103	195	268	318	184	240	278	178	229	262	134	158	175
Optiflux 160NR (polysulfone)	1.5	40	200	45	1064	194	266	308	181	238	264	178	230	275	128	152	157
Optiflux 200NR (polysulfone)	2.0	40	200	56	1321	197	277	330	191	253	289	183	250	290	148	173	189
FX80 (polysulfone)	1.8	35	185	59	1292	197	276	326	189	250	287	185	239	272	148	175	190

19.6 CURRENT MARKET TRENDS

In 2004, more than 100 million dialyzers were produced worldwide. The breakdown is approximately as follows:

- Polysulfone > 65 million/year (60%)
- Cellulose acetate/triacetate > 25 million/year (20–25%)
- Other polymers < 20 million/year (15–20%)

With the planned consolidation of the top four U.S. dialysis providers into two vertically integrated companies, the field of available dialyzers is likely to narrow. Meanwhile, the decrease in dialyzer reuse will continue to drive increased production. Given that the Center for Medicare and Medicaid Services (CMS), formerly known as the Healthcare Financing Administration (HCFA), will continue its downward pressure on costs, and possibly institute a capitated payment system (i.e., a fixed monthly payment to cover all patient costs including hospitalization), only incremental improvements in commercially available dialyzer membranes are expected in the next 5 years.

19.7 FUTURE DIRECTIONS

19.7.1 Middle Molecule Removal

While current therapy is believed to be effective in the clearance of small solutes such as urea, improved removal of middle molecules and protein-bound solutes is desirable. Henderson et al. (2001) point to the importance of quantifying the removal of larger toxic solutes in the light of increasing evidence that shows a positive correlation between survival and middle molecule clearance in hemodialysis patients. A number of studies are underway to enhance the removal of middle molecules. These studies include (1) variations in modes of dialysis, (2) changes in dialyzer design to improve internal filtration, (3) targeted removal of specific molecules, and (4) increased frequency of dialysis.

19.7.1.1 Alternative Modes of Dialysis Various convective therapies to improve the clearance of middle molecules, such as high-flux dialysis (HFD), hemofiltration (HF), and hemodiafiltration (HDF), are being evaluated. High-flux membranes are employed in HFD therapy in which high ultrafiltration rates are counterbalanced by back-filtration. To prevent excessive fluid loss, net ultrafiltration is controlled volumetrically. In this mode, clearances are improved over conventional hemodialysis. Conventional hemodiafiltration utilizes large convective transport with ultrafiltration rates above 70 mL/min. Since such ultrafiltration rates result in total ultrafiltration volumes that exceed the desired weight losses in patients, sterile replacement fluid must be administered. Total replacement fluid required varies between 12 and 22 L/session. Because of the prohibitive cost of prepackaged replacement fluid, on-line generation of sterile fluid has been employed. In this method, the plasma β_2-microglobulin levels were reduced when compared to high-flux HD (Ward et al., 2000). Furthermore, an improvement in survival (35%) has recently been reported using high-efficiency (>15–25 L/session) HDF (Canaud et al., 2003; Jirka et al., 2005). The major drawbacks of this treatment are the complexity of the system and the increased cost of the therapy over conventional hemodialysis.

New dialysis machines with on-line generation of sterile replacement fluid hold promise for overcoming these drawbacks.

A number of studies are being carried out to improve the design of the filter to optimize middle molecule clearances. The OLpūr MD 190 (Nephros Inc., New York), an advanced proprietary filter designed for use in HDF therapy, uniquely combines postdilution and predilution diafiltration into one simple-to-use device. The OLpūr MD 190 has a unique design that achieves the high urea clearance of postdilution hemodiafiltration, while offering improved clearance of toxins in the middle-molecule range (Krieter et al., 2005). Ronco et al. (1998) have used a fixed O-ring on the dialysate side to alter the pressure profile in the dialysate compartment within the dialyzer. This results in an increased rate of filtration and back filtration without affecting the net ultrafiltration rates. Because the filtered fluid is diluted before being back-filtered, it offers improved clearance of middle molecules without the need for replacement fluid. Using a similar philosophy of increasing internal filtration, Mineshima et al. (2000) studied the impact of fiber length and inner diameter on convective solute removal.

Immunoadsorption is another way to remove middle molecules, either specifically or nonspecifically. Adsorptive processes can be carried out either by chemically modifying a hemodialysis membrane to create adsorption sites or by the use of an add-on device during hemodialysis. It should be mentioned that the 1 to 2-m^2 membrane surface area on the hollow-fiber lumen is much smaller than the surface area within the porous membrane structure and may be insufficient to provide significant toxin removal. However, one could argue that adsorptive sites within the membrane wall offer little benefit unless significant back filtration of a toxin is taking place because it makes no difference to the patient whether a toxin is adsorbed within the membrane walls or flushed away with the spent dialysate.

19.7.1.2 Adsorption for Toxin Removal Over the past several years, there have been a number of studies of the use of adsorption for toxin removal. Removal of β_2-microglobulin by both specific and nonspecific methods (Ronco et al., 2001; Kaneko et al., 2001) have been evaluated. The Beta-Sorb column (BetaSorb, Renal Tech International, New York) contains cross-linked divinylbenzene-based resin, with a pore size suitable for removal of middle molecules (Wendler et al., 2003). The Kaneka Lixelle device, composed of cellulosic beads with ligands that trap β_2-microglobulin, was associated with partial regression of dialysis-related amyloidosis (DRA) over 6–13 months (Homma et al., 1995; Hyodo et al., 2001). This device has been approved in Japan, but the high cost prevented more widespread use of this column. As our understanding of uremia has developed, new molecules have been identified that are associated with adverse effects and that are not easily removed by conventional dialysis. These compounds include leptin, β_2-microglobulin, inflammatory mediators, advanced glycosylation end-products (AGE), advanced lipid peroxidation compounds, and protein-bound toxins and are believed to contribute to long-term complications and morbidity.

With advances in surface modification for adsorptive therapy, optimization of device design, and increasing knowledge of uremic toxins, future systems may offer specific removal of toxins within the hemodialyzer.

19.7.1.3 Dialysis Treatment Time and Frequency Dialysis treatment time and frequency are important in the removal of middle molecules and other molecules that transport slowly within the human body. For small solutes such as urea, the transport gradient dissipates within the first few hours, and little is gained by extending treatment times beyond the standard 3.5–4.0 h. Carrying out dialysis more frequently than the usual three times per week schedule results in lower peak concentrations of such small solutes but may not result in substantially greater overall removal each week. Larger-sized solutes are more slowly transported; so equilibrium between blood and dialysate is not generally reached with a typical treatment regimen. Thus, middle molecule removal can be increased by extending the treatment time with nocturnal dialysis, especially if it is performed daily. Short daily dialysis may provide some additional middle molecule removal if "short" is not as short as half the usual treatment time in three times per week dialysis.

Removal of relatively small solutes with substantial compartmental effects, such as inorganic phosphate, may benefit the most from short daily dialysis. In the United States, nocturnal dialysis and short daily dialysis are under investigation in studies funded by the NIH, but are not generally available because Medicare only pays for standard treatment.

19.7.2 Protein-Bound Toxins

Vanholder et al. (2001) reported increases of protein-bound toxins in dialysis patients and their biological impact. Albumin is the principal protein present in blood and is a carrier for a number of solutes that are poorly soluble in free solution. So-called albumin dialysis is a process in which blood is dialyzed against an albumin solution that itself undergoes purification by another method (such as adsorption).

Albumin dialysis is postulated to be effective in removing protein-bound toxins, and evidence of its effectiveness was recently reported (Stange et al., 1993). The MARS Molecular Adsorbents Recirculation system (Teraklin/Gambro Renal Products) and the Prometheus system (Fresenius Medical Care) are specifically designed for treating liver failure, in which concentrations of protein-bound toxins are very high. Ongoing clinical studies are evaluating the use of such systems to treat acute liver failure (Khuroo et al., 2004) and as a bridge to transplant for chronic liver failure (Rifai et al., 2003). In the MARS system, albumin is regenerated and reused, but removal is limited because the membrane characteristics do not change with albumin in the bath, and this is the major barrier to protein-bound toxin removal.

To overcome the need for large amounts of commercial albumin solution as well as the problem of high membrane resistance, the Prometheus system allows the passage of a patient's own albumin through a 250-kDa membrane, then after the albumin-bound solute is removed, the albumin is returned to the patient. While this method is very effective in removing the protein-bound toxins, there have been only a limited number of studies of using albumin solution as a dialysate in hemodialysis patients. Removal of protein-bound solutes by membranes that allow protein leakage or by adsorptive strategies are being investigated (Niwa et al., 1995; Galli et al., 2003), but potentially could worsen the common problem of protein malnutrition in dialysis patients. It is unlikely that alternative time schedules will significantly improve the efficiency of removal of protein-bound solutes in dialysis.

Although current dialysis therapy is life-sustaining, its simple physical transport processes cannot replace the complex transport, metabolic, and endocrine functions of the tubular cells. A bioartificial kidney that includes a conventional dialysis filter and a renal tubule assist device (RenaMed Biologics, formerly Nephros Therapeutics) containing approximately 10^8 renal proximal tubule cells was recently successfully engineered (Tiranathanagul et al., 2005; Humes et al., 1999). In the renal tubule device, cells are grown in a monolayer on the inside of hollow-fiber membranes. This technology is being evaluated to investigate the potential of replacing and maintaining a full range of key functions of the kidney, including endocrine equilibrium, metabolic activity, and immune surveillance in the area of acute renal failure.

The "holy grail" of hemodialysis is the wearable artificial kidney (Kolff et al., 1976). Key challenges in the development of such a device include anticoagulation (avoiding clotting), vascular access (avoiding infection), solute removal, and acid–base balance.

19.8 CONCLUSIONS

Hemodialysis continues to be the largest volume market for polymer membranes. While this is a very mature commodity market, unmet clinical needs continue to drive research along a number of fronts.

REFERENCES

Ariano, R. E., Fine, A., Sitar, D. S., Rexrode, S., and Zelenitsky, S. A. (2005). Adequacy of a vancomycin dosing regimen in patients receiving high-flux hemodialysis. *Am. J. Kidney Dis.* 46, 681.

Bird, R. B., Stewart, W. E., and Lightfoot, E. N. (1960). Velocity distributions in laminar flow. *Transport Phenom.* 34, 34–70.

Bowry, S. K. (2002). Dialysis membranes today. *Int. J. Artif. Organs* 25, 447.

Canaud, B., Bragg-Gresham, J. L., Marshall, M. R., Desmeules, S., Gillespie, B. W., Depner, T. A., Klassen, P., and Port, F. K. (2003). Patients receiving hemodiafiltration or hemofiltration have lower mortality rate than patients receiving hemodialysis without replacement fluid (HD) in Europe: The Dialysis Outcomes and Practice Patterns Study (DOPPS). *Am. Soc. Nephrol.* 14, 31A.

Chenoweth, D. E., Cheung, A. K., and Henderson, L. W. (1983). Anaphylatoxin formation during hemodialysis: Effects of different dialyzer membranes. *Kidney Int.* 24, 764.

Clark, W. R., and Gao, D. (2002). Properties of membranes used for hemodialysis therapy. *Semin. Dial.* 15, 191.

Clark, W. R., Macias, W. L., Molitoris, B. A., and Wang, N. H. (1995). Plasma protein adsorption to highly permeable hemodialysis membranes. *Kidney Int.* 48, 481.

Clinical Performance Measures, (CPM) (2004a). Annual report. End Stage Renal Disease Clinical Performance Measures Project, Centers for Medicare & Medicaid Services, Department of Health and Human Services, Office of Clinical Standards & Quality, Baltimore, MD, p. 22.

Clinical Performance Measures (CPM) (2004b). Annual report. End Stage Renal Disease Clinical Performance Measures Project, Centers for Medicare & Medicaid Services, Department of Health and Human Services, Office of Clinical Standards & Quality, Baltimore, MD, p. 25.

Clinical Performance Measures (CPM) (2004c). Annual report. End Stage Renal Disease Clinical Performance Measures Project, Centers for Medicare & Medicaid Services, Department of Health and Human Services, Office of Clinical Standards & Quality, Baltimore, MD, p. 21.

Code of Federal Regulations (CFR) (2005). 21CFR876, U.S. Government Printing Office, Baltimore, MD.

Daugirdas, J. (1993). Second generation logarithmic estimates of single-pool variable volume Kt/V: An analysis of error. *J. Am. Soc. Nephrol.* 4, 1205.

Deppisch, R., Gohl, H., and Smeby, L. (1998). Microdomain structure of polymeric surfaces—Potential for improving blood treatment procedures. *Nephrol. Dial. Transplant.* 13, 1354.

Dou, L., Bertrand, E., Cerini, C., Faure, V., Sampol, J., Vanholder, R., Berland, Y., and Brunet, P. (2004). The uremic solutes *p*-cresol and indoxyl sulfate inhibit endothelial proliferation and wound repair. *Kidney Int.* 65, 442.

Falkenhagen, D., Strobl, W., Vogt, G., Schrefl, A., Linsberger, I., Gerner, F. J., and Schoenhofen, M. (1999). Fractionated plasma separation and adsorption system: A novel system for blood purification to remove albumin bound substances. *Artif. Organs* 23, 81.

Fujimura, T., Uchi, Y., Fukuda, M., Miyazaki, M., Uezumi, S., and Hiyoshi, T. (2004). Development of a dialyzer with enhanced internal filtration to increase the clearance of low molecular weight proteins. *J. Artif. Organs* 7, 149.

Galli, F., Benedetti, S., Buoncristiani, U., Piroddi, M., Conte, C., Canestrari, F., Buoncristiani, E., and Floridi, A. (2003). The effect of PMMA-based protein-leaking dialyzers on plasma homocysteine levels. *Kidney Int.* 64, 748.

Gohl, H., Buck, R., and Strathmann, H. (1992). Basic features of the polyamide membranes. *Contrib. Nephrol.* 96, 1.

Gotch, F. A., Panlilio, F. M., Buyaki, R. A., Wang, E. X., Folden, T. I., and Levin, N. W. (2004). Mechanisms determining the ratio of conductivity clearance to urea clearance. *Kidney Int. Suppl.* 89, S3.

Grammer, L. C., and Patterson, R. (1987). IgE against ethylene oxide-altered human serum albumin (ETO-HSA) as an etiologic agent in allergic reactions of hemodialysis patients. *Artif. Organs* 11(2), 97–99.

Hakim, R. M., Fearon, D. T., and Lazarus, J. M. (1984). Biocompatibility of dialysis membranes: Effects of chronic complement activation. *Kidney Int.* 26, 194.

Henderson, L. W., Clark, W. R., and Cheung, A. K. (2001). Quantification of middle molecular weight solute removal in dialysis. *Semin. Dial.* 14, 294.

Homma, N., Gejyo, F., Hasegawa, S., Teramura, T., Ei, I., Maruyama, H., and Arakawa, M. (1995). Effects of a new adsorbent column for removing beta-2-microglobulin from circulating blood of dialysis patients. *Contrib. Nephrol.* 112, 164.

Humes, H. D., Buffington, D. A., MacKay, S. M., Funke, A. J., and Weitzel, W. F. (1999). Replacement of renal function in uremic animals with a tissue-engineered kidney. *Nat. Biotechnol.* 17, 451 [see comment].

Hyodo, T., Taira, T., Hiyama, E., Kondo, M., Honma, T., Uchida, T., Baba, S., Sakai, T., and Hidai, H. (2001). The Clinical Study on the Newer Type (Lixelle Type S-15) of the Direct Hemoperfusion B2-MG Adsorption Column. Paper presented at the ASN/ISN World Conference, A1372, San Francisco, CA.

Jirka, T., Cesare, S., Di Benedetto, A., Chang, M., Ponce, P., and Richards, N. (2005). Impact of on-line haemodiafiltration (HDF) on patient survival: Results from a large network database. XLII ERA-EDTA Congress, Istanbul, Turkey.

Kaneko, T., Kudo, M., Okumura, T., Kasiwagi, T., Turuoka, S., Simizu, M., Iino, Y., and Katayama, Y. (2001). Successful treatment of digoxin intoxication by haemoperfusion with specific columns for beta2-microgloblin-adsorption (Lixelle) in a maintenance haemodialysis patient. *Nephrol. Dial. Transplant.* 16, 195.

Khuroo, M. S., Khuroo, M. S., and Farahat, K. L. (2004). Molecular adsorbent recirculating system for acute and acute-on-chronic liver failure: A meta-analysis. *Liver Transplant.* 10, 1099 [see comment].

Koda, Y., Nishi, S., Miyazaki, S., Haginoshita, S., Sakurabayashi, T., Suzuki, M., Sakai, S., Yuasa, Y., Hirasawa, Y., and Nishi, T. (1997). Switch from conventional to high-flux membrane reduces the risk of carpal tunnel syndrome and mortality of hemodialysis patients. *Kidney Int.* 52, 1096.

Kolff, W. J., Jacobsen, S. C., Stephen, R. L., and Ron, D. (1976). Towards a wearable artificial kidney. *Kidney Int.* 10, S300.

Krieter, D. H., Falkenhain, S., Chalabi, L., Collins, G., Lemke, H. D., and Canaud, B. (2005). Clinical cross-over comparison of mid-dilution hemodiafiltration using a novel dialyzer concept and post--dilution hemodiafiltration. *Kidney Int.* 67, 349.

Lavaud, S., Canivet, E., Wuillai, A., Maheut, H., Randoux, C., Bonnet, J.-M., Renaux, J.-L., and Chanard, J. (2003). Optimal anticoagulation strategy in haemodialysis with heparin-coated polyacrylonitrile membrane. *Nephrol. Dial. Transplant.* 18, 2097.

Leypoldt, J. K., Cheung, A. K., Agodoa, L. Y., Daugirdas, J. T., Greene, T., and Keshaviah, P. R. (1997). Hemodialyzer mass transfer-area coefficients for urea increase at high dialysate flow rates. The Hemodialysis (HEMO) Study. *Kidney Int.* 51, 2013.

Lowrie, E. G., Li, Z., Ofsthun, N., and Lazarus, J. M. (2004). Reprocessing dialysers for multiple uses: Recent analysis of death risks for patients. *Nephrol. Dial. Transplant.* 19, 2823.

Lowrie, E. G., Li, Z., Ofsthun, N., and Lazarus, J. M. (2005). The online measurement of hemodialysis dose (Kt): Clinical outcome as a function of body surface area. *Kidney Int.* 68, 1344.

McGill, R. L., Blas, A., Bialkin, S., Sandroni, S. E., and Marcus, R. J. (2005). Clinical consequences of heparin-free hemodialysis. *Hemodial. Int.* 9, 393.

Meyer, T. W., Leeper, E. C., Bartlett, D. W., Depner, T. A., Lit, Y. Z., Robertson, C. R., and Hostetter, T. H. (2004). Increasing dialysate flow and dialyzer mass transfer area coefficient to increase the clearance of protein-bound solutes. *J. Am. Soc. Nephrol.* 15, 1927.

Mineshima, M., Ishimori, I., Ishida, K., Hoshino, T., Kaneko, I., Sato, Y., Agishi, T., Tamamura, N., Sakurai, H., Masuda, T., and Hattori, H. (2000). Effects of internal filtration on the solute removal efficiency of a dialyzer. *ASAIO J.* 46, 456.

Niwa, T., Asada, H., Tsutsui, S., and Miyazaki, T. (1995). Efficient removal of albumin-bound furancarboxylic acid by protein-leaking hemodialysis. *Am. J. Nephrol.* 15, 463.

Niwa, T., Yazawa, T., Kodama, T., Uehara, Y., Maeda, K., and Yamada, K. (1990). Efficient removal of albumin-bound furancarboxylic acid, an inhibitor of erythropoiesis, by continuous ambulatory peritoneal dialysis. *Nephron* 56, 241.

NKF-K/DOQI (2001). NKF-K/DOQI clinical practice guidelines for hemodialysis adequacy: Update 2000. *Am. J. Kidney Dis.* 37, S7.

Ofsthun, N. J., Jenson, J. C., and Kray, M. (1991). Effect of high hematocrit and high blood flow rates on transmembrane pressure and ultrafiltration rate in hemodialysis. *Blood Purif.* 9, 169.

Ofsthun, N. J., and Zydney, A. L. (1994). Importance of convection in artificial kidney treatment. *Contrib. Nephrol.* 108, 53.

Owen, W. F., Lew, N. L., Liu, Y., Lowrie, E. G., and Lazarus, J. M. (1993). The urea reduction ratio and serum albumin concentration as predictors of mortality in patients undergoing hemodialysis. *N. Engl. J. Med.* 329, 1001.

Pitts, R. (1968). *Physiology of the Kidney and Body Fluids*. Year Book Medical Publishers, Chicago, p. 58.

Port, F. K., Orzol, S. M., Held, P. J., and Wolfe, R. A. (1998). Trends in treatment and survival for hemodialysis patients in the United States. *Am. J. Kidney Dis.* 32, S34 [see comment].

Renaux, J. L., Thomas, M., Crost, T., Loughraieb, N., and Vantard, G. (1999). Activation of the kallikrein-kinin system in hemodialysis: Role of membrane electronegativity, blood dilution, and pH. *Kidney Int.* 55, 1097.

Rifai, K., Ernst, T., Kretschmer, U., Bahr, M. J., Schneider, A., Hafer, C., Haller, H., Manns, M. P., and Fliser, D. (2003). Prometheus—A new extracorporeal system for the treatment of liver failure. *J. Hepatol.* 39, 984.

Ronco, C., and Bowry, S. (2001). Nanoscale modulation of the pore dimensions, size distribution and structure of a new polysulfone-based high-flux dialysis membrane. *Int. J. Artif. Organs* 24, 726.

Ronco, C., Brendolan, A., Lupi, A., Metry, G., and Levin, N. W. (2000). Effects of a reduced inner diameter of hollow fibers in hemodialyzers. *Kidney Int.* 58, 809.

Ronco, C., Brendolan, A., Winchester, J. F., Golds, E., Clemmer, J., Polaschegg, H. D., Muller, T. E., Davankov, V., Tsyurupa, M., Pavlova, L., Pavlov, M., La Greca, G., and Levin, N. W. (2001). First clinical experience with an adjunctive hemoperfusion device designed specifically to remove beta 2-microglobulin in hemodialysis. *Contrib. Nephrol.*, 166–173.

Ronco, C., Orlandini, G., Brendolan, A., Lupi, A., and La Greca, G. (1998). Enhancement of convective transport by internal filtration in a modified experimental hemodialyzer: Technical note. *Kidney Int.* 54, 979.

Sato, Y., Mineshima, M., Ishimori, I., Kaneko, I., Akiba, T., and Teraoka, S. (2003). Effect of hollow fiber length on solute removal and quantification of internal filtration rate by Doppler ultrasound. *Int. J. Artif. Organs* 26, 129.

Schaefer, R. M., Fink, E., Schaefer, L., Barkhausen, R., Kulzer, P., and Heidland, A. (1993). Role of bradykinin in anaphylactoid reactions during hemodialysis with AN69 dialyzers. *Am. J. Nephrol.* 13, 473.

Spalding, E. M., Chamney, P. W., and Farrington, K. (2002). Phosphate kinetics during hemodialysis: Evidence for biphasic regulation. *Kidney Int.* 61, 655.

Stange, J., Ramlow, W., Mitzner, S., Schmidt, R., and Klinkmann, H. (1993). Dialysis against a recycled albumin solution enables the removal of albumin-bound toxins. *Artif. Organs* 17, 809.

Sugaya, H., and Sakai, Y. (1999). Polymethylmethacrylate: From polymer to dialyzer. *Contrib. Nephrol.* 125, 1.

Tielemans, C., Madhoun, P., Lenaers, M., Schandene, L., Goldman, M., and Vanherweghem, J. L. (1990). Anaphylactoid reactions during hemodialysis on AN69 membranes in patients receiving ACE inhibitors. *Kidney Int.* 38, 982.

Tiranathanagul, K., Eiam-Ong, S., and Humes, H. D. (2005). The future of renal support: High-flux dialysis to bioartificial kidneys. *Crit. Care Clin.* 21, 379.

U.S. Renal Data System (USRDS) (2005). USRDS annual data report. Available: www.usrds.org.

Vanholder, R., De Smet, R., Glorieux, G., Argiles, A., Baurmeister, U., Brunet, P., Clark, W., Cohen, G., De Deyn, P. P., Deppisch, R., Descamps-Latscha, B., Henle, T., Jorres, A., Lemke, H. D., Massy, Z. A., Passlick-Deetjen, J., Rodriguez, M., Stegmayr, B., Stenvinkel, P., Tetta, C., Wanner, C., Zidek, W., and European Uremic Toxin Work (2003). Review on uremic toxins: Classification, concentration, and interindividual variability. *Kidney Int.* 63, 1934.

Vanholder, R., De Smet, R., and Lameire, N. (2001). Protein-bound uremic solutes: The forgotten toxins. *Kidney Int. Suppl.* 78, S266.

Ward, R. A., Leypoldt, J. K., Clark, W. R., Ronco, C., Mishkin, G. J., and Paganini, E. P. (2001). What clinically important advances in understanding and improving dialyzer function have occurred recently? *Semin. Dial.* 14, 160.

Ward, R. A., Schmidt, B., Hullin, J., Hillebrand, G. F., and Samtleben, W. (2000). A comparison of on-line hemodiafiltration and high-flux hemodialysis: A prospective clinical study. *J. Am. Soc. Nephrol.* 11, 2344.

Wendler, T., Duhr, C., and Bosch, T. (2003). Ex vivo biocompatibility of a new beta2-microglobulin hemoperfusion polymer. *Int. J. Artif. Organs* 26, 467.

Zydney, A. L., and Colton, C. K. (1988). Augmented solute transport in the shear flow of a concentrated suspension. *Physicochem. Hydrodyn.* 10, 77–96.

CHAPTER 20

Tangential-Flow Filtration for Virus Capture

S. RANIL WICKRAMASINGHE

Department of Chemical and Biological Engineering, Colorado State University, Fort Collins, Colorado 80523-1370

20.1 INTRODUCTION

Tangential-flow filtration is frequently used in the biotechnology industry for the purification of proteins. Today, tangential-flow microfiltration (MF) is often the first of the purification operations in the downstream processing of biopharmaceutical products. It is used to separate cells, cell debris, and other insoluble particulate matter in the size range of 0.1–10 μm from the growth medium, which usually contains the product of interest (Kim et al., 2001; Davis, 1992). Microfiltration membranes have their origins in the development of colloidan (nitrocellulose) membranes. In 1926 Membranfilter GmbH was founded in Göttingen, Germany and began commercial manufacture of MF membranes (Baker, 2000). By 1940 other companies such as Sartorius (Göttingen, Germany) and Schleicher and Schuell (Dassel, Germany) were producing similar membranes.

Until the mid-1960s MF membranes were used only in laboratory- and very small scale industrial applications. In the 1960s and 1970s microfiltration became important in biological and pharmaceutical manufacturing. In the 1970s the tangential-flow configuration (also known as cross-flow filtration) was introduced by using the cross-flow technique developed for reverse osmosis (Grandison and Finnigan, 1996). Figure 20.1 shows the tangential-flow configuration. In this chapter we consider only tangential flow as it is commonly used for recovery and purification of products in the biotechnology industry.

Ultrafiltration (UF) uses membranes with pore sizes ranging from 10 to 1000 Å (0.1 μm). Consequently, large-pore UF membranes will have a similar pore size to small-pore MF membranes. The first UF membranes were prepared by Bechhold (1907) from colloidan. By the mid-1920s colloidan UF membranes were available for laboratory use. However, industrial use of UF membranes did not occur till the 1960s after the development of high-flux reverse osmosis (RO) membranes (Loeb and Sourirajan, 1963). Today, ultrafiltration is frequently used in the biotechnology industry for protein concentration and buffer exchange (diafiltration).

Advanced Membrane Technology and Applications. Edited by Norman N. Li, Anthony G. Fane, W. S. Winston Ho, and T. Matsuura
Copyright © 2008 John Wiley & Sons, Inc.

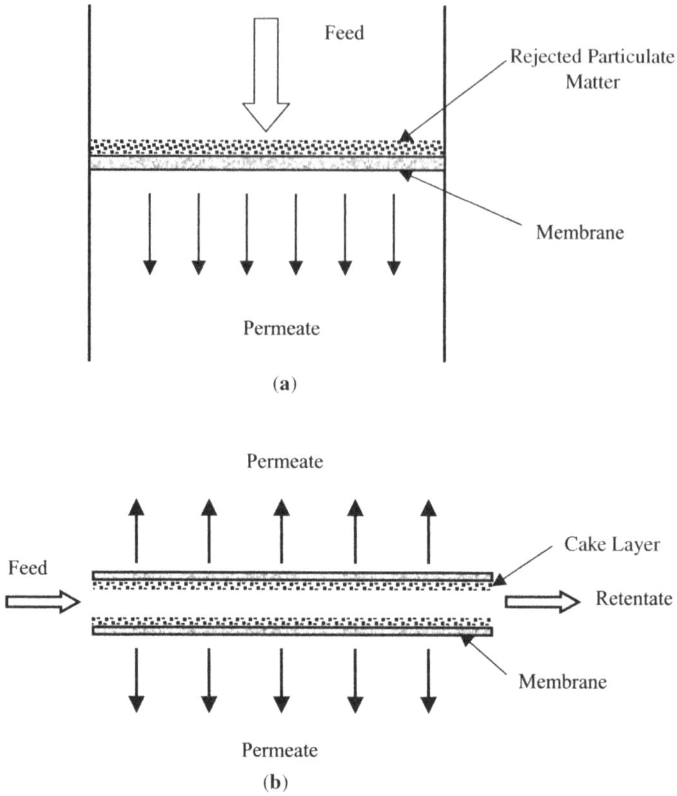

Figure 20.1 Schematic representation of (a) normal flow and (b) tangential-flow filtration.

Virus capture is important in the biotechnology industry for two main reasons. Large-scale production of virus vectors for gene therapy applications and viral vaccines is a major challenge. Morenweiser (2005) estimates that between 10^{11} and 10^{14} viral vectors will be required to satisfy gene therapy regimens. Thus, efficient and scalable processes for producing viral vectors are needed. For example, in the case of human influenza virus, development of a cell-culture-based production system will overcome many of the difficulties of the current process (Genzel et al., 2004). Development of efficient down-stream processing steps for recovery and purification of viral vectors is also essential especially given that the purification steps could account for up to 70% of the production costs (Morenweiser, 2005).

While viral vectors can represent the product of interest, virus capture in order to validate virus clearance from the product of interest is also a major concern in the manufacture of bio-pharmaceuticals (Huang and Peterson, 2001). Validation of virus clearance is a critical component in the purification process. The use of cell lines containing autonomously replicating vectors such as retroviruses and animal-derived cell culture media carries with it the risk of contamination of the final product by viruses and virus like particles associated with the raw materials. Regulatory authorities will not allow clinical testing of purified pharmaceuticals without validation of sufficient removal of these virus particles. Manufacturers of mono-clonal antibodies as therapeutics must demonstrate elimination of 10^3–10^5 more virus particles than is estimated in a single dose equivalent of the unprocessed bulk during purification. In the production of monoclonal antibodies involving the use of cell lines such as Chinese hamster ovary (CHO) cells or murine hybridoma, the presence of type A and type C retrovirus particles indicates the potential risk of retroviral contamination (Huang and Peterson, 2001). Depending on the antibody titre, estimates of the number of viruses in a single dose equivalent could be as high as 10^{10}–10^{15} retrovirus-like particles per milliliter.

This chapter discusses the use of tangential-flow filtration for virus capture to recover, purify, and concentrate virus vectors and for validation of virus clearance. Virus particles may vary is size from about 20 nm (parvoviruses) to more than about 300 nm (van Regenmortel et al., 2000). However, most viruses of significance to the biotechnology industry are no more than 140 nm in size. Consequently, use of MF and large-pore UF membranes could lead to the development of efficient separation processes for virus capture. The chapter begins by briefly reviewing some of the features of the tangential-flow filtration mode of operation. Next tangential-flow filtration for purification of viral vectors for use in viral vaccines and in gene therapy applications is discussed. The final section of the chapter discusses the use of tangential-flow filtration for validation of virus clearance.

20.2 TANGENTIAL-FLOW FILTRATION

In tangential-flow filtration, the feed is pumped tangential to the membrane surface. The permeate flows through the membrane at right angles to the direction of feed flow. In the case of hollow-fiber membranes, the feed is usually pumped through the lumen of the fibers while the permeate flows through the membrane and is collected on the shell side. The driving force for permeate flow is the transmembrane pressure. Microfiltration is usually operated industrially using two pumps, one for the feed and one for the permeate (see Fig. 20.2). By using a permeate pump, the operator can control the permeate flow rate directly (Wickramasinghe et al., 2004a). Ultrafiltration is usually run industrially with only a feed pump.

Tangential-flow filtration is typically operated in one of two modes: concentration or diafiltration (see Fig. 20.2). For microfiltration, in the concentration mode, the product of interest usually passes through the pores of the membrane and is collected in the permeate. However, since MF membranes are nonselective, the concentration of the product of interest will be the same in the retentate and permeate streams.

During tangential-flow microfiltration the imposed pressure drop carries particulate matter to the membrane surface where it is rejected and forms a cake layer. Unlike dead-end or normal flow filtration, the cake layer does not build up indefinitely, as the feed is continually pumped across the cake surface. Deposition of particulate matter leads to a narrowing of the feed channel, which in turn leads to higher average feed velocities and higher shear stresses exerted by the suspension on the surface of the cake layer. Consequently, tangential flow of the feed tends to sweep deposited particles toward the filter exit, thus suppressing growth of the cake layer. Nevertheless, the cake layer does provide an additional resistance to permeate flow. Further it can lead to partial rejection of solutes that would completely pass through the membrane in the absence of particulate matter. If the product of interest is partially rejected, its recovery will be less than expected (Castino and Wickramasinghe, 1996).

Recovery of the solute of interest may be maximized by operating in diafiltration mode (see Fig. 20.2). Here diluent (free of particulate matter) is added to the feed reservoir at the

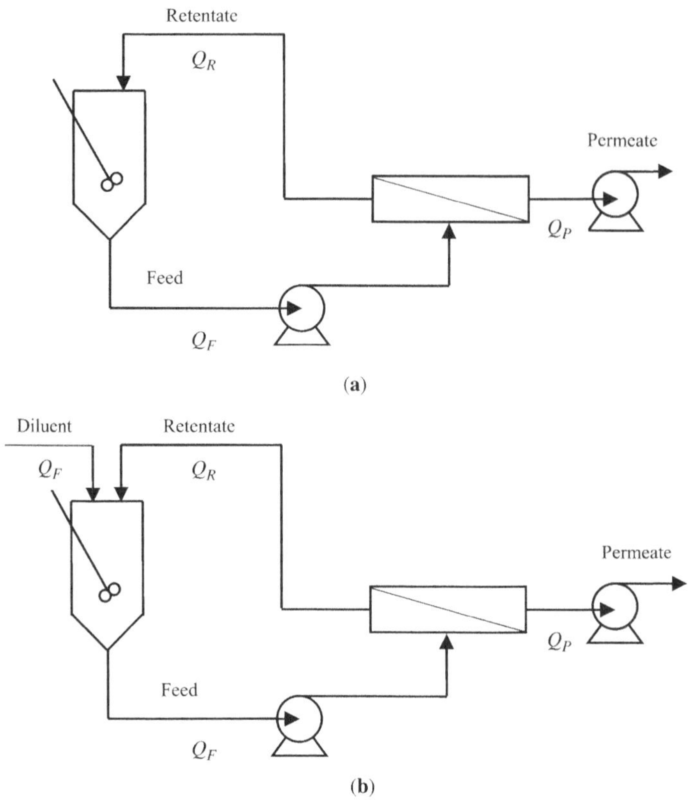

Figure 20.2 Modes of operation of tangential-flow filtration: (a) concentration and (b) diafiltration.

same rate at which permeate is removed. Consequently, the solids concentration in the feed remains constant. While additional product is recovered in the permeate, the product of interest will be diluted.

Ultrafiltration may also be run in concentration and diafiltration modes. Ultrafiltration is usually used for protein concentration and buffer exchange by diafiltration. In addition many studies have explored the possibility of fractionating proteins using ultrafiltration membranes. Though it has generally been assumed that separation of solutes that differ in size by less than an order of magnitude is difficult by tangential-flow filtration, many recent studies indicate that by carefully controlling the operating conditions efficient separation of solutes that differ in size by less than an order of magnitude is possible (Cherkasov and Polotsky, 1996). For example, Pujar and Zydney (1994), have demonstrated the importance of electrostatic and electrokinetic interactions in determining selective protein transport through UF membranes. Saksena and Zydney (1994), showed that the selectivity (defined as the ratio of protein sieving coefficients) for bovine serum albumin (BSA) and immunoglobulin G (IgG) could be increased from 2 at pH 7 to more than 30 by adjusting the pH to 4.7 and lowering the solution ionic strength. Similar improvements in performance have been reported for laboratory-scale filtration of BSA and hemoglobin (van Eijndhoven et al., 1995), BSA and lysozyme (Iritani et al., 1995), and myoglobin and cytochrome (Yang and Tong, 1997).

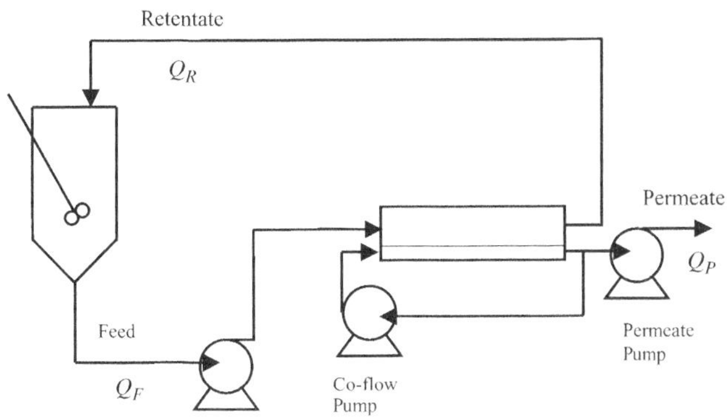

Figure 20.3 Schematic representation of high-performance tangential-flow filtration mode of operation.

Van Reis et al. (1999), have investigated the possibility of protein fractionation using high-performance tangential-flow filtration. In conventional tangential-flow filtration, the feed pressure varies along the feed flow path from the inlet pressure (P_i) to the outlet pressure (P_o). Typically, P_o is close to atmospheric pressure. The average transmembrane pressure (TMP) is given by TMP $= (P_i + P_o)/2 - P_p$, where P_p is the permeate-side pressure. However, since the feed-side pressure varies from P_i to P_o, the TMP will also vary along the feed flow path. This variation in TMP along the feed channel will reduce the resolving power of conventional tangential-flow filtration systems (van Reis et al., 1997). In high-performance tangential-flow filtration, part of the permeate is returned to the permeate side of the module (see Fig. 20.3) such that the permeate flows co-current to the feed. By creating an axial pressure drop along the permeate flow channel, the TMP is more nearly constant throughout the module.

Van Reis et al. (1999) found that high separation factors for solute species with size differences less than an order of magnitude were obtainable using high-performance tangential-flow filtration. Their results indicate that maintaining a constant transmembrane pressure drop across the entire length of the membrane leads to a much finer fractionation of proteins.

The unique features of tangential-flow filtration include the ability to remove cells and cell debris from the growth medium, which contains the product of interest, to concentrate the product of interest and to fractionate solutes of different size. These features have lead to numerous applications of tangential flow filtration in the purification of protein products. These same features could be exploited both for virus purification and validation of virus clearance as described in the following sections.

20.3 TANGENTIAL-FLOW FILTRATION FOR VIRUS CAPTURE

Table 20.1 summarizes some of the important viruses for gene therapy applications and viral vaccines that have been purified using tangential-flow filtration. Viruses can be divided into two groups: enveloped and nonenveloped. Enveloped viruses contain an outer lipoprotein bilayer membrane derived from the membranes of the host cell. Nonenveloped viruses lack this outer lipoprotein membrane.

TABLE 20.1 Examples of Tangential Flow Filtration for Purification of Viruses

Virus	Family	Features	Tangential-Flow Filtration	Reference
Adenovirus	Adenoviridae	Nonenveloped, double-stranded DNA virus, 70–90 nm icosahedral particles	300 and 400 kDa and 0.05-μm membranes used to lyse and concentrate virus	Subramanian et al. (2005)
Aedes aegypti densonucleosisvirus	Parvoviridae	Nonenveloped, single-stranded DNA virus, 18–26 nm icosahedral particles	30-, 50-, 100-, and 300-kDa membranes used to concentrate virus particles	Grzenia et al. (2006)
Human influenza A virus	Orthomyxoviridae	Enveloped, RNA virus, 90–120 nm spherical particles	100- and 300-kDa and 0.1-, 0.2-, and 0.45-μm membranes used to purify, concentrate, and fractionate virus particles	Wickramasinghe et al. (2005)
Lentivirus	Retroviridae	Enveloped, RNA viruses consisting of 80–130 nm spherical particles	100- and 300-kDa membranes used to concentrate virus	Geraerts et al. (2005)
Murine leukemia virus			100-kDa membranes used to concentrate virus	Saha et al. (1994)
			0.45-μm and 300-kDa membranes used in series to purify and concentrate virus	Braas et al. (1996)

Source: Modified from Grzenia et al. (2006).

20.3.1 Nonenveloped viruses

Subramanian et al. (2005) recently described a hollow-fiber-based recovery and concentration process for adenoviral vectors. Adenoviruses are currently used in about 25% of gene therapy protocols (Morenweiser, 2005). Further they appear promising as gene delivery vehicles for vaccination against infectious diseases such as HIV-1 (AIDS) (Shiver and Emini, 2004). Depending on the adenovirus vector used and the processing conditions, up to 50% of the virus particles can be found in the supernatant, the remainder being in the host cells. Consequently, large-scale recovery of adenoviral vectors must be capable of recovering intracellular and extracellular virus particles.

Subramanian et al. (2005) pumped the host cells through the lumen of the hollow fibers at shear rates that lead to cell lysis. The adenoviral vectors that were released were concentrated in the retentate. Since adenoviral vectors are nonenveloped icosahedral particles that are about 80 nm in size, the membrane pore size had to be small enough to reject the virus particles. Membranes with pore sizes of 300 kDa (GE Healthcare, Fairfield, CT), 400 kDa, and 0.05 μm (Spectrum Laboratories, Rancho Dominguez, CA) were tested and found to give good virus rejection.

Condon et al. (2000) describe the use of sequential MF and UF steps to recover virus particles. A 0.65-μm pore size Durapore (Millipore, Bedford, MA) membrane was used in the first MF step to lyse the host cells analogous to the process used by Subramanian et al. (2005). However, since the membrane pore size is larger than the virus particles, the virus particles were recovered in the permeate. To ensure adequate virus recovery, diafiltration of the cell broth was required leading to significant dilution of the recovered virus particles. The second UF step was used to concentrate the virus particles in the permeate from the first MF step.

Grzenia et al. (2006) have investigated the use of 30-, 50-, 100-, and 300-kDa UF membranes for recovery of *Aedes aegypti densonucleosisvirus* (AeDNV). These parvoviruses are similar in size to adeno-associated virus (AAV). AAVs are being extensively investigated as gene therapy vectors since they do not induce an immune response toward viral components; they can integrate into human chromosome 19, and they do not require actively dividing cells for transduction and are nonpathogenic (Morenwiser, 2005). However, large-scale purification of AAV remains a major challenge (Smith et al., 2003; Davidoff et al., 2004; O'Riordan et al., 2000). AeDNV infects the *A. aegypti* mosquito. The *A. aegypti* mosquito is a carrier of viruses that are pathogens that cause dengue and yellow fever. Consequently, development of an AeDNV vector may find important applications in integrated vector-borne disease control programs against human pathogens such as dengue and yellow fever (Carlson et al., 1997).

Grzenia et al. (2006) found that 30-, 50-, and 100-kDa membranes (Sartorius AG, Göttingen, Germany) reject the virus particles while 300-kDa membranes allow some virus particles to pass into the permeate. The decrease in permeate flux during virus concentration for the 300-kDa membrane is much greater than for the 30-, 50-, and 100-kDa membranes indicating possible entrapment of virus particle in the membrane pores. The permeate flux and level of protein rejection is strongly affected by the cell culture growth medium. Their results indicate that when developing a new process, it is essential that the cell culture and purification operations be developed in parallel.

Removal of host cell DNA (deoxyribonucleic acid) was not a major concern for the AeDNV vectors Grazenia et al. (2006) produced for infection of mosquitoes. However, when producing viral vectors for gene therapy applications, high levels of purity are required. For example, AAV vectors have frequently been purified from cell lysates using several rounds of density gradient centrifugation (Morenweiser, 2005). This method, however, has been shown to result in AAV vectors that are contaminated with impurities that can cause local inflammation in vivo (Smith et al., 2003; Davidhoff et al., 2004; O'Riordan et al., 2000).

Grzenia et al. (2006) determined the variation of total protein concentration in the retentate and permeate during concentration of the virus-containing feed solution. Figure 20.4 gives their results for the 30-, 50-, 100-, and 300-kDa membranes tested. As can be

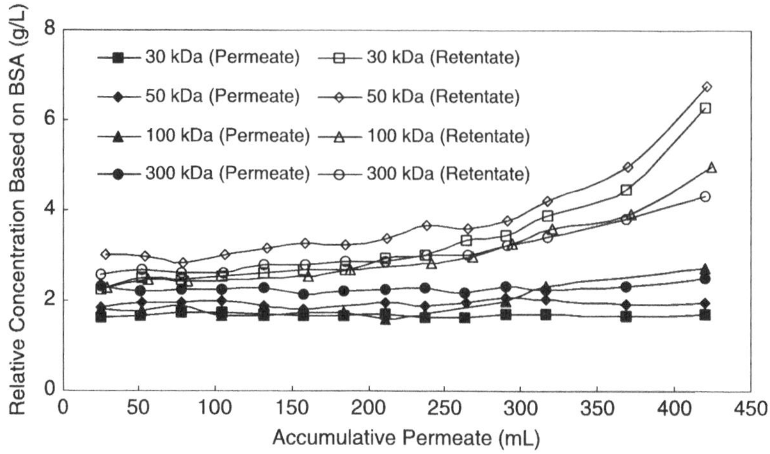

Figure 20.4 Variation of protein concentration in the retentate and permeate during concentration for AeDNV particles (Grzenia et al., 2006).

seen, all four membranes reject protein, the degree of rejection increasing with decreasing membrane molecular weight cutoff of the membrane. A nonzero protein concentration in the permeate for all four membranes indicates that some protein passes through the membrane.

Design of a UF system that concentrates AeDNV particles while minimizing rejection of host cell proteins will require careful selection of the membrane pore size and operating conditions. As described in the previous section, numerous investigators have shown that by manipulating the feed pH, ionic strength, and the TMP, fractionation of proteins that are less than an order of magnitude different in size is feasible. It is likely that in the future a similar method could be used for virus purification. This will be especially important when trying to remove high-molecular-weight contaminants from small nonenveloped viral vectors.

20.3.2 Enveloped Viruses

There have been many more studies on the use of tangential-flow filtration to recover, purify, and concentrate enveloped virus particles than nonenveloped virus particles. Wickramasinghe et al. (2005) have investigated the use of tangential-flow filtration to recover and concentrate human influenza A virus particles grown using a cell culture technique. Ultrafiltration membranes with molecular weight cutoffs of 100 and 300-kDa as well as 0.1-, 0.2-, and 0.45-μm MF membranes from Sartorius AG (Göttingen, Germany) were tested. The two UF membranes displayed complete rejection of the virus particles. The 0.45-μm MF membranes displayed complete passage of the virus particles. However, the 0.1- and 0.2-μm membranes displayed partial rejection of the virus particles.

Wickramasinghe et al. (2005) determined the variation of protein concentration and host cell DNA with permeate volume for the membranes tested. They note that using the 300-kDa membrane is advantageous as significant removal of host cell proteins and DNA is possible while ensuring complete rejection of virus particles. For an industrial process, optimization of the feed pH, ionic strength, and TMP is likely to be necessary in order to maximize removal of high-molecular-weight contaminants.

Wickramasinghe et al. (2005) measured the particle size distribution of the retentate and permeate streams using laser diffraction light scattering. Human influenza A virus particles have a nominal size of 100 nm. Figure 20.5 gives the particle size distribution in the retentate and permeate for 0.1- and 0.2-μm pore size membranes. As can be seen, both membranes are successful in fractionating the virus particles in the feed based on size. For both membranes the retentate contains larger particles than the permeate. Nevertheless there is significant overlap between the particle size distributions in the retentate and permeate.

Though human influenza A virus particle are 80–120 nm in size, Figure 20.5 indicates the existence of much larger and smaller particles. The larger particles are most likely due to virus particle attached to host cell membranes and other host cell fragments. The smaller particles are probably viral fragments. If only free intact virus particles are desired, then fractionation of virus particles could be of practical benefit. Removing unwanted particles during a tangential-flow filtration step will reduce the requirements on the subsequent purification steps such as ion exchange or size exclusion chromatography for whole-virus vaccines.

Comparing Figures 20.5a and 20.5b it can be seen that the sharpness of the virus particle fractionation is better with the 0.1-μm pore size membrane. Since membranes have a pore size distribution, development of a practical membrane-based fractionation step will depend on careful selection of an appropriate membrane pore size and pore size distribution. Further it is likely that operating methods such as high-performance tangential-flow filtration may improve the sharpness of the fractionation.

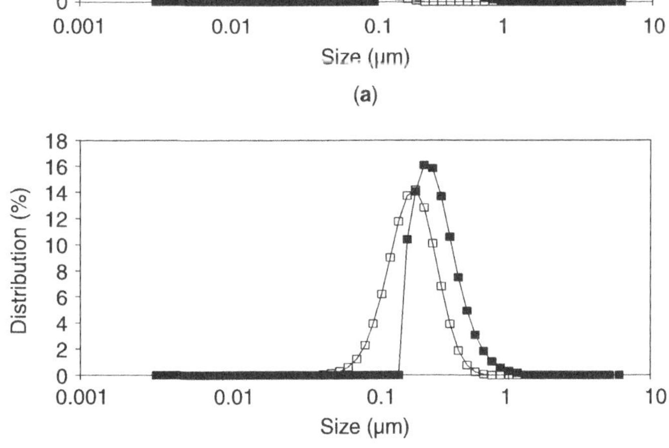

Figure 20.5 Particle size distribution in the retentate and permeate (a) for 0.1- and (b) 0.2-μm pore size membranes (Wickramasinghe et al., 2005).

Geraerts et al. (2005), concentrated lentiviral vectors using 100- and 300-kDa membranes. These viruses are extremely efficient at delivering genes to the brain and hold promise for future gene therapy of neurodegenerative disorders (Baekelandt et al., 2003). Geraets et al. (2005) show that tangential-flow filtration may be used to concentrate these viruses with recoveries between 90 and 100%.

Saha et al. (1994) concentrated murine leukemia virus using 100-kDa membranes. These investigators show that tangential-flow filtration may be used to concentrate enveloped viruses to titers higher than those obtained using a traditional sucrose gradient. Braas et al. (1996) described a two-stage membrane filtration process to recover and purify retrovirus particles. The membrane pore size in the first tangential-flow filtration step is larger than the virus particles; consequently, the virus particles are recovered in the permeate. The membrane pore size in the second filtration step is smaller than the virus particles, thus concentrating the virus in the retentate.

Tangential-flow filtration could find numerous applications in the purification of viral vectors. Membrane pore sizes that are larger than the virus particles may be used to reject host cells and cell debris but allow passage of virus particles. If the membrane pore size is similar to the average virus particle size, fractionation of virus particles may be possible. Membranes with pore sizes smaller than the virus particles may be used to concentrate virus particles. In designing a tangential-flow filtration processes for purification of virus particles, it is important to be aware of the range of virus particle sizes. For smaller virus particles (e.g., nonenveloped viruses) removal of large molecular weight contaminants such as host cell DNA may be challenging. In the case of larger enveloped viruses, careful selection of feed flow rates may be critical in order to prevent disruption of the virus particles.

20.4 TANGENTIAL-FLOW FILTRATION FOR VIRUS CLEARANCE

Virus clearance in excess of 10^4-fold (4 log) is often difficult to validate from a single unit operation. Therefore, virus clearance from a number of different unit operations must be summed to determine the overall level of virus clearance for the purification train. Virus clearance can be by physical removal of the virus particles from the product or by inactivation. Regulatory authorities such as the U.S. Food and Drug Administration (FDA) further specify that reduction factors for two-unit operations with the same mechanism of action may not be added. Furthermore, due to the high variability of infectivity assays, a reduction factor of less than 10-fold should not be included in the purification train.

Large molecular weight cutoff UF membranes usually ($1-3 \times 10^5$ Da; Azari et al., 2000; Ogle and Azari, 2001) have been developed as virus filtration membranes. Virus filtration usually occurs close to the end of the purification train where the feed is relatively pure. Ideally, virus filtration membranes retain all viruses that may be present but show no rejection of the product. Virus filtration is most efficient when the virus particles are at least one order of magnitude larger than the product of interest.

Today most virus filtration is conducted in normal flow mode (Bohonak and Zydney, 2005). The main advantage of tangential-flow filtration is the ability to suppress cake formation and the consequent decrease in permeate flux. This results in much larger feed volumes being filtered (capacity) before the module needs to be cleaned.

During virus filtration, the virus particles are rejected by the membrane. Consequently, reuse of the membrane is not practical as one would need to validate removal of virus particles from the membrane. Sizing of virus filtration membranes depends on the clean membrane permeability and the module capacity or the maximum volume of feed that can be processed before the permeate flux is unacceptably low. The permeability is determined by the pore size, porosity, and overall pore structure of the membrane (Bohonak and Zydney, 2005, Syedain et al., 2006).

If validation of virus clearance is being conducted using test virus particles that are at least one order of magnitude larger than the product of interest, then the product of interest should easily pass through the membrane pores. Further since virus filtration occurs near the end of the purification train, where the feed is free of large molecular weight contaminants and insoluble particulate cell debris, rejection of the product of interest due to cake formation is not a concern. An example of such a situation is the validation of retroviral clearance. Retroviruses are around 100 nm in size, much larger than most protein products. Consequently, due to ease of operation and lower capital costs, virus filtration to validate retroviral clearance is usually conducted in normal flow mode.

Validation of parvovirus clearance is often problematic as these nonenveloped virus particle are small (18–24 nm). Consequently, when the product of interest is large, use of normal flow filtration can lead to significant product rejection. Further validation of parvovirus clearance by low pH or elevated temperature is often not practical due to the high pH and thermal, stability of these virus particles.

Commercially available tangential-flow virus filters for validation of parvovirus clearance include the Millipore Viresolve 70 and 180 modules. Table 20.2 lists several FDA-recommended model viruses together with manufacturers rated level of viral clearance. The Viresolve 70 is rated to pass 90% of proteins with molecular weights less than 70 kDa and reject more than 3.5 log of poliovirus. The Viresolve 180 is rated to pass more than 90% of proteins with molecular weights in excess of 180 kDa and reject more than 2 log of poliovirus. As can be seen from Table 20.2, the larger the average size of the virus particle, the greater the rated clearance. However, the greater the rated clearance of smaller virus particles, the greater is the likelihood of rejection of larger proteins.

Protein passage may be maximized by optimizing the operating conditions. Virus particles are typically negatively charged at neutral pH. If the isoelectric point of the protein and virus are sufficiently far apart, adjustment of the feed pH and ionic strength may lead to improved protein passage through the membrane and hence better recoveries. Careful optimization of the transmembrane pressure and feed flow rates may also be necessary.

TABLE 20.2 Log Clearance of Model Viruses[a]

Virus	Family	Size	Retention by Viresolve 70	Retention by Viresolve 180
Poliovirus	Picornaviridae	22–30 nm, nonenveloped	3.5	2
Simian virus (SV-40)	Polyomaviridae	40 nm, nonenveloped	5.6	Not available
Sinbis	Togaviridae	70 nm, enveloped	7.4	4.5
Reovirus 3	Reoviridae	60–80 nm nonenveloped	7.2	5.5
Murine leukemia virus	Retroviridae	80–130 nm, enveloped	8.5	Not available

[a]As can be seen the log clearance increases with decreasing membrane molecular weight cutoff.

Source: Virus clearance data from Millipore Viresolve 70 and 180 Modules with CorrTest Integrity Test Kits, see: http://www.millipore.com/publications.nsf/docs/dn1180en00 (accessed April 2006).

Validation of virus clearance requires determining the clearance for a number of different unit operations and then summing the individual clearance factors to obtain the overall clearance by the purification train. If the overall clearance is less than the required clearance, additional unit operations must be added to the purification train simply to validate adequate virus clearance. However, addition of new unit operations to validate virus clearance increases the manufacturing cost. Consequently, there is a tremendous need to validate virus clearance in existing unit operations.

Wickramasinghe et al. (2004b) have investigated the feasibility of validating clearance of minute virus of mice (MVM), an FDA-recommended model parvovirus, during microfiltration. Since MVM particles are 18–24 nm in size, they will easily pass through a MF membrane when suspended in buffer. However, since microfiltration is used for bioreactor harvesting, cells and cell debris are present. In addition the authors added a cationic polyelectrolyte to flocculate the feed stream prior to microfiltration. Since mammalian cells are negatively charged at neutral pH, it is expected that the cationic polyelectrolyte will flocculate the biomass leading to a more filterable feed stream. In addition since MVM particles are also negatively charged at neutral pH, it is likely that MVM will be bound in the biomass flocculant aggregate.

Wickramasinghe et al. (2004b) obtained about 4 log clearance of MVM by flocculation of a Chinese hamster ovary cell feed stream spiked with MVM. After addition of the flocculant, microfiltration was conducted using 0.2-μm pore size hollow-fiber membranes (GE Healthcare, Fairfield, CT). In the absence of flocculant, only about 1 log clearance of MVM was obtained in the permeate, while for virus particles suspended in buffer no rejection was observed. In these studies, rejection of MVM is due to the cake layer and not the membrane. Consequently, careful control of the TMP and feed flow rate will be required to ensure similar cake properties.

Akeprathumchai et al. (2004) extended the work by Wickramasinghe et al. (2004b). They investigated the feasibility of retroviral clearance by flocculation and microfiltration of a feed stream consisting of Chinese hamster ovary cells. Clearance of murine leukemia virus (MLV) from flocculated Chinese hamster ovary cell feed streams was investigated. MLV is an FDA-recommended model retrovirus. The virus particle are 80–130 nm in size. Using a 0.1-μm pore size hollow-fiber membranes (GE Healthcare, Fairfield, CT), up to 4 log clearance of MLV was observed in the permeate in the absence of flocculant and cells. Consequently, rejection of MLM was due to the membrane. Using 0.65-μm hollow-fiber membranes (GE Healthcare, Fairfield, CT), little virus clearance was observed in the permeate in the absence of flocculant. However, when a cationic flocculant was added to the Chinese hamster ovary cell feed stream prior to microfiltration, the level of MLV clearance was similar to that obtained using the 0.1-μm pore size membrane.

While virus filtration is generally conducted using normal flow filtration, tangential-flow filtration may be useful especially when the difference in size between the virus particles and product species is less than an order of magnitude. If microfiltration is used for bioreactor harvesting, flocculating the feed prior to microfiltration can lead to significant virus clearance in the permeate.

20.5 CONCLUSIONS

Though tangential-flow filtration is used widely in the manufacture of proteins, there are few reports on the use of tangential-flow filtration for capture of virus particle. In this chapter applications of tangential-flow filtration for purification of virus vectors has been discussed. Both microfiltration and ultrafiltration could find many applications in the purification of virus vectors.

Tangential-flow filtration could also be used to validate virus clearance. Today most virus clearance filters are operated in normal flow mode. However, when the size of the model virus particle for which clearance is being validated, and the desired product are within an order of magnitude of each other, tangential-flow filtration may be the preferred mode of operation.

ACKNOWLEDGMENTS

Dr. Jonathan Carlson, Department of Microbiology, Immunology, and Pathology at Colorado State University is thanked for his assistance. Funding was provided by the National Science Foundation (CAREER program BES 9984095, CTS 0456831) and the National Institutes of Health (N01-AI 25489).

REFERENCES

Akeprathumchai, S., Han, B., Wickramasinghe, S. R., Carlson, J. O., Czermak, P., and Preiss, K. (2004). Murine leukemia virus clearance by flocculation and microfiltration. *Biotechnol. Bioeng.* 88(7), 880.

Azari, M., Boose, J. A., Burhop, K. E., Camacho, T., Catatello, J., Darling, A., Ebeling, A. A., Estep, T. N., Pearson, L., Guzder, S., Herren, J., Ogle, K., Paine, J., Rohn, K., Sarajari, R., Sun, C. S., and Zhang, L. (2000). Evaluation and validation of virus removal by ultrafiltration during the production of diasprin crosslinked hemoglobin (DCLHb). *Biologicals* 28, 81.

Baekelandt, V., Eggermont, K., Michiels, M., Nuttin, B., and Debyser, Z. (2003). Optimized lentiviral vector production and purification procedure prevents immune response after transduction of mouse brain. *Gene Therapy* 10, 1933.

Baker, R. W. (2000). *Membrane Technology and Applications*. McGraw Hill, New York, p. 275.

Bechhold, H. (1907). Kolloidstudien mit der Filtrationsmethode. *Z. Physik. Chem.* 60, 257.

Bohonak, D. M., and Zydney, A. L. (2005). Compaction and permeability effects with virus filtration membranes. *J. Membr. Sci.* 254, 71.

Braas, G., Searle, P. F., Slater, N. K. H., and Lyddiatt, A. (1996). Strategies for the isolation and purification of retroviral vectors for gene therapy. *Bioseparation* 6, 211.

Carlson, J. O., Afansiev, B. N., Higgs, S., Matsubara, T., and Burns, J. C. (1997). Viral transducing vectors for mosquitoes, In J. M. Crampton, C. B. Beard, and C. Louis (Eds.), *The Molecular Biology of Insect Disease Vectors: A Methods Manual*. Chapman and Hall, New York, p. 444.

Castino, F., and Wickramasinghe, S. R. (1996). Washing frozen red blood cell concentrates using hollow fibres. *J. Membr. Sci.* 110, 169.

Cherkasov, A. N., and Polotsky, A. E. (1996). The resolving power of ultrafiltration. *J. Membr. Sci.* 110, 79.

Condon, R. G. G., Connelly, N. V., Frei, A., Glowacki, E., Yabannavar, V., and Batandolo, S. (2000). Methods for cultivating cells and propagating viruses. U.S. Patent 6,146,891.

Davidoff, A. M., Ng. C. Y. C., Sleep, S., Gray, J., Azam, S., Zhao, Y., McIntosh, J. H., Karimipoor, M., and Nathwani, A. C. (2004). Purification of recombinant adeno-associated virus type 8 vectors by ion exchange chromatography generates clinical grade virus stock. *J. Virol. Methods* 121, 209.

Davis, R. H. (1992). Microfiltration. In W. S. W. Ho and K. K. Sirkar (Eds.), *Membrane Handbook*. van Nostrand Reinhold, New York, p. 455.

Genzel, Y., Behrendt, I., König, S., Sann, H., and Reichl, U. (2004). Metabolism of MDCK cells during cell growth and influenza virus production in large scale microcarrier culture. *Vaccine* 22, 2202.

Geraerts, M., Michiels, M., Baekelandt, V., Debyser, Z., and Gijsbers, R. (2005). Upscaling of lentiviral vector production by tangential flow filtration. *J. Gene Med.* 7, 1299.

Grandison, A. S., and Finnigan, T. J. A. (1996). Microfiltration. In A. S. Grandison and M. J. Lewis (Eds.), *Separation Processes in the Food and Biotechnology Industries, Principles and Applications*. Woodhead Publishing, Cambridge, p. 141.

Grzenia, D. L., Carlson, J. O., Czermak, P., Han, B., Specht, R. K., and Wickramasinghe, S. R. (2006). Purification of densonucleosis virus by tangential flow ultrafiltration. *Biotech. Prog.* 22(5), 1346–1353.

Huang, P. Y., and Peterson, J. (2001). Scale up and virus clearance studies on virus filtration in monoclonal antibody manufacture. In W. K. Wang (Ed.), *Membrane Separations in Biotechnology*. Marcel Dekker, New York, p. 327.

Iritani, E., Mukai, Y., and Murase, T. (1995). Upward dead-end ultrafiltration of binary protein mixtures. *Sep. Sci. Technol.* 30, 369.

Kim, J. S., Akeprathumchai, S., and Wickramasinghe, S. R. (2001). Flocculation to enhance microfiltration. *J. Membr. Sci.* 182, 161.

Loeb, S., and Sourirajan, S. (1963). Sea water demineralization by means of an osmosis membrane. In *Saline Water Conversion II*. Advances in Chemistry Series No 38. American Chemical Society, Washington, DC.

Morenweiser, R. (2005). Downstream processing of viral vectors and vaccines. *Gene Therapy* 12, S103.

Ogle, K. F., and Azari, M. R. (2001). Virus removal by ultrafiltration: A case study with diasprin crosslinked hemoglobin (DCLHb). In W. K. Wang (Ed.), *Membrane Separations in Biotechnology*. Marcel Dekker, New York, p. 299.

O'Riordan, C. R., Lachapelle, A. L., Vincent, K. A., and Wadsworth, S. C. (2000). Scaleable chromatographic purification processes for recombinant adeno-associated virus (rAAV). *J. Gene Med.* 2, 444.

Pujar, N. S., and Zydney, A. L. (1994). Electrostatic and electrokinetic interactions during protein transport through narrow pore membranes. *Ind. Eng. Chem. Res.* 33, 2473.

Saha, K., Lin, Y.-C., and Wong, P. K. Y. (1994). Short communication: A simple method for obtaining highly viable virus from culture supernatant. *J. Virol. Methods* 46, 349.

Saksena, S., and Zydney, A. L. (1994). Effect of solute pH and ionic strength on the separation of albumin from immunoglobulins (IgG), by selective membrane filtration. *Biotechnol. Bioeng.* 43, 960.

Shiver, J. W., and Emini, E. A. (2004). Recent advances in the development of HIV-1 vaccines using replication-incompetent adenovirus vectors. *Annu. Rev. Med.* 55, 355.

Smith, R. H., Ding, C., and Kotin, R. M. (2003). Serum free production and column purification of adeno-associated virus type 5. *J. Virol. Methods* 114, 115.

Subramanian, S., Altaras, G. M., Chen, J., Hughes, B. S., Zhou, W., and Altaras, N. E. (2005). Pilot scale adenovirus seed production through concurrent virus release and concentration using hollow fiber membranes. *Biotechnol. Prog.* 21, 851.

Syedain, Z. H., Bohonak, D. M., and Zydney, A. L. (2006). Protein fouling of virus filtration membranes: Effects of membrane orientation and operating conditions. *Biotechnol. Bioeng.* 22(4), 1163.

van Eijndhoven, H. C. M., Saksena, S., and Zydney, A. L. (1995). Protein fractionation using membrane filtration: Role of electrostatic interactions. *Biotechnol. Bioeng.* 48, 406.

van Reis, R., Brake, J. M., Charkoudian, J., Burns, D. B., and Zydney, A. L. (1999). High performance tangential flow filtration using charged membranes. *J. Membr. Sci.* 159, 133.

van Reis, R., Gadam, S., Frautschy, L. N., Orlando, S., Goodrich, E. M., Saksena, S., Kuriyel, R., Simpson, C. M., Pearl, S., and Zydney, A. L. (1997). High performance tangential flow filtration. *Biotechnol. Bioeng.* 56, 71.

van Regenmortel, M. H. V., Fauquet, C. M., Bishop, H. H. L., Carstens, E. B., Estes, M. K., Lemon, S. M., Maniloff, J., Mayo, M. A., McGeoch, D. J., Pringle, C. R., and Wickner, R. B. (2000). *Virus Taxonomy, Seventh Report of the International Committee on Taxonomy of Viruses.* Academic, New York.

Wickramasinghe, S. R., Han, B., Akeprathumchai, S., Chen, V., Neal, P., and Qian, X. (2004a). Improved permeate flux by flocculation of biological feeds: Comparison between theory and experiment. *J. Membr. Sci.* 242(1–2), 57.

Wickramasinghe, S. R., Han, B., Carlson, J. O., and Powers, S. M. (2004b). Clearance of minute virus of mice by flocculation and microfiltration. *Biotechnol. Bioeng.* 86(6), 612.

Wickramasinghe, S. R., Kalbfuss, B., Zimmermann, A., Thom, V., and Reichl, U. (2005). Tangential flow microfiltration and ultrafiltration for human influenza A virus concentration and purification. *Biotechnol. Bioeng.* 92(2), 199.

Yang, M.-C., and Tong, J.-H. (1997). Loose ultrafiltration of proteins using hydrolyzed polyacrylonitrile hollow fiber. *J. Membr. Sci.* 132, 63.

PART III

GAS SEPARATIONS

CHAPTER 21

Vapor and Gas Separation by Membranes

RICHARD W. BAKER

Membrane Technology and Research, Inc., Menlo Park, California 94025

21.1 INTRODUCTION TO MEMBRANES AND MODULES

Gas transport through dense polymer membranes is governed by the expression:

$$j_i = \frac{D_i K_i (p_{i_o} - p_{i_\ell})}{\ell} \tag{21.1}$$

where j_i is the volume (molar) flux expressed as [cm^3 (STP) of component i]/$cm^2 \cdot s$; l is the membrane thickness, p_{io} is the partial pressure of component i on the feed side, and p_{il} is the partial pressure of component i on the permeate side. The diffusion coefficient, D_i, reflects the mobility of the individual molecules in the membrane material; the gas sorption coefficient, K_i, with units [cm^3 (STP) of component i/cm^3 of polymer] $\cdot$ pressure, reflects the number of molecules dissolved in the membrane material. The product D_iK_i can be written as P_i, which is called the membrane permeability and is a measure of the membrane's ability to permeate gas.

The best measure of the ability of a membrane to separate two gases, i and j, is the ratio of their permeabilities, $\alpha_{i/j}$, called the membrane selectivity, which can be written as

$$\alpha_{i/j} = \frac{P_i}{P_j} = \left(\frac{D_i}{D_j}\right)\left(\frac{K_i}{K_j}\right) \tag{21.2}$$

The ratio D_i/D_j is the ratio of the diffusion coefficients of the two gases and can be viewed as the mobility selectivity, reflecting the different sizes of the two molecules. The ratio K_i/K_j is the ratio of the sorption coefficients and can be viewed as the sorption or solubility selectivity, reflecting the relative solubility of the two gases. In polymer materials, the diffusion coefficient decreases with increasing molecular size because large molecules interact with more segments of the polymer chain than do small molecules. Hence, the mobility

Advanced Membrane Technology and Applications. Edited by Norman N. Li, Anthony G. Fane, W. S. Winston Ho, and T. Matsuura
Copyright © 2008 John Wiley & Sons, Inc.

selectivity, D_i/D_j, always favors the permeation of small molecules over large ones. The sorption coefficient of gases and vapors is a measure of the energy required for the gas to be sorbed by the polymer and increases with increasing condensability of the permeant. This dependence on condensability means that the sorption coefficient usually increases with molecular size because large molecules are normally more condensable than smaller ones. Hence the sorption selectivity, K_i/K_j, favors the permeation of larger, more condensable molecules, such as hydrocarbon vapors, over permanent gases, such as oxygen and nitrogen. It follows that the effects of permeant molecular size on the mobility and sorption selectivities are opposed.

Because of the competing effects of the mobility selectivity term and the sorption selectivity term in Eq. (21.2), the selectivity of gas pairs is different in glassy and rubbery polymers. This difference is illustrated by the data in Figure 21.1.

- In glassy polymers, such as polyetherimide, the rigid nature of the polymer chains means the mobility selectivity term in Eq. (2) is dominant. Permeability falls with increasing permeant size, and small molecules permeate preferentially. When used to separate an organic vapor from nitrogen, amorphous glassy membranes preferentially permeate nitrogen.
- In rubbery polymers, such as polyisoprene, the sorption selectivity term is dominant. Permeability increases with increasing permeate size, and large molecules permeate preferentially. When used to separate an organic vapor from nitrogen, rubbery membranes preferentially permeate the organic vapor.

From the discussion above, it appears that vapor–permanent gas separations can be performed with either rubbery vapor-selective membranes or glassy gas-selective membranes.

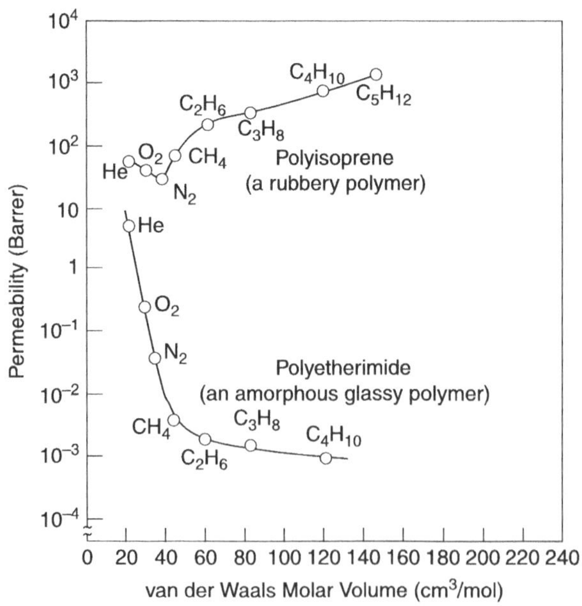

Figure 21.1 Permeability as a function of molar volume for a rubbery and a glassy polymer, illustrating the different balance between sorption and diffusion in these polymer types (Behling et al., 1989).

In practice, almost all commercial plants use rubbery membranes, predominantly membranes made from silicone rubber (polydimethylsiloxane, PDMS). Rubbery membranes are preferred for several reasons:

1. Rubbery polymers tend to have much higher permeability than glassy polymers. (As Fig. 21.1 shows, polyisoprene is over 100 times more permeable than polyetherimide to oxygen.) The higher the permeability, the smaller is the membrane area required to permeate a given volume flow of gas. Therefore, in system design, higher permeability translates to fewer membrane modules and lower capital cost.
2. The vapor–permanent gas selectivity of most glassy polymers is very dependent on organic vapor partial pressure. At low vapor concentrations (or partial pressures), the selectivity of the membrane approaches the selectivity predicted by the ratio of pure gases. However, as the organic vapor concentration (partial pressure) increases, the amount of vapor sorbed in the polymer also increases. The vapor plasticizes the polymer, which becomes rubbery. In the plasticized material, the nitrogen permeability increases, but the organic vapor permeability increases even more. The membrane then switches from being a glassy, permanent-gas-selective membrane to being a rubbery, organic vapor-selective membrane. Examples of the few exceptions to this behavior are the perfluoro polymers Teflon AF (DuPont) and Hyflon AD (Solvay Solexis) (Pinnau et al., 2002). These polymers, because of the inert nature of their perfluoro chemistry, have exceptionally low sorptions for most organic vapors and so retain their glassy nature, even in the presence of high concentrations of organic vapors. Membranes made from these polymers have found a limited use in separating air from gasoline vapor vent streams at retail gas stations.
3. Rubbery membranes provide better selective purge capability. The final reason that rubbery, vapor-permeable membranes are usually preferred is illustrated in Figure 21.2, which shows flow schemes for two types of membrane systems used

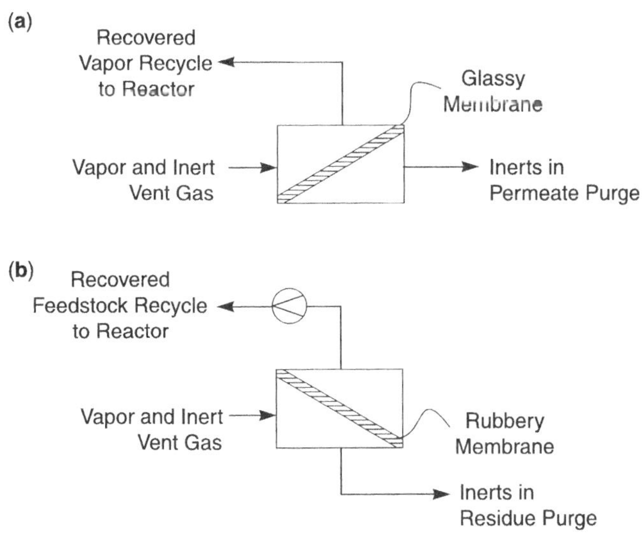

Figure 21.2 Two principal designs for membrane vent-gas systems: (a) vapor retained by a glassy membrane and (b) vapor permeated by a rubbery membrane.

> to recover the vapor component from a vapor–inert gas vent gas stream. The objective of this very common type of process is to produce an inert vent gas purge stripped of as much of the organic vapor component as possible. In the first design shown in Figure 21.2a, a glassy polymer membrane is used to treat the vent gas. The inert gases permeate the membrane, and the vapor-enriched residue gas is recycled to the process. The advantage of this design is that the pressure of the recovered vapor-containing residue gas is only slightly below the pressure of the feed gas. This allows the residue gas to be recycled to the process with minimal recompression. However, the membrane must be very selective to prevent excessive amounts of vapor passing through the membrane. Few glassy polymer membranes are sufficiently selective to make this design practical.

In the design shown in Figure 21.2b, a rubbery membrane is used. The vapor to be recovered permeates the membrane, leaving a residue stream enriched in the inert components. The advantage of this design is that even modestly selective membranes can produce an inert vent gas stream almost completely depleted of vapor. All of the organic vapor component is recycled back to the process. The disadvantage of this design is that the vapor-rich permeate to be recycled to the reactor is at low pressure and must be recompressed.

Silicone rubber composite membranes have been used in almost all of the vapor separation systems installed to date. Silicone rubber is extremely permeable and has adequate vapor–inert gas selectivities for most applications. Polyoctylmethylsiloxane (POMS), closely related to silicone rubber (PDMS), has been suggested as a material with slightly better selectivity but has not been widely used (Schultz and Peinemann, 1996). For a few years, there was also a good deal of interest in the polyacetylene polymers poly(1-trimethylsilyl-1-propyne) (PTMSP), poly(*p-tert*-butyl diphenylacetylene) (PptBDPA) and poly(4-methyl-2-pentyne) (PMP) (Morisato et al., 1999; Pinnau et al., 2004). These glassy polymers, because of the rigidity of their polymer backbones, have extraordinarily high free volumes and very unusual gas permeation properties. In particular, the polymers are even more permeable and more selective to condensable vapors than silicone rubber, the industry benchmark polymer. Unfortunately, the high cost of the polymers and their poor chemical and physical stability have prohibited commercial use, and most researchers have now abandoned work with these materials.

Silicone rubber is soft and elastic. For this reason, silicone rubber membranes are made as composite structures of the type shown in Figure 21.3. A nonwoven support paper provides mechanical strength. The pores of this paper are too large to be coated directly with the silicone rubber selective layer, which is usually less than 5 μm, and sometimes less than 1 μm, thick. Instead, the paper is first coated with a finely microporous ultrafiltration support membrane. This membrane provides a smooth surface on which the very thin silicone rubber layer can be coated. The permeance of the support membrane is usually 10- to 1000-fold higher than that of the selective silicone rubber layer, so the overall resistance to flow is predominantly in the silicone rubber layer.

Composite membranes are made in sheets or rolls and fabricated into large-membrane-area membrane modules for use in industrial systems. GKSS, the main European supplier of vapor–gas separation units, packages its membranes into plate-and-frame modules. Membrane Technology and Research, Inc. (MTR, Menlo Park, CA) the main U.S. producer, uses spiral-wound modules.

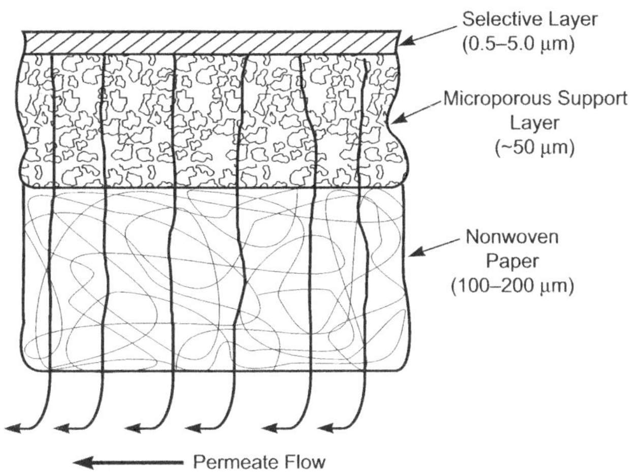

Figure 21.3 Schematic illustration of a composite membrane.

21.2 MEMBRANE PROCESS DESIGN

21.2.1 One-Stage Selective Purge Systems

In a number of processes, the goal of the membrane system is to recover and recycle vapor components previously lost with an inert gas purge (Baker et al., 2000). Figure 21.4 shows the design of a typical reactor purge unit for a reactor that achieves incomplete conversion of feedstock. Typically, the reagents enter the reactor vessel, heat and pressure are applied, and the desired reaction occurs in the presence of a catalyst. The products of reaction then move to a separator and are cooled or scrubbed to remove the product as a liquid. Unreacted components are released as a gas, which is recycled to the reactor.

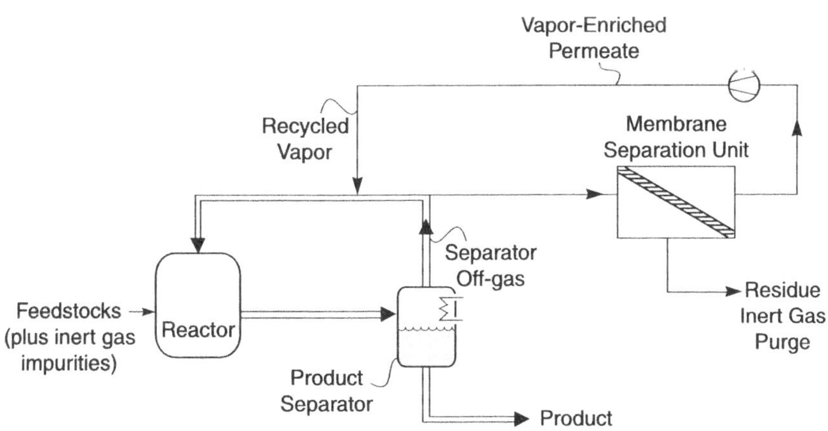

Figure 21.4 Membrane separation system used to recover organic vapor from a reactor vent stream containing vapor components to be recovered and inert gas components to be released.

A common problem with this type of process is the buildup of inert impurities in the reactor–separator–recycle loop. Inert components occupy reactor capacity and progressively diminish the process efficiency. These components can be gases such as carbon dioxide generated by a side reaction in the reactor. More commonly, they enter the reactor with the feedstock. Oxygen is often contaminated with 0.5–1.0% argon, propylene is contaminated with 0.5–5% propane, ethylene with ethane, hydrogen with traces of methane, and so on. To control contaminant buildup, a portion of the recycle gas is purged from the reactor recycle loop. However, the concentration of inerts in the purge may be only a few percent, so for every mole of purged argon or propane, many moles of valuable feedstock may be lost. The opportunity for membranes in such a process is to recover these valuable components.

A simple one-stage membrane system fitted with silicone rubber membranes that preferentially permeate the valuable feedstock vapor and reject the inert gas contaminants is shown in Figure 21.4. The permeate gas, enriched in the vapor components, is compressed and sent back to the reactor; the residue gas, containing only the inert components, is purged. Feedstock losses can be essentially eliminated with such a design.

21.2.2 Multistep and Multistage System Designs

The objective of one-stage membrane systems as shown in Figure 21.4 is to produce an inert gas purge stream stripped of the vapor component. This can be done with great success, but, because the membrane selectivity is modest, the vapor recycle stream recirculated to the reactor is only modestly enriched, perhaps two- or threefold, in the vapor component. In some applications, this is acceptable; in others, a more complete separation is required. The solution is to use two membrane units in series.

One-stage, two-stage, and two-step membrane flow schemes for treating the same feed stream are illustrated in Figure 21.5.* The three designs shown are all based on the same vapor-permeable membranes with a vapor–nitrogen selectivity of 10 and an inert gas permeance, also commonly referred to as pressure-normalized flux, or simply called P/P [based on Eq. (21.1)] of 100 gpu (1 gpu $= 1 \times 10^{-6}$ cm^3 (STP)/$cm^2 \cdot s \cdot cm$ Hg). All these schemes purge the same amount of the inert gas (54.4 scfm). Because the concentration of the hydrocarbon vapors in the permeate gas recycled to the reactor varies, the initial volume of feed gas to the membrane unit (90% inerts, 10% hydrocarbon vapors) is different for each system.

The one-stage system (Fig. 21.5a) divides the feed gas into two roughly equal streams: a residue inert gas purge, from which 90% of the hydrocarbon has been removed; and a permeate recycle stream, enriched about two-fold in hydrocarbons, that is recycled to the reactor. The two-stage system (Fig. 21.5b) also achieves 90% removal of hydrocarbons from the inert gas purge. However, the first-stage permeate gas, after compression to 10 bars, is passed through a second membrane stage. The residue from the second stage is mixed with the feed gas and recycled to the first membrane stage. In the design shown, the second membrane stage is used to produce a recycle residue gas that has the same concentration as the initial feed gas. This is the most efficient design, as separation losses caused by mixing gases of unequal concentrations are avoided. The second-stage permeate

*A system contains a second membrane *stage* when the second membrane unit is placed on the *permeate* gas from the first membrane unit. The system contains a second membrane *step* when the second membrane unit is placed on the *residue* gas from the first membrane unit.

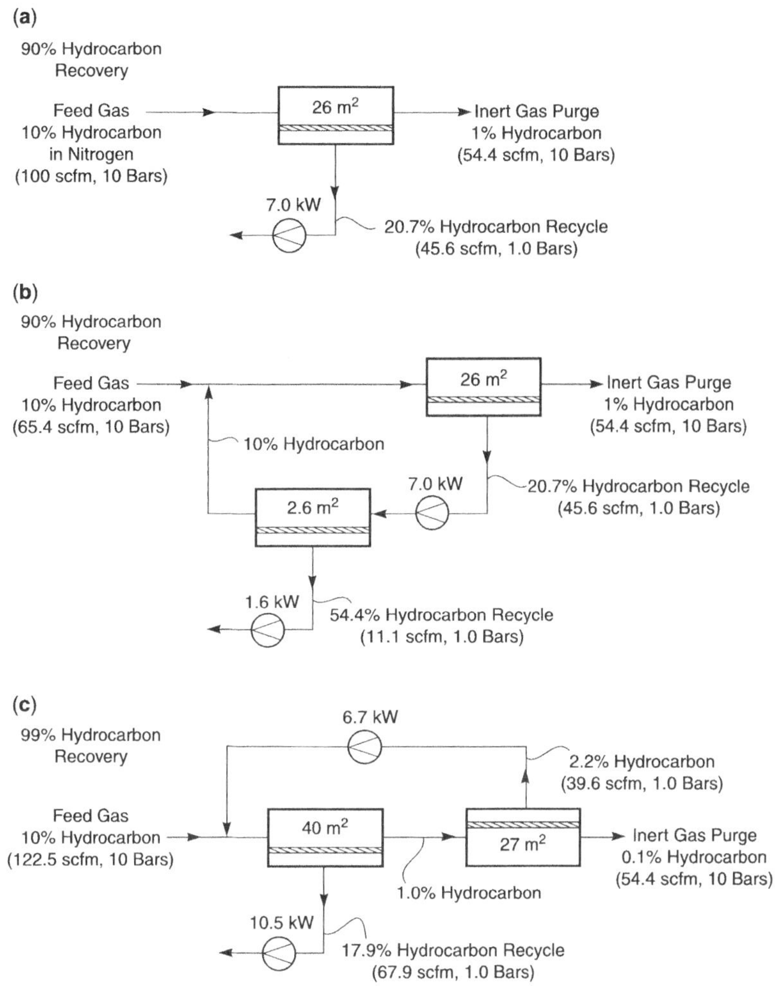

Figure 21.5 Comparison of (a) one-stage, (b) two-stage, and (c) two-step membrane processes, all producing the same volume of inert gas purge (54.4 scfm). These calculations are performed using a computer process simulation package (ChemCAD 5.0, Chemstations, Inc., Houston, TX) modified with code written at MTR for the membrane separation step.

recycled to the reactor is enriched to 54.4% hydrocarbon, compared with about 21% for the one-stage design. The second-stage permeate compressor and membrane unit are small compared to those used in the one-stage system because gas flow to the second-stage membrane is half that to the first stage, and the degree of hydrocarbon removal required is much lower also.

The two-step system design shown in Figure 21.5c is used when 90% recovery of hydrocarbons from the vent gas is not enough, and a higher recovery of hydrocarbons is desired. In the example calculation shown, the recovery is 99%. Two membrane units are used in series. The permeate from the first membrane step is recycled to the reactor, and the permeate from the second membrane step is recycled to the front of the first step. To minimize mixing losses, it is desirable to size each membrane unit so that the second-step permeate recycle gas has the same concentration as the feed gas. Depending on the selectivity of the membrane and the target performance of the process, this may not always be possible. In the design shown, the second-step recycle gas contains only 2.2% hydrocarbon, rather than 10% for the most efficient design. Two-step systems achieve better separations than one-stage units but require significantly more membrane area and higher compression energy. In the example shown in Figure 21.5c, the membrane area and compression power used is 2.5-fold higher than that for the one-stage unit shown in Figure 21.5a.

21.2.3 Hybrid Systems

Multistep and multistage membrane systems of the type shown in Figure 21.5 have been used to treat a number of vapor–gas streams. However, hybrid processes in which membrane separation is combined with another separation process, are attractive because they may enable each unit operation to operate in its preferred range, improving overall process efficiency. The most important of these hybrid processes is the combination of condensation under pressure with membrane separation, illustrated in Figure 21.6. In this process, a vapor-permeable membrane unit is combined with a vapor condensation–flash unit (Baker and Wijmans, 1994; Baker et al., 1998; Wijmans, 1993/1992).

The feed gas is first compressed and then sent to a vapor condenser, operating at −20°C. On cooling the gas, a portion of the vapor contained in the feed gas (in this case propylene), condenses and is removed as liquid. The condenser off-gas, containing uncondensed propylene, is sent to the membrane unit, which preferentially permeates the hydrocarbon vapor, leaving a 99% pure nitrogen residue stream. The propylene-enriched permeate gas is recycled to the incoming feed gas. The gas sent to the membrane unit comes directly from the chilled vapor condenser and is, therefore, saturated with propylene and cold. Under these conditions, the solubility of the hydrocarbon in the membrane is enhanced, and the membrane selectivity is high.

The raw propylene condensate produced in the condenser contains some dissolved nitrogen, so the liquid is flashed at low pressure to remove this gas. The flash step off-gas is recycled to the feed as shown in Figure 21.6. After flashing, the propylene product has a purity of better than 99.5%. The hybrid design takes advantage of the ability of condensation to produce a high-purity liquid and of membrane separation to

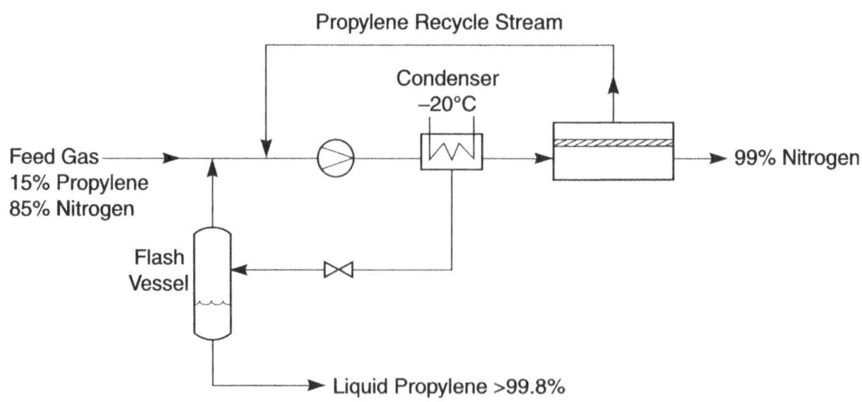

Figure 21.6 Hybrid compression–condensation membrane separation system to recover liquid propylene and nitrogen from a mixed-gas feed.

produce a high-purity residue stream. If condensation alone were used, the only way to avoid loss of propylene in the condenser off-gas would be to cool it to extremely low temperatures, requiring multiple stages of refrigeration. If membrane separation alone were used, it would be almost impossible to achieve both a high-purity residue and a high-purity permeate stream, without resorting to a cascade of many membrane stages. The hybrid design offers flexibility to adjust the operating parameters of each unit for optimized efficiency and product quality.

In the example of Figure 21.6, the inert gas is nitrogen, so there is no possibility of creating flammable gas mixtures in the process. However, in some applications the inert gas is air; so precautions must be taken to eliminate dangers caused by the vapor–air mixture entering the flammable-explosive range. This type of problem exists in the separation of gasoline–air mixtures. The specific system designs used to circumvent this problem are described for the treatment of gasoline vapor–air mixtures in the application section that follows.

21.3 APPLICATIONS

The two principal suppliers of vapor–gas membrane separation systems are MTR and the licensees of GKSS (Borsig, Sihi, and Dalian Eurofilm). The major markets serviced by these companies are described briefly below.

21.3.1 Polyolefin Plant Resin Degassing

Probably the largest single application of vapor separation membranes is in the recovery of hydrocarbon monomers from ethylene and polyethylene and polypropylene plants. These plants make polyolefins, principally from ethylene and propylene. After the polyolefin resin is produced, it contains unreacted monomer and hydrocarbon solvents, dissolved in the resin powder. The dissolved hydrocarbon must be removed before the polymer can be used, and this is done by stripping with hot nitrogen in a column known as a degassing bin. In early polyolefin plants, the vent gas from the degassing bin—containing 10–20 mol% hydrocarbon—was used as boiler fuel. Since the development of vapor separation membranes, most new polyolefin plants have installed hydrocarbon recovery units. In a modern polyolefin plant, the value of the monomer in the nitrogen resin bin off-gas is on the order of \$1–2 million/year; the value of the nitrogen can represent another \$0.5 million/year. Recovery and reuse of these components is well worthwhile.

A process flow stream and a photograph of a typical membrane system fitted to a polyolefin plant resin degassing bin are shown in Figure 21.7 (Baker and Jacobs, 1996). The off-gas from the bin is compressed to 200 psi, dried, and cooled to $-30^{\circ}C$. A portion of the propylene then condenses. The condenser overhead stream (propylene and nitrogen) is sent to the membrane section, which contains two membrane units in series. The first membrane unit produces a permeate stream enriched in propylene and a purified residue stream containing ~97–98% nitrogen. The vapor-enriched permeate stream is recycled to the inlet of the compressor. The nitrogen-rich residue can often be directly recycled to the degassing bin without further treatment. However, in the example shown, the residue gas is passed to a second membrane unit to upgrade the nitrogen to better than 99% purity. The waste hydrocarbon stream from the second membrane unit is sent to flare. The performance of the

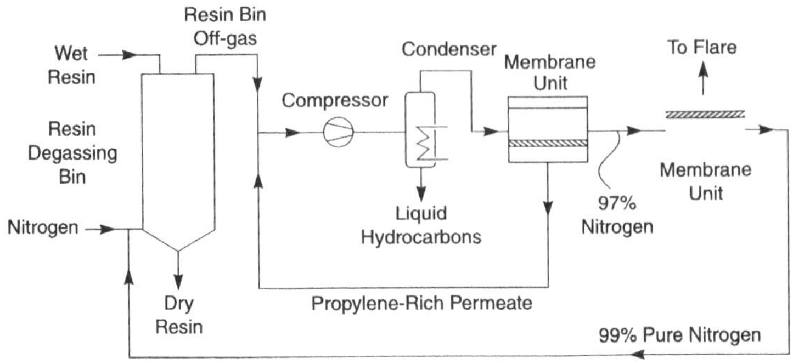

Figure 21.7 Photograph and process flow diagram of a membrane propylene recovery system installed at a modern polypropylene plant. The front portion of the unit is the compressor package. The spiral-wound membrane modules are contained in the horizontal tubes above and behind the compressor. This unit recovers approximately 1000 lb/h of hydrocarbons.

system is summarized in Table 21.1. During the last 10 years, almost 50 of these systems have been installed around the world.

21.3.2 Gasoline Vapor Recovery Systems

21.3.2.1 Large Terminals An important early application of membrane vapor recovery systems was the recovery of gasoline vapors from vent streams produced at large oil and gasoline terminals. During the transfer of hydrocarbons from tankers to holding tanks and then to trucks, off-gases are produced. The off-gas stream volume and vapor concentration vary widely, but the average emissions resulting from each transfer operation are large—in the range of 0.03–0.05% of the hydrocarbon transferred (Ohlrogge et al., 1993; Ohlrogge and Stürken, 2001).

The hydrocarbon concentration of the emitted gas is generally quite high, in the range of 10–30 vol%, depending on the type of hydrocarbon and type of transfer. A typical off-gas composition for gasoline loading or unloading is shown in Table 21.2. Because the off-gas

TABLE 21.1 System Performance of Typical Membrane Resin Degasing Bin Recovery Unit (1999 data)

Feed Flow Rate (lb/h)	5000
Feed Gas Composition (vol%)	
Hydrogen	1.0
Nitrogen	84.4
Propane	0.3
Propylene	14.0
Water	0.3
Hydrocarbon recovery ~1000 lb/h	91%
Nitrogen recovery ~2000 lb/h	50%
Value of recovered hydrocarbons[a]	$1.75 million/year
Value of recovered nitrogen[a]	$0.6 million/year
Capital cost[b]	$2.7 million

[a]Value of recovered nitrogen at $75/ton; hydrocarbons at $400/ton.
[b]Includes cost of the low-temperature refrigeration unit.

is an air–hydrocarbon mixture, the potential for creating an explosive composition has to be considered in the design of the membrane vapor recovery system.

Hydrocarbon vapor–air mixtures containing from 3 to 15% hydrocarbon are in the flammable range. Below 3% hydrocarbon vapor, the mixture is too hydrocarbon-lean to burn. Above 15% hydrocarbon vapor, the mixture does not contain enough oxygen to burn. Problems occur in the intermediate range, where a chance spark can cause an explosion. The usual solution to this problem is to saturate the incoming feed mixture with additional hydrocarbon vapor in a small contactor tower. This ensures that the feed to the compressor needed to operate the membrane unit is always comfortably above the upper explosion limit, regardless of the composition of the feed gas.

As an additional safeguard, liquid-ring compressors are usually chosen. In a liquid-ring compressor, the seal between the rotating vane of the compressor and the compressor chamber is formed by a film of liquid—in this case, liquid gasoline. The liquid seal minimizes metal-to-metal contact and the possibility of sparks. As the gas is compressed, some hydrocarbon vapor is absorbed by the gasoline sealing fluid of the compressor. The fluid leaving the compressor is then a two-phase mixture of gasoline containing dissolved vapors and hydrocarbon-saturated air. A phase separator, after the compressor, separates the hydrocarbon liquid and gas phases. The vapor-saturated gasoline is removed; the saturated vapor then passes to the membrane unit. As with the condensation–membrane

TABLE 21.2 Typical Gasoline Vapor Vent Gas Composition (vol%)

Component	(%)	Component	(%)
Methane	0.01	*i*-Pentane	4.43
Ethane	0.03	Hexane	1.51
Propane	0.69	C_7^+	0.14
Butane	6.66	Benzene	0.29
i-Butane	3.69	Oxygen	16.71
Pentane	2.60	Nitrogen	63.24

Source: From Ohlrogge et al. (1993).

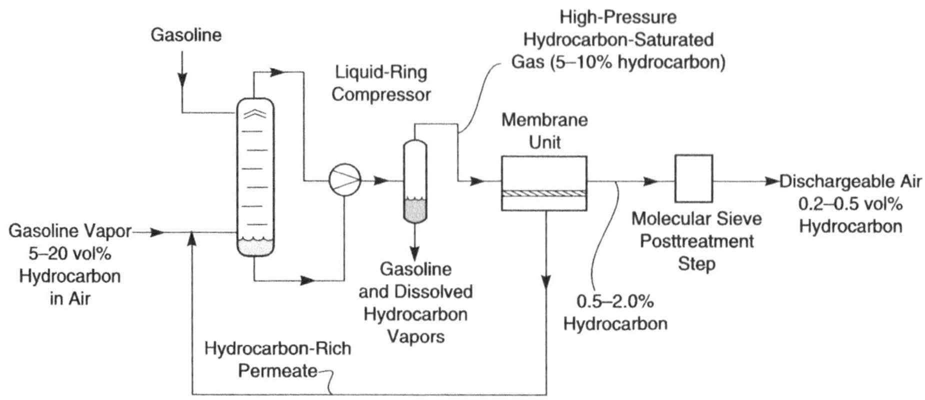

Figure 21.8 Flow schematic of a gasoline vapor recovery system, using a combination of absorption and membrane separation to recover 98 + % of the hydrocarbons in the vent gas, followed by a molecular sieve pressure swing absorption (PSA) unit to remove the final 1–2% hydrocarbon.

separation unit shown in Figure 21.6, hydrocarbon vapors are removed by using a hydrocarbon-selective membrane. The hydrocarbon-enriched permeate is recycled to the front of the feed gas compressor; the hydrocarbon-stripped residue contains 0.5–2% hydrocarbon, mainly the light gases methane, ethane, and propane. To meet air discharge regulations, this gas is usually sent to a final polishing step, most commonly a small, molecular sieve, pressure swing absorption (PSA) unit, which reduces the hydrocarbon level to 0.2–0.5 vol%. As Figure 21.8 shows, the gas under treatment passes through the flammable range from the hydrocarbon-saturated feed (5–10% hydrocarbon) to the hydrocarbon-stripped residue (0.5–2% hydrocarbon) within the membrane module. Since there are no moving parts within the module, the chance of a spark causing an explosion is minimal.

GKSS's licensees have installed about 30 gasoline vapor recovery systems at fuel transfer terminals, mostly in Europe. The alternative technology is to use some sort of thermal oxidizer, and this approach seems to be the most widely used technology, especially in the United States.

21.3.2.2 Retail Gasoline Stations A related gasoline vapor recovery application in which membranes are finding it easier to compete is at retail gasoline stations. Many new gasoline stations are using vacuum-assisted dispensing systems to control the release of hydrocarbon vapors to the atmosphere. These systems use a small pump to draw air and vapors from the gasoline dispensing nozzle. For every liter of gasoline dispensed, as much as 2 L of air and gasoline vapor are returned to the storage tank. The air that builds up in the tank must be vented to the atmosphere. Membrane systems are used to control the vapor emissions.

In the last few years, several hundred retail gasoline stations have installed small membrane systems to treat their tank vents. A flow scheme of this type of system is shown in Figure 21.9. Air from the gas station dispenser is collected and sent to the gasoline storage tank. When the pressure in the tank reaches a preset value, a pressure switch activates a small compressor that draws off excess vapor-laden air. A portion of the hydrocarbon vapors condense and is returned to the tank as a liquid. The remaining hydrocarbons permeate the membrane and are returned to the tank as concentrated

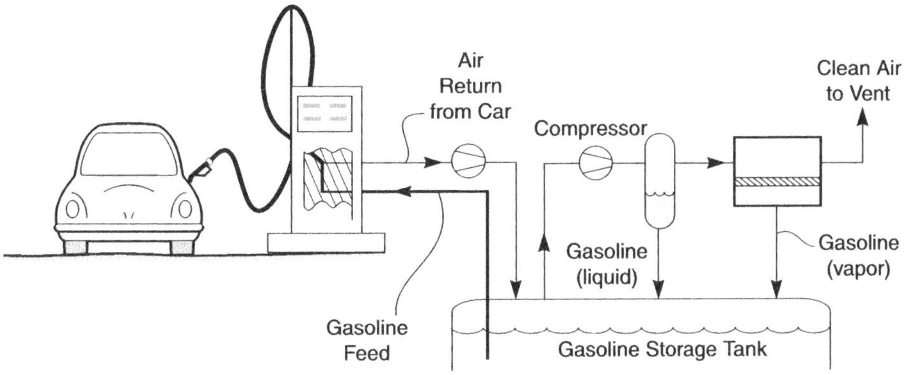

Figure 21.9 Flow diagram of a membrane gasoline-vapor recovery unit suited to a retail gasoline station tank vent. Typical systems are small, containing a single 1- to 2-m^2 membrane module and costing from \$5000 to 15,000. Several hundred, perhaps as many as 1000, of these systems have been installed around the world (Ohlrogge and Wind, 2000).

vapor. Air, stripped of 95–99% of the hydrocarbons, is vented. In addition to eliminating hydrocarbon emissions, the unit essentially pays for itself with the value of the recovered gasoline (Ohlrogge and Wind, 2000).

21.3.3 Polyvinyl Chloride Manufacturing Vent Gas

In the polymerization of polyvinyl chloride, unwanted gas is generated by side reactions, and some small amounts of air leak into the reactors. These inerts must be vented from the process. Because vinyl chloride monomer (VCM) is extremely volatile, the purge gas, although it is typically at 4–5 bars pressure, can contain as much as 50 vol% monomer. As a consequence, the vented gas stream, although small, may contain several hundred thousand dollars worth of monomer values. A typical process flow scheme to recover VCM is shown in Figure 21.10 (Lahiere et al., 1993).

Feed gas containing VCM and air is sent to the membrane system. The VCM-enriched permeate from the membrane system is compressed in a liquid-ring compressor and cooled to liquify the VCM. A liquid-ring compressor is used because of the flammable nature of vinyl chloride. The noncondensable gases are mixed with the feed gas and returned to the membrane section. The residue stream is sent to the incinerator, where the remaining VCM

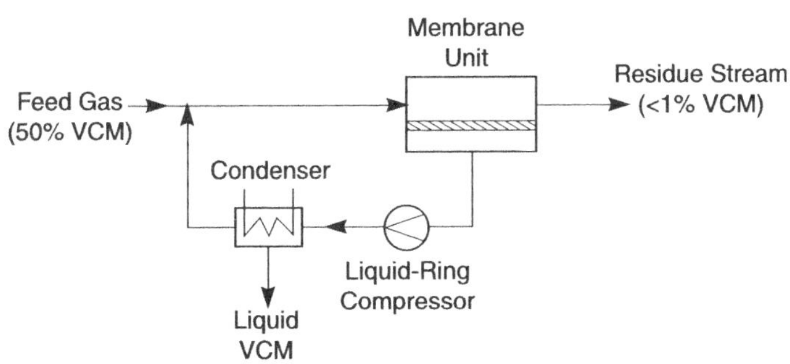

Figure 21.10 Recovery of VCM monomer in a polyvinyl chloride plant.

is destroyed before venting the inerts. VCM recovery is more than 99%. The first unit of this type was installed by MTR in 1992. Since then, about 40 similar systems have been installed.

21.3.4 Ethylene Recovery in Ethylene Oxide and Vinyl Acetate Manufacturing Plants

21.3.4.1 Ethylene Oxide Ethylene oxide (EO) is produced through the catalytic oxidation of ethylene with 99.6 + % pure oxygen. Ethylene, oxygen, and methane are fed into the reactor. Methane is added to moderate the reaction kinetics and keep the gas mixtures outside the explosive range. Ethylene oxide is produced, along with carbon dioxide and water as by-products. The mixture is sent to a water-based scrubber to recover the ethylene oxide. Carbon dioxide is then removed by absorption with hot potassium carbonate, fresh ethylene and oxygen are added to the unreacted gases, and the mixture is recycled back to the reactor. Due to the presence of argon in the incoming oxygen and ethane in the incoming ethylene, a portion of the gases in the reactor loop must be purged to keep the concentration of these inerts under control.

The purge gas for a typical ethylene oxide plant contains approximately 20–30% ethylene, 10–12% argon, 1–10% carbon dioxide, 1–3% ethane, 50% methane, and 4–5% oxygen. This purge gas can be treated in a membrane-based recovery unit, as pictured in Figure 21.11. The purge gas enters the membrane system at approximately 20 bars and 30 EC. Ethylene preferentially permeates the membrane, producing an ethylene-enriched permeate stream and an argon-enriched residue stream. The permeate stream is then recompressed back into the reactor loop via the reclaim compressor. As ethane also preferentially permeates the membrane, there is the potential for buildup of ethane in the system. However, this buildup is mitigated by two factors: the ethane concentration in polymer-grade ethylene is very low; and, as long as the recovery of ethylene is not too high

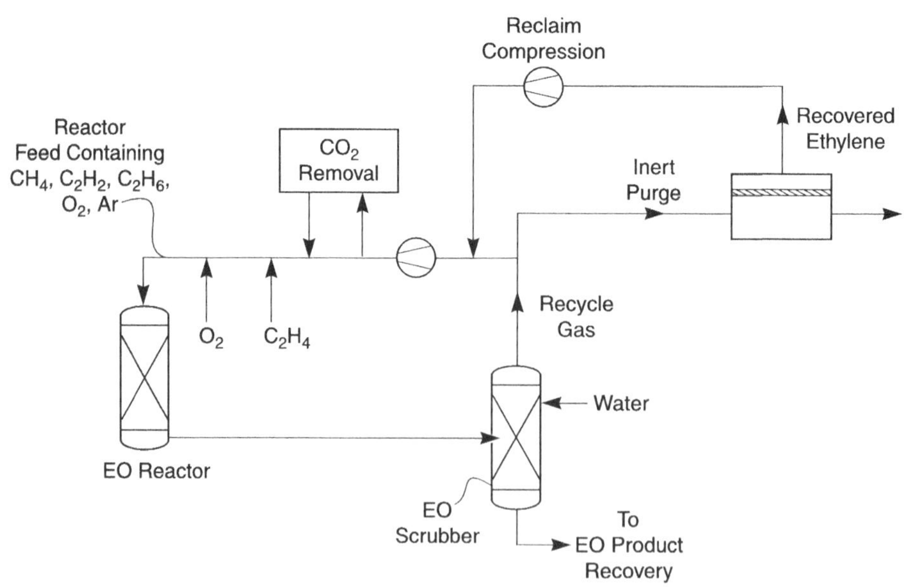

Figure 21.11 Schematic of the ethylene oxide plant ethylene recovery unit.

TABLE 21.3 Performance of Typical Ethylene Oxide Plant Ethylene Recovery Unit

Ethylene recovery (lb/h)	290
Ethylene recovery (%)	75
Methane recovery (lb/h)	150
Value of recovered ethylene[a]	$620,000/year
Value of recovered methane[b]	$215,000/year
System cost	$550,000
Simple payback (months)	<8

[a]Based on $500/ton.
[b]Based on $7.00/1000 ft^3.

(not greater than 90%), the remaining ethane will be removed from the reactor loop through the residue stream. Based on actual field data from operating units, no ethane buildup has been observed. The residue stream, which has been stripped of ethylene, is used in a boiler or incinerator (Baker and Gottschlich, 2000; Jacobs and Billing, 2005).

The performance of a typical ethylene recovery unit is summarized in Table 21.3. The system cost was approximately $550,000, resulting in a simple payback time of less than 8 months.

21.3.4.2 Vinyl Acetate A similar vent gas mixture is created in the production of vinyl acetate. Most vinyl acetate is produced by a catalytic vapor-phase reaction of ethylene and acetic acid in the presence of oxygen. The gases leaving the reactor are cooled, partially condensing the mixture, and the condensed liquid is then purified in a downstream distillation section. The vapor from the condenser is sent to a carbon dioxide removal system, then returned to the reactor, where the unreacted gases are combined with the feed gases. In like manner to the ethylene oxide process, a purge stream must be taken from the reactor loop to remove argon, ethane, and other impurities. The purge gas from vinyl acetate reactors contains a much higher concentration of ethylene (more than 65%), and the remaining components are carbon dioxide (20%), argon (5%), and methane (10%). The feed pressure and temperature are 10 bars and 35 EC, respectively. Table 21.4 summarizes the performance of the ethylene recovery unit, which provides a payback period of less than 9 months.

21.3.5 Natural Gas Processing/Fuel Gas Conditioning

Raw natural gas is often saturated with propane, butane, higher hydrocarbons, and water. Separation of these components is necessary to prevent formation of hydrocarbon liquids

TABLE 21.4 Performance of Typical Vinyl Acetate Plant Ethylene Recovery Unit

Ethylene recovery (lb/h)	460
Ethylene recovery (%)	70
Value of recovered ethylene[a]	$980,000
System cost	$700,000
Simple payback (months)	<9

[a]Based on $500/ton.

and hydrates in the pipeline, as well as to control Btu (British thermal unit) content. In addition, their removal is desirable on economic grounds; the hydrocarbons have more value as recovered natural gas liquids (NGLs).

Membranes can be used to bring raw natural gas to pipeline quality by removing water and higher hydrocarbons. A simple, economical membrane system can lower the dew point of the gas by 80–120 EF (30–50°C). The current alternative technology cools the gas using a refrigeration unit and separates the heavy hydrocarbons by condensation. Thus far, membrane systems have had difficulty replacing refrigeration for removing heavy hydrocarbons from large-volume gas streams, but a number of membrane units have been installed to treat gas used as on-site fuel for remote gas compressor engines (Fenstermaker, 1983; Lokhandwala et al., in press).

Raw unprocessed natural gas is widely used to power field compressor engines and generator sets. Oftentimes this gas has a low octane rating because of the presence of propane and C_{4+} hydrocarbons in the gas. These components lead to predetonation and coking problems, requiring derating of the engines so that they can run smoothly. Engine and turbine manufacturers characterize the quality of natural gas in a number of ways, most commonly by calculating the methane number, or Wobie number, of the gas. These numbers are equivalent to the octane rating used to characterize gasoline. Good gas has a methane number of

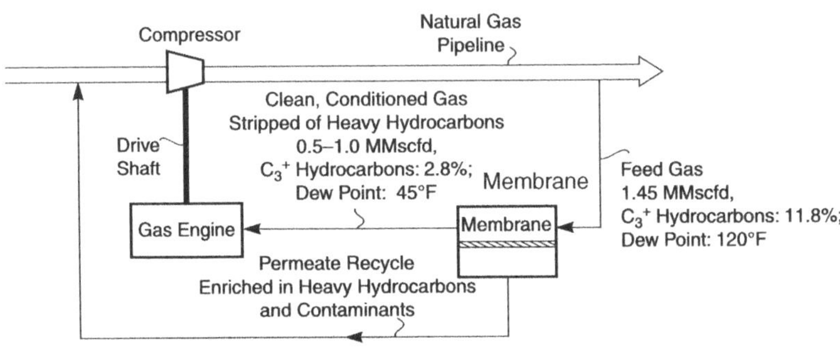

Figure 21.12 Flow diagram and shop photograph of a membrane fuel gas conditioning unit used for a field gas compressor engine. The membrane modules are contained in two horizontal pressure vessels. The unit can produce 0.5 to 1.0 MMscfd of clean gas.

greater than 65; a methane number of 40 or below can be used as engine fuel but will usually require derating of the engine. Another measure of gas quality is its Btu value. Below 600–700 Btu/scf, gas is considered very lean; above 1200 Btu/scf, gas is normally too rich to be used in standard gas-powered equipment. Finally, most engine manufacturers will have a limit on the hydrogen sulfide content of the gas.

The amount of gas used by field engines is usually in the 0.5–2.0-MMscfd (million standard cubic feet per day) range—too small to make treatment of the gas by refrigeration economical. As a consequence, many engine users are forced to live with the problem gas and the resulting low reliability and high maintenance costs.

A membrane-based fuel gas conditioning unit designed to upgrade raw gas used as engine fuel is illustrated in Figure 21.12. In this simple example, the gas to be treated was being used to power a field gas compressor engine. The raw gas was very rich, resulting

TABLE 21.5 Minor and Developing Applications of Membrane Vapor–Gas Separation Technology

Application	Separation	Description
	Early Commercialization Efforts	
CFC, HCFC (Freon) from refrigeration vents and other process streams	CFC, HCFC/air (1–10%)	When the ozone hole in the atmosphere above the Antarctic was discovered, CFC and HCFC fluorocarbons (Freon), widely used as industrial refrigerants, were progressively banned. For a few years a market existed to treat the fluorocarbon/air vents for these units, with approximately 50–100 small systems sold. With the phase-out of CFC and HCFC liquids, this market has disappeared (Wijmans and Baker, 1991).
Semiconductor plasma cleaning gas	C_2F_6/N_2-related separations, including SF_6/N_2, CF_4/N_2	Vent gas from semiconductor plasma cleaning operations contains 0.5–1.0% C_2F_6. C_2F_6 is a high-value chemical ($30/lb) and an egregious global warmer (10,000 times worse than CO_2). This is an interesting application because the best membranes to date are nitrogen-permeable/C_2F_6-rejecting membranes. Pilot systems have been demonstrated, but the market evaporated when the semiconductor industry switched to NF_3 for plasma cleaning (Li et al., 1998/1999; Pinnau et al., 1998; Wijmans et al., 2004).

(*Continued*)

TABLE 21.5 *Continued*

Application	Separation	Description
	Pilot Plant or Small-Scale Commercial Development	
Chlorine recovery from chlor-alkali plant tail gas	Cl_2/air (~20%)	Chlorine produced in chlor-alkali plants is liquefied by compression and cooling. A tail gas containing uncondensed chlorine and air is produced. Silicone rubber membranes can be used to recover the chlorine vapor in a condensation membrane hybrid unit. Pilot plants for the process have been installed (Pinnau et al., 1996; Lokhandwala et al., 1999; Hägg, 2001).
Vacuum pump exhausts	Hydrocarbons, toluene, chlorinated solvents/air	A few systems sold (Baker and Kaschemekat, 1992; Hauk, 1993).
Separation of volatile organic compounds (VOCS) from effluent air streams	0.1–1.0 vol% toluene, hexane, chlorinated solvents, and the like from air	VOC recovery from VOC-containing air streams was a major driver for the early development of vapor separation membranes. However, in this low-concentration range, membranes were not able to compete with incineration, thermal oxidation, carbon adsorption, etc. A few systems have been sold, but the market is now largely abandoned by membrane companies (Baker et al., 1996; Wijmans et al., 1997; Wang et al., 2001). If lower-cost, higher-permeance membranes are developed, this market may be revisited.
	Significant Market Potential	
LPG recovery in refineries	C_{3+} hydrocarbons/CH_4, H_2	Recovery of heavy hydrocarbons or liquefied petroleum gas (LPG) from refinery purge or fuel gas is significantly more profitable than using these components as waste fuel. If the gas contains hydrogen, it can also be recovered. This application is a long-standing area of research; pilot plants for the process are installed. This could become a significant market.

in engine knocking, damage, and frequent shutdown. A portion of the high-pressure compressed gas (450 psia) was diverted from the pipeline and passed across the surface of a rubbery membrane selectively permeable to the heavier components of the gas. Methane and ethane are retained by the membrane; propane, butane, C_{5+} hydrocarbons, and the BTEX (benzene, toluene, ethylbenzene, and xylenes) aromatics all preferentially permeate the membrane. The system removes 80% of the C_{3+} hydrocarbons, lowering the gas hydrocarbon dew point by 75°F.

As membrane-based fuel gas conditioning technology gains credibility, opportunities to compete with low-temperature condensation or to provide membrane-augmented hybrid systems should open up.

21.3.6 Other Applications

The five categories of applications described above cover the bulk of the current membrane vapor–gas separation market. For various reasons, several environmental vent gas applications developed in the 1980s did not materialize commercially, but a number of smaller applications and new applications are currently under development. These are listed in Table 21.5, together with brief descriptions; more detailed discussions can be found in the cited references.

21.4 CONCLUSIONS

The first commercial membrane vapor separation systems were installed in 1988; 4 years later, about 20 systems had been installed, and to date (2006), more than 100 large units and several hundred smaller systems have been installed. Currently, the total membrane vapor separation equipment market is at least $20–30 million/year and growing and diversifying. Although these statistics are unlikely to excite most venture capitalists, the creation of a new market segment in the conservative world of chemical engineering is an unusual achievement. The modest but solid growth of market share for membranes over the last 15 years and the opportunities for development of new product lines are optimistic indicators for a bright future.

21.5 GLOSSARY

Dew Point The temperature at which a vapor-containing gas mixture (such as water vapor in air) becomes saturated and condensation begins.

Permeability Measure of a material's ability to control the rate of permeation of different chemical species. In gas separations, permeability (P) is the product of a diffusion coefficient (D) and a sorption coefficient (K). See Section 21.1 and Baker (2004) for detailed discussion.

Permeance The degree to which a membrane structure admits a gaseous species. Units of permeance are volumetric output per unit membrane area per unit transmembrane pressure. Also referred to as pressure-normalized flux or P/l.

Permeate In gas separation, the portion of a gaseous feed that passes through a membrane when the feed passes across a membrane surface under a transmembrane driving force.

Residue In gas separation, the portion of a gaseous feed that remains (does not pass through) when the feed passes across a membrane surface at elevated pressure.

Selectivity The preference of a membrane to selectively permeate one gaseous species in favor of another. For a given membrane, selectivity is generally expressed as the quotient of the gas permeances for the two species being compared.

Barrer Unit of material permeability. 1 Barrer $= 1 \times 10^{-10}$ $cm^3(STP) \cdot cm/cm^2 \cdot s \cdot cmHg$

BTEX A C_6—C_8 aromatic hydrocarbon stream; the main ingredients are benzene, toluene, ethylbenzene and xylenes.

EO Ethylene oxide

GKSS One of fifteen national research centers that belong to Germany's HGF (Herman Von Helmholz Society of German Research centers). GKSS' main research center is in Geesthacht, Germany.

gpu Gas permeation unit. Unit of membrane permeance. 1 gpu $= 10^{-6}$ $cm^3(STP)/cm^2 \cdot s \cdot cmHg$

LPG Liquid petroleum gas

MMscfd Million standard cubic feet per day. Used to describe flow rates for gaseous hydrocarbon and industrial chemical process or pipeline streams, especially in the natural gas and gas processing industries.

MTR Membrane Technology and Research, Inc. (Menlo Park, CA)

PDMS Polydimethylsiloxane, a silicone rubber

PSA Pressure swing adsorption

VCM Vinyl chloride monomer

All symbols in Eqs. (21.1) and (21.2) are defined in the related discussion.

REFERENCES

Baker, R. W. (2004). *Membrane Technology and Applications*, 2nd ed. Wiley, Chichester.

Baker, R. W., and Gottschlich, D. (2000). Membrane process and apparatus for argon purging from oxidation reactors. U.S. Patent 6,018,060.

Baker, R. W., and Jacobs, M. (1996). Improve monomer recovery from polyolefin resin degassing. *Hydrocarbon Process.* 75(3), 49.

Baker, R. W., and Kaschemekat, J. (1992). Membrane process for treating pump exhausts. U.S. Patent 5,127,926.

Baker, R. W., Kaschemekat, R. W., and Wijmans, J. G. (1996). Membrane systems for profitable VOC recovery. *CHEMTECH* 26, 37.

Baker, R. W., Lokhandwala, K. A., Jacobs, M. L., and Gottschlich, D. E. (2000). Feedstock and product recovery from reactor vent streams. *Chem. Eng. Prog.* 96, 51.

Baker, R. W., and Wijmans, J. G. (1994). Membrane separation of organic vapors from gas streams. In D. R. Paul and Y. P. Yampolskii (Eds.), *Polymeric Gas Separation Membranes*. CRC Press, Boca Raton, FL, pp. 353–397.

Baker, R. W., Wijmans, J. G., and Kaschemekat, J. H. (1998). The design of membrane vapor-gas separation systems. *J. Membr. Sci.* 151 55.

Behling, R. D., Ohlrogge, K., Peinemann, K.-V., and Kyburz, E. (1989). The separation of hydrocarbon from waste vapor streams. In A. E. Fouda, J. D. Hazlett, T. Matsura, and J. Johnson (Eds.),

Membrane Separation in Chemical Engineering. AIChE Symposium Series 272, Vol. 85. AIChE Press, New York, p. 68.

Fenstermaker, R. W. (1983). Engine performance operating on field engine gas as engine fuel. U.S. Patent 4,370,150.

Hägg, M.-B. (2001). Purification of chlorine gas with membranes—An integrated process solution for magnesium production. *Sep. Purif. Technol.* 21, 261.

Hauk, G. (1993). Device for continuously purifying the waste gas from a vacuum unit. U.S. Patent 5,194,074.

Jacobs, M., and Billing, B. (2005). Achieving ethylene efficiency. *Hydrocarbon Eng.* 10(8), 61.

Lahiere, R. J., Hellums, M. W., Wijmans, J. G., and Kaschemekat, J. (1993). Membrane vapor separation: Recovery of vinyl chloride monomer from PVC reactor vents. *Ind. Eng. Chem. Res.* 32, 2236.

Li, Y.-E., Paganessi, J. E., Vassallo, D., and Fleming, G. K. (1998, 1999a, b). Recovery and system for separation and recovery of perfluoro compound gases. U.S. Patents 5,785,741 (1998); 5,858,065 (1999a); and 5,919,285 (1999b).

Lokhandwala, K., Jariwala, A., and Baker, R. W. (in press). Only raw sour gas available for engine fuel? Proven membrane process cleans gas for engines. In The 56th Laurence Reid Gas Conditioning Conference Proceedings, *Norman, OK*, (Feb. 27–Mar. 1, 2006). Also in *Gas Processors' Association (GPA) Annual Meeting Proceedings*, Grapevine, TX, Mar. 5–8, 2006.

Lokhandwala, K. A., Segelke, S., Nguyen, P., Baker, R. W., Su, T. T., and Pinnau, I. (1999). A membrane process to recover chlorine from chloralkali plant tail gas. *Ind. Eng. Chem. Res.* 38, 3606.

Morisato, A., He, Z., and Pinnau, I. (1999). Mixed-gas properties and physical aging of poly(4-methyl-2-pentyne). In B. D. Freeman and I. Pinnau (Eds.), *Polymer Membranes for Gas and Vapor Separation: Chemistry and Materials Science.* ACS Symposium Series 733. American Chemical Society, Washington, DC, pp. 56–67.

Ohlrogge, K., Rockmiller, J. B., Wind, J., and Behling, R. D. (1993). Engineering aspects of the plant design to separate volatile hydrocarbons by membrane vapor separation. *Sep. Sci. Technol.* 28, 227.

Ohlrogge, K., and Stürken, K. (2001). The separation of organic vapors from gas streams by means of membranes. In S. P. Nunes and K.-V. Peinemann (Eds.), *Membrane Technology.* Wiley, Chichester, Part II p. 71.

Ohlrogge, K., and Wind, J. (2000). Method and apparatus for reducing emissions from breather-lined storage tanks. U.S. Patent 6,059,856.

Pinnau, I., He, Z., Da Costa, A. R., Amo, K. D., and Daniels, R. (2002). Gas separation using C_{3+}-hydrocarbon-resistant membranes. U.S. Patents 6,361,582 and 6,361,583.

Pinnau, I., He, Z., Masuda, T., and Sakaguchi, T. (2004). Pure and mixed-gas permeation properties of poly(*p-tert*-butyl diphenylacetylene). In I. Pinnau and B. D. Freeman (Eds.), *Advanced Materials for Membrane Separations.* ACS Symposium Series 876. American Chemical Society Washington, DC, pp. 167–176.

Pinnau, I., Lokhandwala, K. A., Nguyen, P., Toy, L. G., and Jacobs, M. L. (1996). Membrane process for treatment of chlorine-containing gas streams. U.S. Patent 5,538,535.

Pinnau, I., Wijmans, J. G., He, Z., Goakey, S., and Baker, R. W. (1998). Process for recovering semiconductor industry cleaning compounds. U.S. Patent 5,779,763.

Schultz, J., and Peinemann, K.-V. (1996). Membranes for separation of higher hydrocarbons from methane. *J. Membr. Sci.* 110 37.

Wang, X., Daniels, R., and Baker, R. W. (2001). Recovery of VOCs from high-volume low-concentration air streams. *AIChE J.* 47, 1094.

Wijmans, J. G. (1992, 1993). Process for removing condensable components from gas streams. U.S. Patents 5,089,033 (1992) and 5,199,962 (1993).

Wijmans, J. G., and Baker, R. W. (1991). Refrigeration process with purge and recovery of refrigerant. U.S. Patent 5,044,166.

Wijmans, J. G., He, Z., Su, T. T., Baker, R. W., and Pinnau, I. (2004). Recovery of perfluoroethane from chemical deposition operations in the semiconductor industry. *Sep. Purif. Technol.* 35, 203.

Wijmans, J. G., Kamaruddin, H. D., Segelke, S. V., Wessling, M., and Baker, R. W. (1997). Removal of dissolved VOCs from water with an air stripper/membrane vapor separation system. *Sep. Sci. Technol.* 32(14), 2267.

CHAPTER 22

Gas Separation by Polyimide Membranes

YOJI KASE

UBE Industries Ltd., Ichihara, Chiba 290-0045, Japan

22.1 INTRODUCTION

Separation of a gas from the gas mixture is a key issue in the various industrial fields: hydrogen recovery in petroleum refinery process, oxygen removal to prevent flame, air dehumidification to prevent moisture absorption, dehydration of fine chemical products, for example. Although there are various methods to perform gas separation [membrane separation, pressure swing adsorption (PSA) separation, cryogenic separation], the method using membranes attracts much attention.

The gas separation membranes are comprised of either inorganic materials or organic materials (polymers). This chapter discusses the polymeric membranes, and the inorganic membranes are detailed elsewhere.

The gas separation membrane was first commercialized in 1980 in order to recover hydrogen gas produced in the oil refining process. This membrane was a hollow-fiber membrane comprised of polysulfone. The shell of the hollow fiber exhibited an asymmetric cross-sectional structure, which showed the porous structure with pore size varying from one surface to the other surface of the membrane. The asymmetric structure is most preferred for the gas separation membranes due to several reasons described later. A certain number of the hollow-fiber membranes were integrated into a bundle, and the bundle of the hollow-fiber membrane was installed into the piping unit (module) and shipped to the user.

Many types of polymers (polyolefins, polyimides, polysulfones, cellulosics, polycarbonates, etc.) have been explored for fabricating the practical gas separation membranes. In these polymeric membranes, polyimide membranes are the most fascinating due to their excellent properties of:

1. high permselectivity and permeability
2. easy to prepare asymmetric structure

Advanced Membrane Technology and Applications. Edited by Norman N. Li, Anthony G. Fane, W. S. Winston Ho, and T. Matsuura
Copyright © 2008 John Wiley & Sons, Inc.

3. high thermal stability and high chemical stability
4. high mechanical strength
5. long durability

So far, many attempts have been made to develop the polyimide gas separation membranes. At present, the polyimide membranes are commercialized by several manufacturers: Medal (Air Liquide), IMS (Praxair), and UBE Industries Ltd (Baker, 2002).

This chapter details (1) the polyimides used in the membranes, (2) preparation method of the membranes, (3) design of gas separation module, and (4) practical applications of the gas separation membranes.

22.2 PERMEABILITY AND CHEMICAL STRUCTURE OF POLYIMIDES

Recently, a very wide range of polyimide monomers (tetra-carboxylic anhydrides and diamines) has become commercially available, and thereby a rich variety of polyimides can be used in various applications including gas separation membranes. Properties of the polyimides are significantly influenced by the chemical structure of constituent monomers. Therefore, we can modify properties of the polyimide by changing the constituent monomers. For using as gas separation membranes, permeability of the gases, free volume (V_f), glass transition temperature (T_g), solubility of the gases, polarity, mechanical strength, and stiffness are very crucial properties of the polyimides.

Kim et al. (1988), Stern (1984), Stern et al. (1989), and Tanaka et al. (1992) determined the gas permeability for a wide range of polyimides and found that the permeability seems to correlate with the free volume (V_f) of the polyimide. There, the free volume was estimated by Bondi's free-volume method (Bondi, 1964). According to Tanaka et al. (1992) the polymers with higher V_f's tend to show higher gas diffusion coefficients. Although Hirayama et al. also examined the relation between the diffusivity and V_f for several polyimides, they reported the relation was not so good except for the families of polyimides with similar chemical structure (Hirayama et al., 1996a, 1996b, 1999). The chemical structures of polyimides they prepared are shown in Table 22.1.

On the other hand, local motions of polymer segments may influence the gas permeability. As pointed out by Kuroda (1979), elastic modulus E of the polymer chain is a measure of flexibility and mobility of the segments. Hirayama et al. (1996a, 1996b, 1999) investigated dynamic thermal mechanical properties of the polyimides and found the gas diffusion coefficient (D) correlates with the tensile storage modulus (E') of the polyimide. Coefficient D seems to be enhanced by reduction of E' due to an increase in the local segmental motion of the polyimides.

They also suggested that the gas diffusion coefficient seems to relate to the cohesive energy density (CED) as follows:

$$E_d \approx \left(\frac{\pi}{4}\right)\sigma^2\lambda \mathrm{CED}$$

$$D = D_0 \exp\left(\frac{-E_d}{RT}\right)$$

$$\ln(D_0) = aE_d - c$$

TABLE 22.1 Chemical Structure of Polyimide

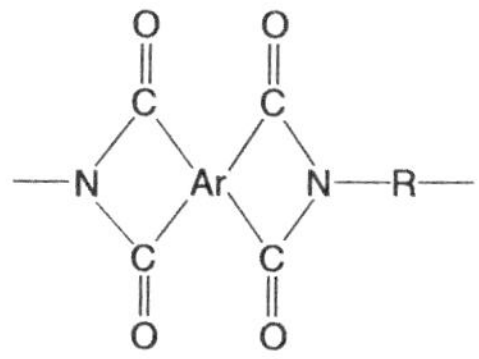

Ar:	BPDA 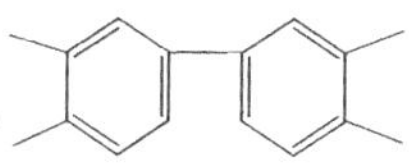	6FDA 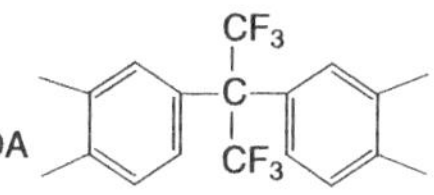
R:	DADE 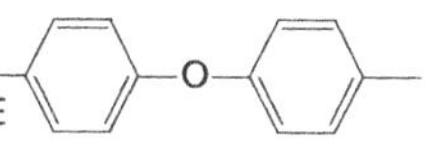	MFA 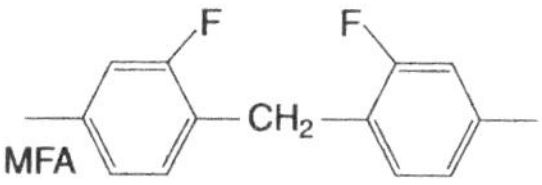
	DADM 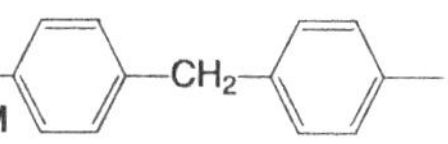	MBA 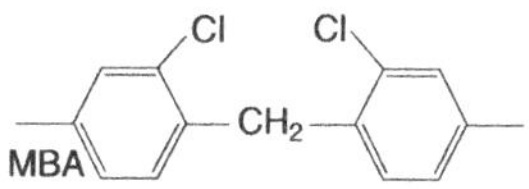
	DADS 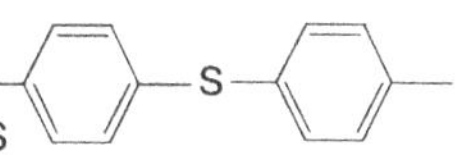	MDX 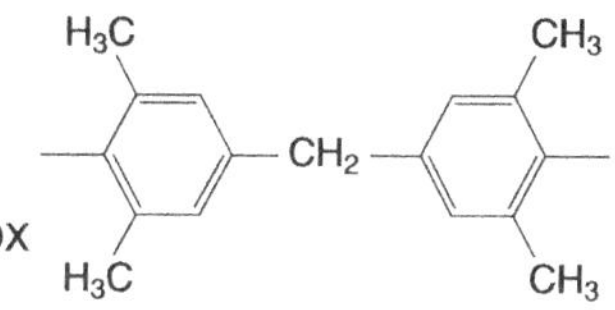
	PASN 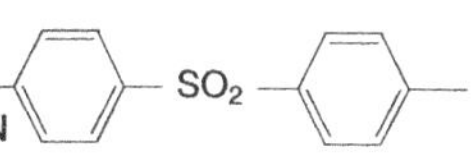	HAB 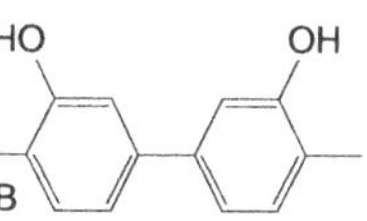
	HFIP 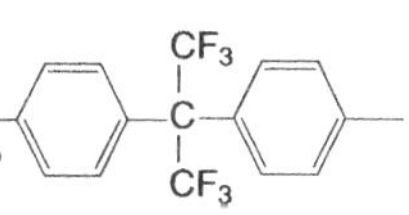	TSN 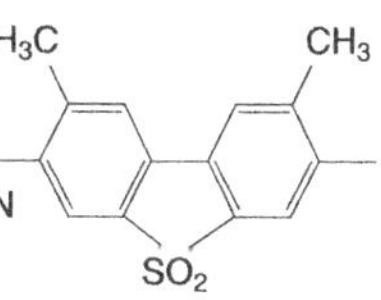
	MDT 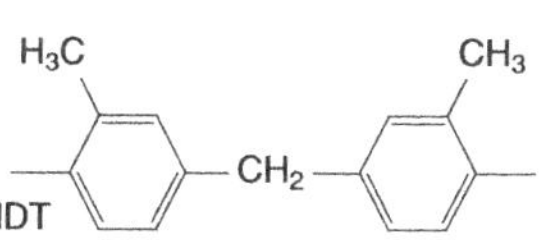	MASN 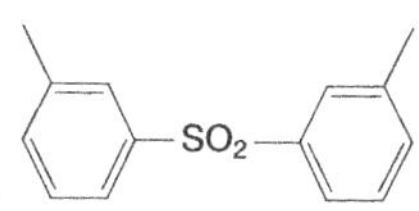
	CDM 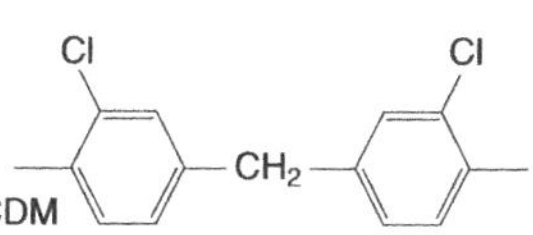	

(Continued)

TABLE 22.1 *Continued*

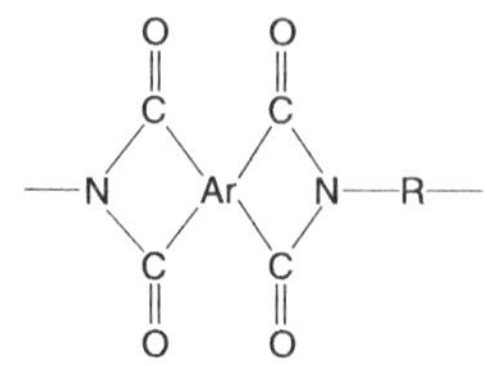

Ar:

BPDA

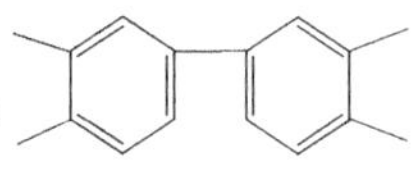

6FDA

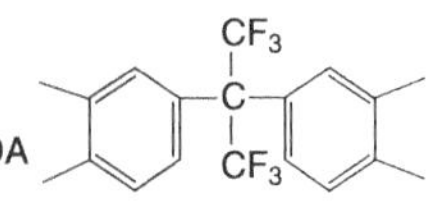

BAPE

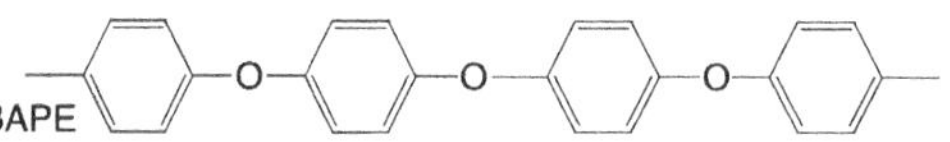

DABA

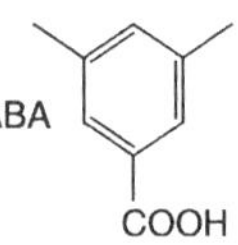

BAPS

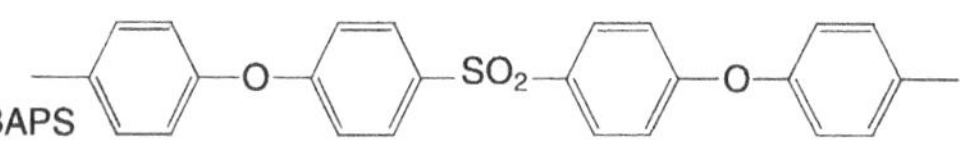

TFDM

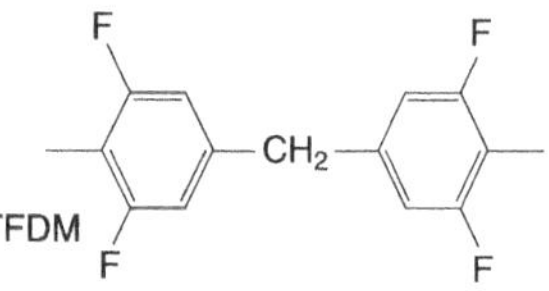

BAPP

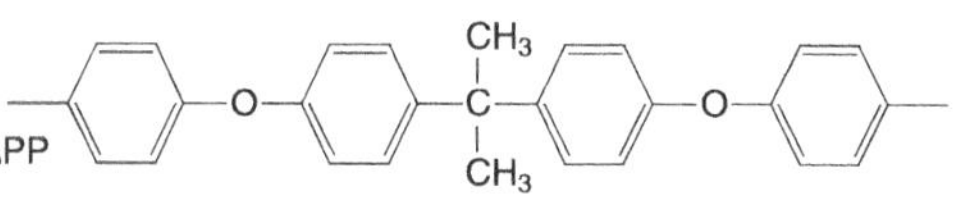

TCDM

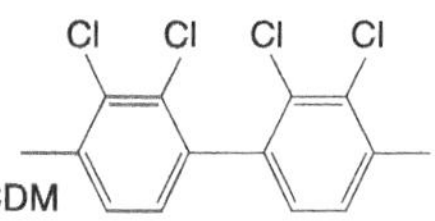

HFBAPP

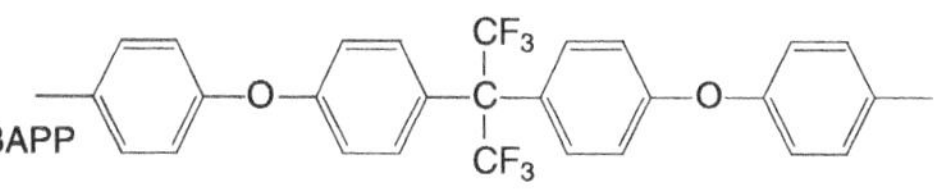

MBHA

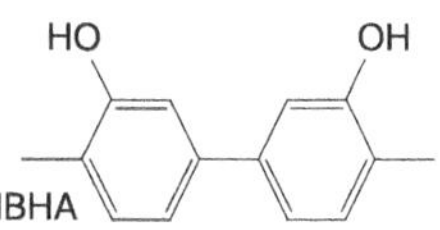

where E_d is the activation energy for the diffusion, σ is the mean size of the penetrant (gas molecule), λ is the length of the diffusional jump, R is the gas constant, T is absolute temperature, a is a constant approximated by $0.001/R$, and c is a phenomenological parameter distinguishing polymers in glassy or rubbery states, which implicitly depends on the tensile modulus of the polyimde. The physical properties of selected aromatic polyimides are given in Table 22.2.

Figure 22.1 shows the diffusivity of CO_2, O_2, CH_4, and N_2 in the polyimide prepared as a function of the CED. The gas molecules are transported through intersegmental gap of the polyimide matrix. Since the energy required to open the intersegmental gaps is larger for the polyimides of large CED, the gas diffusivity seems to increase by decrease in the CED.

On the other hand, Park et al. (1997), Robeson et al. (1997) and Laciak et al. (1999) proposed the method for estimating the gas permeability of the polymer from its chemical structure.

Park et al. (1997) defined 41 fundamental chemical groups that appears in 105 polymers. They assigned empirical factors for each predefined chemical group accounting diffusion of the gas n in the polymer comprising group k. They showed that the gas permeability

TABLE 22.2 Characteristics of Polyimides

No.	Polyimide Sample	*d*-spacing, Å	V_f	E', Gpa	*CED*, J/cm^3
1	BPDA-DADM	5.8	0.110	2.3	990
2	DADS	5.2	0.117	2.8	1050
3	PASN	5.5	0.117	3.5	1050
4	HPIP	6.2	0.163	2.9	740
5	MFA	5.8	0.163	3.1	1040
6	MBA	3.8	0.161	4.2	1030
7	MCA	6.1	0.164	3.0	1050
8	BAPE	4.9	0.151		880
9	BAPS	5.2	0.135	3.2	900
10	BAPP	5.4	0.142	2.7	840
11	HFBAPP	6.0	0.162	2.2	900
12	MDT	5.5	0.173	3.5	840
13	CDM	5.9	0.176	2.6	850
14	MDX	6.2	0.175	2.6	910
15	HAB	5.3	0.145	5.7	1050
16	TSN	5.8	0.118	4.6	850
101	TFDM	5.8	0.135	2.1	1480
102	TCDM	5.8	0.135	2.5	1060
103	MBHA	5.1	0.099	4.4	1090
104	MASN	5.5	0.111	3.7	1050
31	6FDA-DADE	5.8	0.173	2.5	740
32	DADM	5.8	0.155	2.1	720
33	MDT	5.9	0.168	2.7	680
34	MDX	6.2	0.164	2.2	650
35	CDM	5.8	0.182	3.4	770
36	TSN	5.8	0.146	3.1	770
37	BAPE	5.5	0.170	2.2	700
38	HAB	5.9	0.182	4.0	980
105	DABA	5.3	0.153	2.3	850

of the polymer relates to the chemical structure of its repeating unit as follows:

$$P = A \exp\left(\frac{-B}{V_f}\right)$$

$$(V_f)_n = \frac{V - (V_0)_n}{V}$$

$$(V_0)_n = \sum_{k=1}^{K} \gamma_{nk}(V_w)_k$$

$$V = \sum_{k=1}^{K} \beta_k(V_w)_k$$

where A and B are constants depending on the gas n, V is the molar volume of the repeating unit of the polymer, and V_0 is the core volume of the repeating unit. The chemical structure of the repeating unit is divided into the predefined fundamental chemical groups; γ_{nk} and β_k are empirical factors of the chemical group k for gas, and $(V_w)_k$ is the van der Waals volume

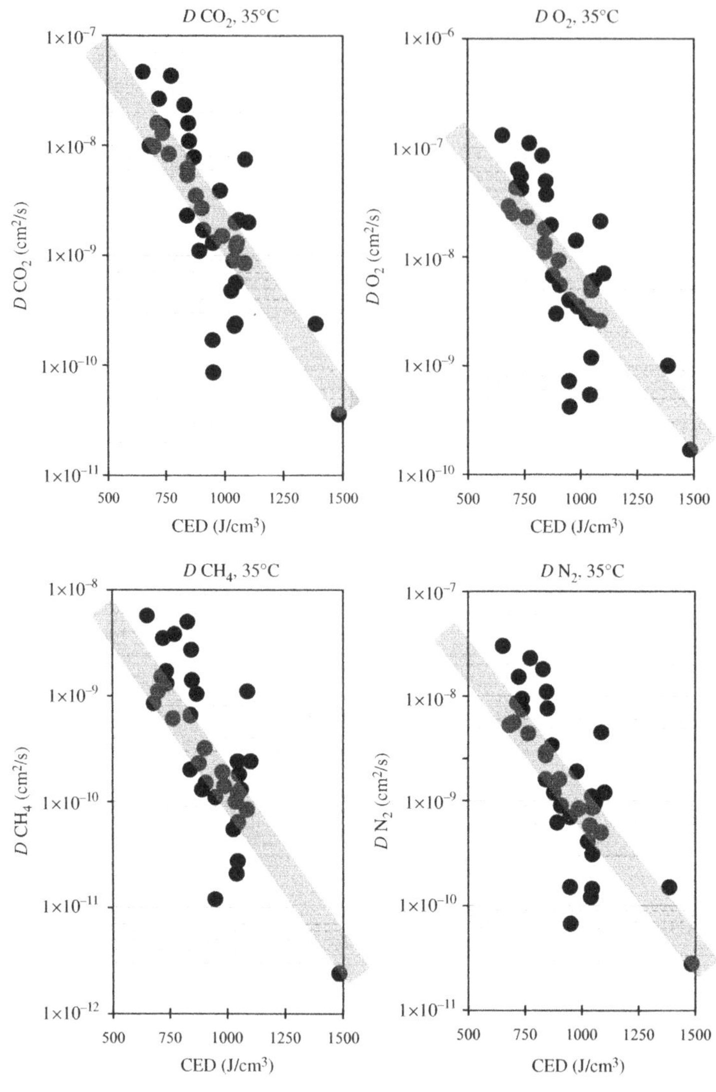

Figure 22.1 Correlation between D and CED at 50°C.

of the group k. They showed excellent agreement of the calculated permeability from the above equation with the experimentally determined ones.

Robeson et al. (1997) and Laciak et al. (1999) proposed another formulation describing the relationship between the permeability and the repeating unit structure. They explored the gas permeability for 65 polymers and defined 24 fundamental chemical groups. They showed the gas permeability of the polymers represented by the following expression:

$$\ln P = \sum_{i=l}^{n} \phi_i \ln P_i$$

where ϕ_i is the volume fraction of a specific group i comprising the repeating unit, and P_i is the permeability contribution of a specific group i. They also showed excellent agreement with the experimental permeability values and the calculated values from this equation.

These findings are helpful to design a new polyimide for the gas separation membrane. Further increase in the simulation accuracy as well as the expansion of the database are expected.

22.3 MANUFACTURE OF ASYMMETRIC MEMBRANE

The preceding section discussed the permeability of the homogeneous films without pores (hereafter, referred to as *symmetric membranes*). However, such symmetric membranes exhibit significantly low permeability, and they are hardly applied to practical uses thereby. In order to increase the apparent permeability of the membrane, the polyimides are usually used in the form of asymmetric membranes. Here, asymmetric membrane denotes the structure consisting of a dense skin layer and a porous support layer, as schematically depicted in Figure 22.2.

In the support layer, the polymer matrix and the pores are co-continuously connected across the layer. The three-dimensionally continuous polymer network exhibits the sufficient mechanical strength, and the three-dimensionally continuous pores allow gases to flow through the layer without significant flow resistance.

On the other hand, the skin layer is a continuous polymer layer without the pores. This layer covers the entire surface of the support layer and is considerably thin (typically less than 100 nm). Although the permeability of the skin layer is the same as that in bulk film, due to the thinness of this layer, the apparent flux of the permeant is significantly enhanced. In order to achieve large permeance, it is crucial to prepare a very thin skin layer without defects.

A number of methods has been developed for the manufacture of asymmetric membranes. However, the asymmetric membranes are mostly produced by the method developed by Loeb and Sourirajan (1962), namely the dry and wet method. This method is comprised of the following four steps: (1) preparation of the polyimide solution (called *dope*), (2) molding (cast or spinning) of the dope, (3) coagulation of the dope by contact with nonsolvent of the polyimide to form the asymmetric structure, and (4) drying of the coagulated membrane.

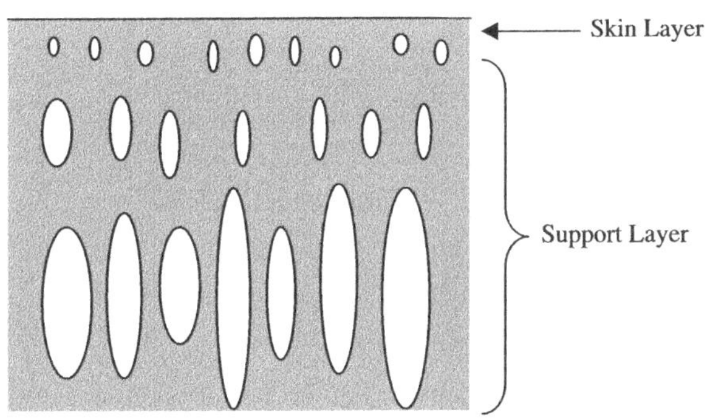

Figure 22.2 Structure of asymmetric membrane.

In the first step, solubility of the polyimide to the solvent, is important. Adequate solvent for the polyimide should be selected. In addition, viscosity of the dope should be adjusted to the range in which the dope is easy to cast or to spin. Further, it is important to select the coagulation solvent since the solvent exchanging rate in the coagulation process affects the structure of the membrane.

Although the asymmetric structure can be formed in flat films, tubular films, and hollow fibers, the hollow fiber is mostly preferred due to its large surface area per unit volume.

22.4 MEMBRANE MODULE

When the gas separation membrane is applied to some practical purposes, the large surface area, that is, large processable capacity for gases is extensively demanded. Since the permeability of each membrane is not so high, the membranes are commonly used by integrating them into bundles. The bundle of the membranes is installed into the piping unit (module) and supplied to the user. For the hollow-fiber membrane, both edges of membrane bundle are finished by epoxy resins, polyurethanes, or silicone resins in order to isolate the inside of the hollow fibers from the outside of the fibers in the module.

Generally, there are two types of modules distinguished by the gas flow direction: The first type feeds the gas mixture into the inside of the hollow fibers, and thereby the permeated gas is collected outside of the fiber bundle (bore-side feed); the second type feeds the gas mixture outside of the fiber bundle, and thereby the permeated gas is collected inside of the hollow fibers (shell-side feed). The suitable type of the module (bore-side feed or shell-side feed) is decided on the basis of feed gas pressure, pressure drop in the module, the composition of the feed gas, and the gas permeability of the membrane.

The hollow-fiber membranes are arranged in a module so that the gas concentration gradient along the gas flow direction is constant and the short pass is inhibited. The membrane module of the bore-side feed case is schematically shown in Figure 22.3.

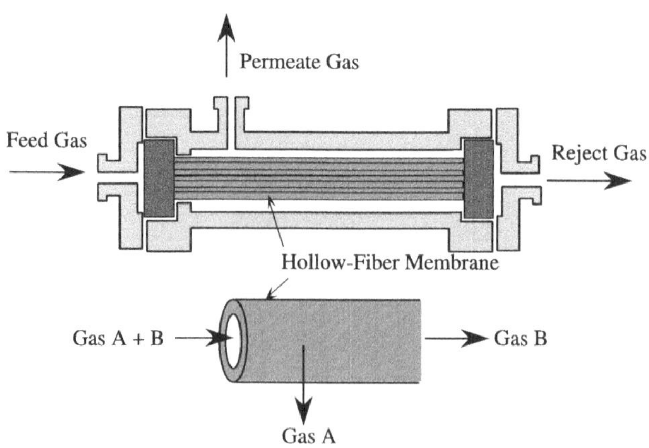

Figure 22.3 Hollow-fiber membrane module.

22.5 APPLICATIONS OF POLYIMIDE GAS SEPARATION MEMBRANES

The gas separation membranes are applied to various fields, as summarized in Table 22.3. This section details some of these applications.

22.5.1 Hydrogen Separation

There are large demands of hydrogen recovery in oil refineries and chemical plants. Long durability at high temperatures and in various chemical substances is requisite for the gas separation membranes to be applied in this area. Polyimides have sufficient stability and, furthermore, exhibit a considerably large permeability of hydrogen compared to that of other gases such as nitrogen. Therefore, the polyimide hollow-fiber membranes have been widely used for hydrogen separation. The hydrogen recovery process in the reformer is represented in Figure 22.4 and Table 22.4.

Since the gas ejected from the reformer contains some amount of the gasoline vapor, the gas is heated in the feed line in order to prevent the condensation of the gasoline. High temperature is convenient for the hydrogen separation since the gas permeability in the polyimide is increased. Typically the hydrogen concentration in the feed gas is 73%, and the hydrogen separation membrane module can concentrate hydrogen up to 98%. In the feed gas 90% of the hydrogen can be recovered by the membrane separation system.

An example of the hydrogen recovering plant using the polyimide membrane module is shown in Figure 22.5. Although conventional hydrogen recovery plants without membrane modules are usually very large sized, this small plant with the polyimide membrane module is able to treat about 10,000 m^3/h of feed gas.

TABLE 22.3 Application of Membrane Separation

Object	Application
H_2	Refinery of petroleum: recovery of H_2 Chemical industry: control the balance of H_2/CO, recovery of H_2, produce of high purity H_2 Recovery of He from natural gas
CO_2	Remove of acid gas from natural gas CO_2 separation from landfil gas
O_2/N_2	Prevention discharge: piping, seal of tank, transportation of powder, tanker for LPG, equipment of washing Prevention for oxidization: molding resin, laser cutting, sorder process Foods: prevention for oxidization, preservation and tranportation of perishable foods Medicines: prevention of oxidization Metal: prevention of oxidization
Dehumidity	Precision machines, measurement of machines, automatic controller, machines with air pressure Ozone generator Analytic equipment, laboratory equipment
Dehydration	Dehydration of organic solvent

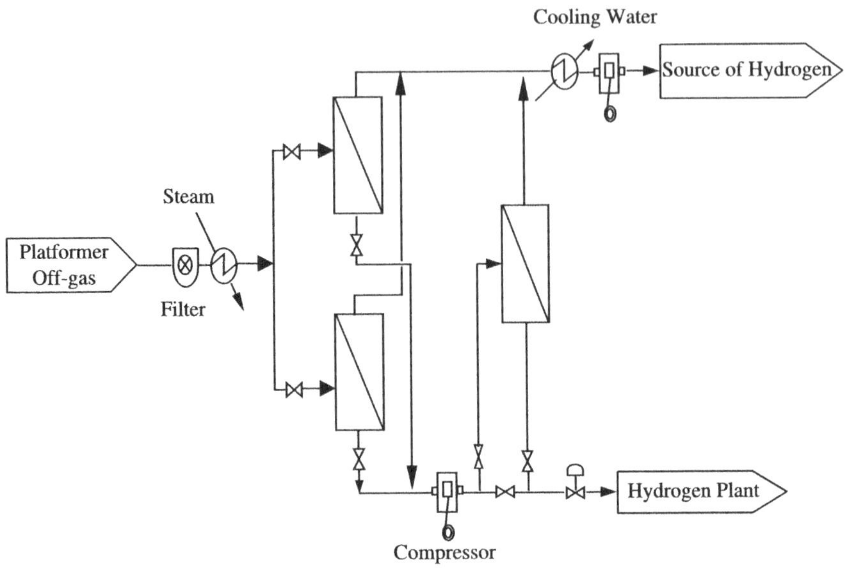

Figure 22.4 Process for recovering hydrogen from offgas of the reformer.

To highly purify hydrogen, a number of modules is frequently combined in series, as shown in Figure 22.6. The production of high-purity hydrogen of 99.999 mol% can be obtained by the system shown in Figure 22.6.

22.5.2 Carbon Dioxide Separation

Most demands of carbon dioxide separation are in the field of natural gases and biogases. In most cases, the carbon dioxide has to be separated from the mixture of CO_2/CH_4. The solubility of CO_2 in the polymers is generally higher than that of other gases. However, the large solubility of carbon dioxide in polymers frequently decreases the selectivity due to occurring swelling of the polymer network at the same time. The polyimide is not significantly plasticized by the carbon dioxide. Therefore, the polyimides are supposed to be the suitable material for the CO_2 separation.

TABLE 22.4 Example of Recovering Hydrogen Data

	Feed	Permeate	Nonpermeate
Flow rate, Nm^3/h	7340	4940	2400
Pressure, kgf/cm^2 G	20	5	35
Temperature, °C	50	—	—
Composition, mol%			
H_2	73.1	98.6	20.4
C_1	10.0	1.1	28.4
C_2	7.9	0.3	23.6
$C_3\sim$	9.0	—	—

Figure 22.5 Hydrogen recovering plant.

22.5.2.1 Natural Gas Most natural gases contain impurity gases, such as carbon dioxide, steam, and hydrogen sulfide. Since some of the impurity gases may cause corrosion of the pipeline, natural gas is first fed to the gas separation line in order to reduce corrosive gas concentration to less than 2%. Polyimide membranes designed for carbon dioxide separation are able to separate steam and hydrogen sulfide as well as carbon dioxide since these gases permeate much more rapidly than hydrocarbons.

22.5.2.2 Landfill Gas At disposal sites, landfill gas is produced by decomposition of organic wastes. This gas typically is composed of CH_4 (50%), CO_2 (40%), and other gases, such as hydrogen sulfide, ammonia, nitrogen, and steam. Since methane in the landfill gas may induce disasters, methane recovery from landfill gas has been explored first. The polyimide membranes showed the possibility of recovering the methane with a high quality suitable for blending in natural gas as a fuel for gas engines. On the other hand, CO_2 recovery by the polyimide membranes has not progressed so much, although reduction of CO_2 emission is crucial for the global environmental problem.

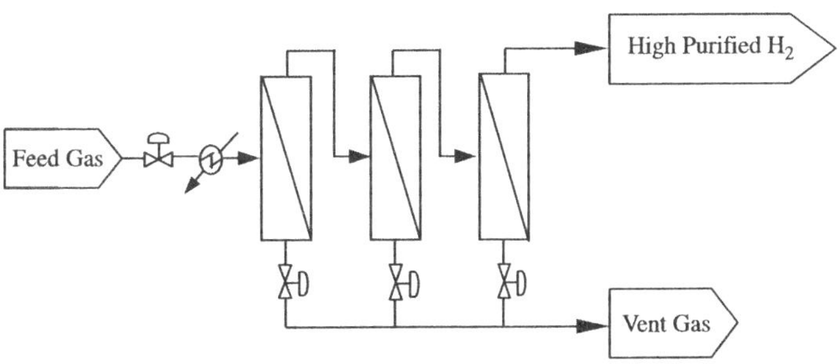

Figure 22.6 High purifying hydrogen process.

22.5.2.3 Exhaust Gases In Japan, the Research Institute of Innovative Technology for the Earth (RITE) is developing a CO_2 recovery system from exhaust gases of stationary CO_2 emission by means of the gas separation membrane (Hirayama et al., 1995; Tokuda et al., 1997; Karashima et al., 1999). This system separates CO_2 from CO_2/N_2. The polyimide membrane used in this system is comprised of a new *cardo*-type polyimide. Here *cardo* designates the polymer structure containning loop-shaped moieties in main chains. Examples of such cardo-type polyimides are represented in Figure 22.7.

The cardo-type polyimide exhibits high gas permeability due to its bulky chemical structure. CO_2 permeability and CO_2/N_2 selectivity were determined for various cardo-type polyimides, and it was found that the cardo polyimide modified with bromine [PI-PMBP64(4Me)-Br] exhibits very high CO_2 permeability and high CO_2/N_2 selectivity. The hollow-fiber membrane of PI-PMBP64(4Me)-Br was successfully prepared by the conventional method as described in Section 22.3. The gas separation module comprising the PI-PMBP64(4Me)-Br membranes showed an excellent ability for CO_2 separation: CO_2 permeability of 1.3×10^{-3} [cm^3(STP)/(cm^2 s cm Hg)], and CO_2/N_2 selectivity of 41 at 25°C. Moreover, the PI-PMBP64 membrane showed a high durability of more than 12,000 h at 50°C without decreasing permselectivity of CO_2/N_2.

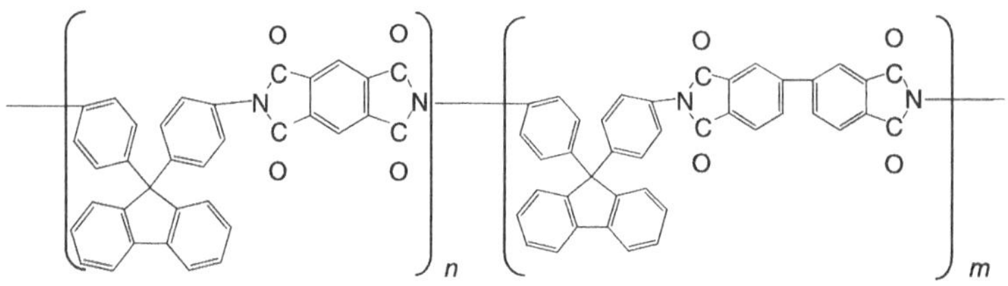

PI-PMBP64 ($n : m = 60 : 40$)

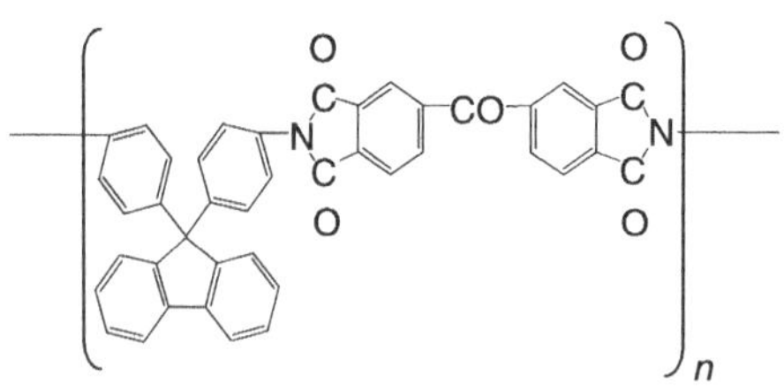

PI-BT-COOMe

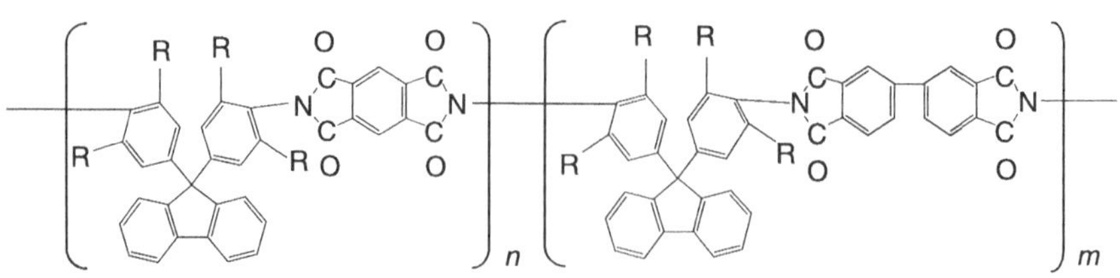

PI-PMBP64(4Me): R = $-CH_3$ ($n : m = 60 : 40$)

PI-PMBP64(4Me)–Br: R = $-CH_2Br$, $-CH_3$, $-CHBr_2$ ($n : m = 60 : 40$)

Figure 22.7 Chemical structures of the cardo-type polyimide.

TABLE 22.5 Nitrogen Products and Module Type of UBE Nitrogen Gas Generator

	Size of Outward Form (mm)			Nitrogen Products (Nm^3/h)				
Module Type	*L*	*D*	*d*	99.9%	99%	98%	97%	95%
NM-B01A	235	55	50	0.025	0.071	0.1	0.14	0.19
NM-B02A	360	55	50	0.06	0.15	0.21	0.27	0.40
NM-B05A	610	55	50	0.12	0.34	0.49	0.63	0.92
NM-B10A	1110	55	50	0.24	0.66	0.95	1.2	1.8
NM-C05A	634	90	70	0.26	0.74	1.1	1.4	2.1
NM-C10A	1110	110	70	0.57	1.6	2.4	3.1	4.5
NM-410A	1110	165	100	1.2	3.6	5.2	6.8	10
NM-615	1545	279	165	3.7	12	17	22	32
NM-815	1554	343	216	4.8	21	31	41	62

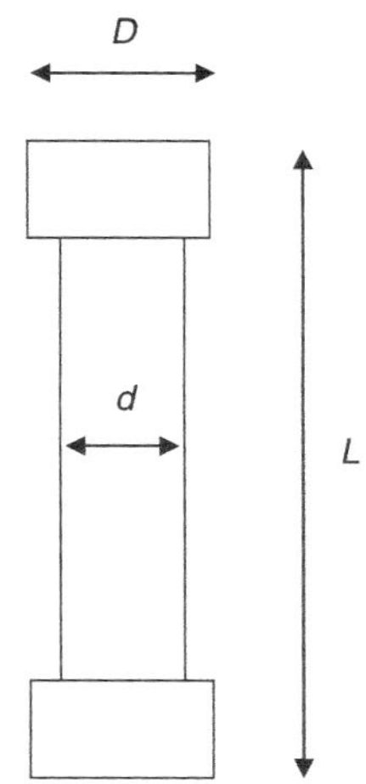

TABLE 22.6 Permeability of H_2O and Organic Solvents

Organic Solvent[a]	$P'_{H_2O} \times 10^3$	$P'org. \times 10^7$	$\alpha H_2O/org.$
MeOH	1.16	427.00	27
EtOH	1.41	31.00	450
IPA	1.38	0.80	17300
Acetone	1.51	1.80	8400
Ethyl acetate	1.33	1.10	12100

Temperature: 130°C; Feed pressure 2 kgf/cm^2 G.
[a]Feed composition: MeOH 80 wt%, EtOH 90 wt%, IPA 95 wt%, Acetone 90 wt%, Ethyl acetate 96 wt%.

22.5.3 Air Separation

Oxygen-enriched air and nitrogen-enriched air can be obtained by membrane separation. When the air is fed to the polyimide membrane module, the oxygen-enriched gas and the nitrogen-enriched gas is obtained on the permeated gas line and the unpermeated gas line, respectively. The nitrogen-enriched air is used as a low-cost inert gas and is widely used for preventing explosion, for suppression of oxidation of industrial materials and foods, and for small-scale chemical experiments. The asymmetrical polyimide hollow-fiber membranes as described in the previous sections are able to be used for this purpose. The performance of the nitrogen separation module shipped from UBE Industries is shown in Table 22.6. The size ($L \times D \times d$) of the module shown in Table 22.5 is in the range of 235–1554 mm × 55–343 mm × 50–216 mm. The process flow scheme for a large amount of nitrogen separation is shown in Figure 22.8.

Since the polyimide membrane installed in the module may be damaged by water, the membrane modules are frequently equipped with a heating unit in order to decrease humidity of the air. Such a heating unit is necessary especially for a large-sized separation system. The compressed air is supplied to the bore side of the hollow fibers, and the (unpermeated) nitrogen-enriched air is obtained inside of the hollow fibers. The unpermeated gas contains very small amounts of gases besides oxygen: argon, carbon dioxide, and water vapor. The

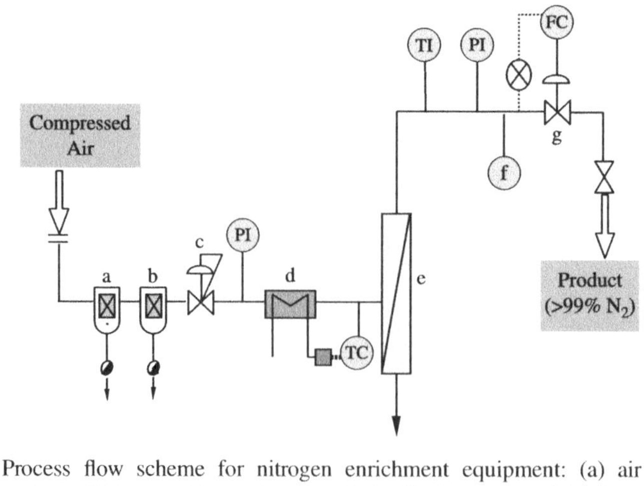

Figure 22.8 Process flow scheme for nitrogen enrichment equipment: (a) air filter, (b) mist separator, (c) regulator, (d) heater, (e) membrane module, (f) O_2 meter, and (g) flow controller value.

concentration of argon and carbon dioxide in the unpermeated gas is about 1% and less than 10–30 ppm, respectively, and the dew point of the unpermeated gas is less than -40°C.

22.5.4 Dehumidification System

The permeability of water vapor is much larger than that of hydrogen for many polymer membranes. In the case of polyimide membrane, the water permeability is significantly large too. Using this high permeability of water vapor in the polyimide, some polyimide membranes have been commercialized in the field of dryers. The membrane dryer showed many advantages over other conventional equipment such as refrigerative dryers and regenerative dryers. The membrane dryer does not need moving parts like a refrigerater and needs no shuttle valve like a regenerative adsorber. Therefore, the membrane dryers are usually maintenance free. In addition, the dried air is sufficiently clean without contaminating with drain and dust. An example of the system for the membrane dryer commercialized by UBE Industries, the UBE Membrane Dryer, is shown in Figure 22.9.

This dryer has a one- or two-stage module in a compact housing. When the wet compressed air is fed to the inner side of the hollow-fiber membrane, water vapor permeates through the membrane toward the outer side of the hollow fiber. As a result, the dry compressed air is obtained from another inner side of the fiber. Further decrease of the dew point of the produced dry air can be achieved by purging the outer side of the hollow-fiber membrane with a part of produced dry air. By adjusting purging airflow rate, the dew point of the produced air can be adjusted optionally.

22.5.5 Dehydration of Organic Compounds

As described above, the water vapor permeability in the polyimide is much larger than other gases. This property of the polyimide makes it possible to dehydrate organic compounds with the polyimide membrane module. Compared to a conventional distillation process,

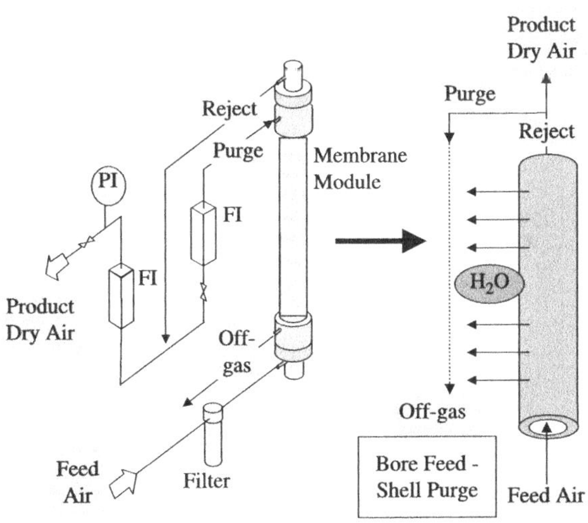

Figure 22.9 Membrane dryer system.

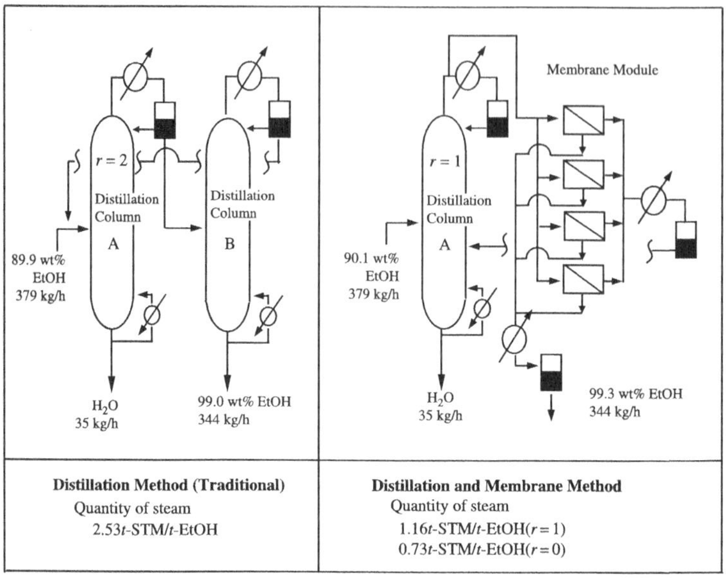

Figure 22.10 Comparison of H_2O/EtOH separation system.

the membrane process proves to be more advantageous in terms of energy savings and overall system size. The excellent heat resistance and chemical stability of the polyimide membrane is preferred in this purpose. The permeability and selectivity of water and organic vapor for polyimide membrane are summarized in Table 22.6.

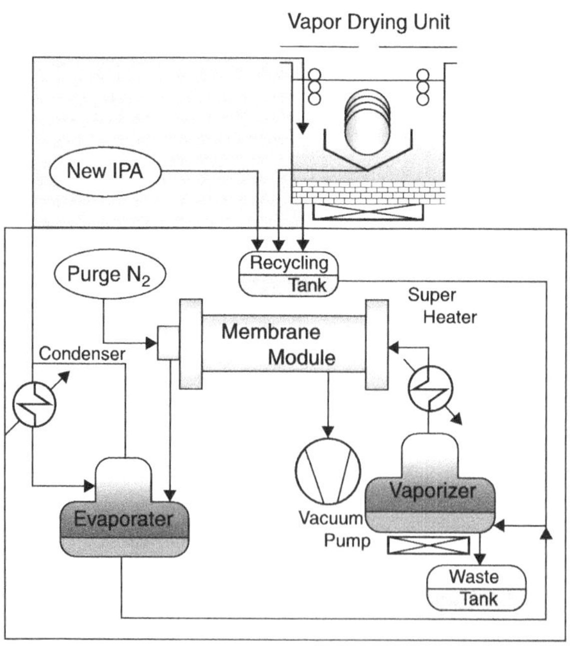

Figure 22.11 IPA circulation purifier.

As shown in Table 22.6, the permeability of water is several hundred or thousand times larger than that of the organic solvents. Hence high dehydration of organic solvent is achieved by the polyimide membranes.

The separation of azeotropic solvents with water have been done by distillation with azeotropic reagent. On the other hand, membrane separation can purify the organic solvent above the azeotropic concentration, and is a effective purification method as a substitution for azeotropic distillation. Some application examples of the dehydration of the organic materials will be presented below.

22.5.5.1 Dehydration of EtOH The azeotropic concentration of ethanol in water is 96%. To obtain more purified EtOH, the azeotropic distillation with benzene or cyclohexane have to be utilized. Figure 22.10 shows the membrane separation system as a substitution for the azeotropic distillation column. When the conventional distillation plant is replaced by the membrane separation system, the steam consumption amount is reduced to one third. The membrane separation system is quite energy saving.

22.5.5.2 Dehydration of Isopropyl Alcohol Isopropyl alcohol (IPA) is widely used as a good solvent for various chemical components, and is utilized as a washing liquid in the production process of large-scale integration circuits (LSI), very large-scale integration circuits (VLSI), and ultra large-scale integration circuits (ULSI), precision machine parts, and optical lenses. In these applications, IPA is required to be of high purity. Hence IPA has been frequently used without recovery and recycle in the factory. However, the IPA circulation purifier, which uses polyimide dehydration membrane as shown in Figure 22.11, enables the user to recover and to recycle IPA to a high dehydration degree. This equipment exhibits dehydration ability of 88% IPA in IPA–water mixture up to 99.995% (water content in IPA–water mixture is 0.005%).

REFERENCES

Baker, R. W. (2002). Future directions of membrane gas separation technology. *Ind. Eng. Chem. Res.* 41, 1393.

Bondi, A. (1964). van der Waals volumes and radii. *J. Phys. Chem.* 68, 441.

Hirayama, Y., Kazama, S., Fujisawa, E., Nakabayashi, M., Matsumiya, N., Takagi, K., Okabe, K., Mano, H., Haraya, K., and Kamizawa, C. (1995). Novel membranes for carbon dioxide separation. *Energy Conserv Mgmt.* 36, 435.

Hirayama, Y., Yoshinaga, T., Kusuki, Y., Ninomiya, K., Sakakibara, T., and Tamari, T. (1996). Relation of gas permeability with struction of aromatic polyimides, I. *J. Membr Sci.* 111, 169.

Hirayama, Y., Yoshinaga, T., Kusuki, Y., Ninomiya, K., Sakakibara, T., and Tamari, T. (1996b). Relation of gas permeability with struction of aromatic polyimides II. *J. Membr. Sci.* 111, 183.

Hirayama, Y., Yoshinaga, T., Nakanishi, S., and Kusuki, Y. (1999). Relation between gas permeabilities and structure of polyimides. In B. D., Freeman and I. Pinnau (Eds.), *Polymer Membranes for Gas and Vapor Separation.* ACS Symposium Series 733. American Chemical Society Washington, DC, p. 194.

Karashima, S., Tokuda, Y., Tachiki, A., Takagi, K., Haraya, K., and Kamizawa, C. (1999). Development of Cardo-type polyimide hollow fiber membranes for CO_2 separation. *Fourth International Conference on Greenhouse Gas Control Technologies (GHGT-4),* 1035.

Kim, T. H., Koros, W. J., Husk, G. R., and O'Brien, K. C. (1988). Relationship between gas separation properties and chemical structure in series of aromatic polyimides. *J. Membr. Sci.* 37, 45.

Kuroda, N. (1979). Permeation of gases through polymer films crystallized by heat treatment. *Sen-i Gakkaishi* 35, T413.

Kusuki, Y. (2002). Polyimide membranes for gas separation. In Y. Imai and R. Yokota (Eds.), *Saishin Polyimide*. NTS, Tokyo, 382.

Laciak, D. V., Robeson, L. M., and Smith, C. D. (1999). Group contribution modeling of gas transport in polymeric membranes. In B. D. Freeman and I. Pinnau (Eds.), *Polymer Membranes for Gas and Vapor Separation*. ACS Symposium Series 733. American Chemical Society, Washington, DC, p. 151.

Loeb, S., and Sourirajan, S. (1962). Sea water demineralization by means of an osmotic membrane. *Adv. Chem. Ser.* 38, 117.

Park, J. Y., and Paul, D. R. (1997). Correlation and prediction of gas permeability in glassy polymer membrane materials via a modified free volume based group contribution method. *J. Membr Sci.* 125, 23.

Robeson, L. M., Smith, C. D., and Langsam, M. (1997). A group contribution approach to predict permeability and permselectivity of aromatic polymers. *J. Membr. Sci.* 132, 33.

Stern, S. A. (1984). Polymers for gas separations: The next decade. *J. Membr. Sci.* 94, 1.

Stern, S. A., Mi, Y., and Yamamoto, H. (1989). Structure/permeability relationships of polyimide membranes. Applications to the separation of gas mixture. *J. Polym. Sci. Part B Polym. Phys.* 27, 1887.

Tanaka, K., Kita, H., Okano, M., and Okamoto, K. (1992). Permeability and permselectivity of gases in fluorinated and non-fluorinated polyimides. *Polymer* 33, 585.

Tokuda, Y., Fujisawa, E., Okabayashi, N., Matsumiya, N., Takagi, K., Mano, H., Haraya, K., and Kamizawa, C. (1997). Development of cardo-type polyimide hollow fiber membranes for CO_2 separation. *Energy Conserv. Mgmt.* 38, S111.

CHAPTER 23

Gas Separation by Carbon Membranes

P. JASON WILLIAMS and WILLIAM J. KOROS

School of Chemical and Biomolecular Engineering, Georgia Institute of Technology, Atlanta, Georgia 30332

23.1 INTRODUCTION

The majority of gas separation membranes currently used in industry are polymeric. Polymer membranes are easily processed into a variety of forms such as asymmetric spiral-wound or hollow fibers while maintaining their separation efficiency; however, the innate separation performance of solution-processable polymers is limited. Robeson (1991) discussed the separation performance of a variety of polymeric membranes for the O_2/N_2 and CO_2/CH_4 separations and formulated the semiempirical upper-bound trade-off line between the permeability and selectivity for solution-processable polymers; see Figure 23.1. Burns and Koros (2003) have shown similar results for the C_3H_6/C_3H_8 separation. Freeman (1999) has also addressed the limitations of polymers in terms of the trade-off line. The logical extension of these earlier analyses is that new membrane materials that exceed this trade-off limit must be developed.

Zeolite and other membranes are of particular interest for gas separation due to their potentially excellent separation efficiency and stability. Carbon membranes have the greatest potential of these nontraditional materials because of the relative ease of their formation.

Several reviews of carbon molecular sieve (CMS) membranes have been presented (Ismail and David, 2001; Saufi and Ismail, 2004). This chapter is a broad overview and considers the modes of transport in carbon membranes, the formation processes used for fabrication, and the separation performance of several membranes, in particular membranes produced in the last 10 years. In addition, emerging efforts to produce industrial-scale carbon membranes will also be reviewed.

23.2 STRUCTURE OF CARBON MEMBRANES

The most common technique used to form carbon membranes is thermal decomposition of polymer precursors. When a polymer is thermally decomposed, it can either form coke or char. Coke can be heated further to form a graphitic structure while char remains in

Advanced Membrane Technology and Applications. Edited by Norman N. Li, Anthony G. Fane, W. S. Winston Ho, and T. Matsuura
Copyright © 2008 John Wiley & Sons, Inc.

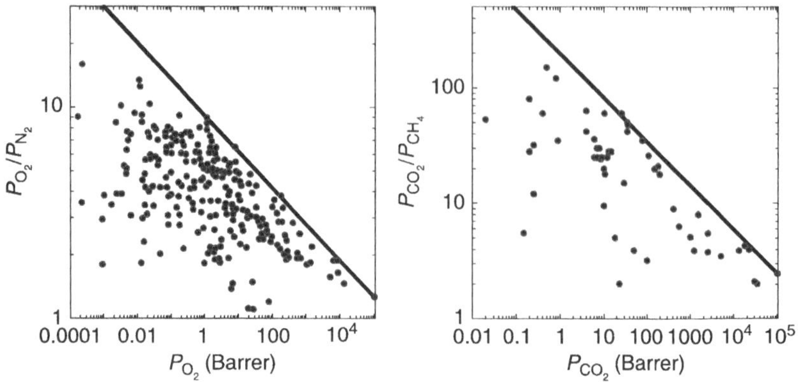

Figure 23.1 Upper-bound curve for O_2/N_2 and CO_2/CH_4 separation. The data used for these plots were obtained from Robeson (1991).

amorphous form (Pierson, 1993). At early stages of decomposition, both cokes and chars have very little long-range order and are composed of aggregates of small crystallites. Therefore, polymers that form coke and char can be used to form CMS membranes in the early stages of decomposition, though polymers that form char are more often used.

Carbon membranes used for gas separation usually have a turbostratic structure (Pierson, 1993, Fig. 23.2a) with very little long-range order and are considered essentially isotropic. Figure 23.2a illustrates that there are also lattice vacancies in the sp^2 hybridized carbon sheets, and pores are formed from packing imperfections between microcrystalline regions in the material, as shown in Figure 23.2b. Therefore, the pore structure in carbon

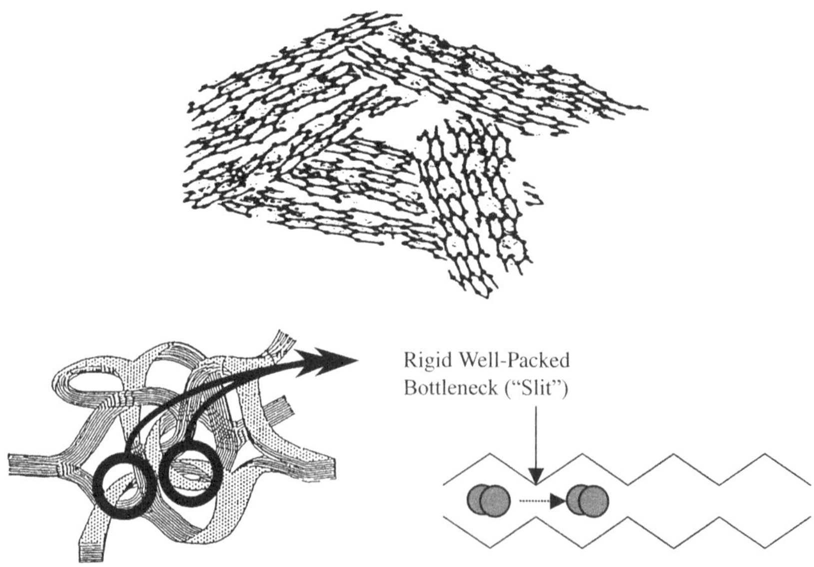

Figure 23.2 (a) Structure of turbostratic carbon (Pierson, 1993). (b) Slitlike pores formed from the disordered packing of microcrystallite regions in turbostratic carbon. This model was provided by Steel (2000).

membranes is said to be slitlike, and an "idealized" cartoon can be seen in Figure 23.2b (Steel, 2000). From the diagram, it is apparent that the pore structure is a function of many parameters including the size of the microcrystalline regions and the degree of order or packing in the material. Each of these factors is related to the polymer precursor and the processing conditions.

23.3 TRANSPORT IN CARBON MEMBRANES

Gas transport can occur in porous membranes by four different idealized mechanisms: Knudsen diffusion, partial condensation/diffusion, selective adsorption/diffusion, and molecular sieving (Rao and Sirkar, 1993a,b). Knudsen diffusion occurs when the mean free path of the molecule is greater than the size of the pore; therefore, the diffusing gas molecule collides more often with the pore wall than with other molecules. Knudsen diffusion can be described using Eq. (23.1) (Hines and Maddox, 1985):

$$D_{A,K} = 97.0r\left(\frac{T}{M_A}\right)^{1/2} \tag{23.1}$$

where $D_{A,K}$ is the Knudsen diffusion coefficient (m^2/s), r is the pore radius (m), T is the temperature (K), and M_A is the molecular weight of component A. Therefore the selectivity of a gas pair based on Knudsen diffusion is

$$\alpha_{A/B,K} = \frac{D_{A,K}}{D_{B,K}} = \left(\frac{M_B}{M_A}\right)^{1/2} \tag{23.2}$$

For gas pairs of similar molecular weight, the selectivity based on Knudsen diffusion is very low.

Selective condensation occurs when one or more of the components in the feed stream condenses in the pores of the membrane (Rao and Sircar, 1993b). Once the pores are filled, the condensed material diffuses across the membrane. The pore size required for selective condensation is generally larger than 30 Å. Very high selectivity can be achieved if one of the components is noncondensible (Sakata and Yamamoto, 1986), but the separation is limited by the condensation partial pressure of the components, which is a function of pore size and shape. Therefore, this mechanism is difficult to control and is rarely used in carbon membranes.

Selective adsorption occurs when one or more components of the feed are preferentially adsorbed into the membrane, followed by diffusion of the adsorbed species across the membrane from sorbed site to sorbed site. The separation efficiency for this mechanism is a function of the physiochemical nature of the pore surface and the pore size. Exclusion of the rejected penetrant from sorption sites in pores is critical for successful application of such materials (Rao and Sircar, 1993b). The driving force for mass transfer across these membranes is the difference in adsorbed phase concentration of the diffusing species. Therefore, only low partial pressure driving forces are needed for strongly adsorbing components. Rao and co-workers (Golden et al., 1998; Naheiri et al., 1997; Rao and Sircar, 1993a,b; Sircar et al., 1999) have developed membranes that rely on this mechanism called Selective Surface Flow (SSF) membranes. High selectivities can be achieved using this method for separation involving highly adsorbing species from low sorbing conponents. Of particular interest has been the separation of low sorbing hydrogen from highly sorbing hydrocarbon streams. Details about this class of membranes will be given later in this review.

Molecular sieving describes the highly restrictive diffusion of a molecule through constrictions that are of similar size to the molecule. For a penetrant to diffuse through the constriction, the penetrant must acquire an activation energy to overcome the repulsion from the walls of the constricted pore. Therefore, smaller penetrants require smaller activation energies and diffuse through much faster than larger penetrants. High selectivities can be obtained using the molecular sieving mechanism, especially when the pore size is very close to the size of the desired penetrant molecule. Molecular sieving is the primary method of transport through carbon membranes used for gas separation and therefore will be discussed in more detail.

Steel (2000) and Steel and Koros (2003) have used a model represented by the cartoon in Figure 23.3 to describe the pore structure in CMS membranes. According to this model, large pores (>5 Å), characterized by dimension d_{tv} are connected by smaller constrictions (<5 Å) characterized by dimension d_c. These two types of pores were dubbed micropores and ultramicropores (Soffer et al., 1987). Diffusion through molecular sieves occurs when molecules adsorbed in micropores obtain the required activation energy to overcome the repulsive forces caused by the walls of the ultramicropores. The diffusive jump length in CMS membranes is characterized by dimension d_λ. Unlike polymeric membranes, which have chain mobility, the structure of CMS membranes are rigid, and therefore the jump length is simply the distance between sorption sites. Diffusion in molecular sieves is an activated process and can therefore be written as

$$D_A = D_{AO} \exp\left(\frac{-E_D}{RT}\right) \tag{23.3}$$

where D_{AO} is the preexponential term, E_D is the activation energy of diffusion, R is the ideal gas constant, and T is the absolute temperature. The activation energy of diffusion depends primarily on the size of the ultramicropores and the size of the penetrant.

Singh (1997) and Singh and Koros (1996) have discussed the importance of *entropic selectivity* in molecular sieving materials such as CMS membranes. These authors analyzed the activation energy of diffusion for oxygen and nitrogen in an upper-bound

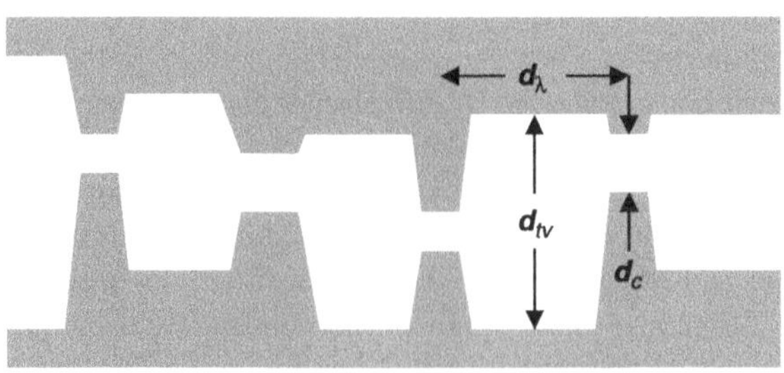

Figure 23.3 Hypothetical model of the pore structure in carbon membranes. (Adapted from Steel, 2000; Steel and Koros, 2003; Singh, 1997; Singh and Koros, 1996; Singh-Ghosal and Koros, 2000.)

polypyrrolone, zeolite 4A, and a CMS membrane. The results showed that the activation energy of diffusion for both O_2 and N_2 was similar for all three materials, while the substantial difference in diffusion coefficients arose from the preexponential factor. From transition state theory, the preexponential factor can be defined as (Stannent, 1968)

$$D_{AO} = e\lambda^2 \frac{kT}{h} \exp\left(\frac{S_D}{R}\right) \tag{23.4}$$

where λ is the average diffusive jump length, S_D is the activation entropy of diffusion, k is the Boltzmann constant, h is Planck's constant, and e is the base of the natural log. The jump length is a strong function of the structure between sorption sites and a weak function of penetrant size. The diffusion coefficient can also be written as (Gladstone et al., 1941)

$$D = \lambda^2 \frac{kT}{h} \frac{F}{F^+} \exp\left(\frac{-E_D}{R}\right) \tag{23.5}$$

where F is the partition function in the normal state and F^+ represents the partition function for the same gas molecule in the transition state. In the case of molecular sieves, the normal sorbed state pertains to the large pores and the transition state pertains to the ultramicropores. The total partition function in either state can be expressed as

$$F = F_{\text{trans}} F_{\text{rot}} F_{\text{vib}} \tag{23.6}$$

If one combines Eqs. (23.4) and (23.5), it is clear that

$$\exp\left(\frac{S_{D,A} - S_{D,B}}{R}\right) = \frac{(F^+/F)_A}{(F^+/F)_B} \tag{23.7}$$

For molecules that have different dimensions, pores can inhibit rotational freedom for one molecule, while allowing it for other molecules. For example, nitrogen can be approximated as spherocylindrical with a length of ~4.07 Å and diameter of ~3.09 Å (Singh, 1997).

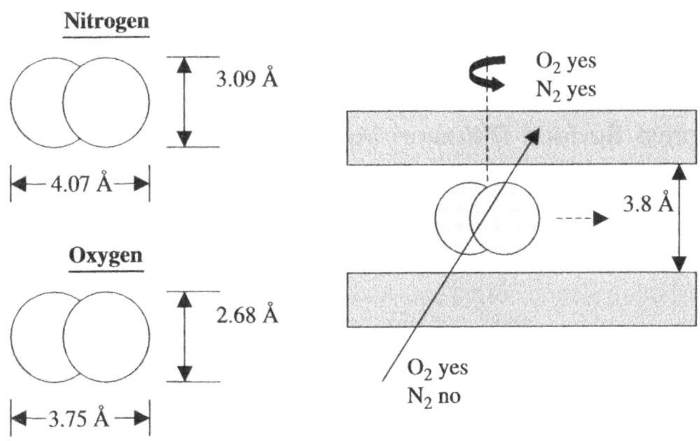

Figure 23.4 View of effect of small pore size on rotational degrees of freedom in CMS pores. Pore size data was calculated by Singh (1997).

Oxygen has a similar shape with diameter of ~2.68 Å and a length of ~3.65 Å. Therefore, if an ultramicropore has a dimension of 3.8 Å, oxygen is allowed to freely rotate, while nitrogen loses rotational freedom, leading to entropic selectivity, as seen in Figure 23.4. Singh (1997) and Singh and Koros (1996) presented a more detailed analysis of this concept and determined the range of entropic selectivity for a 3.8-Å pore varies from 3.7 to 9 for slit-shaped CMS pores. Therefore, entropic selectivity can play a major role in the enhanced separation performance of CMS membranes.

23.4 FORMATION OF CARBON MEMBRANES

Carbon membranes are generally produced from thermal decomposition (pyrolysis) of polymer precursors. Like polymer membranes, a variety of different formats have been produced. The formats can be generalized into two groups, supported and unsupported. Supported membranes are formed by casting a thin polymer film on a surface or wetting a surface with an organic liquid. The supports are generally flat-sheet or tubular porous inorganic materials. Unsupported membranes have been produced in primarily three forms: (1) dense, flat sheet, (2) asymmetric hollow fiber, and (3) asymmetric film.

As mentioned previously, the gas separation performance of a carbon membrane is controlled by the pore size distribution in the material. Several factors have been identified that affect the pore size distribution:

- Polymer precursor composition
- Pyrolysis temperature and ramp rate
- Thermal soak time at maximum pyrolysis temperature
- Pyrolysis atmosphere
- Posttreatment conditions

There are other minor factors that also play a role in the separation performance of carbon membranes, but the factors listed here have been identified as the most significant. In the next sections, these factors will be analyzed in more detail. The syntheses of selective surface flow and CMS membranes follow very similar procedures. First, a review of SSF membranes will be given and then the remainder of the review will focus on CMS membranes.

23.4.1 Selective Surface Diffusion Membranes

Rao, Sircar, and co-workers at Air Products and Chemicals, Inc., have pioneered the production of a specialty type of selective surface diffusion (SSD) membrane, which they refer to as SSF carbon membranes. The development and performance of these membranes have been described in a series of patents and articles (Anand et al., 1995, 1997; Paranjape et al., 1998; Rao and Sircar, 1993a,b, 1996; Rao et al., 1992, 1995a,b; Sircar et al., 1999; Zhou et al., 2003). These membranes usually consist of a thin nanoporous carbon layer (5–7 Å) on the surface of a macroporous support. The mechanism of gas transport through SSF membranes is illustrated in Figure 23.5. The more adsorbable component in a feed stream on the high-pressure side of the membrane selectively adsorbs onto the surface, diffuses across the surface of the membrane to the low-pressure side, and

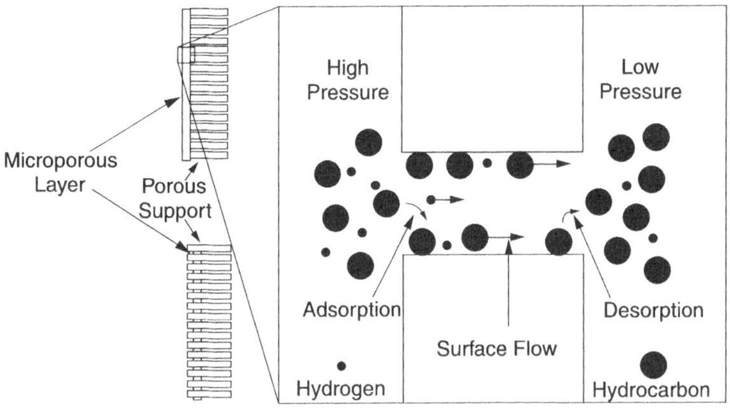

Figure 23.5 Concept of gas transport by SSF membranes. Diagram presented by Sircar et al. (1999).

subsequently desorbs. If the pore size of the selective carbon layer is tuned correctly, then the diffusion of nonadsorbable components in the mixture will be hindered by "plugging" due to the sorption of the desired components. The authors have listed several advantages of SSF membranes.

1. The desired product of effluent streams is often the small, nonadsorbable component such as hydrogen. SSF membranes hinder diffusion of this component and therefore produce streams enriched in the lighter components at higher pressure, reducing recompression costs.
2. The driving force for transport of a component in a gas mixture through SSF membranes is determined by the specific adsorbate loading gradient across the membrane. A large absorbate loading of the selectively adsorbed component can be achieved at a relatively low feed pressure.
3. The energy barrier for surface diffusion is relatively low compared to transport through polymeric membranes, and therefore high flux can be achieved, eliminating the need for very thin membranes.
4. Selectivity is governed by the equilibrium adsorption selectivity at the high-pressure side, the surface diffusion rate of the adsorbed components, and hindered diffusion of the nonadsorbable components through the void space in the pores. Therefore, very high selectivities can be obtained for the preferentially adsorbed component.
5. The adsorptive properties of the SSF membranes can be easily altered by controlling the pore size distribution and surface characteristics.
6. Adsorptive properties increase with decreasing temperature so the membranes can generally be operated near ambient temperature and achieve high selectivity.

Of course, the limitations of this type of membrane are fairly obvious. The membranes are only effective for separating components with substantially different adsorption characteristics. The sorption properties will also be strongly dependent on the feed stream composition. For example, if a small amount of a strongly adsorbing component is present in the feed stream, the adsorption–diffusion of that component will govern the separation ability of the membrane and could cause plugging and shutdown to other desired components. Nevertheless, these membranes have excellent separation performance for some separations, which will be discussed in detail below.

23.4.1.1 Formation of SSF Membranes Rao and co-workers have disclosed several methods for forming supported SSF membranes (Rao and Sircar, 1993a,b; Rao et al., 1993, 1995b; Sircar et al., 1999), but only the preferred method will be discussed. First, a polymer solution or a polymer latex is coated onto the surface of a macroporous support. The support should have a pore size between 0.2 and 50 μm such as macroporous graphite disks (pore size ~0.7 μm) and alumina tubes (pore size <1 μm). A number of polymer precursors for coating have been recommended including polyvinylidene dichloride, polyvinyl chloride, polyacrylonitrile, styrene–divinyl benzene copolymer, and others. The latex is applied to the surface of the support by brushing, spraying, or immersion. In the second step of the process, the latex is heated in an inert environment at a heating rate of 1°C/min to 600–1200°C and subsequently cooled to room temperature at 10°C/min. This process can then be repeated several times to form multiple carbon layers on the surface of the support. Alternate methods include introducing a polymer solution into the pores of a porous support and subsequently pyrolyzing the composite membrane. Oxidative posttreatment (~400°C) has also been used to increase the permeability and, in some cases, the selectivity of the SSF membranes by increasing the average pore size.

23.4.1.2 Examples and Applications An example of the above approaches has been demonstrated for hydrogen separation. An SSF membrane (Sircar et al., 1999) was formed by coating a thin layer of polyvinylidene chloride–acrylate terpolymer latex containing 0.1–0.14 mm polymer beads in an aqueous emulsion on the bore side of a macroporous alumina tube (0.56 cm internal diameter, 0.16 cm wall thickness). The latex was dried at 50°C under N_2 and subsequently heated to 600°C followed by passivation by heating to 200–300°C in an oxidizing atmosphere. The resulting carbon membrane has a pore size of 5–7 Å and a thickness of 2–3 μm.

It is difficult to present the separation performance of SSF membranes in traditional permeability and selectivity format. The separation performance is highly dependent on the feed gas composition and the flow rates on the upstream and downstream side of the membranes. Therefore, it is more helpful to consider applications in which SSF membranes have been illustrated. One such application is the recovery of hydrogen from refinery waste gases (Sircar et al., 1999). The composition of a gas stream from a fluid catalytic cracker off-gas is typically 20% H_2, 20% CH_4, 8% C_2H_6, 29% C_3H_6, and 15% C_3H_8. This stream is generally combusted to recover the heating value. SSF membranes can be used to increase the hydrogen content of the waste stream so that it can be further purified by a pressure swing adsorption (PSA) system and therefore produce high-purity hydrogen (99.99+ mol%). The main advantage of SSF membranes over polymeric membranes for this application is that the enhanced hydrogen stream remains at relatively high pressure and lowers recompression costs. For example, in the stream mentioned above, with a hydrogen recovery of 60%, the high-pressure effluent from the gas stream has a composition of 51.6% H_2, 41.5% CH_4, 2.0% C_2H_6, 2.3% C_3H_6, and 1.5% C_3H_8. Scale-up of this membrane to accommodate a refinery stream has proven successful. Other SSF membrane applications include the separation of H_2 from CO_2 and/or CH_4, H_2 or CH_4 from H_2S, CO_2 from N_2, etc (Sircar et al., 1999).

23.4.2 Carbon Molecular Sieve Membranes

23.4.2.1 Polymer Precursor Composition and Morphology A variety of thermosetting polymers have been used as precursors to produce CMS membranes. Koresh and Soffer (1983) produced the first hollow-fiber CMS membranes by pyrolyzing cellulose hollow fibers. They showed that polymers suitable as precursors for CMS membranes should not flow before they decompose. Several polymers have been used to produce CMS membranes including polyacrylonitrile (David and Ismail, 2003), phenolic resin (Fuertes and Menendez, 2002; Shiflett and Foley, 1999; Shusen et al., 1996), polyfurfuryl alcohol (Acharya and Foley, 1999; Acharya et al., 1997; Shiflett and Foley, 1999, 2000, 2001), poly(vinylidene)-based polymers (Centeno and Fuertes, 2000), cellulose derivatives (Koresh and Soffer, 1983), and polyimides (Barsena et al., 2004; Hatori et al., 1992a,b, 1996; Jones and Koros, 1994a,b; Park et al., 2004; Shao et al., 2004, 2005; Steel and Koros, 2003; Tin et al., 2004c). Polyimides have been the precursor of choice by many researchers likely due to their high T_g, ease of processability, and good separation performance as polymeric membranes. Jones and Koros (1994a,b) have suggested that (based on separation performance and mechanical properties) the best precursors currently available for the production of carbon membranes are polyimides. Table 23.1 lists the precursor and the format of several CMS membranes that have been produced. Figure 23.6 shows the structure of most of the noncommercial polyimides listed in Table 23.1. The majority of the discussion in this review will be on polyimide-based materials because of the breadth of data available in the literature, but initially, alternative precursors will be discussed. There are hundreds of studies on CMS membranes in the literature; therefore, those mentioned in this review are provided to highlight some of the better CMS membranes that have been produced.

Polyfurfuryl Alcohol Polyfurfuryl alcohol (PFA) has been used extensively in the formation of CMS membranes. Because PFA is a liquid at room temperature, all membranes formed from this material have been supported. Chen and Yang (1994) coated PFA onto macroporous graphite supports followed by pyrolysis at 500°C. This process was repeated until CMS layers of 15 μm were formed. The diffusivity ratio for CH_4/C_2H_6 using this process was about 10.

Acharya et al. (1997) have produced CMS membranes by first coating a flat stainless steel support with a solution of PFA followed by subsequent drying. CMS membranes with a selectivity of 2–3 for O_2/N_2 were produced upon subsequent pyrolysis. In another study, Acharya and Foley (1999) used a spray coating technique to coat the stainless steel support. The selectivity of the spray-coated membranes was considerably higher (~4 for O_2/N_2), likely due to the presence of less defects due to improved coating. Shiflett and Foley (1999, 2000, 2001) and Acharya and Foley (1999) have used ultrasonic deposition to coat sintered stainless steel tubes with PFA. Selectivities of up to 30 for O_2/N_2 have been achieved by pyrolyzing and then depositing additional PFA layers followed by additional pyrolysis. These authors have also investigated the use of additives such as TiO_2, silica zeolites, and polyethylene glycol and investigated the effects on separation performance, primarily in an attempt to increase the permeance.

Wang et al. (2000) have used vapor deposition polymerization to coat furfuryl alcohol (FA) onto γ-Al_2O_3/α-Al_2O_3 and glass/α-Al_2O_3 tubes. The tubes were first impregnated with paratoluene sulfonic acid, and then the inside diameter was exposed to FA vapor at 90°C for 15 min, followed by heating to 200°C to polymerize and cross-link. The tubes

TABLE 23.1 Sampling of Carbon Membranes Formed Including the Precursor Material and Morphology

Polymer	Morphology	Reference
Matrimid	Supported film	Fuertes et al. (1999)
Matrimid	Film	Kim et al. (2004)
Matrimid	Film	Steel (2000)
Kapton	Supported film	Fuertes et al. (1999)
Kapton	Film	Suda and Haraya (1995, 1997a,b); Steel (2000); Wang et al. (2003)
Kapton	Asymmetric fiber	Haraya et al. (1995); Peterson et al. (1997)
Kapton	Film	Hatori et al. (2004)
Polyvinylidene chloride-acrylate terpolymer latex	Supported film	Rao and Sircar (1993a)
Phenolic resin	Supported tube	Centeno and Fuertes (2001); Fuertes and Menendez (2002)
6FDA/BPDA-DAM	Film	Singh (1997); Centeno and Fuertes (2000); Singh-Ghosal and Koros (2007)
Poly(vinylidene-chloride-co-vinyl chloride)	Supported film	Centeno and Fuertes (2000)
Phenol-formaldehyde Novolac resin	Asymmetric tubular	Wei et al. (2002)
Sulfonated phenolic resin	Tubular, supported	Centeno and Fuertes (2000); Zhou et al. (2001, 2003)
BPDA/6FDA-TrMPD	Asymmetric fiber	Jones and Koros (1994a,b; 1995a,b); Wei et al. (2002)
BPDA/6FDA-TrMPD	Asymmetric fiber	Geiszler and Koros (1996); Geiszler (1997)
BPDA-DDBT/DABA/CF_3 containing diamine	Asymmetric hollow fiber	Okamoto et al. (1999)
Polyfurfuryl alcohol	Tubular, supported	Shiflet and Foley (1999, 2000, 2001)
Polyfurfuryl alcohol	Supported films	Acharya et al. (1997); Achaya and Foky (1999)
Polyfurfuryl alcohol	Tubular, supported	Wang et al. (2000)
Polyfurfuryl alcohol	Supported films	Chen and Yang (1994)
Phenol-formaldehyde resin	Asymmetric film	Shusen et al. (1996)
Phenolic resin (Novalak type)	Supported films	Centeno and Fuertes (1999)
BPDA-ODA	Tubular, supported	Hayashi et al. (1995, 1996)
P84 polyimide	Film	Tin et al. (2004a,b)
6FDA-durene polyimide	Film	Shao et al. (2005)
6FDA/PMDA(1/6)-TMMDA	Film	Shao et al. (2004)
BTDA-mPDA/BDSA (substituted with Li, Na, or K)	Film	Kim et al. (2003)
NTDA-BAHFDS (sulfonated PI)	Film	Islam et al. (2005)
NTDA-BDSA (sulfonated PI)	Film	Islam et al. (2005)
NTDA-ODADS (sulfonated PI)	Film	Islam et al. (2005)

(*Continued*)

TABLE 23.1 *Continued*

Polymer	Morphology	Reference
NTDA-BAHFDS/BAPF(4/1) (sulfonated PI)	Film	Islam et al. (2005)
NTDA-BDSA/BAPF(4/1) (sulfonated PI)	Film	Islam et al. (2005)
$CH_3COCH_3 + CH_4$ by ICPCVD	Supported film	Wang and Hong (2005b)
Hexamethyldisiloxane by ICPCVD	Supported film	Wang and Hong (2005a,b,c)
BTDA-ODA	Film	Kim et al. (2005b)
BTDA-ODA-polyvinylpyrrolidone blends	Film	Kim et al. (2004, 2005a)
BTDA-ODA/m-PDA(9/1)	Film	Park et al. (2004)
BTDA-ODA/2,4-DAT(9/1)	Film	Park et al. (2004)
BTDA-ODA/m-TMPD(9/1)	Film	Park et al. (2004)

were then heated to carbonize the polymer precursor, and a second polymer coating was added and subsequently carbonized. Using this technique, selectivities of 10–12 were obtained for O_2/N_2 with permeances of around 8×10^{-10} (mol m^{-1} Pa^{-1} s^{-1}). All of these studies show the potential for PFA-based CMS membranes, the primary limitations being that the membranes must be supported and the permeance needs to be increased to produce attractive properties. Shiflett and Foley (2001) have also suggested that the formation of PFA-based CMS membranes is very sensitive to processing conditions. This issue will be discussed in greater detail in subsequent sections.

Phenolic Resin Phenolic resin has also been used to produce CMS membranes. The advantage of phenolic resin over other precursors such as polyimides is low cost. Shusen et al. (1996) have produced asymmetric phenol formaldehyde resin-based CMS membranes. A flat film (0.05–0.10 mm) was formed by thermopressure molding. The membrane was subsequently heated to 850–900°C in nitrogen followed by 0.5–2.0% oxygen. One side was covered by a porous ceramic sheet during this process, and, therefore, the oxidation did not occur symmetrically. The resulting membrane had a reported permeability of 2300 Barrers and an O_2/N_2 selectivity of 10.65.

Fuertes and Menendez (2002) and Centeno and Fuertes (1999, 2001) have published a series of studies using this precursor. Centeno and Fuertes (1999) have spin coated a small amount of a novolak-type phenolic resin on the surface of carbon supports. The membranes were then carbonized in a tubular furnace from 500 to 1000°C in vacuum. The resulting membranes had O_2/N_2 selectivity of around 10 and CO_2/CH_4 selectivity of ~160. This work was later extended and in that case (Centeno and Fuertes, 2001; Fuertes and Menendez, 2002) a novalak-type phenolic resin was deposited on the inner face of a ceramic tubular membrane used for ultrafiltration. The membrane was subsequently pyrolyzed to 700°C. In some cases an oxidative pretreatment was used before pyrolysis or an oxidative posttreatment after pyrolysis. The resulting membranes had O_2 permeabilities around 100 Barrers and O_2/N_2 selectivities around 12 at 25°C. Films dip coated with resin three times had lower permeability and only slightly higher selectivities than those dipped only once. For hydrocarbon mixtures, the separation performance was increased by several treatments: air oxidation of the resin, air oxidation of the carbon, or chemical vapor deposition (CVD) posttreatment of the carbon.

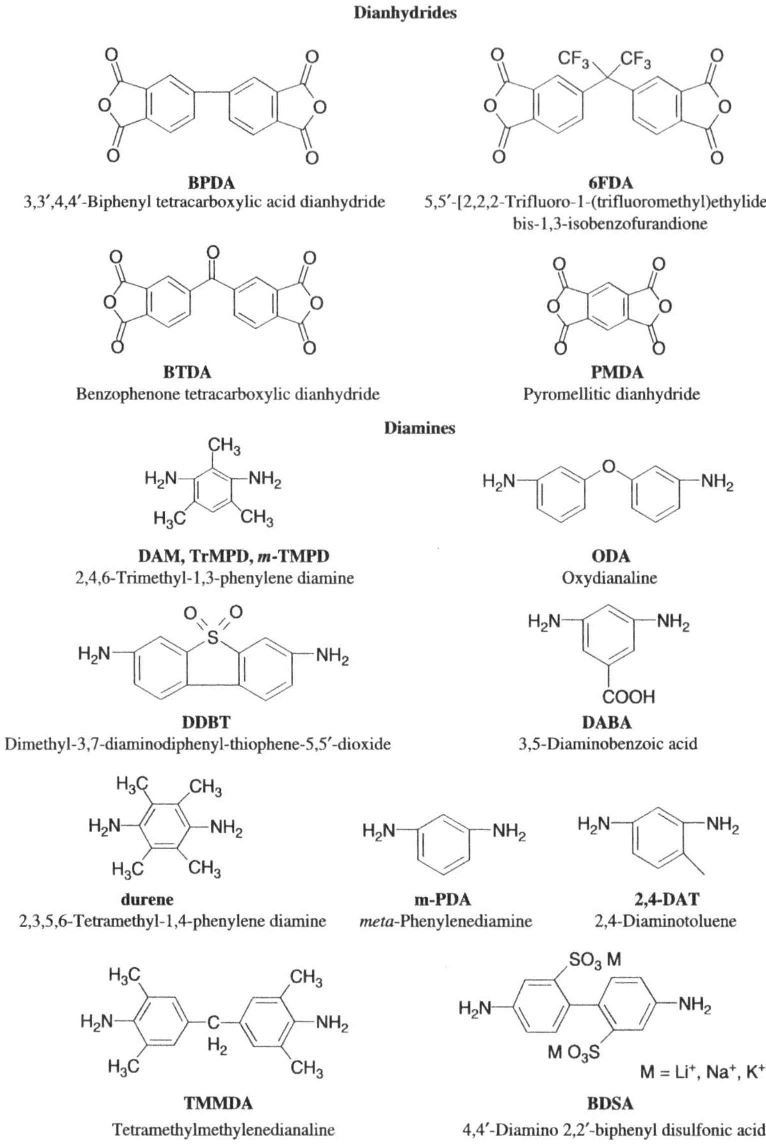

Figure 23.6 Diamines and dianhydrides used to produce many of the polyimide in Table 23.1.

Wei et al. (2002) have dip coated phenol–formaldehyde novolak resin containing a small amount of hexamethylene tetramine onto the surface of green tubular supports of fine phenol–formaldehyde novalac resin particles. During pyrolysis, the hexamethylamine has mobility that reduces the creation of large pores in the surface during pyrolysis. Because the selective layer and support have similar composition, there is also less stress during pyrolysis from shrinkage. The supported membranes were pyrolyzed at 900°C. The concentration of the solution used for the selective layer and the thickness had a considerable effect on the separation performance.

Zhou et al. (2001, 2003) dip coated resol-type phenolic resin and a novalak-type sulfonated phenolic resin/phenolic resin mixture onto porous α-alumina tubes. The membranes were then pyrolyzed at various temperatures in a nitrogen purge. As expected, the separation performance depended on pyrolysis temperature. The sulfonated resin had much better separation performance with both high O_2 permeance (240 GPU* for 500°C CMS) and reasonably attractive selectivity ($O_2/N_2 = 5.2$).

Polyimides Polyimides have become precursors of choice for the production of CMS membranes. As mentioned previously, several factors such as thermal stability and rigidity make polyimides attractive as CMS precursors, in addition to the attractive separation performance of these materials as polymeric membranes. Of particular interest have been commercially available polyimides, which will be the focus of this review. Several noncommercial polyimides have also been used to produce CMS membranes. These are listed in Table 23.1 and the corresponding monomers used to produce these polymers can be found in Figure 23.6.

KAPTON Kapton is a commercially available polyimide made by Dupont. Noncommercially synthesized materials with a similar structure to Kapton (but not the biaxial orientation present in Kapton) are often referred to as PMDA-ODA, since this denotes the monomers used to produce the polyimide (see Fig. 23.6). Kapton has been used extensively to produce CMS membranes (Fuertes et al., 1999; Hatori et al., 2004; Peterson et al., 1997; Suda and Haraya, 1995, 1997a,b). Suda and Haraya (1995, 1997a) produced dense flat CMS membranes from Kapton by pyrolyzing dense film between blocks of graphite at temperature between 773 and 1273 K under a 10^{-5} -Torr vacuum. The CMS film properties were dependent on pyrolysis temperature but had both high permeability and selectivity. He/N_2 selectivities ranged from 6 to 498 ($P_{He} \sim 500-60$ Barrers) while O_2/N_2 selectivities ranged from 3 to 14 ($P_{O_2} \sim 300-1$ Barrers). Suda and Haraya (1997b) also show the pore size distribution can be enlarged with treatment under the presence of water vapor at 623 K. This process will be explained further in a subsequent section. In a later investigation, Suda and Haraya (1997a) perform an extensive study on the formation of CMS membranes from commercially available Kapton film because of the reproducibility of the precursor. The authors investigated a variety of factors such as heating rate, the pyrolysis temperature, and the pyrolysis atmosphere. The results show that for membranes pyrolyzed at 1223 K, permeabilities for O_2, CO_2, and He ranged from 0.92 to 0.15, 3.54 to 0.50, and 26.4 to 11.6 Barrers with selectivities for O_2/N_2, CO_2/N_2, and He/N_2 ranging from 21.6 to 36, 82.7 to 122, and 618 to 2810, respectively, according to the heating rate. The authors also present activation energies for permeation, diffusion, and sorption for a variety of different gases. The results suggest that Kapton is a very good precursor for the production of highly selective CMS membranes.

Peterson et al. (1997) and Suda and Haraya (1995) have also prepared asymmetric capillary PMDA-ODA membranes by first coating a polytetrafluoroethylene (PTFE) tube with polyamic acid, then immersing the capillary in a water and/or ethanol bath, followed by drying and imidization. The PTFE capillary was removed and the asymmetric membranes pyrolyzed at 1223 K. The selectivities for these membranes approached or exceeded

*GPU = gas permeation unit. "GPU" refers to a pressure normalized transmembrane flux, and 1 GPU = 10^{-6} [cc(STP)/(cm^2 cm Hg sec)].

the dense film values when thicker selective layers were formed. In a later study, Peterson et al. (1997) coated the capillary membranes with a thin layer of polydimethlysiloxane, which hindered gas permeation through the defects. The selectivity for He/N_2 reached ~2800. Fuertes et al. (1999) have used a similar technique to produce asymmetric PMDA-ODA CMS membranes supported on a carbon disk. In that study, the authors spin coated a polyamic acid solution onto the carbon disks followed by submersion into a water coagulant bath. The films were then dried, imidized, and subsequently pyrolyzed at 550°C. The resulting asymmetric membranes had an O_2/N_2 selectivity of around 4 with a permeability of 45 Barrers.

Recently, Hatori and co-workers (2004) formed CMS membranes from Kapton film by pyrolyzing in an argon purge from 1173 to 1373 K. The goal was to produce highly selective membranes to remove CO from hydrogen for fuel cell applications. The resulting membranes had selectivities from 200 to 5900 for H_2/CO. Therefore, the study showed that CMS membranes created from Kapton have promising results for hydrogen purification.

MATRIMID 5218 Matrimid 5218, sometimes referred to as BTDA-diaminophenylidene (DAPI), has also been used extensively to produce CMS membranes of varying formats. This polymer is available through Ciba Specialty Chemicals. Fuertes et al. (1999) produced asymmetric CMS membranes from Matrimid supported on carbon disks by a similar method to that used for their PMDA-ODA membranes. Matrimid was dissolved in n-methyl pyrrolidone (NMP) or γ-butyrolactone, spin coated onto carbon disks, and then placed in a coagulant bath that was composed of water or 50/50 mixtures of water–methanol or water–propanol. The membranes were pyrolyzed at temperatures from 450 to 650°C. The structure of the membranes was a strong function of coagulant composition, as expected. The O_2 permeability ranged from 1 to 7 Barrers while the O_2/N_2 selectivity ranged from 5.9 to 2.9 Barrers.

Steel and Koros (2003) produced dense freestanding CMS films from Matrimid. Dense films with thicknesses between 35 and 60 μm were cast from a 2% solution of Matrimid in methylene chloride. The films were then pyrolyzed at 550 and 800°C in a vacuum with a maximum pressure of 0.03 Torr. The CMS membranes had an O_2 permeability ranging from ~300 to 10 Barrers with an O_2/N_2 selectivity of ~6–14.

Several other researchers have also produced CMS membranes from Matrimid. Vu (2001) and Vu et al. (2002, 2003) have produced asymmetric CMS fibers for CO_2/CH_4 separation. Tin et al. (2004a,b) have produced dense flat CMS membranes from Matrimid for a variety of separations and investigated the effect of modification of the precursor on CMS performance. These studies will be discussed in more detail later.

23.4.2.2 Pretreatment of Polymer Precursor

Oxidation Several pretreatments have been used to condition CMS precursors before pyrolysis. The most common pretreatment is preoxidation. Preoxidation can form cross-links in the polymer structure and therefore increase the thermal stability of the precursor. David and Ismail (2003) have shown that the thermal stability of polyacrylonitrile (PAN) hollow-fiber membranes is improved when the precursors were heated to 250°C in air or oxygen for 30 min. The results suggest that stabilization in an inert atmosphere can cyclize PAN, while stabilization in an oxidative environment both cyclizes and oxidizes the structure.

Kusuki and co-workers (1997) and Okamoto et al. (1999) formed CMS membranes from asymmetric polyimide hollow fibers formed from BPDA and aromatic diamines. The authors pretreated the fibers by heating to 400°C in atmospheric air for 30 min. The pretreatment was necessary to maintain the asymmetric structure of the precursors. Otherwise, the precursor softened and the carbon membranes had low separation performance.

Centeno and Fuertes (2000) analyzed the effect of oxidative pretreatment on separation performance of supported poly(vinylidene chloride-*co*-vinyl chloride)-based CMS membranes. The films were oxidized in air at 150 or 200°C for up to 2.5 days before carbonization. Membranes oxidized at 200°C for 6 hours showed a decreased permeability but increase in selectivity, while those treated at 150°C for 2.5 days had increased permeability and decreased selectivity. Therefore, the results show that oxidative pretreatment conditions must be optimized for a given precursor but may improve the CMS membrane properties.

NONSOLVENT PRETREATMENT AND CROSS-LINKING Tin et al. (2004) have introduced two pretreatment methods that can be used to alter the separation performance of CMS membranes. The first treatment involves cross-linking of the polymer before pyrolysis. Dense films of Matrimid 5218 were immersed in a 10% solution of *p*-xylenediamine in methanol at room temperature from 1 to 7 days. The films were then washed with pure methanol and dried in air at room temperature. CMS membranes were then formed by vacuum pyrolysis at 800°C. The permeabilies of the CMS films decrease with increasing cross-linking time, which is likely attributed to a reduction in free volume and chain mobility for increased cross-link density. The selectivity of the membranes also decreased, and therefore it may not generally be a good method to produce high-performance CMS membranes, though different results may be obtained with different polyimide precursors.

Tin et al. (2004a,b) also recognized that the selectivity of one-day immersed films was slightly better than those that were not pretreated. Several studies were then performed to determine the effect of nonsolvent pretreatment on separation performance. Matrimid and polyimide P84 (from Lenzing) were soaked in solutions of methanol, ethanol, propanol, or butanol for one day and air dried. CMS membranes were subsequently pyrolyzed as before. The results show that the permeability decreased but selectivity increased for all of the CMS membranes. The largest enhancement in selectivity was attained from the ethanol soak. The effect of soak time was also tested for methanol and ethanol. The results show that the permeability decreases and selectivity increases with increasing soak time. X-ray diffraction (XRD) and positron annihilation lifetime spectroscopy (PALS) data suggests that the pretreated films have higher packing (smaller pores). Tin et al. (2004a, p. 6482) suggest that the function of nonsolvent pretreatment is "to weaken the intermolecular interactions, allowing structural reorganization of the carbon chains during pyrolysis." These results suggest nonsolvent treatment may be a new method to produce higher performance CMS membranes.

23.4.2.3 Pyrolysis Process

Pyrolysis Temperature The *pyrolysis temperature* is the general term used for the highest temperature to which a precursor is heated during the pyrolysis process. Several researchers have used a series of ramps and dwell times during the decomposition. Dwell times are often used at lower temperatures (<300°C) to remove residual solvent and water. Dwell times above the glass transition temperature of the polymer are sometimes used to remove the thermal history of the polymer precursor.

The pyrolysis temperature is generally chosen to be above the decomposition point of the polymer but below the graphitization temperature; therefore, the pyrolysis temperature is strongly linked to the precursor properties. As a general rule, an increase in pyrolysis temperature (above the decomposition point) will lower the permeability of the CMS and increase the selectivity. Higher pyrolysis temperatures produce increased crystallinity, increased density, and lower average interplanar spacing in the membranes.

As an example, Shoa et al. (2004, 2005) have produced CMS membranes from the decomposition of 6FDA-durene. This polyimide decomposes at ~496°C under the conditions used in the study. The separation performance of membranes pyrolyzed at temperature from 250 to 800°C was evaluated. The results show that the permeability of the films generally increases from 325 to 475°C for CO_2, O_2, N_2, and CH_4. From 475 to 525°C, the permeability of the smaller gases remains fairly constant but decreases for the larger gases. There is another increase in permeability at 550°C and then a steady decrease to 800°C. The authors suggest the second increase in permeability at 550°C may be due to the accelerated decomposition at that point because the maximum decomposition rate occurs at 531°C. The selectivity is also greatly affected by temperature. There is a general increase in selectivity as the pyrolysis temperature is increased, except for two local minima at 450 and 550°C. Above the decomposition temperature the selectivity increases with temperature. This trend has also been observed by a variety of other researchers.

Ramp Rate Carbon molecular sieve membranes have been produced with ramp rates up to ~13.3°C/min. Increasing the ramp rate tends to increase the permeability and lower the selectivity of the membrane. Suda and Haraya (1997a) produced CMS membranes from Kapton polyimide in an argon atmosphere. Heating rates were varied from 13.3 to 1.33 K/min. The permeability of the membranes to all gases decreased with lower heating rates. The authors suggest this could be due to (1) the rate of evolution of by-products or (2) an increased pyrolysis time that allows pore sintering to occur. The first explanation agrees with the study of Hatori et al. (1996), which showed that thicker films produce different amounts of by-products due to further reaction of the by-products before they diffuse out of the membrane. Increasing the pyrolysis rate may have a similar effect as increasing the film thickness. As a general rule, most pyrolysis processes now reported in the literature have slow ramp rates ($\leq$5 K/min) to increase CMS selectivity.

Thermal Soak Time The so-called thermal soak time is the time held at the pyrolysis temperature. The thermal soak time is also used to tune the microstructure of the carbon membrane. Of course, the final properties of the material are a complex function of both the thermal soak time and the pyrolysis temperature. Several researchers have shown that increasing the thermal soak time increases the selectivity and decreases the permeability of the CMS membrane. Steel (2000) showed that increasing the soak time, when pyrolyzing at 550°C (~100°C above the decomposition temperature), had a considerable effect on permeability and selectivity, which could be attributed to both pore sintering and further decomposition. At 800°C, thermal soak time had little effect on selectivity, but permeability was decreased significantly, probably caused by sintering of the pores and narrowing of the pore size distribution. It should be noted that the thermal soak times in that study (2–8 h) were rather long in comparison to those commonly found in the literature ($\leq$2 h). In a separate study, Kim et al. (2005b) investigated shorter thermal soak times (0–60 min) for a CMS pyrolyzed at 700°C. The results showed that the oxygen permeability decreased by ~80% with a corresponding 66% increase in the selectivity of O_2/N_2 when the thermal soak time was increase from 0 to 60 min. Therefore, as mentioned previously, the effect of thermal soak time is a strong function of polymer composition and pyrolysis temperature.

Pyrolysis Atmosphere Geiszler and Koros (1996) have studied the effect of pyrolysis atmosphere on the separation performance of asymmetric CMS hollow-fiber membranes created from 6FDA/BPDA-DAM polyimide. The fibers were pyrolyzed under three different atmospheres including argon, helium, and vacuum (0.01–0.03 Torr) with gas flow rates between 20 and 200 sccm. The results showed that inert purge pyrolysis at 200 sccm resulted in much higher permeance but lower selectivity than vacuum pyrolysis for the O_2/N_2 separation (Geiszler, 1997; Geiszler and Koros, 1996). A similar phenomenon was observed by Vu (2001) for the CO_2/CH_4 separation. The CMS hollow fibers in Vu's study were created from Matrimid precursors. When Geiszler (1997) reduced the flow rate to 20 sccm, the permeance was reduced to below that of membranes formed in vacuum. The selectivity of these fibers remained unchanged within the error of the tests.

Geiszler (1997, p. 61) gave several qualitative explanations for the above phenomena. First, it was suggested that inert purge pyrolysis changes the mechanism of the carbonization reaction by increasing the gas-phase heat and mass transfer over vacuum pyrolysis. Further, "by accelerating the carbonization reaction, the inert gas molecules appeared to produce a more 'open' porous matrix in the CMS membranes, resulting in a higher permeability and less selective pore structure." Geiszler (1997) also suggested that at high flow rates (200 sccm), convective mass transport may enhance decomposition, at least to the point where the inert gases enhance the transfer of large by-products of decomposition away from the carbon membrane. At low flow rates (20 sccm), by-products may have further decomposed and deposited in the pores or on the surface of the membrane, resulting in lower permeate flux.

Suda and Haraya (1997a) have formed flat CMS films from Kapton polyimide by pyrolysis at 1000°C in argon and in vacuum. The results from this study have shown that there was little difference in the permeation properties between membranes formed in argon purge and vacuum. Therefore, it is likely that the effect of pyrolysis atmosphere may be related to the properties of the precursor and the temperature profile used during pyrolysis.

23.4.2.4 Posttreatment

Oxidation Most posttreatment processes are used to alter the pore size distribution in the CMS membrane. The most common method used to increase the pore size in CMS membranes is low-temperature oxidation. Researchers have used several different techniques. Soffer et al. (1987) treated hollow-fiber CMS membranes derived from cellulose precursors using 400°C air for 15 min to increase the permeance. Kusakabe et al. (1998) produced supported CMS membranes from BPDA-ODA polyimide. The membranes were then posttreated with oxygen at 300°C for 3 h. The posttreatment increased the permeance by an order of magnitude but had little effect on the selectivity. The authors attributed this to an increase in pore volume without broadening the pore size distribution.

Fuertes (2001) has used postoxidation to increase the pore size in CMS membranes derived from phenolic resin. The results show that oxidation at temperature from 100 to

450°C can increase the pore distribution to such an extent that the diffusion mechanism goes from molecular sieving to adsorption selective. Therefore, when the correct oxidation pretreatment is used, heavier hydrocarbons such as *n*-butane diffuse faster than nonadsorbing materials, and therefore very good selectivities are obtained. The results also show that the selectivity of adsorbable/nonadsorbable decreases with high temperature (>450°C) and oxidation times longer than a few hours. These results show the magnitude by which postoxidation can be used to effect the pore size distribution in CMS membranes.

23.5 CURRENT SEPARATION PERFORMANCE

As noted earlier, polymeric membranes currently dominate the commercial market for gas separations since they are inexpensive, can be produced as asymmetric fibers (which have high surface-to-volume ratio), and have very good mechanical properties. The drawbacks to polymer membranes are that they lack good separation performance at high temperatures and are strongly affected by highly aggressive components, especially for molecules with similar size. The cost to produce carbon membranes is estimated to be 1–2 orders of magnitude higher than polymeric membranes due to the difficulty of scaling up to rapid production rates for modules. So they will presumably be used only in applications where polymer membranes have some serious difficulty. One method commonly used to compare membrane performance is to plot permeability versus selectivity, the so-called upper-bound plot (Fig. 23.1), to enable comparing performance of different materials. Table 23.2 gives the separation performance for several CMS membranes discussed in the literature. These studies are only a sample of the CMS membranes that have been produced and were chosen for the following reasons:

1. Most are formed from commercially available precursors and have shown promise for CMS formation.
2. Most of the studies detail several membranes produced under different conditions and the effect of those conditions on separation performance.
3. Permeability and selectivity data are either given or can be determined using the information in the publications, especially for the O_2 and N_2.

The precursors selected were Kapton, Matrimid, a phenolic resin, PVD-PVDC, PFA, and polyimide P84. Two laboratory-synthesized polyimides were also chosen for comparison, BTDA-ODA and 6FDA:BPDA-DAM. Figure 23.7 compares the performance of some of these CMS membranes to that of polymer membranes listed in the literature for O_2/N_2 separation. The polymer data and the upper bound were obtained from Robeson (1991). The results show that almost all of the CMS membranes discussed here have properties above the polymeric upper bound. The "best" separation performance, in comparison to the polymer upper bound were from the CMS membranes produced from PFA and the polyimides Kapton, BTDA-ODA, and 6FDA:BPDA-DAM.

The PFA membranes presented here were presented by Shiflett and Foley (1999, 2000, 2001). These membranes were synthesized by ultrasonically depositing a layer of PFA on a stainless steel support followed by drying and subsequent pyrolysis. This process was then repeated several times to form the resulting supported membranes. This process may result in an asymmetric membrane, and authors have also shown that producing reproducible properties for this type of membrane is difficult (Shiflett and Foley, 2001). The

TABLE 23.2 Separation Performance of Selected CMS Membranes

Precursor	Temp (K)	Atmosphere	Heat Rate (K/min)	Soak (min)	P_{O_2}	P_{CO_2}	P_{He}	O_2/N_2	CO_2/CH_4	CO_2/N_2	He/N_2	Reference
Kapton	873	Vacuum	10	120	383	1820	534	4.7		22.2	6.54	Suda and Haraya (1997a)
Kapton	1073	Vacuum	10	120	34.8	128	248	11.5		42.2	81.4	
Kapton	1273	Vacuum	10	120	0.96	4.15	27.9	23.4		101	680	
Kapton	1223	Argon	13.3	120	0.92	3.54	26.4	21.6		82.7	618	
Kapton	1223	Argon	4.5	120	0.66	2.51	21.1	23		87.7	739	
Kapton	1223	Argon	1.33	120	0.15	0.5	11.6	36		122	2810	
Matrimid	923	Vacuum	0.5	60	6	10	80	2.6	5.5	4.5	30	Fuertes et al. (1999)
	748	Vacuum	0.5	60	3	9	25	4.5	15	15	35	
	723	Vacuum	0.5	60	1	2.5	10	5.9	15	15	60	
Matrimid	823	Vacuum	4	120	323	1264		8	64	30		Steel (2000)
	1073	Vacuum	4	120	24	66		13	220	35		
	823	Vacuum	4	540	166	439		9	81	24		
	1073	Vacuum	4	540	1	3		16	300	38		
Phenolic Resin	973	Vacuum	0.5		13.4	46.1	158.7	10.7	85.3	36.9	127	Centeno and Fuertes (2001)
	973	Vacuum	0.5		8.96	12	53.8	10	87	18.2	90	Centeno and Fuertes (1999)
PVDC-PVC	773	Vacuum	1	0	0.24	4.8	7	2.3	8.0		10	Centeno and Fuertes (2000)
	873	Vacuum	1	0	5.97	10.7	21	4.2	11.8		15	

(*Continued*)

TABLE 23.2 *Continued*

Precursor	Temp (K)	Atmosphere	Heat Rate (K/min)	Soak (min)	P_{O_2}	P_{CO_2}	P_{He}	O_2/N_2	CO_2/CH_4	CO_2/N_2	He/N_2	Reference
	973	Vacuum	1	0	9.56	19.1	35	3.8	16.7		14.3	
	1073	Vacuum	1	0	2.39		14	2.0			12	
	973	Vacuum	1	0	6.69	17.9	38	7.4	70.0	18		
	973	Vacuum	1	0	0.96	2.2	23	14.0	75.0	25	300	
PFA (6 coats)	873	Helium	5	120	848.9		1329	2.1			3.3	Shiflet and Foley (2000)
PFA (3 coats)	723	Helium	5	120	1.8		10.4	30.4			178.0	
PFA (3 coats)	573	Helium	5	120	0.4		4.3	13.5			160.0	
PFA (4 coats)	723	Helium	5	60	5.1		13.7	6.3			16.9	
PFA (5 coats)	723	Helium	5	0	2.7		8.9	6.9			22.4	
PFA (3 coats)	723	Helium	5	120	5.5		18.5	8.0			26.9	
PFA(4 coats)	423,723	Helium	5	60,120	8.7		12.7	5.0			7.3	
PFA (3 coats)	800,723,650	Helium	5	120	4.1		10.5	28.3			189.0	
P84 Polyimide	550	Vacuum	Varies	120		1808			22			Kim et al. (2004c)
	650	Vacuum	Varies	120		738			37			
	800	Vacuum	Varies	120	158	499		9	89	28		

P84 Polyimide	500	Nitrogen	1	0	3.9	17.5	26.5	5.3		24.0	36.3	Barsema (2004)
	600	Nitrogen	1	0	72.2	276.0	447	9.3		35.4	57.3	
	700	Nitrogen	1	0	16.8	64.1	166	8.0		30.5	79.0	
	750	Nitrogen	1	0	8.3	30.6	178	10.4		38.3	222.5	
	800	Nitrogen	1	0	3	3.3	58.5	1.1		1.2	20.9	
	900	Nitrogen	1	0	0.32	0.5	17.4	8.0		12.8	435.0	
	700	Nitrogen	1	120	0.46	2.4	37	3.3		17.1	264.3	
	500	Nitrogen	1	0	3.9	17.5	26.5	5.3		24.0		
6fda-bpda-dam	823	Vacuum	4	120	1052	4864		7	53	33		Steel (2000)
	1073	Vacuum	4	120	24	94		11	118	45		
	823	Vacuum	4	540	368	1305		9	70	32		
	1073	Vacuum	4	540	16	95		10	95	63		
BTDA-ODA	773	Ar, 200 cm^3(STP)/min	3	30	509	1535	2134	9		27	38	Kim et al. (2005b)
	973	Ar, 200 cm^3(STP)/min	3	0	632	1989	2707	9		28	38	
	973	Ar, 200 cm^3(STP)/min	3	30	204	505	1593	13		32	100	
	973	Ar, 200 cm^3(STP)/min	3	60	136	350	1106	15		39	123	
	1073	Ar, 200 cm^3(STP)/min	3	30	61	176	872	15		44	218	

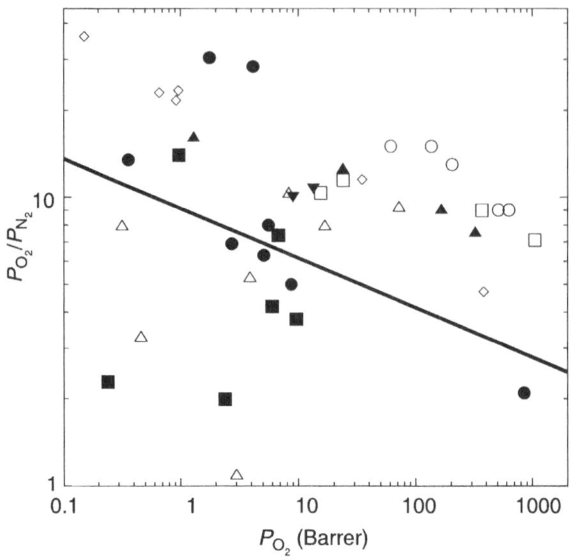

Figure 23.7 Comparison of various CMS membranes with polymeric membrane upper bound. Upper bound (Robeson, 1991), • are supported tubular PFA CMS (Shiflett and Foley, 2000), □ are dense flat 6FDA-BPDA-DAM CMS (Steel, 2000), ■ are PVDC-PDV CMS (Centeno and Fuertes, 2000), ○ are dense flat BTDA-ODA CMS (Kim et al., 2005b), ▲ are dense flat CMS from Matrimid (Steel, 2000), ◇ are dense Kapton CMS (Suda and Haraya, 1997a), ▼ are flat phenolic-resin-based CMS supported on carbon disks (Centeno and Fuertes, 1999, 2001), and Δ are dense flat P84 (Barsema, 2004).

permeabilities listed here are based on the entire thickness of the membrane and therefore may be overestimated; nevertheless, the selectivity is very high, and, thus, these membranes are interesting. Also, the results from the PFA study show that the processing conditions have a drastic effect on the separation performance. The separation performance ranges from $P_{O_2} = 850$ Barrers ($O_2/N_2 = 2.1$) to $P_{O_2} = 1.9$ Barrers ($O_2/N_2 = 30.4$) with fairly similar pyrolysis procedures. The remainder of the "high-performance" CMS membranes from Figure 23.6 were produced from polyimides. Polyimides still appear to be the most attractive polymer for the production of CMS membranes, as mentioned by Jones and Koros (1994a). One advantage of polyimide precursors is that they can be used unsupported, and several studies have shown that they can be used to produce asymmetric hollow-fiber CMS membranes (Geiszler and Koros, 1996; Jones and Koros, 1994a; Okamoto et al., 1999; Vu, 2001). Also, each of the polyimide-based CMS membranes shown here have a range of separation properties. Therefore, for commercial application, producing high-performance CMS membranes for multiple applications may only require altering the pyrolysis process, without changing the polymer precursor. This is a major advantage of CMS membranes over other materials.

23.6 PRODUCTION OF CMS MODULES

For commercial applications, it is desirable to fabricate membranes into modular form that maximizes both productivity and selectivity. The most common commercial membrane geometries are flat sheet and tubular. There have been limited studies investigating the production of CMS membrane modules, and most have focused on laboratory-scale modules. The main difficulty in the production of high-density, industrial-scale CMS membrane modules comes from the poor mechanical properties of the CMS membranes.

Soffer et al. (1987, 1996, 1997) described the production of the first hollow-fiber CMS membranes in a series of patents. The majority of the membranes discussed by Soffer and co-workers were produced by first pyrolyzing cellulose fibers followed by posttreatment to improve the performance of the membranes. Several others have subsequently described the development of laboratory-scale membrane modules. For example, Jones and Koros (1994a,b), Vu (2001) and Vu et al. (2002, 2003) have constructed modules based on Figure 23.8.

To date, there have been few industrial-scale CMS membrane modules discussed in the literature. Lagorssee et al. (2004) recently discussed the production and characterization of membranes and membrane modules from Carbon Membranes Ltd. This was the only company in the world to produce large-scale CMS membrane modules, but it closed in 2001. The modules were based on the patents of Soffer et al. (1987, 1996, 1997). CMS membranes were produced in hollow-fiber configuration by first pyrolyzing cellulose hollow fibers. The structure formed from this step was dubbed the "primary" microporosity. Bundles of these fibers were then placed in membrane modules similar to those described by Jones and Koros (1994a) but on a much larger scale. Carbon CVD (using propylene as the source) was then used on the bore side of the fiber to fill the existing pores on the bore side and produce a thin carbon layer. The layer was then oxidized to open up the pores and obtain the desired pore size in the selective layer. The resulting membrane had an asymmetric structure. The packing density reported for these modules was 2000 m^2/m^3 (10,000 fibers and 4 m^2/module). Soffer et al. (1996) have also described a method for clogging the pores of defective or broken CMS fibers in a module, but no information could be found on whether or not this process was used by Carbon Membranes Ltd.

Recently, Lagorssee et al. (2005) have discussed a new CMS membrane module being developed by Blue Membranes GmbH (Germany). This module is constructed from a flat membrane precursor in a "honeycomb" configuration. The packing density is up to 2500 m^2/m^3 in a 10-m^3 module. The support is an industrial-grade paper with ceramic support. The support is then sheeted with a polymer precursor layer using an imprinting

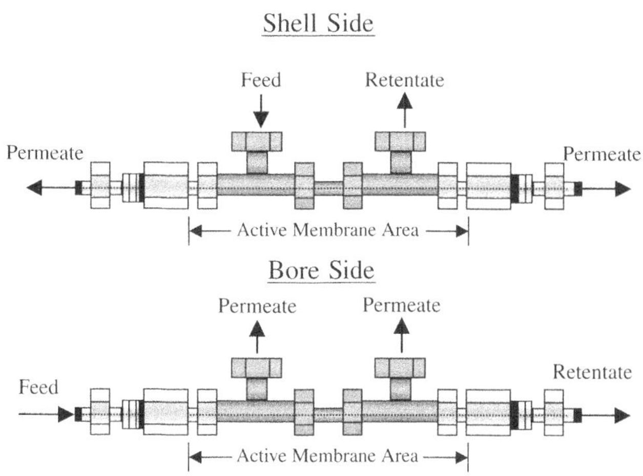

Figure 23.8 Minimodule used for single fiber testing (Vu, 2001).

technique. The precursor described was composed of a commercially available resol-type phenolic resin, Phenodur PR 515, and an epoxy resin, Beckopox EP309.

The flat precursor sheet is used to produce a corrugated sheet using a stamping procedure. The pattern is diagonal to the sheet's direction. The sheet is then pleated. Overlapping the pleated corrugated segments creates corrugated cross-flow channels. The edges are then sealed creating the upstream and downstream side of the module. The module is subsequently pyrolyzed. The properties are tuned with a procedure similar to that described by Soffer et al. (1997) and Fuertes and Menendez (2002). First, a CVD layer is deposited on the surface, and then the final porosity tuned by subsequent oxidation. These steps also repair any defects formed during pyrolysis of the original layer. The module is then sealed in a housing. The results show that by using different activation procedures, membranes with O_2/N_2 selectivities as high as 13.2, with good permeances have been obtained. These membranes also have good properties for other separations such as CO_2/N_2 and may be the next-generation membranes for gas separations. The processing technique is almost entirely automated and provides high surface area.

23.7 CHALLENGES AND DISADVANTAGES OF CMS MEMBRANES

23.7.1 Reproducibility

One of the major problems that must be overcome when producing CMS membranes is to be able to easily produce membranes with reproducible separation properties. Several groups have shown that CMS membranes produced under what appear to be the same pyrolysis conditions tend to have a range of separation performance, especially selectivity. Singh-Ghosal and Koros (2000) produced membranes at three different temperatures (535.550, and 800°C). As expected, temperature had a major effect on separation performance. Interestingly, the O_2/N_2 selectivity of films produced at the same temperature varied by as much a 5. Similar results were obtained by Steel (2000). Strano and Foley (2003) have studied the variability in separation performance in great detail. As mentioned previously, this group has produced supported CMS membranes from PFA by a variety of different techniques including brush coating of a PFA solution onto the support, spray coating the supports, and ultrasonic deposition (Acharya and Foley, 1999; Acharya et al., 1997; Shiflett and Foley, 1999, 2000, 2001; Strano and Foley, 2002, 2003). Though fairly similar pyrolysis techniques have been used to produce all of the membranes, the permeance and selectivity have varied considerably. Strano and Foley (2003) have attributed the variations in separation performance to defective pores formed during the pyrolysis process.

Strano and Foley (2003) discussed the following observations from permeance data for supported CMS membranes: (1) The flux versus pressure for weakly adsorbing species is linear for ideal selectivity ratios significantly above Knudsen values, while those with selectivity ratios near Knudsen values show a nonlinear dependence. The authors attribute the first effect to adsorption within the nanopores in the carbon in the Henry's law regime, while the latter is attributed to viscous and Knudsen flow in large pores. (2) The separation performance of CMS adsorbents are fairly repeatable and almost directly related to carbon synthesis conditions, while the performance of CMS membranes sometimes varies when similar synthesis conditions are used. (3) Increasing the selectivity with respect to one gas pair parallels increasing selectivity for all weakly adsorbing gas pairs. (4) The activation energy for permeation for CMS membranes varies from author to author, even when similar synthesis conditions are used. The heats of adsorption for bulk carbon samples are sometimes much lower than for CMS membranes formed under similar conditions. The authors suggest that these observations are consistent with defective nanopores in the membrane, which could be created during synthesis such as polymer contamination, crack formation during cooling, or thermal property mismatch between carbon and support.

In support of these observations, the authors have suggested the following parallel resistance model to represent the permeability of a weakly adsorbing gas in a CMS membrane:

$$\pi_i = (1 - f)\pi_{np,i} + f\,\pi_{d,i} \tag{23.8}$$

where π_i is the permeability of component i, $\pi_{np,i}$ is the pressure independent of intrinsic permeability of the carbon membrane, $\pi_{d,i}$ is the permeability through the defects, and f is the surface area fraction of defective pores (the authors originally used α). The authors suggest the following model to represent the flow through the defective pores, which takes into account Knudsen diffusion and viscous flow:

$$\pi_{d,i} = \frac{\varepsilon}{\tau RT}\left(\frac{4}{3}\beta d_m\sqrt{\frac{8RT}{M_i}} + \frac{d_m^2\bar{p}}{\gamma\eta_i}\right) \tag{23.9}$$

where β and γ are 0.8 and 2.5, respectively, for membranes. The combined porosity and tortuosity factor (ε/τ) is a function of membrane structure but is equal to 1 for a theoretical bundle of straight pores. Therefore, the pore size, d_m, completely describes the separation performance of the defective nanopores. The authors also modeled the flux versus pressure at various pore sizes and have shown that as the pore size increases from 1 to 200 nm, there is a much more pronounced quadratic dependence of flux versus pressure, which is consistent with observation (1) above. The authors have assumed that the permeability in the bulk carbon is constant and independent of synthesis conditions. This may not be a very valid assumption since several authors have shown considerable differences in performance with pyrolysis temperature, which cannot be entirely attributed to nanopores. Nonetheless, the authors suggest that any variation in bulk carbon permeability is insignificant in comparison to variations due to nanopores.

Using this model, the authors have determined the best-fit volume fraction of defects, intrinsic carbon permeability, and film thickness (which is determined by the actual thickness and the area fraction of defects) for all membranes using a regression technique for pores of 1, 50, and 200 nm. The authors have shown that a single 200-nm pore in a 10-cm^2 film could have a significant effect on separation performance but be too small to detect by adsorption or microscopic techniques. Using the model parameters, and a 50-nm pore size, the authors show that their data fits experimental data well and accurately predicts pressure-dependent flux in several membranes. The model was also used to examine the variability in the data of Singh-Ghosal and Koros (2000), though different intrinsic permeabilies were estimated at each pyrolysis temperature. The model is useful and estimates intrinsic permeabilities of the carbon without any defects and stresses the importance of the production of defect-free membranes, which are likely the major factor that leads to irreproducible membrane performance.

23.7.2 Organic Exposure

Jones and Koros (1994b) have studied the effect of organic exposure on air separation performance for hollow-fiber CMS membranes. Membranes were first tested using dry air feeds to determine a baseline permeance and selectivity. The membranes were then exposed to various hydrocarbons, and then the permeance and selectivity to air was measured over time. For C_6 and heavier hydrocarbons, concentrations as low as 0.1 ppm, decreased the permeability and selectivity by over 90% in less than one day. Though the use of these membranes would normally be limited to streams that do not have organics, exposure may occur in certain situations. It is evident from these results that exposure to organics could greatly affect the separation performance.

Traditional molecular sieves for adsorption are regenerated using thermal techniques. Due to limitations of the epoxy used to form the modules in this study, only temperatures below 90°C were investigated but showed very little regeneration. Jones and Koros though, have shown that regeneration can be achieved, at least to some extent, by exposing CMS membranes to propylene at near unity activity (~150 psia). Other hydrocarbons such as ethylene and 1,3-butadiene were also tested. 1,3-Butadiene at near unity activity had little regenerative effect. Ethylene was only tested at pressure up to ~150 psi, and also had little effect, though the authors suggest ethylene at near unity activity may have similar effects than propylene, though the pressure is very high and not as practical as propylene.

Vu et al. (2002) have investigated the effect of impurities in 90%/10% CH_4/CO_2 streams on the separation performance of CMS hollow-fiber membranes. The results are similar to that of Jones and Koros (1991b) for air streams though not as drastic. When exposed to a stream with 100 ppm of heptane, the CO_2 permeance dropped from 42 to 38 GPU in about 10 min but remained constant for 5 h. There was very little effect on selectivity. Similar results were noted when exposed to 70.4 ppm of toluene. Increasing the concentration of either impurity decreased the permeance but had little effect on selectivity. The membrane properties were almost completely regenerated by heating the membrane to 90°C in N_2 for more than 12 h. These results show that CO_2 may not be as greatly affected by contaminants as weakly sorbing gases, likely due to the strength at which CO_2 can adsorb and compete with organic contaminants.

Menendez and Fuertes (2001) have studied aging effects in supported CMS membranes stored under propylene for 30 days. This study was performed for several reasons, but the most important was to test the usefulness of these CMS membranes to separate hydrocarbons. The permeance of the membranes were tested periodically with a variety of gases with different kinetic diameter. The results show that the permeance of the membranes increases with time for almost all gases and the permselectivity decreases slightly. Therefore, these results agree with the work of Jones and Koros (1994b), which shows that propylene can "open up" the pore structure of CMS membranes and also that CMS membranes may be useful for separation of light hydrocarbons, as long as the streams are free of heavier hydrocarbons such as C_6 and above.

Therefore, Jones and Koros (1994b) have shown that CMS membranes are particularly susceptible to organic exposure, but propylene exposure may be a regenerative technique that would allow CMS membranes to continue to be used if they are inadvertently exposed to organic feeds.

23.7.3 Air and Humidified Feeds

Jones and Koros (1995a,b) have also investigated the effect of water vapor on the separation performance of CMS membranes. The separation performance of CMS hollow-fiber membranes for humidified air streams was investigated. Even though CMS membranes are for the most part hydrophobic, the authors suggest that oxygen-containing sites in the CMS structure can act as nucleation points for water droplets to form. This has also been observed in carbon adsorbents. The results show that when the humidity levels are below ~45%, there is very little effect on the separation performance. At higher humidities, though, the permeance can be greatly diminished with little change in selectivity. Of course, one solution is to pretreat streams before exposing the membranes, which the authors have also investigated. The authors coated the CMS hollow-fiber membranes with poly(4-methyl-1-pentene), Teflon AF 1600, or Teflon AF 2400. The membranes coated with Teflon AF 2400 showed very little change in separation performance (~11%) after exposure to 24 h of an 85% relative humidity (RH) stream, though the coating lowered the selectivity from ~10.5 to ~7.8. Uncoated films showed a 58% loss in permeance. Therefore, this method shows promise for hindering the negative effect that high humidity levels have on CMS separation performance.

Menendez and Fuertes (2001) have also investigated the effects of aging in air for supported CMS membranes. CMS membranes were stored in four different environments: nitrogen, laboratory air, dry air, and air with 100% RH. The results show that the N_2 permeability and O_2/N_2 selectivity of the films stored in nitrogen changed by about 30 and 20%, respectively, over 25 days. The original selectivity of this membrane was very low, though (O_2/N_2 ~3.3). When the membrane was stored in laboratory air or dry air, there was a substantial decrease in the membrane performance. The N_2 permeance dropped by ~50% in one day and 90% after about 50 days for the dry feed. There are no data available between 6 and ~140 days for the laboratory-stored film; so it is difficult to determine the rate at which the permeance changed with respect to time. When stored in 100% RH air, Menendez and Fuertes (2001) note that the rate of permeance change is decreased. There is a considerable difference in the selectivity change of the films stored in dry air and laboratory air, though. When exposed to laboratory air, the selectivity for O_2/N_2 first drops to ~80% of the original after 6 days, and then increases by about 50% over the original after ~140 days. The membrane exposed to the dry stream shows a fairly constant increase in selectivity over 132 days (~50% increase after 132 days).

These results suggest that oxygen is likely chemisorbed to the surface of the carbon over time and narrows the pore size of the membrane, thereby increasing selectivity but decreasing permeability. It appears that water may actually slow the rate of chemisorption of oxygen in the membrane. Menendez and Fuertes (2001) suggest that changes in the permeance for membranes stored in nitrogen are likely attributed to exposure to air and oxygen when periodically testing the membrane, though a 30% drop over 30 days is substantial.

There are some issues that should be addressed with this study, though. The (O_2/N_2) selectivity of the membranes stored in laboratory air and dry air were 9 and 7, respectively. Therefore, these membranes presumably have a pore size very similar to that of oxygen and nitrogen. The (O_2/N_2) selectivity of the membrane stored in nitrogen was only 3.3, therefore, the intrinsic pore size of these membranes is fairly large in comparison to the samples stored in dry air. Because the permeance in molecular sieving media is very sensitive to pore size, comparison of these two membranes may be difficult because changes in the pore structure of the latter membrane will not have as significant of an effect as for the highly selectivity membranes. Also, to more accurately test the effect of oxygen chemisorption, membranes should be completely free of oxygen for the duration of the test. Second, the original selectivity of the CMS exposed to 100% RH in the Menendez and Fuertes (2001) study is not provided; so speculation about the original pore size in the membrane cannot be made. Therefore, any conclusions about the effect of humidity level from this study should be viewed with caution.

It is difficult to draw any conclusions about the effect of humidity on separation performance based on the results of the two studies presented here. Jones and Koros (1995a,b) have shown that increasing the humidity in the feed stream causes increased reduction in the permeance to O_2 and N_2, while Menendez and Fuertes (2001) suggest that exposure to high humidity levels hinders chemisorption of oxygen and reduces the rate of aging over time. It is possible that both of these results are correct and that the membranes produced by Jones and Koros (1995a,b) have already aged before the test by being exposed to air during the module making process. The stabilized (aged) membranes already have chemisorbed oxygen. In that case, higher humidity levels during testing may have an adverse effect on permeance. The membranes produced by Menendez and Fuertes (2001) are very young at the beginning of the test and have not formed a stabilized structure (with adsorbed oxygen), and therefore the effect of oxygen is very pronounced, and humidity levels might have the opposite effect, hindering the aging process.

23.8 DIRECTION OF CARBON MEMBRANE DEVELOPMENT

As demonstrated, CMS membranes have the potential to technically compete with polymer membranes and to advance the use of membranes in new and challenging separations. Traditionally, the majority of CMS research has dealt with synthesizing new CMS materials using (1) new polymer precursors, (2) new pyrolysis techniques, or (3) modifying existing precursors and/or CMS membranes. Unfortunately, there are still only a limited number of studies that address the relationship between the polymer precursor or formation parameters and the resulting CMS membranes. This will be very important for systematic development of these materials for new separations such as olefin/paraffin.

The other area still in need of technical CMS membrane research is in the scale-up of these materials to industrial size. The main drawback of CMS membranes is that they have poor mechanical properties. For optimum surface-to-volume ratios, membrane modules consisting of asymmetric hollow fibers are desired. Unfortunately, this geometry may not be feasible for CMS membranes based on the work attempted to date. New membrane geometries such as the "honeycomb" configuration discussed by Lagorosee et al. (2005) need to be investigated or techniques to produce and maintain CMS hollow-fiber modules. Of particular importance will also be investigations of how the performance of different CMS geometries compare. Since CMS formation involves the decomposition of polymer precursors, the release of small molecules during decomposition and heat transfer throughout the membranes may also play a major role in scale-up considerations.

ACKNOWLEDGMENTS

The authors gratefully acknowledge the financial support of the Department of Energy under grant number DE-FG02-04ER15510 and the Georgia Research Alliance. The authors would also like to thank John Perry and Melissa Zubris-Williams for their helpful comments.

REFERENCES

Acharya, M., Raich, B. A., Foley, H. C., Harold, M. P., and Lerou, J. J. (1997). Metal-supported carbogenic molecular sieve membranes: Synthesis and applications. *Ind. Eng. Chem. Res.* 36, 2924–2930.

Acharya, M., and Foley, H. C. (1999). Spray-coating of nanoporous carbon membranes for air separation. *J. Membr. Sci.* 161, 1–5.

Anand, M., Langsam, M., Rao, M. B., and Sircar, S. (1997). Multicomponent gas separation by selective surface flow (SSF) and poly-trimethylpropyne (PTMSP) membranes. *J. Membr. Sci.* 123, 17–25.

Anand, M., Rao, M. B., and Sircar, S. (1995). Hydrogen recovery by adsorbent membranes. U.S. Patent 5,435,836.

Barsema, J. (2004). *Carbon Membranes: Precursor, Preparation, and Functionalization.*, University of Twente, Twente.

Barsema, J. N., Klijnstra, S. D., van der Vegt, N. F. A., Koops, G. H., and Wessling, M. (2004). Intermediate polymer to carbon gas separation membranes based on Matrimid PI. *J. Membr. Sci.* 328, 93–102.

Burns, R. L., and Koros, W. J. (2003). Defining the challenges for C_3H_6/C_3H_8 separation using polymeric membranes. *J. Membr. Sci.* 211(2), 299–309.

Centeno, T. A., and Fuertes, A. B. (1999). Supported carbon molecular sieve membranes based on a phenolic resin. *J. Membr. Sci.* 160, 201–211.

Centeno, T. A. and Fuertes, A. B. (2000). Carbon molecular sieve gas separation membranes based on poly(vinylidene chloride-*co*-vinyl chloride). *Carbon* 38, 1067–1073.

Centeno, T. A., and Fuertes, A. B. (2001). Carbon molecular sieve membranes derived from a phenolic resin supported on porous ceramic tubes. *Sep. Purif. Technol.* 25, 379–384.

Chen, Y. D., and Yang, R. T. (1994). Preparation of carbon molecular sieve membrane and diffusion of binary mixtures in the membrane. *Ind. Eng. Chem. Res.* 33, 3146–3153.

David, L. I. B., and Ismail, A. F. (2003). Influence of the thermostabilization process and soak time during pyrolysis on the polyacrylonitrile carbon membranes for O_2/N_2 separation. *J. Membr. Sci.* 213, 285–291.

Freeman, B. D. (1999). Basis of permeability/selectivity tradeoff relations in polymeric gas separation membranes. *Macromolecules* 32(2), 375–380.

Fuertes, A. B. (2001). Effect of air oxidation on gas separation performance of adsorption-selective carbon membranes. *Carbon* 39, 697–706.

Fuertes, A. B., and Menendez, I. (2002). Separation of hydrocarbon gas mixtures using phenolic resin-based carbon membranes. *Sep. Purif. Technol.* 28, 29–41.

Fuertes, A. B., Nevskaia, D. M., and Centeno, T. A. (1999). Carbon composite membranes from Matrimid(r) and Kapton(r) polyimides for gas separation. *Microporous Mesoporous Mater.* 33, 115–125.

Geiszler, V. (1997). *Polyimide Precursors for Carbon Molecular Sieve Membranes.* University of Texas, Austin, TX.

Geiszler, V. C., and Koros, W. J. (1996). Effects of polyimide pyrolysis conditions on carbon molecular sieve membrane properties. *Ind. Eng. Chem. Res.* 35, 2999–3003.

Glasstone, S., Laidler, K. J., and Eyring, H. (1941). *The Theory of Rate Processes.* McGraw-Hill, New York.

Golden, C. M. A. T., Rao, M. B., and Sircar, S. (1998). Selective surface flow carbon membranes for gas separation. Presented at the 6th Conference on Fundamentals of Adsorbtion, Giens, France.

Haraya, K., Suda, H., Yanagishita, H., and Matsuda, S. (1995). Asymmetric capillary membrane of a carbon molecular sieve. *J. Chem. Soc. Chem. Commun.* 17, 1781–1782.

Hatori, H., Takagi, H., and Yamada, Y. (2004). Gas separation properties of molecular sieving carbon membranes with nanopore channels. *Carbon* 42, 1169–1173.

Hatori, H., Yamada, Y., and Shiraishi, M. (1992a). Preparation of macroporous carbon films from polyimide by phase inversion method. *Carbon* 30, 303–304.

Hatori, H., Yamada, Y., Shiraishi, M., Nakata, H., and Yoshitomi, S. (1992b). Carbon molecular sieve films from polyimide. *Carbon* 30, 305–306.

Hatori, H., Yamada, Y., Shiraishi, M., Yoshihara, M., and Kimura, T. (1996). The mechanism of polyimide pyrolysis in the early stage. *Carbon* 34(2), 201–208.

Hayashi, J.-I., Mizuta, H., Yamamoto, M., Kusuki, K., Morooka, S., and Suh, S.-H. (1996). Separation of ethane/ethylene and propane/propylene systems with a carbonized BPDA-pp'ODA polyimide membrane. *Ind. Eng. Chem. Res.* 35, 4167–4181.

Hayashi, J.-I., Yamamoto, M., Kusakabe, K., and Morooka, S. (1995). Simultaneous improvement of permeance and permselectivity of 3,3′-4,4′-biphenyltetracarboxylic dianhydride-4,4′-oxydianiline polyimide membrane by carbonization. *Ind. Eng. Chem. Res.* 34, 4364–4370.

Hines, A. L., and Maddox, R. N. (1985). *Mass Transfer Fundamentals and Applications.* Prentice-Hall, Englewood Cliffs, NJ.

Islam, M. N., Zhou, W., Honda, T., Tanaka, K., Kita, H., and Okamoto, K.-I. (2005). Preparation and gas separation performance of flexible pyrolytic membranes by low-temperature pyrolysis of sulfonated polyimides. *J. Membr. Sci.* 261, 14–26.

Ismail, A. F., and David, L. I. B. (2001). A review of the latest development of carbon membranes for gas separation. *J. Membr. Sci.* 193, 1–19.

Jones, C. W., and Koros, W. J. (1994a). Carbon molecular sieve gas separation membranes—I. Preparation and characterization based on polyimide precursors. *Carbon* 32(8), 1419–1425.

Jones, C. W., and Koros, W. J. (1994b). Caron molecular sieve gas separation membranes—II. Regeneration following organic exposure. *Carbon* 32(8), 1427–1432.

Jones, C. W., and Koros, W. J. (1995a). Characterization of ultramicroporous carbon membranes with humidified feeds. *Ind. Eng. Chem. Res.* 34, 158–163.

Jones, C. W., and Koros, W. J. (1995b). Carbon composite membranes: A solution to adverse humidity effects. *Ind. Eng. Chem. Res.* 34, 164–167.

Kim, Y. K., Park, H. B., and Lee, Y. M. (2003). Carbon molecular sieve membranes derived from metal-substituted sulfonated polyimide and their gas separation properties. *J. Membr. Sci.* 226, 145–158.

Kim, Y. K., Park, H. B., and Lee, Y. M. (2004). Carbon molecular sieve membranes derived from thermally labile polymer containing blend polymers and their separation properties. *J. Membr. Sci.* 243, 9–17.

Kim, T. K., Park, H. B., and Lee, Y. M. (2005a). Gas separation properties of carbon molecular sieve membranes derived from polyimide/polyvinylpyrrolidone blends: Effect of the molecular weight of polyvinylpyrrolidone. *J. Membr. Sci.* 251, 159–167.

Kim, Y. K., Park, H. B., and Lee, Y. M. (2005b). Preparation and characterization of carbon molecular sieve membranes derived from BTDA-ODA polyimide and their separation properties. *J. Membr. Sci.* 255, 365–273.

Koresh, J. E., and Soffer, A. (1983). Molecular sieve permselective membrane. Part I. Presentation of a new device for gas mixture separation. *Sep. Sci. Technol.* 18(8), 723–734.

Kusakabe, K., Yamamoto, M., and Morooka, S. (1989). Gas permeation and micropore structure of carbon molecular sieving membranes modified by oxidation. *J. Membr. Sci.* 149, 59–67.

Kusuki, Y., Shimazaki, H., Tanihara, N., Nakanishi, S., and Yoshinaga, T. (1997). Gas permeation properties and characterization of asymmetric carbon membranes by pyrolyzing asymmetric polyimide hollow fiber membrane. *J. Membr. Sci.* 134, 245–253.

Lagorsse, S., Leite, A., Magalhaes, F. D., Bischofberger, N., Rathenow, J., and Mendez, A. (2005). Novel carbon molecular sieve honeycomb membrane module: Configuration and membrane characterization. *Carbon* 43, 809–819.

Lagorsse, S., Magalhaes, F. D., and Mendez, A. (2004). Carbon molecular sieve membranes sorption, kinetic and structural characterization. *J. Membr. Sci.* 241, 275–287.

Menendez, I., and Fuertes, A. B. (2001). Aging of carbon membranes under different environments. *Carbon* 39, 733–740.

Naheiri, T., Ludwig, K. A., Anand, M., Rao, M. B., and Sircar, S. (1997). Scale-up of selective surface flow membrane for gas separation. *Sep. Sci. Technol.* 32(9), 1589–1602.

Okamoto, K.-I., Kawamura, S., Yoshino, M., Kita, H., Hirayama, Y., Tanihara, N., and Kusuki, Y. (1999). Olefin/paraffin separation through carbonized membranes derived from an asymmetric polyimide hollow fiber membrane. *Ind. Eng. Chem. Res.* 38, 4424–4432.

Paranjape, M., Clarke, P. F., Pruden, B. B., Parrillo, D. J., Thaeron, C., and Sircar, S. (1998). Separation of bulk carbon dioxide-hydrogen mixtures by selective surface flow membranes. *Adsorption* 4, 355–360.

Park, H. B., Kim, T. K., Lee, J. M., Lee, S. Y., and Lee, Y. M. (2004). Relationship between chemical structure of aromatic polyimides and gas permeation properties of their carbon molecular sieve membranes. *J. Membr. Sci.* 229, 117–127.

Peterson, J., Matsuda, M., and Haraya, K. (1997). Capillary carbon molecular sieve membranes derived from Kapton for high temperature gas separation. *J. Membr. Sci.* 131, 85–94.

Pierson, H. O. (1993). *Handbook of Carbon, Graphite, Diamond, and Fullerenes.* Noyes Publication, Park Ridge, NJ.

Rao, M. B., and Sircar, S. (1993a). Nanoporous carbon membrane for gas separation. *Gas Sep. Purif.* 7(4), 279–284.

Rao, M. B., and Sircar, S. (1993b). Nanoporous carbon membranes for separation of gas mixtures by selective surface flow. *J. Membr. Sci.* 85, 253–264.

Rao, M. B., and Sircar, S. (1996). Performance and pore characterization of nanoporous carbon membranes for gas separation. *J. Membr. Sci.* 110, 109–118.

Rao, M. B., Sircar, S., Abradaro, J. M., and Baade, W. F. (1995a). Hydrogen recovery by adsorbent membranes. U.S. Patent 5,447,559.

Rao, M. B., Sircar, S., and Golden, T. C. (1992). Gas separation by adsorbent membranes. U.S. Patent 5,104,425.

Rao, M. B., Sircar, S., and Golden, T. C. (1995b). Method of making composite porous carbonaceous membranes. U.S. Patent 5,431,864.

Robeson, L. M. (1991). Correlation of separation factor versus permeability for polymeric membranes. *J. Membr. Sci.* 62(2), 165–185.

Sakata, J., and Yamamoto, M. (1986). Apparatus for separating condensable gas. U.S. Patent 4,583,996.

Saufi, S. M., and Ismail, A. F. (2004). Fabrication of carbon membranes for gas separation—A review. *Carbon* 42, 241–259.

Shao, L., Chung, T.-S., and Pramoda, K. P. (2005). The evolution of physiochemical and transport properties of 6FDA-durene toward carbon membranes; from polymer, intermediate to carbon. *Microporous Mesoporous Mater.* 84, 59–68.

Shao, L., Chung, T.-S., Wensley, G., Goh, S. H., and Pramoda, K. P. (2004). Casting solvent effects on morphologies, gas transport properties of a novel 6FDA/PMDA-TMMDA copolyimide membrane and its derived carbon membranes. *J. Membr. Sci.* 244, 77–87.

Shiflett, M. B., and Foley, H. C. (1999). Ultrasonic deposition of high-selectivity nanoporous carbon membranes. *Science* 285(5435), 1902–1905.

Shiflett, M. B., and Foley, H. C. (2000). On the preparation of supported nanoporous carbon membranes. *J. Membr. Sci.* 179, 275–282.

Shiflett, M. B., and Foley, H. C. (2001). Reproducible production of nanoporous carbon membranes. *Carbon* 39, 1421–1446.

Shusen, W., Meiyun, Z., and Zhizhong, W. (1996). Asymmetric molecular sieve carbon membranes. *J. Membr. Sci.* 109, 267–270.

Singh, A. (1997). Membrane materials with enhanced selectivity; an entropic interpretation. Ph.D. Dissertation, University of Texas, Austin, TX.

Singh, A., and Koros, W. J. (1996). Significance of entropic selectivity for advanced gas separation membranes. *Ind. Eng. Chem. Res.* 35, 1231–1234.

Singh-Ghosal, A., and Koros, W. J. (2000). Air separation of flat sheet homogeneous pyrolytic carbon membranes. *J. Membr. Sci.* 174, 177–188.

Sircar, S., Rao, M. B., and Thaeron, C. M. A. (1999). Selective surface flow membrane for gas separation. *Sep. Sci. Technol.* 34(10), 2081–2093.

Soffer, A., Azariah, M., Amar, A., Golub, D., Saguee, S., and Tobias, H. (1997). Method of improving the selectivity of carbon membranes by chemical vapor deposition. U.S. Patent 5,695,818.

Soffer, A., Koresh, J. E., and Saggy, S. (1987). Separation device. U.S. Patent 4,685,940.

Soffer, A., Saguee, S., Golub, D., Cohen, H., and Azariah, M. (1996). Selective clogging of failed fibers. U.S. Patent 5,575,963.

Stannent, V. (1968). Simple gases. In J. Crank (Ed.), *Diffusion in Polymers*. Academic, New York.

Steel, K. M. (2000). Carbon membranes for challenging separations. University of Texas, Austin, TX.

Steel, K. M., and Koros, W. J. (2003). Investigation of porosity of carbon materials and related effects on gas separation properties. *Carbon* 41, 253–266.

Strano, M. S., and Foley, H. C. (2002). Temperature- and pressure-dependent transient analysis of single component permeation through nanoporous carbon membranes. *Carbon* 40, 1029–1041.

Strano, M. S. and Foley, H. C. (2003). Modeling ideal selectivity variation in nanoporous membranes. *Chemi. Eng. Sci.* 58, 2745–2758.

Suda, H., and Haraya, K. (1995). Molecular sieving effect of carbonized polyimide membrane. *J. Chem. Soc. Chem. Commun.* 17, 1179–1180.

Suda, H., and Haraya, K. (1997a). Gas permeation through micropores of carbon molecular sieve membranes derived from Kapton polyimide. *J. Phys. Chem. B* 101, 3988–3994.

Suda, H., and Haraya, K. (1997b). Alkane/alkene permselectivities of a carbon molecular sieve membrane. *Chem. Commun.* 1, 93–94.

Tin, P. S., Chung, T.-S., and Hill, A. J. (2004a). Advanced fabrication of carbon molecular sieve membranes by nonsolvent pretreatment of precursor polymers. *Ind. Eng. Chem. Res.* 43, 6476–6483.

Tin, P. S., Chung, T.-S., Kawi, S., and Guiver, M. D. (2004b). Novel approaches to fabricate carbon molecular sieve membranes based on chemical modified and solvent treated polyimides. *Microporous Mesoporous Mater.* 73, 151–160.

Tin, P. S., Chung, T.-S., Liu, Y., and Wang, R. (2004c). Separation of CO_2/CH_4 through carbon molecular sieve membranes derived from P84 polyimide. *Carbon* 42, 3123–3131.

Vu, D. Q. (2001). Formation and characterization of asymmetric carbon molecular sieve and mixed matrix membranes for natural gas purification. University of Texas, Austin, TX.

Vu, D. Q., Koros, W. J., and Miller, S. J. (2002). High Pressure CO_2/CH_4 separation using carbon molecular sieve hollow fiber membranes. *Ind. Eng. Chem. Res.* 41, 367–380.

Vu, D. Q., Koros, W. J., and Miller, S. J. (2003). Effect of condensable impurities in CO_2/CH_4 gas feeds on carbon molecular sieve hollow-fiber membranes. *Ind. Eng. Chem. Res.* 42, 1064–1075.

Wang, L.-J., and Hong, F. C.-N. (2005a). Effects of surface treatments and annealing on carbon-based molecular sieve membranes for gas separation. *Appl. Surf. Sci.* 240, 161–174.

Wang, L.-J., and Hong, F.C.-N. (2005b). Surface structure modification on the gas separation performance of carbon molecular sieve membranes. *Vacuum* 78, 1–12.

Wang, L.-J., and Hong, F.C.-N. (2005c). Carbon-based molecular sieve membranes for gas separation by inductively-coupled-plasma chemical vapor deposition. *Microporous Mesoporous Materi.* 77, 167–174.

Wang, K., Suda, H., and Haraya, K. (2003). The characterization of CO_2 permeation in a CMSM derived from polyimide. *Sep. Purif. Technol.* 21, 61–69.

Wang, H., Zhang, L., and Gavalas, G. R. (2000). Preparation of supported carbon membrane from furfuryl alcohol by vapor deposition polymerization. *J. Membr. Sci.* 177, 25–31.

Wei, W., Hu, H., You, L., and Chen, G. (2002). Preparation of carbon molecular sieve membrane from phenol-formaldehyde Novolac resin. *Carbon* 40, 445–467.

Zhou, W., Yoshino, M., Kita, H., and Okamoto, K.-I. (2001). Carbon molecular sieve membranes derived from phenolic resin with a pendant sulfonic acid group. *Ind. Eng. Chem. Res.* 40, 4801–4807.

Zhou, W., Yoshino, M., Kita, H., and Okamoto, K.-I. (2003). Preparation and gas permeation properties of carbon molecular sieve membranes based on sulfonated phenolic resin. *J. Membr. Sci.* 217, 55–67.

CHAPTER 24

Polymeric Membrane Materials and Potential Use in Gas Separation

HO BUM PARK and YOUNG MOO LEE

School of Chemical Engineering, Hanyang University, Seoul, South Korea

24.1 INTRODUCTION

To date, polymer membranes have served as key elements in many significant scientific and technological areas such as tissue repair (Magnusson et al., 1990), protective garments (Gregor et al., 1988), pharmaceuticals production (Mueller et al., 2003), food and beverage packaging (Alves et al., 2004), microelectronics (Nagel and Will, 1999), sensors (Srivastava et al., 2000), fuel cells (Smitha et al., 2005), water purifications (Krzysztof et al., 2002), and gas and vapor separations (Koros and Fleming, 1993; Stern, 1994; Pratibha and Chauhan, 2001). The vital property for which membranes are utilized is their ability to adjust the permeation rate of a chemical component through the membrane with driving forces such as pressure, concentration, and electrical potential. Among these driving forces, pressure difference across the membrane is mostly used in ultrafiltration (UF), microfiltration (MF), reverse osmosis (RO), pervaporation (PV), and gas separation (GS) (Strathmann, 1981; Mulder, 2000).

Membrane-based gas separation is actually developing industrial membrane separation technologies, and the market size and number of applications served are expanding because it offers a number of advantages in terms of energy and capital cost. In principle, all films made of polymers can selectively separate gases from gaseous mixtures by the differential permeation of the components. The concept of separating gases using polymer membranes dates back 175 years. The first publications related to the issue of gas separation using polymer films were by Mitchell (1830, 1833) and Graham (1866). Mitchell observed that natural rubber balloons filled with hydrogen gas deflated over time. He proposed that this phenomenon is due to the release or diffusion of gas through the balloon wall. Graham repeated Mitchell's experiments with natural rubber films and reported the first quantitative measurements of the gas permeation rate through natural rubber films.

Advanced Membrane Technology and Applications. Edited by Norman N. Li, Anthony G. Fane, W. S. Winston Ho, and T. Matsuura
Copyright © 2008 John Wiley & Sons, Inc.

Despite a long history, it was not until 150 years later that gas separations using polymer membranes became a reality in the gas separation industry. Until the 1970s, the dense polymer membranes available were too thick to obtain the high permeation flux required for practical applications. The development of the Loeb–Sourirajan process for making defect-free, high-flux, ultrathin asymmetric membranes for RO was a milestone in membrane history (Loeb and Sourirajan, 1963). Since that time, many polymer membranes based upon asymmetric membranes and their modules have been developed and evaluated in pilot-scale for MF, UF, and RO applications (Baker, 2000).

In 1980, the first polymer membrane system for commercial gas separations appeared on the gas separation market (Koros and Fleming, 1993). The Prism membrane system developed of hydrogen from purge-gas streams in ammonia synthesis plants. Conventionally, ammonia is produced by the reaction of hydrogen and nitrogen at high pressure and temperature over a catalyst. Hydrogen feedstock is obtained by steam reforming of natural gas. Because the conversion rate in ammonia synthesis is limited to 18–20%, unreacted gases need to be recycled back to the feed to improve production yields. To avoid buildup of inert gases and unreacted methane, gas is purged off. This purge gas contains valuable hydrogen, but the conventional separation methods were deemed too expensive at the time (Spillman, 1989). Membrane-based gas separation was considered a promising process if feed was available at high pressure and permeate could be tolerated at lower pressures. Since then, gas separation processes using polymeric membranes have steadily competed with other separation processes such as adsorption and cryogenic distillation in niche application areas. Membrane-based gas separations remain competitive because of their easy operation, small units, low energy cost, portability, reliability, and space efficiency (Spillman, 1995).

Current membrane-based gas processes and potential applications include hydrogen separation and recovery from ammonia purge gas, refinery, and syngas stream in the petrochemical industry. Other applications include carbon-dioxide-enhanced oil recovery, natural gas processing, landfill gas upgrading, air separation (oxygen enrichment or nitrogen production), air dehydration, and helium recovery (Paul and Yampolskii, 1994). The market for polymer membranes for gas separation has expanded annually and sales of membrane gas separation systems have become an approximately $500 million per year business. The primary reasons for the fast growth of membrane markets and sales include consumer demand for higher quality products, increased regulatory pressures, deteriorating natural resources, and the need for environmental and economic sustainability. The worldwide sales of all synthetic membranes are estimated at over $2 billion. Since membranes account for only 40% of the total investment of a membrane separation system, the total annual turnover for the membrane-based industry can be estimated around $5 billion. The annual growth rate in sales of all membrane products has been estimated as approximately 12–15%. The gas separation market is a $455 million per year business, totaling 24% of the whole membrane market. The future market will expand and further growth of this technology is expected for the next 10 years or so.

In this chapter, current research trends of polymer membranes for gas separation are discussed. This review reports on significant polymer membrane materials highlighted in academia and industry, together with existing and potential gas separation applications. Many commercial polymer membranes such as polysulfones, cellulose acetates, polycarbonates, and polyimides are omitted because they were already discussed in previous reviews. Polymeric membranes receiving increased attention, such as siloxane polymers, amorphous fluoropolymers, substituted polyacetylenes, and high-temperature polymers are mainly dealt with in specific and potential applications emerging from these polymeric membrane materials.

24.2 BASIC PRINCIPLES OF GAS SEPARATION IN POLYMER MEMBRANES

24.2.1 Permeability and Selectivity

Gas permeation through a nonporous dense polymer membrane is described using a solution–diffusion model in which the permeability coefficient (P_A) [cm^3 standard temperature and pressure (STP) cm/cm^2 s cm Hg] of gas molecule A is the product of a diffusion coefficient (D_A) (cm^2/s) and a solubility coefficient (S_A) (cm^3-gas/cm^3-polymer cm Hg):

$$P_A = D_A S_A \tag{24.1}$$

Equation (24.1), first postulated by Graham (1866), is a basic expression representing the solution–diffusion mechanism of gases through homogeneous dense polymer membranes. Diffusion is the process by which a small molecule (of gas, vapor, etc.) is moved in a system due to random molecular motions. Diffusion in a polymer matrix is classified into three categories depending on the relative mobility of the polymer and penetrant gas (Crank, 1975; Rogers, 1985). These categories are (1) Fickian diffusion (case I), (2) non-Fickian diffusion (case II), and (3) anomalous diffusion (Table 24.1). Solubility is related to the local concentration (C) of the gas dissolved in the polymer at a given temperature and the solubility coefficient $S(C)$ is expressed as a function of pressure (p):

$$S(C) = \frac{C}{p} \tag{24.2}$$

TABLE 24.1 Different Diffusion Mechanisms of Gas through a Polymer Membrane

Diffusion Mode	Description
Case I: Fickian diffusion	The diffusion process has a rate much smaller than that of the relaxation modes of the polymeric matrix. The sorption equilibrium is rapidly reached. The boundary conditions are independent of time and swelling kinetics.
Case II: non-Fickian diffusion	This mode relates to a fast diffusion process compared to the simultaneous relaxation processes of the polymer. The sorption phenomena are complicated by a strong dependence with the swelling kinetics. These deviations from the Fickian behavior are generally found in the case of the sorption of organic vapors by solid polymers.
Anomalous diffusion	This mode relates to a process when the diffusion and the polymer relaxation rates are comparable. The sorption and the transport of molecules are affected by the presence of preexisting microvoids in the matrix and the penetrant motion is affected by polymer microstructure.

Coefficient $S(C)$ depends on gas condensability as well as polymer–penetrant interactions. Equation (24.2) can be expressed differently using various sorption modes in polymers. These modes, shown in Figure 24.1, are Henry's law sorption, Langmuir-mode sorption, Flory–Huggins mode sorption, dual-mode sorption, and Brunauer-Emmett-Teller (BET) mode sorption.

In general, gas transport phenomenon in polymer membranes progresses through five successive steps as follows (Crank and Park, 1968) (illustrated in Fig. 24.2).

1. Diffusion through the boundary layer at the upstream side.
2. Relative sorption of the gases by the polymer membrane.
3. Diffusion of the gases inside the polymer membrane. The diffusion step is the slowest and becomes the rate-determining step in gas permeation.
4. Desorption of gases at the permeate side (lower partial pressure).
5. Diffusion out of the boundary layer of the downstream side.

In membrane gas separation, the effect of concentration polarization on the feed side is negligible, and so the first and last steps are frequently excluded. However, membrane performance in the system generally decreases with time due to concentration polarization occurring as a result of the limited permeation of certain species. Therefore, in practical applications, membrane module and system designs must be thoroughly considered to prevent concentration polarization, fouling, or both. Generally, the permeation rate is a dominating factor in affecting concentration polarization, while the influences of separation

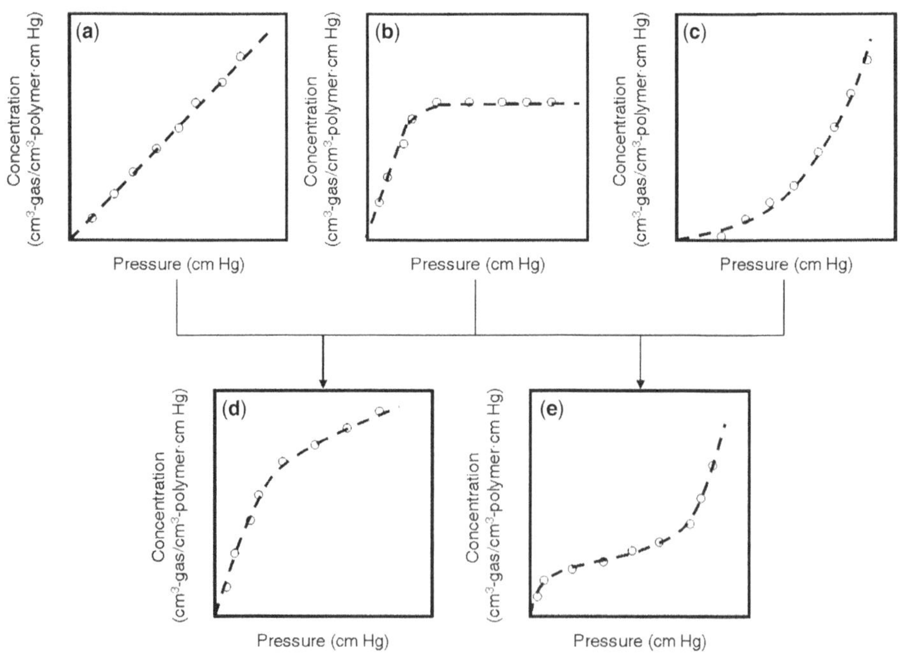

Figure 24.1 Representative sorption phenomena of gases in polymer membranes: (a) Henry's law sorption, (b) Langmuir-mode sorption, (c) Flory–Huggins mode sorption, (d) dual-mode sorption, and (e) BET mode sorption.

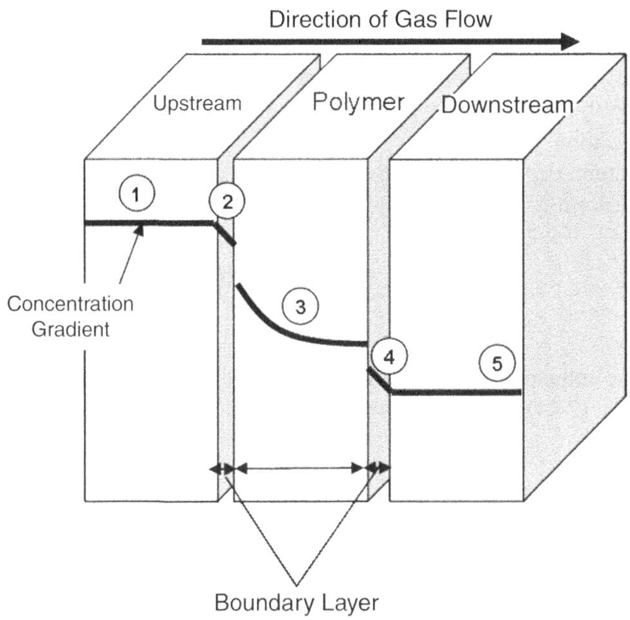

Figure 24.2 Schematic representation of gas permeation steps across polymer membranes according to the solution–diffusion mechanism.

factor to be significant initially and to level off gradually. In addition, increasing feed gas velocity leads to a decrease in the concentration polarization, operation pressure effect is very limited, and the composition of feed gas is less significant.

Selectivity ($\alpha_{A/B}$) of gas A to gas B is either due to differences in the solubility coefficient (solubility selectivity, S_A/S_B) or to differences in the diffusion coefficient (diffusivity selectivity, D_A/D_B):

$$\alpha_{A/B} = \frac{P_A}{P_B} = \left[\frac{D_A}{D_B}\right]\left[\frac{S_A}{S_B}\right] \tag{24.3}$$

Here separation occurs because of the difference in solubilities and mobilities of different penetrants in the membrane.

24.2.2 Temperature Dependence on Gas Transport

Generally, at a given pressure, diffusion of gas through a polymer membrane increases with temperature. The temperature dependence of diffusion can be expressed by an activated process following a relationship similar to the Arrhenius equation:

$$D_A = D_{A0}\ \exp\left[\frac{-E_D}{RT}\right] \tag{24.4}$$

where D_A is the diffusion coefficient (cm^2/s), D_{A0} is the preexponential factor, E_D is the activation energy of diffusion (kJ/mol), R is a universal gas constant (8.31 J/mol · K), and T is the Kelvin temperature (K). The activation energy is the energy used against cohesive forces between polymer segments in making the gaps through which diffusion occurs. Typically, gas diffusion coefficients increase over the temperature range of interest if the polymer does not undergo thermally induced structural rearrangements such as crystallization.

The temperature dependence of S_A of a given polymer–penetrant system can be described in the form of a van't Hoff relationship:

$$S_A = S_{A0} \exp\left[\frac{-\Delta H_S}{RT}\right] \tag{24.5}$$

where ΔH_S is the enthalpy of solution of the penetrant and S_{A0} is the preexponential factor. From Eqs. (24.1), (24.4), (24.5) it follows that over a narrow range of temperatures,

$$\begin{aligned} P_A &= D_{A0} \exp\left[\frac{-E_D}{RT}\right] S_{A0} \exp\left[\frac{-\Delta H_S}{RT}\right] \\ &= D_{A0} S_{A0} \exp\left[\frac{-(E_D + \Delta H_S)}{RT}\right] \\ &= P_{A0} \exp\left[\frac{-E_P}{RT}\right] \end{aligned} \tag{24.6}$$

where E_P, the activation energy of permeation, is

$$E_P = E_D + \Delta H_S \tag{24.7}$$

Values of E_p, E_d, and ΔH_S depend upon the polymer morphology and thermal properties such as the glass transition temperature (T_g). For many polymer and gas pairs, these parameters can be found in the literature (Brandrup et al., 1999). According to Gee (1947), the heat of the solution, ΔH_S may be expressed as

$$\Delta H_S = \Delta H_{Co} + \Delta H_{Pm} \tag{24.8}$$

where ΔH_{Co} is the molar heat of condensation (always <0 and small for gases) (Crank and Park, 1968) and ΔH_{Pm} is the partial molar heat of mixing (always >0). These values can be estimated from the cohesive energy densities of the penetrant and the polymer using Hildebrand's theory (Hildebrand and Scott, 1950).

24.2.3 Pressure Dependence of Gas Transport

Gas permeability of a polymer is affected by feed pressure variations. Koros and Chern (1987) classified patterns observed in permeability versus pressure relationships into four characteristic patterns, illustrated in Figure 24.3. They are: (i) linear, with a slope close to zero, representing the ideal case that meets the assumption of diffusion and solution being independent of gas pressure (e.g., low sorbing penetrants such as He or N_2 in rubbery or glassy polymers), (ii) a nearly linear increase of permeability with increasing

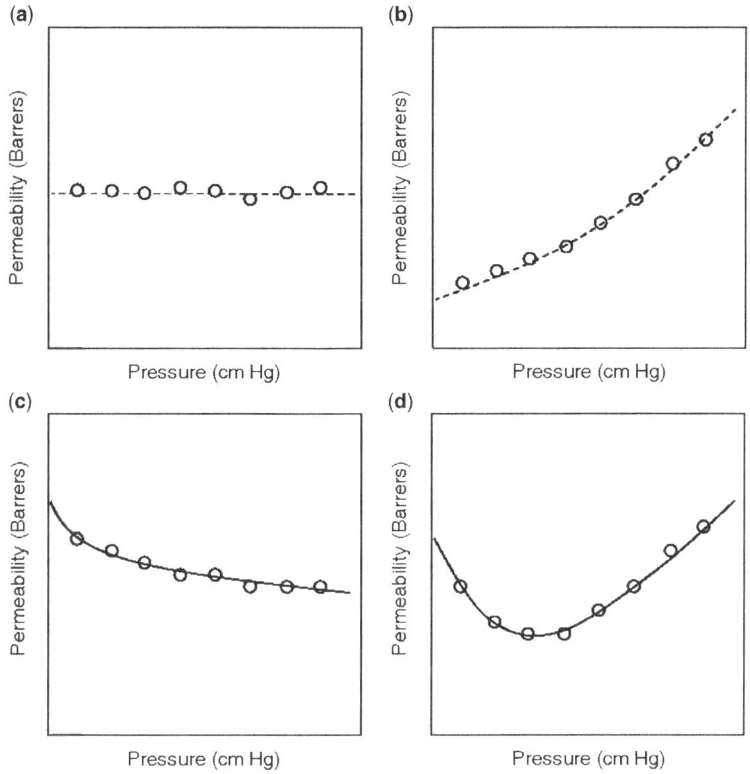

Figure 24.3 Changes in gas permeability of a polymer membrane as a function of pressure.

pressure, which often describes the permeability of an organic vapor into a rubbery polymer, (iii) a decreasing trend of permeability with increasing pressure, which is typically observed for highly soluble gases such as CO_2 in glassy polymers, and (iv) concave upward, which can be perceived as a combination of (ii) and (iii) and is typical of a plasticizing penetrant such as an organic vapor in a glassy polymer.

24.2.4 Unusual Sorption Behavior of Glassy Polymers

Polymers are divided into rubbery and glassy polymers. Rubbery polymers have more flexible chains than glassy polymers. Rubbery polymers are at equilibrium at room temperature because their glass transition temperatures commonly exist below room temperature. Therefore, gas transport behavior in rubbery polymers can be sufficiently explained using a simple solution–diffusion mechanism as long as the rubbery polymers do not contain any crystalline phase that acts as a complete barrier against the penetrant. On the other hand, gas transport through glassy polymer membranes is more complicated because the sorption sites of gases in glassy polymers are unique. In glassy polymers, changes of gas permeability as a function of pressure are often elucidated by the dual-mode and partial immobilization model (Petropoulos, 1970; Koros et al., 1977; Vieth and Sladek, 1965). The basic assumptions applied to these models are: (i) Henry's and Langmuir's sorption modes coexist in the glassy polymers, (ii) local equilibrium between the two modes is maintained throughout the glassy membranes, (iii) the gas sorbed on the Langmuir's sorption sites is completely immobilized, and (iv) diffusion occurs only in Henry's sorption sites. The pressure dependence of the solubility coefficient for a single pure component is represented by

$$S = \frac{C}{p} = k_D + \frac{C'_H b}{1 + bp} \tag{24.9}$$

where k_D is Henry's law constant, C'_H is the Langmuir capacity constant, and b is the Langmuir affinity constant. The C'_H term indicates the amount of unrelaxed free volume in the glassy polymer. This term is used to determine the nonequilibrium nature of glassy polymers. The affinity constant b describes the specific nature of a given penetrant to sorb into the excess unrelaxed volume in the nonequilibrium glassy polymer matrix. In these models, the true diffusion coefficient of a given penetrant is assumed to be constant regardless of concentration or position in the glassy membrane. The nonequilibrium state of glassy polymers can be affected considerably by process and operation parameters such as preparation conditions, pressure, temperature, and interactions between the polymer and penetrant.

24.2.5 Criteria for Membrane Material Selection

To select a suitable polymer membrane and design for each gas separation application, the nature of the polymer membrane (glassy or rubbery) must be considered as well as the target separating mixtures (permanent gas/permanent gas or condensable vapor/permanent gas). The main factors contributing to selectivity (solubility-selective or diffusivity-selective materials) can then be determined and used to improve separation efficiency with minimal loss of productivity. The most ideal membrane material design is to improve both diffusivity-selectivity and solubility-selectivity, but it is difficult to achieve both simultaneously. Moreover, for a given polymer membrane, both diffusivity and solubility depend strongly on process parameters such as pressure difference, feed composition, and temperature, as well as the intrinsic nature of the polymer. Each selectivity contribution (diffusivity or solubility) to total selectivity can be varied by such factors, but research has mainly focused on improving selectivity by modifying the polymer chain structure. More information on the effects of process parameters on selectivity contribution should be thoroughly considered for identifying membrane materials suitable for each application.

24.2.5.1 Diffusivity-Selective Polymer Membranes Molecular sieves are materials that can separate gas molecules by the size and shape on a molecular level. Most of molecular sieves are inorganic materials having well-defined pore sizes and shapes. For example, zeolite molecular sieves (Kusakabe et al., 1996) are well-known catalysts and inorganic membranes. Carbon molecular sieves (Hayashi et al., 1997; Suda and Haraya, 1997; Park et al., 2004; Park and Lee, 2005), known as amorphous carbons, in contrast to carbons, diamonds, or graphite, can be prepared by pyrolyzing polyimides that have a *d*-spacing around 3.65 Å, a value between the kinetic diameters of oxygen (3.46 Å) and nitrogen (3.64 Å). High gas selectivity [i.e., $\alpha(O_2/N_2) > 10$] of these molecular sieving materials is ascribed to high diffusivity-selectivity caused by the differences of molecular sizes and shapes.

Diffusivity (or mobility)–selectivity reflects the ability of the polymer matrix to be selective for the shape and size of penetrant molecules. Diffusivity-selectivity is governed by relative penetrant mobilities and by structural factors such as polymer chain stiffness and intersegmental polymer packing. However, in contrast to inorganic or carbon molecular sieves with a controlled molecular size, polymer membranes are seldom considered as molecular sieves. If an organic molecular sieve with high selectivity can be prepared, it will have vast advantages over zeolites or carbon molecular sieves because of its easy processability, thin-film formability, and other significant reasons. Generally, improving gas separation performance by increasing diffusivity-selectivity is more effective for small gas separations that use rigid glassy polymer membranes such as polysulfones, polycarbonates, polyimides, and polypyrrolones with narrow free volume distributions. For small gas separations such as O_2/N_2 and CO_2/CH_4, the diffusivity-selectivity needs significant advances in its permeability-selectivity properties.

When as-cast polyaniline membranes are exposed to different gases, the gas permeabilities are dependent on the size of the penetrant gas (Anderson et al., 1991). The O_2/N_2 selectivity in as-cast polyaniline membranes is as good as conventional polymers, such as polyimides, and is better than polymers typically used in commercial gas separations such as polysulfones (Koros et al., 1992). Lee et al. (1999) showed that the gas permeability of polyaniline membranes is dependent on the acidic-doping level. A doping–dedoping process using acidic dopants considerably enhanced oxygen permeability. After redoping membranes for several hours in an acidic aqueous solution, oxygen–nitrogen selectivity drastically increased (Fig. 24.4). The increase in selectivity reflects the increase in the diffusion selectivity caused by the change in d-spacing via the doping–dedoping process.

So far, strong interest in highly selective glassy polymer membranes has been in academic laboratories. Despite abundant reports about newly synthesized polymers showing remarkable permselectivity, most of the polymers are not yet practical for gas separation applications. Their complicated synthetic routes and poor solubilities in

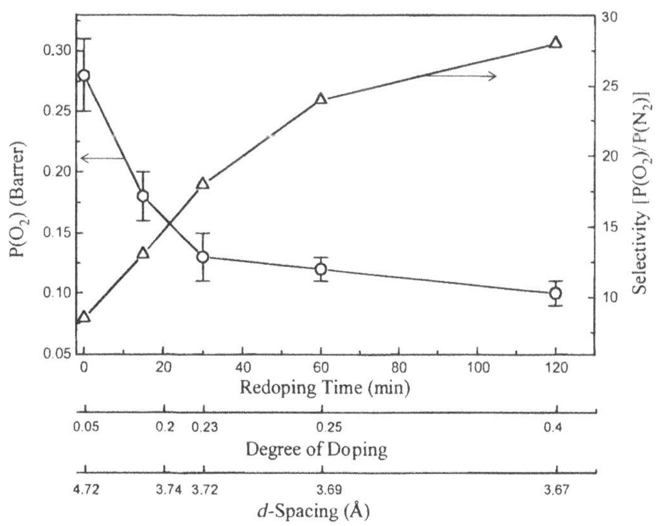

Figure 24.4 Changes gas permeability and selectivity of polyaniline composite membranes by a doping–dedoping–redoping process (Lee et al., 1999).

common organic solvents do not meet commercial criteria for economic membrane modules and systems. In addition, increasing segmental motion at elevated temperatures undermines the ability of polymer membranes to discriminate between penetrants of different dimensions, thereby resulting in low diffusivity-selectivity. For high-temperature gas separations, further improvements in total selectivity (or separation factor) made by decreasing the diffusivity-selectivity may not be possible.

24.2.5.2 *Solubility-Selective Membrane* The solubility coefficient reflects how many gas molecules can be sorbed in polymer membranes. It depends on the condensability as well as the physical interactions of the penetrants with the polymer membrane. Solubility is determined by the concentration of the sorbed gas per unit polymer volume. Generally, the concentration as a function of pressure at constant temperature shows a sorption isotherm with a characteristic shape that is concave to the pressure axis. The solubility-selectivity term in Eq. (24.3) is thermodynamic in nature and is governed by the relative polymer–penetrant interactions and the relative condensability of the penetrants. Solubility-selectivity terms contribute significantly to separations of condensable vapors and polar molecules.

In the absence of specific interactions of gases with the polymer, solubility can be determined by the critical temperature (T_c) of the gas. Although the unrelaxed volume in between the polymer chain segments, together with polymer–penetrant interactions, determine the concentration value, the effectiveness of these factors is reduced by the condensability of vapor, which is related to their critical temperatures. Thus, the solubility-selectivity of gases that do not interact with the polymer is primarily determined by the ratio of their critical temperatures. On the other hand, the critical volume (V_c) or kinetic diameter of the gas is more significant in the size-exclusive mechanism (diffusivity-selectivity). It is important to know the differences in polymer–penetrant interactions for a more exact evaluation of the contribution of solubility-selectivity to the permselectivity. Typically, the permselectivity of rubbery polymers is governed by solubility-selectivity. In this respect, there are no large differences in permanent gas pairs nor in glassy polymers while, for vapor–gas or CO_2–permanent gas pairs, the solubility-selectivity is large enough to lead to a high permselectivity. Table 24.2 summarizes some physical properties of significant gases in membrane gas separation.

Vapor–gas separation membranes or CO_2-selective membranes, such as silicone- and poly(ethylene oxide)-cross-linked rubbery polymer membranes, are the best examples representing high solubility-selectivity contributions to permselectivity. Recently, Lin and Freeman (2005) suggested a material selection guideline for CO_2 selective membranes. To achieve high CO_2 permeability and high CO_2–light gas (i.e., CH_4, N_2, and H_2) selectivity, many polar groups such as ether oxygens (—O—), nitriles (—CN), carbonyls (—C=O—), acetates (—COO—), and amides (—NHCO—) were introduced into the polymers. These polar groups contribute to high CO_2 solubility and CO_2–light gas solubility-selectivity, but polar groups in a polymer decrease CO_2 diffusion coefficients due to the strong affinity between CO_2 and the polar groups. It is well-known that poly (ethylene oxide) (PEO) with ether oxygen units appears to provide excellent CO_2 separation and permeation properties. However, pure PEO materials tend to crystallize easily, making it difficult to improve CO_2 separation and permeation properties. To solve this problem, cross-linked poly(ethylene glycol) arylates (XLPEO) containing branches with methoxy end groups that do not crystallize at typical operation temperatures were prepared. At 35°C, the CO_2 permeability and CO_2/H_2 gas selectivity reached up to 570 Barrers [1 Barrer $= 10^{-10}$ cm^3 (STP) $\cdot$ cm/(cm$^2 \cdot$ s $\cdot$ cm Hg)] and 12 Barrers, respectively.

TABLE 24.2 Physical Properties of Simple Gases

Gas	T_b^a (K)	T_c^b (K)	ε/k^c (K)	σ^d (nm)	d_k^e (nm)
He	4.3	5.3	10.2	0.255	0.260
H_2	20	33	60	0.283	0.289
N_2	77	126	71	0.380	0.364
CO	82	133	92	0.369	0.376
O_2	90	155	107	0.347	0.346
CH_4	112	191	149	0.376	0.380
C_2H_4	175	283	225	0.416	0.390
C_2H_6	185	305	216	0.444	—
CO_2	195	304	195	0.394	0.330
H_2S	212	373	301	0.362	0.360
C_3H_8	231	370	237	0.512	0.430
NH_3	240	406	558	0.290	—
SO_2	263	431	335	0.411	—
n-C_4H_{10}	272	425	331	0.469	0.430
H_2O	373	647	809	0.264	0.265

[a]Boiling temperature.
[b]Critical temperature.
[c]Lennard-Jones temperature.
[d]Potential length constant or collision diameter of the molecule.
[e]Kinetic diameter (Breck, 1974).

Furthermore, the CO_2/H_2 pure gas selectivity of XLPEO reached a value of 40 at $-20°C$ (i.e., CO_2 permeability = 52 Barrers). This is due to the increase of CO_2 solubility and CO_2/H_2 solubility-selectivity. It indicates that solubility-selectivity can be strengthened by increasing the condensability of gas at lower temperatures, leading to high selectivity. However, considerations about economic process conditions (e.g., pressure and temperature) in practical applications must first be evaluated.

24.3 LIMITATIONS OF GAS SEPARATIONS USING POLYMER MEMBRANES

"Tradeoff relations of gas permeability and selectivity—there exists an empirical rule that more permeable polymers are statistically less selective and vice versa." Robeson (1991).

An ideal gas separation membrane must have high permeability and high selectivity. The gas permeation properties of polymer membranes have been extensively studied, and a wide variety of polymers have been synthesized to be more permeable and selective. Nevertheless, there are still strong trade-off relations of gas permeability and selectivity in polymer membranes that will not easily be broken in the future. Moreover, this trade-off behavior of polymer membranes is not yet fully comprehended theoretically. However, it is believed that this trade-off behavior for specific gas pairs is unique and related to parameters of the gas molecules.

Based on an enormous survey of approximately numerous references, including journals, books, and patents, Robeson (1991) proposed the trade-off relations of polymer membranes for various gas pairs (i.e., O_2/N_2, H_2/CH_4, CO_2/CH_4, H_2/N_2, He/CH_4, He/N_2, He/H_2, He/O_2, and H_2/O_2). The transport parameters, selectivity α_{AB} and

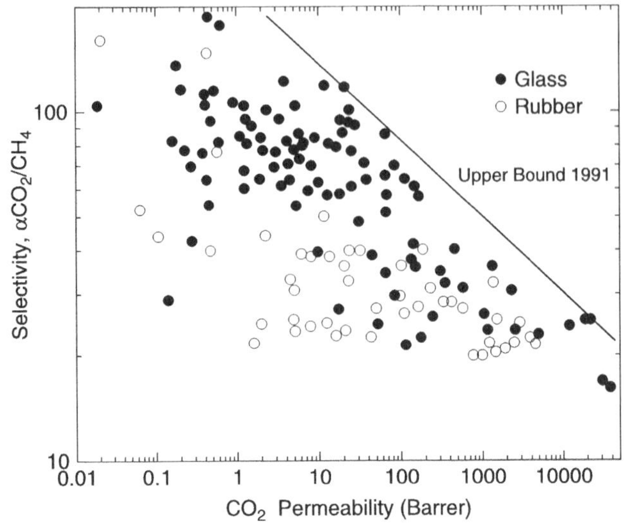

Figure 24.5 Upper bound line of CO_2 permeability and CO_2/CH_4 permselectivity (Robeson, 1991).

permeability P_A, make a cloud on this diagram located below an upper-bound line. This line represents the best combination of α_{AB} and P_A values (Fig. 24.5).

The following empirical equation was proposed to explain the upper-bound lines for given gas pairs:

$$\log \alpha_{AB} = \log \beta_{AB} - \lambda_{AB} \log P_A \tag{24.10}$$

According to Robeson (1991), the transport parameters, β_{AB} and λ_{AB}, for various gas pairs correlate with the Lennard-Jones kinetic diameters of gases:

$$\lambda_{AB} = d_B - d_A \tag{24.11}$$

where d_A and d_B are the kinetic diameters (angstros) of the gas molecules A and B, respectively.

Freeman (1999) expounded upon these trade-off relations. Using well-known correlations such as the compensation effect and dependence of the activation energy of diffusion on penetrant molecular size, Eq. (24.12), identical to Eq. (24.10), was obtained. Equation (24.12) was derived from (i) the Arrhenius equation for the diffusion of small molecules [Eq. (24.13)], (ii) the correlation between the front factor and activation energy [Eq. (24.14)], and (iii) the correlation of the effect of penetrant size on activation [Eq. (24.15)]:

$$\begin{aligned}\ln \alpha_{A/B} = {} & -\left[\left(\frac{d_B}{d_A}\right)^2 - 1\right] \ln P_A \\ & + \left\{\ln\left(\frac{S_A}{S_B}\right) - \left[\left(\frac{d_B}{d_A}\right)^2 - 1\right] \times \left[b - f\left(\frac{1-a}{Rt}\right) - \ln S_A\right]\right\}\end{aligned} \tag{24.12}$$

$$D_A = D_{0A} \exp\left(-\frac{E_{DA}}{RT}\right) \tag{24.13}$$

$$\ln D_{0A} = a\frac{E_{DA}}{RT} - b \tag{24.14}$$

In these equations, a and b are independent of gas type; a is a universal value (0.64) independent of polymer type (Barrer and Skirrow, 1948) and b has a value of 9.2 for rubbery polymers and 11.5 for glassy polymers (Van Krevelen, 1990).

$$E_{DA} = cd_A^2 - f \tag{24.15}$$

c and f are constants that depend on the polymer and d_A is the penetrant diameter (angstroms).

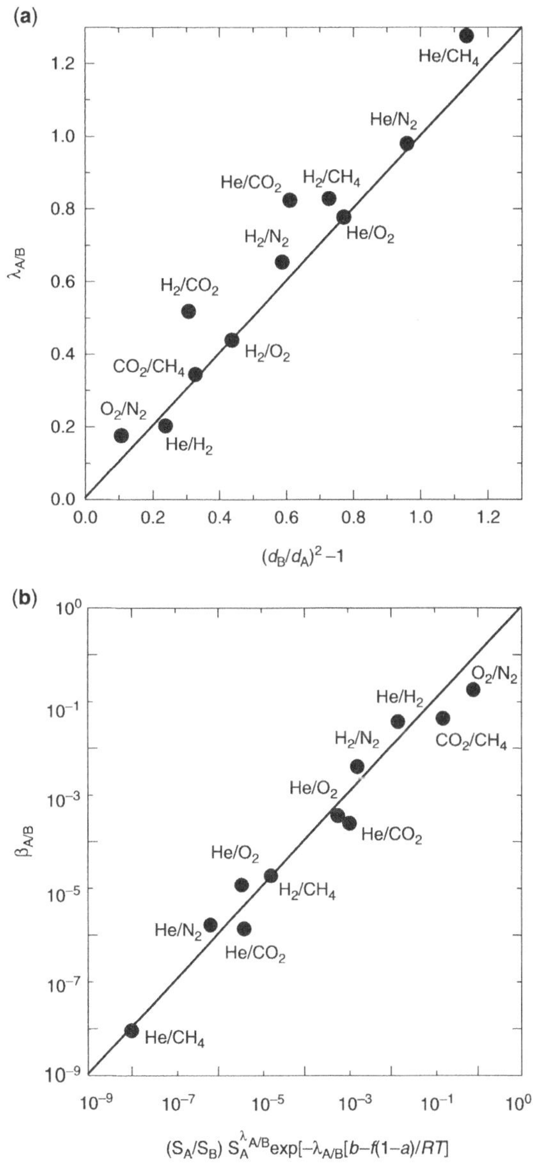

Figure 24.6 Comparison of slopes of ln $\alpha_{A/B}$ vs. ln P_A plots: (a) $\lambda_{A/B}$ and (b) $\beta_{A/B}$ reported by Robeson in 1991 (Fig. 24.6a) and in 1994 (Fig. 24.6b) with the theoretical prediction (solid line) Freeman and cited by (1999).

Equation (24.12) has the same mathematical form as the empirical relation by Robeson (1991) if β_{AB} and λ_{AB} are identified as follows (Fig. 24.6):

$$\lambda_{A/B} = \left(\frac{d_B}{d_A}\right)^2 - 1 = \left(\frac{d_B + d_A}{d_A^2}\right)(d_B - d_A) = k(d_B - d_A) \tag{24.16}$$

$$\beta_{A/B} = \frac{S_A}{S_B} S_A^{\lambda_{A/B}} \exp\left\{-\lambda_{A/B}\left[b - f\left(\frac{1-a}{RT}\right)\right]\right\} \tag{24.17}$$

If k in Eq. (24.16) is constant, Eq. (24.13) correlates well with Eq. (24.11). The value of k is approximately 0.8 and shows little variation.

The trade-off relations of gas separation properties of polymer membranes were repeatedly considered using a free volume model and linear correlations (Alentiev and Yampolskii, 2000). Here the equation identical to Eq. (24.10) was obtained from direct consequences of the free volume model for glassy polymers:

$$\log \alpha_{AB} = -x - (y - 1) \log P_A \tag{24.18}$$

These parameters determine the average lines of experimental correlations of permeability and selectivity and not the upper-bound lines as suggested by Robeson (1991) and Freeman (1999). They lead to the same recommendations of sorption selectivity enhancement for moving above the upper-bound lines of Robeson's correlation. Trade-off relations of gas permeability and selectivity were well correlated with data obtained at room temperature.

24.4 POLYMER MEMBRANE MATERIALS

24.4.1 Siloxane Polymers

Silicon-based polymers are one of the most studied polymer membranes because of their versatile properties. Siloxane-based elastomers are high-performance materials with excellent oxidative and thermal stability, biocompatibility, and high gas and vapor permeability (Zeigler and Fearon, 1990). Siloxane-based polymers are typically prepared by a ring-opening polymerization of a trimer or tetramer (Ivin and Saegusa, 1984), as shown in Scheme 1.

The effects of polysiloxane structures on transport properties are reviewed in detail (Stern et al., 1987). In particular, polydimethylsiloxane (PDMS) is the most important siloxane polymer (Rochow, 1987). Particularly, it is a useful membrane material for gas and vapor separation. PDMS is known as the most permeable rubbery polymer. Due to

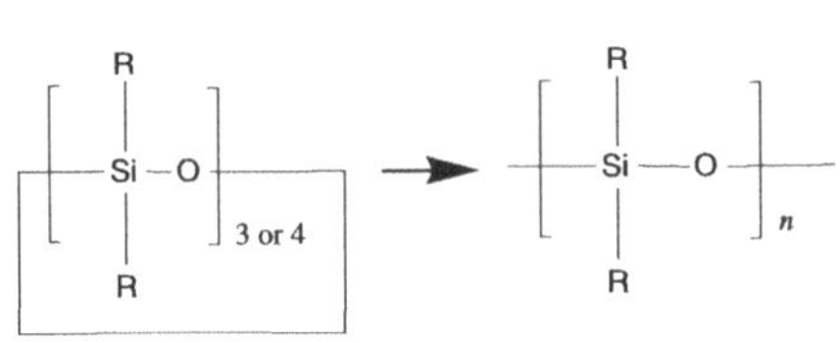

Scheme 24.1

chain flexibility, rotational mobility and large free volume, the low glass transition temperature (-123°C) of PDMS explains its high gas permeability. The extreme chain mobility originates from both dynamic and equilibrium flexibilities. High dynamic flexibility of PDMS is due to the molecule's ability to change spatial arrangements rotating around its skeletal bonds. In addition, the equilibrium flexibility of PDMS is the ability of a chain to be compact in the form of a random coil. In the absence of excluded volume effects, equilibrium flexibility is measured indirectly by the mean-square end-to-end distance or the radius of gyration of the chain (Flory, 1969).

Although siloxane polymers show much higher gas permeability than many other rubbery polymers, the gas selectivity is lower than that of glassy polymers such as polysulfones and polyimides. For instance, oxygen–nitrogen selectivity of PDMS is approximately 2.1 (Mulder, 2000). There have, therefore, been many efforts to improve selectivity without decreasing permeability by varying the basic siloxane structure. Modified siloxane polymers were investigated (Zeigler and Fearon, 1990), but these structural changes did not greatly improve gas separation performance. Therefore, for permanent gas separation, such as O_2/N_2 separation, siloxane polymers have been largely used as a protective coating to avoid the significant decrease of separation performance caused by the probable formation of cracks or defects of asymmetric glassy polymer membranes in bulk production of hollow-fiber or flat-sheet membranes.

The most promising application of siloxane polymer membranes is the separation of organic vapors from permanent gases, such as the recovery of high-value organic vapors, ethylene, propylene, gasoline, and vinyl chloride from industrial waste gas streams (Schultz and Peinemann, 1996; Baker, 2000). So far, the permeation properties of PDMS membranes have been reported for various pure gases and organic vapors or mixed gases (Barrer and Chio, 1965; Baker et al., 1987; Stern et al., 1987; Lee et al., 1988; Jordan and Koros, 1990; Bhide and Stern, 1991; Leemann et al., 1996; Singh et al., 1998; Hagg, 2000; Merkel et al., 2000; Yeom et al., 2002). Although the poly(1-trimethylsilyl-1-propyne) (PTMSP) membrane, a stiff chain and ultrahigh free volume glassy polymer, shows the highest hydrocarbon permeability and excellent selectivity, PDMS is still the most practical polymer membrane. This is because PTMSP exhibits poor chemical resistance toward aliphatic and aromatic hydrocarbons (Pinnau et al., 1997).

Usually, PDMS membranes easily sorb condensable gases and vapors and are swollen by them. Preferable swelling by these penetrants leads to a decrease of selectivity of a vapor or gas over a smaller penetrant in a mixed-gas separation. However, a recent study by Pinnau and He (2004) showed that vapor-induced swelling of PDMS membrane results in improvements in both permeability and selectivity, particularly for hydrocarbon–methane and hydrocarbon–hydrogen separation. They explained that vapor-induced swelling of PDMS results in increased permeability and leads to higher diffusion coefficients of all penetrants. In addition, the relative permeability increase becomes greater as the size of the components in the mixture increase. This is a good example of improving vapor–gas selectivity by increasing vapor–gas diffusivity selectivity. Accordingly, the vapor–gas selectivity increases by increasing vapor concentration in the feed. Pinnau and He (2004) found that a decrease in temperature results in a significant increase in hydrocarbon–methane and hydrocarbon–hydrogen selectivity (Fig. 24.7). Permeation properties of all mixture components are considerably influenced by the swelling of PDMS because the solubility coefficients of condensable mixture components are very high at low temperatures. This might be due to the increase of vapor–gas solubility

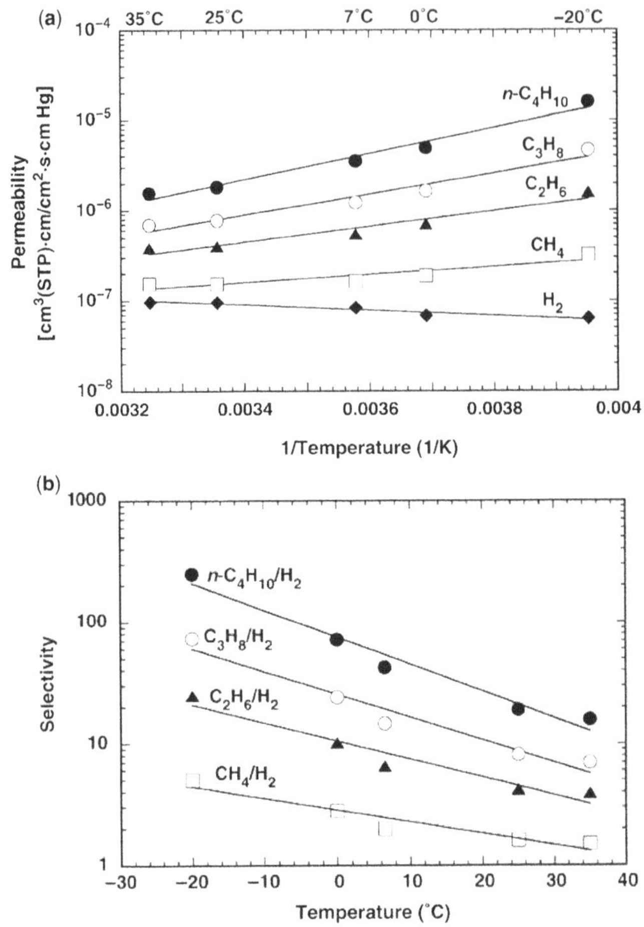

Figure 24.7 For a PDMS membrane: (a) mixed-gas hydrogen, methane, ethane, propane, and *n*-butane permeability vs. inverse feed temperature, and (b) *n*-butane/hydrogen, propane/hydrogen, ethane/hydrogen, and methane/hydrogen selectivity vs. temperature (Pinnau and He, 2004).

selectivity. As a result, vapor-induced swelling of PDMS membranes is promising for industrial applications such as VOC/air, CH_4/H_2, CH_4/N_2, and CO_2/H_2 separation.

24.4.2 Substituted Polyacetylenes

Glassy polymers have long been considered low permeable materials as compared to rubbery ones. Glassy polymers have rigid aromatic chains that may retard the penetration of gas molecules between intersegmental polymer chains. However, the investigation of mono- and disubstituted acetylene polymers and other glassy polymers has increased interest in the microstructures of glassy polymers.

Polyacetylene-based polymers have been evaluated for use in gas separation applications because these amorphous glassy polymers are characterized by high glass transition temperatures above 200°C, low densities, and high gas permeabilites. In particular, PTMSP, a substituted polyacetylene containing bulky substituents, is one of the most studied polymers for membrane use because of its unique structure and excellent gas and vapor permeation properties. A large number of publications about PTMSP and related polymers have appeared since the first report of the synthesis of PTMPS in 1983 (Masuda et al., 1983). The chemical structures of several polyacetylenes are shown in Figure 24.8. The synthesis of high-molecular-weight PTMSP was accelerated by the discovery of WCl_6 analogous catalysts ($NbCl_5$ and $TaCl_5$). PTMSP is a glassy polymer with high gas permeability and low selectivity. It is soluble in nonpolar solvents such as toluene, cyclohexane, and carbon tetrachloride, thus permitting an easy film formation by a solution casting method. The potential applications of PTMSP include oxygen enrichment from air applicable for use in car engines, respiration-aiding apparatuses, and combustion furnaces.

Nagai et al. (2001) reported a detailed review on the synthesis and characterization of substituted polyacetylenes and their use in membrane applications. Discovery of the PTMSP membrane with high gas permeability (Table 24.3) prompted membrane scientists to explore the relationship between gas transport behavior and PTMSP microstructure. High gas or vapor permeability in the PTMSP membrane is due to the ultrahigh free volume of this polymer. This polymer's free volume can be two or three orders of magnitude higher than free volumes of conventional glassy polymers such as polyimides, poly(phenylene oxide)s, and polysulfones. The fractional free volume (FFV) of PTMSP is about 0.34. Typically, it is believed that the large amount of free volume ($\sim$30%) in freshly prepared PTMSP is interconnected, thus permitting the rapid diffusion of gas or vapor (Srinivasan et al., 1994). After the synthesis of PTMSP, other substituted polyacetylenes such as poly(4-methyl-2-pentene) (PMP) and poly(*tert*-butylacetylene) (PTBA) were

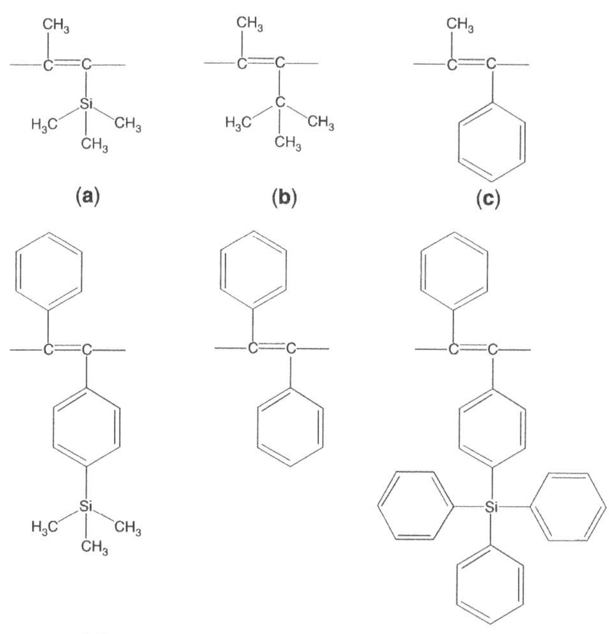

Figure 24.8 Chemical structures of disubstituted polyacetylenes: (a) PTMSP, (b) PMP, (c) PPP, (d) PTMSDPA, (e) PDPA, and (f) PTPSDPA.

TABLE 24.3 Gas Permeability and Selectivity of Disubstituted Polyacetylene Polymers

	Permeability (Barrer)			Selectivity to N_2		
	PTMSP	PMP	PTMSDPA	PTMSP	PMP	PTMSDPA
He	6500	2600	1100	1.03	2.00	1.96
H_2	17,000	5800	2600	2.70	4.46	4.64
N_2	6300	1300	560	1.00	1.00	1.00
O_2	9700	2700	1200	1.54	2.08	2.14
CH_4	15,000	2900	1600	2.38	2.23	2.86
CO_2	34,000	11,000	4900	5.40	8.46	8.75
C_2H_6	26,000	37,000	2700	4.13	28.46	4.82
C_3H_8	32,000	7300	4400	5.08	5.62	7.86
n-C_4H_{10}	102,000	26,000	20,000	16.19	20.00	35.71

synthesized. These polymers, analogous to PTMSP structures, exhibit similar physical properties (densities, free volumes, glass transition temperature, etc.) and are classified as *superglassy polymers.*

Moreover, PTMSP shows unusual permeation properties for large, condensable gases. PTMSP is significantly more permeable to large, condensable gases than small, penetrant gases, in contrast to conventional glassy polymers such as polysulfones. Particularly, PTMSP exhibits both the highest C_{3+} hydrocarbon permeability and the highest C_{3+}/methane and C_{3+}/hydrogen selectivity of any known polymers (Pinnau and Toy, 1996a). For that reason, PTMSP is often called a *reverse selective polymer membrane.* This reverse selectivity makes PTMSP membranes potentially useful for several industrial gas separations such as the removal of higher hydrocarbons from hydrogen streams and the recovery of organic vapors from process streams. Gas permeabilities of permanent gases and condensable gases in superglassy polymers, such as PTMSP and PMP, decrease with increasing temperature. The activation energies of permeation in PTMSP and PMP are negative, contrary to the behavior of conventional glassy polymers.

Despite excellent gas permeation properties of PTMSP, potential application of PTMSP for gas separation is limited because of its rapid loss of free volume by physical aging or relaxation and its lack of chemical stability in heat, radiation, or ultraviolet light in the presence of oxygen. The accessible free volume of PTMSP seems to be very dependent upon the method of preparation or sample history. Therefore, these polymers require improving their long-term stability (Pinnau, 1997) using further cross-linking or mixing with inorganic nanoparticles.

24.4.3 Amorphous Fluoropolymers

A number of fluoropolymers have been extensively investigated since Roy Plunkett discovered Teflon in 1937. Polytetrafluoroethylene (PTFE) exhibits remarkable electric, chemical, thermal, and mechanical properties. Most PTFE-based fluoropolymers are either crystalline or semicrystalline. However, the Teflon AF series is a family of amorphous polymers that was reported in the literature (Resnick, 1976) because these polymers also showed the desirable electric, chemical, thermal, and mechanical properties similar to semicrystalline fluoropolymers. These amorphous fluoropolymers have unique physical properties such as high glass transition temperatures and high free volumes similar to the well-known high free volume glassy polymer, PTMSP.

Fluorinated polymers have many features different from common hydrocarbon polymers. Since they have high free volume and low cohesive energy density (CED), these fluoropolymers can be dissolved in selected fluorinated solvents. They show extraordinarily high gas and vapor permeability (Pinnau and Toy, 1996b; Merkel et al., 1999; Polyakov et al., 2003, 2004). Generally, fluoropolymers have lower CED than any other hydrocarbon polymers, resulting in both enhanced gas solubility and reduced activation energy of diffusion for small molecules.

Copolymers of the Teflon AF series are prepared in four steps starting with hexafluoroacetone (HFA) and ethylene oxide (EO). Condensation reactions of HFA and EO result in 2,2-bis-trifluoromethyl-1,3-dioxolane, which is successively chlorinated, fluorinated, and dechlorinated to give the 2,2-bis-trifluoromethyl-4,5-difluoro-1,3-dioxole (TFMDFD) monomer (Hung, 1993; Resnick, 1976). This monomer copolymerizes with tetrafluoroethylene (TFE). The physical properties of these amorphous copolymers vary according to the relative amounts of the co-monomers, TFMDFD and TFE. Currently, DuPont is producing two commercial grades, AF-1600 and AF-2400. Teflon AF-2400 and AF-1600 are the names of copolymers for which $n = 0.87$ and 0.65, respectively, where n is the percentage of the TFMDFD monomer (see Fig. 24.9). Table 24.4 summarizes physical properties of the two fluoropolymers.

Several recent studies report gas permeation, sorption, and free volume characteristics of these two amorphous fluoropolymers. Nemser and Roman (1991; Nemser, 1993) reported permeability properties of Teflon AF polymer membranes prepared by a melt-press method. Similar to PTMSP, these polymers are permeable to permanent gases. For example, oxygen

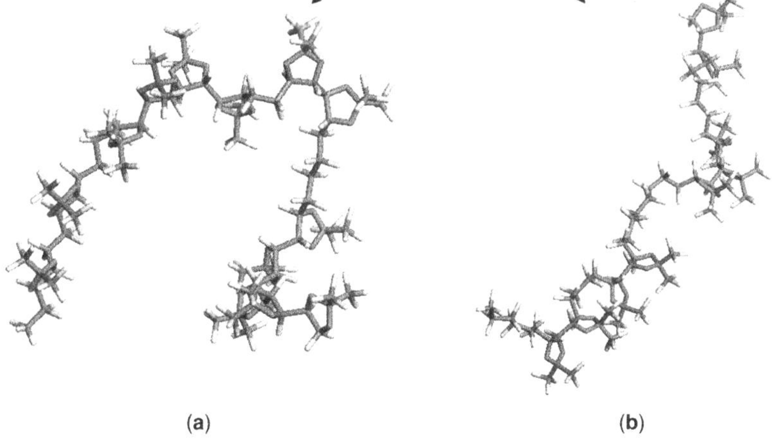

(a) (b)

Figure 24.9 Chemical structures of Teflon (a) AF-2400 and (b) AF-1600.

TABLE 24.4 Physical Properties of Teflon AF-1600 and AF-2400

Trade Name	Fractional Free Volume	TFMDFD (mol%)	Glass Transition Temperature (°C)	Solvent
AF-1600	0.30	65	160	Fluorinert Hostinert Flutec
AF-2400	0.33	87	240	Galden Liquids

and nitrogen permeability of a Teflon AF-2400 membrane is 990 and 480 Barrers, respectively. Pinnau and Toy (1996a) compared gas permeation properties of an AF-2400 membrane prepared by a solution-casting method and a PTMSP membrane. PTMSP and AF-2400 are both ultrahigh free volume glassy polymers, but their gas permeation properties are somewhat different. Permeabilities of small gas molecules in AF-2400 membranes are higher than those of large, condensable gases. PTMSP membranes exhibit the opposite permeation behaviors. The main contribution to selectivity of AF-2400 is due to diffusivity-selectivity, while the selectivity of PTMSP is controlled by large differences in solubility. Also, AF-2400, like PTMSP, has negative activation energies of permeation for small gases but exhibits weak temperature dependence of gas permeability as compared to PTMSP. Because the vapor–gas selectivity of AF-2400 membrane is sensitive to vapor activity, vapor–permanent gas selectivity varies depending upon feed pressure. Therefore, AF-2400 membranes are not suitable for vapor–gas separations, such as the separation of air from organic vapors. On the other hand, AF-2400 membrane can be used as an effective coating material for composite and asymmetric polymer membranes such as siloxane polymers because of its high gas permeability.

24.4.4 High-Temperature Polymers

Polymer membranes are severely limited in their uses in extremely high-temperature and corrosive environments. At temperatures above 100°C at oxidative atmospheres, most of organic polymers may be thermally decomposed or deformed. For that reason, microporous inorganic membranes such as zeolite (Min et al., 2003), silica (Maene et al., 1998), modified alumina (Cho et al., 1995), and carbon molecular sieve membranes (Vu et al., 2002) have been regarded as potential membrane materials able to survive extreme operating conditions such as high temperature, high pressure, and oxidative streams. These microporous inorganic membranes exhibit high selectivity and high permeability as well as thermal, chemical, and mechanical stability. They can be used in specific applications in which polymer membranes fail to make the separation.

However, the cost of inorganic membranes or their modules is still too expensive to be competitive against current polymer membranes or modules. Also, careful attention should be needed to prevent the formation of undesirable pinholes or cracks during fabrication procedures of these inorganic membranes (Fain, 1991). Since they cannot be employed as free-standing thin membranes due to extensive brittleness, inorganic membranes such as zeolite, silica, and carbon molecular sieve require mechanically stable microporous support membranes such as porous steel, glass, or alumina tubes. According to Baker's recent review (2002), it is hard to expect a positive future for ceramic or carbon fiber

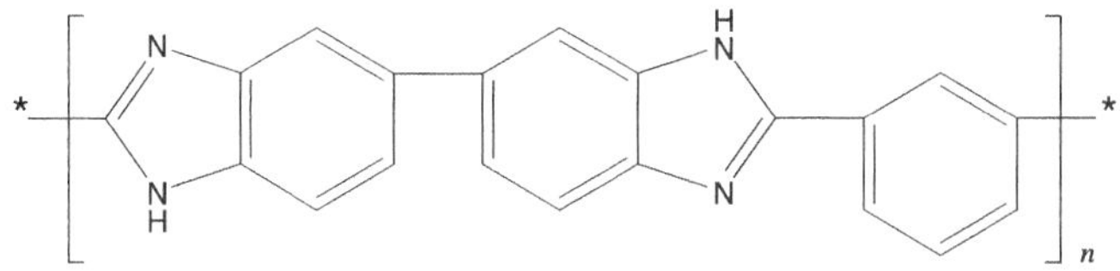

Figure 24.10 Chemical structure of poly[2,2′-(*m*-phenylene)-5,5′-bibenzimidazole] (PBI).

membranes in permanent gas separation applications despite their excellent selectivity and high productivity. At present, it is known that these membranes cost 10–100 times as much as the equivalent polymer membranes.

Therefore, high-temperature polymer membranes for gas separation applications have gradually received much more attention. Candidate polymer membrane materials suitable to high-temperature gas separation applications must not thermally degrade and retain chemically resistant properties. Recently, Pesiri et al. (2003) reported the availability of polybenzimidazole (PBI) for gas separations at high temperatures. PBI was chosen because of its high glass transition temperature above 427°C and its onset of thermal decomposition near 600°C. Figure 24.10 shows the chemical structure of poly[2,2′-(*m*-phenylene)-5,5′-bibenzimidazole] (PBI). It can be synthesized from condensation polymerization of 3,3′,4,4′-tetraaminobiphenyl (TAB) and diphenyl isophthalate (DPIP) in poly(phosphoric) acid or in a hot molten nonsolvent such as sulfolane or diphenyl sulfone (Neuse, 1982). In this experiment, gas permeation tests were carried out at 300–350°C, near the temperature of reformate streams used in hydrogen separations. Poor gas separation performance of PBI at room temperature is due to its rigid-rod structure. However, higher separation performance can be achieved by using operational temperatures above 100°C. For instance, hydrogen permeability of the PBI membrane at room

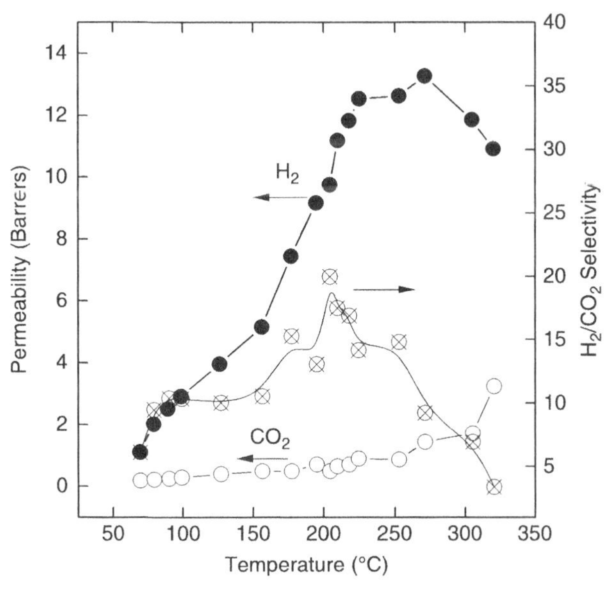

Figure 24.11 Mixed-gas permeation and separation data as a function of temperature for H_2/CO_2 separations using PBI membranes (Pesiri et al., 2003).

temperature is low (0.09 Barrer) but increases up to 18 Barrers at 340°C. Carbon dioxide permeability also increases from 0.01 Barrer at 25°C to 4 Barrers at 340°C. In contrast, H_2/CO_2 selectivity decreased from 9 to 4.5. Interestingly, H_2/CO_2 selectivity at temperature ranges between 200 and 270°C increased to about 20 (Fig. 24.11). It is interesting that hydrogen permeability of PBI membrane shows non-Arrhenius behavior within the temperature ranges used in the experiments. At higher temperature ranges, carbon dioxide permeability might be more dominant, resulting in unexpected changes in selectivities.

Although this study provides insight into polymer membrane materials for separating valuable gases from high-temperature gas streams, more information is required to understand gas transport and the mutual interactions that affect gas selectivity at temperatures above 100°C. There are very few reports so far, and the single or mixed-gas permeation data (including diffusivity and/or solubility) is not sufficient for analyzing gas transport behavior at high temperature ranges. Although H_2/CO_2 selectivity of PBI is potentially useful for commercial hydrogen production applications, the H_2 permeability of PBI is only about 100 Barrers at 300°C. If PBI membrane could be considered for practical applications, such as hydrogen purification or as a water gas shift membrane reactor, its properties must be tailored to achieve high H_2 flux (productivity) in a practical module and system.

Other known high-temperature polymers, such as polyimide, polybenzoxazole, polybenzithiazole, and polypyrrolone (Fig. 24.12) should also be investigated as potential H_2 separation membranes, microporous substrate materials in a thin-film composite membrane structure, or matrix (base) materials for making organic-inorganic hybrid or composite membranes with high H_2 permeability and selectivity. Polyimide has been used in membranes for commercial gas separation. However, the selectivity of polyimides for H_2 over other synthesis gas species is much lower than that of PBI. For instance, the H_2/CO_2 selectivity of polyimide is in the range of 5–13 at 200–300°C. Thus, polyimides are considered mainly as a substrate for thin-film composite membranes or as hybrid or composite membrane matrix materials.

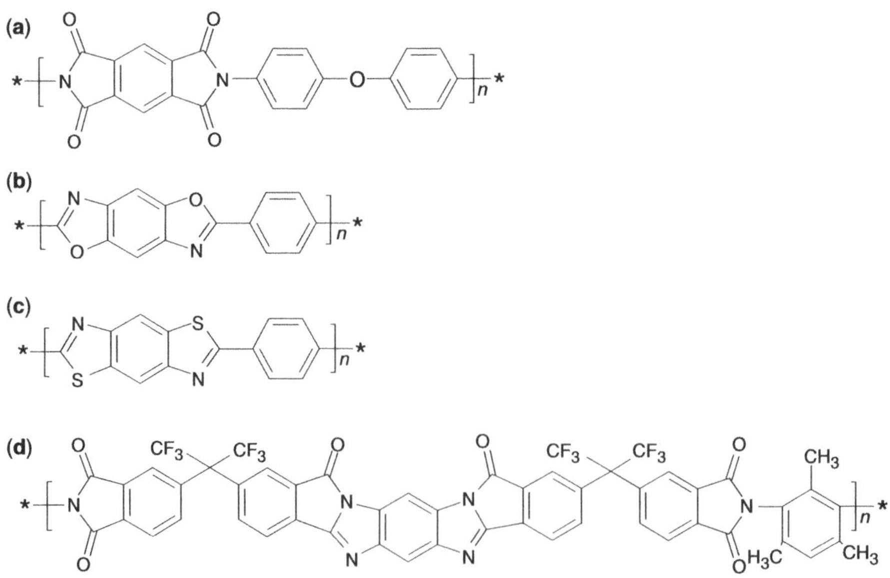

Figure 24.12 Chemical structures of promising high-temperature polymers: (a) polyimide, (b) polybenzoxazole, (c) polybenzithiazole and (d) polypyrrolone copolymer.

In the near future, it is a great challenge to extend the applicable range of polymer membranes by developing better polymer membranes for gas separation at high temperatures. In this regard, many questions still remain for membrane scientists and engineers because numerous studies on permeability and permselectivity of polymer membranes have been carried out at ambient temperature, but there are only a few experimental data and theoretical approaches at high temperature ranges.

24.4.5 Nanocomposites and Mixed-Matrix Membranes

Over the last several decades, the introduction of inorganic nanoparticles into a continuous polymer matrix has been tried to improve mechanical and thermal stability of conventional polymer membranes and to improve gas separation performance. These membrane materials are called *mixed-matrix membranes* or *hybrids* or *composite membranes*. Inorganic particles mixed with organic polymer matrix can be classified into two categories: porous and nonporous inorganic particles. Colloidal silica (from in situ sol–gel reactions) and zeolite belong to porous inorganic particles, while fumed silica and clay belong to nonporous inorganic particles.

Although there are many publications dealing with mixed-matrix membrane materials in the literature in the last 10 years, the original concept dates back to the past decades. Acting as physical cross-linkers, inorganic fillers such as fumed silica or carbon black were already used for improving the mechanical properties of natural rubbers or synthetic rubbery polymers. However, with the advent of nanotechnology, the controlled combination of polymer and nanosized inorganic components becomes useful beyond the usual mixing for the improvement of physical and chemical properties. It is also a promising technology for tailoring polymer microstructures that might affect gas transport behavior in the membrane-based gas separation field.

Typically, the addition of inorganic particles may improve the gas selectivity by increasing the diffusional pathway for gases and can also strengthen the antiplasticization effect by restricting the thermal motion of polymer chains. More recently, unexpected gas transport behavior of polymer nanocomposite membranes was reported (Merkel et al., 2002). Polymer nanocomposites consisting of poly(4-methyl-2-pentyne) (PMP) and hydrophobic-surface-treated, nanosized fumed silica were prepared in the wide range of fumed silica content. The fumed silica is so nonporous that gas permeability was expected to decrease with the increase of silica content. Classically, the incorporation of impermeable filler particles into a polymer leads to a systematic reduction in molecular transport (Maxwell, 1873; Neogi, 1996). The original idea of this work might be to prevent the severe decline of gas permeabilities caused by the progress of physical aging in ultrahigh free volume glassy polymers such as PTMSP and PMP, and simultaneously to improve the gas selectivity by increasing the diffusion pathway because nonporous inorganic fillers can act as both a physical hardener and barrier. In contrast to this initial expectation, it was reported that the gradual addition of nanostructured fumed silica to several glassy high free volume polymers results in an increase in permeability (Merkel et al., 2002; Gomes et al., 2005), as shown in Figure 24.13. The unexpected improvement of gas permeability is supposed to be due to the disruption of polymer chain packing by the silica particles, leading to an increase in the free volume available for molecular transport.

When the high free volume polymer PMP is combined with inorganic nanoparticles, an increase in vapor–gas selectivity was also reported (Merkel et al., 2003a). For PTMSP combined with inorganic nanoparticles, different results on selectivity have been published

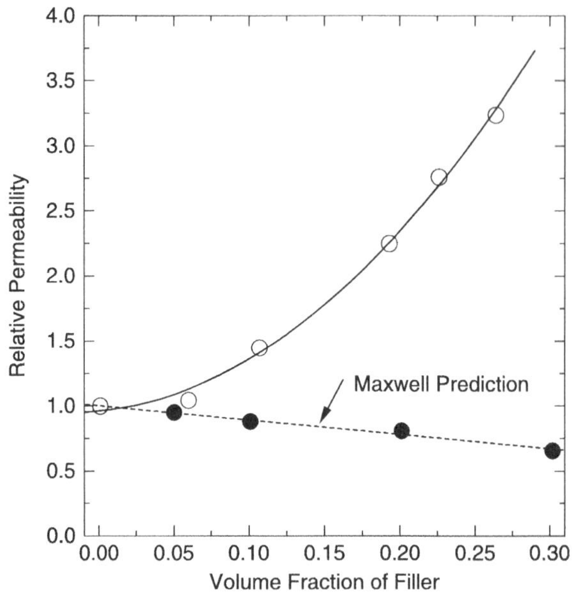

Figure 24.13 Ratio of penetrant permeability in the nanocomposite to that in the pure polymer as a function of filler volume fraction (Merkel et al., 2002).

(Merkel et al., 2003a, 2003b, 2003c). The incorporation of fumed silica can increase the free volume of PTMSP without creating nonselective defects. However, it can also create free volume elements large enough to allow nonselective Knudsen transport (Gomes et al., 2005). To fully understand these phenomena, more systematic studies of important factors (e.g., the dispersion state of inorganic nanoparticles into the polymer matrix, the effect of different kinds of polymers, particle sizes, particle shapes, and kinds of particles) are needed. Control of the complex nanoscale interfaces between the two major phases represents a significant technical matter to be overcome in transitioning from small scale to large scale.

Hybrid membranes of organic polymers and inorganic particles may be promising materials for improving the weaknesses of organic polymer membranes. Hybrid membranes can be prepared from inorganic particles coupled with an organic polymer matrix by covalent bonding. These couplings between different phases can prevent undesirable reductions of selectivity due to large interfaces between the inorganic particle and the organic polymer. The incorporation of inorganic particles or networks into organic polymer matrices can significantly reduce the loss of selectivity at elevated temperatures by retarding thermal motion of the polymer chain. Park et al. (2003) reported the gas permeation properties of imide-siloxane copolymer–silica hybrid membranes. They used silane-coupling agents to prepare imide-siloxane copolymer–silica hybrid membranes. Even at high silica content (up to 50 wt%), the hybrid membranes did not experience macroscopic phase separation due to the existence of coupling agents. This might be due to the improvement of compatibility between organic polymer matrix and inorganic particles. Changes of selectivity of the hybrid membranes at low and high temperatures depends on silica content (Fig. 24.14). High silica content in hybrid membranes decreases the falling-off ratio of gas selectivity at an elevated temperature in comparison with organic polymer membranes.

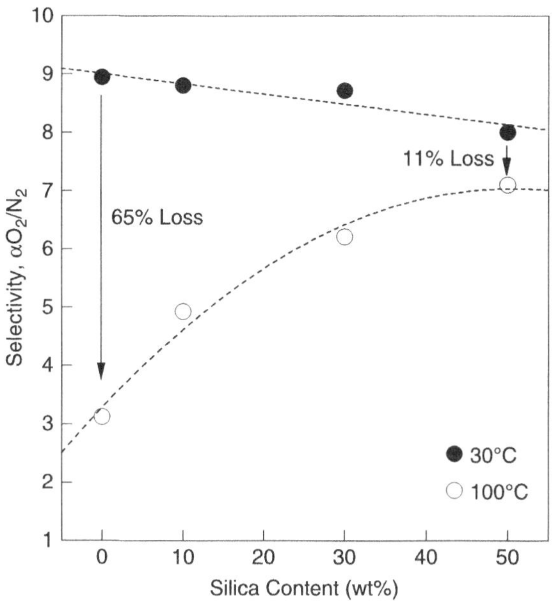

Figure 24.14 Falling-off ratio of O_2/N_2 selectivity at an elevated temperature in polyimide–silica hybrid membranes (Park et al., 2003).

24.4.6 Other Promising Polymers

24.4.6.1 Semicrystalline Polymer Many efforts to improve permselectivity by increasing solubility-selectivity in the polymer matrix have failed. This is due to molecules sorbing favorably in the polymer and acting as plasticizers, resulting in a significant decrease of permselectivity by increasing the permeability coefficient of undesirable molecules. Poly(amide-6-*b*-ethylene oxide) (PEBAX) may be a good choice to overcome this problem. PEBAX is a block copolymer consisting of a polyamide (PA) block as a hard segment and a polyether (PE) block as a soft segment. A crystalline amide block in PEBAX functions as an impermeable phase, whereas an ether block acts as a permeable phase because of its high chain flexibility.

A couple of studies with gas permeation results of PEBAX were reported in the literature. Rezac et al. (1997) reported the sorption and diffusion characteristics of water and methanol in a series of PEBAX copolymers. The study reveals that those polymers can be used to separate methanol selectively from air, rather than separating methanol from water. Bondar et al. (2000) found that in applications such as the removal of CO_2 from mixtures of hydrogen in syngas, PEBAX shows high selectivity in polar or quadrupolar/nonpolar systems (e.g., CO_2/H_2 or CO_2/N_2). The sorption and permeation results indicate strong interactions between the polar gas CO_2 and the PE blocks in the copolymer. Kim et al. (2001) and Kim and Lee (2001) found that strong affinities of polar species to the PE block is attributable to the high permeability and permselectivity of polarizable gases through the PEBAX copolymer. This leads to high CO_2/N_2 selectivity up to 61 and SO_2/N_2 selectivity up to 500. The permeability of CO_2 in PEBAX ranged between 110 and 140 Barrers. The CO_2 permeability in PEBAX decreased with an increase of hard PA content from 20 to 46% in PEBAX, as shown in Figure 24.15. In the presence of

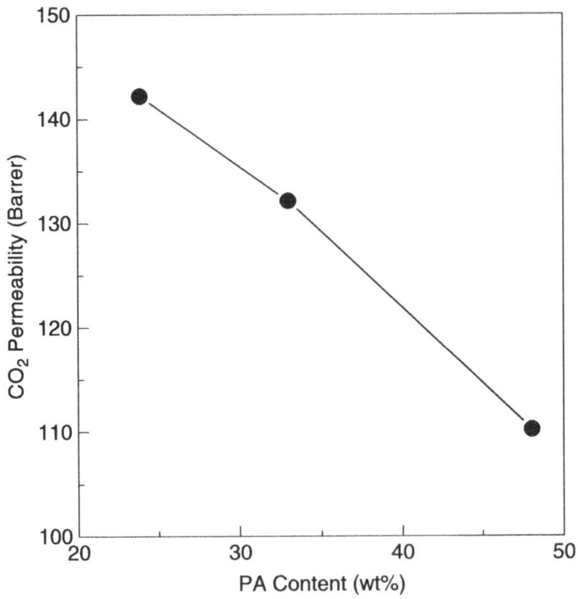

Figure 24.15 Change of CO_2 permeability with the amount of crystalline region in PEBAX membranes (Kim et al., 2001).

uniform and rigid chain packing, a crystalline polymer is lacking in both the sorption sites as well as the mobility of the chains to allow transport of gas molecules.

24.4.6.2 Polymers with Intrinsic Microporosity Dense polymer membranes are classified as nonporous materials because most amorphous polymers have flexible backbones enabling them to pack together efficiently. However, all polymers contain microvoids or free volume. Highly rigid glassy polymers have a free volume up to 20%. Physical aging such as annealing and quenching or the use of different solvents can change this free volume.

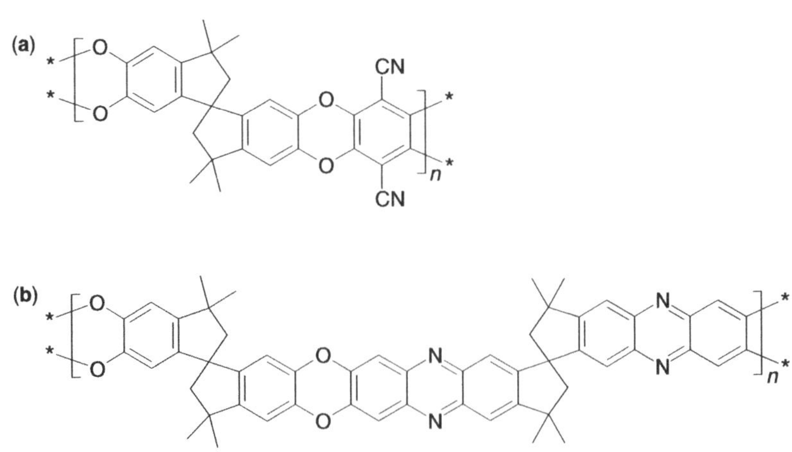

Figure 24.16 Chemical structures of PIM: (a) PIM-1 and (b) PIM-7 (Budd et al., 2005b).

Recently, British researchers (Budd et al., 2004, 2005a, 2005b; McKeown et al., 2005) reported novel polymer membranes with intrinsic microporosity, PIM, shown in Figure 24.16. These polymers show exceptional microporous structures, high gas permeability, and high permselectivity. Their rigid but contorted molecular structures hinder polymer chain packing and create free volume coupled with chemical functionality, providing strong intermolecular interaction. These PIM polymers are unique in that they have high surface areas (500–1000 $m^2\ g^{-1}$) with micropore diameters in the range of 0.6–0.8 nm from the fact that microporous materials containing large surface areas of 300–1500 $m^2\ g^{-1}$ are mostly inorganic materials such as crystalline zeolites (aluminosilicates) and activated carbons. The oxygen and carbon dioxide permeabilities of the PIMs are in the range of 190–370 Barrers and 1100–2300 Barrers, respectively, and the PIMs exhibit selectivities significantly higher than any other high free volume glassy polymer membranes such as PTMSP and Teflon AF series polymer membranes. These materials are good candidates for applications such as the generation of oxygen-enriched air for enhanced combustion and fermentation processes or for the removal of carbon dioxide from methane.

24.5 MEMBRANE GAS SEPARATION APPLICATIONS AND CONCLUSIONS

24.5.1 Air Separation

Gas separation membranes for removing nitrogen from air have been available since the mid-1980s. The first membranes were spiral wound, but, today, hollow-fiber membranes have been used. For the main gases involved in air separation, the following applies: water–vapor permeates fastest, followed by CO_2, O_2, and finally N_2. This order of gas permeation is utilized in nitrogen separation since oxygen can be favorably extracted from compressed air. Today, polymer membrane systems are commonly used if a residual oxygen content of 1% is acceptable. If greater degrees of purity are needed, pressure swing adsorption (PSA) systems are more cost effective.

Air separation membrane units can be operated in either pressure or vacuum mode, as shown in Figure 24.17. In the pressure mode, feed air is typically pressurized to several bars, while permeate is maintained near atmospheric pressure ($\sim$1 bar). The differential pressures are higher than those of the vacuum mode, leading to reduced membrane area requirements. In the vacuum mode, feed air is pressurized slightly above atmospheric pressure, and a vacuum is maintained on the permeate side of the membrane. The retentate is vented at atmospheric pressure. The vacuum mode is typically more energy efficient than the pressure mode because a vacuum is applied only to the permeate stream.

For oxygen enrichment air applications, the vacuum mode is more energy efficient and therefore more suitable. However, the vacuum mode requires a larger membrane area than the pressure mode because of the limited differential pressure. The pressure mode is more suitable for nitrogen production because the retentate can be obtained at conditions closer to feed conditions. In some cases, vacuum and pressure modes can be used together for a synergetic effect. In this mixed mode of operation, the feed air is pressurized and a partial vacuum is maintained on the permeate side to increase both the feed-to-permeate pressure ratio and the feed-to-retentate pressure ratio. The oxygen-enriched permeate is then mixed with an ambient airstream to obtain the desired airflow and oxygen concentration (Fig. 24.18). Significant applications of air separation are the production of

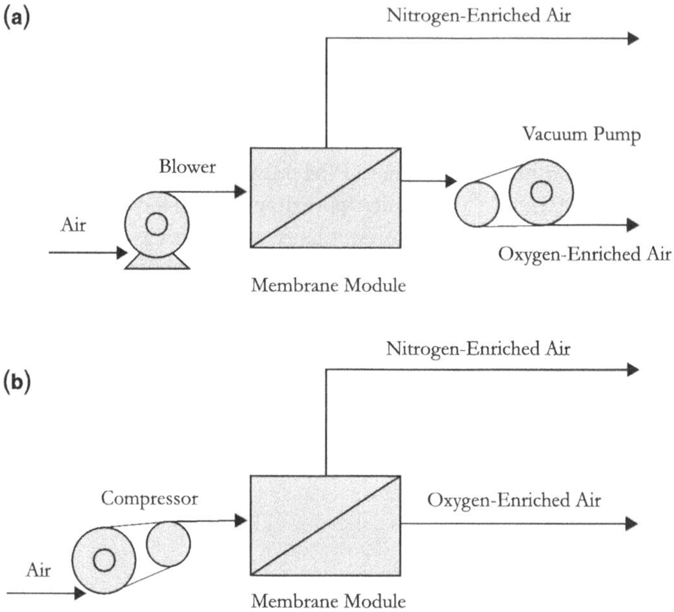

Figure 24.17 Schematic representation of (a) vacuum and (b) pressure mode for air separations.

oxygen and water-depleted air for oil and gas drilling, controlled atmosphere, maritime transport, inerting, beverage dispensing, tire inflation, and laboratory gas supply.

24.5.2 Hydrogen Separation

Hydrogen separation was the first commercial usage for membrane-based gas separations, and it is of great importance in many petrochemical processes. Polysulfone

Figure 24.18 Oxygen generator operated at both pressure and vacuum mode: (a) polysulfone hollow-fiber module, (b) oxygen sensor, (c) compressor and vacuum, (d) vacuum gauge, and (e) feed pressure gauge.

hollow-fiber membranes were used for hydrogen separation in 1980 when H_2 demands were rising. Because hydrogen is highly permeable, selectivities to the other gases are typically high. Hydrogen separation by membranes can be used for ammonia purge gas streams (H_2/N_2, H_2/CH_4, and H_2/Ar), refineries, and syngas production. Another application includes extraction of hydrogen from petroleum naphtha cracking streams. Spillman and Grace (1989) compared the economic figures for a membrane-based system with a cryogenic separation. The investment costs were comparable for both options ($1.35 million for 450,000 m^3/day separation), and the hydrogen recovery rates were similar. Even though this comparison is from 1989, the membrane advantages are still clear.

The main processes for hydrogen upgrading in refineries are the PSA, cryogenic separation, and membrane separation. The increased need for hydrogen in refineries is due to increased environmental regulation and heavier crude. Product purity can be maintained for small changes in feed composition by adjusting the feed-to-permeate pressure ratio. But, the relatively strong impact of product purity on recovery for these systems means that hydrogen recovery can be significantly affected. In most refinery membrane applications, however, the major impurity is methane. This can be allowed to increase in the product slightly without major downstream impact.

In the near future, the hydrogen economy will expand, and hydrogen production will need to be increased from what it is today. Hydrogen comes from multiple sources. These include renewable sources such as solar, wind, hydropower, and biomass, and nonrenewable sources such as fossil fuels and nuclear energy. Hydrogen-based power generation systems, such as fuel cells, have the potential to eliminate emission of all pollutants, including nitrogen oxides, carbon monoxide, volatile organic chemicals, and particulates. Fuel cells are also potentially useful for clean power generation at small, distributed facilities as well as for emergency backup and portable power generation. The availability of inexpensive yet highly pure hydrogen is a prerequisite for the transition to fuel-cell-based power generation. It would be ideal to produce pure H_2 from electrolysis of H_2O using solar energy generated by solar cells. However, for now, the inevitable choice is to obtain H_2 from fossil and biomass fuels. Hydrogen can be recovered from several kinds of gas streams. Many of them include low levels of hydrogen and undesirable impurities. One significant separation is the separation of CO_2/H_2. H_2 is produced via steam reforming of hydrocarbons to make CO and H_2 and then a water–gas shift is used to make CO_2 and H_2. Syngas (or synthesis gas) can be prepared from methane as an intermediate step in the production of basic chemicals, such as methanol, ammonia, and oxo-alcohols. Carbon dioxide is produced in the syngas reaction and must be removed before further catalytic conversion of CO and H_2 can proceed toward the desired chemical. Depending upon the kinds of basic chemicals to be produced, the H_2/CO syngas ratio may need to be adjusted. Water–gas shift is an old and established technology that may find new applications in fuel cells and coal-processing systems. Water–gas shift is a reversible, exothermic reaction that is thermodynamically unfavorable at elevated temperatures. Membrane technology may potentially provide opportunities for new applications of water–gas shift. By combining the water–gas shift reaction with selective removal of one of the reaction products, a single reactor can operate simultaneously at high temperature and high conversion, without the requirement of excess steam. Such a membrane reactor needs a membrane that is highly permselective for either H_2 or CO_2.

24.5.3 Hydrocarbon–Hydrocarbon Separation

A potential and challenging membrane-based gas separation process is olefin–paraffin separation. The existing technology of cryogenic distillation consumes a lot of energy because the olefin–paraffin pairs of interest (ethylene–ethane, propylene–propane) have similar boiling and evaporation properties. Olefin–paraffin separations are a heavy expense for the petrochemical industry. With a growing awareness of the significance of conserving natural resources, companies are dedicated to finding techniques that reduce energy consumption and that recycle purge or waste streams. Conventionally, cryogenic distillation at elevated pressures in trayed fractionators has been used to separate olefins and paraffins. However, this distillation system is too expensive to build and operate and is currently only economically feasible for streams containing high-quality olefins.

Unfortunately, current polymer membranes show poor selectivities in olefin–paraffin separations. The effect of plasticization and temperature on the selectivity must be further improved. The most promising membranes for olefin–paraffin separations are facilitated-transport polymer membranes. Olefin compounds having a π-bond can penetrate through the membranes with transition metals such as cuprous and silver ions by a *facilitated transport mechanism*. An active transport of specific molecules through a membrane is achieved by utilizing a carrier species. The carrier shuttles a specific penetrant across membrane boundaries. The penetrant is quickly transported from the side with higher penetrant concentration to the side with lower penetrant concentration. When a feed mixture contains only one species with which the carrier can react, only that species is transported through the membrane. The driving force in facilitated transport is a concentration gradient of penetrant–carrier complexes across the membrane. Unfortunately, these facilitated transport membranes have severe problems such as poor long-term physical stability and chemical instability caused by degradation of the carrier ion.

24.5.4 Carbon Dioxide Separation

Carbon-dioxide-selective separation is becoming an important issue in areas such as petrochemical engineering (e.g., CO_2 removal from natural gas), environment (e.g., CO_2 removal from flue gas), agriculture (e.g., control of CO_2 concentration), and other related industries. Membranes can selectively separate CO_2 from industrial processes.

24.5.4.1 Carbon Dioxide Removal from Natural Gas

The production of natural gas requires pretreatment before delivery to the pipeline. Typically, raw natural gas contains 75–90% methane. It also contains ethane, propane, butane, and 1–3% of other hydrocarbons. Impurities such as nitrogen, water, carbon dioxide, and hydrogen sulfide may be present and should be removed to upgrade to the required natural gas purity. Membrane technology competes with the widely used amine absorption technique in small-scale applications (<5 million scfd). Although amine absorption suffers from high capital costs and complex operations, membrane systems cannot be a substitute for the amine absorption techniques because current membranes still exhibit low gas permeability and selectivity. Nonplasticized high-CO_2-permable polymeric membranes should be developed for practical applications.

24.5.4.2 Natural Gas Sweetening Processes

These processes remove CO_2 from high-pressure methane stemming from natural gas walls. Many natural gas streams are available in the pressure range of 330–1200 psig. In this case, membrane systems are frequently used in combination with conventional gas separation processes. The bulk of CO_2 can be separated from raw natural gas by using membrane systems. The final purification to pipeline-quality gas can be carried out by the amine absorption process. The capacity of an existing amine plant can be increased by integrating a membrane unit.

24.5.4.3 Carbon Dioxide Recovery from Landfill Gas Landfill gas is produced at atmospheric pressures from the decomposition of organic materials under anaerobic conditions. This biogas contains 40–45 mol% CO_2, 54–59 mol% CH_4, 4% nitrogen, 1% oxygen, 1% water vapor, and traces of hydrogen sulfide and halogenated hydrocarbons (CFC). This biogas can be recovered for potential use when the landfill is covered. The upgrading of CH_4 for use in the local gas distribution system is possible by applying a membrane system. For this application, the toxic trace gases of H_2S and CFC are removed first by an adsorption process. The remaining gas stream is fed to a membrane system and is often compressed to higher pressure, up to 500 psig, to enhance membrane efficiency.

24.5.4.4 Enhanced Oil Recovery In enhanced oil recovery (EOR), the gas stream exits the field at pressures frequently as high as 2000 psig. CO_2 is used as injection medium in existing oil fields to increase the oil production. Because its viscosity is lowered by the dissolved CO_2, the oil is easily driven to the surface. Along with the oil, a casing head gas is produced. This gas contains a variety of hydrocarbons including substantial amounts of methane. The casing head gas is mixed with CO_2 at the well. Concentration levels of 40–90 mol% carbon dioxide are achieved. The CO_2 must be removed to utilize the hydrocarbons. In this application, the natural gas as well as the carbon dioxide is the desired product. The natural gas is to be used as fuel and the CO_2 is to be used for reinjection. In the latter case, a purity of 95 mol% is often sufficient to maintain the solvent power of CO_2.

24.5.4.5 Further Natural Gas Purification Further expansion of the natural gas purification market to include CO_2 removal, dehydration, and removal of higher hydrocarbons to control heating values and dew points will be possible using membrane systems.

24.5.5 Vapor–Gas Separation

In this application, organic vapor removal from air or nitrogen streams, such as ethylene, or propylene recovery from polyolefin production, and gasoline vapor from air are representative membrane processes. MTR (Membrane Technology Research) installed the first propylene recovery membrane system (VaporSep) in 1996. Since then, over 20 systems that recover propylene and nitrogen from resin degasing bins have been installed. Typically, during the production of polypropylene, a considerable amount of the propylene feedstock is lost. The value of the lost feedstock is substantial, ranging from $1 million to $3 million per year for a typical polypropylene plant. Usually, for vapor–gas separation, rubbery polymers such as cross-linked PDMS or reverse-selective glassy polymers such as PTMSP or PMP can be used. These polymers are mostly solubility-selective membranes.

Usually, the emission of gasoline vapors at gas stations generates the formation of ground-level ozone due to high emissions of hydrocarbons. Membrane-based gas separations can recover and recycle gasoline vapors back to the underground storage tanks. For gasoline vapor recovery, a silicone-based membrane system was first considered.

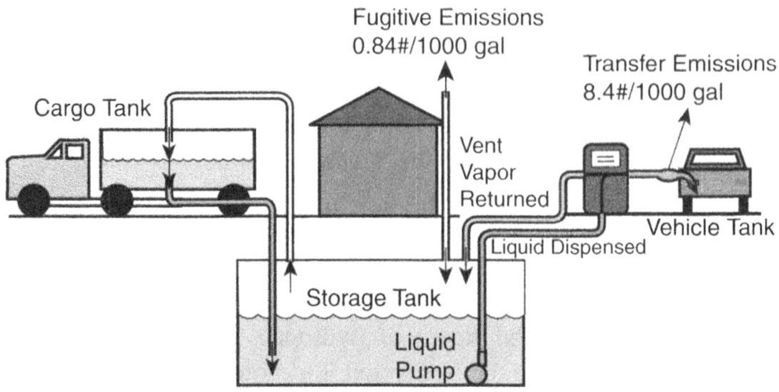

Figure 24.19 Gasoline vapor separation system using fluoropolymer membranes (Koch, 2001).

Silicone membranes for vapor recovery and recycling separate the hydrocarbon portion of the gasoline vapor, and the retained air is vented. However, silicone rubbers exhibit a weak resistance to gasoline vapors. If the system failed, further safeguards must be present to avoid uncontrolled release of gasoline vapors into the atmosphere. In addition, silicone-based membrane systems require many components such as pumps, condensers, valving, fans, and membrane cartridges to operate properly. For that reason, CMS (Compact Membrane Systems, Inc.) used fluoropolymer membranes to permeate and vent relatively clean air to the atmosphere while retaining and recycling gasoline vapors to the tank (Fig. 24.19). Amorphous fluoropolymers have high free volume and high permanent gas permeability but selectively prevent large vapor penetration across the membrane. Accordingly, no gasoline vapor emissions would result if power or system failures occurred. Because complex mechanical systems such as a condenser are unnecessary and fluoropolymers have excellent chemical stability in gasoline vapor, the recovery of gasoline vapor at gas stations is more economic and reliable when fluoropolymers are used.

REFERENCES

Alentiev, A. Y., and Yampolskii, Y. P. (2000). Free volume model and tradeoff relations of gas permeability and selectivity in glassy polymers. *J. Membr. Sci.* 165, 201.

Alves, V. D., Koroknai, B., Belafi-Bako, K., and Coelhoso, I. M. (2004). Using membrane contactors for fruit juice concentration. *Desalination* 162(1–3), 263.

Anderson, M. R., Mattes, B. R., Reiss, H., and Kaner, R. B. (1991). Conjugated polymer films for gas separations. *Science* 252, 1412.

Baker, R. W. (2000). *Membrane Technology and Applications*. McGraw-Hill, New York.

Baker, R. W. (2002). Future directions of membrane gas separation technology. *Ind. Eng. Chem. Res.* 41, 1393.

Baker, R. W., Yoshioka, N., Mohr, J. M., and Khan, A. J. (1987). Separation of organic vapors from air. *J. Membr. Sci.* 31, 259.

Barrer, R. M., and Chio, H. T. (1965). Solution and diffusion of gases and vapors in silicone rubber membranes. *J. Polym. Sci. C* 10, 111.

Barrer, R. M., and Skirrow, G. J. (1948). Transport and equilibrium phenomena in gas-elastomer systems. I. Kinetic phenomena. *J. Polym. Sci.* 3, 549.

Bhide, B. D., and Stern, S. A. (1991). Permeability of silicone polymers to hydrogen. *J. Appl. Polym. Sci.* 42, 2397.

Bondar, V., Freeman, B. D., and Pinnau, I. (2000). Gas transport properties of poly(ether-*b*-amide) segmented block copolymers. *J. Polym. Sci. Polym. Phys.* 38, 2051.

Brandrup, J., Immergut, E. H., and Grulke, E. A. (1999). *Polymer Handbook*, 4th ed. Wiley-Interscience, New York.

Breck, D. W. (1974). *Zeolite Molecular Sieves: Structure, Chemistry and Use*. Wiley, New York, pp. 593–724.

Budd, P. M., Elabas, E. S., Ghanem, B. S., Makhseed, S., McKeown, N. B., Msayib, K. J., Tattershall, C. E., and Wang, D. (2004). Solution-processed, organophilic membrane derived from a polymer of intrinsic microporosity. *Adv. Mater.* 16, 456.

Budd, P. M., McKeown, N. B., and Fritsch, D. (2005a). Free volume and intrinsic microporosity in polymers. *J. Mater. Chem.* 15, 1977.

Budd, P. M., Msayib, K. J., Tattershall, C. E., Ghanem, B. S., Reynolds, K. J., McKeown, N. B., and Fritsch, D. (2005b). Gas separation membranes from polymers of intrinsic microporosity. *J. Membr. Sci.* 251, 263.

Cho, Y.-K., Han, K., and Lee, K.-H. (1995). Separation of CO_2 by modified-Al_2O_3 membranes at high temperature. *J. Membr. Sci.* 104, 219.

Crank, J. (1975). *The Mathematics of Diffusion*, 2nd ed. Oxford University Press, 438 pp.

Crank, J., and Park, G. S. (Eds.) (1968). *Diffusion in Polymers*, 2nd ed. Clarendon, Oxford.

Fain, D. E. (1991). Technical and economic aspects and prospects for gas separation with inorganic membranes. *Key Eng. Mater.* 61–62, 327.

Flory, P. J. (1969). *Statistical Mechanics of Chain Molecules*. Wiley-Interscience, New York.

Freeman, B. D. (1999). Basis of permeability/selectivity tradeoff relations in polymeric gas separation membranes. *Macromolecules* 32, 375.

Gee, G. (1947). Equilibrium properties of high-polymer solutions and gels. *J. Chem. Soc.* 280–288.

Gomes, D., Nunes, S. P., and Peinemann, K.-V. (2005). Membrane for gas separation based on poly(1-trimethylsilyl-1-propyne)-silica nanocomposites. *J. Membr. Sci.* 246, 13.

Graham, T. (1866). On the absorption and dialytic separation of gases by colloid septa. *Philos. Mag.* 32, 401.

Gregor, E. C., Tanny, G. B., Shchori, E., and Kenigsberg, Y. (1988). Sunbeam Process microporous membranes. A high-performance barrier for protective clothing. *J. Coated Fabrics* 18, 26.

Hagg, M. B. (2000). Membrane purification of Cl_2 gas. I. Permeabilites as a function of temperature for Cl_2, O_2, N_2, H_2 in two types of PDMS membranes. *J. Membr. Sci.* 170, 173.

Hayashi, J., Mizuta, H., Yamamoto, M., Kusakabe, K., and Morooka, S. (1997). Pore size control of carbonized BPDA-pp'ODA polyimide membrane by chemical vapor deposition of carbon. *J. Membr. Sci.* 124, 243.

Hildebrand, J. H., and Scott, R. L. (1950). *The Solubility of Non Electrolytes*, 3rd ed. Reinhold, New York.

Hung, M. H. (1993). Structure-property relationship of fluorinated dioxole polymers. *Macromolecules* 26, 5829.

Ivin, K. J., and Saegusa, T. (1969). In K. C. Frisch and S. L. Reegen (Eds.), *Ring-Opening Polymerization*. Dekker, New York.

Ivin, K. J., and Saegusa, T. (1984). Ring-opening polymerization. Chapter 14. Cyclic Siloxane. Elsevier. pp. 1055–1133.

Jordan, S. M., and Koros, W. J. (1990). Permeability of pure and mixed gases in silicone rubber at elevated pressures. *J. Polym. Sci. B: Polym. Phys.* 28, 795.

Kim, J. H., Ha, S. Y., and Lee, Y. M. (2001). Gas permeation of poly(amide-6-*b*-ethylene oxide) copolymer. *J. Membr. Sci.* 190, 179.

Kim, J. H., and Lee, Y. M. (2001). Gas permeation properties of poly(amide-6-*b*-ethylene oxide)-silica hybrid membranes. *J. Membr. Sci.* 193, 209.

Koch, W. H. (2001). Developing technology for enhanced vapor recovery: Part 1—Vent processors. *Petrol. Equip. Technol.* 16.

Koros, W. J., and Chern, R. T. (1987). Separation of gaseous mixtures using polymer membranes. In: R. W. Rousseau (Ed.). *Handbook of Separation Process Technology.* Wiley, New York, pp. 862–953.

Koros, W. J., Coleman, M. R., and Walker, D. R. B. (1992). Controlled permeability polymer membranes. *Ann. Rev. Mater. Sci.* 22, 47.

Koros, W. J., and Fleming, G. K. (1993). Membrane-based gas separation. *J. Membr. Sci.* 83, 1.

Koros, W. J., Chan, A. H., and Paul, D. R. (1977). Sorption and transport of various gases in polycarbonate. *J. Membr. Sci.* 2, 165.

Krzysztof, K., Marek, G., and Antoni, M. (2002). Membrane processes used for potable water quality improvement. *Desalination* 145, 315.

Kusakabe, K., Yoneshige, S., Murata, A., and Morooka, S. (1996). Morphology and gas permeance of ZSM-5-type zeolite membrane formed on a porous α-alumina support tube. *J. Membr. Sci.* 116, 39.

Lee, C. L., Chapman, H. L., Cifuentes, M. E., Lee, K. M., Merrill, L. D., Ulman, K. L., and Venkataraman, K. (1988). Effects of structure on the gas permeability of silicone membranes. *J. Membr. Sci.* 38, 55.

Lee, Y. M., Ha, S. Y., Lee, Y. K., Suh, D. H., and Hong, S. Y. (1999). Gas separation through conductive polymer membranes. 2. Polyaniline membranes with high oxygen selectivity. *Ing. Eng. Chem. Res.* 38, 1917.

Leemann, M., Eigenberger, G., and Strathmann, H. (1996). Vapor permeation for the recovery of organic solvents from waste air streams: Separation capacities and process optimization. *J. Membr. Sci.* 113, 313.

Lin, H., and Freeman, B. D. (2005). Material selection guidelines for membranes that remove CO_2 from gas mixtures. *J. Mol. Struct.* 739, 2005.

Loeb, S., and Sourirajan, S. (1963). *ACS Advances in Chemistry Series 38*, American Chemical Society Washington, DC, p. 117.

Maene, N., Nair, B. N., D'hooghe, P., Nakao, S.-I., and Keizer, K. (1998). Silica-polymers for processing gas separation membranes: High temperature growth of fractal structure. *J. Sol-Gel Sci. Technol.* 2, 117.

Magnusson, I., Stenberg, W. V., Batich, C., and Egelberg, J. (1990). Connective tissue repair in circumferential periodontal defects in dogs following use of a biodegradable membrane. *J. Clin Periodontol* 17(4), 243.

Masuda, T., Isobe, E., Higashimura, T., and Takada, K. (1983). Poly[1-(trimethylsilyl)-1-propyne]: A new high polymer synthesized with transition-metal catalysts and characterized by extremely high gas permeability. *J. Am. Chem. Soc.* 105, 7473.

Maxwell, C. (1873). *Treatise on Electricity and Magnetism*, Vol. 1. Oxford University Press, London.

McKeown, N. B., Budd, P. M., Msayib, K. J., Ghanem, B. S., Kingston, H. J., Tattershall, C. E., Makhseed, S., Reynolds, K. J., and Fritsch, D. (2005). Polymers of intrinsic microporosity (PIMs): Bridging the void between microporous and polymeric materials. *Chem. A Eur. J.* 11, 26.

Merkel, T. C., Bondar, V., Nagai, K., and Freeman, B. D. (1999). Hydrocarbon and perfluorocarbon gas sorption in poly(dimethylsiloxane), poly(1-trimethylsilyl-1-propyne), and copolymers of tetrafluoroethylene and 2,2-bis(trifluoromethyl)-4,5-difluoro-1,3-dioxole. *Macromolecules* 32, 370.

Merkel, T. C., Bondar, V. I., Nagai, K., Freeman, B. D., and Pinnau, I. (2000). Gas sorption, diffusion, and permeation in poly(dimethylsiloxane). *J. Polym. Sci. B: Polym. Phys.* 38, 415.

Merkel, T. C., Freeman, B. D., Spontak, R. J., He, Z., Pinnau, I., Meakin, P., and Hill, A. J. (2002). Ultrapermeable, reverse-selective nanocomposites membranes. *Science* 296, 519.

Merkel, T. C., Freeman, B. D., Spontak, R. J., He, Z., Pinnau, I., Meakin, P., and Hill, A. J. (2003a). Sorption, transport, and structural evidence for enhanced free volume in poly(4-methyl-2-pentyne)/fumed silica nanocomposite membranes. *Chem. Mater.* 15, 109.

Merkel, T. C., He, Z., Pinnau, I., Freeman, B. D., Meakin, P., and Hill, A. J. (2003b). Nanoparticles on gas sorption and transport in poly(1-trimethylsilyl-1-propyne). *Macromolecules* 36, 6844.

Merkel, T. C., He, Z., Pinnau, I., Freeman, B. D., Meakin, P., and Hill, A. J. (2003c). Sorption and transport in poly(2,2-bis(trifluoromethyl)-4,5-difluoro-1,3-dioxole-cotetrafluoroethylene) containing nanoscale fumed silica. *Macromolecules* 36, 8406.

Min, J.-S., Kiyozumi, Y., and Itoh, N. (2003). A sealant-free preparation technique for high-temperature use of a composite zeolite membrane. *Ind. Eng. Chem. Res.* 42, 80.

Mitchell, J. K. (1830). On the penetrativeness of fluids. *Am. J. Med.* 13, 36.

Mitchell, J. K. (1833). On the pentration of gases. *Am. J. Med. Sci.* 25, 100.

Mueller, M., Oehr, C., Malthaner, H., Goehl, H., Deppisch, R., and Storr, M. (2003). A process for production of a regioselective polymeric membrane for purification of biological fluids. *PCT Int. Appl.* 33 pp.

Mulder, M. (2000). *Basic Principles of Membrane Technology*. Kluwer Academic, Dordrecht.

Nagai, K., Masuda, T., Nakagawa, T., Freeman, B., and Pinnau, I. (2001). Poly[1-(trimethylsilyl)-1-propyne] and related polymers: Synthesis, properties and functions. *Prog. Polym. Sci.* 26, 721.

Nagel, R., and Will, T. (1999). Microelectronics. Membrane processes for water treatment in the semiconductor industry. *Ultrapure Water* 16(8), 35.

Neogi, P. (1996). Transport phenomena in polymer membranes. In P. Neogi (Ed.), *Diffusion in Polymers*. Marcel Dekker, Inc., New York, pp. 173–209.

Nemser, S. M. (1993). Applications of membranes in industry—Glassy fluoropolymer membranes. Paper presented at the 21st Aharon Katzir-Katchalsky Conference, Rehovot, Israel.

Nemser, S. M., and Roman, I. C. (1991). Perfluorinated membranes. U.S. Patent 5,051,114.

Neuse, E. W. (1982). Aromatic polybenzimidazoles. Syntheses, properties, and applications. *Adv. Polym. Sci.* 47, 1.

Park, H. B., Kim, Y. K., Lee, J. M., and Lee, Y. M. (2004). Relationship between chemical structure of aromatic polyimides and gas permeation properties of their carbon molecular sieve membranes. *J. Membr. Sci.* 229, 117.

Park, H. B., Kim, J. K., Nam, S. Y., and Lee, Y. M. (2003). Imide-siloxane copolymer/silica hybrid membranes: Preparation, characterization and gas separation properties. *J. Membr. Sci.* 220, 59.

Park, H. B., and Lee, Y. M. (2005). Fabrication and characterization of nanoporous carbon/silica membranes. *Adv. Mater.* 17, 477.

Paul, D. R., and Yampolskii, Y. P. (1994). *Polymeric Gas Separation Membranes*. CRC Press, London.

Pesiri, D. R., Jorgensen, B., and Dye, R. C. (2003). Thermal optimization of polybenzimidazole meniscus membranes for the separation of hydrogen, methane, and carbon dioxide. *J. Membr. Sci.* 218, 11.

Petropoulos, J. H. (1970). Quantitative analysis of gaseous diffusion in glassy polymers. *J. Polym. Sci., Polym. Phys. Ed.* 8, 1797.

Pinnau, I., Casillas, C. G., Morisato, A., and Freeman, B. D. (1997). Long-term permeation properties of poly(1-trimethylsilyl-1-propyne) membranes in hydrocarbon-vapor environment. *J. Polym. Sci. B: Polym. Phys.* 35, 1483.

Pinnau, I., and He, Z. (2004). Pure- and mixed-gas permeation properties of polydimethylsiloxane for hydrocarbon/methane and hydrocarbon/hydrogen separation. *J. Membr. Sci.* 244, 227.

Pinnau, I., and Toy, L. G. (1996a). Transport of organic vapors through poly(1-trimethylsilyl-1-propyne). *J. Membr. Sci.* 116, 199.

Pinnau, I., and Toy, L. G. (1996b). Gas and vapor transport properties of amorphous perfluorinated copolymer membranes based on 2,2-bistrifluoromethyl-4,5-difluoro-1,3-dioxole/tetrafluoroethylene. *J. Membr. Sci.* 109, 125.

Polyakov, A. M., Starannikova, L. E., and Yampolskii, Yu. P. (2003). Amorphous Teflons AF as organophilic pervaporation materials. Transport of individual components. *J. Membr. Sci.* 216, 241.

Polyakov, A. M., Starannikova, L. E., and Yampolskii, Yu. P. (2004). Amorphous Teflons AF as organophilic pervaporation materials. Separation of mixtures of chloromethanes. *J. Membr. Sci.* 238, 21.

Pratibha, P., and Chauhan, R. S. (2001). Membranes for gas separation. *Prog. Polym. Sci.* 26, 853.

Resnick, P. R. (1976). Polymers of fluorinated dioxoles. U.S. Patent 3,978,030.

Rezac, M. E., John, T., and Pfromm, P. H. (1997). Effect of copolymer composition on the solubility and diffusivity of water and methanol in a series of polyether amides. *J. Appl. Polym. Sci.* 65, 1983.

Robeson, L. M. (1991). Correlation of separation factor versus permeability for polymeric membranes. *J. Membr. Sci.* 62, 165.

Robeson, L. M., Borgoyne, W. F., Langsam, M., Savoca, A. C., and Tien, C. F. (1994). High performance polymers for membrane separation. *Polymer* 35, 4970.

Rochow, E. G. (1987). *Silicon and Silicones*. Springer-Verlag, Berlin.

Rogers, C. E. (1985). Permeation of gases and vapors in polymers. *Polym. Permeability* 11–73.

Schultz, J., and Peinemann, K. V. (1996). Membranes for separation of higher hydrocarbons from methane. *J. Membr. Sci.* 110, 37.

Singh, A., Freeman, B. D., and Pinnau, I. (1998). Pure and mixed gas acetone/nitrogen permeation properties of polydimethylsiloxane (PDMS). *J. Polym. Sci. B: Polym. Phys.* 36, 289.

Smitha, B., Sridhar, S., and Khan, A. A. (2005). Solid polymer electrolyte membranes for fuel cell applications—a review. *J. Membr. Sci.* 259(1–2), 10.

Spillman, R. W., and Grace, W. R. (1989). Economics of gas separation membranes. *Chem. Eng. Prog.* 85(1), 41–62.

Spillman, R. W. (1995). Economics of gas separation membrane processes. In R. D. Noble and S. A. Stern (Eds.), *Membrane Separation Technology. Principles and Applications*. Elsevier, Amsterdam.

Srinivasan, R., Auvil, S. R., and Burban, P. M. (1994). Elucidating the mechanism(s) of gas transport in poly[1-(trimethylsilyl)-1-propyne] (PTMSP) membranes. *J. Membr. Sci.* 86, 67.

Srivastava, R. C., Sahney, R., Upadhyay, S., and Gupta, R. L. (2000). Membrane permeability based cholesterol sensor—A new possibility. *J. Membr. Sci.* 164(1–2), 45.

Stern, S. A. (1994). Polymers for gas separations: The next decade. *J. Membr. Sci.* 94, 1.

Stern, S. A., Shah, V. M., and Hardy, B. J. (1987). Structure-permeability relationships in silicone polymers. *J. Polym. Sci. B: Polym. Phys.* 25, 1263.

Strathmann, H. (1981). Membrane separation processes. *J. Membr. Sci.* 9, 121.

Suda, H., and Haraya, K. (1997). Gas permeation through microporous of carbon molecular sieve membranes derived from Kapton polyimide. *J. Phys. Chem. B.* 101, 3988.

Van Krevelen, D. W. (1990). *Properties of Polymers: Their Correlation with Chemical Structure; Their Numerical Estimation and Prediction from Additive Group Contribution*. Elsevier, Amsterdam.

Vieth, W. R., and Sladek, K. J. (1965). Model for diffusion in a glassy polymer. *J. Colloid Sci.* 20, 1014.

Vu, D. Q., Koros, W. J., and Miller, S. J. (2002). High pressure CO_2/CH_4 separation by using carbon molecular sieve hollow fiber membranes. *Ind. Eng. Chem. Res.* 41, 367.

Yeom, C. K., Lee, S. H., and Lee, Y. M. (2002). Vapor permeations of a series of $VOCs/N_2$ mixtures through PDMS membranes. *J. Membr. Sci.* 198, 129.

Zeigler, J. M., and Fearon, F. W. G. (1990). *Silicon-Based Polymer Science: A Comprehensive Resource*. American Chemical Society, Washington, DC.

CHAPTER 25

Hydrogen Separation Membranes

YI HUA MA

Center for Inorganic Membrane Studies, Department of Chemical Engineering, Worcester Polytechnic Institute, Worcester, Massachusetts 01609

25.1 INTRODUCTION

Hydrogen is among one of the top industrial gases produced yearly in the world. Its production and importance as one of the key industrial gases will further increase due to the additional need for upgrading the low-grade crude feedstock and our entering into the twenty-first century of the hydrogen economy. Over 90% of the hydrogen in the United States is produced by steam reforming of natural gas, whereas the worldwide hydrogen production from natural gas is around 45–50%. Currently, the separation of hydrogen from the reformate gases is primarily done by pressure swing adsorption (PSA). However, there are plenty of opportunities for using membranes to obtain hydrogen with various purity. Membrane separations have several advantages including low capital costs, ease of operation (low operation costs), and modular arrangement. One of the major shortcomings of membrane separations is the low-pressure product gas hydrogen, which requires recompression for process uses. On the other hand, the separation of hydrogen by membranes becomes even more attractive when the hydrogen separation membranes are used as a membrane reactor for continuous hydrogen removal during reaction.

Membrane reactors offer an inherent ability to combine reaction, product concentration, and separation in a single unit operation and are especially suited for reactions such as water gas shift reaction with conversions limited by the thermodynamic equilibrium of the reaction. The use of a membrane reactor for the continuous separation of hydrogen from the reformate allows the reformer to be run at lower temperatures and, at the same time, provides process intensification by eliminating the shift reactors, PreOx (preferential oxidation) reactor and hydrogen separator.

Although both porous and dense inorganic membranes are suited for gas separations, dense inorganic membranes have the advantage of high separation factors (theoretically infinite) unsurpassed by any of their porous counterparts. However, one major drawback of dense inorganic membranes is their low permeation fluxes. Operating the membrane at high temperatures and using a thin membrane layer supported on a porous structure

Advanced Membrane Technology and Applications. Edited by Norman N. Li, Anthony G. Fane, W. S. Winston Ho, and T. Matsuura
Copyright © 2008 John Wiley & Sons, Inc.

can increase the flux to a commercially viable level. Dense metal membranes are especially well suited for high-temperature hydrogen separations and productions. Palladium and its alloys are the most commonly used materials for the preparation of dense metal membranes for hydrogen separations because of their high hydrogen permeability relative to other hydrogen-permeable inorganic materials (e.g., silica, tantalum, and nickel) and excellent chemical resistance to oxidation and corrosion.

This chapter will describe membranes for hydrogen separations for both high- and low-temperature applications, with a special emphasis on Pd and Pd–alloy supported on porous metal substrates for high-temperature and high-pressure membrane reactor applications.

25.2 POROUS NONMETALLIC MEMBRANES FOR HYDROGEN SEPARATIONS

The required hydrogen purity generally depends on a specific application. For example, high-purity hydrogen is not required for refinery applications, whereas high-purity hydrogen, especially low-CO content, is necessary for proton exchange membrane (PEM) fuel cell applications. Because of these varied applications, materials ranging from polymeric hollow fibers to dense metal can all be used as hydrogen separation membranes for specific applications. The discussion in this section will cover some of these hydrogen separation membranes.

25.2.1 Hollow-Fiber Membranes

The first widespread commercial application of membrane separations is to use hollow-fiber membrane separators for hydrogen recovery from processes, ammonia plants and petrochemical purge streams, and H_2/CO ratio adjustment in synthesis gas (Gardner et al., 1977; Bollinger et al., 1984; Koros and Mahajan, 2000). The reported ideal hydrogen selectivities with respect to different gases, such as CO_2, CO, N_2, and CH_4, appear to be reasonable, ranging from 170 for H_2/CH_4 to 6.75 for H_2/CO_2. The operation of these permeators, however, is restricted to low temperatures because of the polymeric material used for the synthesis of hollow fibers.

Membrane permeators have to compete against pressure swing adsorption (PSA) or cryogenic separations. In general, membrane separations are competitive in a relatively wide range of conditions (Poffenbarger and Gastinne, 1989). Since hydrogen molecule is the second smallest molecule in the periodic table, it generally appears at the low-pressure side of the separator. Recompression of the low-pressure product gas hydrogen can represent a substantial additional cost for the separation process.

25.2.2 Molecular Sieving Membranes

25.2.2.1 Molecular Sieve Carbon Membranes The use of molecular sieve carbon membranes (MSCM) for gas separations was first reported by Koresh and Soffer (1983), who based their research on molecular sieve carbon adsorbents, and developed MSCM by controlled pyrolysis of thermosetting polymer membranes. They demonstrated that the permeation characteristics of the membrane could be controlled by mild stepwise thermochemical treatments. The permeability for both hydrogen and methane reached a maximum as the heat treatment temperature varied from 400 to 800°C (Koresh and

Soffer, 1986). For light gases, such as helium, hydrogen, argon, and oxygen, at relatively low pressures, the permeability is essentially independent of pressure, a consequence of free molecules in the pores or a liner adsorption isotherm. They further demonstrated that the separation factor for H_2/CH_4 appeared to be independent of the mixture composition and decreased from 57 at 200°C to 35 at 500°C (Koresh, 1987). Although these values of separation factors appear to be reasonable, the suitability of such membranes for high-pressure applications requires careful study because of the possible loss of separation efficiency due to the possible change from free molecules in the pores to Knudsen or even molecular diffusion at high pressures.

Surface-selective flow membranes made of nanoporous carbon, which is a variation of molecular sieving membranes, were developed by Rao et al. (1992) and Rao and Sircar (1993). The membrane can be produced by coating poly(vinylidene chloride) on the inside of a macroporous alumina tube followed by carbonization to form a thin membrane layer. The mechanism of separation is by adsorption–surface-diffusion–desorption. Certain gas components in the feed are selectively adsorbed, permeated through the membrane by surface diffusion, and desorbed at the low-pressure side of the membrane. This type of membrane was used to separate H_2 from a mixture of H_2 and CO_2 (Sircar and Rao, 2000), and their main advantage is that the product hydrogen is at the high-pressure side eliminating the need for recompression. The membrane, however, is not industrially viable because of its low overall separation selectivity. In addition, since the separation mechanism involves physical adsorption, operation at low temperatures is required.

25.2.2.2 Zeolite Membranes Zeolites are porous aluminosilicate crystals composed of SiO_4 and AlO_4 tetrahedra arranged in various geometric patterns. The tetrahedra are linked together at the corners by shared oxygen ions to form ordered lattices. These lattices can be visualized as three-dimensional combinations of ions, layers, and polyhedra. Cations are present in the porous aluminosilicates, and some of these cations can be readily exchanged by placing zeolite crystals containing a given ion in a solution containing another ion. The sorptive and catalytic properties of zeolites may be altered by the presence of different cations. Therefore, zeolite are microporous crystalline materials with a uniform pore size distribution on a molecular scale and with high thermal and chemical stability. A membrane separation system that takes advantage of the adsorptive and molecular sieving properties of zeolites requires a continuous zeolite membrane layer, which is deposited on a porous support to provide the needed mechanical strength for practical applications.

At the present, industrial applications of zeolite membranes are limited to the pervaporation for the dehydration from mixtures of alcohols (Morigami et al., 2001). One of the most difficult tasks for making supported zeolite membranes is the elimination of interstitial pores, which is extremely critical for gas separations. Methods for repairing the interstitial pores include repeated zeolite synthesis to close the interstitial pores by forming additional thin layers (Kumakiri et al., 1999; Ma et al., 2001), modification (pretreatment) of support material and/or its microstructure (Zhu et al., 2005), and modification by silylation in the nonzeolitic pores (Hong et al., 2005). Hong et al. 2005 applied silylation by the catalytic cracking of methyldiethoxysilane to modify the pore size of boron-substituted ZSM-5 and SAPO-34 to increase the hydrogen selectivity over light gases. The H_2 permeance of zeolite membranes is, in general, around 1.0 m^3/m^2 h atm in the temperature range between 400 and 700 K with H_2 selectivity over light gases such as CO_2, CH_4, and N_2 under 100. Recent experimental work on permeation of gas mixtures by Nernoff's group at Sandia National Laboratories showed higher H_2 selectivity than single gas measurements for ZSM-5 and silicalite-1 (Welk and Nernoff, 2004).

Zeolite membranes have long been recognized to have great potential for gas separations. However, considerable challenges, such as elimination of intercrystalline pores and reduction of pore size to increase the hydrogen selectivity, remain to be resolved.

25.2.3 Mixed-Matrix Membranes

One way to achieve high permselectivity is to use the so-called mixed-matrix membranes, which can be formed by embedding the high-permselectivity molecular sieving materials into the structure of polymeric materials (Koros and Mahajan, 2000; Kulprathipanja, 2003). Since polymeric materials are commonly used in the formation of a mixed-matrix membrane, it is more appropriate for low-temperature applications.

25.3 HIGH-TEMPERATURE HYDROGEN SEPARATION MEMBRANES

25.3.1 Composite Dense (Nonporous) Pd and Pd–Alloy Membranes

The permeation of hydrogen through a palladium layer is by the solution and diffusion mechanism. For the case of low H concentrations in the Pd–H phase, no α-to-β phase transformation, and the permeation controlled by the hydrogen diffusion through the bulk of palladium layer, the hydrogen flux can be expressed as

$$J = kD\frac{p_{\mathrm{HP}}{}^{1/2} - p_{\mathrm{LP}}{}^{1/2}}{\delta} \tag{25.1}$$

where J is the hydrogen flux, $\mathrm{m^3/m^2\ s}$ or $\mathrm{kmol/m^2\ s}$, k the absorption or Henry's law constant, $\mathrm{kmol/m^3\ Pa^{1/2}}$, D the diffusivity, $\mathrm{m^2/s}$, and δ the thickness of the membrane, m. Subscripts HP and LP designate the high- and low-pressure side of the membrane, respectively, and kD is the permeability, which is temperature dependent. Since the membrane thickness is often unknown or difficult to determine accurately, $\mathscr{P} = kD/\delta$, defined as permeance, is frequently used to characterize the permeation property of the membrane. Furthermore, Eq. (25.1) can be expressed in a more general form as

$$J = \mathscr{P}(p_{\mathrm{HP}}{}^{n} - p_{\mathrm{LP}}{}^{n}) \tag{25.2}$$

where $1 \geq n \geq ½$. When $n = ½$, Eq. (25.2) reduces to Eq. (25.1), which is known as the Sieverts law. A typical Sieverts plot is shown in Figure 25.1 (Ma et al., 2003).

Work on dense metallic membranes for hydrogen separations has been mostly on Pd and Pd–alloy, with early work concentrated on Pd and Pd–alloy foils and discussed in detail by Lewis (1967). Palladium foils are not suitable for large chemical process applications, especially membrane reactor applications, because of the difficulties of using plane geometry in large chemical processes. In the case of tubular geometry, in order to have sufficient mechanical strength for high-pressure applications, the diameter of the thin pure Pd tube has to be small in order to have the needed mechanical strength. The large pressure drop resulted from the small-diameter tubes, prohibiting them from large-scale applications.

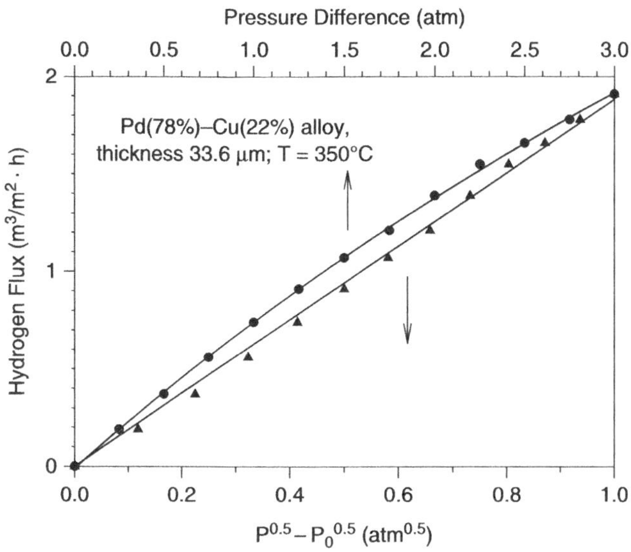

Figure 25.1 Sieverts' plot for a Pd–Cu/porous stainless steel (PSS) composite membrane at 350°C. Pd: 78 wt%, Cu: 22 wt%, thickness: 33.6 μm, support: Mott PSS grade 0.2 μm (Ma et al., 2003).

In addition, although foils as thin as 25 μm or less are available commercially, the Pd cost for the tube is too high for large-scale chemical process applications because there are limits on how thin the tube wall can be practically made. One way to overcome the lack of mechanical strength and cost problem is to form a thin Pd layer on a porous support. The benefits of decreasing the Pd layer thickness are twofold: decrease the Pd cost and at the same time increase the hydrogen flux. Therefore, a decrease of the membrane thickness by a factor of 2 would decrease the total Pd cost by a factor of 4.

One of the problems associated with using a pure palladium membrane layer is the hydrogen embrittlement caused by the phase transition between α and β phase of palladium hydride. Figure 25.2 is the phase diagram of the palladium–hydrogen system and shows the existence of a two-phase region below the critical temperature of ~300°C (Guazzone, 2006). One effective way of minimizing the hydrogen embrittlement problem is to form palladium-rich alloys such as Pd/Ag and Pd/Cu to lower the critical temperature. Furthermore, a number of palladium-rich alloys has higher hydrogen fluxes than that of pure palladium, as shown in Table 25.1 (Knapton, 1977). Therefore, another added benefit of using Pd–alloy membranes is the reduction of the needed amount of Pd, thereby further reducing the total cost of the membrane. However, the benefit of using palladium-rich alloys to improve the hydrogen permeation flux may vary with temperature (Gryaznov, 2000).

Both porous ceramic and metallic tubes or plates can be used as the support for Pd and Pd–alloy membrane layers. Porous ceramic supports have the advantages of small pore sizes, uniform pore size distributions, and excellent chemical stability. The small pore size and uniform pore size distribution allow the formation of thinner and uniform membrane layers. However, ceramic supports are brittle and prone to cracking. In addition, connecting ceramic material to metallic elements in a process presents a considerable challenge to both material scientists and process engineers. On the other hand, porous metallic supports, such as porous stainless steel and other porous specialty alloys, can

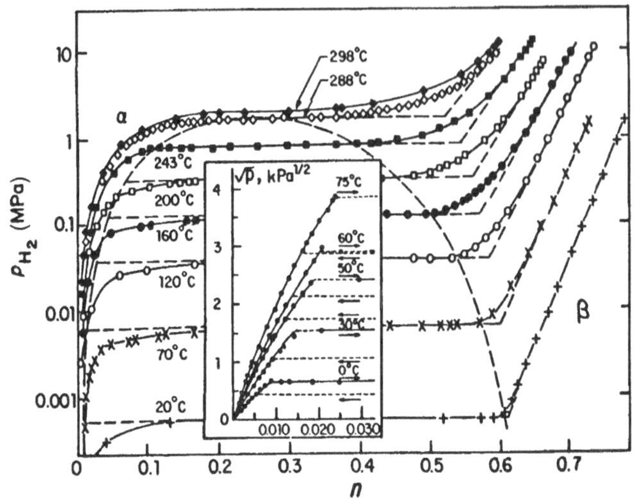

Figure 25.2 Isotherm Pd phase diagram for Pd–H system where $n = H/Pd$ (Guazzone, 2006).

easily be connected to other metallic elements in the process, making the process of integrating the membrane module to a specific process relatively easy.

Although porous metallic supports have good mechanical strength, their large pore sizes and wide pore size distributions make the formation of thin and uniform membrane layers a real challenge. In addition, intermetallic diffusion between the metal elements in the support and the membrane layer can cause the hydrogen flux to deteriorate, especially at high temperatures. One way to alleviate or minimize the problem of intermetallic diffusion is to insert an intermetallic diffusion barrier layer between the support and the membrane layer.

Methods for making an intermetallic diffusion barrier reported in the literature include placing a γ-alumina layer sandwiched between a vanadium foil and Pd foil (Pd/γ-Al_2O_3/V) (Edlund and McCarthy, 1995), formation of a TiN layer by sputtering titanium in a nitrogen atmosphere (Shu et al., 1996), application of nickel powder followed by sintering and application of a thin γ-alumina layer by the sol–gel method (Nam and Lee, 2001), and a controlled in situ oxidation of the porous metal at temperatures greater than

TABLE 25.1 Improvement in Hydrogen Permeability of Various Binary Pd Alloys at 350°C

Alloy Metal	wt% for Maximum Permeability	Normalized Permeability (P_{alloy}/P_{Pd})
Y	10	3.8
Ag	23	1.7
Ce	7.7	1.6
Cu	40	1.1
Au	5	1.1
Pure Pd	—	1.0

Source: After Knapton (1977).

400°C to form an in situ intermetallic barrier layer (Ma et al., 2000) and coating with a thin layer of silver (Tong et al., 2005). Although long-time hydrogen permeation data have been reported in the literature, the length of the long-term stability tests varied widely and most were in the hundreds of hours or less, except the work of Ma et al. (1998), who showed their membrane with intermetallic diffusion barrier prepared by the in situ oxidation method to be stable for over 6000 h in the temperature range of 350–450°C. In order for membranes to be viable for large-scale industrial applications, the membrane must be tested for long-term stability in thousands of hours. Ayturk et al. (2004) and Ma et al. (2004a) modified their in situ oxidation technology to extend the temperature range to above 500°C by using the bimetal multilayer (BMML) deposition technique with the formation of a porous Pd–Ag layer by consecutive deposition of Pd and Ag without intermediate surface activation. Further improvement of their intermetallic barrier layer is covered in several additional patent applications (Ma et al., 2004b,c,d).

25.3.1.1 Electroless Deposition of Pd and Pd–Alloy on Porous Substrates Some of the methods using more sophisticated equipment for the deposition of a Pd layer on porous supports include spray pyrolysis to deposit a Pd–Ag alloy on a porous alumina support (Li et al., 1993), metal–organic chemical vapor deposition (MOCVD) technique by decomposing palladium(II) acetate in argon under reduced pressure (Yan et al., 1994; Morooka et al., 1995), supercritical fluid transport-chemical deposition (SFTCD) method using the metal β-diketonate complex-(2,2,7-trimethyl-3,5-octanedionato) palladium(II) (Hybertson et al., 1991), sputter deposition technique (Konno et al., 1988; Jayaraman and Lin, 1995; Mardilovich et al., 1996), electron beam evaporation and ion beam sputtering (Peachey et al., 1996), and magnetic sputtering (Gryaznov et al., 1993; Bryden and Ying, 1995). These are powerful techniques, especially for the deposition of palladium alloy, but more research may be required for using them for large-scale membrane preparation.

One of the most commonly used techniques for depositing Pd on porous substrates is the electroless plating, which is relatively simple and easy to scaleup and can provide uniform deposition on complex shapes and large substrate areas. In addition, the deposited film has good adhesion and hardness. A variety of porous supports has been used as substrates for the deposition of Pd and Pd–alloy including tantalum and niobium by Buxbaum and his co-workers (Buxbaum and Marker, 1993; Buxbaum and Kinney, 1996), porous silver by Govind and Atnoor (1991), porous glass by Uemiya et al. (1988, 1991), porous alumina by Kikuchi and Uemiya (1991), Collins and Way (1993), Huang et al. (2001), and Paglieri et al. (1999), and porous stainless steel (PSS) by Mardilovich et al. (1998) and Shu et al. (1993). Electroless plating assisted by osmotic pressure to manipulate the microstructure, porosity, and thickness of the deposited film was used to plate Pd on alumina (Yeung and Varma, 1995), Pd on Vycor (Yeung et al., 1999), and Pd on PSS (Souleimanova et al., 2002).

An electroless plating solution generally consists of a palladium ion source, a complexant, a reducing agent, stabilizers, and accelerators. The plating procedure includes (1) complete removal of foreign contaminants (grease, oil, dirt, corrosion products); (2) activation by seeding the support surface with palladium nuclei, which during the electroless plating initiate an autocatalytic process of the reduction of a metastable Pd salt complex on the target surface; and (3) electroless plating. A typical plating solution composition, pH, and temperature are shown in Table 25.2 (Ma et al., 1998).

TABLE 25.2 Composition of a Typical Electroless Palladium Plating Solution

$Pd(NH_3)_4Cl_2 \cdot H_2O$, g/L	4.0
NH_4OH (28%), mL/L	198
Na_2EDTA, g/L	40.1
H_2NNH_2 (91 M), mL/L	5.6–7.6
pH	~10.4
Temperature, °C	60

Source: Ma et al. (1998).

The chemical reaction taking place during the plating is

$$2Pd(NH_3)_4Cl_2 + H_2NNH_2 + 4NH_4OH \rightarrow 2Pd^0 + N_2 + 8NH_3 + 4NH_4Cl + H_2O$$

or

$$2Pd^{2+} + H_2NNH_2 + 4OH^- \rightarrow 2Pd^0 + N_2 + 4H_2O$$

The plating rate with the conditions shown in Table 25.2 is around 1 μm per hour depending on the condition and type of surface. For porous metal supports, an intermetallic diffusion barrier layer needs to be inserted before plating, and the various ways of inserting this layer were discussed at the beginning of this section. This barrier layer is necessary for a membrane to have long-term stability.

Palladium–alloy membranes not only have lower critical temperatures for α-β phase transformation (hydrogen embrittlement) but also give higher hydrogen permeation fluxes (see Table 25.1). Although only binary alloys are shown in Table 25.1, some tertiary systems have also been shown to have higher hydrogen permeation fluxes (Shu et al., 1991; Gryaznov et al., 1993; Gryaznov, 2000). Since it is easier to form Pd alloy in bulk samples such as foils and tubes, most early fundamental work on hydrogen interaction with Pd alloys has been done with bulk samples. For deposited thin films, Pd alloys are generally formed by sequentially depositing alternating layers of Pd and the second metal followed by high-temperature annealing, referred to as the coating and diffusion method. One of the difficulties for using the coating and diffusion method is the required high annealing temperatures, which may cause cracking of the membrane layer. The problem becomes even more serious for membranes with porous metal substrates because of the high rate of intermetallic diffusion at high temperatures. Detailed discussion on Pd–alloy membranes and other aspects of composite Pd membranes can be found in Ma et al. (2003).

25.3.1.2 Desired Membrane Characteristics The required qualities for a hydrogen separation membrane to be commercially viable for reforming reactors are high permeation flux and good long-term chemical, thermal, and mechanical stability under high temperatures and pressures. The draft technical targets set by the U.S. Department of Energy are shown in Table 25.3. Of special importance is the required long-term stability (durability) which ranges from 3 years at 2007 to greater than 10 years by 2015. Up to now, the longest tested time for composite Pd and Pd–alloy membranes reported in the literature is around 6000 h (slightly over 8 months) carried out both at 350–450°C under pure hydrogen (Ma et al., 1998) and 500°C under steam reforming reaction conditions (Ma and

TABLE 25.3 U.S. DOE Hydrogen Separation—Technical Targets

Performance Criteria	2007 Target	2010 Target	2015 Target
Flux sccm/cm^2 @100 psi ΔP H_2 partial pressure	50	100	150
Operating Temp, °C	400–700	300–600	250–500
S tolerance	Yes	Yes	Yes
Cost, \$/ft^2	1000	500	<250
WGS activity	Yes	Yes	Yes
ΔP operating capability, system pressure, psi	100	Up to 400	Up to 800–1000
CO tolerance	Yes	Yes	Yes
Hydrogen purity	95%	99.5%	99.99%
Stability/durability (years)	1	3	>5

Source: From Office of Fossil Energy *Hydrogen from Coal RD&D Plan*, June 10, 2004—Draft.

Matzakos, 2005). For commercial applications, it is imperative that such long-time tests be carried out to evaluate the long-term stability of the membrane.

For Pd foils, Knapton (1977) reported a hydrogen permeability of 3.55×10^{-4} m^3 m/m^2 h atm$^{1/2}$ for a 25-μm-thick foil at 350°C, while others reported similar but different values. Based on the various values for the hydrogen permeability on Pd foils and activation energy reported in the literature, Guazzone (2006) provided a generalized equation to estimate the hydrogen permeability in the temperature range of 200–500°C. On the other hand, the hydrogen permeability in composite Pd and Pd–alloy membranes reported in the literature from different investigators varied widely. A detailed summary of permeability obtained by various investigators for different composite Pd membranes was also presented by Guazzone (2006).

Hydrogen flux deterioration due to the presence of certain gas components, such as CO, CO_2, sulfur-containing compounds, and the like, can also be problematic for Pd and Pd–alloy membranes. The contamination of the membrane by CO and CO_2 is thought to be caused by physical adsorption that reduces the area for hydrogen absorption, but the membrane can be readily regenerated by passing pure hydrogen over the membrane surface at a high temperature for a short period of time. On the other hand, for both Pd and Pd–Cu membranes, the sulfur-containing compounds, such as H_2S, could block the surface sites for hydrogen absorption, causing the hydrogen flux to deteriorate at low H_2S concentrations, while for high H_2S concentrations, membrane failure occured due to the formation of micron-size pores resulting from the sulfidation of Pd and Cu (Kulprathipanja et al., 2005).

Although somewhat misleading, the cost of palladium has been one of the major concerns for large-scale process applications of composite Pd and Pd–alloy membranes in the past. Of course, the Pd cost is a factor necessary to consider, but it is not the dominant element in the overall cost of the Pd membrane reactors or separators as long as a thin layer of Pd membrane is used (e.g., ≈5 μm). On the other hand, since it is difficult to make small and uniform metallic powders, which are necessary for the formation of the metallic porous support with small and uniform pores required for forming thin membrane layers, the cost of the support may be high and may contribute significantly to the overall cost of the membrane module. However, when a large quantity is needed, the cost of the support is expected to drop. In addition, the process intensification afforded by the membrane reactor will further reduce the capital cost, making the membrane reactor process competitive to the current conventional steam-reforming process. The high-pressure CO_2 produced from the membrane reactor will not only add economical value to the process but also is beneficial to the environment.

25.3.2 Composite Silica Membranes

Due to their chemical stability, silica membranes have been considered as attractive hydrogen separation membranes with high permeability and selectivity for hydrogen. Silica membranes formed by the high-temperature CVD of tetraethylorthosilicate (TEOS) in an inert gas atmosphere showed high hydrogen permeances with essentially 100% hydrogen selectivities over CH_4, CO, and CO_2 (Prabhu and Oyama, 2000). Furthermore, the membrane was stable to hydrothermal stresses with 10% H_2O at 873 K and 1 bar for over 150 h. Composite silica–alumina membranes synthesized by CVD of TOES showed reasonable hydrogen permeances with hydrogen selectivity over CH_4, CO, and CO_2 exceeding 1000 (Lee et al., 2004). Recent work by Galuszka et al. (2006) showed that the He/N_2 selectivity of their silica membrane prepared by CVD of TOES on an asymmetric porous α-alumina support declined approximately 50% after heating and cooling between 500 and 580°C. The long-term hydrothermal stability, especially at high temperatures and pressures, has always been a major concern for silica membranes. Therefore, data for long-term stability of silica membranes are essential for the membrane to be considered commercially viable as reformers.

25.3.3 Dense Cermet (Ceramic Metal) Composite Membranes

Another class of dense membranes for hydrogen separation is the dense ceramic metal composite membranes (cermet) (Balachandran et al., 2005). In order to increase the electronic conductivity of the metal–ceramic composite, 40–50 vol% of a metal (typically Pd) or its alloy can be dispersed in a stable oxide matrix such as Al_2O_3 or ZrO_2. Pd as a hydrogen transport metal can not only increase the electronic conductivity of the composite but also provide an additional hydrogen transport path. Balachandran et al. (2005) reported reasonable hydrogen fluxes for a thin ($\approx$15 μm thick) cermet membrane with a hydrogen transport metal dispersed in a thermodynamically and mechanically stable ceramic matrix. The cermet membrane was stable for 100–120 h when it was exposed to hydrogen–helium mixture streams containing 100 and 400 ppm of H_2S at 900°C but deteriorate rapidly when the H_2S concentration was increased to 1000 ppm.

25.4 CONCLUDING REMARKS

The demand for hydrogen will certainly continue to grow, and there exist ample exciting opportunities for developing membrane technologies for both large- and small-scale hydrogen production from steam reforming reactions. The membrane technologies afford the possibility of producing pure hydrogen with simultaneous production of the high-pressure by-product greenhouse gas CO_2 suitable for sequestration or industrial applications such as enhanced oil recovery.

Technologies must be developed to produce membranes for high-temperature separation and reaction applications. Composite dense metallic membranes, such as Pd and its alloys, supported on metal substrates are especially suited for such applications due, in part, to their high separation selectivity, good permeability at high temperatures, and ease of module fabrication and process integration. Other inorganic membranes such as zeolite, silica, and cermet also have potential for high-temperature hydrogen separation applications, although considerable research is still needed to provide reliable ways for connecting inorganic material to metal parts normally used in industrial processes.

The total cost of the membrane, support, and module fabrication must be reduced in order for the membrane to be economically viable for industrial applications, especially for large-scale applications. The development of technologies for making thin and durable membranes at high temperatures and pressures should be explored with interdisciplinary approaches. Long-term (in thousands of hours) permeation data under reaction conditions should be obtained. Significant technical barriers still exist but overcoming these barriers will bring membrane technologies to the next level of high-temperature and high-pressure applications in large-scale chemical and petrochemical industry.

REFERENCES

Ayturk, M. E., Mardilovich, I. P., Engwall, E. E., and Ma, Y. H. (2004). Microstructure analysis of Pd/Ag alloy membranes supported on porous stainless steel. In *Proceedings of the AIChE 2004 Annual Meeting*, Austin, TX, Nov. 7–12, 2004.

Balachandran, U. (Balu), Lee, T. H., Chen, L., Song, S. J., Picciolo, J. J., and Dorris, S. E. (2005). Hydrogen permeation and chemical stability of dense hydrogen separation membranes. In *Proceedings of the 22nd Annual International Pittsburgh Coal Conference*, Pittsburgh, PA, Sept. 12–15, 2005.

Bollinger, W. A., Long, S. P., and Metzger, T. R. (1984). Optimizing hydrocracker hydrogen. *Chem. Eng. Prog.* 80(5), 51.

Bryden, K. J., and Ying, J. Y. (1995). Nanostructured palladium membranes synthesis by magnetron sputtering. *Mater. Sci. Eng.* A204, 140.

Buxbaum, R. E., and Kinney, A. B. (1996). Hydrogen transport through tubular membranes of palladium-coated tantalum and niobium. *Ind. Eng. Chem. Res.* 35, 530.

Buxbaum, R. E., and Marker, T. L. (1993). Hydrogen transport through non-porous membranes of palladium-coated niobium, tantalum and vanadium. *J. Membr. Sci.* 85, 29.

Collins, J. P., and Way. J. D. (1993). Preparation and characterization of a composite palladium-ceramic membrane. *Ind. Eng. Chem. Res.* 32, 3006.

Edlund, D. J., and McCarthy, J. (1995). The relationship between intermetallic diffusion and flux decline in composite-metal membranes: Implications for achieving long membrane lifetime. *J. Membr. Sci.* 107, 147.

Galuszka, J., Giddings, T., and Clelland, I. (2006). Silica H-membrane thermal and hydrothermal stability. In R. Bredsen and H. Reader (Eds.) *Proceedings of the 9th International Conference on Inorganic Membranes*, Lillehammer, Norway, June 25–26, 2006, p. 124.

Gardner, R. J., Crane, R. A., and Hannan, J. F. (1977). Hollow fiber permeator for separating gases. *Chem. Eng. Prog.* 73(10), 76.

Govind, R., and Atnoor, D. (1991). Development of a composite palladium membrane for selective hydrogen separation at high temperature. *Ind. Eng. Chem. Res.* 30, 591.

Gryaznov, V. (2000). Metal containing membranes for the production of ultrapure hydrogen and the recovery of hydrogen isotopes. *Sep. Purif. Methods* 29(2), 171.

Gryaznov, V. M., Serebryannikova, O. S., Serov, M. Yu., Ermilova, M. M., Karavanov, A. N., Mischenko, A. P., and Orekhova, N. V. (1993). Preparation and catalysis over palladium composite membranes. *Appl. Catal. A: General* 96, 15.

Guazzone, F. (2006). Engineering of substrate surface for the synthesis of ultra-thin Pd and Pd-Cu membranes for H_2 Separation. Ph.D. Dissertation, Worcester Polytechnic Institute, Worcester, MA.

Hong, M., Falconer, J. L., and Noble, R. D. (2005). Modification of zeolite membranes for H_2 separation by catalytic cracking of methyldiethoxysilane. *Ind. Eng. Chem. Res.* 44, 4035.

Huang, T. C., Wei, M. C., and Chen, H. I. (2001). Permeation of hydrogen through palladium/alumina composite membranes. *Sep. Sci. Technol.* 36, 199.

Hybertson, B. M., Hansen, B. N., Barkley, R. M., and Sievers, R. E. (1991). Deposition of palladium films by a novel supercritical fluid transport-chemical deposition process. *Mater. Res. Bull.* 26, 1127.

Jayaraman, V., and Lin, Y. S. (1995). Synthesis and hydrogen permeation properties of ultrathin palladium-silver alloy membranes. *J. Membr. Sci.* 99, 89.

Kikuchi, E., and Uemiya, S. (1991). Preparation of supported thin palladium-silver alloy membranes and their characteristics for hydrogen separation. *Gas Sep. Purif.* 5, 261.

Knapton, A. G. (1977). Palladium alloys for hydrogen diffusion membranes—A review of high permeability materials. *Plat. Met. Rev.* 21, 44.

Konno, M., Shindo, M., Sugawara, S., and Saito, S. (1988). A composite palladium and porous aluminum oxide membrane for hydrogen gas separation. *J. Membr. Sci.* 37, 503.

Koresh, J. (1987). Carbon molecular sieve membranes: General properties and the permeability of CH_4/H_2 mixtures. *Sep. Sci. Technol.* 22, 973.

Koresh, J. E., and Soffer, A. (1983). Molecular sieve carbon membranes for gas separation. *Sep. Sci. Technol.* 18, 723.

Koresh, J. E., and Soffer, A. (1986). Mechanism of permeation through molecular-sieve carbon membrane. *J. Chem. Soc. Faraday Trans.* I82, 2057.

Koros, W. J., and Mahajan, R. (2000). Pushing the limits on possibilities for large scale gas separation: Which strategies? *J. Membr. Sci.* 175, 181.

Kulprathipanja, A., Alptekin, G. O., Falconer, J. L., and Way, J. D. (2005). Pd and Pd-Cu membranes: Inhibition of H_2 permeation by H_2S. *J. Membr. Sci.* 254, 49.

Kulprathipanja, S. (2003). Mixed matrix membrane development. *Ann. N. Y. Acad. Sci.* 984, 361.

Kumakiri, I., Yamaguchi, T., and Nakao, S. (1999). Preparation of zeolites A and faujasite membranes from a clear solution. *Ind. Eng. Chem. Res.* 38, 4282.

Lee, D.-W., Nam, S.-E., Sea, Ihm, S.-K., and Lee, K.-H. (2004). Permeation behavior of a H_2/CO gaseous mixture through Pt-included composite membranes. *J. Membr. Sci.* 243, 243.

Lewis, F. A. (1967). *The Palladium Hydrogen System*. Academic, London.

Li, Z. Y., Maeda, H., Kusakabe, K., Morooka, S., Anzai, H., and Akiyama, S. (1993). Preparation of palladium-silver alloy membranes for hydrogen separation by the spray pyrolysis method. *J. Membr. Sci.* 78, 247.

Ma, Y. H., Mardilovich, I. P., and Engwall, E. E. (2003). Thin composite palladium and palladium/alloy membranes for hydrogen separation. *Ann. N. Y. Acad. Sci.* 984, 346.

Ma, Y. H., Mardilovich, I. P., and Engwall, E. E. (2004a). Method for fabricating composite gas separation modules. U.S. Patent application 20040237780.

Ma, Y. H., Mardilovich, I. P., and Engwall, E. E. (2004b). Composite gas separation modules having high Tamman temperature intermediate layers. U.S. Patent application 20040244590.

Ma, Y. H., Mardilovich, I. P., and Engwall, E. E. (2004c). Method for curing defects in the fabrication of a composite gas separation module. U.S. Patent application 20040244583.

Ma, Y. H., Mardilovich, I. P., and Engwall, E. E. (2004d). Composite gas separation modules having intermediate porous metal layers. U.S. Patent application 20040237779.

Ma, Y. H., Mardilovich, P. P., and She, Y. (1998). Stability of hydrogen flux through Pd/porous stainless composite membranes. In *Proceedings of the ICIM*$_6$, Nagoya, Japan, June 22–26, 1998, pp. 246–249.

Ma, Y. H., Mardilovich P. P., and She, Y. (2000). Hydrogen gas-extraction module and method of fabrication. U.S. Patent 6,152,987 (2000).

Ma, Y. H., and Matzakos, A. (2005). Composite palladium porous stainless membranes for hydrogen separation and reaction applications: An example of university-industry collaboration. In *Proceedings of the North American Membrane Society Annual Meeting*, Providence, RI, June 11–15, 2005.

Ma, Y. H., Zhou, Y., Poladi, R., and Engwall, E. (2001). The synthesis and characterization of zeolite A membranes. *Sep. Purif. Technol.* 25, 235.

Mardilovich, P. P., She, Y., Ma, Y. H., and Rei, M. H. (1998). Defect-free palladium membranes on porous stainless-steel support. *AIChE J.* 44, 310.

Morigami, Y., Kondo, M., Abe, J., Kita, H., and Okamoto, K. (2001). The first large-scale pervaporation plant using tubular type module with zeolite NaA membrane. *Sep. Purif. Technol.* 25, 251.

Morooka, S., Yan, S., Yokoyama, S., and Kusakabe, K. (1995). Palladium membrane formed in macropores of support tube by chemical vapor deposition with crossflow through a porous wall. *Sep. Sci. Technol.* 30, 2877.

Nam, S. E., and Lee, K. H. (2001). Hydrogen separation by Pd alloy composite membranes: Introduction of diffusion barrier. *J. Membr. Sci.* 192, 177.

Paglieri, S. N., Foo, K. Y., Way, J. D., Collins, J. P., and Harper-Nixon, D. L. (1999). A new preparation technique for Pt/alumina membranes with enhanced high-temperature stability. *Ind. Eng. Chem. Res.* 38, 1925.

Peachey, N. M., Snow, R. C., and Dye, R. C. (1996). Composite Pd/Ta metal membranes for hydrogen separation. *J. Membr. Sci.* 111, 123.

Poffenbarger, G. L., and Gastinne, P. (1989). Hydrogen applications and design considerations. In *Proceedings of the AIChE Spring National Meeting and Petrochemical Refining Exposition*, Houston, TX, Apr. 3, 1989.

Prabhu, A. K., and Oyama, S. T. (2000). Highly hydrogen selective ceramic membranes. *J. Membr. Sci.* 176, 243.

Rao, M. B., and Sircar, S. (1993). Nanoporous carbon memb ranes for separation of gas mixtures by selective surface flow. *J. Membr. Sci.* 85, 253.

Rao, M. B., Sircar, S., and Golden, T. C. (1992). Gas separation by adsorbent membranes. U.S. Patent 5,104,425.

Shu, J., Adnot, A., Grandjean, B. P. A., and Kaliaguine, S. (1996). Structurally stable composite Pd-Ag alloy membranes: Introduction of a diffusion barrier. *Thin Solid Films* 286, 72.

Shu, J., Grandjean, B. P. A., Ghali, E., and Kaliaguine, S. (1993). Simultaneous deposition of Pd and Ag on porous stainless steel by electroless plating. *J. Membr. Sci.* 77, 181.

Shu, J., Grandjean, B. P. A., Van Neste, A., and Kaliaguine, S. (1991). Catalytic palladium-based membrane reactors: A review. *Can. J. Chem. Eng.* 69, 1036.

Sircar, S., and Rao, M. B. (2000). Nano-porous carbon membranes for gas separation. In N. Kanellopoulos (Ed.), *Recent Advances on Gas Separation by Microporous Membranes*. Elsevier, Amsterdam, The Netherlands, p. 473.

Souleimanova, R. S., Mukasyan, A. S., and Varma, A. (2002). Pd membranes formed by electroless plating with osmosis: H_2 permeation studies. *AIChE J.* 48, 262.

Tong, J., Shirai, R., Kashima, Y., and Matsumura, Y. (2005). Preparation of a pinhole-free Pd-Ag membrane on a porous metal support for pure hydrogen separation. *J. Membr. Sci.* 260, 84.

Uemiya, S., Kude, Y., Sugino, K., Sato, N., Matsuda, T., and Kikuchi E. (1988). A palladium/porous-glass composite membrane for hydrogen separation. *Chem. Lett.* 10, 1687.

Uemiya, S., Sato, N., Ando, H., Kude, Y., Matsuda, T., and Kikuch, E. (1991). Separation of hydrogen through palladium thin film supported on a porous glass tube. *J. Membr. Sci.* 56, 303.

Welk, M., and Nernoff, T. M. (2004). H_2 separation through zeolite thin film membranes. *Prepr. Pap. Am. Chem. Soc. Div. Fuel Chem.* 49, 889.

Yan, S., Maeda, H., Kusakabe, K., and Morooka, S. (1994). Thin palladium membrane formed in support pores by metal-organic chemical vapor deposition method and application to hydrogen separation. *Ind. Eng. Chem. Res.* 33, 616.

Yeung, K. L., Christiansen, S. C., and Varma, A. (1999). Palladium composite membranes by electroless plating technique, relationship between plating kinetics, film microstructure and membrane performance. *J. Membr. Sci.* 159, 107.

Yeung, K. L., and Varma, A. (1995). Novel preparation techniques for thin metal-ceramic composite membranes. *AIChE J.* 41, 2131.

Zhu, W., Gora, L., van den Berg, A. W. C., Kapteijn, F., Jansen, J. A., and Moulin, J. A. (2005). Water vapour separation from permanent gases by a zeolite-4A membrane. *J. Membr. Sci.* 253, 57.

PART IV

MEMBRANE CONTACTORS AND REACTORS

CHAPTER 26

Membrane Contactors

KAMALESH K. SIRKAR

Otto H. York Department of Chemical Engineering, Center for Membrane Technologies, New Jersey Institute of Technology, Newark, New Jersey 07102

26.1 INTRODUCTION

Separation and purification processes and techniques may be broadly classified into the following three categories:

1. Processes and techniques that need selective partitioning between or creation of two immiscible phases.
2. An external force field is an essential basis for separation.
3. Membrane separation processes.

Examples of common processes/techniques in the first category are: absorption, adsorption, chromatography, crystallization, distillation, extraction (solvent-based or supercritical fluid based), ion exchange, and so forth. In many such processes/techniques, two immiscible phases are brought together in a variety of geometrical and flow configurations, such as packed columns, spray columns, mixer-settlers, bubble-tray columns, chromatographic columns, and the like. If the two immiscible phases are fluids, one of the phases is generally dispersed in the other phase as drops or bubbles. Distribution of the liquid phase as thin films over packings is also widely practiced in gas–liquid contactors where the gas phase flows continuously. (Packings that are in the form of thin long wires are also known for such applications in gas–liquid and liquid–liquid contacting.) A common use of such a flow configuration takes place in solid–fluid contacting applications where the solid phase in particulate form allows the fluid phase to contact the solid phase by flowing around the particles. Dispersion/suspension of solid particles is another option.

Separation in such systems takes place as different species are transferred between the two phases by different amounts. Some species concentrate more in one phase while others concentrate in the other immiscible phase due to different partitioning tendencies. The concentrations of different species to be separated are therefore different in the two

Advanced Membrane Technology and Applications. Edited by Norman N. Li, Anthony G. Fane, W. S. Winston Ho, and T. Matsuura
Copyright © 2008 John Wiley & Sons, Inc.

phases at equilibrium leading to separation. Research and development in these separation processes/techniques are generally geared toward the following goals:

1. What material and chemical properties/compositions of the two phases and operating conditions are needed to create a partitioning behavior that leads to higher selectivity between two species and higher loading of a given species in a given phase?
2. How can one increase the rate at which this process is implemented?
3. How does one mitigate/eliminate the operational problems and reduce operational cost in these devices/processes?

The solution of problems of the first type is attempted via synthesis of new absorbents, adsorbents, chromatographic media, extractants, and the like. Problems of the second type are resolved by creating very small sized particles (adsorbent, chromatographic medium), bubbles and drops, thin liquid film, and so forth that which can generate very high interfacial area per unit device volume. Problems of the third type encountered are: high-pressure drop, foaming, emulsion-formation, flooding, loading, limits on the ratio of the two-phase flow rates, poor utilization of solid adsorbent sites for partitioning, lower values of volumetric mass transfer coefficient, and the like.

Membrane contactors provide a novel approach to the solution of many such problems (especially of the second and third kind) of contacting two different phases, one of which must be a fluid. Essentially, a porous membrane, most often in hollow-fiber form, is the basic element in such a device. Any membrane in flat or spiral-wound or hollow-fiber or any other form has two interfaces since it has two sides. However, conventional separation processes involve usually one interface in a two-phase system, for example, gas–liquid, vapor–liquid, liquid–liquid, liquid–supercritical fluid, gas–solid, liquid–solid, and the like. Membrane contactors allow the creation of one immobilized phase interface between two phases participating in separation via the porous membrane. Three types of *immobilized phase interfaces* in two-phase configurations are relevant:

1. Two fluid phases in contact: gas–liquid, liquid–liquid, vapor–liquid, supercritical fluid–fluid (fluid phase membrane contactors) (Fig. 26.1a).
2. One fluid phase in contact with one solid phase: liquid–solid; gas/vapor–solid; supercritical fluid–solid (solid–fluid phase membrane contactors) (Fig. 26.2).

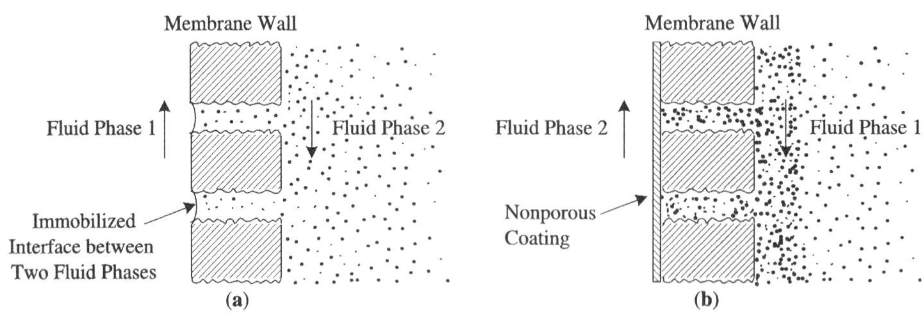

Figure 26.1 Membrane contactor allowing two fluid phases to contact each other.

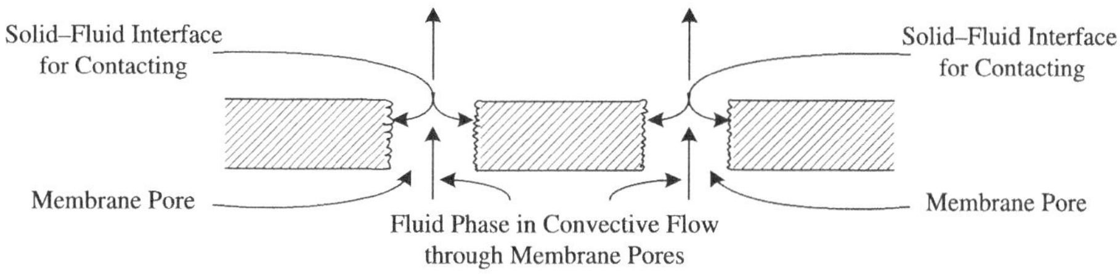

Figure 26.2 Membrane contactor for solid–fluid phase contacting.

3. The membrane system has two immobilized phase interfaces; each interface is between two immiscible fluid phases, for example, gas–liquid (1) and liquid–gas (2), liquid–liquid (1) and liquid–liquid (2) (multiple-phase-interface-based membrane contactor) (Fig. 26.3).

In all of the two-phase configurations identified above, phase interfaces are not created by dispersing one phase as drops or bubbles or particles into the other phase. One form of membrane contactor, however, helps in creating a dispersion through the membrane structure, and then the two phases continue to contact each other as they flow through the membrane device flow channels. This *fourth form of membrane contactors* has been utilized in liquid–liquid systems via emulsion formation and in gas–liquid systems via bubble generation (Fig. 26.4). Although this configuration is a radical departure from the practice of membrane contactors in the first three categories and is more in line with conventional chemical engineering contactor devices, it has its own utility.

The membrane contactors in all four categories were primarily developed for separation and purification. However, they are being increasingly exploited for the purposes of reaction-separation, separation-reaction, and so forth. Membrane contactors can therefore function as separator-reactors as well as reactor-separators.

What we will do now is briefly introduce each of the four separation and purification applications of membrane contactors, describe their operational modes and constraints (if any), provide illustrative examples, and identify, if possible, problem areas and potential needs. We will not provide a detailed review of the literature. The following selected references provide a chronological list of introductions, reviews, and overviews of various types of membrane contactors: Sirkar and Prasad (1992), Sirkar et al. (1992), Sirkar (1992), Reed et al. (1995), Gableman and Hwang (1999), Joscelyne and Trägård (2000), Klaassen and Jansen (2001), Kovvali and Sirkar (2003a, b), and Curcio and Drioli (2005).

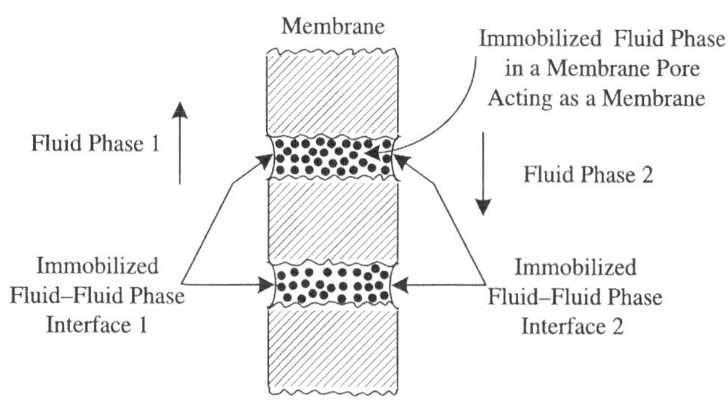

Figure 26.3 Membrane contactor system with multiple immobilized fluid phase interfaces.

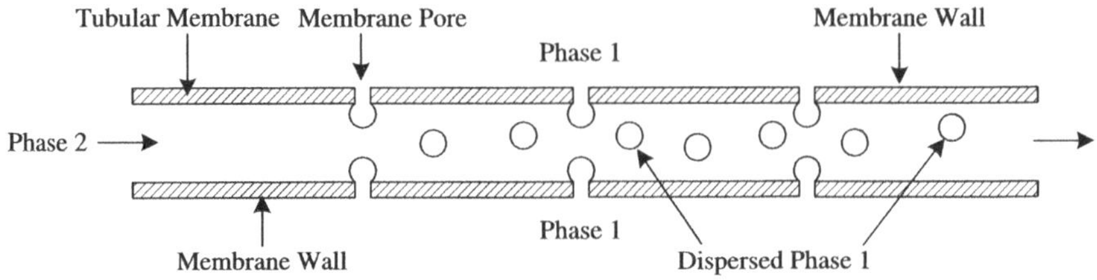

Figure 26.4 Dispersion creation and phase contacting in a dispersion-based membrane contactor.

26.2 MEMBRANE-BASED CONTACTING OF TWO FLUID PHASES

This is currently the most important area of large-scale use of membrane contactors. Three general classes of two fluid phase systems are relevant: gas–liquid (also vapor–liquid), liquid–liquid, and supercritical fluid–liquid. We will describe/touch upon the following aspects of each of such two-phase systems: basis of contacting, membranes used for contacting, mass transfer issues, and applications.

26.2.1 Gas–Liquid (Vapor–Liquid) Systems

26.2.1.1 Phase Contacting Conditions Contacting of a gas stream and a liquid stream through a porous membrane for the purpose of *gas absorption or stripping* usually employs a *porous hydrophobic membrane* and a *nonwetting liquid phase*, which is more often than not, an aqueous phase. As shown in Figure 26.1a, the fluid phase 2 (a gas stream) flows on one side of the membrane; a liquid stream (fluid phase 1) flows in the countercurrent direction (as shown) or co-current direction on the other side of the membrane. The pores are filled with the gas; the liquid flows at a pressure equal to or higher than that of the gas pressure. The gas–liquid interface is immobilized on the liquid side of the membrane pore. The excess liquid phase pressure must be less than the value of the breakthrough pressure, ΔP_{br}:

$$\Delta P_{\mathrm{br}} = \frac{2\gamma \cos\theta}{r_{\mathrm{pore}}} \tag{26.1}$$

where γ is the surface tension of the liquid, θ is the contact angle between the liquid and the membrane polymer surface, and r_{pore} is the radius of the pores in the membrane. Otherwise, the liquid will enter the pores and be dispersed as drops in the gas phase on the other side. Alternately, if the gas pressure is higher than that of the liquid, it will be dispersed as bubbles into the liquid (Sirkar, 1992).

Dispersion of either phase into the other in gas–liquid systems is undesirable. By having the pores filled with gas, the resistance to species transfer through the membrane pores is kept at a low level (Qi and Cussler, 1985a). For nondispersive operation, the liquid phase must not spontaneously wet the membrane pores. For porous hydrophobic membrane-based contactors, aqueous solutions are, therefore, preferred liquid phases; for any nonwetting liquid, the value of θ is $\geq 90°$. The surface tension of the liquid should be greater than the critical surface tension of the polymer, γ_c. However, dissolved substances, especially surface-active ones, in water can reduce the value of γ and can lead to wetting. Adsorption of particular substances at the membrane pore mouth can also lead to spontaneous pore wetting.

Polar volatile organic species/solutes present in aqueous solutions can desorb from the solution and get adsorbed on the hydrophobic polymer surface of the pore wall/mouth while the hydrophilic part of the solute sticks out making the membrane surface more hydrophilic and more susceptible to wetting (Kosaraju et al., 2005). An example of such a volatile solute is monoethanolamine used for absorption of CO_2. To avoid such volatile solute adsorption and subsequent surface modification in CO_2 absorption, nonvolatile absorbents containing amine functionalities, such as amino acids (Kumar et al., 2002) and polyamidoamine (PAMAM) dendrimers (Kosaraju et al., 2005) are being explored. For example, no wetting was observed in runs using generation 0 PAMAM dendrimer in an aqueous solution for periods as long as 55 days with porous hydrophobic polypropylene hollow fibers. This run employed continuous CO_2 absorption in one membrane module and CO_2 stripping in another module with the absorbent solution being circulated between the two modules.

The discussion presented above is valid whether absorption of a gas into a liquid or desorption of a gas from a liquid is carried out. If a vapor is absorbed into a liquid, the same considerations are valid. If stripping is implemented via vacuum in a nondispersive fashion, the pressure conditions for the two-phase system and the wetting criterion remain unchanged to ensure that the hydrophobic porous membrane remains nonwetted. If, however, the liquid phase exists inside the pores, then to achieve nondispersive operation, the gas-phase pressure should be equal to or greater than that of the liquid phase (Karoor and Sirkar, 1993). Vacuum stripping is not possible with liquid in the membrane pores, that is, wetted pores; it is necessary to have a nonporous coating on the side of the membrane where the gas–liquid interface would have been located otherwise (Sirkar, 1992).

One can employ a porous membrane with a nonporous coating or an asymmetric membrane having a nonporous skin (Fig. 26.1b). In case of nonwetted hydrophobic pores, conventional operating conditions may be maintained with aqueous solutions, namely aqueous solution at a higher pressure than the gas phase unless there is significant permeation of the aqueous phase through the coating; therefore, the coating/skin is also hydrophobic (Kosaraju et al., 2005). For wetted pores, the liquid phase is also maintained at a higher pressure than the gas phase with hydrophobic nonporous coatings even though the liquid phase may be nonpolar organic (as shown in Fig. 26.1b). Such a condition allows vacuum stripping of the liquid phase to be implemented without any problem provided the liquid does not permeate to any degree. However, it is desirable to have the pore nonwetted to allow high transfer rates in vacuum stripping or even gas absorption (Poddar et al., 1996a, b).

An additional case of gas–liquid/vapor–liquid contacting through a porous hydrophobic membrane (Fig. 26.1a) happens in the case of *vacuum membrane distillation*. In this case, the aqueous/nonwetting solution is at a higher temperature; volatile components that may be major components (e.g., water in desalination) or minor components (e.g., alcohol present in water) are evaporated and removed by vacuum on the other side. Operationally, it is similar to stripping using nonwetted pores. However, there are different transport rate considerations here. Specifically, heat transfer rate in the feed solution is very important in vacuum membrane distillation (Li and Sirkar, 2005).

26.2.1.2 Membranes Membranes used for gas–liquid contacting include porous polypropylene, porous tetrafluoroethylene, porous poly-4-methyl-1-pentene. The latter is asymmetric with a dense skin (Kosaraju et al., 2005). Polysulfone membranes (mildly hydrophobic) have also been used with a nonporous hydrophobic high free volume fluoropolymer coating (Kovvali and Sirkar, 2003a). (The same reference identified porous contacting membranes built completely out of a fluoropolymer.) However, most of the membranes employed are symmetric. The geometrical form of the membrane employed most often is hollow fine fibers. Commercially, they are utilized generally in a mat form (Celgard Inc., Charlotte, NC). The porosity of the membranes varies between 0.3 and 0.8. The pore size of the membranes in membrane contactors varies depending on the membrane, the application and the presence/absence of an integral or composite skin. Where a dense skin or a dense coating is employed, one must seek a skin having as high a permeance as possible (e.g., silicone rubber) without allowing the liquid phase to permeate through. At the same time the porous support should be as porous as possible (Albrecht et al., 2005) to reduce the transport resistance; an asymmetry in the porous structure is useful for applying a nonporous coating successfully on the side of the smaller pore size. If the liquid phase is kept out of the pores in such a structure, the porous substrate may be hydrophilic.

26.2.1.3 Mass Transfer in Gas–Liquid Systems

As in conventional contactors, mass transfer rates in membrane contactors for gas–liquid systems are generally described by means of an overall mass transfer coefficient, K, and the gas–liquid interfacial area per unit device volume, a_v. The overall mass transfer coefficient based on the liquid phase for any species i, K_{il}, is usually described via the principle of the following resistances in series: liquid film resistance ($1/k_{il}$), membrane resistance (H_i/k_{im}), and the gas film resistance (H_i/k_{ig}) for the gas-filled membrane pore case in series leading to the overall resistance ($1/K_{il}$):

$$\frac{1}{K_{il}} = \frac{1}{k_{il}} + \frac{H_i}{k_{im}} + \frac{H_i}{k_{ig}} \tag{26.2}$$

where H_i is Henry's law constant for species i defined as

$$C_{il} = H_i p_{ig} \tag{26.3}$$

Here C_{il} is the liquid-phase concentration of species i in equilibrium with the partial pressure p_{ig} of species i in the gas phase. The objective in any membrane contactor design is to make sure that all three resistances are as small as possible. For the liquid-phase resistance, minimization requires employing cross flow on the shell side around hollow fibers (Sirkar, 1992; Reed et al., 1995; Gableman and Hwang, 1999). In gas–liquid contacting generally the liquid-phase resistance dominates; however, for highly reactive systems, the gas-phase resistance may contribute substantially. There will be an additional coating resistance in the case of membranes having a coating; this resistance can be substantial (Poddar et al., 1996a).

The overall species transfer rate is proportional to the product of K_{il} and a_v, the gas–liquid interfacial area per unit device volume. Usually, the k_{im} term in Eq. (26.2) incorporates the membrane porosity effect, so that the value of a_v used is the total membrane surface area per unit device volume. The typical range of a_v in hollow-fiber membrane devices is 16–65 cm^{-1} (500–2000 ft^{-1}), which is orders of magnitude larger than conventional devices (Reed et al., 1995). In general, the product K_{il} a_v is around 5–40 times larger than those in most conventional contacting devices. An important aspect of membrane contactor operation and mass transfer is that there can be a very wide variation of the gas–liquid flow rate ratio. There is no flooding or loading-related phenomena in a membrane contactor as long as the correct phase pressure conditions are maintained. Foaming is also avoided. The phase flow rates on each side of the membrane wall are such that the Reynolds numbers are not very high so that pressure drops are limited. Yet the mass transfer coefficients especially on the shell side are quite high if cross flow is employed.

26.2.1.4 Applications The following are some of the widely used and emerging applications of gas–liquid membrane contactors: oxygenation of aqueous solutions, degassing of water/aqueous solutions for ultrapure water production, carbonation of aqueous solutions/beverages, absorption of NH_3, SO_2, and so forth, stripping of volatile organic compounds (VOCs), and humidification/dehumidification of airstreams.

26.2.2 Liquid–Liquid Systems

26.2.2.1 Phase Contacting Conditions The contacting of one liquid phase with another immiscible liquid phase in a nondispersive membrane contactor may be carried out in the fashion illustrated in Figure 26.1a for a porous membrane. If the porous membrane is *hydrophobic* and is spontaneously wetted by an organic nonpolar phase, then an aqueous phase (or any other nonwetting liquid phase) is brought on the other side of the membrane at a pressure equal to or higher than that of the wetting nonpolar organic phase in the pores (Kiani et al., 1984). The interface between the two immiscible phases is at the membrane pore mouth on the side of the aqueous (nonwetting) phase. As long as the excess phase pressure of the aqueous (nonwetting) phase over that of the wetting phase in the pores is less than the value of a critical pressure, ΔP_{cr}, the interface of the two phases at the pore mouth is stable:

$$\Delta P_{cr} = \frac{2\gamma_{in}\cos\theta}{r_{pore}} \tag{26.4}$$

Here γ_{in} is the interfacial tension between the two liquids, unlike the surface tension, γ, in Eq. (26.1) for gas–liquid contacting. When the excess phase pressure difference exceeds ΔP_{cr}, the aqueous (nonwetting) phase is dispersed as drops in the organic (wetting) phase. Otherwise, the two immiscible phases will flow on two sides of the membrane without any dispersion regardless of their density differences; mass transfer or solvent extraction or back extraction will take place between the two phases.

If the porous membrane is hydrophilic, then it can be spontaneously wetted by an aqueous solution or an organic solution, polar or nonpolar (Prasad et al., 1990). Generally, an aqueous solution is used as the wetting solution in the pores of a hydrophilic membrane; the organic solution on the other side of the membrane is maintained at the same pressure or higher pressure than that of the aqueous phase. Unless the organic phase pressure exceeds that of the aqueous phase flowing on the other side by an amount exceeding ΔP_{cr}, the phase interface is immobilized at the pore mouth of the membrane on the organic side; nondispersive solvent extraction/back extraction can be implemented. Experimental estimates of the magnitudes of the value of ΔP_{cr} for a variety of membranes and solvent extraction systems are provided in Sirkar and Prasad (1992). One can also introduce an aqueous nonwetting phase in the pores of a hydrophobic membrane; however, an organic solution present at the interface will gradually displace the aqueous solution from the pores. Stable interfaces have been formed and extractions studied in the case of a nonpolar organic and a polar organic system (Prasad and Sirkar, 1987).

26.2.2.2 *Membranes* The membranes employed have to possess considerable solvent resistance. The membrane most commonly employed for solvent extraction from aqueous solutions is of porous hydrophobic polypropylene invariably in hollow-fiber form (Celgard Inc., Charlotte, NC). It can be used with alcohols, alkanes, aromatics, and the like. The swelling increases with ketones; chlorinated solvents are not used. Fluorinated copolymers (Mykrolis Inc., Billerica, MA) are good candidates for use with most solvents (exceptions include perflourinated solvents) as hydrophobic membranes. Membranes made of nylon, regenerated cellulose, or polyacrylonitrile have been successfully employed with aqueous phase in pores. Recently, porous ceramic membranes having a polymeric coating have been introduced (Kühni AG, Allschwil, Switzerland).

26.2.2.3 *Mass Transfer* The rate of mass transfer in a liquid–liquid extraction system implemented in a nondispersive membrane contactor is analyzed in the manner followed in conventional contactor analysis. The overall organic phase-based mass transfer coefficient K_{io} of a species i being extracted (or back extracted) from an aqueous solution into a solvent wetting the pores of a *hydrophobic membrane* is described via the resistances-in-series model:

$$\frac{1}{K_{io}} = \frac{m_i}{k_{iw}} + \frac{1}{k_{imo}} + \frac{1}{k_{io}} \tag{26.5}$$

Here m_i is the distribution coefficient of solute i between the organic phase and the aqueous phase:

$$m_i = \frac{C_{io}}{C_{iw}} \tag{26.6}$$

where C_{io} is the organic phase solute concentration in equilibrium with the aqueous phase solute concentration C_{iw}. The three resistances in series are m_i/k_{iw}, the aqueous phase resistances; $1/k_{imo}$, the membrane phase resistance; and $1/k_{io}$, the organic phase resistance. When the solute strongly prefers the organic phase, $m_i >> 1$, the resistance of the aqueous phase controls the transfer rate. When $m_i << 1$, the organic phase along with the membrane resistance control the extraction rate. These considerations are valid whether there is solute extraction or back extraction. The resistance-in-series model for a hydrophilic membrane with the aqueous phase in pore leads to [instead of Eq. (26.5)]

$$\frac{1}{K_{io}} = \frac{1}{k_{io}} + \frac{m_i}{k_{imw}} + \frac{m_i}{k_{iw}} \tag{26.7}$$

Detailed expressions for a variety of configurations are available in Sirkar and Prasad (1992). Design of such contactors using the number of transfer units (NTU) and height of a transfer unit (HTU) has been illustrated in Sirkar and Prasad (1992). A more detailed design approach has been presented by Karabelas and Asimakopoulou (2006).

The very high value of a_v in polymeric hollow-fiber-based extractors leads to high values of $K_{io}\, a_v$. Further the HTU [or LTU (length of transfer unit)] values can be as low as 2 cm (Prasad and Sirkar, 1990). As a result very small and compact hollow-fiber extractors can replace conventional dispersion-based devices that are 10–50 times larger.

26.2.2.4 Applications This technique has been applied to the following extraction applications: metal extraction from aqueous solution; metal back extraction from organic solvents into an aqueous solution; organic pollutant extraction; aromatics extraction from a nonpolar organic to a polar organic; pharmaceuticals extraction; and fermentation product extraction (Sirkar and Prasad, 1992). Large-scale commercial applications are reported in Klaassen and Jansen (2001) for recovery of an aromatic compound from wastewater back into a reactor using porous hydrophobic polypropylene hollow-fiber membranes and in Lopez and Matson (1997) for extraction of an enzymatic reaction product in a multiphase/extractive enzyme membrane reactor for the industrial production of diltiazem chiral intermediate using an acrylonitrile-based hydrophilic asymmetric hollow-fiber membrane.

26.2.3 Supercritical Fluid–Liquid Systems

When a subcritical fluid becomes a supercritical fluid (SCF), its physical properties change drastically, for example, density increases considerably and the solubilities of solutes also increase correspondingly; hence, supercritical fluids are often called extractants as if they were solvents. Supercritical fluid formed from CO_2 has been used to contact aqueous solutions using porous hydrophobic polypropylene hollow-fiber devices in a nondispersive fashion (Fig. 26.1a) (Robinson and Sims, 1996). Dense CO_2 gas at a high pressure [e.g., 124 bars Gableman et al. (2005)] has been used to study extraction from an aqueous solution. Subcritical propane gas at 34.5 atm and 25°C has also be used to extract ethanol/acetone from an aqueous solution (Bocquet et al., 2005). Although Robinson and Sims (1996) have claimed that "the process is conducted with the pressure on both sides of the membrane in the module being essentially the same," (lines 12–14, Abstract) such cannot be the case for countercurrent flow of two phases on two sides of the membrane. Gableman et al. (2005) have maintained the aqueous solution at a higher pressure when using dense CO_2 gas and porous hydrophobic polypropylene membrane to prevent leakage of CO_2 from the membrane pores into the aqueous liquid solution.

Membranes employed so far for the extraction of species from an aqueous solution by a supercritical fluid or a subcritical dense gas consist of porous polypropylene hollow fibers. These fibers do not have high enough strength to resist the high pressures employed with such extractants (34.5 to >124 bars). Therefore the housing of the membrane module must withstand the pressure of the supercritical/subcritical fluid. This is needed regardless of the membrane's capacity to withstand high pressure since the pressure difference needed to maintain a stable phase interface at the pore mouth is not too high.

The conventional method of describing mass transfer in porous hollow-fiber-based membrane contactors (Sirkar and Prasad, 1992) has also been found to be adequate for describing transport in SCF–liquid solution systems. However, special attention has to be paid to buoyancy-induced flow due to low kinematic viscosity of dense gas systems in vertical configuration where gravitational force is important (Gableman et al., 2005). This last reference provides a comprehensive transport analysis in dense gas extraction for an aqueous solution. Bocquet et al. (2005) have also provided considerable data and analysis for dense gas extraction and suggest that a high value of the ratio of membrane porosity to tortuosity is needed for obtaining high flux and extraction efficiency.

Towsley et al. (1999) have described a number of applications of porous membrane-based supercritical extraction involving extraction of flavors from wines and juices, recovering aromas from vegetables, and marine oils and nuts as well as fermentation broths, recovering and recycling industrial organic solvents.

26.3 MEMBRANE-BASED SOLID–FLUID CONTACTING

There are two general modes of membrane-based solid–fluid contacting: (1) membrane pore surface provides sites for solid–fluid contacting as in ion exchange/adsorption/complexation of ionic/nonionic species; (2) as the fluid flows through the flow channel on one side of the membrane wall, the membrane pores allow species movement and efficient contact with adsorbent particles on the other side of the membrane. An additional contacting mode is (3) two adjacent but different membrane device flow channels allow fluid to flow across from one to the other through short bed lengths of adsorbent beads.

26.3.1 Membrane Adsorption, Membrane Chromatography: Liquid–Solid Contacting

Consider Figure 26.2 in the context of liquid–solid adsorption/ion exchange processes for macrosolutes/proteins and the like. In conventional ion exchange chromatography, porous resin beads are employed to separate the protein by adsorption-based processes. Considerable diffusional resistance is encountered by protein molecules in reaching the sites in the inside surfaces of the porous beads employed in chromatographic columns. As a result, the fractional utilization of the large site density in the porous resin beads is low. In processes identified as *membrane adsorption/membrane chromatography*, these deficiencies of conventional porous resin beads are eliminated. Feed solution containing the macrosolute/protein flows on one side of the membrane wall. The membranes are usually microfiltration membranes having pore sizes in the range of 0.02–10 μm. The excess pressure of the liquid solution drives the solution through the membrane pores. This convective motion brings protein molecules into contact with individual sites/functional groups on the surfaces of the pore wall. The utilization of the active sites/ligands on the pore surfaces is very high, if not complete.

However, since the total site capacity of a membrane pore surface has a finite/limited value, within a short period of time, the site capacity of a membrane pore (thickness of the membrane wall is small) is exhausted. In the case of flat membranes, the approach adopted is to stack a number of flat membranes one over the other. Therefore much greater pore surface area is available as the liquid flows through the pores of consecutive flat membranes. In the case of hollow fibers, the solution flows through the bore of microfiltration membrane-based hollow fibers. As the liquid solution flows out through the pores, proteins get bound with the ligands on the pore surface. Since there is limited site capacity in the pore in the wall of a hollow fiber, site saturation is achieved relatively quickly. So one must stop the convective liquid motion of feed solution through the microfiltration membrane pore prior to pore exhaustion and start passing the pore surface regenerating solution in the same manner. After the pores are regenerated, one begins the cycle of feed solution introduction, pore surface adsorption, pore surface regeneration, and so forth in a cyclic fashion. The same approach is followed for a stack of flat membranes except the cycle time is longer.

This method of fluid–solid contacting is now practiced commercially in the biotechnology industry. An early introduction to this technique is available in Brandt et al. (1988). A number of reviews are available (Thommes and Kula, 1995; Roper and Lightfoot, 1995). A book illustrating this technique was published sometime back (Klein, 1991).

This technique has now been successfully adopted to adsorption of heavy metals from contaminated waters with or without disposable membranes (Hestekin et al., 2001).

26.3.2 Gas–Solid Membrane Contactor

In adsorption and chromatographic processes, there is an increasing tendency toward smaller particles to provide large adsorbent surface area. However, flow in beds of such particles will lead to high pressure drops. An approach to avoid the high pressure drops in beds made of small-size particles is to provide a porous hollow fiber (or a spiral-wound module) whose shell side is packed with small-size beads. The feed gas stream flows through the fiber bore. Gaseous species diffuse rapidly through the pores in the wall and contact the fine particles (as small as 10 μm). Feed gas undergoes a limited pressure drop characteristic of conventional hollow-fiber-based flow. At the same time, gas–solid contact with fine adsorbent particles having a very high surface area is achieved without the feed gas undergoing a high pressure drop. The shell side in a hollow-fiber device does not have any inlet or outlet open for any gas to come in or go out. A similar arrangement can be practiced with a spiral-wound membrane contactor. A summary of such devices is provided in Feng and Ivory (2002). One could employ a liquid instead of gas also in this configuration. Such an arrangement is also suitable for carrying out catalytic reactions with catalyst particles kept on the shell side in a trickle bed format (Yang and Cussler, 1987) where, however, liquid is introduced on the shell side as the gas flows through the tube side.

26.3.3 Short-Bed-Length Fluid–Solid Contactor

If direct-flow-based contacting of a fluid phase with a solid adsorbent is desirable without excessive pressure drop, one can take recourse to short bed lengths. This short bed length is created by adsorbents/ion exchange resin beds located between two adjacent sets of porous hollow fibers. The feed liquid flows through the bore of one set of hollow fibers. As the liquid permeates through the wall of the first set of hollow-fiber membranes, it encounters resin particles on the shell side. There is a second set of porous hollow fibers immediately adjacent acting as a receiver channel of the liquid permeating through the resin bed. More details is provided in a U.S. Patent (Asher, 1997). One problem in such an arrangement is maldistribution of flow from the top to the bottom of the bed.

26.4 TWO IMMOBILIZED PHASE INTERFACES

The phenomenon of a critical pressure ΔP_{cr} or a breakthrough pressure ΔP_{br} [Eqs. (26.4) and (26.1), respectively] needed to destroy immobilized interfaces between two fluid phases at the mouth of a membrane pore led to the possibility of one fluid phase being contained within the pores of a porous solid membrane and contacting two different fluid phases on two sides of the solid membrane (Fig. 26.3). The fluid phase inside the pores is identified as a membrane since it acts as a membrane with two phase interfaces. In case the fluid phase is liquid, it is identified as either an immobilized liquid membrane (ILM) or a supported liquid membrane (SLM). The following phase contacting configurations have been extensively studied using an SLM/ILM:

SLM: Flowing liquid phase 1: SLM: Flowing liquid phase 2
ILM: Flowing gas phase 1: ILM: Flowing gas phase 2

When a gaseous phase exists inside the membrane pore and two different liquids flow on two sides of the membrane, we have a supported gas membrane (SGM) inside the membrane pores:

SGM: Flowing liquid phase 1: SGM: Flowing liquid phase 2

An essential prerequisite for each such case is that the phase present in the pore is not spontaneously displaced by the contacting fluid on each side of the pore. In the case of an SGM, the porous membrane must be hydrophobic for aqueous (or nonwetting) solutions flowing on two sides of the membrane (Imai et al., 1982; Qi and Cussler, 1985b, c). In the case of an SLM based on an organic phase in the membrane pores, hydrophobic membranes are conventionally used with two flowing aqueous phases on two sides; hydrophilic membranes may be used also. However, aqueous phases in the pores of a hydrophobic membrane cannot be used with two organic phases flowing on two sides. In the case of an ILM, however, with two gas phases flowing on two sides of the membrane, either organic or aqueous phase may act as the immobilized liquid membranes in the pores. The gas pressure differential between the two sides must not exceed the value of ΔP_{br} [Eq. (26.1)]. Hydrophilic as well as hydrophobic membranes may be used to imbibe the liquid into the pores.

Successful operation of any such membrane—SLM, ILM, or SGM—requires that the phase in the pore remains intact and does not disappear with time: Liquid membrane stability is of utmost importance. In the case of an ILM with two gas phases flowing on two sides, there are three requirements for stability: (1) The liquid membrane must not disappear by evaporation into the gas phase; (2) gas-phase condensation should not flood the liquid membrane; and (3) the sorption of gas-phase constituents and their reaction (if any) in the liquid membrane must not damage the liquid membrane. Considerable progress toward the solution of requirements (1) and (2) vis-à-vis extended stability has been summarized in Kovvali and Sirkar (2003b). In the case of SLM, stability requires no dissolution of liquid membrane constituents in the two flowing liquid phases and no displacements of the SLM by constituents of the flowing liquid phases. A variety of techniques have been tried especially with organic SLMs and two flowing aqueous phases. The goal of an extended life time remains elusive (see Kovvali and Sirkar, 2003b) due to leaching of membrane constituents and potential wetting by flowing liquid phases. Supported gas membranes appear to be quite stable (Muttoo and Dean, 2001; Qi and Cussler, 1985c) unless the two nonwetting flowing phases slowly seep into the pores.

Membranes employed for such SGM studies generally are of polypropylene, polytetrafluoroethylene, and the like. The membrane configuration employed span from hollow fibers (Qi and Cussler, 1985b, c) to stacks of flat membranes (Muttoo and Dean, 2001) in large scale.

A variation of SGM-based separation involves the processes of *osmotic distillation* (OD) and *direct-contact membrane distillation* (DCMD). They are not usually identified as SGM processes. In OD, two aqueous solutions generally at the same temperature, that do not wet the pores of a hydrophobic membrane, are on two sides with gas phase in the pores. However, the osmotic pressure of one solution is much higher than the other; this causes water vapor to be transported through the pores to the more concentrated solution having a higher osmotic pressure. In direct-contact membrane distillation, hot nonwetting solution flows on one side of the SGM: vapor from the solution is evaporated, diffuses through the pores, and is condensed in the cold distillate flowing on the other side of the membrane.

Large-scale studies of OD have been conducted (Sirkar, 1995). Highly successful studies on DCMD for brine solutions have also been implemented by Li and Sirkar (2004). Pilot-plant studies have been recently completed successfully by the same group. For a recent review of both OD and MD, see Curcio and Drioli (2005).

26.5 DISPERSIVE CONTACTING IN A MEMBRANE CONTACTOR

All topics on fluid–fluid contacting considered so far employed a porous membrane (generally without any nonporous coating) in which at the pore mouths, the phase interfaces between the fluid phases were immobilized. Further the contacting was nondispersive. Over the last 17 years, fluid–fluid contacting in a dispersive format has been carried out in porous membrane devices. The porous membrane has been used as a device to create a dispersion of drops of one liquid in another or a dispersion of gas bubbles in a liquid (Fig. 26.4). Next, in some cases, the membrane device channel may allow continued dispersive contact between the two phases if so designed. Thus, the membrane device can become more than an efficient emulsifier/sparger/homogenizer.

One of the earliest studies was carried out by Nakashima et al. (1991): *Membrane emulsification* was achieved by forcing an oil phase through a microporous glass membrane, dispersing the oil phase into the water phase, and creating a narrow drop size distribution whose average size was around three times the size of the membrane pores. Matsuoka et al. (1992) employed hydrophobic porous polyethylene hollow-fiber membranes with one bore end closed and gas introduced into the bore through the other end and sparged into the liquid on the shell side as tiny bubbles. The power consumption of this method of *bubble generation* for an equivalent amount of transfer of, say, oxygen into the surrounding liquid was the lowest of all sparging techniques. Downstream this dispersion of bubbles could be used in a fermentation system; emulsified drops are useful for two-phase reaction systems. An additional variation on gas dispersion via bubbles is *membrane foaming* (Bals and Kulozik, 2003) where due to the presence of surface-active agents, proteins, and the like in a solution flowing at a low rate on one side of the membrane, gas sparging leads to the formation of a foam. Very fine size bubbles in a foam were produced via small pores in a membrane.

In these processes, the contacting between the two phases takes place during the formation of the bubbles and drops. As the bubbles and drops next move/flow with the liquid, the membrane device will act as a traditional dispersive phase contactor. On the other hand, if the membrane function is limited to the generation of dispersion, which moves next in a non-membrane-based device, then the membrane functions as a *supersparger*; contacting function is limited unless most of the contacting takes place during sparging.

The dispersion creation requires application of a pressure difference, ΔP, between the phase to be dispersed and the phase in which dispersion takes place through the pores. This ΔP has two contributions: (1) capillary pressure ΔP_{cr} defined by Eq. (26.4) and (2) viscous flow pressure drop ΔP_{flow} through the capillary, which is the membrane pore (usually calculated from Hagen–Poiseuille law). The wetting characteristics of the pore material as well as the membrane channel are important for the emulsification process, especially, the nature of the emulsion produced. An earlier review of membrane-based emulsification is available in Joscelyne and Trägårdh (2000). A comparison of membrane emulsification with that through porous microchannel plate is provided in Lambrich and Schubert (2005).

When two miscible phases have to be contacted by dispersing one in the other, traditionally conventional stirred mixers are employed. One can also use porous membranes to mix two miscible phases. Zarkadas and Sirkar (2006) have employed porous hydrophobic hollow-fiber membranes to force either an isopropanol phase or an aqueous solution of an amino acid into the other phase. This was done to carry out antisolvent-based crystallization. Extraordinarily high rates of crystallization were achieved with much smaller crystal size.

26.6 CONCLUDING REMARKS

Membrane contactors provide an elegant opportunity for contacting two fluid phases in a nondispersive fashion. Membrane contactors provide also a window into efficient contacting of a solid phase with a fluid phase. Dispersive contacting can be efficiently carried out in addition by porous membrane contactor devices during the dispersion process. Large-scale applications of membrane contactors are an increasingly reality. Many potential applications are waiting to be realized [see Yang et al. (2006) for membrane distillation of organic solvents].

REFERENCES

Albrecht, W., Hilke, R., Kneifel, K., Weigel, Th., and Peinemann, K.-V. (2005). Selection of microporous hydrophobic membranes for use in gas/liquid contactors: An experimental approach. *J. Membr. Sci.* 263, 66.

Asher, W. J. (1997). Hollow fiber contactor and process. U.S. Patent 5,693,230.

Bals, A., and Kulozik, U. (2003). The influence of the pore size, the foaming temperature and the viscosity of the continuous phase on the properties of foams produced by membrane foaming. *J. Membr. Sci.* 220, 5.

Bocquet, S., Torres, A., Sanchez, J., and Rios, G. J. (2005). Modeling the mass transfer in solvent-extraction processes with hollow-fiber membranes. *AIChE J.* 51, 1067.

Brandt, S., Goffe, R. K., Kessler, S. B., O'Connor, J. L., and Zale, S. E. (1988). Membrane-based affinity technology for commercial purifications. *Biotechnology* 6, 779.

Curcio, E., and Drioli, E. (2005). Membrane distillation and related operations—A review. *Sep. Purifi. Rev.* 34, 35.

Feng, X., and Ivory, J. (2002). Hollow fiber and spiral wound contactors for fluid/particle contact and interaction. *Chem. Eng. Commun.* 189, 247.

Gableman, A., and Hwang, S.-T. (1999). Hollow fiber membrane contactors. *J. Membr. Sci.* 159, 61.

Gableman, A., Hwang, S.-T., and Krantz, W. B. (2005). Dense gas extraction using a hollow fiber membrane contactor: Experimental results versus model predictions. *J. Membr. Sci.* 257, 11.

Hestekin, J. A., Bachas, L. G., and Bhattacharyya, D. (2001). Poly(amino acid) functionalized cellulosic membranes: Metal sorption mechanisms and results. *Ind. Eng. Chem. Res.* 40, 2668.

Imai, M., Furusaki, S., and Miyauchi, T. (1982). Separation of volatile materials by gas membranes. *Ind. Eng. Chem. Process Des. Dev.* 21, 421.

Joscelyne, S. M., and Trägård, G. (2000). Membrane emulsification—A literature review. *J. Membr. Sci.* 169, 107.

Karabelas, A. J., and Asimakopoulou, A. G. (2006). Process and equipment design for membrane-based extraction: Basic considerations. *J. Membr. Sci.* 272, 78.

Karoor, S., and Sirkar, K. K. (1993). Gas absorption studies in microporous hollow fiber membrane modules. *Ind. Eng. Chem. Res.* 32, 674.

Kiani, A., Bhave, R. R., and Sirkar, K. K. (1984). Solvent extraction with immobilized interfaces in a microporous hydrophobic membrane. *J. Membr. Sci.* 20, 125.

Klaassen, R., and Jansen, A. E. (2001). The membrane contactor: Environmental applications and possibilities. *Environ. Prog.* 20, 37.

Klein, E. (1991). *Affinity Membranes: Their Chemistry and Performance in Adsorption Separation Processes*. Wiley, Hoboken, NJ.

Kosaraju, P., Kovvali, A. S., Korikov, A., and Sirkar, K. K. (2005). Hollow fiber membrane contactor based CO_2 absorption-stripping using novel solvents and membranes. *Ind. Eng. Chem. Res.* 44, 1250.

Kovvali, A. S., and Sirkar, K. K. (2003a). Membrane contactors: Recent developments. In D. Bhattacharyya and D. Allan Butterfield (Eds.), *New Insights into Membrane Science and Technology: Polymeric and Biofunctional Membranes*. Membrane Science and Technology Series 8. Elsevier, New York, p. 147.

Kovvali, A. S., and Sirkar, K. K. (2003b). Stable liquid membranes: Recent developments and future directions. *Ann. N. Y. Acad. Sci.* 984, p. 279.

Kumar, P. S., Hogendoorn, J. A., Feron, P. H. M., and Versteeg, G. F. (2002). New absorption liquids for removal of CO_2 from dilute gas streams using membrane contactors. *Chem. Eng. Sci.* 57, 1639.

Lambrich, U., and Schubert, H. (2005). Emulsification using microporous systems. *J. Membr. Sci.* 257, 76.

Li, B., and Sirkar, K. K (2004). Novel membrane and device for direct contact membrane distillation-based desalination process. *I&EC Res.* 43, 5300.

Li, B., and Sirkar, K. K. (2005). Novel membrane and device for vacuum membrane distillation-based desalination process. *J. Membr. Sci.* 257, 60.

Lopez, J. L., and Matson, S. L. (1997). A multi-phase/extractive enzyme membrane reactor for production of diltiazem chiral intermediate. *J. Membr. Sci.* 125, 189.

Matsuoka, H., Fukada, S., and Toda, K. (1992). High oxygen transfer rate in a new aeration system using hollow fiber membrane. *Biotechnol. Bioeng.* 40, 346.

Muttoo, T., and Dean, D. (2001). Development of chlorine dioxide gas transport contactor (ERCOR101 Generator) for use in drinking water disinfection. Paper presented at the North American Membrane Society 12th Annual Meeting, Lexington, KY, May 15–20, 2001.

Nakashima, T., Shimizu, M., and Kukizaki, M. (1991). Membrane emulsification by microporous glass. Paper presented at the Second International Conference on Inorganic Membranes, Montpellier, France, July 1–4, 1991.

Poddar, T. K., Majumdar, S., and Sirkar, K. K. (1996a). Membrane-based absorption of VOCs from a gas stream. *AIChE J.* 42, 3267.

Poddar, T. K., Majumdar, S., and Sirkar, K. K. (1996b). Removal of VOCs from air by membrane-based absorption and stripping. *J. Membr. Sci.* 120, 221.

Prasad, R., Khare, S., Sengupta, A., and Sirkar, K. K. (1990). Novel liquid-in-pore configurations in membrane solvent extraction. *AIChE J.* 36, 1592.

Prasad, R., and Sirkar, K. K. (1987). Microporous membrane solvent extraction. *Sep. Sci. Technol.* 22, 619.

Prasad, R., and Sirkar, K. K. (1990). Hollow fiber solvent extraction: Performances and design. *J. Membr. Sci.* 50, 153.

Qi, Z., and Cussler, E. L. (1985a). Microporous hollow fibers for gas absorption, I. Mass transfer in liquid, II. Mass transfer across the membrane. *J. Membr. Sci.* 23, 321.

Qi, Z., and Cussler, E. L. (1985b). Bromine recovery with hollow fiber gas membranes. *J. Membr. Sci.* 24, 43.

Qi, Z., and Cussler, E. L. (1985c). Hollow fiber gas membranes. *AIChE J.* 31, 1548.

Reed, B. W., Semmens, M. J., and Cussler, E. L. (1995). Membrane contactors. In R. D. Noble and S. A. Stern (Eds.), *Membrane Separations Technology, Principles and Applications*. Elsevier Science, New York, p. 468.

Robinson, J. R., and Sims, M. J. (1996). Method and system for extracting a solute from a fluid using dense gas and a porous membrane. U.S. Patent 5,490,884.

Roper, D. K., and Lightfoot, E. N. (1995). Separation of biomolecules using adsorptive membranes. *J. Chromatogr. A* 702, 3.

Sirkar, K. K. (1992). Other new membrane processes. In W. S. W. Ho and K. K. Sirkar (Eds.), *Membrane Handbook*. Van Nostrand Reinhold, New York, Chapter 46. Reprinted 2001, Kluwer Academic, Boston, MA.

Sirkar, K. K. (1995). Membrane separations: Newer concepts and applications for the food industry. In R. K. Singh and S. S. H. Rizvi (Eds.), *Bioseparation Processes in Foods*. Marcel Dekker, New York, p. 351.

Sirkar, K. K., and Prasad, R. (1992). Membrane-based solvent extraction. In W. S. W. Ho and K. K. Sirkar (Eds.), *Membrane Handbook*. Van Nostrand Reinhold, New York, p. 727. Reprinted 2001, Kluwer Academic, Boston, MA.

Sirkar, K. K., Sengupta, A., and Majumdar, S. (1992). Hollow fiber contained liquid membrane. In W. S. W. Ho and K. K. Sirkar (Eds.), *Membrane Handbook*. Van Nostrand Reinhold, New York, p. 764. Reprinted 2001, Kluwer Academic, Boston, MA.

Thommes, J., and Kula, M. R. (1995). Membrane chromatography—An integrative concept in the downstream processing of proteins. *Biotechnol. Prog.* 11, 357.

Towsley, R. W., Turpin, J. A., Sims, M. J., and Robinson, J. R. (1999). Porocritical fluid extraction using CO_2 for industrial organic solvent recovery and recycle. Paper presented at the CISF-99 Fifth Conference on Supercritical Fluids and Their Applications, Garda (Verona), Italy, June 13–19, 1999.

Yang, D., Barbero, R. S., Delvin, D. J., Colling, C. W., Cussler, E. L., and Carrera, M. E. (2006). Hollow fibers as structured packing for olefin/paraffin separation. *J. Membr. Sci.* 279, 61.

Yang, M. C., and Cussler, E. L. (1987). A hollow fiber trickle-bed reactor. *AIChE J.* 33, 1754.

Zarkadas, D., and Sirkar, K. K. (2006). Antisolvent crystallization in porous hollow fiber devices. *Chem. Eng. Sci.* 61, 5030.

CHAPTER 27

Membrane Reactors

ENRICO DRIOLI and ENRICA FONTANANOVA

Institute on Membrane Technology of the National Council Research (ITM-CNR), and Department of Chemical Engineering and Materials, University of Calabria, Rende (CS), Italy

27.1 STATE-OF-THE-ART ON CATALYTIC MEMBRANE REACTORS

Catalytic reactions are intensively used in the chemical industry, energy conversion, wastewater treatment, and in many other processes; however, the necessity to realize a *sustainable growth* calls for additional and substantial developments in this field.

In this perspective, the application of membrane technology offers interesting opportunities, from both an environmental and an industrial point of view (Drioli and Fontananova, 2004). The combination of advanced molecular separation and chemical conversion realized in a catalytic membrane reactor (CMR) has many advantages in comparison to traditional systems. The catalyst can be easily recycled; moreover, the selective transport properties of the membranes can be used to shift the equilibrium conversion (e.g., esterification reaction), to remove selectively products and by-products from the reaction mixture, and to supply selectively the reagent (e.g., oxygen for partial oxidation reactions). The main limitation still existing for the large-scale applications of CMRs is related to the manufacturing cost of the membranes and modules and the limited membrane lifetime. Considerable technical complexity of the process that makes modeling and prediction more difficult is also to be considered.

A special category of membrane reactors exists when the membrane defines the reaction volume, for example, by providing a contacting zone for two immiscible phases (phase transfer catalysis), excluding solvents and thus making the process environmentally more attractive. The first application of CMRs has concerned high-temperature reactions. However, the use of inorganic membranes, characterized by higher chemical and thermal stability with respect to polymeric membranes, still today suffers from some important drawbacks: high cost, limited lifetime, difficulties in reactor manufacturing (e.g., delamination of the membrane top layer from the support due to the different thermal expansion coefficients).

On the other hand, the use of polymeric membranes in CMRs is of increasing interest (Vankelecom and Jacobs, 2000). The cost of polymeric membranes is generally low, and

Advanced Membrane Technology and Applications. Edited by Norman N. Li, Anthony G. Fane, W. S. Winston Ho, and T. Matsuura
Copyright © 2008 John Wiley & Sons, Inc.

the preparation protocols allow a better reproducibility; moreover, the relatively low operating temperatures are associated with a less stringent demand for materials used in the reactor construction. In general, polymeric membranes are less resistant to high temperatures and aggressive chemicals than inorganic or metallic membranes. However, polymeric materials resistant under rather harsh conditions—for example, polydimethylsiloxane, Nafion, Hyflon, polyvinylydene fluoride—are today available. Moreover, many reactions of relevant interest in fine chemical synthesis or in water treatment take place under mild conditions.

Biocatalytic membrane reactors can also be used in production, processing, and treatment operations. The trend toward environmentally friendly technologies makes these units particularly attractive because of their ability to operate at moderate temperature and pressure and to reduce the formation of by-products. Enzymes, compared to inorganic catalysts, generally permit greater stereospecificity and higher reaction rates under milder reaction conditions. Relevant applications of biocatalytic membrane reactors include: production of new or better foodstuffs, in which desired nutrients are not lost during thermal treatment; novel pharmaceutical products with well-defined enantiomeric compositions; and wastewater processes.

An interesting example of industrial application of CMRs is the consortium of BP/Amoco, Praxair, Statoil, SASOL, and Phillips Petroleum, called the Oxygen Transport Membrane Alliance (OTM) (http://www.eere.energy.gov/hydrogenandfuelcells/), working on solid oxide membrane reactor-based synthesis for syngas production. Another important case study in commercializing membrane reactors was the membrane reactor facility for the drug diltiazem or Cardizem production installed by Sepracor Inc. in Japan (www.sepracor.com).

In the following sections some cases study in wastewater treatments, oxidation reactions, H_2 production, and stereolective synthesys will be presented.

27.2 ADVANCED OXIDATION PROCESSES FOR WASTEWATER TREATMENTS

The development of new effective and environmental benign methods for wastewater treatments is an important research area. Most traditional processes simply transfer the pollutants present in the water from one medium to another [liquid–liquid extraction (Liu et al., 2001; Pandit and Basu, 2004), absorption onto activated carbon (Krishnan and Anirudhan, 2002; Lin and Wang, 2003), air stripping (Heggemann et al., 2001)] or generate waste that require further treatment and disposal [biological oxidation (Beltran et al., 2001) and classical chemical treatment such as addition of chlorine or potassium permanganate (Battistoni et al., 2001; Rajkumar and Palanivelu, 2003; Wang and Lemley, 2003)]. Advanced oxidation processes (AOP), are very promising methods for wastewater treatments (Gogate and Pandit, 2004a,b). These methods, mainly based on the photoinduced generation of hydroxyl radicals, can lead to complete mineralization of organic pollutants.

Early transition metals (V, Nb, Ta, Mo, W) in their highest oxidation states, can form metal–oxygen cluster anions, commonly named polyoxoanions or polyoxometalates (POMs). POMs have been largely used as efficient photocatalyst for the complete degradation of organic pollutants in water (Antonaraki et al., 2002; Kormali et al., 2004; Mylonas and Constantinou, 1996). These compounds have interesting redox properties for catalytic applications. In particular, POMs in the solid state are particularly suitable for catalyst design at the atomic and molecular level (Misono, 2001). There are several large-scale industrial applications of POMs as oxidation and acid catalysts, both in the solid state and in solution (Misono, 1987; Nojiri and Misono, 1993; Okuhara et al., 1996; Risono and Nojiri, 1990).

However, POMs are characterized by a low surface area, which in many cases limits their applications (Misono, 2001). In order to overcome this problem, numerous attempts have been made to disperse POMs on various supporting materials such as silica (Guo et al., 2002, 2003; Li et al., 2002; Maldotti et al., 2002) and organic polymers (Feng et al., 2002a,b; Moriguchi and Fendler, 1998). The heterogenization of catalysts has also interesting implications in both environmental and industrial sectors, allowing the reuse several times of the same catalyst, in many cases with higher catalytic activity.

Among the different heterogenization routes, the entrapping of catalysts in membranes offers new possibility in catalysis. In recent works (Bonchio et al., 2003; Fontananova et al., 2006b), we reported about the heterogenization of decatungstate ($W_{10}O_{32}^{4-}$), an interesting photocatalyst used for the partial oxidation of organic substrates and for the complete degradation of organic pollutants in polymeric membranes made of polyvinilidene fluoride (PVDF), a polymer characterized by a high chemical, thermal, and ultraviolet (UV) stability. Decatungstate (W_{10}) exhibits especially interesting properties for the photocatalytic detoxification of water since its absorption spectrum overlaps the UV solar emission spectrum, opening the potential route for environmentally benign solar-photoassisted applications (Texier et al., 1997). However, W_{10} is characterized by low quantum yields, poor selectivity, and limited stability at pH higher than 2.5 (Mylonas and Constantinou, 1996). Membrane technology could offer interesting possibilities to overcome these limitations.

The heterogenization of a lipophilic salt of decatungstate, $(^{n}Bu_4N)_4W_{10}O_{32}$ (not soluble in water), in PVDF membranes (PVDF-W_{10}), prepared by phase inversion technique induced by a nonsolvent (Kimmerle and Strathmann, 1990), at different catalyst loading (14.3, 25.0, and 33.3 wt% with respect to polymer mass) was investigated (Bonchio et al., 2003; Fontananova et al., 2006b).

Solid-state characterization techniques have been used to verify that the heterogenization procedure did not damage the catalyst. Fourier transform–infrared (FT-IR) spectra give useful indication about the structure of the catalyst heterogenized in the membrane. FT-IR spectra confirmed that the catalyst structure is preserved within the polymeric membrane (Fig. 27.1). The infrared spectrum of the catalytic membranes show the characteristic bands of W_{10} units (595, 803, 895, 958 cm^{-1}), as well as those typical of the employed alkylammonium cation (2870 cm^{-1}).

Also spectroscopic properties of the catalyst heterogenized in PVDF membranes are preserved as confirmed by the presence of the CT band at 324 nm in the DR-UV spectra of the catalytic membranes (Fig. 27.2). Moreover, it is important to note that the PVDF membrane prepared without the catalyst (PVDF reference) does not show any absorption in the region of interest for the catalyst.

The catalytic membranes were successfully applied in the aerobic photooxidation of phenol, one of the main organic pollutants in wastewater, providing stable and recyclable photocatalytic systems. The catalytic tests were carried out in a photocatalytic membrane reactor operating with flow-through at different transmembrane pressures (TMP).

From the experimental tests a dependence of the phenol degradation rate from the catalyst loading in the membrane was observed (Fig. 27.3). The better results were obtained at 25.0 wt% catalyst loading; a further increase of catalyst loading (33.3 wt%) reduces the reaction rate probably because of catalyst aggregation phenomena. Also the transmembrane

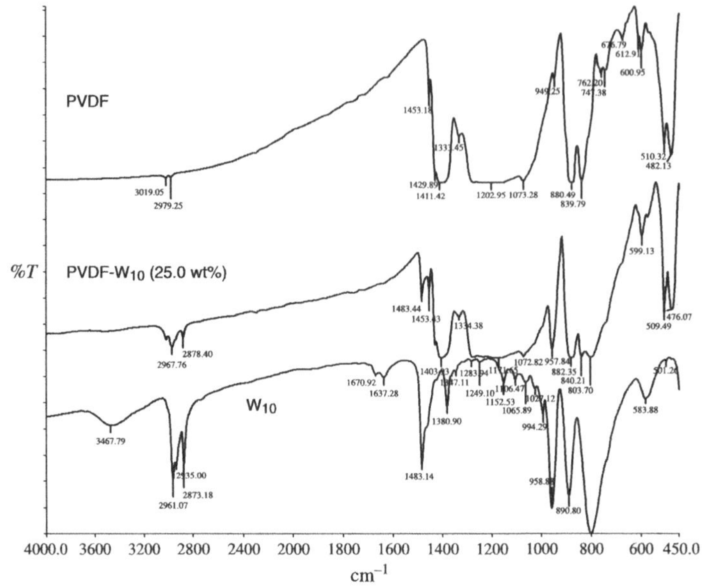

Figure 27.1 FT-IR spectra of the native PVDF membrane, the catalytic PVDF-W_{10} (25.0 wt%) membrane and decatungstate (W_{10} in KBr). (Adapted from Fontananova et al., 2006b.)

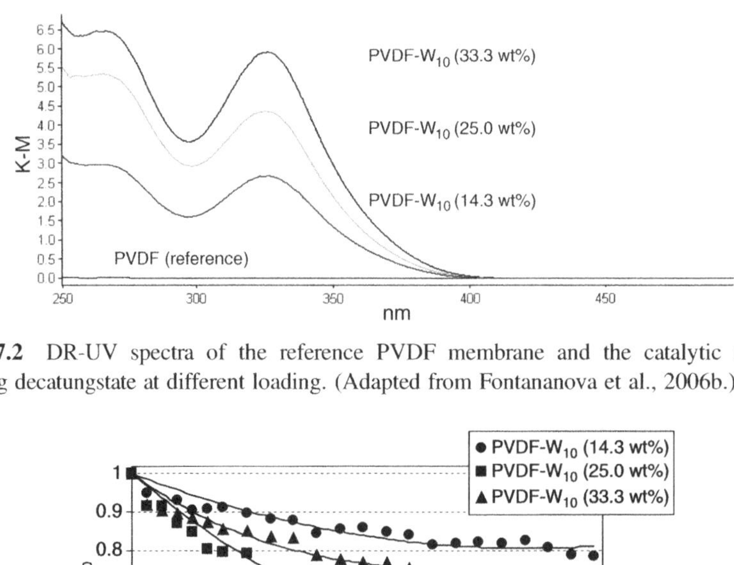

Figure 27.2 DR-UV spectra of the reference PVDF membrane and the catalytic membranes containing decatungstate at different loading. (Adapted from Fontananova et al., 2006b.)

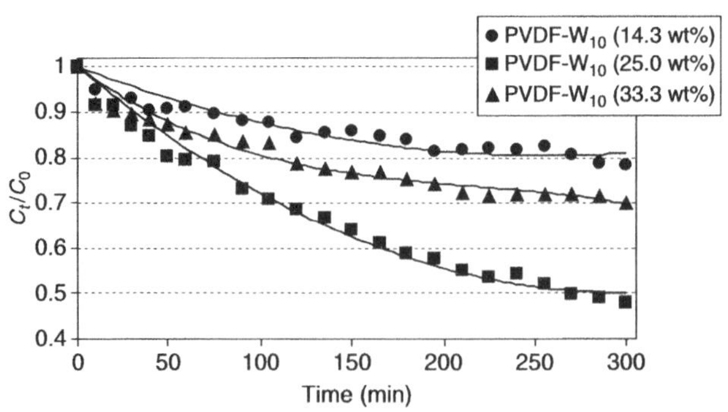

Figure 27.3 Phenol degradation by PVDF-W_{10} membranes at different catalyst loading (C_t is the phenol concentration at the time t and C_0 the initial concentration; TMP = 1 bar). (Adapted from Fontananova et al., 2006b.)

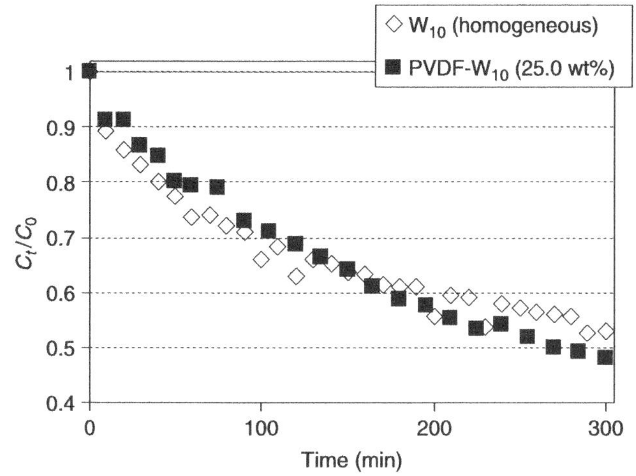

Figure 27.4 Comparison between homogenous ($Na_4W_{10}O_{32}$; 4×10^{-4} M) and heterogeneous (PVDF-W_{10}; 25.0 wt%) phenol degradation reaction. (Adapted from Fontananova et al., 2006b.)

pressure influences the catalytic activity of the membranes and the faster phenol disappearance, due to an optimal catalyst–substrate contact time, was observed operating at TMP 1 bar (Fontananova et al., 2006b).

The rates of phenol degradation with homogenous $Na_4W_{10}O_{32}$ (4×10^{-4} M) and heterogeneous PVDF-W_{10} (25.0 wt%) were similar (Fig. 27.4). However, in the case of the homogenous reaction, several persistent intermediates were individuated (benzoquinone, hydroquinone, and catechol). On the contrary, photodegradation performed by PVDF-W_{10} leads to nearly complete decomposition of phenol to CO_2 and H_2O as confirmed by total organic carbon (TOC) analyses (Fontananova et al., 2006b).

The interesting catalytic activity of the PVDF-W_{10} membranes can be ascribed to the selective absorption of the organic substrate from water on the PVDF polymer membrane that increases the effective phenol concentration and retains also intermediate products around the catalytic sites until their complete mineralization. Moreover, the polymeric hydrophobic environment protects the decatungstate from the conversion to an isomer that absorbs only light under 280 nm (Texier et al., 1999). This phenomenon, known to occur in homogeneous solution at pH >2.5 (Mylonas and Constantinou, 1996), was monitored also in our homogeneous catalytic test (pH $= 6$) by the reduction of intensity of the CT band (324 nm) in the UV–visible (UV–Vis) spectra of the homogenous $Na_4W_{10}O_{32}$. As a consequence the reduction of the catalyst efficiency, compared to the heterogeneous system, over a long time (>150 min) was observed (Fig. 27.4). The photocatalytic membranes prepared were recycled without loss of activity in successive catalytic runs.

The use of plasma techniques for the heterogenization of POMs on the membrane surface was also investigated (Fontananova et al., 2006a; Lopez et al., 2006). Novel catalytic membranes have been prepared by linking phosphotungstic acid $H_3PW_{12}O_{40}$(W_{12}), a polyoxometalate having interesting properties as photocatalyst, on the surface of plasma-modified PVDF membranes (PVDF-NH_2-W_{12}).

The application of plasma technique for the heterogenizazion of a catalyst on the surface of membranes is very interesting because, generally, when a catalyst is heterogenized on solid support, it results only partially active due to diffusion limitations and aggregation phenomena.

Plasma treatments represent a versatile tool to modify only the few topmost layers of solid surfaces, thus leaving the bulk properties totally unaltered (Ratner et al., 1990;

Wilson et al., 2001). By plasma techniques it is possible to graft, in a controlled way, different functional groups on solid surfaces (e.g., —COOH, —OH, —NH_2, etc.) (Favia et al., 1997). These functionalities can be conveniently used as anchor groups for catalyst binding, which can be well dispersed on the membrane surface, overcoming diffusion and aggregation limits. In particular, in our experiments we have used ammonia-fed discharges (Arefi-Khnosari et al., 1997; Creatore et al., 2000; Favia et al., 1996; Tatoulian et al., 1995), because the grafting of basic groups on membrane surfaces can allow the formation of ionic interactions with the acid W_{12} catalyst; moreover, the formation of hydrogen bonds with other polar groups grafted on the surface is also possible.

Polar chemical groups (we hypothesize principally NH_2, together with OH, CN, NH, and CO) have been grafted by an NH_3 plasma discharge on the upper surface of a PVDF membrane and pretreated with Ar in order to control hydrophobic recovery (Favia et al., 2005), The groups grafted on the surface can act as anchor sites for immobilizing the acid catalyst W_{12}.

The plasma-activated membrane has been immersed in an aqueous solution of W_{12} for 24 h; after it has been washed three times with water and finally dried at room temperature. Principally, ionic interaction between W_{12} and the NH_2 groups, together with hydrogen bonding and different interactions with the other polar groups also grafted by plasma treatments, link the catalyst on the membrane surface.

Surface diagnostics techniques such as X-ray photoelectron spectroscopy (XPS), contact angle (CA) measurements, and RX maps, have been used to attest the surface modification by plasma treatment.

Tungsten-based catalyst binding can be monitored by XPS atomic-composition results. The W_4f peak shows two major components, and this behavior is typical of tungsten oxides (Khyzhun, 2000; Martin-Litas et al., 2002) (Fig. 27.5).

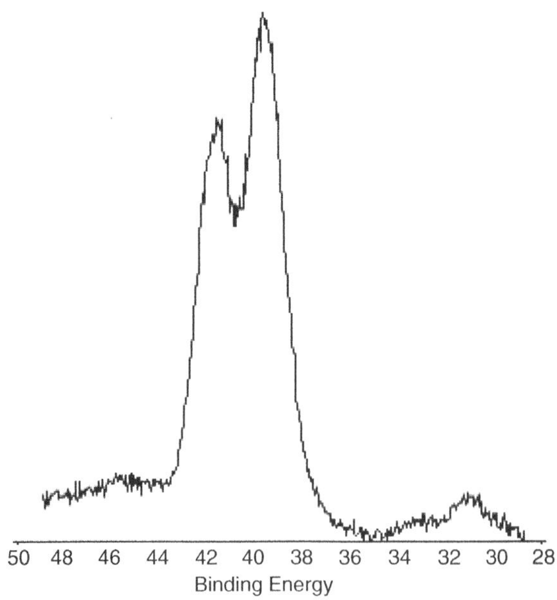

Figure 27.5 W4f component in the XPS spectra of the PVDF-NH_2-W_{12} catalytic membrane (From Fontananova et al., 2006a.)

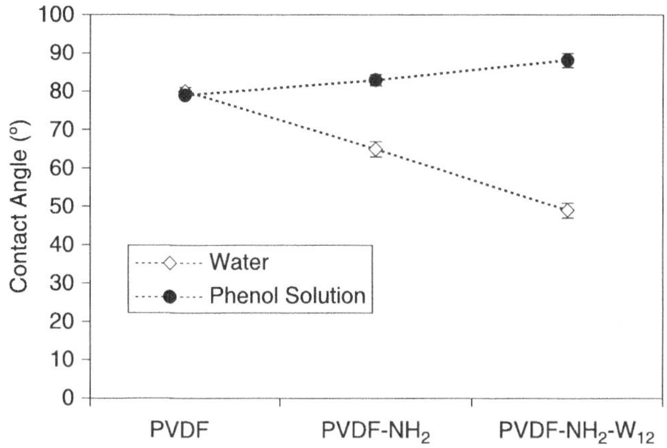

Figure 27.6 Contact angles (± standard deviation) with water and a phenol solution (0.002 M) on the upper surface of the PVDF-based membranes. (Adapted from Fontananova et al., 2006a.)

RX maps on the upper surface of the PVDF-NH_2-W_{12} membrane underline a good dispersion of the catalyst with a homogeneous W and F surface distribution. CA measurements on the upper surface of the membranes confirm the presence of polar groups by the decreasing of the water contact angle (WCA) with respect to the native membrane (Fig. 27.6). The WCA value is further decreased after the immobilization of W_{12} occurred, probably due to the presence of many oxygen moieties.

Interesting results were obtained considering the contact angle with a 0.002 M phenol solution (Fig. 27.6). As expected, the surface tension of the phenol solution was lower than the value of the pure water (respectively, 73.47 and 75.24 mN/m); consequently, the plasma-modified membranes have a higher contact angle with phenol solution, compared to the native PVDF membrane, but the values are still lower than 90°.

Catalytic activity of the PVDF-NH_2-W_{12} membranes was also evaluated in the aerobic degradation reaction of phenol in water. By comparison with homogeneous reactions, the catalyst heterogenized on the membrane surface was more efficient concerning the rate of phenol photodegradation (Table 27.1), also if the amount of catalyst present on the membrane surface is several orders of magnitude lower than those used in homogeneous phase (7.0×10^{-4} mol/L). First-order rate constant for both homogeneous and heterogeneous systems were observed (Table 27.1).

A better dispersion of the catalyst on the membrane surface and an increase of the catalyst stability, with respect to the homogeneous solution, probably contribute to increase the catalyst efficiency both in term of percent active catalyst and superficial area of the catalyst.

TABLE 27.1 First-Order Rate Constant (k) for Phenol Photodegradation and Correlation Coefficient (R^2) for the First-Order Plot[a]

Process	k (min^{-1})	R^2
W_{12} (homogeneous)	0.00110	0.99
PVDF-NH_2-W_{12}	0.00313	0.98

[a]For the heterogeneous reaction TMP = 1 bar.

Further studies are in progress with other catalysts belonging to the polyoxometalates group, for their heterogenization in fluorinated membranes (not only PVDF but also Hyflon) by the two techniques described: immobilization in the membrane starting from homogeneous solution of the catalyst and the polymer and linking of the catalyst on the surface of plasma-modified membranes.

27.3 SELECTIVE OXIDATIONS

One of the most interesting aspects of catalyst heterogenization in membranes is the effect of the membrane environment on the catalyst activity. Membrane composition (hydrophobic or hydrophilic characteristics of the membrane material, presence of chemical groups with acid or basic properties, etc.) and membrane structure (dense or porous, symmetric or asymmetric) can positively influence the catalyst performance not only by the selective sorption and diffusion of reagents and/or products but also influencing the catalyst activity by electronic (electron-donating and electron-withdrawing groups) and conformational effects (stabilization of the transition states). These effects are similar to those occurring in biological membranes.

A membrane-induced structure–reactivity trend, which may be exploited to achieve selective processes, was observed in polymeric catalytic membranes prepared by embedding decatungstate within porous membranes made of PVDF or dense polydimethylsiloxane (PDMS) membranes. These photocatalytic systems are characterized by different and tunable properties depending on the nature of the polymeric microenvironment (Bonchio et al., 2003). The polymeric catalytic membranes prepared were used for the batch-selective photooxidation of water-soluble alcohols. Membrane-induced discrimination of the substrate results from the oxidations of a series of alcohols with different polarity, through comparison with the homogeneous reactions (Fig. 27.7).

The catalyst activity is influenced by the selective sorption and diffusion of the substrate in the membrane. Generally, substrates that are preferentially adsorbed, and therefore are more concentrated around the catalytic sites, increase the reaction rate; on the other hand, diffusion is also an important factor for the catalytic reactions, particularly in dense membranes. Cycloheptanol is the more hydrophobic alcohol in this series and, consequently, can better interact with the PDMS by Van der Waals interactions; however, it is also the more sterically constrainted, and the reaction rate is lower than for *n*-cyclopentanol (Fig. 27.8).

In all cases, the alcohols oxidation occurs following a degradation pathway in which the aldehyde is formed as an intermediate. This reactivity behavior is known for photooxidation promoted by POMs in water, where an efficient flow of hydroxyl radicals is generated from the solvent. Control experiments on the reaction mixture and on the recovered membrane (UV–Vis, FT-IR) allow to exclude the occurrence of the catalyst leaching and, as a consequence, of a competing homogeneous pathway.

The incorporation of homogenous Ti(IV)/trialkanolamine catalyst in polymeric membranes has been investigated as heterogeneous catalysts for the selective oxidation of secondary amines to nitrones by alkyl hydroperoxides (Buonomenna et al., 2004, 2006). Three polymers, PVDF, a modified polyetherketone (PEEK-WC), and polyacrilonitrile (PAN) with different functional groups and chemical-physical properties were used to tune the reactivity of the catalytic polymeric membranes.

The catalytic activity of these systems was investigated in the oxidation to nitrone of dybenzylamine and in the stereoselective oxidation to sulfoxide by alkyl hidroperoxides,

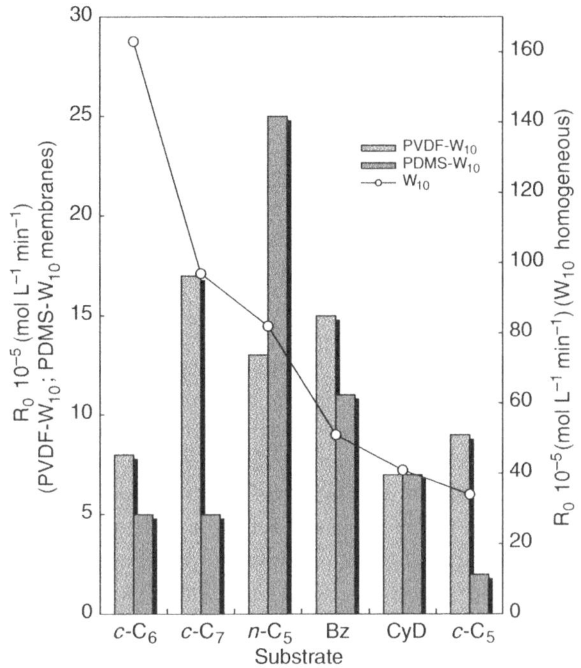

Figure 27.7 Structure–reactivity trend observed in the photoxidation of alcohols in water by heterogeneous PDMS-W_{10}, PVDF-W_{10}, and homogeneous $Na_4W_{10}O_{32}$ (W_{10}). In all reactions: substrate (0.04 mmol), H_2O (2 mL, pH = 7), $pO_2 = 1$ atm, $\lambda > 345$ nm, $T = 25°C$. W_{10} (0.6 μmol), PDMS-W_{10} (6.3% loading, 0.6 μmol); PVDF-W_{10} (25% loading, 0.6 μmol). Substrate: cis-1,2-cyclohexandiol (CyD); cycloheptanol (*c*-C_7); cyclohexanol (*c*-C_6); cyclopentanol (*c*-C_5); *n*-pentanol (*n*-C_5); benzyl alcohol (Bz). (Adapted from Bonchio et al., 2003.)

under batch conditions (membranes cut in small pieces and introduced in the reaction mixture) using cumyl hydroperoxide (CHP) as oxidant. PVDF-based catalytic membranes gave the best results affording products in short reaction times, high yields, and selectivity comparable with the performances of the corresponding homogeneous system (Buonomenna et al., 2004, 2006).

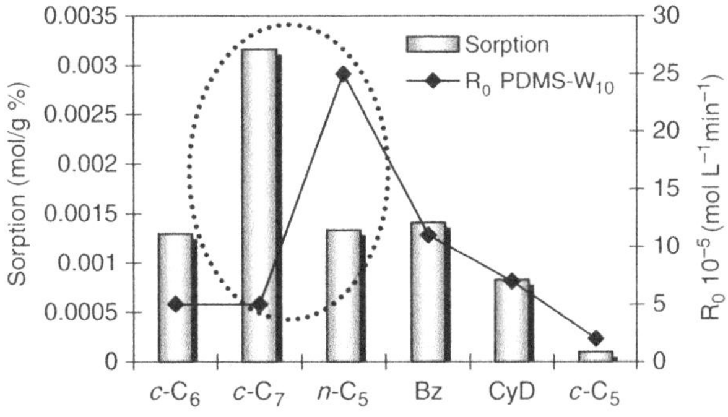

Figure 27.8 Sorption of alcohols (0.02 M; 25°C) in PDMS membranes and photooxidation rates (R_0) of the same alcohols in water by PDMS-W_{10} (6.3% catalyst loading). (Adapted from Bonchio et al., 2006.)

27.4 BIOCATALYTIC MEMBRANE REACTORS

Membrane reactors using biological catalysts have been largely used in enantioselective processes. Methodologies for the preparation of emulsions (submicron) of oil in water have been developed, and such emulsions have been used for kinetic resolutions in phase transfer biocatalysis (Giorno and Drioli, 2000; Li et al., 2003; Sakaki et al., 2001). The potentiality of these systems has been already confirmed in the production of significantly high enantiomeric excess of the (S)-naproxen acid (antiinflammatory drug) from racemic mixtures of (R, S)-naproxen methyl ester in emulsion enzyme membrane reactors in which lipase is entrapped by physical methods in polymeric membranes (Giorno and Drioli, 2000; Li et al., 2003; Sakaki et al., 2001). In this reactor the substrate has been fed as an emulsion. The distribution of the water–organic interface at the level of the immobilized enzyme has remarkably improved the property of transport, kinetic, and selectivity of the immobilized biocatalyst. These results are due to the improvements of the mass transfer efficiency in the emulsion enzyme membrane reactor, where monodispersed emulsion well reproduced the organic–water interface within the membrane pores. In these conditions, each pore worked as a continuous microtank reactor where the immobilized lipase could be activated by the organic–water interface. As a consequence the enzyme stability, as well as activity and enantioselectivity, were preserved. The performance of the biocatalytic system for the hydrolysis of triglycerides to obtain fatty acids and glycerol was also studied (Giorno et al., 2007).

27.5 CATALYTIC CRYSTALS

An innovative potential application of membrane technology in catalysis and in catalytic membrane reactors is the possibility to produce catalytic crystals with a well-defined size, size distribution, and shape by membrane crystallization (Fig. 27.9) (Di Profio et al., 2003, 2005). This innovative technology makes use of the evaporative mass transfer of volatile solvents through microporous hydrophobic membranes in order to concentrate feed solutions above their saturation limit, thus attaining a supersaturated environment where crystals may nucleate and grow. In addition, the presence of a polymeric membrane increases the probability of nucleation with respect to other locations in the system (heterogeneous nucleation).

Two different methodologies are employed, depending on the strategy used to activate the driving force (a partial pressure gradient), for the selective extraction of the solvent. In osmotic membrane crystallization, the difference in solute concentration, and, consequently, in water activity of the two solutions, induces at the vapor–liquid interfaces, a vapor pressure difference that generates the mass transfer from the dilute solution (containing the organic or inorganic molecule of interest) toward the stripping solution.

In thermal membrane crystallization, the membrane is in contact with the warm crystallizing solution at the retentate side and with a cold condensing solution at the distillate side, so that the driving force is generated by a temperature difference. Membrane crystallization is particularly attractive for the preparation of heat-sensitive catalysts such as enzymes.

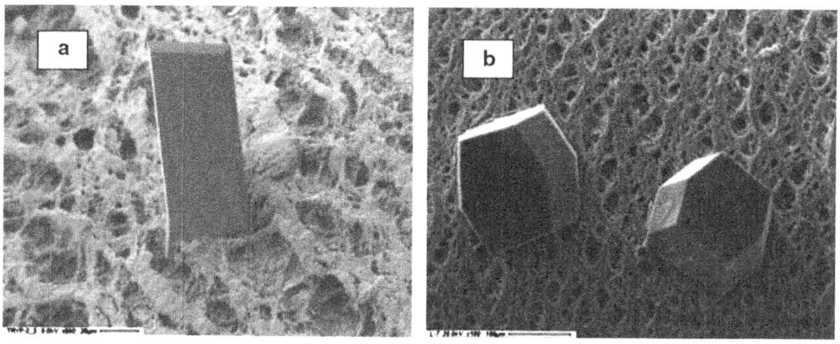

Figure 27.9 Trypsin (A, from Drioli et al., 2005) and lysozyme (B, from Di Profio et al., 2003) crystals grown on hydrophobic polypropylene (PP) membranes.

27.6 INORGANIC MEMBRANE REACTORS

Zeolite membranes are able to separate continuously mixtures of substances on the basis of differences in molecular size and shape and also on the basis of different adsorption properties (Coronas and Santamaria, 2004a). Due to their good selectivity and high thermal and chemical resistance, zeolite membranes are also interesting for their use in membrane reactors. Moreover, the possibility to host ions, atoms, or complexes having specific catalytic activity opened new opportunities as advanced materials for catalytic applications.

Recently, thin FAU (NaY) zeolite membranes were synthesized on porous α-Al_2O_3 tubular supports by a new secondary growth method (Algieri et al., 2003). FAU zeolite membranes were loaded with Pt by ionic exchange (Hasegawa et al., 2001). These catalytic (PtY) membranes were used in membrane reactors for CO-selective oxidation in H_2-rich mixtures (Bernardo et al., 2006). Zeolite membrane reactors allow to combine the selective permeation with the Pt catalytic activity. H_2 is more permeable than CO through the zeolite membrane; therefore, its contact time with the catalytic phase is minor than that of the CO. A flow-through configuration was used in the membrane reactor, feeding a mixture composed of CO (10.5%), O_2 (8.7%), and H_2 (80.8%) at 200°C; a CO conversion of 98% and a CO selectivity of 63% were observed. These PtY catalytic membranes are very promising for the possible application in the petrochemical industry. For example, the large H_2 amount produced in ethylene plants can be upgraded, giving to it an added value and opening its use to other applications such as fuel cells and hydrogenation reactions.

27.7 MICROREACTORS

Microreactors have demonstrated their clear advantages in catalysis, and microreaction engineering is rapidly developing. Zeolites are also excellent candidates for microreactor technology; an interesting overview in this field was reported by Coronas and Santamaria (2004a,b). The main advantages in microreactors compared to conventional reactors are the reduction of length scale, the decreasing of heat and mass transfer distances, the realization of specific interfaces for multiphase flow and the smaller reagent volumes, and correspondingly lower costs (Table 27.2).

TABLE 27.2 Comparison between Conventional and Microreactors

Properties	Conventional Reactor	Microreactor
Internal volume	$\gg$1 mL	1–10 μL
Specific surface	100 m^2/m^3	10000 m^2/m^3
Phase boundary surfaces for liquid/liquid mixtures	100 m^2/m^3	5000–50,000 m^2/m^3
Flows	Turbulent	Laminar character

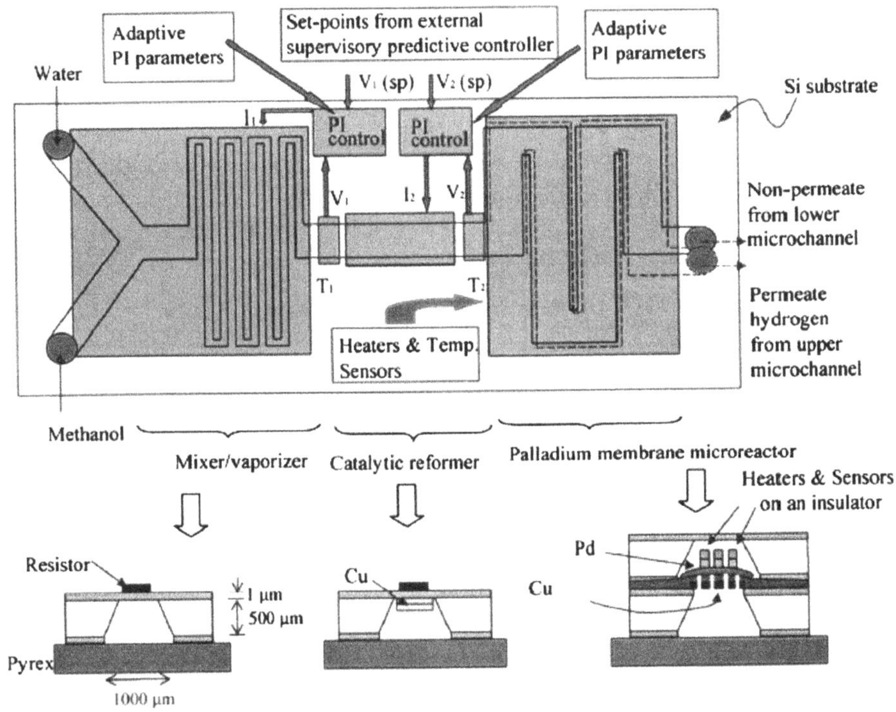

Figure 27.10 Scheme of the membrane microreactor. (From Karnik et al., © 2003 IEEE.)

Micromembranes are today fabricated using the standard microelectronic fabrication processes (Hessel and Lowe, 2003). A Pd-based micromembrane was successfully used for the water gas shift reaction (WGSR) in a miniature fuel cell. The micromembrane structure was built in a silicon substrate, using the MEMS microfabrication processes (Karnick et al., 2003). The composite membrane was composed by four layers: copper, aluminum, spin-on-glass (SOG), and Pd (Fig. 27.10). The first three layers provide a structutral support for the Pd film; moreover, the copper acts as a catalyst in the WGSR, and Pd is used to separate H_2 from the other gases present with high selectivity.

27.8 CONCLUSIONS

Catalytic membrane reactors are an interesting example of integrated systems in which molecular separation and chemical conversions are combined in one unit. The heterogenization of catalysts in membrane is particularly suitable for catalyst design at the atomic and molecular level.

Catalytic membrane reactors have been successfully applied in small-scale operations and can really be considered as an emerging technology. For further progress in the catalytic membrane reactors the understanding of the influence of the membrane in the mechanism of the chemical reaction and on its kinetic aspects is fundamental. Moreover, the understanding of properties, behavior, and synthesis of innovative materials is important in order to exploit the potential use of novel highly complex composite systems, molecules, and new multifunctional materials derived from them. Computational strategies, theoretical simulations, and modeling are key tools to be considered in catalysis.

ACKNOWLEDGMENTS

Financial support from the Ministero dell'Istruzione dell'Università e della Ricerca (MIUR) (CEMIF.CAL -CLAB01TYEF and CAMERE-RBNE03JCR5) is gratefully acknowledged.

REFERENCES

Algieri, C., Bernardo, P., Golemme, G., Barbieri, G., and Drioli, E. (2003). Permeation properties of a thin silicalite-1 (MFI) membrane. *J. Membr. Sci.* 222, 181–190.

Antonaraki, S., Androulaki, E., Dimotikali, D., Hiskia, A., and Papaconstantinou, E. (2002). *J. Photochem. Photobios. A.-C.* 148, 191.

Arefi-Khnosari, F., Tatoulian, M., Placenta, G. J., Kurdi, J., and Amoroux. (1997). In. R. d'Agostino and P. Favia (Eds.), *Plasma Processing of Polymers.* Kluwer Academic, Dordrecht, p. 165.

Battistoni, P., Boccadoro, R., Bolzonella, D., and Pezzoli, S. (2001). *Ind. Eng. Chem. Res.* 40(21), 4506.

Beltran, F. J., Alvarez, P. M., Rodriguez, E. M., Garcia-Araya, J. F., and Rivas, J. (2001). *Biotechnol. Prog.* 17(3), 462.

Bernardo, P., Algieri, C., Barbieri, G., and Drioli, E. (2006). Cataltic (PtY) membranes for the purification of H_2-rich streams. *Catal. Today* 118, 90–97.

Bonchio, M., Carraro, M., Scorrano, G., Fontananova, E., and Drioli, E. (2003). Heterogeneous photooxidation of alchols in water by photocatalytic membranes incorporating decatungstate. *Adv. Synthesis Catal.* 345(2003), 1119–1126.

Bonchio, M., Carraro, M., Gardan, M., Scorrano, G., Drioli, E., Fontananova, E. (2006). Hybrid photocatalytic membranes embedding decatungstate for heterogeneous photooxygenation. *Top. Catal.* 40, 133–140.

Buonomenna, M. G., Drioli, E., Bertoncello, R., Milanese, L., Prins, L. J., Scrimin, S., and Licini, G. (2006). Ti(IV)/trialkanolamine catalytic polymeric membranes: Preparation, characterization, and use in oxygen transfer reactions. *J. Catal.* 238, 221–231.

Buonomenna, M. G., Drioli, E., Nugent, W. A., Prins, L. J., Scrimin, P., and Licini, G. (2004). Ti(IV)-based catalytic membranes for efficient and selective oxidation of secondary amines. *Tetrahedr. Lett.* 45, 7515–7518.

Coronas, J., and Santamaria, J. (2004a). State-of-the-art in zeolite membrane reactors. *Top. Catal.* 29, 29–44.

Coronas, J., and Santamaria, J. (2004b). The use of zeolite films in small-scale and micro-scale applications. *Chem. Eng. Sci.* 59, 4879–4885.

Creatore, M., Favia, P., Tenuto, G., Valentini, A., and d'Agostino, R. (2000). *Plasmas and Polymers* 5(3/4), 201.

Di Profio, G., Curcio, E., and Drioli, E. (2003). Membrane crystallization of lysozyme: Kinetic aspects. *J. Crystal Growth* 257, 359–369.

Di Profio, G., Curcio, E., and Drioli, E. (2005). Trypsin crystallization by membrane-based techniques. *J. Struct. Biol.* 150, 41–49.

Drioli, E., and Fontananova, E. (2004). Membrane technology and sustainable growth. *Chem. Eng. Res. Des.* 82(A12), 1557–1562.

Drioli, E., Curcio, E., and Di Profio, G. (2005). State of the art and recent progresses in membrane contactors. *Trans IChemE, Part A, Chem. Eng. Res. Des.* 83(A3), 223–233.

Favia, P., d'Agostino, R., and Palumbo, F. (1997). *J. Phys. IV* 7, 199.

Favia, P., Milella, A., Iacobelli, L., and d'Agostino, R. (2005). In d'Agostino, R., Favia, P., Oehr, C., and Wertheimer, M. R., (Eds.), *Plasma Processes and Polymers*. Wiley, VCH, Berlin, p. 271.

Favia, P., Stendardo, M., and d'Agostino, R. (1996). *Plasmas and Polymers* 1(2), 91.

Feng, W., Zhang, T. R., Liu, Y., Lu, R., Zhao, Y. Y., Li, T. J., and Yao, J. N. (2002a). *J. Solid State Chem.* 169, 1.

Feng, W., Zhang, T. R., Wei, L., Lu, R., Bai, Y. B., Li, T. J., Zhao, Y. Y., and Yao, J. N. (2002b). *Mater. Lett.* 54, 309.

Fontananova, E., Donato, L., Drioli, E., Lopez, L. C., Favia, P., and d'Agostino, R. (2006a). Heterogenization of polyoxometalates on the surface of plasma modified polymeric membranes. *Chem. Mater.* 18, 1561–1568.

Fontananova, E., Drioli, E., Donato, L., Bonchio, M., Carraro, M., and Scorrano, G. (2006b). Heterogeneous photooxidation of phenol by catalytic membranes. *Chin. J. Process Eng.* 6, 54–59.

Giorno, L., D'Amore, E., Mazzei, R., Piacentini, E., Zhang, J., Drioli, E., Cassano, R., and Picci, N. (2007). An innovative approach to improve the performance of a two separate phase enzyme membrane reactor by immobilizing lipase in presence of emulsion. *J. Membr. Sci.* 295, 95–101.

Giorno, L., and Drioli, E. (2000). Biocatalytic membrane reactors: Applications and perspectives. *Trends Biotechnol.* 18, 339–348.

Gogate, P. G., and Pandit, A. B. (2004a). *Adv. Environ. Res.* 8, 501.

Gogate, P. R., and Pandit, A. B. (2004b). *Adv. Environ. Res.* 8, 553.

Guo, Y., Hu, C., Jiang, S., Guo, C., Yang, Y., and Wang, E. (2002). *Appl. Catal. B Environ.* 36, 9.

Guo, Y.-H., and Hu, C.-W. (2003). *J. Clust. Sci.* 14, 505–527.

Hasegawa, Y., Kusakabe, K., and Morooka, S. (2001). Selective oxidation of carbon monoxide in hydrogen rich mixture by permeation through a platinum-loaded Y-type zeolite membrane. *J. Membr. Sci.* 190, 1.

Heggemann, M. H., Warnecke, H.-J., and Viljoen, H. J. (2001). *Ind. Eng. Chem. Res.* 40(15), 3361.

Hessel, V., and Lowe, H. (2003). Microchemical engineering: Components, plant concept user acceptance. *Chem. Eng. Technol.* 26, 1–24.

Karnik, S. V., Hatalis, M. K., and Kothare, M. V. (2003). Towards a palladium micro-membrane for the water gas shift reaction: Microfabrication approach and hydrogen purification results. *J. Microelectromechanical Sys.* 12, 93–100.

Khyzhun, O. Y. (2000). *J. Alloys Compounds* 305, 1.

Kimmerle, K., and Strathmann, H. (1990). *Desalination* 79, 283.

Kormali, P., Dimoticali, D., Tsipi, D., Hiskia, A., and Papaconstantinou, E. (2004). *Appl. Catal. B Environ.* 48, 175.

Krishnan, K. A., and Anirudhan, T. S. (2002). *J. Hazard. Mater.* B92, 161.

Li, D., Guo, Y., Hu, C., Mao, L., and Wang, E. (2002). *Appl. Catal. A Gen.* 235, 11.

Li, N., Giorno, L., and Drioli, E. (2003). Effect of immobilization site and membrane materials on multiphasic enantiocatalytic enzyme membrane reactors. *Ann. N. Y. Acad. Sci.* 984, 436–452.

Lin, S. H., and Wang, C. H. (2003). *Ind. Eng. Chem. Res.* 42(8), 1648.

Liu, W., Howell, J. A., Arnot, T. C., and Scott, J. A. (2001). *J. Membr. Sci.* 181, 127.

Lopez, L. C., Buonomenna, M. G., Fontananova, E., Iacoviello, G., Drioli, E., d'Agostino, R., and Favia, P. (2006). New generation of catalytic PVDF membranes: Coupling plasma treatments with chemical immobilization of W-based catalysts. *Adv. Funct. Mater.* 16, 1417–1424.

Maldotti, A., Molinari, A., Varani, G., Lenarda, M., Storaro, L., Bigi, F., Maggi, R., Mazzacani, A., and Sartori, G. (2002). *J. Catal.* 209, 210.

Martin-Litas, I., Vinatier, P., Levasseur, A., Dupin, J. C., Gonbeau, D., and Weill, F. (2002). *Thin Solid Films* 416, 1–9.

Misono, M. (1987). *Catal. Rev. Sci. Eng.* 29, 269.

Misono, M. (2001). *Chem. Commun.* 13, 1141.

Moriguchi, I., and Fendler, J. H. (1998). *Chem. Mater.* 10, 2205.

Mylonas, A., and Constantinou, E. (1996). Photodegradation of phenol and *p*-cresol by polyoxotungstates. Mechanistic implications. *Polyhedron* 15, 3211–3217.

Nojiri, N., and Misono, M. (1993). *Appl. Catal.* 93, 103.

Okuhara, T., Mizuno, N., and Misono, M. (1996). *Adv. Catal.* 41, 113.

Pandit, P., and Basu, S. (2004). *Ind. Eng. Chem. Res.* 43(24), 7861.

Rajkumar, D., and Palanivelu, K. (2003). *Ind. Eng. Chem. Res.* 42(9), 1833.

Ratner, B. D., Chilkoti, A., and Lopez, G. P. (1990). In R. d'Agostino (Ed.), *Plasma Deposition, Treatment and Etching of Polymers*, Academic, New York.

Risono, M., and Nojiri, N. (1990). *Appl. Catal.* 64, 1.

Sakaki, K., Giorno, L., and Drioli, E. (2001). Lipase-catalyzed optical resolution of racemic naproxen in biphasic enzyme membrane reactors. *J. Membr. Sci.* 184, 27–38.

Tatoulian, M., Arefi-Khnosari, F., Mabille-Rouger, I., Gheorgiu, M., Amoroux, J., and Bouchier, D. (1995). *J. Adhes. Sci. Technol.* 9, 923.

Texier, I., Giannotti, C., Malato, S., Richter, C., and Delaire, J. (1999). Solar photodegradation of pesticides in water by sodium decatungstate. *Catal. Today* 54, 297–307.

Vankelecom, I. F. J., and Jacobs, P. A. (2000). Polymeric membranes in catalytic reactors. *Catal. Today* 56, 147–157.

Wang, Q., and Lemley, A. T. (2003). *J. Agric. Food Chem.* 51(18), 5382.

Wilson, D. J., Williams, R. L., and Pond, R. C. (2001). *Surf. Interface Anal.* 31, 385.

PART V

ENVIRONMENTAL AND ENERGY APPLICATIONS

CHAPTER 28

Facilitated Transport Membranes for Environmental, Energy, and Biochemical Applications

JIAN ZOU, JIN HUANG, and W.S. WINSTON HO

Department of Chemical and Biomolecular Engineering, Department of Materials Science and Engineering, The Ohio State University, Columbus, Ohio 43210-1180

28.1 INTRODUCTION

Separations using synthetic membranes have been widely adopted for environmental, energy, and biochemical applications since the 1960s (Ho and Sirkar, 1992). In most current membrane processes, the separation mechanisms are based on solution, diffusion, and/or sieving. With these types of membranes, an increase in selectivity is often accompanied by a decrease in flux and vice versa (Gottschlich et al., 1988; Ho and Sirkar, 1992). Facilitated transport membranes offer an attractive method of achieving high selectivity while also maintaining high flux. This type of membrane is based on facilitated transport, which combines diffusion with the reversible reaction of a targeted component with reactive carriers inside the membrane.

There are two main types of reactive carriers for facilitated transport: (1) the mobile carrier, which can move freely across the membrane, and (2) the fixed carrier, which only has limited mobility around its equilibrium position. Figure 28.1 is a schematic diagram of the membranes with these two types of carriers. In a mobile-carrier membrane, the mobile carriers react with the targeted component on the feed side of the membrane, move across the membrane, and release this component on the permeate side. The carrier–component complex diffuses in parallel with the molecular diffusion of the component. As a result, the transport of this component is augmented or facilitated. The other components, which do not react with the reactive carriers, diffuse across the membrane down their concentration gradients via the solution–diffusion mechanism. Their transport is not affected by the facilitated transport. In a fixed-carrier membrane, the targeted component reacts at one carrier site and then hops to the next unreacted carrier site along the direction of the concentration gradient via the "hopping" mechanism (Cussler et al., 1989).

Advanced Membrane Technology and Applications. Edited by Norman N. Li, Anthony G. Fane, W. S. Winston Ho, and T. Matsuura
Copyright © 2008 John Wiley & Sons, Inc.

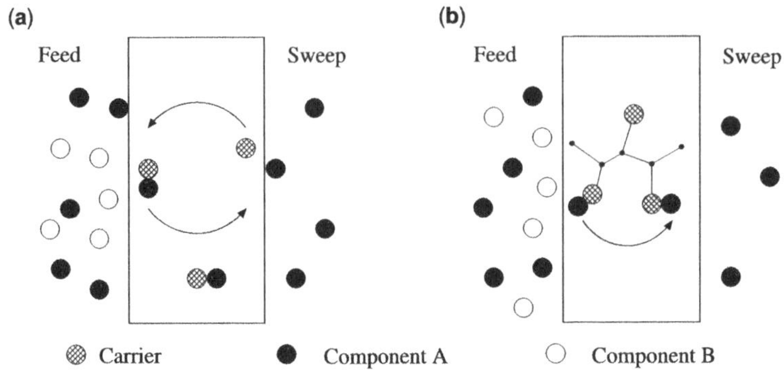

Figure 28.1 Schematic of facilitated membranes: (a) membrane with mobile carriers and (b) membrane with fixed carriers.

Compared with the conventional membranes based on the solution–diffusion mechanism, facilitated transport membranes have several advantages: (1) they are often highly selective, especially at low concentration driving forces; (2) high permeability can be achieved when the concentration driving force is low; and (3) they can maintain both high permeability and high selectivity at the same time.

Facilitated transport membranes have been studied for a long time. The earliest report was on the facilitated transport of oxygen through the hemoglobin solutions (Scholander, 1960). Since then, new membranes and new applications have emerged in many areas. Today, the potential applications of facilitated transport membranes include metal recovery, acid gas removal, bioseparations, olefin/paraffin separation, and O_2/N_2 separation.

This chapter reviews the recent developments of two types of facilitated transport membranes: (1) supported liquid membranes (SLMs) with strip dispersion and (2) carbon-dioxide-selective polymeric membranes, for environmental, energy, and biochemical applications.

28.1.1 Supported Liquid Membranes with Strip Dispersion

Facilitated transport can take place in both liquid and polymer membranes. Liquid membranes combine extraction and stripping, which are generally carried out in two separate steps in conventional processes such as solvent extraction, into one step. A one-step liquid membrane process provides the maximum driving force for the separation of a target species, leading to the best cleanup and recovery of the species (Ho and Sirkar, 1992). There are two types of liquid membranes: (1) emulsion liquid membranes (ELMs) and (2) supported liquid membranes (SLMs). ELMs, first invented by Li (1968), have been reviewed extensively (Ho and Li, 1992; Ho and Sirkar, 1992).

28.1.1.1 Supported Liquid Membranes In SLMs, the liquid membrane phase is the organic liquid imbedded in pores of a microporous support, for example, microporous polypropylene hollow fibers (Ho and Sirkar, 1992). When the organic liquid contacts the microporous support, it readily wets the pores of the support, and the SLM is formed.

For the extraction of a target species from an aqueous feed solution, the organic-based SLM is placed between two aqueous solutions, the feed solution and the strip solution, where the SLM acts as a semipermeable membrane for the transport of the target species from the feed solution to the strip solution. The organic in the SLM is immiscible in the aqueous feed and strip streams and contains an extractant, a diluent that is generally an inert organic solvent, and sometimes a modifier. In both types of liquid membranes, facilitated transport is the mass transfer mechanism for the target species to go from the feed solution to the strip solution (Ho and Li, 1992; Ho and Sirkar, 1992; Way and Noble, 1992). Quinn (Otto and Quinn, 1971; Smith and Quinn, 1979) and Li (Matuleviclus and Li, 1975; Li, 1978) pioneered the facilitated transport mechanism and laid the solid foundation for this mechanism.

The use of SLMs for the removal of metals, including chromium (Molinari et al., 1989; Ho and Sirkar, 1992; Huang et al., 1998; Yang et al., 2001; Ho, 2003), copper (Largman and Sifniades, 1978; Danesi, 1984; Ho and Sirkar, 1992; Yang and Fane, 1999; Gherrou et al., 2002; Ho, 2003), zinc (Danesi et al., 1983; Tanigaki et al., 1988; Ho and Sirkar, 1992; Gherrou et al., 2002; Ho, 2003), cobalt (Danesi et al., 1984; Chaudry et al., 1990; Ho and Sirkar, 1992; Gega et al., 2001; Alguacil, 2002; Ho, 2003), and strontium (Buchalter et al., 1987; Ho and Sirkar, 1992; Ramadan and Danesi, 1988; Dozol et al., 1994; Mackova and Mikulaj, 1996; Ho, 2003), from aqueous solutions and wastewaters has long been pursued by the scientific and industrial community. The use of SLMs for the recovery of benzylpenicillin (Marchese et al., 1989), penicillin G (Tsikas et al., 1992; Lee et al., 1994; Juang et al., 1998; Ho, 2003), phenylalanine (Bryjak et al., 1993; Campbell et al., 1994; Dzygiel et al., 1999), lactic acid (Sirman et al., 1991; Juang et al., 1997; Yahaya, 2001), citric acid (Friesen et al., 1991; Juang et al., 1997; Rockman et al., 1997), propionic acid (Nuchnoi et al., 1987; Shen et al., 1994), and butanoic acid (Nuchnoi et al., 1987; Shen et al., 1994) from aqueous solutions and fermentation broths has also long been pursued by the scientific and industrial community.

Although an SLM process is very effective for the removal of trace contaminants to very low levels due to its ability to circumvent equilibrium limitations (Ho and Sirkar, 1992), its use has been hampered by its stability. The traditional SLM suffers from a gradual loss of the organic membrane phase to the aqueous feed and strip solutions, due to emulsification (e.g., resulting from lateral shear forces) at the membrane–aqueous interfaces and to the osmotic pressure difference across the membrane (Kemperman et al., 1996; Hill et al., 1996; Dreher and Stevens, 1998). The osmotic pressure difference displaces the organic membrane phase from the micropores of the support. Displacement of the organic membrane phase from the pores can ultimately allow mixing of the feed and strip solutions, leading to complete failure of the separation unit.

This chapter reviews recent advances in SLMs with strip dispersion for removal and recovery of metals, including chromium, copper, zinc, cobalt, and strontium, from wastewaters and aqueous streams. This chapter also discusses potential applications of SLMs for removal and recovery of antibiotics including penicillin G and others.

28.1.1.2 Supported Liquid Membrane with Strip Dispersion The SLM system with strip dispersion developed recently is shown schematically in Figure 28.2 (Ho, 2001a, 2001b, 2003; Ho and Poddar, 2001; Ho and Wang, 2002). As shown in Figure 28.2, an aqueous strip solution is dispersed in an organic membrane solution containing an extractant or extractants in a mixer, and the water-in-oil dispersion formed is then pumped into a membrane module to contact with one side of a microporous

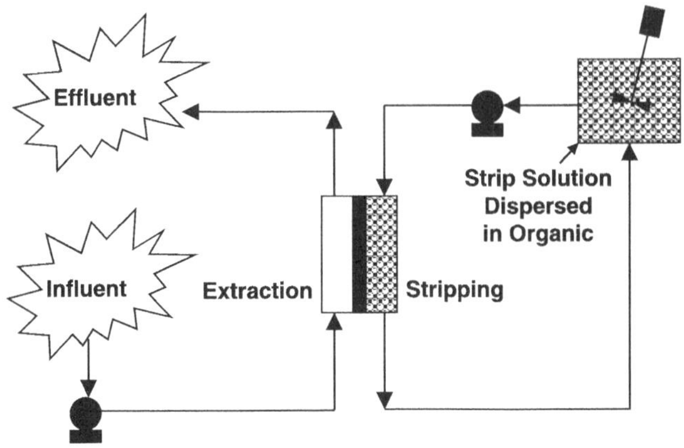

Figure 28.2 Schematic of supported liquid membrane with strip dispersion.

support (e.g., which is passed through the shell side of a microporous polypropylene hollow-fiber module). The aqueous feed solution containing a target species to be extracted is on the other side of the support (e.g., which is passed through the other side of the fibers, the tube side). The continuous organic phase of the dispersion readily wets the pores of a hydrophobic microporous support (e.g., microporous polypropylene hollow fibers in the module), and a stable liquid membrane (the organic phase) supported in the pores of the microporous support is formed.

Figure 28.3 shows an enlarged view of the SLM with strip dispersion. A low-pressure differential [about 13.8 kPa (2 psi)] between the aqueous feed solution side (P_a) and the strip dispersion side (P_o) is applied to prevent the organic solution of the strip dispersion from passing through the pores to come into the feed solution side. The dispersed droplets of the aqueous strip solution in a typical size of 80–800 μm are orders of magnitude larger than the pore size of the microporous polypropylene support employed for the SLM, which is 0.03 μm. Thus, these droplets are retained on the strip dispersion side and cannot pass through the pores to go to the feed solution side.

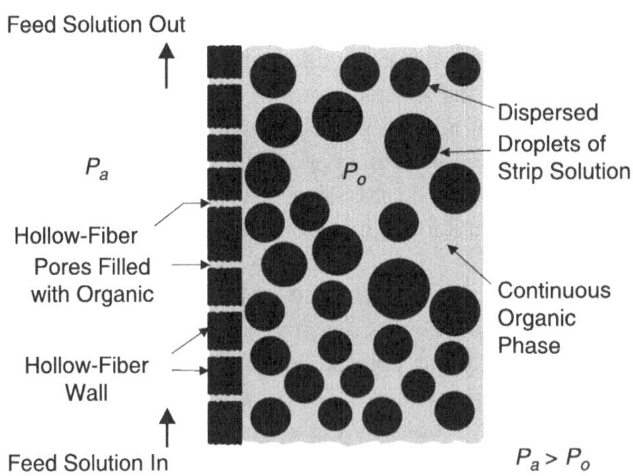

Figure 28.3 Enlarged view of the supported liquid membrane with strip dispersion.

This SLM may be considered as a SLM with a constant supply of the organic membrane solution into the pores. This ensures a stable and continuous operation. Following the approach of including organic membrane solution in the strip side, Teramoto et al. (2000) have stabilized their SLM by adding their organic membrane solution in the strip side. In addition, for the SLM with strip dispersion, the direct contact between the organic and strip phases (with high-shear mixing if necessary) on the strip dispersion side provides an additional mass transfer area between these two phases for stripping in addition to the hollow-fiber surface area, resulting in very efficient mass transfer for stripping. The stripping is more efficient than that with the conventional SLM.

The SLM supported in a hollow-fiber module has a very large mass transfer area. For example, using microporous polypropylene hollow fibers of 300-μm outside diameter, a commercial-size module of 25.4 cm (10 inches) in diameter by 71.1 cm (28 inches) in length has a surface area of 135 m^2, a pilot plant module of 10.2 cm (4 inches) in diameter by 71.1 cm (28 inches) in length possesses a surface area of 19 m^2, and a laboratory module of 6.35 cm (2.5 inches) in diameter by 20.3 cm (8 inches) in length provides a surface area of 1.4 m^2.

A typical SLM system consists of an SLM hollow-fiber module (or a series of SLM modules), a feed solution vessel, a feed pump, a strip dispersion vessel, a mixer for the strip dispersion vessel (for dispersing an aqueous strip solution in an organic solution; this may be omitted under a pumping and flow condition), and a strip dispersion pump. The configuration of the SLM module is similar to that of a shell-and-tube heat exchanger. Upon the completion of the removal of the target species, the mixer for the strip dispersion is turned off, and the dispersion separates into the two phases, the organic solution and the concentrated strip solution, upon standing. The phase separation is very fast (less than about 1 min), and there is no formation of an emulsion. The concentrated strip solution can be the product of this process.

28.1.2 Carbon-Dioxide-Selective Polymeric Membranes

Facilitated transport membranes are very attractive for the removal of CO_2 from synthesis gas containing H_2, CO, and CO_2. For membranes based on the solution–diffusion mechanism, it is challenging to achieve a high CO_2/H_2 selectivity since H_2 usually exhibits an unfavorably higher diffusivity coefficient than CO_2 does. As a result, in most conventional polymeric membranes, the CO_2/H_2 selectivities are usually less than one (Lin and Freeman, 2005). On the other hand, facilitated transport membranes can achieve very high CO_2/H_2 selectivities. Ho et al. (Tee et al., 2006; Zou and Ho, 2006) recently reported a CO_2/H_2 selectivity of 450 using membranes containing both mobile and fixed carriers.

This chapter reviews and discusses recent advances in carbon-dioxide-selective polymer membranes for hydrogen purification and carbon dioxide removal.

28.1.2.1 Facilitated Transport Membranes for CO_2 Removal The separation of CO_2 from other gases, such as H_2, N_2, CO, and CH_4, has various potential applications, including the purification of synthesis gas to obtain high-purity hydrogen for fuel cells, the separation of CO_2 from flue gas for the greenhouse gas sequestration, the removal of CO_2 from natural gas for natural gas sweetening, and the removal of CO_2 from breathing air in a spacecraft or confined space. For fuel cells, the need to obtain high-purity hydrogen as the main fuel is critical

(Song, 2002). CO_2-selective membranes have the potential to obtain high-purity H_2 by separating CO_2 from the synthesis gas to enhance the water gas shift (WGS) reaction (Huang et al., 2005a) or to be followed by methanation (Ledjeff-Hey et al., 2000), thus producing high-purity H_2 on the high-pressure feed side. Conventional CO_2 separation processes, like amine scrubbing, are highly energy consuming and require regeneration steps and additional chemicals. Their applications were also limited in certain areas, such as on-board hydrogen production, off-shore natural gas sweetening, and life support system in a spacecraft, due to their system complexity and large sizes/weights. Membrane separation provides one prospective approach to capturing and concentrating CO_2 with reduced energy consumption, enhanced weight and space efficiency, and operational simplicity.

Facilitated transport CO_2-selective membranes reported in the literature include SLMs, ion exchange membranes, and membranes with reactive carriers bonded in the matrices of membranes. For SLMs, Ward and Robb (1967) immobilized an aqueous bicarbonate–carbonate solution into a porous support and obtained a CO_2/O_2 separation factor of 1500. Meldon et al., (1977) investigated the facilitated transport of CO_2 through an immobilized alkaline liquid film. Their experimental results confirmed that weak acid buffers significantly increased the CO_2 transport. However, such SLMs have two major problems: loss of solvent and loss or degradation of carriers. The loss of solvent is caused by its evaporation, especially at a high temperature and/or its permeation through support under a high transmembrane pressure. The loss of carrier occurs when the carrier solution is forced to permeate through support ("washout"), and the degradation of carriers is led by the irreversible reaction of the carrier with impurities in the feed gas stream (LeBlanc et al., 1980; Way et al., 1987).

Ion-exchange-facilitated transport membranes were first reported by LeBlanc et al. (1980) to address the instability issue of SLMs. In the ion exchange membranes, ionic carriers were retained inside the membranes by electrostatic forces; therefore, minimizing the washout of carriers. Way et al. (1987) and Yamaguchi et al. (1996) used perfluorosulfonic acid ionomer cation exchange membranes containing amines as the carriers. The ion exchange membrane used by Langevin et al. (1993) was sulfonated styrene–divinylbenzene in a fluorinated matrix, and the transport model based on the Nernst–Planck equation was developed. Matsuyama et al. (1994, 1996) grafted acrylic acid and methacrylic acid on different substrates and used various diamines, diethlylenetriamine, and triethylenetetramine as the carriers. They also blended poly(acrylic acid) with poly(vinyl alcohol) to prepare membranes and introduced monoprotonated ethylenediamine into the membranes by ion exchange and used it as the carrier (Matsuyama et al., 2001). Other approaches were also proposed to solve the instability problems of SLMs. Quinn et al. (1995) developed membranes consisting of molten salt hydrates, which were nonvolatile and immobilized in microporous polypropylene supports. Teramoto et al. (2001, 2002, 2004) developed a "bulk flow liquid membrane," in which a carrier solution was forced to permeate through the membrane and then was recycled continuously.

Membranes with reactive carriers bonded in the membrane matrices were reported by several researchers and were believed to have better stability than SLMs. Yamaguchi et al. (1995) developed membranes with poly(allylamine) and compared them with ion exchange membranes containing amines as the counterions. Matsuyama et al. (1999) heat-treated poly(vinyl alcohol)–polyethylenimine membranes to improve their stability and to increase the amount of polyethylenimine retained inside the membranes, which increased the water content, therefore increasing the diffusivity of the carrier complex. Quinn and Laciak (1997) developed polyelectrolyte membranes based on poly(vinylbenzyltrimethylammonium fluoride) (PVBTAF) and achieved a CO_2/H_2 selectivity of 87 at 23°C. They also blended fluoride-containing organic and inorganic salts, like CsF, into the PVBTAF membranes and obtained a CO_2 permeance more than four times that of PVBTAF (Quinn et al., 1997). Ho and his co-workers synthesized cross-linked poly(vinyl alcohol) membranes containing polyamines as fixed carriers and amino-acid salts as mobile carriers (Ho, 1997, 2000; Tee et al., 2006; Zou and Ho, 2006). They reported membranes containing dimethylglycine (DMG) salts and polyethylenimine, and found that both CO_2 permeability and CO_2/H_2 selectivity of the membranes increased as temperature increased in the range of 50–100°C (Tee et al., 2006). Recently, they reported on membranes containing amines up to about 180°C (Zou and Ho, 2006).

Models for facilitated transport on different systems have been developed and studied. Analytical and numerical methods were used to do the modeling and to compare with experimental results (Cussler et al., 1989; Donaldson and Lapinas, 1982; Gottschlich et al., 1988; Ho and Dalrymple, 1994; Hong et al., 1996; Kang et al., 1996; Noble, 1990, 1991, 1992; Smith and Quinn, 1979; Ward, 1970).

There are two parameters to characterize the separation performance of a membrane. One is the selectivity (or the separation factor), which is defined as

$$\alpha_{ij} = \frac{y_i/y_j}{x_i/x_j} \tag{28.1}$$

Another parameter is the permeability P_i, which is defined as

$$P_i = \frac{N_i}{\Delta p_i/l} \tag{28.2}$$

The common unit of P_i is the Barrer, which is 10^{-10} cm^3 (standard temperature and pressure or STP) $cm/(cm^2\ s\ cm\ Hg)$; P_i/l is referred to as the permeance, and its common unit is the gas permeation unit (GPU), which is 10^{-6} cm^3 (STP)/(cm^2 s cm Hg) (Ho and Sirkar, 1992).

28.1.2.2 Modeling of CO_2-Selective WGS Membrane Reactor for Fuel Cells Fuel cells are regarded as a promising energy conversion approach in the twenty-first century. Hydrogen is the preferred fuel for fuel cells in most cases, especially for proton exchange membrane fuel cells (PEMFCs). But, for PEMFC applications, one key issue is how to obtain high-purity hydrogen. In the United States, most hydrogen is produced by steam reforming of natural gas followed by the WGS reaction [Eq. (28.3)] (Kroschwitz and Howe-Grant, 1995). The resulting synthesis gas (syngas) mainly contains hydrogen, carbon monoxide, and carbon dioxide. For PEMFCs, the CO concentration in synthesis gas needs to be reduced to less than 10 ppm, since even a very small amount of CO deteriorates fuel cell performance by poisoning platinum, the electrocatalyst in PEMFCs (Ahmed and Krumpelt, 2001; Song, 2002). There are several options for the CO cleanup. One option is to develop a WGS membrane reactor in which a membrane removes either H_2 or CO_2 from the reactor to shift the reversible WGS reaction forward so that the CO concentration can be further converted/reduced:

$$CO + H_2O \Leftrightarrow CO_2 + H_2 \qquad (\Delta H = -41.16\ kJ/mol) \tag{28.3}$$

Several researchers have achieved high CO conversions beyond the equilibrium conversions or even close to 100% by using palladium or other inorganic H_2-selective WGS membrane reactors (Basile et al., 1996; Criscuoli et al., 2000; Giessler et al., 2003; Tosti et al., 2003; Uemiya et al., 1991; Xue et al., 1996). However, preparing thin, flawless, and durable H_2-selective membranes remains a challenge for the commercial application of this type of membrane reactor (Armor, 1998). With CO_2-selective membranes (Ho, 1997, 2000; Tee et al., 2006; Zou and Ho, 2006) removing CO_2 continuously, a CO_2-selective WGS membrane reactor provides a promising approach of decreasing CO concentration and enhancing hydrogen production at relatively low temperatures ($\sim 150°C$). In comparison with the H_2-selective membrane reactor, the CO_2-selective WGS membrane reactor is more advantageous because (1) an H_2-rich product is recovered at a high pressure (feed gas pressure), and (2) air and/or steam can be used to sweep the permeate, CO_2, out of the reactor to obtain a high driving force for the separation.

Based on the experimental data of the CO_2-selective facilitated transport membranes described earlier in this chapter, we proposed the concept of CO_2-selective WGS membrane reactor and developed a one-dimensional nonisothermal model to simulate the reaction and transport process (Huang et al., 2005a). The modeling results have shown that H_2 enhancement and CO reduction to 10 ppm or lower are achievable for autothermal reforming synthesis gas via CO_2 removal. With this model, we have elucidated the effects of system parameters on the membrane reactor performance. Using the membrane synthesized and the commercial WGS catalyst, we have obtained a CO concentration of less than 10 ppm in the H_2 product in WGS membrane reactor experiments and verified the model developed (Huang et al., 2005b; Zou et al., 2007).

28.1.2.3 CO_2 Capture from Flue Gas and Synthesis Gas In recent years, the increasing public concern over global warming has concentrated on the man-made emissions of greenhouse gases. In the United States, 82.4% of the total greenhouse emissions consisted of CO_2 from the combustion of fossil fuels [U.S. Department of Energy (DOE), 2005]. Currently, the principle CO_2 capture technologies include chemical and/or physical absorption, physical adsorption, cryogenic separation, and membrane separation (Aaron and Tsouris, 2005). Compared to other technologies, membrane-based CO_2 capture has the advantages of low energy consumption, simplicity of operation, and absence of moving parts. However, today's commercial polymeric membranes, such as cellulose acetate, suffer from low-CO_2 permeability and selectivity since they are based on the solution–diffusion mechanism and primarily rely on the subtle size differences of the penetrants to achieve separation (Koros and Mahajan, 2000; Meisen and Shuai, 1997; Wind et al., 2004). Also, these membranes are not suitable for high-temperature operations, which are required in the direct CO_2 capture from the flue gas in power plants (Meisen and Shuai, 1997). For the application of membranes in CO_2 capture from the flue gas, Hirayama et al. (1999) reported that a CO_2/N_2 selectivity of greater than 70 and a minimum CO_2 permeability of 100 Barrers were required for an economic operation.

As an alternative to conventional polymeric membranes, facilitated transport membranes have shown to be more promising in satisfying these goals. By using the novel polymeric facilitated transport membrane that we synthesized, CO_2 capture was studied with a feed gas simulating the flue gas.

28.2 SUPPORTED LIQUID MEMBRANES WITH STRIP DISPERSION

28.2.1 Chromium Removal and Recovery

28.2.1.1 Chromium and Sulfuric Acid Fluxes For chromium removal and recovery, the organic membrane solution of the SLM used for both laboratory experiments and field demonstration was composed of 10 wt% (0.213 M) of *N*-lauryl-*N*-trialkylmethylamine with a molecular weight of 372 (a total number of 25.3 carbon atoms per amine molecule, e.g., Amberlite LA-2), 1 wt% of 1-dodecanol (modifier for the extractant), and 3 wt% of PLURONIC L31 (a block copolymer of ethylene oxide and propylene oxide additive to enhance the phase separation of the aqueous strip solution from the organic membrane solution in the absence of agitation for the recovery of the strip solution) in Isopar L [an isoparaffinic hydrocarbon solvent with a flash point of 62°C, a boiling point of 207°C, a viscosity of 1.5 cP (at 25°C), and a density of 0.767 g/mL (at 15.6°C)] (Ho, 2001a, 2003; Ho and Poddar, 2001).

The flux of Cr(VI) through the SLM as a function of Cr(VI) concentration in the feed is shown in Figure 28.4 (Ho, 2001a, 2003; Ho and Poddar, 2001). As shown in Figure 28.4, a carrier saturation phenomenon with a constant flux for the facilitated transport mechanism becomes evident when the feed Cr(VI) concentration is 1500 ppm or greater. This flux was about 12.2 $g/m^2/h$, which is quite high. The flux predicted from the model developed as a function of the chromium concentration in the aqueous feed solution is shown in Figure 28.4 in comparison with the experimental data (Ho, 2001a, 2003; Ho and Poddar, 2001). As shown in this figure, the model prediction is in good agreement with the data for a very wide range of aqueous feed Cr(VI) concentrations from approximately 20–6000 ppm.

Sulfuric acid was used for the pH adjustment of the aqueous feed solution (pH 1.5). However, the supported liquid membrane also removes sulfuric acid in the similar mechanism. Thus, the sulfuric acid competes with the chromic acid in the feed solution for complexation with the amine in the membrane phase. The extraction of sulfuric acid into the strip solution is not desirable since it degrades the quality of chromate recovered in

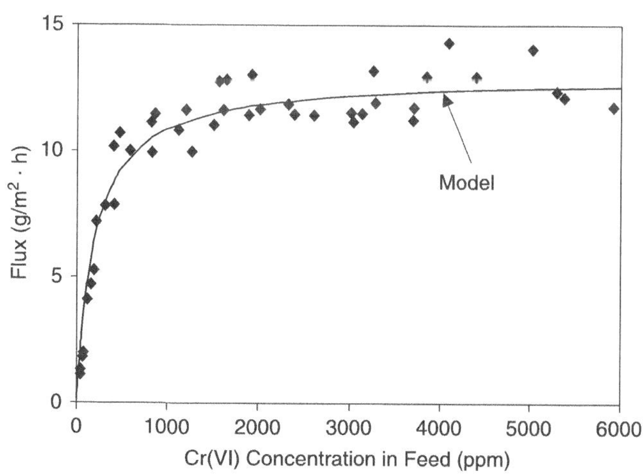

Figure 28.4 Chromium(VI) flux as a function of chromium(VI) concentration in the feed (Ho and Poddar, 2001).

the strip solution. However, Ho and his co-workers have unexpectedly found that the flux of sulfuric acid depends on the Cr(VI) concentration in the feed (Ho, 2001a, 2003; Ho and Poddar, 2001). The concentration of sulfuric acid in the feed was approximately 1500 ppm at feed pH 1.5. The sulfuric acid flux reduces very significantly as the Cr(VI) concentration in the feed increases. The sulfuric acid flux for a feed Cr(VI) concentration of greater than about 100 ppm is much less than that for a lower concentration. In other words, the chromium–sulfuric acid selectivity for a feed Cr(VI) concentration of greater than about 100 ppm is much higher than that for a lower concentration.

28.2.1.2 Chromium Removal and Recovery Process Ho and co-workers have developed a novel chromium removal and recovery process based on the new findings with unexpected results described above, where the sulfuric acid flux for a feed Cr(VI) concentration of greater than about 100 ppm is much less than that for a lower concentration, that is, the chromium–sulfuric acid selectivity for a feed Cr(VI) concentration of greater than about 100 ppm is much higher than that for a lower concentration (Ho, 2001a, 2003; Ho and Poddar, 2001). The novel process comprises two SLM steps: (1) A feed solution containing hexavalent chromium [Cr(VI)] is treated to decrease the chromium to an acceptable level [0.05 ppm Cr(VI) or lower] for discharge or recycle with an aqueous strip solution (strip 1), which results in a moderately concentrated Cr(VI) solution (1000–6000 ppm), and (2) the aqueous strip solution, after its phase separation from the continuous, organic phase and its pH adjustment to pH 1.5, is then processed to decrease Cr(VI) to about 100 ppm or greater, a concentration similar to that in the feed solution in step 1, for recycling back to the feed solution in step 1, by the use of a new aqueous strip solution (strip 2), which results in a highly concentrated Cr(VI) solution (about 200,000 ppm). This two-step process allows for the production of a highly concentrated Cr(VI) solution with low sulfate concentration from treating a moderately concentrated Cr(VI) solution with a Cr(VI) concentration of about 100 ppm or greater during the second step of this process.

In this process, the highly concentrated Cr(VI) solution with low sulfate concentration is a product that is suitable for reuse or resale. Therefore, this process not only solves the environmental problem but also recovers the chromium. In each SLM step of this process, the organic phase after its phase separation from the aqueous strip solution is reused for making the strip dispersion with the fresh aqueous strip solution.

28.2.2 Copper Removal and Recovery

In the removal of the metal ions such as copper, zinc, cobalt, and strontium, the net result of the metal ion removal from the aqueous feed solution into the aqueous strip solution is the transfer of proton ions from the strip solution into the feed solution. Thus, the pH of the feed solution reduces as the metal is removed. This has a significant effect on module utilization efficiency for SLM systems, and an SLM system operating at a low feed pH of about 2 is highly desirable in order to have a module utilization efficiency of 100% (Ho et al., 2001; Ho, 2003).

We have recently developed an SLM system with strip dispersion effective at the low-feed pH of about 2 for copper removal and recovery from wastewaters and process streams (Ho, 2001b, 2002a, 2003; Ho et al., 2001, 2002). The organic membrane solution in the strip dispersion comprised 15 wt% LIX 973 N (about 46% nonylsalicyl aldoxime, 18% ketoxime, 6% nonylphenol, and 30% diluent), 2 wt% 1-dodecanol, and 83 wt%

n-dodecane. The aqueous strip solution was 3 M sulfuric acid. A high volume ratio of 19 between the organic solution and the strip solution was used for greatly concentrating the copper in the strip solution.

The SLM system has been recently scaled up in a pilot plant unit using two 10.2-cm (4-inch) diameter microporous polypropylene hollow-fiber modules in parallel with 16.7 L/min (4.4 gal/min) total feed flow, that is, 8.3 L/min (2.2 gal/min) per module, for the selective removal and recovery of copper from the Berkeley Pit water containing copper, zinc, and the other metals after iron removal with pH (~4.3) adjustment (Ho, 2003). In a 3-h pilot plant run with a feed volume of 129.8 L (34.3 gal), the copper in the feed was removed from 152 to 0.08 ppm for a run time of 2 h and to less than 0.07 ppm for the 3-h run time, whereas it was concentrated to 3700 ppm in the aqueous strip solution for a run time of 2 h. Figure 28.5 shows the copper concentrations in the aqueous feed and strip solutions as a function of time in the recycle operation (Ho et al., 2002; Ho, 2003).

The scaleup results from pilot plant runs agreed reasonably well with laboratory results, particularly for the copper concentrations in the treated feed solutions—less than 0.1 ppm achieved from both the laboratory and pilot plant runs (Ho, 2003). The treated feed solutions with such a low copper concentration would be allowed for discharge in terms of the copper concentration. The copper concentration in the aqueous strip solution recovered from the pilot plant run was in line with that from the laboratory experiment under their operating conditions. The copper concentration from the pilot plant run was about 3700 ppm, and it was high enough for reuse, including its use in electrowinning for copper metal recovery. Thus, the goals of zero discharge and no sludge would be achievable.

In addition to the analyses of the copper concentrations in both the aqueous feed and strip solutions, the solutions were also analyzed for the other metals, that is, zinc, cobalt, nickel, aluminum, manganese, and cadmium. There were no significant changes in the concentrations of the other metals in both the aqueous feed and strip solutions before and after the pilot plant run. In other words, there were no significant contaminants in the concentrated strip solution recovered, that is, the strip solution product had high quality.

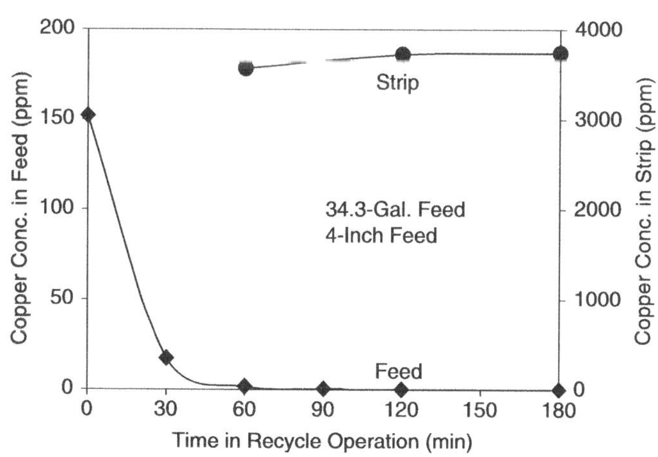

Figure 28.5 Copper concentrations in feed and strip solutions as a function of time in recycle operation from the pilot plant/scaleup run (Ho et al., 2002).

This also demonstrated the selective removal and recovery of copper from the feed solution containing the contaminant metals. This also reinforced the reuse opportunity of the strip solution recovered.

In both the laboratory and pilot runs, the phase separation for the strip dispersion was very satisfactory. Upon the completion of the removal of the target species, copper, the mixer for the strip dispersion was turned off, and the dispersion separated into the two phases, the organic solution and the concentrated strip solution, upon standing. The phase separation was very fast (less than about 1 min), and there was no formation of an emulsion. There was no physical loss of the organic membrane solution except its solubility in the aqueous solution that was very low due to the use of the isoparaffinic hydrocarbon solvent. Thus, the stability of the SLM with strip dispersion appeared to be satisfactory for a period of several weeks.

28.2.3 Zinc Removal and Recovery

We have also developed an SLM system with strip dispersion effective at the low feed pH of also about 2 for zinc removal and recovery from wastewaters and process streams (Ho, 2001b, 2002a, 2003; Ho et al., 2001, 2002). The organic membrane solution in the strip dispersion consisted of 8 wt% di(2,4,4-trimethylpentyl) dithiophosphinic acid (Cyanex 301), 2 wt% dodecanol, and 90 wt% Isopar L. The aqueous strip solution was 3 M sulfuric acid.

The treated Berkeley Pit water remained at a pH value of about 2 after the copper removal, and it was used for the removal and recovery of zinc at this pH. The pilot plant runs for zinc used one 10.2-cm (4-inch) diameter module with the feed flow rate of 3.8 L/min (1 gal/min). In a 5-h pilot plant run with a feed volume of 121.1 L (32 gal), the zinc in the feed was removed from 526 to 0.65 ppm for a run time of 4 h and to less than 0.62 ppm for the 5-h run time, whereas it was concentrated to 20,000 ppm in the strip for a run time of 4 h. Figure 28.6 shows the zinc concentrations in the aqueous feed and strip solutions as a function of time in recycle operation (Ho et al., 2002; Ho, 2003).

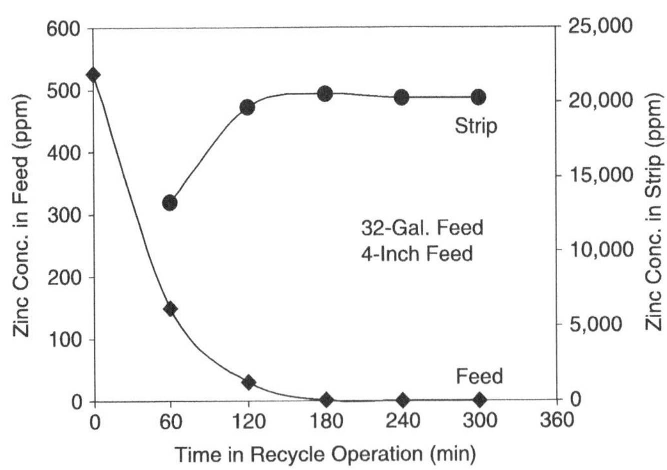

Figure 28.6 Zinc concentrations in feed and strip solutions as a function of time in recycle operation from the pilot plant/scaleup run (Ho et al., 2002).

The scaleup results from pilot plant runs agreed reasonably well with laboratory results in terms of the zinc concentrations in the treated feed and concentrated strip solutions (Ho, 2003). The zinc concentration in the aqueous strip solution recovered from the pilot plant run was in good agreement with that from the laboratory experiment. The zinc concentration was more than 17,000 ppm, and it was high enough for reuse, including its use in electrowinning for zinc metal recovery. The zinc concentration in the treated feed solution from the pilot plant run was in line with that from the laboratory experiment under their operating conditions. The treated feed solutions with such a low zinc concentration (0.3–0.65 ppm) would be allowed for discharge in terms of the zinc concentration. Thus, the goals of zero discharge and no sludge would be achievable.

In addition to the analyses of the zinc concentrations in both the aqueous feed and strip solutions, the solutions were also analyzed for the other metals, that is, cobalt, nickel, aluminum, manganese, and cadmium. There were no significant changes in the concentrations of aluminum, manganese, and cadmium in both the aqueous feed and strip solutions before and after the pilot plant run. However, about 35% of the cobalt and 25% of the nickel extracted from the feed solution were transferred to the aqueous strip solution. The cobalt and nickel were about 0.1 and 0.04 wt% of the zinc concentration in the aqueous strip solution recovered, respectively. Thus, there were no significant contaminants in the concentrated strip solution recovered, that is, the strip solution product had high quality. This also demonstrated the selective removal and recovery of zinc from the feed solution containing the contaminant metals. This also reinforced the reuse opportunity of the strip solution recovered.

The phase separation for the strip dispersion and the stability of the SLM with strip dispersion in both the laboratory and pilot runs for zinc were similar to those for copper described earlier. Both the phase separation and the stability were satisfactory.

28.2.4 Cobalt Removal and Recovery

We have identified an SLM system with strip dispersion effective at the low feed pH of about 2 for cobalt removal and recovery from wastewaters and process streams (Ho, 2001b, 2002a, 2003). This SLM system can have a feed pH not to reduce significantly due to proton transfer during extraction in a hollow-fiber module in order to utilize the entire length of the module (Ho et al., 2001). The organic membrane solution in the strip dispersion consisted of 8 wt% di(2,4,4-trimethylpentyl) dithiophosphinic acid (Cyanex 301), 2 wt% 1-dodecanol, and 90 wt% Isopar L (described earlier). The aqueous strip solution was 5 M hydrochloric acid. A relatively high volume ratio of about 13.3 between the organic solution (800 mL) and the strip solution (60 mL) was used for greatly concentrating the cobalt in the strip solution.

In the laboratory, the SLM system treated the 1-L feed solution containing about 525 ppm cobalt in the recycle mode of operation. The SLM removed the cobalt to 0.7 ppm in the treated feed solution in just 15 min (Ho, 2001b, 2002a, 2003). It also concentrated cobalt to about 30,000 ppm in the aqueous strip solution (from the strip solution of 5 M hydrochloric acid preloaded with about 21,250 ppm cobalt). The cobalt flux at the cobalt concentration of about 380 ppm in the feed solution was about 5 $g/m^2/h$, which is very high.

In a pilot plant run, the SLM using the same organic membrane solution but with an aqueous strip solution of 6.5 M hydrochloric acid was used to remove cobalt from a 40-L feed solution containing 492 ppm cobalt and concentrate it in the strip solution.

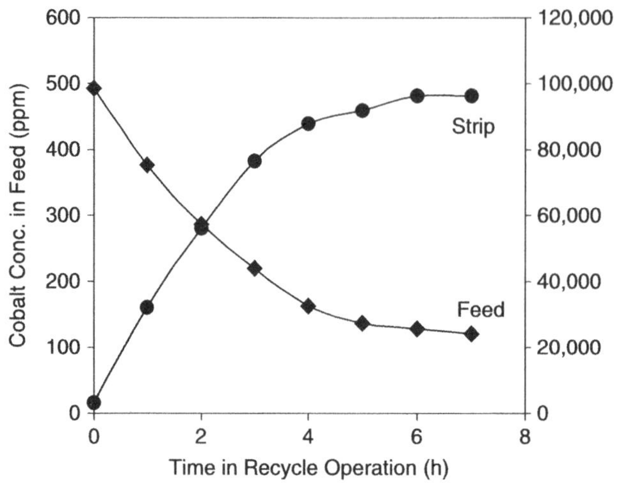

Figure 28.7 Cobalt concentrations in the feed and strip solutions as a function of time in recycle operation for concentrating to about 100,000 ppm in the final strip (Ho, 2003).

For concentrating the cobalt in the strip solution from an initial concentration of about 3100 ppm to a final concentration of about 100,000 ppm, a relatively high volume ratio of about 8.6 between the organic solution and the strip solution was used (a very high volume ratio of about 260 between the feed solution and the strip solution). Again, the SLM system treated the 40-L feed solution in the recycle mode of operation. Figure 28.7 shows the results. As shown in this figure, the SLM concentrated the cobalt effectively to about 100,000 ppm in the strip solution in about 7 h in the recycle operation (Ho, 2003).

28.2.5 Strontium Removal

28.2.5.1 Non-radioactive Sr-87 Removal We have synthesized a family of new extractants, alkyl phenylphosphonic acids, for strontium removal by SLMs with strip dispersion (Ho, 2001b, 2003; Ho and Wang, 2002). The extractants synthesized were 2-butyloctyl phenylphosphonic acid (C12 BOPPA), 2-hexyldecyl phenylphosphonic acid (C16 HDPPA), 2-octyldecyl/2-hexyldodecyl phenylphosphonic acid (C18 ODPPA/HDPPA), and 2-octyldodecyl phenylphosphonic acid (C20 ODPPA).

Laboratory SLM experiments were carried out in the recycle mode of operation for both a feed solution and a strip dispersion. The strip dispersion was prepared by dispersing an aqueous strip solution of 1 M HCl in the organic membrane solution at an organic-to-strip solution volume ratio of 3. Each of the organic membrane solutions of the SLMs used for the experiments comprised 8 wt% of one of the four alkyl phenylphosphonic acids and 2 wt% of 1-dodecanol (modifier for the extractant) in *n*-dodecane.

In the experiments, all the four extractants synthesized were used for strontium removal from aqueous feed solutions containing about 5.5 ppm of nonradioactive Sr-87 at pH 3. Figure 28.8 shows the results for all of these four extractants by the use of 1 M HCl as the stripping solutions (Ho, 2001b, 2003; Ho and Wang, 2002). As shown in this figure, all of these four extractants removed strontium from the aqueous feed solutions and recovered it in the aqueous strip solutions. The strontium concentration in the aqueous strip solutions was about 40 ppm, which was nearly the maximum concentration achievable for a feed-to-strip solution volume ratio of 8. Thus, the stripping with hydrochloric acid

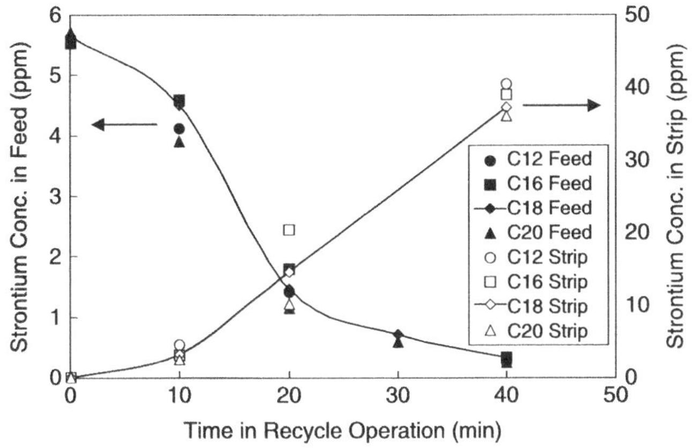

Figure 28.8 Strontium concentrations in feed and strip solutions as a function of time in recycle operation for four extractants (C12 feed: concentration in the feed solution for the C12 BOPPA extractant; C12 strip: concentration in the strip solution for the C12 BOPPA extractant; in the same way for the C16 HDPPA, C18 ODPPA/HDPPA, and C20 ODPPA extractants; the aqueous feed solutions also contained about 80 ppm calcium, 20 ppm magnesium, and 50 ppm zinc in addition to Sr-87) (Ho and Wang, 2002).

was very satisfactory and nearly quantitative. It should be noted that the aqueous feed solutions also contained about 80 ppm calcium, 20 ppm magnesium, and 50 ppm zinc in addition to strontium to simulate groundwater.

28.2.5.2 Radioactive Sr-90 Removal After we demonstrated the new extractants effective for nonradioactive Sr-87 removal, we pursued the experiments using radioactive Sr-90. In view of the fact that the four extractants removed Sr-87 equally well as described before, C20 ODPPA (with the longest alkyl chain for the lowest solubility in water) was used for the Sr-90 removal in order to increase the probability of achieving the goal of 8 pCi/L or lower. Table 28.1 shows the results for the Sr-90 removal from aqueous feed

TABLE 28.1 Radioactive Sr-90 Removal Using the C20 ODPPA Extractant

Run	Sr-90 Concentration in Feed (pCi/L)	Time (min)	Sr-90 Concentration in Treated Feed (pCi/L)
1	317	120	3.3
2	317	120	3.5
3	317	120	3.3
4	317[a]	240	4.0
5	1000[a]	240	5.5
6	1000[a]	360	1.0
7	30,000	60	1171.0
		120	352
8	30,000[a]	300	84

Source: Ho and Wang (2002).
[a]The feed solution also contained about 80 ppm calcium, 20 ppm magnesium, and 50 ppm zinc.

solutions at pH 3 with 1 M HCl as the stripping solution. As shown in this table, the SLM containing C20 ODPPA removed Sr-90 to the target concentration of 8 pCi/L (the drinking water standard) or lower from feed solutions containing 317–1000 pCi/L Sr-90 (Ho and Wang, 2002; Ho, 2003). Especially, this target concentration was also achieved from a feed solution containing 1000 pCi/L Sr-90 with the presence of about 80 ppm calcium, 20 ppm magnesium, and 50 ppm zinc.

This SLM also removed Sr-90 to a concentration of 352 pCi/L or lower from feed solutions containing 30,000 pCi/L Sr-90 (also shown in Table 28.1) with and without the presence of about 80 ppm calcium, 20 ppm magnesium, and 50 ppm zinc. The strip concentration was concentrated to more than 263,000 pCi/L (with an aqueous strip solution volume of about 220 mL, that is, a feed-to-strip solution volume ratio of about 9). Based on the results for the feed solutions of 317–1000 pCi/L, the treated concentration of 352 pCi/L or lower (in the treated solutions), which is lower than 1000 pCi/L, can be reduced to meet the target concentration of 8 pCi/L or lower by the use of a second SLM step.

28.2.6 Antibiotics Removal and Recovery

28.2.6.1 Penicillin G Removal and Recovery We have developed an SLM system with strip dispersion effective at the low feed pH of about 3 for penicillin G removal and recovery from aqueous solutions (Ho, 2002b, 2003). The organic membrane solution in the strip dispersion consisted of 10 wt% *N*-lauryl-*N*-trialkylmethylamine with a molecular weight of 372 (a total number of 25.3 carbon atoms per amine molecule, e.g., Amberlite LA-2), 1 wt% *o*-nitrophenyl octyl ether (*o*-NPOE), and 89 wt% Isopar L (described earlier). The aqueous strip solution was 1.2 M sodium carbonate (Na_2CO_3).

A strip dispersion was prepared by mixing together 200 mL of the 1.2 M sodium carbonate (Na_2CO_3) solution and 800 mL of the organic solution. The strip dispersion was fed into the shell side of a 2.5-inch polypropylene hollow-fiber module (2.5 inches in diameter by 8 inches in length). A 1-L feed solution containing a penicillin G concentration of 8840 ppm was passed into the tube side of the hollow-fiber module countercurrently to the flow of the strip dispersion in the recycle mode of operation. The pH of the feed solution was maintained at 3 ± 0.1 by adding 3 M sulfuric acid as needed. Samples of the feed and strip solutions were collected at timed intervals and analyzed by ultraviolet (UV) light.

Figure 28.9 shows the results (Ho, 2002b, 2003). The SLM removed the penicillin G from a high concentration of 8840 ppm to a relatively low concentration of 877 ppm in the feed solution in 3 h. On the other hand, the penicillin G was recovered and concentrated to a high concentration of 41,011 ppm in the aqueous strip solution at the same time. This represented a high recovery efficiency of about 93%. The penicillin G flux at the penicillin G concentration of about 5500 ppm in the feed solution was about 9 $g/m^2/h$, which is quite high.

28.2.6.2 Potential for Other Antibiotics The SLMs also have the potential for the removal and recovery of other antibiotics, including Cephalexin, Ampicillin, and Amoxicillin. Cephalexin is in the cephalosporin class of antibiotics, and it is one of the most prescribed antibiotics, approximately 3000 tons/year worldwide (Sheldon, 2004). The cephalosporin molecules contain acid and amine functional groups, which are amendable for complexation with extractants incorporated in liquid membranes. Studies on the

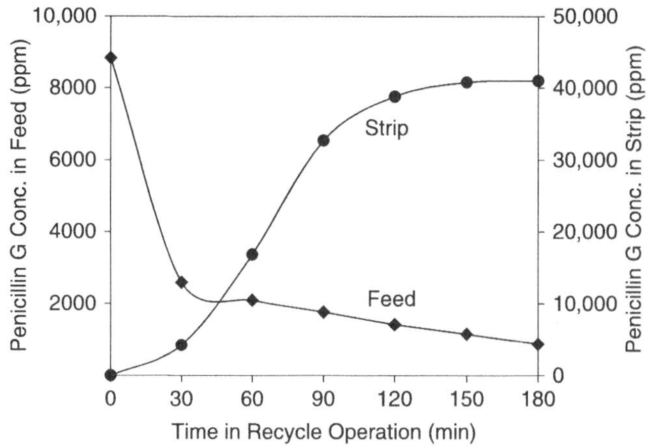

Figure 28.9 Penicillin G concentrations in feed and strip solutions as a function of time in recycle operation (Ho, 2003).

recovery of the antibiotics using liquid membranes have been rather limited, and there are no investigations using SLMs with strip dispersion (Ho, 2003; Dutta and Sahoo, 2002). Due to their good membrane stability, SLMs with strip dispersion have the potential to be an economical and efficient way to separate antibiotics from their fermentation reaction mixtures.

28.3 CARBON-DIOXIDE-SELECTIVE MEMBRANES

This section reviews our recent work on new facilitated transport CO_2-selective membranes for CO_2 removal, WGS membrane reactor, and CO_2 capture from flue gas/synthesis gas.

28.3.1 CO_2-Selective Membranes for CO_2 Removal

A review of our recent work on the synthesis and transport properties of new CO_2-selective membranes follows. Transport properties of the membranes synthesized, including CO_2 permeability and flux, H_2 flux, CO_2/H_2, CO_2/N_2, and CO_2/CO selectivities, were studied. The effects of feed pressure, water concentration, and temperature on transport properties were investigated.

28.3.1.1 Synthesis of CO_2-Selective Membranes CO_2-selective polymeric membranes with the thin-film composite structure were prepared by casting an aqueous solution onto microporous BHA Teflon supports (thickness: 60 μm, average pore size: 0.2 μm, BHA Technologies, Kansas City, MO), or GE E500A microporous polysulfone supports (thickness: about 60 μm excluding nonwoven fabric support, average pore size: 0.05 μm, GE Infrastructure, Vista, CA). The aqueous solution was prepared from poly(vinyl alcohol) (PVA), formaldehyde (cross-linking agent), potassium hydroxide, 2-aminoisobutyric acid (AIBA) potassium salt, and poly(allylamine) (Zou and Ho, 2007). The active layer was dense and about 20–80 μm thick. The thickness of a membrane mentioned hereafter all refers to the thickness of the active layer.

The membranes used in the present study contained 50 wt% PVA, 18.3 wt% KOH, 20.7 wt% AIBA-K, and 11.0 wt% poly(allylamine), unless otherwise indicated. Formaldehyde equivalent to a 60-mol% cross-linking degree was added into the casting solution (Zou and Ho, 2006).

28.3.1.2 Transport Measurement and Mechanism The gas permeation tests were conducted by using a gas permeation apparatus (Tee et al., 2006; Zou and Ho, 2006). Two feed gases were used for the gas permeation tests: one consisting of 20% CO_2, 40% H_2, and 40% N_2, and the other consisting of 17% CO_2, 1.0% CO, 45% H_2, and 37% N_2 (all on dry basis). The second composition was used to simulate the composition of the synthesis gas from autothermal reforming of gasoline with air. Argon was used as the sweep gas for the ease of gas chromatography analysis.

The membranes synthesized contained both AIBA-K and $KHCO_3$-K_2CO_3 (converted from KOH) as the mobile carriers and poly(allylamine) as the fixed carrier for CO_2 transport. 2-Aminoisobutyric acid is a sterically hindered amine, and its reaction with CO_2 is depicted in Eq. (28.4) (Sartori et al., 1987). Poly(allylamine) contains unhindered, primary amino groups, and the reaction of these groups is shown in Eq. (28.5) (Sartori et al., 1987). The reaction mechanism of the CO_2 with $KHCO_3$-K_2CO_3 was presumably similar to that of potassium carbonate promoted by hindered amine described in Eq. (28.6) (Sartori and Savage, 1983; Shulik et al., 1996):

$$R{-}NH_2 + CO_2 + H_2O \Leftrightarrow R{-}NH_3^+ + HCO_3^- \tag{28.4}$$

$$2R{-}NH_2 + CO_2 \Leftrightarrow R{-}NH{-}COO^- + R{-}NH_3^+ \tag{28.5}$$

$$CO_3^{2-} + CO_2 + H_2O \Leftrightarrow 2HCO_3^- \tag{28.6}$$

The transfer of CO_2 across the membrane is enhanced by the facilitated transport with the reactions mentioned above, and the total flux equation for the CO_2 transport in the membranes can be expressed as (Ho and Dalrymple, 1994)

$$N_A = D_A(C_{A|p1} - C_{A|p2})/l + D_{AB}(C_{AB|p1m} - C_{AB|p2m})/l \tag{28.7}$$

In Eq. (28.7) the first term on the right-hand side is the flux due to the physical solution–diffusion mechanism, while the second term is contributed by the chemical reaction. The nonreacting gases, such as H_2, N_2, CO, and CH_4, do not have chemical association with the carriers and therefore can only be transported by the physical mechanism, which is limited by their low solubility on the highly polar sites in the membranes (Ho and Sirkar, 1992; Quinn et al., 1995; Kim et al., 2004). For these nonreacting gases, the flux equation in the membrane is the first term on the right-hand side of Eq. (28.7) only.

28.3.1.3 Effects of Feed Pressure on Separation Performance The effects of feed pressure on CO_2 flux and permeability, H_2 flux, and CO_2/H_2 selectivity were investigated using a membrane with a thickness of $\sim$60 μm on the BHA microporous Teflon support. The feed gas consisted of 20% CO_2, 40% H_2, and 40% N_2. Figure 28.10 illustrates the effects of feed pressure on CO_2 flux and permeability. As illustrated in this figure, CO_2 flux increased first linearly with the feed pressure and then approached a constant value. This can be explained with the carrier saturation phenomenon. As described by Ho and

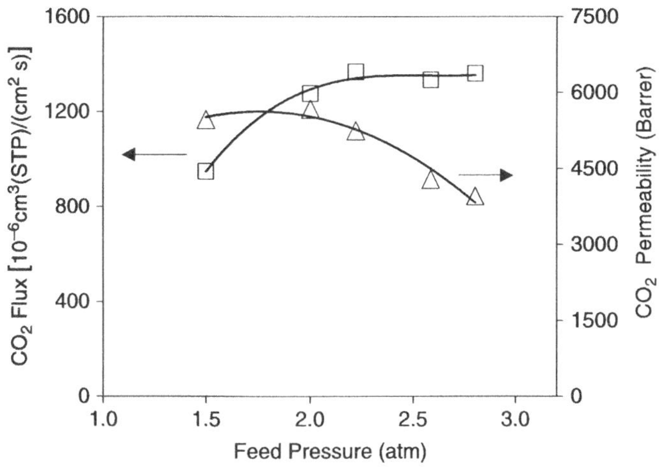

Figure 28.10 Effect of feed pressure on CO_2 flux and permeability: (□) CO_2 flux; (△) CO_2 permeability; at 110°C with water rates = 0.03/0.03 cm^3/min (feed/sweep) (Zou and Ho, 2006).

Dalrymple (1994), when the partial pressure of CO_2 is equal to or higher than a critical CO_2 partial pressure, p_{1c}, the carrier saturation occurs, in which the concentration of CO_2–carrier reaction product attains its maximum value, $C_{AB,max}$ and becomes constant. In other words, further increase in the partial pressure of CO_2 will not increase the concentration of CO_2–carrier reaction product. This can be expressed as follows:

$$C_{AB|p1} = H_{AB|p1}\, p_1 = C_{AB,max} = \text{const.} \quad \text{when } p_1 \geq p_{1c} \tag{28.8}$$

Compared with the facilitated transport flux, the flux due to the physical solution and diffusion of CO_2 in the membrane is negligible. Therefore, the total CO_2 flux becomes constant eventually as the feed pressure increases.

As demonstrated in Figure 28.10, CO_2 permeability decreased when the feed pressure increased. This can also be explained by using the carrier saturation phenomenon. In Eq. (28.2), in order to maintain the equality, increasing the CO_2 partial pressure, p_1, will not further increase the CO_2 flux since all the carriers have already reacted with CO_2 and attained the maximum capacities. Thus, CO_2 permeability will decrease.

However, unlike CO_2 flux, H_2 flux increased linearly with the feed pressure (Zou and Ho, 2006). This is because H_2 has no chemical association with carriers. Its sorption in the membrane can be described by Henry's law, and the permeability is usually independent of the feed pressure (Zolandz and Fleming, 1992). Therefore, the flux increased linearly with the feed pressure.

Figure 28.11 depicts the effect of feed pressure on CO_2/H_2 selectivity. As depicted in this figure, the CO_2/H_2 selectivity dropped as the pressure increased. Again, this can be explained using the carrier saturation phenomenon described earlier that CO_2 permeability reduces as the pressure increases. As a result, the selectivity decreases as the pressure increases since H_2 permeability usually does not change with pressure significantly.

28.3.1.4 Effects of Water Content on Separation Performance In this study, both the feed and the sweep gases were fed with controlled amounts of water before they

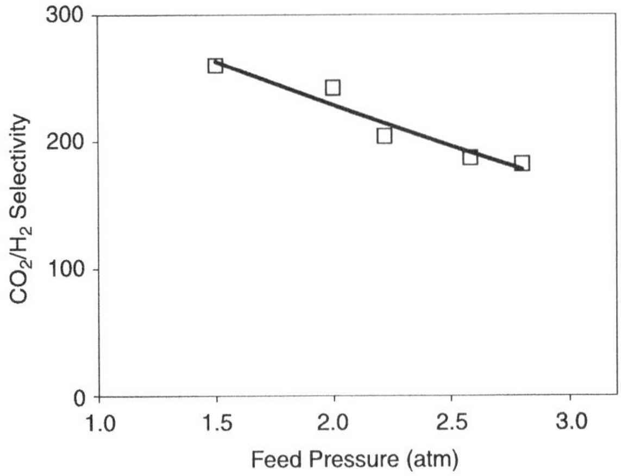

Figure 28.11 Effect of feed pressure on CO_2/H_2 selectivity. At 110°C with water rates = 0.03/0.03 cm^3/min (feed/sweep) (Zou and Ho, 2006).

entered the permeation cell. The effects of water content on the membrane separation performance at 120 and 150°C were investigated. The feed gas consisted of 20% CO_2, 40% H_2, and 40% N_2. Figure 28.12 depicts the CO_2 permeability as a function of the water concentration on the sweep side. As the water concentration on the sweep side increased, CO_2 permeability increased almost linearly. When the water content was increased from 58 to 93% (mol), the CO_2 permeability at 120°C increased from 3700 to as high as 8200 Barrers, while the permeability at 150°C increasing from 920 to 2700 Barrers. These increases were presumably due to two reasons: (1) Higher water content on the sweep side raised the water retention inside the membrane, thus resulting in increasing the mobility of both mobile and fixed carriers and the reaction rates of CO_2 with the carriers [Eqs. (28.4)–(28.6)], and (2) higher water content on the sweep side diluted the permeated CO_2 concentration in the sweep side, thus resulting in increasing the driving

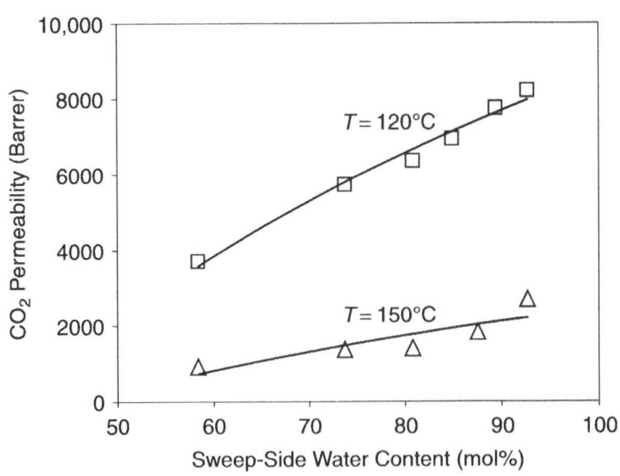

Figure 28.12 CO_2 permeability vs. water content on the sweep side. (□) T = 120°C; (△) T = 150°C; feed water content = 41 mol% and feed pressure = 2.0 atm (Zou and Ho, 2006).

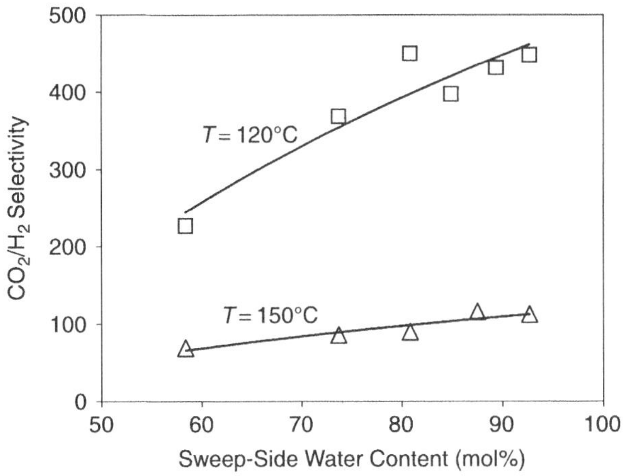

Figure 28.13 CO_2/H_2 selectivity vs. water content on the sweep side. (□) $T = 120°C$; (△) $T = 150°C$; feed water content = 41 mol% and feed pressure =2.0 atm (Zou and Ho, 2006).

force for the CO_2 transport. The increase of CO_2 permeability with increasing gas–water content was also reported by Quinn et al. (Quinn and Laciak, 1997; Quinn et al., 1997).

Figure 28.13 shows the CO_2/H_2 selectivity as a function of the water concentration on the sweep side at 120 and 150°C. The CO_2/H_2 selectivities at both temperatures increased as the water content on the sweep side increased. When the sweep water content was 93 mol%, the CO_2/H_2 selectivities at 120 and 150°C reached 450 and 270, respectively. This increase could be explained by the rise of CO_2 permeation rate while the transport of H_2 was not significantly affected by the increase of the water content.

The water content on the feed side also had significant effects on CO_2 permeability and CO_2/H_2 selectivity (Zou and Ho, 2006). Both CO_2 permeability and CO_2/H_2 selectivity increased as the water content in the feed stream increased. This might be explained by the case that the higher water content on the feed side raised the water retention inside the membrane, therefore increasing the mobility of both mobile and fixed carriers and the reaction rates of CO_2 with the carriers. As a result, the CO_2 transport was enhanced, while the transport of H_2 was not significantly affected.

28.3.1.5 *Effects of Temperature on Separation Performance*

The effects of temperature on CO_2 permeability, CO_2/H_2, CO_2/N_2, and CO_2/CO selectivity were studied in the temperatures range of 100–180°C (Zou and Ho, 2006). Figure 28.14 illustrates the effects of temperature on CO_2 permeability and CO_2/H_2 selectivity using a 26-μm membrane containing 40.0 wt% PVA, 20.0 wt% KOH, 20.0 wt% AIBA-K, and 20.0 wt% poly(allylamine) (Zou and Ho, 2006). As shown in this figure, the CO_2 permeability and CO_2/H_2 selectivity all decreased as temperature increased. This was due to the fact that CO_2 permeability decreased as temperature increased, while the transport of both H_2 was not affected by temperature significantly. However, the membrane showed good separation performance from 110 to 160°C. Even at 150°C, the CO_2 permeability and CO_2/H_2 selectivity were 2500 Barrers and 80, respectively. The CO_2/H_2 selectivity was reduced slightly as the temperature increased to 170°C, and it decreased significantly to about 10 at 180°C mainly due to the significant swelling of the membrane, thus resulting in a sharp increase of H_2 permeability at this high temperature.

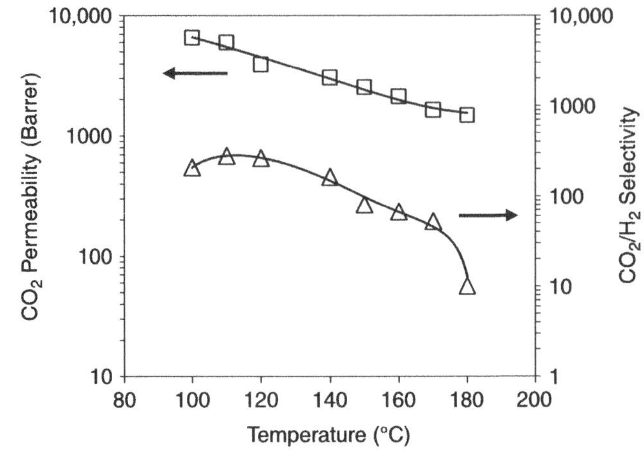

Figure 28.14 Effects of temperature on CO_2 permeability and CO_2/H_2 selectivity. (□) CO_2 permeability; (△) CO_2/H_2 selectivity; feed pressure = 2.1 atm; with increasing water rates at elevated temperatures (Zou and Ho, 2006).

The CO_2/CO selectivity reduced as temperature increased, which can be explained by the decrease of CO_2 transport rate at elevated temperatures. However, even at 170°C, the CO_2/CO selectivity was 160, which is very high (Zou and Ho, 2006).

The results from this study were compared with the results reported in the literature (Quinn and Laciak, 1997; Quinn et al., 1997) and our previous results (Tee et al., 2006) in Table 28.2. As we can see from the comparison, the results from this work are gratifying. The improvement in the working temperature over our previous work could be mostly attributed to better cross-linking of PVA as the polymer matrix, which was confirmed by Fourier transform infrared spectroscopy (FTIR). The high permeability and selectivity were presumably attributable to better carriers since the membranes contained both AIBA-K and $KHCO_3$-K_2CO_3 as the mobile carriers and poly(allylamine) as the fixed carrier.

TABLE 28.2 Comparison of Membrane Performance with Literature Results

Membrane	T (°C)	CO_2 Partial Pressure (atm)	P_{CO_2} (Barrer)	α (CO_2/H_2)	Reference
Poly(vinylbenzyl-trimethylammonium fluoride)	23	0.42	~120	87	Quinn and Laciak (1997)
Poly(vinylbenzyl-trimethylammonium fluoride)-CsF	23	0.40	~510	127	Quinn et al. (1997)
Dimethylglycine-Li/PVA	90	0.24	1700	50	Tee et al. (2006)
AIBA-K/polyallylamine/PVA	100	0.24	6500	210	Zou and Ho (2006) (Fig. 28.14)
	150	0.24	2500	80	

Gas permeation results presented in this work, especially above 150°C, showed that the polymeric membranes that we prepared were capable of more applications at elevated temperatures, such as the WGS reaction membrane reactors, which incorporate both CO_2 removal and the WGS reaction to produce high-purity H_2 (Huang et al., 2005a, b; Zou et al., 2007).

28.3.2 CO_2-Selective WGS Membrane Reactor

28.3.2.1 Model Description As one of the two main types of commercialized membrane modules, the hollow-fiber membrane module has shown excellent mass transfer performance because of its large surface area per unit volume (about 3000 ft^2/ft^3 for gas separation). In the modeling work by Huang et al. (2005a), the WGS membrane reactor was configured to be a hollow-fiber membrane module with catalyst particles packed inside the fibers. The catalyst packed was assumed to be the commercial $Cu/ZnO/Al_2O_3$ catalyst for the lower-temperature WGS reaction. The reaction rate reported in the literature (Keiski et al., 1993; Moe, 1962) was used in the model, incorporating the molar and energy balances on both feed (lumen) and sweep (shell) sides.

Although hydrogen is the preferred fuel for fuel cells, currently there are issues with its storage and distribution (Brown, 2001). As a more practical way, hydrogen used in automotive fuel cells is suggested to be produced by on-site or on-board reforming reactions of the hydrocarbon fuels, such as natural gas, gasoline, and diesel. In the modeling work by Huang et al. (2005a), the syngas from autothermal reforming of hydrocarbon fuel with air was chosen as the feed gas, and the feed molar flow rate, n_{t0} was 1 mol/s. With the composition of the autothermal reforming syngas given in Table 28.3, this flow rate was chosen because a sufficient H_2 molar flow rate would hence be provided to generate a power of 50 kW via the fuel cell for a five-passenger car (Brown, 2001). Heated air was used as the sweep gas. The concentrations of hydrogen and carbon dioxide in the inlet air were set as 0.5 ppm and 370 ppm, respectively.

28.3.2.2 Modeling Results In the modeling work by Huang et al. (2005a), a reference case was chosen with a CO_2/H_2 selectivity of 40, a CO_2 permeability of 4000 Barrers, an inlet sweep-to-feed molar flow rate ratio of 1, a membrane thickness of 5 μm, and a total of 52,500 hollow fibers (a length of 61 cm, an inner diameter of 0.1 cm, and a porous support with a porosity of 50% and a thickness of 30 μm). Both the feed inlet and the sweep inlet temperatures were kept at 140°C, and the feed and the sweep pressures were 3 and 1 atm, respectively. Figure 28.15 shows the profiles of the CO and CO_2 mole fractions on the feed side along the length of the membrane reactor. The modeling results demonstrated that this membrane reactor could decrease CO concentration from 1% to 9.82 ppm along with the removal of almost all the CO_2.

For the autothermal reforming syngas, the membrane reactor could enhance H_2 concentration from 45.30 to 54.95% on the dry basis, that is, from 41 to 49.32% on the wet basis

TABLE 28.3 Composition of Autothermal Reforming Syngas

	CO	H_2O	H_2	CO_2	N_2
Molar fraction (%)	1	9.5	41	15	33.5

Source: Huang et al. (2005a).

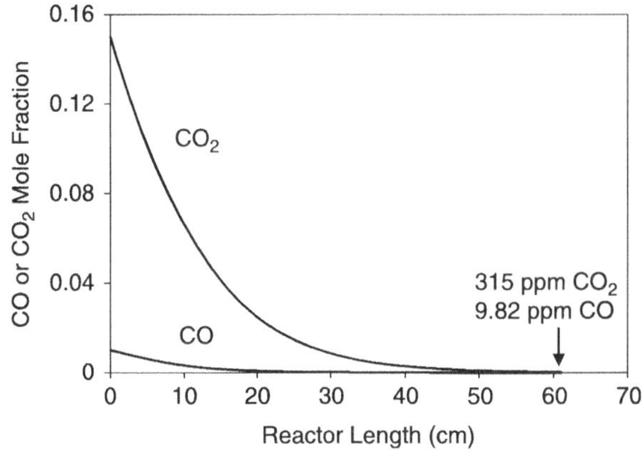

Figure 28.15 Feed-side CO and CO_2 mole fraction profiles along the length of membrane reactor for autothermal reforming syngas (Huang et al., 2005a).

(Huang et al., 2005a). The H_2 recovery calculated from the model was 97.38%. With the advancement of the high-temperature PEMFC (120–160°C), it is expected that the constraint of CO concentration can be relaxed to about 50 ppm in the near future. Then, the required hollow-fiber number could be reduced significantly to 39,000 based on the modeling results.

The temperature profiles for both feed and sweep sides are shown in Figure 28.16 (Huang et al., 2005a). Since the overall module was adiabatic, the feed gas was heated by the exothermic WGS reaction. The highest feed side temperature was 158°C at about $z = 15$ cm. Beyond that, the feed side temperature decreased, and it became very close to the sweep side temperature at the end of membrane reactor. This was due to the efficient heat transfer provided by the hollow-fiber configuration. Higher temperatures enhance WGS reaction rates but reduce CO conversion. Thus, it is important to use air at an appropriate temperature, that is, 140°C, as the sweep gas to keep the feed gas within 150 ± 10°C.

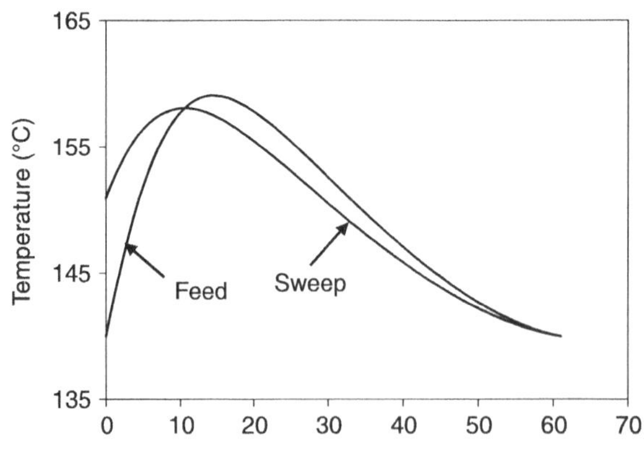

Figure 28.16 Feed-side and sweep-side temperature profiles along the length of membrane reactor for autothermal reforming syngas (Huang et al., 2005a).

With respect to this case, the effects of CO_2/H_2 selectivity, CO_2 permeability, sweep-to-feed ratio, inlet feed temperature, inlet sweep temperature, feed pressure, and catalyst activity on the reactor behavior were then investigated (Huang et al., 2005a). As the CO_2/H_2 selectivity increased, the recovery of H_2 increased without affecting the membrane area requirement and the low CO attainment significantly. Higher membrane permeability resulted in the reduction of the required membrane area. Increasing sweep-to-feed ratio enhanced the permeation driving force but decreased the feed-side temperature and thus the reaction rate, resulting in a net effect balanced between them and an optimal ratio of about 1. As either of the inlet feed and sweep temperatures increased, the required membrane area decreased. However, the temperatures greater than about 170°C would be unfavorable to the exothermic and reversible WGS reaction. Increasing feed pressure decreased the required membrane area significantly, particularly from 2 to 4 atm. Increasing catalyst activity enhanced WGS reaction and CO_2 permeation.

28.3.2.3 Water Gas Shift Membrane Reactor Experiments Ho and his students set up a rectangular flat-sheet membrane reactor using their CO_2-selective membrane and the commercial $Cu/ZnO/Al_2O_3$ catalyst to conduct WGS membrane reactor experiments and to demonstrate the WGS membrane reactor concept (Huang et al., 2005b; Zou et al., 2007). This reactor had well-defined gas flow and velocity for both the feed and sweep sides, and it was suitable for modeling and scaleup work. It had a width of about 7 inches (17.8 cm) and a length of about 7.5 inches (19.1 cm), with an effective membrane area of about 340 cm^2. The gas mixture of 1% CO, 45% H_2, 17% CO_2, and 37% N_2 was used as the feed gas, and argon was used as the sweep gas for the ease of gas chromatography (GC) analysis. The rationales for this CO level are twofold: (1) It can be readily produced from commercial WGS reactors, and (2) it requires CO_2 removal for its reduction via WGS reaction.

By varying the feed flow rate, the performance of the membrane reactor was investigated. Figure 28.17 shows the results obtained from this membrane reactor for the syngas at 150°C and 2 atm. As shown in this figure, the CO concentration in the exit

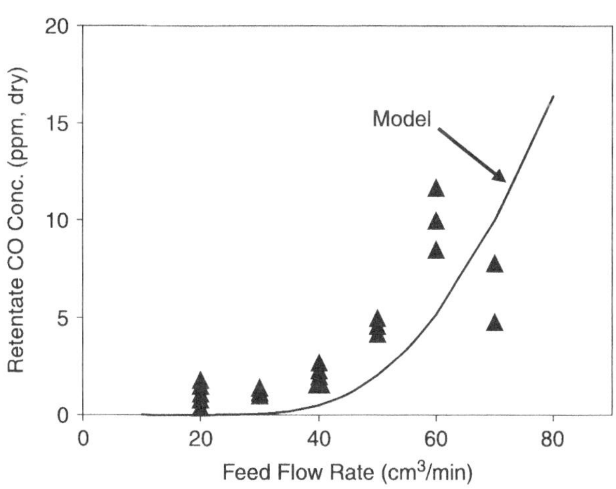

Figure 28.17 Exit dry CO concentration vs. feed flow rate in the rectangular WGS membrane reactor. [Feed gas: 1% CO, 17% CO_2, 45% H_2, 37% N_2, $T = 150°C$, $p_f = 2.0$ atm, $ps = 1.0$ atm, feed/sweep flow rates $= 1/1$ (dry basis)] (Zou et al., 2007).

stream, that is, the H_2 product, was <10 ppm (on the dry basis) for the various feed rates of the syngas at 20, 30, 40, 50, 60, and 70 cm^3/min (with ~20% steam in the syngas). The nonisothermal model was modified for the flat-sheet membrane configuration of the reactor. As shown in this figure, the data agreed reasonably with the prediction by the model.

28.3.3 CO_2 Capture from Flue Gas/Synthesis Gas

28.3.3.1 CO_2 Capture Experiments The experiments of CO_2 capture from a flue gas or synthesis gas were conducted in the gas permeation apparatus described earlier (Tee et al., 2006; Zou and Ho, 2006). The gas used contained 20% CO_2, 40% N_2, and 40% H_2. Steam was applied as the sweep gas since water condenses at ambient condition and a high-purity CO_2 can be obtained readily from the permeate side. The gas permeation was performed at 110°C to achieve the best membrane separation performance. Feed-side and sweep-side pressures were set at 2 and 1 atm, respectively. A circular gas permeation cell with a membrane area of 45.6 cm^2 was used. Both the retentate and permeate streams leaving the gas permeation apparatus were cooled down to ambient temperature in their respective water knockout vessels, which removed the water condensed.

28.3.3.2 CO_2 Capture Results Although flue gases from different sources might have various compositions, N_2, CO_2, and water vapor are generally the major components accounting for over 90% of the gas stream. After mixing the dry feed gas with the water vapor in the vessel in the experimental setup, the total feed gas composition used was about 12% CO_2, 24% H_2, 24% N_2, and 40% H_2O, which would provide a close estimation to the CO_2 capture from a flue gas or synthesis gas.

As shown in Figure 28.18, a CO_2 concentration of greater than 98% in the permeate was achieved for a feed flow rate as high as 100 cm^3/min (on dry basis, 20°C and 1 atm). Furthermore, if we deducted the H_2 portion from the permeate side, the dry CO_2 concentration would be greater than 99%. The high CO_2/N_2 selectivity of the membrane clearly accounted for this high CO_2 purity on the permeate side. Slightly increased permeate dry CO_2 concentration was observed with the increase of the feed flow rate. This might be

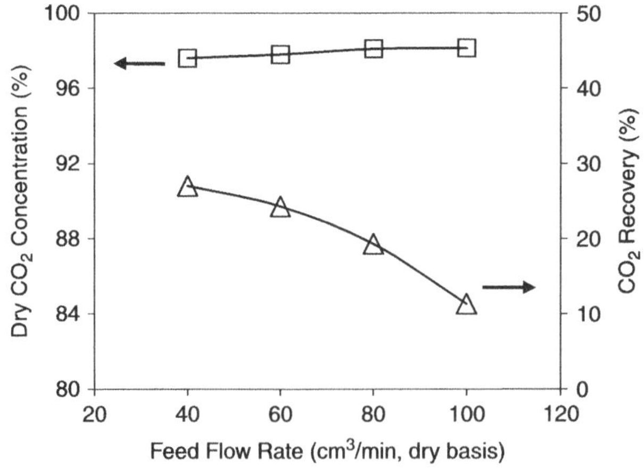

Figure 28.18 Permeate dry CO_2 concentration and CO_2 recovery vs. feed flow rate. (□) Permeate dry CO_2 concentration; (△) CO_2 recovery. Feed-to-sweep molar ratio = 1 (web basis).

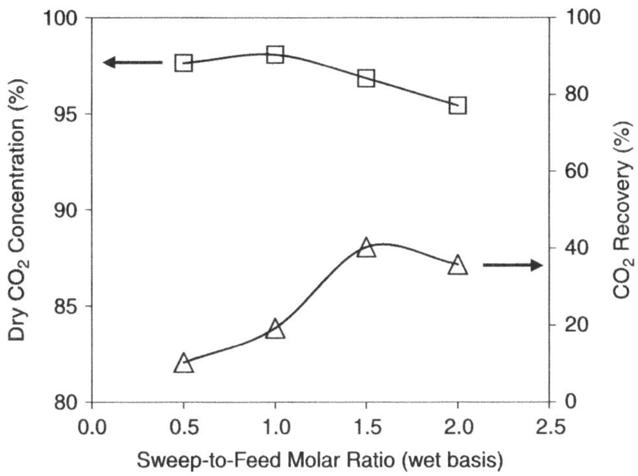

Figure 28.19 Permeate dry CO_2 concentration and CO_2 recovery vs. sweep-to-feed molar ratio. (□) Permeate dry CO_2 concentration; (△) CO_2 recovery. Feed flow rate = 80 cm^3/min (dry basis, 20°C and 1 atm).

due to the case that a higher feed flow rate brought more steam onto the membrane, which could enhance the facilitated transport for the reasons discussed earlier. The CO_2 recovery dropped as the feed flow rate increased, which could be explained by the fact that the residence time of the gas mixture in the permeation cell reduced as the feed flow rate increased.

Figure 28.19 illustrates the effect of the sweep-to-feed ratio on permeate dry CO_2 concentration and CO_2 recovery. As illustrated in this figure, the permeate CO_2 concentrations (on the dry basis) were all above 95% and did not change significantly with the increase of the sweep-to-feed ratio. However, the CO_2 recovery increased significantly as the sweep-to-feed ratio increased from 0.5 to 1.5, which could be explained by the fact that a higher sweep-to-feed ratio resulted in a lower CO_2 concentration on the sweep side and then a higher driving force for CO_2 permeation.

28.4 CONCLUSIONS

We have reviewed the recent developments of two types of facilitated transport membranes: (1) supported liquid membranes (SLMs) with strip dispersion and (2) carbon-dioxide-selective polymeric membranes, for environmental, energy, and biochemical applications. SLMs with strip dispersion have been developed for removal and recovery of metals, including chromium, copper, zinc, cobalt, and strontium, from wastewaters and process streams. The SLM system for chromium not only has removed the hexavalent chromium from about 100–1000 ppm to less than 0.05 ppm in the treated effluent allowable for discharge or recycle but also has recovered the chromium product at a high concentration of about 20% Cr(VI) (62.3% Na_2CrO_4) suitable for resale or reuse. In other words, the goals of zero discharge and no sludge have been achieved.

The copper SLM system not only has removed the copper from 150 ppm in the inlet feed to less than 0.15 ppm in the treated feed but also recovered the copper at a high concentration of greater than 10,000 ppm in the strip solution. For the zinc SLM system, the zinc at the inlet feed concentration of 550 ppm has been removed to less than 0.3 ppm in the treated feed while a high zinc concentration of more than 17,000 ppm has been recovered in the strip solution. The cobalt SLM system has removed it from about 525 to 0.7 ppm in the treated feed solution while concentrating it to at least 30,000 ppm in the aqueous strip solution. For strontium removal, the SLM has removed radioactive Sr-90 to the target of 8 pCi/L or lower from feed solutions of 300–1000 pCi/L. The stability of the SLM has been ensured by a modified SLM with strip dispersion, where the aqueous strip solution is dispersed in the organic membrane solution in a mixer. The strip dispersion formed is circulated from the mixer to the membrane module to provide a constant supply of the organic solution to the membrane pores.

Supported liquid membranes with strip dispersion have great potential for removal and recovery of antibiotics and other bioproducts from aqueous solutions and fermentation broths. On penicillin G recovery, the SLM removed it from a feed of 8840 ppm and concentrated it to a high concentration of 41,011 ppm in the aqueous strip solution with a high recovery of about 93%.

CO_2-selective membranes containing both mobile and fixed carriers in cross-linked poly(vinyl alcohol) have been synthesized. The membranes have shown high CO_2/H_2, CO_2/N_2, and CO_2/CO selectivities and high CO_2 permeability up to 170°C. The CO_2 permeability and CO_2/H_2 selectivity decreased with increasing feed pressure, which could be explained with the carrier saturation phenomenon. Both the permeability and selectivity increased significantly with increasing water contents in both feed and sweep.

A one-dimensional nonisothermal model has been developed to predict the performance of the novel CO_2-selective WGS membrane reactor. The modeling results have shown that H_2 enhancement and CO reduction to <10 ppm via CO_2 removal are achievable. Results from WGS membrane reactor experiments have shown carbon monoxide reduction to 10 ppm as well as significant hydrogen enhancement via CO_2 removal. The data have been in good agreement with modeling prediction.

On CO_2 capture experiments from the gas mixture with N_2 and H_2, the permeate CO_2 dry concentration of greater than 98% was obtained by using steam as the sweep gas. The permeate CO_2 concentration did not change significantly with the increase of the sweep-to-feed ratio. However, the CO_2 recovery increased significantly as the sweep-to-feed ratio increased from 0.5 to 1.5.

Nomenclature

C_A	CO_2 concentration (mol/cm^3)
C_{AB}	CO_2–carrier reaction product concentration (mol/cm^3)
D_A	diffusivity coefficient for CO_2 (cm^2/s)
D_{AB}	diffusivity coefficient for CO_2–carrier reaction product (cm^2/s)
H_{AB}	Henry's law constant for CO_2–carrier reaction product (mol/cm^3/atm)
ΔH	heat of reaction (J/mol)
l	membrane thickness (cm)
N_i	steady-state flux of component i (mol/cm^2/s)
n	molar flow rate (mol/s)
P_i	permeability of component i (Barrer)
p	pressure (atm)
p_1	CO_2 partial pressure on the high-pressure side of membrane (atm)
p_2	CO_2 partial pressure on the low-pressure side of membrane (atm)

p_{1c}	critical CO_2 partial pressure at which carrier saturation occurs (atm)
p_{1m}	CO_2 partial pressure in the membrane on the high-pressure side of membrane (atm)
p_{2m}	CO_2 partial pressure in the membrane on the low-pressure side of membrane (atm)
Δp_i	partial pressure difference of component i
x_i	mole fractions of component i in the retentate stream
y_i	mole fractions of component i in the permeate stream

Greek letter

α_{ij}	selectivity of component i over component j

Subscripts

0	initial
f	feed side
s	sweep
t	total

ACKNOWLEDGMENT

We would like to thank Debbie de la Cruz and GE Infrastructure and Chris Plotz and BHA Technologies for giving us the GE E500A microporous polysulfone support and the BHA microporous Teflon support, respectively. We would also like to thank the National Science Foundation, the Office of Naval Research, and The Ohio State University for the financial support. Part of this material is based upon work supported by the National Science Foundation under Grant No. 0625758.

REFERENCES

Aaron, D., and Tsouris, C. (2005). Separation of CO_2 from flue gas: A review. *Sep. Sci. Technol.* 40, 321.

Ahmed, S., and Krumpelt, M. (2001). Hydrogen from hydrocarbon fuels for fuel cells. *Int. J. Hydrogen Energy* 26, 291.

Alguacil, F. J. (2002). Facilitated transport and separation of manganese and cobalt by a supported liquid membrane using DP-8R as a mobile carrier. *Hydrometallurgy* 65, 9.

Armor, J. N. (1998). Applications of catalytic inorganic membrane reactors to refinery products. *J. Membr. Sci.* 147, 217.

Basile, A., Criscuoli, A., Santella, F., and Drioli, E. (1996). Membrane reactor for water gas shift reaction. *Gas Sep. Purif.* 10, 243.

Brown, L. F. (2001). A comparative study of fuels for on-board hydrogen production for fuel-cell-powered automobiles. *Int. J. Hydrogen Energy* 26, 381.

Bryjak, M., Kozlowski, J., Wieczorek, P., and Kafarski, P. (1993). Enantioselective transport of amino acid through supported chiral liquid membranes. *J. Membr. Sci.* 85, 221.

Buchalter, E. M., Hofman, D. L., Craig, W. M., Birkill, R. S., and Smit, J. J. (1987). Supported liquid membrane technology applied to the recovery of useful isotopes from reactor pool water. *Chem. Eng. Res. Des.* 65, 381.

Campbell, M. J., Walter, R. P., Singleton, R., and Knowles, C. J. (1994). Investigation of the stability and selectivity of phenylalanine transport across a supported liquid membrane. *J. Chem. Technol. Biotechnol.* 60, 263.

Chaudry, M. A., Malik, M. T., and Ali, A. (1990). Transport of cobalt (II) ions through di(2-ethylhexyl) phosphoric acid–carbon tetrachloride supported liquid membranes. *Sep. Sci. Technol.* 25, 1161.

Criscuoli, A., Basile, A., and Drioli, E. (2000). An analysis of the performance of membrane reactors for the water-gas shift reaction using gas feed mixtures. *Catal. Today* 56, 53.

Cussler, E. L., Aris, R., and Bhown, A. (1989). On the limit of facilitated diffusion. *J. Membr. Sci.* 43, 149.

Danesi, P. R. (1984). A simplified model for the coupled transport of metal ions through hollow-fiber supported liquid membranes. *J. Membr. Sci.* 20, 231.

Danesi, P. R., Chiarizia, R., and Castagnola, A. (1983). Transfer rate and separation of cadmium (II) and zinc (II) chloride species by a trilaurylammonium chloride–triethylbenzene supported liquid membrane. *J. Membr. Sci.* 14, 161.

Danesi, P. R., Reichley-Yinger, L., Cianetti, C., and Rickert, P. G. (1984). Separation of cobalt and nickel by liquid-liquid extraction and supported liquid membranes with bis(2,4,4-trimethylpentyl)-phosphinic acid [Cyanex 272]. *Solvent Extr. Ion Exch.* 2, 781.

Donaldson, T. L., and Lapinas, A. T. (1982). Secondary flux enhancement in two-carrier facilitated transport. *Chem. Eng. Sci.* 37, 715.

Dozol, J. F., Casas, J., and Sastre, A. M. (1994). Influence of the extractant on strontium transport from reprocessing concentrate solutions through flat-sheet supported liquid membranes. *Sep. Sci. Technol.* 29, 1999.

Dreher, T. M., and Stevens, G. W. (1998). Instability mechanisms of supported liquid membranes. *Sep. Sci. Technol.* 33, 835.

Dutta, N. N., and Sahoo, G. C. (2002). Perspectives in liquid membrane extraction of Cephalosporin antibotics. *Adv. Biochem. Eng. Biotechnol.* 75, 209.

Dzygiel, P., Wieczorek, P., Jonsson, J. A., Milewska, M., and Kafarski, P. (1999). Separation of amino acid enantiomers using supported liquid membrane extraction with chiral phosphates and phosphonates. *Tetrahedron* 55, 9923.

Friesen, D. T., Babcock, W. C., Brose, D. J., and Chambers, A. R. (1991). Recovery of citric acid from fermentation beer using supported-liquid membranes. *J. Membr. Sci.* 56, 127.

Gega, J., Walkowiak, W., and Gajda, B. (2001). Separation of Co (II) and Ni (II) ions by supported and hybrid liquid membranes. *Sep. Purif. Technol.* 22/23, 551.

Gherrou, A., Kerdjoudj, H., Molinari, R., and Drioli, E. (2002). Removal of silver and copper ions from acidic thiourea solutions with a supported liquid membrane containing D2EHPA as carrier. *Sep. Sci. Technol.* 28, 235.

Giessler, S., Jordan, L., Diniz da Costa, J. C., and Lu, G. Q. (2003). Performance of hydrophobic and hydrophilic silica membrane reactors for the water gas shift reaction. *Sep. Purif. Technol.* 32, 255.

Gottschlich, D. E., Roberts, D. L., and Way, J. D. (1988). A theoretical comparison of facilitated transport and solution-diffusion membrane modules for gas separation. *Gas Sep. Purif.* 2, 65.

Hill, C., Dozol, J.-F., Rouquette, H., Eymard, S., and Tournois, B. (1996). Study of the stability of some supported liquid membranes. *J. Membr. Sci.* 114, 73.

Hirayama, Y., Kase, Y., Tanihara, N., Sumiyama, Y., Kusuki, Y., and Haraya, K. (1999). Permeation properties to CO_2 and N_2 of poly(ethylene oxide)-containing and crosslinked polymer films. *J. Membr. Sci.* 160, 87.

Ho, W. S. W. (1997). Membranes comprising salts of aminoacids in hydrophilic polymers. U.S. Patent 5,611,843.

Ho, W. S. W. (2000). Membranes comprising aminoacid salts in polyamine polymers and blends. U.S. Patent 6,099,621.

Ho, W. S. W. (2001a). Supported liquid membrane process for chromium removal and recovery. U.S. Patent 6,171,563.

Ho, W. S. W. (2001b). Combined supported liquid membrane/strip dispersion process for the removal and recovery of radionuclides and metals. U.S. Patent 6,328,782.

Ho, W. S. W. (2002a). Combined supported liquid membrane/strip dispersion process for the removal and recovery of metals. U.S. Patent 6,350,419.

Ho, W. S. W., (2002b). Combined supported liquid membrane/strip dispersion process for the removal and recovery of penicillin and organic acids. U.S. Patent 6,433,163.

Ho, W. S. W. (2003). Removal and recovery of metals and other materials by supported liquid membranes with strip dispersion. In N. N. Li, E. Drioli, W. S. W. Ho, and G. G. Lipscomb (Eds.), *Advanced Membrane Technology*. New York Academy of Sciences, New York, p. 97.

Ho, W. S. W., and Dalrymple, D. C. (1994). Facilitated transport of olefins in Ag^+-containing polymer membranes. *J. Membr. Sci.* 91, 13.

Ho, W. S. W., and Li, N. N. (1992). Definitions and theory for emulsion liquid membranes. In W. S. W. S. W. Ho and K. K. Sirkar (Eds.), *Membrane Handbook*. Chapman & Hall, New York, p. 595.

Ho, W. S. W., and Poddar, T. K. (2001). New membrane technology for removal and recovery of chromium from waste waters. *Environ. Prog.* 20, 44.

Ho, W. S. W., Poddar, T. K., and Neumuller, T. E. (2002). Removal and recovery of copper and zinc by supported liquid membranes with strip dispersion. *J. Chin. Inst. Chem. Eng.* 33, 67.

Ho, W. S. W., and Sirkar, K. K. (Eds.) (1992). *Membrane Handbook*. Chapman & Hall, New York.

Ho, W. S. W., and Wang, B. (2002). Strontium removal by new alkyl phenylphosphonic acids in supported liquid membranes with strip dispersion. *Ind. Eng. Chem. Res.* 41, 381.

Ho, W. S. W., Wang, B., Neumuller, T. E., and Roller, J. (2001). Supported liquid membranes for removal and recovery of metals from waste waters and process streams. *Environ. Prog.* 20, 117.

Hong, J., Kang, Y. S., Jang, J., and Kim, U. Y. (1996). Analysis of facilitated transport in polymeric membrane with fixed site carrier 2. Series RC circuit model. *J. Membr. Sci.* 109, 159.

Huang, J., El-Azzami, L., and Ho, W. S. W. (2005a). Modeling of CO_2-selective water-gas-shift membrane reactor for fuel cell. *J. Membr. Sci.* 261, 67.

Huang, J., Zou, J., and Ho, W. S. W. (2005b). A modeling and experimental study of CO_2-selective water-gas-shift membrane reactor for fuel cells. Paper 501b. In *Proceedings of Topical Conference on Fuel Cells Technology at AIChE Annual Meeting*, Cincinnati, OH, Oct. 30–Nov. 4, 2005.

Huang, T. C., Huang, C. C., and Chen, D. H. (1998). Transport of chromium (VI) through a supported liquid membrane containing a tri-*n*-octylphosphine oxide. *Sep. Sci. Technol.* 33, 1919.

Juang, R.-S., Huang, R.-H., and Wu, R.-T. (1997). Separation of citric and lactic acids in aqueous solutions by solvent extraction and liquid membrane processes. *J. Membr. Sci.* 136, 89.

Juang, R.-S., Lee, S.-H., and Shiau, R.-C. (1998). Carrier-facilitated liquid membrane extraction of penicillin G from aqueous streams. *J. Membr. Sci.* 146, 95.

Kang, Y. S., Hong, J., Jang, J., and Kim, U. Y. (1996). Analysis of facilitated transport in solid membranes with fixed site carriers 1. Single RC circuit model. *J. Membr. Sci.* 109, 149.

Keiski, R. L., Desponds, O., Chang, Y. F., and Somorjai, G. A. (1993). Kinetics of the water-gas-shift reaction over several alkane activation and water-gas-shift catalysts. *Appl. Catal. A* 101, 317.

Kemperman, A. J. B., Bargeman, D., van den Boomgaard, Th., and Strathmann, H. (1996). Stability of supported liquid membranes: State of the art. *Sep. Sci. Technol.* 31, 2733.

Kim, T., Li, B., and Hagg, M. (2004). Novel fixed-site-carrier polyvinylamine membrane for carbon dioxide capture. *J. Polym. Sci. Part B: Polym. Phys.* 42, 4326.

Koros, W. J., and Mahajan, R. (2000). Pushing the limits on possibilities for large scale gas separation: Which strategies. *J. Membr. Sci.* 175, 181.

Kroschwitz, J. I., and Howe-Grant, M. (Eds.) (1995). *Encyclopedia of Chemical Technology*, 4th ed., Vol. 13, Wiley, New York, p. 927.

Langevin, D., Pinoche, M., Selegny, E., Metayer, M., and Roux, R. (1993). CO_2 facilitated transport through functionalized cation-exchange membranes. *J. Membr. Sci.* 82, 51.

Largman, T., and Sifniades, S. (1978). Recovery of copper (II) from aqueous solutions by means of supported liquid membranes. *Hydrometallurgy* 3, 153.

LeBlanc, O. H., Ward, W. J., Matson, S. L., and Kimura, S. G. (1980). Facilitated transport in ion-exchange membranes. *J. Membr. Sci.* 6, 339.

Ledjeff-Hey, K., Roes, J., and Wolters, R. (2000). CO_2-scrubbing and methanation as purification system for PEFC. *J. Power Sources* 86, 556.

Lee, C. J., Yeh, H. J., Yang, W. J., and Kan, C. R. (1994). Separation of penicillin G from phenyl-acetic acid in a supported liquid membrane system. *Biotechnol. Bioeng.* 43, 309.

Li, N. N. (1968). Separating hydrocarbons with liquid membranes. U.S. Patent 3,410,794.

Li, N. N. (1978). Facilitated transport through liquid membranes—An extended abstract. *J. Membr. Sci.* 3, 265.

Lin, H., and Freeman, B. D. (2005). Materials selection guidelines for membranes that remove CO_2 from gas mixture. *J. Mol. Struct.* 739, 57.

Mackova, J., and Mikulaj, V. (1996). Transport of strontium cation through a hollow fiber supported dichlorobenzene membrane using 18-C-6 crown ether. Nitrate and anion of dinonylnaphthalene sulfonic acid. *J. Radioanal. Nucl. Chem.* 208, 111.

Marchese, J., Lopez, J. L., and Quinn, J. A. (1989). Facilitated transport of benzylpenicillin through immobilized liquid membrane. *J. Chem. Technol. Biotechnol.* 46, 149.

Matsuyama, H., Terada, A., Nakagawara, T., Kitamura, Y., and Teramoto, M. (1999). Facilitated transport of CO_2 through polyethylenimine/poly(vinyl alcohol) blend membrane. *J. Membr. Sci.* 163, 221.

Matsuyama, H., Teramoto, M., and Iwai, K. (1994). Development of a new functional cation-exchange membrane and its application to facilitated transport of CO_2. *J. Membr. Sci.* 93, 237.

Matsuyama, H., Teramoto, M., Matsui, K., and Kitamura, Y. (2001). Preparation of poly(acrylic acid)/poly(vinyl alcohol) membrane for the facilitated transport of CO_2. *J. Appl. Polym. Sci.* 81, 936.

Matsuyama, H., Teramoto, M., Sakakura, H., and Iwai, K. (1996). Facilitated transport of CO_2 through various ion exchange membranes prepared by plasma graft polymerization. *J. Membr. Sci.* 117, 251.

Matuleviclus, E. S., and Li, N. N. (1975). Facilitated transport through liquid membranes. *Sep. Purif. Methods* 4, 73.

Meisen, A., and Shuai, X. (1997). Research and development issues in CO_2 capture. *Energy Convers. Manag.* 38(Suppl.), S37.

Meldon, J. H., Smith, K. A. and Colton, C. K. (1977). The effect of weak acids upon the transport of carbon dioxide in alkaline solutions. *Chem. Eng. Sci.* 32, 939.

Moe, J. M. (1962). Design of water-gas-shift reactors. *Chem. Eng. Prog.* 58, 33.

Molinari, R., Drioli, E., and Pantano, G. (1989). Stability and effect of diluents in supported liquid membranes for chromium (III), chromium (VI), and cadmium (II) recovery. *Sep. Sci. Technol.* 24, 1015.

Noble, R. D. (1990). Analysis of facilitated transport with fixed site carrier membranes. *J. Membr. Sci.* 50, 207.

Noble, R. D. (1991). Facilitated transport mechanism in fixed site carrier membranes. *J. Membr. Sci.* 60, 297.

Noble, R. D. (1992). Generalized microscopic mechanism of facilitated transport in fixed site carrier membranes. *J. Membr. Sci.* 75, 121.

Nuchnoi, P., Yano, T., Nishio, N., and Nagai, S. (1987). Extraction of volatile fatty acids from diluted aqueous solution using a supported liquid membrane. *J. Ferment. Technol.* 65, 301.

Otto, N. C., and Quinn, J. A. (1971). Facilitated transport of carbon dioxide through bicarbonate solutions. *Chem. Eng. Sci.* 26, 949.

Quinn, R., Appleby, J. B., and Pez, G. P. (1995). New facilitated transport membranes for the separation of carbon dioxide from hydrogen and methane. *J. Membr. Sci.* 104, 139.

Quinn, R., and Laciak, D. V. (1997). Polyelectrolyte membranes for acid gas separations. *J. Membr. Sci.* 131, 49.

Quinn, R., Laciak, D. V., and Pez, G. P. (1997). Polyelectrolyte-salt blend membranes for acid gas separations. *J. Membr. Sci.* 131, 61.

Ramadan, A., and Danesi, P. R. (1988). Transfer rate and separation of strontium (2+) and cesium (1+) by supported liquid membranes utilizing synergized crown ether carriers. *Solvent Extr. Ion Exch.* 6, 157.

Rockman, J. T., Kehat, E., and Lavie, R. (1997). Thermally enhanced extraction of citric acid through supported liquid membrane. *AIChE J.* 43, 2376.

Sartori, G., Ho, W. S. W., Savage, D. W., Chludzinski, G. R., and Wiechert, S. (1987). Sterically-hindered amines for acid-gas absorption. *Sep. Purif. Methods* 16, 171.

Sartori, G., and Savage, D. W. (1983). Sterically hindered amines for CO_2 removal from gases. *Ind. Eng. Chem. Fund.* 22, 239.

Scholander, P. F. (1960). Oxygen transport through hemoglobin solutions. *Science* 131, 585.

Sheldon, R. (2004). Biocatalysis for sustainable organic synthesis. *Austral. J. Chem.* 57, 281.

Shen, Y., Groenberg, L., and Joensson, J. A. (1994). Experimental studies on the enrichment of carboxylic acids with tri-*n*-octylphosphine oxide as extractant in a supported liquid membrane. *Anal. Chim. Acta* 292, 31.

Shulik, L. J., Sartori, G., Ho, W. S. W., Thaler, W. A., Milliman, G. E.., and Wilbur, J. C. (1996). A novel, V^{+5}-stable K_2CO_3 promoter for CO_2 absorption. *Sep. Sci. Technol.* 31, 1663.

Sirman, T., Pyle, L., and Grandison, A. S. (1991). Extraction of organic acids using a supported liquid membrane. *Biochem. Soc. Trans.* 19, 274S.

Smith, D. R., and Quinn, J. A. (1979). The prediction of facilitation factors for reaction augmented membrane transport. *AIChE J.* 25, 197.

Song, C. (2002). Fuel processing for low-temperature and high-temperature fuel cells: Challenges, and opportunities for sustainable development in the 21st century. *Catal. Today* 77, 17.

Tanigaki, M., Shiode, T., Okumi, S., and Eguchi, W. (1988). Facilitated transport of zinc chloride through hollow fiber supported liquid membrane. Part 3. Module operation. *Sep. Sci. Technol.* 23, 1171.

Tee, Y. H., Zou, J., and Ho, W. S. W. (2006). CO_2-selective membranes containing dimethylglycine mobile carriers and polyethylenimine fixed carrier. *J. Chin. Inst. Chem. Eng.* 37, 37.

Teramoto, M., Kitada, S., Ohnishi, N., Matsuyama, H. and Matsumiya, N. (2004). Separation and concentration of CO_2 by capillary-type facilitated transport membrane module with permeation of carrier solution. *J. Membr. Sci.* 234, 83.

Teramoto, M., Sakaida, Y., Fu, S. S., Ohnishia, N., Matsuyama, H., Makia, T., Fukuib, T., and Arai, K. (2000). An attempt for the stabilization of supported liquid membrane. *Sep. Purif. Technol.* 21, 137.

Teramoto, M., Takeuchi, N., Maki, T., and Matsuyama, H. (2001). Gas separation by liquid membrane accompanied by permeation of membrane liquid through membrane physical transport. *Sep. Purif. Technol.* 24, 101.

Teramoto, M., Takeuchi, N., Maki, T., and Matsuyama, H. (2002). Facilitated transport of CO_2 through liquid membrane accompanied by permeation of carrier solution. *Sep. Purif. Technol.* 27, 25.

Tosti, S., Basile, A., Chiappetta, G., Rizzello, C., and Violante, V. (2003). Pd-Ag membrane reactors for water gas shift reaction. *Chem. Eng. J.* 93, 23.

Tsikas, D., Kaltsidou-Schottelius, E., and Brunner, G. (1992). Hollow fiber-supported liquid membranes for the extraction of penicillins and the synthesis of 6-aminopenicillanic acid. *Chem. Ing. Tech.* 64, 545.

Uemiya, S., Sato, N., Ando, H., and Kikuchi, E. (1991). The water gas shift reaction assisted by a palladium membrane reactor. *Ind. Eng. Chem. Res.* 30, 585.

U.S. Department of Energy (DOE) (2005). Emissions of greenhouse gases in the United States 2004. Available: ftp://ftp.eia.doe.gov/pub/oiaf/1605/cdrom/pdf/ggrpt/057304.pdf, accessed Mar. 2006.

Ward, W. J. (1970). Analytical and experimental studies of facilitated transport. *AIChE J.* 16, 405.

Ward, W. J., and Robb, W. L. (1967). Carbon dioxide-oxygen separation: Facilitated transport of carbon dioxide across a liquid film. *Science* 156, 1481.

Way, J. D., and Noble, R. D. (1992). Facilitated transport. In W. S. W. Ho, and K. K. Sirkar (Eds.), *Membrane Handbook*. Chapman & Hall, New York, p. 834.

Way, J. D., Noble, R. D., Reed, D. L., Ginley, G. M., and Jarr, L. A. (1987). Facilitated transport of CO_2 in ion exchange membranes. *AIChE J.* 33, 480.

Wind, J. D., Paul, D. R., and Koros, W. J. (2004). Natural gas permeation in polyimide membranes. *J. Membr. Sci.* 228, 227.

Xue, E., O'Keeffe, M., and Ross, J. R. H. (1996). Water-gas shift conversion using a feed with a low steam to carbon monoxide ratio and containing sulphur. *Catal. Today* 30, 107.

Yahaya, G. O. (2001). Kinetic studies on organic acid extraction by a supported liquid membrane using functionalized polyorganosiloxanes as mobile and fixed-site carriers. *Sep. Sci. Technol.* 36, 3563.

Yamaguchi, T., Boetje, L. M., Koval, C. A., Noble, R. D., and Brown, C. N. (1995). Transport properties of carbon dioxide through amine functionalized carrier membranes. *Ind. Eng. Chem. Res.* 34, 4071.

Yamaguchi, T., Koval, C. A., Noble, R. D., and Bowman, C. N. (1996). Transport mechanism of carbon dioxide through perfluorosulfonate ionomer membranes containing an amine carrier. *Chem. Eng. Sci.* 51, 4781.

Yang, X. J., and Fane, A. G. (1999). Performance and stability of supported liquid membranes using LIX 984 N for copper transport. *J. Membr. Sci.* 156, 251.

Yang, X. J., Fane, A. G., and MacNaughton, S. (2001). Removal and recovery of heavy metals from wastewaters by supported liquid membranes. *Water Sci. Technol.* 43, 341.

Zolandz, R. R., and Fleming, G. K. (1992). Gas permeation. In W. S. W. Ho, and K. K. Sirkar (Eds.), *Membrane Handbook*. Chapman & Hall, New York, p. 27.

Zou, J., and Ho, W. S. W. (2006). CO_2-selective polymeric membranes containing amines in crosslinked poly(vinyl alcohol). *J. Membr. Sci.* 286, 310.

Zou, J., Huang, J., and Ho, W. S. W. (2007). CO_2-selective water gas shift membrane reactor for fuel cell hydrogen processing. *Ind. Eng. Chem. Res.* 46, 2272.

CHAPTER 29

Fuel Cell Membranes

PETER N. PINTAURO and RYSZARD WYCISK

Department of Chemical Engineering, Case Western Reserve University, Cleveland, Ohio 44106-7217

29.1 INTRODUCTION TO FUEL CELLS

Fuel cells are electrochemical devices that convert the energy liberated in a chemical reaction directly into electricity. The basic building blocks of a fuel cell are an ionically conducting electrolyte that separates two electrodes, an anode where oxidation reactions occur, and a cathode for reduction reactions. During operation, fuel (e.g., H_2 gas or methanol) is fed continuously to the anode, and an oxidant (normally oxygen from air) is supplied to the cathode. Fuel oxidation and oxygen reduction reactions occur spontaneously on the electrodes with the production and consumption of electrons (at the anode and cathode, respectively) and the generation of heat (because the electrical energy conversion process is not 100% efficient). Electrical energy is extracted from the electrons as they flow from the anode to the cathode via an external circuit. A fuel cell, although having components and characteristics similar to those of a battery, differs in one primary way. The battery is an energy storage device, where the maximum energy available is determined by the amount of chemical reactant stored within the battery itself. The fuel cell, on the other hand, is an energy conversion device with separate (external) storage of fuel and oxidant. Thus, a fuel cell has the capability of producing electrical energy for as long as reactants are supplied to the electrodes.

There are a variety of different fuel cells that are normally classified according to the electrolyte type (see Table 29.1). The amount of power produced by a fuel cell depends upon several factors, such as fuel cell type, cell size, the temperature at which the cell operates, and the pressure at which the gases are supplied to the cell. A single fuel cell element (one anode and cathode) produces enough electricity for only the smallest applications. Therefore, individual fuel cells are typically combined into a stack configuration.

Two types of fuel cells that employ polymeric cation exchange membranes as the electrolyte material are the proton exchange membrane fuel cell (PEMFC) and the direct methanol fuel cell (DMFC). The membrane has a multifunctional role in these devices: (1) It physically separates the anode and cathode to prevent an electrical short circuit, (2) it separates the fuel and oxidant to eliminate a chemical short circuit, and (3) it provides

Advanced Membrane Technology and Applications. Edited by Norman N. Li, Anthony G. Fane, W. S. Winston Ho, and T. Matsuura
Copyright © 2008 John Wiley & Sons, Inc.

TABLE 29.1 Types of Fuel Cells

Fuel Cell Type	Electrolyte	Anode Fuel	Cathode Gas	Temperature	Efficiency
Molten carbonate	Alkali-carbonates	Hydrogen	Atmospheric oxygen	650°C (1200°F)	40–55%
Phosphoric acid	Phosphoric acid	Hydrogen	Atmospheric oxygen	210°C (400°F)	35–50%
H_2/O_2(air) proton exchange membrane	Solid polymer membrane	Hydrogen	Atmospheric oxygen	75°C (180°F)	35–60%
Direct methanol	Solid polymer membrane	Methanol solution in water	Atmospheric oxygen	75°C (180°F)	35–40%
Alkaline	Potassium hydroxide	Hydrogen	Pure oxygen	Below 80°C	50–70%
Solid oxide	Ceramic oxide	Hydrogen, reformed methane	Atmospheric oxygen	800–1000°C (1500–1800°F)	45–60%

a facile pathway for the flow (transport) of ionic species (protons) from the anode to the cathode. Further details on PEMFCs and DMFCs follow.

29.1.1 Proton Exchange Membrane Fuel Cell

The PEMFC, also known as the solid polymer or polymer electrolyte fuel cell, delivers high power density and offers the advantage of low weight and volume, as compared to other fuel cell types. A PEMFC uses a thin polymer membrane layer as the electrolyte, with electrodes composed of platinum on carbon powder. This type of fuel cell operates at a temperature of about 80°C and requires only hydrogen, oxygen from the air, and water to function. These fuel cells are typically fueled with pure hydrogen supplied from storage tanks or onboard reformers. A schematic diagram of a hydrogen–air PEMFC fuel cell, with the relevant electrode reactions, is shown in Figure 29.1. The key component of this type of fuel cell is a *membrane–electrode assembly* (MEA) composed of a thin (typically 50–200 μm) cation (proton) exchange membrane (which is usually hydrated to promote proton transport) with precious metal catalyst powder electrodes pressed directly onto the opposing membrane surfaces. Immediately adjacent to the electrodes are carbon paper or carbon cloth gas diffusion layers (GDLs), which assist in distributing gas to the catalyst and removing products from the electrodes (e.g., water from the air cathode). An MEA is sandwiched between two metal or graphite composite flow field plates to create a fuel cell element. These plates contain grooves to supply fuel and air to the electrodes and to remove water from the cathode. The plates themselves conduct electrons out of the assembly (see Fig. 29.2). The thermodynamic potential for each H_2–air cell (MEA) is 1.23 V, although lower voltages are obtained when current is flowing (typically about 0.5–0.7 V per cell). In order to generate a higher overall voltage and to increase power output, a number of individual cells are combined in series to form a fuel cell stack.

Proton exchange membrane fuel cells have a number of attributes that make them ideal candidates for use in automotive applications, as stationary distributed power plants (including remote off-grid applications), and in small domestic/portable applications (i.e., replacements for rechargeable batteries). They operate at a relatively low temperature, which allows them to start up rapidly from cold, and they have a high power density, which makes them relatively compact. In addition, PEMFCs work at high efficiencies, producing around 40–50% of the maximum theoretical voltage and can vary their output quickly to meet shifts in power demand. Since 50–60% of the available energy of the fuel is lost as heat, a PEMFC may require cooling for proper operation. Water management (removal of

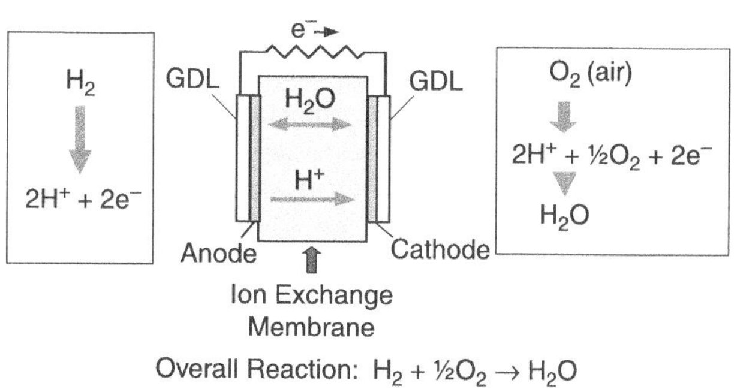

Figure 29.1 Schematic diagram of an H_2–air PEMFC.

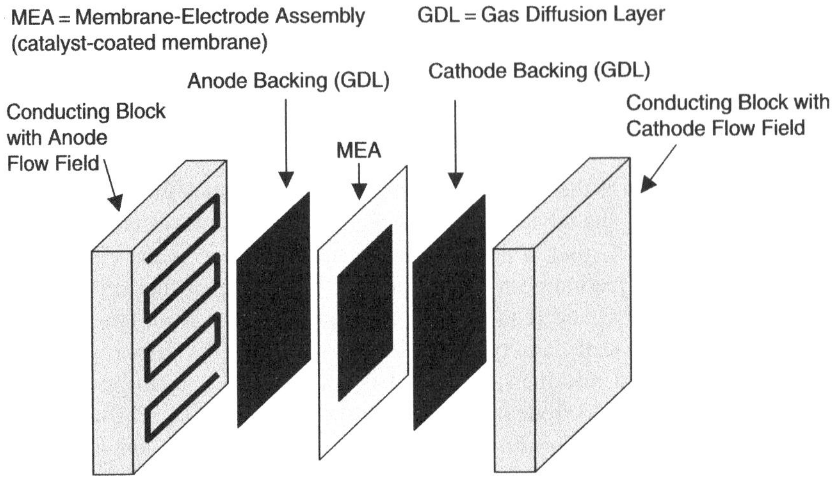

Figure 29.2 Schematic of a single MEA fuel cell test apparatus.

product water in the cathode air exhaust, humidification of the anode feed hydrogen gas, and adequate water content in the proton exchange membrane) is also critical for a PEMFC to function properly.

There are a number of barriers that need to be overcome before there is widespread commercialization of PEMFC devices. The main issue is MEA cost. The membrane and catalysts are expensive, but ongoing research and development has resulted in significant cost reductions, and economies of scale will take effect once these cells are mass produced. The other drawback of PEMFCs is that they require pure hydrogen to operate and are susceptible to poisoning by carbon monoxide and other gaseous impurities. This is largely due to the low operating temperature of the cell, which necessitates the use of highly sensitive precious metal electrode catalysts. Again, work is being carried out to produce more tolerant catalyst systems along with membranes capable of operating at elevated temperatures where catalyst poisoning is less of a problem.

29.1.2 Direct Methanol Fuel Cell

Direct methanol fuel cell technology is relatively new as compared to that of PEMFCs. The direct methanol fuel cell (DMFC) is similar to the PEMFC in that it employs an MEA with a polymeric proton exchange membrane as the electrolyte that is positioned between two precious metal catalyst electrodes. During fuel cell operation, liquid methanol (from an aqueous solution) is oxidized to CO_2, protons, and electrons, with the electrons exiting the cell and the protons migrating through the membrane to the cathode where they react with oxygen (from air) and electrons to form water. While potentially a very attractive solution to the issues of hydrogen storage (particularly for portable applications), the principal problem facing the commercialization of DMFCs stems from their relatively low performance (low power output) in comparison to hydrogen-fueled PEMFCs.

A schematic diagram of a DMFC and the anode–cathode reactions during fuel cell operation are shown in Figure 29.3. As is the case of a PEMFC, the proton exchange membrane is hydrated to ensure facile proton transport during current flow. Typical operating temperatures are in the range of about 60–140°C with an energy conversion efficiency

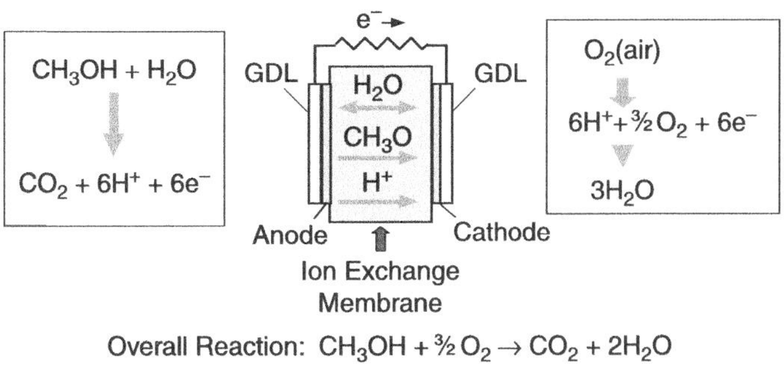

Figure 29.3 Schematic diagram of a direct methanol fuel cell.

of about 40%. While the cathode catalyst is either platinum black or Pt on carbon, the anode (methanol oxidation) catalyst is a Pt–Ru black alloy (Ru is used to minimize catalyst poisoning by CO, an intermediate product of methanol electrooxidation), and catalyst loadings for both the anode and cathode are higher than those used in an H_2–air PEMFC.

Direct methanol fuel cells do not have the fuel storage problems typical of hydrogen–air fuel cells. Liquid methanol has a higher volumetric energy density, although less than that of gasoline or diesel fuel. Methanol is easy to transport and supply to the public using our current infrastructure since it is a liquid. The principal technical problems associated with a DMFC are: (1) the low electrochemical reaction rates for methanol oxidation reaction as compared to hydrogen oxidation, which gives rise to lower cell voltages, less power output, and lower energy conversion efficiencies, and (2) methanol leakage (crossover) across the proton exchange membrane, which leads to a number of deleterious side effects (as will be discussed in the following section).

Direct methanol fuel cells are being considered as the power supply for a number of applications such as automotive, portable electronics including cellular phones and laptop computers, auxiliary power for instrumentation, and as a battery replacement for combat personnel and other military needs.

29.2 BACKGROUND ON FUEL CELL MEMBRANES

29.2.1 General Requirements

The general requirements of a proton-conducting membrane for PEMFC and DMFC applications are as follows: (1) high ionic conductivity (with zero electronic conductivity) under cell operating conditions, (2) long-term chemical and mechanical stability at elevated temperatures in hydrated oxidizing and reducing environments, (3) good mechanical strength, preferably with resistance to swelling, (4) low fuel (H_2 or methanol) and O_2 gas crossover and pinhole free, (5) interfacial chemical/mechanical compatibility with catalyst layers in an MEA, (6) easy processing into an MEA, and (7) low cost.

29.2.2 Current Use of Perfluorosulfonic Acid Membranes

For PEMFC operation under high relative humidity (RH) conditions at $T \leq 80°C$, perfluorosulfonic acid (PFSA) proton conductors (e.g., Nafion®, whose chemical structure is shown in Fig. 29.4) are the membrane material of choice due to their high conductivity

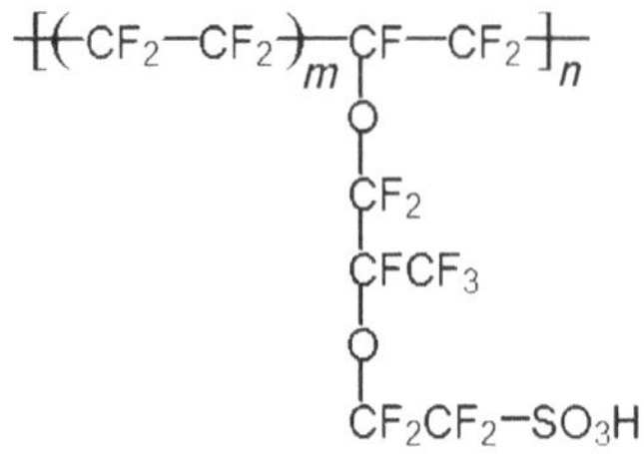

Figure 29.4 Chemical structure of Nafion®.

and chemical/mechanical stability (Banerjee and Curtin, 2004). There are numerous studies in the literature that detail the performance of Nafion-based MEAs in an H_2–air fuel cell (see, e.g., Gottesfeld and Zawodzinski, 1997). Early studies on DMFCs with liquid methanol feeds operating at 60–80°C utilized, for the most part, Nafion as the proton-conducting solid polymer electrolyte because of its prior use and acceptable performance in a PEMFC. Unfortunately, methanol leakage (crossover from the anode to the cathode) through Nafion during DMFC operation is unacceptably high. When methanol contacts the air cathode in a DMFC, it will be oxidized chemically (with oxygen from the cathode feed air) to form CO_2 and water, with a number of unwanted consequences: (1) There is cathode depolarization (a loss in voltage) due to a mixed potential phenomenon (where two reactions, the electrochemical reduction of O_2 and the chemical oxidation of methanol, occur simultaneously on the same electrode), (2) CO intermediate, generated on the cathode during the oxidation of methanol, poisons the cathode catalyst, (3) oxygen is consumed during methanol oxidation and is not available for electrochemical reduction, (4) there is excess water generation at the cathode and flooding of the electrode occurs (which will lower O_2 access to catalytic sites), and (5) there is consumption of methanol without electricity generation, thus lowering the overall fuel efficiency of the fuel cell.

29.2.3 Challenges for Membrane Scientists

There are two major challenges for membrane scientists with regard to proton exchange membrane fuel cells: (1) PEMFC membranes that conduct protons under high-temperature and low-humidity conditions and (2) DMFC membranes that are highly conductive but act as barriers to methanol.

Proton exchange membrane fuel cell operation with lightly humidified or dry gases at 120°C would be highly advantageous with regard to heat rejection from a fuel cell stack, compatibility with automotive radiators, catalyst tolerance to CO impurities in the hydrogen gas stream, and faster electrode kinetics. Unfortunately, the conductivity of PFSA membranes drops dramatically at $T > 100°C$ under low-humidity conditions due to an insufficient number of membrane-phase water molecules for proton separation from sulfonate fixed charge sites, a loss of percolation pathways for proton movement due to low membrane swelling, and structural changes in the polymer that cause membrane pores to collapse (the glass transition temperature of Nafion is only about 100°C). At high temperatures and low humidity, PFSA membranes also exhibit a loss in mechanical strength (e.g., creep and pinhole formation), increased gas permeability, and higher rates of oxidative degradation.

For a DMFC, the membrane properties of utmost interest are the proton conductivity and methanol crossover. Ideally, the membrane should be a good proton conductor (with a conductivity >0.1 S/cm) and a good methanol barrier. Since proton conductivity in polymeric ion exchange membranes is highly dependent on membrane water content (as noted above for PEMFCs) and since methanol and water are highly miscible, it is difficult to block methanol from absorbing in (and eventually moving through) a hydrated DMFC membrane. Nonetheless, researchers have made progress in this area, as will be described below.

Specific targeted properties of membranes for use in high-temperature, low relative humidity PEMFCs and in DMFCs are listed in Table 29.2 (Gasteiger and Mathias, 2005), along with references for general procedures as to how the properties are measured.

TABLE 29.2 Targeted Membrane Properties for High-Temperature PEMFCs and Low Crossover DMFCs

Property	Target Membrane Properties for PEMFC (H_2/air fuel cell)	Target Membrane Properties for DMFC	Method(s) to Determine Property
Proton conductivity	>0.1 S/cm at 120°C and 25% RH >0.03 S/cm at 25°C and 25% RH	Minimum of 0.01–0.03 S/cm, with a membrane-area-specific (areal) resistance of $0.2\ \Omega \cdot cm^2$, in water at 25°C	2- or 4-point method (Zawodzinski et al., 1991)
Fuel permeability (H_2 or methanol)	$<1.2 \times 10^{-12}$ at 80°C (mol cm)/(cm^2 s kPa); see Gasteiger and Mathias (2005) for more permeabilities at other temperatures	No established target, but methanol flux should be <5.0 μmol/cm^2-min at 60°C	GC method (Broka and Ekdunge, 1997) for H_2; limiting current method (Ren et al., 2000); CO_2 sensor method (Wycisk et al., 2005); or methanol diffusion cell (Wycisk et al., 2005)
O_2 permeability	$<4 \times 10^{-12}$ at 80°C (mol cm)/(cm^2 s kPa); see Gasteiger and Mathias (2005) for permeabilities at other temperatures	$<4 \times 10^{-12}$ at 80°C (mol cm)/(cm^2 s kPa); see Gasteiger and Mathias (2005) for permeabilities at other temperatures	GC method (Broka and Ekdunge, 1997)
Solubility in H_2O	<1% at 150°C for 24 h	<1% at 150°C for 24 h	Autoclave in H_2O
Swelling in H_2O	<100% H_2O uptake in boiling water	<100% H_2O uptake in boiling water and 1–3 M methanol solutions	Weight-gain measurement (Wycisk et al., 2005)
Chemical stability	Stable in presence of peroxyl species	Stable in presence of peroxyl species	Test with H_2O_2 and/or Fenton's reagent (Guo et al., 1999)
Mechanical stability	Critical	Critical	Specifications and requirements not yet determined

TABLE 29.3 DOE Technical Targets for High-Temperature, Low-Relative Humidity PEMFC Membranes

		Calendar Year			
Characteristic	Units	2004 Status	2005	2010	2015
Membrane conductivity at					
Operating temperature	S/cm	0.1	0.1	0.1	0.1
Room temperature	S/cm	0.07	0.07	0.07	0.07
−20°C	S/cm	0.01	0.01	0.01	0.01
Operating temperature	°C	≤80	≤120	≤120	≤120
Inlet water vapor partial pressure	kPa	50	25	1.5	1.5
Oxygen crossover[a]	mA/cm^2	5	5	2	2
Hydrogen crossover[a]	mA/cm^2	5	5	2	2
Cost	$\$/m^2$	65[b]	200	40	40
Durability with cycling					
At operating temp. of ≤80°C	h	1000[c]	2000[d]	5000[d]	5000[d]
At operating temp. of >80°C	h	Not available[e]		2000	5000
Low-temperature survivability	°C	−20	−30	−40	−40
Thermal cyclability in presence of condensed water		Yes	Yes	Yes	Yes

[a]Tested in MEA at 1 atm O_2 or H_2 at nominal stack operating temperature.
[b]Based on 2004 TIAX cost analysis and will be periodically updated.
[c]Durability is being evaluated. Steady-state durability is 9000 h.
[d]Includes typical driving cycles.
[e]High-temperature membranes are still in a development stage and durability data are not available.

The Hydrogen, Fuel Cells, and Infrastructure Technologies Program at the U.S. Department of Energy (DOE) (http://www.eere.energy.gov/hydrogenandfuelcells/tech_teams.html#targets) has also compiled a list of membrane properties for high-temperature, low relative humidity (RH) PEMFC operation. Target goals for membrane

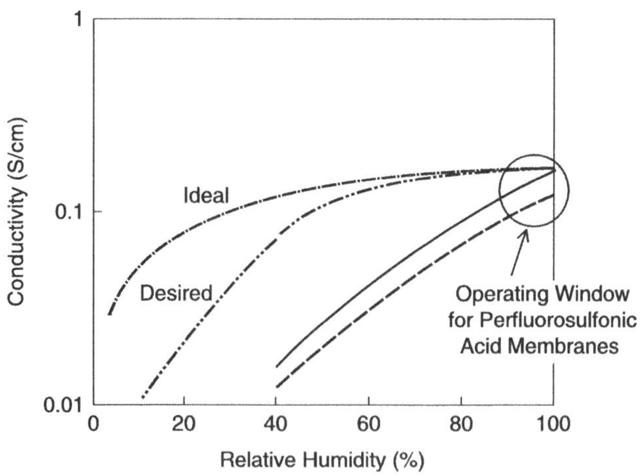

Figure 29.5 Effect of relative humidity on the ideal, desired, and actual proton conductivity of fuel cell membranes (Gasteiger and Mathias, 2005).

properties have been proposed for the next 10 years, as shown in Table 29.3. Proton conductivity targets at low humidity conditions are particularly ambitious. In this regard, Gasteiger and Mathias (2005) have plotted the desired, ideal, and current (Nafion) conductivity/humidity behavior for PEMFC membranes, where the current–actual data were collected for a Nafion 117 membrane at 80 and 120°C (see Fig. 29.5). Rapid water loss from the membrane with decreasing relative humidity leads to the observed loss in proton conductivity.

With regards to the proton conductivity data in Tables 29.2 and 29.3, it is important to note that the membrane resistance loss in an operating PEMFC or DMFC is a function of both the proton conductivity and the membrane thickness (i.e., one can tolerate a low proton

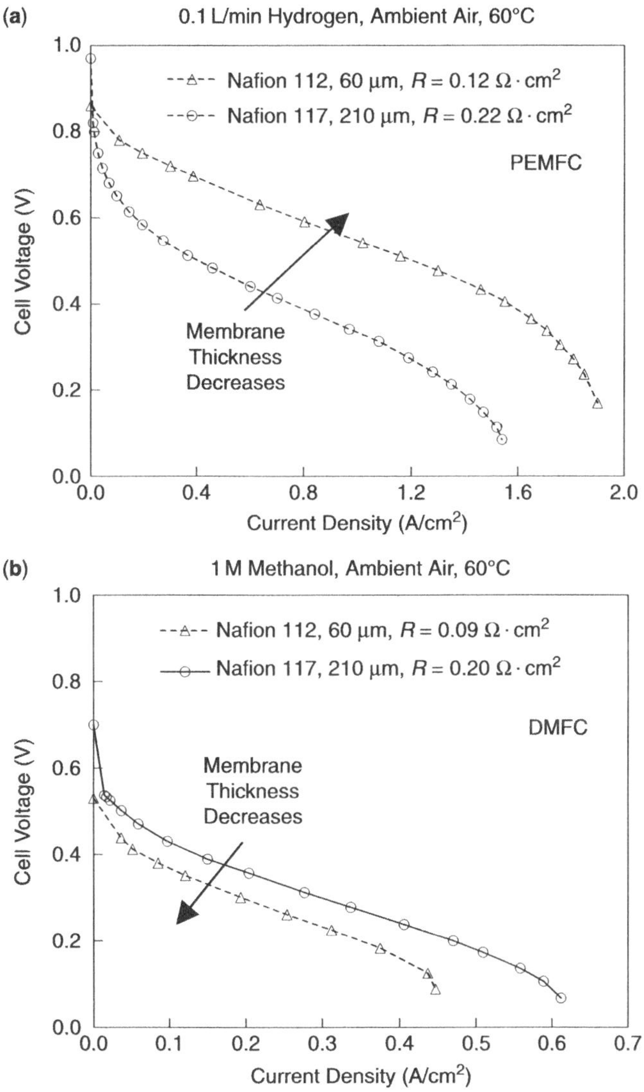

Figure 29.6 Effect of membrane thickness on full cell performance with two commercial Nafion membranes. *R* denotes the membrane areal resistance: (a) PEMFC and (b) DMFC.

conductivity if the membrane is made sufficiently thin). The membrane property parameter that takes into account both film thickness and conductivity is the area-specific (areal) resistance (with units of $\Omega \cdot cm^2$), defined as the ratio of membrane thickness to conductivity. For a water-equilibrated Nafion 117 membrane (200 μm wet thickness), the areal resistance at 60°C is about 0.2 $\Omega \cdot cm^2$. For a high temperature (e.g., 120°C) and low relative humidity (25–50%) PEMFC, the U.S. DOE has established a long-term target of $<0.05\ \Omega \cdot cm^2$ for the areal resistance of the membrane material.

Membrane thickness also plays an important role in DMFC operation. Here one must balance an increase in ohmic loss (due to finite membrane conductivity) with a decrease in methanol crossover as the membrane thickness is increased. If the membrane is too thick, ohmic losses dominate; but if the membrane is too thin, methanol crossover will be too high. In both cases the DMFC power output will be low. This situation is not the case for a PEMFC, where decreasing the membrane thickness improves fuel cell performance (until the membrane becomes so thin that H_2 and/or O_2 gas crossover is a concern). Figure 29.6 shows the fundamental difference in thickness behavior of a proton-conducting membrane (commercial Nafion samples) in a PEMFC and DMFC (Pintauro, unpublished data). The voltage–current density plots in Figure 29.6a show an improvement in PEMFC performance when the thickness of Nafion is decreased from 210 μm (Nafion 117, wet) to 60 μm (Nafion 112, wet) due to a decrease in the areal resistance. Here better performance is quantified in terms of power density (defined as the product of voltage and current density). The situation is much different for a DMFC (Fig. 29.6b), where the fuel cell performance plot with Nafion 112 lies below that for Nafion 117. Methanol crossover is the dominant factor with regard to fuel cell operation, in which case a thicker membrane is a better methanol barrier, even though its areal resistance is large.

29.3 RECENT WORK ON NEW FUEL CELL MEMBRANES

Recent fuel cell membrane research and development (R&D) efforts are summarized with a focus on: (1) membranes for high-temperature, low-humidity PEMFC operation, (2) low-cost alternatives to PFSA membranes, and (3) direct methanol fuel cell membranes. A listing of fuel cell membrane performance data is given in Tables 29.4–29.6 for each membrane subcategory. The review material is by no means exhaustive, but it is representative of the kinds of fuel cell membranes previously/currently under investigation. For different viewpoints on the evolutionary development of various fuel cell membranes, the reader is directed to other review articles in the open literature (Rikukawa and Sanui, 2000; Li et al., 2003; Haile, 2003; Jannasch, 2003; Savadogo, 2004; Hickner et al., 2004; Hogarth et al., 2005; Smitha et al., 2005).

29.3.1 Membranes for Low-Humidity PEMFC Operation

The membranes described here are designed to work in a hydrogen–air fuel cell operating at a temperature of 100–200°C with minimal or no humidification of the reactant gases.

29.3.1.1 All-Polymeric Membranes—Immobilized Imidazole System Presently, there is no purely polymeric membrane available with a sufficiently high proton conductivity (>0.1 S/cm) for fuel cell operation under anhydrous or low water vapor pressure

TABLE 29.4 Examples of High-Temperature PEMFC Membranes

Membrane Type	Conductivity[a] (S/cm)	Fuel Cell Characteristics	Comments	Reference
Nafion 112	0.10	635 mW/cm^2 at 1500 mA/cm^2	60°C, hydrogen at 0.1 L/min., 0.5 L/min. air at 1 atm, RH 100%, 0.4 mg Pt/cm^2	Pintauro (unpublished data)
Nafion 117	0.10	340 mW/cm^2 at 1000 mA/cm^2	60°C, hydrogen at 0.1 L/min., 0.5 L/min. air at 1 atm, RH 100%, 0.4 mg Pt/cm^2	Pintauro (unpublished data)
Fluorinated styrene copolymer	0.22 at 60°C	At 0.4 V: 46 mA/cm^2	50°C, 20% RH, membrane IEC = 1.5 mmol/g, thickness 190 μm. Nafion 117 reference was 8 mA/cm^2	Zhou et al. (2005)
PBI/phosphoric acid		At 0.5 V: 1900 mA/cm^2 for hydrogen–oxygen; 760 mA/cm^2 for hydrogen air	200°C, 3 atm hydrogen, 3 atm oxygen (air), no humidification	Li et al. (2004)
Nafion 115 + 6% SiO_2	0.78 $\Omega \cdot cm^2$ at 130°C (measured in MEA)	At 0.4 V: 470 mA/cm^2 (>50 h lifetime); as compared to Nafion 115, 217 mA/cm^2 (<1 h lifetime)	130°C, 3 atm hydrogen (humidifier temp. 130°C), 3 atm air (humidifier temp. 130°C), 0.4 mg Pt/cm^2	Adjemian et al. (2002a)
Microporous polyester filled with zirconium phosphate–sulphophenyl phosphonate	0.03 at 150°C and 95% RH		Membrane thickness = 43 μm, matrix pore diameter = 0.8 μm	Alberti et al. (2005)
Nafion–silicotungstic acid	0.07–0.10	At 0.6 V: 500 mA/cm^2 (350 mA/cm^2 for a reference recast Nafion without filler)	80°C, 1 atm hydrogen/oxygen, gas humidifier temperature = 95°C, 1 mg Pt/cm^2	Tian and Savadogo (2005)
Nafion 112 + arsenotungstic acid		At 0.5 V: 600 mA/cm^2	120°C, hydrogen/oxygen, 25% humidity, anode and cathode loading: 0.5 mg Pt/cm^2; 170 mA/cm^2 for Nafion 112 with no HPA	Herring et al. (2004)
CsH_2PO_4 (25 μm in thickness)	0.022 at 240°C	415 mW/cm^2 at 1200 mA/cm^2	240°C, 3 atm hydrogen (humidifier temp. = 72°C), 1 atm oxygen (humidifier temp. 72°C), 7.7 mg Pt/cm^2	Uda and Haile (2005)

[a]At 25°C in water, unless otherwise noted.

TABLE 29.5 Low-Cost Substitutes for Perfluorosulfonic Acid Polymers in PEMFCs

Membrane Type	Conductivity[a] (S/cm)	Fuel Cell Characteristics	Comments	Reference
DAIS membrane: sulfonated poly(styrene–ethylene–butylene)	0.09–0.10	At 0.6 V: 260 mA/cm^2; at 0.4 V: 600 mA/cm^2	40°C, hydrogen and air at 1 atm, no humidification, anode and cathode loading: 0.5 mg/cm^2	Wnek et al. (1999)
Photocross-linked DAIS polymer	0.25 at 80°C and 100% RH	At 0.6 V: 620 mA/cm^2 (initial); 500 mA/cm^2 (after 85 h); at 0.4 V: 1300 mA/cm^2 (initial) 970 mA/cm^2 (85 h)	80°C, 1 atm hydrogen, humidifier temp. = 88°C, 1 atm oxygen, humidifier temp. = 86°C, 0.6 mg Pt/cm^2, membrane thickness 90 μm	Chen et al. (2005)
Disulfonated polysulfone with hexafluoroisopropy lidene groups		At 0.6 V: 950 mA/cm^2 (initial); 780 mA/cm^2 (after 1500 h)	80°C, 1.4 atm; hydrogen and air, anode 8 mg Pt/cm^2, cathode 6 mg Pt/cm^2, membrane thickness 58 μm	Kim et al. (2005)
Sulfonated poly(ether ether ketone)	0.035 at 80°C 100% RH	At 0.6 V: 990 mA/cm^2 at 100% RH, 110 mA/cm^2 at 66% RH	80°C, hydrogen and oxygen at 1 atm, 0.7 mg/cm^2 thickness = 25 μm	Jiang et al. (2005)
Sulfonated polyimide (branched and cross-linked)	0.11 at 90°C 100% RH	At 0.6 V: 1450 mA/cm^2 (as compared to Nafion 112: 1500 mA/cm^2)	90°C, 3 atm hydrogen, humidifier temp. = 88°C, 3 atm oxygen, humidifier temp. = 85°C, 0.5 mg Pt/cm^2, membrane thickness = 35 μm, membrane IEC = 2.3 mmol/g	Yin et al. (2005)
Sulfonated FEP-*g*-acrylic acid	Conductance = 4.0 S/cm^2	At 0.6 V: 2190 mA/cm^2; 100 h stable operation at 1 A constant current	50°C, 1 atm hydrogen and oxygen, humidifier temp. = 60°C, 0.3 mg Pt/cm^2	Patri et al. (2004)

[a] At 25°C in water, unless otherwise noted.

TABLE 29.6 Examples of DMFC Membranes

Membrane Type	Conductivity[a] (S/cm)	Fuel Cell Characteristics	Comments	Reference
Nafion 112	0.10	70 mW/cm^2 at 300 mA/cm^2 (1.4 times smaller than with Nafion 117)	60°C, 1.0 M methanol at 2 mL/min, 0.5 L/min air at 1 atm, 4 mg Pt/cm^2 (crossover 1.74 at open circuit)[b]	Wycisk et al. (2005)
Nafion 117	0.10	97 mW/cm^2 at 400 mA/cm^2	60°C, 1.0 M methanol at 2 mL/min, 0.5 L/min air at 1 atm, 4 mg Pt/cm^2 (crossover 1.00 at open circuit)	Wycisk et al. (2005)
Disulfonated poly(arylene sulfone) with phosphine oxide units	0.05	700 h performance loss: 21 mA/cm^2	Methanol permeability 0.7×10^{-6} cm^2/s	Bashyam et al. (2005)
SPEEK	0.04	125 mW/cm^2 at 500 mA/cm^2 (1.3 times greater than with Nafion 117)	130-μm-thick membrane; 60°C, 1.0 M methanol at 2 mL/min, 0.5 L/min air at 1 atm, 4 mg Pt/cm^2 (crossover 0.47 at open circuit)	Pavlak et al. (2007)
SPEKK	0.06	11 mW/cm^2 at 60 mA/cm^2	90°C, 1.5 M methanol at 26 mL/min, air at 100% RH and 0.6 L/min, 0.5 mg Pt/cm^2 (crossover 0.83)	Vetter et al. (2005)
Nafion–polyfurfuryl alcohol	0.07	8.4 mW/cm^2 at 36 mA/cm^2 (1.8 times greater than with Nafion 115)	7% PFA, 60°C, 1.0 M methanol at 10 mL/min, air at 1 atm, 2 mg Pt/cm^2 (crossover 0.45 at room temp.)	Liu et al. (2005)
Nafion-FEP (50%)	0.03	97 mW/cm^2 at 400 mA/cm^2 (equivalent to Nafion 117)	50% FEP, 60°C, 1.0 M methanol at 2 mL/min, 0.5 L/min ambient air, 4 mg Pt/cm^2 (crossover 0.65 at open circuit)	Lin et al. (2005)
Nafion-PBI	0.06	97 mW/cm^2 at 400 mA/cm^2 (equivalent to Nafion 117)	5% PBI, 60°C, 1.0 M methanol at 2 mL/min, 0.5 L/min air at 1 atm, 4 mg Pt/cm^2 (crossover 0.72 at open circuit)	Wycisk et al. (2005)

(*Continued*)

TABLE 29.6 *Continued*

Membrane Type	Conductivity[a] (S/cm)	Fuel Cell Characteristics	Comments	Reference
Poly[bis(phenoxy) phosphazene (1.2 mmol/g) + 3 wt% PBI	0.03	89 mW/cm^2 at 390 mA/cm^2	3% PBI, 60°C, 1.0 M methanol, 2 mL/min, 0.5 L/min ambient air, 4 mg Pt/cm^2 (crossover 0.39 at open circuit)	Wycisk et al. (2005)
SPEEK + PBI + aminated polysulfone	0.03	151 mW/cm^2 at 302 mA/cm^2 (1.3 times greater than with Nafion 117)	IEC = 1.18 mmol/g, 110°C, 1.0 M methanol, 4 mL/min, 1.5 L/min 4 atm air, 5.2/5.3 mg Pt/cm^2 (crossover 0.42 at open circuit)	Jorissen et al. (2002)
Polystyrene sulfonic acid + poly(vinylidene fluoride)	0.08	52 mW/cm^2 at 170 mA/cm^2 (1.2 times greater than with Nafion 117)	55°C, 0.50 M methanol, 0.1 L/min ambient air, 8 mg Pt/cm^2 (crossover 0.36 at open circuit)	Prakash et al. (2004)
SPEEK + zirconium phosphate +PBI	0.01	4.8 mW/cm^2 at 24 mA/cm^2 (46% of that with unmodified SPEEK)	SPEEK sulfonation degree = 42%, 110°C, 1.5 M methanol at 4 mL/min and 2.5 atm, 0.6 L/min 3 atm air, 1/0.4 mg Pt/cm^2 anode/cathode loading (crossover 0.93 at open circuit as compared to unmodified SPEEK)	Silva et al. (2005b)
Porous polyethylene–poly(acrylamide *tert*-butyl sulfonic acid)	0.05	82 mW/cm^2 at 400 mA/cm^2 (1.1 times greater than with Nafion 117)	50°C, 8 wt% methanol at 10 mL/min, 0.5–1.0 L/min air at 1 atm, 2.0/1.0 mg Pt/cm^2 anode/cathode loading (crossover 1.2, measured in a diffusion cell at 25°C and 1.0 M methanol)	Yamaguchi et al. (2005)
Porous polyethylene–poly(acrylamide *tert*-butyl sulfonic acid)	0.02	48 mW/cm^2 at 300 mA/cm^2	50°C, 32 wt% methanol at 10 mL/min, 0.5–1.0 L/min ambient air, 2.0/1.0 mg Pt/cm^2 anode/cathode loading (crossover 0.97, from diffusion cell at 25°C and 1.0 M methanol)	Yamaguchi et al. (2005)

[a] At 25°C in water unless otherwise noted.
[b] Methanol crossover relative to that in Nafion 117.

condition above 100°C. There are, however, promising advances in concepts and early-phase model compound studies in this area. One example is given here: the immobilized imidazole system.

The role of water in a proton exchange membrane is twofold: It is needed for the dissociation of H^+ from sulfonic acid sites, and it provides a network for proton conduction. Two ways to circumvent the problem of high-temperature conductivity loss by membrane dehydration is to replace water with a higher boiling solvent or to immobilize a proton solvent within the membrane. Both approaches have been proposed and studied by Kreuer and co-workers (Kreuer, 2001). Imidazole was used as a water substitute in the proton exchange membranes because it possesses similar amphoteric character and hydrogen bonding and autoprotolysis capabilities. Additionally, in contrast to water, imidazole moieties can be covalently immobilized within an appropriate polymeric scaffold to form an anhydrous, medium/high-temperature proton-conducting membrane. Operational feasibility of this concept has been demonstrated with several model systems (Schuster et al., 2001; Kreuer et al., 2004). For example, polystyrene with imidazole-terminated flexible side chains (see Fig. 29.7) and benzimidazole covalently bonded to an inorganic SiO_2 network by a flexible spacer were fabricated and tested by Herz et al. (2003). Moderately high proton conductivities of up to 7×10^{-4} S/cm at 200°C were obtained in the absence of water. This conductivity is still too low for use in a fuel cell.

29.3.1.2 Hybrid Membranes All examples given below relate to hybrid systems, consisting of a nonconducting polymeric scaffold combined with an inorganic proton conductor (polybenzimidazole–phosphoric acid) or a proton-conducting polymer filled with inorganic hygroscopic particles, either Nafion with nonconducting silica or sulfonated poly(ether ether ketone) (SPEEK) with proton-conducting mixed-zirconium phosphate–phosphonate.

Acid-Doped Polybenzimidazole Phosphoric-acid-doped polybenzimidazole (PBI; see Fig. 29.8a) fuel cell membranes were developed some 13 years ago at Case Western Reserve University (Wainright et al., 1995). Researchers found that a significant amount of phosphoric acid could be absorbed into a PBI film without scarifying the mechanical strength of the membrane. The sorbed acid species conduct protons in the absence of

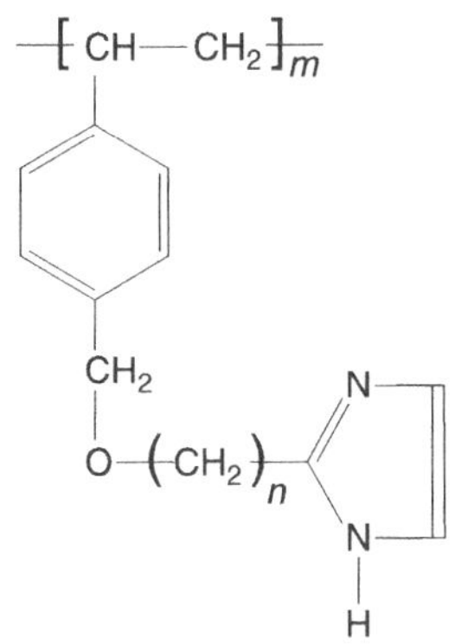

Figure 29.7 Example of an all-polymeric membrane material with covalently immobilized imidazole.

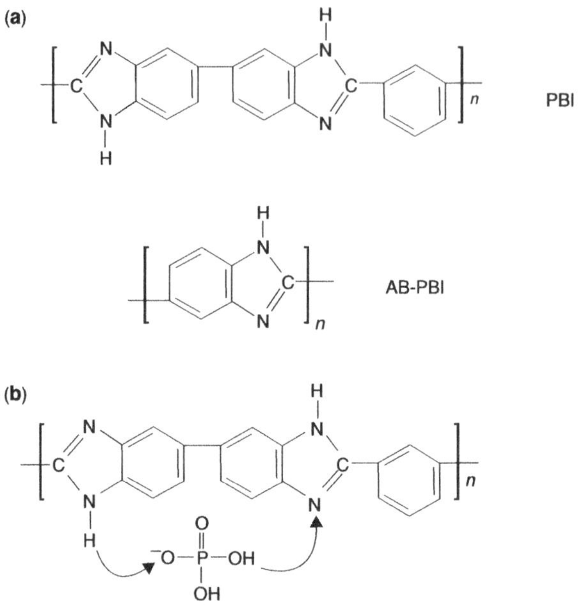

Figure 29.8 Structures of two basic polybenzimidazoles: (a) PBI and poly(2,5-benzimidazole), also known as AB-PBI and (b) schematic description of polymer matrix-assisted proton transport in phosphoric acid doped PBI.

water at temperatures in the 140–200°C range. Under anhydrous conditions, phosphoric-acid-doped PBI membranes exhibited good fuel cell performance at elevated temperatures that has not been matched by any other PEM material (Li et al., 2004). A fuel cell employing this type of membrane is very much like a phosphoric acid fuel cell, the only difference being the use of PBI to contain the acid. The presence of the PBI scaffold, however, has an important function; it promotes the formation of protonic defects through acid–base interactions at the imidazole sites. Although PBI–phosphoric acid membranes are commercially available in Celtec MEAs from PEMEAS Fuel Cell Technologies, Inc., developmental research aimed at improving the properties of these materials is continuing (Xiao et al., 2005). Due to the competing requirements of increasing proton conductivity, which requires higher doping level, and improving mechanical properties, which calls for lower doping, serious research efforts are still required. One obvious need is to increase the molecular weight of the PBI polymer. Other PBI structures, for example, the so-called AB-PBI material, poly(2,5-benzimidazole) shown in Figure 29.8 (Asensio and Gomez-Romero, 2005), are also being explored. There is also concern about the loss of phosphoric acid during long-term fuel cell operation (by slow evaporation and/or leaching into condensed water within the MEA structure). Nonetheless, this membrane is presently the only PEM that allows continuous fuel cell operation at temperatures of near 200°C.

Composite Membranes Containing Inorganic Particles or Heteropolyacids Doping a proton-conducting organic polymer with inorganic particles can help in achieving several goals. When hydrophilic particles are embedded into a proton-conducting polymer, the water retention capability of the composite may significantly increase, which is an important characteristic for membrane self-humidification in a fuel cell. In many instances, the addition of inorganic fillers also improves the mechanical/thermal properties and dimensional stability of the resultant membrane. When the added hygroscopic particles are also good proton conductors (as is the case for heteropolyacids such as phosphotungstic acid), then the combined benefits of composite membrane are, in principle, even greater, especially at high-temperature, low-RH fuel cell operating conditions.

Addition of hydroscopic metal oxides such as silica, zirconia, or titania to a proton-conducting polymer is the most obvious way to improve water retention at elevated temperatures (Aparicio et al., 2003). Unfortunately, due to the negligible proton conductivity of these oxides, an increase in the overall resistance of the composite membrane is observed, especially at low temperatures. However, as the temperature is increased, the conductivity gain due to better hydration offsets the loss due to the excluded conducting volume, and the net fuel cell performance is improved, as compared to an unmodified membrane (Adjemian et al., 2002a,b). It should be stressed that there are limits to the water sorption capability of the oxides. While these membranes retain more water than traditional PEM materials such as Nafion at high temperature and low RH, water uptake is insufficient and ohmic losses are still unacceptably high for PEMFC applications.

Layered metal mixed phosphonates–sulfoarylphosphonates were shown to possess a high proton conductivity and high hydrophilicity (Alberti and Casciola, 1997, 2003; Alberti et al., 1996, 2000, 2001), properties strongly desired in fuel cell applications. Two types of composite polymer–inorganic conductor systems have been investigated. In one, the polymer served merely as a binder leading to virtually an "all-inorganic" conductor where phosphate–phosphonate was the major component of the membrane. Unfortunately, problems with brittleness have retarded the development of this type of hybrid. In the second type, the high-water-retention capability of the phosphate–phosphonate particles was utilized by admixing exfoliated particles into a proton-conducting polymer. Loadings of up to 30% proved to be optimal. The undesirable loss of proton conductivity with particle loading at full humidification, as seen in metal-oxide-doped membranes, was avoided because the proton conductivity of zirconium phosphate–phosphonate platelets (see Fig. 29.9) was comparable to that of the matrix polymer. This route proved to be the most successful utilization of the hybrid membrane concept, especially for medium temperature fuel cell applications. In most literature studies, Nafion or sulfonated poly(arylene ethers) were used as the base polymers (Alberti et al., 2004, 2005). Because the proton conductivity of the phosphate–phosphonate component is strongly dependent on hydration conditions, fuel cell performance of this type of PEM is still not satisfactory at low humidification.

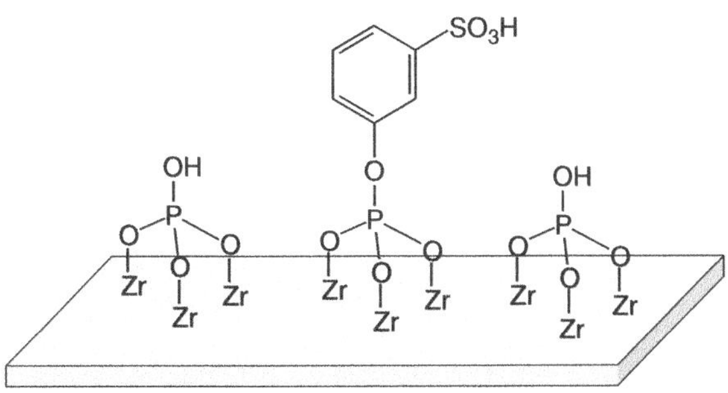

Figure 29.9 Schematic structure of a zirconium phosphate–sulfophenyl phosphonate platelet.

Heteropolyacids (HPAs) are inorganic hydrates known for their high ambient temperature proton conductivity, approaching 0.2 S/cm. Compounds with a "Keggin unit" structure, $[PM_{12}O_{40}]^{+3}$, have received the most attention. Silicotungstic acid and phosphotungstic acid are examples of commercially available HPAs. Heteropolyacids have been shown to improve electrode kinetics in fuel cells (Limoges et al., 2005). The biggest hurdle to overcome in employing HPAs in fuel cell membranes is their high water solubility. This problem was addressed via immobilization (embedding) of HPA in silica (Staiti et al., 1999) or by partial exchange of available protons with large cations (Ramani et al., 2005). Reports have shown that HPA-based membranes perform well in a hydrogen fuel cell at elevated temperatures (100–120°C) and moderately low (50–70%) humidification (Smitha et al., 2005; Tian and Savadogo, 2005; Herring et al., 2004).

29.3.1.3 Inorganic Solid Acid Membranes Inorganic solid acids such as cesium hydrogen sulfate ($CsHSO_4$) or cesium dihydrogen phosphate (CsH_2PO_4) have been shown to offer the advantages of anhydrous proton transport and high-temperature stability (up to 250°C) (Chisholm and Haile, 2000; Chisholm et al., 2002; Merle et al., 2003). Cesium phosphate, when heated above 141°C, undergoes an abrupt "superprotonic" phase transition, which is accompanied by an increase in proton conductivity by 2–3 orders of magnitude, to a value of 10^{-3}–10^{-2} S/cm. Significant progress on this material has been made by researchers from Caltech. A maximum power density of 415 mW/cm^2 has been obtained in a weakly humidified hydrogen–oxygen fuel cell at 240°C (Boysen et al., 2004; Uda and Haile, 2005). Water solubility and the high-temperature ductility of all solid acid membranes are the main obstacles limiting development of solid-acid-based fuel cell membranes.

29.3.2 Cost-Effective PEM Materials for Low-Temperature Operation

The high cost of perfluorosulfonic acid membranes, such as Nafion ($800–$1000/m^2) continues to be an issue. There are several sulfonated hydrocarbon polymers that work well in PEMFCs at 60–80°C and full humidification, but there is concern regarding their long-term oxidative stability (i.e., the possibility of polymer degradation by peroxide and hydroxyl radicals generated during fuel cell operation). Sulfonated and cross-linked polystyrene is a classic example of an early hydrocarbon-based PEM (used in the Gemini space program) that exhibited oxidative degradation. Recently, the concept of using a sulfonated styrene copolymer for cost-sensitive applications has reemerged via the Dais Analytic Corporation membrane. Sulfonated poly(arylene ethers), sulfonated polyimides, and sulfonated fluoropolymer grafts–interpolymers also belong to this group of low-cost PFSA alternatives. The main advantage of these materials is their very competitive fabrication costs, combined with reasonable fuel cell lifetimes at low operating temperatures (which makes them attractive in fuel cells for certain applications, such as battery substitutes for portable power).

29.3.2.1 Sulfonated Styrenic Block Copolymers Two kinds of styrene-based membranes were introduced by Dais Analytic Corporation (see Fig. 29.10). The first was fabricated from a commercially available styrene–ethylene–butylene–styrene triblock polymer (Kraton) by postsulfonation of the styrenic units (Ehrenberg et al., 1995). Later, a new Dais Analytic membrane was developed based on an ethylene–styrene pseudorandom interpolymer from Dow Chemical Corporation (Serpico et al., 2002). Using a

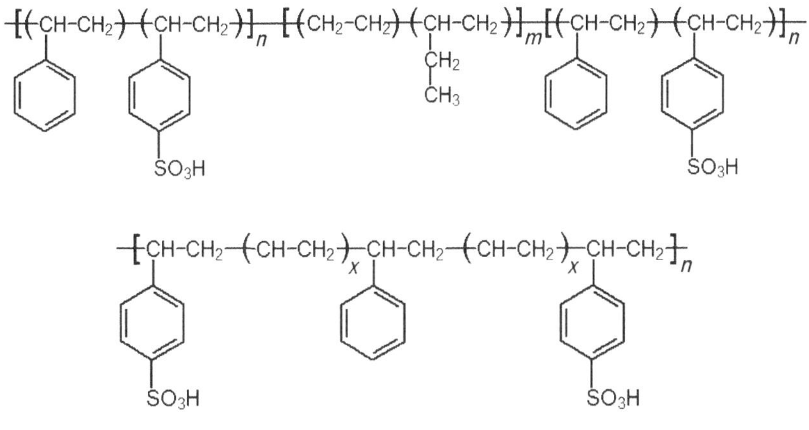

Figure 29.10 Dais analytic membrane polymers.

proprietary process, 30–60% of the aromatic rings were sulfonated. The resultant films exhibited a proton conductivity approaching 0.1 S/cm at full hydration. The mechanical properties of the membranes were acceptable for low-temperature fuel cell operation. Membranes from this polymer are currently being used to dehydrate air in commercial ventilation systems (ConsERV). Other researchers have studied similar styrene-based PEM materials and found proton conductivities similar to that of the Dais Analytic material (Mokrini et al., 2004; Chen et al., 2005; Shim and Weiss, 2005).

29.3.2.2 Sulfonated Poly(arylene ethers) Poly(arylene ethers) (PAEs), comprising, among others, polysulfones and polyketones (see Fig. 29.11), belong to a group of high-performance thermoplastics that possess excellent thermal, mechanical, and chemical properties. Sulfonated derivatives of these polymers can be obtained by postsulfonation of a given PAE (Lufrano et al., 2000; Iojoiu et al., 2005). McGrath and co-workers at Virginia Tech University used the copolymerization of sulfonated and nonsulfonated monomers to synthesize a family of disulfonated poly(arylene ether sulfones) (Wang et al., 2002; Harrison et al., 2003, 2005; Summer et al., 2004) (see Fig. 29.12). PEMFC performance up to 2000 h has been reported at 80°C using membrane–electrode assemblies fabricated from a fluorinated PAE derivative (Kim et al., 2005). Sulfonated aromatic polyketones have also been investigated as PEM materials (Gao et al., 2003; Kaliaguine et al., 2003;

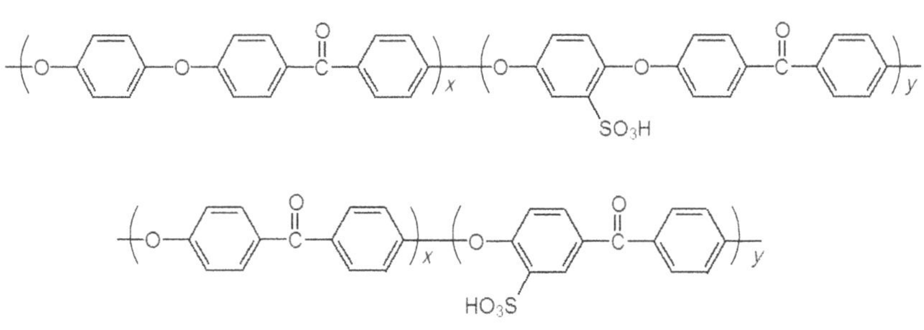

Figure 29.11 Sulfonated poly(ether ether ketone) and sulfonated poly(ether ketone).

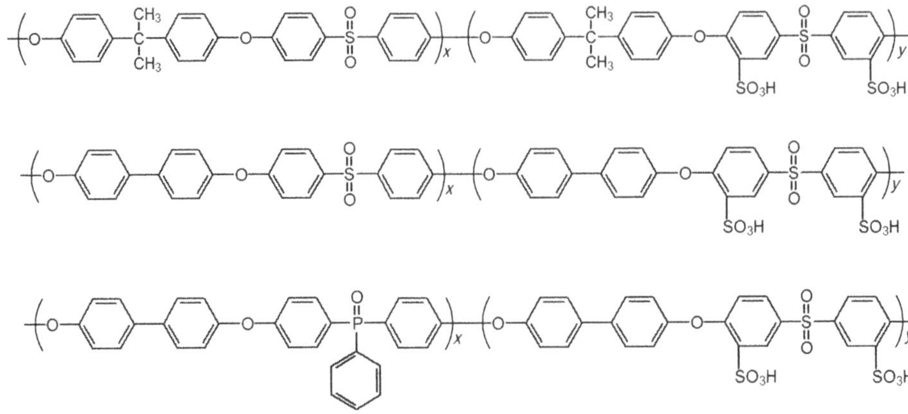

Figure 29.12 Three different disulfonated polysulfone polymers.

Xing et al., 2004b; Jiang et al., 2005), with conductivities in the range of 0.01–0.2 S/cm (see Table 29.5). In general, sulfonated PAEs suffer from dehydration and a loss of conductivity at low humidities and temperatures exceeding 80°C. There is also concern as to the long-term oxidative stability of these polymers.

29.3.2.3 Sulfonated Polyimides Sulfonated polyimides (SPI) have attracted considerable attention due to their high proton conductivity, good thermal stability, and excellent mechanical strength. These materials are generally terpolymers, synthesized from a suitable dianhydride, sulfonated diamine, and a nonsulfonated diamine. Two main classes of sulfonated derivatives have been studied: phthalic polyimides (with five-member rings) and naphthalenic polyimides (containing six-member rings) (see Fig. 29.13). It was found that sulfonated phthalic polyimides degrade hydrolytically in water so the naphthalenic type became the preferred membrane material (Genies et al., 2001a,b). Due to their lower ring strain, naphthalenic derivatives have superior chemical and thermal stability that renders them suitable for medium-temperature fuel cells. Lifetime tests showed good hydrogen–air fuel cell performance for more than 3000 h at 60°C (Roziere and Jones, 2003). Direct copolymerization of sulfonic acid monomers has been reported to give better stability of the protogenic sites and better control of the final membrane morphology (Einsla et al., 2004, 2005; Zhou et al., 2005; Sundar et al., 2005;

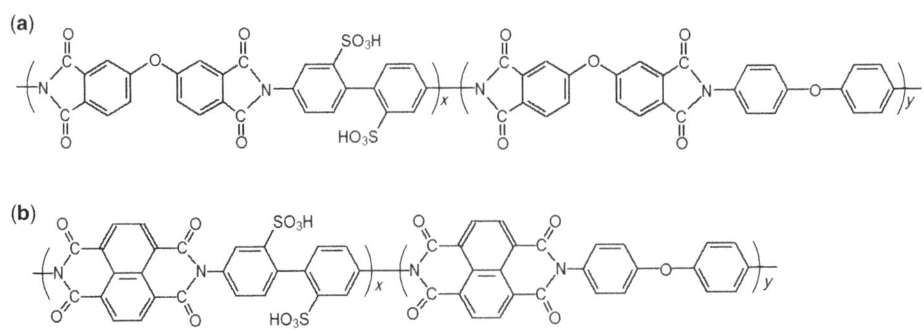

Figure 29.13 (a) Five-member ring and (b) six-member ring sulfonated polyimides.

Yamada et al., 2005; Yin et al., 2005). Proper selection of the nonsulfonated diamine monomer allowed for control of important membrane properties, including water sorption and mechanical strength. For example, it was shown that the incorporation of bulky or angled co-monomers (aromatic diamines) improved the water retention capability of SPIs, leading to very high proton conductivities (Zhang and Litt, 1999), but the mechanical properties (brittleness) and hydrolytic stability of these polymer were unattractive. They have not yet been fabricated in MEAs and tested in a fuel cell.

29.3.2.4 Fluoropolymer Grafts Membranes prepared from graft copolymers (Dargaville et al., 2003; Jannasch, 2005) have been shown to offer interesting properties for fuel cell applications. Radiation-induced grafting, which was developed in Finland (Kallio et al., 2002; Gode et al., 2003) and Switzerland (Gubler et al., 2004, 2005), involves the polymerization of a monomer (styrene) or monomer mixture in a preformed polymer film (usually fluoropolymer). The preferred matrix polymer is either poly(vinylidene fluoride), ethylene-tetrafluoroethylene copolymer, or Teflon FEP. There are two variants of the process: (1) the *preirradiation method* where the polymer film is irradiated first and then exposed to the monomer (here the radicals are generated on the polymer only) and (2) the *postirradiation method* when the monomer is first allowed to be absorbed into the film and then both the monomer and matrix are irradiated simultaneously (where radicals form on both the monomer and the matrix polymers). Postsulfonation of the styrenic component of the copolymer films is usually carried out with chlorosulfonic acid. Numerous studies on radiation-grafted membranes showed that this type of material can be successfully used in fuel cell MEAs (Horsfall and Lovell, 2001; Kabanov, 2004; Patri et al., 2004). At low grafting degrees, membranes exhibited acceptable swelling, good mechanical properties, and high proton conductivity (as high as 0.1 S/cm at room temperature). Although the oxidative stability of styrenic proton conductors is questionable, optimized styrene-divinylbenzene-grafted membranes exhibited durabilities of several thousand hours in a hydrogen fuel cell at temperatures up to 80°C (Gubler et al., 2005).

29.3.3 Direct Methanol Fuel Cell Membranes

Current research activities are focused on the identification of membrane materials and structures with high proton conductivity and low methanol permeability. Ultimately, one would like membranes that work well in a DMFC at 10–20 M methanol, but the focus of most research is on much lower methanol feed concentrations (0.5 and 1.0 M). Since DMFCs are designed primarily for the portable power/electronics market, operating temperatures in the 25–80°C range are usually considered. Extensive data is available in the literature on new membrane materials with reduced methanol permeability, including sulfonated or phosphonated copolymers, phosphoric-acid-doped polybenzimidazole, and various blends and composites (see Table 29.6 for a listing of DMFC properties).

29.3.3.1 Membranes from Sulfonated Copolymers Polymers containing aromatic units can usually be converted into proton conductors by postsulfonation. Even if the starting material is a homopolymer, partial sulfonation leads to a copolymer containing sulfonated and nonsulfonated repeat units. Alternatively, copolymerization can be performed using sulfonic acid and uncharged monomers. Two polymer groups, namely sulfonated polyphosphazenes and sulfonated polyarylenes, will be discussed in greater detail below.

Sulfonated Polyphosphazenes Polyphosphazenes (Allcock, 2002; Wycisk and Pintauro, 2003) belong to the class of inorganic polymers that possess a heteroatomic backbone consisting of alternating phosphorus and nitrogen atoms, with two side groups attached to each phosphorus (see Fig. 29.14). The most attractive feature of polyphosphazene chemistry is the very broad choice of groups that can be easily incorporated into the macromolecular chain, resulting in nearly limitless possibilities as to how one might control the macromolecular architecture and fine tune the properties of the resultant polymer. The first polyphosphazene-based proton conducting membranes were reported by Pintauro and co-workers (Wycisk and Pintauro, 1996) from sulfonated poly(aryloxy phosphazenes). Later the same group showed that films from sulfonated and ultraviolet (UV)-light crosslinked poly[bis(3-methylphenoxy)phosphazene] possessed high proton conductivity, low methanol diffusivity, and good thermooxidative stability (Guo et al., 1999; Tang et al., 1999, 2001). Phosphonated poly(aryloxyphosphazenes) were synthesized by Allcock and co-workers (Allcock et al., 2002; Fedkin et al., 2002), who found that the methanol diffusion coefficient of a phenyl phosphonic acid functionalized polyphosphazene membrane in a 50% aqueous methanol solution was 40 times lower than that of Nafion 117. No fuel cell data has been reported for phosphonated polyphosphazene membranes.

Sulfonated Polyarylenes Poly(arylene ethers), including polysulfones and polyketones, were briefly introduced earlier in this review as potential membrane materials for low-temperature hydrogen fuel cells. Proton-conducting sulfonated derivatives are obtained by either postsulfonation of a given PAE or by direct copolymerization of sulfonated monomers. Numerous studies on the DMFC performance of these membranes are available in the literature. The most significant progress has been reported by the cooperative efforts of Los Alamos National Laboratory and Virginia Tech University. A recent report (Bashyam et al., 2005) showed that a PAE membrane prepared from disulfonated polysulfone having a conductivity of 0.05 S/cm and methanol permeability of 0.7×10^{-6} cm^2/s performed well in a DMFC at 80°C and 0.5 M methanol, with a current loss of 21 mA/cm^2 (at a cell voltage of 0.5 V) over a 700-h test period.

Various sulfonated polyketones have also been examined for use in a DMFC. For example, the 60°C fuel cell performance of an MEA prepared from postsulfonated poly(ether ether ketone) with a room temperature conductivity of 0.04 S/cm was superior to that of Nafion 117 (Pavlak et al., 2007). The methanol permeability of the SPEEK

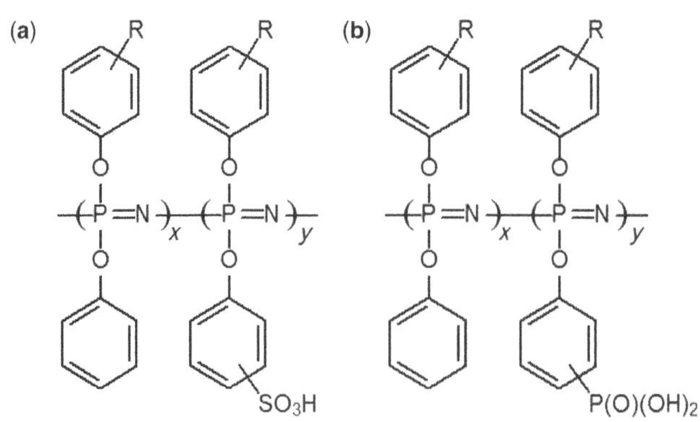

Figure 29.14 (a) Sulfonated and (b) phosphonated polyaryloxyphosphazenes.

membrane was 2.1 times lower than that of Nafion 117. Other types of polyketones have also been investigated as DMFC membrane materials. The power output from a DMFC with MEAs fabricated from sulfonated poly(ether ketone ketone) (SPEKK) was comparable to that of Nafion, but with lower methanol crossover (Vetter et al., 2005). In order to better control the nanomorphology of a sulfonated polyketone, the direct copolymerization of sulfonated and nonsulfonated aromatic monomers was studied. The resultant polymers possessed good mechanical and proton conductivity properties, but no fuel cell data were reported (Xing et al., 2004a,b; Gil et al., 2004).

29.3.3.2 Membranes from Blends and Composites The modification of various proton-conducting polymers by blending or doping with inorganic particles has been reported. Several examples, with low methanol permeability (relative to Nafion), are described next.

Modified Nafion Numerous modifications of Nafion have been attempted in order to improve its methanol barrier properties. These include: partial substitution of the sulfonic acid groups with Cs^+ ions (Tricoli, 1998), Pd coating (Hejze et al., 2005; Tang et al., 2005), sandwiching a Pd foil between two Nafion 115 films (Pu et al., 1995), incorporating inorganic nanoparticles into the polymer (Antonucci et al., 1999; Dimitrova et al., 2002; Jung et al., 2003; Baglio et al., 2005; Silva et al., 2005a–c; Rhee et al., 2005; Park and Yamazaki, 2005a,b; Lee et al., 2005; Sauk et al., 2005), interpolymerization (Jia et al., 2000; Liu et al., 2000), blending with vinylidene fluoride–hexafluoropropylene copolymer (Lin et al., 1998), blending with Teflon FEP (Lin et al., 2006), doping with PBI (Wycisk et al., 2006), filling a porous matrix with Nafion (Shim et al., 2005), and Nafion surface treatments (Walker et al., 1999). Unfortunately, a decrease in methanol permeability was always accompanied by a loss in proton conductivity. The Nafion-FEP system, in particular, is attractive because it performs well in a DMFC with low methanol crossover, and the total Nafion content is reduced by a factor of 10 as compared to Nafion 117 due to a reduction in membrane thickness and Nafion content.

Polyphosphazene Blends Blends of sulfonated polyphosphazene, for example, sulfonated poly[bis(3-methylphenoxy)phosphazene] or poly[(bisphenoxy)phosphazene] (see Fig. 29.14) with either an inert organic polymer such as poly(vinylidene fluoride) (Pintauro and Wycisk, 2004; Wycisk et al., 2002) or polyacrylonitrile (Carter et al., 2002) or a "reactive" polymer (e.g., polybenzimidazole) (Wycisk et al., 2005) have been investigated. The resultant membranes had conductivities of 0.01–0.06 S/cm (in water at 25°C) and equilibrium water swelling from 20 to 60% (at 25°C). Blends of poly[(bisphenoxy)phosphazene] and polybenzimidazole (where acid–base complexation occurred between the sulfonic acid and the imidazole nitrogen) exhibited good mechanical properties and low methanol permeability. MEAs with this membrane material outperformed Nafion 117 in a DMFC at 60°C with concentrated (5–10 M) methanol feeds. With 1.0 M methanol and 0.5 L/min ambient air at 60°C, the maximum power density was 97 mW/cm^2 and the methanol crossover was ≈2.5 times lower than that with Nafion 117 (Wycisk et al., 2005).

Polyarylene Acid–Base Blends Kerres (2001) utilized the concept of acid–base cross-linking to improve the osmotic (swelling) and mechanical properties of proton-conducting membranes. He showed that blending a sulfonated polymer, for example, SPEEK,

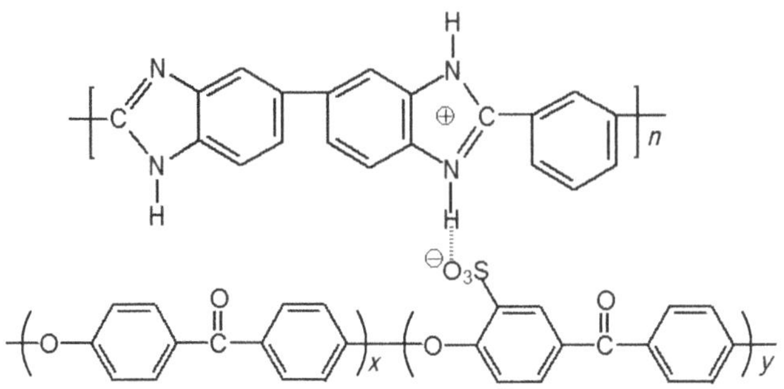

Figure 29.15 Cross-link formation through acid–base interaction between a sulfonic acid group and the benzimidazole nitrogen atom.

sulfonated poly(ether ketone) (SPEK) or sulfonated polysulfone with a basic polymer (in the Bronsted sense), for example, polybenzimidazole or aminated polysulfone, resulted in proton transfer (partial or complete) between sulfonic acid sites and the polymeric basic moieties in the blend (see Fig. 29.15). Such complexation effectively cross-linked the polymer. This allowed for the control of membrane swelling and also reduced membrane brittleness upon drying. In general, the membranes performed well in a DMFC up to 130°C, reaching a peak power density of nearly 300 mW/cm^2. The ionic cross-links, however, decomposed during extended heating at a temperature exceeding 70–80°C (Kerres, 2005).

Poly(styrenesulfonic Acid)–Poly(vinylidene Fluoride) Composite Membranes Semi-interpenetrating polymer network composite membranes consisting of poly (styrenesulfonic acid) and poly(vinylidene fluoride) (PVDF) were developed jointly at the University of Southern California and the Jet Propulsion Laboratory (Prakash et al., 2004). The methodology made use of the thermally initiated radical interpolymerization of styrene that was absorbed into a PVDF film with subsequent sulfonation of the resultant polystyrene. Membranes had a proton conductivity of 0.06–0.09 S/cm with three times lower methanol crossover in a DMFC at 55°C and 0.5 M methanol.

Sulfonated Polymers with Inorganic Barrier Fillers The basic premise is that the incorporation of methanol-impermeable inorganic particles into a polymeric proton conductor will result in a composite film of improved selectivity (where selectivity is defined as the ratio of proton conductivity to methanol permeability) and better fuel cell performance. Recast or extruded Nafion nanocomposites were investigated most often, but other polymer systems were also examined, such as sulfonated poly(arylene ethers) and sulfonated polyimides with silica, titania, exfoliated layered clays, phosphates, or phosphonates. Although no significant breakthroughs (in terms of conductivity and methanol permeability) have been achieved, several interesting results have been published. For example, a significant reduction in methanol and water permeability was achieved by inorganic modification of SPEK and SPEEK (Nunes et al., 2002). The modification was performed by in situ hydrolysis of different alkoxides of Si, Ti, and Zr. Silanes covalently bound to the polymer chain improved the incorporation of SiO_2 in the polymer matrix. Additional improvements in membrane properties were obtained by using organically modified silanes with basic groups. Membranes composed of sulfonated SPEEK and zirconium phosphate pretreated with *n*-propylamine and polybenzimidazole were employed in moderate temperature (90–110°C) DMFCs (Silva et al., 2005c), with lower methanol crossover and higher thermal/chemical stability, as compared to unmodified SPEEK. The membranes were also tested in a DMFC operating at 130°C and pure oxygen as the cathode feed, with a maximum power density of 50 mW/cm^2 at 250 mA/cm^2.

Pore-Filling Membranes These membrane are composed of a porous supporting substrate with the pores filled with a proton-conducting polymer. The mechanically stable substrate restricts swelling of the conducting phase, resulting in a mechanically and osmotically robust membrane with reduced methanol permeability. This type of membrane was developed by researchers at the University of Tokyo (Yamaguchi et al., 2003) over the last several years. Several substrates and various pore-filling polyelectrolytes were explored, such as PTFE/poly(vinylsulfonic acid) or polyethylene/poly(acrylamide sulfonic acid). With the latter material, excellent DMFC performance was reported with 10 M methanol feeds at 50°C, where the maximum power density was 10 times higher than that with Nafion 117 (Yamaguchi et al., 2005).

29.4 CONCLUSIONS

The discovery of new membrane materials and morphologies is a high-priority subset of PEM fuel cell R&D. An aggressive approach is needed to overcome the deficiencies of present-day membranes, in terms of performance, cost, and durability. There is a particularly compelling need for membranes that are good proton conductors at high-temperature and low-humidification conditions for H_2–air fuel cells. Similarly, proton-conducting membranes that are good methanol barriers are needed for direct methanol fuel cells. This review chapter has touched on some, but by no means all, of the recent work on fuel cell membranes. As fuel cell commercialization approaches, there are exciting opportunities for membrane scientists and engineers.

REFERENCES

Adjemian, K. T., Lee, S. J., Srinivasan, S., Benziger, J., and Bocarsly, A. B. (2002a). Silicon oxide Nafion composite membranes for proton-exchange membrane fuel cell operation at 80–140°C. *J. Electrochem. Soc.* 149(3), A256.

Adjemian, K. T., Lee, S. J., Srinivasan, S., Benziger, J., and Bocarsly, A. B. (2002b). Investigation of PEMFC operation above 100°C employing perfluorosulfonic acid silicon oxide composite membranes. *J. Power Sources* 109, 356.

Alberti, G., Boccali, L., Casciola, M., Massinelli, L., and Monotneri, E. (1996). Protonic conductivity of layered zirconium phosphonates containing $-SO_3H$ groups. III. Preparation and characterization of γ-zirconium sulfoaryl phosphonates. *Solid State Ionics* 84, 97.

Alberti, G., and Casciola, M. (1997). Layered metal IV phosphonates, a large class of inorgano-organic proton conductors. *Solid State Ionics* 97, 177.

Alberti, G., and Casciola, M. (2003). Composite membranes for medium-temperature PEM fuel cells. *Annu. Rev. Mater. Res.* 33, 129.

Alberti, G., Casciola, M., D'Alessandro, E., and Pica, M. (2004). Preparation and proton conductivity of composite ionomeric membranes obtained from gels of amorphous zirconium phosphate sulfophenylphosphonates in organic solvents. *J. Mater. Chem.* 14, 1910.

Alberti, G., Casciola, M., and Palombari, R. (2000). Inorganic-organic proton conducting membranes for fuel cells and sensors at medium temperatures. *J. Membr. Sci.* 172, 233.

Alberti, G., Casciola, M., Pica, M., Tarpanelli, T., and Sganappa, M. (2005). New preparation methods for composite membranes for medium temperature fuel cells based on precursor solutions of soluble inorganic compounds. *Fuel Cells* 5(3), 366.

Alberti, G., Constantino, U., Casciola, M., Ferroni, S., Massinelli, L., and Staiti, P. (2001). Preparation, characterization and proton conductivity of titanium phosphate sulfophenylphosphonate. *Solid State Ionics* 145, 249.

Allcock, H. R. (2002). *Chemistry and Applications of Polyphosphazenes.* Wiley, Hoboken, NJ.

Allcock, H. R., Hofmann, M. A., Ambler, C. M., Lvov, S. N., Zhou, X. Y., Chalkova, E., and Weston, J. (2002). Phenyl phosphonic acid functionalized poly[aryloxyphosphazenes] as proton-conducting membranes for direct methanol fuel cells. *J. Membr. Sci.* 201, 47.

Antonucci, P. L., Aroco, A. S., Creti, P., Ramunni, E., and Antonucci, V. (1999). Investigation of a direct methanol fuel cell based on a composite Nafion®-silica electrolyte for high temperature operation. *Solid State Ionics* 125, 431.

Aparicio, M., Damay, F., and Klein, L. C. (2003). Characterization of SiO_2-P_2O_5-ZrO_2 sol-gel/Nafion™ composite membranes. *J. Sol-Gel Sci. Technol.* 26, 1055.

Asensio, J. A., and Gomez-Romero, P. (2005). Recent developments on proton conducting poly(2,5-benzimidazole) (ABPBI) membranes for high temperature polymer electrolyte membrane fuel cells. *Fuel Cells* 5(3), 336.

Baglio, V., Arico, A. S., Di Blasi, A., Antonucci, V., Antonucci, P. L., Licoccia, S., Traversa, E., and Serraino Fiory, F. (2005). Nafion-TiO_2 composite membranes: Physico-chemical properties of the filler versus electrochemical performance. *Electrochim. Acta* 50, 1241.

Banerjee, S., and Curtin, D. E. (2004). Nafion® perfluorinated membranes in fuel cells. *J. Fluorine Chem.* 125, 1211.

Bashyam, R., Brosha, E., Hamon, Ch., Kim, Y. S., Choi, J.-H., Neergat, M., Piela, B., Piela, P., Pivovar, B., Purdy, G., Ramsey, J., Wilson, M., and Zelenay, P. (2005). Direct methanol fuel cells. In *Hydrogen, Fuel Cells & Infrastructure Technologies Program, 2005 Annual Review*, Washington, DC, May 23–27, 2005. Available: http://www.hydrogen.energy.gov.

Borka, K., and Ekdunge, P. (1997). Oxygen and hydrogen permeation properties and water uptake of Nafion 117 membrane and recast film for PEM fuel cell. *J. Appl. Electrochem.* 27, 117.

Boysen, D. A., Uda, T., Chisholm, C. R. I., and Haile, S. M. (2004). High-performance solid acid fuel cell through humidity stabilization. *Science* 303, 68.

Carter, R., Wycisk, R., Yoo, H., and Pintauro, P. N. (2002). Blended polyphosphazene/polyacrylonitrile membranes for direct methanol fuel cells. *Electrochem. Solid-State Lett.* 5, A195.

Chen, S.-L., Benziger, J.B., Bocarsly, A. B., and Zhang, T. (2005). Photo-cross-linking of sulfonated styrene-ethylene-butylene copolymer membranes for fuel cells. *Ind. Eng. Chem. Res.* 44, 7701.

Chisholm, C. R. I., and Haile, S. M. (2000). Superprotonic behavior of $Cs_2(HSO_4)(H_2PO_4)$—A new solid acid in the $CsHSO_4$-CsH_2PO_4 system. *Solid State Ionics* 136/137, 229.

Chisholm, C. R. I., Merle, R. B., Boysen, D. A., and Haile, S. M. (2002). Superprotonic phase transition in $CsH(PO_3H)$. *Chem. Mater.* 14, 3889.

Dargaville, T. R., George, G. A., Hill, D. J. T., and Whittaker, A. K. (2003). High energy radiation grafting of fluoropolymers. *Prog. Polym. Sci.* 28, 1355.

Dimitrova, P., Friedrich, K. A., Vogt, B., and Stimming, U. (2002). Transport properties of ionomer composite membranes for direct methanol fuel cells. *J. Electroanal. Chem.* 532, 75.

Ehrenberg, S. G., Serpico, J. K., Wnek, G. E., and Rider, J. N. (1995). Fuel cell incorporating novel ion-conducting membrane. U.S. Patent 5,468,574.

Einsla, B. R., Hong, Y.-T., Kim, Y. S., Wang, F., Gunduz, N., and McGrath, J. E. (2004). Sulfonated naphthalene dianhydride based polyimide copolymers for proton-exchange-membrane fuel cells. I. Monomer and copolymer synthesis. *J. Polym. Sci., Part A: Polym. Chem.* 42, 862.

Einsla, B. R., Kim, Y. S., Hickner, M. A., Hong, Y.-T., Hill, M. L., Pivovar, B. S., and McGrath, J. E. (2005). Sulfonated naphthalene dianhydride based polyimide copolymers for proton-exchange-membrane fuel cells. II. Membrane properties and fuel cell performance. *J. Membr. Sci.* 255, 141.

Fedkin, M. V., Zhou, X., Hofmann, M. A., Chalkova, E., Weston, J. A., Allcock, H. R., and Lvov, S. N. (2002). Evaluation of methanol crossover in proton-conducting polyphosphazene membranes. *Mater. Lett.* 52, 192.

Gao, Y., Robertson, G. P., Guiver, M. D., and Jian, X. (2003). Synthesis and characterization of sulfonated poly(phthalazinone ether ketone) for proton exchange membrane materials. *J. Polym. Sci. Part A: Polym. Chem.* 41, 497.

Gasteiger, H. A., and Mathias, M. F. (2005). Fundamental research and development challenges in polymer electrolyte fuel cell technology. In M. Murthy, T. F. Fuller, J. W. Van Zee, and S. Gottesfeld (Eds.), *Proton Conducting Membrane Fuel Cells*, Vol. 3, Electrochemical Society, Pennington, NJ, pp. 1–24.

Genies, C., Mercier, R., Sillion, B., Cornet, N., Gebel, G., and Pineri, M. (2001a). Soluble sulfonated naphthalenic polyimides as materials for proton exchange membranes. *Polymer* 42, 359.

Genies, C., Mercier, R., Sillion, B., Cornet, N., Gebel, G., and Pineri, M. (2001b). Stability study of sulfonated phthalic and naphthalenic polyimide structures in aqueous medium. *Polymer* 42, 5097.

Gil, M., Ji, X. L., Li, X. F., Na, H. Hampsey, J. E., and Lu, Y. F. (2004). Direct synthesis of sulfonated aromatic poly(ether ether ketone) proton exchange membranes for fuel cell applications. *J. Membr. Sci.* 234, 75.

Gode, P., Ihonen, J., Strandroth, A., Ericson, H., Lindberg, G., Paronen, M., Sundholm, F., Sundholm, G., and Walsby, N. (2003). Membrane durability in a PEM fuel cell studied using PVDF based radiation grafted membranes. *Fuel Cells* 3(1–2), 21.

Gottesfeld, S., and Zawodzinski, T. A. (1997). Polymer electrolyte fuel cells. In R. Alkire, H. Gerischer, D. Kolb, and C. Tobias (Eds.), *Advances in Electrochemical Science and Engineering*, Vol. 5, VCH, New York, NY p. 195.

Gubler, L., Gursel, S. A., and Scherer, G. G. (2005). Radiation grafted membranes for polymer electrolyte fuel cells. *Fuel Cells* 5(3), 317.

Gubler, L., Kuhn, H., Schmidt, T. J., Scherer, G. G., Brack, H. P., and Simbeck, K. (2004). Performance and durability of membrane electrode assemblies based on radiation-grafted FEP-*g*-polystyrene membranes. *Fuel Cells* 4(3), 196.

Guo, Q., Pintauro, P. N., Tang, H., and O'Connor, S. (1999). Sulfonated and crosslinked polyphosphazene-based proton-exchange membranes. *J. Membr. Sci.* 154, 175.

Haile, S. M. (2003). Fuel cell materials and components. *Acta Materialia* 51, 5981.

Harrison, W. L., Hickner, M. A., Kim, Y. S., and McGrath, J. E. (2005). Poly(arylene ether sulfone) copolymers and related systems form disulfonated monomer building blocks: Synthesis, characterization, and performance—A topical review. *Fuel Cells* 5(2), 201.

Harrison, W. L., Wang, F., Mecham, J. B., Bhanu, V. A., Hill, M., Kim, Y. S., and McGrath, J. E. (2003). Influence of the bisphenol structure on the direct synthesis of sulfonated poly(arylene ether) copolymers. *J. Polym. Sci. Part A: Polym. Chem.* 41, 2264.

Hejze, T., Gollas, B. R., Sauerbrey, R. K., Schmied, M., Hofer, F., and Besenhard, J. O. (2005). Preparation of Pd-coated polymer electrolyte membranes and their application in direct methanol fuel cells. *J. Power Sources* 140, 21.

Herring, A. M., Turner, J. A., Dec, S. F., Sweikart, M. A., Malers, J. L., Meng, F., Pern, F. J., Horan, J., and Vernon, D. (2004). The use of heteropoly acids in composite membranes for elevated temperature PEM fuel cell operation; lessons learnt from three different approaches. Paper presented at the 2004 Fuel Cell Seminar; Technology, Markets, and Commercialization. San Antonio, TX, Nov. 2–5, 2004. Available: www.fuelcellseminar.com.

Herz, H. G., Kreuer, K. D., Maier, J., Scharfenberger, G., Schuster, M. F. H., and Meyer, W. H. (2003). New fully polymeric proton solvents with high proton mobility. *Electrochim. Acta* 48, 2165.

Hickner, M. A., Ghassemi, H., Kim, Y. S., Einsla, B. R., and McGrath, J. E. (2004). Alternative polymer systems for proton exchange membranes (PEMs). *Chem. Rev.* 104, 4587.

Hogarth, W. H. J., Diniz da Costa, J. C., and Lu, G. Q. (2005). Solid acid membranes for high temperature (>140°C) proton exchange membrane fuel cells. *J. Power Sources* 142, 223.

Horsfall, J. A., and Lovell, K. V. (2001). Fuel cell performance of radiation grafted sulphonic acid membranes. *Fuel Cells* 1(3–4), 186.

Iojoiu, C., Marechal, M., Chabert, F., and Sanchez, J.-Y. (2005). Mastering sulfonation of aromatic polysulfones: Crucial for membranes for fuel cell application. *Fuel Cells* 5(3), 344.

Jannasch, P. (2003). Recent developments in high-temperature proton conducting polymer electrolyte membranes. *Curr. Opin. Colloid Interface Sci.* 8, 96.

Jannasch, P. (2005). Fuel cell membrane materials by chemical grafting of aromatic main-chain polymers. *Fuel Cells* 5(2), 248.

Jia, N., Lefebvre, M. C., Halfyard, J., Qi, Z., and Pickup, P. G. (2000). Modification of Nafion proton exchange membranes to reduce methanol crossover in PEM fuel cells. *Electrochem. Solid-State Lett.* 3, 529.

Jiang, R., Kunz, H. R., and Fenton, J. M. (2005). Investigation of membrane property and fuel cell behavior with sulfonated poly(ether ether ketone) electrolyte: Temperature and relative humidity effects. *J. Power Sources* 150, 120.

Jorissen, L., Gogel, V., Kerres, J., and Garche, J. (2002). New membranes for direct methanol fuel cells. *J. Power Sources* 105, 267.

Jung, D. H., Cho, S. Y., Peck, D. H., Shin, D. R., and Kim, J. S. (2003). Preparation and performance of a Nafion®/montmorillonite nanocomposite membrane for direct methanol fuel cell. *J. Power Sources* 118, 205.

Kabanov, V. Ya. (2004). Preparation of polymer membranes for fuel cells by radiation graft polymerization. *High Energy Chem.* 38(2), 57.

Kaliaguine, S., Mikhailenko, S. D., Wang, K. P., Xing, P., Robertson, G., and Guiver, M. (2003). Properties of SPEEK based PEMs for fuel cell application. *Catal. Today* 82, 213.

Kallio, T., Lundstrom, M., Sundholm, G., Walsby, N., and Sundholm, F. (2002). Electrochemical characterization of radiation-grafted ion-exchange membranes based on different matrix polymers. *J. Appl. Electrochem.* 32, 11.

Kerres, J. A. (2001). Development of ionomer membranes for fuel cells. *J. Membr. Sci.* 185, 3.

Kerres, J. A. (2005). Blended and cross-linked ionomer membranes for application in membrane fuel cells. *Fuel Cells* 5, 230.

Kim, Y. S., Siu, A., and Pivovar, B. (2005). Non-Nafion membrane electrode assemblies. In *Hydrogen, Fuel Cells & Infrastructure Technologies Program, 2005 Annual Review*, Washington, DC, May 23–27, 2005. Available: http://www.hydrogen.energy.gov.

Kreuer, K. D. (2001). On the development of proton conducting polymer membranes for hydrogen and methanol fuel cells. *J. Membr. Sci.* 185, 29.

Kreuer, K.-D., Paddison, S. J., Spohr, E., and Schuster, M. (2004). Transport in proton conductors for fuel-cell applications: Simulations, elementary reactions, and phenomenology. *Chem. Rev.* 104, 4637.

Lee, K., Nam, J.-H., Lee, J. H., Lee, Y., Cho, S. M., Jung, C. H., Choi, H. G., Chang, Y.-Y., Kwon, Y.-U., and Nam, J.-D. (2005). Methanol and proton transport control by using layered double hydroxide nanoplatelets for direct methanol fuel cell. *Electrochem. Commun.* 7, 113.

Li, Q., He, R., Jensen, J. O., and Bjerrum, N. J. (2003). Approaches and recent development of polymer electrolyte membranes for fuel cells operating above 100°C. *Chem. Mater.* 15, 4896.

Li, Q., He, R., Jensen, J. O., and Bjerrum, N. J. (2004). PBI-based polymer membranes for high temperature fuel cells—Preparation, characterization and fuel cell demonstration. *Fuel Cells* 4(3), 147.

Limoges, B. R., Stanis, R. J., Turner, J. A., and Herring, A. M. (2005). Electrocatalyst materials for fuel cells based on the polyoxometalates $[PMo_{(12-n)}V_nO_{40}]^{(3+n)-}$ ($n = 0-3$). *Electrochim. Acta* 50, 1169.

Lin, J., Lee, J. K., Kellner, M., Wycisk, R., and Pintauro, P. N. (2006). Nafion-fluorinated ethylene-propylene resin membrane blends for direct methanol fuel cells. *J. Electrochem. Soc.* 153, A1325.

Lin, J.-Ch., Ouyang, M., Fenton, J. M., Kunz, H. R., Koberstein, T., and Cutlip, M. B. (1998). Study of blend membranes consisting of Nafion® and vinylidene fluoride-hexafluoropropylene copolymer. *J. Appl. Polym. Sci.* 70, 121.

Liu, J., Wang, H., Cheng, S., and Chan, K.-Y. (2005). Nafion-polyfurfuryl alcohol nanocomposite membranes for direct methanol fuel cells. *J. Membr. Sci.* 246, 95.

Lufrano, F., Squadrito, G., Patti, A., and Passalacqua, E. (2000). Sulfonated polysulfone as promising membranes for polymer electrolyte fuel cells. *J. Appl. Polym. Sci.* 77, 1250.

Merle, R. B., Chisholm, C. R. I., Boysen, D. A., and Haile, S. M. (2003). Instability of sulfate and selenate solid acids in fuel cell environments. *Energy & Fuels* 17, 210.

Mokrini, A., Del Rio, C., and Acosta, J. L. (2004). Synthesis and characterization of new ion conductors based on butadiene styrene copolymers. *Solid State Ionics* 166, 375.

Nunes, S. P., Ruffmann, B., Rikowski, E., Vetter, S., and Richau, K. (2002). Inorganic modification of proton conductive polymer membranes for direct methanol fuel cells. *J. Membr. Sci.* 203, 215.

Park, Y.-S., and Yamazaki, Y. (2005a). Low methanol permeable and high proton-conducting Nafion/calcium phosphate composite membrane for DMFC. *Solid State Ionics* 176, 1079.

Park, Y.-S., and Yamazaki, Y. (2005b). Novel Nafion/hydroxyapatite composite membrane with high crystallinity and low methanol crossover for DMFCs. *Polym. Bull.* 53, 181.

Patri, M., Hande, V. R., Phadnis, S., Somaiah, B., Roychoudhury, S., and Deb, P. C. (2004). Synthesis and characterization of SPE membrane based on sulfonated FEP-*g*-acrylic acid by radiation induced graft copolymerization for PEM fuel cell. *Polym. Adv. Technol.* 15, 270.

Pavlak, C., Wycisk, R., Lin, J., and Pintauro, P. N. (2007). Effect of membrane transport characteristics on direct methanol fuel cell performance. *The Ohio Journal of Science* 107(1), A21.

Pintauro, P. N., and Wycisk, R. (2004). Sulfonated polyphosphazene membranes for direct methanol fuel cells. In M. Gleria and R. DeJaeger (Eds.), *Phosphazenes: A Worldwide Insight*, Nova Science Publishers, Hauppauge, NY, p. 591.

Prakash, G. K. S., Smart, M. C., Wang, Q.-J., Atti, A., Pleynet, V., Yang, B., McGrath, K., Olah, G. A., Narayanan, S. R., Chun, W., Valdez, T., and Surampudi, S. (2004). High efficiency direct methanol fuel cell based on poly(styrenesulfonic) acid (PSSA)-poly(vinylidene fluoride) (PVDF) composite membranes. *J. Fluorine Chem.* 125, 1217.

Pu, C., Huang, W., Ley, K. L., and Smotkin, E. S. (1995). A methanol impermeable proton conducting composite electrolyte system. *J. Electrochem. Soc.* 142, L119.

Ramani, V., Kunz, H. R., and Fenton, J. M. (2005). Stabilized heteropolyacid/Nafion® composite membranes for elevated temperature/low relative humidity PEFC operation. *Electrochim. Acta* 50, 1181.

Ren, X., Springer, T. E., Zawodzinski, T. A., and Gottesfeld, S. (2000). Methanol transport through Nafion membranes. Electro-osmotic drag effects on potential step measurements. *J. Electrochem. Soc.* 147, 466.

Rhee, C. H., Kim, H. K., Chang, H., and, Lee, J. S. (2005). Nafion/sulfonated montmorillonite composite: A new concept electrolyte membrane for direct methanol fuel cells. *Chem. Mater.* 17, 1691.

Rikukawa, M., and Sanui, K. (2000). Proton-conducting polymer electrolyte membranes based on hydrocarbon polymers. *Prog. Polym. Sci.* 25, 1463.

Roziere, J., and Jones, D. J. (2003). Non-fluorinated polymer materials for proton exchange membrane fuel cells. *Annu. Rev. Mater. Res.* 33, 503.

Sauk, J., Byun, J., and Kim, H. (2005). Composite Nafion/polyphenylene oxide (PPO) membranes with phosphomolybdic acid (PMA) for direct methanol fuel cells. *J. Power Sources* 143, 136.

Savadogo, O. (2004). Emerging membranes for electrochemical systems. Part II. High temperature composite membranes for polymer electrolyte fuel cell (PEFC) applications. *J. Power Sources* 127, 135.

Schuster, M., Meyer, W. H., Wegner, G., Herz, H. G., Ise, M., Schuster, M., Kreuer, K. D., and Maier, J. (2001). Proton mobility in oligomer-bound proton solvents: Imidazole immobilization via flexible spacers. *Solid State Ionics* 145, 85.

Serpico, J. M., Ehrenberg, S. G., Fontanella, J. J., Jiao, X., Perahia, D., McGrady, K. A., Sanders, E. H., Kellogg, G. E., and Wnek, G. E. (2002). Transport and structural studies of sulfonated styrene-ethylene copolymer membranes. *Macromolecules* 35, 5916.

Shim, J. H., Koo, I. G., and Lee, W. M. (2005). Nafion-impregnated polyethylene-terephthalate film used as the electrolyte for direct methanol fuel cells. *Electrochim. Acta* 50, 2385.

Shim, S.-Y., and Weiss, R. A. (2005). Sulfonated poly(ethylene-ran-styrene) ionomers. *Polym. Int.* 54, 1220.

Silva, R. F., Passerini, S., and Pozio, A. (2005a). Solution-cast Nafion®/montmorillonite composite membrane with low methanol permeability. *Electrochim. Acta* 50, 2639.

Silva, V. S., Ruffmann, B., Vetter, S., Mendes, A., Madeira, L. M., and Nunes, S. P. (2005b). Characterization and application of composite membranes in DMFC. *Catal. Today* 104, 205.

Silva, V. S., Weisshaar, S., Reissner, R., Ruffmann, B., Vetter, S., Mendes, A., Madeira, L. M., and Nunes, S. (2005c). Performance and efficiency of a DMFC using non-fluorinated composite membranes operating at low/medium temperatures. *J. Power Sources* 145, 485.

Smitha, B., Sridhar, S., and Khan, A. A. (2005). Solid polymer electrolyte membranes for fuel cell applications—A review. *J. Membr. Sci.* 259, 10.

Staiti, P., Freni, S., and Hocevar, S. (1999). Synthesis and characterization of proton-conducting materials containing dodecatungstophosphoric acid and dodecatungstosilicic acid supported on silica. *J. Power Sources* 79, 250.

Summer, M. J., Harrison, W. L., Weyers, R. M., Kim, Y. S., McGrath, J. E., Riffle, J. S., Brink, A., and Brink, M. H. (2004). Novel proton conducting sulfonated poly(arylene ether) copolymers containing aromatic nitriles. *J. Membr. Sci.* 239, 199.

Sundar, S., Jang, W., Lee, C., Shul, Y., and Han, H. (2005). Crosslinked sulfonated polyimide networks as polymer electrolyte membranes in fuel cells. *J. Polym. Sci. Part B: Polym. Phys.* 43, 2370.

Tang, H., Pan, M., Jiang, S., Wan, Z., and Yuan, R. (2005). Self-assembling multi-layer Pd nanoparticles onto Nafion™ membrane to reduce methanol crossover. *Colloids Surfaces Physicochem. Eng. Aspects* 262(2005), 65.

Tang, H., and Pintauro, P. N. (2001). Polyphosphazene membranes. IV. Polymer morphology and proton conductivity in sulfonated poly[bis(3-methylphenoxy)phosphazene] films. *J. Appl. Polym. Sci.* 79, 49.

Tang, H., Pintauro, P. N., Guo, Q., and O'Connor, S. (1999). Polyphosphazene membranes. III. Solid-state characterization and properties of sulfonated poly[bis(3-methylphenoxy)phosphazene]. *J. Appl. Polym. Sci.* 71, 387.

Tian, H., and Savadogo, O. (2005) Effect of silicotungstic acid (STA) on the performance of a polymer electrolyte fuel cell (PEMFC) based on Nafion cast in dimethylformamide. *Fuel Cells* 5(3), 375.

Tricoli, V. (1998). Proton and methanol transport in poly(perfluorosulfonate) membranes containing Cs^+ and H^+ cations. *J. Electrochem. Soc.* 145, 3798.

Uda, T., and Haile, S. M. (2005). Thin-membrane solid-acid fuel cell. *Electrochem. Solid-State Lett.* 8(5), A245.

Vetter, S., Ruffmann, B., Buder, I., and Nunes, S. P. (2005). Proton conductive membranes of sulfonated poly(ether ketone ketone). *J. Membr. Sci.* 260, 181.

Wainright, J. S., Wang, J.-T., Weng, D., Savinell, R. F., and Litt, M. (1995). Acid-doped polybenzimidazoles: A new polymer electrolyte. *J. Electrochem. Soc.* 142, L121.

Walker, M., Baumgartner, K.-M., Feichtinger, J., Kaiser, M., Rauchle, E., and Kerres, J. (1999). Barrier properties of plasma-polymerized thin films. *Surf. Coat. Technol.* 116–119, 996.

Wang, F., Hickner, M., Kim, Y. S., Zawodzinski, T. A., and McGrath, J. E. (2002). Direct polymerization of sulfonated poly(arylene ether sulfone) random (statistical) copolymers: Candidates for new proton exchange membranes. *J. Membr. Sci.* 197, 231.

Wnek, G., Ehrenberg, S., Serpico, J., Tangredi, T., Doell, G., and Zador, E. (1999). Hydrocarbon PEMs revisited. *Fuel Cell Bull.* 2(4), 6.

Wycisk, R., Carter, R., Pintauro, P. N., and Byrne, C. (2002). Ion-exchange membranes from blends of sulfonated polyphosphazene and Kynar FLEX PVDF. In I. Pinnau and B. D. Freeman (Eds.), *Advanced Materials for Membrane Separations*. ACS Symposium Series 876. American Chemical Society, Washington, DC, Chapter 23, p. 335.

Wycisk, R., Chisholm, J., Lee, J., Lin, J., and Pintauro, P. N. (2006). Direct methanol fuel cell membranes from Nafion-polybenzimidazole blends. *J. Power Sources* 163, 9.

Wycisk, R., Lee, J. K., and Pintauro, P. N. (2005). Sulfonated polyphosphazene-polybenzimidazole membranes for direct methanol fuel cells. *J. Electrochem Soc.* 152, A892.

Wycisk, R., and Pintauro, P. N. (1996). Sulfonated polyphosphazene ion-exchange membranes. *J. Membr. Sci.* 119, 155.

Wycisk, R., and Pintauro, P. N. (2003). Polyphosphazenes. In J. I. Kroschwitz (Ed.), *Encyclopedia of Polymer Science and Technology*, Vol. 7, 3rd ed. Wiley, Hoboken, NJ, pp. 603–625.

Xiao, L., Zhang, H., Jana, T., Scanlon, E., Chen, R., Choe, E.-W., Ramanathan, L. S., Yu, S., and Benicewicz, B. C. (2005). Synthesis and characterization of pyridine-based polybenzimidazoles for high temperature polymer electrolyte membrane fuel cell applications. *Fuel Cells* 5(2), 287.

Xing, P., Robertson, G. P., Guiver, M. D., Mikhailenko, S. D., and Kalliaguine, S. (2004a). Sulfonated poly(aryl ether ketone)s containing naphthalene moieties obtained by direct copolymerization as novel polymers for proton exchange membranes. *J. Polym. Sci. Part A: Polym. Chem.* 42, 2866.

Xing, P., Robertson, G. P., Guiver, M. D., Mikhailenko, S. D., Wang, K., and Kaliaguine, S. (2004b). Synthesis and characterization of sulfonated poly(ether ether ketone) for proton exchange membranes. *J. Membr. Sci.* 229, 95.

Yamada, O., Yin, Y., Tanaka, K., Kita, H., and Okamoto, K.-I. (2005). Polymer electrolyte fuel cells based on main-chain-type sulfonated polyimides. *Electrochim. Acta* 50, 2655.

Yamaguchi, T., Kuroki, H., and Miyata, F. (2005). DMFC performances using a pore-filling polymer electrolyte membrane for portable usages. *Electrochem. Commun.* 7, 730.

Yamaguchi, T., Miyata, F., and Nako, S. (2003). Pore-filling type polymer electrolyte membranes for a direct methanol fuel cell. *J. Membr. Sci.* 214, 283.

Yin, Y., Hayashi, S., Yamada, O., Kita, H., and Okamoto, K.-I. (2005). Branched/crosslinked sulfonated polyimide membranes for polymer electrolyte fuel cells. *Macromol. Rapid Commun.* 26, 696.

Zawodzinski, T., Neeman, M., Sillerud, L. O., and Gottesfeld, S. (1991). Determination of water diffusion coefficient in perfluorosulfonate ionomeric membranes. *J. Phys. Chem.* 95, 6040.

Zhang, Y., and Litt, M. H. (1999). Polyimides based on naphthalene tetracarboxylic dianhydride and benzidinedisulfonic acid. *ACS Polym. Preprints* 40(2), 480.

Zhou, H., Miyatake, K., and Watanabe, M. (2005). Polyimide electrolyte membranes having fluorenyl and sulfopropoxy groups for high temperature PEFCs. *Fuel Cells* 5(2), 296.

PART VI

MEMBRANE MATERIALS AND CHARACTERIZATION

CHAPTER 30

Recent Progress in Mixed-Matrix Membranes

CHUNQING LIU and SANTI KULPRATHIPANJA
UOP LLC, 25 East Algonquin Road, Des Plaines, Illinois, 60017

ALEXIS M. W. HILLOCK, SHABBIR HUSAIN, and WILLIAM J. KOROS
School of Chemical & Biomolecular Engineering, Georgia Institute of Technology, Atlanta, Georgia, 30332

30.1 INTRODUCTION

30.1.1 Historical Review

The drive toward greater economic and environmental efficient separation processes has resulted in the rapid development of membrane processes in recent years. The adoption of membranes for separation and purification has resulted from the progress made in membrane materials, membrane structure, and large-scale membrane production methods. The breakthrough development of Loeb–Sourirajan process[1] to make defect-free ultrathin cellulose acetate membranes in the 1960s led to initial commercialization of reverse osmosis membranes in the 1970s followed by intense research activity and further commercialization in the 1980s. Today, polymer-membrane-based separation processes are widely used in the petrochemical, semiconductor, food, water, pharmaceutical, and biotechnology industries as well as a wide range of environmental applications.

To be useful for separation or purification processes, membranes must exhibit a number of characteristics such as high flux, high selectivity, mechanical stability, resistance to fouling, and low cost.

30.1.2 Current State-of-the-Art Polymeric Membranes

The membranes most commonly used in membrane separation processes are polymeric and nonporous. The separation is based on a solution–diffusion mechanism. This mechanism involves molecular-scale interactions of the permeating molecule with the membrane polymer. The mechanism assumes that each molecule is sorbed by the membrane at one interface, transported by diffusion across the membrane through the voids between

Advanced Membrane Technology and Applications. Edited by Norman N. Li, Anthony G. Fane, W. S. Winston Ho, and T. Matsuura
Copyright © 2008 John Wiley & Sons, Inc.

the polymeric chains (or called free volume), and desorbed at the other interface. According to the solution–diffusion model, the permeation of molecules through membranes is controlled by two major parameters: diffusivity coefficient (D) and solubility coefficient (S). Diffusivity is a measure of the mobility of individual molecules passing through the voids between the polymeric chains in a membrane material. The solubility coefficient equals the ratio of sorption uptake normalized by some measure of uptake potential, such as partial pressure. Permeability (P), defined in Eq. (30.1), represents the ability of molecules to permeate a membrane:

$$P = DS \tag{30.1}$$

The ability of a membrane to separate two molecules, for example, A and B, is the ratio of their permeabilities, called the membrane selectivity, α_{AB}:

$$\alpha_{AB} = \frac{P_A}{P_B} \tag{30.2}$$

Since P is the product of D and S, Eq. (30.2) can be rewritten as

$$\alpha_{AB} = \left(\frac{D_A}{D_B}\right)\left(\frac{S_A}{S_B}\right) \tag{30.3}$$

where D_A/D_B is the ratio of the diffusion coefficients of the two molecules and can be viewed as the mobility or diffusivity selectivity, reflecting the different sizes of the two molecules; S_A / S_B is the ratio of the Henry's law sorption coefficients of the two molecules and can be viewed as the sorption or solubility selectivity of the two molecules. The balance between the solubility selectivity and the diffusivity selectivity determines whether a membrane material is selective for molecule A or molecule B in a feed mixture. Gases can have high permeability coefficients because of high solubility coefficients, high diffusion coefficients, or both. Generally speaking, the diffusion coefficient decreases and the solubility coefficient increases with an increase in the molecular size of the gas. For high-performance polymer membranes, both high permeability and selectivity are desirable. The higher the permeability, the less the amount of membrane area required to treat a given amount of gas, and the higher the selectivity, the higher the purity of the product gas under equivalent feed and permeate conditions.

Polymers provide a range of desirable properties that are important for gas separation processes including low cost, high permeability, good mechanical stability, and ease of processability. A polymer material with a high glass transition temperature (T_g), high melting point, and high crystallinity is generally preferred.[2] Glassy polymers (i.e., polymers below their T_g) have stiffer polymer backbones and therefore let smaller molecules such as H_2 and He pass more quickly, and larger molecules such as hydrocarbons permeate the membrane more slowly.[3–5] To increase the membrane selectivity, either the diffusivity or the solubility needs to be enhanced; however, polymers that are more permeable are generally less selective and vice versa.[6] A rather general trade-off exists between permeability and selectivity (so-called polymer upper-bound limit).[7,8] This is illustrated in Figure 30.1, which shows an upper bound in the relationships between the CO_2/CH_4 selectivity and the permeability of CO_2 for various glassy and rubbery polymers. A substantial research effort has been directed to overcoming the limit imposed by the upper bound,

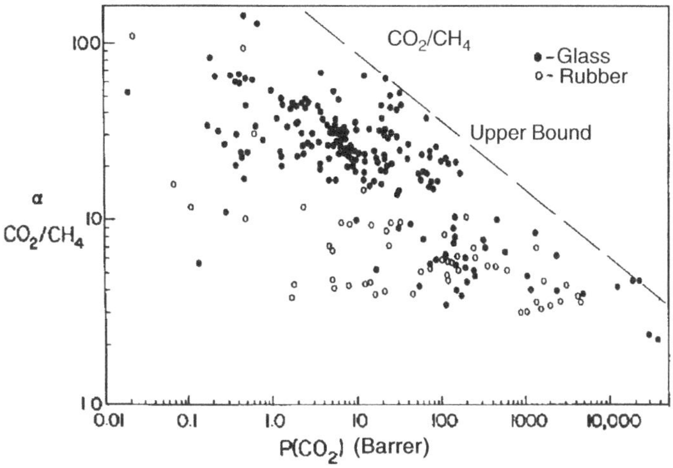

Figure 30.1 Literature data for CO_2/CH_4 separation factor versus CO_2 permeability.[7]

TABLE 30.1 Kinetic Diameters for Various Molecules

Molecule	Kinetic Diameter (A)
H_2	2.89
O_2	3.46
N_2	3.64
CO	3.76
CO_2	3.3
H_2O	2.65
NH_3	2.6
SO_2	3.6
CH_4	3.8
C_2H_2	3.3
C_2H_4	3.9

and despite many attempts, success has been rather limited.[9,10] This is due to the small difference in the kinetic diameters of each molecule to be separated (see Table 30.1).[11]

Cellulose acetate (CA) glassy polymer membranes[12] are used extensively in gas separation. Although CA membranes have many advantages, they are limited in terms of selectivity and chemical, thermal, and mechanical stabilities. One of the immediate challenges facing CA polymer membranes is achieving higher selectivity with equal or greater permeability to the current generation of these materials.

30.1.3 Current State-of-the-Art Inorganic Membranes

Inorganic membranes, such as purely molecular sieving zeolite membranes,[13–38] carbon membranes,[39,40] alumina membranes,[41] and silica membranes,[42] have high thermal and chemical stabilities. Over the past 25 years, extensive work has been reported on the synthesis, characterization, and application of inorganic membranes.

Membranes of various zeolites, such as deca-dodecasil 3R (DDR) type,[13,14] aluminophosphate ($AlPO_4$) type,[15] silicoaluminophosphate,[16–19] ZSM-5 (MFI),[20–35] Y type,[36] A type,[37] P type,[38] and modernite,[31] have been synthesized on porous supports. Some of the membranes have good selectivity. For example, DDR-type zeolite membrane[13,14] has shown much higher CO_2 permeability and CO_2 over CH_4 selectivity compared to CA polymer membrane. MFI-type membrane[34,35] has shown good selectivity toward *p*-xylene over other C8-aromatics. Although good selectivity is achievable using zeolite membranes in some specific separations, their permeability is relatively low. Attempts were made to increase their permeability by developing a very thin zeolite membrane.[32,33] It is also noticed that these zeolite membranes, however, have poor processability and cannot be fabricated in an economically feasible way with current membrane manufacturing techniques for large-scale applications.

30.1.4 Concept of Mixed-Matrix Membranes

Despite all the advantages, polymeric membranes cannot overcome the polymer upper-bound limit between permeability and selectivity. On the other hand, some inorganic membranes such as zeolite and carbon molecular sieve membranes offer much higher permeability and selectivity than polymeric membranes but are expensive and difficult for large-scale manufacture. Therefore, it is highly desirable to provide an alternate cost-effective membrane in a position above the trade-off curves between permeability and selectivity.

Based on the need of a more efficient membrane than polymer and inorganic membranes, a new type of membranes, mixed-matrix membranes, has been developed recently. Mixed-matrix membranes are hybrid membranes containing solid, liquid, or both solid and liquid fillers embedded in a polymer matrix.[43–193] The various material combinations possible with mixed-matrix technology are represented in Figure 30.2. All of these combinations, with the exception of supported liquids, will be covered in this chapter.

Mixed-matrix membranes have the potential to achieve higher selectivity with equal or greater permeability compared to existing polymer membranes while maintaining their advantages. Enhanced separation properties are accomplished by adding a dispersed phase to the processable polymer matrix, the net result of which should be

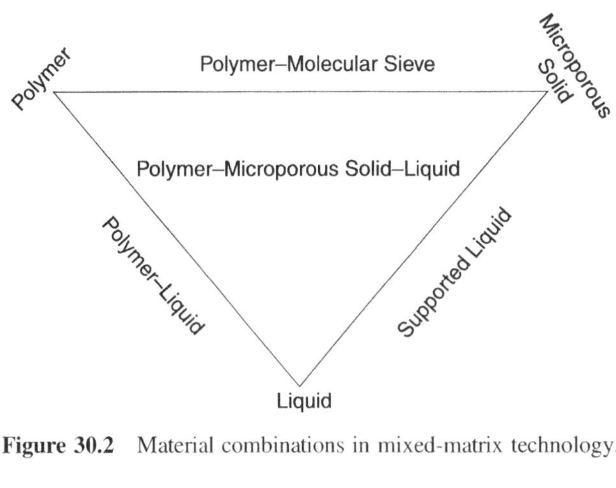

Figure 30.2 Material combinations in mixed-matrix technology.

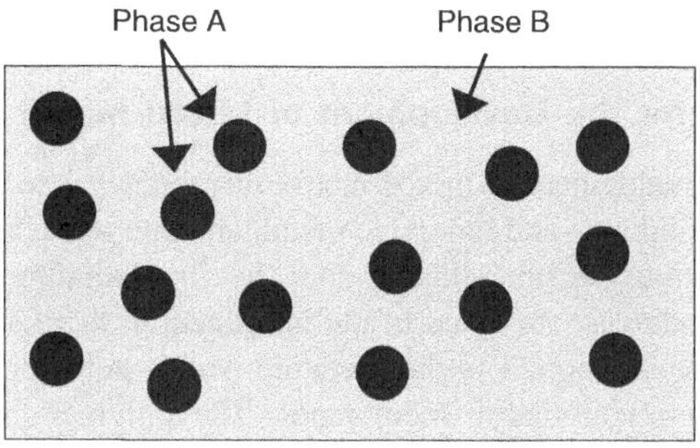

Figure 30.3 Mixed matrix example: phase A dispersed in phase B matrix.

an improvement in the separation properties of the overall membrane (Fig. 30.3). While the bulk phase (phase B in Fig. 30.3) is typically a polymer or a ceramic support, the dispersed phase (phase A in Fig. 30.3) can represent molecular sieves, liquids, or liquid-impregnated sieves.

Whatever combination is chosen, the resulting mixed-matrix membrane properties can be estimated to a first approximation through use of the so-called Maxwell model.[43] This model is well understood and accepted as a simple, but effective, tool for estimating mixed-matrix membrane properties. The Maxwell model equation is as follows:

$$P_{\mathrm{MM}} = P_M \left[\frac{P_D + 2P_M - 2\Phi_D(P_M - P_D)}{P_D + 2P_M + \Phi_D(P_M - P_D)} \right] \tag{30.4}$$

In this equation, P is permeability, Φ_D is the volume fraction of the dispersed phase, the MM subscript refers to the mixed-matrix membrane, the M subscript refers to the continuous matrix, and the D subscript refers to the dispersed phase. Provided a researcher knows the volume fraction of the dispersed phase and the permeability through the two pure materials, the calculation is uncomplicated. A vast majority of researchers compared their mixed-matrix membrane results to either the Maxwell model or some extension of this equation.

The concept of mixed-matrix membranes has been demonstrated at UOP[103] in the mid-1980s using silicalite–cellulose acetate mixed-matrix membranes for CO_2/H_2 separation. In the demonstration, a feed mixture of 50/50 CO_2/H_2 with a differential pressure of 50 psi was used. The calculated separation factor for CO_2/H_2 was found to be 5.15 ± 2.2. In contrast, a CO_2/H_2 separation factor of 0.77 ± 0.06 was found for cellulose acetate membrane. This indicates that silicalite in the membrane phase reversed the selectivity from H_2 to CO_2. Experimental results and modeling predictions indicate that mixed-matrix membranes with the incorporation of fillers within polymeric substrates provide potential possibilities to achieve enhanced membrane performance, which will open up new opportunities for the separation and purification processes. Highlighted applications for mixed-matrix membranes include separation and purification of gas mixtures such as separation of N_2 from air,[59,68,103,135] CO_2 removal from natural gas,[59,65,129,130] and separation of *n*-pentane from *i*-pentane,[69] as well as separation of liquid mixtures[81,84,86,87,131–134,183] such as pervaporation of ethanol–water mixtures[86,131–133] and toluene–ethanol mixtures.[134] This chapter reviews several key advances in the design, preparation, and applications of mixed-matrix membranes presented in the literature.

30.2 RECENT PROGRESS IN MIXED-MATRIX MEMBRANES

30.2.1 Strategies for the Development of Mixed-Matrix Membranes

The strategy for the development of mixed-matrix membranes is to combine the advanced features of polymer membrane and inorganic membrane into one composite membrane. As discussed in the previous section, this is done by incorporating dispersed fillers into continuous polymer matrices. As noted in the introduction, there are three main types of mixed-matrix membranes reported in the literature: solid–polymer, liquid–polymer, and solid–liquid–polymer mixed-matrix membranes. The polymer matrices providing low cost and easy processability are selected from either glassy polymers (e.g., polyimide, polysulfone, polyethersulfone, or cellulose acetate) or rubbery polymers (e.g., silicone rubber). The dispersed fillers include solid, liquid, or both solid and liquid.

For solid–polymer mixed-matrix membranes, the solid fillers dispersed in the polymer matrix include microporous molecular sieves (e.g., zeolites),[58–70,73,81–88,105,108,121,140–151,161,162,167,179,181,184] mesoporous molecular sieves,[102,155] metal–organic frameworks,[171] activated carbon,[45,78,154] carbon molecular sieves,[46,76,77,163] layered materials,[98,152] silicas,[74,100,180,182,183] C60,[80] TiO_2,[99] and microporous organic host.[156] To date, almost all of the studies on zeolite–polymer mixed-matrix membranes use commercially available large zeolite particles with particle sizes in the micron range as solid fillers. Most recently, increasing activities in using nanosized zeolites and other solid materials in mixed-matrix membrane applications have been reported.[157,158]

For the liquid–polymer mixed-matrix membranes, the physical state of the fillers incorporated into the continuous polymer matrix is liquid such as poly(ethylene glycol) (PEG).[135–139] Because of the long-term stability concern of liquid polymer encapsulated in the continuous polymer matrix, a new type of mixed-matrix membranes, solid–liquid–polymer mixed-matrix membrane, has been developed recently.[160,193] The solid such as activated carbon impregnated with liquid polymer such as PEG is functioned as stabilizer of the liquid polymer in the continuous polymer phase. Besides, activated carbon enhances the mixed-matrix membrane performance.

To enhance the commercial applicability of membrane separation processes, UOP LLC developed two types of mixed-matrix membranes in the mid-1980s[103,135–139] The first type is solid–polymer mixed-matrix membranes with solid inorganic adsorbent embedded in the polymer phase.[103,135,160] The polymers can be CA, polysulfone, polyethersulfone, polyelectrolyte complex polymer, or silicone rubber. The solid adsorbent can be zeolites such as NaX, AgX, NaY, NaA, and silicalite, silica gel, alumina, or activated carbon. The second type is liquid–polymer mixed-matrix membranes produced by casting PEG or other liquid filler and silicone rubber on a porous polysulfone or CA support.[136–139] Both types of mixed-matrix membranes were evaluated for the separation of polar gas from nonpolar gas, carbon dioxide from nitrogen and methane, and light paraffins from light olefins. Most recently, UOP LLC developed a third type of solid–liquid–polymer mixed-matrix membranes.[160,193] For this type of mixed-matrix membranes, the solid such as activated carbon was impregnated with liquid polymer such as PEG. The impregnated activated carbon was then dispersed in the continuous silicone rubber polymer matrix. The resulting mixture was then coated on top of a porous polymer membrane to form a mixed-matrix membrane. This hybrid mixed-matrix membrane combines the properties of the continuous polymer phase, the dispersed solid filler phase, and the impregnated liquid phase.

30.2.2 Solid–Polymer Mixed-Matrix Membranes

Mixed-matrix materials are excellent candidates to overcome the "upper-bound" constraint, as shown in Figure 30.1. Much of the research conducted to date on mixed-matrix membranes has focused on the combination of a solid molecular sieving phase, such as zeolites or carbon molecular sieves, with a processable bulk polymer matrix. The sieving phase in a solid–polymer mixed-matrix scenario can have a selectivity that is significantly larger than the pure polymer. Addition of a small volume fraction of sieves to the polymer matrix, therefore, increases the overall separation efficiency significantly. While the upper-bound curve has been surpassed using these solid–polymer mixed-matrix membrane,[43–47] there are still many aspects of this new technology that require more detailed investigation.

The first known article concerning mixed-matrix membranes was published in 1960 by Barrer and James. It examined several different zeolites dispersed in an inert polymer resin to create ion exchange membranes.[48,49] Voids were noted at the interface of the two media (i.e., where polymer and sieve meet). These voids resulted in a degradation of mixed-matrix performance. Today, researchers still struggle with material compatibility issues that exist at the solid–polymer interface, as well as other nonideal morphologies introduced by the solid–polymer mixed-matrix technology.[50]

Research has shown that the interfacial region, which is a transition phase between the continuous polymer and dispersed sieve phases, is of particular importance in successful mixed-matrix membrane formation.[51–55] The type of morphology that forms at the interfacial region has a direct impact on a membrane's separation properties, and its ability to reach the predicted Maxwell model properties. As shown in Figure 30.4, the ideal mixed-matrix membrane will exhibit both an increase in selectivity and permeability as the solid-phase volume fraction is increased, and the Maxwell model can be used to estimate these separation properties (as discussed earlier).

As mentioned in the publications by Barrer and James,[48,49] poor interfacial adhesion can result in interfacial voids that are much larger than the penetrating molecules. These voids are nonselective and permit the transport of both the slow and fast penetrant indiscriminately. Such voids destroy the selectivity enhancement due to addition of the sieving

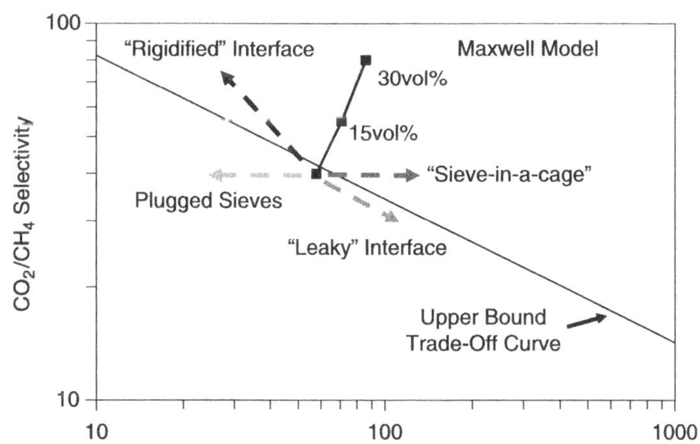

Figure 30.4 Possible interfacial morphologies for mixed-matrix membranes and their effect on CO_2/CH_4 gas transport properties; center point represents a polymer membrane with no sieving phase.

phase, reducing the overall selectivity of the mixed-matrix membrane to the pure polymer value, and increasing the permeability due to transport through the interfacial voids (see Fig. 30.4). This ruins any opportunity for the mixed-matrix membrane to achieve Maxwell model enhancement. The formation of these voids is usually termed *sieve-in-a-cage* morphology. The sieve-in-a-cage phenomenon is shown pictorially in the scanning electron micrograph (SEM) in Figure 30.5.[54] Likewise, an interface with molecular-scale or submolecular-scale extra free volume between segments can also occur, and this produces a small decrease in selectivity below that of the pure polymer while still demonstrating an increase in permeability. This situation is often termed a *leaky* interface for convenience of discussion.[56]

If sieve–polymer interfacial adhesion is good, but a reduction in free volume is believed to occur near the sieve surface,[47,53] the result is termed *matrix rigidification*. It is believed that the layer of rigidified polymer that surrounds the sieves in this morphology displays a lower permeability value than the bulk polymer matrix, resulting in a lower overall membrane permeability[47,50] (Fig. 30.4). The enhancement in selectivity caused by the sieving phase should not be affected significantly by matrix rigidification, unless the rigidified polymer permeability is so low that it effectively starves the zeolites. In this latter case, permeation proceeds around the sieve as in the case of the truly *plugged sieves*. Provided the rigidified polymer permeability is not exceedingly low, the resulting separation properties should only match the Maxwell model selectivity estimate, as permeability will be lower than predicted.

Plugging of zeolite pores with solvents, contaminates, or other transport-limiting entities is always a concern for mixed-matrix membranes. Research has been conducted on how contaminates and other sorbing penetrants affect mixed-matrix membranes,[57,58] but more detailed studies are required before large-scale application in industry will be possible. Partial zeolite pore blockage decreases permeability through the sieving phase and can, in some cases, affect the selectivity as well, based on the new pore diameter after discounting the plugged portion of the pore.[47,53] If the pores are completely blocked, which is often the case, no enhancement in selectivity over the pure polymer is seen for the mixed-matrix membrane (plugged sieves in Fig. 30.4). Obviously, Maxwell model enhancement is not possible when plugging of zeolite pores occurs. To deal with potential pore blockage via

Figure 30.5 SEM of zeolite particles exhibiting sieve-in-a-cage morphology.

water, hydrophobic zeolites have been studied[59–61] in mixed-matrix scenarios. Also, the addition of pretreatment steps to the membrane separation process has been examined in order to remove contaminates that may cause pore blockage or membrane fouling before they reach the membrane module.[62,63]

30.2.2.1 Gas Separation Using Solid–Polymer Mixed-Matrix Membranes Common industrial gas separations that utilize membrane technology include: air separation (i.e., O_2/N_2), hydrogen recovery (i.e., H_2/CH_4, H_2/CO, H_2/N_2), and acid gas removal from natural gas (i.e., CO_2/CH_4, SO_2/CH_4). Researchers have investigated many polymers for use in solid–polymer mixed-matrix membranes, including: cellulose acetate, polyvinyl acetate, polyetherimide (commercially Ultem), polysulfone (commercially Udel), polydimethylsiloxane, polyethersulfone, and several polyimides (including commercial Matrimid). Typical sieving phases include various zeolites, carbon molecular sieves, and traditional silica. As mentioned in the previous section, a large amount of work has focused on material compatibility and adhesion at the solid–polymer interface in order to achieve separation property enhancements over traditional polymers with mixed-matrix membranes.

A number of polymers have been embedded with zeolites to form mixed-matrix membranes; however, like Barrer, many researchers note poor adhesion at the solid–polymer interface. The use of silane coupling agents to chemically link polymer to the zeolite particles, membrane formation under high-temperature conditions, and membrane annealing are common tools used to "heal" a poorly adhered, sieve-in-a-cage, interface.[64,65]

Permeation results for several different zeolites embedded in a polyethersulfone matrix give improvements over neat polymer properties, but only at zeolite loadings above 40 wt%.[66] Theoretically, addition of even 10 wt% of a sieving material with superior permeation properties to the pure polymer should result in an enhanced membrane. Polyethersulfone has also been investigated with 3A, 4A, and 5A zeolites.[67] These are similar Linde Type A, hydrophilic zeolites with differing pore sizes. The zeolite with the largest pore size, 5A, produced the most selective polyethersulfone mixed-matrix membrane. The 4A and 5A membrane permeabilities decreased with increasing zeolite loading, contradicting known models.[67] This is most likely the result of matrix rigidification at the solid–polymer interface, discussed in Section 30.2.2.1.

Polydimethylsiloxane (PDMS) embedded with zeolite particles displayed permeation improvements compared to the original polymer, but only when zeolite loadings of 40 wt% or larger are implemented.[59,68–70] Certain target separations, such as *n*-pentane/*i*-pentane, did not improve relative to neat PDMS permeation properties.[69] Fundamental transport of gas molecules through solid–rubbery polymer systems has been studied using zeolite 5A–silicone rubber membranes.[71,72]

Polyvinyl acetate (PVAc) is another rubbery polymer used in mixed-matrix research. Its flexible nature helps to prevent void formation at the solid–polymer interface. Although it may not have practical industrial applications, PVAc aids in developing proof-of-concept associated with mixed-matrix membranes. Zeolite 4A–PVAc membranes have been proven to enhance membrane selectivity in mixed-matrix membranes with only 15 vol% zeolite;[43] however, the permeability is lower than predicted presumably due to matrix rigidification. The rubbery nature of PVAc allows for more polymer relaxation at the solid–polymer interface as compared to the case with traditional, glassy polymers.[43,55]

Several researchers have successfully created glassy polyimide mixed-matrix membranes with separation properties that fall in line with accepted models, if one accounts for matrix rigidification effects. Zeolite 4A–polyetherimide mixed-matrix membranes resulted in separation property improvements over the neat polymer with only 15 vol% zeolite and corresponded with accepted models.[44] If the zeolite loading is increased to 35 vol%, the influence of matrix rigidification at the interface is seen, and the mixed-matrix permeability decreases while the selectivity continues to increase.[44,55]

Poly[2,2′-bis-(3,4-dicarboxyphenyl)hexafluoropropane dianhydride-4,4′-hexafluoroiso-propyl-idene dianiline/2,3,5,6-tetramethy1-1,4-phenylene diamine/3,5-diaminobenzoic acid)] (6FDA-6FpDA/4MPD/DABA), (2 : 2 : 1), 2,2′-BAPB + BPADA, and Matrimid, each embedded with zeolite 4A,[43,44,61] have shown improved separation properties with an increase in selectivity over the neat polymer and a corresponding decrease in permeability.[43,44] Again, the permeability loss can be attributed to matrix rigidification at the solid–polymer interface.

Using silane coupling agents with zeolite–polyimide pairs has succeeded for certain cases, such as with zeolite 4A–Ultem[44] but falls short for others, such as a faujasite-type zeolite ZSM-2/poly[2,2′-bis-(3,4-dicarboxyphenyl) hexafluoropropane dianhydride-4,4′-hexafluoroisoproply-idene dianiline-3,5-diaminobenzoic acid)] (6FDA-6FpDA-DABA) ZSM-2/6FDA-6FpDA-DABA.[73] In almost all cases where silane coupling agents are employed, imaging techniques such as scanning and transmission electron microscopy (SEM and TEM) suggest that a well-adhered interface exists, even when this is not strictly the case at a molecular level based on deviation from models that assume good adherence.[54,65,73] Adhesion through hydrogen bonds at the zeolite–polyimide interface provides an alternative to covalent coupling, particularly for CO_2/CH_4 separation;[61] however, in the presence of humidity these hydrogen bonds may be less desirable than covalent links.

Solid–polymer mixed-matrix scenarios using traditional silica, as opposed to zeolites, have also been examined. Polyimides, in particular, have been the focus of this mixed-matrix technology. 6FDA-6FpDA-DABA and 6FDA-6FpDA mixed-matrix membranes were formed with various silica compounds acting as the dispersed phase. Interfacial voids were an issue, as with other researchers, but the DABA-based polyimide showed favorable silica–polymer adhesion and improved permeation properties compared to the neat polymer. This is most likely due to interaction between the reactive DABA monomer and the functionalized silica. Annealing these mixed-matrix films increased permeability but lowered membrane selectivity.[47,52,74]

Besides zeolites and traditional silica, carbon molecular sieves have been investigated as dispersed phases. Mixed-matrix membranes have been created using carbon molecular sieves dispersed in polyetherimide (Ultem) and Matrimid, separately. These mixed-matrix membranes displayed an increase in both permeability and selectivity over their neat polymer counterparts.[46,75,76] The effect of trace amounts of toluene impurity in the feed stream of these carbon–polymer membranes was tested, and the membranes showed promising stability over time against the impurity.[77] Zeolite–carbon mixed-matrix membranes have recently been developed where the carbonized polymer matrix is derived from a pure Matrimid membrane.[45] These mixed-matrix membranes double the CO_2/CH_4 selectivity of the pure carbonized Matrimid membranes tested but lose over half of their productivity in the process. While these properties are well above Robeson's upper bound, other researchers have achieved better separation properties using only pure carbonized Matrimid membranes.[78]

Activated carbon dispersed in acrylonitrile–butadiene–styrene (ABS) polymer provided significant enhancement to both the neat polymer permeability and selectivity. Good interfacial contact was observed via SEM, thorough agglomeration of the activated carbon was also noted.[79] C_{60}, fullerene, was embedded in Matrimid; however, this mixed-matrix membrane resulted in decreases in membrane permeability with increased fullerene concentration.[80] This is most likely due to impermeability of the solid fullerene particles. As with other researchers, the two phases appeared compatible when characterized, even though favorable permeation properties did not result.

30.2.2.2 Pervaporation Using Solid–Polymer Mixed-Matrix Membranes The separation of various alcohols from water is the primary pervaporation application of solid–polymer mixed-matrix membranes. Researchers have investigated PDMS embedded with several different sieving materials. ZSM-5 proved to be a viable sieving phase, resulting in both selectivity and flux enhancement for the separation of isopropyl alcohol from water when embedded in a PDMS matrix. The separation properties continued to improve as the weight fraction of ZSM-5 was increased, but selectivity dropped as the percentage of isopropyl alcohol in the feed was increased.[81] Modified ZSM-5 and Y zeolite were each dispersed in a PDMS matrix and sorption experiments showed an increase in ethanol solubility for both mixed-matrix scenarios, though no permeation data was reported.[82] Carbon–PDMS mixed-matrix membranes gave increased selectivity, compared to neat PDMS membranes for the ethanol–water separation but suffered from reduced fluxes.[83]

Poly(vinyl alcohol) (PVA) has also been applied to alcohol–water pervaporation. ZSM-5 was again utilized, and mixed-matrix enhancements resulted. The ZSM-5/PVA mixed-matrix membranes demonstrated increased selectivity and flux, compared to pure PVA, for the water–isopropyl alcohol separation. Membrane swelling and fluxes increase as the water concentration in the feed increases.[84] Sorption studies of ZSM-5/PVA systems indicate that this hybrid material is a good candidate for the extraction of water from alcohol.[85] Modified poly(vinyl chloride) embedded with NaA zeolite has demonstrated both flux and selectivity enhancements for the ethanol–water separation at high zeolite loadings. Voids at the solid–polymer interface prevented selectivity enhancement at low zeolite loadings.[86]

Several polymers other than PDMS and PVA have been examined for solid–polymer mixed-matrix pervaporation. An activated carbon–block polyether–amide mixed-matrix membrane was investigated for the separation of volatile organic compounds (VOCs). An enhancement in both trichloroethane and water flux was noted, as well as increase in trichloroethane–water selectivity compared to the neat polymer.[87] Zeolite–polyimide mixed-matrix membranes were explored for xylene isomer separation. Solid–polymer adhesion was poor, resulting in a sieve-in-a-cage morphology; therefore, no separation properties that exceeded neat polymer membranes were identified.[88] Methanol–toluene separation was successfully performed with NaX–Viton mixed-matrix membranes, though other solid–polymer combinations were attempted and failed to form an enhanced membrane.

30.2.2.3 Specialty Solid–Polymer Mixed-Matrix Membranes Ion exchange membranes composed of organic polymer and ion-conducting ceramics have been researched for specialty gas separation applications. The most relevant of these are proton exchange membranes, which are being examined mostly for fuel cell applications.[89] Ion exchange mixed-matrix membranes for protein recovery have been studied using both cation and anion exchange particles. These membranes are composed of ion exchange particles incorporated into an ethylene vinyl alcohol copolymer for impressive protein separation with industrially important qualities (i.e., membrane flexibility, ease of scaleup).[90–92] Researchers examining traditional gas separations with ion exchange membranes have seen a range of results.[93–95] Many of these ion exchange mixed-matrix membranes rely on polyelectrolytes or other conductive polymers as the dispersed phase, with an insulating polymer as the matrix. These "polymer-only" mixed-matrix membranes have been able to enhance the selective ability of a membrane through addition of conductive polymer particles in several studies.[93–96] Ion exchange mixed-matrix membranes continue to develop, and should provide many promising industrial applications for mixed-matrix technology.

Incorporation of selective flakes into mixed-matrix membranes promises lowered permeability and enhanced selectivity compared to the pure polymer. This concept was discussed by Cussler,[97] based on existing theory concerning impermeable flakes. Data exists for aluminophosphate flakes dispersed in a polyimide matrix, and the separation trends from Cussler's theory hold for all gases examined.[98] This technology is applicable to highly permeable polymers that require selectivity enhancement to meet industrial needs. Flakes are a promising mixed-matrix material, but loss of membrane productivity, in the end, may limit the application of this technology unless adequate intrinsic flake permeability can be achieved.

Nanocomposite mixed-matrix membranes have been investigated for close to a decade. TiO_2–poly(amide-imide) membranes showed selectivity improvement but suffered loss of productivity when TiO_2 was added.[99] Nonporous, nanoscale, fumed silica was embedded in a glassy, amorphous polymer, poly(4-methyl-2-pentyne), which resulted in enhancements in both permeability and selectivity for the mixed-matrix membrane.[100] These membranes were discovered to be *reverse selective*, so the membrane is selective for the larger penetrant. This phenomenon is attributed to increased free volume in the bulk polymer from chain packing disruption, which occurs when the filler is added.[101]

While most solid–polymer mixed-matrix membranes rely on sieve micropores to generate high selectivities within the membrane by discriminating between gas penetrants, mesoporous molecular sieves have been studied as well. These mesoporous sieves are used to increase permeability in mixed-matrix membranes, instead of enhancing selectivity. Mesoporous MCM-41 incorporated into a polysulfone matrix improved membrane permeability for all gases tested. The selectivity of pure polysulfone was maintained, as expected.[102] Mesoporous sieves are an excellent material for highly selective polymer membranes that require increased productivity; however, nanoscopic-scale crystals will probably be required to enable accommodation within the submicron selective layer typical for high-performance asymmetric membranes in use today.

30.2.2.4 How to Formulate and Process Solid–Polymer Mixed-Matrix Membranes

Dense Films Considerable differences exist between formulations (dopes) for forming dense films versus asymmetric membranes. Typically, dense film polymer solutions consist of fewer components as compared to casting dopes for asymmetric structures. Dense films are made via the gradual removal of the solvent from the cast film, while asymmetric films are made through a phase separation mechanism. The difference in formation mechanisms reflects the difference in dope compositions for the two types of membranes. While in dense films, homogeneous mixing of solids and stresses at the film–support interface are the major concerns, asymmetric membrane formation also includes complications introduced by phase separation, especially at the solid–polymer interface.

Most reported mixed-matrix research focuses on dense films. Emphasis on procedures to homogeneously disperse solids in polymer solutions[103] followed by casting under controlled evaporation regimes, with various annealing procedures.[53,65,104] Initial mixing procedures have varied considerably. Addition of powdered solid dispersed phase (DP) to the polymer with processing in an extruder,[105] and the use of high-speed mixers and sonicators[103,106] to mix the DP dispersion with a solvent prior to polymer addition is common to disperse and stabilize that particle suspension. Multistep mixing procedures in which a small quantity of polymer is sonicated in a dispersion solvent and dispersed phase (a so-called priming step) followed by the addition of the bulk portion of polymer to increase viscosity for casting is also common. This multistep process produces the desired polymer-to-inorganic or carbon solid ratio in the final sample.[43,46] The trade-off between excessive mixture viscosity (casting difficulty) and inadequate viscosity (sedimentation of the denser DP) is especially important in the formulation concentrations. The experience gained from dense film work has been used as a first step in processing mixed-matrix dopes for asymmetric membranes since the selection of such asymmetric structures must also allow the dense selective skin layer to function properly.

Dense mixed-matrix film formulations typically include the polymer, solvent, and the DP. As discussed in Section 30.2.2.1, for an enhancement in the selectivity (mixed-matrix effect) it is necessary that the sieve have good contact with the polymer to prevent nonselective bypass of the gas molecules. Although the objective is to study as simple a system as possible, additives are typically included in the formulation to improve polymer and DP adhesion. For example, plasticizers have been used to increase the flexibility of the matrix polymer. Although the presence of the plasticizer typically improves the adhesion between the polymer and zeolite, this usually results in poor transport properties of the membrane due to undesirable softening of the matrix, with loss in intrinsic size and shape discriminating ability.[53] Silane coupling agents have also been used with varying success to improve the adhesion between the polymer and the DP,[54,65,107,108] and using polymers with reactive functionality that can interact with the DP.[44]

Casting procedures generally include using a doctor's knife to spread the mixture on a support, Teflon, or glass among others, from which the film can be easily removed after drying. The initial solvent evaporation is generally done under controlled conditions and temperatures, followed by a drying step at higher temperatures to remove any remaining solvent.[109]

Asymmetric Mixed-Matrix Membranes Practical membranes used in film or hollow-fiber geometry for separations consist of an asymmetric structure, where a thin skin layer performs the actual separation and the porous layer underneath acts as a support to the skin. This structure greatly increases the productivity of the membrane, which is inversely proportional to the separating layer thickness, with the support layer ideally providing little or no resistance to permeate flow. As the skin layer performs the separation function, the highly selective DP need only be dispersed in the skin region of the asymmetric membrane. Dual-layer membranes, where the DP is included only in the outer layer with the inner layer forming the porous support, lower the use of expensive DP in mixed-matrix membranes. This structure is depicted in Figure 30.6.

The dope formulation for dense mixed-matrix films must be revisited for making asymmetric membranes as the structure of the membrane is controlled by phase separation kinetics. Different solvents used in dense film formulation may be used in asymmetric formulations to control the phase separation rate of the membrane within the nonsolvent (quench) bath and obtain membranes of varying porosities.[110] Further, nonsolvents and additives can be added to fine tune the morphology of the membrane to control pore sizes and suppress the formation of large voids within the membrane.[111–116] Asymmetric skinned membranes can also contain a second more volatile solvent that can evaporate and form a high solids concentration layer (skin) on the exposed surface

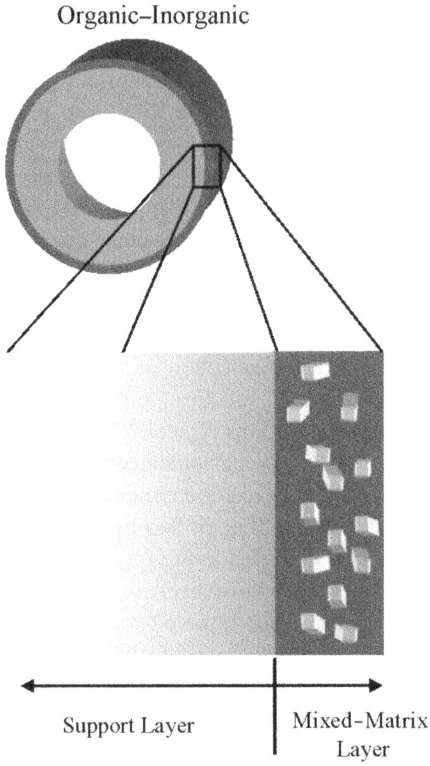

Figure 30.6 Graphic depicting dual-layer mixed-matrix hollow-fiber membrane with zeolite particles embedded in the outer layer.

of the membrane.[117] A final difference between films and fibers is that fibers can include a higher viscosity dope for spinning to prevent capillary instability during the spinning process.[118]

Once the formulation is established, asymmetric membranes can be cast as flat sheets or spun (extruded through a spinneret) into fiber form, followed either by an evaporation step or directly submerged into the quench bath. In this quenching process, the nonsolvent penetrates the membrane and the solvent diffuses out into the quench bath. The entering nonsolvent causes the phase separation of the membrane with the porosity being formed by the domains of polymer lean phase, which are washed out of the final membrane structure. For mixed-matrix membranes, it is believed that defects (sieve-in-cage) can form due to the nucleation of nonsolvent–polymer lean phase around the DP during the phase separation.[119] One successful approach to combat this problem describes the modification of the DP to increase surface hydrophobicity leading to a hypothesized suppression of the nucleation of the hydrophilic polymer lean phase by the aqueous quench medium.

Successful spinning of mixed-matrix hollow-fiber membranes for gas separation has so far been only demonstrated in a patent by Ekiner et al.[120] A major hurdle to the commercial implementation of mixed-matrix membranes has been the lack of reproducibility in forming successful mixed-matrix membranes. Challenges with poor polymer–sieve interaction, variability in molecular sieve transport, surface characteristics, and effects of contaminants on molecular sieve performance have been identified in dense mixed-matrix fields. These factors are relevant to asymmetric membranes in addition to the control of phase separation issues noted above.[50]

A recent publication raises the issue of possible particle migration within the mixed-matrix hollow-fiber membranes.[121] Comparing with earlier evidence of particle migration in sheared viscoelastic suspensions,[122–124] such a development seems possible in mixed-matrix hollow-fiber spinning, however, only if the particle concentration in the spinning dope is high (generally greater than 10% by volume) or the elasticity of the dope is very large.[123,125] Additionally, experimental verification of these suggested trends is challenging since the use of energy dispersive spectroscopy (EDS) line scans across a membrane morphology filled with voids used by the authors is difficult to interpret unambiguously. Such methods, depending on the beam voltage, analyze extensive regions below the visible surface of the sample and can easily lead to erroneous results brought about by embedded large particle agglomerates. This difficulty in probing the behavior of solid particles and their subsequent effect on membrane properties promises to provide considerable experimental opportunities for researchers. These issues must be understood in a general framework across multiple molecular sieve and polymer combinations before the technology of mixed-matrix hollow fibers can be applied successfully in the field.

While porous defects between the solid and polymer are undesirable in microporous solid–polymer gas separation membranes, the opposite is true for ultrafiltration[126] and ion exchange mixed-matrix membranes.[127] The presence of such voids enhances membrane performance as the solid acts only in the adsorbent capacity and not as a molecular sieve as required in the case of the microporous solid.

Mesoporous solid–polymer membranes can be easily formed in asymmetric form using carbons and mesoporous ion exchange resins embedded in a polymer matrix. Although carbons and activated carbons have traditionally been used for their adsorbent properties, increasing attention is being paid to their use in gas separation.[58,128] The research on carbon–polymer and carbon–zeolite mixed-matrix membranes is still in its infancy, and challenges of membrane brittleness seen in carbon–polymer membranes with relevant carbon loadings must be overcome before their general application.

30.2.3 Liquid–Polymer Mixed-Matrix Membranes

For liquid–polymer mixed-matrix membranes, the physical state of the fillers incorporated into the continuous polymer matrix is liquid such as PEG and amines. The following section provides an introduction to this less common mixed-matrix technology. Since the existing literature on this new type of membrane is less developed, more detail is provided here.

As noted in the introduction with regard to Figure 30.2, research in polymer–microporous solid mixed-matrix materials has been able to advance by applying the framework provided by Eq. (30.4). Similarly, supported liquid membranes have benefited from the application of a theoretical framework. The other variants of the mixed-matrix paradigm (polymer–liquid and polymer–microporous solid–liquid) have been impeded by the lack of such a framework. In fact, the effects of the liquid component on the polymer matrix complicate the development, but it does not prevent it in principle. If one applied free volume theory and carried out characterizations of the effects of miscible liquids in the "matrix," a generalization of Eq. (30.4) should be possible. In this case, information regarding partition coefficients and DP volume fractions would enable the development of a more quantitative framework for these interesting systems.

The following section gives examples of these types of mixed-matrix materials derived from liquid components. In many cases, the detailed DP volumes are not known or specified directly. Moreover, in virtually no cases are partition coefficients known between the liquid and polymer matrix. Nevertheless, these cases allow some vision of the possibilities of these types of systems.

30.2.3.1 Preparation of Liquid–Polymer Mixed-Matrix Membranes In one approach,[136–139] a liquid–polymer mixed-matrix membrane was prepared by dissolving silicone rubber polymer (e.g., RTV-615A + RTV-615B) in a nonpolar solvent such as cyclohexane or Freon T F. To this solution, a certain amount of liquid filler (e.g., PEG) was added to form an emulsified solution. The emulsified solution was blended by vigorous shaking for a period of time and then degased by using a vacuum pump. The degased solution was coated on a porous support membrane (e.g., polysulfone) under vacuum. After coating, the liquid–polymer mixed-matrix membrane was cured for a period of 1–2 h at a temperature of 85°C.

In another approach,[136] a liquid–polymer mixed-matrix membrane was prepared by first soaking a wet porous support membrane (e.g., polysulfone) in a solution containing liquid fillers (e.g., glycerol) for a period of 2 h, followed by drying for a period of 10 h at room temperature. The treated porous support membrane was then coated with the emulsified solution described above to form a liquid–polymer mixed-matrix membrane.

30.2.3.2 Poly(ethylene glycol) (PEG)–Polymer Mixed-Matrix Membranes (MMM_{PEG}) PEG can dissolve acidic gases,[185,186] therefore, the gas diffusivity in the flexible PEG segment may be high. However, a thin film of PEG with good mechanical and thermal stability is difficult to obtain.

In recent years, PEG–polymer mixed-matrix membranes prepared by blending PEG with other polymers have been reported in the literature.[136–139,187–189] These membranes combined the high permeability coefficient and high selectivity of the PEG segment with the good processability and good mechanical stability of other polymers.

In 1982, Kawakami et al. reported PEG–cellulose nitrate mixed-matrix membranes having up to 50 wt% of PEG.[187] These mixed-matrix membranes exhibited CO_2 permeabilities of 1.4–8.2 Barrers and CO_2 selectivities of 29–38. The PEG–cellulose nitrate mixed-matrix membranes showed significantly increased CO_2 permeabilities and selectivities with increasing PEG loading. The CO_2 permeability improvement was attributed to the increments of both diffusivity and solubility of CO_2.

In the mid-1980s, Kulprathipanja et al. at UOP LLC developed a new type of MMM_{PEG} by casting PEG or other liquid filler and silicone rubber on a porous polysulfone or CA support.[136–139] They prepared MMM_{PEG} by adding PEG (400–600 MW) to a silicone rubber solution to form an emulsified solution. The emulsified solution was coated on a porous polysulfone membrane under vacuum. After coating, the membrane was further cured. They also prepared a gly-MMM_{PEG} membrane by first soaking a wet porous polysulfone in a solution containing 15% glycerol by weight for a period of 2 h, followed by drying for a period of 10 h at room temperature. The treated polysulfone support was then coated with the emulsified solution described above.

Both MMM_{PEG} and gly-MMM_{PEG} were tested for the separation of gases such as SO_2, NH_3, H_2S, CO_2, H_2, N_2, CH_4, C_2H_4, and C_2H_6. The gas to be tested was passed through the membrane under a pressure of 5–50 psig at ambient temperature. The permeation rate and selectivity for each pure gas, both polar and nonpolar, was measured and calculated. For comparison purposes, a reference membrane composed of silicone rubber coated on a porous polysulfone (SR-PS) was also prepared and tested. The results are summarized in Tables 30.2–30.6.

TABLE 30.2 Mixed-Matrix Membranes for SO_2 Separation[a]

Membrane	P_{SO_2}/P_{N_2}	P_{SO_2}/P_{CO_2}	$P_{SO_2}/d \times 10^3$ cm^3 (STP)/cm^2-s-cm Hg
MMM_{PEG}	4350	74	10
Gly-MMM_{PEG}	30,000	300	42
SR-PS	85	7.5	4.6

[a] MMM_{PEG} = PEG–silicone rubber coated on porous polysulfone
Gly-MMM_{PEG} = PEG–silicone rubber coated on glycerol-treated porous polysulfone
SR-PS = silicone rubber coated on porous polysulfone
d = thickness of membrane

TABLE 30.3 Mixed-Matrix Membrane for NH_3 Separation

Membrane	P_{NH_3}/P_{N_2}	P_{NH_3}/P_{H_2}	$P_{NH_3}/d \times 10^4$ cm^3 (STP)/cm^2-s-cm Hg
MMM_{PEG}	1270	80	2.25
Gly-MMM_{PEG}	7200	510	2.89
SR-PS	35	12	1.86

TABLE 30.4 Mixed-Matrix Membrane for H_2S Separation

Membrane	P_{H_2S}/P_{CH_4}	P_{H_2S}/P_{H_2}	P_{H_2S}/P_{CO_2}	$P_{H_2S}/d \times 10^5$ cm^3 (STP)/cm^2-s-cm Hg
MMM_{PEG}	136	21	6.7	7.8
Gly-MMM_{PEG}	119	74	8.1	10.7
SR-PS	10	9	2.3	14.0

TABLE 30.5 Mixed-Matrix Membrane for CO_2 Separation

Membrane	P_{CO_2}/P_{CH_4}	P_{CO_2}/P_{H_2}	P_{CO_2}/P_{N_2}	$P_{CO_2}/d \times 10^6$ cm^3 (STP)/cm^2-s-cm Hg
MMM_{PEG}	24	4.8	67	13.6
Gly-MMM_{PEG}	22	9.1	100	13.2
Sil-PS	4.3	3.8	11.4	61

TABLE 30.6 Mixed-Matrix Membrane for Ethylene Separation

Membrane	$P_{C_2H_4}/P_{C_2H_6}$	$P_{C_2H_4}/d \times 10^6$ cm^3 (STP)/cm^2-s-cm Hg	$P_{C_2H_6}/d \times 10^6$ cm^3 (STP)/cm^2-s-cm Hg
MMM_{PEG}	1.7	0.79 ± 0.010	0.47 ± 0.003
Gly-MMM_{PEG}	2.28	0.40 ± 0.020	0.17 ± 0.002
SR-PS	0.86	7.19 ± 0.152	8.34 ± 0.070

The results in Tables 30.2–30.6 indicate that PEG enhanced the permeation rates of SO_2 and NH_3. PEG, however, hindered the permeation rates of the other gases. The effect of PEG on lowering the gas permeation rates increased as $H_2S < CO_2 < H_2 < CH_4 < N_2$. This enhanced the selectivity of polar gas over nonpolar gas substantially. Table 30.6 also shows the effect of PEG on ethylene and ethane.[188] PEG slowed down the permeation rate of ethylene and ethane in the membrane phase. However, the effect is more prevalent for ethane than ethylene. This promotes the ethylene–ethane separation.

Vijitjunya[191] also studied the effect of some diol isomers as the liquid fillers in mixed-matrix membranes for gas separation. They incorporated 1,4-butanediol, 1,2-butanediol, 1,3-butanediol, and 2,3-butanediol liquid fillers into silicone rubber coating on top of the porous polysulfone support membrane. Similar to PEG liquid filler, it was found that 1,2-butanediol and 2,3-butanediol could improve propylene over propane selectivity. However, the selectivity of propylene over propane was not enhanced when 1,3-butanediol and 1,4-butanediol were added to silicone rubber. In PEG, the hydroxyl group is attached to each carbon atom on carbon backbones. A portion of 1,2- and 2,3-butanediol molecules are similar to PEG in which the hydroxyl groups are attached to the adjacent carbon atoms. In contrast, the hydroxyl groups in 1,3- and 1,4-butanediol molecules are not attached to the adjacent carbon atoms. Therefore, it was proposed that the position of hydroxyl groups of the butanediols plays an important role for the improvement of propylene over propane selectivity.

The stability of MMM_{PEG} was carried out at a temperature range of 40–75°C. CO_2 and CH_4 gases were saturated with water and then were passed through the MMM_{GEG} at pressures of 25 and 50 psi, respectively. The stability of MMM_{PEG} was excellent over the period studied, namely 200 days on stream, with no apparent deterioration in CO_2 and CH_4 flux or CO_2/CH_4 selectivity.

A plot of selectivity versus critical temperature for these gases is presented in Figure 30.7. The plot shows that the permeabilities of the molecules increase as the critical temperatures increase. This indicates that the solubility is the controlling mechanism for mixed-matrix membrane selectivity. However, for lighter gases such as H_2 and N_2 the permeability decreases as the molecular weight increases. The inverse relationship between the diffusion coefficient and molecular weight implies the importance of molecular diffusion coefficients.

Recently, Li et al. reported the preparation of PEG/CA mixed-matrix membranes.[189] They investigated the effect of PEGs with different molecular weights on the permeabilities and selectivities of a series of gases including CO_2, H_2, O_2, CH_4, and N_2.

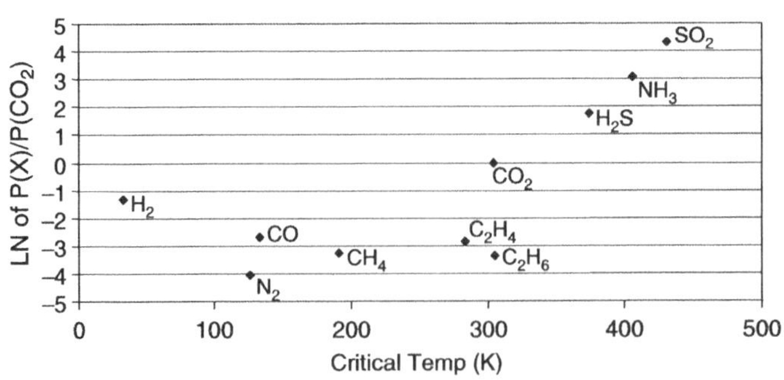

Figure 30.7 Critical temperature vs. selectivity.

It was determined that the PEG/CA mixed-matrix membrane have 10 wt% PEG20000 exhibited higher permeability for CO_2 and higher selectivity for CO_2 over N_2 and CH_4 than those of the PEG/CA mixed-matrix membranes containing 10 wt% PEG of the molecular weight in the range 200–6000. Favorable attractions between the sorbed CO_2 and the ether units in the main chain of PEG20000 in the amorphous domains in the PEG/CA mixed-matrix membranes should enable the CO_2 penetrant to permeate much more rapidly through the mixed-matrix membranes as compared to the much less strongly sorbing N_2 even under relatively high temperatures. On the other hand, Li et al.,[189] suggested that the higher permselectivity was due to a higher diffusion coefficient for CO_2 vs. N_2 in the PEG20000. This seems highly unlikely, and a careful study of the PEG20000 system vs. temperature would be useful to resolve this issue.

30.2.4 Solid–Liquid–Polymer Mixed-Matrix Membranes

All liquid–polymer mixed-matrix membranes have liquid polymer encapsulated in the continuous polymer matrix. The long-term stability of these membranes for industrial gas separation processes is still a critical issue because of the undesirable leakage of the liquid from the membrane. To stabilize the liquid in the polymer membrane, a new type of mixed-matrix membranes, solid–liquid–polymer mixed-matrix membranes, has been developed most recently.[160,191–193] The solid, such as activated carbon impregnated with liquid polymer such as PEG, functions as a stabilizer of the liquid polymer in the continuous polymer phase. These hybrid solid–liquid–polymer mixed-matrix membranes combine the properties of the continuous polymer phase, the dispersed solid filler phase, and the impregnated liquid phase.

30.2.4.1 Preparation of Solid–Liquid–Polymer Mixed-Matrix Membranes A solid/liquid–polymer mixed-matrix membrane was prepared by dispersing different types of liquid fillers (e.g., PEG and amines) stabilized by adsorbing into porous solid fillers (e.g., activated carbons or molecular sieves) in polymer membrane matrix.[160,192,193] First, the solid filler-stabilized liquid fillers were prepared by dissolving the liquid fillers such as PEG in an organic solvent and then adding the solid fillers such as activated carbon into the solution. The mixed solution was stirred until a homogeneous slurry solution was formed. The organic solvent was then removed under vacuum to produce the solid filler-stabilized liquid fillers, which were then dispersed into silicone rubber solution. The solution was degased under vacuum and then the bubble-free solution was coated on a porous support membrane, such as polysulfone or cellulose acetate membrane under vacuum. After coating, the solid–liquid–polymer mixed-matrix membrane was cured for a period of 1–2 h at a temperature of 85°C.

30.2.4.2 Poly(ethylene glycol)–Solid–Polymer Mixed-Matrix Membranes In 2002, Kulprathipanja and Charoenphol[160,192] studied PEG-activated carbon-polymer mixed-matrix membranes. These membranes were prepared by impregnating activated carbon with PEG first, then adding the impregnated activated carbon to a silicone rubber solution to form a suspended solution, and finally coating the suspended solution on a porous polysulfone membrane. After coating, the membrane was further cured to form the SR + (PEG–activated carbon)–PS mixed-matrix membranes. Activated carbon was

TABLE 30.7 Permeabilities and Selectivities of Gases Through SR/PS, SR + PEG/PS, SR + Activated Carbon PS and SR + [PEG-activated carbon] PS Membranes

Membrane	P_{CO_2}/P_{N_2}	$P_{CO_2}/d \times 10^6$ cm^3 (STP)/cm^2-s-cm Hg	$P_{N_2}/d \times 10^6$ cm^3 (STP)/cm^2-s-cm Hg
SR/PS	10.92	14.03	1.29
SR + PEG/PS	34.96	8.13	0.23
SR + activated carbon/PS	15.83	34.78	2.20
SR+ (PEG–activated carbon)–PS	34.64	26.23	0.75

used to prevent the leak out of PEG. Pure gas permeation results (Table 30.7) show that the SR + (PEG–activated carbon)–PS membrane is less permeable to both CO_2 and N_2 than the SR + activated carbon–PS membrane without PEG. It is proposed that the relatively lower permeability is due to the presence of PEG on the surface of activated carbon, which hindered the path of gas diffusion. Table 30.7 also indicates that the SR + (PEG–activated carbon)–PS membrane has much higher selectivity of CO_2/N_2 than those of both SR/PS and SR + activated carbon–PS membranes, which is due to the higher solubility of CO_2 than that of N_2 in PEG.

30.2.4.3 Amine–Solid–Polymer Mixed-Matrix Membranes In order to stabilize the amine liquid in the polymer membrane, Kulprathipanja et al.[193] designed a new type of diethanolamine (DEA)–NaX–polymer mixed-matrix membrane in 2004. The membranes were prepared using similar approach to that they used for the PEG–NaX–polymer mixed-matrix membranes,[191] except the use of DEA rather than PEG as the liquid filler and the use of CA rather than polysulfone as the support membrane. These new membranes were tested for the separation of CO_2 from CH_4. It can be seen from Table 30.8 that the incorporation of DEA-impregnated NaX into the silicone rubber coating on top of the porous CA support membrane improved both CO_2/CH_4 and CO_2/N_2 selectivities. In addition, the results shown in Table 30.8 and Figure 30.8 also indicate that the selectivity of CO_2/CH_4 increases continuously as the concentration of DEA increases. The permeability of CO_2, however, decreases continuously as the concentration of DEA increases. It is proposed that since DEA is a chemical adsorption solvent, it will enhance the selectivity of both CO_2/CH_4 and CO_2/N_2 by facilitating the mechanism and decreasing the permeability of CO_2.

TABLE 30.8 Permeabilities and Selectivities of Gases Through DEA–NaX-Polymer Mixed-Matrix Membranes

Membrane	P_{CO_2}/P_{CH_4}	P_{CO_2}/P_{N_2}	$P_{CO_2}/d \times 10^6$ cm^3 (STP)/cm^2-s-cm Hg
CA	8.37	8.34	17.69
SR/CA	13.74	15.81	3.94
SR + (5%DEA-10%NaX)/CA	11.53	19.34	3.38
SR + (10%DEA-10%NaX)/CA	13.25	18.70	3.04
SR + (20%DEA-10%NaX)/CA	17.89	22.39	2.42
SR + (30%DEA-10%NaX)/CA	22.31	26.67	2.28

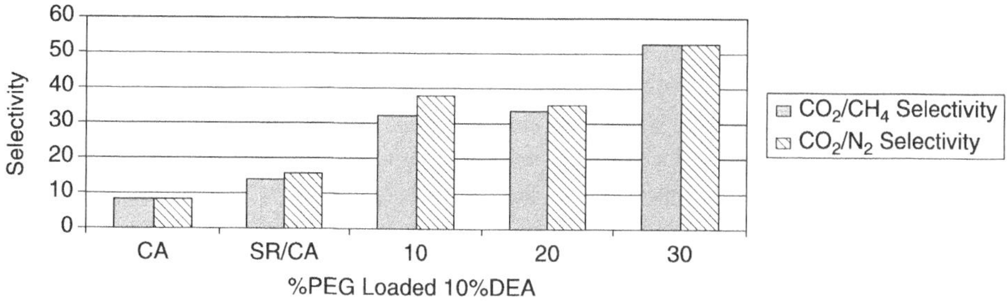

Figure 30.8 Selectivity vs. DEA loading for DEA–NaX–polymer mixed-matrix membranes.

30.3 SUMMARY AND FUTURE OPPORTUNITIES

Mixed matrix membrane research activity is increasing both in academia and industry after the concept was proved in the mid 1980s. The research activity is aimed to overcoming the limit imposed by the so-called polymer upper bound limit by imbedding solid, liquid or a combine liquid/solid into the polymer phase. Although enormous technical progress has been reported to date, no commercial mixed-matrix membrane application is found. The lack of commercial mixed-matrix membranes could be due to current insufficient information in membrane processability, stability, and cost. More importantly, it requires a better strategy approach to develop a commercial mixed-matrix membrane for a particular separation. This, in turn, requires a more in-depth understanding of molecular transport mechanism of transport species through the mixed phase of solid–polymer, liquid–polymer, or solid–liquid–polymer of the mixed-matrix membrane.

Especially for the liquid-containing systems, development of an engineering model to generalize Eq. (30.4) is needed badly to assist in material design and optimization. Considering partition coefficients for liquid components into the polymer would be a valuable step forward. Also, including effects of sorbed liquid on polymer matrix permeabilities should be included in any extension, since Eq. (30.4) uses pure permeation properties of the dispersed and continuous phases in its standard form.

However, with the current research activity, it is believed that the parameters mentioned above will be identified and understood in the near future. When that happens, one would expect to find commercial mixed-matrix membranes for some specific applications.

REFERENCES

1. Loeb, S., and Sourirajan, S. (1963). Sea water demineralization by means of an osmotic membrane. *Adv. Chem. Ser.* 38, 117–132.
2. Singh, R. (1998). Industrial membrane separation processes. *CHEMTECH* 28(4), 33–44.
3. Matsuura, T. (1994). *Synthetic Membranes and Membrane Separation Processes*. CRC Press, Boca Raton, FL, Chapter 2.
4. Costello, L. M., Walker, D. R. B., and Koros, W. J. (1994). Analysis of a thermally stable polypyrrolone for high temperature membrane-based gas separations. *J. Membr. Sci.* 90(1–2), 117–130.
5. Kawakami, H., Mikawa, M., Nagaoka, S. (1996). Gas transport properties in thermally cured aromatic polyimide membranes. *J. Membr. Sci.* 118(2), 223–230.

6. Stern, S. A. (1994). Polymers for gas separations: The next decade. *J. Membr. Sci.* 94(1), 1–65.
7. Robeson, L. M. (1991). Correlation of separation factor versus permeability for polymeric membranes. *J. Membr. Sci.* 62(2), 165–185.
8. Robeson, L. M., Burgoyne, W. F., Langsam, M., Savoca, A. C., and Tien, C. F. (1994). High performance polymers for membrane separation. *Polymer* 35(23), 4970–4978.
9. Abelson, P. H. (1989). Synthetic membranes. *Science* 244(4911) 1421.
10. Liu, C., and Martin, C. R. (1991). Composite membranes from photochemical synthesis of ultrathin polymer films. *Nature* 352, 50–52.
11. Break, D. W. (1974). *Zeolite Molecular Sieves*, Wiley, New York, p. 636.
12. Puleo, A. C., Paul, D. R., and Kelley, S. S. (1989). The effect of degree of acetylation on gas sorption and transport behavior in cellulose acetate. *J. Membr. Sci.* 47(3), 301–332.
13. Tomita, T., Nakayama, K., and Sakai, H. (2004). Gas separation characteristics of DDR type zeolite membrane. *Micropor. Mesopor. Mater.* 68, 71–75.
14. Nakayama, K., Suzuki, K., Yoshida, M., Yajima, K., and Tomita, T. (2006). Method for preparing DDR type zeolite film, DDR type zeolite membrane, and composite DDR type zeolite membrane, and method for preparation thereof. U.S. Patent 7,014,680B2.
15. Guan, G., Tanaka, T., Kusakabe, K., Sotowa, K.-I., and Morooka, S. (2003). Characterization of $AlPO_4$-type molecular sieving membranes formed on a porous α-alumina tube. *J. Membr. Sci.* 214(2), 191–198.
16. Poshusta, J. C., Tuan, V. A., Falconer, J. L., and Noble, R. D. (1998). Synthesis and permeation properties of SAPO-34 tubular membranes. *Ind. Eng. Chem. Res.* 37, 3924–3929.
17. Poshusta, J. C., Tuan, V. A., Pape, E. A., Noble, R. D., and Falconer, J. L. (2000). Separation of light gas mixtures using SAPO-34 membranes. *AIChE J.* 46(4) 779.
18. Poshusta, J. C., Noble, R. D., Falconer, J. L. (2001). Characterization of SAPO-34 membrane by water adsorption. *J. Membr. Sci.* 186(1), 25–40.
19. Li, S., Alvarado, G., Noble, R. D., and Falconer, J. L. (2005). Effects of impurities on CO_2/CH_4 separations through SAPO-34 membranes. *J. Membr. Sci.* 251(1–2), 59–66.
20. Bakker, W. J. W., van den Broeke, L. J. P., Kapteijn, F., and Moulijn, J. A. (1997). Temperature dependence of one-component permeation through a silicalite-1 membrane. *AIChE J.* 43(9), 2203–2214.
21. van den Broeke, L. J. P., Bakker, W. J. W., Kapteijn, F., and Moulijn, J. A. (1999). Transport and separation properties of a silicalite-1 membrane—I. Operating conditions. *Chem. Eng. Sci.* 54(2), 245–258.
22. van den Broeke, L. J. P., Kapteijn, F., and Moulijn, J. A. (1999). Transport and separation properties of a silicalite-1 membrane—II. Variable separation factor. *Chem. Eng. Sci.* 54(2), 259–269.
23. Funke, H. H., Kovalchick, M. G., Falconer, J. L., and Noble, R. D. (1996). Separation of hydrocarbon isomer vapors with silicalite zeolite membranes. *Ind. Eng. Chem. Res.* 35, 1575–1582.
24. Funke, H. H., Argo, A. M., Falconer, J. L., and Noble, R. D. (1997). Separations of cyclic, branched, and linear hydrocarbon mixtures through silicalite membranes. *Ind. Eng. Chem. Res.* 36, 137–143.
25. Lin, X., Kita, H., and Okamoto, K.-I. (2000). Silicalite membrane preparation, characterization, and separation performance. *Ind. Eng. Chem. Res.* 40, 4069–4078.
26. Nomura, M., Yamaguchi, T., and Nakao, S.-I. (1997). Silicalite membranes modified by counterdiffusion CVD technique. *Ind. Eng. Chem. Res.* 36, 4217–4223.
27. Mabande, G. T. P., Pradhan, G., Schwieger, W., Hanebuth, M., Dittmeyer, R., Selvam, T., Zampieri, A., Baser, H., and Herrmann, R. (2004). A study of silicalite-1 and Al-ZSM-5 membrane synthesis on stainless steel supports. *Micropo. Mesopor. Mater.* 75, 209–220.

28. Raileanu, M., Popa, M., Moreno, J. M. C., Stanciu, L., Bordeianu, L., and Zaharescu, M. (2002). Preparation and characterization of alumina supported silicalite membranes by sol-gel hydrothermal method. *J. Membr. Sci.* 210(2), 197–207.

29. Yuan, W., Lin, Y. S., and Yang, W. (2004). Molecular sieving MFI-type zeolite membranes for pervaporation separation of xylene isomers. *J. Am. Chem. Soc.* 126, 4776–4777.

30. Aoki, K., Tuan, V. A., Falconer, J. L., and Noble, R. D. (2000). Gas permeation properties of ion-exchanged ZSM-5 zeolite membranes. *Micropor. Mesopor. Mater.* 39(3), 485–492.

31. Bernal, M. P., Piera, E., Coronas, J., Menendez, M., Santamaria, J. (2000). Mordenite and ZSM-5 hydrophilic tubular membranes for the separation of gas phase mixtures. *Catal. Today* 56, 221–227.

32. Hedlund, J., Sterte, J., Anthonis, M., Bons, A., Carstensen, B., Corcoran, N., Cox, D., Deckman, H., Gijnst, W., de Moor, P., Lai, F., McHenry, J., Mortier, W., Reinoso, J., and Peters, J. (2002). High-flux MFI membranes. *Micropor. Mesopor. Mater.* 52, 179–189.

33. Chau, C., Sicard, M., and Dred, R. L. (2004). Thin zeolite membrane, its preparation and its use in separation. U.S. Patent 6,818,333 B2.

34. Takahashi, T., Sakai, H., and Ogawa, N. (2002). Method for separation of p-xylene. *U.S.* Patent 2002/0045792 A1.

35. Xomeritakis, G., Lai, Z., and Tsapatsis M. (2001). Separation of xylene isomer vapors with oriented MFI membranes made by seeded growth. *Ind. Eng. Chem. Res.* 40, 544–552.

36. Weh, K., Noack, M., Sieber, I., and Caro, J. (2002). Permeation of single gases and gas mixtures through faujasite-type molecular sieve membranes. *Micropor. Mesopor. Mater.* 54, 27–36.

37. Aoki, K., Kusakabe, K., and Morooka, S. (2000). Separation of gases with an A-type zeolite membrane. *Ind. Eng. Chem. Res.* 39, 2245–2251.

38. Dong, J., and Lin, Y. S. (1998). In situ synthesis of P-type zeolite membranes on porous α-alumina supports. *Ind. Eng. Chem. Res.* 37, 2404–2409.

39. Jones, C. W., and Koros, W. J. (1994). Carbon molecular sieve gas separation membranes—I. Preparation and characterization based on polyimide precursors. *Carbon* 32(8), 1419–1425.

40. Geiszler, V. C., and Koros, W. J. (1996). Effect of polyimide pyrolysis conditions on carbon molecular sieve membrane properties. *Ind. Eng. Chem. Res.* 35, 2999–3003.

41. Cho, Y.-K., Han, K., and Lee, K.-H. (1995). Separation of CO_2 by modified γ-Al_2O_3 membranes at high temperature. *J. Membr. Sci.* 104(3), 219–230.

42. Kusakabe, K., Sakamoto, S., Saie, T., and Morooka, S. (1999). Pore structure of silica membranes formed by a sol–gel technique using tetraethoxysilane and alkyltriethoxysilanes. *Sep. Purif. Technol.* 16, 139–146.

43. Mahajan, R., and Koros, W. J. (2000). Factors controlling successful formation of mixed-matrix gas separation materials. *Ind. Eng. Chem. Res.* 39(8), 2692–2696.

44. Mahajan, R., and Koros, W. J. (2002). Mixed matrix membrane materials with glassy polymers. Part 2. *Polym. Eng. Sci.* 42(7), 1432–1441.

45. Tin, P. S., Chung, T.-S., Jiang, L., and Kulprathipanja, S. (2005). Carbon-zeolite composite membranes for gas separation. *Carbon* 43(9), 2025–2027.

46. Vu, D. Q., Koros, W. J., and Miller, S. J. (2003). Mixed matrix membranes using carbon molecular sieves. I. Preparation and experimental results. *J. Membr. Sci.* 211(2), 311–334.

47. Hibshman, C., Cornelius, C. J., and Marand, E. (2003). The gas separation effects of annealing polyimide-organosilicate hybrid membranes. *J. Membr. Sci.* 211(1), 25–40.

48. Barrer, R. M., and James, S. D. (1960). Electrochemistry of crystal-polymer membranes. I. Resistance measurements. *J. Phys. Chem.* 64, 417–421.

49. Barrer, R. M., and James, S. D. (1960). Electrochemistry of crystal-polymer membranes. II. Membrane potentials. *J. Phyis. Chem.* 64, 421–427.

50. Moore, T. T., and Koros, W. J. (2005). Non-ideal effects in organic-inorganic materials for gas separation membranes. *J. Mol. Struct.* 739(1–3), 87–98.
51. Bruggemann, D. A. G. (1935). Berechnung verschiedener physikalischer Konstanten von heterogenen Substanzen. I. Dielektrizitatskonstanten und Leitfahigkeiten der Mischkorper aus isotropen Substanzen [The calculation of various physical constants of hetereogeneous substances. I. The dielectric constants and conductivities of mixtures composed of isotropic substances]. *Annal. Phys.V* 24, 636.
52. Cornelius, C. J., and Marand, E. (2002). Hybrid silica-polyimide composite membranes: Gas transport properties. *J. Membr. Sci.* 202(1–2), 97–118.
53. Mahajan, R., Burns, R., Schaefer, M., and Koros, W. (2002). Challenges in forming successful mixed-matrix membranes with rigid polymeric materials. *J. Appl. Polym. Sci.* 86, 881–90.
54. Mahajan, R., Koros, W. J. (2002). Mixed matrix membrane materials with glassy polymers. Part 1. *Polym. Eng. Sci.* 42(7), 1420–1431.
55. Moore, T. T., Mahajan, R., Vu, D. Q., and Koros, W. J. (2004). Hybrid membrane materials comprising organic polymers with rigid dispersed phases. *AIChE J.* 50(2), 311–321.
56. Kaeger, J., and Ruthven, D. M. (1992). *Diffusion in Zeolites and Other Microporous Solids.* Wiley, New York.
57. Manson, J. A., Sperling, L. H. (1976). *Polymer Blends and Composites.* Plenum, New York p. 513.
58. Zimmerman, C. M., Singh, A., and Koros, W. J. (1997). Tailoring mixed-matrix composite membranes for gas separations. *J. Membr. Sci.* 137(1–2), 145–154.
59. Duval, J. M., Folkers, B., Mulder, M. H. V., Desgrandchamps, G., and Smolders, C. A. (1993). Adsorbent-filled membranes for gas separation. Part 1. Improvement of the gas separation properties of polymeric membranes by incorporation of microporous adsorbents. *J. Membr. Sci.* 80, 189–198.
60. Moore, T. T., Vo, T., Mahajan, R., Kulkarni, S., Hasse, D., and Koros, W. J. (2003). Effect of humidified feeds on oxygen permeability of mixed-matrix membranes. *J. Appl. Polym. Sci.* 90(6), 1574–1580.
61. Yong, H. H., Park, H. C., Kang, Y. S., Won, J., and Kim, W. N. (2001). Zeolite-filled polyimide membrane containing 2,4,6-triaminopyrimidine. *J. Membr. Sci.* 188(2), 151–163.
62. Hale, P., and Lokhandwala, K. (2004). Advances in membrane materials provide new solutions in the gas business. *Paper presented at the 83rd Annual Convention Proceedings—Gas Processors Association*, pp. 138–153, March 14–17, 2004, San Antonio, Texas.
63. Tabe-Mohammadi, A. (1999). A review of the applications of membrane separation technology in natural gas treatment. *Sep. Sci. Technol.* 34(10), 2095–2111.
64. Duval, J. M. (1993). Adsorbent filled polymeric membranes; Application to pervaporation and gas separation. Ph.D. Dissertation, University of Twente, The Netherlands.
65. Duval, J. M., Kemperman, A. J. B., Folkers, B., Mulder, M. H. V., Desgrandchamps, G., and Smolders, C. A. (1994). Preparation of zeolite filled glassy polymer membranes. *J. Appl. Polym. Sci.* 54(4), 409–418.
66. Suer, M. G., Ba, N., and Yilmaz, L. (1994). Gas permeation characteristics of polymer-zeolite mixed-matrix membranes. *J. Membr. Sci.* 91(1–2), 77–86.
67. Li, Y., Chung, T.-S., Cao, C., and Kulprathipanja, S. (2005). The effects of polymer chain rigidification, zeolite pore size and pore blockage on polyethersulfone (PES)-zeolite A mixed-matrix membranes. *J. Membr. Sci.* 260(1–2), 45–55.
68. Tantekin-Ersolmaz, S. B., Atalay-Oral, C., Tatlier, M., Erdem-Senatalar, A., Schoeman, B., and Sterte, J. (2000). Effect of zeolite particle size on the performance of polymer-zeolite mixed-matrix membranes. *J. Membr. Sci.* 175(2), 285–288.

69. Tantekin-Ersolmaz, S. B., Senorkyan, L., Kalaonra, N., Tatlier, M., and Erdem-Senatalar, A. (2001). *n*-Pentane/*i*-pentane separation by using zeolite-PDMS mixed-matrix membranes. *J. Membr. Sci.* 189(1), 59–67.

70. Jia, M., Peinemann, and K. V., Behling, R. D. (1991). Molecular sieving effect of the zeolite-filled silicone rubber membranes in gas permeation. *J. Membr. Sci.* 578(2–3), 289–296.

71. Paul, D. R., and Kemp, D. R. (1973). Diffusion time lag in polymer membranes containing adsorptive fillers. *J. Polym. Sci. Polym. Symp.* 41, 79–93.

72. Kemp, D. R., Paul, D. R. (1974). Gas sorption in polymer membranes containing adsorptive fillers. *Journal of Polymer Science, Polymer Physics Edition* 12(3), 485–500.

73. Pechar, T. W., Tsapatsis, M., Marand, E., and Davis, R. (2002). Preparation and characterization of a glassy fluorinated polyimide zeolite-mixed-matrix membrane. *Desalination* 146(1–3), 3–9.

74. Cornelius, C. J., and Marand, E. (2002). Hybrid inorganic-organic materials based on a 6FDA-6FpDA-DABA polyimide and silica: Physical characterization studies. *Polymer* 43(8), 2385–2400.

75. Vu, D. Q., Koros, W. J., and Miller, S. J. (2002). High pressure CO_2/CH_4 separation by using carbon molecular sieve hollow fiber membranes. *Ind. Eng. Chem. Res.* 41(3), 367–380.

76. Vu, D. Q., Koros, W. J., Miller, S. J. (2003). Mixed matrix membranes using carbon molecular sieves. II. Modeling permeation behavior. *J. Membr. Sci.* 211(2), 335–348.

77. Vu, D. Q., Koros, W. J., and Miller, S. J. (2003). Effect of condensable impurity in CO_2/CH_4 gas feeds on performance of mixed-matrix membranes using carbon molecular sieves. *J. Membr. Sci.* 221(1–2), 233–239.

78. Steel, K. M., and Koros, W. J. (2005). An investigation of the effects of pyrolysis parameters on gas separation properties of carbon materials. *Carbon* 43(9), 1843–1856.

79. Anson, M., Marchese, J., Garis, E., Ochoa, N., and Pagliero, C. (2004). ABS copolymer-activated carbon mixed-matrix membranes for CO_2/CH_4 separation. *J. Membr. Sci.* 243(1–2), 19–28.

80. Chung, T.-S., Chan, S. S., Wang, R., Lu, Z., and He, C. (2003). Characterization of permeability and sorption in Matrimid/C60 mixed-matrix membranes. *J. Membr. Sci.* 211(1), 91–99.

81. Kittur, A. A., Kariduraganavar, M. Y., Kulkarni, S. S., and Aralaguppi, M. I. (2005). Preparation of zeolite-incorporated poly(dimethyl siloxane) membranes for the pervaporation separation of isopropyl alcohol/water mixtures. *J. Appl. Polym. Sci.* 96(4), 1377–1387.

82. Chandak, M. V., Lin, Y. S., Ji, W., and Higgins, R. J. (1997). Sorption and diffusion of VOCs in DAY zeolite and silicalite-filled PDMS membranes. *J. Membr. Sci.* 133(2), 231–243.

83. Vankelecom, I. F. J., De Kinderen, J., Dewitte, B. M., and Uytterhoeven, J. B. (1997). Incorporation of hydrophobic porous fillers in PDMS membranes for use in pervaporation. *J. Phys. Chem. B* 101(26), 5182–5185.

84. Kittur, A. A., Kariduraganavar, M. Y., Toti, U. S., Ramesh, K., Aminabhavi, T. M. (2003). Pervaporation separation of water-isopropanol mixtures using ZSM-5 zeolite incorporated poly(vinyl alcohol) membranes. *J. Appl. Polym. Sci.* 90(9), 2441–2448.

85. Bartels-Caspers, C., Tusel-Langer, E., and Lichtenthaler, R. N. (1992). Sorption isotherms of alcohols in zeolite-filled silicone rubber and in PVA-composite membranes. *J. Membr. Sci.* 70(1), 75–83.

86. Goldman, M., Fraenkel, D., and Levin, G. (1989). A zeolite/polymer membrane for separation of ethanol-water azeotrope. *J. Appl. Polym. Sci.* 37(7), 1791–800.

87. Ji, W., Sikdar, S. K., and Hwang, S.-T. (1995). Sorption, diffusion and permeation of 1,1,1-trichloroethane through adsorbent-filled polymeric membranes. *J. Membr. Sci.* 103(3), 243–255.

88. Vankelecom, I. F. J., Merckx, E., Luts, M., and Uytterhoeven, J. B. (1995). Incorporation of zeolites in polyimide membranes. *J. Phys. Chem.* 99(35), 13187–13192.

89. Lee, E. K., and Koros, W. J. (2002). Membranes, synthetic, applications. In Mark, H. F. (Ed.), *Encyclopedia of Polymer Science and Technology*, 3rd ed., Vol. 9. Academic, New York, pp. 279–344.

90. Avramescu, M.-E., Borneman, Z., and Wessling, M. (2003). Dynamic behavior of adsorber membranes for protein recovery. *Biotechnol. Bioeng.* 84(5), 564–572.

91. Avramescu, M.-E., Borneman, Z., and Wessling, M. (2003). Mixed-matrix membrane adsorbers for protein separation. *J. Chromatogr. A* 1006(1–2), 171–183.

92. Avramescu, M.-E., Girones, M., Borneman, Z., and Wessling, M. (2003). Preparation of mixed-matrix adsorber membranes for protein recovery. *J. Membr. Sci.* 218(1–2), 219–233.

93. Gulsen, D., Hacarloglu, P., Toppare, L., and Yilmaz, L. (2001). Effect of preparation parameters on the performance of conductive composite gas separation membranes. *J. Membr. Sci.* 182(1–2), 29–39.

94. Hacarlioglu, P., Toppare, L., and Yilmaz, L. (2003). Polycarbonate-polypyrrole mixed-matrix gas separation membranes. *J. Membr. Sci.* 225(1–2), 51–62.

95. Kim, J. H., Won, J., and Kang, Y. S. (2004). Pi-complexes of polystyrene with silver salts and their use as facilitated olefin transport membranes. *J. Polym. Sci. Part B: Polym. Phys.* 42(12), 2263–2269.

96. Lebrun, L., Da Silva, E., and Metayer, M. (2002). Elaboration of ion-exchange membranes with semi-interpenetrating polymer networks containing poly(vinyl alcohol) as polymer matrix. *J. Appl. Polym. Sci.* 84(8), 1572–1580.

97. Cussler, E. L. (1990). Membranes containing selective flakes. *J. Membr. Sci.* 52(3), 275–288.

98. Jeong, H.-K., Krych, W., Ramanan, H., Nair, S., Marand, E., and Tsapatsis, M. (2004). Fabrication of polymer/selective-flake nanocomposite membranes and their use in gas separation. *Chem. Mater.* 16(20), 3838–3845.

99. Hu, Q., Marand, E., Dhingra, S., Fritsch, D., Wen, J., and Wilkes, G. (1997). Poly(amide-imide)/TiO_2 nano-composite gas separation membranes: Fabrication and characterization. *J. Membr. Sci.* 135(1), 65–79.

100. He, Z., Pinnau, I., and Morisato, A. (2002). Nanostructured poly(4-methyl-2-pentyne)/silica hybrid membranes for gas separation. *Desalination* 146(1–3), 11–15.

101. Merkel, T. C., Freeman, B. D., Spontak, R. J., He, Z., Pinnau, I., Meakin, P., and Hill, A. J. (2002). Ultrapermeable, reverse-selective nanocomposite membranes. *Science* 296(5567), 519–522.

102. Reid, B. D., Ruiz-Trevino, F. A., Musselman, I. H., Balkus, K. J., Jr.; and Ferraris, J. P. (2001). Gas permeability properties of polysulfone membranes containing the mesoporous molecular sieve MCM-41. *Chem. Mater.* 13(7), 2366–2373.

103. Kulprathipanja, S., Neuzil, R. W., and Li, N. N. (1988). Separation of fluids by means of mixed-matrix membranes. U.S. Patent 4,740,219.

104. Cornelius, C., Hibshman, C., and Marand, E. (2001). Hybrid organic-inorganic membranes. *Sep. Purif. Technol.* 25(1–3), 181–193.

105. Guer, T. M. (1994). Permselectivity of zeolite filled polysulfone gas separation membranes. *J. Membr. Sci.* 93(3), 283–289.

106. Rojey, A., Deschamps, A., Grehier, A., and Robert, E. (1990). Process for separation of the constituents of a mixture in the gas phase using a composite membrane. U.S. Patent 4,925,459.

107. Vankelecom, I. F. J., Van den Broeck, S., Merckx, E., Geerts, H., Grobet, P., and Uytterhoeven, J. B. (1996). Silylation to improve incorporation of zeolites in polyimide films. *J. Phys. Chem.* 100(9), 3753–3758.

108. Kulkarni, S. S., Hasse, D. J., Corbin, D. R., and Patel, A. N. (2003). Gas separation membrane with organosilicon-treated molecular sieve. U.S. Patent 6,508,860.

109. Moore, T. T. (2004). Effects of materials, processing, and operating conditions on the morphology and gas transport properties of mixed-matrix membranes. Ph.D. Dissertation, University of Texas, Austin, TX.

110. Frommer, M. A., Feiner, I., Kedem, O., and Bloch, R. (1970). The mechanism for formation of "skinned" membranes: II. Equilibrium properties and osmotic flows determining membrane structure. *Desalination* 7(3), 393–402.

111. Pinnau, I., and Koros, W. J. (1991). Structures and gas separation properties of asymmetric polysulfone membranes made by dry, wet, and dry/wet phase inversion. *J. Appl. Polym. Sci.* 43(8), 1491–1502.

112. Fritzsche, A. K., Cruse, C. A., Kesting, R. E., and Murphy, M. K. (1990). Polysulfone hollow-fiber membranes spun from Lewis acid:base complexes. II. The effect of Lewis acid-to-base ratio on membrane structure. *J. Appl. Polym. Sci.* 39(9), 1949–1956.

113. Kesting, R. E., Fritzsche, A. K., Cruse, C. A., and Moore, M. D. (1990). The second-generation polysulfone gas-separation membrane. II. The relationship between sol properties, gel macrovoids, and fiber selectivity. *J. Appl. Polym. Sci.* 40(9–10), 1575–1582.

114. Cabasso, I., Klein, E., and Smith, J. K. (1977). Polysulfone hollow fibers. II. Morphology. *J. Appl. Polym. Sci.* 21(1), 165–180.

115. Ekiner, O. M., and Vassilatos, G. (1990). Polyaramide hollow fibers for hydrogen/methane separation—spinning and properties. *J. Membr. Sci.* 53(3), 259–273.

116. Pesek, S. C., and Koros, W. J. (1993). Aqueous quenched asymmetric polysulfone membranes prepared by dry/wet phase separation. *J. Membr. Sci.* 81(1/2), 71–88.

117. Pinnau, I. (1991). Skin formation of integral-asymmetric gas separation membranes made by dry/wet phase inversion (gas separation). Ph.D. Dissertation, University of Texas, Austin, TX.

118. Wallace, D. W. (2004). Crosslinked hollow fiber membranes for natural gas purification and their manufacture from novel polymers. Ph.D. Dissertation, University of Texas, Austin, TX.

119. Husain, S. (2006). Mixed matrix dual layer hollow fiber membranes for natural gas separation. Ph.D. Dissertation, Georgia Institute of Technology, Atlanta, GA.

120. Ekiner, O. M., and Kulkarni, S. S. (2003). Process for making mixed-matrix hollow fiber membranes for gas separation. U.S. Patent 6,663,805.

121. Jiang, L. Y., Chung, T. S., Cao, C., Huang, Z., and Kulprathipanja, S. (2005). Fundamental understanding of nano-sized zeolite distribution in the formation of the mixed-matrix single- and dual-layer asymmetric hollow fiber membranes. *J. Membr. Sci.* 252(1–2), 89–100.

122. Karnis, A., and Mason, S. G. (1966). Particle motions in sheared suspensions. XIX. Viscoelastics media. *Trans. Soc. Rheol.* 10(2), 571–592.

123. Tehrani, M. A. (1996). An experimental study of particle migration in pipe flow of viscoelastic fluids. *J. Rheol. (N.Y.)* 40(6), 1057–1077.

124. Gauthier, F., Goldsmith, H. L., and Mason, S. G. (1971). Particle motions in non-Newtonian media. II. Poiseuille flow. *Trans. Soc. Rheol.* 15(2), 297–330.

125. Leighton, D., and Acrivos, A. (1987). The shear-induced migration of particles in concentrated suspensions. *J. Fluid Mechan.* 181, 415–439.

126. Polotskaya, G., Biryulin, Y., Rozanov, V. (2004). Asymmetric membranes based on fullerene-containing polyphenylene oxide. *Fullerenes, Nanotubes, and Carbon Nanostructures* 12(1, 2), 371–376.

127. Kiyono, R., Koops, G. H., Wessling, M., and Strathmann, H. (2004). Mixed matrix microporous hollow fibers with ion-exchange functionality. *J. Membr. Sci.* 231(1–2), 109–115.

128. Vu, D. Q. (2001). Formation and characterization of asymmetric carbon molecular sieve and mixed-matrix membranes for natural gas purification. Ph.D. Dissertation, University of Texas, Austin, TX.

129. Miller, S. J., Kuperman, A., and Vu, D. Q. (2005). Mixed matrix membranes with small pore molecular sieves and methods for making and using the membranes. U.S. Patent 2005139065.

130. Buttal, T., Bac, N., and Yilmaz, L. (1995). Effect of feed composition on the performance of polymer-zeolite mixed-matrix gas separation membranes. *Sep. Sci. Technol.* 30(11), 2365–2384.

131. Chen, X., Yang, H., Gu, Z., and Shao, Z. (2001). Preparation and characterization of HY zeolite-filled chitosan membranes for pervaporation separation. *J. Appl. Polym. Sci.* 79(6), 1144–1149.

132. Shah, D., Kissick, K., Ghorpade, A., Hannah, R., and Bhattacharyya, D. (2000). Pervaporation of alcohol–water and dimethylformamide–water mixtures using hydrophilic zeolite NaA Membranes: Mechanisms and experimental results. *J. Membr. Sci.* 179(1–2), 185–205.

133. Okumus, E., Gurkan, T., and Yilmaz, L. (1994). Development of a mixed-matrix membrane for pervaporation. *Sep. Sci. Technol.* 29(18), 2451–2473.

134. Duval, J. M., Folkers, B., Mulder, M. H. V., Desgrandchamps, G., and Smolders, C. A. (1994). Separation of a toluene/ethanol mixture by pervaporation using active carbon-filled polymeric membranes. *Sep. Sci. Technol.* 29(3), 357–373.

135. Kulprathipanja, S., Neuzil, R. W., and Li, N. N. (1992). Separation of gases by means of mixed-matrix membranes. U.S. Patent 5,127,925.

136. Kulprathipanja, S. (1986). Separation of polar gases from nonpolar gases. U.S. Patent 4,606,740.

137. Kulprathipanja, S., and Kulkarni, S. S. (1986). Separation of polar gases from nonpolar gases. U.S. Patent 4,608,060,

138. Kulprathipanja, S., Kulkarni, S. S., and Funk, E. W. (1988). Multicomponent membranes. U.S. Patent 4,737,165.

139. Kulprathipanja, S., Kulkarni, S. S., and Funk, E. W. (1988). Preparation of gas selective membranes. U.S. Patent 4,751,104.

140. Zimmerman, C. M., Mahajan, R., and Koros, W. J. (1997). Fundamental and practical aspects of mixed-matrix gas separation membranes. *Polym. Mater. Sci. Eng.* 77, 328–329.

141. Rattanawong, W., Osuwam, S., Risksomboen, T., and Kulprathipanja, S. (2001). Zeolite-cellulose acetate mixed-matrix membrane for olefin/paraffin separation. *Am. Chem. Soc. Div. Petrochem.* 46, 166.

142. Rattanawong, W. (2001). Zeolite/cellulose acetate mixed matrix membranes for olefin/paraffin separations. M.S. Thesis, The Petroleum and Petrochemical College, Chulalongkovn University, Bangkok, Thailand.

143. Boom, J. P., Bargeman, D., and Strathmann, H. (1994). Zeolite-filled membranes for gas separation and pervaporation. *Stud. Surf. Sci. Catal.* 84, 1167–1174.

144. Kerres, J., and Haering, T. (2001). Organic-inorganic membranes. WO Patent 2001083092.

145. Süer, M. G., Bac, N., Yilmaz, L., Gurkan, T., and Sacco, A., Jr. (1994). In E. F. Vansant (Ed.), *Gas Separation with Zeolite Based Polyethersulfone Membranes. Separation Technology.* Elsevier, Amsterdam; New York, p. 661.

146. Haering, T., Kerres, J., and Ullrich, A. (2000). Silicate-supported proton-conducting composite membranes containing polymeric ionomer blends. WO Patent 2000074827.

147. Tatlier, M., Tantekin- Ersolmaz, S., Birguil, S., Ataly-Oral, C., and Erdem-Senatalar, A. (2001). Power-law scaling behavior of membranes. *J. Membr. Sci.* 182(1–2), 183–193.

148. Boom, J. P., Punt, I. G. M., Zwijnenberg, H., de Boer, R., Bargeman, D., Smolders, C. A., and Strathmann, H. (1998). Transport through zeolite filled polymeric membranes. *J. Membr. Sci.* 138(2), 237–258.

149. Sukapintha, W. (2000). Mixed matrix membrane for olefin/paraffin separation. M.S. Thesis, The Petroleum and Petrochemical College, Chulalongkorn University, Bangkok, Thailand.

150. Jiang, L., Chung, T. S., Li, D. F., Cao, C., and Kulprathipanja, S. (2004). Fabrication of Matrimid/polyethersulfone dual-layer hollow fiber membranes for gas separation. *J. Membr. Sci.* 240(1–2), 91–103.

151. Miller, S. J., Munson, C. L., Kulkani, S. S., and Hasse, D. J. (2002). Purification of *p*-xylene using composite mixed-matrix membranes. U.S. Patent 6,500,233 B1.

152. Tsapatsis, M., Jeong, H.-K., and Nair, S. (2005). Layered silicate material and applications of layered materials with porous layers. U.S. Patent 6,863,983.

153. Koros, W. J., and Mahajan, R. (2000). Pushing the limits on possibilities for large scale gas separation: Which strategies. *J. Membr. Sci.* 175(2), 181–196.

154. Corbin, D. R., Foley, H. C., and Shiffett, M. B. (2004). Mixed matrix nanoporous carbon membranes. U.S. Patent 6,740,143 B2.

155. Rallabandi, P. S., and Ford, D. M. (2000). Permeation of small molecules through polymers confined in mesoporous media. *J. Membr. Sci.* 171(2), 239–252.

156. Yamasaki, A., Iwatsubo, T., Masuoka, T., Mizoguchi, K. (1994). Pervaporation of ethanol/water through a poly(vinyl alcohol)/cyclodextrin (PVA/CD) membrane. *J. Membr. Sci.* 89, 111–117.

157. Moermans, B., Beuckelaer, W. D., Vankelecom, I. F. J., Ravishankar, R., Martens, J. A., and Jacobs, P. A. (2000). Incorporation of nano-sized zeolites in membranes. *Chem. Commun.* 2467–2468.

158. Wang, H., Holmberg, B. A., Yan, Y. (2002). Homogeneous polymer-zeolite nanocomposite membranes by incorporating dispersible template-removed zeolite nanocrystals. *J. Mater. Chem.* 12, 3640–3643.

159. Kulprathipanja, S., Funk, E. W., Kulkarni, S. S., and Chang, Y. A. (1988). Separation of a monosaccharide with mixed-matrix membranes. U.S. Patent 4,735,193.

160. Kulprathipanja, S., and Charoenphol, J. (2004). Mixed matrix membrane for separation of gases. U.S. Patent 6,726,744.

161. Miller, S. J., Kuperman, Al., and Vu, D. Q. (2005). Mixed matrix membranes with small pore molecular sieves and methods for making and using the membranes. U.S. Patent 2005139066.

162. Hasse, D. J., Kulkarni, S. S., Corbin, D. R., and Patel, A. N. (2003). Mixed matrix membranes incorporating chabazite type molecular sieves. U.S. Patent 6626980.

163. Koros, W. J., Vu, D. Q., Mahajan, R., and Miller, S. J. (2003). Gas separations using mixed-matrix membranes. U.S. Patent 6,503,295.

164. Koros, W. J., Vu, D. Q., Mahajan, R., and Miller, S. J. (2002). Mixed matrix membranes and manufacture and purification of methane from a gas stream. U.S. Patent 2002056369.

165. Koros, W. J., Vu, D. Q., Mahajan, R., and Miller, S. J. (2002). Carbon molecular sieves and manufacture for separation membranes. U.S. Patent 2002053284.

166. Koros, W. J., Wallace, D., Wind, J. D., Miller, S. J., Staudt-Bickel, C., and Vu, D. Q. (2003). Crosslinked and crosslinkable hollow fiber mixed matrix membrane and method of making same. U.S. Patent 2003140789 A1.

167. Miller, S. J., and Yuen, L.-T. (2005). Mixed matrix membrane with super water washed silica containing molecular sieves and methods for making and using the same. U.S. Patent 2005043167.

168. Figoli, A., Sager, W., and Wessling, M. (2002). Synthesis of novel nanostructured mixed-matrix membranes. *Desalination* 148, 401–405.

169. Zhang, X., Wang, T., and Liu, H. (2004). Preparation of composite carbon-zeolite membranes using a simple method. *J. Mater. Sci.* 39(16–17), 5603–5606.

170. Koros, W. J. (2004). Evolving beyond the thermal age of separation processes: Membranes can lead the way. *AIChE J.* 50(10), 2326–2334.

171. Yehia, H., Pisklak, T. J., Ferraris, J. P., Balkus, K. J., Jr., and Musselman, I. H. (2004). Methane facilitated transport using copper (II) biphenyl dicarboxylate-triethylenediamine poly(3-acetoxyethylthiophene) mixed-matrix membranes. *Polym. Prepr.* 45, 35–36.

172. Mahajan, R., Vu, D. Q., and Koros, W. J. (2002). Mixed matrix membrane materials: An answer to the challenges faced by membrane based gas separations today. *J. Chin. Inst. Chem. Eng.* 33(1), 77–86.

173. Mahajan, R., Zimmerman, C. M., and Koros, W. J. (1999). Fundamental and practical aspects of mixed-matrix gas separation membranes. *ACS Symp. Ser.* 733, 277–286.

174. Rezac, M. E., and Koros, W. J. (1992). Preparation of polymer-ceramic composite membranes with thin defect-free separating layers. *J. Appl. Polym. Sci.* 46, 1927–1938.

175. Sforca, M. L., Yoshida, I. V. P., and Nunes, S. P. (1999). Organic–inorganic membranes prepared from polyether diamine and epoxy silane. *J. Membr. Sci.* 159(1–2), 197–207.

176. Kulprathipanja, S. (2003). Mixed matrix membrane development. *Ann. N.Y. Acad. Sci.* 984, 361–369.

177. Bouma, R. H. B., Checchetti, A., Chidichimo, G., and Drioli, E. (1997). Permeation through a heterogeneous membrane: The effect of the dispersed phase. *J. Membr. Sci.* 128(2), 141–149.

178. Ahmad, Z., and Mark, J. E. (2001). Polyimide-ceramic hybrid composites by the sol-gel route. *Chem. Mater.* 13(10), 3320–3330.

179. Lee, J.-Y., Shim, M.-J., and Kim, S.-W. (1999). Effect of natural zeolite on the mechanical properties of epoxy matrix. *Polym. Eng. Sci.* 39(10), 1993–1997.

180. Tsagaropoulos, G., Eisenburg, A. (1995). Direct observation of two glass transitions in silica-filled polymers. Implications to the morphology of random ionomers. *Macromolecules* 28(1), 396–398.

181. Vankelecom, I. F. J., Beukelaer, S. D., and Uytterhoeven, J. B. (1997). Sorption and pervaporation of aroma compounds using zeolite-filled PDMS membranes. *J. Phys. Chem. B* 101, 5186–5190.

182. Xenopoulos, C., Mascia, L., and Shaw, S. J. (2002). Polyimide–silica hybrids derived from an isoimide oligomer precursor. *J. Mater. Chem.* 12(2), 213–218.

183. Moaddeb, M., and Koros, W. J. (1995). Silica-treated ceramic substrates for formation of polymer-ceramic composite membranes. *Ind. Eng. Chem. Res.* 34, 263–274.

184. Marand, E., and Pechar, T. W. (2003). Zeolite-polyimide mixed-matrix membranes for gas separations. U.S. Patent 2003220188.

185. Matsumoto, K., Xu, P., and Nishikimi, T. (1993). Gas permeation of aromatic polyimides. I. Relationship between gas permeabilities and dielectric constants. *J. Membr. Sci.* 81(1–2), 15–22.

186. Matsumoto, K., Xu, P. (1993). Gas permeation of aromatic polyimides. II. Influence of chemical structure. *J. Membr. Sci.* 81(1–2), 23–30.

187. Kawakami, M., Iwanaga, H., Hara, Y., Iwamoto, M., and Kagawa, S. (1982). Gas permeabilities of cellulose nitrate/poly(ethylene glycol) blend membranes. *J. Appl. Polym. Sci.* 27, 2387–2393.

188. Sukapintha, W. (2000). Mixed matrix membrane for olefin/paraffin separation. M.S. Thesis, The Petroleum and Petrochemical College, Chulalongkorn University, Bangkok, Thailand.

189. Li, J., Wang, S., Nagai, K., Nakagawa, T., and Mau, A. W-H. (1998). Effect of polyethyleneglycol (PEG) on gas permeabilities and permselectivities in its cellulose acetate (CA) blend membranes. *J. Membr. Sci.* 138(2), 143–152.

190. *Gas Conditioning Fact Book.* (1962). The Dow Chemical Company, Midland, MI.

191. Vijitjunya, P. (2001). Dispersed liquid/polymer mixed matrix membrane for olefin/paraffin separation. M.S. Thesis, The Petroleum and Petrochemical College, Chulalongkorn University, Bangkok, Thailand.

192. Charoenphol, J. (2002). Mixed matrix membranes for CO_2/N_2 separation. M.S. Thesis, The Petroleum and Petrochemical College, Chulalongkorn University, Bangkok, Thailand.

193. Kulprathipanja, S., Soontraratpong, J., and Chiou, J. J. (2005). Mixed matrix membrane for gas separation. U.S. Patent 11/216823.

CHAPTER 31

Fabrication of Hollow-Fiber Membranes by Phase Inversion

TAI-SHUNG NEAL CHUNG

Department of Chemical and Biomolecular Engineering, National University of Singapore, Singapore 119260

31.1 INTRODUCTION

Since Mahon 1966 proposed using hollow-fiber membranes as separation apparatus in his patents 4 decades ago, synthetic polymeric hollow-fiber membranes have advanced as a key player in separation technologies. Compared to the flat-sheet membrane, hollow-fiber configuration has the following advantages: (1) a much larger membrane area per unit volume of membrane module and, hence, resulting in a higher productivity; (2) self-mechanical support that can be back flushed for liquid separation; and (3) good flexibility and easy handling during module fabrication, membrane reparation, and system operation (Kesting, 1985; Matsuura, 1994; Ho and Sirkar, 1992; Paul and Yampol'skii, 1994). Nowadays, hollow fibers are widely used in gas separation, ultrafiltration, pervaporation, dialysis and supported liquid membrane extraction.

Three key elements determine the potential and applications of a hollow-fiber membrane: (1) pore size and pore size distribution, (2) selective layer thickness, and (3) inherent properties (chemistry and physics) of the membrane material. Pore size and its distribution usually determine membrane applications, separation factor, or selectivity. The selective layer thickness determines the membrane flux or productivity. Material chemistry and physics govern the intrinsic permselectivity for gas separation and pervaporation, fouling characteristics for RO (reverse osmosis), UF (ultrafiltration), and MF (microfiltration) membranes, chemical resistance for membranes used in harsh environments, protein and drug separation, as well as biocompatibility for biomedical membranes used in dialysis and biomedical and tissue engineering.

In the last 4 decades, a significant effort has been given by membrane scientists to develop hollow fibers with desirable pore structures and thin selective layers. However, even today, most membrane scientists admit that we understand hollow-fiber forming mechanisms qualitatively not quantitatively. New hollow-fiber membranes are invented mostly

Advanced Membrane Technology and Applications. Edited by Norman N. Li, Anthony G. Fane, W. S. Winston Ho, and T. Matsuura
Copyright © 2008 John Wiley & Sons, Inc.

based on our experience, empirical data, limited qualitatively scientific understanding, trial and error, and luck.

The objective of this chapter is to review the fundamental understanding of the asymmetric hollow-fiber formation and the technology developed for asymmetric hollow fibers with desired skin and cross-section morphology.

31.2 BASIC UNDERSTANDING

Phase inversion is one of the most important means to prepare asymmetric hollow-fiber membranes. The resultant membranes have a dense skin layer that is integrally bonded in series with a thick porous substructure. The skin and the substructure are composed of the same material. The skin layer, which contains the effective separating layer, is one of the key elements in determining membrane flux and separation factor for liquid separation, as well as permeance and selectivity for gas separation.

Since Loeb and Sourirajan (1963) invented cellulose acetate membranes for RO in the late 1950s, the principles of asymmetric membrane formation have been reviewed by many authors (Sourirajan, 1970; Kesting, 1985; Ho and Sirkar, 1992; Koros and Pinnau, 1994; Matsuura, 1994; Wienk et al., 1996). However, most attention has been given to phase-inversion and thermal-inversion mechanisms of asymmetric flat membranes. Flat membranes are usually prepared by casting a polymeric solution on a substrate and dried for a short period of time before immersing in a nonsolvent coagulation bath. The process of fabricating flat membranes is much simpler than that of hollow fibers. However, the controlling factors for hollow-fiber morphology are quite different from those for flat membranes (Chung et al., 1992; Chung, 1997). Therefore, it is a well-known fact in membrane scientific community that it is difficult to simulate the hollow-fiber spinning process by adopting the process conditions developed for asymmetric flat membranes.

The formation of modern asymmetric hollow-fiber membranes consists of seven steps: (1) dope preparation, (2) degas, (3) metering, (4) spinning, (5) evaporation (in the air gap region), (6) coagulation, and (7) solvent exchange. Most commercially available hollow fibers are spun from a hot spinneret with an air gap distance and a moderate speed in order to increase the fiber production as well as to reduce fiber diameter (in order to increase fiber packing density per module). It is a nonisothermal process, and there are at least three forces (stresses) applied upon the as-spun polymeric spinning solution: (1) shear and elongation stresses within the spinneret, (2) gravity induced by the fiber's own weight, and (3) stresses induced by the take-up unit and coagulation bath (Chung, 1997; Wang et al., 2004a). There are two coagulations taking place at the internal and external surfaces during hollow-fiber spinning. Properly choosing bore fluid chemistry, flow rate, and controlling the internal coagulation, one can reasonably control the internal skin morphology. Similarly, one can control the outer skin morphology by adjusting the outer coagulant chemistry and coagulation conditions. If liquids are used as bore fluids, the internal coagulation process starts immediately after extrusion from a spinneret, and then the fiber goes through the external coagulation. Usually, water is the preferred external coagulant because of low cost and being environmentally friendly. The proper choice of the internal coagulant is, therefore, becoming very important because the rate of demixing (i.e., phase separation) and the resultant inner skin morphology strongly depend on the chemistry and compositions of the internal coagulant (Chung et al., 1992, 1994, 1997).

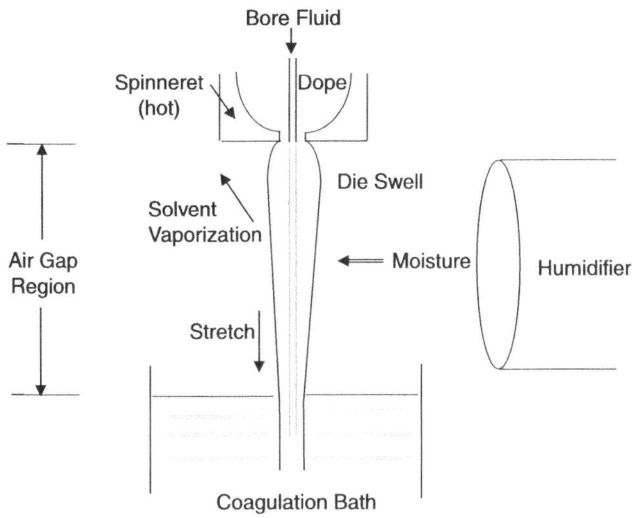

Figure 31.1 Schematic of dry-jet wet-spinning of hollow-fiber membranes.

If the spinneret is hot, a fraction of low-boiling dope solvents may evaporate and form the skin as shown in Figure 31.1. Humidity surrounding the fiber may affect the skin morphology because high humidity may induce early precipitation and increase the selective skin thickness (Chung et al., 1992). However, the effects of humidity on phase inversion and membrane performance depend on the duration of nascent fiber traveling through the air gap as well as if there is strong solvent evaporation occurring at the outer layer of nascent membranes. If the duration is very short or if there is a highly volatile solvent evaporation, the adsorption of water moisture upon the nascent hollow fiber may be diminished. Under these circumstances, the humidity effects on membrane separation performance may be negligible. The molecular sizes and solubility parameters of solvents and internal and external coagulants play important roles on membrane morphology (Kesting et al., 1990; Kesting and Fritzsche, 1993). Large-size solvents may have difficulties to leach out during the precipitation. The residual solvents may redissolve the skin and densify the selective layer. The difference in solubility parameter between spinning solution and internal or external coagulant affect the coagulation rate (Kesting, 1985; Ho and Sirkar, 1992; Chung et al., 1992, 1994, 1997; Matsuura, 1994). In addition, the ratio (k) of solvent outflow to coagulant influx determines the macroscopic membrane porosity (Yilmaz and McHugh, 1986, 1988; Matsuura, 1994), as illustrated in Figure 31.2. Depending on the initial dope composition and k value, the precipitation path may occur via nuclei growth or spinodal decomposition. Because k is not a constant across the membrane thickness and k is also a function of temperature and dope viscosity, it becomes difficult to microscopically predict local membrane porosity. Therefore, Figure 31.2 can only be used as a qualitative understanding of phase inversion.

In addition, the spinning dope suitable for fabricating hollow fibers generally has a much greater viscosity and elasticity than that for flat membranes. To produce hollow-fiber membranes for gas separation, Chung et al. (1992) in their early work had hypothesized that a dope exhibiting significant chain entanglement is one of the key requirements to produce hollow fibers with minimum defects. A critical concentration extrapolated from the viscosity versus polymer concentration relationship, as illustrated in Figure 31.3, is

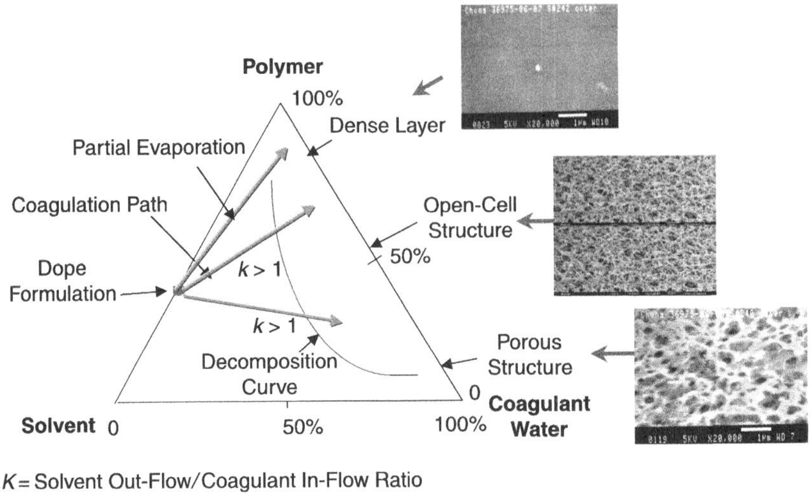

Figure 31.2 Relationships among dope composition, precipitation kinetics, and membrane morphology.

recommended for the spinning solution (Chung et al., 1992). Below the critical concentration, the resultant hollow fibers may have too many defects and cannot be properly repaired by the silicone rubber coating (Henis and Tripodi, 1981). Above the critical concentration, the resultant hollow fibers may have a thick dense selectivity layer that reduces the permeance. As a rule of thumb, the optimal polymer concentration for gas separation hollow-fiber membranes may be located at or 1–2 wt% above the critical concentration. This practice has been widely used and proved to be valid for gas separation and pervaporation membranes (Chung et al., 1997; Kapantaidakis et al., 2002; Liu et al., 2005; Ren et al., 2002; Barsema et al., 2003).

Rheological data suggest that high-viscosity non-Newtonian fluids behave differently from low-viscosity Newtonian fluids in the flow chamber of the spinneret. Non-Newtonian fluids memorize the shear and elongation stresses imposed on them, while Newtonian fluids do not. Furthermore, the hollow-fiber spinning process is a dynamic process not an equilibrium process. Therefore, it may not be appropriate to predict the phase-inversion process by the traditional Flory–Huggins theory (Chung, 1997). Gravity

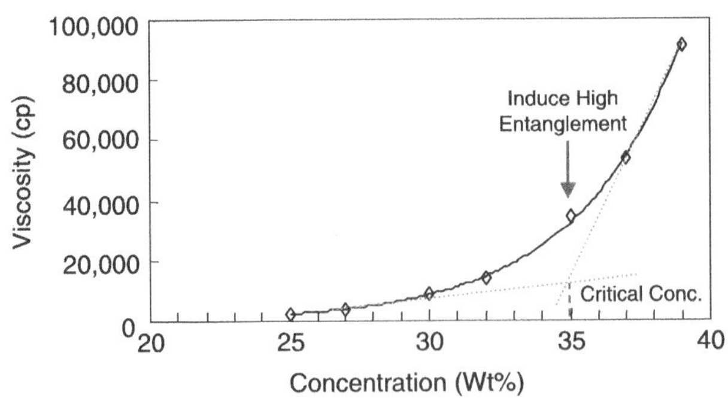

Figure 31.3 Viscosity vs. polymer concentration of PES–NMP dopes and critical concentration (Chung et al., 1997).

force must sometimes be considered (Chung and Teoh, 1997). For a high-speed spinning process, the effects of spinning-line tension, convective, and drag flow that occurred at the membrane surface in the coagulation bath may not be ignored. One must take these factors into consideration to develop hollow-fiber membranes with a desirable structure and separation performance.

31.3 RECENT PROGRESSES ON SINGLE-LAYER ASYMMETRIC HOLLOW-FIBER MEMBRANES

During the last 4 decades, many researchers have studied the effects of dope concentration, internal and external coagulants' chemistry, and spinning conditions on membrane formation and performance. Most of their works have been summarized elsewhere (Mulder, 1996; Chung and Kafchinski, 1997; Freeman and Pinnau, 1999; Sharpe et al., 1999; Clausi and Koros, 2000; Nunes and Peinemann, 2001; Carruthers et al., 2003; Ho, 2003; Baker, 2004). Here we only summarize key progress in the last 10 years.

31.3.1 Effects of Spinneret Design and Flow Behavior within the Spinneret on Hollow-Fiber Membranes

The systematic study of the effects of spinneret design and flow behavior within the spinneret on membrane performance started about one decade ago (Chung et al., 1998; Qin and Chung, 1999; Cao et al., 2004; Wang et al., 2004b; Widjojo and Chung, 2006). When fabricating UF and gas separation membranes using straight annular spinnerets (i.e., mainly shear stress exists), it was found that an increase in shear rate would result in an increase in the orientation of polymeric chains and chain packing. The increasing shear rate may elongate and reduce pore sizes (Wang and Chung, 2001). As a consequence, flux or permeance may decrease, while separation factor or selectivity may increase.

The influence of shear rate on membrane structure is illustrated in Figures 31.4 (Ren et al., 2002) and 31.5 (Chung et al., 2002). Figure 31.4 exhibits the cross-section morphology of hollow-fiber membranes spun at shear rates of 812 and 2436 s^{-1}. Some macrovoids can be observed near the inner skin of fibers spun at a low shear rate (812 s^{-1}), while these macrovoids are apparently eliminated or suppressed when the shear rate increases to 2436 s^{-1}. Clearly, high shear rates modify the precipitation path and retard the formation of macrovoids. In addition, with an increase in shear rate, the membrane structure becomes

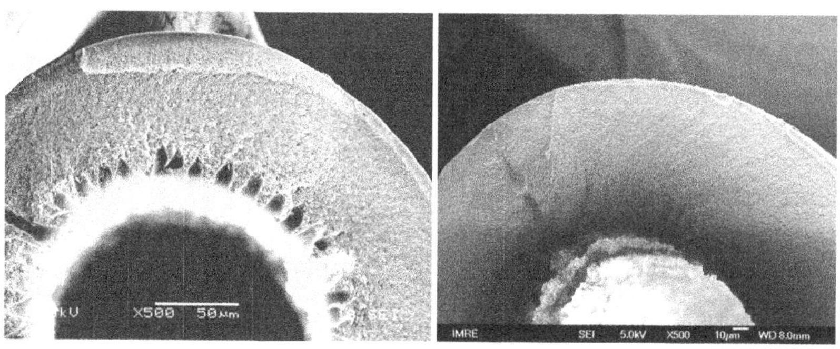

Figure 31.4 Influence of shear rate on the cross-section and outer skin morphology of hollow-fiber membranes spun from a polyimide solution [(*left*) 812 s^{-1} and (*right*) 2436 s^{-1}] (Ren et al., 2002).

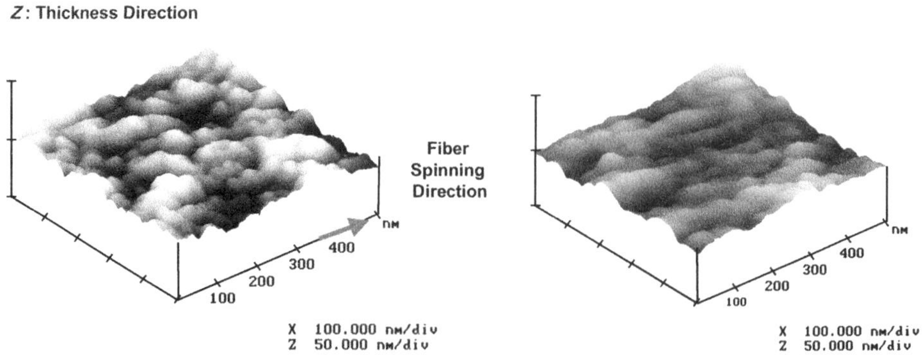

Figure 31.5 Three-dimensional AFM image of outer surface of hollow-fiber UF membrane spun with a shear rate of (*left*) 1305 s^{-1} and (*right*) 11,066 s^{-1} (Chung et al., 2002).

more compact. Furthermore, high shear rates affect the outer surface morphology. Figure 31.5 displays three-dimensional atomic force microscopy (AFM) images of the outer surface of hollow-fiber UF membranes spun with different shear rates [1305/s (*left*) vs. 11,066/s (*right*)] (Chung et al., 2002). It is revealed that nodules in the outer skin changing from random arrangement to obviously tidier align along the direction of dope extrusion when the shear rate is increased. In addition, both nodule sizes in the fiber spinning and transversal directions decrease with increasing shear rate, possibly because of chain disentanglement and thermodynamically favored. Furthermore, the roughness of the outer surface of hollow-fiber UF membranes decreases with an increase in shear rate.

The combined influence of elongation and shear rate induced by the geometry of spinnerets on membrane performance for gas separation has been studied as illustrated in Figure 31.6 (Cao et al., 2004). The flow profiles of dope solution and the elongation and

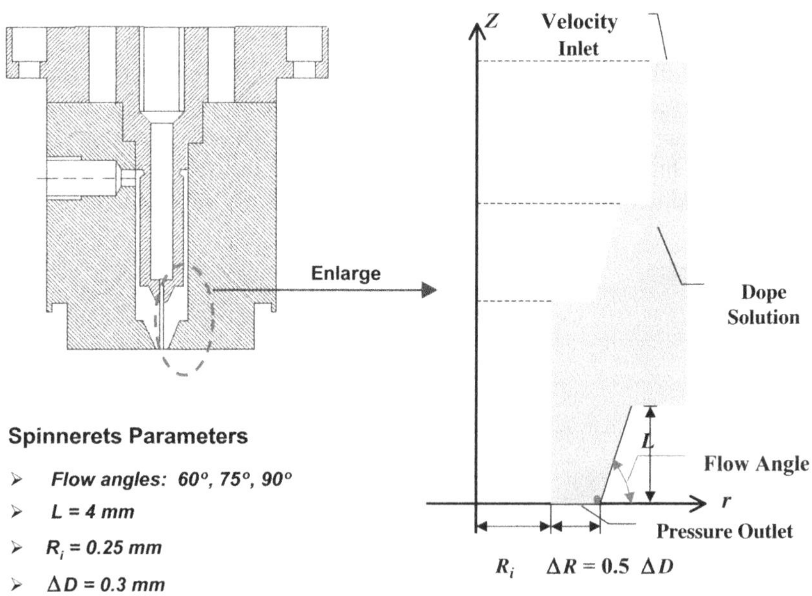

Figure 31.6 Spinnerets with various flow angles (Cao et al., 2004).

shear rates at the outermost point of the outlet of spinnerets can be simulated by the computational fluid dynamics (CFD) model. The preliminary conclusion indicates that the elongation rate has more contribution portion in permselectivity than in permeance, while the shear rate has more contribution portion in permeance than in permselectivity.

For UF membranes, experimental results suggest hollow fibers spun from a conical spinneret have smaller mean pore sizes with larger geometric standard deviations, thus exhibiting lower water flux and greater solute separation than hollow fibers spun from a traditional straight spinneret (Wang et al., 2004b). In addition, scanning election microscopy (SEM) studies indicate macrovoids respond differently for the 90° straight and 60° conical spinnerets with an increase in shear rate (i.e., dope flow rate). Macrovoids can be significantly suppressed and almost disappear in the cross section of the fibers spun by using a 90° spinneret at high dope flow rates. This phenomenon cannot be observed for the 60° conical spinneret.

31.3.2 Effects of Air Gap on Hollow-Fiber Membranes

Table 31.1 summarizes the effect of air gap distance on the permeance of hollow fibers spun from 30 wt% polyerthersulfone (PES) (Chung and Hu, 1997). It clearly shows that air gap distance plays a very important role on nascent fiber performance. An increase in air gap distance results in a significant decrease in permeance. This interesting phenomenon may arise not only from the fact that different precipitation paths take place during the wet-spinning and dry-jet wet-spinning process, but also from elongation-induced chain orientation and packing. However, if the air gap is too big, it may also create defects because of gravity and elongational stresses.

Many scientists have used the Flory–Huggins equation of solutions, derived in 1942, to study the phase-inversion process during the formation of asymmetric flat membranes. However, the Flory–Huggins equation of solutions may not be fully capable of describing the Gibbs free energy for the state of as-spun hollow-fiber solutions (nascent fibers) spun nonisothermally with tension (Chung, 1997). At least two additional terms have to be included in this equation if the fiber is spun isothermally; one is the work, done by the external stresses to the as-spun nascent fiber and the other is an extra entropy change, $\Delta S^{\mathrm{Extra}}/RT$, induced by these stresses.

Figure 31.7 displays the SEM examples of the external fiber surface structures spun with and without air gap from a 4,4′-hexafluoroisopropylidene diphthalic anhydride (6FDA) (6FDA)–polyimide solution (Chung, 1997). Their morphology is completely different. These pictures indicate that external stresses (work) probably have three effects on the states of solutions: (1) creating extra phase instability, (2) facilitating phase separation, and (3) inducing orientation. The first two will either shorten the time for a solution

TABLE 31.1 Permeance Change with Different Air Gap Distances

Air Gap Distance (cm)	O_2 Permeance (GPU)[a]	N_2 Permeance (GPU)	He Permeance (GPU)	O_2/N_2 Selectivity
14.4	22.87	25.23	55.94	0.91
6.1	24.87	26.23	76.67	0.95
1.2	42.93	40.01	117.1	0.93
0	75.79	74.90	165.2	0.99

[a]GPU = 1 × 10 − 6 cm^3 (STP)/cm^2-s-cm Hg.

Source: Chung and Hu (1997).

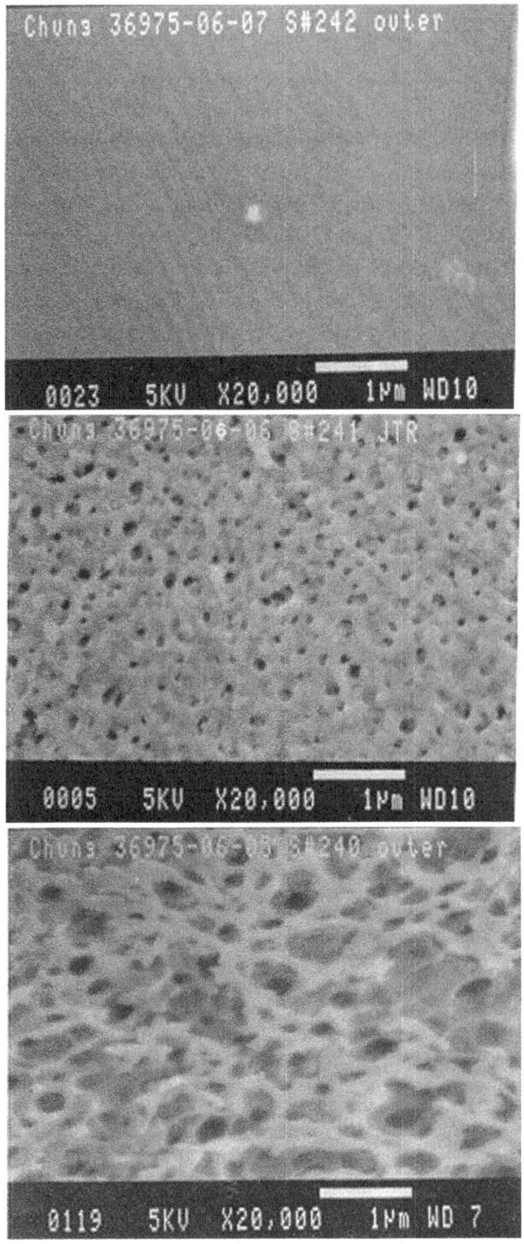

Figure 31.7 Effects of air gap distance on external fiber morphology: [(*top*) no air gap, (*middle*) 2.54 cm air gap, and (*bottom*) 12.7 cm] (Chung, 1997).

moving from the binodal boundary to the spinodal boundary or reduce the distance of precipitation path between binodal and spinodal boundaries, while the last one results in an oriented fiber structure. Experimental results also suggest that the effectiveness of external stresses (work) on the phase stability of a spinning solution depends on the chemistry of coagulants. The wet-spun hollow fiber has a tight external surface morphology, while the dry-jet wet-spun fiber with a long air gap may have a three-dimensional open-cell structure.

31.3.3 Macrovoid and Elongational Drawing

For most cases, macrovoids in asymmetric hollow fibers may be undesirable because they are the weak mechanical points that usually result in membrane failure at high pressures. The origins of macrovoid formation in the cross section of phase-inversion membranes have been often studied and seriously debated. Several researchers believe it originates from thermodynamic aspects of chemical potential gradient (Broens et al., 1980; Reuvers et al., 1987; Yao et al., 1988; Smolders et al., 1992). Others consider it starts from local surface instability and material and stress imbalance, which induce solvent intrusion and capillary flow (Matz, 1972; Strathmann et al., 1975; Strathmann and Kock, 1977; Chung and Kafchinski, 1997; Wang et al., 2004a; Widjojo and Chung, 2006). Other mechanisms such as Marangoni effects (Levich and Krylov, 1969; Shojaie et al., 1994), osmosis pressure (McKelvey and Koros, 1996), and moisture effects (Menut et al., 2002; Tsai et al., 2005; Khare et al., 2005) have also been proposed.

Ways to reduce macrovoids or modify the phase-inversion conditions to yield a sponge-like structure have been studied; they may consist of (1) using high polymer concentration solutions (Kesting et al., 1990; Kesting and Fritzsche, 1993), (2) using high viscosity spinning solutions (Chung et al., 1994), (3) spinning at high shear rates (Ren et al., 2002), (4) the induction of delayed demixing (Kim et al., 2001) or gelation (Lin et al., 2002), (5) the addition of surfactants (Tsai et al., 2001), (6) the addition of high-viscosity components (Li et al., 2002), or (7) the reduction of air gap distance and flow channel thickness in the spinneret (Widjojo and Chung, 2006).

Recently, it was reported that macrovoids in asymmetric hollow-fiber membranes can be completely eliminated at high elongational draws (Wang et al., 2004a; Xiao et al., 2006). The number of macrovoids and the number of macrovoid layer decrease with an increase in elongational draw ratio, while the dimension of macrovoids varies with increasing elongational draw ratio until the macrovoids are fully eliminated. As shown in Figure 31.8, for the free fall (i.e., no elongational draw) hollow fibers, a clear double-layer structure of

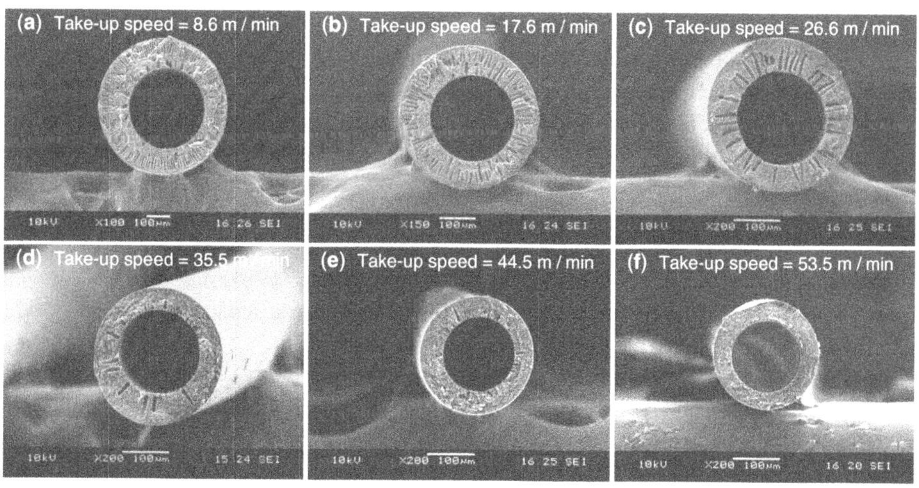

Figure 31.8 Effects of elongational draw ratio on the morphology of dry-jet wet-spin PES single-layer hollow-fiber membranes. Scale bar: 100 μm. Draw ratio: (a) $\phi = 2.42$; (b) $\phi = 4.55$; (c) $\phi = 7.31$; (d) $\phi = 9.78$; (e) $\phi = 12.00$; (f) $\phi = 15.20$ (Wang et al., 2004a).

macrovoids can be observed (Fig. 31.8a). One layer near the outer skin edge is finger-like, whereas the other near the inner skin edge is teardrop-like. When doubling the takeup speed, the number of finger-like macrovoids decreases. When tripling the takeup speed, only single layers of finger-like macrovoids can be observed. However, the size and dimension of macrovoids become larger and longer and their total number is clearly reduced. Further increasing the takeup speed, Figure 31.8e shows that the finger-like macrovoids disappear at 53.5 m/min, which is about 6 times of the free fall speed. Here the elongational draw ratio, ϕ, is defined as the ratio of the cross-section area of the spinneret for dope flow to the solid cross-section area of the precipitated hollow-fiber membrane as follows:

$$\phi = \frac{(\mathrm{OD}^2 - \mathrm{ID}^2)_{\mathrm{spinneret}}}{(\mathrm{OD}^2 - \mathrm{ID}^2)_{\mathrm{hollow\ fiber}}} \tag{31.1}$$

Compared to other methods to eliminate macrovoids, using high elongational draw ratios appears to have many advantages. Not only does it reduce fiber dimension but it also increases fiber production rate. In addition, the whole process becomes easier and cheaper when comparing with the addition of other materials in the spinning dope. The cause of this dramatic morphology change may likely have arisen from the rapid shrinkage of fiber diameter during the elongational stretch, which may induce radial outflow (which is inward to the bore fluid side, while outward to the external coagulation side). As a consequence, it hinders the capillary intrusion or diffusion of coagulants from both internal and external coagulations and eliminates the chance of forming macrovoids (Wang et al., 2004a; Xiao et al., 2006).

31.3.4 Posttreatment and Additional Coating

Usually the as-spun hollow fibers have to go through solvent exchange and posttreatments in order to (1) remove residual solvents, (2) remove residual stress, (3) improve chain packing, and/or (4) eliminate defects. Conventional solvent exchange approaches are to immerse as-spun fibers in water, then in methanol, and/or then in hexane (Clausi and Koros, 2000; Chung et al., 2002; Ren et al., 2002; Cao et al., 2004). The hexane treatment is highly recommended because of its low surface tension, which may eliminate pore collapse during drying. Without proper solvent exchange, it has been reported that the apparent dense-selective layer thickness may increase from 730 Å (after water, methanol, and hexane treatments) to 4400 Å (after both water and methanol treatments) and 18,800 Å (only after water treatment) (Clausi and Koros, 2000).

The silicone rubber coating invented by Henis and Tripodi (1981) may be the most well-known method to seal hollow-fiber defects and recover intrinsic separation properties. This technique has been well practiced in academia and industry.

Heat treatment has been considered as an effective method to (1) control pores size, (2) reduce pore sizes, and (3) remove membrane defects of as-spun hollow fibers since the birth of RO (Sourirajan and Matsuura, 1985). Its effects on microporous (Chung, 1996), gas separation (Chung et al., 2003), and pervaporation (Chung et al., 2006; Liu et al., 2005; Qiao and Chung, 2006) membranes have been summarized elsewhere. Basically, heat treatment induces molecule relaxation and microscopically repackages the polymeric chains, which tend to perfect and densify the selective skins and minimize the surface defects.

31.4 DUAL-LAYER HOLLOW FIBERS

Compared with single-layer asymmetric hollow-fiber membranes, the dual-layer ones spun from a simultaneous extrusion process are more attractive because of the following reasons:

1. Not only do they keep all the advantages of asymmetric membranes, but they also would greatly save the cost of high-performance materials as much as 95% or even more, depending on the ratio of the inner to the outer layer thicknesses.
2. By means of the inner supporting layer and co-extrusion process, it is possible to deploy brittle but highly selective and permeable materials as the selective layer to form a composite membrane.
3. The simultaneous co-extrusion approach eliminates the secondary step of depositing a selective layer upon hollow-fiber membranes, thus making the production of dual-layer asymmetric hollow-fiber membranes much more straightforward and cost effective compared to the preparation of other types of composite membranes.

The study of dual-layer asymmetric hollow-fiber membranes formed by the phase-inversion process started in the late 1980s. In 1987, Yanagimoto invented dual-layer asymmetric flat-sheet and hollow-fiber membranes to improve the antifouling properties of membranes for ultrafiltration and microfiltration (Yanagimoto, 1987, 1988). Since then, Kuzumoto and Nitta (1989) simultaneously extruded inner and outer dopes containing the same polymer but different solvents and additives to improve water permeability. Ekiner et al. (1992) disclosed the procedures for the fabrication of dual-layer hollow fibers for gas separation. Li et al. (2002) developed a delamination-free dual-layer asymmetric

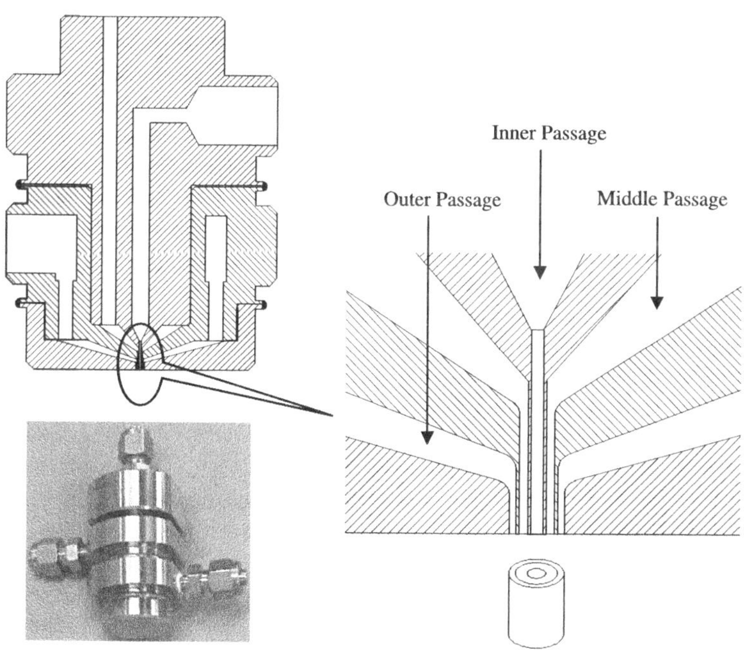

Figure 31.9 Structure and flow channels of a dual-layer spinneret (Li et al., 2002).

hollow-fiber membrane that possessed the same permselectivity as the single-layer ones but saved as much as 90% of material costs. Jiang et al. (2004) and Li et al. (2004a) studied the spinning conditions on dual-layer membranes and produced various dual-layer hollow-fiber membranes with selective layer thickness of 0.8 μm and 400–500 Å, respectively.

Figure 31.9 illustrates the dual-layer spinneret design in our lab (Li et al., 2002). The outer-layer dope, inner-layer dope, and bore fluid are delivered to the orifice by passing through three independent channels. The most difficult parts of spinneret design are precision and channel alignment. To run the dual-layer hollow-fiber spinneret, two metering pumps were employed to deliver inner and outer dopes.

31.4.1 Layer Shrinkage, Delamination, and Membrane Morphology

To prepare dual-layer asymmetric hollow-fiber membranes using the co-extrusion approach, the delamination phenomenon is critical and cannot be overlooked since it will directly affect the integrity of membranes. Figure 31.10 demonstrates how different shrinkage percentages affect the membrane structure. Delamination-free structure may be obtained when the shrinkage percentages of both layers are roughly same (Fig. 31.10a). A gap may form between the inner and outer layers if the shrinkage percentage of the inner layer is more than that of the outer layer (Fig. 31.10b). There are three possibilities if the outer layer shrinks more than the inner layer. The first situation is that the mechanical strength of the inner layer is stronger than the outer layer and the outer material layer is brittle; then the weak outer

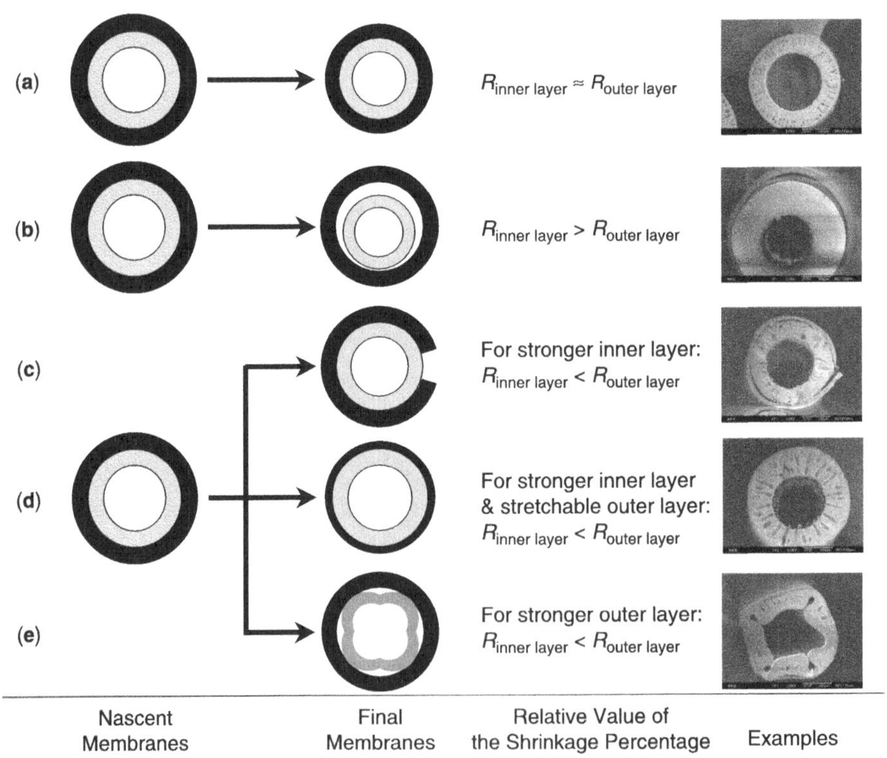

Figure 31.10 Schematic diagram of the influence of shrinkage percentage on dual-layer hollow-fiber structure (Li et al., 2004a).

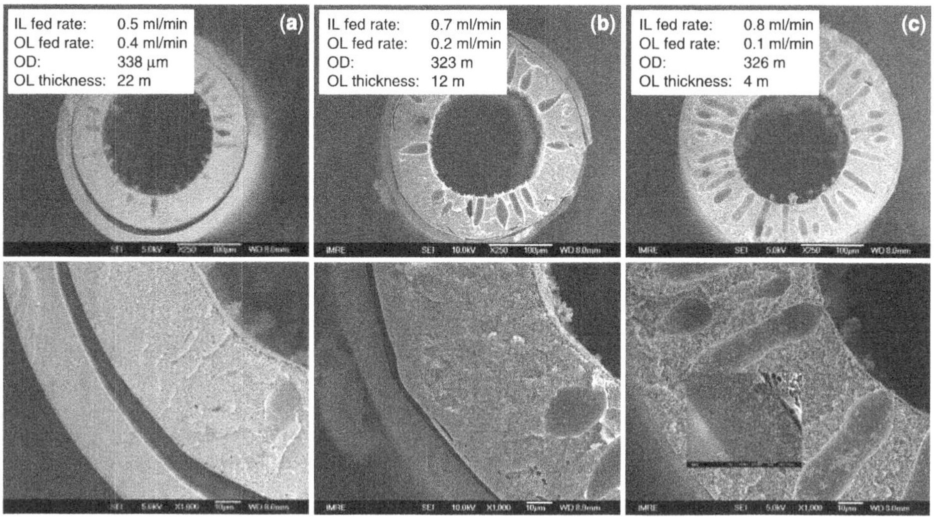

Figure 31.11 Effect of ratio of outer to inner dope flow rates on the shrinkage percentage and delamination. Ratio of outer to inner dope flow rates: (*left*) 0.83, (*middle*) 0.285, and (*right*) 0.125 (Li et al., 2004a).

layer may burst due to the accumulated internal stress (Fig. 31.10c). If the outer layer is stretchable, an apparent delamination-free structure may be achievable because the outer layer may tighten around the inner layer (Fig. 31.10d).

Figure 31.11 illustrates an example by adjusting the ratio of outer to inner dope flow rates. The delamination is significantly reduced with a decrease in the ratio and is totally eliminated when the ratio is 0.125 (Fig. 31.11c). Because the diffusions of solvents and nonsolvents in or out of a membrane are reciprocally proportional to the square of membrane thickness, a decrease in the ratio of outer-to-inner dope flow rates results in an increase in the shrinkage ratio of outer to inner layers. Consequently, the outer layer gradually tightens the inner layer. In addition, a thick membrane tends to have less shrinkage because its outer skins (i.e., outer contours) may gel or solidify well before completing the solvent exchange. The third one is that the outer layer is much stronger than the inner layer, which results in a fully deformed inner layer (Fig. 31.10e).

31.4.2 Inter-layer Diffusion in Dual-Layer Membranes

Figure 31.12 displays the desirable skin morphologies at various locations of dual-layer membranes. Briefly, both the inner and outer skins of the inner layer are porous with intimate adhesion between them. When using two different materials, one can control the inter-layer diffusion as a function of spinning conditions. Table 31.2 summarizes the elemental analysis data of the interfacial layers of hollow fibers spun from PES as the inner layer and Matrimid polyimide as the outer layer at different spinneret temperatures (Jiang et al., 2004). After being extruded from the spinneret, the two dopes should make contact with each other under normal conditions. When the spinneret temperature is low (i.e., 25°C), their viscosities are high, the diffusion rate of the polymer molecules between the two layers will not be fast; therefore, no sulfur can be found in the outer layer, as shown in this table. When the spinneret temperature is increased to 60°C, the interlayer diffusion between the two polymers apparently occurs, as evidenced in Table 31.2, in which it is

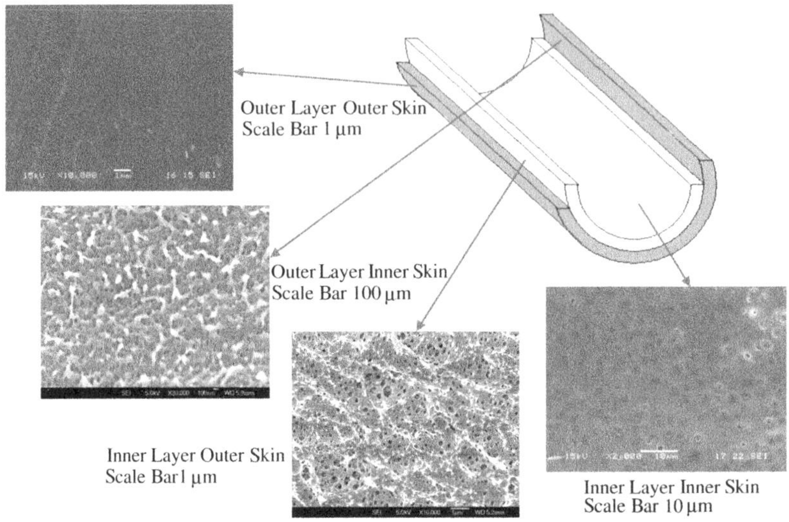

Figure 31.12 SEM morphology of different skins of the dual-layer hollow fibers (Jiang et al., 2004).

TABLE 31.2 Elemental Analysis of Interface of Dual-Layer Hollow Fibers Spun with an Air Gap of 1.5 cm at 25°C of Coagulation[a]

	Sulfur (%)	
	A (25°C)	B (60°C)
Outer layer inner skin	0	0.19
Inner layer outer skin	5.18	2.97

[a]A: spinneret temperature 25°C; B: spinneret temperature 60°C.
Source: Jiang et al. (2004).

found that the inner skin of the outer layer contains the sulfur diffused from the PES inner layer. Clearly, the spinneret temperature plays an important role in the degree of interlayer diffusion. A high spinneret temperature favors the interlayer diffusion because of low dope viscosities and high molecular diffusion rates.

31.4.3 Effects of Posttreatment on Dual-Layer Hollow-Fiber Membranes

Different posttreatment protocols have been employed to seal the defects on the outer surface of dual-layer hollow fibers (Li et al., 2004a). Figure 31.13 displays the cross-section morphology near the outer edge of the outer layer after various heat treatments. They are approximately 40 and 110 nm for membranes heat treated at 75 and 150°C, respectively (Li et al., 2004b). Clearly, the apparent dense-selective layer thickness of dual-layer hollow-fiber membranes increases with an increase in heat treatment temperature. This is due to the fact that heat treatment induces relaxation of the stresses imposed in hollow fibers when they were fabricated. Hollow fibers shrink inward gradually (i.e., reduced outer diameter), leading to higher packing density of polymer chains. Therefore, the heat treatment not only reduces the surface defects but also the bulk porosity. This phenomenon is similar to that observed in single-layer hollow-fiber membranes.

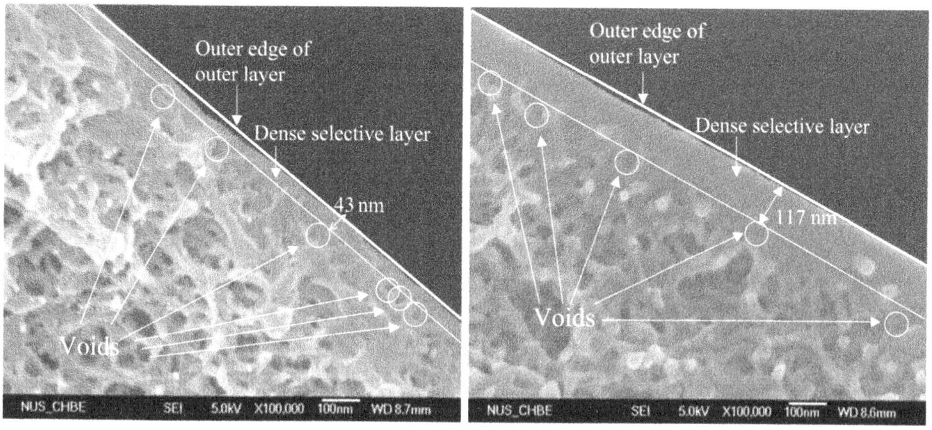

Figure 31.13 Visual estimation of the dense-selective layer thickness of dual-layer hollow-fiber membranes with different heat treatment methods: (*left*) 75°C for 3 h and (*right*) 150°C for 1 h. (Li et al., 2004b).

31.5 CONCLUDING REMARKS

A significant advancement has been made by membrane scientists in the last 4 decades to qualitatively understand membrane formation via the trial-and-error method. However, more efforts must be given to investigate hollow-fiber formation and understand the effects of spinning conditions on membrane morphology quantitatively in order to build the capabilities of developing hollow fibers with designed pore structure and ultrathin selective layer. Dual-layer hollow fibers apparently have many advantages over single-layer hollow fibers. They should gain more attention in order to develop functional, high-performance, and cost-effective hollow-fiber membranes.

ACKNOWLEDGMENTS

The authors would like to thank NUS for funding this research under grant numbers of R-279-000-165-112 and R-279-000-184-112. Special thanks are given to Yohannes Ervan Santoso and Natalia Widjojo for their useful comments on this chapter. Thanks also are due to my former and current staff and students D. F. Li, J. J. Shieh, C. Cao, R. Wang, J. Z. Ren, K. Y. Wang, L. Y. Jiang, Y. Li, X. Y. Qiao, Y. C. Xiao, M. L. Chng, R. X. Liu, and N. Peng for their wonderful work.

REFERENCES

Baker, R. W. (2004). *Membrane Technology and Applications*. Wiley, New York.

Barsema, J. N., Kapantaidakis, G. C., van der Vegt, N. F. A., Koops, G. H., and Wessling, M. (2003). Preparation and characterization of highly selective dense and hollow fiber asymmetric membranes based on BTDA-TDI/MDI co-polyimide. *J. Membr. Sci.* 216, 195.

Broens, L., Altena, F. W., Smolders, C. A., and Koenhen, D. M. (1980). Asymmetric membrane structures as a result of phase separation phenomena. *Desalination* 32, 33.

Cao, C., Chung, T. S., Chen, S. B., and Dong, Z. J. (2004). The study of elongation and shear rates in spinning process and its effect on gas separation performance of poly(ether sulfone) (PES) hollow fiber membranes. *Chem. Eng. Sci.* 59, 1053.

Carruthers, S. B., Ramos, G. L., and Koros, W. J. (2003). Morphology of integral-skin layers in hollow-fiber gas-separation membranes. *J. Appl. Polym. Sci.* 90, 399.

Chung, T. S. (1996). A review of microporous composite polymeric membrane technology for air-separation. *Polym. Polym. Composites* 4, 269.

Chung, T. S. (1997). The limitations of using Flory–Huggins equation for the states of solutions during asymmetric hollow fiber formation. *J. Membr. Sci.* 126, 19.

Chung, T. S., Guo, W. F., and Liu, Y. (2006). Enhanced Matrimid membranes for pervaporation by homogenous blends with polybenzimidazole (PBI). *J. Membr. Sci.* 271, 221.

Chung, T. S., and Hu, X. (1997). The effect of air-gap distance on the morphology and thermal properties of polyethersulfone hollow fibers. *J. Appl. Polym. Sci.* 66, 1067.

Chung, T. S., and Kafchinski, E. R. (1997). The effects of spinning conditions on asymmetric 6FDA/6FDAM polyimide hollow fibers for air-separation. *J. Appl. Polym. Sci.* 65, 1555.

Chung, T. S., Kafchinski, E. R., and Foley, P. (1992). Development of asymmetric hollow fibers from polyamides for air separation. *J. Membr. Sci.* 75, 181.

Chung, T. S., Kafchinski, E. R., and Vora, R. (1994). Development of a defect-free 6FDA-durene asymmetric hollow fiber and its composite hollow fibers. *J. Membr. Sci.* 88, 21.

Chung, T. S., Qin, J. J., Huan, A., and Toh, K. C. (2002). Study for the effect of dope shear rate on the outer surface morphology of hollow fiber ultrafiltration membranes by the atomic force microscope (AFM). *J. Membr. Sci.* 196, 251.

Chung, T. S., Ren, J. Z., Wang, R., Li, D. F., Liu, Y., Pramoda, K. P., Cao, C., and Loh, W. W. (2003). Development of asymmetric 6FDA-2,6DAT hollow fiber membranes for CO_2/CH_4 separation: Part 2. Suppression of plasticization. *J. Membr. Sci.* 214, 57.

Chung, T. S., and Teoh, S. K. (1997). Breaking the limitation of composition change during isothermal mass-transfer processes at the spinodal. *J. Membr. Sci.* 130, 141.

Chung, T. S., Teoh, S. K., and Hu, X. D. (1997). Formation of ultra-thin high-performance polyethersulfone hollow fiber membranes. *J. Membr. Sci.* 133, 161.

Chung, T. S., Teoh, S. K., Lau, W. Y. W., and Srinivasan, M. P. (1998). Effect of shear stress within the spinneret on hollow fiber membrane morphology, separation performance, and thermal and mechanical properties. *Ind. Eng. Chem. Res.* 37, 3930, 4903.

Clausi, D. T., and Koros, W. J. (2000). Formation of defect-free polyimide hollow-fiber membranes for gas separation. *J. Membr. Sci.* 167, 79.

Ekiner, O. M., Hayes, R. A., and Manos, P. (1992). Novel multicomponent fluid separation membranes. U.S. Patent 5,085,676.

Freeman, B. D., and Pinnau, I. (1999). *Polymer Membranes for Gas and Vapor Separation: Chemistry and Materials Science*. American Chemical Society, Washington, DC.

Henis, J. M. S., and Tripodi, M. K. (1981). Composite hollow fiber membranes for gas separation: The resistance model approach. *J. Membr. Sci.* 8, 233.

Ho, W. S. W. (2003). Recent developments and applications for hollow fiber membranes. *J. Chin. Inst. Chem. Eng.* 34, 75.

Ho, W. S. W., and Sirkar, K. K. (Eds.) (1992). *Membrane Handbook*. Chapman & Hall, New York.

Jiang, L. Y., Chung, T. S., Li, D. F., Cao, C., and Kulprathipanja, S. (2004). Fabrication of Matrimid/polyethersulfone dual-layer hollow fiber membranes for gas separation. *J. Membr. Sci.* 240, 91.

Kapantaidakis, G. C., Koops, G. H., and Wessling, M. (2002). Effect of spinning conditions on the structure and the gas permeation properties of high flux polyethersulfone/polyimide blend hollow fibers. *Desalination* 144, 121.

Kesting, R. E. (1985). *Synthetic Polymeric Membranes: A Structural Perspective.* Wiley, New York.

Kesting, R. E., and Fritzsche, A. K. (1993). *Polymeric Gas Separation Membranes.* Wiley, New York.

Kesting, R. E., Fritzsche, A. K., Cruse, A., Murphy, M. K., Handermann, A. C., Malon, R. F., and Moore, M. D. (1990). The second-generation polysulfone gas-separation membranes. I: The use of Lewis acid:base complexes as transient templates to increase free volume. *J. Appl. Polym. Sci.* 40, 1557.

Khare, V. P., Greenberg, A. R., and Krantz, W. B. (2005). Vapor-induced phase separation—Effect of the humid air exposure step on membrane morphology. Part I. Insights from mathematical modeling. *J. Membr. Sci.* 258, 140.

Kim, J. H., Min, B. R., Won, J., Park, H. C., and Kang, Y. S. (2001). Phase behavior and mechanism of membrane formation for polyimide/DMSO/water system. *J. Membr. Sci.* 187, 47.

Koros, W. J., and Pinnau, I. (1994). Membrane formation for gas separation process. In D. R. Paul and Y. P. Yampol'skii (Eds.), *Polymeric Gas Separation Membranes.* CRC Press, Boca Raton, FL, Chapter 5.

Kuzumoto, E., and Nitta, K. (1989). Manufacture of permselective hollow membranes. Japanese Patent 01,015,104.

Levich, V. G., and Krylov, V. S. (1969). Surface-tension-driven phenomena. *Annu. Rev. Fluid Mech.* 1, 293.

Li, D. F., Chung, T. S., Wang, R., and Liu, Y. (2002). Fabrication of fluoropolyimide/ polyethersulfone dual-layer asymmetric hollow fiber membranes for gas separation. *J. Membr. Sci.* 198, 211.

Li, D. F., Chung, T. S., and Wang, R. (2004a). Morphological aspects and structure control of dual-layer asymmetric hollow fiber membranes formed by a simultaneous co-extrusion approach. *J. Membr. Sci.* 243, 155.

Li, Y., Cao, C., Chung, T. S., and Pramoda, K. P. (2004b). Fabrication of dual-layer polyethersulfone (PES) hollow fiber membranes with an ultrathin dense selective layer for gas separation. *J. Membr. Sci.* 245, 53.

Lin, K. Y., Wang, D. M., and Lai, J. Y. (2002). Nonsolvent-induced gelation and its effect on membrane morphology. *Macromolecules* 35, 6697.

Liu, R. X., Qiao, X. Y., and Chung, T. S. (2005). The development of high performance P84 co-polyimide hollow fibers for pervaporation dehydration of isopropanol. *Chem. Eng. Sci.* 60, 6674.

Loeb, S., and Sourirajan, S. (1963). Sea water demineralization by means of an osmotic membrane. *Adv. Chem. Ser. ACS* 38, 117.

Mahon, H. I. (1966). Permeability separatory apparatus and process using hollow fibers. U.S. Patent 3,228,877.

Matsuura, T. (1994). *Synthetic Membranes and Membrane Separation Process.* CRC Press, Boca Raton, FL, Chapter 5.

Matz, R. (1972). The structure of cellulose acetate membranes 1. The development of porous structures in anisotropic membranes. *Desalination* 10, 1.

McKelvey, S. A., and Koros, W. J. (1996). Phase separation, vitrification, and the manifestation of macrovoids in polymeric asymmetric membranes. *J. Membr. Sci.* 112, 29.

Menut, P., Bohatier, C. P., Deratani, A., Dupuy, C., and Guilbert, S. (2002). Structure formation of poly(ether-imide) films using nonsolvent vapor induced phase separation: Relationship between mass transfer and relative humidity. *Desalination* 145, 11.

Mulder, M. (1996). *Basic Principles of Membrane Technology.* Kluwer Academic, The Netherlands.

Nunes, S. P., and Peinemann, K. V. (2001). *Membrane Technology in the Chemical Industry*. Wiley-VCH, Singapore.

Paul, D. R., and Yampol'skii, Y. P. (1994). *Polymeric Gas Separation Membranes*. CRC Press, Boca Raton, FL.

Qiao, X. Y., and Chung, T. S. (2006). Pervaporation dehydration of isopropanol through cross-linked P84 co-polyimide membranes. *AIChE J.* 52, 3462.

Qin, J. J., and Chung, T. S. (1999). Effect of dope flow rate on the morphology, separation performance, thermal and mechanical properties of ultrafiltration hollow fibre membranes. *J. Membr. Sci.* 157, 35.

Ren, J. Z., Chung, T. S., Li, D. F., Wang, R, and Liu, Y. (2002). Development of asymmetric 6FDA-2,6 DAT hollow fiber membranes for CO2/CH4 separation—1. The influence of dope composition and rheology on membrane morphology and separation performance. *J. Membr. Sci.* 207, 227.

Reuvers, A. J., Van den Berg, J. W. A., and Smolders, C. A. (1987). Formation of membranes by means of immersion precipitation. Part I: A model to describe mass transfer during immersion precipitation. *J. Membr. Sci.* 34, 45.

Sharpe, I. D., Ismail, A. F., and Shilton, S. J. (1999). A study of extrusion shear and forced convection residence time in the spinning of polysulfone hollow fiber membranes for gas separation. *Sep. Purif. Technol.* 17, 101.

Shojaie, S. S., Krantz, W. B., and Greenberg, A. R. (1994). Dense polymer film and membrane formation via the dry-cast process. 1. Model validation and morphological study. *J. Membr. Sci.* 94, 281.

Smolders, C. A., Reuvers, A. J., Boom, R. M., and Wienk, I. M. (1992). Microstructures in phase-inversion membranes. 1. Formation of macrovoids. *J. Membr. Sci.* 73, 259.

Sourirajan, S. (1970). *Reverse Osmosis*. Logos, London.

Sourirajan, S., and Matsuura, T. (1985). *Reverse Osmosis and Ultrafiltration*. American Chemical Society, Washington, DC.

Strathmann, H., and Kock, K. (1977). The formation mechanism of phase inversion membranes. *Desalination* 21, 241.

Strathmann, H., Kock, K., Amar, P., and Baker, R. W. (1975). The formation mechanism of asymmetric membranes. *Desalination* 16, 179.

Tsai, H. A., Huang, D. H., Ruaan, R. C., and Lai, J. Y. (2001). Mechanical properties of asymmetric polysulfone membranes containing surfactant as additives. *Ind. Eng. Chem. Res.* 40, 5917.

Tsai, H. A., Kuo, C. Y., Lin, J. H., Wang, D. M., Deratani, A., Pochat-Bohatier, C., Lee, K. R., and Lai, J. Y. (2005). Morphology control of polysulfone hollow fiber membranes via water vapor induced phase separation. *J. Membr. Sci.* 278, 390.

Wang, R., and Chung, T. S. (2001). Determination of pore sizes and surface porosity and the effect of shear stress within a spinneret on asymmetric hollow fiber membranes. *J. Membr. Sci.* 188, 29.

Wang, K. Y., Li, D. F., Chung, T. S., and Chen, S. B. (2004a). The observation of elongation dependent macrovoid evolution in single- and dual-layer asymmetric hollow fiber membranes. *Chem. Eng. Sci.* 59, 4657.

Wang, K. Y., Matsuura, T., Chung, T. S., and Guo, W. F. (2004b). The effects of flow angle and shear rate within the spinneret on the separation performance of poly(ethersulfone) (PES) ultrafiltration hollow fiber membranes. *J. Membr. Sci.* 240, 67.

Widjojo, N., and Chung, T. S. (2006). The thickness and air-gap dependence of macrovoid evolution in phase-inversion asymmetric hollow fiber membranes. *Ind. Eng. Chem. Res.* 45, 7618.

Wienk, I. M., Boom, R. M., Beerlage, M. A. M., Bulte, A. M. W., Smolders, C. A., and Strathmann, H. (1996). Recent advances in the formation of phase inversion membranes made from amorphous or semi-crystalline polymers. *J. Membr. Sci.* 113, 361.

Xiao, Y. C., Wang, K. Y., Chung, T. S., and Ta, J. N. (2006). Evolution of nano-particle distribution during the fabrication of mixed matrix TiO_2-polyimide hollow fiber membranes. *Chem. Eng. Sci.* 61, 6228.

Yanagimoto, T. (1987). Manufacture of ultrafiltration membranes. Japanese Patent 62,019,205.

Yanagimoto, T. (1988). Method for manufacture of hollow-fiber porous membranes. Japanese Patent 63,092,712.

Yao, C. W., Burford, R. P., Fane, A. G., and Fell, C. J. D. (1988). Effect of coagulation conditions on structure and properties of membranes from aliphatic polyamides. *J. Membr. Sci.* 38, 113.

Yilmaz, L., and McHugh, A. J. (1986). Analysis of non-solvent-solvent-polymer phase diagrams and their relevance to membrane formation modeling. *J. Appl. Polym. Sci.* 31, 997.

Yilmaz, L., and McHugh, A. J. (1988). Modeling of asymmetric membrane formation. II. The effects of surface boundary conditions. *J. Appl. Polym. Sci.* 35, 1969.

CHAPTER 32

Membrane Surface Characterization

M. KALLIOINEN and M. NYSTRÖM

Laboratory of Membrane Technology and Technical Polymer Chemistry, Department of Chemical Technology, Lappeenranta University of Technology (LUT), Lappeenranta, Finland

32.1 INTRODUCTION

Membranes are utilized widely in very different applications, and the requirements for an optimal membrane might differ strongly. There are several commercial membranes on the market from which to choose a suitable membrane. However, fundamental knowledge of the membrane characteristics affecting its performance in a specific application is needed to ensure a good choice. Yet, the information given in the data sheets of the membrane manufacturers is often insufficient. Thus, different membrane surface characterization methods are needed to obtain enough information on the properties of the membranes.

The most important characteristics of membranes affecting their performance and stability in a specific application are their chemical composition, hydrophilicity/hydrophobicity, charge, and morphology. These properties can be characterized using several different methods. By combining the results of these methods, it is possible to characterize a membrane almost perfectly. However, perfect characterization is not always needed for an optimal membrane selection: information on the most crucial properties is generally enough. Therefore, the aim of this chapter is to give an idea of which methods can be utilized to obtain the needed information. There are many different characterization methods, and, thus, it is not possible to introduce all of them here. However, the most commonly used methods in surface characterization at this moment have all been included in the chapter. More information can be found from some review articles recently published in the field (Chan and Chen, 2004; Pontié et al., 2005). Because the methods suitable for the characterization of the morphology of porous membranes (the bubble point method, mercury intrusion porometry, gas adsorption-desorption, thermoporometry, permporometry, liquid displacement, permeation measurements, and fractional rejection measurements) are comprehensively presented elsewhere (Mulder, 1996), this chapter introduces the suitability of the microscopical methods (optical microscopy, confocal scanning electron microscopy, electron microscopy, and atomic force microscopy) for the characterization of membrane morphology.

Advanced Membrane Technology and Applications. Edited by Norman N. Li, Anthony G. Fane, W. S. Winston Ho, and T. Matsuura
Copyright © 2008 John Wiley & Sons, Inc.

32.2 CHARACTERIZATION OF THE CHEMICAL STRUCTURE OF A MEMBRANE

Information on the chemical structure of a membrane surface and on its hydrophilicity and charge is needed for a better understanding of membrane stability under different conditions. The knowledge about the surface chemistry also helps in the determination of fouling mechanisms and optimization of cleaning procedures. The membrane manufacturers usually give information about the most commonly used membrane materials, but analytical methods can still give crucial further information about, for example, the additives used in the membrane manufacturing, the orientation of the polymer chains, and the crystallinity of the polymer in the membrane.

The analytical methods generally used in the examination of the chemical structure of the membrane are Fourier transform infrared (FTIR) spectroscopy, Raman spectroscopy, energy-dispersive X-ray spectroscopy (EDS), electron spectroscopy for chemical analysis (ESCA, XPS), Auger electron spectroscopy (AES), and secondary ion mass spectrometry (SIMS). The operation principles of these methods differ, and, thus, the results of the different methods complement each other. The applicability of the methods is presented in Table 32.1. The analysis of the data in Table 32.1 reveals that XPS, AES, and SIMS are the most suitable methods for analysis of the thin skin layers of the membranes, while with FTIR, Raman and EDS information about both the skin layer and the bulk of the membranes is obtained. It can be concluded from the lateral resolution values that in mapping of the chemical composition of a membrane surface, Raman, EDS, XPS, AES and SIMS give more detailed information on the surface composition than FTIR. Membranes are often used in their wet state, and therefore, in order to get the best knowledge on their characteristics in such applications, the membranes should also be characterized in their wet state. However, in most of the above methods the samples have to be analyzed in their dry state, either water produces such strong signals that it covers the information from the other compounds or because the samples have to be stable under vacuum conditions. Thus, a great advantage of Raman spectroscopy is the possibility of analyzing samples in their wet state.

Other methods, for instance, nuclear magnetic resonance (NMR) spectroscopy and matrix-assisted laser desorption ionization mass spectrometry (MALDI-MS), can also be used in the characterization of the chemical composition. Information on these methods can be found elsewhere (Chan and Chen, 2004).

32.2.1 Infrared Spectroscopy (IR)

In IR spectroscopy (IR, Table 32.1) IR radiation is focused on the sample. When the frequency of the IR radiation is equal to the specific vibration of the sample molecules, the molecules absorb the radiation. The IR radiation passing through the sample is detected, and the obtained spectrum shows the changes in IR radiation intensity as a function of frequency. Usually, the positions of the IR absorption bands are presented in the spectrum as wavenumbers (the number of waves per unit length), which are directly proportional to frequency. The intensity of the absorption band depends on the change in dipole moment of the molecule caused by the absorption. Thus, functional groups containing polar bonds, for instance, O—H, C═O, and $—NH_2$, can easily be detected with IR while analysis of groups containing nonpolar bonds is much more difficult. IR spectroscopy enables both qualitative

TABLE 32.1 Summary of Spectroscopic Methods Used in Membrane Surface Characterisation

	Infrared Spectroscopy (IR)	Raman Spectroscopy	Energy Dispersive X-ray Spectroscopy (EDS)	X-ray Photoelectron Spectroscopy (XPS)	Auger Electron Spectroscopy (AES)	Time-of Flight Secondary Ion Mass Spectroscopy (TOF-SIMS)
Detection of elements	No	Yes	Yes, Be-U	Yes, H-U	Yes, Li-U	Yes, H-U
Detection of functional groups	Yes	Yes	No	No	No	No
Detection of chemical bonds	Yes	Yes	No	Yes	Yes	Yes
Determination of crystallinity	Yes	Yes	No	No	Yes	No
Qualitative analysis	Yes	Yes	Yes	Yes	Yes	Yes
Quantitative analysis	Yes	Yes	Yes	Yes	Yes	Yes
Mapping/imaging of the chemical composition possible	Yes	Yes	Yes	Yes	Yes	Yes
Depth profiling of the chemical composition possible	Yes	Yes	No	Yes	Yes	Yes
Analysis in vacuum	No	No	Yes	Yes	Yes	Yes
Wet samples can be analyzed	No	Yes	No	No	No	No
Special sample pretreatment	No	No	Yes	No	No	No
Sample destructive technique	No	Yes/No	Yes	No	Yes	Yes/No
Analysis depth	1–100 μm (PAS) 1–10 μm (ATR)	1–5 μm	0.5–5 μm	1–3 nm	Few monolayers	0.2–1 nm
Spatial resolution	0.5–1 μm (ATR, PAS)	0.5–1 μm (confocal Raman)	0.5–5 μm	5–50 μm	10 nm	1–2 μm
Detection limit	1 atom% (ATR)	[a]	0.1 atom%	1 atom%	0.1–1 atom%	0.1 atom%

[a] A general value is not available because the detection limit depends strongly on the equipment and the measuring conditions.

and quantitative analysis, and it is applicable for both inorganic and organic membrane samples (Chalmers and Everall, 1993; Colthup et al., 1975; Holmbom and Stenius, 2000; Socrates, 1980; Wartewig and Neubert, 2005).

Infrared spectroscopy is often utilized in the determination of the chemical composition of membrane samples and in the localization of different compounds on the sample surface (Belfer et al., 2000; Bottino et al., 2001; Fontyn et al., 1987; Kallioinen et al., 2003; Pihlajamäki et al., 1998). Considering the example in Figure 32.1, the FTIR-ATR (attenuated total reflectance) spectra show that the commercial nanofiltration membranes NF45, NF270, and NTR7450 contain a layer of Udel polysulfone and not polyethersulfone. The bands at 1586 and 1488 cm^{-1} indicate C=C stretching vibration of the aromatic ring, which is typical for polysulfone. The weak bands at 1385 and 1365 cm^{-1} reveal methyl groups (C—H bending in alkane, gem-dimethyl), which are not found in polyethersulfone. The doublet 1323/1295 cm^{-1} indicates asymmetric SO_2 stretching and the band at 1151 cm^{-1} symmetric SO_2 stretching. The aromatic ether band is at 1243 cm^{-1} (Puro et al., 2006). IR spectroscopy can additionally be used in the determination of the crystallinity of polymeric membranes (Boccaccio et al., 2002).

Nowadays, FTIR spectrometers are mainly used due to their better performance compared to the dispersive instruments. In an FTIR spectrometer, the interferograms with and without a sample in the focus of a beam are measured, and these interferograms are transformed mathematically with Fourier transformation into spectra of a sample and a background. With an FTIR instrument the whole spectrum is obtained with one measurement while the dispersive instrument measures only limited wavenumber bands, one at a time (Griffits and de Haseth, 1986; Stenius and Vuorinen, 1999; Wartewig and Neubert, 2005).

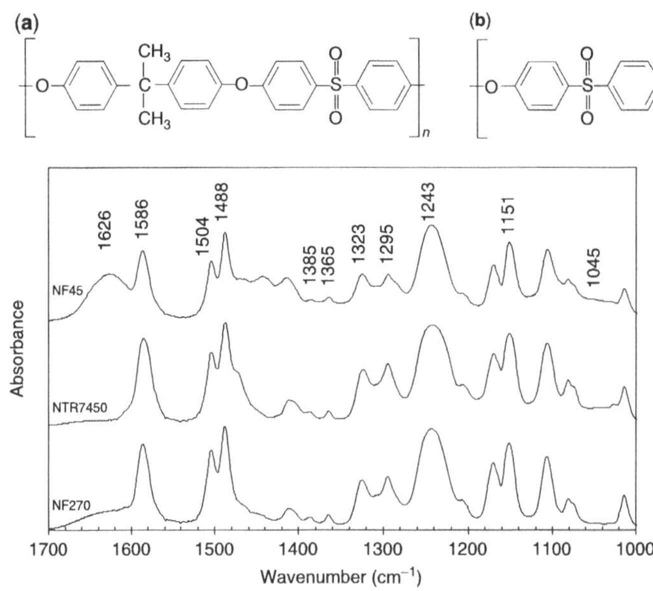

Figure 32.1 Chemical structures of (a) polysulfone and (b) polyethersulfone and FTIR-ATR spectra of the commercial nanofiltration membranes NF45, NF270, and NTR7450. The spectra reveal that the membranes contain a polysulfone layer (Puro et al., 2006, with permission of IChemE's journals).

The most common way to perform membrane surface analysis is to combine FTIR spectrometry with an ATR technique, also known as internal reflection spectroscopy (IRS) (Belfer et al., 2000; Bottino et al., 2001; Kallioinen et al., 2003; Pihlajamäki et al., 1998). In the ATR technique the sample is placed against an ATR crystal, which is made from IR transparent material having a high refractive index. The IR radiation is focused on one end of the crystal, where the beam undergoes several total internal reflections before exiting from the crystal and going to the detector. The beam penetrates slightly into the sample at each reflection, and thus, a spectrum providing information about the surface chemical structure can be obtained. FTIR-ATR is a nondestructive analyzing technique that requires no pretreatment (Chalmers and Everall, 1993; Garbassi et al., 1994; Wartewig and Neubert, 2005). However, because the OH group has very strong IR absorption, its absorption bands cover and obscure other interesting bands if the samples are analyzed in their wet state. For this reason the membrane samples must be analyzed in their dry state (Holmbom and Stenius, 2000).

The penetration of the IR radiation into the sample during the FTIR-ATR measurements can be controlled by changing the incident angle and by using crystals having different refractive indices. Commonly used crystal materials are KRS 5 (thallium–bromide–iodide), zinc selenide, and germanium (Katon, 1996). The penetration depth of IR radiation also depends on the wavelength. Thus, the FTIR-ATR technique is suitable for depth profiling of membrane surfaces (Boccaccio et al., 2002; Kim et al., 1997). Information on the chemical structure of a sample can be obtained from a depth of 1 μm up to 10 μm (Garbassi et al., 1994). The FTIR-ATR method is at its best in the characterization of dense and smooth selective sides of membranes because surface roughness and high porosity decrease the contact between the crystal and the membrane surface, thus, disturbing the measurements (Boccaccio et al., 2002). The spatial resolution in FTIR-ATR measurements is 0.5–1 μm at best and the detection limit is 1 atom-% (Holmbom and Stenius, 2000). An IR microscope can also be equipped with an ATR objective (an objective into which the ATR crystal is incorporated). In that case, the crystal is in the form of a small hemisphere, against which the sample is pressed. The most commonly used hemisphere materials are germanium and zinc selenide but also diamond can be applied (Katon, 1996).

Other IR-spectroscopy-based techniques suitable for membrane surface characterization are FTIR microspectroscopy (Katon, 1996) and FTIR photoacoustic spectroscopy (PAS) (Boccaccio et al., 2002). With FTIR microspectroscopy, a map of the chemical composition of the sample surface can be composed by measuring the IR spectrum in a narrow frequency band while the focal spot is moved across the surface. The location or concentration of different functional groups can be determined by repeating the mapping with another frequency band (Katon, 1996). This method is very suitable, for instance, in the detection of defects or impurities on the membrane surface. In PAS the sample, which is placed in a closed chamber, is irradiated with short pulses of IR radiation. The absorbed radiation is dissipated from the sample as heat. The heat dissipation creates pressure waves in the gas surrounding the sample. These pressure changes are detected with a sensitive microphone and the signal is transformed to a spectrum. Depth profiling is also possible with the PAS method because the sampling depth can be varied from 1 to 100 μm by changing the modulation frequency of the radiation. The FTIR-PAS spectrum contains information both from the membrane skin layer and the membrane porous layer (Boccaccio et al., 2002). The analyzed sample must be dry, but no other sample pretreatment is needed (Garbassi et al., 1994; Holmbom and Stenius, 2000; Stenius and Vuorinen, 1999).

32.2.2 Raman Spectroscopy

In Raman spectroscopy (Table 32.1) the sample is irradiated with an intense source of monochromatic light. The laser used as the light source can operate either in the ultraviolet, visible, or near-infrared regions. The incident light energy causes changes in the molecular vibrational energy levels of the sample. Raman scattering, which is detected in Raman spectroscopy, means that the molecular vibrational state does not return to the initial state but remains at a higher vibrational energy level (Stokes scattering) or at a lower vibrational energy level (anti-Stokes scattering). The intensity of Raman scattering is strong when the change in the molecular vibrational energy causes a change in molecular polarizability (Stevenson and Vo-Dinh, 1996; Wartewig, 2003). Therefore, groups like double and triple carbon–carbon bonds, disulfide bonds, and carbon–hydrogen bonds are easily detectable with a Raman spectrometer, and the Raman bands for water are weak. This insensitivity to water is an advantage of Raman spectroscopy compared to IR spectroscopy because also wet membrane samples can be analyzed without the disturbing effect of water in the spectrum (Holmbom and Stenius, 2000).

In a Raman spectrum the intensity of scattered light versus Raman shift is shown. The Raman shift is the energy difference between the incident photon and the Raman scattered photon. Raman spectroscopy can give information on chemical structure, morphology, intermolecular interactions, and crystallinity (Stuart, 1996) of a sample, and it can be used for both qualitative and quantitative characterization. However, quantitative information is very difficult to obtain because besides the concentration of the molecules in the sample, several different factors, including laser power and other equipment-related factors, have an influence on the band intensity (Smith and Dent, 2005). Raman spectroscopy is applicable for both polymeric and inorganic membranes, and a good spectrum can usually be obtained with little or no sample pretreatment. Fluorescence, optical clarity, and sample degradation under high-energy laser radiation can, however, disturb the analysis, although the influence of these phenomena can, to some extent, be diminished with advanced analyzing techniques. IR and Raman spectroscopy complement each other, and often bands that are strong in an IR spectrum are weak in the Raman spectrum and vice versa. This can also be noticed in Figure 32.2, where the ATR-IR and FT-Raman spectra of Nafion 117 membrane and highly crystalline polytetrafluoroethylene (PTFE) are shown (Gruger et al., 2001). Thus, the best information about the chemical structure of a sample is obtained, when both of these methods are utilized (Khulbe and Matsuura, 2000; Smith and Dent, 2005; Wartewig and Neubert, 2005). The acquisition of a Raman spectrum is relatively fast; for instance, in micro-Raman spectroscopy a good-quality spectrum is obtained in a few minutes. However, the seeking of the optimal analysis conditions and the spectra interpretation might be time consuming.

At least two Raman spectroscopic methods are applicable in membrane surface characterization; Fourier transform Raman spectroscopy (FT-Raman), and micro-Raman spectroscopy (Boccaccio et al., 2002; Gruger et al., 2001; Khulbe and Matsuura, 2000). In FT-Raman the sample molecules are excited with a near-infrared (NIR) laser and appropriately configured Michelson interferometers, and Fourier transform processes are used in the collecting of scattered light and the analysis of the collected light. Almost the only requirement is that the sample to be analyzed with FT-Raman must not be black (Hendra, 2005). FT-Raman is a valuable tool in determining the overall chemical structure of a membrane, but it cannot characterize the asymmetric structure of a porous membrane matrix (Boccaccio et al., 2002).

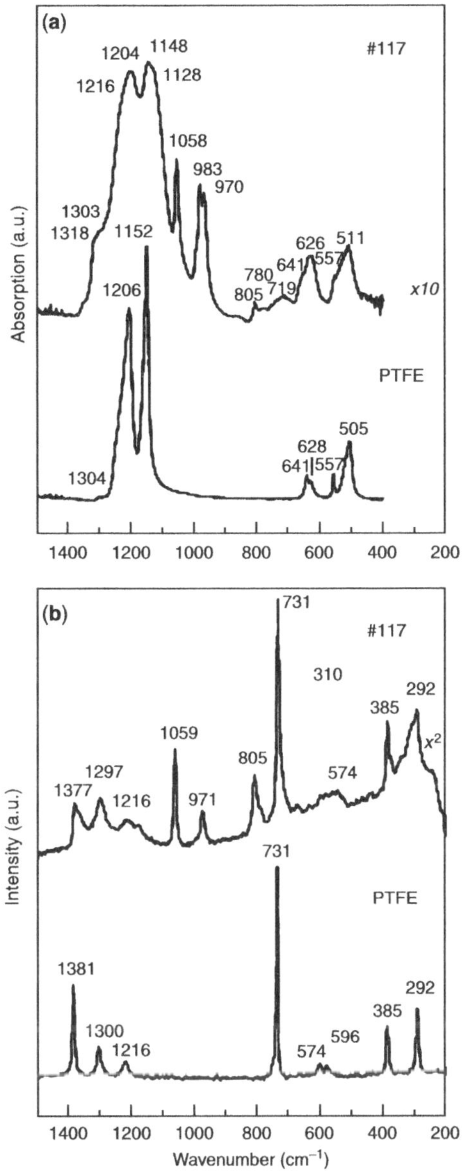

Figure 32.2 (a) IR-ATR and (b) FT-Raman spectra of Nafion® 117 membrane and highly crystalline PTFE (Gruger et al., 2001, with permission from Elsevier).

In micro-Raman spectroscopy, the spectrometer is attached to a microscope and the backscattered Raman signal is collected within the cone defined by the same objective through which the laser beam is focused on the sample for excitation of the sample molecules. This system also enables chemical mapping and imaging of a sample surface. A microscope can also be attached to an FT-Raman spectrometer. When a confocal microscope is used, Raman scattering originating from a small specific region of a sample can be detected and optical sectioning of a transparent sample is possible. Thus, with confocal micro-Raman spectroscopy a three-dimensional map of the chemical composition of a sample is possible to obtain by combining depth profiling and *x-y* mapping results. The main advantage of this is that cross-sectioning of a sample is not needed. Micro-Raman spectroscopy also enables the detection of very small amounts of material, even though its Raman scattering is weak, because the Raman signal is collected from a small specific area. The resolution in micro-Raman spectroscopy depends on the microscope type and the laser used as the incident light source. With confocal microscopy both lateral and depth resolution can be as good as a few micrometers, which is better than with a light microscope (Fleming et al., 2005; Khulbe and Matsuura, 2000; Pastorczak et al., 2005; Smith and Dent, 2005; Thibault et al., 2002; Turrel and Dhamelincourt, 2005).

32.2.3 Energy-Dispersive X-ray Spectroscopy (EDS)

In an electron microscope a focused electron beam interacts with the atoms in a sample and element-specific X-rays are generated. These X-rays can be detected with an energy-dispersive spectrometer (EDS, Table 32.1) coupled to a scanning electron microscope (SEM) or to a transmission electron microscope (TEM). Due to the vacuum conditions in the sample chamber, the samples must be analyzed in their dry state. EDS is useful both in qualitative and quantitative analysis of all elements above atomic number 4 (Be) at typical beam currents used for secondary electron imaging in SEM. The detection limit of the EDS is about 0.1 atom-% and the spatial resolution is 0.5–5 μm (Holmbom and Stenius, 2000). An advantage of the use of EDS with SEM is the possibility of elemental mapping and line scanning, which can show an intensity profile of an element in the sample (Goldstein et al., 2003). Figure 32.3 presents an example of the use of EDS and intensity profiles in the examination of ceramic membranes. From the figure it can be seen that the titanium-containing sol for making the membrane has not penetrated into

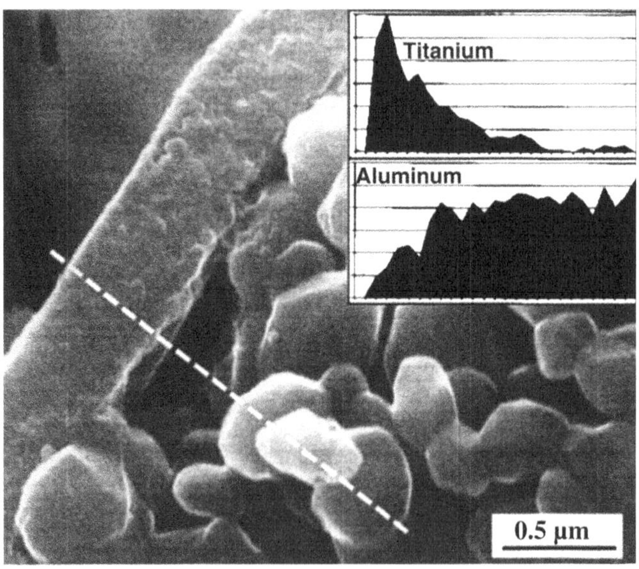

Figure 32.3 Intensity profiles of Ti and Al in a ceramic titanium nitride (TiN) membrane having an Al_2O_3 support. The profiles are scanned along the dotted line (Tomandl et al., 2000, with permission from Elsevier).

the supporting ceramic (Tomandl et al., 2000). EDS is a relatively rapid analysis method; a qualitative analysis can be performed in a few minutes. However, polymeric membranes and some other nonconducting samples have to be coated with conductive material before analysis. Thus, the sample preparation is time consuming. The coating layer might also disturb the analysis. The depth of analysis depends on the atomic number of the sample and on the energy of the incident electron beam. Information about the elements of the sample can be obtained even from a depth of 0.5 μm (Holmbom and Stenius, 2000). Geometric factors, such as surface roughness and sample thickness, also affect the analysis depth. A further problem is that in the analysis of polymeric membranes the sample composition might alter under the electron beam. Coating of the sample might also damage the surface (Goldstein et al., 2003).

32.2.4 X-ray Photoelectron Spectroscopy (XPS, ESCA)

In X-ray photoelectron spectroscopy (XPS, Table 32.1), known also as elemental spectroscopy for chemical analysis (ESCA), X-rays are focused on the sample, which is located under ultrahigh vacuum. Due to the vacuum conditions the sample must be dry. The X-rays interact with the sample, causing different photoemissions, but the primary event is photoemission of core electrons. The binding energies of the core electrons are characteristic for the elements, and because the binding energy can be determined when the kinetic energy is known, the detection of the emitted electrons and their kinetic energies enable an identification of the elements of the sample (Briggs, 1994; Holmbom and Stenius, 2000; Munro and Singh, 1993).

With XPS all the elements except hydrogen can be detected. The peak intensities in the energy spectrum are proportional to the number of atoms in the sample. Thus, both qualitative and quantitative analysis is possible with XPS. The detection limit varies with different sample types, but, in general, the detection limit is approximately 1 atom-% (Holmbom and Stenius, 2000). The local environment of the atom affects the binding energies of the core-level electrons, and, therefore, the chemical bonding of the atoms can be seen as a chemical shift in the XPS spectrum (Briggs, 1994; Holmbom and Stenius, 2000; Munro and Singh, 1993).

The analysis depth is typically 1–3 nm, but it depends on the electron energy and can even be 10 nm. Due to this surface sensitivity, XPS is especially useful for analysis of thin membrane skin layers, nanofiltration membrane structures, and modifications of membrane surfaces (Momtaz et al., 2005; Mukherjee et al., 1996; Song et al., 2004; Vrijenhoek et al., 2001). Figure 32.4. shows XPS spectra of a phenolphthalein poly(ether ether ketone) (PEK-C) ultrafiltration (UF) and PEK-C nanofiltration (NF) membranes. The NF membrane was made by forming a skin layer containing polyamidoamine (PAMAM) dendrimers over a microporous PEK-C UF membrane in an in situ interfacial polymerization process. The XPS spectra verified the chemical changes occurring in the polymerization process; the spectra of the PEK-C NF membrane contains a nitrogen peak, which is not seen in the spectrum of the PEK-C UF membrane (Lianchao et al., 2006).

The advantages of XPS analysis are due to the low energy of the incident X-rays, meaning that the analysis is nondestructive, and nonconducting samples (polymeric membranes) can be analyzed without conducting coating. Sample charging can be controlled in modern XPS equipment, but it can still occur to some extent and cause artefacts in the spectra. The analysis time, which is typically several minutes, should be kept as short as possible since long exposure times might increase the risk of sample destruction

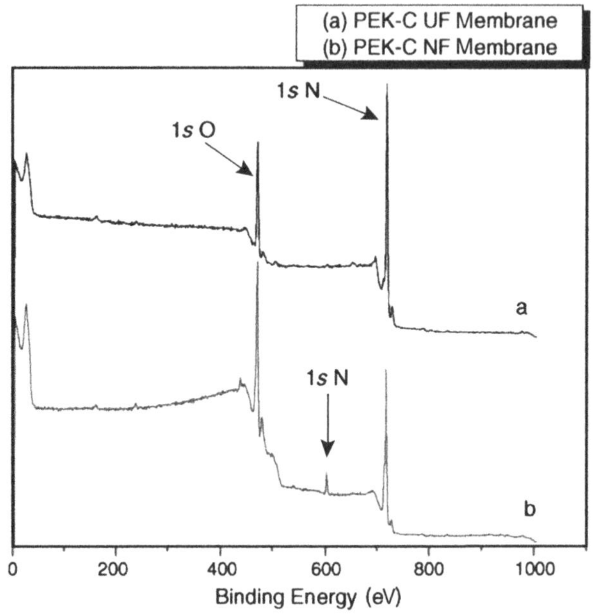

Figure 32.4 XPS spectra of (a) PEK-C UF and (b) PEK-C NF membranes (Lianchao et al., 2006, with permission from Elsevier).

(Briggs, 1994; Fulghum, 1999). For instance, it is known that poly(vinylidene difluoride) (PVDF) membranes lose fluorine-containing fragments during long irradiation times (Flösch, 1992). In general, modern XPS instruments enable spectra acquisition with high or moderate energy resolution from areas having diameters of 15–100 μm in 5–40 min (Fulghum, 1999).

X-ray photoelectron spectroscopy enables detailed imaging of compositional maps of a sample surface because its lateral resolution can reach even 3 μm (Reniers and Tewell, 2005; Fulghum, 1999; Holmbom and Stenius, 2000; Kai et al., 2000). The spatial resolution of XPS is 5–50 μm (Holmbom and Stenius, 2000). Depth profiling of the sample in order to obtain information from subsurface layers is also possible. The nondestructive way to perform depth profiling is to alter the electron take-off angle so that electrons escaping from different depths of the sample are detected. High surface roughness might, however, disturb this analysis (Briggs, 1994). Another depth-profiling method also utilized in analysis of membranes is ion sputtering for material removal (Kim et al., 2005; Sah et al., 2004). In this method spectral information is obtained from increasingly bigger depths as material is removed layer by layer. Ion sputtering enables depth analysis down to some 100 nm, but it might also cause several artefacts in the results and destruct the sample (Oswald and Reiche, 2001).

32.2.5 Auger Electron Spectroscopy (AES)

In Auger electron spectroscopy (AES, Table 32.1) the sample is bombarded with a high-energy electron beam. This causes ionization of a sample atom by removing a core electron. Consequently, an upper level electron falls to a lower level, and a third electron (Auger electron) is excited by the energy given off. The Auger electrons are detected. The energies of

Auger electrons are characteristic for the elements, and, thus, the elements of the sample and the chemical states of the elements can be identified from the obtained spectrum. The crystalline nature of a sample can affect the Auger signal. Due to this AES can also be used in examination the crystalline order of the sample. All elements except hydrogen (H) and helium (He) can be detected, and both qualitative and quantitative analysis is possible to perform with AES (Bishop, 1990). AES is very surface sensitive: The depth of analysis is a few monolayers, and, thus, the information is obtained only from the upmost surface layers (Reniers and Tewell, 2005). For this reason AES is better at identifying the fine structure of surfaces than XPS. The spot size of the incident electron beam can be focused to a very small area. Thus, the spatial resolution of AES can be as good as 0.5 μm, and the lateral resolution can even be lower than 10 nm (Reniers and Tewell, 2005). The detection limit of AES is 0.1–1 atom-% (Bishop, 1990). Scanning Auger microscopy (SAM) in which Auger electrons are detected with SEM enables chemical mapping of sample surfaces. Depth profiling is also possible, as with XPS, by removing the sample surface, layer by layer, by sputtering with an energetic ion beam. However, if the needed information is located deeper in the sample than 2 μm, the sputter depth profiling becomes very time consuming and the resolution suffers due to uneven sputtering effects in the sample. As in XPS, also in AES the samples are placed in a vacuum chamber. Thus, the samples must be analyzed in their dry state (Bishop, 1990; Walls, 1990).

Nonconducting samples, such as polymeric membranes, are damaged during AES analysis because they become charged due to the bombarding with high-energy electron beams. This limits the utilization of AES in membrane characterization. However, the charging problems can be avoided when AES analysis of polymer films is performed with X-ray photoelectron spectrometry (Otsuka et al., 2002). Nevertheless, AES is still not used very much in membrane surface characterization.

32.2.6 Secondary Ion Mass Spectrometry (SIMS)

In secondary ion mass spectrometry (SIMS) a beam of primary ions (e.g., gases He^+, Ne^+, Ar^+, and Xe^+ or liquid metals Ga^+ and Cs^+) is focused to the sample surface. In consequence, some material is sputtered from the surface. A small fraction of the sputtered material contains negatively and positively charged atoms (elemental ions), which are detected with a mass spectrometer. Thus, as a result, positive and negative secondary ion mass spectra showing the ion mass/charge versus the number of ions detected at each mass/charge are obtained (Belu et al., 2003; Munro and Singh, 1993).

The SIMS method suitable for surface characterization is called static SIMS (SSIMS). In this method the total number of primary ions in the analysis area is kept low enough to ensure that the probability of a particular region being struck by a primary ion more than once is negligible. Thus, sensitivity to the uppermost monolayers is obtained while sample damage is minimized. SSIMS is applicable also to organic materials. In dynamic SIMS (DSIMS) a higher flux of primary ions is focused on the surface, which causes a rapid erosion of the sample surface. DSIMS is applicable in depth profiling (Garbassi et al., 1994; Munro and Singh, 1993).

The mass analyzers most commonly used in the SSIMS technique are magnetic sector, quadrupole, and time-of-flight (TOF) analyzers. The TOF analyzer is based on the principle that ions with different mass/charge ratios will travel a fixed path length in different times, when they are accelerated through the same potential field. The lighter ions travel at a faster velocity and are thus detected before the heavier ions. Compared to the other mass analyzers, the advantages of TOF analyzers are their high transmission capability and simultaneous detection of all masses. Thus, other mass analyzers require higher ion doses to acquire the same number of secondary ions in the spectrum (Belu et al., 2003; Munro and Singh, 1993). In addition, a TOF analyzer is also faster than other analyzers. The mass resolution of a TOF analyzer is high, typically 7000 $M/\Delta M$ at a mass/charge ratio of 28 and higher than 10,000 $M/\Delta M$ at a mass/charge ratio higher than 100. This means that ions of nominally the same mass can be resolved (Briggs, 1994).

Time-of-flight SIMS (Table 32.1) can be used in the determination of the chemical structure and the composition of a surface, and it detects all the elements from hydrogen to uranium (Holmbom and Stenius, 2000). The advantage of SIMS compared to XPS is that with SIMS precise molecular information of polymers can be obtained. *Fingerprint spectra* can also be found from literature for a series of polymers from low-density polyethylene to polyethyleneterephtalate (PET) (Vickerman, 1989). It is also possible to analyze the molar masses of polymer samples up to 10,000 g/mol (Belu et al., 2003). Several sample types can be analyzed (powders, liquids, and solids) and special pretreatment is not needed. However, the analysis is performed under ultrahigh vacuum, which means that to ensure stability under vacuum conditions membrane samples must be analyzed in their dry state (Belu et al., 2003; Hagenhoff, 1995).

The surface sensitivity of TOF-SIMS analysis is high; the detection limit is 0.1–0.5 atom-% (Holmbom and Stenius, 2000), and the depth of analysis can be as small as 1 nm (Delcorte et al., 1996; Hagenhoff, 1995). In surface characterization also TOF-SIMS imaging can be utilized. Modern TOF-SIMS equipment enables the imaging of a sample surface under microprobe and microscope mode with a 1–2 μm lateral resolution and with a 10,000 $M/\Delta M$ resolution (Hoshi and Kudo, 2003). Thus, SIMS is very applicable for the characterization of both clean and fouled membrane surfaces as well as in the examination of adsorbate–membrane interactions (Fontyn et al., 1987; Kaeselev et al., 2002; Spevack and Deslandes, 1996). A high surface sensitivity requires a high cleanliness of the sample surface. Samples should also be analyzed immediately after preparation in order to avoid surface diffusion, molecular reorientation, or surface contamination before the analysis. In addition, the stability of the SIMS spectrum depends on the thickness and the morphology of the sample (Belu et al., 2003). Nonconducting samples become charged under the incident beam of primary ions. In modern SIMS equipments the charging can be avoided with electron flood sources, and, thus, charging does not disturb the analysis (Briggs, 1994).

Quantitative analysis with TOF-SIMS is challenging because the yield of secondary ions depends on several factors and is not directly proportional to concentration. The ion yields vary, for instance, due to the surface composition of the sample. This means that the ion yields of the same analyte differ from each other when the chemical environment of the analyte changes. Also sputtering of the sample can cause chemical environment changes during the analysis. Therefore, comparison between samples is difficult. This problem is called *matrix effects*, and it cannot be totally avoided even in modern SIMS systems (Belu et al., 2003; Hagenhoff, 1995).

32.3 CHARACTERIZATION OF MEMBRANE HYDROPHILICITY

Membrane hydrophilicity is a crucial factor affecting membrane performance when organic molecules are separated from aqueous solutions, for instance, in membrane distillation processes (Franken et al., 1987), in the separation of dispersions (Katoh et al., 1996; Vigo et al., 1985), in filtration of process waters from food (Goulas et al., 2002; Shaban, 1996), and in wastewater filtration in the pulp and paper industry (Carlsson et al., 1998; Dal-Cin et al., 1995; Nuortila-Jokinen and Nyström, 1996). Therefore, the determination of membrane hydrophilicity is an important part of membrane surface characterization.

The most common method utilized in the determination of hydrophilicity is measurement of the contact angle, which describes the edge of the two-phase boundary, where it ends at a third phase. The contact angle depends on the interfacial tensions of the interfaces involved (Palacio et al., 1999). Thus, when a drop of water is put on a solid surface under air, the shape of the drop is modified under the pressure of the different surface–interfacial tensions until an equilibrium state is obtained (Good, 1979). If a surface is very hydrophilic, a water droplet placed on the surface spreads along the surface and the contact angle is small. For instance, the contact angle between a cellulose surface and water is less than 30°. A hydrophobic surface repels water and tends, thus, to form as small a common surface with the water droplet as possible. Consequently, the contact angle between a hydrophobic surface and water is high. For example, the contact angle between a polypropylene surface and water is higher than 100° (Cheryan, 1998). In addition to the determination of hydrophilicity, contact angle measurements can also be utilized in the characterization of the surface tension of a membrane (Geens et al., 2004; Hiemenz, 1986).

The contact angle determination is based on Young's equation in thermodynamic equilibrium:

$$\cos\theta = \frac{\gamma_{SV} - \gamma_{SL}}{\gamma_{LV}} \tag{32.1}$$

where γ_{SV} and γ_{LV} are the surface tensions of the solid and liquid against the vapor, γ_{SL} is the tension of the solid–liquid interface and θ is the equilibrium (Young) contact angle (Good, 1979). However, this equation assumes the surface to be smooth, rigid, and homogeneous, and because a membrane surface is a nonideal surface, it is commonly observed that a droplet of water on a membrane surface exhibits a range of contact angles instead of just one. This phenomenon is called *contact angle hysteresis*, and the range of observed angles is often determined by measurement of the advancing angle (maximum angle) and the receding angle (minimum angle). The advancing angle represents mainly the hydrophobic parts (Zhang and Hallström, 1990) and the receding angle mainly the hydrophilic parts of a membrane surface (Zhang et al., 1989). Chemical heterogeneity of a surface, surface roughness, and porosity affect the amplitude of the hysteresis phenomenon (Cain et al., 1983; Extrand, 2002; Gekas et al., 1992; Zhang and Hallström, 1990).

The easiest way to measure the contact angle between water and a membrane surface is the sessile drop method. In this method, a water droplet is placed on the membrane surface with a syringe or a micropipette. The droplet is observed with a microscope and the contact angle is measured with a goniometer. The advancing contact angle can be measured with the sessile drop method by increasing the drop volume and the receding angle by decreasing the drop volume (Garbassi et al., 1994; Neumann and Good, 1979). Usually the membrane is analyzed with the sessile drop method in its dry state. However, a membrane contact angle might vary owing to changes in the surrounding medium, and, thus, the contact angle of a dry membrane surface is not necessarily the same as the contact angle of a wet membrane surface. Membranes are used in their wet states, and, therefore, in order to obtain better knowledge of membrane hydrophilicity under working conditions, contact angle measurements of membranes should be performed in their wet states (Zhang et al., 1989). The contact angle of a wet membrane surface can be measured by using a captive bubble/drop method in which an air or liquid droplet is put with a U-shaped needle on the sample surface immersed in a liquid medium. In this method the angle is measured with a goniometer (Zhang et al., 1989). If a liquid drop is used, the liquids of the experimental setup must be immiscible. The advancing contact angle can be measured by decreasing the volume of the bubble, while the receding angle is obtained by increasing the bubble volume (Garbassi et al., 1994; Neumann and Good, 1979; Zhang et al., 1989). In both sessile drop and captive bubble methods vibrations and distortions of the droplet/bubble disturb the measurements. It must be noted that the angle obtained in simple measurements with sessile drop and captive bubble methods is a static contact angle, and if contact angle hysteresis can be expected, the static equilibrium angle measurement does not give enough information about the membrane hydrophilicity (Garbassi et al., 1994).

The contact angle can also be measured with the Wilhelmy rod method (a modification of the Wilhelmy plate method using a rod of a diameter of at least 20 mm), in which the membrane piece is fastened around a vertical rod partially immersed in the liquid. The static contact angle is obtained when the rod is at rest, and the dynamic contact angle is obtained when the rod is in motion. The advancing and receding angles can be measured by immersing and withdrawing the rod, respectively. The advantages of the Wilhelmy method are that the contact angle is measured with wet membrane samples and the method is easy to apply also in fouling studies (Palacio et al., 1999; Väisänen and Nyström, 1997). The equipment used in the measurements of the contact angle of membrane samples with the Wilhelmy method is presented in Figure 32.5 (Palacio et al., 1999).

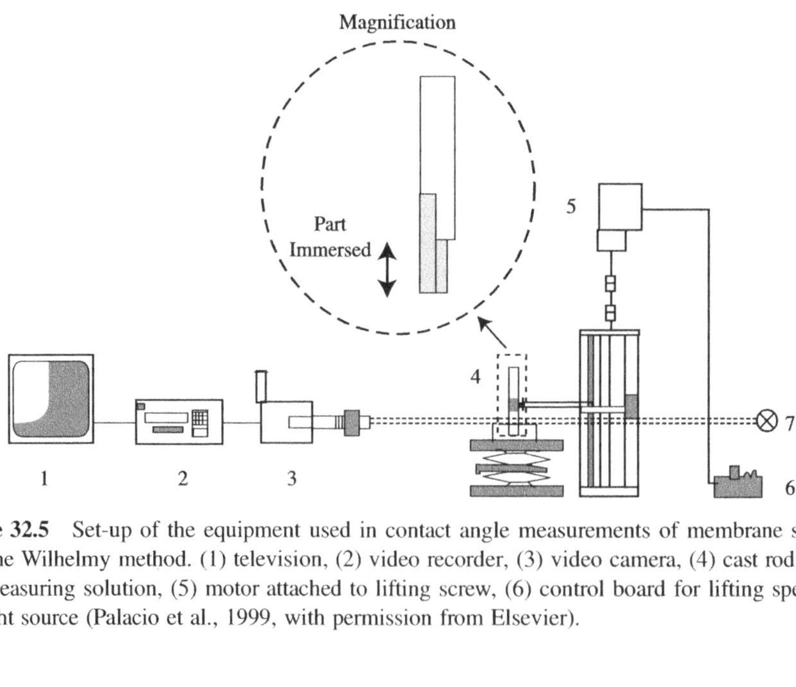

Figure 32.5 Set-up of the equipment used in contact angle measurements of membrane samples with the Wilhelmy method. (1) television, (2) video recorder, (3) video camera, (4) cast rod dipped into measuring solution, (5) motor attached to lifting screw, (6) control board for lifting speed and (7) light source (Palacio et al., 1999, with permission from Elsevier).

TABLE 32.2 Contact Angles of Commercial Nanofiltration and Tight Ultrafiltration Membranes Measured with the Sessile Drop Method

Membrane	Contact Angle (deg)	
PVD-1	20	
Desal-5 DK	31	
SR1	32	
ATF-60	38	Hydrophilicity decreases
Desal-5 DL	42	
XN-40	46	
NF-PES-10	54	
NTR 7450	58	

Source: Adapted from Mänttäri et al. (2002).

The contact angle method is widely utilized in the evaluation of membrane surface hydrophilicity owing to the simplicity of the method. Only small pieces of membranes are needed for the measurements and the measurements are also fast to perform. Although accurate contact angle values of membrane surfaces are difficult to measure due to the non-ideality of the surface, the contact angle can be utilized as a relative index of membrane hydrophilicity/hydrophobicity (Cho et al., 2000; Combe et al., 1999; Drelich et al., 1996; Keurentjes et al., 1989). For instance, Mänttäri et al. (2002) have measured contact angles of commercial nanofiltration and tight ultrafiltration membranes with sessile drop technique. The results showed that (Table 32.2) the membrane PVD-1 was the most hydrophilic and the membrane NTR 7450 was the most hydrophobic (Mänttäri et al., 2002).

32.4 CHARACTERIZATION OF MEMBRANE CHARGE

The membrane surface becomes charged when it is brought into contact with an aqueous electrolyte solution. The charging occurs due to, for instance, dissociation of functional groups, adsorption of ions from solution, and adsorption of polyelectrolytes, ionic surfactants, and charged macromolecules. In addition to the membrane surface, also the interior membrane pore surface can become charged. Membrane charging strongly affects the filtration properties of the membrane, and, thus, information on the electrical characteristics is needed. Although it is possible to evaluate the membrane charge when the chemical structure of the membrane is known, in practice, experiments are required for determination of the charge sign and density (Elimelech et al., 1994; Schaep and Vandecasteele, 2001; Zhao et al., 2005). The surface charge properties of membranes are often characterized by the zeta (ζ) potential, which is estimated using streaming potential measurements. Electrical characterization of a membrane surface can also be performed with impedance spectroscopy, membrane potential measurements, and titration. These methods show the membrane charge both qualitatively and quantitatively. However, the different approaches of the methods must be taken into account when results from different types of experiments are compared quantitatively (Schaep and Vandecasteele, 2001).

32.4.1 Streaming Potential Measurements

Streaming potential measurements can be used to determine the ζ potential of a membrane. The ζ potential is the potential at the plane of shear between a charged surface and a liquid that move in relation to each other. In streaming potential measurements, a pressure difference is applied across the membrane, which is in an electrolyte solution. As a consequence, the flow of fluid causes the electrical double layer to shear and the electric charges of the diffuse part of the double layer to displace. As a result, a potential is generated between the ends of the measurement channel. The potential difference is proportional to the charge on the surface and to the pressure used. The streaming potential is determined as the slope of potential versus pressure ($\Delta E/\Delta p$), and it should end at zero to ensure no asymmetric potentials are present (Kim et al., 1996; Nyström et al., 1989). The streaming potential can also be determined by applying a pressure pulse to one side of a membrane and measuring the instantaneous potential difference that results. This method, however, is valid only when the electrical potential difference is linearly dependent on the transmembrane pressure (Fievet et al., 2000; Szymczyk et al., 1999). The membrane sample is not destroyed during the streaming potential measurements.

The electrodes are located on the same side of the membrane, when the streaming potential is measured along the surface (tangential streaming potential), and on both sides of the membrane, when the streaming potential is measured through the pores of the membrane. The streaming potential measurements are performed in a specified pH range depending on electrode materials, and the ζ potentials can be calculated from the results at different pH using different modifications of the Helmholz–Smolukowski equation. Zeta potentials are usually shown as functions of pH (Elimelech et al., 1994; Kim et al., 1996; Nyström et al., 1989; Pihlajamäki, 1998; Szymczyk et al., 1998).

The surface charges evaluated with streaming potential measurements are averages of the charges of the skin layer and the support layer if the measurements have been done through the pores of the membranes. The asymmetric nature of the membranes, the membrane porosity, and pore geometry must be taken into account, as well as the fact that the nature of the ions of the electrolyte solution affect the streaming potential results. Thus, the sign and changes of the streaming potential can be determined for comparison purposes, but most streaming potential measurement systems cannot give accurate, absolute numerical values for the charge of a surface free from adsorbed ions (Chan and Chen, 2004; Kim et al., 1996; Nyström et al., 1994).

The internal membrane structure has a greater effect on the ζ potential results, when the measurements are performed through the pores. This can be seen in Figure 32.6. (Ho et al., 1999), where ζ potentials of polycarbonate track-etched membranes having different pore sizes, measured both along the surface and through the pores, are shown. The results show that the ζ potential measured along the surface is the same for all the membranes, while a pore size dependence can be seen in the results of the measurements through the pores. For the membranes having bigger pore size, the magnitude of the apparent ζ potential is greater. This phenomenon has been found also in earlier studies (Kim et al., 1997), and it can be explained by the differences in the chemical structure of the membrane surface and pore surface owing to the manufacturing process, where the material has been more oxidized when bigger pores were made (Ho et al., 1999).

The ζ potential values depend on the chemical structure of the membrane materials. If the membrane contains strongly acidic groups, the dissociation of the groups occurs immediately at a low pH and the ζ potential can be expected to be strongly negative even at low pH values (pH 2–3). When the membrane contains weakly acidic groups, the ζ potential can be expected to become more negative from the point the groups start to dissociate to the point where the groups are totally dissociated. Similarly, strongly

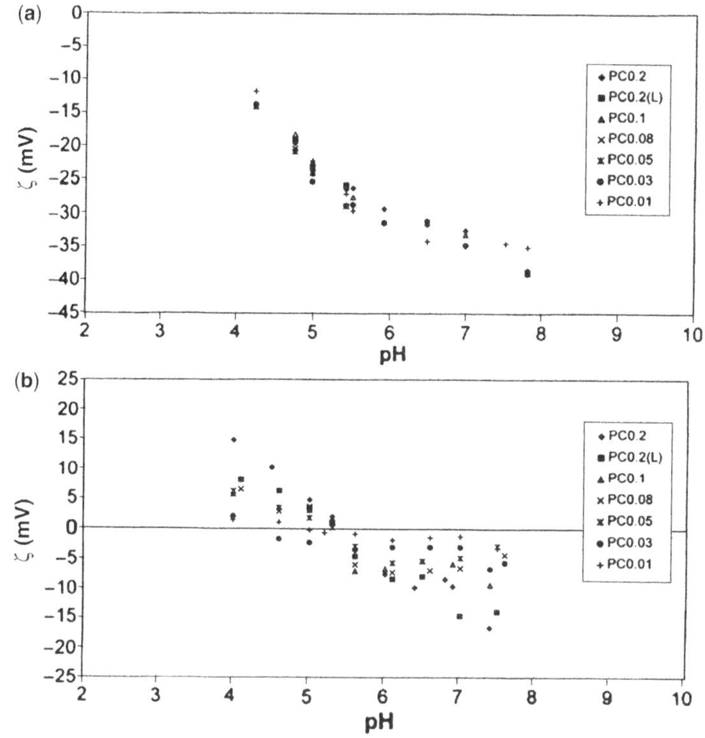

Figure 32.6 Apparent ζ potentials of track-etched polycarbonate membranes vs. pH measured (a) along the membrane surface and (b) through the membrane pores. The membrane PC0.2 has the biggest pore size and PC0.01 the smallest pore size. (Adapted from Ho et al., 1999.)

basic groups give positive potentials in most of the pH range, while weakly basic groups have no positive charges at pH values higher than 8 (Kim et al., 1996). The isoelectric point of the membrane can be evaluated from the pH dependence of the ζ potential (Kim et al., 1996; Martín et al., 2003). The isoelectric point does not vary with measurement method but can be modified by adsorption of specific salts, modification agents, or foulants (Kim et al., 1997).

32.4.2 Electrochemical Impedance Spectroscopy (EIS)

In electrochemical impedance spectroscopy (EIS) measurements a membrane is placed in an electrolyte solution of specified concentration. Two electrodes are located on the membrane surface, and a small alternating current of known frequency and with small amplitude is applied to the system. The phase difference and the amplitude of the simultaneous electrical potential that develops across the membrane are measured. A wide frequency range is used in the measurements, and the membrane is assumed to be a plane capacitor. As a result, the resistances and capacitances of the membrane are obtained and the thickness of the membrane can be evaluated. By using knowledge of the dielectric constant of water and the membrane material, the porosity of the membrane can be evaluated from the capacitance values. Impedance can be calculated from the values for resistance and capacitance; by plotting the impedance values against the frequency values used in the measurements, information on the molecular structure and layer structure of the membrane can be obtained. The resistance and capacitance values, and thus porosities and thicknesses, can be obtained for each sublayer separately. The membrane is not damaged during EIS measurements (Benavente et al., 2005; Cañas et al., 2001; Cañas and Benavente, 2002; Coster et al., 1996).

32.4.3 Membrane Potential

In membrane potential measurements, a membrane is positioned between two half-cells. The cells are filled with electrolyte solutions; the solution of higher electrolyte concentration is in contact with the active side of the membrane, and the solution of lower electrolyte concentration is in contact with the membrane support layer. The electrical potential difference is measured by inserting electrodes directly into the bulk solutions (Fig. 32.7a). The membrane potential measurements evaluate the amount of charge inside the membrane (Cañas and Benavente, 2002; Nakagaki and Takagi, 1986; Schaep and Vandecasteele, 2001; Takagi et al., 2000).

The membrane potential is the electrical potential difference between the two sides of the membrane when the membrane is separating two solutions of the same electrolyte but at different concentrations. The membrane potential is the sum of three contributors: two Donnan potentials (ΔE_{D1}, ΔE_{D2}, Fig. 32.7b) and the diffusion potential (ΔE_M, Fig. 32.7b). The Donnan potentials are the potential differences occurring at the interfaces between the membrane and the solution due to the existence of a fixed membrane charge causing an ion distribution between the membrane phase and the bulk solution. The magnitude of the Donnan potential, and thus also the membrane potential, depends on the membrane pore radius. The diffusion potential originates from the difference in transport velocity of anions and cations through the membrane (Cañas and Benavente, 2002; Nakagaki and Takagi, 1986; Schaep and Vandecasteele, 2001; Takagi et al., 2000).

32.4.4 Titration

In titration, the positively and negatively charged groups on the membrane surface are determined separately. The principle of titration is simple. A membrane is immersed in a solution having a high concentration of ions. It is assumed that because of the high ion concentration of the immersion solution, the original counterions of the membrane surface are exchanged to ions from the immersion solution. Thus, all charged groups on the surface have an ion from the immersion solution as a counterion. Then the membrane is rinsed with water, and it is immersed in another solution having a lower concentration of another type of ions. The ions of this second immersion solution take the places of the ions from the first solution as counterions, and the ions of the first solution are set free in the solution. This solution is analyzed and the number of ions from the first immersion solution is set equal to the number of negatively or positively charged groups on the membrane surface. In the determination of negatively charged groups, the first immersion solution can be, for instance, CsCl and the second solution $MgCl_2$, while in the determination of positively charged groups NaF and Na_2SO_4 can be used (Schaep and Vandecasteele, 2001). The charge of a membrane surface can also be determined with potentiometric titration (Bowen and Mukhtar, 1993; Jimbo et al., 1998).

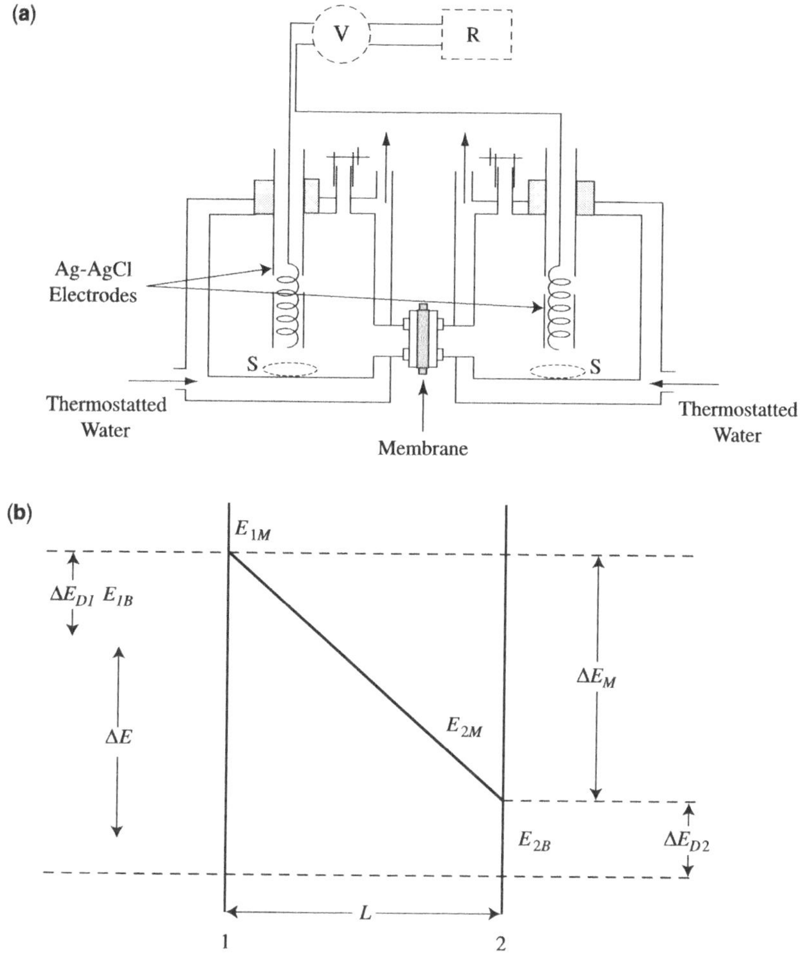

Figure 32.7 (a) Apparatus for membrane potential measurements: potentiometer (V), recorder (R) and stirring bar (S) (Takagi et al., 2000, with permission from Elsevier). (b) Schematic diagram of asymmetric membrane potential: diffusion potential within the membrane (ΔE_M), surface potential at interfaces 1 and 2 (ΔE_{D1}, ΔE_{D2}), membrane potential (ΔE), thickness of the membrane (L), bulk (B); $\Delta E = \Delta E_D + \Delta E_M$ and $\Delta E_D = \Delta E_{D2} - \Delta E_{D1}$ (Adapted from Nakagaki and Takagi, 1986.) Reproduced with permission from Chemical & Pharmaceutical Bulletin, Vol. 34, No. 3. Copyright [1986] Pharmaceutical Society of Japan.

32.5 CHARACTERIZATION OF MEMBRANE MORPHOLOGY

The structural features of membranes vary strongly depending on their ultimate function [microfiltration (MF), reverse osmosis (RO), and pervaporation (PV)]. Thus, membrane morphology can be classified in many ways. For instance, membranes can be categorized into porous and nonporous membranes, and the porous membranes can be further grouped into subclasses based on pore size. The characterization of membrane morphology becomes more difficult with decreasing pore size (Mulder, 1996).

Several methods can be used in the characterization of the morphology of porous membranes, like the bubble point method, mercury intrusion porometry, gas adsorption–desorption, thermoporometry, permporometry, liquid displacement, permeation measurements, and fractional rejection measurements. These methods are mainly used to characterize microfiltration and ultrafiltration membranes and details on them are presented comprehensively elsewhere (Mulder, 1996). In addition to these methods, also microscopical methods are applicable in the characterization of membrane morphology. The microscopical methods will be considered in more detail in the following chapters, because among these methods, suitable methods for characterization of both porous and nonporous membranes can be found. Optical microscopy, confocal scanning laser microscopy (CSLM), scanning electron microscopy (SEM), transmission electron microscopy (TEM), and atomic force microscopy (AFM) visualize the membrane morphology directly. CSLM (Ferrando et al., 2005), SEM (Goldstein et al., 2003), and TEM (Freger et al., 2002) can be utilized in the characterization of the chemical composition of the membranes, and AFM can be used in the examination of the forces (London–van der Waals and the electrical double-layer forces) affecting the interaction between the membrane surface and the colloids in the process feed (Bowen et al., 2002). A summary of the applicability of the microscopical methods is given in Table 32.3. It should be noted that the resolution is defined in this chapter as the smallest distance, which can be noticed between two details in a magnified image.

32.5.1 Optical Microscopy

In optical microscopy (Table 32.3) light transmitted through the sample or reflected from the sample surface forms an image, which is magnified with a lens system. The advantages of optical microscopy compared to electron microscopy are that it is less expensive to buy and use and that the interactions between a sample and the used radiation cause almost no distortions in the sample. No sample pretreatment is needed, and the samples can be analyzed in their wet state as well as in their dry state because the sample is not exposed to a vacuum during imaging (Hemsley, 1989). However, the resolution of an optical microscope is poor compared to the other microscopical characterization methods; only in the order of 1 μm (Vaughan, 1993). Thus, the use of optical microscopy in surface characterization is limited to the characterization of surface macrostructure. For instance, a quick examination to ensure successful coating of a surface can be performed with an optical microscope (Bessarabov and Sandersson, 2004).

32.5.2 Confocal Scanning Laser Microscopy (CSLM)

In confocal scanning laser microscopy (CSLM, Table 32.3) the incident light source is a laser. The laser beam is focused on a limited spot in the sample and the light reflected, backscattered, or emitted from the spot in focus is detected. A pinhole in front of the detector obstructs the light reflected, backscattered, or emitted from the illuminated regions of the sample below and above the in-focus point to reach the detector. Thus, only one point of the sample is observed at a time. In order to obtain an image of the sample the laser beam is scanned over the sample surface and the detected signal is recorded point by point to produce the image. The measurements can be carried out perpendicular to the optical axis of the microscope, in the x-y plane, and also parallel to the optical axis of the microscope, in the x-z or the y-z plane. By combining the series of images recorded at different depths by changing the position of the focalization plane a three-dimensional image can be built up (Charcosset et al., 2000; Charcosset and Bernengo, 2000; Ferrando et al., 2005).

TABLE 32.3 **Microscopical Methods Used in Membrane Surface Characterization**

	Optical Microscopy	Confocal Scanning Laser Microscopy (CSLM)	Scanning Electron Microscopy (SEM)	Transmission Electron Microscopy (TEM)	Atomic Force Microscopy (AFM)
High resolution	No	No	Yes	Yes	Yes
Analysis of MF membranes (pore size, fine structure)	Yes	Yes	Yes	Yes	Yes
Analysis of UF membranes (pore size, fine structure)	No	No	Yes	Yes	Yes
Analysis of NF membranes (pore size, fine structure)	No	No	Yes	Yes	Yes
Analysis of RO membranes (pore size, fine structure)	No	No	No	Yes	Yes
Vacuum required	No	No	Yes	Yes	No
Sample pretreatment required	No	No	Yes	Yes	No
Wet samples can be analyzed	Yes	Yes	No	No	Yes
Sample destructive technique	No	No	Yes	Yes	Yes/No
3D image obtained	No	Yes	No	No	Yes
Depth profiling possibility	No	Yes	No	No	No
Examination of porous structure	Yes	Yes	Yes	Yes	Yes
Examination of chemical composition	No	Yes	Yes	Yes	No
Examination of roughness	No	No	No	No	Yes
Examination of surface forces	No	No	No	No	Yes
Visualization of membrane damages	Yes	Yes	Yes	Yes	Yes
Visualization of fouling layer/ coating layer	Yes	Yes	Yes	Yes	Yes

Confocal scanning laser microscopy can be used in three different modes. In the reflection mode the light backscattered from the sample forms an echo image describing the topography of the sample. In the fluorescence mode the image gives information on the location of different compounds in the sample because the image is composed of the fluorescent light, which the compounds in the sample emit. CSLM can also be used in the transmission mode as a conventional optical microscope. In addition, it is possible to measure in two modes at the same time in order to get a connected transmittance and intensity image (Charcosset et al., 2000; Charcosset and Bernengo, 2000; Ferrando et al., 2005).

If the membrane sample is transparent, the reflection mode of CSLM makes it possible to obtain three-dimensional surface porosity data without physical sample slicing. An example is presented in Figure 32.8, where the CSLM images visualize the differences in membrane morphology of a cellulose acetate membrane at different depths. Black areas in the images represent membrane pores and white areas represent membrane material. Thus, the images show that the membrane structure is denser at a depth of 4 μm than on the membrane surface (Charcosset and Bernengo, 2000). An advantage of the CSLM is that the membrane samples can be analyzed in their wet state, in their dry state, or mounted in immersion oil (Charcosset et al., 2000; Charcosset and Bernengo, 2000). However, in the best case CSLM

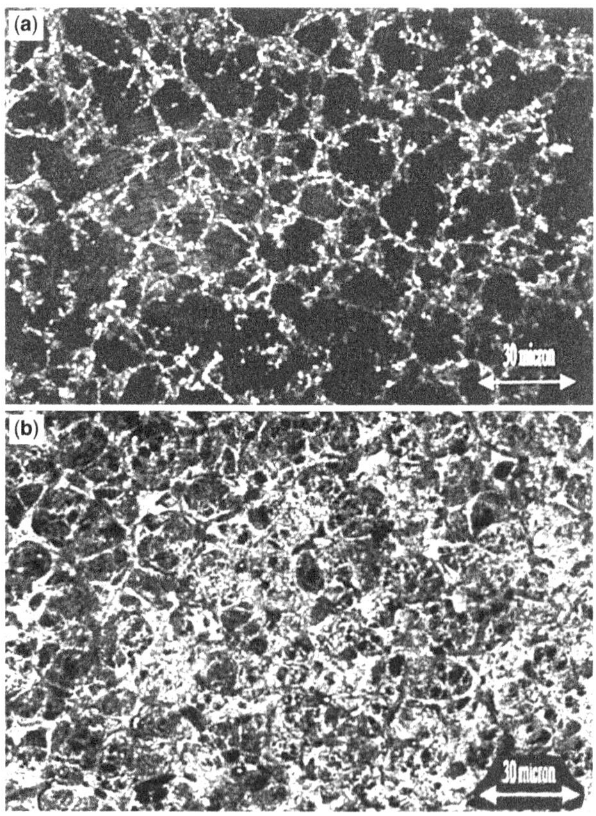

Figure 32.8 CSLM images from different depths of a cellulose acetate membrane: (a) $z = 0$ μm, (b) $z = 4$ μm. Membrane sample was mounted in immersion oil during the imaging. (Adapted from Charcosset and Bernengo, 2000, with permission from Elsevier.)

resolution is 180 nm in the focal plane (x, y) and only 500–800 nm along the optical axis (z), and, thus, porosity characterization is limited to microfiltration membranes (Ferrando et al., 2005). The fluorescent mode is useful in on-line visualization of the adsorption–desorption processes occurring at different depths in the membrane, and, thus, it has been used especially in studies on membrane fouling caused by proteins (Hayama et al., 2003; Ferrando et al., 2005; Reichert et al., 2002). The fluorescence mode can also be used in the characterization of membranes made from multiphase polymer blends. If the compounds of the analyzed sample do not present autofluorescence, the fluorescence can be induced by chemical treatment, for instance, by treating the sample with specific stains (Ferrando et al., 2005).

32.5.3 Scanning Electron Microscopy (SEM)

In a scanning electron microscope (SEM, Table 32.3) a fine beam of electrons scans the membrane surface. This causes several kinds of interactions generating different signals, of which secondary electrons (SE) and backscattered electrons (BSE) are used in the image forming (Goldstein et al., 2003).

Secondary electron images can be used in membrane characterization in imaging of membrane morphology, for example, pore geometry, pore size, pore size distribution, and surface porosity (Schossig-Tiedemann and Paul, 2001). Due to the large depth of field, the SE images visualize the membrane surface morphology three-dimensionally (Goldstein et al., 2003). The lateral resolution of a conventional SEM is 10–50 nm, which enables determination of the fine morphology of microfiltration membranes (Chahboun et al., 1992; Goldstein et al., 2003). However, with field emission scanning electron microscopy (FESEM), which has a resolution of 1–5 nm, also visualization of the fine morphology of ultrafiltration membranes is possible (Kim et al., 1991; Masselin et al., 2001). The depth resolution in SE imaging is 1–10 nm and in BSE imaging 10–1000 nm (Goldstein et al., 2003). To observe cross sections of flat-sheet membranes or the inner surfaces of hollow-fiber membranes, the samples are usually briefly frozen in liquid nitrogen and then broken manually (Charcosset and Bernengo, 2000; Khatib et al., 1997; Zhao et al., 2000). In addition to the visualization of the sample surface, SE and BSE images provide information on sample topography and chemical composition of the sample (BSE images) (Goldstein et al., 2003).

Vacuum conditions are required in conventional SEM, which means that the sample has to be analyzed in its dry state. Thus, in the examination of membrane morphology with SEM, it has to be noted that sample drying might alter the pore structure (Bowen et al., 1996b; Fritzsche et al., 1992). Polymers are generally insulators, therefore, polymeric membrane samples (and other nonconducting samples) have to be coated with a conductive coating (carbon, gold, platinum, or palladium) to eliminate surface charging and to minimize sample damage caused by the electron beam (Charcosset et al., 2000; Fritzsche et al., 1992; Mulder, 1996; Schossig-Tiedemann and Paul, 2001). However, the electron beam might still damage the polymeric membrane sample (Chahboun et al., 1992), and the coating process might cause artificial changes to the membrane surface (Schossig-Tiedemann and Paul, 2001). Furthermore, the coating layer can lead to an underestimation of the pore size (Bowen et al., 1996b; Fritzsche et al., 1992). The charging problem and beam damage to the sample can be diminished by applying the FESEM technique, which uses a lower accelerating voltage compared to conventional electron microscopy (Kim et al., 1991; Kim and Fane, 1994).

Environmental scanning electron microscopy (ESEM) has been developed for studies of wet or nonconducting materials, and it enables the characterization of membrane samples also without coating (Yu et al., 2005). However, the resolution of ESEM is limited compared to conventional SEM and FESEM (Chan and Chen, 2004; Schossig-Tiedemann and Paul, 2001). In addition to morphology characterization, ESEM can also be used in the characterization of the hydrophilicity/hydrophobicity of a surface and in the determination of the swelling of a membrane (Yu et al., 2005, 2006).

32.5.4 Transmission Electron Microscopy (TEM)

In transmission electron microscopy (TEM, Table 32.3) an electron beam is focused on the sample, and the electrons passing through the sample are detected for image forming (Hernández et al., 1999). Thus, the darker areas of the image represent the sample areas through which fewer electrons are transmitted, and the lighter areas represent the areas through which more electrons are transmitted. TEM analysis occurs in a vacuum, and, thus, the sample has to be analyzed in its dry state. The maximum resolution of TEM is 0.3–0.5 nm (Zhao et al., 2000), and, therefore, it can be used even in the characterization of nano- and RO membranes (Freger et al., 2002, 2005; Sheldon, 1991). TEM visualizes the pore size of the membrane, and a pore size distribution can be achieved with image analysis (Tomandl et al., 2000). TEM is also useful in the characterization of multiphase morphologies because it provides information on the inner structure of the particles and gives the possibility to show the differences in chemical structure in the sample as contrasts in the image (Freger et al., 2002; Wallheinke et al., 1998). An example is given in Figure 32.9, where a thin section containing a thin-film composite membrane, which has

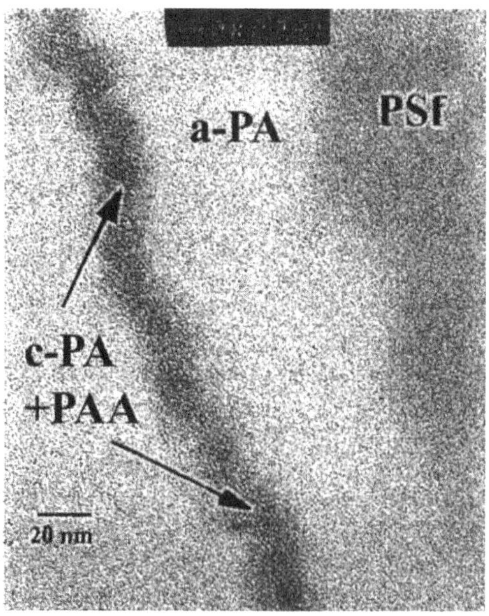

Figure 32.9 Thin section of an NF-270 membrane after modification with acrylic acids. Magnification of image is 320 K. Sample was stained with uranyl nitrate to achieve better contrast for seeing the double-layer structure of the skin of the membrane. The thin, topmost layer of the membrane contains carboxylic groups (c-PA + PAA) and the thicker and brighter part of the skin layer is carboxylic free (a-PA). (Adapted from Freger et al., 2002, with permission from Elsevier.)

a semiaromatic piperazine-based polyamide (PA) layer modified with acrylic acids on top of a polysulfone (PSf) layer is visualized with TEM. The TEM image clearly shows the double-layer structure of the skin layer of the membrane: PA containing uranyl-stained carboxylic groups can be seen as a thin dark layer on the top of the membrane (c-PA), and the carboxyl-free polymer layer is seen (a-PA) between the supporting PSf layer and the top c-PA layer (Freger et al., 2002).

In TEM analysis the analyzed sample has to be thin enough for electrons to penetrate (less than 50 nm). Thus, the depth of the TEM analysis depends on the sample thickness (Chan and Chen, 2004; Zhao et al., 2000). Preparation of a thin section of this kind is difficult and time consuming and might lead to artefacts in the resulting image. In addition to this, some materials need to be stained using heavy-metal salts to increase the image contrast (Sheldon, 1991; Wallheinke et al., 1998). A sample for TEM analysis can also be prepared by making a freeze–fracture of the sample and then replicating the fracture surface using carbon and carbon/platinum. The replica is characterized with TEM. An overall membrane structure is seen when using the thin-section method, while rapid freezing and replica preparation with the deep-etching method allows the fine structure of the membrane to be characterized. Thus, the best knowledge about membrane structure is obtained, when samples prepared with both of these methods are analyzed (Sheldon, 1991). Due to the demanding sample preparation process causing artefacts and the analysis in the dry state of the membrane, the porosity evaluated with TEM analysis might be smaller than in reality (Bowen et al., 1999).

32.5.5 Atomic Force Microscopy (AFM)

In atomic force microscopy (AFM, Table 32.3), also known as scanning force microscopy (SFM), a small, sharp tip scans the surface of a sample. The tip is located at the free end of a cantilever. Interaction forces between the tip and the sample surface cause the cantilever to bend or deflect. A detector measures the cantilever deflections, and a three-dimensional map of the surface topography of the sample is generated based on the measured deflections (Bowen et al., 1999; Reich et al., 2001; Yalamanchili et al., 1998).

In contact AFM (C-AFM) mode, also known as the repulsive mode, the tip is less than a nanometer from the surface being imaged, and it responds to very short range repulsive interactions (Born repulsion) with the sample. When imaging with noncontact mode (NC-AFM), the tip is generally vibrated at a 5- to 10-nm distance away from the surface, and it responds to attractive van der Waals interactions with the sample. In NC-AFM the forces used for imaging are lower, and, thus, it is an especially suitable imaging mode for polymeric membrane materials, which are soft or liable to mechanical damage (Bowen et al., 1996b, 1999). An image of a rigid sample characterized both with C-AFM and NC-AFM look the same. However, the images look different when there are a few monolayers of condensed water lying on the surface of the rigid sample. The C-AFM will penetrate the liquid layer to image the underlying surface, whereas the NC-AFM will image the surface of the liquid layer. An intermittent contact mode (IC-AFM), known also as the tapping mode, is a hybrid of C-AFM and NC-AFM. As in NC-AFM, a cantilever is vibrated, but the cantilever is held at a tip-sample distance closer to the region of contact imaging. Thus, at the lower limit of the cantilever movement, the tip just taps the sample surface. Owing to the intermittent contact of the tip with the sample surface, IC-AFM is less likely to damage the sample surface than the C-AFM (Bowen et al., 1999).

Atomic force microscopy is widely used in the characterization of membrane surface morphology, from microfiltration membranes to reverse osmosis membranes and to gas separation membranes (Bottino et al., 1994; Bowen et al., 1996a; Dietz et al., 1992; Khulbe et al., 1996; Stamatialis et al., 1999; Wilbert et al., 1998). One of its advantages is that both the lateral resolution and the depth resolution (z direction) can reach the subnanometer range (Yalamanchili et al., 1998). Other advantages of AFM measurements compared to electron microscopy are also that special sample preparation is not needed and that the imaging can be performed directly in air or in liquid (Bowen et al., 1996b). The pore size, surface porosity (a ratio between porous area and total area), and pore density (number of pores per surface unit) can be obtained directly from the AFM images. The AFM method also allows determination of pore size distribution (Hilal et al., 2005; Ochoa et al., 2001) and surface roughness (Boussu et al., 2005; Vrijenhoek et al., 2001). AFM images describing the surface roughness of the nanofiltration membranes NF-PES10 and Desal 5 DL are presented in Figure 32.10 (Boussu et al., 2005). The measurements were performed with the NC-AFM technique. According to the authors, the roughness values measured with the IC-AFM technique were of the same

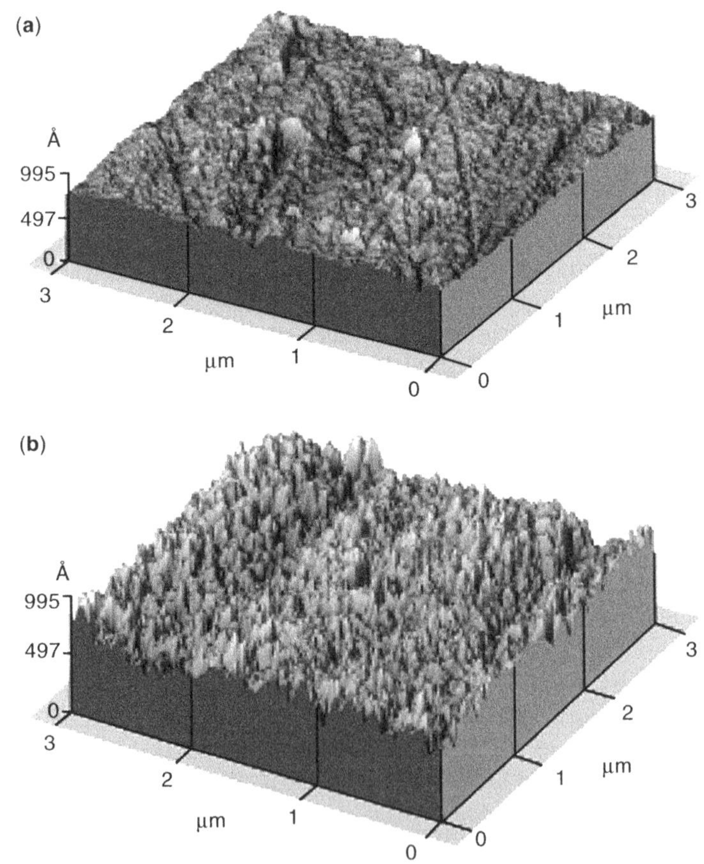

Figure 32.10 AFM images describing the roughness of nanofiltration membranes (a) NF-PES-10 and (b) Desal 5 DL. The measurements were performed with NC-AFM (Boussu et al., 2005, with permission from Elsevier).

order as the values measured with NC-AFM. However, it was found that a comparison of the roughness values is only valid when the same surface area is scanned in the measurements (Boussu et al., 2005).

In the characterization of membrane fine morphology with AFM, it has to be noted that if the pores are smaller than the tip AFM underestimates the pore dimensions. This effect is expected to be less significant for NC-AFM than for C-AFM (Bowen et al., 1996b). The AFM images might also contain artefacts resulting from physical distortion or damage to the sample caused by the tip (C-AFM) or a convolution of the tip and sample shapes, especially when the sizes of the tip and the surface pores are similar. Artefacts can also occur due to the inaccuracy of the scanning mechanism of AFM (Bowen et al., 1996b; Bowen and Doneva, 2000). A high surface roughness might also disturb the characterization of membrane fine morphology (Ochoa et al., 2001).

In addition to morphological characterization, AFM can quantify surface forces by using a colloid probe method. In this method, a colloid particle is glued to the end of the cantilever instead of the tip. Measuring adhesion on the membrane surface enables determination of the fouling tendency of a membrane surface by particles without full-scale filtration experiments, which facilitates membrane development processes (Bowen et al., 1998). Furthermore, by using the colloid probe method with AFM measurements in aqueous solution, the electrical properties of the membrane surface can be quantified (Bowen et al., 2002). AFM can also be used in the examination of membrane swelling (Freger et al., 2005).

32.6 CONCLUSIONS

Membrane surface characteristics affect membrane performance in the filtration process, and, thus, knowledge of surface characteristics is needed. Information given by membrane manufacturers on membrane material, cut-off value, and sometimes even on membrane charge often has to be supplemented with information from the characterization methods presented in this chapter.

The chemical structure of the membrane can be analyzed with spectroscopic methods, of which the Fourier transform infrared spectroscopy attenuated total reflectance (FTIR-ATR) method is the most utilized. However, Raman spectroscopy and IR spectroscopy complement each other, and, thus, if both methods are used, the information obtained on the membrane chemical structure is quite comprehensive. If only information from the membrane skin layer is needed, the most surface-sensitive methods—X-ray photoelectron spectroscopy (XPS) and time-of-flight secondary ion mass spectroscopy (TOF-SIMS)—have to be used.

Membrane charge, as well as hydrophilicity, can be predicted based on known membrane chemical structure. Often, however, more accurate information is needed. Contact angle measurements offer a fast and simple way to get the needed information on hydrophilicity. Several methods can be applied in the characterization of the electrical properties of the membrane. The most utilized method is the determination of the ζ potential from streaming potential measurements. The ζ potential values give information about the overall membrane surface charge, while the charge inside the membrane can be determined with membrane potential measurements. Thus, the ζ potential is more useful when knowledge on the membrane surface charge affecting the interaction with the macromolecules of the feed in the filtration process is needed, whereas membrane potential measurement results increase knowledge on the mobility of ions in the membrane material and on its Donnan properties. If information about the electrical properties of different sublayers of the membrane is needed, electrochemical impedance spectroscopy can be used. Information on the negative and positive groups in the membrane can also be determined with titration.

Direct information on membrane porous structure and sublayer structure is obtained with microscopical methods. The most commonly applied methods are SEM and AFM because the resolution of the microscopes is good enough for characterization of ultra- and nanofiltration membranes and even RO membranes. In rough surface characterization conventional optical microscopy can also be used. The resolution of CSLM is sufficient only for characterization of microfiltration membranes. However, the advantage of CSLM is that information on the membrane bulk structure can be obtained without physical sample cross sectioning.

The choice of characterization method is generally made based on the problem to which an answer is required and on the time, money, and resources available. However, the best knowledge is always obtained by combining results from different characterization methods.

Nomenclature

AES	Auger electron spectroscopy
AFM	atomic force microscopy
ATR	attenuated total reflectance
a-PA	amine-dominated polyamide
B	bulk
BSE	backscattering electrons
Be	beryllium
C-AFM	contact mode in atomic force microscopy
CSLM	confocal scanning laser microscopy
c-PA	carboxylic groups containing polyamide
EDS	energy-dispersive X-ray spectroscopy
EIS	electrochemical impedance spectroscopy
ESCA	elemental spectroscopy for chemical analysis
ESEM	environmental scanning electron microscopy
FESEM	field-emission scanning electron microscopy
FTIR	Fourier transform infrared spectroscopy
FT-Raman	Fourier transform Raman spectroscopy
H	hydrogen
IC-AFM	intermittent contact mode in atomic force microscopy
Li	litium
$M/\Delta M$	molar mass/change in molar mass
MALDI-MS	matrix-assisted laser desorption ionization mass spectrometry
NMR	nuclear magnetic resonance spectroscopy
NF	nanofiltration
NC-AFM	noncontact mode in atomic force microscopy
PA	polyamide
PAA	polyacrylic acid
PAMAM	polyamidoamine

PAS	photoacustic spectrometry
PEK-C	phenolphthalein poly(ether ether ketone)
PET	polyethyleneterephtalate
PSf	polysulfone
PTFE	polytetrafluoroethylene
PV	pervaporation
R	recorder
RO	reverse osmosis
S	stirring bar
SE	secondary electrons
SEM	scanning electron microscopy
SIMS	secondary ion mass spectrometry
TEM	transmission electron microscopy
TOF	time of flight
U	uranium
UF	ultrafiltration
V	potentiometer
XPS	X-ray photoelectron spectroscopy

Symbols

ΔE	potential difference
ΔE_D	Donnan potential difference
ΔE_{D1}	Donnan potential at the interfaces between the membrane and the solution
ΔE_{D2}	potential at the interfaces between the membrane and the solution
ΔE_M	diffusion potential within the membrane
L	thickness of the membrane
Δp	pressure difference
γ_{LV}	surface tension of liquid against vapor
γ_{SL}	surface tension of solid–liquid interface
γ_{SV}	surface tension of solid against vapor
Θ	equilibrium (Young) contact angle

ACKNOWLEDGMENT

The authors are grateful to the Graduate School on Functional Surfaces, the Academy of Finland (projects SA/206064 and SA/208292), Walter Ahlström Foundation, and the Gustav Komppa Fund of the Foundation of Alfred Kordelin for financial support.

REFERENCES

Belfer, S., Fainchtain, R., Purinson, Y., and Kedem, O. (2000). Surface characterization by FTIR-ATR spectroscopy of polyethersulfone membranes: Unmodified, modified and protein fouled. *J. Membr. Sci.* 172, 113.

Belu, A. M., Graham, D. J., and Castner, D. G. (2003). Time-of-flight secondary ion mass spectrometry: Techniques and applications for the characterisation of biomaterial surfaces. *Biomaterials* 24, 3635.

Benavente, J., Zhang, X., and Valls, R. G. (2005). Modification of polysulfone membranes with polyethylene glycol and lignosulfate: Electrical characterization by impedance spectroscopy measurements. *J. Colloid Interface Sci.* 285, 273.

Bessarabov, D., and Sandersson, S. (2004). Solid polyelectrolyte (SPE) membranes with textured surface. *J. Membr. Sci.* 244, 69.

Bishop, H. E. (1990). Auger electron spectroscopy. In J. M. Walls (Ed.), *Methods of Surface Analysis, Techniques and Applications*. Cambridge University Press, Cambridge, pp. 87–126.

Boccaccio, T., Bottino, A., Capannelli, G., and Piaggio, P. (2002). Characterization of PVDF membranes by vibrational spectroscopy. *J. Membr. Sci.* 210, 315.

Bottino, A., Capanelli, G., D'Asti V., and Piaggio, P. (2001). Preparation and properties of novel organic-inorganic porous membranes. *Sep. Purif. Technol.* 22/23, 269.

Bottino, A., Capanelli, G., Grosso, A., Monticelli, O., Cavalerri, O., Rolandi, R., and Soria, R. (1994). Surface characterization of ceramic membranes by atomic force microscopy. *J. Membr. Sci.* 95, 289.

Boussu, K., Van der Bruggen, B., Volodin, A., Snauwaert, J., Van Haesendonck C., and Vandecasteele, C. (2005). Roughness and hydrophobicity studies of nanofiltration membranes using different modes of AFM. *J. Colloid Interface Sci.* 286, 632.

Bowen, W. R., and Doneva, T. A. (2000). Artefacts in AFM studies of membranes: Correcting pore images using fast fourier transform filtering. *J. Membr. Sci.* 171, 141.

Bowen, R. W., Doneva, T. A., and Stoton, J. A. G. (2002). The use of atomic force microscopy to quantify membrane surface electrical properties. *Colloids Surf. A* 201, 73.

Bowen, W.-B., Hilal, N., Lovitt, R. W., and Williams, P. M. (1996a). Atomic force microscope studies of membranes: Surface pore structures of Cyclopore and Anopore membranes. *J. Membr. Sci.* 110, 233.

Bowen, W. R., Hilal, N., Lovitt, R. W., and Williams, P. M. (1996b). Atomic force microscopy studies of membranes: Surface pore structures of Diaflo ultrafiltration membranes. *J. Colloid Interface Sci.* 180, 350.

Bowen, W. R., Hilal, N., Lovitt, R. W., and Wright, C. J. (1998). A new technique for membrane characterisation: Direct measurement of the force adhesion of a single particle using an atomic force microscope. *J. Membr. Sci.* 139, 269.

Bowen, W. R., Hilal, N., Lovitt, R. W., and Wright, C. J. (1999). Atomic force microscope studies of membrane surface, In T. S. Sørensen (Ed.), *Surface Chemistry and Electrochemistry of Membranes.* Surfactant Science Series Vol. 79. Marcel Dekker, New York, pp. 1–37.

Bowen, R., and Mukhtar, H. (1993). Properties of microfiltration membranes: The surface electrochemistry of mixed oxide membranes. *Colloids Surf. A* 81, 93.

Briggs, D. (1994). Polymer surface characterisation by XPS and SIMS. In S. J. Spells (Ed.), *Characterisation of Solid Polymers, New Techniques and Developments*. Chapman & Hall, London, pp. 312–358.

Cain, J. B., Francis, W., Venter, R. D., and Neumann, A. W. (1983). Dynamic contact angles on smooth and rough surfaces. *J. Colloid Interface Sci.* 94, 123.

Cañas, A., Ariza, M. J., and Benavente, J. (2001). Characterization of active and porous sublayers of a composite reverse osmosis membrane by impedance spectroscopy, streaming and membrane potentials, salt diffusion and X-ray photoelectron spectroscopy measurements. *J. Membr. Sci.* 183, 135.

Cañas, A., and Benavente, J. (2002). Electrochemical characterisation of an asymmetric nanofiltration membrane with NaCl and KCl solutions: Influence of membrane asymmetry on transport parameters. *J. Colloid Interface Sci.* 246, 328.

Carlsson, D. J., Dal-Cin, M. M., Black, P., and Lick, C. N. (1998). A surface study of membranes fouled by pulp mill effluent. *J. Membr. Sci.* 142, 1.

Chahboun, A., Coratger, R., Ajustron, F., and Beauvillain, J. (1992). Comparative study of micro- and ultrafiltration membranes using STM, AFM and SEM. *Ultramicroscopy* 41, 235.

Chalmers, J. M., and Everall, N. J. (1993). Vibrational spectroscopy. In B. J. Hunt and M. I. James (Eds.), *Polymer Characterisation*. Blackie Academic & Professional, Suffolk, England, pp. 69–106.

Chan, R., and Chen, V. (2004). Characterisation of protein fouling on membranes: Opportunities and challenges. *J. Membr. Sci.* 242, 169.

Charcosset, C., and Bernengo, J.-C. (2000). Comparison of microporous membrane morphologies using confocal scanning laser microscopy. *J. Membr. Sci.* 168, 53.

Charcosset, C., Cherfi, A., and Bernengo, J.-C. (2000). Characterization of microporous membrane morphology using confocal scanning laser microscopy. *Chem. Eng. Sci.* 55, 5351.

Cheryan, M. (1998). *Ultrafiltration and Microfiltration Handbook*. Technomic Company, Lancaster, pp. 245–247.

Cho, J., Amy, G., and Pellegrino, J. (2000). Membrane filtration of natural organic matter: Factors and mechanism affecting rejection and flux decline with charged (UF) membrane. *J. Membr. Sci.* 164, 89.

Colthup, N. B., Daly, L. H., and Wiberley, S. E. (1975). *Introduction to Infrared and Raman Spectroscopy*. Academic, New York, Chapter I, p. 57.

Combe, C., Molis, E., Lucas, P., Riley, R., and Clark, M. M. (1999). The effect of CA membrane properties on adsorptive fouling by humic acid. *J. Membr. Sci.* 154, 73.

Coster, H. G. L., Chilcott, T. C., and Coster, A. C. F. (1996). Impedance spectroscopy of interfaces, membranes and ultrastructures, a review. *Bioelectrochem. Bioenerget.* 40, 79.

Dal-Cin, M. M., Strietz, C.-N., Tweddle, T. A., Capes, C. E., McLellan, F., and Buisson, H. (1995). Effect of adsorptive fouling on membrane performance: Case study with pulp mill effluent. *Desalination* 101, 155.

Delcorte, A., Bertrand, P., Arys, X., Jonas, A., Wischerhoff, E., Mayer, B., and Laschewsky, A. (1996). Tof-SIMS study of alternate polyelectrolyte thin films: Chemical surface characterisation and molecular secondary ions sampling depth. *Surface Sci.* 366, 149.

Dietz, P., Hansma, P. K., Inacker, O., Lehmann, H.-D., and Herrmann, K.-H. (1992). Surface pore structures of micro- and ultrafiltration membranes imaged with atomic force microscope. *J. Membr. Sci.* 65, 101.

Drelich, J., Miller, J. D., and Good, R. J. (1996). The effect of drop (bubble) size on advancing and receding contact angles for heterogeneous and rough solid surfaces as observed with sessile-drop and captive-bubble techniques. *J. Colloid Interface Sci.* 179, 37.

Elimelech, M., Chen, W. H., and Waypa, J. J. (1994). Measuring the zeta (electrokinetic) potential of reverse osmosis membranes by a streaming potential analyzer. *Desalination* 95, 269.

Extrand, C. W. (2002). Water contact angles and hysteresis of polyamide surfaces. *J. Colloid Interface Sci.* 248, 136.

Ferrando, M., Rŏžek, A., Zator, M., López F., and Güell, C. (2005). An approach to membrane fouling characterisation by confocal scanning laser microscopy. *J. Membr. Sci.* 250, 283.

Fievet, P., Szymczyk, A., Aoubiza, B., and Pagetti, J. (2000). Evaluation of three methods for the characterisation of the membrane-solution interface: Streaming potential, membrane potential and electrolyte conductivity inside pores. *J. Membr. Sci.* 168, 87.

Fleming, O. S., Kazarian, S. G., Bach, E., and Schollmeyer, E. (2005). Confocal Raman study of poly(ethylene terephthalate) fibres dyed in supercritical carbon dioxide: Dye diffusion and polymer morphology. *Polymer* 46, 2943.

Flösch, D. (1992). Surface analysis of PVDF membranes. *J. Membr. Sci.* 70, 53.

Fontyn, M., Bijsterbosch, B. H., and Van't Riet, K. (1987). Chemical characterization of ultrafiltration membranes by spectroscopic techniques. *J. Membr. Sci.* 36, 141.

Franken, A. C. M., Nolten, J. A. M., Mulder, M. H. V., Bargeman, D., and Smolders, C. A. (1987). Wetting criteria for the applicability of membrane distillation. *J. Membr. Sci.* 33, 315.

Freger, V., Bottino, A., Capanelli, G., Perry, M., Gitis, V., and Belfer, C. (2005). Characterization of novel acid-stable NF membranes before and after exposure to acid using ATR-FTIR, TEM and AFM. *J. Membr. Sci.* 256, 134.

Freger, V., Gilron, J., and Belfer, S. (2002). TFC polyamide membranes modified by grafting of hydrophilic polymers: An FT-IR/AFM/TEM study. *J. Membr. Sci.* 209, 283.

Fritzsche, A. K., Arevalo, A. R., Connolly, A. F., Moore, M. D., Elings, V., and Wu, C. M. (1992). The structure and morphology of the skin of polyethersulfone ultrafiltration membranes: A comparative atomic force microscope and scanning electron microscope study. *J. Appl. Polym. Sci.* 45, 1945.

Fulghum, J. (1999). Recent developments in high energy and spatial resolution analysis of polymers by XPS. *J. Electron Spectrosc. Relat. Phenom.* 100, 331.

Garbassi, F., Morra, M., and Occhiello, E. (1994). *Polymer Surfaces; from Physics to Technology.* Wiley West Sussex, England, Chapter 3, pp. 69–150.

Geens, J., Van der Bruggen, B., and Vandecasteele, C. (2004). Characterisation of the solvent stability of polymeric nanofiltration membranes by measurement of contact angles and swelling. *Chem. Eng. Sci.* 59, 1161.

Gekas, V., Persson, K. M., Wahlgren, M., and Sivik, B. (1992). Contact angles of ultrafiltration membranes and their possible correlation to membrane performance. *J. Membr. Sci.* 72, 293.

Goldstein, J., Newbury, D., Joy, D., Lyman, C., Echlin, P., Lifshin, E., Sawyer, L., and Michael, J. (2003). *Scanning Electron Microscopy and X-ray Microscopy,* Kluwer Academic/Plenum, New York, Chapters 1, 9, and 10, pp. 1–20, 391–449, 453–535.

Good, R. J. (1979). Contact angles and the surface free energy of solids. In R. J. Good and R. R. Stromberg (Eds.), *Surface and Colloid Science, Volume II, Experimental Methods.* Plenum, New York, Chapter 1, pp. 1–27.

Goulas, A. K., Kapasakalidis, P. G., Sinclair, H. R., Rastall, R. A., and Grandison, A. S. (2002). Purification of oligosaccharides by nanofiltration. *J. Membr. Sci.* 209, 321.

Griffits, P. R., and de Haseth, J. A. (1986). *Fourier Transform Infared Spectrometry.* Wiley, New York.

Gruger, A., Régis, A., Schmatko, T., and Colomban, P. (2001). Nanostructure of Nafion® membranes at different states of hydration, an IR and Raman study. *Vib. Spectrosc.* 26, 215.

Hagenhoff, B. (1995). Surface mass spectroscopy—Application to biosensor characterisation. *Biosens. Bioelectron.* 10, 885.

Hayama, M., Miyasaka, T., Mochizuki, S., Asahara, H., Yamamoto, K.-I., Kohori, F., Tsujioka, K., and Sakai, K. (2003). Optimum dialysis membrane for endotoxin blocking. *J. Membr. Sci.* 219, 15.

Hemsley, D. A. (1989). Basic light microscopy and phase contrast microscopy. In D. A. Hemsley (Ed.), *Applied Polymer Light Microscopy.* Elsevier Science Publishers, Essex, England, p. 43.

Hendra, P. J. (2005). Fourier transform Raman Spectroscopy. In J. J. Laserna (Ed.), *Modern Techniques in Raman Spectroscopy.* Wiley, Eastbourne, England, pp. 73–106.

Hernández, A., Calvo, J. I., Prádanos, P., and Palacio, L. (1999). A multidisciplinary approach towards pore size distributions of microporous and mesoporous membranes. In T. S. Sørensen (Ed.),

Surface Chemistry and Electrochemistry of Membranes. Surfactant Science Series, Vol. 79. Marcel Dekker, New York, pp. 40–90.

Hiemenz, P. C. (1986). *Principles of Colloid and Surface Chemistry*, 2nd ed. Marcel Dekker, New York, pp. 287–345.

Hilal, N., Al-Khatib, L., Al-Zoubi, R., and Nigmatullin, R. (2005). Atomic force microscopy study of membranes modified by surface grafting of cationic polyelectrolyte. *Desalination* 184, 45.

Ho, A. K., Pereira, J. M., Dunstan, D. E., Stevens, G. W., and Nyström, M. (1999). Measurement and theoretical modeling of protein mobility through the membranes. *AIChe J. Chem. Eng. Res. Dev.* 45, 1343.

Holmbom, B., and Stenius, P. (2000). Analytical methods. In P. Stenius (Ed.), *Forest Products Chemistry*. Fapet Oy, Jyväskylä, Finland, pp. 107–172.

Hoshi, T., and Kudo, M. (2003). High resolution static imaging by time of flight SIMS. *Appl. Surf. Sci.* 203/204, 818.

Jimbo, T., Tanioka, A., and Minoura, N. (1998). Comparison of surface charge and net charge densities of poly(acrylonitrile) membrane grafted with ionic monomers: Hydrophilic and hydrophobic effects of graft chain. *J. Colloid Interface Sci.* 204, 336.

Kaeselev, B., Kingshott, P., and Jonsson, G. (2002). Influence of the surface structure on the filtration performance of UV-modified PES membranes. *Desalination* 146, 265.

Kai, T., Tsuru, T., Nakao, S.-I., and Kimura, S. (2000). Preparation of hollow-fiber membranes by plasma-graft filling polymerization for organic-liquid separation. *J. Membr. Sci.* 170, 61.

Kallioinen, M., Reinikainen, S.-P., and Nuortila-Jokinen, J. (2003). Foulant characterisation in membrane filtration in pulp and paper mill applications. In R. Chan and P. R. Neal (Eds.), *Papers and Abstracts IMSTEC'03 5th International Membrane Science and Technology Conference*. UNESCO Centre for Membrane Science and Technology, Sydney, Australia.

Katoh, R., Asano, Y., Furuya, A., Sotoyama, K., and Tomita, M. (1996). Preparation of food emulsions using a membrane emulsification system. *J. Membr. Sci.* 113, 131.

Katon, J. E. (1996). Infrared microspectroscopy. A Review of fundamentals and applications. *Micron* 27, 303.

Keurentjes, J. T. F., Harbrecht, J. G., Brinkman, D., Hanemaaijer, J. H., Cohen Stuart, M. A., and van't Riet, K. (1989). Hydrophobicity measurements of microfiltration and ultrafiltration membranes. *J. Membr. Sci.* 47, 333.

Khatib, K., Rose, J., Barres, O., Stone, W., Bottero, J.-Y., and Anselme, C. (1997). Physico chemical study of fouling mechanisms of ultrafiltration membrane on Biwa lake (Japan). *J. Membr. Sci.* 130, 53.

Khulbe, K. C., Kruczek, B., Chowdury, G., Gagné, S., and Matsuura, T. (1996). Surface morphology of homogenous and asymmetric membranes made from poly(phenylene oxide) by tapping mode atomic force microscopy. *J. Appl. Polym. Sci.* 59, 1151.

Khulbe, K. C., and Matsuura, T. (2000). Characterization of synthetic membranes by Raman spectroscopy, electron spin resonance, and atomic force microscopy; a review. *Polymer* 41, 1917.

Kim, K. J., Dickson, M. R., Fane, A. G., and Fell, C. J. D. (1991). Electron microscopy in synthetic polymer membrane research. *J. Microsc.* 162, 403.

Kim, K.-J., and Fane, A. G. (1994). Low voltage scanning electron microscopy in membrane research. *J. Membr. Sci.* 88, 103.

Kim, K. J., Fane, A. G., Nyström, M., Pihlajamäki, A., Bowen, W. R., and Mukhtar, H. (1996). Evaluation of electroosmosis and streaming potential for measurement of electric charges of polymeric membranes. *J. Membr. Sci.* 116, 149.

Kim, K.-J., Fane, A. G., Nyström, M., and Pihlajamäki, A. (1997). Chemical and electrical characterization of virgin and protein-fouled polycarbonate track-etched membranes by FTIR and streaming-potential measurements. *J. Membr. Sci.* 134, 199.

Kim, Y.-W., Ahn, W. S., Kim, J.-J., and Kim, Y. H. (2005). In situ fabrication of self-transformable and hydrophilic poly(ethylene glycol) derivative-modified polysulfone membranes. *Biomaterials* 26, 2867.

Lianchao, L., Baoguo, W., Huimin, T., Tianlu, C., and Jiping, X. (2006). A novel nanofiltration membrane prepared with PAMAM and TMC by in situ interfacial polymerization on PEK-C ultrafiltration membrane. *J. Membr. Sci.* 269, 84.

Mänttäri, M., Pihlajamäki, A., and Nyström, M. (2002). Comparison of nanofiltration and tight ultrafiltration membranes in the filtration of paper mill process water. *Desalination* 149, 131.

Martín, A., Martinez, F., Malfeito, J., Palacio, L., Prádanos, P., and Hernández, A. (2003). Zeta potential of membranes as a function of pH: Optimization of isoelectric point evaluation. *J. Membr. Sci.* 213, 225.

Masselin, I, Durand-Bourlier, L., Laine, J.-M., Sizaret, P. Y., Chasseray, X., and Lemordant, D. (2001). Membrane characterisation using microscopic image analysis. *J. Membr. Sci.* 186, 85.

Momtaz, M., Dewez, J. L., and Marchand-Brynaert, J. (2005). Chemical reactivity of a poly(vinylidene fluoride) microfiltration membrane ("Durapore DVPP"). *J. Membr. Sci.* 250, 29.

Mukherjee, D., Kulkarni, A., and Gill, W. N. (1996). Chemical treatment for improved performance of reverse osmosis membranes. *Desalination* 104, 239.

Mulder, M. (1996). *Basic Principles of Membrane Technology*, 2nd ed. Kluwer Academic Publishers, Dordrecht, The Netherlands, 1996, pp. 158–204.

Munro, H. S., and Singh, S. (1993). The characterisation of polymer surfaces by XPS and SIMS. In B. J. Hunt and M. I. James (Eds.), *Polymer Characterisation*. Blackie Academic & Professional, Suffolk, Great Britain, pp. 333–354.

Nakagaki, M., and Takagi, R. (1986). Experimental verification of the theory of membrane potential for collodion membranes with asymmetric charge distribution. *Chem. Pharm. Bull.* 34, 957.

Neumann, A. W., and Good, R. J. (1979). Techniques of measuring contact angles. In R. J. Good and R. R. Stromberg (Eds.), *Surface and Colloid Science, Vol. II, Experimental Methods*. Plenum, New York, Chapter 2., pp. 31–87.

Nuortila-Jokinen, J., and Nyström, M. (1996). Comparison of membrane separation processes in the internal purification of paper mill water. *J. Membr. Sci.* 119, 99.

Nyström, M., Lindström, M., and Matthiasson, E. (1989). Streaming potential as a tool in the characterization of ultrafiltration membranes. *Colloids Surf.* 36, 297.

Nyström, M., Pihlajamäki, A., and Ehsani, N. (1994). Characterization of ultrafiltration membranes by simultaneous streaming potential and flux measurements. *J. Membr. Sci.* 87, 245.

Ochoa, N. A., Prádanos, P., Palacio, L., Pagliero, C., Marchese, J., and Hernández, A. (2001). Pore size distributions based on AFM imaging and retention of multidisperse polymer solutes, characterization of polyethersulfone UF membranes with dopes containing different PVP. *J. Membr. Sci.* 187, 227.

Oswald, S., and Reiche, R. (2001). Binding state information from XPS depth profiling: Capabilities and limits. *Appl. Surface Sci.* 179, 307.

Otsuka, T., Koizumi, S., Endo, K., and Chong, D. P. (2002). Theoretical X-ray photoelectron and Auger electron spectra of polymers by density functional theory calculations using model dimmers. *J. Mol. Struct. (Theochem.)* 619, 241.

Palacio, L., Calvo, J. I., Prádanos, P., Hernández, A., Väisänen P., and Nyström, M. (1999). Contact angles and external protein adsorption onto UF membranes. *J. Membr. Sci.* 152, 189.

Pastorczak, M., Wiatrowski, M., Kozanecki, M., Lodzinski, M., and Ulanski, J. (2005). Confocal Raman microscopy in 3-dimensional shape and composition determination of heterogeneous systems. *J. Mol. Struct.* 744–747, 997.

Pihlajamäki, A. (1998). Electrochemical charactersation of filter media properties and their exploitation in enhanced filtration. Ph.D. thesis, Lappeenranta University of Technology, Lappeenranta, Finland, Chapter 2, pp. 2–53.

Pihlajamäki, A., Väisänen, P., and Nyström, M. (1998). Characterization of clean and fouled polymeric ultrafiltration membranes by Fourier transform IR spectroscopy—attenuated total reflection. *Colloids Surf. A* 138, 323.

Pontié., M., Rapenne, S., Thekkedath, A., Duchesne, J., Jacquemet, V., Leparc, J., and Suty, H. (2005). Tools for membrane autopsies and antifouling strategies in seawater feeds: A review. *Desalination* 181, 79.

Puro, L., Mänttäri, M., Pihlajamäki, A., and Nyström, M. (2006). Characterization of modified nanofiltration membranes by octanoic acid permeation and FTIR analysis. *Trans IChemE, Part A,Chem. Eng. Res. Des.* 84(A2), 87.

Reich, Z., Kapon, R., Nevo, R., Pilpel, Y., Zmora, S., and Scolnik, Y. (2001). Scanning force microscopy in the applied biological sciences. *Biotechnol. Adv.* 19, 451.

Reichert, U., Linden, T., Belfort, G., Kula, M.-R., and Thömmes, J. (2002). Visualising protein adsorption to ion-exchange membranes by confocal microscopy. *J. Membr. Sci.* 199, 161.

Reniers, F., and Tewell, C. (2005). New improvements in energy and spatial (x, y, z) resolution in AES and XPS application. *J. Electron. Spectrosc. Relat. Phenom.* 142, 1.

Sah, A., Castricum, H. L., Bliek, A., Blank, D. H. A., and ten Elshof, J. E. (2004). Hydrophobic modification of γ-alumina membranes with organochlorosilanes. *J. Membr. Sci.* 243, 125.

Schaep, J., and Vandecasteele, C. (2001). Evaluating the charge of nanofiltration membranes. *J. Membr. Sci.* 188, 129.

Schossig-Tiedemann, M., and Paul, D. (2001). Improved preparation of membrane surfaces for field-emission scanning electron microscopy. *J. Membr. Sci.* 187, 85.

Shaban, H. (1996). Removal of water from aroma aqueous mixtures using pervaporation processes. *Sep. Technol.* 6, 69.

Sheldon, J. (1991). The fine-structure ultrafiltration membranes. I. Clean membranes. *J. Membr. Sci.* 62, 75.

Smith, E., and Dent, G. (2005). *Modern Raman Spectroscopy—A Practical Approach*, Wiley, West Sussex, England, Chapters 1 and 2, pp. 1–67.

Socrates, G. (1980). *Infrared Characteristics: Group Frequencies, Tables and Charts.* Wiley, West Sussex, England.

Song, W., Ravindran, V., Koel B. E., and Pirbazari, M. (2004). Nanofiltration of natural organic matter with H_2O_2/UV pre-treatment: Fouling mitigation and membrane surface characterization. *J. Membr. Sci.* 241, 143.

Spevack, P., and Deslandes, Y. (1996). ToF-SIMS analysis of adsorbate/membrane interactions. I. Adsorption of dehydroabietic acid on poly(vinylidene fluoride). *Appl. Surf. Sci.* 99, 41.

Stamatialis, D. F., Dias, C. R., and Norberta de Pinho, M. (1999). Atomic force microscopy of dense and asymmetric cellulose-based membranes. *J. Membr. Sci.* 160, 235.

Stenius, P., and T. Vuorinen (1999). Direct characterization of chemical properties of fibers. In E. Sjöström, and R. Alén (Eds.), *Analytical Methods in Wood Chemistry, Pulping and Papermaking*. Springer-Verlag, Berlin and Heidelberg, Germany, pp. 149–189.

Stevenson, C. L., and Vo-Dinh, T. (1996). Signal expressions in Raman spectroscopy. In J. J. Laserna (Ed.), *Modern Techniques in Raman Spectroscopy*. Wiley, Eastbourne, England, pp. 1–38.

Stuart, B. H. (1996). Polymer crystallinity studied using Raman spectroscopy. *Vib. Spectrosc.* 10, 79.

Szymczyk, A., Fievet, P., Aoubiza, B., Simon, C., and Pagetti, J. (1999). An application of the shape charge model to the electrolyte conductivity inside a charged microporous membrane. *J. Membr. Sci.* 161, 275.

Szymczyk, A., Fievet, P., Mullet, M., Reggiani, J. C., and Pagetti, J. (1998). Comparison of two electrokinetic methods—electroosmosis and streaming potential—to determine the zeta-potential of plane ceramic membranes. *J. Membr. Sci* 143, 189.

Takagi, R., Hori, M., Gotoh, K., Tagawa, M., and Nakagaki, M. (2000). Donnan potential and ζ-potential of cellulose acetate membrane in aqueous sodium chloride solutions. *J. Membr. Sci.* 170, 19.

Thibault, C., Huguet, P., Sistat, P., and Pourcelly, G. (2002). Confocal Raman microspectroscopy and electrochemical investigation of anion transport through ion-exchange membranes. *Desalination* 49, 149.

Tomandl, T., Mangler, M., Pippel, E., and Woltersdorf, J. (2000). Evidence of nanopores in sol-gel TiO_2 and TiN ultrafiltration membranes. *Mater. Chem. Phys.* 63, 139.

Turrel, G., and Dhamelincourt, P. (2005). Micro-Raman spectroscopy. In J. J. Laserna, (Ed.), *Modern Techniques in Raman Spectroscopy*. Wiley, Eastbourne, England, pp. 109–140.

Väisänen, P., and Nyström, M. (1997). Comparison of polysulfone membranes and polysulfone films. *Acta Polytech. Scand.*, Ch. 247, pp. 25–34.

Vaughan, A. S. (1993). Polymer microscopy. In B. J. Hunt and M. I. James (Eds.), *Polymer Characterisation*. Blackie Academic & Professional, Suffolk, England, p. 298.

Vickerman, J. C. (1989). Static secondary ion mass spectrometry. In J. M. Walls (Ed.), *Methods of Surface Analysis, Techniques and Applications*. Cambridge University Press, Cambridge, pp. 169–215.

Vigo, F., Uliana, C., and Lupino, P. (1985). The performance of a rotating module in oily emulsions ultrafiltration. *Sep. Sci. Technol.* 20, 213.

Vrijenhoek, E. M., Hong, S., and Elimelech, M. (2001). Influence of membrane surface properties on initial rate of colloidal fouling of reverse osmosis and nanofiltration membranes. *J. Membr. Sci.* 188, 115.

Wallheinke, K., Heckmann, W., Potschke, P., and Stutz, H. (1998). Localizing compatibilizers in immiscible blends by SEM. *Polym. Test.* 17, 247.

Walls, J. M. (1990). Methods for surface analysis. In J. M. Walls (Ed.), *Methods of Surface Analysis, Techniques and Applications*. Cambridge University Press, Cambridge, pp. 1–19.

Wartewig, S. (2003). *IR and Raman Spectroscopy, Fundamental Processing*, Wiley-VCH Verlag GmbH & Co. KGaA, Weinheim, Germany, pp. 30–32.

Wartewig, S., and Neubert, R. H. H. (2005). Pharmaceutical applications of mid-IR and Raman spectroscopy. *Adv. Drug Deliv. Rev.* 57, 1144.

Wilbert, M. C., Pellegrino, J., and Zydney, A. (1998). Bench-scale testing of surfactant-modified reverse osmosis/nanofiltration membranes. *Desalination* 115, 15.

Yalamanchili, M. R., Veeramasuneni, S., Azevedo, M. A. D., and Miller, J. D. (1998). Use of atomic force microscopy in particle science and technology research. *Colloids Surf. A* 133, 77.

Yu, H. M., Schumacher, J. O., Zobel, M., and Hebling, C. (2005). Analysis of membrane electrode assembly (MEA) by environmental scanning electron microscope (ESEM). *J. Power Source* 145, 216.

Yu, H. M., Ziegler, C., Oszcipok, M., Zobel, M., and Hebling, C. (2006). Hydrophilicity and hydrophobicity study of catalyst layers in proton exchange membrane fuel cells. *Electrochim. Acta* 51, 1199.

Zhang, W. and Hallström, B. (1990). Membrane characterisation using the contact angle technique, I. Methodology of the captive bubble technique. *Desalination* 79, 1.

Zhang, W., Wahlgren, M., and Sivik, B. (1989). Membrane characterisation by the contact angle technique, II. Characterization of UF-membranes and comparison between the captive bubble and sessile drop methods to obtain water contact angles. *Desalination* 72, 263.

Zhao, Y., Xing, W., Xu, N., and Wong, F.-S. (2005). Effects of inorganic electrolytes on zeta potentials of ceramic microfiltration membranes. *Sep. Purif. Technol.* 42, 117.

Zhao, C., Zhou, X., and Yue, Y. (2000). Determination of pore size distribution on the surface of hollow-fiber filtration membranes: A review of methods. *Desalination* 129, 107.

CHAPTER 33

Membrane Characterization by Ultrasonic Time-Domain Reflectometry

WILLIAM B. KRANTZ

Department of Chemical and Biomolecular Engineering, National University of Singapore, The Republic of Singapore, 117576

ALAN R. GREENBERG

Department of Mechanical Engineering, University of Colorado, Boulder, Colorado 80309-0427

33.1 INTRODUCTION

There is a continuing need for better ways to characterize membranes and membrane processes. In particular, noninvasive, real-time techniques are needed for characterizing membrane fouling, cleaning, module hydrodynamics, compaction, formation, morphology, and quality control. There is a particular need for characterization techniques that can be readily adapted to commercial membrane modules and processes.

Acoustic wave transmission and reflection have been used extensively throughout the twentieth century for nondestructive testing. In the latter part of the twentieth century ultrasound techniques were developed for the noninvasive imaging of fetuses and internal organs in the health-care industry. In 1995 the first application of ultrasonic time-domain reflectometry (UTDR) for characterizing membrane processes was reported (Bond et al., 1995). This chapter will provide an overview on the developments in applying UTDR for membrane and membrane process characterization.

This chapter is organized as follows. Section 33.2 provides an overview of the UTDR measurement principle. Section 33.3 considers the application of UTDR to inorganic membrane fouling and module cleaning. Section 33.4 discusses preliminary studies in which scanning acoustic microscopy has been applied to characterizing biofouling. Section 33.5 reviews how UTDR has been used to measure membrane compaction and to develop compaction-resistant membranes. Section 33.6 applies UTDR to studying membrane formation processes. Section 33.7 reviews recent work on applying UTDR for detecting defects and characterizing membrane morphology with potential application to on-line quality control. Section 33.8 provides a prognosis for the use of UTDR characterization in commercial applications that could help develop new membranes and membrane processes.

Advanced Membrane Technology and Applications. Edited by Norman N. Li, Anthony G. Fane, W. S. Winston Ho, and T. Matsuura
Copyright © 2008 John Wiley & Sons, Inc.

33.2 PRINCIPLE OF UTDR MEASUREMENT

This section is intended to provide an overview of the physics of UTDR measurement. The interested reader seeking a more in-depth treatment of this subject is referred to comprehensive reviews such as that of Krautkramer and Krautkramer (1990) and others. UTDR analysis is based on the physical principle that a wave is affected by the media through which it travels. Thus, changes in the transit time, attenuation, scattering, and frequency content of a waveform can provide information on the properties of the media. Ultrasonic waves are created by electrically exciting a piezoelectric transducer that is acoustically coupled in a position that enables wave transmission to the object of interest; *acoustically coupled* implies that the media between the transducer and the object of interest transmit the waveform without undue attenuation. The resulting waveform is characterized by its peak frequency and bandwidth. This waveform is reflected and transmitted at each interface (with appropriate properties) that is encountered within the material being analyzed. Ultrasonic waveforms characterized by a higher peak frequency are more responsive to changes in the media being analyzed but also are more strongly attenuated, which makes resolving their reflections more difficult. Hence, the choice of the peak frequency of the UTDR transducer is a compromise between sensitivity and attenuation. Current UTDR technology for membrane applications uses transducers in the 1- to 100-MHz frequency range that permits noninvasive, real-time characterization with micron-scale resolution.

Because an ultrasonic wave propagates via compression and rarefaction, its velocity c is a property of the material in which it is being transmitted. In general, c increases with the density of the material. For example, $c \cong 1500$ m/s for water, whereas $c \cong 2000$ m/s for a typical polymeric membrane material.

When an ultrasonic wave encounters an interface between two media, reflection, transmission, and mode conversion (e.g., change of phase) can occur. The magnitude of the reflected and transmitted waves is dictated by the difference between the acoustic impedance of the media, $Z_2 - Z_1$, where $Z_i \equiv \rho_i c_i$, and the subscripts 1 and 2 denote the media from which and into which, respectively, the wave is propagating. For an incident wave perpendicular to an interface, the amplitude A of the reflected relative to the incident wave is given by

$$A = \frac{Z_2 - Z_1}{Z_2 + Z_1} \tag{33.1}$$

If $Z_2 > Z_1$ corresponding to $\rho_2 c_2 > \rho_1 c_1$, the reflected wave will be in phase with the incident wave, whereas if $Z_1 < Z_2$ corresponding to $\rho_2 c_2 < \rho_1 c_1$, it will be 180° out-of-phase.

Implementation of the UTDR technique involves appropriately placing one or more transducers such that a transmission path to the object being analyzed is created as shown in the schematic cross section of a membrane module in Figure 33.1a. A viscous coupling agent is applied between the transducer and the surface that it contacts in order to ensure good ultrasonic transmission. UTDR applications in membrane technology thus far have used the pulse–echo mode, whereby the transducer alternates rapidly between transmitting an ultrasonic waveform and receiving its reflection from the various interfaces in the object being analyzed. A storage oscilloscope permits determining the time between transmission and reflection of a waveform. The UTDR data are plotted in terms of the instantaneous voltage of the reflected waveform as a function of time as

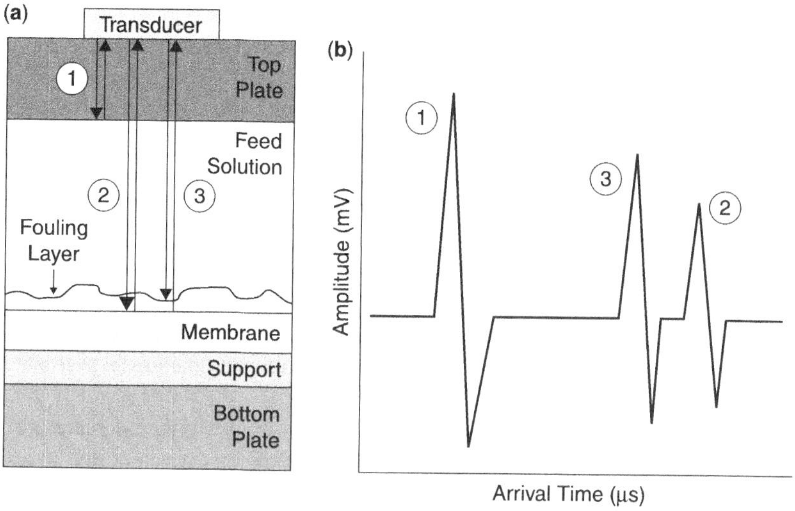

Figure 33.1 (a) Cross section of a liquid-filled membrane cell showing a UTDR transducer and reflected waveforms from the top plate (1), membrane (2), and fouling layer (3). (b) Reflected waveform amplitude as a function of its arrival time showing the relative displacements of reflections (1), (2), and (3).

shown in Figure 33.1b. Information can be obtained from the arrival time, amplitude, and phase modulation of the reflected waves. For example, reflection 1 from the interface between the top plate of the membrane cell and the feed solution will be detected at a shorter time than reflection 2 from the interface between the feed solution and the membrane as shown in Figure 33.1b. However, fouling deposits on the membrane surface will decrease the arrival time as shown by reflection 3 in Figure 33.1b. Hard fouling deposits will increase the amplitude of the reflected wave as shown in the right panel, whereas soft deposits or rough surfaces that cause diffuse reflection will decrease the amplitude. The arrival time T for any primary reflection is related to Δd, the distance between the reflecting interface and the transducer, as follows:

$$T = \frac{2\Delta d}{c} \tag{33.2}$$

Since reflection 1 provides a stationary reference point, the arrival times of reflections 2 and 3 can be used to determine the thickness of a fouling layer or the amount of membrane compaction.

One must select the sampling rate and total display time for the storage oscilloscope; the former determines how often data are taken, whereas the latter is the total sampling time. The display time is dictated by the thickness and speed of sound in the various media within the sample thickness of interest. The accuracy of UTDR depends on the ability to resolve the reflected signal, which in turn depends on the sampling rate and signal amplitude. The latter depends on the peak frequency of the transducer, the attenuation of the acoustic waveform, the thickness and physical properties of the material being analyzed, and the type of UTDR transducer being employed. Increased resolution can also be obtained by displaying the reflected signal in the frequency or Fourier transform domain (ultrasonic frequency-domain reflectometry, UFDR) that separates adjacent peaks.

33.3 CHARACTERIZATION OF INORGANIC MEMBRANE FOULING

Fouling refers to the deposition of dissolved solutes, particulates, and colloids on a membrane. Fouling usually is inferred from a decline in permeate flux and quality. However, permeate flux and quality can decline for reasons other than fouling, such as concentration polarization or membrane compaction. Moreover, permeate flux and quality do not provide any information on local conditions in a membrane module; that is, one might want to arrest fouling at its inception, which usually occurs before significant changes in permeate flux or quality occur.

A major breakthrough in the analysis of membrane fouling was the adaptation of UTDR for noninvasively studying membrane processes in real time that was reported in a series of studies that emanated from the NSF Center for Membrane Applied Science and Technology (MAST) at the University of Colorado (Mairal et al., 1999, 2000; Chai et al., 2001; Chai et al., 2003; Greenberg and Krantz, 2003; Zhang et al. 2003). Other research groups soon recognized the power of UTDR for studying membrane fouling (Sanderson et al., 2002; Li and Sanderson, 2002; Li et al., 2002a,b,c, 2003). However, most of these studies involved flat-sheet membranes; only Chai et al. (2001) and Zhang et al. (2003) have applied UTDR to the more industrially important spiral-wound modules.

Applying UTDR to a spiral-wound module is complicated by multiple reflections that occur from each of the interfaces associated with the wrapping material and concentric membrane, membrane support, and channel-spacer layers as shown in the schematic cross section in Figure 33.2. Two methods that have been developed to address this

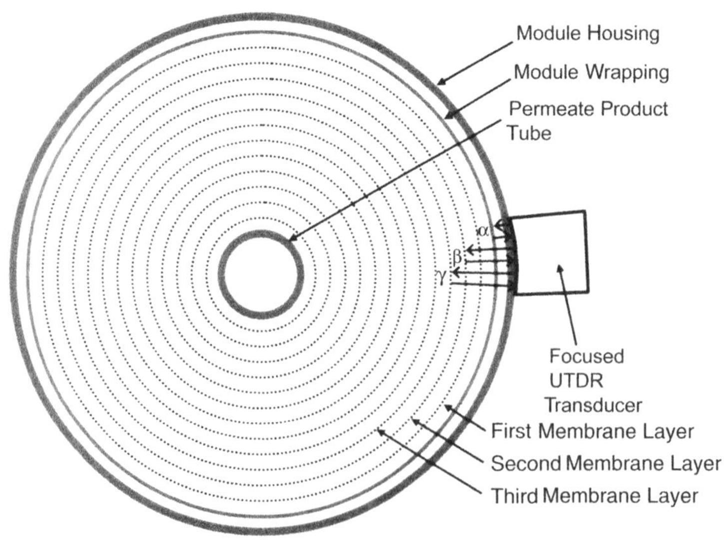

Figure 33.2 Cross section of a spiral-wound membrane module having 12 membrane layers and an outer wrapping that is encased in a housing onto which a UTDR transducer is attached; schematic shows reflections from the outer wrapping and second and third membrane layers.

problem will be discussed, namely analyzing the reflections from only the outermost layers and using the acoustic signature for the entire waveform response. Each method offers particular advantages for characterizing membranes and membrane processes.

Chai et al. (2001) used UTDR to study calcium-sulfate fouling from aqueous solutions in a 2521 Koch spiral-wound reverse osmosis (RO) membrane module. Only the reflections from the outer wrapping and the second and third membrane layers denoted by α, β, and γ in Figure 33.2 were studied. The UTDR system consisted of a 3.5-MHz transducer having a focal length of 7 cm (Research Institute of Acoustics, Chinese Academy of Science), pulsar receiver (Panametrics 5052 PRX), and digital oscilloscope (Nicolet Pro 50).

Figure 33.3 shows typical permeate flux and UTDR response data for operation at 0.68 MPa, 20°C, and a feed concentration that was varied between 0.8 and 1.6 g/L in order to accelerate fouling. The permeate flux in Figure 33.3a displays a monotonic decline associated with progressive membrane fouling. The markedly different information provided by UTDR is shown in panel Figure 33.3b in which the arrival time for peak β from the second membrane layer in Figure 33.2 is seen to decrease and then level off; its amplitude is seen to increase and then level off. The permeate flux provides an integral

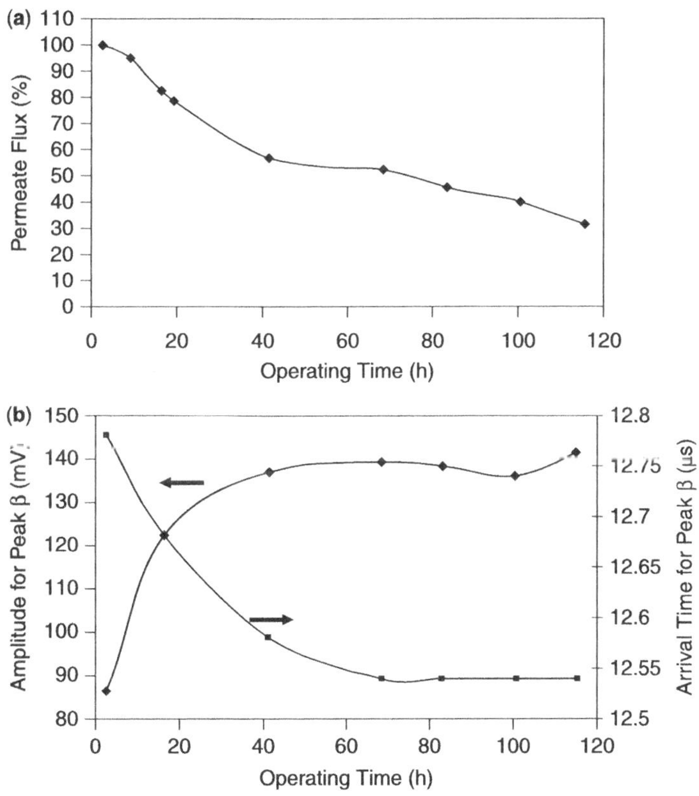

Figure 33.3 Comparison between (a) permeate flux and (b) UTDR signal amplitude and arrival time for peak β in Figure 33.2 as a function of operating time for calcium–sulfate fouling in a Koch 2521 spiral-wound membrane module operating at 0.68 MPa, 20°C, and feed concentration of 0.8–1.6 g/L.

measure of fouling throughout the membrane module. In contrast, UTDR provides a point measurement of local fouling. In this case the UTDR transducer was located near the downstream end of the 2521 module where concentration polarization and fouling would be most pronounced. The UTDR data indicate that the thickness of the local fouling layer increased for approximately 60 h, at which time it had reached its steady-state thickness. However, fouling was occurring elsewhere in the module for the entire 120-h operating time. The spiral-wound module was subsequently unwrapped in order to weigh the fouling deposits on small sections throughout the module. This gravimetric analysis confirmed the UTDR measurements that indicated fouling was more pronounced in the direction of the feed flow and also closer to the permeate-product tube. This study illustrates the advantage of analyzing specific reflection peaks in the UTDR response waveform; that is, this analysis provides information on specific locations within a membrane module. As such, UTDR analysis of specific peak reflections can be used to optimally design membrane modules; for example, to improve the spacer design in the feed channels. These pioneering studies of Chai et al. (2001) led to the invention of a fouling meter that can be installed on commercial flat-sheet, spiral-wound, or hollow-fiber membrane modules to provide real-time, noninvasive, characterization of membrane fouling and cleaning (Bond et al., 2000).

Zhang et al. (2003) developed an alternative method for coping with the complex multiple reflections in spiral-wound membrane modules. They defined an amplitude shift factor Q_A and an arrival time shift factor Q_T that characterize the entire reflected waveform response or selected regions of it as follows:

$$Q_A \equiv \sqrt{\frac{\sum_{j=1}^{j=N} \left(V_{ju} - V_{jr}\right)^2}{N}} \qquad Q_T \equiv \sqrt{\frac{\sum_{j=1}^{j=N} \left(T_{ju} - T_{jr}\right)^2}{N}} \tag{33.3}$$

where V_{ju} and T_{ju} are the UTDR amplitude and arrival time of peak j, respectively, in any reflected waveform of interest, and V_{jr} and T_{jr} are the UTDR amplitude and arrival time of peak j in the reference waveform, which is usually that for the unfouled membrane module. When either of these shift factors is significantly different from zero, it indicates that the membrane module has experienced some change such as might be due to fouling.

Figure 33.4 shows a plot of the percent relative permeation flux (i.e., relative to an unfouled membrane) and the arrival time shift factor as a function of operating time for the fouling and cleaning of a Koch 2521 spiral-wound RO membrane module. The UTDR analysis employed the same apparatus as the earlier work of Chai et al. (2001); however, this study employed two 3.5-MHz transducers on the module housing located at the midpoint and downstream end of the module. The superimposed lines for the UTDR response illustrate the variation in Q_T that results from taking the sum in Eq. (33.3) over different arrival time intervals of the waveform. The storage oscilloscope in combination with a personal computer permitted determining the arrival time and amplitude shift factors in real time. Figure 33.4 shows a time sequence during which pure water was run through the module for 30 h, after which the feed was shifted to a 1.6-g/L aqueous solution of calcium sulfate for 100 h, after which the feed was shifted back to pure water in order to clean the module of fouling deposits. The permeate flux displays the expected behavior of decreasing and leveling-off during the fouling phase and then progressively recovering during the cleaning phase. The arrival time shift factor also tracks the fouling and cleaning phases. However, there is a time delay for the arrival

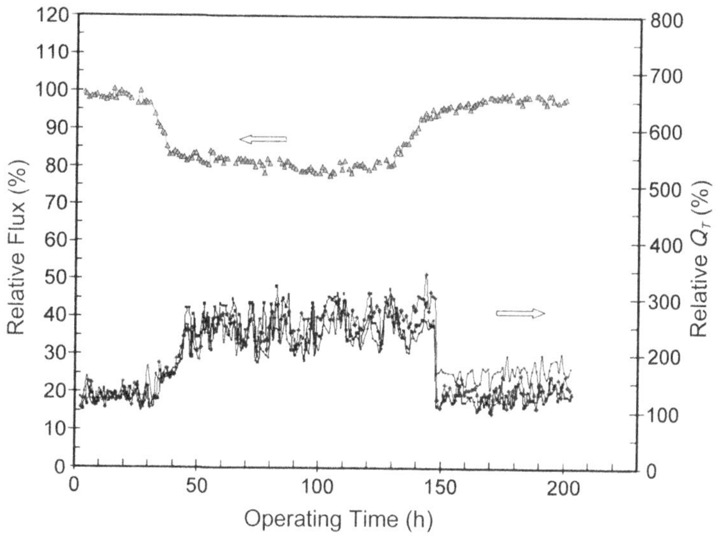

Figure 33.4 Comparison between the permeation flux relative to that for an unfouled membrane (upper data and left ordinate) and the normalized arrival time shift factor (lower data and right ordinate) as a function of operating time for a sequence consisting of pure water feed (0–30 h), fouling with a 1.6-g/L aqueous solution of calcium–sulfate (30–100 h), and cleaning with a pure water feed (100–171 h) in a Koch 2521 spiral-wound membrane module operating at 0.7 MPa and 20°C.

time shift factor response owing to the fact that UTDR provides a local measurement rather than an integral response over the volume of the module. We also see that the module was completely cleaned after 145 h at the axial location where the UTDR data were taken; however, the permeate flux indicates that the entire module was not clean until approximately 170 h. The amplitude shift factor also confirmed similar response behavior.

The arrival time and amplitude shift factors provide convenient metrics for use in combination with the fouling meter patented by Bond et al. (2000). That is, a microprocessor can be incorporated into the fouling meter to determine these two metrics in order to indicate the condition of a membrane module.

33.4 CHARACTERIZATION OF MEMBRANE BIOFOULING

Biofouling is a generic term referring to the flux decline caused by biofilms growing in and on membrane surfaces; this is a serious and extensive problem in the separations industry. The detection of biofilm on membrane surfaces represents a significant challenge for the application of ultrasonic reflectometry due to the lack of impedance contrast at the biofilm–membrane and biofilm–feed solution interfaces. Initial work involving the use of scanning acoustic microscopy to characterize biofilm growth on membrane coupons exposed in a bioreactor was first described by Fonseca (2002). Scanning acoustic microscopy, which employs an immersion transducer, will be described in more detail in Section 33.7, that discusses its use to characterize membrane morphology.

Evans et al. (2005) recently reported significant improvements in the scanning acoustic microscopy technique as well as in the extension of ultrasonic reflectometry to the in situ monitoring of biofilm growth on microfiltration membranes in a cross-flow module. Results suggest that ultrasonic reflectometry can detect biofilm mass on the order of 100 $\mu g/cm^2$, which corresponds to early-stage formation when the biofilm distribution is decidedly nonuniform. In addition, this study shows that the response time for changes in the flux and acoustic measurements are of the same order of magnitude. While these initial studies regarding the use of UTDR are encouraging, much additional work is required in order to statistically relate changes in the ultrasonic signals with biofilm development and membrane performance parameters.

33.5 CHARACTERIZATION OF MEMBRANE COMPACTION

The first application of UTDR to membrane processes was that of Bond et al. (1995) who used it to characterize the compaction of Dow SW-30 membranes. Peterson (1996) and Peterson et al. (1998) used UTDR to study the compaction of both cellulose acetate and Dow BW-30 membranes. The characteristic UTDR response for membrane compaction is the opposite of that for peaks 2 and 3 in Figure 33.1; that is, compaction causes an increase in the time required for the reflection from the membrane surface, since compression moves the membrane–fluid interface away from the transducer.

Compaction data are usually reported in terms of the compressive strain, which is the change in membrane thickness divided by its initial thickness. Figure 33.5 shows a plot of the compressive strain and permeation rate as a function of the time that a constant pressure of 4.1 MPa was applied to a Dow BW-30 flat-sheet membrane through which pure water was permeating. One sees a monotonic decrease in the permeation rate that corroborates with a monotonic increase in the compressive strain over the duration of the test. Of interest is the elastic compressive strain $\cong 0.02$ that occurs instantaneously. This implies that the inherent permeability of the unstressed membrane might not be obtained accurately by extrapolating the permeation rate back to zero time. In the case of a thin-film composite membrane such as the BW-30, most of the compaction is occurring in the thick support layer. Nonetheless, the strong relation between the compaction and permeation data in Figure 33.5 suggests that the high pressure is progressively changing either the functional layer of the membrane or the functional support interface region.

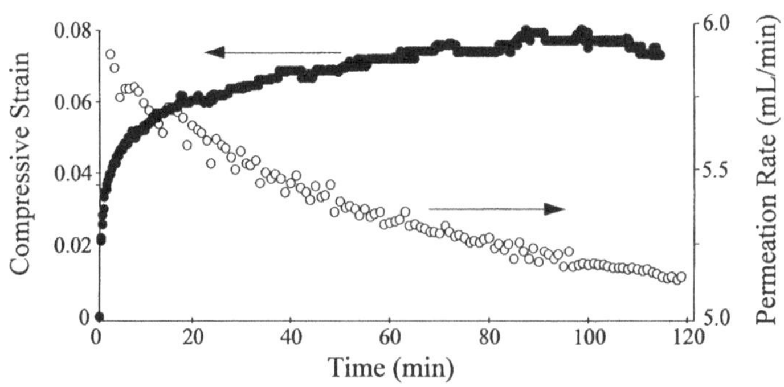

Figure 33.5 Compressive strain (left ordinate) and permeation rate (right ordinate) as a function of operating time for pure water permeation through a Dow BW-30 RO flat-sheet membrane at 4.1 MPa and 23°C.

Peterson et al. (1998) also used UTDR to show that the compressive strain for integral cellulose–acetate membranes has both elastic and inelastic components, only some of which are recoverable; that is, membranes display some degree of nonrecoverable compressive strain when subjected to high pressures. UTDR also revealed that the recovery of these membranes is viscoelastic or time dependent, which possibly explains the hysteresis effects observed when membranes are subjected to pressure cycling.

A potentially very useful application of UTDR is to design membranes that offer improved resistance to compaction. Kelley et al. (2002) studied the effect of cross-linking on the compaction resistance of cellulose–acetate membranes. Figure 33.6 shows the compressive strain as a function of time for pure water permeation through a cellulose–acetate membrane at 4.1 MHz that has been exposed for different periods of time at 23°C to a titanium–isopropoxide cross-linking agent. Sufficient cross-linking time can reduce the compressive strain by 65% and nearly totally eliminate the elastic compaction.

Aerts et al. (2001) used UTDR to study the effect of the filler loading on the compaction characteristics of Zirfon® membranes; the latter are asymmetric polysulfone membranes (~0.2 μm dense layer and ~340 μm substrate) containing filler particles (~0.9 μm) of zirconia. The effect of filler loading on the permeability and compressive strain for pure water permeation through these membranes at 6.3 MPa is shown in Figure 33.7. The permeability of the pure polysulfone polymer in these Zirfon® membranes is vanishingly small but increases dramatically with filler content owing to an increase in the porosity of the functional layer as inferred via field emission scanning electron microscopy (FESEM). Increasing the filler content, which has a significantly higher Young's modulus relative to the polymer matrix, reduces the elastic as well as the long-term compressive strain. However, it also imparts more pronounced viscoelastic time-dependent behavior. The reduction in strain occurs over a much longer time scale than does the decrease in permeability. This is caused by the effect of the transmembrane pressure on progressively collapsing the voids between the filler and polymer matrix. The permeability on the other hand is determined primarily by the functional layer that is too thin to contain any filler particles; hence, it responds more quickly to the applied pressure. These studies demonstrate that UTDR in combination with FESEM can be used to ascertain the effects

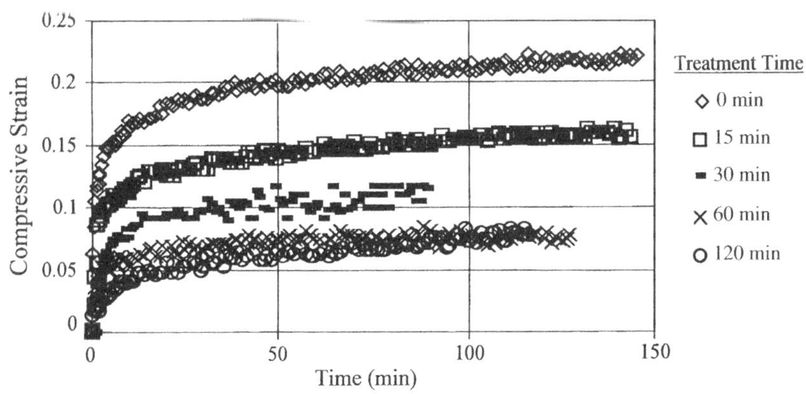

Figure 33.6 Compressive strain as a function of time for pure water permeation through a cellulose acetate membrane at 4.1 MHz that has been exposed for different periods of time at 23°C to a titanium–isopropoxide cross-linking agent.

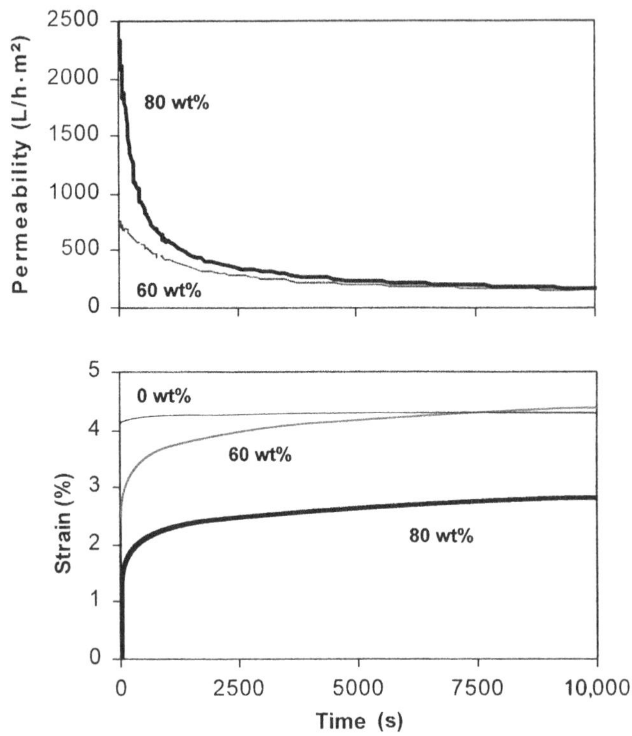

Figure 33.7 Permeability and compressive strain as a function of time for the permeation of pure water through a Zirfon® composite UF membrane at 6.3 MPa and 21°C showing the effect of the zirconia filler content expressed as wt%.

of filler content on both the functional layer and substrate. As such, UTDR provides a valuable tool for designing improved and new membranes with properties optimized for specific applications.

Compaction of polymeric membranes also occurs during gas separation. Reinsch et al. (2000) described the use of UTDR to measure compaction of 175-μm-thick (with backing) asymmetric cellulose–acetate gas separation membranes provided by Grace Davison (Littleton, CO). Figure 33.8 shows a schematic of the membrane cell used in these characterization studies of membrane compaction during gas separation and the primary reflections of acoustic waves A and B, which correspond to the cell top-plate–gas interface and the gas–membrane interface, respectively. Compaction was studied as a function of feed gas pressure and composition. Figure 33.9 shows a plot of the membrane strain as a function of time for compaction at a transmembrane nitrogen gas pressure difference 2.8 MPa followed by a recovery cycle at atmospheric pressure for a commercial asymmetric cellulose–acetate membrane. An instantaneous strain of approximately 13% is observed followed by a small time-dependent strain.

Interestingly, after the transmembrane pressure is reduced to zero, a permanent strain of approximately 6% is observed at a recovery time of 150 min. These studies also indicated that membrane compaction increases with an increasing concentration of CO_2 in the feed stream. The ability to characterize membrane mechanical response to CO_2 is of particular interest owing to the plasticizing nature of this gas.

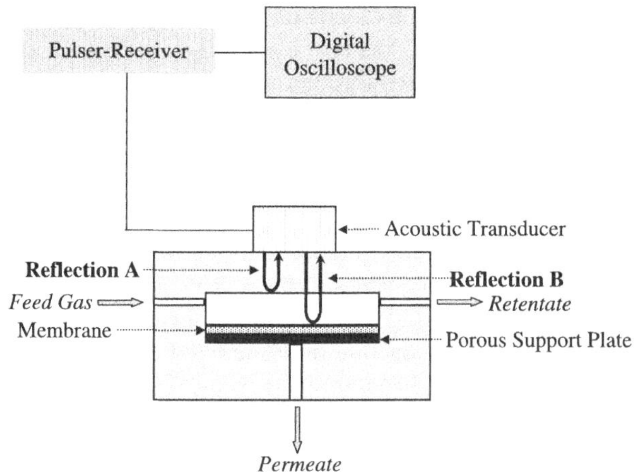

Figure 33.8 Schematic (not-to-scale) of a membrane cell with the externally mounted acoustic transducer used to characterize membrane compaction during gas separation showing the primary reflections of acoustic waves A and B, which correspond to the cell top-plate–gas interface and the gas–membrane interface, respectively.

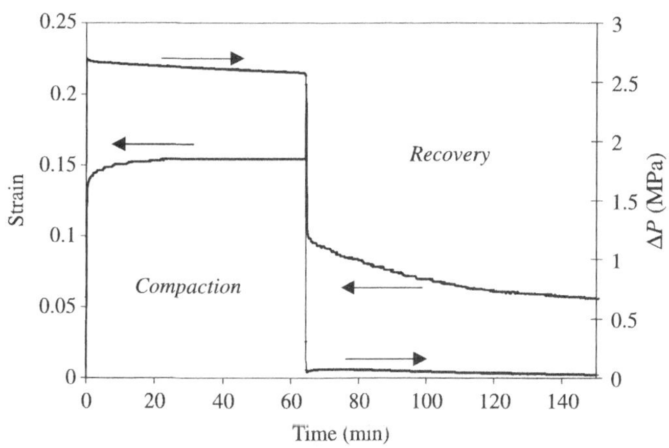

Figure 33.9 Membrane strain and transmembrane pressure difference during a compaction (at a transmembrane pressure difference of 2.8 MPa) and recovery cycle for a commercial asymmetric cellulose–acetate membrane with nitrogen as the feed gas.

33.6 CHARACTERIZATION OF MEMBRANE FORMATION

Characterizing polymeric membrane formation processes is particularly challenging since the casting solutions are very thin (~100–500 μm), and many of the functional properties are typically imparted within a few seconds. The ability of UTDR to provide high-resolution, noninvasive, real-time characterization data is particularly advantageous for this application.

Kools et al. (1998) published the first use of UTDR to study a membrane formation process, namely the evaporative casting of cellulose–acetate from an acetone solution containing water as the pore former. Evaporation of the volatile acetone solvent increases the nonsolvent water concentration, thereby eventually causing phase separation. The phase separation front propagates downward through the casting solution, whose thickness simultaneously decreases owing to evaporation and densification. The manner in which UTDR was implemented to study this evaporative casting process in real time is shown in Figure 33.10. A UTDR transducer was placed on the underside of the aluminum support plate for the casting solution in order to track both the overall casting solution thickness and the phase separation boundary. Figure 33.11 shows the amplitude of a portion of the reflected waveform at evaporation times of 0, 70, 165, and 450 s for a casting solution initially at 23°C and a composition of 10 : 65 : 25 wt% of cellulose acetate, acetone, and water, respectively. Peak A is the reflection from the fixed reference plane at the interface between the casting solution and the support plate; peak B is the reflection from the interface between the phase-separated region and the underlying homogeneous solution; peak C is the reflection from the liquid–gas interface. This application of UTDR again illustrates the advantage of being able to track specific reflections in the waveform response. UTDR is able to track the location of both the upper interface, peak C, as well as the phase separation boundary, peak B, in real time. One observes the progressive decrease in arrival times for both peaks, corresponding to overall thinning of the casting solution and penetration of the phase separation boundary.

The UTDR technique offers considerable promise for obtaining real-time characterization data on other membrane formation processes as well. In particular, UTDR might well provide invaluable characterization data for very rapid membrane formation processes such as wet-casting, thermally induced phase separation (TIPS), and vapor-induced phase separation (VIPS). It might also be used to characterize membrane formation via interfacial polymerization (IP). In this application the thickness of the IP layer is too thin to be followed via the arrival time of the UTDR waves. However, the amplitude of the reflected waves might be used to infer the progressive polymerization and cross-linking that occurs during the IP process.

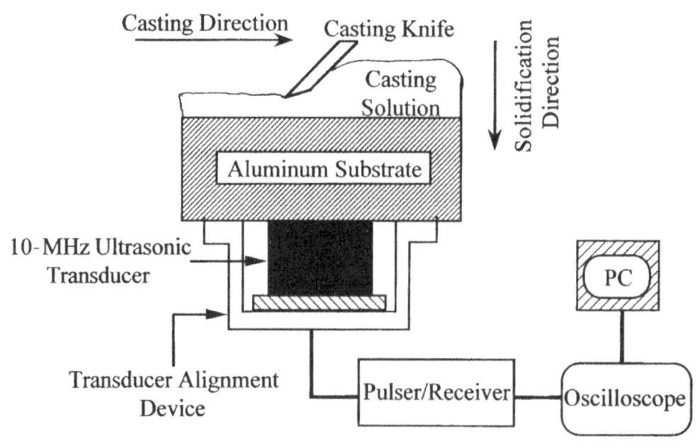

Figure 33.10 Schematic of an apparatus for the evaporative casting of polymeric membranes from a casting solution consisting of cellulose acetate, acetone, and water; the simultaneous downward movement of the phase separation front and liquid–gas interface was monitored with a 10-MHz UTDR transducer placed on the underside of the aluminum support surface.

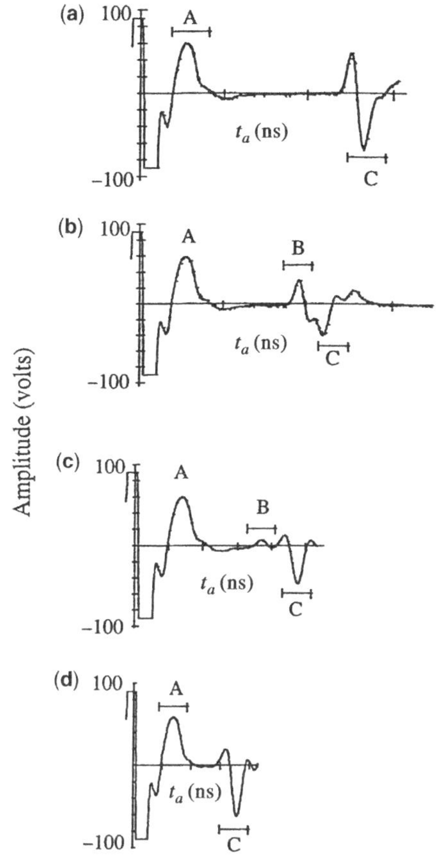

Figure 33.11 UTDR amplitude as a function of arrival time for a portion of the reflected waveform during the evaporative casting of polymeric membranes from a casting solution initially at 23°C and a composition of 10 : 65 : 25 wt% of cellulose–acetate, acetone, and water, respectively, showing the effect of evaporation time: (a) 0 s, (b) 70 s, (c) 165 s, and (d) 450 s; peaks A, B, and C correspond to reflections from the fixed support plate, phase separation boundary, and liquid–gas interface, respectively.

33.7 CHARACTERIZATION OF MEMBRANE MORPHOLOGY

Quality control in membrane manufacturing is of considerable interest. Most commercial membranes are made via continuous casting that involves several processing steps. Ideally, one would like to detect the effects of malfunctioning in any of these steps noninvasively in real time. Clearly, UTDR offers the potential for achieving this goal. The first publication of an attempt to use UTDR to characterize membrane morphology was that of Gómez Álvarez-Arenas (2003). This investigator used highly sensitive air-coupled piezoelectric transducers to relate the acoustic attenuation and velocity to the properties of polymeric microfiltration membranes. However, it was not possible to establish a significant relationship between the acoustic measurements and membrane structure, probably owing to the low-frequency transducers that were employed in this study. A major advancement toward characterizing membrane morphology noninvasively in real time was made by

Ramaswamy et al. (2004) who were the first investigators to use immersion UTDR for membrane characterization. This promising characterization technique will be described in more detail here.

Figure 33.12 illustrates the principles of immersion UTDR. In this characterization technique both the UTDR transducer and the sample are immersed in a water bath that provides uniform coupling to all points. This is particularly advantageous for quality control applications since the signals are not compromised by artifacts introduced by different media between the transducer and points on the sample. Immersion UTDR typically tethers the transducer to a motorized arm that permits accurately translating it in all three directions. Two characteristic reflections are shown in Figure 33.12. Reflection A from the top of the sample provides information on the roughness and pore size at the surface. Reflection B, which passes through the sample twice, provides information on the internal structure of the membrane that is of primary interest for characterizing the membrane morphology. If the transducer frequency is too high, reflection B will be "swamped" by reflection A. Hence, the transducer frequency has to be chosen appropriate to the sample thickness. When one is seeking information on pore size rather than thickness changes such as in fouling or compaction, it is convenient to capitalize on the acoustic scattering that can occur from surfaces. In particular, Rayleigh scattering becomes significant for $D > 0.01\lambda$, where D is the diameter of the scattering site and λ is the acoustic wavelength. Note that the acoustic wavelength λ and frequency ω are related to the acoustic velocity c in the media by the equation

$$\omega\lambda = c \tag{33.4}$$

Hence, to obtain good Rayleigh scattering for microfiltration membranes having characteristic pore sizes in the range of 0.1–1 μm via immersion UTDR, it is necessary to use an acoustic transducer with a frequency in the range of 15–150 MHz (since $c = 1500$ m/s for water). The Rayleigh scattering increases as the fourth power of the diameter of the

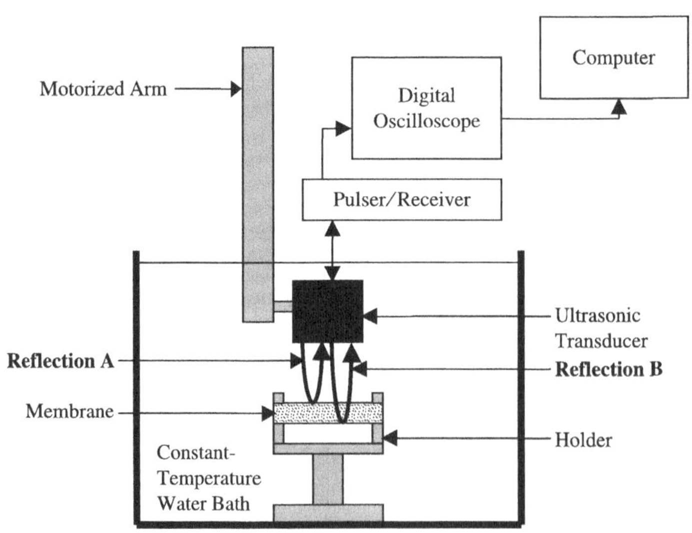

Figure 33.12 Schematic of immersion UTDR whereby both the transducer and sample are coupled directly through water; reflection A characterizes the surface morphology of the sample, whereas reflection B is sensitive to its internal structure.

scattering site. Hence, larger pores will cause greater signal attenuation. This principle then can be used to relate acoustic wave attenuation to pore size distribution in a membrane. Clearly, choosing the proper transducer for immersion UTDR requires a compromise between a sufficiently low frequency to separate the desired waveform reflection (i.e., reflection B from reflection A in Fig. 33.12) and an appropriately high frequency to obtain the required sensitivity.

Ramaswamy (2002) first applied scanning acoustic microscopy for determining the location of pinholes with diameters greater than 100 μm inserted into commercial microporous poly(vinylidene fluoride) (PVDF) membranes as well as for identifying partially penetrating defects with diameters ranging from 100 to 300 μm. Subsequent work also successfully detected the presence of macrovoid defects in laboratory-cast cellulose–acetate membranes.

The goal of later studies of Ramaswamy et al. (2004) was to ascertain whether UTDR could distinguish between the morphology of symmetric microfiltration membranes. They studied Durapore PVDF membranes with nominal pore sizes of 0.1, 0.45, and 0.6 μm and mixed cellulose–ester membranes having pore sizes of 0.1, 0.2, and 0.45 μm, all of which were obtained from the Millipore Corporation. Ramaswamy et al. (2004) used a 90-MHz transducer having a 13-mm spherical focal length (Panametrics, Inc., Model No. V3512), which achieved a reasonable compromise between separating the waveform reflections and obtaining adequate sensitivity.

Figure 33.13 shows the amplitude of reflection B in Figure 33.12 as a function of frequency for the three PVDF microfiltration membranes. One observes that the amplitude decreases (i.e., attenuation increases) as the nominal pore size increases. A similar trend was observed for the mixed cellulose–ester membranes. Ramaswamy et al. (2004) corroborated their UTDR measurements with SEM and porometry analyses that confirmed that there were substantive differences in the structure of these membranes. They also developed a simple neural network model in which five characteristic frequencies in the UTDR waveform response were used to determine the mean pore size of a membrane. Hence, the

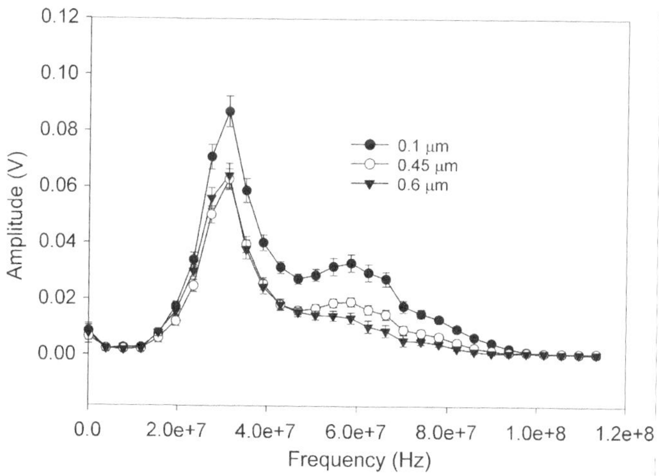

Figure 33.13 Amplitude of reflection B (shown schematically in Fig. 33.12) as a function of frequency in the waveform response for Duropore® PVDF membranes having nominal pore diameters of 0.1, 0.45, and 0.6 μm.

UTDR waveform can be used to detect differences in membrane morphology. Moreover, in combination with a technique for data analysis such as neural networks, it might be used for on-line noninvasive quality control during membrane manufacture.

33.8 SUMMARY AND RECOMMENDATIONS

The prior sections of this chapter have focused on background information on the UTDR technique and on how it has been applied for characterizing membranes and membrane processes. In this section we will suggest applications for which UTDR might provide characterization data of particular value to the membrane industry. We also will indicate developments that will permit UTDR to become an even more discriminating and useful characterization tool. Finally, we will provide our recommendations as to what is needed to make it possible for the membrane industry to capitalize on the power of UTDR to provide valuable characterization data to improve its products and processes.

Hopefully, this chapter has convinced the reader that UTDR offers unique advantages relative to other characterization techniques for membranes and membrane processes. In particular, it is a noninvasive technique that can provide micron-scale resolution of spatial features in real time. Moreover, it is a relatively low-cost technology that can be readily adapted to commercial membrane modules.

Section 33.3 should have provided convincing evidence that UTDR is capable of noninvasive, real-time characterization of membrane fouling and cleaning. This capability should now be put to use in developing optimal permeation and cleaning cycles for commercial membrane modules and systems involving liquid separations. It also can be used to develop improved and new fouling remediation methods. UTDR is particularly useful in this context since it provides local or point information on the inception of fouling rather than a metric such as permeation flux that is a measure of the overall module performance. The studies of Chai et al. (2001) established that UTDR can determine where fouling begins and develops more rapidly in a membrane module. As such, UTDR can be used to optimize membrane module construction via better spacer design and improved hydrodynamics.

Applying UTDR to characterize biofouling is particularly challenging owing to the small differences in acoustic impedance generally encountered in this application. However, the recent studies of Evans et al. (2005) summarized in Section 33.4 have demonstrated that this indeed can be done. Clearly, more effort needs to be devoted to this important application of UTDR.

The studies of Aerts et al. (2001) and Kelley et al. (2002) discussed in Section 33.5 demonstrated that UTDR can be used to determine the effectiveness of fillers and cross-linking to improve the compaction resistance of membranes. This application needs to be explored by membrane manufacturers in order to create a new generation of more robust membranes.

Section 33.6 reviewed the work of Kools et al. (1998) who published the only study to date of the use of UTDR to characterize membrane formation processes. This powerful application of UTDR should be used to study other membrane formation processes such as wet-casting, thermally induced phase separation, vapor-induced phase separation, and interfacial polymerization. Indeed, UTDR characterization can provide invaluable insight into precisely how membranes acquire their unique permselective properties during the fabrication process. This knowledge in turn will help develop improved and new membranes.

It is surprising in view of the classical use of UTDR for nondestructive testing and quality control for many years that it has not found use for this purpose in membrane technology until recently. The pioneering studies of Ramaswamy (2002) and Ramaswamy et al. (2004) reviewed in Section 33.7 demonstrated that UTDR can detect pinhole defects and can discriminate between the structures of microfiltration membranes having different nominal pore sizes. More effort needs to be devoted to adapting UTDR as an on-line quality control tool and to using powerful techniques such as neural networks for relating the acoustic waveform response to membrane structural characteristics.

Rapid advances continue to be made in both the hardware and software technology that is used in UTDR. Transducers are now available with a wide range of focal lengths and frequencies ranging from kilohertz to gigahertz. Moreover, materials are being developed that permit using these acoustic transducers in hostile environments. Progressively faster storage oscilloscopes are being developed that permit significantly improved data acquisition and analysis. Improved signal analysis software lends itself well for use in microprocessors that will permit deconvoluting and extracting more information from the complex UTDR waveforms. Clearly, rapid advances in the supporting technology for UTDR will enhance its utility for the applications discussed in this chapter and suggest new applications for it in the membrane industry.

It is interesting to note that the open literature indicates that the application of UTDR characterization to membrane science and technology has taken place entirely in university laboratories. Relatively few of these studies have applied UTDR characterization to commercial membrane modules. Moreover, UTDR has been used on only small-scale commercial modules such as the 2521 spiral-wound module, which can be studied in the laboratory. Clearly, only industry has the capability to explore the potential of UTDR for large-scale membrane modules and processes. Hence, the authors strongly recommend cooperative efforts between industry and university researchers in order to utilize the full power of UTDR as a characterization tool for developing improved membranes and membrane processes.

Symbols List

A	UTDR waveform amplitude, V
c	phase velocity of acoustic wave, m/s
D	pore diameter, m
N	number of peaks included in summation to determine shift factor, dimensionless
Q_A	amplitude shift factor normalized with respect to the amplitude shift factor for an unfouled membrane module, dimensionless
Q_T	arrival time shift factor normalized with respect to the arrival time shift factor for an unfouled membrane module, dimensionless
T	arrival time, s
V	voltage, V
Z	acoustic impedance, $kg/m^2 \cdot s$
Δd	distance traversed to a reflecting surface by an acoustic wave, m
λ	wavelength of acoustic wave, m
ρ	mass density, kg/m^3
ω	frequency of acoustic wave, s^{-1}

Subscripts	
i	property of material i
j	peak j in UTDR waveform
r	reference waveform, usually that for the unfouled membrane
1	refers to property of material 1
2	refers to property of material 2

ACKNOWLEDGMENTS

The authors gratefully acknowledge support for much of the UTDR research reported here from the NSF Industry/University Cooperative Research Center (I/U CRC) for Membrane Applied Science and Technology (MAST) at the University of Colorado and the University of Cincinnati (NSF grants ECD-9103095, EEC-9632722 and EEC-0120823). The authors acknowledge Chengbin Guo of the Research Institute of Acoustics, Chinese Academy of Science, who designed and supplied the specialized focused UTDR transducers used in some of this research.

REFERENCES

Aerts, P., Greenberg, A. R., Leysen, R., Krantz, W. B., Reinsch, V. E., and Jacobs, P. A. (2001). The influence of filler concentration on the compaction and filtration properties of Zirfon® composite ultrafiltration membranes. *Sep. Purif. Technol.* 22–23, 663.

Bond, L. J., Chai, G.-Y., Greenberg, A. R., and Krantz, W. B. (2000). Method and apparatus for determining the state of fouling/cleaning of membrane modules. U.S. Patent 6,161,435.

Bond, L. J., Greenberg, A. R., Mairal, A. P., Loest, G., Brewster, J. H., and Krantz, W. B. (1995). Real-time nondestructive characterization of membrane compaction and fouling. In D. O. Thompson and D. E. Chimenti (Eds.), *Review of Progress in Quantitative Nondestructive Evaluation*, Vol. 14. Plenum, New York, p. 1167.

Chai, G.-Y., Greenberg, A. R., and Krantz, W. B. (2003). Study of using in-situ ultrasonic measurement technique in membrane separation. *J. Membr. Sci. Technol. (China)* 23, 135.

Chai, G.-Y., Krantz, W. B., and Greenberg, A. R. (2001). Measurement of membrane fouling in spiral-wound modules. Paper presented at the 6th World Congress of Chemical Engineering, Melbourne, Australia, Sept. 23–27, 2001.

Evans, E. A., Greenberg, A. R., Hernandez, M., and Peterson, M. (2005). Early-stage biofilm detection using acoustic methods. Paper presented at the International Congress on Membranes and Membrane Processes, Seoul, Korea, Aug. 21–26, 2005.

Fonseca, A. C. (2002). Biofouling of nanofiltration membranes and in-situ biofilm detection. Ph.D. Dissertation, University of Colorado, Boulder, CO.

Gómez Álvarez-Arenas, T. E. (2003). Air-coupled ultrasonic spectroscopy for the study of membrane filters. *J. Membr. Sci.* 213, 195.

Greenberg, A. R., and Krantz, W. B. (2003). UTDR-based sensors for real-time quantification of membrane module fouling and cleaning. *Fluid/Particle Sep. J.* 15, 43.

Kelley, S. S., Greenberg, A. R., Filley, J., Peterson, R., and Krantz, W. B. (2002). Chemical modification of cellulose acetate with titanium isopropoxide. *Int. J. Polym. Anal. Characterization* 7, 162.

Kools, W. F. C., Konagurthu, S., Greenberg, A. R., Bond, L. J., Krantz, W.B., van den Boomgaard, T., and Strathmann, H. (1998). Use of ultrasonic time-domain reflectometry for real-time measurement of thickness changes during evaporative casting of polymeric films. *J. Appl. Polym. Sci.* 69, 2013.

Krautkramer, J., and Krautkramer, H. (1990). *Ultrasonic Testing of Materials*, 4th ed. Springer-Verlag, Berlin.

Li, J. X., Hallbauer, D. K., and Sanderson, R. D. (2003). Direct monitoring of membrane fouling and cleaning during ultrafiltration using a non-invasive ultrasonic technique. *J. Membr. Sci.* 215, 33.

Li, J. X., Hallbauer-Zadorozhnaya, V. Y., Hallbauer, D. K., and Sanderson, R.D. (2002a). Cake-layer deposition, growth and compressibility during microfiltration measured and modeled using a non-invasive ultrasonic technique. *Ind. Eng. Chem. Res.* 41, 4106.

Li, J. X., and Sanderson, R. D. (2002). In situ measurement of particle deposition and its removal in microfiltration by ultrasonic time-domain reflectometry. *Desalination* 146, 169.

Li, J. X., Sanderson, R. D., Hallbauer, D. K., and Hallbauer-Zadorozhnaya, V. Y. (2002b). Measurement and modeling of organic fouling deposition in ultrafiltration by ultrasonic transfer signals and reflections. *Desalination* 146, 177.

Li, J. X., Sanderson, R. D., and Jacobs, E. P. (2002c). Non-invasive visualization of the fouling of microfiltration membranes by ultrasonic time-domain reflectometry. *J. Membr. Sci.* 201, 17.

Mairal, A. P., Greenberg, A. R., and Krantz, W. B. (2000). Investigation of membrane fouling and cleaning using ultrasonic time-domain reflectometry. *Desalination* 130, 45.

Mairal, A. P., Greenberg, A. R., Krantz, W. B., and Bond, L. J. (1999). Real-time measurement of inorganic membrane fouling in RO desalination using ultrasonic time-domain reflectometry. *J. Membr. Sci.* 159, 185.

Peterson, R. A. (1996). Use of acoustic TDR to assess the effect of crosslinking on membrane compaction. M.S. Dissertation, University of Colorado, Boulder, CO.

Peterson, R. A., Greenberg, A. R., Bond, L. J., and Krantz, W. B. (1998). Use of ultrasonic TDR for real-time noninvasive measurement of compressive strain during membrane compaction. *Desalination* 116, 115.

Ramaswamy, S. (2002). Development of an ultrasonic technique for the non-invasive characterization of membrane morphology. Ph.D. Dissertation, University of Colorado, Boulder, CO.

Ramaswamy, S., Greenberg, A. R., and Peterson, M. L. (2004). Non-invasive measurement of membrane morphology via UFDR: Pore-size characterization. *J. Membr. Sci.* 239, 143.

Reinsch, V. E., Greenberg, A. R., Kelley, S. S., Perterson, R., and Bond, L. J. (2000). A new technique for the simultaneous, real-time measurement of membrane compaction and performance during exposure to high-pressure gas. *J. Membr. Sci.* 171, 217.

Sanderson, R. D., Li, J. X., Koen, L. J., and Lorenzen, L. (2002). Ultrasonic time-domain reflectometry as a non-destructive visualization technique to monitor fouling and cleaning on reverse osmosis membranes. *J. Membr. Sci.* 207, 105.

Zhang, Zh.-X., Greenberg, A. R., Krantz, W. B., and Chai, G.-Y. (2003). Study of membrane fouling and cleaning in spiral wound modules using ultrasonic time-domain reflectometry. In A. A. Butterfield and D. Bhattacharyya (Eds.), *New Insights into Membrane Science and Technology: Polymeric, Inorganic and Biofunctional Membranes*. Elsevier, Amsterdam, The Netherlands, p. 65.

CHAPTER 34

Microstructural Optimization of Thin Supported Inorganic Membranes for Gas and Water Purification

M. L. MOTTERN, J. Y. SHI, K. SHQAU, D. YU, and HENK VERWEIJ

Department of Materials Science & Engineering, The Ohio State University, Columbus, Ohio 43210-1178

34.1 INTRODUCTION

Inorganic or "ceramic" membranes are generally considered and used because of their thermochemical stability under harsh conditions, their incompressible pore structure, and the possibility of achieving very high fluxes and selectivities in specific processes. Typical applications include high-temperature, high-pressure gas separation, treatment of aggressive liquids, and use in high-temperature reactors with an improved space yield.

Transport of a species through a 100% selective, ideally permeable membrane from the feed to the permeate side at the same chemical potential occurs without energy dissipation. Molecules in the gaseous state have in that case about the same partial pressure at both sides of the membrane; aqueous solutions experience isostatic pressures that differ by the osmotic value. Real membranes are dissipative and often not 100% selective for one species. The selectivity of membranes is generally optimized for a target separation by choosing a certain microstructure and chemical composition. The resistance of the supported structure can generally be diminished by making the membranes as thin as possible on a smooth and strong but permeable porous support structure. However, decreasing the membrane thickness increases the risk of loss of selectivity by pinholes. In addition, the occurrence of surface transfer limitations may result in a characteristic thickness below which further reductions have little effect.

The viability of any membrane type depends critically on its transport characteristics as well as its stability at operational conditions. Inorganic membranes are often considered for their energy-saving potential. Since inorganic membranes are traditionally made via ceramic routes, they tend to have a higher cost than polymer membranes. This has the

Advanced Membrane Technology and Applications. Edited by Norman N. Li, Anthony G. Fane, W. S. Winston Ho, and T. Matsuura
Copyright © 2008 John Wiley & Sons, Inc.

consequence that there is a major emphasis on cost-effective design and implementation of inorganic membranes for practical applications, such as follows:

- Separation of O_2 from air at high temperatures as a reactant in the production of H_2 and chemicals via partial oxidation of (hydro)carbons
- Membrane reactors with improved space yield for dehydrogenation, steam reforming, and water gas shift by selective removal of H_2 from the reaction mixture
- Particulate filtration of aggressive (waste) streams
- Removal of dissolved salts from contaminated, hard and brackish water and possibly seawater by nanofiltration

One of the earliest applications of supported inorganic membranes was developed more than 60 years ago for the separation of $^{235}UF_6$ from $^{238}UF_6$ (Bhave, 1991). Those membranes consisted of a thin layer of partially sintered γ-alumina membranes with $\sim$4-nm connected pores on a coarse porous α-Al_2O_3 multilayer support, as shown in Figure 34.1, left. The isotope separation was based on very small differences in Knudsen gas diffusion rates in the mesoporous membrane structure. While the small Knudsen selectivity was acceptable for the high value-added UF_6 separation, it was certainly not acceptable for more common gas separations. This led, some 20 years ago, to a rapidly growing research activity that resulted in the identification of several promising inorganic membrane materials and concepts (Verweij, 2002). New inorganic membranes were realized as either stand-alone or supported tubular or disk-shaped structures. Current stand-alone membrane compositions are generally dense structures such as thin-walled cylinders of H_2 semipermeable Pd alloys (Lin and Baxbaum, 2000) and dense perovskite ceramic plates and tubes (Bouwmeester and Burggraaf, 1997). To preserve their structural integrity at operational conditions, stand-alone membranes have a minimum thickness of $\sim$20 μm for Pd alloys and up to several millimeters for the dense ceramics.

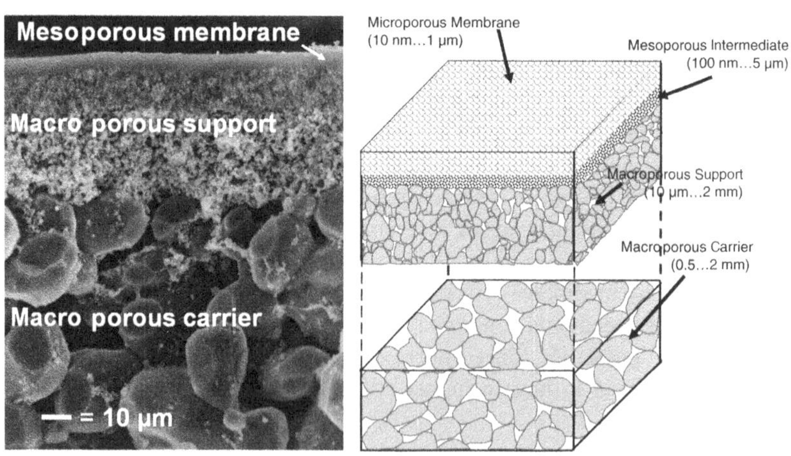

Figure 34.1 (*Left*) Scanning electron microscopy (SEM) microstructure of state-of-the-art supported mesoporous γ-alumina membrane structure. (*Right*) Idealized microstructure of supported zeolite membrane structure; the intermediate layer is a reaction barrier and facilitates thin zeolite membrane deposition. Only part of the support is shown; its structure coarsens slightly toward the permeate side.

Supported membranes are made in a wide variety of compositions with a selective thickness of <10 μm. They are present on a sufficiently strong and permeable porous structure that is compatible with the targeted process conditions. The support structure often consists of more than one layer with, starting from the membrane side, increasing thickness X and pore diameter $Ø_p$, as shown in Figure 34.1, right. The layer that carries a highly selective membrane is often identified as the intermediate layer. That layer is deposited on a support layer that can be thick and stand alone or thinner and supported by a thick, coarse-porous carrier. The membrane–intermediate layer–support/carrier sequence is very common for experimental supported membranes, but other combinations and continuously graded structures are very well possible. The simpler structures are often used for fundamental research purposes. Stand-alone and supported membranes can be made as flat plates and tubular shapes and, more recently, as hollow fibers (Liu and Hughes, 2004; Li et al., 2006) and honeycomb structures (Ceramem, 2006).

Supported membrane concepts that were studied in the past 10 years include 5-μm-thick dense Pd on supported γ-alumina (Pacheco Tanaka et al., 2006), 10-μm-thick dense $La_{0.5}Sr_{0.5}CoO_{3-\delta}$ for O_2 purification (van der Haar, 2001), 60-nm thin amorphous silica for small gas molecule separation and water pervaporation (de Vos and Verweij, 1998), >1 μm-thick MFI zeolite membranes for *para/ortho*-xylene isomer separation (Caro et al., 2000), 0.4–2-μm thick mesoporous γ-alumina (Yu et al., 2006), and 4 μm mesoporous $CoAl_2O_4$ for nanofiltration of liquids (Condom et al., 2006).

Membrane gas transport parameters are obtained for individual layers or as an overall quantity for multilayer stacks. The molar flux j is generally used for gases and expressed as moles per second per unit area A. Permeance quantities are often used for structures where the effective membrane layer thickness is unknown. In addition, they can be useful in comparisons of the transport resistance of individual layers. The molar permeance (or permeation) f is the stationary flux normalized for pressure difference Δp. The molar permeability k is the permeance normalized for thickness. Most dense membrane fluxes scale nonlinearly with $p^f - p^p$ so that f and k are not very useful as a figure of merit. The actual molecular selectivity α_{l_1,l_2} of layers is the ratio of gas permeances, measured for mixtures, present at the feed, f, and the permeate, p, side. Practical values of α_{l_1,l_2} are often calculated from steady-state average composition ratios. However, this value can deviate from the true membrane selectivity because of mass transfer limitation in the boundary layers near the feed and permeate surfaces. It is for this reason that molecular selectivity is often expressed as permselectivity (or ideal selectivity), $\alpha^s_{l_1,l_2}$. This is the ratio of the single gas permeances of l_1 with respect to (w.r.t.) l_2 measured under the same conditions. The selectivity $\alpha^s_{l_1,l_2}$ can be a membrane material's constant but does not account for interaction between species in the membrane separation of mixtures. The volumetric liquid flux j_ℓ is expressed as cubic meters per second per unit area. The mechanical permeance f_ℓ is the stationary flux normalized for dynamic viscosity, η_ℓ, and pressure difference. The mechanical permeability k_ℓ is the mechanical permeance normalized for thickness. The selectivity in liquid separation is expressed in terms of retention or rejection, $R_l = 1 - c_p/c_f$. These definitions can be used in the formulation of a number of interrelated, very generic long-term targets for supported inorganic membranes:

1. $f_l^{tot} > 10^{-5}$ mol/($m^2 \cdot$ s $\cdot$ Pa) and $\alpha^{tot}_{l_1,l_2} > 100$; $f_\ell^{tot} > 10^{-12}$ m and $R_l > 99\%$
2. <50% unfavorable change in these properties after >10.000 h operation with >10 start/stop cycles

3. Regeneration and/or in situ repair capability to restore >90% of the original transport properties after degradation
4. Manufacturing yield >90% with zero hours transport properties, reproducible within 10%
5. Membrane surface per volume >100 m^2/m^3
6. Cost <\$500/m^2

Improved estimates of the quantitative goals may become available for specific applications in cooperative projects. Such projects should have the common goals of high definition, reproducible membrane synthesis routes, accurate descriptions of membrane transport and operational stability, and membrane concepts and designs that are to be optimized for one specific process. Several groups worldwide have made substantial progress in the past 10 years for target 1 in particular and more recently for targets 2, 4, and 5. Target 4 feeds back into almost all the other targets; the inability to realize high-quality supported membrane structures reproducibly has hindered progress for several years. This made us decide to refocus our research program to synthesis and manufacturing routes for multilayer membrane and support structures with a much improved definition of the predesigned microstructure. This chapter provides an overview of related concepts in supported membrane morphology and transport and highlights recent progress in properties and characterization of supported structures.

34.2 MORPHOLOGY, POROSITY, AND DEFECTS

Inorganic membranes and supporting layers are classified into four different groups according to the presence and characteristic scale of connected porosity. Transport of liquids occurs only in connected meso- and macroporous structures with $Ø_p > 2$ nm; liquid transport mechanisms in such structures are described in terms of (tortuous) laminar viscous flow and retention of (charged) species by electrosteric effects. Transport of gas molecules can occur in dense structures as well as porous structures with $Ø_p > 0.5$ nm. These $Ø_p$ effectively define the minimum connected void space that is considered a pore as opposed to "dense" structures. Gas transport of molecules through dense structures often occurs as atomic and electronic species that form from molecules in a surface transfer reaction. Molecular gas transport in porous structures occurs by at least three largely different mechanisms, depending on the characteristic length scale of $Ø_p$. Morphological characteristics and typical properties of the four different groups are discussed in more detail in the next four sections.

34.2.1 Dense Structures

In dense structures there is no intentional "void space" with dimensions much larger than the atomic building blocks. Zeolites that may have a large structural void space in their crystal structure are generally not considered dense structures. Dense membranes generally have an ideal selectivity in which only one species is transported. The separative layer in supported polymer membranes is also considered dense, even though it may not be 100% selective for one component. In dense inorganic membrane transport, molecules such as H_2 and O_2 are converted at the surface into (smaller) atomic, ionic, and/or

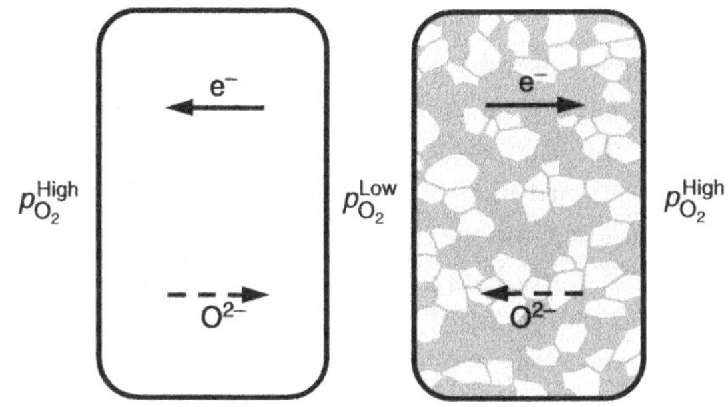

Figure 34.2 Net O_2 transport through dense and composite membranes in response to gradient in O_2 chemical potential expressed in terms of effective p_{O_2}. Upon entering/exiting the solid structure O_2 is converted into/formed from O^{2-} and electrons. These are transported simultaneously in a single mixed-conducting phase or a dual-phase bipercolative composite. Transport may be governed by either bulk phase or surface transfer limitations.

electronic species that have sufficient mobility in the membrane; see Figure 34.2. The most important groups of dense inorganic membrane materials are as follows:

- Palladium and Pd alloys that exhibit a selective net H_2 transport that occurs internally by interstitial diffusion of H atoms (Lin and Baxbaum, 2000); j_{H_2} of 20-μm thick Pd membranes exceeds 10^{-6} mol/m$^2 \cdot$ s at $T > 300°C$ with $p^f = 700$ kPa and $p^p \rightarrow 0$.
- Perovskites, based on the $SrCeO_{3-\delta}$ composition, in which overall H_2 transport occurs by interstitial vacancy diffusion of H^+ in combination with electronic carrier transport on the Ce site sublattice (Gupta and Lin, 2006).
- Perovskites, based on the $LaCoO_{3-\delta}$ composition, in which O_2 is transported by substitutional vacancy diffusion of O^{2-} coupled to electronic carrier transport on the Co site sublattice (Bouwmeester and Burggraaf, 1997); j_{O_2} of 10-μm thick supported $La_{0.5}Sr_{0.5}CoO_{3-\delta}$ membranes exceeds 10^{-6} mol/m$^2 \cdot$ s at $T > 750°C$ with $p^f = 200$ kPa and $p^p \rightarrow 0$ (van der Haar, 2001).
- Dual-phase mixtures of an electronic and an ionic conductor. These materials have primarily been explored for O_2 separation, in which case the coupled transport takes place with each species in a different phase (Chen et al., 1996). However, the microstructures so far have been very inhomogeneous, and much work is yet to be done to further develop membrane protoypes based on these materials.

The perovskite compositions contain several dopants to obtain an optimum for electronic and ionic conductivity and thermochemical stability at operational conditions.

34.2.2 Microporous Structures

Microporous structures have an internal network of connected porosity with a typical $0.5 < Ø_p < 2$ nm. This implies that more than one type of molecule fits in the pores. Flux and selectivity are the result of a combined effect of sorption selectivity and mobility of individual molecules. Permeance values, corrected for support resistance, of up to 10^{-5} mol/(m$^2 \cdot$ s $\cdot$ Pa) have been reported (Benes et al., 2000) for microporous membranes. These membranes can have a substantial selectivity for, generally, smaller molecules such as H_2, CO_2, CO, H_2O, N_2, O_2, and light hydrocarbons. Separation of mixtures can be described in terms of a competitive process of diffusion hopping on the connected micropore network. Since the molecules enter the microstructure without breaking any intramolecular bonds, their interfacial transfer rates are likely higher than for dense membranes. At this point there is no conclusive evidence for the occurrence of surface transfer rate limitations in microporous membrane transport. This offers perspectives on achieving very high fluxes by realizing a homogeneous membrane thickness of < 100 nm. The inability of larger molecules to enter the structure at all is generally referred to as "size exclusion." Examples of microporous membrane materials are as follows:

- Sol–gel amorphous silica (de Vos and Verweij, 1998) in which micropores form around templating water or alcohol molecules (Lu et al., 1999). The templating molecules are removed by thermal evaporation, leaving a connected pore system in the silica network with $Ø_p \approx 0.5$ nm. The micropore surface may have a substantial amount of terminating Si–OH or Si–OCH_3 groups.
- Zeolites that have a well-defined crystal structure, in which the structural microporosity can also be formed around templating molecules (Caro et al., 2000). The most common zeolites that have been applied in membranes are sodalite (A), fauyasite (Y), and silicalite and ZSM-5 with respectively the SOD, FAU, and MFI structure types (IZA, 2006).
- Microporous carbon that is generally made by pyrolysis of cross-linked polymers, made as hollow fibers (Saufi and Ismail, 2004).

A comprehensive review of the literature on this subject until 2002 is provided in Lin et al. (2002). A more recent review on the application of silica and zeolite membranes for H_2 separation is provided in Verweij et al. (2006).

34.2.3 Mesoporous Structures

Mesoporous structures have an intentional connected porosity with $2 < Ø_p < 50$ nm. They are used as an intermediate and buffer layer to deposit thin microporous membranes. In addition, mesoporous membranes can be used for gas separation and nanofiltration:

- High-flux, limited-selectivity gas separations rely on (small) differences in Knudsen diffusion fluxes. Mesoporous γ-alumina membranes have also been reported to exhibit a substantial CO_2 flux by surface diffusion on the internal pore surface (Uhlhorn et al., 1992). This mechanism might be of use for the separation of CO_2 and other polar molecules.
- Substantial ion retention can be obtained in pressure-driven permeation of aqueous (polar) solutions through supported mesoporous oxide membranes (Yu et al., 2006). The retention mechanism requires that the internal membrane surface is selectively charged by chemisorption of either positive or negative ions such as H^+, Ca^{2+}, and HSO_4^-. The net surface charge, in turn, results in a space charge of mobile counterions in the pore entrance area that hinders ion entry to a certain extent. Charge-neutral water molecules on the other hand are still transported by pressure-driven viscous flow in the connected mesopore system. The retention increases with the

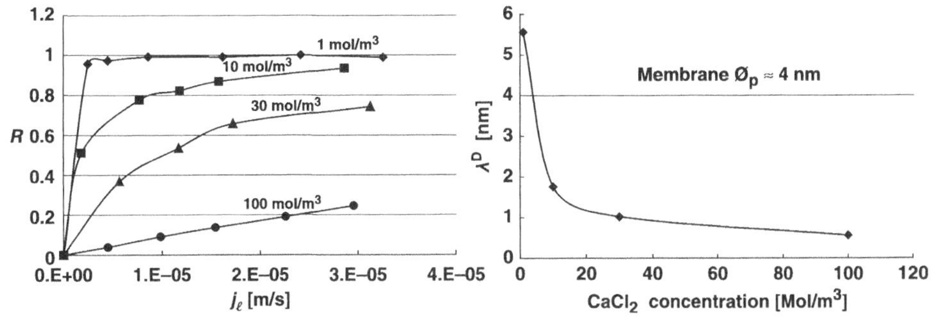

Figure 34.3 (*Left*) $CaCl_2$ rejection of aqueous solution with $p_H = 4.6$ by quasi-homogeneous 800 nm-thick γ-alumina membranes with $\O_p \approx 4$ nm on AKP15 α-Al_2O_3 supports; see Figure 34.11. (*Right*) Debye length λ^D of same solutions versus c_{CaCl_2}.

charge of the co-ions (with the same charge as the pore surface). The change of retention with pressure for these membranes is remarkable in that it appears to increase asymptotically with the permeation flux provided external mass transfer limitations in the laminar boundary layer can be minimized. The retention for a certain type of ion increases with $\lambda_\ell^{D,f}/\O_p$, in which $\lambda_\ell^{D,f}$ is the Debye length in the feed solution. The above-mentioned retention behavior and the effect of the Debye length are demonstrated in Figure 34.3.

- Larger molecules and particulates, dissolved or dispersed in liquid, may exhibit a substantial retention based on a combination of charge *and* steric-hydrodynamic effects. The steric effects may become noticeable at molecule or particle size $>\O_p/10$.

34.2.4 Macroporous Structures

Macroporous structures have a connected porosity with $\O_p > 50$ nm. They are applied as supporting layers or nonselective reaction barriers (Sloot et al., 1990) and in (hot) particulate filtration.

34.2.5 Organized and Random Packing Structures

Structures at all relevant length scales, as described in Sections 34.2.1–34.2.4, can be classified further into organized and random packing structures. For dense materials, structural organization is expressed in the presence (or absence) of a periodic crystal lattice. For microporous materials there is a clear distinction between crystalline zeolites and amorphous silica with a very short range order. Zeolite membranes may consist of a three-dimensional mosaic of crystallites that may be either randomly orientated with respect to each other or possess a certain preferred orientation or texture (Lai et al., 2003). The polycrystalline nature of and presence of texture in zeolite membranes can have important consequences for flux and separation behavior.

Mesoporous structures are made by random packing of initially dispersed nanoparticles followed by thermal processing. In addition, they can be made into periodic structures by surfactant-assisted (sol–gel) formation of an oxide from solutions (Lu et al., 1997). Random-packed macroporous structures are commonly made by ceramic green processing

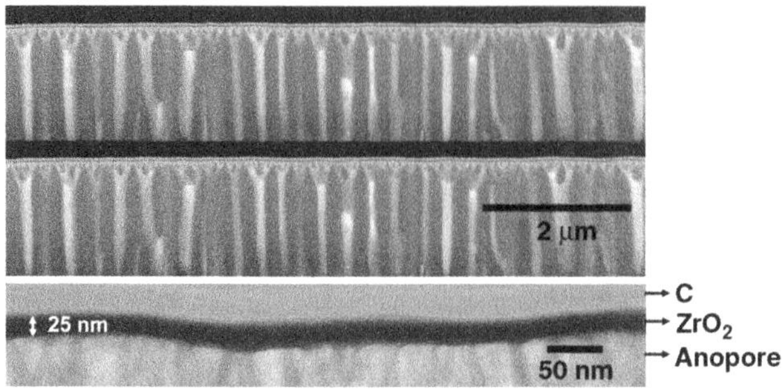

Figure 34.4 Focused ion beam (FIB) TEM cross section of Wattman Anopore support with $Ø_p =$ 200 nm in bulk structure. At the top surface the structure branches out to a finer porosity and carries a ~25 nm-thick ZrO_2 membrane. The C layer was deposited as part of the TEM sample preparation procedure.

of particles in the size range of 100 nm–10 μm. Colloidal dispersions of monosized spheres of similar dimensions can also be packed in periodic hexagonal and cubic close packings (HCP and CCP) (Philipse et al., 1990). In addition, highly organized structures, such as thin flat microsieves, can be produced by modern microfabrication techniques (Kuiper et al., 1998). Controlled anodization of aluminum metal followed by removal of the metal (Hoenicke and Dietzsch, 2002) is used to produce thin disks with a structure as shown in Figure 34.4. Such structures have no tortuosity, which may result in a very low transport resistance and minimal blocking of the internal structure by particulates.

34.2.6 Homogeneity and Defects

All solid materials and structures are inhomogeneous by nature, either at the atomic level or at larger length scales. It is for this reason that for the description of the homogeneity of porous structures we introduced the term "quasi-homogeneous" (Verweij, 2003). A quasi-homogeneous structure has a homogeneous average density in which the averaging length is chosen as

- the lattice constant of periodic structures in the three main directions or
- 3 times the typical particle diameter, $Ø_s$, in random-packed structures.

One of the major challenges in the technology of supported membranes is to avoid deviations from quasi-homogeneity or, more specific, the occurrence of "pinholes" in the selective membrane layer. Pinholes are pores that connect the membrane feed side with the permeate side with a characteristic $Ø_p$ much larger than that of the quasi-homogeneous structure. Pinholes can reduce the overall selectivity in gas separation to the Knudsen value or even less and reduce retention in water purification. The effect of pinholes on dense and microporous membrane gas separation can be very outspoken since the pinhole can reduce the driving force for the selective species very effectively. Figure 34.5 shows how the difference in chemical potential of a species between the feed and the permeate side is effectively removed over an area that is much larger than that of the pinhole. The effect of pinholes in water purification by mesoporous membranes is expected

Figure 34.5 A surface bubble in the support causes a leak in the eventual deposited membrane, which effectively removes H_2 driving force and selectivity over a much larger area, in particular when the support's resistance is similar to that of the membrane.

to be less pronounced since in that case ion retention is induced by flow. The pinholes in an otherwise very flat membrane surface may quickly become plugged by particulates.

In general, any deviation of the quasi-homogeneous structure is identified as a meso- or macrodefect, with a characteristic length scales of <50 or >50 nm, respectively. The defects can consist of void space, material with a different composition and/or deviating porosity. Surface defects in intermediate and supporting layers are easily replicated in thin layers prepared on top of such structures. Deviations of the quasi-homogeneous structure can also be the major cause of thermochemical degradation of supported membranes when operated at target conditions. Unfortunately, the occurrence of pinholes, (surface) defects, and inhomogeneous and disconnected microstructures is very common in state-of-the-art inorganic membrane technology. Figure 34.6 shows an example of how a surface defect in a fairly homogeneous support structure leads to a significant pinhole in a dense membrane. Inspection by TEM of focused ion beam (FIB) cross sections of mesoporous γ-alumina structures that were state-of-the-art in 2000 revealed the presence of low-density structures and other defects, as shown in Figure 34.7, upper. These low-density structures limit the minimal membrane thickness that can be made defect free and cause excessive shrinkage and grain growth during calcination. The shrinkage, in turn, causes interfacial microcracks (shown in Fig. 34.7, lower) and leads to poor adhesion to the support (Nijmeijer et al., 2001).

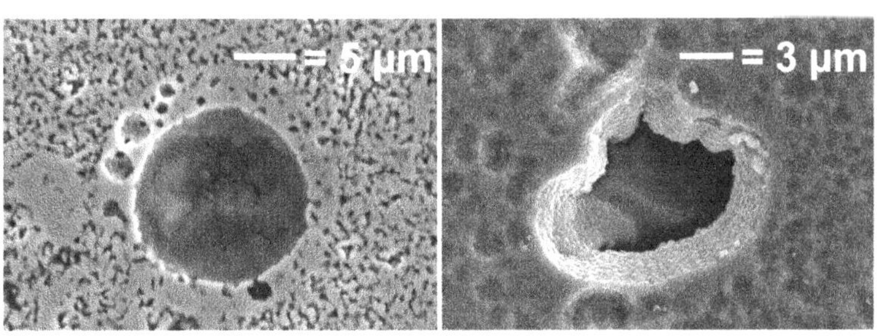

Figure 34.6 (*Left*) Surface SEM of polished $La_{0.5}Sr_{0.5}CoO_{3-\delta}$ support prepared by colloidal filtration of untreated commercial powder followed by sintering at 1020°C. This cross section clearly demonstrates the presence of solid inhomogeneity and a hollow sphere that was formed in spray roasting of the powder. (*Right*) Defect in 10-μm-thick $La_{0.5}Sr_{0.5}CoO_{3-\delta}$ membrane prepared by pulsed laser deposition on support as shown right. A surface defect has caused a large, nonselective pinhole.

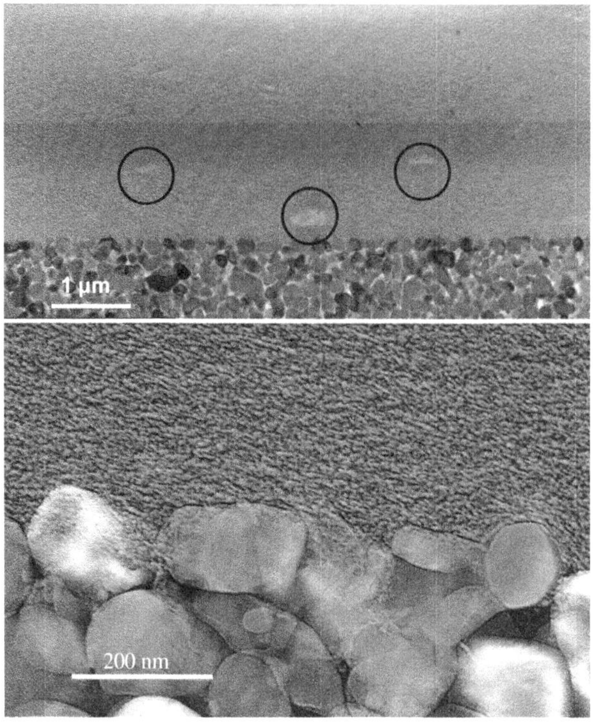

Figure 34.7 (*Upper*) Low-density areas (circles) in transmission electron microscopy (TEM) FIB mesoporous γ-alumina sample with $\varnothing_p = 4$ nm. (*Lower*) Microcracking at support interface caused by excessive shrinkage and poor adhesion. These effects diminish operational lifetime significantly.

34.3 OPTIMIZATION OF SUPPORTED MEMBRANE STRUCTURES

The individual layers in a supported membrane structure may have different and sometimes overlapping functions:

- The thin membrane top layer provides the desired selectivity.
- One or more intermediate layers provide a pore bridge between the dense or fine porous membrane film and the underlying support layer. The intermediate layer may also have the function of a chemical buffer layer to minimize reaction between the membrane and the support.
- The support layer provides strength to the overall structure.

While the membrane is expected to provide most of the selectivity, a fine-porous intermediate-layer structure may also contribute to separation, if it shows Knudsen selectivity for gases, or selective sorption of ions in liquids. This effect may obscure the performance of low-selectivity membranes. Pore bridging is important in minimizing penetration during deposition of membrane material in the support structure. A fine-porous deposition surface also diminishes the risk of spontaneous, surface-tension-induced break-up of the membrane film during operation. The support strength must be such that it allows for handling, sealing, and application of significant pressure differences, up to 10 MPa. Since the membrane and intermediate layers play no role in providing strength to the structure, those layers can be very thin to minimize their transport resistance as long as they maintain their functionality. In gas separation, intermediate layers may not contribute much to the overall resistance, in which case they can be made at a thickness such that pinholes do not form. However, that thickness should remain below 1 μm to minimize mechanochemical degradation and adhesion problems due to thermal expansion mismatches. The support or carrier layer, on the other hand, has primarily a strengthening function that can only be realized if it has a certain thickness, often on the order of millimeters.

The surface decomposition and recombination processes at the gas–solid interface of dense membranes may become rate limiting at a thickness well in excess of 10 μm; see Figure 34.8. As a consequence, thinner dense membranes have received very little attention. The surface transfer mechanism for microporous membranes involves the entire molecule so that microporous membranes may already show selectivity and a very high flux at a thickness on the order of 1 nm. The current minimum reported thickness of microporous membranes is on the order of 20 nm (Lee et al., 2004). However, the apparent layer thickness of membranes and intermediate layers can be increased substantially if part of the material is deposited in the larger pores of the supporting layer. This penetration effect can be controlled by selecting a proper pore size sequence and adjusting deposition conditions (Shi et al., 2006).

The full microstructural functionality and transport properties of quasi-homogeneous random-packed structures are established on a length scale on the order of Ø_s. This implies that these structures may bridge much larger surface pores in a layer that is just a few Ø_s thick. If the layers are made by deposition of particle dispersions, bridges can still form spontaneously if the Ø_s of those particles is larger than one-third of the surface Ø_p of the supporting structure. The formation of homogeneous layers (of smaller particles) may be promoted by the following:

- Pretreatment of the porous support with polymers that (temporarily) block the support pores or modify the surface chemistry during deposition
- Adjusting the deposition rheology by the addition of polymer thickeners; see Figure 34.9

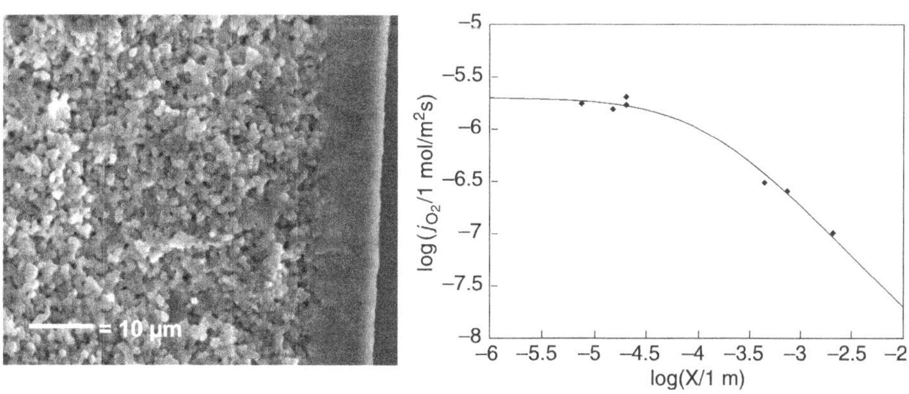

Figure 34.8 (*Left*) A 5 μm-thick $La_{0.5}Sr_{0.5}CoO_{3-\delta}$ membrane prepared by pulsed laser deposition on porous support with ϕ_p as described in Figure 34.6. (*Right*) Membrane thickness dependence of O_2 flux at 900°C, $p^{\text{f}}_{O_2} = 20$ kPa, $p^{\text{p}}_{O_2} = 3$ kPa. The drawn line is based on a transport model with surface transfer becoming as resistive as bulk transport at $X = 50$ μm (van der Haar, 2001).

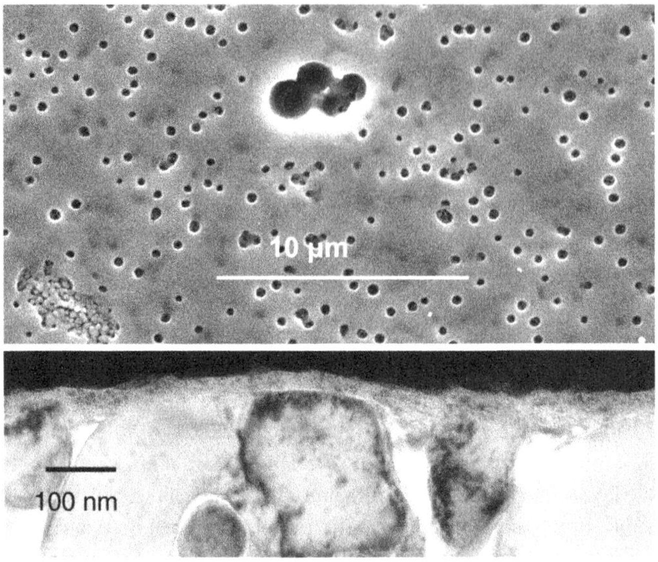

Figure 34.9 (*Upper*) SEM of defects in thin supported γ-alumina layer; the major defect in the top middle is caused by a surface defect in the support; in the lower left is an adhesion defect; the smaller defects are related to a less than optimum deposition viscosity. (*Lower*) FIB TEM of continuous film formed by deposition of $\text{Ø}_p = 10$ nm particles. Layer formation was promoted by the addition of polyvinyl alcohol (PVA) to the aqueous dispersion medium.

A typical example for a thin membrane–intermediate layer–support structure is the microporous silica membrane reported in de Vos and Verweij (1998). The microporous silica membrane layers in that study had $\text{Ø}_p \approx 0.5$ nm and a thickness $X \approx <100$ nm. The intermediate layer consisted of partially sintered γ-alumina with $\text{Ø}_p \approx 4$ nm, $X \approx$ 1 μm, and porosity $\phi_p \approx 0.5$. The macroporous support was made by dry pressing 20 μm Baikowski CR6 alumina granules and sintering at 1250°C to $\text{Ø}_p \approx 160$ nm and $\phi_p \approx 0.4$. The supports were mechanically ground and polished to $X = 2$ mm followed by ultrasonic removal of surface debris. The microstructure of the support–intermediate layer combination was fairly inhomogeneous. However, the observation of a significant H_2 selectivity at a permeance of 10^{-6} mol/(Pa · m^2 · s) indicated that the intermediate layer covered most of the surface defects in the support. The permeance of the most permeable membrane, corrected for support resistance, approached the 10^{-5} mol/(m^2 · s) goal mentioned in the introduction. However, sufficiently strong supporting structures that would result in an overall permeance of 10^{-5} mol/(m^2 · s) are yet to be developed. A recent homogeneous supported amorphous silica structure (on a too-resistive support) is shown in Figure 34.10, left. Comparison of Figure 34.10 left with Figure 34.1 demonstrates a remarkable improvement in microstructural definition in the past 15 years.

34.3.1 Transport Properties

Understanding transport mechanisms in supported membrane structures is important for interpretation of flux and separation mechanism and for the design of structures that are optimized for a particular application. While a highly separative membrane is expected

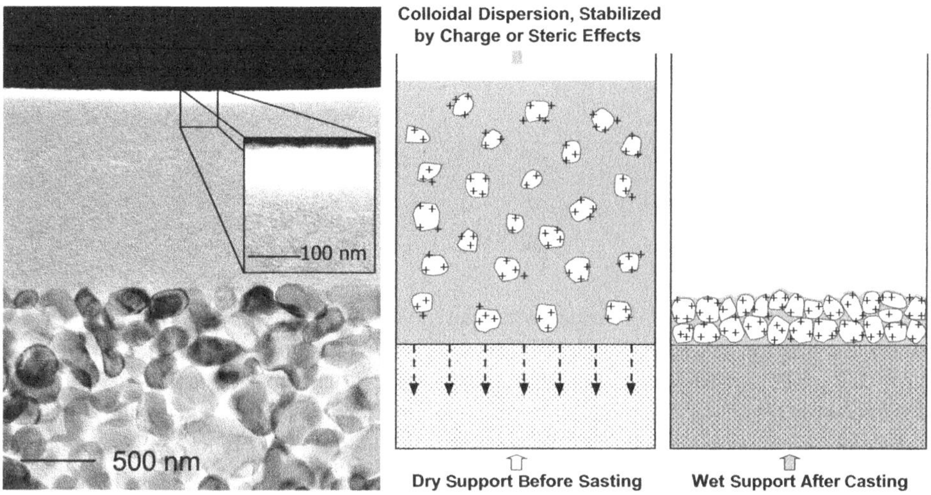

Figure 34.10 (*Left*) FIB TEM of recent supported amorphous silica structure; the silica is just visible as a 100 nm (light gray) layer on top of a 1-mm (darker gray) mesoporous γ-alumina layer. The support is made by partial sintering of dense-packed 300-nm α-Al_2O_3 particles. (*Right*) Three layers (left) are all formed from a colloidally stable dispersion by the filtration as shown.

to be resistive, thick support and carrier structures may provide an equally significant or even dominant contribution to overall transport resistance. The transport mechanisms in dense and microporous layers can be very specific for the membrane material and the separation process (Verweij, 2003). The membrane flux for a species, ℓ, in such structures is commonly described in terms of its thermodynamic driving force, affinity for the membrane structure, and mechanical mobility. The thermodynamic driving force is provided by differences of the chemical potential of species, l, at the feed and the permeate side. For gases, the ideal chemical potential can be used to a good approximation:

$$\mu_{l,g} = \mu_{l,g}^0 + RT \ln\left(\frac{p_l}{p^0}\right) \tag{34.1}$$

The simplest descriptions for dense and microporous membranes predict a scaling of resistance with X. This type of scaling also applies to transport in the pore volume of meso- and macroporous structures. The most important transport mechanisms for different structure types are described in the next sections.

34.3.1.1 Microporous Membrane and Surface Transport A full treatment of microporous membrane transport by site hopping is found in the literature (Verweij, 2003; Verweij et al., 2006) and is summarized below. This treatment is also applicable to the case of surface transport by diffusion hopping on the pore walls of meso- and macroporous structures.

The limited microstructural definition of current membranes justifies the use of a simple competitive Langmuir chemical potential for species l in the membrane phase:

$$\mu_{l,m} = \mu_{l,m}^0 + RT \ln\left(\frac{\theta_l}{1-\theta}\right) \tag{34.2}$$

Since there is little evidence for surface transfer limitations, membrane concentrations at the feed and permeate side can be derived from $\mu_{l,g} = \mu_{l,m}$ to obtain the Langmuir isotherm:

$$\frac{\theta_{l,f|p}}{1-\theta_{f|p}} = \frac{p_{l,f|p}}{p^0}\exp\left(\frac{\mu_{l,g}^0 - \mu_{l,m}^0}{RT}\right) = \frac{p_{l,f|p}}{p^0}\exp\left(-\frac{\Delta s_l^0}{R}\right)\exp\left(\frac{\Delta h_l^0}{RT}\right) \tag{34.3}$$

A high sorption affinity is obtained when the molecule actually fits in the structure, and if that is the case, it has an energetically favorable interaction with the membrane material ($\Delta h_l^0 < 0$). The occurrence of *size exclusion* is mathematically equivalent to $\Delta h_l^0 < 0$. The molecule's mobility, b, is determined by the hopping activation barrier and the availability of vacant neighboring sites. Membrane separation for a binary mixture can be obtained by calculating individual fluxes from diagonal Onsager transport equations. The most generic expression for a quasi-homogeneous Langmuir lattice is

$$j_{l_1} = -\tilde{f}_{L,l_1} b_{l_1}^0 c^{\mathrm{tot}} RT[(1-\theta_{l_2})\nabla\theta_{l_1} + \theta_{l_1}\nabla\theta_{l_2}] \tag{34.4}$$

Expressions, similar to (34.4), can be obtained for other (numbers of) components. Effects of grain boundaries and orientational tortuosity in polycrystalline zeolites may require additional correction factors. In the absence of correlation effects, b_l can be written as

$$b_l = b_l^0(1-\theta) \qquad b_l^0 = b_l^{\mathrm{a}} \exp\left(\frac{-u_l^{\mathrm{h}}}{RT}\right) \tag{34.5}$$

The nonequilibrium correlation factor, $0 < \tilde{f}_{L,l_1} < 1$, provides a correction for the oversimplification in (34.5) that θ of neighboring sites is always statistically distributed. Generally $\tilde{f}_{L,l_1}$ is close to 1 but can become very small for mobile molecules with low θ that percolate on a micropore network that is also occupied by slower molecules with high θ (Benes et al., 2002). This leads to the distinction of two types of microporous or surface diffusion separation:

Type 1 Both molecules have little affinity and hence $\theta \to 0$. This type of behavior is often found for small gas molecules, low pressures, and high temperatures. It results in significant simplification of (34.4) so that $\alpha_{l_1,l_2} = \alpha_{l_1,l_2}^{\mathrm{s}}$ in the absence of external mass transfer limitations.

Type 2 One of the molecules has a much higher affinity (and hence lower mobility). It may occupy >50% of the sites and form a percolative network, nearly impermeable for the other molecule.

Most studies of microporous amorphous silica membranes assume, implicitly or supported by experiments, the occurrence of type 1 behavior. Studies of zeolite membranes, however, often consider separation of hydrocarbons with a much higher affinity and take the possibility of type 2 behavior into account. Type 2 behavior can also be expected with CO_2 and H_2O separation with amorphous silica membranes and may actually lead to preferential transport of these molecules. For single gas and type 1 permeation on a Langmuir lattice, any θ_f, and normal gradients

$$j_l = -b_l^0 c^{\text{tot}} RT \, \nabla\theta_l = -\tilde{D}_l \nabla c_l \tag{34.6}$$

which is Fick's first law with the chemical diffusion coefficient $\tilde{D}_l = b_l^0 RT$, independent of θ. For type 1 behavior $\theta \to 0$ so that (34.3) simplifies to the Henry isotherm, which means that it can easily be combined with (34.5) and (34.6) to obtain for each component,

$$f_j^1 = \frac{j_l}{p_{l,f} - p_{l,p}} = \frac{b_l^a c^{\text{tot}} RT}{p^0 X} \exp\left(-\frac{\Delta s_l^0}{R}\right) \exp\left(\frac{\Delta h_l^0 - u_l^h}{RT}\right) \tag{34.7}$$

Molecules in micropores can be considered as being captured in a potential well with depth $\approx \Delta h_l^0$. This makes that the activation energy u_l^h that is needed for hopping to a neighboring site has a magnitude similar to Δh_l^0. The result is that, contrary to what is the case for dense membranes, single gas and type 1 microporous membranes generally exhibit a fairly weak temperature dependence of (34.7). A slight increase with T of permeance at low p is found for most light gases. On the other hand, f_{CO_2}, for which the Henry approximation may not apply, shows a slight decrease with T. The apparent thermal activation, $\exp(-u^{\text{tot}}/RT)$, of these permeances has little direct physical meaning. A positive u^{tot} is an indication of dominant microporous membrane transport. A negative u^{tot}, however, may be difficult to distinguish from similar behavior for transport in meso- and macropores, as discussed below.

34.3.1.2 Meso- and Macroporous Membrane Gas Transport

Transport descriptions of gases in meso- and macropores are used to establish the effects of pinhole defects in thin microporous membranes and to estimate transport resistance in the supporting structure. The relevant transport mechanisms for gases are Knudsen diffusion, viscous flow, and molecular diffusion. The occurrence of these mechanisms depends on the Knudsen number,

$$\text{Kn}_l = \frac{\lambda_l}{\emptyset_\text{p}} \qquad \lambda_l = \frac{k_B T}{\sqrt{2}\sigma_l p} \tag{34.8}$$

The values of λ_l for simple gases and a wide range of conditions cover all relevant length scales of practical supported membrane structures, as shown in Figure 34.10, left. Knudsen diffusion dominates at $\text{Kn} > 1$ with intrinsic behavior for individual quasi-homogeneous layers,

$$f_l^{\text{Kn}} = \frac{\phi_p \emptyset_\text{p}}{3\tau X} \sqrt{\frac{8}{\pi RTM_l}} \tag{34.9}$$

The Knudsen mechanism is slightly separative with a theoretical separation factor $\alpha_{l_1,l_2}^{\text{Kn}} = \sqrt{M_{l_2}/M_{l_2}}$ equal to 4.7 for H_2/CO_2. This often makes it difficult to distinguish poor microporous separation from Knudsen separation in membrane defects or a resistive support.

For $\text{Kn} < 1$, molecular *viscous* flow and diffusion become the most important mechanisms. While molecular diffusion may contribute to overall transport and separation, it is not considered separately in most analytical treatments. Viscous flow is nonseparative and of particular importance for estimates of support and carrier resistance using

$$f_l^{\mathrm{mf}} \approx \frac{\phi_p \O_{\mathrm{p}}^2}{32\tau X \eta_l} \frac{\bar{p}_l}{RT} \qquad \bar{p}_l = \frac{p_{l,f} + p_{l,p}}{2} \tag{34.10}$$

The permeance of porous structures is often expressed as $f = \alpha + \beta p$ to represent a combination of Knudsen and viscous flow behavior. While such an expression may have little physical significance, it often describes experimental permeance data very well and can be used to analyze membrane defects and the dominant flow mechanism. The dusty gas model (DGM) (Mason, 1983) provides a fairly accurate and comprehensive treatment for the transport of mixtures in meso- and macroporous structures for all flow regimes and mechanisms. Solutions of DGM equations, however, require finite-element methods (Benes et al., 1999).

From (34.9) f_l^{Kn} varies as $\sqrt{1/T}$, and since η of dilute gases varies as $\sqrt{T}$, f_l^{mf} varies as $\sqrt{T^3}$. This implies that either decrease or increase of the experimental values with T may be found. This variation of T can be of the same order as what is found for microporous membranes and much smaller that the generally strong increase of j with T in dense membranes.

34.3.1.3 Meso- and Macroporous Membrane Liquid Transport The volumetric mechanical permeance for pure liquids is obtained as a variation of Darcy's law for both mesoporous and macroporous structures as

$$f_\ell = \frac{\phi_p \O_{\mathrm{p}}^2}{32\tau X} \tag{34.11}$$

This expression assumes ideal sticking of the liquid to the pore wall, and for small mesopores, a correction may be necessary for the occurrence of a <1-nm thin immobile layer. Experimental values of (34.11) are best obtained with pure liquids from the linear relation between j_ℓ and $p^{\mathrm{f}} - p^{\mathrm{p}}$; for solutions that exhibit salt retention, $p^{\mathrm{f}} - p^{\mathrm{p}}$ may need to be corrected for osmotic effects. Calculations of ion retention require a numerical solution involving an irreversible thermodynamic description of multicomponent transport, ion sorption equilibria at the pore wall, and the Navier–Stokes and Poisson equations. An example of such a calculation is provided in de Lint and Benes (2005). More work is needed to properly describe phenomena at the pore entrance and to account for osmotic effects.

34.3.2 Design Considerations for Multilayer Structures

The pressure-independent permeances of (34.7) and (34.9) can be used to calculate the overall permeance for gases in a multilayer sequence with type 1 or Knudsen transport with a simple series resistance expression:

$$\frac{1}{f_l^{\mathrm{tot}}} = \frac{1}{\phi_{m/u_1} f_{l,m}^{1}} + \frac{1}{\phi_{u_1/u_2} f_{l,u_1}^{\mathrm{Kn}}} + \cdots + \frac{1}{f_{l,u_k}^{\mathrm{Kn}}} \tag{34.12}$$

Since in this case different gas species move independent of each other for both type 1 as well as Knudsen transport, (34.12) can be used for single gases as well as mixtures. Because the gas viscous flow permeance (34.10) contains a pressure-dependent term ($\bar{p}_l$), it is not possible to use it in expressions as simple as (34.12). However, for incompressible liquids, an expression similar to (34.12) can also be written for a meso/macroporous multilayer structure independent of η_ℓ:

$$\frac{1}{f_\ell^{\text{tot}}} = \frac{1}{\phi_{m/u_1} f_{\ell,m}} + \frac{1}{\phi_{u_1/u_2} f_{\ell,u_1}} + \cdots + \frac{1}{f_{\ell,u_k}} \tag{34.13}$$

The pore constriction resistance factors ϕ_{m/u_1}, ϕ_{u_1/u_2}, ... account for the fact that all quasi-homogeneous transport in a layer becomes confined to the surface porosity. The actual values of parameters such as ϕ_{m/u_1} can be obtained from finite element effective resistance calculations for specific combinations of ϕ_m, ϕ_{u_1}, and $\O_{u_1}/X_m$. If, for instance, all the membrane material forms as a flat sheet on top of the intermediate layer, $\phi_{m/u} \approx \phi_{u_1}$ for $X_m < X_{u_1}$ while $\phi_{m/u} \rightarrow 1$ for $X_m >_{u_1}$. Penetration of membrane material in the supporting structure may result in a significantly smaller $\phi_{m/u}$. To maximize $\phi_{m/u}$, it is important to apply intermediate structures with a large ϕ_p. Pore constriction effects can also become important for the other layer interfaces when very thin intermediate and support layers are applied.

The differences in scaling behavior for normal gases and liquids at small $\O_p$ have important consequences for the distribution of transport resistance in subsequent layers. This is demonstrated in the next two paragraphs where two typical examples are discussed for the initial optimization of multilayer structures: The resistance of the same mesoporous structure is found to be relatively unimportant for the gas separation application while it provides a major contribution for liquid permeation.

34.3.2.1 Supported Membranes for H_2 Purification It is assumed that a thin microporous or dense membrane can be found that is sufficiently permeable and selective for H_2. From (34.9), a typical intermediate layer with $\phi_p = 0.5$, $\O_p = 4$ nm, $\tau_p = 3$, and $X = 1$ μm is expected to have $f_{H_2}^{Kn} \cong 10^{-4}$ mol/(m$^2 \cdot$ s $\cdot$ Pa) at 200°C so that the resistance of such layers is small compared to the $f_{H_2}^{1} \cong 10^{-5}$ mol/(m$^2 \cdot$ s $\cdot$ Pa) benchmark. A fine porous support layer with $\phi_p = 0.3$, $\O_p = 100$ nm, and $\tau_p = 3$ at 200°C reaches the benchmark permeance at $X = 150$ μm. This explains why the flux and separation factor of very thin supported membranes is often affected by support resistance and this aspect must be considered in the design of optimized support structures. Since a support with $X <$ 150 μm will not have sufficient strength for practical application, a second support or carrier layer must be applied. Assuming that $\O_p$ for such a carrier will be large enough for transport by viscous flow, (34.10) shows that the benchmark permeance will be reached at $\bar{p}_l/X = 1.5 \times 10^8$ Pa/m ($X = 1.5$ mm at $\bar{p}_l = 1$ bar) for $\eta_{H_2} = 12.2 \times 10^{-6}$ Pa $\cdot$ s, $\phi_p =$ 0.3, $\O_p = 1$ μm, $\tau_p = 3$, and $T = 200$°C. This implies that carriers with larger $\O_p$ and/or graded structures must be considered.

34.3.2.2 Supported Membranes for Water Purification Water purification is assumed to be carried out with a supported mesoporous membrane with characteristics similar to that of the intermediate layer of the previous example. Using (34.11), such a membrane is expected to reach the benchmark $f_\ell = 10^{-12}$ m at $X = 80$ nm. The fine porous support layer from the H_2 case reaches the benchmark permeance at $X = 30$ μm. This implies that such a layer can only be applied as a 1 μm bridging layer on a carrier with much larger $\O_p$. Such a carrier would reach the benchmark permeance at $X =$ 30 μm with $\O_p = 1$ μm. And that pore diameter would be sufficiently small to be bridged by the support layer.

34.3.3 Design Considerations for Supporting Structures

The examples in the previous paragraphs show that improving membrane permeance goes hand in hand with improving the overall permeance of the supporting structure. The thick and strong support–carrier structure that is needed to reach the benchmark goals has large pores with transport resistance determined by viscous flow, scaling as $Ø_p^{-2}$. This implies that the most effective approach to improving the permeability of the supported membrane structure is making all other layers as thin as needed or possible and increasing $Ø_p$ of the carrier as much as needed. However, increasing $Ø_p$ has negative effects on the following:

- Mechanical strength of the carrier (Shqau et al., 2006)
- Ability to bridge the carrier pores with a defect-free layer

These considerations easily lead to the conclusion that the ideal membrane support structure has one smooth deposition surface with a homogeneous fine porosity and a graded structure that continuously coarsens in the direction perpendicular to the membrane surface. A design optimization study of such a structure would consider qualitative aspects of membrane deposition and quantitative aspects of permeability and strength under application conditions. An example of an early study for the simpler case of one quasi-homogeneous support with transport resistance in the viscous flow regime can be found in Biesheuvel and Verweij (1999). One interesting conclusion was that, given a certain $Ø_p$, the maximum flux was obtained at $\phi_p = 0.3$ on the condition that no predefined limits were set to $p^f - p^p$. However, the study relied to a large extent on estimates of the porous tensile strength from empirical models and scarcely available experimental data. The internal consistency of these data is only beginning to improve with the recent availability of more detailed studies of the relation between strength and porosity (Ji et al., 2006) and processing protocols for defect-free quasi-homogeneous porous structures. Figure 34.11 shows data for flexural strength, gas and water permeability, and $Ø_p$ of porous α-Al_2O_3 disks, prepared by colloidal filtration of Sumitomo AKP30 and AKP15 powders, followed by drying and sintering (Shqau et al., 2006). Those powders have typical particle sizes of 300 and 600 nm, respectively, which after slight sintering lead to

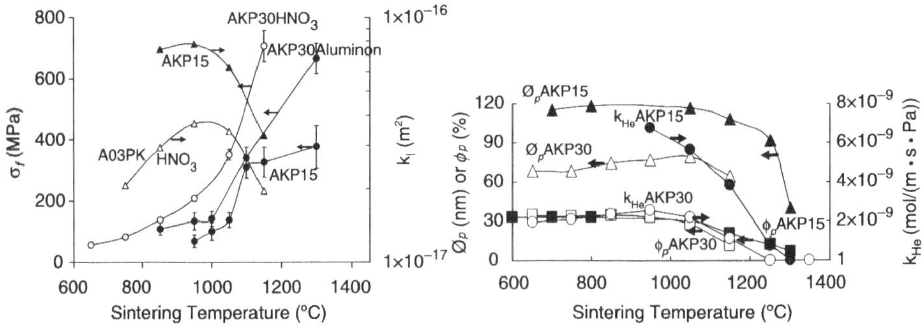

Figure 34.11 (*Left*) σ_f and k_ℓ of porous α-Al_2O_3 disks with 42 mm diameter, prepared by colloidal filtration of aqueous Sumitomo Chemical AKP15 and AKP30 powder dispersions as a function of sintering temperature. HNO_3 at pH 2 and Aluminon at pH 9.5 were used for initial colloidal stabilization of AKP30 and Aluminon only for AKP15. (*Right*) k_{He}, ϕ_p by Hg pycnometry, and $Ø_p$ by Hg intrusion of the same porous α-Al_2O_3 supports as a function of sintering temperature.

$\phi \approx 30\%$ and $Ø_p \approx 80$ and 110 nm, respectively. Both support types show very high flexure strength values that enable water purification at pressure differences of >1.7 and >1.1 MPa, respectively. However, the higher mechanical permeability of AKP15 supports results in better operational characteristics, despite their strength being lower than that of AKP30 supports. Further optimization studies require detailed information for the geometries of the supported membrane and its sealing/seating construction.

Preparation of porous supports is frequently done by consolidating powders (particles) into a predefined shape followed by sintering. With increasing sintering T, the initial porous strength develops by neck formation between the particles followed by gradual decrease of the connected porosity $\phi_{p,0}$ to zero at $\phi_p < 0.05$. Pore diameter $Ø_p$ shows a maximum with sintering T, while σ_f gradually increases to a limiting value as long as $\phi_p > 0.05$. Many practical consolidation methods result in an initial ϕ_p well above the random close-packing limit and hence an inhomogeneous microstructure by nature. Such an initial microstructure is likely to result, after sintering, in less than optimum combinations of permeability and flexure strength, σ_f. Colloidal consolidation methods, discussed in the next chapter, have the potential of realizing homogeneous packing structures at or slightly above the random close-packing limit. Such structures can be sintered to acceptable strength with a slight decrease of ϕ_p. This implies that practical optimized values of $\phi_{p,0}$ are likely to occur just above the random close-packing limit of 0.63.

34.4 SYNTHESIS AND MANUFACTURING

The large-scale introduction of inorganic membranes has been hindered by their high cost, in which important contributing factors include the following:

- Lack of optimized designs for specific processes and poor zero-hour yield and reproducibility in pilot-scale manufacturing. These problems are related to the limited definition of supported microstructures as discussed in the previous sections.
- Use of time-consuming, conventional ceramic routes for forming and thermal processing. Macroporous carriers are traditionally shaped with methods such as dry pressing and extrusion. These methods employ mixtures of particles and additives with a low volumetric particle fraction between 0.2 and 0.5. This makes it that the sintered microstructures are likely to deviate substantially from the target quasi homogeneous structure, which impacts their strength and surface quality. As a result surface defects must be repaired by application of relatively thick supporting and intermediate layers by colloidal deposition, as shown in Figure 34.12. The fact that these layers are thicker than needed may have negative implications for the overall transport resistance and thermochemical stability.

In the past 5 years substantial efforts were made in our laboratories to realize random-packed structures with the best possible homogeneity at a wide range of length scales. Recent examples of the results of these efforts are shown in Figures 34.4; 34.10, left; and 34.12. These results were obtained by using several variations of the method of colloidal filtration and deposition, as illustrated in Figure 34.10, right. The method of colloidal deposition starts with fairly dilute dispersions of particles with a nearly isometric shape and a narrow particle size distribution. Any initially present inhomogeneity in the dispersions can be removed by ultrasound-assisted screening for 10 nm $< Ø_s <$ 1 μm or centrifugation

Figure 34.12 (*Left to right*) Improvement in visual appearance over 4 years development of thin γ-alumina membranes prepared on optically smooth, 42-mm-Ø macroporous α-Al_2O_3 supports. The colors are due to an optical interference effect used as an indication of thickness variations. The improvements are due to removal of defects as in Figure 34.4.

for $Ø_s < 50$ nm. Examples of such inhomogeneity are large agglomerates and particles, residual hollow shells from production by spray drying, organic binder residuals, and airborne contamination. Any subsequent particulate (surface) contamination is avoided by performing all consolidation and deposition processing at class 100 clean-room conditions and working as clean as possible. To make the thermal processing steps less time consuming, all conventional ceramic drying, burnout, and sintering of membrane and intermediate layers might be replaced by more appropriate methods as are used in the semiconductor industry. Examples of such methods are ozonation and reactive ion etching for the removal of organic additives and rapid thermal processing for drying and sintering of thin layers. These methods can have a process cycle time of seconds, can target a specific layer, and as a consequence are much less contaminating.

34.5 CHARACTERIZATION

Further development of supported membrane structures will benefit from the availability of suitable characterization tools. The acquisition of accurate fundamental transport parameters, reproducibility, and lifetime behavior requires that several membrane parameters are obtained, preferably in a nondestructive way. Development of membrane properties is best followed for the same membrane. Membranes and intermediate layers can be very thin and often show considerable penetration in their supporting structure. This hinders unambiguous determination of their thickness and accurate determination of the parameters $\phi_{m/u_1}, \phi_{u_1,u_2}, \ldots$. In particular, for amorphous silica the actual membrane phase may differ from "unsupported" bulk materials that are formed under simulated conditions. This is related to the fact that deposition kinetics and drying play an important role in the formation of the nonequilibrium silica structure.

In addition to properties of the quasi-homogeneous structure, detailed information is needed for the morphology of typical defects and statistical information for the occurrence of a certain defect type. However, with state-of-the-art supported membranes it is already a challenge to obtain accurate quasi-homogeneous transport numbers and information about the nature and occurrence of defects. Detection of pinhole defects can possibly be done by scanning a probe across the membrane surface that senses local deviations in (transport) properties in the quasi-homogeneous structure. To the authors' knowledge such techniques are not readily available. However, the advent of very strong and optically smooth supports, as used for Figure 34.10, left, has enabled a wide range of characterization methods that provide either statistically averaged or local information. Examples of techniques particularly suited for supported membranes are discussed in the next sections, classified according to their level of nondestructiveness.

34.5.1 Fully Nondestructive Techniques

Fully nondestructive techniques do not touch the membrane surface and may be used during operation. These requirements make it that such techniques are likely limited to in-line transport measurements and optical reflection measurements. The transport measurements are discussed under quasi nondestructive techniques. Spectroscopic ellipsometry provides the thickness and composition for layers with $\text{Ø}_p < 50$ nm and $X < 5$ μm, mutually independent, for the optically smooth supports and membranes. The composition is in that case obtained from interpretation of the refractive index and may include information about ϕ_p and the amount of adsorbed molecules. The use of this method was demonstrated first in Benes et al. (2001) for the layer thickness and CO_2 sorption in supported amorphous silica membranes, as shown in Figure 34.10, left. Spectroscopic ellipsometry is now used routinely in our laboratory for layer thickness of supported (γ-alumina) membranes. This analysis involves determination of the optical constants of uncoated macroporous AKP15 and AKP30 α-Al_2O_3 supports, described in Figure 34.11. As the γ-alumina membranes are optically transparent, the dispersion in refractive index n can be described as a "Cauchy type" material with the form

$$n^2 = C_1 + \frac{C_2}{\lambda^2} + C_3\lambda^2 \tag{34.14}$$

where λ is the wavelength of the light and C_1, C_2, and C_3 are fitting parameters. The Cauchy model of the γ-alumina membrane is further modified with a Bruggeman effective media approximation (Aspnes, 1982), which allows for the determination of ϕ_p. Effective media approximations (EMAs) are similarly used to obtain the membrane surface roughness and the thickness of the penetration zone between layers. Such an analysis results for current supported γ-alumina membranes in a surface roughness of 8 nm, a membrane thickness on the order of 800 nm with $\phi_p = 0.36$, and a membrane/support penetration of 49 nm. The surface roughness is much larger than the local value of <1 nm, as is observed by TEM inspection of FIB cross sections, and is likely due to the thickness variations in the colloidal filtration support that occur on a length scale of >1 μm.

34.5.2 Quasi-Nondestructive Techniques

Quasi-nondestructive techniques include several transport measurements that are used to test specific membrane properties. They require sample mounting by compression sealing or glass solders that rarely leave the delicate membrane surface intact. Gas transport properties of dense and microporous membranes are tested by measuring single gas j_l as a function of p^f and p^p and by obtaining fluxes and α_{l_1,l_2} from the stationary composition and flow rate of gas mixtures at the membrane feed and permeate side. To use the results of these measurements for comparison and optimized membrane designs, substantial interpretation is needed in terms of fundamental material parameters (Verweij, 2003). Examples of such parameters are b^{a}, c^{tot}, Δs_l^0, Δh_l^0, and u^{h} in (34.7). In addition, important corrections must be made for contributions of the supporting structure and external mass transfer limitations in mixtures. Last but not least, interpretation of experimental data is significantly affected by the presence of pinholes in the membrane. This, together with the fact that supported membrane morphology is often poorly defined or simply unknown, has historically hindered quantitative comparison of literature data.

The transport properties of meso- and macroporous structures are determined by measuring viscous (liquid or gas) permeability, as shown in Figure 34.11, left, and Knudsen (gas) permeability, as shown in Figure 34.11, right. The ion separation of aqueous solutions by nanofiltration is done preferably by analyzing feed and permeate compositions by ion chromatography. An example of a benchmarking test is measuring the salt rejection of NaCl, $CaCl_2$, or $AlCl_3$ solutions by positively charged membranes. The salt rejection is in that case expected to approach 100% with $\lambda_\ell^D/\text{Ø}_p > 1$ at limiting j_ℓ; see Figure 34.3, left. A deviation from 100% might in that case be tentatively interpreted as a surface fraction of inactive pinholes. However, a quantitative treatment for the interpretation of the effect of pinholes is yet to be carried out. The tentative interpretation, however, indicates that, contrary to what is the case for highly selective gas separation, the salt retention by nanofitration is not affected much by minor pinhole concentrations.

34.5.2.1 Permeation Porometry The transport measurements, mentioned above, are fast and provide averaged information about the quasi-homogeneous pore structure and the effect of pinholes. Application of more than one method such as Knudsen gas permeation and viscous liquid permeation may result in more detailed structural information, but it will generally be difficult to quantify the morphology and occurrence of defects. Single gas permeation methods are easier than separation measurements and are often preferred because of their unambiguous interpretation. However, single gas permeation does not provide any information about the interaction of fluxes of different species in the membrane structure. A more comprehensive and complete picture for meso- and macroporous structures with $\text{Ø}_p < 100$ nm may be obtained by permeation porometry (Mottern et al., 2005). In this method, connected pores larger than a certain Ø_p^{min} are blocked by subjecting the (multilayer) structure to unsaturated cyclohexane (CH) vapor. The actual value of Ø_p^{min} depends on the relative CH vapor pressure p_{CH}^{rel} through the Kelvin equation

$$\ln(p_{CH}^{rel}) = -\frac{2f\,\gamma_{\ell g} V^{mol} \cos(\theta^{wet})}{\text{Ø}_K^{min} RT} \tag{34.15}$$

were f is a geometric constant equal to 2 or 1 for desorption or adsorption of the condensate, respectively; $\gamma_{\ell g}$ is the liquid–vapor surface tension, V^{mol} is the molar volume of the condensate, and θ^{wet} is the contact angle between the condensate and the solid (generally taken to be zero). The Kelvin diameter Ø_K is slightly smaller than Ø_p due to physisorption of a CH monolayer (t layer) on the pore surface. The remaining connected porosity is determined by measuring the permeability of a probe gas in the partly saturated structure. The probe gas is commonly O_2, but recent studies (Mottern et al., 2006) have shown that chemical diffusion in CH-saturated top layers causes the superposition of membrane and support data, which effectively masks the effects of any pinholes in the membrane.

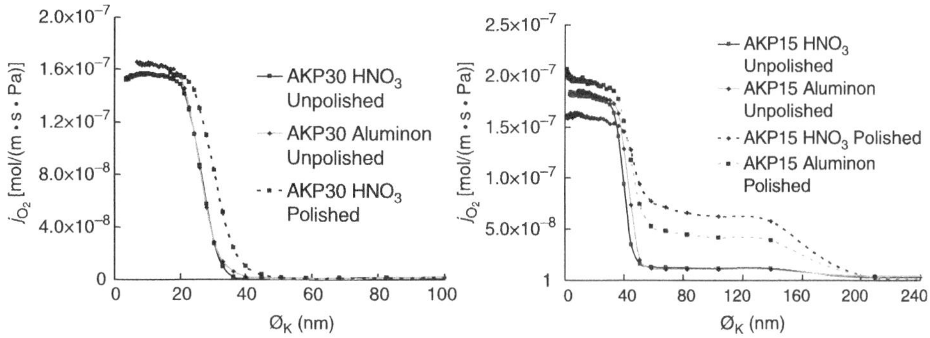

Figure 34.13 Permeation porometry results for polished and unpolished supports made by colloidal filtration: (*left*) AKP30 and (*right*) AKP15 with HNO_3 or aluminon stabilization and sintering at 950 and 1050°C, respectively. The inflections in the curves correspond to maxima in connected pore size distribution.

Permeation porometry of AKP30 α-Al_2O_3 supports prepared by colloidal filtration and sintering at 950°C revealed a connected pore size distribution (CPSD) centered on $Ø_K \approx$ 26 nm; see Figure 34.13, left. The connected pores are found to be substantially smaller than the ~80 nm bulk pores as determined by mercury intrusion porometry; see Figure 34.11. Since permeation porometry measures the smallest pore constrictions, it provides an indicator of "bottleneck" behavior. Polishing the AKP30 α-Al_2O_3 support resulted in a slight shift of the CPSD to an average $Ø_K \approx$ 28 nm. This increase in $Ø_K$ suggests the formation of a slightly deviating surface structure due to preferential deposition of small particles in the surface formed in colloidal filtration. The narrow-particle-size distribution of AKP30 makes substantial surface film formation unlikely.

AKP15 α-Al_2O_3 supports prepared by colloidal filtration and sintering (Shqau et al., 2006) have a CPSD with an average $Ø_K \approx$ 42 nm. There is a slight but noticeable variation in the $Ø_K$ due to the choice of colloidal stabilizer; see Figure 34.13, left. The AKP15 desorption curves also reveal a small number of large pores with an average $Ø_K \approx$ 160 nm. Although these large pores are small in number, they may have a detrimental impact on flexure strength and deposition of thin membranes through the formation of pinholes. The $CaCl_2$ retention results shown in Figure 34.3 were obtained with 800-nm-thick γ-alumina on an unpolished AKP15 support. The 99.9% maximum retention for 1 mol/m^3 indicates that the γ-alumina membrane was virtually pinhole free, which indicates, in turn, that the 160 nm-Ø surface pores in the AKP15 may have been plugged with γ-alumina. The 160 nm Ø, however, indicates that pinholes become more likely for a membrane thickness less than 200 nm. Polishing the AKP15 supports obtained with HNO_3^- stabilization results in a substantial increase in the amount of larger pores such that a bimodal CPSD is generated with approximately equal contributions to the O_2 permeation of large and small pores. The large pores have a similar $Ø_K \approx$ 160 nm, but the small pores have a larger $Ø_K \approx$ 100 nm. The dramatic change in the observed CPSD suggests the formation of a significant surface film on the AKP15 supports composed primarily of small AKP30-like particles. The structure appears to be graded such that a fine porosity exists at the surface of the support, which is considered favorable for colloidal membrane deposition. Below the surface film, the pore structure coarsens, which results in less transport resistance. For polished AKP15 supports prepared with Aluminon stabilization, the bulk pore structure

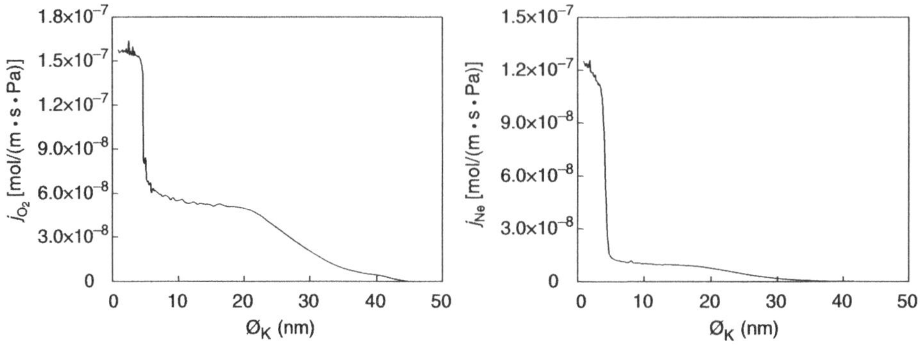

Figure 34.14 Permeation porometry desorption isotherm for ~1 μm state-of-the-art supported γ-alumina membrane on an AKP30 α-Al_2O_3 support prepared by colloidal filtration with HNO_3 stabilization and sintering at 950°C; see Figure 34.11. (*Left*) O_2 probe. (*Right*) Ne probe.

does not have as many large pores contributing to the total amount of O_2 permeation. In fact, the large pores contribute for only one-fourth of the total O_2 permeation compared to about one-half for AKP15 supports obtained with HNO_3 stabilization. This indicates that the use of Aluminon results in a more homogeneous packing structure for AKP15 α-Al_2O_3, which made us decide to standardize on that colloidal stabilizer.

The individual layer permeance $f_{l,\ell}$ due to chemical diffusion of probe l in liquid ℓ can be estimated by

$$f_{l,\ell} = \frac{j_l}{\Delta p_l} = \tilde{D}_{l,\ell} K_{l,\ell} \frac{\phi}{\tau X} \tag{34.16}$$

in which $\tilde{D}$ is the coefficient for chemical diffusion and K the Henry solubility constant. Inspection of these quantities for several gases in CH (Krieger et al., 1967) shows that $\tilde{D}_{Ne,CH} K_{Ne,CH}$ is 5 times smaller than $\tilde{D}_{O_2,CH} K_{O_2,CH}$, which indicates that the Ne probe will be more selective for the membrane top layer. Use of Ne as a probe gas has the disadvantage that it is very expensive since it is very rare. The latter, however, was also found to be an advantage because it results in a very low background concentration, enabling reliable detection of extremely low Ne fluxes in thick saturated support structures. Permeation porometry desorption isotherms for a state-of-the-art 1 μm-thick supported γ-alumina membrane with O_2 and Ne as the probe gas are shown in Figure 34.14. This figure clearly demonstrates the ~4 nm $Ø_K$ of the γ-alumina and the 5 times lower permeability of Ne in the saturated top layer. The "plateau" at $10 < Ø_K < 20$ nm, where the support is fully unsaturated, is easily misinterpreted in terms of the occurrence of pinhole defects in the membrane top layer.

One drawback of the permeation porometry method is that it is very time consuming, up to one week for one sample. As a consequence, it will typically be used as a research tool and for calibration of faster but less comprehensive methods.

34.5.3 Destructive Techniques

Destructive techniques naturally destroy membrane structure and functionality. Typical examples include mechanical strength measurements, accelerated lifetime tests, and inspection of cross sections by microscopy. The σ_f values shown in Figure 34.11 were obtained in biaxial flexure mode of AKP15 and AKP30 α-Al_2O_3 disks supported by three equally spaced steel balls concentrically located about the applied load (Shqau et al., 2006). This method does not require any sample machining so that introduction of stress and damage by such a treatment is avoided. The advent of dual-beam FIB techniques for TEM sample preparation (Giannuzzi and Stevie, 1999) has enabled the detection of structural defects on a much smaller scale than what is shown in Figure 34.6 and with much more detail than what can be obtained with fracture surfaces. We routinely use this technique to prepare $5 \times 15 \times 0.1$ μm electron-transparent supported membrane samples with the 5-μm dimension perpendicular to the membrane surface. Inspection by TEM of these samples provides a local, nearly two-dimensional cross section of subsequent layers and their connection. The FIB method is particularly suitable for membranes that are deposited on smooth, strong, and quasi-homogeneous AKP30 and AKP15 α-Al_2O_3 supports that are developed in our laboratory (Shqau et al., 2006). Samples made from membranes on these supports are easy to handle and generally stay intact during transfer to the TEM grid. Several examples are provided in this chapter in Figures 34.4; 34.7; 34.9, lower; and 34.10, left. The FIB method typically provides excellent information about X, surface roughness, interlayer penetration and adhesion, and the occurrence of defects. However, the FIB method is rather time consuming and elaborate so it is not very suitable to obtain statistical information.

34.6 CONCLUSIONS

Application of thin supported inorganic membranes offers interesting perspectives on energy savings in the production of industrial chemicals and water purification and on energy conversion with improved yield. However, a large breakthrough in applications has not yet been realized due to limited reproducibility of prototypes, lack of fundamental transport data, and cost perspectives. On the other hand, much progress has been made in the development of new membrane concepts and material synthesis methods and more recently in the synthesis of defect-free thin membranes and smooth, strong supports by high-definition colloidal deposition techniques. These synthesis methods, in turn, have enabled new, essential characterization techniques such as spectroscopic ellipsometry and TEM of FIB cross sections. The status of current membranes is summarized as follows:

- *Concepts for Multilayer Transport* More systematic work is needed for surface and bulk transport of dense membranes, correlation coefficients in multicomponent microporous membrane transport, and descriptions of ion retention by nanofiltration. However, the most common supported membrane cases can be addressed very well with what is currently available.
- *Accurate Equilibrium and Transport Characterization* There is a real need to obtain such data with improved accuracy and internal consistency measured on very well defined (multilayer) structures. In particular, high-temperature, high-pressure data are very scarce.
- *Thermochemical Stability* Systematic data on thermochemical stability are particularly scarce. Since testing of operational stability is always destructive for the membrane, such data can be obtained best with a series of samples that are reproducible within 10%. This leads us back to the availability of well-defined samples.

- *Thin Membranes and Intermediate Layers* There has been much progress in this area, but excellent performance can only be obtained with membranes that are deposited on optimized support structures. Little is known about the actual formation mechanism of porous layers by colloidal consolidation (Biesheuvel and Lyklema, 2005). Future synthesis and manufacturing of thin supported layers should occur with fast and noncontaminating methods as used in the semiconductor industry to reduce cycle times per layer and hence cost. These techniques will also enable viable manufacturing routes in which several layers of the same composition are deposited on top of each other to minimize the influence of surface defects. Protocols for mesoporous membranes appear to stabilize for $X > 500$ nm, but more work is needed for $X < 500$ nm. Problems with adhesion to the support have been solved by improving mesoporous homogeneity and the application of adhesion agents (Nijmeijer et al., 2001).
- *Design Optimization for Target Processes* Not much has been done in this area because of the lack of systematic design parameters, as mentioned before.
- *Developments of Smooth, Strong, and Permeable Supports and Carriers* Optimization studies with commercially available AKP15 and AKP30 α-Al_2O_3 have resulted in optimized microstructures with $50 < Ø_p < 150$ nm. More evolutionary development is certainly needed for this size range, but optimization of homogeneity for $Ø_p > 1$ μm likely needs more attention. The larger $Ø_p$ requires very high sintering temperatures in conventional processing and may result in poor mechanical strength and reliability. This problem might be addressed by application of wet-chemical techniques such as phosphate bonding (Toy and Whittemore, 1989), which provides thermochemical stability up to 900°C.

Nomenclature

Glossary

CH	Cyclohexane
CPSD	Connected pore size distribution
FIB	Focused ion beam
SEM	Scanning electron microscopy
TEM	Transmission electron microscopy

Symbols

$Ø$	Diameter
$Ø_K$	Kelvin diameter
$Ø_p$	Typical pore diameter
$Ø_s$	Typical particle/grain diameter
ℓ	Fluid phase
α_{l_1,l_2}	Membrane selectivity from mixture measurement, $\alpha_{l_1,l_2} = f_{l_1}/f_{l_2}$
$\alpha^s_{l_1,l_2}$	Membrane permselectivity from single gas measurement, $\alpha^s_{l_1,l_2} = f_{l_1}/f_{l_2}$
ϕ_p	Relative porosity ($0 < \phi_p < 1$)
$\phi_{u_k/u_{k+1}}$	Constriction resistance at transition between layers u_k and u_{k+1}
η	Dynamic viscosity

λ	Mean free path length of gas molecules
λ^D	Debye length
μ	Molar chemical potential
μ^0	Part of μ exclusive of partial entropy due to distribution over locations or kinetic states
θ	Time-averaged occupation of site in dense or microporous structures that are available to accommodate one mobile species
θ^{wet}	Wetting angle
σ	Collision cross section of gas molecules
τ	Tortuosity: ratio of actual transport length and projected parallel length
f	Membrane feed-side superscript
g	Gas phase
m	Membrane ($m = u_0$)
min	Minimum
p	Membrane permeate-side superscript
rel	Relative quantity
A	Apparent exposed membrane surface
b	Particle/molar mechanical mobility
b^0	Unconditional b (all neighboring sites available for diffusion hopping transport)
b^a	Preexponential term for T dependence of b^0
c	Molar concentration
c^{tot}	Total c of micropore sites
$\tilde{D}$	Chemical diffusion coefficient
f	Molar permeance of single layer, $f = j/\Delta p$
f^{tot}	Overall f of multilayer structure
f^1	f for type 1 transport (34.7)
f^{Kn}	Knudsen f
f^{mf}	Molecular flow f
f_ℓ	Volumetric liquid (mechanical) permeance, $f_\ell = -j_\ell \eta_\ell / \Delta p$
$\tilde{f}_L$	Nonequilibrium correlation coefficient for chemical diffusion on lattice L; $0.5 < \tilde{f}_L < 1$ for $\theta \rightarrow 1$ in single-component mixtures, $0 < \tilde{f}_{l,L} < 1$ in multicomponent mixtures; $\tilde{f}_{l,L}$ tends toward unity with decreasing θ
h	Partial molar enthalpy
j	Molar gas flux
j_ℓ	Volumetric liquid flux
k_ℓ	Mechanical fluid permeability, $k_\ell = j_\ell X \eta_\ell / \Delta p_\ell$
K	Henry equilibrium constant
Kn	Knudsen number
l	Gas-phase and mobile membrane species such as H_2 and CO_2
p	Gas pressure
R_l	Rejection of dissolved species l, $R_l = 1 - c_l^p / c_l^f$
s	Partial molar entropy
s^0	Part of s exclusive of contribution due to distribution over locations or kinetic states
T	Temperature
u^h	Energy barrier for species hopping to adjacent vacant micropore
u^{tot}	Apparent molar activation energy of f as in $\exp(-u^{tot}/RT)$

V^{mol} Molar volume

X Apparent layer thickness

ACKNOWLEDGMENT

The authors acknowledge the support received from the National Science Foundation IGERT and WaterCAMPwS STC programs under grant numbers 0221678 and 0120978, the Ohio Department of Development, and the Ohio State University.

REFERENCES

Aspnes, D. E. (1982). Optical properties of thin films. *Thin Solid Films* 89, 249.

Benes, N. E., Bouwmeester, H. J. M., and Verweij, H. (2002). Multi-component lattice gas diffusion. *Chem. Eng. Sci.* 57(14), 2673.

Benes, N. E., Nijmeijer, A., and Verweij, H. (2000). Microporous Silica Membranes. In N. K. Kanellopoulos (Ed.), *Recent Advances in Gas Separation by Microporous Ceramic Membranes*. Elsevier, Amsterdam, p. 335.

Benes, N. E., Spijksma, G., Verweij, H., Wormeester, H., and Poelsema, B. (2001). CO_2 sorption of a thin silica layer determined by spectroscopic ellipsometry. *AIChE J.* 47(5), 1212.

Benes, N. E., Verzijl, R., and Verweij, H. (1999). Numerical scheme for simulating multicomponent mass transport accompanied by reversible chemical reactions in porous media. *Computers Chem. Eng.* 23, 975.

Bhave, R. R. (1991). *Inorganic Membranes, Synthesis, Characteristics and Applications.* Van Nostrand Reinhold, New York.

Biesheuvel, P. M., and Lyklema, J. (2005). Sedimentation–diffusion equilibrium of binary mixtures of charged colloids including volume effects. *J. Phys. Condens. Matter* 17, 6337.

Biesheuvel, P. M., and Verweij, H. (1999). Design of ceramic membrane supports: Permeability, tensile strength and stress. *J. Membr. Sci.* 156(1), 141.

Bouwmeester, H. J. M., and Burggraaf, A. J. (1997). Dense ceramic membranes for oxygen separation. In P. J. Gellings and H. J. M. Bouwmeester (Eds.), *The CRC Handbook of Solid State Electrochemistry*. CRC Press, Boca Raton, FL, p. 481.

Caro, J., Noack, M., Koelsch, P., and Schaefer, R. (2000). Zeolite membranes state of their development and perspective. *Micropor. Mesopor. Mater.* 38(1), 3.

Ceramem (2006). Technology brief: Ceramic membrane modules. Available: http://www.ceramem.com/techbrief.pdf, accessed Jan. 2006.

Chen, C. S., Kruidhof, H., Bouwmeester, H. J. M., Verweij, H, and Burggraaf, A. J. (1996). Oxygen permeation through oxygen ion-conducting oxide-noble metal dual phase composites. *Solid State Ionics* 86/88(1), 569.

Condom, S., Chemlal, S., Larbot, A., and Persin, M. (2006). Behaviour of a cobalt spinel ultrafiltration membrane during salt filtration with different ionic strengths. *J. Membr. Sci.* 268(2), 175.

de Lint, W. B. S., and Benes, N. E. (2005). Separation properties of gamma-alumina nanofiltration membranes compared to charge regulation model predictions. *J. Membr. Sci.* 248(1–2), 149.

de Vos, R. M., and Verweij, H. (1998). High-selectivity, high-flux silica membranes for gas separation. *Science* 279(5357), 1710.

Giannuzzi, A. L., and Stevie, F. A., (1999). A review of focused ion beam milling techniques for TEM specimen preparation. *Micron* 30(3), 197.

Gupta, V. K., and Lin, Y. S. (2006). Ceramic proton conductors. In A. F. Sammells and M. V. Mundschau (Eds.), *Nonporous Inorganic Membranes for Chemical Processing*. Wiley VCH, Weinheim, Germany.

Hoenicke, D., and Dietzsch, E. (2002). Anodic alumina. In F. Schueth, K. S. W. Sing, and J. Weitkamp (Eds.), *Handbook of Porous Solids*. Wiley VCH, Weinheim, Germany, p. 1395.

IZA (2006). Structure Commission website, Database of zeolite structures. Available: http://www.iza-structure.org/databases/, accessed Jan. 2006.

Ji, S. C., Gu, Q., and Xia, B. (2006). Porosity dependence of mechanical properties of solid materials. *J. Mater. Sci.* 41(6), 1757.

Krieger, I. M., Mulholland, G. W., and Dickey, C. S. (1967). Diffusion coefficients for gases from the rates of solution of small gas bubbles. *J. Phys. Chem.* 71(4), 1123.

Kuiper, K., van Rijn, C. J. M., Nijdam, W., and Elwenspoek, M. C. (1998). Microsieves made with laser interference lithography for micro-filtration applications. *J. Membr. Sci.* 150(1), 1.

Lai, Z. P., Bonilla, G., Diaz, I., Nery, J. G., Sujaoti, K., Amat, A. M., Kokkoli, E., Terasaki, O., Thompson, R. W., Tsapatsis, M., and Vlachos, D. G. (2003). Microstructural optimization of a zeolite membrane for organic vapor separation. *Science* 300(5618), 456.

Lee, D., Zhang, L., Oyama, S. T., Niu, S., and Saraf, R. F. (2004). Synthesis, characterization, and gas permeation properties of a hydrogen permeable silica membrane supported on porous alumina. *J. Membr. Sci.* 231, 117.

Li, K., Tan, X. Y., and Liu, Y. T, (2006). Single-step fabrication of ceramic hollow fibers for oxygen permeation. *J. Membr. Sci.* 272(1), 1.

Lin, Y. S., and Baxbaum, R. E. (2000). Metal membranes: Membrane separation. In I. D. Wilson, T. R. Adlard, C. F. Poole, and M. Cooke (Eds.), *Encyclopedia of Separation Science*. Academic, San Diego, CA, p. 3365.

Lin, Y. S., Kumakiri, I., Nair, B. N., and Alsyouri, H. (2002). Microporous inorganic membranes. *Sep. Purif. Meth.* 31(2), 229.

Liu, S., Li, K., and Hughes, R. (2004). Preparation of $SrCe_{0.95}Yb_{0.05}O_{3-\alpha}$ perovskite for use as a membrane material in hollow fiber fabrication. *Mater. Res. Bull.* 39, 119.

Lu, Y. F., Cao, G. Z., Kale, R. P., Prabakar, S., Lopez, G. P., and Brinker, C. J. (1999). Microporous silica prepared by organic templating: Relationship between the molecular template and pore structure. *Chem. Mater.* 11(5), 1223.

Lu, Y. F., Ganguli, R., Drewien, C. A., Anderson, M. T., Brinker, C. J., Gong, W. L., Guo, Y. X., Soyez, H., Dunn, B., Huang, M. H., and Zink, J. I. (1997). Continuous formation of supported cubic and hexagonal mesoporous films by sol-gel dip coating. *Nature* 389(6649), 364.

Mason, E. A., and Malinauskas, A. P. (1983). *Gas Transport in Porous Media: The Dusty Gas Model*. Elsevier, New York.

Mottern, M. L., Oyola, M., Shqau, K., Yu, D., and Verweij, H. (2005). Permeation porometry of membrane supports and thin supported γ-alumina membranes. Paper presented at PCM 2005, Brugge, Belgium, Session MP-4, Oct. 20–21, 2005.

Mottern, M. L., Shqau, K., Yu, D., and Verweij, H. (2006). A closer examination of permeation porometry. In R. Bredesen and H. Ræder (Eds.), *Proc. 9th Int. Conf. Inorganic Membranes*, Lillehammer, Norway, Session 3B2, June 25–29, 2006.

Nijmeijer, A., Kruidhof, H., Bredesen, R., and Verweij, H. (2001). Preparation and properties of hydrothermally stable γ-alumina membranes. *J. Am. Ceram. Soc.* 84(1), 136.

Pacheco Tanaka, D. A., Llosa Tanco, M. A., Nagase, T., Okazaki, J., Wakui, Y., Mizukami, F., and Suzuki, T. M. (2006). Fabrication of hydrogen-permeable composite membranes packed with palladium nanoparticles. *Adv. Mater.* 18(5), 630.

Philipse, A. P., Bonekamp, B. C., and Veringa, H. J. (1990). Colloidal filtrations and (simultaneous) sedimentations of alumina and silica suspensions. *J. Am. Ceram. Soc.* 73(9), 2720.

Saufi, S. M., and Ismail, A. F. (2004). Fabrication of carbon membranes for gas separation—A review. *Carbon* 42, 241.

Shi, J. Y., Shqau, K., Verweij, H., Revur, R., Sengupta, S., and Schorr, J. R. (2006). Preparation of well-defined thin YSZ and SSZ membranes with a narrow pore size distribution. In R. Bredesen and H. Ræder (Eds.), *Proc. 9th Int. Conf. Inorganic Membranes*, Lillehammer, Norway, June 25–29, 2006, p. 69.

Shqau, K., Mottern, M. L., Yu, D., and Verweij, H. (2006). Preparation and properties of porous α-Al_2O_3 membrane supports. *J. Am. Ceram. Soc.* 89(6), 1790.

Sloot, H. J., Versteeg, G. F., and van Swaaij, W. P. M. (1990). A non permselective membrane reactor for chemical processes normally requiring strict stoichiometric feed rates of reactants. *Chem. Eng. Sci.* 45(8), 2415.

Toy, C., and Whittemore, O. J. (1989). Phosphate bonding with several calcined aluminas. *Ceram. Int.* 15, 167.

Uhlhorn, R. J. R., Huis in 't Veld, M. H. B. J., Keizer, K., and Burggraaf, A. J. (1992). Synthesis of ceramic membranes. Part I. Synthesis of non-supported and supported γ-alumina membranes without defects. *J. Mater. Sci.* 27(2), 527.

van der Haar, L. M. (2001). Mixed-conducting perovskite membranes for oxygen separation—Towards the development of a supported thin film membrane. *Thesis*, University of Twente, Enschede, The Netherlands.

Verweij, H. (2002). Novel synthesis of ceramic membranes. In P. Vincenzini (Ed.), *Proc. 10th Int. Ceramics Congres & 3rd Forum on New Materials*, Session J-4, Florence, Italy, July 14–18, 2002.

Verweij, H. (2003). Ceramic membranes: Morphology and transport. *J. Mater. Sci.* 38(23), 4677.

Verweij, H., Lin, Y. S., and Dong, J. H. (2006). Micro-porous silica and zeolite membranes for hydrogen purification. *MRS Bull.* 31(10), 756.

Yu, D., Mottern, M. L., Shqau, K., and Verweij, H. (2006). Synthesis and Optimization of supported γ-alumina membranes for water purification. In R. Bredesen and H. Ræder (Eds.), *Proc. 9th Int. Conf. Inorganic Membranes*, Lillehammer, Norway, June 25–29, 2006, p. 691.

CHAPTER 35

Structure/Property Characteristics of Polar Rubbery Membranes for Carbon Dioxide Removal

VICTOR A. KUSUMA, BENNY D. FREEMAN and MIGUEL JOSE-YACAMAN

Department of Chemical Engineering, University of Texas at Austin, Austin, Texas 78712

HAIQING LIN

Membrane Technology and Research, Inc., Menlo Park, California 94025

SUMOD KALAKKUNNATH and DOUGLASS S. KALIKA

Department of Chemical and Materials Engineering and Center for Manufacturing, University of Kentucky, Lexington, Kentucky 40506-0046

35.1 INTRODUCTION AND BACKGROUND

Selective carbon dioxide removal from mixtures with light gases such as H_2, N_2, CH_4, and O_2 is of significant industrial interest. Examples include separation of CO_2 from hydrogen following steam reforming of hydrocarbons (Kohl and Nielson, 1997), separation of CO_2 from oxygen for food packaging purposes (Paul and Clarke, 2002), and removal of CO_2 from natural gas streams (Baker, 2004). While traditional methods such as pressure swing adsorption and amine absorption have been widely used in industry for CO_2 separation in chemical and petrochemical applications, membrane-based options have been explored as alternatives due to their simplicity, efficiency, and small footprint (Kohl and Nielson, 1997). Membranes could be more widely utilized for CO_2 removal if their separation performance was improved. From a practical viewpoint, CO_2 must be permeated to the low-pressure side of the membrane to retain the desired light gases (e.g., H_2 and CH_4) at or near feed pressure for subsequent transport and use. Therefore, membrane materials for such applications must possess high CO_2/light gas selectivity and high CO_2 permeability to produce a product stream with satisfactory flow rate and purity (Kalakkunnath et al., 2005).

Separation of CO_2 from larger light gases is currently achieved with glassy polymer membranes, which rely on the size-sieving ability of the polymer for effective separation (Koros et al., 1992), for example, removal of CO_2 from CH_4 in natural gas. Carbon dioxide is a major impurity in natural gas streams, and it often must be removed to meet

Advanced Membrane Technology and Applications. Edited by Norman N. Li, Anthony G. Fane, W. S. Winston Ho, and T. Matsuura
Copyright © 2008 John Wiley & Sons, Inc.

pipeline specifications (Baker, 2004). Due to the large volume of natural gas produced annually, even small improvements in CO_2 removal efficiency could lead to considerable cost reduction. Carbon dioxide has a smaller kinetic diameter than CH_4 (3.3 vs. 3.8 Å) (Breck, 1974), which leads to higher diffusion coefficients for CO_2 than for CH_4. Additionally, CO_2 is more condensable than CH_4 (as characterized by its higher critical temperature), leading to greater solubility for CO_2 than for CH_4 (Freeman and Pinnau, 1999). Therefore, CO_2 is inherently more permeable than CH_4 owing to both favorable diffusivity selectivity and favorable solubility selectivity. Most materials design efforts to improve CO_2/CH_4 separation performance aim to increase the size-sieving ability of the polymer (Freeman, 1999; Koros et al., 1992) because penetrant diffusivity is typically more sensitive than penetrant solubility to changes in polymer structure. In practice, however, highly condensable CO_2 can be strongly sorbed in the polymer matrix, plasticizing the polymer and decreasing the size-sieving ability of the membrane, leading to significant reductions in CO_2/CH_4 selectivity (Staudt-Bickel and Koros, 1999; White et al., 1995).

Another emerging application involving CO_2 removal from a light gas stream is the separation of CO_2 from oxygen. Food-packaging applications are increasingly reliant upon packaging materials with carefully designed permeation rates to gases such as CO_2 and O_2 to increase the shelf life of various types of produce (Paul and Clarke, 2002). For instance, there is a need to maintain an optimum level of oxygen for fresh fruit and vegetable storage, as these items continue to respire after harvesting: aerobic respiration at high O_2 concentration breaks down starch, sugars, and organic acids into CO_2 and H_2O plus heat and metabolic energy, and this loss of substrate affects shelf life; on the other hand, anaerobic respiration at low O_2 concentration results in the formation of ethanol, acetaldehyde, and organic acids (fermentation) that may cause undesirable odors and flavors and may lead to growth of anaerobic pathogens that can cause food poisoning (Paul and Clarke, 2002). For these reasons, it is of interest to control the relative transport rates of CO_2 and O_2 in food packaging, and the CO_2/O_2 selectivity of the packaging material is one variable that can influence the relative transport rates of CO_2 and O_2. However, controlling the relative permeation properties of CO_2 and O_2 by size sieving can be limited since the kinetic diameters of CO_2 and O_2 (3.3 vs. 3.46 Å, respectively) (Breck, 1974) are similar. Accordingly, alternative membrane materials need to be explored to improve separation performance for this application.

Hydrogen gas is becoming an increasingly important commodity, not only for its current use in the fertilizer and refinery industries, but also for its potential application as an energy carrier for future fuel cell technologies (Service, 1999). Hydrogen is mostly produced by steam reforming of hydrocarbons, a process leading to CO_2/H_2 gas mixtures with H_2 as the major component (Kohl and Nielson, 1997). The separation properties of conventional polymeric membranes for the removal of CO_2 from H_2 are unfavorable, since H_2 has a smaller kinetic diameter than CO_2 (2.89 Å as opposed to 3.3 Å) (Breck, 1974) and, therefore, strongly size-sieving polymeric membranes preferentially permeate H_2 over CO_2. As a result, H_2 is produced in the permeate side at low pressure, even though it is required to be at high pressure for further use. The subsequent recompression of this stream diminishes the advantage of using membranes to perform the separation. Potentially, this separation could benefit from a reverse-selective membrane; that is, a membrane that selectively permeates the larger component (CO_2) while retaining the smaller component (H_2) at elevated pressure (Lin and Freeman, 2005b).

Our approach for CO_2 separations from light gases focuses on rubbery materials with high gas permeability (Lin et al., 2006b). Rubbery materials have weak size-sieving ability due to the flexible nature of rubbery polymer chains. As such, solubility selectivity is very important in these materials. One way to increase CO_2 solubility in rubbery materials is to exploit the quadrupolar character of CO_2 and its favorable interaction with polar groups present in polymeric membranes (Koros, 1985; Ghosal et al., 1996). In particular, the presence of ether oxygen groups in polymers increases CO_2 permeability, with more pronounced effects at higher ether oxygen concentrations (Bondar et al., 2000). An appropriate model material, therefore, is poly(ethylene oxide) (PEO), and the gas transport properties of PEO for CO_2 removal have recently been reported (Lin and Freeman, 2004). One potential advantage of rubbery polymers for these separations is that their separation properties may be less likely to deteriorate catastrophically when strongly sorbing gases or vapors, such as CO_2, dissolve into the polymer matrix and plasticize it. Rubbery membranes do not rely as heavily on size discrimination to achieve separation as conventional glassy polymers, and the loss of size selectivity that typically accompanies plasticization may not be as problematic in rubbery materials as it is in glassy membrane materials. In fact, for reverse-selective separations such as the removal of CO_2 from H_2, plasticization can actually enhance the CO_2/H_2 selectivity as it further reduces the relatively modest size-sieving ability of the rubbery matrix, sieving that tends to favor permeation of H_2 (Lin et al., 2006a).

Several studies have focused on using PEO or PEO-containing materials for reverse-selective CO_2/H_2 and CO_2/N_2 separations, and some of this literature has been summarized in a recent review (Lin and Freeman, 2005b). The primary disadvantage of pure, high-molecular-weight PEO is its strong tendency to crystallize, which is deleterious for gas permeability (Lin and Freeman, 2005b). Several methods have been explored to reduce crystallinity in PEO: using low-molecular-weight liquid PEO as an additive to other polymers (Li et al., 1995), designing phase-separated block copolymers containing ethylene oxide segments that are too short to crystallize (Bondar et al., 1999), and preparing highly cross-linked networks with large concentrations of PEO (Lin and Freeman, 2005b). For the latter case, when the molecular weight of PEO segments between cross-links is lower than 1500, crystallization can be avoided, and wholly amorphous networks are obtained (Graham, 1987).

Our research focuses on the structure and transport properties of cross-linked PEO (XLPEO) networks prepared with varying cross-link density and network architecture. Networks were synthesized via photopolymerization of poly(ethylene glycol) diacrylate cross-linker in the presence of water as a diluent or via copolymerization with acrylate monomers, leading to network structures as depicted schematically in Figure 35.1. Dynamic mechanical analysis was used to characterize the viscoelastic properties of the networks, and the static and dynamic characteristics of these materials were related to their gas separation performance.

35.2 THEORY AND EXPERIMENT

35.2.1 Permeability and Selectivity

Gas transport through a dense or nonporous polymeric film is often described by the solution diffusion mechanism (Graham, 1866). The permeability of a polymer to a gas

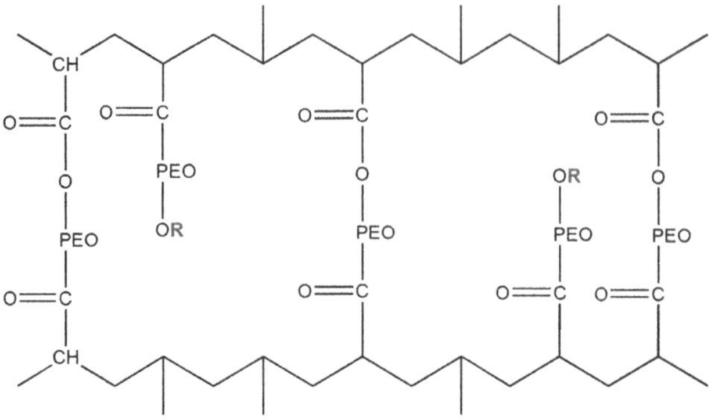

Figure 35.1 Scheme of representative network elements in XLPEO, with R representing either —H (when PEGA is being copolymerized with PEGDA) or —CH_3 (when PEGMEA is being copolymerized with PEGDA) and PEO representing —$(CH_2—CH_2—O)_n$—. Structures of the monomers are shown later in Table 35.1.

A, P_A, is defined as (Petropoulos, 1994)

$$P_A \equiv \frac{N_A l}{f_2 - f_1} \tag{35.1}$$

where N_A is the steady-state gas flux through the film, l is the film thickness, and f_2 and f_1 are the upstream and downstream fugacities of gas A, respectively. Fugacity is used instead of pressure to account for nonideal behavior in the gas phase, which can be significant for compressible penetrants (such as CO_2) at low temperature and high pressure, particularly in mixtures with other gases. Fugacity can be calculated from the virial equation of state, with parameters available in the literature (Stern et al., 1969; Dymond and Smith, 1980; Lin and Freeman, 2005a). If the diffusion process obeys Fick's law and the downstream pressure is much less than the upstream pressure (as is the case for our experiments), then the permeability is given by (Wijmans and Baker, 1995)

$$P_A = D_A S_A \tag{35.2}$$

where S_A is the solubility coefficient and D_A is the average effective diffusivity through the film. The ideal selectivity of a membrane for gas A over gas B is defined as the ratio of their pure gas permeabilities (Freeman and Pinnau, 1999):

$$\alpha_{A/B} = \frac{P_A}{P_B} = \left[\frac{D_A}{D_B}\right]\left[\frac{S_A}{S_B}\right] \tag{35.3}$$

where D_A/D_B is the diffusivity selectivity and S_A/S_B is the solubility selectivity. Diffusivity selectivity is strongly influenced by the size difference between the penetrant molecules and the size-sieving ability of the polymer matrix, while solubility selectivity is controlled by the relative condensability of the penetrants and the relative affinity of the penetrants for the polymer matrix (Freeman and Pinnau, 1999). Enhanced gas selectivity, therefore, can be achieved by modifying either diffusivity or solubility selectivity or both (Ghosal and Freeman, 1994).

Gas solubility in rubbery polymers depends on gas condensability (as represented by critical temperature), interaction of the gas with the polymer matrix, and also weakly on the free volume of the polymer (Freeman and Pinnau, 1999). Sorption of gases at low pressure usually obeys Henry's law (Freeman, 1992):

$$C = k_D p \tag{35.4}$$

where k_D is the solubility coefficient or Henry's law constant.

The average effective diffusivity through the film can be defined in terms of local diffusion coefficients by (Lin and Freeman, 2006a)

$$D_A = \frac{1}{C_2 - C_1} \int_{C_1}^{C_2} \frac{D_{\text{loc}}}{1 - w_2} dC = \frac{1}{C_2 - C_1} \int_{C_1}^{C_2} D_{\text{eff}}\, dC \tag{35.5}$$

where D_{loc} is the local concentration-dependent diffusion coefficient, C_2 and C_1 are the dissolved penetrant concentrations at the upstream and downstream faces of the polymer, respectively, w_2 is the penetrant mass fraction in the polymer, and D_{eff} is the local effective diffusion coefficient that characterizes the penetrant mobility in the polymer. The coefficient D_{eff} can be expressed using Eqs. (35.1), (35.2), and (35.5) as (Koros et al., 1977)

$$D_{\text{eff}}(C_2) = \left[P_A + f\frac{dP_A}{df}\right]_{f_2} \left(\frac{df}{dC_2}\right)_{f_2} \tag{35.6}$$

Diffusion of small molecules in a polymer is often quantitatively related to free volume by (Cohen and Turnbull, 1959; Fujita, 1961)

$$D_A = A_D \exp\left(-\frac{B}{\text{FFV}}\right) \tag{35.7}$$

where A_D is a pre-exponential factor, B is a constant proportional to penetrant size, and FFV is the fractional free volume in the polymer. The diffusion coefficient appearing in Eq. (35.7) is, strictly speaking, the gas self-diffusion coefficient and not the binary mutual diffusion, D_{loc}. However, we have recently shown that for the cases of interest to this work there is no significant error in replacing the self-diffusion coefficient with D_A (Lin and Freeman, 2006a). Gas diffusivity is also determined by polymer chain flexibility, often characterized by polymer glass transition temperature T_g (Freeman and Pinnau, 1999). Combining Eqs. (35.2) and (35.7) yields an expression for permeability:

$$P_A = A_P \exp\left(-\frac{B}{\text{FFV}}\right) \tag{35.8}$$

where A_P is a pre-exponential factor that is equal to A_D multiplied by the solubility of penetrant A in the polymer, S_A. The FFV is often estimated using the following expression (Lee, 1980):

$$\text{FFV} = \frac{V - V_0}{V} \tag{35.9}$$

where V is the specific volume of the amorphous polymer at the temperature of interest as obtained from density measurements, and V_0 is the specific occupied volume at 0 K, estimated as 1.3 times the van der Waals volume, which in turn can be calculated using Bondi's group contribution method (van Krevelen, 1990). However, in the case of a strongly sorbing penetrant such as CO_2 in XLPEO, FFV is a function of not only temperature but also the partial pressure of CO_2, since the sorption of large amounts of CO_2 into the polymer will swell it, which increases free volume. This effect is accounted for in the modified free volume model discussed below (Lin and Freeman, 2006a).

For the permanent gases (H_2, N_2, and O_2), permeability coefficients are essentially independent of pressure, whereas the permeability coefficients of CO_2 and the hydrocarbons increase with increasing pressure. This behavior is consistent with gas permeation properties in rubbery polymers (Merkel et al., 2000). Strongly sorbing penetrants, such as CO_2, can plasticize the polymer matrix by increasing polymer local segmental motion, thus enhancing their own diffusivity and corresponding permeability (Koros and Hellums, 1989). Therefore, all permeability values reported in this work, including those plotted in the figures, have been extrapolated to infinite dilution from several permeability measurements performed at different pressures up to 15 atm. The permeability is often empirically related to fugacity (instead of pressure, to account for possible nonideal behavior of CO_2, e.g., at lower temperature) to obtain infinite-dilution permeability ($P_{A,0}$) as follows (Stern et al., 1971):

$$P_A = P_{A,0} \exp(m_{P,E}\,\Delta f) = P_{A,0} \exp(m_{P,E} f_2) \tag{35.10}$$

Here, $m_{P,E}$ is an adjustable constant at a given temperature and Δf is the difference between the upstream and downstream fugacity, $\Delta f = f_2 - f_1$. In this study, the downstream fugacity f_1 is practically zero, so Δf can be replaced by f_2. Therefore, from Eq. (35.10), $P_{A,0}$ is the permeability coefficient when $\Delta f = 0$ (this state is often achieved by allowing both f_1 and f_2 to go to zero, hence the name "infinite-dilution permeability"). For permanent gases, the permeability typically changes little with fugacity and $m_{P,E} f_2$ is very small. In this case, Eq. (35.10) reduces to a linear form (Merkel et al., 2000; Lin and Freeman, 2004):

$$P_A = P_{A,0}(1 + m_{P,E}\,\Delta f) = P_{A,0}(1 + m_{P,E} f_2) \tag{35.11}$$

Infinite-dilution solubility and diffusivity can be calculated using similar relationships.

35.2.2 Dynamic Relaxation Characteristics of Polymer Networks

The dynamic relaxation characteristics of cross-linked polymer networks are sensitive to the architecture and topography of the network as well as the overall cross-link density. Dynamic thermal analysis techniques such as dynamic mechanical analysis and broadband dielectric spectroscopy can be used to characterize the subglass and glass–rubber relaxation characteristics of polymer networks across a wide range of temperatures and timescales, and a number of experimental studies have appeared that examine the influence of cross-link density on the time–temperature characteristics of the underlying motional processes (Glatz-Reichenbach et al., 1994; Roland, 1994; Kannurpatti et al., 1997, 1998; Kannurpatti and Bowman, 1998; Fitz and Mijovic, 1999; Litvinov and Dias, 2001; Schroeder and Roland, 2002; Alves et al., 2004). These studies provide valuable insight as to the influence of cross-links on the relaxation environment of the polymer chain segments as related to network composition, structure, and local free volume. The presence of covalent cross-links typically results in a reduction in segmental mobility in the vicinity of the cross-link junctions, with the corresponding constraint manifested by an increase in glass transition temperature T_g; this effect is most pronounced at high cross-link densities, where the average distance between cross-links approaches the length-scale characteristic of cooperative segmental motion. The presence of cross-links also leads to an inhomogeneous broadening of the glass transition owing to the range of relaxation environments experienced by the chain segments and their relative proximity to the cross-link points. For the cross-linked networks described here, the maximum effective cross-link density is established by the molecular weight of the PEGDA cross-linker, which has a monomeric repeat length (n) corresponding to 14 ethylene oxide units. Copolymerization with PEG acrylates (i.e., PEGA, PEGMEA) or photopolymerization in the presence of diluent can be used to produce networks with lower effective cross-link density.

Measurement of the rubbery plateau modulus (E_R) provides a direct means by which to assess cross-link density in polymer networks. Classical rubber elasticity theory (Treloar, 1975; Mark, 1982; Hill, 1997) relates the elastic modulus in the rubbery region to effective cross-link density (ν_e) as shown, for example, in the equation suggested by Hill (1997):

$$\nu_e = \frac{E_R}{3RT} \tag{35.12}$$

where R is the gas constant and T is absolute temperature. For the networks described here, time–temperature superposition methods (Ferry, 1980) can be applied across the glass transition to construct master curves of storage modulus versus ωa_T, where ω is the applied test frequency ($\omega = 2\pi f$, with f expressed in hertz) and a_T is the dimensionless shift factor. The glass–rubber relaxation is satisfactorily described using the Kohlrausch–Williams–Watts (KWW) (Williams et al., 1971) stretched exponential relaxation time distribution function:

$$\phi(t) = \exp\left[-\left(\frac{t}{\tau_0}\right)^{\beta}\right] \tag{35.13}$$

where τ_0 is the central relaxation time and β is the relaxation time distribution parameter, which is used to characterize the breadth of the relaxation. β ranges from 0 to 1, with lower values indicating increased intermolecular coupling and broadening of the relaxation owing to the influence of the cross-links. The time–temperature characteristics of the various networks can be compared via the construction of cooperativity or fragility plots, normalized Arrhenius plots wherein $\log(a_T)$ is plotted versus T_g/T in the vicinity of the glass transition (Angell, 1991; Roland, 1994). These plots provide an objective basis to examine the intermolecular cooperativity characteristics of the networks and the apparent activation energy associated with segmental motion.

35.2.3 Experimental Details

Fundamental insight into the relationship between the structural and dynamic characteristics of the polymer networks and their corresponding transport properties is required to design cross-linked PEO-based membranes with optimum gas separation performance. Our studies are focused on copolymer networks prepared from mixtures of PEGDA cross-linker (ethylene oxide chain length $n = 14$) with either PEGA ($n = 7$) or PEGMEA ($n = 8$). The corresponding monomer structures are shown in Table 35.1. Reaction mixtures were prepared

TABLE 35.1 Chemical Structures of the Cross-linker and Monomers Used in This Study

Chemical	Structure
Poly(ethylene glycol) diacrylate (PEGDA)	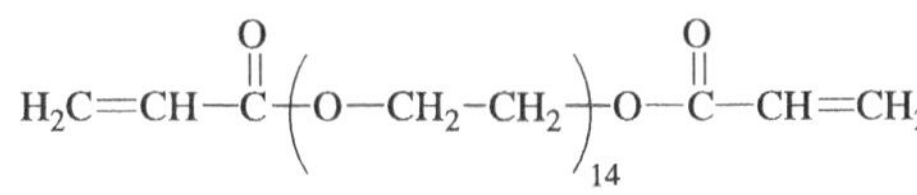
Poly(ethylene glycol) methyl ether acrylate (PEGMEA)	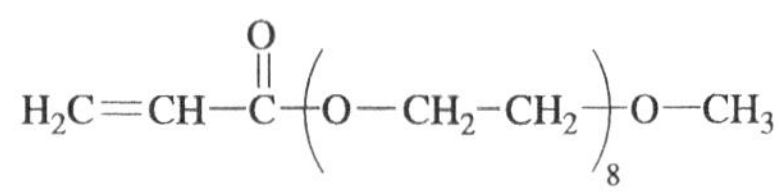
Poly(ethylene glycol) acrylate (PEGA)	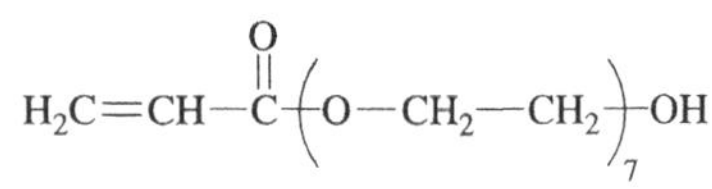

by combining appropriate monomer ratios with 0.1 wt% 1-hydroxyl-cyclohexyl phenyl ketone (HCPK) photoinitiator followed by magnetic stirring for at least 1 h. The prepolymer solution was sandwiched between quartz plates with controlled spacing and photopolymerized using an ultraviolet (UV) cross-linker (model FB-UVXL-1000, Fisher Scientific) with 312-nm UV light for 90 s at 3 mW/cm^2. The resulting solid film was subsequently immersed in ultrapure water to remove any unbound low-molecular-weight "sol." Attenuated total reflection Fourier transform infrared spectrometry (FTIR-ATR, Thermo Nicolet Nexus 470, Madison, WI) indicated virtually complete conversion of the acrylate groups for the network compositions described here. Additional details of the film preparations were given in an earlier paper (Lin et al., 2005).

The ethylene oxide (EO) chain lengths in the cross-linker and monomers were carefully chosen to keep the concentration of ethylene oxide in the resulting networks constant (~82 wt%), so that property differences among the networks could be attributed solely to changes in network structure and would not be complicated by variations in EO concentration. For networks prepared via copolymerization of PEGA or PEGMEA with the PEGDA cross-linker, the cross-link density can be varied directly by varying the relative amount of cross-linker (PEGDA) in the prepolymer formulation, with virtually no change in the overall chemical composition of the resulting network. The presence of the PEGA or PEGMEA comonomer leads to the insertion of nonreactive pendant branches throughout the network, which is cross-linked by reaction of the bifunctional PEGDA; see Figure 35.1. Alternately, photopolymerization of PEGDA cross-linker in the presence of water, which acts as a diluent, can be used to produce networks of varying cross-link density, as the water in the prepolymerization reaction mixture leads to the introduction of "wasted" cross-links and reduced effective cross-link density (Lin et al., 2005; Kalakkunnath et al., 2006). For the PEO-based materials described here, dynamic mechanical analysis was used to characterize the viscoelastic relaxation properties of the rubbery networks in relation to gas separation performance. Permeability, diffusivity, and solubility measurements were performed using a selection of relevant gases over a wide range of temperature and pressure. Experimental details associated with the dynamic mechanical and gas transport studies have been provided elsewhere (Kalakkunnath et al., 2005; Lin and Freeman, 2006b).

35.3 RESULTS AND DISCUSSION

35.3.1 Viscoelastic Properties of Cross-linked PEO Networks

The glass–rubber relaxation characteristics of the cross-linked PEO networks were studied using dynamic mechanical analysis. Figure 35.2 shows representative results (storage modulus vs. temperature) for cross-linked PEGDA ($n = 14$) and the PEGDA/PEGA copolymers. The modulus curves indicate a modest decrease in glass transition temperature (T_α; 1 Hz) with increasing acrylate content that reflects insertion of the nonreactive PEGA branches into the polymer network; a similar effect, with an even stronger negative shift in glass transition temperature, was observed for the PEGDA/PEGMEA copolymers (see summary of transition temperatures in Table 35.2). The rubbery plateau modulus displays a systematic decrease with increasing PEGA (i.e., decreasing cross-linker content). Rubbery modulus (E_R) is plotted versus PEGDA content in Figure 35.3 for the PEGDA/PEGA and PEGDA/PEGMEA copolymer networks as well as for networks formed via the photopolymerization of PEGDA in the presence of water diluent. For all three series, a single relationship is observed between rubbery modulus and cross-linker content, implying that effective cross-link density (ν_e) depends solely on the amount of PEGDA cross-linker present in the reaction mixture, even though the structural details of the resulting networks may differ. Notably, for the PEGDA/water networks, no change in glass transition temperature is observed with varying effective cross-link density (Lin et al., 2005). For this particular cross-linker ($n = 14$), the modification in network structure that occurs with reaction in the presence of diluent apparently does not lead to a significant change in the constraint experienced by the chain segments at the length scale associated with the glass transition. While the observed variation in rubbery modulus clearly indicates a reduction in effective cross-link density, the formation of loops or wasted cross-links in the presence of water has little influence on the resulting value of T_g (Kalakkunnath et al., 2006).

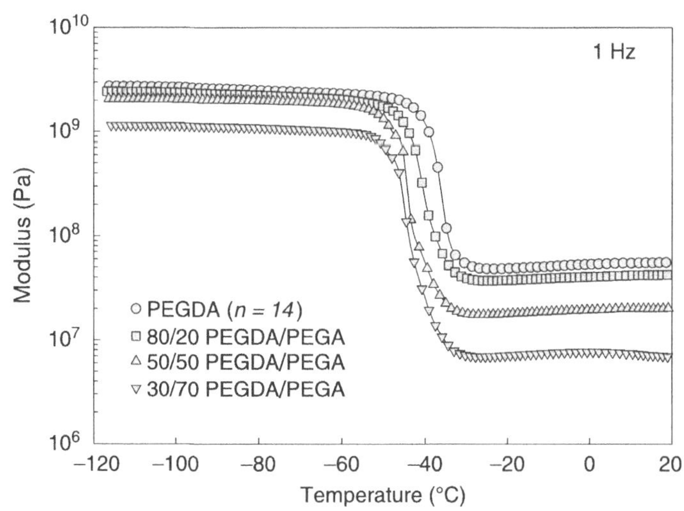

Figure 35.2 Storage modulus (E') versus temperature for PEGDA/PEGA copolymer networks with varying weight fraction PEGDA cross-linker. Frequency of 1 Hz; heating rate of 1°C/min. [Reprinted with permission from Kalakkunnath et al. (2005). Copyright 2005 American Chemical Society.]

TABLE 35.2 Characteristics of Cross-linked PEGDA (n = 14) and Copolymer Networks

	T_α (1 Hz)	T_g (DSC)	β_{KWW}	FFV
PEGDA	−35°C	−40°C	0.30	0.118
PEGDA/PEGMEA				
80/20	−41°C	−44°C	0.33	0.122
50/50	−47°C	−52°C	0.34	0.127
30/70	−52°C	−57°C	—	0.128
PEGDA/PEGA				
80/20	−38°C	−40°C	0.34	0.112
50/50	−41°C	−42°C	0.35	0.112
30/70	−42°C	−44°C	0.38	0.110

Note: T_α, dynamic mechanical peak temperature for glass transition (1 Hz); T_g, calorimetric glass transition temperature; β_{KWW}, KWW distribution parameter for glass–rubber relaxation; FFV, fractional free volume based on density measurements; DSC, differential scanning calorimetry.

Time–temperature superposition was used to establish master curves of storage modulus versus ωa_T, as shown in Figure 35.4 (PEGDA/PEGA copolymer networks); the shift of the curves to higher frequency with PEGA content reflects a progressive decrease in relaxation time consistent with the observed trend in glass transition temperature. The KWW curve fits (solid curves in Fig. 35.4) indicate narrowing of the glass–rubber relaxation with decreasing overall cross-link density, as reflected in the value of the distribution parameter β, which increases from 0.30 to 0.38 across the series; see Table 35.2. The relaxation narrowing suggests a reduction in the elastic constraint imposed by the cross-link junctions, leading to a more homogeneous relaxation environment at lower cross-link density. This result, which was also observed for the PEGDA/PEGMEA and PEGDA/water series, is

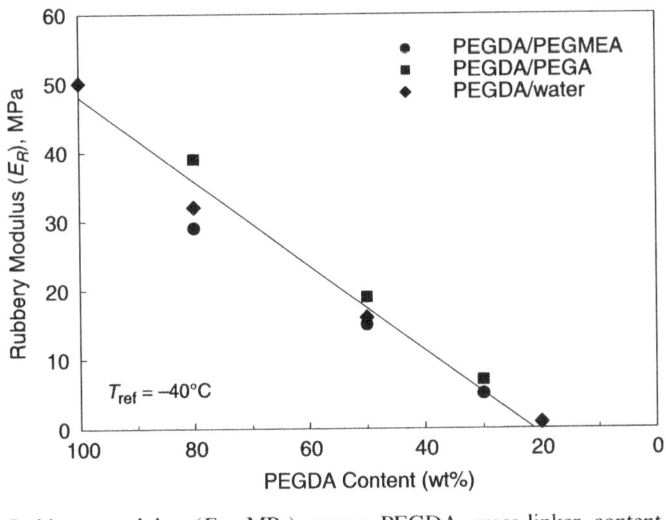

Figure 35.3 Rubbery modulus (E_R; MPa) versus PEGDA cross-linker content for PEGDA/PEGMEA, PEGDA/PEGA, and PEGDA/water networks, where E_R is evaluated from time–temperature master curves at −40°C. [Reprinted with permission from Kalakkunnath et al. (2005). Copyright 2005 American Chemical Society.]

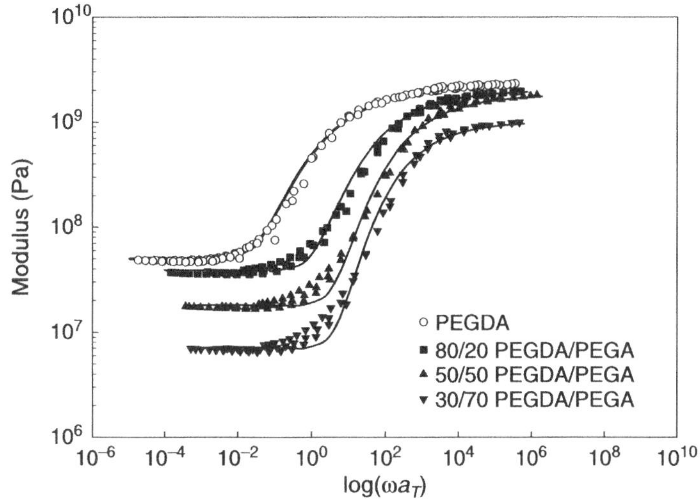

Figure 35.4 Time–temperature master curves for PEGDA/PEGA copolymer networks; $T_{ref} = -40^{\circ}C$. Solid curves are KWW best fits. [Reprinted with permission from Kalakkunnath et al. (2005). Copyright 2005 American Chemical Society.]

consistent with observations reported for comparable polymer networks of varying cross-link density (Kannurpatti et al., 1997, 1998; Schroeder and Roland, 2002; Alves et al., 2004).

An effective method to compare the time–temperature relaxation characteristics of the networks is the construction of cooperativity plots (Roland, 1994), wherein the time–temperature shift factor is plotted as $\log(a_T)$ versus reciprocal temperature, expressed as T_α/T, where T_α is the dynamic mechanical glass transition temperature at 1 Hz. The cooperativity plot for the PEGDA/PEGA series is shown in Figure 35.5. The local slope of the individual curves reflects the activation energy associated with the glass transition process and provides an indication of the relative intermolecular cooperativity inherent to the relaxation in each network. For the PEGDA copolymer networks, decreasing cross-link density leads to a gradual reduction in apparent activation energy (i.e., fragility), which is consistent with the relaxation of a less constrained network in which the underlying motional processes encompass a lower degree of intermolecular cooperativity (Angell, 1991; Ngai and Roland, 1993). The observed behavior, which is characteristic of many homopolymer network systems with varying cross-link density, suggests that the glass–rubber relaxation process in the PEGDA copolymer networks is not substantially altered or impeded by the introduction of the pendant branches. In fact, the segmental motions occurring along the ethylene oxide segments in the pendant branches are likely to be very similar in character to those occurring across the PEGDA cross-link bridges and, as such, the presence of the branches does not significantly change the time–temperature character of the glass transition. The decrease in cooperativity that is evident in Figure 35.5 is primarily due to the net decrease in cross-link density with increasing branch content.

The presence of the pendant branches does, however, lead to a decrease in the glass transition temperature for the copolymer networks as measured by both dynamic mechanical and calorimetric methods (refer to Table 35.2), as the nonreactive chain ends introduce

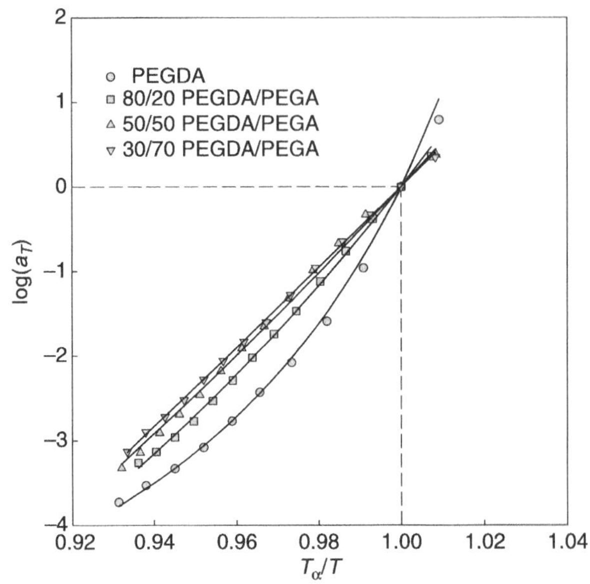

Figure 35.5 Cooperativity plots [log(a_T) vs. T_α/T] for PEGDA/PEGA copolymer networks. Solid curves are Williams-Landel-Ferry (WLF) fits, as defined by Ferry (1980). [Reprinted with permission from Kalakkunnath et al. (2005). Copyright 2005 American Chemical Society.]

defects into the network structure. This effect is more pronounced for the PEGDA/PEGMEA copolymers (—OCH_3 branch ends) as compared to the PEGDA/PEGA copolymers (—OH branch ends). For the PEGDA/PEGMEA copolymers, the decrease in T_g is accompanied by a progressive increase in FFV as determined according to Eq. (35.9); see Table 35.2. Interestingly, for the PEGDA/PEGA copolymers, a small *decrease* in FFV is observed with increasing PEGA branch content. As will be shown, this relatively minor difference in network structure (—OCH_3 vs. —OH end group) has a strong influence on the gas transport properties of these rubbery networks.

35.3.2 Correlation of Structural Detail of the Membrane with Gas Transport Properties

To analyze the effect of cross-link density and structural details on the gas transport properties of XLPEO, samples with varying degrees of cross-linking were subjected to permeability measurements using selected gases. Figure 35.6a presents the CO_2 permeability at infinite dilution (Lin et al., 2006c). The infinite-dilution permeability of the PEGDA/PEGMEA network increased by more than 400%, from 110 to 580 Barrers [where 1 Barrer = 1×10^{-10} cm^3 at standard temperature and pressure (STP) cm/cm^2 s cm Hg], as PEGDA content was decreased from 100 to 1.3 wt%. In contrast, the infinite-dilution permeability of the PEGDA/PEGA network remained relatively constant (110 Barrers) with varying PEGDA content. For comparison, samples of PEGDA diluted before photopolymerization with different amounts of water were tested: The dilution reduced the effective cross-link density of the network, as shown in Figure 35.3. In this case, the permeability increased from 110 to 140 Barrers as water content increased

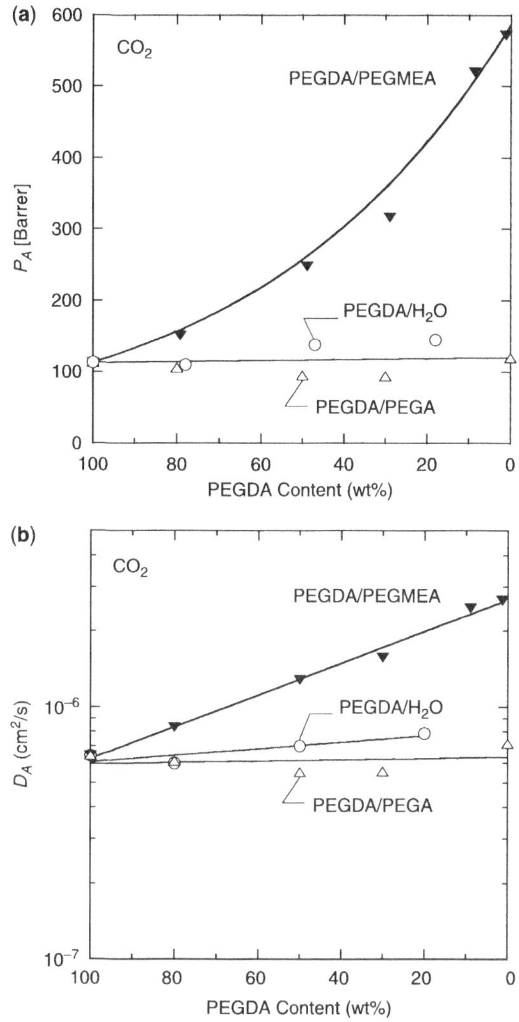

Figure 35.6 (a) CO_2 permeability and (b) CO_2 diffusivity (both extrapolated to infinite dilution) through XLPEO to downstream vacuum as a function of PEGDA content in prepolymer solution; 35°C. Here, D_A is calculated from P_A based on Eq. (35.2). [Reprinted with permission from Lin et al. (2006c). Copyright 2005 Elsevier.]

from 0 to 80 wt%, so cross-link density per se had only a minimal effect on the permeability of the polymer membranes.

To understand the effect of network structure on permeability, the diffusion coefficients for each system were calculated based on independent sorption studies, and the result is presented in Figure 35.6b. Comparison with Figure 35.6a. reveals similar trends such that the observed permeability variation can be attributed mostly to a change in penetrant diffusivity in the networks. In a homogeneous blend, gas transport properties (X) such as permeability or diffusivity are often modeled empirically as (Paul, 1984)

$$\ln X = \Phi_1 \ln X_1 + \Phi_2 \ln X_2 \tag{35.14}$$

where Φ_i is the volume fraction of component i and the subscripts 1 and 2 correspond to components 1 and 2, respectively. This equation, represented as the solid curves in Figure 35.6a, is able to satisfactorily predict the change in gas permeability with varying PEGDA content; the equation only takes into account the correlation between permeability and composition of the polymer blends and copolymers but does not account for cross-link density. Therefore, the changes in CO_2 permeability can be attributed mainly to the influence of the methoxy (—OCH_3) chain end groups on bulk density and not to variations in cross-link density. The presence of the methoxy chain end groups increases the polymer FFV and, therefore, the diffusivity of gases through the network (see FFV values in Table 35.2); a detailed analysis of this result has been published separately (Lin et al., 2006c).

Since permeability of gases in these systems depends primarily on diffusivity, Eq. (35.7) can be used to correlate the selectivity of the networks (calculated based on measured pure gas permeabilities) with FFV using previously calculated adjustable constants (Lin and Freeman, 2005a, 2006a); the selectivity values for several CO_2/light gas systems are presented in Figure 35.7 and compared with model calculations as described below. The experimental FFVs were determined from bulk density measurements according to Eq. (35.9). The lines through the data points were calculated using the following free-volume model derived from Eqs. (35.3) and (35.8):

$$\alpha_{A/B} = \frac{A_{P,A}}{A_{P,B}} \exp\left(\frac{B_B - B_A}{\text{FFV}}\right) \tag{35.15}$$

which predicts that CO_2/light gas selectivity increases with increasing FFV if the light gas is smaller than CO_2 and CO_2/light gas selectivity decreases with increasing FFV if the light

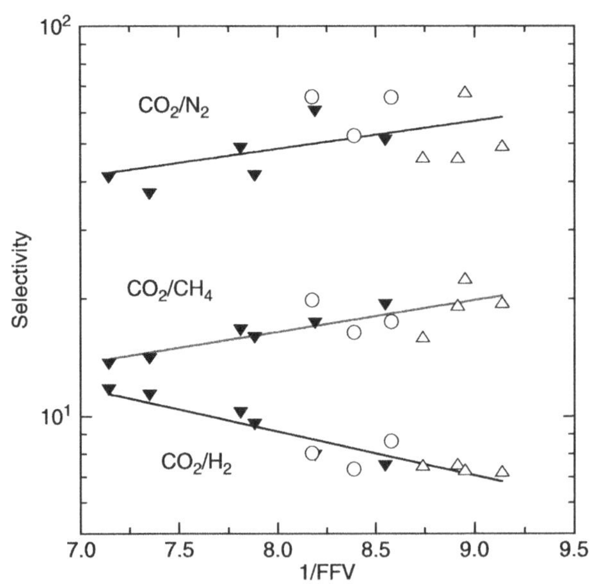

Figure 35.7 Correlation between pure gas selectivity and FFV for various XLPEO networks. Symbols correspond to those used in Figure 35.6. Permeability data used were as extrapolated to infinite dilution. [Reprinted with permission from Lin et al. (2006c). Copyright 2006 Elsevier.]

gas is larger than CO_2. For the gas pairs shown in Figure 35.7, H_2 is smaller than CO_2, while N_2 and CH_4 have larger kinetic diameters. For each gas pair considered, the selectivity values indicate preferential permeation of CO_2 ($\alpha > 1$); in the case of CO_2 and H_2, CO_2/H_2 selectivity ranges from 7 to 12, showing that these networks have high permeability and high reverse selectivity.

35.3.3 Temperature and Pressure Effects on the Transport Properties of XLPEO

Temperature plays an important role in determining the transport performance of any membrane. Usually, with increasing temperature, permeability of all penetrants increases, as described by the Van't Hoff–Arrhenius expression (Baker, 2004; Freeman and Pinnau, 1999):

$$P_A = P_{A,0}, \exp\left(\frac{-E_P}{RT}\right) \tag{35.16}$$

This behavior is demonstrated in Figure 35.8a, where CO_2 permeability data for selected PEGDA/PEGMEA copolymers are plotted against reciprocal temperature according to Eq. (35.16). Higher temperatures and greater PEGMEA content lead to higher permeability.

The influence of temperature on gas diffusion coefficients is presented in Figure 35.8b, which shows pure gas CO_2 diffusivity at various temperatures as a function of CO_2 fugacity for cross-linked PEO prepared by photopolymerization of PEGDA cross-linker in the presence of 20 wt% water. As indicated by these data, diffusivity increases as temperature increases.

In this series of experiments, the diffusivity of CO_2 increases much more strongly with fugacity at lower temperature than at higher temperature. For example, from Figure 35.8b, CO_2 diffusivity increases from 6×10^{-7} to 9×10^{-7} cm^2/s as fugacity increases from 0 to 10 atm at 308 K; in contrast, diffusivity increases by a full order of magnitude, from 6×10^{-9} to 6×10^{-8} cm^2/s, over a similar fugacity range at 253 K. This difference can be attributed to the plasticization effect of sorbed CO_2 in the polymer, leading to an increase in FFV and, in turn, gas diffusivity (Lin et al., 2006a). Carbon dioxide sorption increases as temperature decreases (Lin and Freeman, 2005a), so the plasticization effect of CO_2 on diffusivity becomes more pronounced at lower temperatures.

Figure 35.8c presents the influence of CO_2 partial pressure on CH_4 permeability in CO_2/CH_4 mixed-gas permeation experiments. In this case, the presence of CO_2 increases the permeability of CH_4 strongly. This effect is believed to be linked to plasticization of the polymer matrix by CO_2, which would act to increase the FFV and, in turn, CH_4 permeability. This effect is more pronounced at lower temperature since, at a given partial pressure of CO_2, the thermodynamic activity of CO_2 (and, therefore, the CO_2 concentration in the polymer) is higher. Methane is much less soluble than CO_2, so the permeation properties of both CO_2 and CH_4 are essentially independent of CH_4 partial pressure (Lin and Freeman, 2005a). For reverse-selective separations such as the removal of CO_2 from H_2, plasticization by CO_2 actually *improves* separations performance by increasing gas permeability without loss in selectivity in rubbery XLPEO (Lin et al., 2006a).

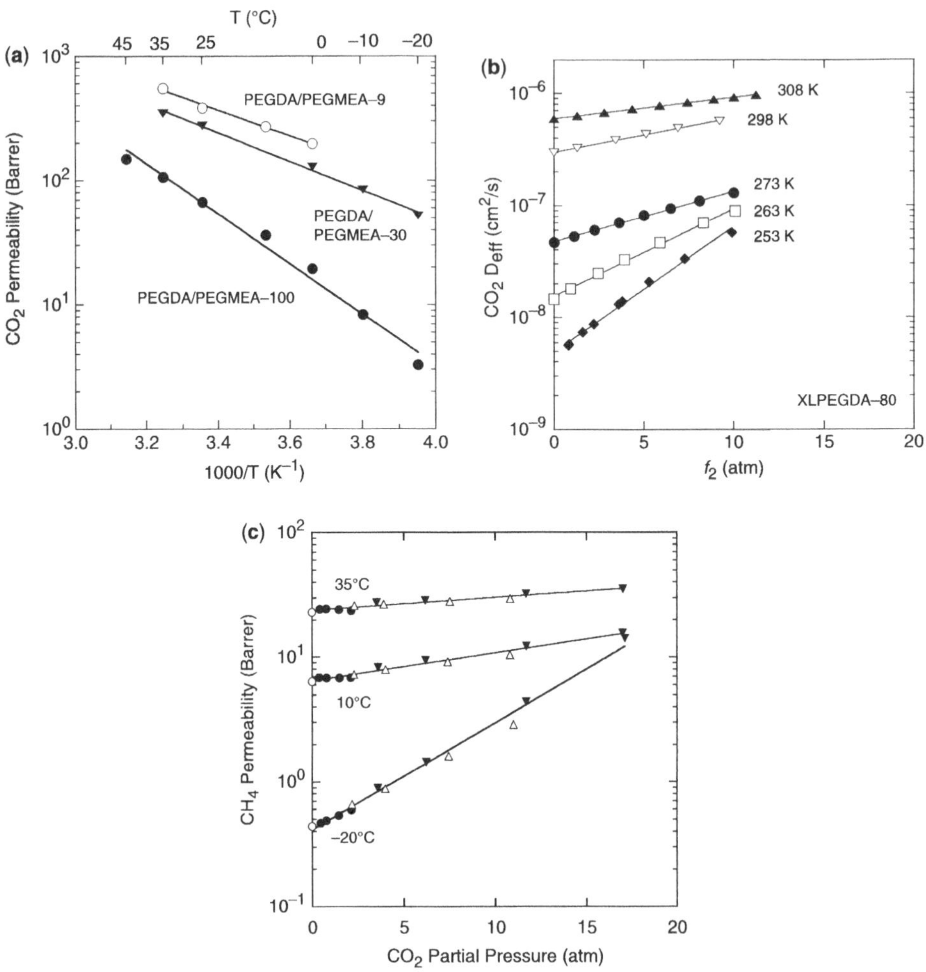

Figure 35.8 (a) Effect of temperature and polymer composition on CO_2 permeability for various PEGDA/PEGMEA networks. (b) Effect of pure CO_2 upstream fugacity and temperature on CO_2 diffusion coefficient in XLPEGDA-80. (c) Effect of CO_2 partial pressure on CH_4 permeability in PEGDA/PEGMEA-30 from CO_2/CH_4 gas mixtures. Symbols denote different CO_2/CH_4 mixed-gas compositions: (○) pure CO_2; (●) 10% CO_2; (△) 50% CO_2; (▼) 80% CO_2). XX in PEGDA/PEGMEA-XX refers to the weight percent of PEGDA in the prepolymer mixture with PEGMEA, while XX in XLPEGDA-XX refers to the weight percent of PEGDA in the prepolymer mixture with water.

35.3.4 Modified Free-Volume Model

The general relationship between permeability and fugacity, as given in Eq. (35.10), can be substituted into Eq. (35.6) to estimate diffusion coefficients as a function of temperature and penetrant concentration (Merkel et al., 2000; Lin and Freeman, 2004). However, values of $P_{A,0}$ and $m_{P,E}$ are required for each gas at each temperature of interest, leading to a large number of empirical parameters needed to characterize diffusivity in a particular material (Lin and Freeman, 2006a). To address this shortcoming, a model based on free-volume theory was used to correlate the data, as described below.

An increase in the concentration of methoxy (—OCH_3) chain end groups in the PEO copolymers leads to an increase in diffusivity and permeability due to enhanced polymer FFV. This result is confirmed by positron annihilation lifetime spectroscopy (PALS) measurements; as the concentration of methoxy chain end groups in the PEGDA/PEGMEA networks is increased, the FFV as probed by PALS increases (Lin and Freeman, 2006a). In addition, the glass transition temperature as measured by the calorimetric method (T_g) in these polymers decreases as the methoxy chain end-group concentration (i.e., PEGMEA concentration) increases. Here, T_g has been correlated with FFV in rubbery PEO according to the following expression (Sperling, 2001):

$$\mathrm{FFV} = \mathrm{FFV}(T_g) + \alpha_r\left(T - T_g\right) \tag{35.17}$$

where $\mathrm{FFV}(T_g)$ is the apparent FFV at T_g and α_r is the thermal expansion coefficient of the FFV. Figure 35.9 presents this correlation for a series of PEGDA/PEGMEA copolymers with varying concentrations of methoxy chain end groups. Assuming α_r to be independent of copolymer composition, Eq. (35.17) can be used to fit the data, as per the straight line shown in Figure 35.9. Based on this result, $\mathrm{FFV}(T_g) = 0.055 \pm 0.001$ and $\alpha_r = (8.4 \pm 2.6) \times 10^{-4}\ \mathrm{K}^{-1}$ (Lin et al., 2006a). The value of $\mathrm{FFV}(T_g)$ compares favorably to values obtained for other materials in previous studies (Sperling, 2001). Likewise, the value of α_r is similar to the magnitude of the thermal expansion coefficient of the entire polymer, which is determined from dilatometry measurements on PEO; the reported value is $7.8 \times 10^{-4}\ \mathrm{K}^{-1}$ (Zoller and Walsh, 1995).

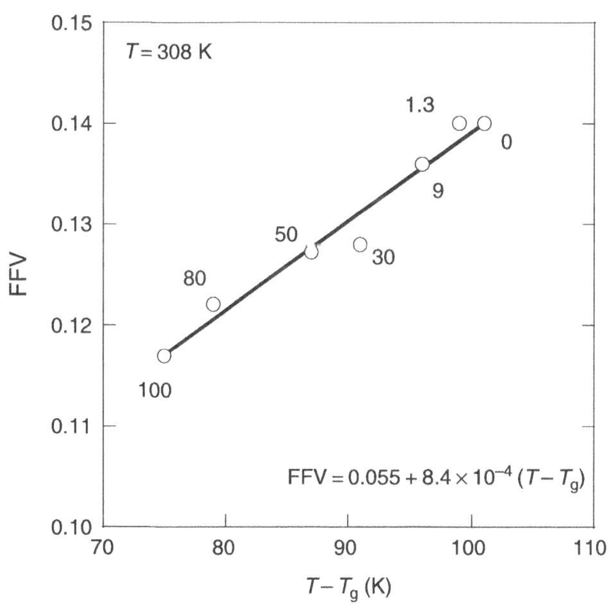

Figure 35.9 Correlation of FFV and $T - T_g$ based on Eq. (35.17) for various PEGDA/PEGMEA networks. Numbers above each data point indicate weight percent of PEGDA (i.e., cross-linker) in prepolymerization mixture with PEGMEA. This figure is based on Lin and Freeman (2005b).

The correlation between FFV and $T - T_g$ as defined in Eq. (35.17) can be combined with the Fujita-modified Cohen–Turnbull model given by Eq. (35.7) to obtain an equation relating diffusivity to temperature and FFV(T_g) (Lin and Freeman, 2006a):

$$D_{\text{eff}} = A_D \exp\left(\frac{-B}{0.055 + 8.4 \times 10^{-4}(T - T_g)}\right) \tag{35.18}$$

This model is then used in Figure 35.10a to correlate CO_2 infinite-dilution diffusivity at 35°C in various PEGDA/PEGMEA networks; the experimental results are provided to illustrate the fit. The fit appears to provide a reasonable description of these PEO-containing materials at 35°C. In Figure 35.10b, the model is used to describe infinite-dilution diffusivity of CO_2 in similar systems at different temperatures. Thus, the model can be generalized to correlate CO_2 diffusivity within similar PEO-containing systems across a range of temperatures.

The free-volume model can be extended beyond infinite dilution in the case of strongly sorbing components like CO_2. As CO_2 sorbs into the polymer matrix, it swells the polymer, resulting in enhanced segmental mobility and a depression in T_g and, therefore, an increase in diffusivity as indicated by Eq. (35.18) (Ismail and Lorna, 2002). Chow (1980) proposed an expression relating T_g depression to diluent concentration:

$$\ln\frac{T_g}{T_{g0}} = \beta[(1 - \theta)\ \ln(1 - \theta) + \theta\ \ln\ \theta] \tag{35.19}$$

where

$$\theta = \frac{M_p}{zM_d}\frac{w_2}{1 - w_2} \qquad w_2 = \frac{C_2 M_d}{C_2 M_d + 22414\rho_p} \qquad \beta = \frac{zR}{M_p \Delta C_{\text{pp}}}$$

Here T_{g0} and T_g are the glass transition temperatures of pure polymer and the polymer–diluent mixture, respectively, when the diluent weight fraction is w_2; M_p and M_d are the molecular weights of the polymer repeat unit (which is taken as 44 g/mol in this study since that is the molecular weight of an ethylene oxide moiety) and diluent, respectively; ρ_p is the polymer density; ΔC_{pp} is the change in heat capacity of the pure polymer at its glass transition, 0.99 J/(g · °C), which was obtained from DSC experiments (Lin and Freeman, 2006a). Equation (35.19) can be combined with the modified Cohen–Turnbull model [Eq. (35.18)] to quantitatively predict the effect of CO_2 concentration on diffusivity as a function of temperature and CO_2 partial pressure. This model uses a single parameter (FFV) and two adjustable parameters, A_D and B in Eq. (35.18), to describe the effect of temperature and pressure on gas diffusion coefficients.

Based on this model, Figure 35.11a presents the effect of temperature and pressure on CO_2 diffusivity in two different members of the family of cross-linked PEO materials. The permeation properties of CH_4, a co-permeant in gas mixtures with CO_2, may also be correlated with FFV using this model. Unlike CO_2, CH_4 solubility changes little with temperature (Lin and Freeman, 2005a), so CH_4 permeability can be directly plotted against 1/FFV at different temperatures and partial pressures of CO_2, as shown in Figure 35.11b. Again, there is good agreement between the model and the experimental results.

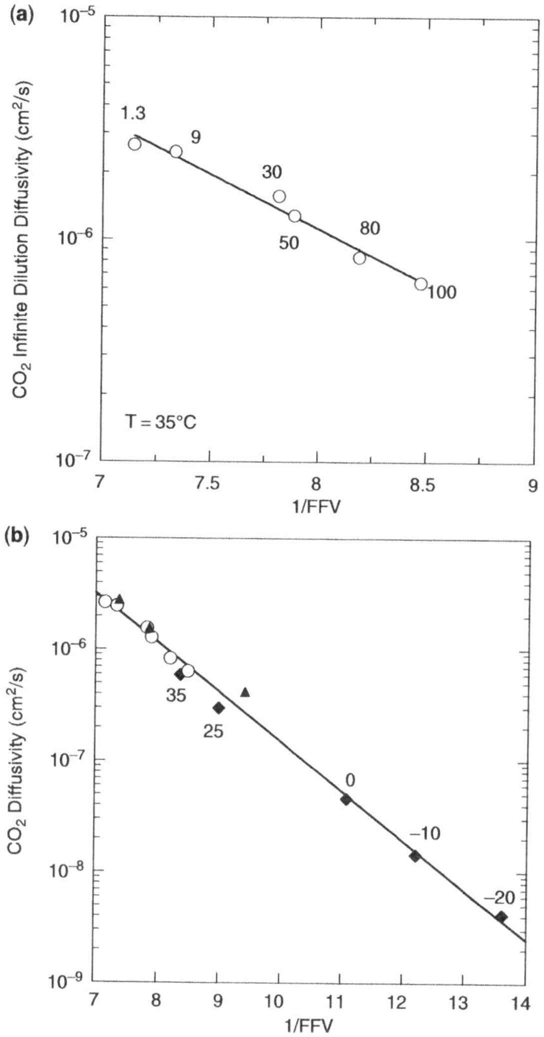

Figure 35.10 Infinite-dilution diffusivity of CO_2 as function of 1/FFV for several PEGDA/PEGMEA networks at (a) 35°C and (b) several temperatures showing the general application of Eq. (35.18). Numbers above each data point in (a) indicate weight percent of PEGDA (i.e., cross-linker) in prepolymerization mixture with PEGMEA. In (b): (○) PEGDA/PEGMEA at 35°C with various compositions; (▲) PEGDA/PEGMEA-9 (contained 9% PEGDA in prepolymer mixture) at various temperatures; (◆) XLPEGDA-80 (prepared using 80% by weight PEGDA in prepolymer mixture with water) at various temperatures, denoted above data points in degrees Celsius.

35.3.5 Performance of XLPEO Relative to the Upper Bound

The performance of XLPEO for CO_2 removal compares favorably to other polymers in terms of permeability and selectivity, quantities that are often plotted against each other as a basis of comparison between various materials. There is a characteristic trade-off between permeability and selectivity observed for polymeric membranes: More permeable polymers are generally less selective and vice versa. This trade-off was first empirically

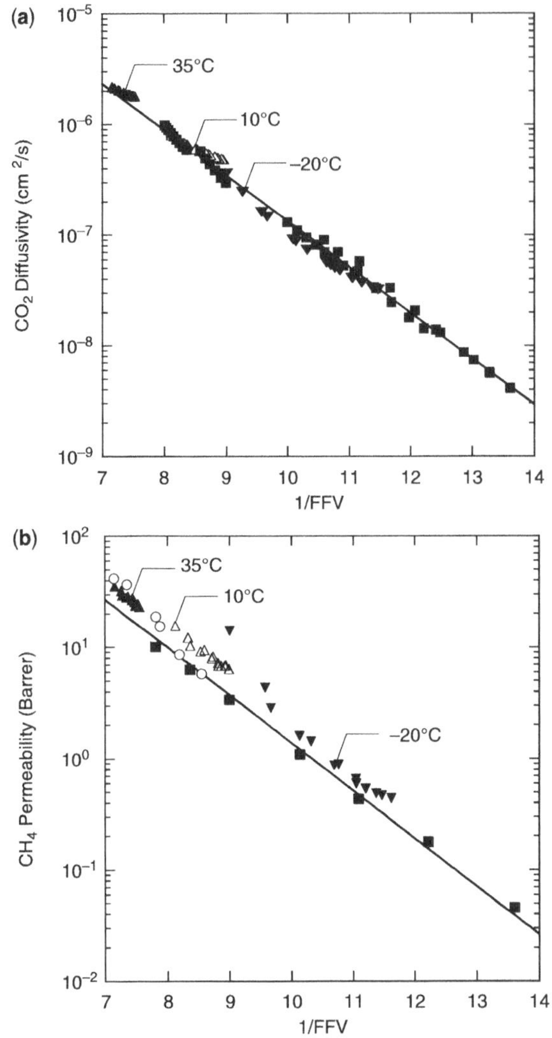

Figure 35.11 (a) Diffusivity of CO_2 as function of 1/FFV for PEGDA/PEGMEA networks and (b) mixed-gas permeability of CH_4 as function of 1/FFV for several PEGDA/PEGMEA networks at several temperatures and partial pressures of CO_2 showing general application of Eq. (35.18): (■) PEGDA/PEGMEA-100; (▲, ▼, △) PEGDA/PEGMEA-30 at various designated temperatures; (○) PEGDA/PEGMEA-1.3.

described by Robeson (1991) in examining the selectivity of several gas pairs in glassy polymers and was later determined to be a function of factors such as the ratio of the kinetic diameters of the penetrants, as shown below (Freeman, 1999):

$$\ln \alpha_{A/B} = -\left[\left(\frac{d_B}{d_A}\right)^2 - 1\right] \ln P_A + \left\{\ln\left(\frac{S_A}{S_B}\right) - \left[\left(\frac{d_B}{d_A}\right)^2 - 1\right] \times \left(b - f\left(\frac{1-a}{RT}\right) - \ln S_A\right)\right\} \quad (35.20)$$

where d_A and d_B are the kinetic diameters of the penetrants, a and b are gas species-independent parameters where a has a universal value of 0.64 and b has a value of $-\ln(10^{-4}\ \mathrm{cm}^2/\mathrm{s}) = 9.2$ for all rubbery polymers (Barrer, 1942; van Amerongen, 1946), and f is a polymer-dependent parameter related to the average distance between polymer chains. For rubbery polymers, f may be set to zero as a first approximation (Freeman, 1999). The selectivity of CO_2 relative to various gases of interest is plotted against CO_2 permeability in the cross-linked PEO family of materials in Figure 35.12 and compared to other previously studied materials.

In Figure 35.12a the separation characteristics of XLPEO are compared to the performance of cellulose acetate (CA) (Lin et al., 2006b). Mixed-gas CO_2/CH_4 selectivity values in XLPEO are similar to those in CA, but XLPEO is more than an order of magnitude more permeable to CO_2. At 253 K, the separation performance of XLPEO at high CO_2 partial pressure exceeds the Robeson (1991) upper bound limit. Similar performance can be observed in Figures 35.12b and c for CO_2/N_2 and CO_2/O_2, respectively. In these cases,

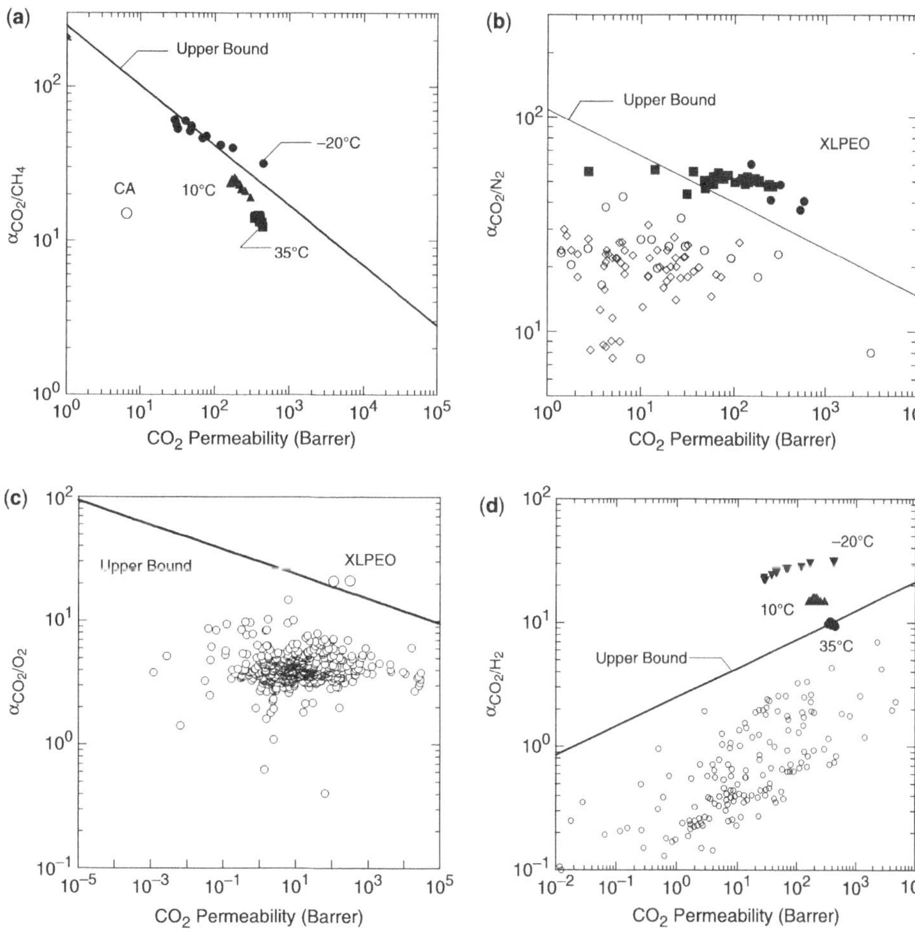

Figure 35.12 Permeability/selectivity performance map of XLPEO for (a) CO_2/CH_4, (b) CO_2/N_2, (c) CO_2/O_2, and (d) CO_2/H_2 showing upper bound calculated by Eq. (35.20) (Freeman, 1999). Comparisons with cellulose acetate (CA) and other previously studied systems are provided wherever available (Li et al., 1987; Robeson, 1991). PEO-based materials are depicted as solid symbols.

the performance of cross-linked PEO generally exceeds the upper bound within the limits of our studies. In Figure 35.12d, CO_2/H_2 separation performance, based on mixed-gas studies in the case of XLPEO, is compared with the performance of other materials. Since the kinetic diameter of H_2 is smaller than that of CO_2 (2.89 Å as opposed to 3.3 Å; see Section 35.2), the upper bound line constructed in Figure 35.12d has a positive slope, as per Eq. (35.20). Therefore, unlike separation based on strong size-sieving ability (which always results in a distinct trade-off between permeability and selectivity), high CO_2 permeability and high CO_2/H_2 selectivity can be realized simultaneously with reverse-selective membranes: Increased permeability in the rubbery polymer networks is achieved by increasing FFV and corresponding diffusivity for all penetrants, thus reducing sieving effects and unfavorable selectivity based on relative size. Indeed, the materials explored in our study exhibit excellent separation performance for CO_2/H_2 mixtures, and lowering the temperature moves the separation performance above the upper bound.

35.4 CONCLUSIONS

The separation properties of polar rubbery materials for the removal of CO_2 from mixtures with light gases have been examined. Cross-linked PEO has separation characteristics that are essentially independent of cross-link density but highly dependent on FFV; this result has been confirmed by a combination of dynamic mechanical analysis and comparative permeability measurements on XLPEO with different network structures. Plasticization of the polymer network by highly condensable penetrants actually improves the separation achieved in these rubbery membranes for reverse-selective separations such as the removal of CO_2 from H_2. A simple, modified free-volume model can accurately describe the effects of temperature and CO_2 partial pressure on the performance of the cross-linked PEO systems. The CO_2/light gas separation performance of these materials was compared to the performance of other polymers via permeability/selectivity trade-off maps, and the cross-linked PEO materials approached or exceeded the established upper bound in a number of cases. Such materials may be the basis of new design strategies for preparing high-performance membranes to remove acid gases from mixtures with light gases.

ACKNOWLEDGMENTS

We gratefully acknowledge partial support of this work from the National Science Foundation under grant number CTS-0515425. Activities at the University of Kentucky were supported by a grant from the Kentucky Science and Engineering Foundation as per grant agreement KSEF-148-502-05-130 with the Kentucky Science and Technology Corporation.

REFERENCES

Alves, N. M., Gomez Ribelles, J. L., Gomez Tejedor, J. A., and Mano, J. F. (2004). Viscoelastic behavior of poly(methyl methacrylate) networks with different cross-linking degrees. *Macromolecules* 37, 3735.

Angell, C. A. (1991). Relaxation in liquids, polymers and plastic crystals—Strong/fragile patterns and problems. *J. Non-Crystalline Solids* 13, 131–133.

Baker, R. W. (2004). *Membrane Technology and Applications*, 2nd ed. Wiley, New York.

Barrer, R. M. (1942). Permeability in relation to viscosity and structure of rubber. *Trans. Faraday Soc.* 38, 322.

Bondar, V. I., Freeman, B. D., and Pinnau, I. (1999). Gas sorption and characterization of poly(ether-*b*-amide) segmented block copolymers. *J. Polym. Sci. Part B: Polym. Phys.* 37, 2463.

Bondar, V. I., Freeman, B. D., and Pinnau, I. (2000). Gas transport properties of poly(ether-*b*-amide) segmented block copolymers. *J. Polym. Sci. Part B: Polym. Phys.* 38, 2051.

Breck, D. W. (1974). *Zeolite Molecular Sieves*. Krieger Publishing, Melbourne, FL.

Chow, T. S. (1980). Molecular interpretation of the glass transition temperature of polymer-diluent systems. *Macromolecules* 13, 362.

Cohen, M. H., and Turnbull, D. (1959). Molecular transport in liquids and glasses. *J. Chem. Phys.* 31, 1164.

Dymond, J. H., and Smith, E. B. (1980). *The Virial Coefficients of Pure Gases and Mixtures: A Critical Compilation*. Oxford University Press, New York.

Ferry, J. D. (1980). *Viscoelastic Properties of Polymers*, 3rd ed. Wiley, New York.

Fitz, B. D., and Mijovic, J. (1999). Segmental dynamics in poly(methylphenylsiloxane) networks by dielectric relaxation spectroscopy. *Macromolecules* 32, 3518.

Freeman, B. D. (1992). Mutual diffusion in polymeric systems. In S. L. Aggarwal and S. Russo (Eds.), *Comprehensive Polymer Science: First Supplement*. Pergamon, New York p. 167.

Freeman, B. D. (1999). Basis of permeability/selectivity tradeoff relations in polymeric gas separation membranes. *Macromolecules* 32, 375.

Freeman, B. D., and Pinnau, I. (1999). Polymeric materials for gas separations. In B. D. Freeman and I. Pinnau (Eds.), *Polymer Membranes for Gas and Vapor Separation*, Vol. 733. American Chemical Society, Washington, DC.

Fujita, H. (1961). Diffusion in polymer-diluent systems. *Fortschritte der Hochpolymeren-Forschung* 3, 1.

Ghosal, K., Chern, R. T., Freeman, B. D., Daly, W. H., and Negulescu, I. I. (1996). Effect of basic substituents on gas sorption and permeation in polysulfone. *Macromolecules* 29, 4360.

Ghosal, K., and Freeman, B. D. (1994). Gas separation using polymer membranes: An overview. *Polym. Adv. Technol.* 5, 673.

Glatz-Reichenbach, J. K. W., Sorriero, L. J., and Fitzgerald, J. J. (1994). Influence of cross-linking on the molecular relaxation of an amorphous copolymer near its glass-transition temperature. *Macromolecules* 27, 1338.

Graham, N. B. (1987). Poly(ethylene oxide) and related hydrogels. In N. A. Peppas (Ed.), *Hydrogels in Medicine and Pharmacy*, Vol. II. CRC Press, Boca Raton, FL, p. 95.

Graham, T. (1866). On the absorption and dialytic separation of gases by colloid septa part I: Action of a septum of caoutchouc. *Philos. Mag.* 32, 401.

Hill, L. W. (1997). Calculation of crosslink density in short chain networks. *Prog. Orga. Coat.* 31, 235.

Ismail, A. F., and Lorna, W. (2002). Penetrant-induced plasticization phenomenon in glassy polymers for gas separation membrane. *Sep. Purif. Technol.* 27, 173.

Kalakkunnath, S., Kalika, D. S., Lin, H., and Freeman, B. D. (2005). Segmental relaxation characteristics of crosslinked poly(ethylene oxide) copolymer networks. *Macromolecules* 38, 9679.

Kalakkunnath, S., Kalika, D. S., Lin, H., and Freeman, B. D. (2006). Viscoelastic characteristics of U.V. polymerized poly(ethylene glycol) diacrylate networks with varying extents of crosslinking. *J. Polym. Sci. Part B: Polym. Phys.* 44, 2058.

Kannurpatti, A. R., Anderson, K. J., Anseth, J. W., and Bowman, C. N. (1997). Use of living radical polymerizations to study the structural evolution and properties of highly crosslinked polymer networks. *J. Polym. Sci. Part B: Polym. Phys.* 35, 2297.

Kannurpatti, A. R., Anseth, J. W., and Bowman, C. N. (1998). A study of the evolution of mechanical properties and structural heterogeneity of polymer networks formed by photopolymerizations of multifunctional (meth)acrylates. *Polymer* 39, 2507.

Kannurpatti, A. R., and Bowman, C. N. (1998). Structural evolution of dimethacrylate networks studied by dielectric spectroscopy. *Macromolecules* 31, 3311.

Kohl, A. L., and Nielson, R. (1997). *Gas Purification*, 5th ed. Gulf Publishing, Houston, TX.

Koros, W. J. (1985). Simplified analysis of gas/polymer selective solubility behavior. *J. Polym. Sci. Polym. Phys. Ed.* 23, 1611.

Koros, W. J., Chan, A. H., and Paul, D. R. (1977). Sorption and transport of various gases in polycarbonate. *J. Membr. Sci.* 2, 165.

Koros, W. J., Coleman, M. R., and Walker, D. R. B. (1992). Controlled permeability polymer membranes. *Annu. Rev. Mater. Sci.* 22, 47.

Koros, W. J., and Hellums, M. W. (1989). Gas separation membrane material selection criteria: Differences for weakly and strongly interacting feed components. *Fluid Phase Equilibr.* 53, 339.

Lee, W. M. (1980). Selection of barrier materials from molecular structure. *Polym. Eng. Sci.* 20, 65.

Li, J., Nagai, K., Nakagawa, T., and Wang, S. (1995). Preparation of polyethyleneglycol (PEG) and cellulose acetate (CA) blend membranes and their gas permeabilities. *J. Appl. Polym. Sci.* 58, 1455.

Li, N. N., Funk, E. W., Chang, Y. A., Kulkarn, S. S., Swamikannu, A. X., and White, L. S. (1987). Membrane separation processes in the petrochemical industry: Phase II, final report for U.S. Department of Energy, ID/12422-2, Allied-Signal http://www.osti.gov/energycitations/product.biblio.jsp?osti_id=6290984.

Lin, H., and Freeman, B. D. (2004). Gas solubility, diffusivity and permeability in poly(ethylene oxide). *J. Membr. Sci.* 239, 105.

Lin, H., and Freeman, B. D. (2005a). Gas and vapor solubility in crosslinked poly(ethylene glycol diacrylate). *Macromolecules* 38, 8394.

Lin, H., and Freeman, B. D. (2005b). Materials selection guidelines for membranes that remove CO_2 from gas mixtures. *J. Mol. Struct.* 739, 57.

Lin, H., and Freeman, B. D. (2006a). Gas permeation and diffusion in crosslinked poly(ethylene glycol) diacrylate. *Macromolecules* 39, 3568.

Lin, H., and Freeman, B. D. (2006b). Permeation and diffusion. In H. Czichos, L. E. Smith, and T. Saito (Eds.), *Springer Handbook of Materials Measurement Methods*. Springer Science + Business Media, New York, p. 371.

Lin, H., Kai, T., Freeman, B. D., Kalakkunnath, S., and Kalika, D. S. (2005). The effect of cross-linking on gas permeability in crosslinked poly(ethylene glycol) diacrylate. *Macromolecules* 38, 8381.

Lin, H., Van Wagner, E., Freeman, B. D., Toy, L. G., and Gupta, R. P. (2006a). Plasticization-enhanced H_2 purification using polymeric membranes. *Science* 311, 639.

Lin, H., Van Wagner, E., Raharjo, R., Freeman, B. D., and Roman, I. (2006b). High performance polymer membranes for natural gas sweetening. *Adv. Mater.* 18, 39.

Lin, H., Van Wagner, E., Swinnea, J. S., Freeman, B. D., Pas, S. J., Hill, A. J., Kalakkunnath, S., and Kalika, D. S. (2006c). Transport and structural characteristics of crosslinked poly(ethylene oxide) rubbers. *J. Membr. Sci.* 276, 145.

Litvinov, V. M., and Dias, A. A. (2001). Analysis of network structure of UV-cured acrylates by ^1H nmr relaxation, ^{13}C NMR spectroscopy, and dynamic mechanical experiments. *Macromolecules* 34, 4051.

Mark, J. E. (1982). Experimental determinations of crosslink densities. *Rubber Chem. Technol.* 55, 762.

Merkel, T. C., Bondar, V. I., Nagai, K., Freeman, B. D., and Pinnau, I. (2000). Gas sorption, diffusion, and permeation in poly(dimethylsiloxane). *J. Polym. Sci. Part B: Polym. Phys.* 38, 415.

Ngai, K. L., and Roland, C. M. (1993). Chemical structure and intermolecular cooperativity: Dielectric relaxation results. *Macromolecules* 26, 6824.

Paul, D. R. (1984). Gas transport in homogeneous multicomponent polymers. *J. Membr. Sci.* 18, 75.

Paul, D. R., and Clarke, R. (2002). Modeling of modified atmosphere packaging based on designs with a membrane and perforations. *J. Membr. Sci.* 208, 269.

Petropoulos, J. H. (1994). Mechanisms and theories for sorption and diffusion of gases in polymers. In D. R. Paul and Y. P. Yampolskii (Eds.), *Polymeric Gas Separation Membranes*. CRC Press, Boca Raton, FL, p. 17.

Robeson, L. M. (1991). Correlation of separation factor versus permeability for polymeric membranes. *J. Membr. Sci.* 62, 165.

Roland, C. M. (1994). Constraints on local segmental motion in poly(vinylethylene) networks. *Macromolecules* 27, 4242.

Schroeder, M. J., and Roland, C. M. (2002). Segmental relaxation in end-linked poly(dimethylsiloxane) networks. *Macromolecules* 35, 2676.

Service, R. F. (1999). Hydrogen power: Bringing fuel cells down to earth. *Science* 285, 682.

Sperling, L. H. (2001). *Introduction to Physical Polymer Science*, 3rd ed. Wiley, New York.

Staudt-Bickel, C., and Koros, W. J. (1999). Improvement of CO_2/CH_4 separation characteristics of polyimides by chemical crosslinking. *J. Membr. Sci.* 155, 145.

Stern, S. A., Fang, S.-M., and Jobbins, R. M. (1971). Permeation of gases at high pressures. *J. Macromol. Sci. Phys.* B5, 41.

Stern, S. A., Mullhaupt, J. T., and Gareis, P. J. (1969). The effect of pressure on the permeation of gases and vapors through polyethylene. Usefulness of the corresponding states principle. *AIChE J.* 15, 64.

Treloar, L. R. G. (1975). *The Physics of Rubber Elasticity*, 3rd ed. Oxford University Press, New York.

van Amerongen, G. J. (1946). Permeability of different rubbers to gases and its relation to diffusivity and solubility. *J. Appl. Phys.* 17, 972.

van Krevelen, D. W. (1990). *Properties of Polymers: Their Correlation with Chemical Structure: Their Numerical Estimation and Prediction from Additive Group Contributions*. Elsevier, Amsterdam.

White, L. S., Blinka, T. A., Kloczewski, H. A., and Wang, I.-F. (1995). Properties of a polyimide gas separation membrane in natural gas streams. *J. Membr. Sci.* 103, 73.

Wijmans, J. G., and Baker, R. W. (1995). The solution-diffusion model: A review. *J. Membr. Sci.* 107, 1.

Williams, G., Watts, D. C., Dev, S. B., and North, A. M. (1971). Further considerations of non symmetrical dielectric relaxation behaviour arising from a simple empirical decay function. *Trans. Faraday Soc.* 67, 1323.

Zoller, P., and Walsh, D. (1995). *Standard Pressure-Volume-Temperature Data for Polymers*, 1st ed. Technomic Publishing, Lancaster, PA.

INDEX

Advanced Membrane Technology and Applications. Edited by Norman N. Li, Anthony G. Fane, W. S. Winston Ho, and T. Matsuura
Copyright © 2008 John Wiley & Sons, Inc.

Printed and bound by CPI Group (UK) Ltd, Croydon, CR0 4YY
16/09/2026
14963451-0001